Ocean Currents

T0228946

Editor-in-Chief

John H. Steele
Marine Policy Center, Woods Hole Oceanographic Institution, Mail Stop 41, Woods Hole, Massachusetts 02543, USA

Editors

Steve A. Thorpe
National Oceanography Centre, University of Southampton, Waterfront Campus, European Way, Southampton, SO14 3ZH, UK
School of Ocean Sciences, Bangor University, Menai Bridge, Anglesey, LL59 5AB, UK

Karl K. Turekian
Yale University, Department of Geology and Geophysics, New Haven, Connecticut, 06520-8109, USA

Subject Area Volumes from the Second Edition

Climate & Oceans edited by Karl K. Turekian
Elements of Physical Oceanography edited by Steve A. Thorpe
Marine Biology edited by John H. Steele
Marine Chemistry & Geochemistry edited by Karl K. Turekian
Marine Ecological Processes edited by John H. Steele
Marine Geology & Geophysics edited by Karl K. Turekian
Marine Policy & Economics Guest Edited by Porter Hoagland, Marine Policy Center, Woods Hole Oceanographic Institution, Woods Hole, Massachusetts
Measurement Techniques, Sensors & Platforms edited by Steve A. Thorpe
Ocean Currents edited by Steve A. Thorpe

OCEAN CURRENTS

A DERIVATIVE OF ENCYCLOPEDIA OF OCEAN SCIENCES, 2ND EDITION

Editor

STEVE A. THORPE

ELSEVIER

Boston • Heidelberg • London • New York • Oxford
Paris • San Diego • San Francisco • Singapore • Sydney • Tokyo
Academic Press is an imprint of Elsevier

ACADEMIC
PRESS

Academic Press is an imprint of Elsevier
32 Jamestown Road, London NW1 7BY, UK
30 Corporate Drive, Suite 400, Burlington, MA 01803, USA
525 B Street, Suite 1900, San Diego, CA 92101-4495, USA

Material in the work originally appeared in *Encyclopedia of Ocean Sciences* (Elsevier Ltd., 2001) and *Encyclopedia of Ocean Sciences*, 2nd Edition (Elsevier Ltd., 2009), edited by John H. Steele, Steve A. Thorpe and Karl K. Turekian.

British Library Cataloguing in Publication Data
A catalogue record for this book is available from the British Library

Library of Congress Control Number: 2009907114

ISBN: 978-0-08-096486-7

For information on all Elsevier publications
visit our website at www.elsevierdirect.com

CONTENTS

THE ABYSSAL CIRCULATION

ENCLOSED OR SEMI-ENCLOSED SEAS, FJORDS, ESTUARIES AND RIVERS

GRAVITY AND TURBIDITY CURRENTS, AND FLOWS IN CHANNELS

EDDIES AND WAVES

CIRCULATION AND RELATED MODELS

APPENDICES

OCEAN CURRENTS: INTRODUCTION

This volume is a collection of articles published in the *Encyclopedia of Ocean Sciences* on the subject of ocean currents and the circulation of the ocean. The articles summarize the state of knowledge at a particular time, in about 2006.

The causes of ocean currents and ocean circulation have fascinated scientists for centuries, and several of the articles describe how the understanding of particular ocean currents has developed over time. The existence of the Gulf Stream, for example, was known in the early 16th century and mariners took advantage of its drift to cross the Atlantic from west to east, but a convincing explanation of the current based on hydrodynamic theory was not developed until four centuries later. Henry Stommel's book on the subject (Stommel, 1965) is still well worth reading. Based largely on navigation and ship drift in the early days, the methods of observation of oceanic flow have undergone radical change, particularly in the 20th century. Eulerian point measurements using moored current meters or those based on geostrophic estimates became possible in the 1950s, the former demanding robust instruments and moorings and the latter the precise measurement of density (and both depending on the ability of sea-going scientists to deploy moorings and make measurements, often in the most extreme weather imaginable). The development of acoustically-tracked neutrally buoyant Lagrangian floats in the late 1950s provided information about the previously unimagined variability of ocean currents, the mesoscale eddies, and the meandering paths of water motions. The acoustic tracking of floats has now been largely superseded by satellite tracking of drifters or floats that return periodically to the surface to communicate their position and other data back to shore. *Measurement Techniques, Platforms and Sensors* are the subjects of a companion volume in this series of articles extracted from the complete *Encyclopedia of Ocean Sciences*.

Precisely how the ocean circulation is driven, how the circulation is related to the mixing of density stratified water within the ocean (a topic pursued further in articles of the companion volume *Elements of Physical Oceanography*) and how the circulation may respond to future changes in climate (see the *Climate and Oceans* volume), are still matters of debate. The relation of circulation to climate change presently makes the study of ocean currents one of particularly great importance.

This volume is laid out in a series of sections or topic areas, beginning with one in which the currents in the major oceans, the Atlantic, Pacific, Indian, Southern and Arctic Oceans, and the major Indonesian connection between the Pacific and Indian Ocean, are described. Next, the circulation of deep water, the Abyssal Circulation, is addressed. The articles in the third section discuss the (contrasting) circulations of the Seas, such as the Mediterranean, Baltic and Red Seas, as well as the circulation in fjords, estuaries and the effects of rivers. The intermittency and variability of the oceans is the theme of the fourth and fifth sections the former describing flows that occur near ocean boundaries: cascades, gravity flows and turbidity currents, and the latter, large-scale waves and eddies. Such waves and mesoscale eddies are now known to dominate the kinetic energy of the oceans at time scales of a few months or less; a reminder that, as Walter Munk put it, 'the ocean is an AC system, not DC'. The final section is more general, theoretical and less 'observational' than the earlier. It sets out knowledge of how the ocean currents are driven, and describes numerical models of the ocean. A course on ocean currents might well begin with information selected from articles in this section, the earlier sections being used to illustrate the consequences of the driving forces. The selected order reflects a preference for theory to be driven by the challenge to explain data, to develop theories on the basis of observations, but the order is not intended to diminish the enormous importance of the powerful theories and theoreticians whose work has led to the present-day understanding of the dynamics of ocean currents; a complete appreciation of the ocean currents cannot be obtained without an understanding of the physics of the motion and mixing of the stratified water of the ocean on a rotating Earth, and of the methods used and the care required to obtain reliable measurements. Articles on other topics intimately related to ocean currents and circulation, such as 'Water Types and Water Masses' are to be found in the full Encyclopedia and in the companion special topic volume, *Elements of Physical Oceanography*.

Although the subject of ocean currents is a small part of 'oceanography,' it is vitally important because of its connections to other topics, most especially because currents are agents of advection and transport of water properties, including heat, around the world; the topic is therefore related to climate and to the spread of solutes, dissolved matter, pollutants, algae and larvae. The demands of recent years for a better and more complete understanding of the ocean have blurred the boundaries between disciplines. Whilst believing this collection of articles, from the limited disciplinary field of ocean currents, will make subjects more accessible (if not more affordable) for readers, the Editors acknowledge and emphasise the connections with other articles within the Encyclopedia, references to which are provided at the end of each article.

The authors of the articles are experts who have themselves undertaken research to reveal the main motion of water in the particular region addressed in their article. They are distinguished researchers who have given time to write concisely and lucidly about their subjects, and the Editors are indebted to them all for their care in preparing these accounts.

The articles in this volume would not have been produced without the considerable help of members of the Encyclopedia's Editorial Advisory Board, particularly Harry Bryden, John Gould, Gerold Siedler and Bruce Warren. In addition to thanking the authors of the articles in this volume, the Editors wish to thank the members of the Editorial Board for the time they gave in recommending subjects for review, in identifying and encouraging the authors, and in reading (and sometimes suggesting improvements to) their initial drafts – indeed in making this venture possible. This collection of articles about the currents of the world's oceans is perhaps, as a single volume, unique in its breadth and coverage.

The volume is a tribute to those oceanographers who spent many tiring and often uncomfortable hours, day and night, on swaying research vessels in many parts of the world's ocean and seas to collect data, and also to those who argued for, or funded, the international measurement programmes, notably WOCE, the World Ocean Circulation Experiment, and so made it possible to obtain some of the data on which many of the articles in this volume are based.

Steve A. Thorpe
Editor

REFERENCES

Stommel H (1965) *The Gulf Stream*, Second Edition. Berkley: University of California Press, London: Cambridge University Press, 248.

THE OCEANS: THE ATLANTIC OCEAN

CURRENT SYSTEMS IN THE ATLANTIC OCEAN

L. Stramma, University of Kiel, Kiel, Germany

Introduction

By the late nineteenth century our present view of the Atlantic Ocean surface circulation had already been largely worked out. The voyages of discovery brought startling observations of many of the important surface currents. During the twentieth century the focus turned to a detailed description of the surface currents and the investigation of the subsurface currents. Recently, much attention has been focused on climate research as it became clear that climate goes through long period variability and can affect our lives and prosperity. The physical climate system is controlled by the interaction of atmosphere, ocean, land and sea ice, and land surfaces. To understand the influence of the ocean on climate, the physical processes and especially the ocean currents storing and transporting heat need to be thoroughly investigated. Since the end of the twentieth century, the general circulation of the Atlantic Ocean has been considered in a climatological context. A new picture emerged with the North Atlantic Ocean being seen not only as relevant to the climate of Europe, but for its influence on the entire globe due to its unique thermohaline circulation. Its warm upper-ocean currents transport mass and heat, originating in part from the Pacific and Indian Oceans, towards the north in the South-Atlantic and the North Atlantic. Cold deep waters of the North Atlantic flow southward, cross the South Atlantic, and are exported into the Indian and even the Pacific Oceans. The research focus on ocean currents in the Atlantic Ocean at the beginning of the new millennium will be further investigation of the mean surface and deep currents of the Atlantic and its variability on short-term to decadal timescales, so that the ocean's role in climate change can be better understood.

Basin Structure

The Atlantic Ocean extends both into the Arctic and Antarctic regions, giving it the largest meridional extent of all oceans. The north–south extent from Bering Strait to the Antarctic continent is more than 21 000 km, while the largest zonal distance from the Gulf of Mexico to the coast of north-west Africa is only about 8300 km. Here the Atlantic Ocean is described between the northern polar circle and the southern tip of South America at 55°S (see **Figure 1**). In the north, the Davis Strait between northern Canada and west Greenland separates the Labrador Sea from Baffin Bay to the north of Davis Strait. The Denmark Strait between east Greenland and Iceland, and the ridges between Iceland and Scotland separate the Irminger Basin and the Iceland Basin from the Greenland Sea and Norwegian Sea. In the south a line from the southern tip of South America to the southern tip of South Africa separates the South Atlantic Ocean from the Southern Ocean. Although there is no topographic justification for this separation, defining a southern ocean around Antarctica allows the separate investigation of the Antarctic Circumpolar Current and the processes near Antarctica as a whole. The Atlantic Ocean has the largest number of adjacent seas, and the larger ones are discussed in other articles (*see* Baltic Sea Circulation, North Sea Circulation.)

The Mid-Atlantic Ridge, which in many parts rises to <2000 m depth and reaches the 3000 m depth contour nearly everywhere, is located zonally in most places near the middle of the Atlantic and divides the Atlantic Ocean into a series of eastern and western basins (**Figure 1**). The basin names presented in **Figure 1** are only the major ones; in more detailed investigations of special regions many more topographically identified structures exist with their related names. The major topographic features of the Atlantic strongly affect the deep currents of the deep and bottom water masses, either by blocking or guiding the flow. The Walvis Ridge off south-west Africa limits the northward flow of Antarctic Bottom Water (AABW) in the Cape Basin and consequently the major northward flow of AABW takes place in the western basins. Although the Rio Grande Rise between the Argentina Abyssal Plain and the Brazil Basin disturbs a smooth northward spreading of the AABW in the western basins, the Vema and Hunter Channels within the Rio Grande Rise are deep and wide enough to allow a continuous northward flow. The Romanche Fracture Zone at the equator allows part of the AABW to enter the eastern basins of the Atlantic, where the AABW spreads poleward in both hemispheres.

Once a water mass is formed, there is a conservation of its angular momentum, or rather potential

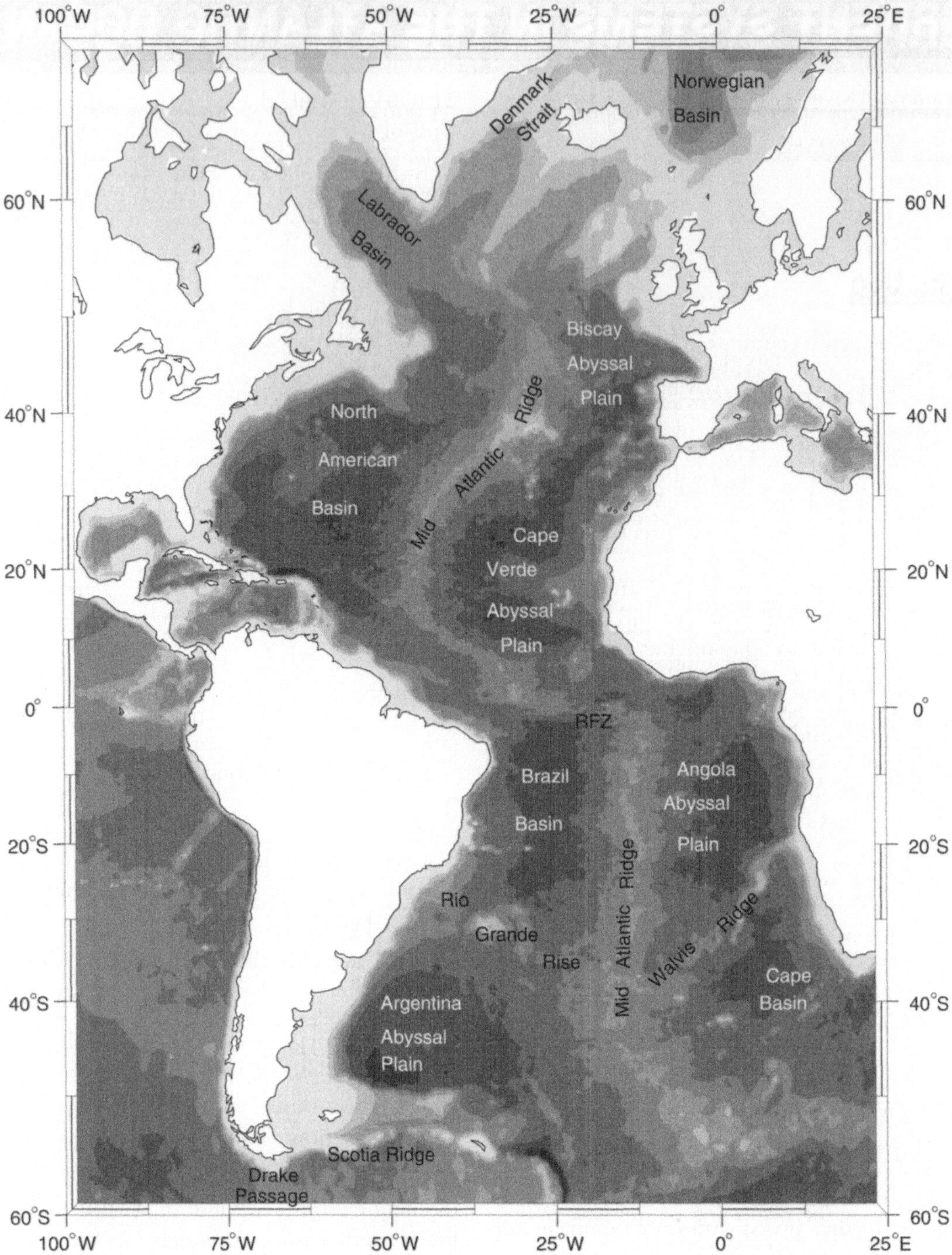

Figure 1 Topography of the Atlantic Ocean with large-scale depth contours in 1000 m steps. RFZ, Romanche Fracture Zone.

vorticity. For large-scale motions, in the interior of the ocean, potential vorticity reduces to $f/h = \text{constant}$ (where f is the Coriolis parameter and h the water depth). From this expression we can predict which way a current will swing on passing over bottom irregularities – equatorward over ridges and poleward over troughs in both hemispheres. A prominent example is the interaction between the relatively narrow Drake Passage south of the South American continent and the Scotia Ridge, which connects Antarctica with South America and contains numerous islands, located about 2000 km east of Drake Passage. The Antarctic Circumpolar Current accelerates to pass through the Drake Passage, meets the

Scotia Ridge with increased speed, and shifts sharply northward. In subpolar and polar regions density variations with depth are small and the pressure gradient force is more evenly distributed over the water column than in the tropical and subtropical regions. As a result, the currents in the subpolar and polar regions extend to great depth. In the case of the subpolar gyre of the North Atlantic, the currents have a strong barotropic component (with little vertical velocity shear) and hence tend to follow f/h contours.

Historical Developments

Charts of ocean currents from the late nineteenth century show that by then the patterns of surface circulation in regions away from the equator and polar latitudes were already well understood. This fundamental knowledge accumulated gradually through centuries of sea travel and had reached a state of near correctness by the time dedicated research cruises, full depth measurements, and the practical application of the dynamical method were begun.

By the fifth century AD, mariners had probably acquired intimate knowledge of coastal currents in the Mediterranean, but little information about them is reported in Classical writing. Following the dark and Middle Ages when little progress was made, the voyages of discovery brought startling observations of many of the Atlantic's most important ocean currents, such as the North and South Equatorial Currents, the Gulf Stream, the Agulhas Current, and others. The Gulf Stream appears to have been mapped as early as 1525 (by Ribeiro) on the basis of Spanish pilot charts. The fifteenth to seventeenth centuries were marked by attainments of knowledge that increasingly taxed the abilities of science writers to reconcile new information with accepted doctrine.

Significant advances in determining the global ocean circulation beyond local mapping of currents came only after the routine determination of longitude at sea was instituted. The introduction of the marine chronometer in the late eighteenth century made this possible. Largely because of the marine chronometer, a wealth of unprecedentedly accurate information about zonal, as well as meridional, surface currents began to accumulate in various hydrographic offices. In the early nineteenth century data from the Atlantic were collected and reduced in a systematic fashion (by James Rennell), to produce the first detailed description of the major circulation patterns at the surface for the entire mid- and low-latitude Atlantic, along with evidence for cross-equatorial flow. This work provided a foundation for the assemblage of a global data set (by Humboldt and Berghaus) that yielded worldwide charts of the nonpolar currents by the late 1830s. Heuristic and often incorrect theories of what causes the circulation in the atmosphere and oceans were popularized in the 1850s and 1860s and led to a precipitous decline in the quality of charts intended for the public (Maury; Gareis and Becker). However, errors in popular theories provided motivation for the adoption of analytical methods, which in turn led directly to the discovery of the full effect of Earth's rotation on relatively large-scale motion and the realization of how that effect produces flow perpendicular to horizontal pressure gradients (Ferrel). The precedents for modern dedicated research cruises came in the 1870s (e.g. the *Challenger* cruise), as well as mounting evidence for the existence of a deep and global thermohaline circulation (Carpenter, Prestwich). With the ever-increasing numbers of observations made at and near the surface, the upper-layer circulation in nonpolar latitudes was approximately described by the late 1880s. A current map by Krümmel (1887) nicely described the surface currents of the entire Atlantic. This figure is not reproducible; however, **Figure 2** shows as an example an earlier and slightly less accurate map from Krümmel (1882), but only for the South Atlantic Ocean.

Currents of the Atlantic Ocean Warmwatersphere

The warmwatersphere, consisting of the warmer upper waters of the ocean, is the most climatologically important part of the ocean due to its direct interaction with the atmosphere. The transition from the warm- to the cold-watersphere takes place in a relatively thin layer at temperatures between 8° and 10°C. The warm watersphere reaches to 500–1000 m depth in the Atlantic's subtropics and rapidly rises towards the ocean surface poleward of about 40° latitude.

The near-surface circulation is driven primarily by the wind and forced into closed circulation cells by the continental boundaries. The circulation of the Atlantic is governed by the subtropical gyres of the North and South Atlantic (**Figure 3**). The subtropical gyre of the North Atlantic includes the Florida Current and Gulf Stream as western boundary currents, the North Atlantic Current, the Azores and Canary Currents in the eastern Atlantic, the North Equatorial Current, and the Caribbean, Cayman and Loop Current in the Caribbean and Gulf of Mexico. The subtropical gyre of the South Atlantic includes

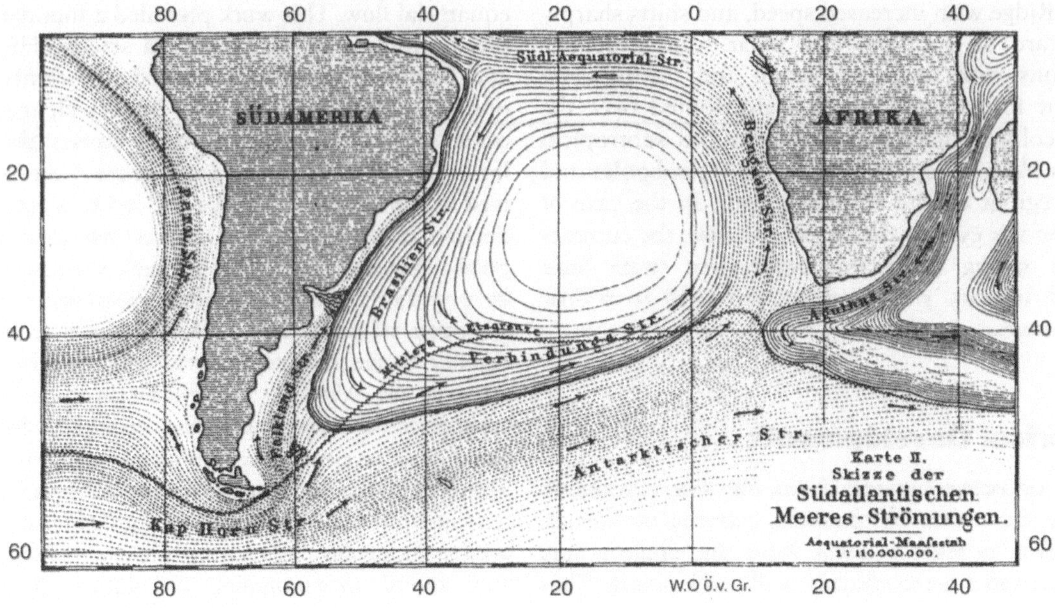

Figure 2 Chart of surface currents in the South Atlantic Ocean by Krümmel (1882).

the poleward-directed Brazil Current as western boundary current, which turns eastward at the Brazil/Falkland (Malvinas) confluence region as eastward flow near 40°S named South Atlantic Current. The South Atlantic Current in part continues to the Indian Ocean and in part adds to water from the Agulhas retroflection to the northward-flowing Benguela Current and the westward-flowing South Equatorial Current. Near the coast of north Brazil the South Equatorial Current contributes in part to the Brazil Current, but in part also to the subsurface intensified North Brazil Undercurrent, responsible for the warm water flow from the Southern to the Northern Hemisphere. The two anticyclonic subtropical gyres, clockwise in the Northern and counterclockwise in the Southern Hemisphere, reach through the entire warmwatersphere and show only weak seasonal changes. Subtropical gyres, although existing longitudinally to basin-scale, also tend to have sub-basin-scale recirculation gyres in their western reaches (**Figure 3**). The northward extent of the South Atlantic subtropical gyre decreases with increasing depth. It is located near Brazil at 16°S in the near-surface layer and at 26°S in the layer of Antarctic Intermediate Water.

The preferential north–south orientation of the continents bounding the Atlantic Ocean lead to meridional eastern and western boundary currents which together with the wind-induced zonal currents, westward flow under the trade winds, and eastward flow under the midlatitude westerly winds, form the closed gyres. The western ocean boundary regions are associated with an intensification of the currents. The consequent energetic western boundary currents, the Florida Current and the Gulf Stream in the North Atlantic Ocean and the Brazil Current in the South Atlantic Ocean, have large transports and typical width scales of ~100 km. The western boundary currents are intensified because the strength of the Coriolis effect varies with latitude. The western boundary currents of the Atlantic Ocean are generally so deep that they are constrained against the continental shelf edge and do not reach the shore (*see* Brazil and Falklands (Malvinas) Currents and Florida Current, Gulf Stream and Labrador Current).

In the Tropics the two subtropical gyres are connected via a complicated tropical circulation system. The tropical circulation shows a north-westward cross-equatorial flow at the western boundary and several zonal current and countercurrent bands (**Figure 4**) of smaller meridional and vertical extent and a lot of vertical diversification. The north-westward flow along the western boundary starts as a subsurface flow, the North Brazil Undercurrent, which becomes surface intensified north of the northeastern tip of Brazil by near-surface inflow from the South Equatorial Current and is named North Brazil Current. The North Brazil Current crosses the equator north-westwards and retroflects eastward at about 8°N. In this North Brazil Current retroflection zone, eddies detach from the current and progress north-westward towards the Caribbean. In northern spring, when the North Equatorial Countercurrent is

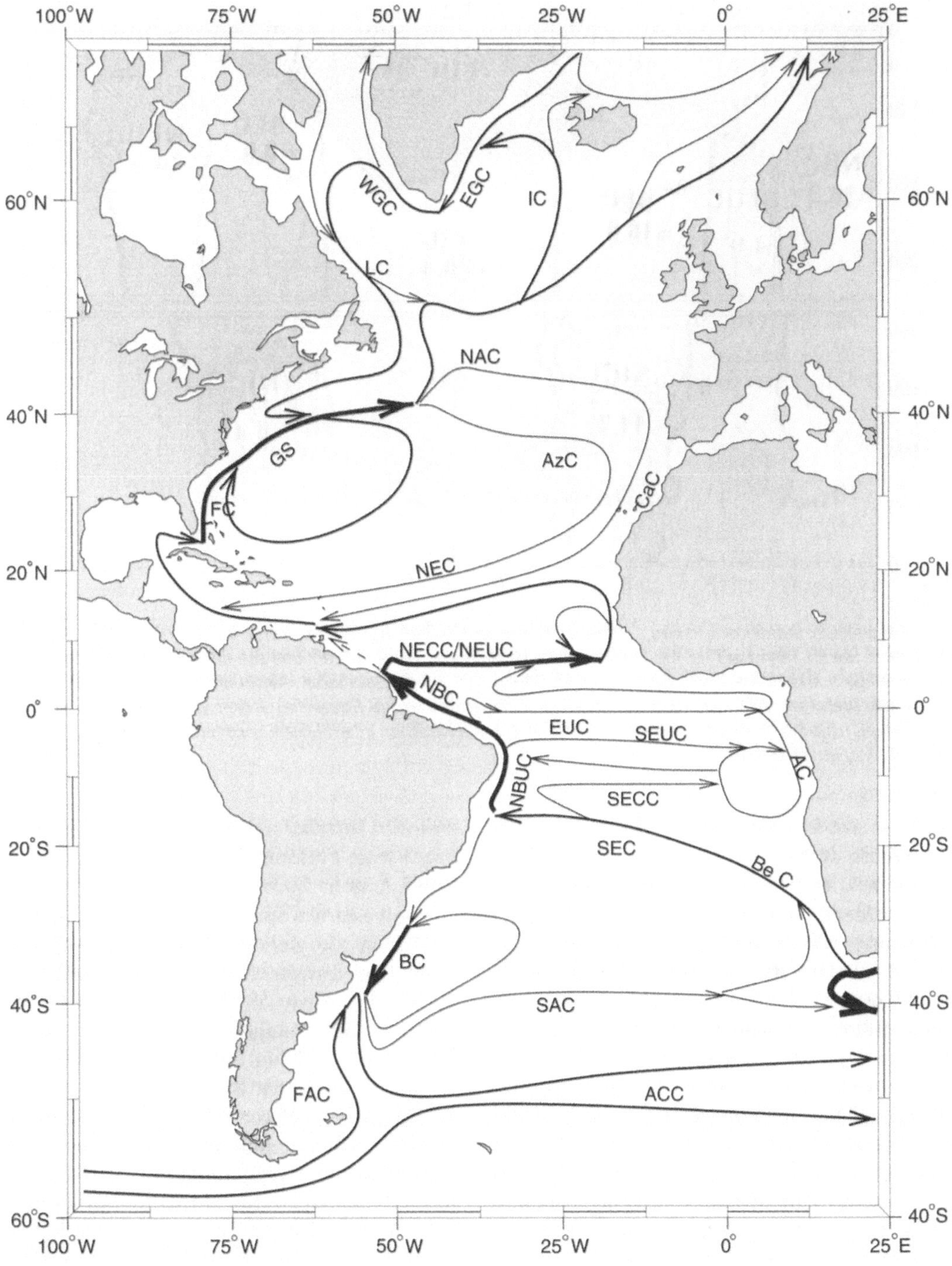

Figure 3 Schematic representation of upper-ocean currents in the North and South Atlantic Oceans in northern fall. For abbreviations of current bands see **Table 1**.

weak, there seems to be a continuous flow of about 10 Sv towards the Caribbean called Guyana Current. The westward flows are regarded as different bands of the South Equatorial Current, the northern one even crossing the equator. The eastward subsurface flows are named the Equatorial Undercurrent at the equator, and the North and South Equatorial Undercurrents at about 5° latitude. The eastward surface intensified flows at about 9° latitude are the North and South Equatorial Countercurrents. In northern fall the North Equatorial Countercurrent and the North Equatorial Undercurrent override one another and it is difficult to distinguish between the two current bands. In the Antarctic Intermediate

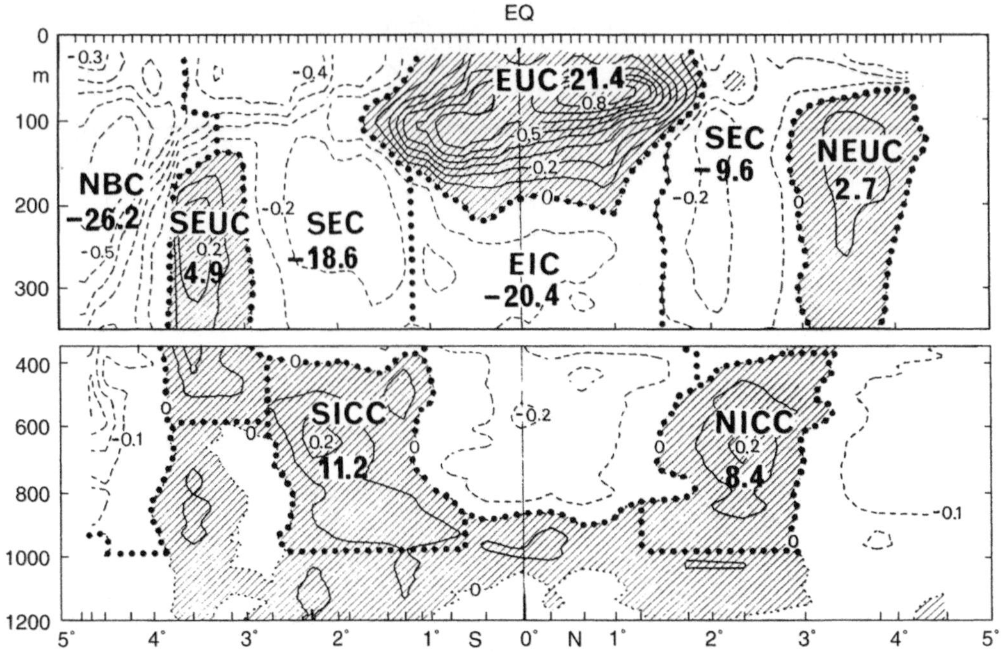

Figure 4 Zonal velocity component in m s^{-1} (eastward flow is shaded) from direct velocity measurements (ADCP) across the equator at 35°W in March 1994 north of the north-eastern tip of Brazil. Current branches are indicated and transport numbers are given in Sv. The figure shows the North Brazil Current (NBC), the South Equatorial Undercurrent (SEUC), the South Equatorial Current (SEC) with branches north and south of the equator separated by the Equatorial Undercurrent (EUC), the North Equatorial Undercurrent (NEUC), the Equatorial Intermediate Current (EIC), the Northern Intermediate Countercurrent (NICC), and the Southern Intermediate Countercurrent (SICC).

Water layer at about 700 m depth there are intermediate currents at the equator (Equatorial Intermediate Current), as well as north and south of the equator (Northern and Southern Intermediate Countercurrents) which flow in the opposite direction to the currents above (**Figure 4**). The Intertropical Convergence Zone in the Atlantic, where the trade winds of both hemispheres converge, is located north of the equator throughout the year, and reaches the South Atlantic only in southern summer and then only at the north coast of Brazil. Seasonal changes of the wind field lead obvious variations in the tropical near-surface currents; however, with different strengths. The strongest seasonal signal is observed in the North Equatorial Countercurrent. The eastward-flowing North Equatorial Countercurrent is strongest in August, when the Intertropical Convergence Zone is located at its northernmost position. At that time the North Equatorial Countercurrent crosses the entire Atlantic basins zonally, but in late boreal winter it becomes weak or even reverses to westward in the western domain. South of the Cape Verde Islands at 9°N, 25°W, there is a cyclonic feature named Guinea Dome throughout the year, but it is weaker in northern winter. The Southern Hemispheric counterpart is the Angola Dome at 10°S 9°E. The Angola Dome is seen only in southern summer and it is imbedded in a permanent larger-scale cyclonic feature centered near 13°S, 5°E called Angola Gyre.

The western tropical Atlantic is a region of special interest in the global ocean circulation. The meridional heat transport across the equator is accomplished by warm surface water, central water, and subpolar intermediate water from the Southern Hemisphere moving northward in the upper 900 m mainly in the North Brazil Current, and cold North Atlantic Deep Water (NADW) moving southward between 1200 m and 4000 m. These reversed and compensating water spreading paths are often referred to as part of the global thermohaline conveyor belt. A clear distinction has to be made between the cross-equatorial flow at the western boundary and the interhemispheric water mass exchange. The latter is the amount of transfer from the Southern to the Northern Hemisphere of about 17 Sv in the upper ocean, and to a small degree in the Antarctic Bottom Water, compensated by the transfer from the Northern to the Southern Hemisphere of about 17 Sv by the North Atlantic Deep Water. The cross-equatorial flow within the North Brazil Current with about 35 Sv (**Table 1**) is much larger, since part of this cross-equatorial flow originates from the zonal equatorial circulation, retroflects north of the

Table 1 Major upper-ocean currents of the Atlantic Ocean and transport in Sv (1 Sv = 10^6 m^3 s^{-1})

Current name	*Abbreviation in* ***Figure 3***	*Transport in Sv*
Subpolar gyre		
East Greenland Current	EGC	40–45
West Greenland Current	WGC	40–45
Irminger Current	IC	16
Labrador Current	LC	40–45
North Atlantic subtropical gyre		
Gulf Stream	GS	90–130
North Atlantic Current	NAC	35
Azores Current	AzC	12
Canary Current	CaC	5
North Equatorial Current	NEC	20
Florida Current	FC	32
Equatorial currents		
North Equatorial Countercurrent	NECC	40
North Equatorial Undercurrent	NEUC	19 (mean)
Equatorial Undercurrent	EUC	20–30
North Brazil Current	NBC	35
North Brazil Undercurrent	NBUC	25
South Equatorial Undercurrent	SEUC	5–23
Angola Current	AC	5
South Equatorial Countercurrent	SECC	7
South Atlantic subtropical gyre		
Brazil Current	BC	5–22
South Atlantic Current	SAC	15–30
Benguela Current	BeC	25
South Equatorial Current	SEC	20 (southern band)
Southern South Atlantic		
Falkland (Malvinas) Current	FAC	up to 70
Antarctic Circumpolar Current	ACC	110–150

equator, and returns into the equatorial circulation system.

Poleward of the subtropical gyre the current field of the North and South Atlantic are completely different. In the North Atlantic a cyclonic subpolar gyre is present, driven in part by the wind stress curl associated with the atmospheric Icelandic low pressure system and in part by the fresh water from the subarctic. This subpolar gyre includes the northern part of the North Atlantic Current, the Irminger Current, the East and West Greenland Currents and the Labrador Current off north-eastern North America. In the South Atlantic the continents terminate and an eastward flow of water all around the globe within the Antarctic Circumpolar Current is driven mainly by the midlatitudes westerlies. The South Atlantic counterpart of the Labrador Current is the Falkland (Malvinas) Current, which flows equatorward along the south-eastern South American shelf edge to about 38°S. However, this current differs in origin as it is essentially a meander of a branch of the Antarctic Circumpolar Current. In the Brazil/Falkland (Malvinas) confluence region the Falkland (Malvinas) Current is retroflected southward to join the Antarctic Circumpolar Current. The South Atlantic Current as southern current band of the South Atlantic subtropical gyre and the Antarctic Circumpolar Current can be distinguished as separate current bands, nevertheless mass and heat exchange between the subtropics and subpolar region takes place in this region.

The currents of the North Atlantic subpolar gyre have a strong barotropic flow component, which lead to large water mass transports (**Table 1**). As the major method of estimating transport is by geostrophy, which provides only the baroclinic component, early transport estimates for this region with strong barotropic flow fields largely underestimated the real transports. Another prominent example is the Falkland (Malvinas) Current in the South Atlantic, where estimates including the barotropic component lead to transports of up to 70 Sv, while earlier geostrophic computations resulted in transports of about 10 Sv. Differences between transports presented in **Table 1** and transport values presented elsewhere might also arise from the location where the transport is estimated, as the mass transport changes along the flow path, or from different definitions of the boundaries of the current bands. For example, the Gulf Stream is measured to the deepest depth reached by the northward flow, while the southward-flowing Brazil Current is typically estimated only for the transport in the warmwatersphere, while the southward flow underneath is estimated separately as Deep Western Boundary Current.

Currents of the Deep Atlantic Ocean

The deep-ocean circulation depends heavily on the changes in density imposed by air–sea interaction. The flow in the deep ocean is driven by the equator-to-pole differences in ocean density. This thermohaline circulation, driven by temperature and salinity gradients, provides global-scale transport of heat and salt. The forcing of this flow is concentrated in a few areas of intense production of dense water in the far North Atlantic, the Labrador Sea, and along the margin of Antarctica within so-called convection areas.

The water formed at the Antarctic continent is the densest water mass in the Atlantic and, once it has crossed the Antarctic Circumpolar Current (ACC), spreads as Antarctic Bottom Water through the South Atlantic western basins northward into the North Atlantic (**Figure** 5), where it can usually be found near the seafloor even north of 40°N. Actually, the real Antarctic water masses are so dense that they can be followed only to about 4.5°S, and Lower Circumpolar Deep Water spreads to the Northern

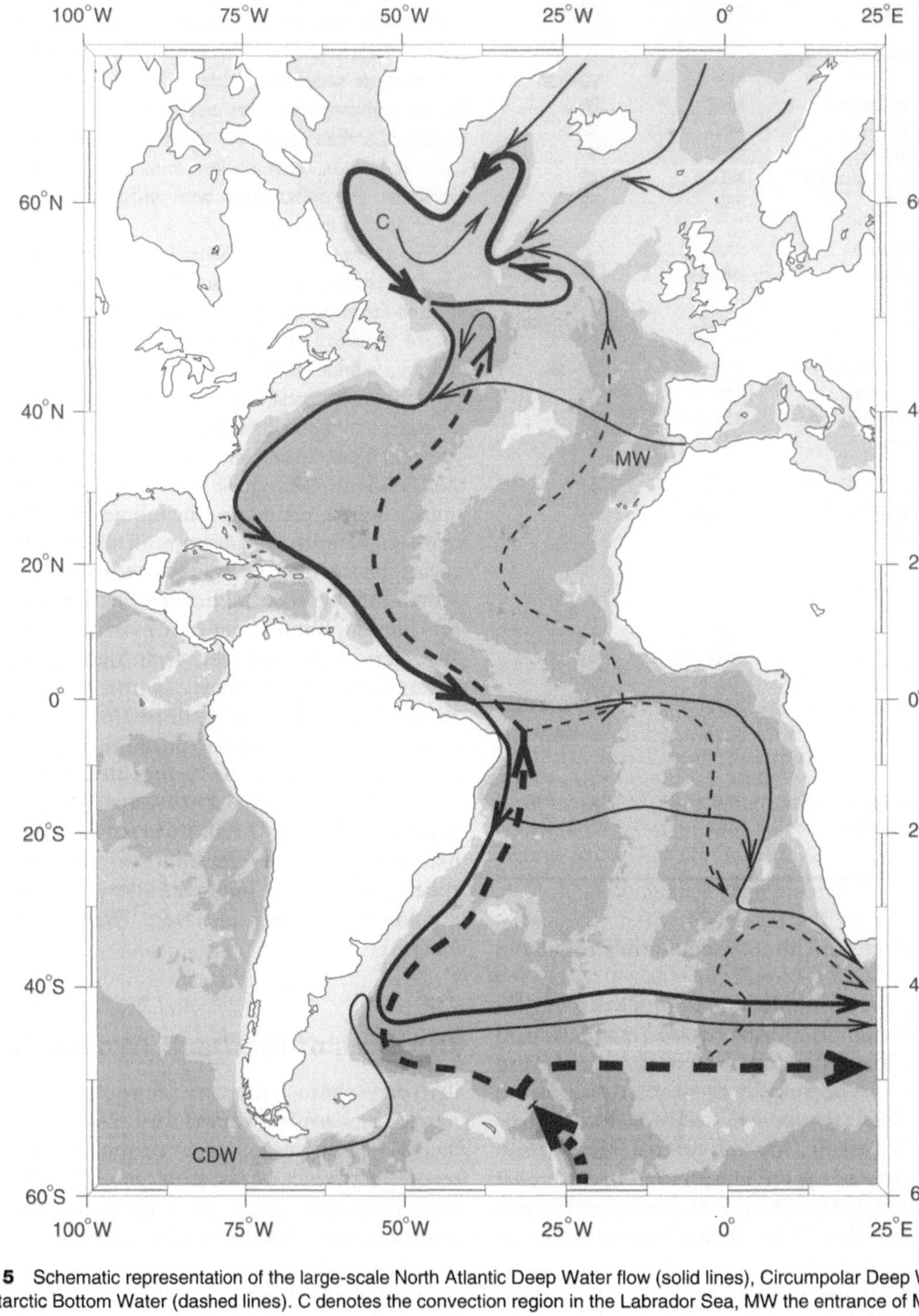

Figure 5 Schematic representation of the large-scale North Atlantic Deep Water flow (solid lines), Circumpolar Deep Water (CDW), and Antarctic Bottom Water (dashed lines). C denotes the convection region in the Labrador Sea, MW the entrance of Mediterranean Water to the North Atlantic. For readability of the figure no recirculation cells are drawn. Bottom topography in 2000 m steps.

Hemisphere. However, for historical reasons the name Antarctic Bottom Water is used generally for this water mass and is used here for consistency. In the north, Antarctic Bottom Water is modified by mixing and contributes to the North Atlantic deep water formation. The deep water of the North Atlantic (NADW) is composed of different sources in the northern Atlantic. The source for the deepest NADW layer is dense water from the Greenland Sea which overflows the Denmark Strait and is called Denmark Strait Overflow Water or lower NADW. South of the Denmark Strait the lower NADW entrains surrounding water, which in part contains modified Antarctic Bottom Water. The middle layer of NADW is a combination of overflow across the Iceland–Scotland Ridge with a light component of modified Antarctic Bottom Water. The upper layer of the NADW is caused by open-ocean convection in the Labrador Sea and is called Labrador Sea Water or upper NADW. A closer investigation of the Labrador Sea Water shows that it has two different sources. Mediterranean Water entering over the Strait of Gibraltar spreads westward in the North Atlantic and contributes saline water mainly to the upper NADW.

The NADW is trapped for some years within the deep-reaching North Atlantic subpolar gyre before it enters the Deep Western Boundary Current. Then the NADW spreads southward as Deep Western Boundary Current in the western ocean basins with recirculation cells to the east. When the NADW crosses the equator towards the South Atlantic part of the NADW flows eastward along the equator and then southward within the eastern basins. However, the major portion of the NADW continues to flow southward at the Brazilian continental margin as Deep Western Boundary Current. When the NADW reaches the latitude of the ACC the NADW is modified by mixing as it is carried eastward with the ACC around the Antarctic continent. Branches of the modified NADW, now often referred to as Circumpolar Deep Water, move northward again into the Indian and Pacific Oceans.

Future Aspects

Ocean research is always influenced by political and economic interests. The improvement in understanding of the surface currents at the time of the voyages of discovery was caused by the need for good and safe sailing routes. The more detailed look at the currents of the surface as well as the deep Atlantic were influenced by the interest in the resources of the sea for food, and the search for economic sources. The research focus on ocean currents in the Atlantic Ocean at the beginning of the new millennium will be improvement in understanding the mean surface and deep currents of the Atlantic, and its variability on short-term to decadal timescales to clarify the ocean's role in climate changes. These investigations are driven by the need to protect and manage the environment and the living conditions of all countries and are managed in large international research programs.

To improve the climate prediction models it is necessary to understand the ocean's role in climate changes better. International programs like Climate Variability and Predictability (CLIVAR) started to describe and understand the physical processes responsible for climate and predictability on seasonal, interannual, decadal, and centennial timescales, through the collection and analysis of observations and the development and application of models of the coupled climate system.

Another new important focus will be to describe and understand the interactive physical, chemical, and biological processes that regulate the total Earth system. This is also the overall objective of the International Geosphere-Biosphere Program (IGBP). One core project of IGBP is GLOBEC, which is now changing from a planning to a research status with the goal of advancing understanding of the structure and functioning of the global ocean ecosystem, its major subsystems, and its response to physical forcing so that a capability can be developed to forecast the responses of the marine ecosystem to global change.

Despite the future focus on the Atlantic's role in climate changes as well as interactive processes, and although the major components of the near-surface circulation from ship drift observations have been known for more than 100 years, there is still also the need to investigate details of the Atlantic Ocean subsurface and abyssal circulation and its physical processes, which so far are unrevealed.

See also

Arctic Ocean Circulation. Atlantic Ocean Equatorial Currents. Benguela Current; Current Systems in the Southern Ocean. Florida Current, Gulf Stream and Labrador Current.

Further Reading

Krauss W (ed.) (1996) *The Warmwatersphere of the North Atlantic Ocean*. Berlin: Gebrüder Borntraeger.

Peterson RG, Stramma L, and Kortum G (1996) Early concepts and charts of ocean circulation. *Progress in Oceanography* 37: 1–115.

Robinson AR and Brink KH (eds.) (1998) *The Sea, Ideas and Observations on Progress in the Study of the Seas The Global Coastal Ocean, Regional Studies and Syntheses*, vol. 11. New York: Wiley.

Segar DA (1998) *Introduction to Ocean Sciences*. London: Wadsworth Publishing Company.

Tomczak M and Godfrey JS (1994) *Regional Oceanography An Introduction*. Oxford: Elsevier.

Wefer G, Berger WH, Siedler G, and Webb DJ (eds.) (1996) *The South Atlantic Present and Past Circulation*. Berlin: Springer Verlag.

Zenk W, Peterson RG and Lutjeharms JRE (eds.) (1999)New view of the Atlantic: A tribute to Gerold Siedler. *Deep-Sea Research II* 46: 527.

FLORIDA CURRENT, GULF STREAM AND LABRADOR CURRENT

P. L. Richardson, Woods Hole Oceanographic Institution, Woods Hole, MA, USA

Introduction

The swiftest oceanic currents in the North Atlantic are located near its western boundary along the coasts of North and South America. The major western boundary currents are (1) the Gulf Stream, which is the north-western part of the clockwise flowing subtropical gyre located between 10°N and 50°N (roughly); (2) the North Brazil Current, the western portion of the equatorial gyre located between the equator and 5°N; (3) the Labrador Current, the western portion of the counter-clockwise-flowing subpolar gyre located between 45°N and 65°N; and (4) a deep, swift current known as the Deep Western Boundary Current, which flows southward along the whole western boundary of the North Atlantic from the Labrador Sea to the equator at depths of around 1000–4000 m.

The swift western boundary currents are connected in the sense that a net flow of warmer upper ocean water (0–1000 m very roughly) passes northward through the Atlantic to the farthest reaches of the North Atlantic where the water is converted to colder, denser deep water that flows back southward through the Atlantic. This meridional overturning circulation, or thermohaline circulation as it is also known, occurs in a vertical plane and is the focus of much recent research that is resulting in new ideas about how water, heat, and salt are transported by ocean currents. The combination of northward flow of warm water and southward flow of cold water transports large amounts of heat northward, which is important for North Atlantic weather and climate.

History

The Florida Current, the part of the Gulf Stream flowing off Florida, was described by Ponce de León in 1513 when his ships were frequently unable to stem the current as they sailed southward. The first good chart of the Gulf Stream was published in 1769–1770 by Benjamin Franklin and Timothy Folger, summarizing the Nantucket ship captain's knowledge gained in their pursuit of the sperm whale along the edges of the Stream (**Figure 1**). By the early nineteenth century the major circulation patterns at the surface were charted and relatively well known. During that century, deep hydrographic and current meter measurements began to reveal aspects of the subsurface Gulf Stream. The first detailed series of hydrographic sections across the Stream were begun in the 1930s, which led to a much-improved description of its water masses and circulation. During World War II the development of Loran improved navigation and enabled scientists to identify and follow Gulf Stream meanders for the first time. Shipboard surveys revealed how narrow, swift, and convoluted the Gulf Stream was. Stimulated by these new observations, Henry Stommel in 1948 explained that the western intensification of wind-driven ocean currents was related to the meridional variation of the Coriolis parameter. Ten years later, Stommel suggested that cold, dense water formed in the North Atlantic in late winter does not flow southward along the seafloor in the mid-Atlantic but instead is constrained to flow southward as a Deep Western Boundary Current (DWBC). This prediction was soon verified by tracking some of the first subsurface acoustic floats in the DWBC off South Carolina. In the 1960s and 1970s, deep floats, and moored current meters began to provide some details of the complicated velocity fields in the Gulf Stream and DWBC. Satellite infrared measurements of sea surface temperature provided much information about the near-surface currents including temporal variations and eddies. In the 1980s and 1990s, surface drifters, subsurface floats, moored current meters, satellite altimetry, and various kinds of profilers were used to obtain long (few years) time series. These have given us our present understanding of western boundary currents, their transports, temporal variations, and role in transporting mass, heat and salt. Models of ocean circulation are playing an important role in helping to explore ocean processes and the dynamics that drive western boundary currents.

Generating Forces

Two main forces generate large-scale ocean currents including western boundary currents. The first is wind stress, which generates the large-scale oceanic gyres like the clockwise-flowing subtropical gyre

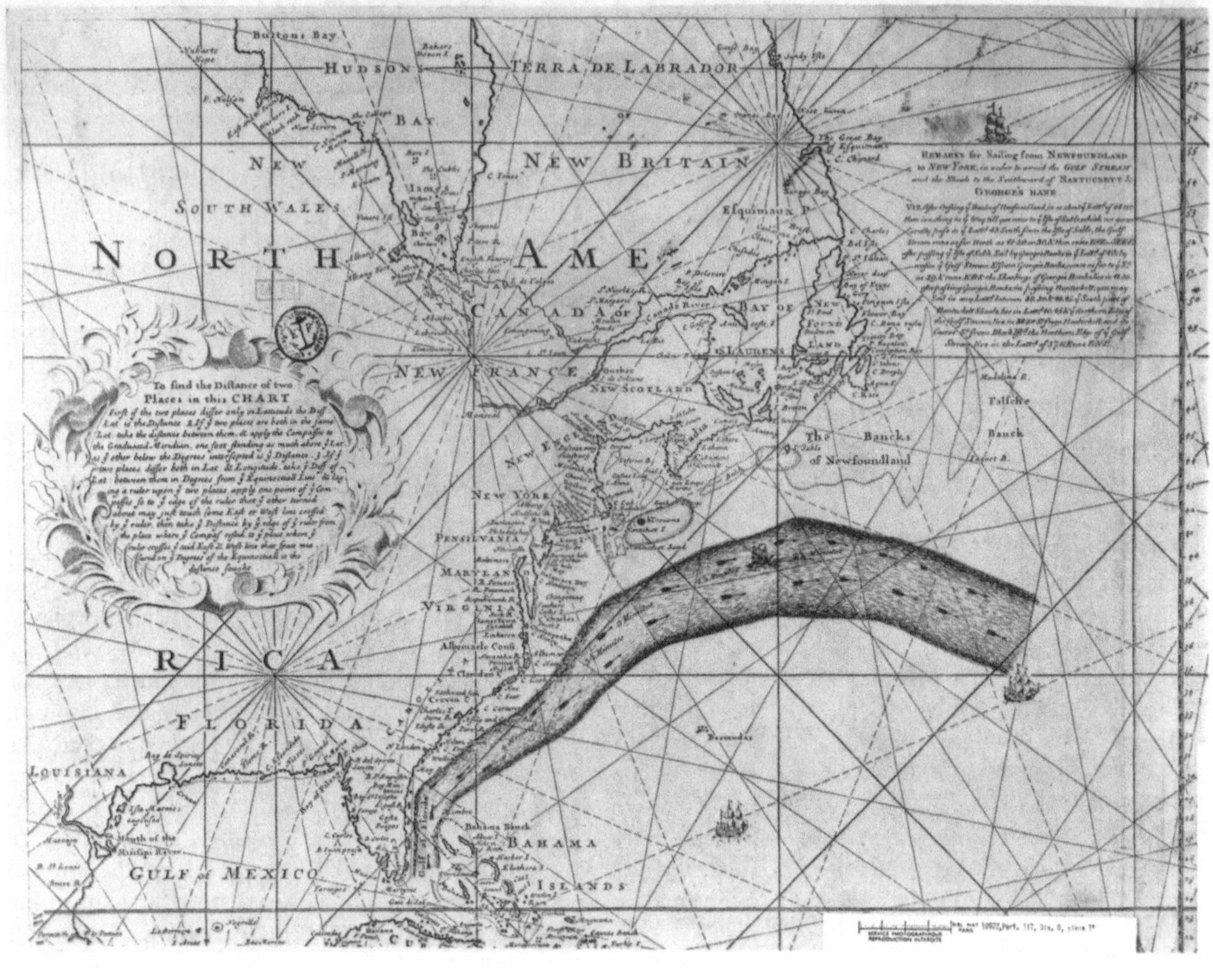

Figure 1 The Franklin–Folger chart of the Gulf Stream printed circa 1769–1770. This was the first good chart of the Stream and continues today to be a good summary of its mean speed, course, and width (assuming the charted width is the limit of the Stream's meanders). (See Richardson PL (1980) Benjamin Franklin and Timothy Folger's first printed chart of the Gulf Stream. *Science* 207: 643–645.)

centered in the upper part of the Atlantic (upper 1000 m roughly). The second is buoyancy forcing, which generates differences in water density by means of heating, cooling, precipitation, and evaporation and causes the large-scale meridional overturning circulation (MOC) – the northward flow of warmer less dense water and the southward flow of colder, denser water underneath. The rotation and spherical shape of the earth intensify currents along the western margins of ocean basins. The westward intensification is a consequence of the meridional poleward increase in magnitude of the vertical component of Earth's rotation vector, the Coriolis parameter. The vertical component is important because the oceans are stratified and much wider than they are deep.

In the Gulf Stream, the northward wind-driven gyre circulation and the northward-flowing upper part of the buoyancy-forced MOC are in the same direction and are additive. The relative amounts of transport in the Gulf Stream due to wind forcing and buoyancy forcing are thought to be roughly 2:1, although this subdivision is an oversimplification because the forcing is complicated and the Gulf Stream is highly nonlinear. In the vicinity of the Guyana Current (~10°N) the upper layer MOC is counter to and seems to overpower the wind-driven tropical gyre circulation, resulting in northward flow, where southward flow would be expected from wind stress patterns alone. Farther south in the North Brazil Current (0–5°N), the western boundary part of the clockwise-flowing wind-driven equatorial gyre is in the same direction as the MOC, and their transports add. Along the western boundary of the Labrador Sea, the southward-flowing part of the subpolar gyre, the Labrador Current, is in the same direction as the

DWBC, resulting in a top–bottom southward-flowing western boundary current.

Gulf Stream System

The Gulf Stream System is an energetic system of swift fluctuating currents, recirculations, and eddies. The swiftest surface currents of ~5 knots or 250 cm s^{-1} are located where the Stream is confined between Florida and the Bahamas. In this region and sometimes farther downstream, the Stream is known as the Florida Current. Off Florida, the Gulf Stream extends to the seafloor in depths of 700 m and also farther north along the Blake Plateau in depths of 1000 m. Near Cape Hatteras, North Carolina (35°N), the Stream leaves the western boundary and flows into deep water (4000–5000 m) as an eastward jet (**Figure 2**). The Stream's departure point from the coast is thought to be determined by the large-scale wind stress pattern, interactions with the southward-flowing DWBC, and inertial effects. The Gulf Stream flows eastward near 40°N toward the Grand Banks of Newfoundland, where part of the Stream divides into two main branches that continue eastward and part returns westward in recirculating gyres located north and south of the mean Gulf Stream axis.

The first branch, the North Atlantic Current, flows northward along the eastern side of Grand Banks as a western boundary current reaching 50°N, where it meets the Labrador Current, turns more eastward, and crosses the mid-Atlantic Ridge. Part of this flow circulates counterclockwise around the subpolar gyre and part enters the Nordic Seas. This latter part eventually returns to the Atlantic as cold, dense overflows that merge with less dense intermediate and deep water in the Labrador Sea and flow southward as the DWBC.

The second branch of the Gulf Stream flows southeastward from the region of the Grand Banks, crosses the mid-Atlantic Ridge, and flows eastward near 34°N as the Azores Current. Water from this and other more diffuse flows circulates clockwise around the eastern side of the subtropical gyre and returns westward to eventually reform into the Gulf Stream. A north-westward flow along the eastern side of the Antilles Islands is known as the Antilles Current. It is predominantly a subsurface thermocline flow.

Water flowing in the Gulf Stream comes from both the westward-flowing return current of the subtropical gyre and from the South Atlantic. Roughly half of the transport of the Florida Current is South Atlantic water that has been advected through the Gulf of Mexico and the Caribbean from the North Brazil Current.

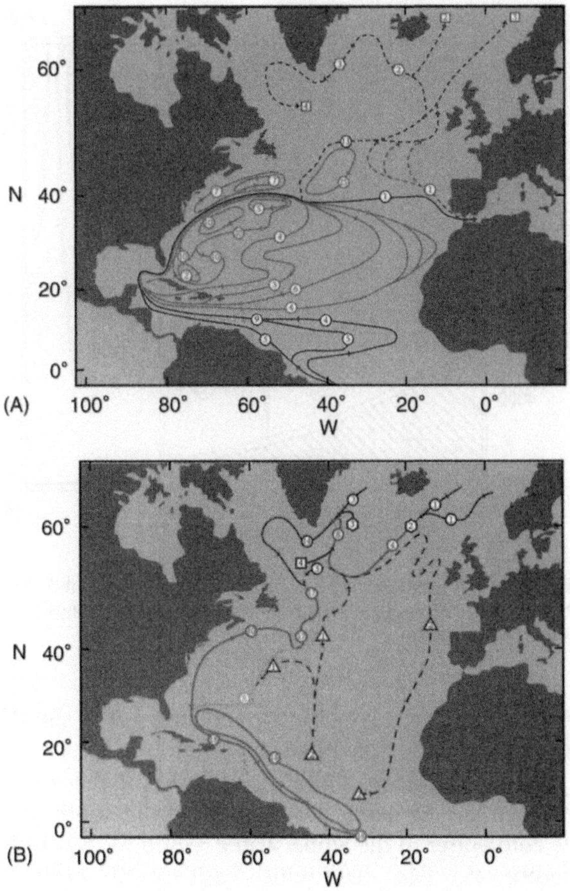

Figure 2 (A) Schematic showing the transport of the upper layer (temperatures greater than around 7°C) North Atlantic circulation (in Sverdrups, 1 Sv = 10^6 $m^3 s^{-1}$). Transports in squares denote sinking and those in hexagons denote entrainment. Red lines show the upper layer flow of the meridional overturning circulation. The blue boxes attached to dashed lines indicate that significant cooling may occur. Solid green lines characterize the subtropical gyre and recirculations as well as the Newfoundland Basin Eddy. Dashed blue lines indicate the addition of Mediterranean Water to the North Atlantic. (B) Schematic circulation showing the transport in Sverdrups of North Atlantic Deep Water (NADW). Green lines indicate transports of NADW, dark blue lines and symbols denote bottom water, and red lines upper layer replacement flows. Light blue line indicates a separate mid-latitude path for lower NADW. The square represents sinking, hexagons entrainment of upper layer water, and triangles with dashed lines water mass modification of Antarctic Bottom Water. (Reproduced from Schmitz (1996).)

Current Structure

The Gulf Stream is a semicontinuous, narrow (~100 km wide) current jet with fastest speeds of 200–250 cm s^{-1} located at the surface decreasing to around 20 cm s^{-1} near 1000 m depth (**Figure 3**). The mean velocity of the Stream extends to the seafloor even in depths below 4000 m with a mean velocity of a few centimeters per second. The deep flow field

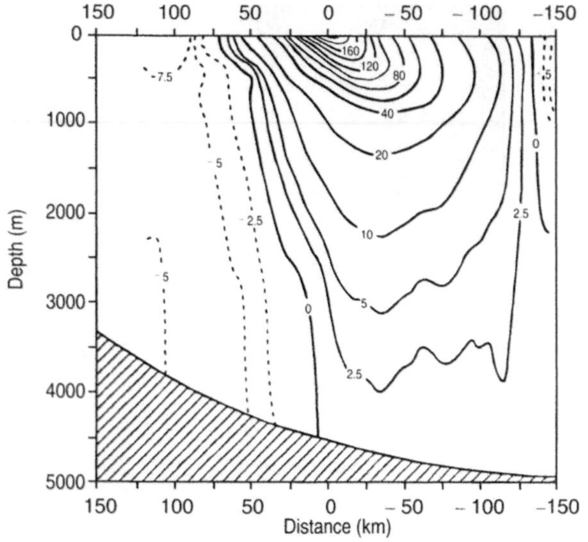

Figure 3 Downstream speed (positive) (cm s $^{-1}$) of the Gulf Stream at 68 W measured with current meters. (Adapted from Johns *et al.* (1995).)

under the Stream is complex and not just a simple deep extension of the upper layer jet.

The Stream flows along the juncture of the warm water in the Sargasso Sea located to the south and the cold water in the Slope Water region to the north (**Figure 4**). The maximum temperature gradient across the Stream is located near a depth of 500 m and amounts to around 10°C. Maximum surface temperature gradients occur in late winter when the Slope Water is coolest and the Stream advects warm tropical water northward. This relatively warm water is clearly visible in infrared images of the Stream (**Figure 5**). The sea surface slopes down northward across the Gulf Stream by around 1 m due to the different densities of water in the Sargasso Sea and Slope Water region. The surface slope has been measured by satellite altimeters and used to map the path and fluctuations of the Stream.

Variability

The Gulf Stream jet is unstable. It meanders north and south of its mean position forming convoluted paths that seem to have a maximum amplitude of around 200 km near 55°W (**Figure 6**). The most energetic meanders propagate downstream with a period of around 46 days and wavelength near 430 km. Frequently the edges of individual large meanders merge and coherent pieces of the Stream separate from the main current in the form of large eddies or current rings. Gulf Stream rings are typically 100–300 km in diameter. Those north of the Stream rotate clockwise and those south of the Stream rotate counterclockwise. As rings form on one side of the Stream, they trap in their centers, water from the opposite side of the Stream. Cold core rings are located to the south and warm core rings to the north. Once separate from the Stream, rings tend to translate westward at a few centimeters per second. Often they coalesce with the Stream after lifetimes of several months to years. The formation of rings is an energy sink for the swift Gulf Stream jet and also acts to decrease the mean temperature gradient across the Stream.

Infrared images and other data show that the Gulf Stream region is filled with meanders, rings, and smaller-scale current filaments and eddies that appear to be interacting with each other (**Figure 5**). The picture of the Gulf Stream as a continuous current jet is probably an oversimplification. Often various measurements of the Stream are combined and schematic pictures are drawn of its 'mean' characteristics. Although no such thing as a 'mean' occurs in reality, the schematics are very useful for simplifying very complex phenomena and for showing conceptual models of the Stream. One of the first such schematics by Franklin and Folger is shown in **Figure 1**.

The swift current speeds in the Gulf Stream in combination with its energetic meanders, rings, and other eddies result in very large temporal current fluctuations and very high levels of eddy kinetic energy (EKE) that coincide with the Gulf Stream between 55°W and 65°W (**Figure 7A**). Roughly 2/3 of the EKE is a result of the meandering Stream. Peak values of EKE near the surface are over 2000 cm^2 s^{-2}. The region of high EKE extends down to the sea floor underneath the Stream where values are in excess of 100 cm^2 s^{-2}, roughly 100 times larger than values in the deep Sargasso Sea (**Figure 7B**). EKE is a measure of how energetic time variations and eddies are in the ocean; EKE is usually much larger than the energy of the mean currents which suggests that eddies are important to ocean physics at least where the largest values of EKE are found.

The Gulf Stream varies seasonally and interannually, but because of energetic eddy motions and the required long time series, the low-frequency variability has only been documented in a very few places. The amplitude of the seasonal variation of the Florida Current transport is relatively small, around 8% of the mean transport, with maximum flow occurring in July–August. There is little evidence for longer-term variations in the Florida Current. Interannual variations in the wind patterns over the North Atlantic have been observed as well as variations in hydrographic properties and amounts of Labrador Sea Water formed. Present research is

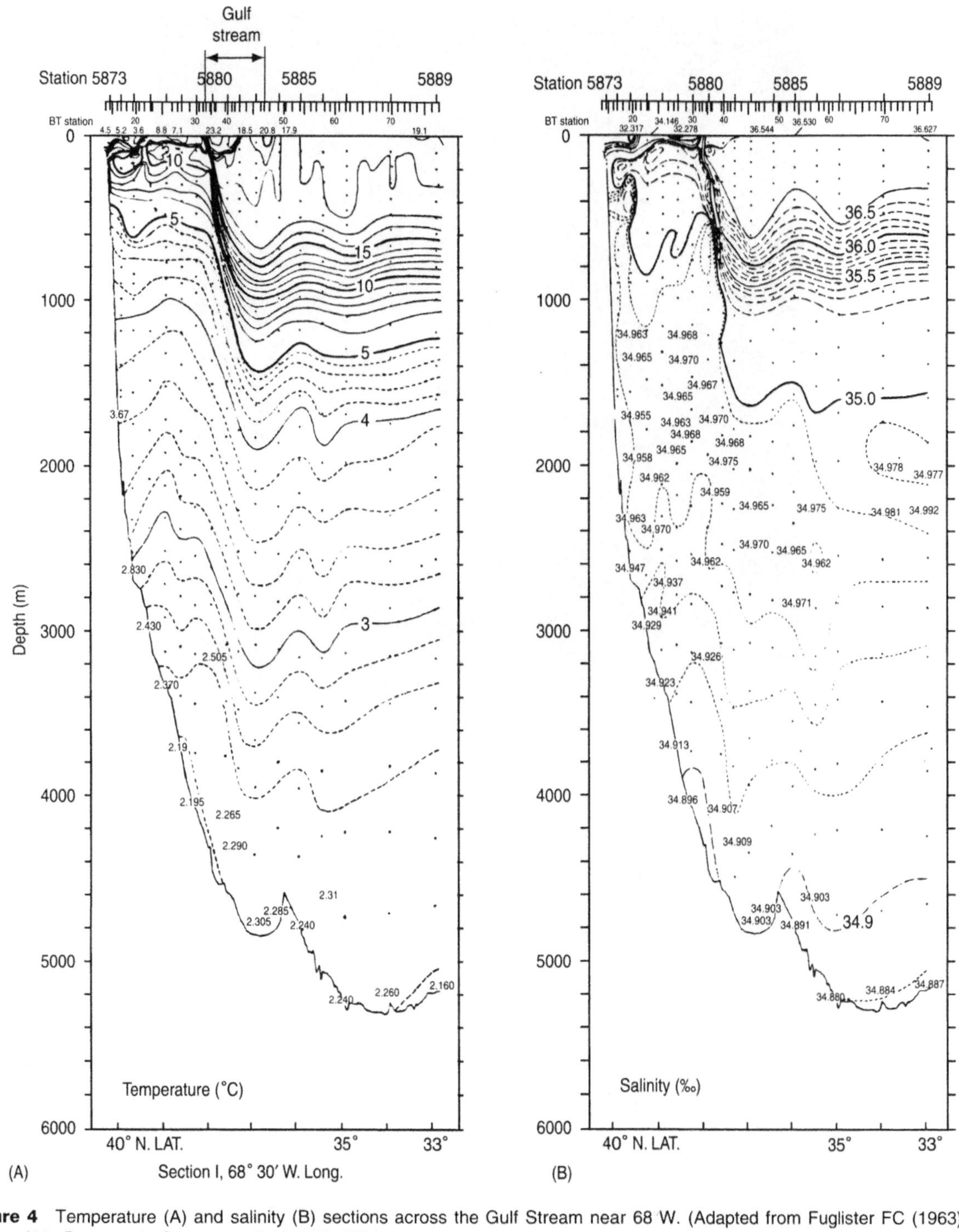

Figure 4 Temperature (A) and salinity (B) sections across the Gulf Stream near 68 W. (Adapted from Fuglister FC (1963) Gulf Stream '60. *Progress in Oceanography* 1: 265–373.)

investigating these climate variations of the ocean and atmosphere.

Transport

The volume transport of the Florida Current or amount of water flowing in it has been measured to be around 30 Sv where 1 Sverdrup (Sv) $= 10^6\,m^3\,s^{-1}$. This transport is around two thousand times the annual average transport of the Mississippi River into the Gulf of Mexico. As the Stream flows northward, its transport increases 5-fold to around 150 Sv located south of Nova Scotia. Most of the large transport is recirculated westward in recirculating gyres located

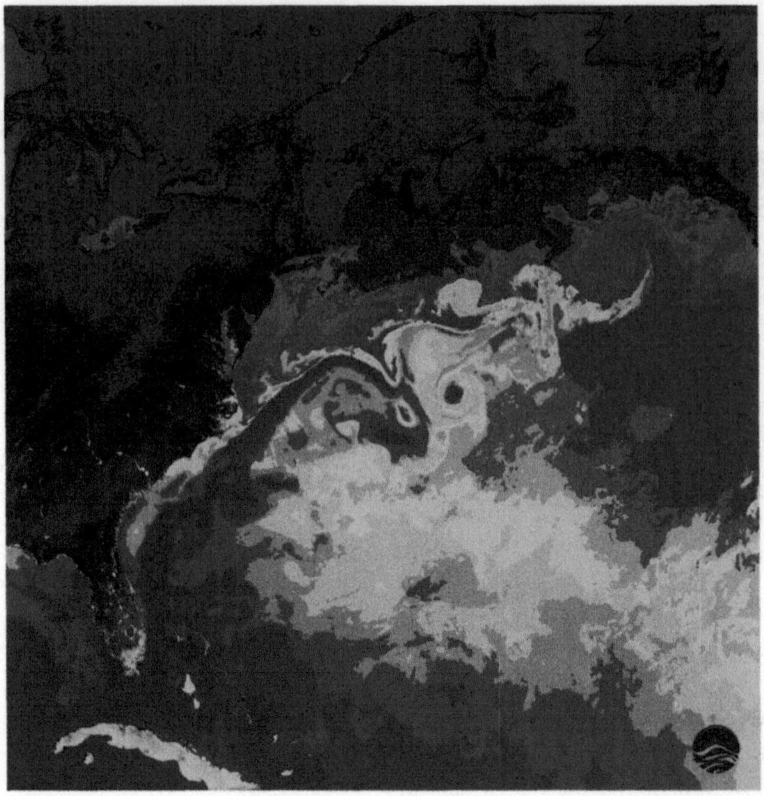

Figure 5 Satellite infrared image showing sea surface temperature distribution. Warm water (orange-red) is advected northward in the Gulf Stream. Meanders and eddies can be inferred from the convoluted temperature patterns. Image courtesy of O. Brown, R. Evans and M. Carle, University of Miami, Rosenstiel School of Marine and Atmospheric Science.

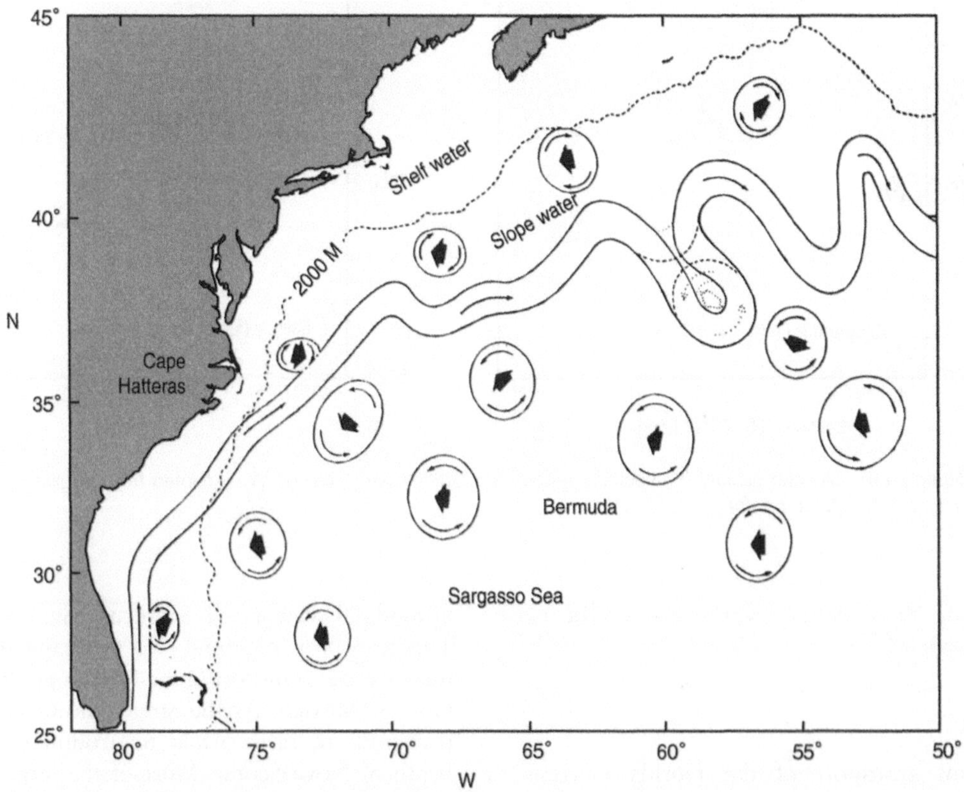

Figure 6 Schematic representation of the instantaneous path of the Gulf Stream and the distribution and movement of rings. Each year approximately 22 warm core rings form north of the Gulf Stream and 35 cold core rings form south of the Stream. (Reproduced from Richardson PL (1976) Gulf Stream rings. *Oceanus* 19(3): 65–68.)

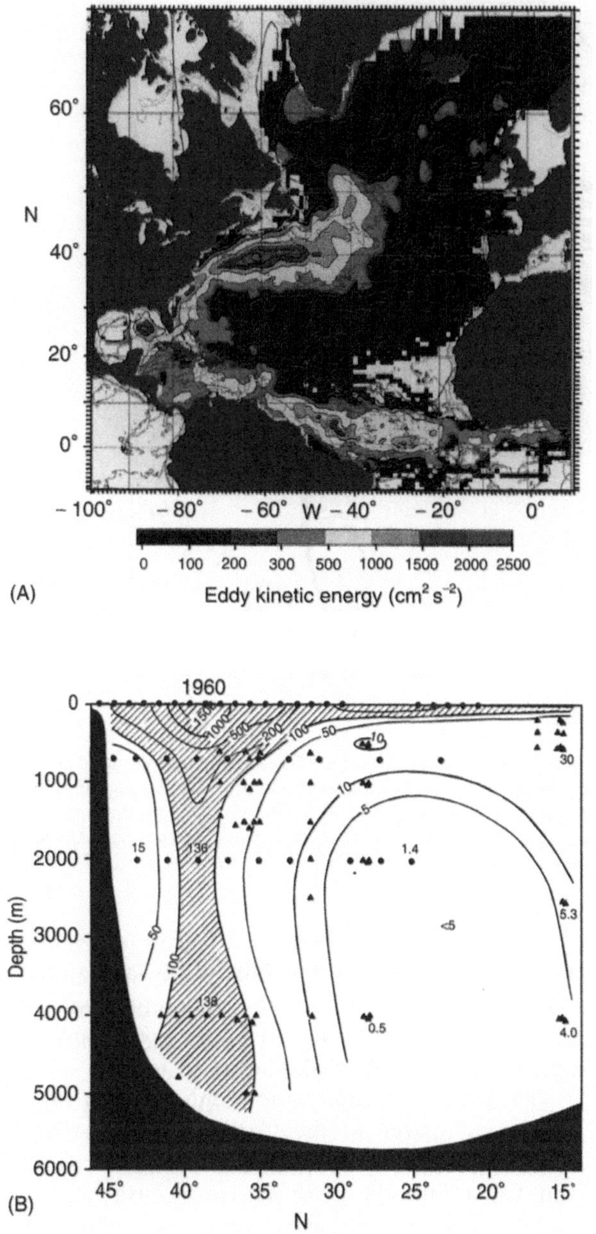

Figure 7 (A) Horizontal distribution of eddy kinetic energy (EKE) in the North Atlantic based on a recent compilation of surface drifter data. EKE is equal to $\frac{1}{2}(\overline{u'^2} + \overline{v'^2})$ where u' and v' are velocity fluctuations from the mean velocity calculated by grouping drifter velocities into small geographical bins. High values of EKE ($cm^2\ s^{-2}$) coinciding with the Gulf Stream are 10 times larger than background values in the Sargasso Sea. (Courtesy of D. Fratantoni, WHOI.) (B) Vertical section of EKE ($cm^2\ s^{-2}$) across the Gulf Stream system and subtropical gyre near 55°W based on surface drifters, SOFAR floats at 700 m and 2000 m (dots) and current meters (triangles). High values of EKE coincide with the mean Gulf Stream axis located near 40°N (roughly). (Adapted from Richardson PL (1983) A vertical section of eddy kinetic energy through the Gulf Stream system. *Journal of Geophysical Research* 88: 2705–2709.)

north and south of the Gulf Stream axis (**Figure 8**). The 150 Sv is much larger than that estimated for the subtropical gyre from wind stress (~30–40 Sv) and the net MOC (~15 Sv). Numerical models of the Gulf Stream suggest that the large increase in transport and the recirculating gyres are at least partially generated by the Stream's energetic velocity fluctuations. The region of largest EKE coincides with the maximum transport and recirculating gyres both in the ocean and in models. Inertial effects and buoyancy forcing are also thought to contribute to the recirculating gyres.

The velocity and transport of the Stream can be subdivided into two parts, a vertically sheared or baroclinic part consisting of fast speeds near the surface decreasing to zero velocity at 1000 m, and a constant or barotropic part without vertical shear that extends to the sea floor underneath the Stream. The baroclinic part of the Stream remains nearly constant (~50 Sv) with respect to distance downstream from Cape Hatteras, but the barotropic part more than doubles, from around 50 Sv to 100 Sv, causing the large increase in transport. The lateral meandering of the Stream and the north and south recirculating gyres are also highly barotropic and extend to the seafloor.

Estimates of the transport of the North Atlantic Current east of Newfoundland are also as large as 150 Sv. A recirculating gyre, known as the Newfoundland Basin Eddy, is observed to the east of this current.

North Brazil Current

The North Brazil Current (NBC) is the major western boundary current in the equatorial Atlantic. The NBC is the northward-flowing western portion of the clockwise equatorial gyre that straddles the equator. Near 4°N the mean transport of the NBC is around 26 Sv, which includes 3–5 Sv of flow over the Brazilian shelf. Maximum near-surface velocities are over $100\ cm\ s^{-1}$.

Large seasonal changes in near-equatorial currents are forced by the annual meridional migration of the Intertropical Convergence Zone that marks the boundary of the north-east tradewinds located to the north and the south-east trades to the south. The meridional migration of the winds causes large annual variations of wind stress over the tropical Atlantic. During summer and fall most of the NBC turns offshore or retroflects near 6°N and flows eastward between 5 to 10°N in the North Equatorial Countercurrent. During July and August the transport of the NBC increases to 36 Sv; most of the transport lies above 300 m, but the northward

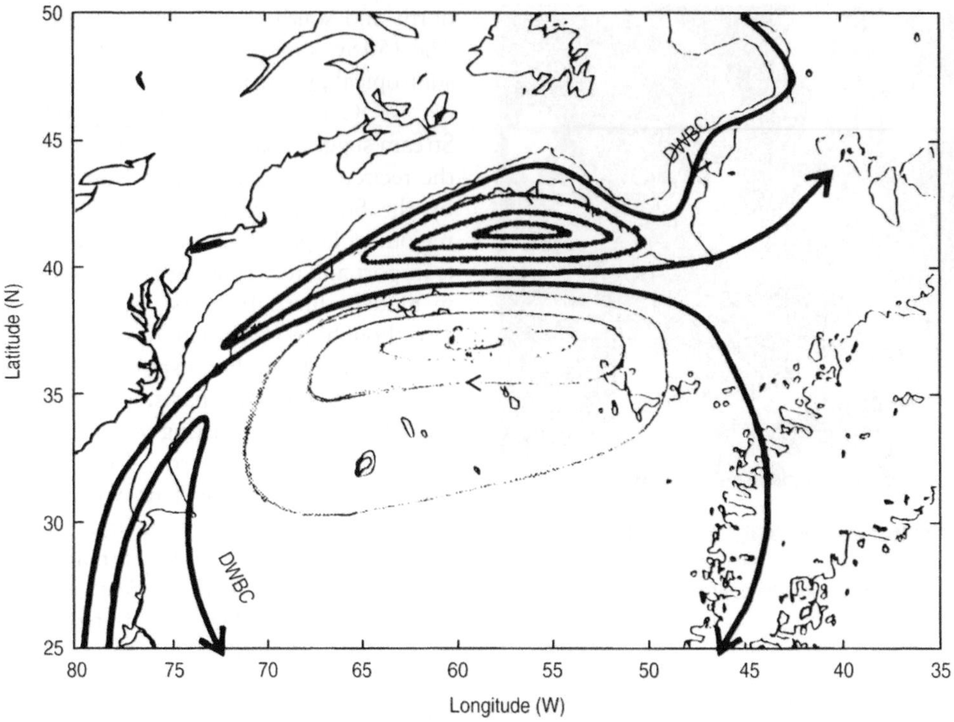

Figure 8 Schematic circulation diagram showing average surface to bottom transport of the Gulf Stream, the northern and southern recirculating gyres, and the DWBC. Each streamline represents approximately 15 Sv. Because of the vertical averaging, the DWBC looks discontinuous in the region where it crosses under or through the Gulf Stream near Cape Hatteras. (Adapted from Hogg NG (1992) On the transport of the Gulf Stream between Cape Hatteras and the Grand Banks. *Deep-Sea Research* 39: 1231–1246.)

current extends down to roughly 800 m and includes Antarctic Intermediate Water. In April and May the transport decreases to 13 Sv and the NBC becomes weaker, more trapped to the coast, and more continuous as a boundary current. At this time the countercurrent also weakens.

Occasionally, 3–6 times per year, large 400-km diameter pieces of the retroflection pinch off in the form of clockwise-rotating current rings. NBC rings translate northward along the western boundary 10–20 cm s^{-1} toward the Caribbean, where they collide with the Antilles Islands. The northward transport carried by each ring is around 1 Sv. During fall–spring the western boundary current region north of the retroflection is dominated by energetic rings.

Amazon River Water debouches into the Atlantic near the equator, is entrained into the western side of the NBC, and is carried up the coast. Some Amazon Water is advected around the retroflection into the countercurrent and some is caught in rings as they pinch off. Farther north near 10°N the Orinoco River adds more fresh water to the north-westward-flowing currents.

The path of South Atlantic water toward the Gulf Stream is complicated by the swift, fluctuating currents. Some northward transport occurs in alongshore flows, some in NBC rings, and some by first flowing eastward in the countercurrent, then counterclockwise in the tropical gyre, and then westward in the North Equatorial Current.

The South Atlantic Water and Gulf Stream recirculation merge in the Caribbean. Water in the Caribbean Current is funneled between Yucatan and Cuba into the swift narrow (~100 km wide) Yucatan Current. The flow continues in the Gulf of Mexico as the Loop Current, and then exits through the Straits of Florida as the Florida Current. Roughly once per year a clockwise rotating current ring pinches off from the Loop Current in the Gulf of Mexico and translates westward to the western boundary, where a weak western boundary current is observed.

Labrador Current

The Labrador Current is formed by very cold −1.5°C water from the Baffin Island Current and a branch of the West Greenland Current, which merge on the western side of the Labrador Sea. The current flows southward from Hudson Strait to the

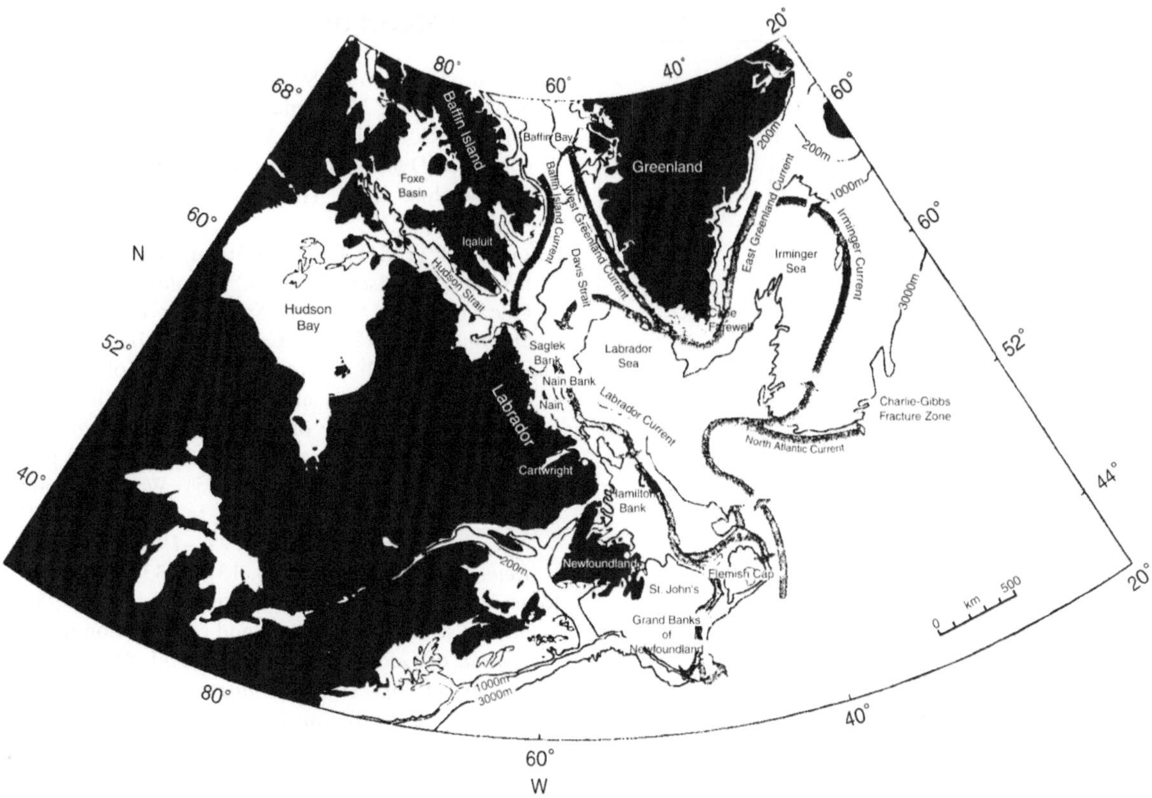

Figure 9 Schematic figure showing the North Atlantic subpolar gyre circulation and the Labrador Current. (Reprodued from Lazier and Wright (1993).)

southern edge of the Grand Banks of Newfoundland (**Figure 9**). The Labrador Current consists of two parts. The first ~ 11 Sv is located over the shelf and upper slope and is concentrated in a main branch over the shelf break. This part is highly baroclinic and is thought to be primarily buoyancy-driven by fresh water input from the north. The second part, the deep Labrador Current, lies farther seaward over the lower continental slope, is more barotropic, and extends over the full water depth down to around 2500 m. Below roughly 2000 m is Nordic Seas overflow water, which also flows southward along the western boundary. The deep Labrador Current is the western portion of the subpolar gyre, which has a transport of around 40 Sv.

Most of the Labrador Current water leaves the boundary near the Grand Banks, part entering a narrow northward-flowing recirculation and part flowing eastward in the subpolar gyre. Very sharp horizontal gradients in temperature and salinity occur where cold, fresh Labrador Current Water is entrained into the edge of the North Atlantic Current. Some Labrador Current Water passes around the Grand Banks and continues westward along the shelf and slope south of New England and north of the Gulf Stream (**Figure 10**).

Deep Western Boundary Current (DWBC)

The DWBC flows southward along the western boundary from the Labrador Sea to the equator. Typical velocities are 10–20 cm s^{-1} and its width is around 100–200 km. It is comprised of North Atlantic Deep Water that originates from very cold, dense Nordic Seas overflows and from less cold and less dense intermediate water formed convectively in the Labrador Sea in late winter. The DWBC is continuous in that distinctive water properties like the high freon content in the overflow water and in Labrador Sea Water have been tracked southward to the equator as plumes lying adjacent to the boundary. However, the DWBC is discontinuous in that it is flanked by relatively narrow sub-basin-scale recirculating counterflows that exchange water with the DWBC and recirculate DWBC water back northward. The net southward transport of the DWBC is thought to be around 15 Sv, but the measured southward transport is often two or three times that, the excess over 15 Sv being recirculated locally.

Near the Grand Banks of Newfoundland, part of the DWBC water leaves the boundary and divides

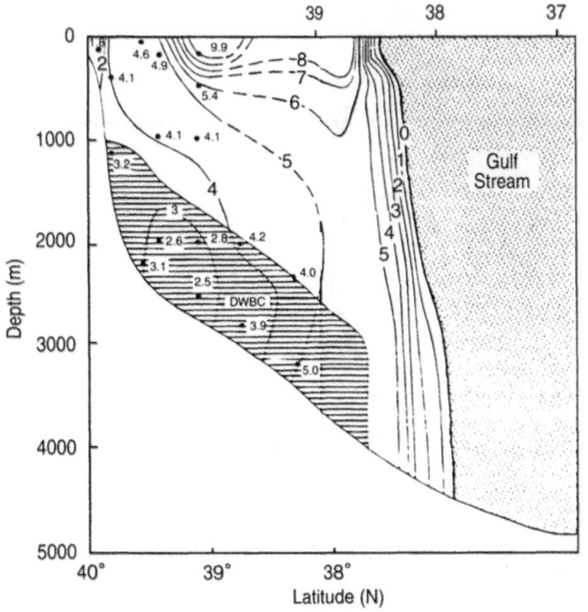

Figure 10 Vertical section near 68°W of mean zonal currents in the Slope Water region north of the Gulf Stream. Plotted values are westward velocity in cm s^{-1}. The region of the DWBC is indicated schematically by light lines. (Adapted from Johns *et al.* (1995).)

into a northward-flowing recirculation and the eastward-flowing subpolar gyre circulation. The other part flows around the Grand Banks and westward inshore of the Gulf Stream. Between the Grand Banks and Cape Hatteras, the DWBC coincides with the northern recirculating gyre (or Slope Water gyre), which also flows westward there (**Figure 10**). Near Cape Hatteras the DWBC encounters the deep Gulf Stream. Most of the deeper part of the DWBC seems to flow continuously south-westward underneath the mean axis of the Stream, but most of the water in the upper part of the DWBC appears to be entrained into the deep Gulf Stream and flows eastward. Some of this water recirculates in the northern recirculating gyre, and some crosses the mean axis of the Stream, recirculates westward south of the Stream toward the western boundary, and continues southward. The formation of energetic Gulf Stream meanders and pinched-off current rings is probably an important mechanism by which DWBC water passes across or through the mean axis of the Stream.

See also

Benguela Current. Brazil and Falklands (Malvinas) Currents. Mesoscale Eddies. Wind Driven Circulation.

Further Reading

Bane JM (1994) Gulf Stream system: an observational perspective. In: Majumdar SK, Miller EW, Forbes GS, Schmalz RF, and Panah AA (eds.) *The Oceans: Physical-Chemical Dynamics and Human Impact*, pp. 99–107. Philadelphia, PA: The Pennsylvania Academy of Science.

Fofonoff NP (1981) The Gulf Stream system. In: Warren BA and Wunsch C (eds.) *Evolution of Physical Oceanography, Scientific Surveys in Honor of Henry Stommel*, pp. 112–139. Cambridge, MA: MIT Press.

Fuglister FC (1960) Atlantic Ocean atlas of temperature and salinity profiles and data from the International Geophysical Year of 1957–1958. *Woods Hole Oceanographic Institution Atlas Series* 1: 209 pp.

Hogg NG and Johns WE (1995) Western boundary currents. *U.S. National Report to International Union of Geodesy and Geophysics 1991–1994*. Supplement to *Reviews of Geophysics*, pp. 1311–1334.

Johns WE, Shay TJ, Bane JM, and Watts DR (1995) Gulf Stream structure, transport and recirculation near 68W. *Journal of Geophysical Research* 100: 817–838.

Krauss W (1996) *The Warmwatersphere of the North Atlantic Ocean*. Berlin: Gebrüder Borntraeger.

Lazier JRN and Wright DG (1993) Annual velocity variations in the Labrador Current. *Journal of Physical Oceanography* 23: 659–678.

Lozier MS, Owens WB, and Curry RG (1995) The climatology of the North Atlantic. *Progress in Oceanography* 36: 1–44.

Peterson RG, Stramma L, and Korturn G (1996) Early concepts and charts of ocean circulation. *Progress in Oceanography* 37: 1–15.

Reid JL (1994) On the total geostrophic circulation of the North Atlantic Ocean: flow patterns, tracers, and transports. *Progress in Oceanography* 33: 1–92.

Robinson AR (ed.) (1983) *Eddies in Marine Science*. Berlin: Springer-Verlag.

Schmitz WJ and McCartney MS (1993) On the North Atlantic circulation. *Reviews of Geophysics* 31: 29–49.

Schmitz WJ Jr. (1996) *On the World Ocean Circulation*, vol. I: *Some Global Features/North Atlantic Circulation*. Woods Hole Oceanographic Institution Technical Report, WHOI-96-03, 141 pp.

Stommel H (1965) *The Gulf Stream, A Physical and Dynamical Description*, 2nd edn. Berkeley, CA: University of California Press.

Worthington LB (1976) On the North Atlantic circulation. *The Johns Hopkins Oceanographic Studies*, 6. Baltimore, MD: The Johns Hopkins University Press.

BENGUELA CURRENT

L. V. Shannon, University of Cape Town, Cape Town, South Africa

Introduction

The Benguela, which shares its name with a town in Angola, is one of four major current systems situated at the eastern boundaries of the world oceans, and the oceanography of the region is in many respects similar to that of the Canary Current off north-west Africa, the California Current off the west coast of the USA, and the Humboldt Current off Peru and Chile. The Benguela is, however, unique in that it is bounded at both equatorward and poleward ends by warm-water systems. Eastern boundary current systems are characterized by wind-driven upwelling along the coast of cold, nutrient-rich water which supports high biological productivity.

Different interpretations exist as to what constitutes the Benguela Current, its ecosystem and its boundaries. In this article, the broader definition of the Benguela will be used and the currents and physical processes which occur in that part of the South Atlantic between 5°S and 38°S and east of the 0° meridian will be considered. This encompasses the coastal upwelling regime, the eastern part of the South Atlantic gyre, and a series of fronts and transitional zones, overlying some complex bathymetry. The continental shelf along the west coast of southern Africa is variable in width, being narrow off southern Angola, near Lüderitz in Namibia, and near Cape Town, and widest off the Orange River and in the extreme south where the Agulhas Bank protrudes polewards (**Figure 1**). The shelf-break (edge of the continental shelf) lies at depths between 200 m and 500 m, from which a steep continental slope descends to about 5000 m where it meets the abyssal plains of the Cape and Angola Basins which, in turn are separated by an extensive submarine mountain chain, the Walvis Ridge (**Figure 1**).

The earliest physical measurements in the area were those necessary for the safe and efficient passage of sailing ships along the trade routes between Europe and the East – wind, currents, waves. The records of the early navigators contain a remarkable amount of information about the physical oceanography of the region, and charts of currents published in the 1700s display features which were 'discovered' by scientists during the past half century! However, the application of late twentieth century technology to the study of the Benguela – moored instruments, surface and submerged floats and in particular satellite technology – has highlighted the complexity and variability of the Benguela. Like the adjacent African continent it is an area of contrasts.

Winds

Winds significantly influence the oceanography of the Benguela region on various time- and space-scales from events of only a few hours duration to basin-wide seasonal and longer period processes. The prevailing winds along the west coast of southern Africa are controlled by the anticyclonic (anticlockwise in the Southern Hemisphere) motion around the South Atlantic High (SAH) pressure system, the seasonal pressure field over the land, and eastward moving cyclones which cross the southern part of the continent. The SAH, part of a discontinuous belt of high pressure which encircles the Southern Hemisphere, is maintained throughout the year but migrates seasonally, being further north in the austral winter. The pressure over the land alternates between a well-developed low during summer and a weak high in winter, and so winds are seasonally variable. The coastal plain acts as a thermal barrier to cross-flow, and hence winds tend to be longshore southerly over most of the Benguela region, steered by the coast. It is these winds which produce coastal upwelling. The essential seasonal and longshore differences in the intensity of the upwelling-producing winds are illustrated in **Figure 2**.

The principal area of perennially strong southerly winds lies near Lüderitz (27°S). In winter the northward shift of the pressure systems has its strongest influence south of 31°S, where there is a relaxation of the southerly winds and increased frequency of westerlies. Off northern Namibia the southerly winds are strongest during autumn and spring, while north of 15°S (southern Angola) winds are weak throughout the year, although still strongest during winter. Hot dry, dusty downslope winds ('Berg' winds) – analogous to the Santa Ana in California – are common during autumn. Coastal winds are pulsed on timescales of 3–10 days over much of the Benguela – particularly in the south (during summer) by the passage of easterly moving cyclones associated with the belt of westerlies which

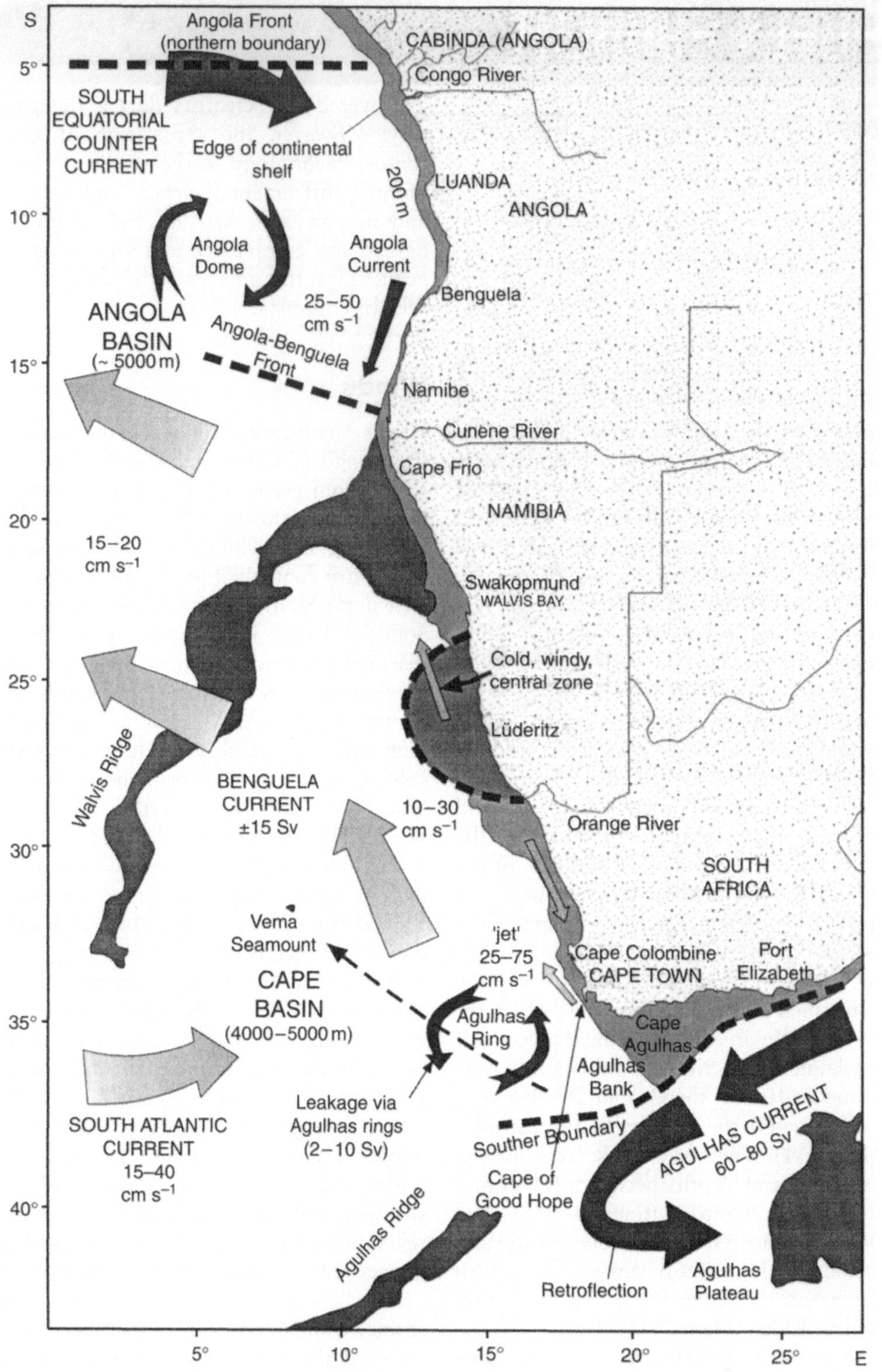

Figure 1 Schematic of the Benguela Current region showing general bathymetry, currents in the upper layer, and system boundaries.

lies south of Africa. This pulsing of the winds and the associated changes in atmospheric pressure are important in terms of the dynamics of coastal upwelling and shelf processes. Diurnal changes in coastal winds are common throughout the Benguela north of 33°S – a classical land–sea breeze effect.

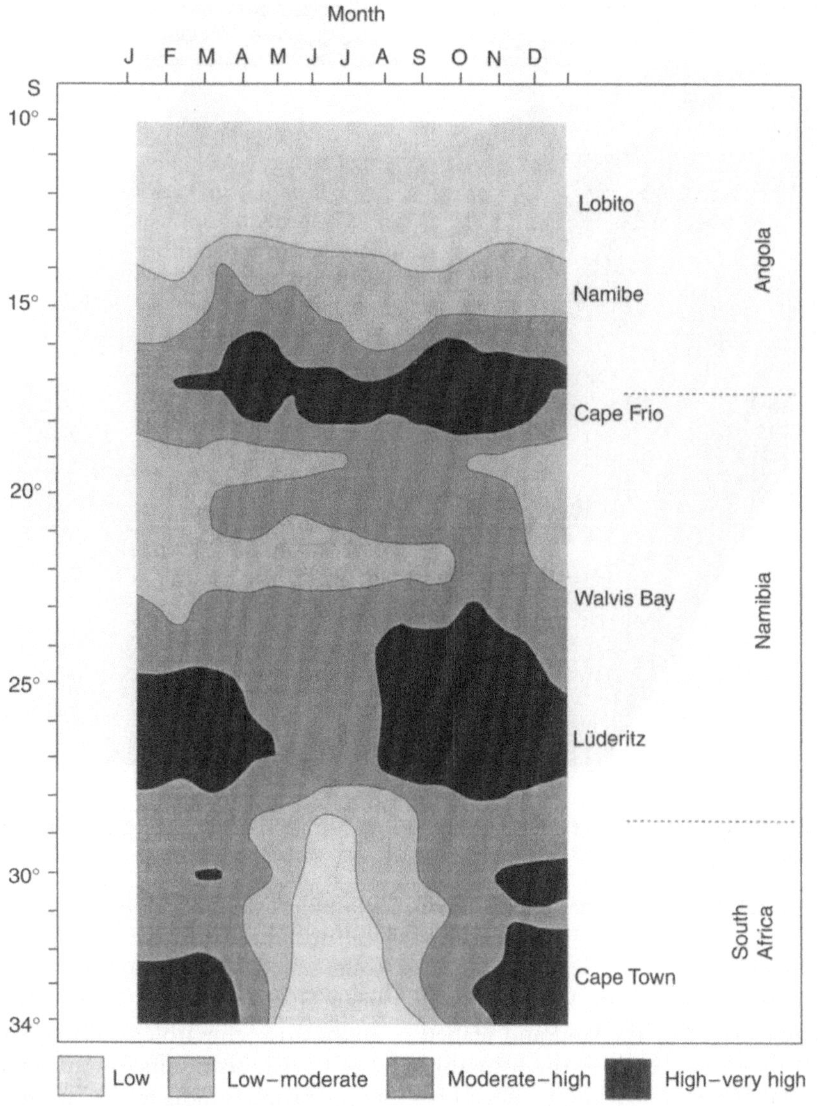

Figure 2 Monthly wind stress along the west coast of southern Africa 75 km offshore (according to A.J. Boyd, 1987). Note the zones of high wind stress near Lüderitz, Cape Frio and also the seasonality in the south.

Upwelling

The process whereby cold subsurface water is brought to the surface near the coast as a consequence of longshore equatorward winds is termed coastal upwelling. In simple terms, longshore winds displace warm surface water equatorwards and, as a consequence of the Earth's rotation, offshore, resulting in a drop in sea level against the coast, and an uplift of water from below (and alongshore) to correct the imbalance. In a simple one cell system, the thermocline (layer where there is a strong vertical temperature gradient) is tilted upwards, and may result in a front (horizontal gradient) between the cool upwelled water and the warmer oceanic water, with water moving at depth over the shelf and upwards, and sinking at the front. Slow poleward undercurrents and fast equatorward jet currents (near the front or shelf-break) are often characteristic of coastal upwelling systems. The process is illustrated schematically in **Figure 3**.

This description is really an over-simplification, as two or three longshore fronts may develop with complex circulations in between, while the actual degree of upwelling and the intensity and direction of shelf currents will be significantly influenced by coastal trapped waves. (The latter is a type of internal wave in the ocean over the shelf which propagates polewards at speeds of around 4–8 cm s^{-1}, displacing the thermocline and trapped or guided by the coast.) The existence of these coastal

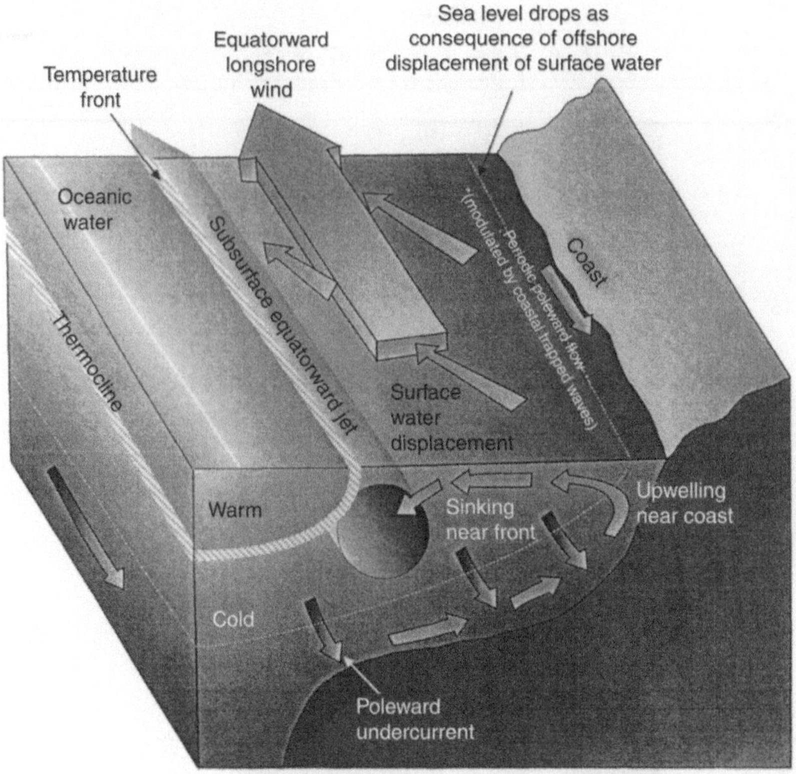

Figure 3 Schematic showing the structure of coastal upwelling and associated circulation in the Southern hemisphere.

trapped waves can result in enhanced or reduced upwelling and larger sea level changes than might be inferred from the wind.

The wind field (in particular the curl of wind stress), topographic features on the land and seabed, and orientation of the coast, result in the formation of a number of areas where upwelling is more pronounced. The principal upwelling cell in the Benguela is near Lüderitz (27°S) and strong upwelling occurs there throughout the year. This cell effectively divides the system into two quasi-independent subsystems. Other cells exist north of Cape Frio (at about 18°S), off northern Namibia, and at 31°S, 33°S, and 34°S (Cape Town), the last two being seasonal. In the southern area maximum upwelling occurs between September and March, whereas off northern and central Namibia and southern Angola upwelling is more perennial with a late winter-spring maximum.

In the northern Benguela, upwelling and insolation (solar heating) are out of phase and inshore sea surface temperatures have a clear seasonal cycle. In the south, upwelling and insolation are in phase and surface temperatures vary little seasonally (in places by only 1°C). The main upwelling areas of the Benguela, i.e. south of 15°S, is a major heat sink, and negative climatological sea surface temperature anomalies of 5–6°C occur off Lüderitz. In sharp contrast the ocean off Angola in summer is a heat source, as is the region south of Africa which is influenced by the warm Agulhas Current where there are positive climatological temperature anomalies of 2–4°C. All this results in some spectacular horizontal temperature gradients or fronts in the south-east Atlantic. These, the complexity of the area and the extent of coastal upwelling are illustrated in **Figure 4**.

Water Masses and Large-scale Circulation

Overviews of the water masses and general circulation in the Atlantic Ocean are given in Current Systems in the Atlantic Ocean: respectively. In the broadly defined Benguela region the principal water masses are Tropical and Subtropical Surface Waters, Central (Thermocline) Waters, Antarctic Intermediate Water (AAIW), North Atlantic Deep Water (NADW), and Antarctic Bottom Water (AABW). Central water which corresponds to the linear part of the temperature–salinity curve applicable to the south-east Atlantic (approximately 6°C, 34.5 – 16°C, 35.5; see **Figure 5**), lies below the surface layer, and it is this water which upwells along the west coast of

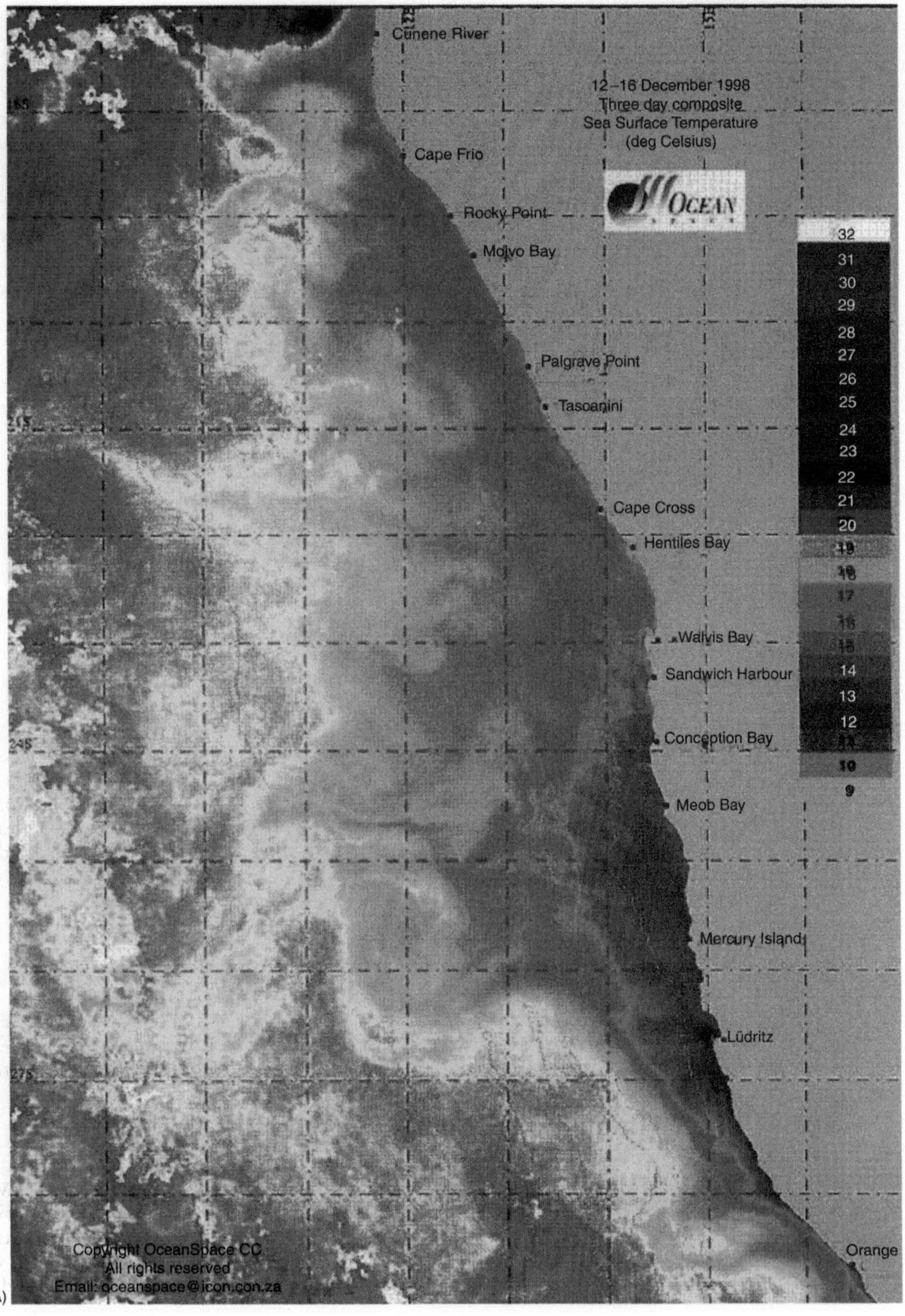

(A)

Figure 4 Satellite sea surface temperature images (°C): (A) off Namibia, (B) South Africa. (Courtesy of OceanSpace CC.)

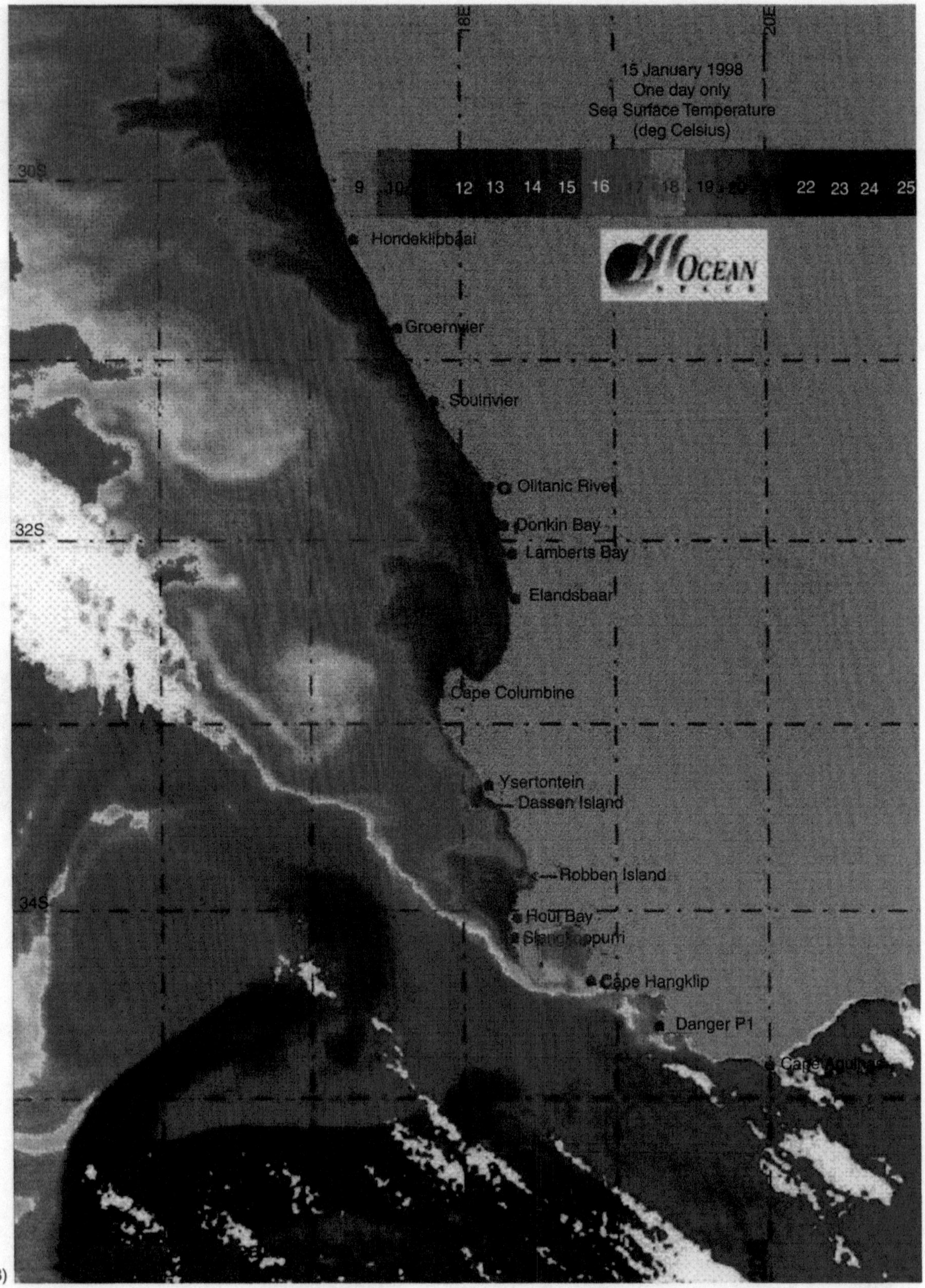

(B)

Figure 4 *Continued*

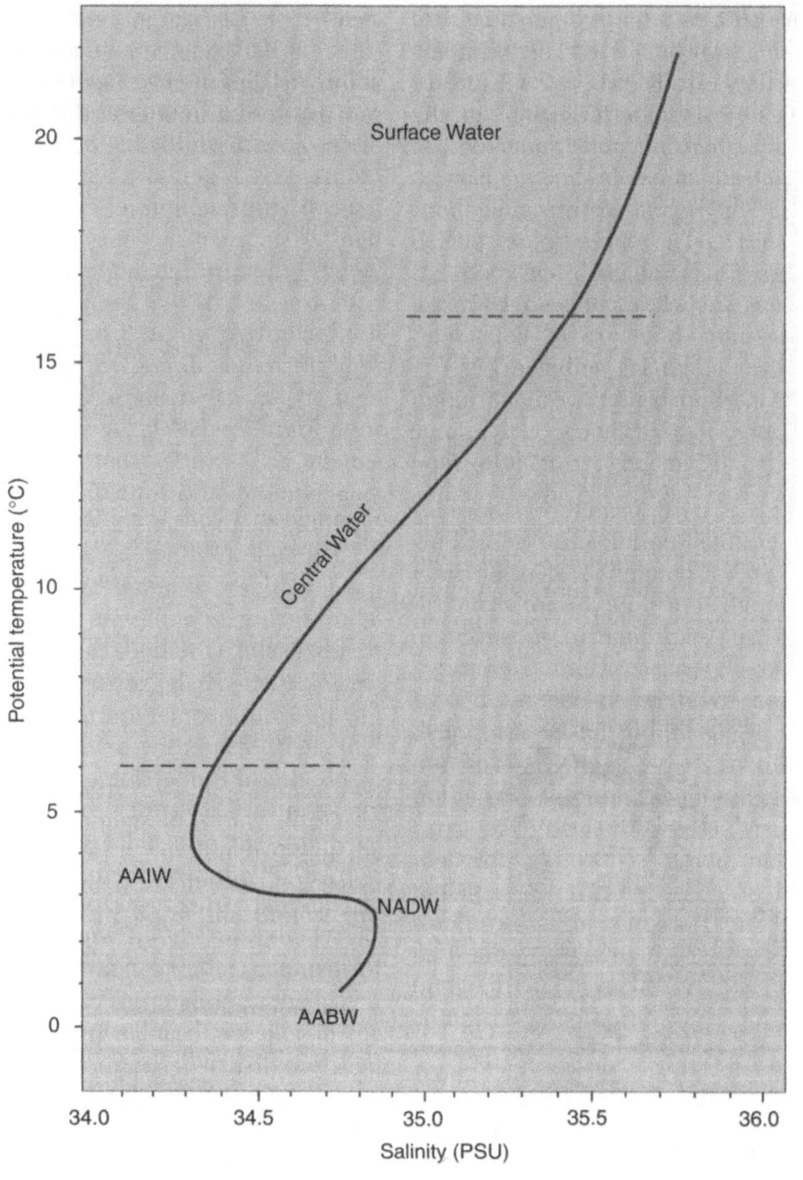

Figure 5 Characteristic temperature–salinity relationship for the south-east Atlantic between 30° and 35°S.

southern Africa. The characteristics of this 'water mass' in the region are quite variable, reflecting different origins, i.e. from the 'open' South Atlantic, the Indian Ocean, and the tropical Atlantic. Except over, or adjacent to, the shelf the flow of central water(s) is similar to that of the overlying surface water(s). A generalized description of the circulation in the upper layer is provided in **Figure 1**. The broad gray arrows between 15°S and 35°S represent the Benguela Current, which is defined as the integrated equatorward flow in the upper layers in the south-east Atlantic east of the 0° meridian. Speeds are in the range 10–30 $cm\,s^{-1}$. North of 30°S, the Benguela Current has a pronounced westward component. The total equatorward transport in the Benguela of surface, central and AAIW is thought to be 15–25 Sv (1 Sv is $10^6\,m^3\,s^{-1}$). In the south the Benguela Current is fed by the South Atlantic Current and by leakage from the Indian Ocean/Agulhas Current. This leakage of 2–10 Sv takes place mainly via the shedding of Agulhas rings at the retroflection of the Agulhas – typically six rings per year which have a translation speed of 5–8 $cm\,s^{-1}$ (*see* Agulhas Current) and which result in a net equatorward transport of heat in the south-east Atlantic. North of 15°S, the region is dominated by the influence of the South Equatorial Counter Current/Under Current (*see* Atlantic Ocean Equatorial Currents), the circulation around the quasi-seasonal Angola Dome and the coastal poleward-flowing Angola Current.

Circulation in the area between 5°S and 15°S has to be considered an integral part of the Benguela Current proper, particularly in terms of subsurface and also biochemical processes. It is here in the vicinity of the Angola Dome and in the area to the east of it that the main oxygen minimum layer in the South Atlantic forms. **Figure 6** illustrates conceptually the linkages between this area and the Benguela

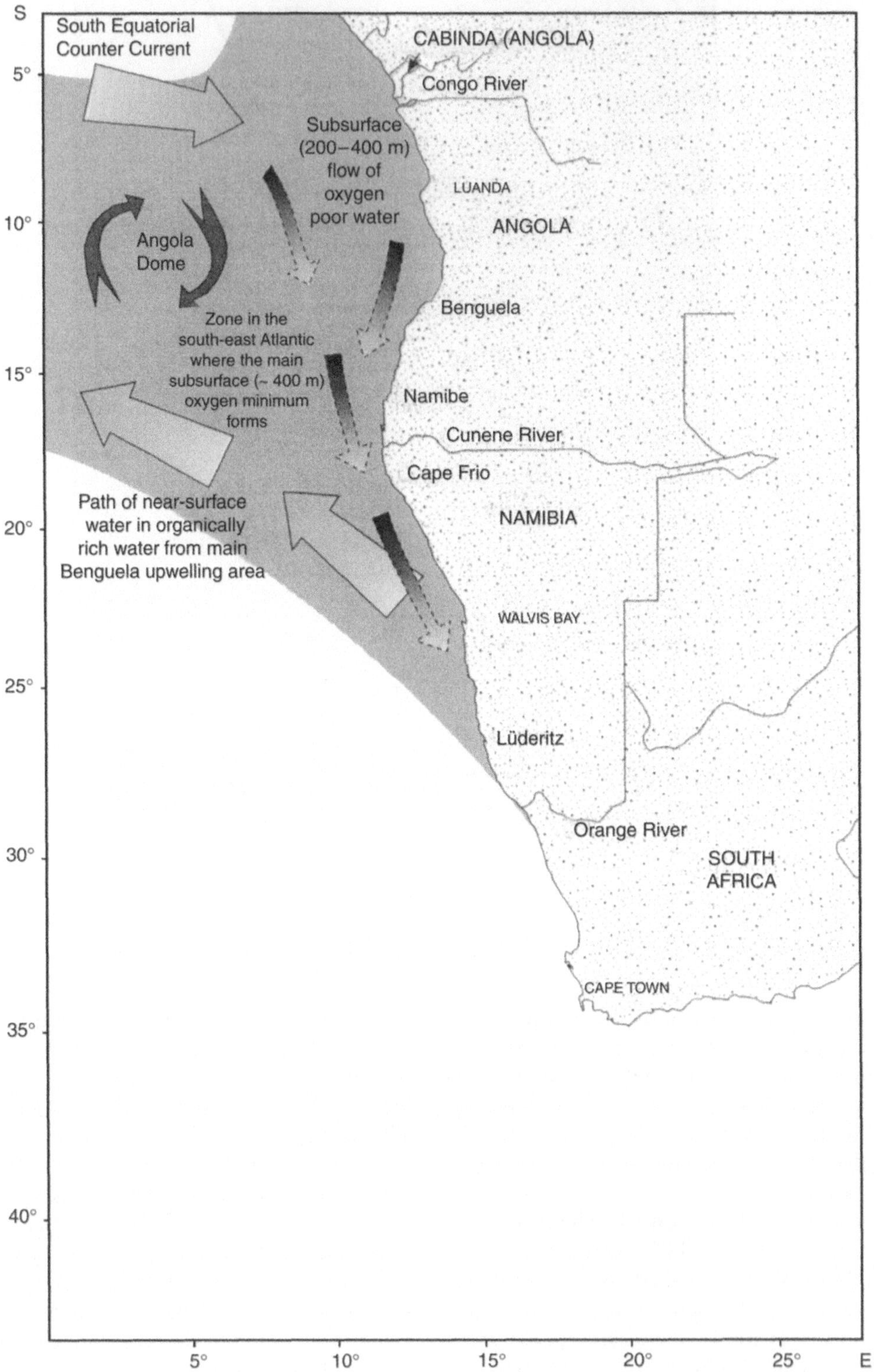

Figure 6 Schematic showing linkage between the shelf system off Namibia and processes off Angola.

shelf system, with organically rich water advected from the highly productive Namibian upwelling area northwards and westwards. The sinking and decay of organic particulates form a marked oxygen minimum at around 400 m. This oxygen poor water subsequently advects slowly polewards adjacent to the continental shelf at depths of 200–400 m. It must be emphasized that **Figure 6** is somewhat speculative; it is only intended to convey a perceived concept.

AAIW which has a core temperature and salinity of 4–5°C and 34.2–34-5 and corresponds to the salinity minimum is present throughout the region at an average depth of 700–800 m, and in volume accounts for about 50% of the water present above 1500 m. It is much fresher in the south west than in the south east and in the north, again reflecting different origins, i.e. temperate South Atlantic, Indian Ocean, and tropical Atlantic. A schematic of the origins and mean flow of AAIW based on analysis of water properties is given in **Figure 7**. (A recent international experiment using subsurface 'floats' which can track the movement of water masses such as the AAIW has shown that 'snapshots' of the flow of AAIW around southern Africa is much more complex than the mean pattern given in **Figure 7**.)

Below the AAIW lies NADW, corresponding to the deep salinity maximum (typically >34.8 PSU) that comprises a thick layer between 1000 m and 3500 m of relatively warm saline water. West of southern Africa its flow is generally polewards, becoming diluted *en route*. Volumetrically NADW is the main water mass present in the region. The deepest water mass present in the Cape Basin is AABW, and on average it circulates slowly in a clockwise direction over the abyssal plain at depths below 4000 m. The Walvis Ridge forms an almost impenetrable barrier to the northward movement of AABW into the Angola Basin, as a consequence it is not significantly present there (*see* Abyssal Currents).

Shelf Circulation

Over much of the Benguela shelf, surface currents are largely influenced by prevailing winds. Poleward flow often occurs close inshore near the surface (see **Figure 1**). In the extreme south, there is a convergent flow of surface water from the Agulhas bank funnelling northwards into a shelf-edge frontal jet around the Cape of Good Hope with characteristic speeds of 25–75 cm s^{-1}. The surface current then separates into two components near 33°S (Cape Columbine). Over the Namibian shelf surface currents are usually in an equatorward direction, aligned to the prevailing wind, but periodic and episodic reversals do occur. Off Angola the coastal poleward-flowing Angola Current is detectable between the surface and 200 m, and current speeds of 70 cm s^{-1} at the surface and 88 cm s^{-1} subsurface have been reported during late summer. The dynamics of the Angola Current appear to be linked with the Angola Dome and the South Equatorward Counter Current. At the surface there is not much evidence for significant continuity of poleward flow of this current into Namibian waters, as the Angola Current tends to turn westwards just north of the Angola Benguela front. At 400 m the poleward flow does seem to be more continuous.

As in other coastal upwelling systems, a poleward undercurrent is a dominant characteristic of near bottom water over the shelf, extending from the coast to west of the shelf-break. The subtidal currents over the shelf are dominated by coastal trapped waves with periods of 3–10 days. The net poleward flow in the undercurrent is 5–8 cm s^{-1} or about 5 km d^{-1}. On occasions this southwards-moving current can reach the surface, resulting in episodes of poleward flow at the surface, i.e. when the zero flow intersects the sea surface. Tidal currents in the Benguela are of relatively small amplitude.

The characteristics of shelf circulation in the Benguela region can be summarized as follows:

- Wind-driven equatorward flow in the upper mixed layer between about 35°S and 15°S.
- Poleward-flowing coastal Angola Current north of 15°S
- Shelf-edge/frontal jets associated with the upwelling system in the southern Benguela.
- Poleward-propagating coastal trapped waves resulting in reversals with a period of 3–10 days.
- Small amplitude tidal and inertial high-frequency oscillations (10–15 cm s^{-1}).
- Poleward undercurrent over the shelf with net flow of about 6 km d^{-1}.

Viewed simplistically, much of the region can be viewed as having an equatorward-moving surface layer overlying water moving slowly polewards.

System Boundaries, Fronts, and Filaments

The physical boundaries of, and within, the Benguela are usually associated with fronts (see **Figure 1**). These tend to form barriers to the horizontal movement of water and small particles. Over much of the area between Cape Frio (18°S) and the Cape of Good Hope (34°S) there is a meandering (wave-like) longshore temperature front which approximately follows the shelf-break (see **Figure 4**). This is the boundary between the area influenced by coastal upwelling and the warmer oceanic water lying further west. The greater part of the total equatorward flow in the south-east Atlantic, i.e. the Benguela Current proper, takes place between the upwelling oceanic front and the 0° meridian. Major perturbations of the front produce filament-like features. These filaments are characteristic of upwelling systems, and have life spans of days to several weeks. Usually orientated perpendicular to the coast they

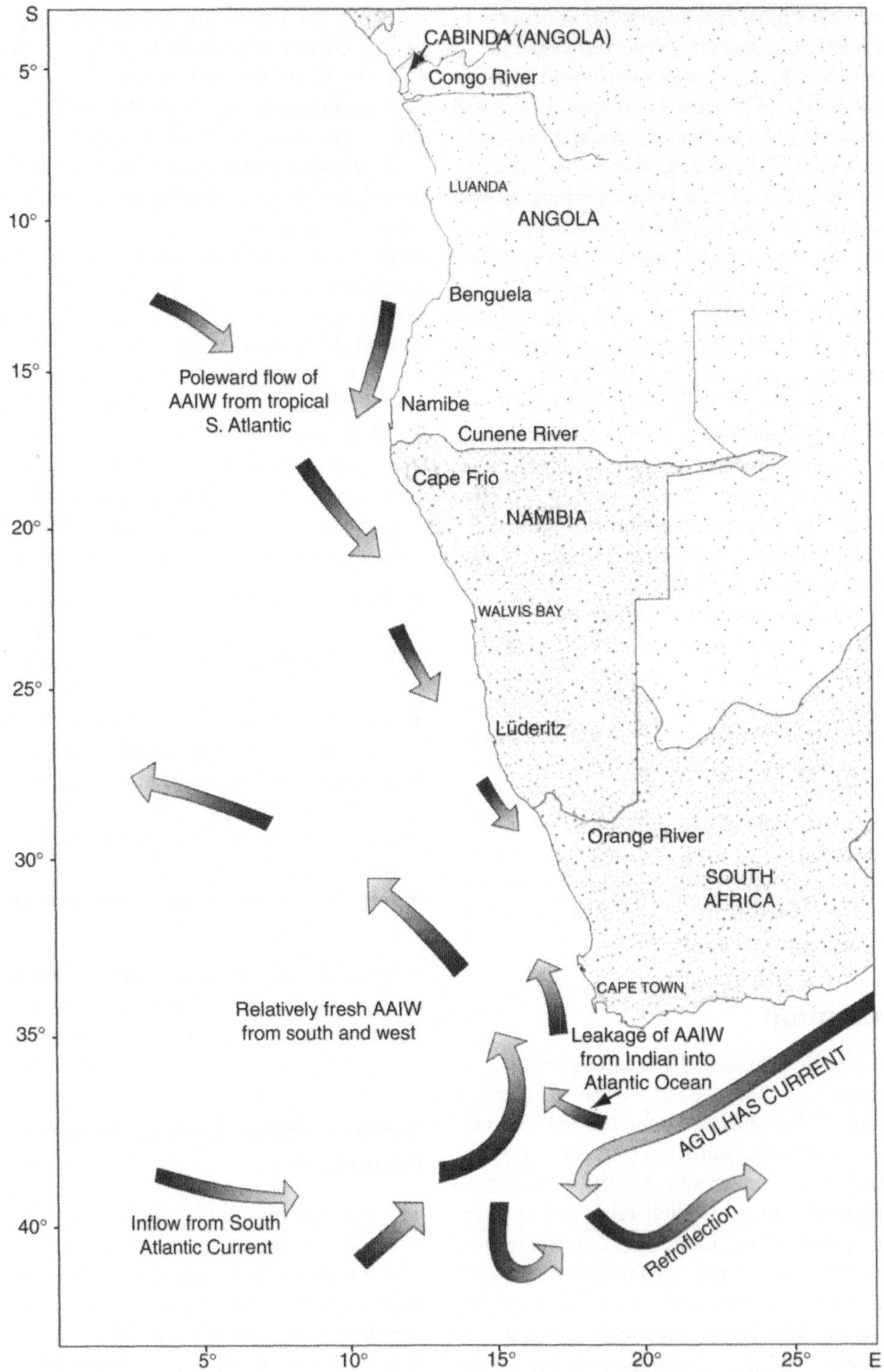

Figure 7 Deduced circulation of Antarctic Intermediate Water in the Benguela region.

cause the front to be highly convoluted. Benguela filaments are generally (but not always) site specific. Water circulation tends to follow the filament/front, so filaments may not be a significant source of transfer of water between coastal and oceanic systems.

The southern boundary of the Benguela system is the Agulhas retroflection area, characteristically between 36° and 38°S. This boundary is highly variable. Rings shed from the retroflecting Agulhas Current every 2 months on average result in periodic bursts of water leaking from the Indian Ocean into the Atlantic (discussed previously and in Agulhas Current). The main path of these rings is west-northwest. When a major ring is shed this can appear like a flooding of the south-east Atlantic by tropical surface water, as evident in **Figure 4B**. On occasion, Agulhas rings may interact with the Benguela shelf and draw shelf water around the ring in the form of a large curved filament. In addition to rings, there is a small almost continuous leakage of Agulhas water around the Cape of Good Hope and this tends to strengthen the shelf edge/upwelling front there (**Figure 4B**).

At about 5°S there is a pronounced front between the system of equatorward current and the Benguela region, i.e. the Angola Front, which is particularly marked at subsurface depths. There is differing opinion as to whether the Angola Front is the northern boundary of the Benguela or whether the boundary should rather be taken as the Angola-Benguela Front off southern Angola which is the northern extent of pronounced coastal upwelling. The latter front is most evident as a sharp horizontal surface temperature gradient, maintained throughout the year between 14° and 17°S. The Angola-Benguela Front (ABF) is orientated perpendicular to the coast and is in reality a collection of two or three fronts, at times rather convoluted. On average the ABF lies furthest south in summer. It is detectable to a depth of 200 m and appears to be maintained primarily by a contribution of bathymetry, coastline orientation, and wind stress.

System Variability

Processes on many orders of temporal and spatial scales impact on the Benguela system. Of the physical features, the distribution of the wind field in time and in space is of overarching importance, significantly influencing coastal trapped waves, upwelling, frontal dynamics, currents in the upper mixed layer, stratification, etc. Seasonal changes in insolation exert an important influence on the surface waters. (One third of the upwelling area and the entire Angolan system lie within the tropics and receive high levels of thermal radiation and light.) Tides and tidal currents are only really important in near-shore environments.

Event-scale Variability

The processes which occur on timescales of hours to several days and space scales of meters to tens of kilometers (i.e. event-scale) are characteristic of upwelling events. The southern Benguela displays more event-scale variability than the central and northern parts of the region as a consequence of the influence of eastward-moving cyclones which pass south of the continent with periods of 3–10 days. Further north, upwelling tends to be more consistent than in the south, although the diurnal changes in wind forcing become more important (land–sea breeze influence). There is a substantial body of literature on event-scale oceanography and biology of the Benguela.

Seasonal Variability

The seasonal migration of the weather systems and changes in insolation is manifest in changes in the upper mixed layer – intensity and distribution of upwelling, stratification, and intrusion of tropical water from the south and from the north into the upwelling area. Off Angola and Namibia there is a distinct seasonal cycle in near surface temperature with a maximum in March and a minimum in August/September with a range of 4–6°C. In the south the range in seasonal average temperature is only 1–2°C. The seasonal signal in the south-east Atlantic is in general larger than that in the eastern Pacific.

Interannual Variability and Episodic Events

While the seasonal signal is large in comparison with some eastern boundary current systems, the interannual signal in the Benguela is smaller. Nevertheless, interannual changes in the wind field do significantly impact on the Benguela. This interannual variability manifests itself in two main ways, i.e. through system (or subsystem)-wide change in upwelling as a consequence of large-scale changes in wind speed and/or direction – resulting in cool and warm years – and through major perturbations at the boundaries of the system which are a result of basin-wide or global changes in the ocean–atmosphere system.

In the south, extreme disturbances in the Agulhas retroflection can be manifest as a major incursion of Agulhas Current water penetrating into the Benguela – either in the form of shallow filaments or occasionally shed rings which take a more northerly path than usual. There is evidence that major events occurred in 1957, 1964, 1986, and 1997/98. Occasionally these are followed by equatorward pulses of subAntarctic water, e.g. 1987.

Major perturbations of the Angola-Benguela Front – termed *Benguela Niños* – occur in the south-east Atlantic. These events have a character similar to their Pacific *El Niño* counterpart, but are not necessarily linked to the latter. Every few years the tropical eastern Atlantic becomes anomalously warm as a consequence of relaxation of trade winds, deepening of the thermocline, and reduced loss of heat from the ocean. Occasionally, every 10 years on average, this warming is more extreme, as a consequence of sudden relaxation of winds off Brazil, and when this happens a Kelvin wave is generated and there is an apparent intensification in the South Equatorial Current with more warm water than usual pushing eastwards and southwards, displacing the Angola-Benguela Front and flooding the northern Benguela shelf with tropical water. The most recent Atlantic or *Benguela Niño* occurred in 1995. Previous events were recorded in 1934, 1949, 1963, and 1984, while others may have occurred in 1910, the mid-1920s and in 1972–74. Although not necessarily in phase with Pacific *El Niños*, they do appear to be a regional response to changes in the global ocean–atmosphere system.

Concluding Remarks

While there is a fairly substantial body of knowledge about the structure and functioning of the Benguela, there are still several notable gaps. These include a proper understanding of the linkages between shelf and offshore processes, processes occurring at the northern and southern boundaries of the system, teleconnections between the Benguela and the global ocean–atmospheric system, and the effect of global climate change on the region. Research priorities for the future must focus on modeling to improve predictability of system variability and response to climate change.

See also

Abyssal Currents. Agulhas Current. Atlantic Ocean Equatorial Currents. Canary and Portugal Currents. Current Systems in the Atlantic Ocean. Current Systems in the Southern Ocean.

Further Reading

Bakun A and Nelson CS (1991) The seasonal cycle of wind-stress curl in subtropical eastern boundary current regions. *Journal of Physical Oceanography* 21: 1815–1834.

Hart TJ and Currie RI (1960) The Benguela Current. *Discovery Report* 31: 123–297.

Hill AE, Hickey BM, Shillington FA, *et al.* (1998) Eastern boundariesc: oastal segment (E). In: Robinson AR and Brink KH (eds.) *The Sea*, 11, pp. 583–604. New York: John Wiley Sons.

Moroskkin KV, Bubnov VA, and Bulatov RP (1970) Water circulation in the eastern South Atlantic Ocean. *Oceanology* 10: 27–34.

Shannon LV (1985) The Benguela ecosystem. 1. Evolution of the Benguela, physical features and processes. In: Barnes M (ed.) *Oceanography and Marine Biology: An Annual Review*, 23, pp. 105–182. Aberdeen: Aberdeen University Press.

Shannon LV and Nelson G (1996) The Benguelal: arge scale features and processes and system variability. In: Wafer G, Berger WH, Siedler G, and Webb DJ (eds.) *The South Atlantic: Past and Present Circulation*, pp. 163–210. Berlin: Springer Verlag.

Shillington FA (1998) The Benguela upwelling system off southwestern Africa. Coastal segment (16,E). In: Robinson AR and Brink KH (eds.) *The Sea*, 11, pp. 583–604. John Wiley Sons.

Stramma L and Peterson RG (1989) Geostrophic transport in the Benguela Current region. *Journal of Physical Oceanography* 19: 1440–1448.

Wefer G, Berger WH, Siedler G, and Webb DJ (eds.) (1996) *The South Atlantic: Past and Present Circulation*. Berlin: Springer Verlag.

Yamagata T and Iizuka S (1995) Simulation of the tropical thermal domes in the Atlantic: a seasonal cycle. *Journal of Physical Oceanography* 25: 2129–2139.

BRAZIL AND FALKLANDS (MALVINAS) CURRENTS

A. R. Piola, Universidad de Buenos Aires, Buenos Aires, Argentina
R. P. Matano, Oregon State University, Corvallis, OR, USA

Introduction

The zonal component of the mean prevailing winds, low latitude easterlies and mid-latitude westerlies induce anticyclonic[1] upper ocean circulation patterns referred to as subtropical gyres. The latitudinal rate of change of the Earth's rotation induces a zonal asymmetry in these gyres, and intensifies the flow near the western boundaries. The Brazil Current is the western limb of the subtropical gyre that carries warm and salty waters poleward along the continental slope of South America (**Figure 1**). Near 39°S the Brazil Current collides with a northward branch of the Antarctic Circumpolar Current (ACC), the Malvinas (Falkland) Current, which transports cold and relatively fresh subAntarctic waters equatorward. The collision between these distinct water masses generates one of the most energetic regions of the world ocean: the Brazil/Malvinas Confluence (BMC). This article reviews *in situ* and remote observations and the results of numerical simulations that describe the mean structure and time variability of the Brazil and Malvinas Currents and the frontal region that they generate.

Water Masses

The western South Atlantic has been referred to as the 'cross-roads of the world ocean circulation', because it hosts water formed in remote areas of the world, and brought into this region by the large-scale ocean circulation. This meeting of water masses generates a highly complex vertical stratification structure. In the upper ocean, this structure is dominated by the confluence of subtropical and subAntarctic waters associated with the opposing flows of the Brazil and Malvinas Currents. In the deep ocean, the vertical stratification structure is dominated by contributions from deep and bottom waters from the North Atlantic, South Pacific, and Antarctic regions.

To illustrate the water mass structure of the upper layer of the western South Atlantic, **Figure 2** shows a diagram of potential temperature[2] versus salinity (θ–S) from summer stations collected within the cores of the Brazil and Malvinas Currents (1000–2000 m depth range). From 20°S to 35°S, **Figure 2** illustrates the θ/S characteristics associated with the water masses advected by the Brazil Current and, from 55°S to 40°S, with those advected by the Malvinas Current. In addition, **Figure 2** also shows the θ–S diagram of a hydrographic station collected downstream from the separation of both boundary currents from the continental margin (e.g. within the core of the BMC).

Upper Ocean

The upper portion of the water mass carried poleward by the Brazil Current is referred to as Tropical Waters (TW), and is characterized by high potential temperature ($\theta > 20$°C) and salinity (S > 36 PSU, **Figure 2**). The high temperatures of the TW are due to heat gained from the atmosphere at low latitudes, while the high salinities are due to freshwater losses at mid-latitudes. The upper portion of the Brazil Current is also characterized by the presence of relatively thin low salinity layers capping the TW structure (e.g. the 35°S curve in **Figure 2**). These low salinity layers are thought to be caused by mixing between TW and shelf and river waters. Below the TW, but still within the Brazil Current, there is a sharp thermocline and halocline (see the quasilinear θ–S relation in the 20–10°C temperature range) that is referred to as South Atlantic Central Water (SACW). The SACW shows a very stable θ–S pattern with only minor variations induced by winter sea–air interactions near the southern limit of the Brazil Current.

The upper layer of the Malvinas Current (i.e. the curves corresponding to 40° and 50°S in **Figure 2**) is substantially colder ($\theta < 15$°C) and fresher (S < 34.2 PSU) than the corresponding layer of the Brazil Current. These properties reflect the subAntarctic origin of the Malvinas waters. In the northern portion of the Drake Passage, the source for the Malvinas transport, the surface temperature is close to

[1] Clockwise in the northern hemisphere and counter-clockwise in the southern hemisphere.

[2] θ is the temperature of a water parcel raised adiabatically to the sea surface, thus removing the effect of pressure.

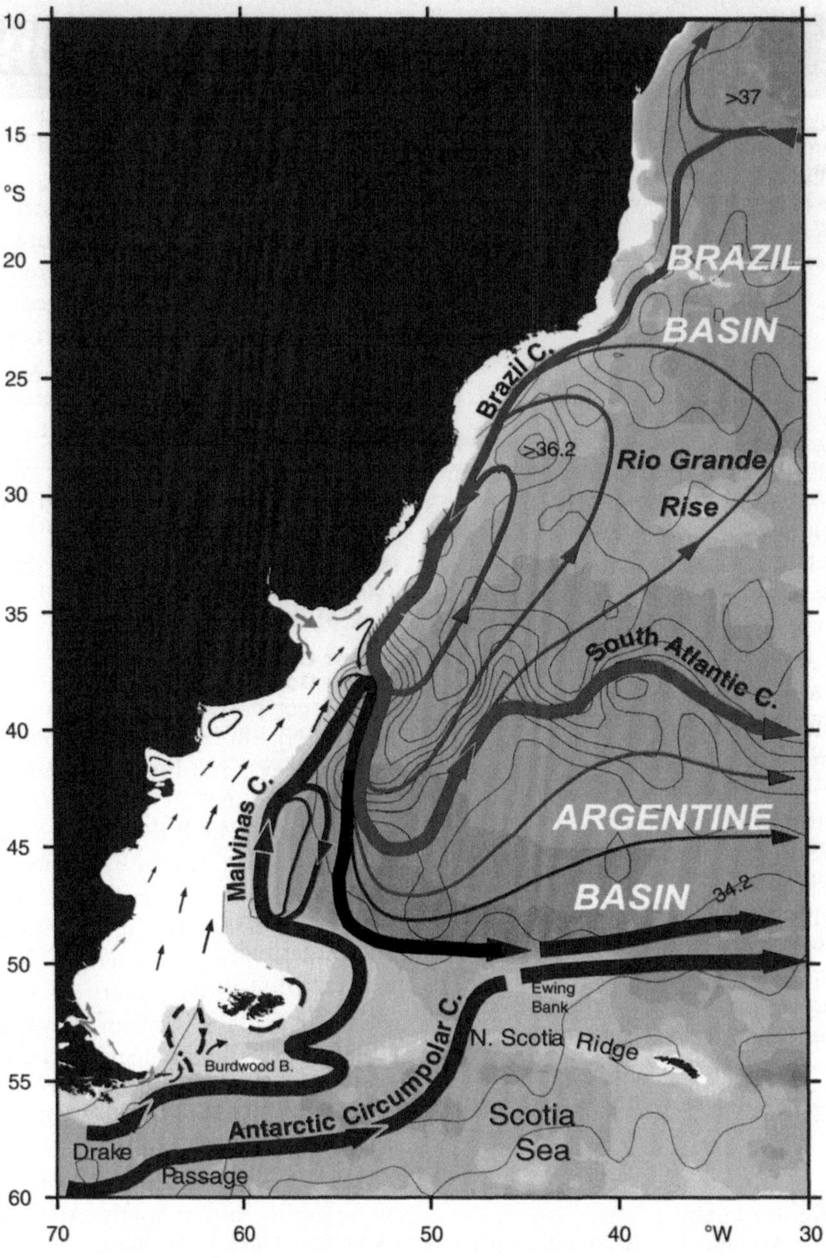

Figure 1 Schematic diagram of the upper layer circulation of the South Atlantic western boundary currents. Black lines are used for the Antarctic and subAntarctic water flows, associated with the Antarctic Circumpolar Current and the Malvinas Current. Red lines are used for the flow of the subtropical waters carried by the Brazil Current. Over the Patagonian continental shelf the arrows represent the mean surface currents. The thin contour lines show the salinity field at 200 m depth that was used to infer part of the circulation scheme. The salinity at 200 m ranges from 34.2 south of the Confluence to 37 near 15°S. A sharp salinity front develops at the Brazil/Malvinas Confluence and extends in a meandering fashion towards the ocean interior where it marks the South Atlantic Current. The background shading represents the bottom topography with darker shading corresponding to deeper waters. The deepest area in the southern Argentine Basin is the Argentine Abyssal Plain where depth is greater than 6000 m. Major topographic features and currents cited in the text are labeled.

4°C and increases northward up to 16°C at the latitude where the Malvinas separates from the continental boundary (~40°S). Although the sub-Antarctic waters of the Malvinas Current and the SACW of the Brazil Current thermocline occupy the same density range ($\sigma_\theta \sim 25.5$–$27.0\,\mathrm{kg\,m^{-3}}$) they have very different thermohaline characteristics and the convergence of these water masses, in the BMC, leads to the formation of alternate layers of sub-Antarctic and subtropical water. These intrusions are referred to as interleaving or fine-structure (see station at BMC, **Figure 2**).

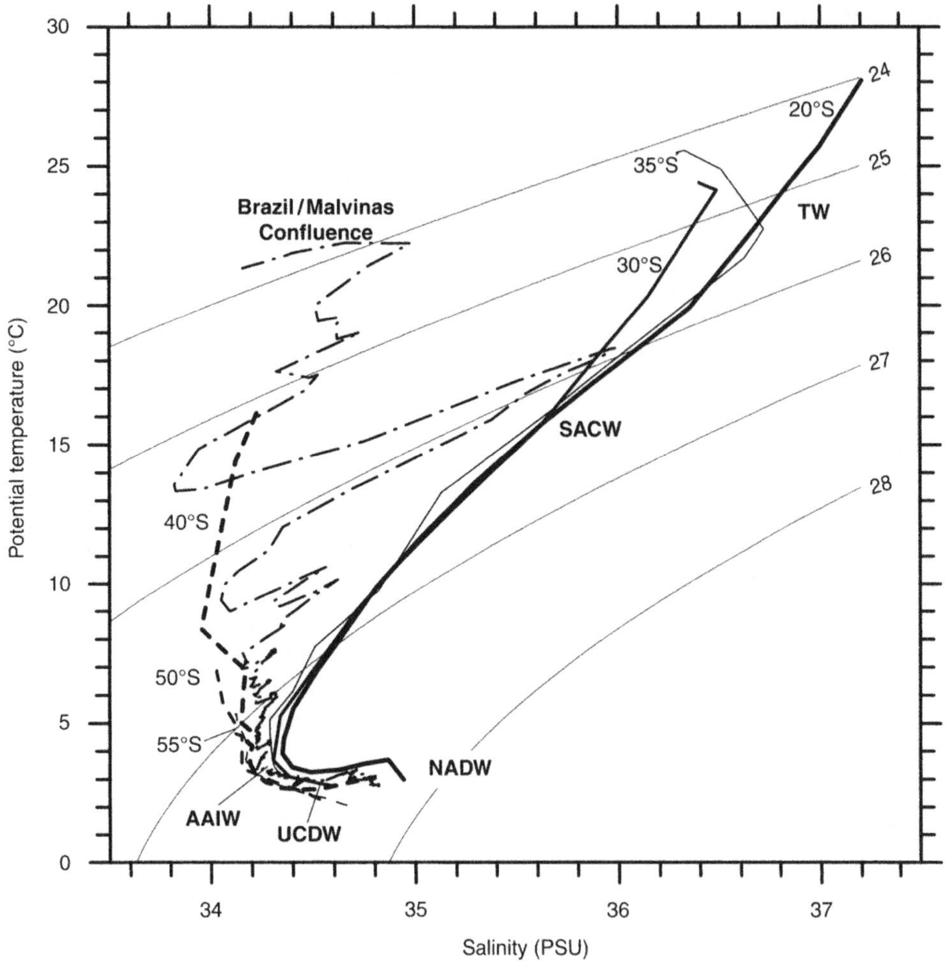

Figure 2 Potential temperature–salinity diagrams from hydrographic stations collected during austral summer along the paths of the Brazil Current (from 20°S in the Brazil Basin to 35°S, solid lines) and along the Malvinas Current (from 55°S in the northern Drake Passage and 40°S, dashed lines). These stations are located between the 1000 m and 2000 m isobaths near the cores of the western boundary currents. Also included is a station from the Brazil/Malvinas Confluence after separation from the western boundary (dashed-dotted line). Constant density anomaly (σ_θ in units of km m^{-3}) lines are included. See **Figure 3** for abbreviations.

Antarctic Intermediate Water

The water mass structure of the Brazil Current at intermediate depths (700–1000 m) is dominated by the presence of Antarctic Intermediate Water (AAIW). The AAIW, characterized by a salinity minimum ($S < 34.3$ PSU), has contributions from the coldest and densest ($\sigma_\theta \sim 27.3\,\mathrm{kg\,m^{-3}}$) member of the southern hemisphere Subpolar Mode Water or Sub-Antarctic Mode Water (SAMW), which originates from deep winter convection along the SubAntarctic Zone. The Malvinas Current carries newly formed AAIW and SAMW into the Argentine Basin. Data collected during the austral winter show that, as the AAIW/SAMW enter into the Argentine Basin from the south, they are exposed to the atmosphere and are subject to further modification by local air–sea interactions. South of the BMC, the AAIW/SAMW are less salty ($S < 34.1$ PSU) than within the Brazil Current (**Figure 2**) and these lateral property gradients across the BMC induce interleaving. Similarly to the upper layer flow, the temperature of the AAIW core increases from 3°C, at the Drake Passage, to 3.5°C at 40°S.

It is interesting to note that although on average the AAIW must spread northward (away from the region of formation), direct current observations at 28°S, and close to the continental margin indicate that, in the subtropical basin, the AAIW follows the upper ocean anticyclonic gyre, and flows southward below the Brazil Current. After leaving the continental boundary the AAIW turns into the subtropical gyre, where vertical and lateral mixing increase its salinity and decrease its dissolved oxygen concentration. The water mass resulting from this recirculation process is known as recirculated AAIW.

Deep and Abyssal Water

The deep layers of the western South Atlantic show a variety of water masses which are depicted by their properties in **Figure 3**. Below the Brazil Current, there is the poleward flow of North Atlantic Deep Water (NADW), which is the primary source of ventilation underneath the main thermocline. The NADW originates at the high latitudes of the North Atlantic Ocean, from where it spreads southward along the continental slope of the American continent. At 30°S the NADW is characterized by relatively high potential temperature ($\theta \sim 3^\circ C$), salinity ($S \sim 34.8$ PSU), and dissolved oxygen ($O_2 \sim 250\,\mu mol\,kg^{-1}$). Below 800–1000 m Circumpolar Deep Water (CDW) flows northward within the Malvinas Current. Although the nutrient-rich CDW originates from NADW, mixing along its path around the Antarctic Continent leads to decreased concentrations of oxygen and salinity. Consequently, although CDW is still identified by a relative salinity maximum, its salinity at the core still is lower than that of NADW. In the western Argentine Basin the NADW splits the CDW into two layers: the upper CDW (UCDW) and the lower CDW (LCDW). The latter are identified by two minima in dissolved oxygen above and below the high salinity, oxygen-rich NADW. The existence of two separate oxygen minimum layers is apparent north of 50°S (**Figure 3**). From the Drake Passage the UCDW flows into the Argentine Basin closely following the 1000–1500 m isobaths. At 40°S the UCDW is characterized by deep ($\sim$1400 m) temperature ($\theta < 2.9^\circ C$) and dissolved oxygen minima ($O_2 < 200\,\mu mol\,kg^{-1}$). The LCDW is the densest water flowing eastward through Drake Passage. It enters the Argentine Basin over the Falkland Plateau and primarily east of Ewing Bank, flows westward along the escarpment located at 49°S and continues northward along the continental slope of the Argentine Basin at 3000–3500 m depth (**Figure 3**).

The abyssal waters of the southern hemisphere oceans are derived from southern high latitudes and are generally referred to as Antarctic Bottom Water. In the western South Atlantic the bottom waters are cold ($\theta < 0^\circ C$), oxygen-rich ($O_2 \sim 225\,\mu mol\,kg^{-1}$), and nutrient-rich. These abyssal waters are denser and colder than the densest water found in the Drake Passage and must derive from the Weddell Sea. Underneath the continental ice shelves of the southern Weddell Sea the densest water mass of the world ocean is formed, but it is the Weddell Sea Deep Water (WSDW), a product of mixing between the CDW and the Weddell Sea Bottom Water, which flows northward around the Scotia Trench and enters into the Argentine Basin as an abyssal western boundary current.

Circulation

Brazil Current

The Brazil Current originates along the continental slope of South America, between 10° and 15°S, through a branching of the westward-flowing South Equatorial Current. The northern branch of the South Equatorial Current forms the North Brazil Current, and represents a loss of upper layer mass from the South Atlantic to the North Atlantic. The southern branch forms the Brazil Current, the western boundary current of the subtropical South Atlantic Ocean. A substantial amount of the southward upper ocean flow occurs on the outer continental shelf of Brazil. Although the term Brazil Current usually refers to the flow within the upper 1500 m, there is evidence that the current may extend well beyond that depth. In fact, hydrographic observations suggest that the AAIW layer is also part of the southward-flowing western boundary current. Direct current measurements off southern Brazil also reveal that although the upper layer flow of the South Equatorial Current reaches South America near 15°S, at intermediate depths the bifurcation shifts south of 24°S. The addition of re-circulated AAIW to the southward flow would contribute to the increase of volume transport of the Brazil Current observed south of approximately 28°S.

Geostrophic calculations, and a few direct current measurements of the Brazil Current transport, yield a value of only about 4–6 Sv, between 10 and 20°S, and this increases to about 20 Sv at 38°S, near the BMC. The rate of transport increase for the Brazil Current is comparable to that of the Gulf Stream. The increase of the Brazil Current's transport is partially associated with a tight recirculation cell near the western boundary and, perhaps, the addition of intermediate waters near 25°S. *In situ* observations, between 20° and 28°S, have shown that the poleward increase of volume transport of the Brazil Current to 16 Sv is associated with a deepening of the current from 100 m to 600 m. While the discussion on the Brazil Current's transport has focused on the upper 1000 m, there are also important poleward, western boundary undercurrents below the thermocline. At 27°S, for example, the core of the southward-flowing NADW ($S > 34.94$ PSU) is found at approximately 2000 m and east of the upper ocean jet. If this undercurrent is included in the transport calculation then the southward volume transport relative to a deep reference level is close to 11 Sv at 27°S, and increases southward to 70–80 Sv at 36°S. Although this estimate may include some southward recirculation of subAntarctic water and CDW from the Malvinas Current it is, nevertheless, much larger than previous values.

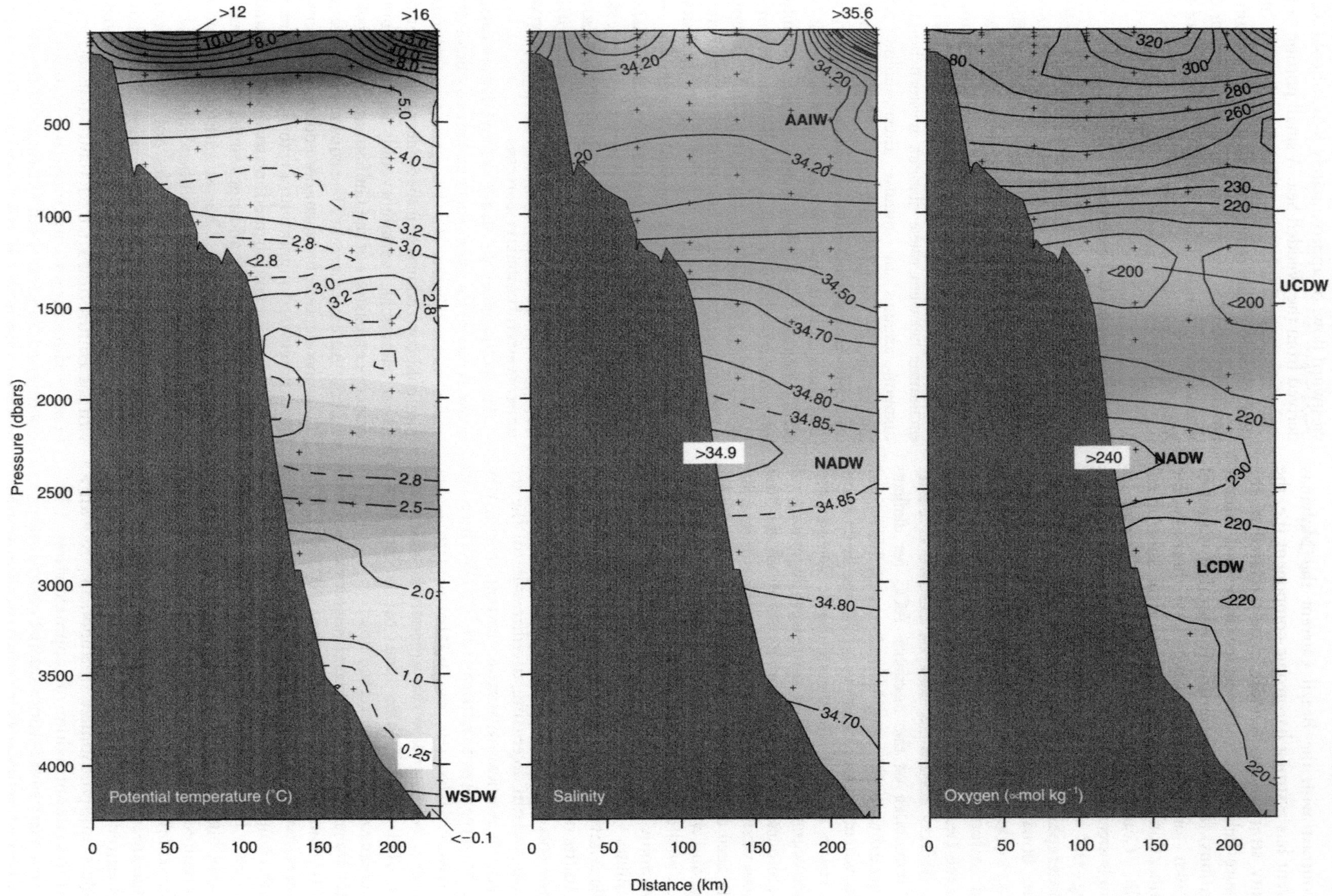

Figure 3 Late winter vertical potential temperature (°C), salinity (PSU), and dissolved oxygen (μmol kg^{-1}) sections from the Brazil/Malvinas Confluence near 38°S. Water masses identified by property extrema are labeled as follows: TW, Tropical Water; SACW, South Atlantic Central Water; AAIW, Antarctic Intermediate Water; UCDW, Upper Circumpolar Deep Water; NADW, North Atlantic Deep Water; LCDW, Lower Circumpolar Deep Water; WSDW, Weddell Sea Deep Water.

Malvinas Current

In contrast with the Brazil Current the Malvinas Current has a strong barotropic component (i.e. the density stratification is more closely related to the pressure field than to the temperature and salinity variations). This characteristic is typical of waters of subpolar origin, which have less thermohaline stratification than waters of tropical or subtropical origin. Consequently the Malvinas Current is strongly steered by the bottom topography as it flows along the continental slope of South America. Hydrographic observations suggest that most of the water flowing eastward along the SubAntarctic Front, in the northern Drake Passage, loops northward to form the Malvinas Current. Downstream from the Drake Passage, a portion of the upper layer flow deflects northward west of Burdwood Bank. The remainder of the northern ACC jet deflects northward through a gap located east of Burdwood Bank, where bottom depth is >1700 m. Both branches rejoin north of Burdwood Bank. Deeper and denser water can only flow northward depending on the complex bottom topography. Most of the flow deflects westward following the bottom topography of the deep chasm, which separates the north Scotia Ridge and the Falkland Plateau. Near 48°S the current is well organized, closely following the bottom topography, and appears to have little spatial variability.

The Malvinas Current extends from the sea surface to the ocean floor. From 50° to 40°S maximum surface speeds ($>0.7\,\mathrm{m\,s^{-1}}$) are observed over the 1000 m isobath, decreasing at either side of the jet. Within the northward-flowing current, the AAIW, and the oxygen-poor/nutrient-rich UCDW core, are also observed close to the 1000 m isobath – suggesting a coherent flow throughout the water column. This confirms the idea of a substantial barotropic contribution to the flow.

Relative geostrophic volume transport estimates of the Malvinas Current vary between 10 and 12 Sv. However, mass conservation arguments for the cross-isobath component in the BMC suggest that the total flow (e.g. barotropic + baroclinic) must be substantially higher (~70 Sv). These high estimates are also required if most of the waters within and north of the SubAntarctic Front of the northern Drake Passage are included in the Malvinas Current. Recent direct current observations near 41°S lead to a mean volume transport estimate of about 35 Sv, with a barotropic contribution of approximately 50%. However, these observations may be north enough to miss part of the flow that recirculates southward as the Malvinas Return Current (**Figure 1**). Along the southern edge of the Argentine Basin there is a westward flow, of approximately 8 Sv, of cold water derived from the Weddell Sea. A small portion of the Antarctic contribution (~ 2 Sv) is relatively new WSDW, while the remaining value of 6 Sv corresponds to WSDW recirculated within a bottom cell whose western branch is observed to the east of the Malvinas upper layer jet (**Figure 3**).

Brazil/Malvinas Confluence

Near 38°S, close to the region of highest volume transport, the Brazil Current meets the northward-flowing Malvinas Current. This head-on collision causes the current systems to separate from the western boundary forming a large, quasi-stationary, meander that extends southward to about 45°S. The encounter between these distinct western boundary currents creates an intense thermohaline front known as the Brazil/Malvinas Confluence. Surface velocities along the frontal jet exceed $1\,\mathrm{m\,s^{-1}}$, and there are indications that the current extends vertically beyond the 4000 m depth. Quasi-continuous temperature-salinity profiles reveal signatures of intense mixing of subAntarctic waters and subtropical waters along the front. This intensive mixing extends to the deep waters where NADW-CDW interleaving is also observed. Observations of relatively high salinity ($S > 34.8$ PSU), oxygen-rich ($O_2 > 210\,\mu\mathrm{mol\,kg^{-1}}$) deep waters below the Malvinas Current, show that the NADW found underneath the Brazil Current flows poleward beyond the separation point of the upper layer, suggesting a decoupling of the deep and upper western boundary currents.

The collision of the Brazil and Malvinas currents spawns one of the most spectacular eddy fields of the global ocean (**Figure 4**). The generation of warm- and cold-core eddies at either side of the front have led to mesoscale variability only matched by the offshore extensions of the Gulf Stream, the Kuroshio, and the Agulhas Current. The conspicuous precursor to the production of warm-core eddies is the anomalous poleward migration of the Brazil Current, which forms a complicated intrusive pattern leading to a set of meanders and rings sometimes referred to as the intrusion eddy. The sea surface temperature anomalies of these rings can be as large as 10°C and they occur over timescales of approximately 2 months. Hydrographic observations show that the eddies in the BMC region are vertically coherent, with signatures down to several thousand meters. These eddies are an important mechanism for the meridional transfer of salt and heat. Warm-core rings detached from the Brazil Current retroflection are frequently entrained into the subantarctic gyre.

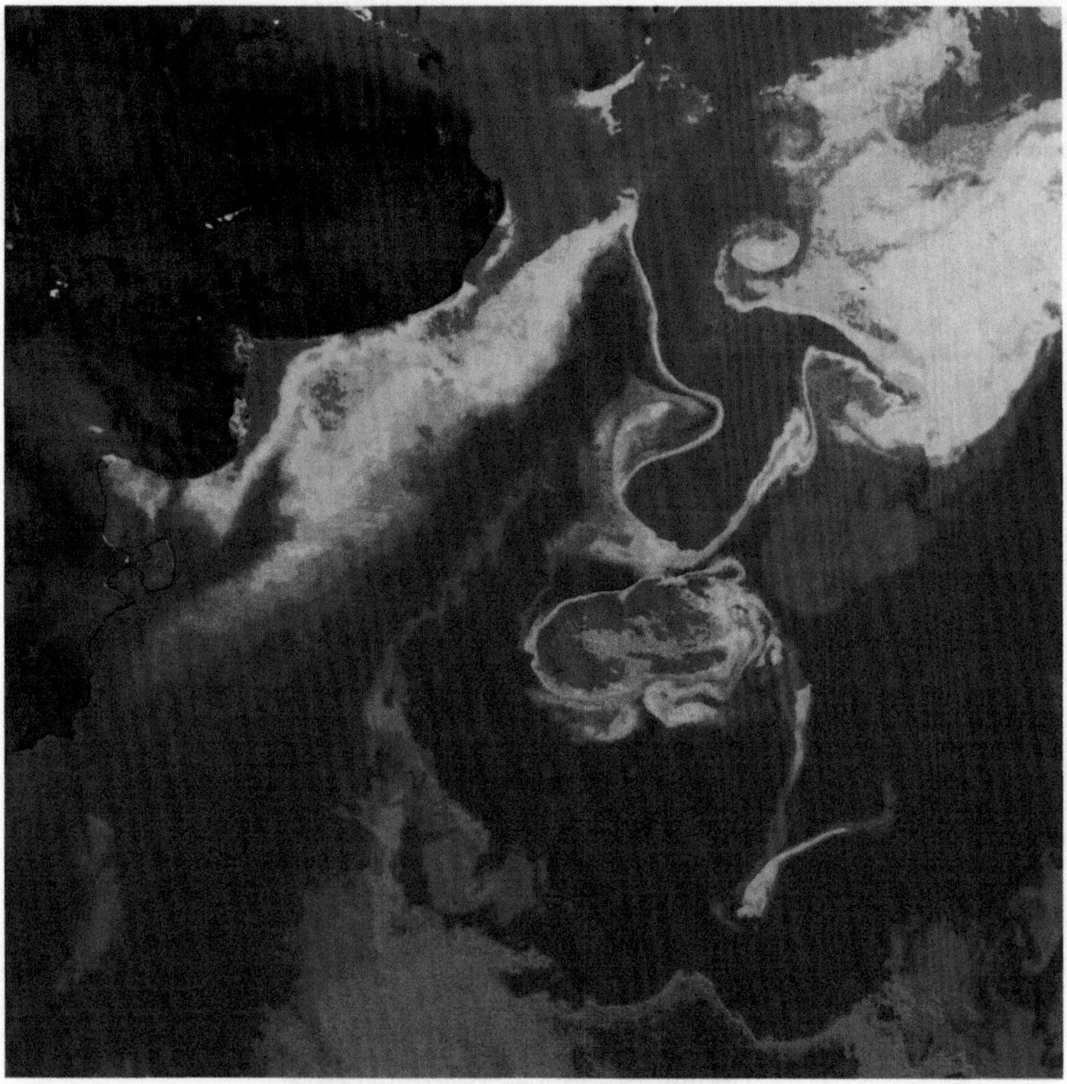

Figure 4 Advanced very high resolution radiometer image of the Brazil/Malvinas Confluence at a time when the Brazil Current was in its southward extension and a large anticyclonic (warm) eddy was being formed. Warmest surface waters (approximately 25°C) are coded in red. The surface temperature decreases are color-coded through yellow and green; the dark blue areas show the coldest waters advected northward by the Malvinas Current (approximately 9°C). (Figure courtesy of O. Brown, R. Evans, and G. Podestá, Rosenstiel School of Marine and Atmospheric Sciences, University of Miami and Estación HRPT Alta Resolución, Servicio Meteorológico Nacional, Argentina.)

Likewise, cold filaments or rings detached from the Malvinas Current are frequently driven onto the recirculation cell that dominates the south-western portion of the subtropical gyre.

Besides the mesoscale variability associated with the formation of eddies and meanders, the variability of the BMC system also has distinctive peaks at semi-annual and annual periods. While the semi-annual variability in the Brazil Current is very small it increases to nearly half the magnitude of the annual signal in the Malvinas Current. It is thought that the semi-annual component of the Malvinas Current variability relates to a similar component in the wind forcing over the Southern Ocean. The annual component of the variability in the south-western Atlantic is dominated by the large meridional excursions of the BMC front. Satellite-derived sea surface temperature and sea surface height anomalies revealed that the latitude of the location of the BMC front has a tendency to move north during the austral winter (July–September) and south during the austral summer (January–March).

Since the location of the BMC is thought to depend on the mass transports of the two western boundary currents the seasonal displacements of the confluence may reflect relative variations of those transports. Numerical simulations of this region (e.g. **Figure 5**) indicate that the transport of the Brazil Current follows the annual evolution of the wind stress curl over the subtropical basin, reaching a maximum during the summer (when the BMC is at its southernmost position) and a minimum during the winter (when the confluence moves farther north). It is not known whether the Malvinas Current makes any significant contribution to these seasonal oscillations. Although it is known that the

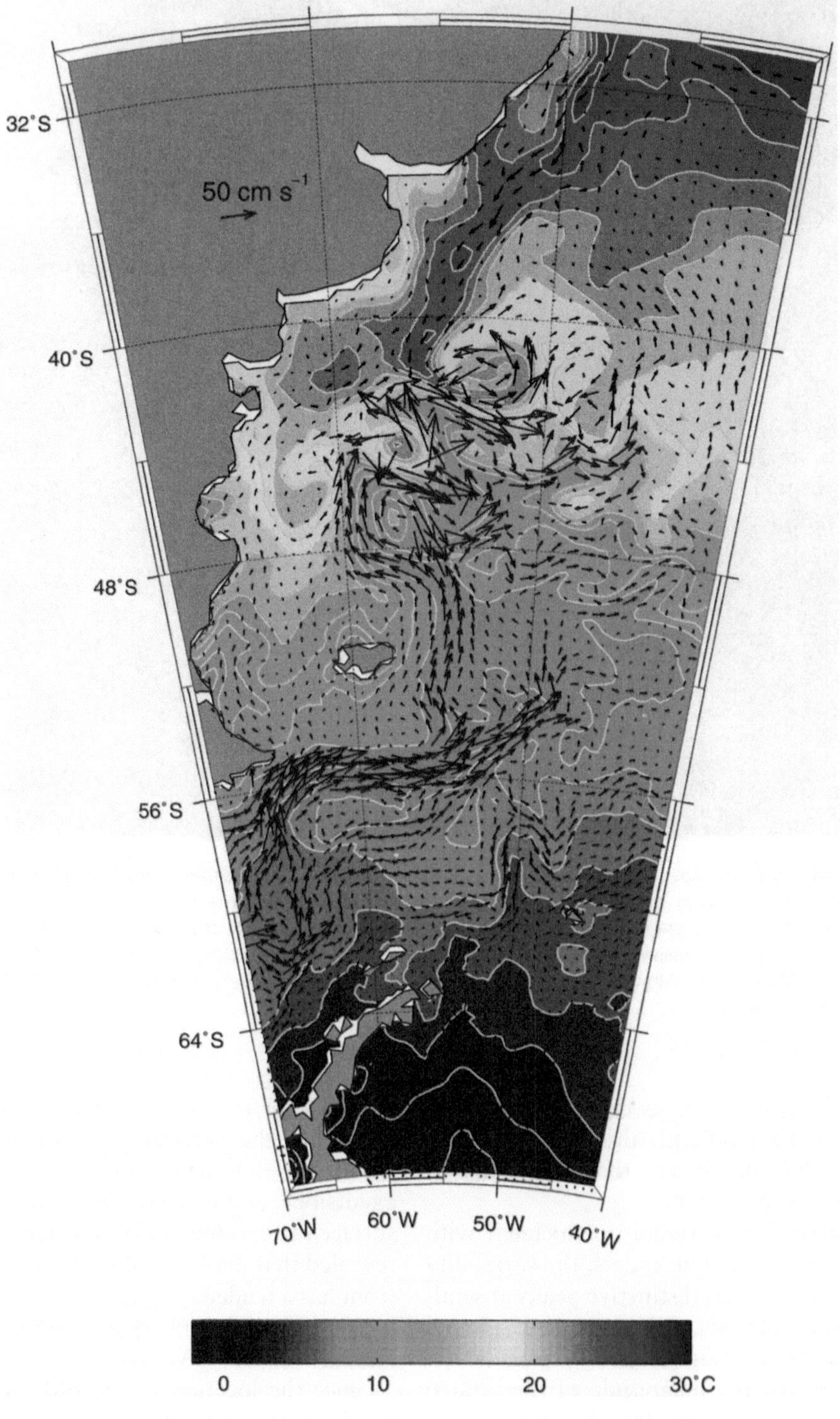

Figure 5 Snapshot of the upper ocean circulation in the western South Atlantic as simulated by the Parallel Ocean Climate Model (POCM). The color fields correspond to sea surface temperatures and the arrows to velocity vectors. The velocity scale is included in the figure. (Figure prepared courtesy of E. Beier.)

northern branch of the ACC, which feeds the Malvinas transport, has a clearly defined annual signal at the Drake Passage, *in situ* and remote observations have failed to identify any clear seasonal variation at mid-latitudes.

Acknowledgments

A. Piola acknowledges the support of the Inter-American Institute for Global Change Research and the Agencia Nacional de Promoción Científica y Tecnológica (Argentina), and R. Matano the support from the National Science Foundation (U.S.A.) grant OPP 9527695, OCE 9819223 and the Jet Propulsion Laboratory contact 1206714. We thank E. Beier for preparing **Fig. 5**.

See also

Abyssal Currents. Antarctic Circumpolar Current. Arctic Ocean Circulation. Current Systems in the Atlantic Ocean. Current Systems in the Southern Ocean. Mesoscale Eddies. Regional and Shelf Models.

Further Reading

Gordon AL (1981) South Atlantic thermocline ventilation. *Deep-Sea Research* 28A(11): 1239–1264.

Peterson RG and Stramma L (1991) Upper-level circulation in the South Atlantic Ocean. *Progress in Oceanography* 26(1): 1–73.

Reid JL, Nowlin WD, and Patzert WC (1977) On the characteristics and circulation of the southwestern Atlantic Ocean. *Journal of Physical Oceanography* 7(1): 62–91.

Reid JL (1989) On the total geostrophic transport of the South Atlantic Oceanf: low patterns, tracers and transports. *Progress in Oceanography* 23(3): 149–244.

Stramma L and England M (1999) On the water masses and mean circulation of the South Atlantic Ocean. *Journal of Geophysical Research* 104(C9): 20 863–20 883.

Tomczak M and Godfrey JS (1994) *Regional Oceanography An Introduction*. London: Pergamon.

CANARY AND PORTUGAL CURRENTS

E. D. Barton, University of Wales, Bangor, UK

Introduction

The Canary and Portugal Currents form the eastern limb of the North Atlantic Subtropical Gyre. Detailed knowledge of the currents is still surprisingly sparse in some respects and is largely based on indirect methods of estimating the flow. The entire eastern boundary of the gyre is affected by the process of coastal upwelling, driven by the seasonally varying Trade Winds. Upwelling is intimately related to the currents on the continental shelf, and varies on timescales from several days upwards. This phenomenon has been studied intensively at different places and times and long-term sampling has only begun in recent years.

Large-scale Circulation

The low to midlatitudes of the North Atlantic Ocean are occupied by the clockwise rotating subtropical gyre (**Figure 1**). The western boundary of this system is made up by the Gulf Stream, which feeds into the North Atlantic Current and the Azores Current. The latter flows eastward to supply the eastern subtropical boundary region. Branches of the Azores Current loop gently into the Portugal Current and further south into the Canary Current. The latter separates from the African coast at around 20°N to become the North Equatorial Current, which eventually feeds into the Caribbean Current and back to the Gulf Stream.

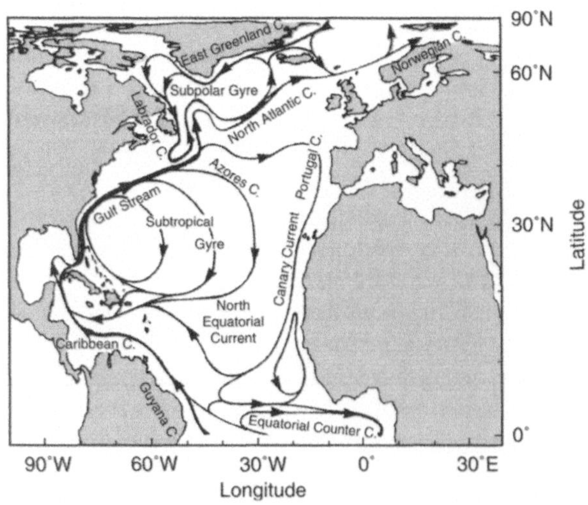

Figure 1 Sketch of the general near-surface circulation of the North Atlantic Ocean.

Using all available hydrographic data, the long-term average geostrophic flow for the region has been calculated assuming a level of no motion near 1200 m. Scarcity of data limits the resolution of the analysis to a grid of $3^\circ \times 3^\circ$; near the coast, where deep data are even fewer, results are more uncertain. Where the eastward-flowing Azores Current turns south as it nears the eastern boundary, two branches of the Canary Current are formed (**Figure 2A**) separated by Madeira. West of Iberia only weak near-surface flow toward the eastern boundary is indicated, while nearer to shore the Portugal Current carries about 2×10^6 $m^3 s^{-1}$ equatorward in the layers above 200 m depth. Only some of this continues southward into the Canary Current, while the rest apparently enters the Mediterranean in a shallow surface layer. The total amount of water carried equatorward above 200 m in the Canary Current, including input from the Portugal Current, was estimated at about 4×10^6 $m^3 s^{-1}$ between 35°W and the African coast. Near 20°N the Canary Current breaks away from the African coast to turn westward as the North Equatorial Current near 15°N. South of this separation point, a recirculation cell around the 'Guinea Dome' lies east of the Cape Verde Islands between the coast and the equatorward flow.

The Canary Current varies seasonally by slight changes in position but not greatly in transport (**Figure 3**). Areas of larger uncertainty, shaded in the figure, correspond to few or no hydrographic samples in that study. The near-shore branch of the current migrates seasonally across the Canary Islands, closer to Africa in summer and farther offshore in winter. As it does so, the Azores Current oscillates south in summer and north in winter, so that the eastern part of the gyre has an annual 'wobble.' Some streamlines intersect the African coast in the northern half of the area and leave in the south, suggestive of a narrow, intense equatorward flow in the under-sampled coastal band between 20° and 30°N, particularly in spring and summer. Recirculation south of 20°N is clearest in winter and spring, but indications of northward flow are also seen there during other seasons. In autumn, near-shore northward flow extends as far as the Canaries, although again in the coastal under-sampled band.

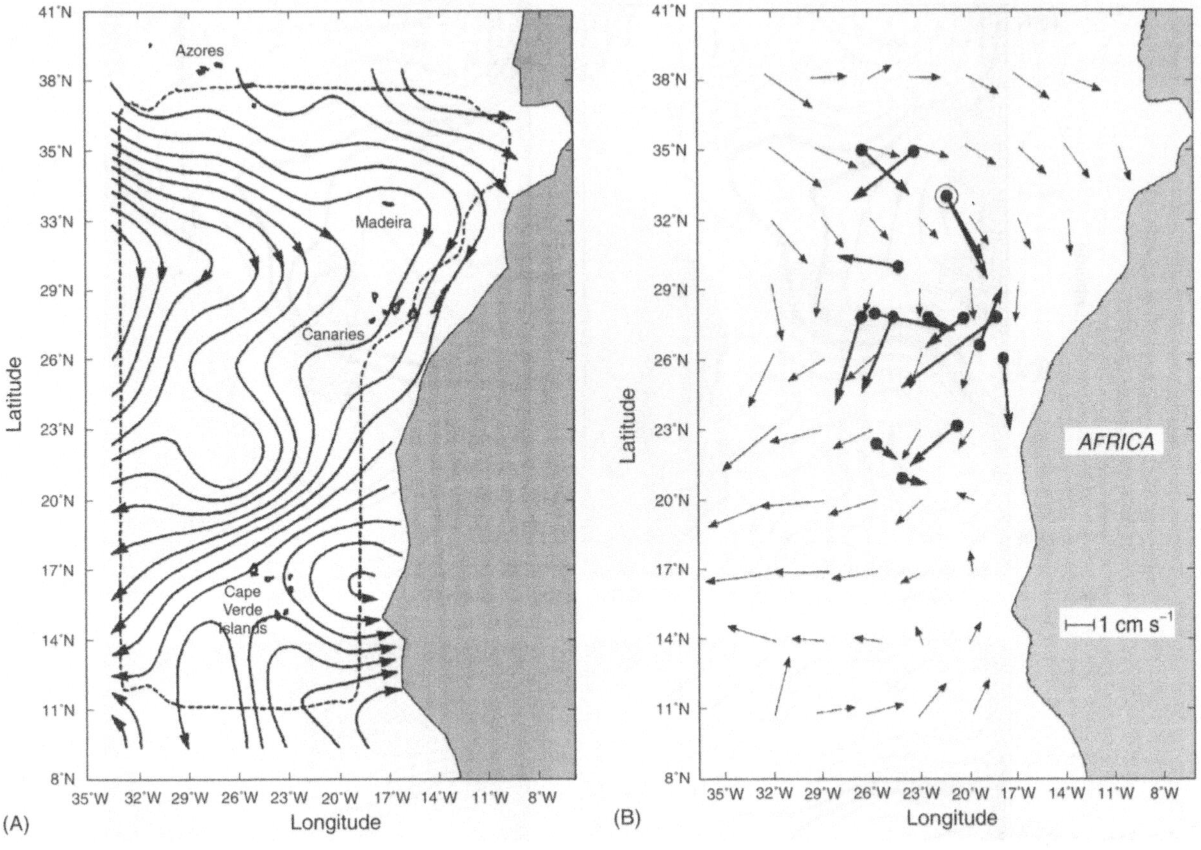

Figure 2 (A) Total transport of volume calculated from the long-term mean density field with the geostrophic assumption and summed from 200 m depth to the sea surface. Between any pair of streamlines the volume transport is $0.5 \times 10^6\ m^3\ s^{-1}$. The calculations have less uncertainty inside the dashed line. (B) Geostrophic current vectors (thin arrows) at 200 m depth calculated from the long-term mean density field and mean observed currents (thick arrows) near 200 m at the sites marked by dots. The circled dot indicates the longest record, Kiel276.

The few available long-term moored observations are mainly well away from shore and the shallowest records are at about 200 m depth. Most lasted 1–2 years, but one mooring (Kiel276) is continuous since 1980 on the southern edge of the Azores Current (33°N, 22°W). In general, fluctuations were more than 5 times more energetic than the mean flow and so most records do not provide a reliable statistical estimate of the average. Nevertheless, the measured mean currents agree broadly with the large-scale geostrophic flow (**Figure 2B**) but indicate that the transport may be >50% more than indicated by the geostrophic calculations. No subsurface record indicated significant seasonal variability. However, the Kiel276 record demonstrated the dominance of meanders and eddies in the Azores Current and variability on the scale of a decade. Since the Azores Current feeds into the eastern boundary system, similar long-term variability likely occurs in the Canary and Portugal Currents as well, although no long-term current monitoring has yet been achieved there.

Several year-long moorings, deployed recently between the Canary Islands and the African shelf, showed highly variable flow, with maximum velocities near 0.3–0.4 m s^{-1}, but seasonal mean currents 10 times smaller. In mid-channel the upper 200 m layer flowed southward with variable strength. The bulk of the upper 600 m flowed mainly southward in winter and spring, but northward in summer and autumn. Despite increased Trade Winds in summer the Canary Current decreased east of the Canary Islands, at the same time as the flow through the western islands increased. These observations are compatible with the suggested seasonal variation in the gyre and the extension of near-surface poleward flow to the latitude of the Canaries, although the timing is out of phase with the long-term seasonal average results. Of course, any single year is not necessarily representative of the long-term mean. There was no evidence of strong equatorward flow along the continental margin in spring and summer, though measurements did not extend over the upper slope and shelf.

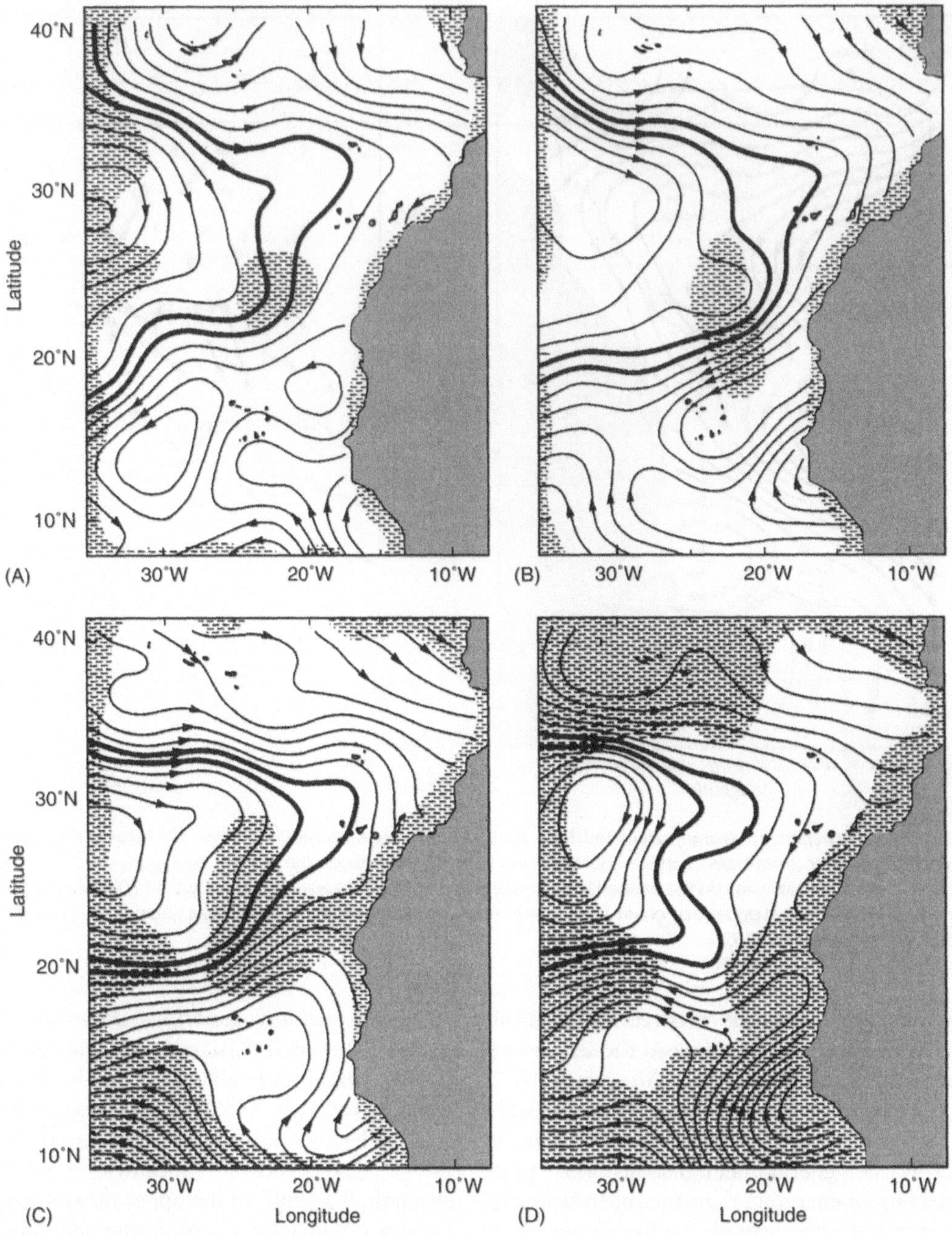

Figure 3 Seasonal variation of the geostrophic volume transport above 200 m depth in the eastern boundary region, calculated as before for (A) spring, (B) summer, (C) autumn, and (D) winter. Areas of larger uncertainty are shaded. Flow between any pair of streamlines is $0.5 \times 10^6\ m^3\ s^{-1}$.

Coastal Upwelling

For some part of the year the north-east Trade Winds blow along every part of the subtropical Eastern boundary with a strong alongshore component that produces offshore Ekman transport in the surface layers and therefore upwelling at the coast. The strength of upwelling is conventionally expressed in terms of the upwelling (or Bakun) index, which is simply the Ekman transport $T_E = (\tau/\rho f)$ where τ is the component of wind stress parallel to shore, ρ is the density of sea water, and f is the Coriolis parameter. The index is routinely calculated from coastal surface wind data on a daily or monthly basis. It represents the rate at which water is removed from shore in the surface layer and, in a two-dimensional system, the amount of water upwelled to replace it.

The annual cycle of upwelling for the region (**Figure 4**) shows the coastal temperature anomaly with respect to central ocean alongside the monthly mean upwelling index. Summer Trade Winds affect the Iberian and African coasts north of 20°N. Off the West Coast of Iberia, upwelling generally starts in May or June and lasts only until September. The southern, Algarve, coast of Portugal and north coast of Morocco (33–37°N) are oriented at a large angle to the Trade Winds and so upwelling there is intermittent and short-lived. In winter the Trades shift southward to provoke upwelling between 30° and 12°N. South of 20°N upwelling starts in December and lasts until April or May. Between 20° and 30°N the coast is subject to year round upwelling that peaks in July and August.

Wind forcing and strength of upwelling, as represented by coastal temperature anomaly, show variations up to a factor of 2 between years and decades. The 1960s were typified by upwelling of about half

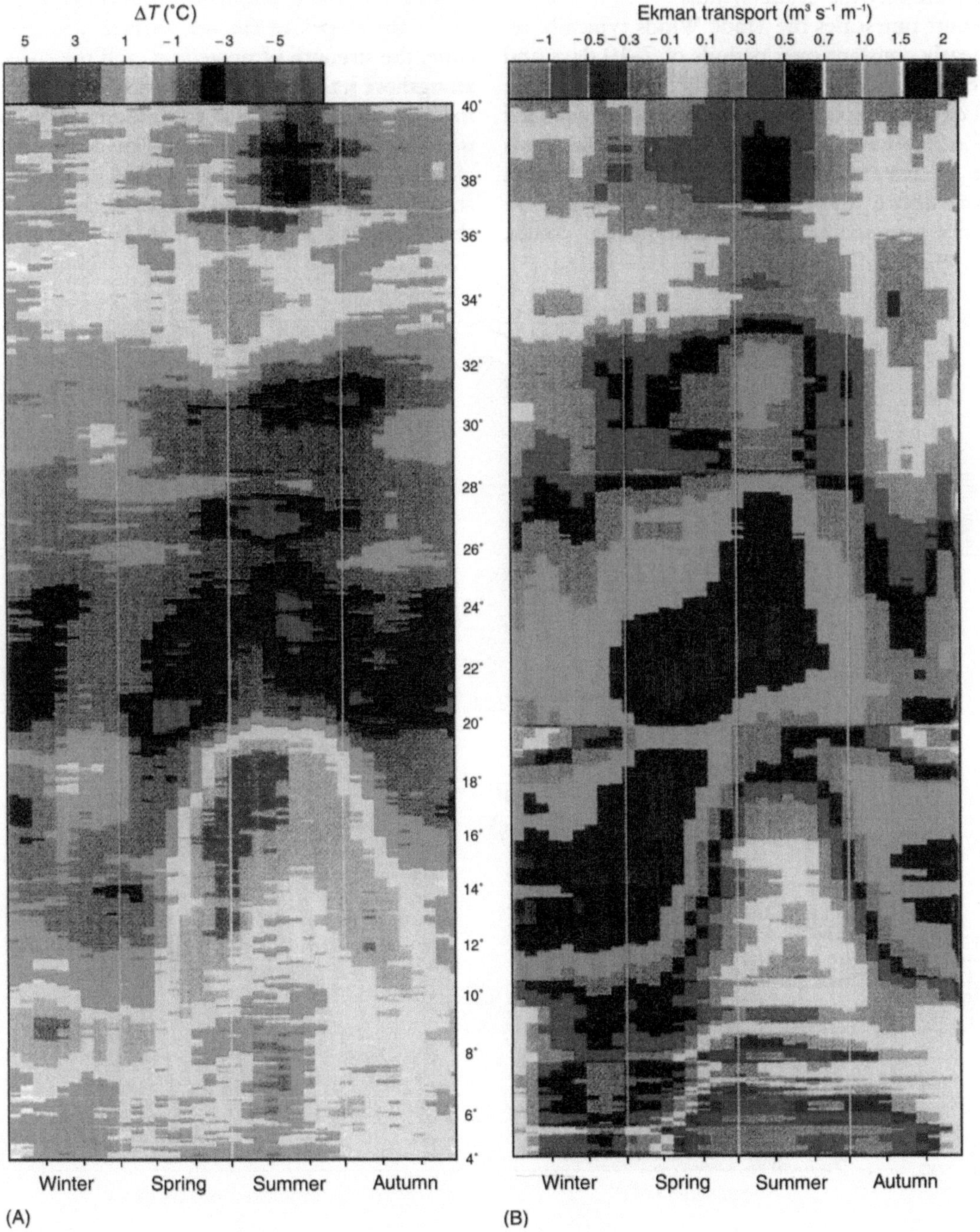

Figure 4 Annual cycle of upwelling represented by (A) temperature difference between the coast and midocean determined by satellite sea surface temperature estimates, and (B) the upwelling index calculated from surface atmospheric pressure analyses.

of the average intensity off West Africa. Upwelling increased through the 1970s, only to weaken again in the 1980s and increase in the 1990s. An intriguing finding is that the strength of upwelling appears to be increasing in the long term – a trend common to all the upwelling regions of the world over the last 40 years. This may be linked to global warming, because increased summer heating deepens the continental low-pressure systems so increasing the atmospheric pressure difference with the oceanic highs to intensify the Trade Winds.

On short timescales the Trade Winds typically remain nearly constant over periods of 7–10 days and then relax to near zero or weakly northward for several days. Observations over the continental shelf of NW Africa have shown how the system responds (**Figure 5**). Favorable winds drive offshore Ekman transport above about 30 m depth and raise the pycnocline near shore so that within one day colder, less salty subsurface water breaches the sea surface. The upwelled water, which can originate from as deep as 200 m, is denser, and there is a slight downward slope of the sea surface toward the coast because of the offshore transport. These give rise to an equatorward geostrophic flow that weakens with depth over the shelf. The jet is strongest where the pycnocline intersects the surface as a boundary between the denser upwelled waters and the less dense oceanic waters, where it reaches velocities of up to $0.8\ \mathrm{m\ s^{-1}}$ in the example shown. As the wind varies, so do the slopes of the sea surface and the pycnocline, the strength of upwelling, and the speed of the alongshore jet. The scale of the upwelling region, i.e., the distance within which the isosurfaces are uplifted is given by the Rossby radius of deformation $\lambda = \sqrt{(g\Delta\rho/\rho h)}/f$ where g is the gravitational acceleration, h is the undisturbed depth of the surface layer above the pycnocline, ρ is the density of the deep

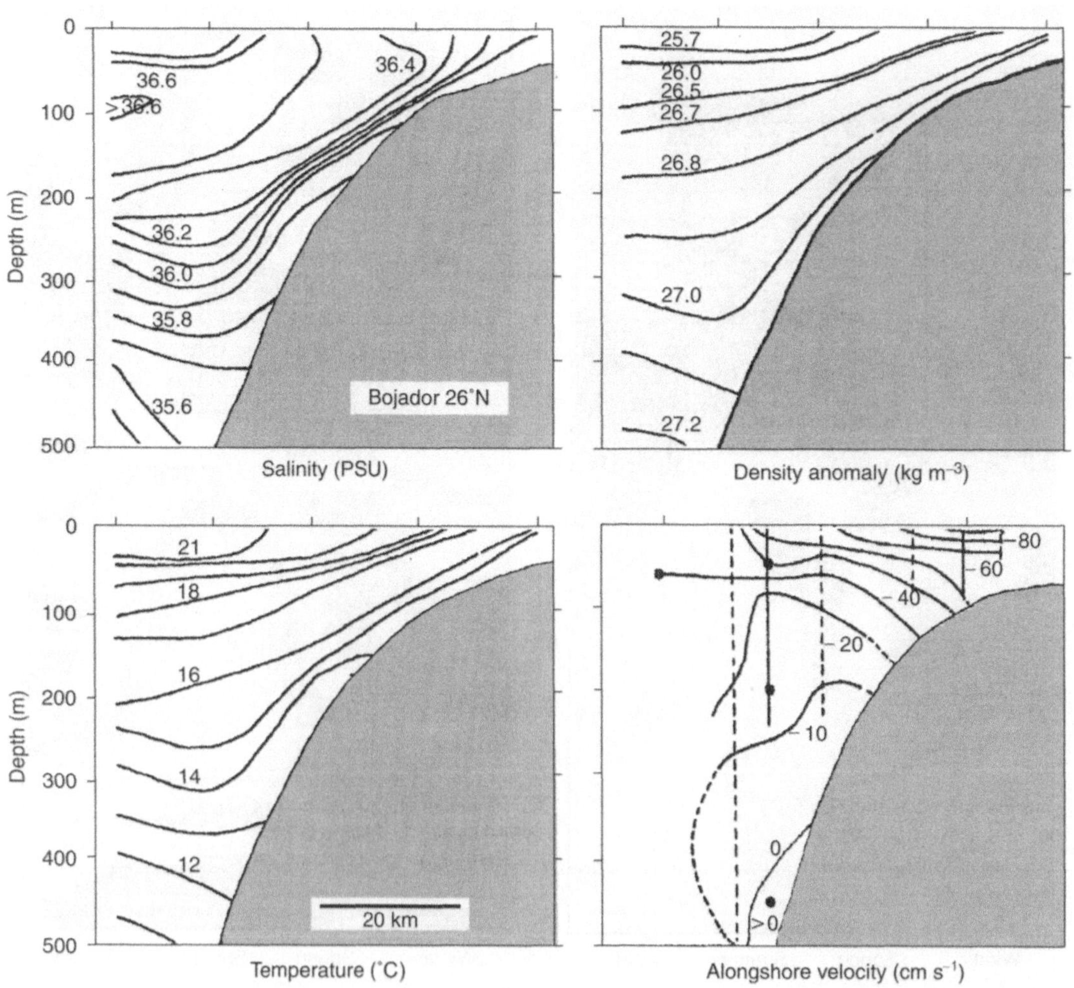

Figure 5 Sections of temperature, salinity, density, and alongshore current off north-west Africa during favorable winds in August. Contours are elevated from ~200 m to the sea surface. The current field is based on a combination of moored current meter (solid dots), profiling current meter (solid vertical lines), and geostrophic estimates (dashed vertical lines).

water, and $\Delta\rho$ is the density contrast between the surface and deep layers. The characteristic upwelling velocity, w, is given by the upward transport divided by the width of the upwelling zone $w = T/\lambda$. For a typical situation $\lambda \sim 10$–20 km, and $w \sim 10\,\mathrm{m\,d^{-1}}$.

The onshore–offshore flow observed off NW Africa conforms to the classic picture of upwelling (**Figure 6**) where the surface layer offshore flow is compensated by deeper onshore flow. When the wind relaxes, the whole water column on the shelf moves shoreward, carrying with it warm oceanic waters. Detailed comparisons of the time-varying offshore and onshore transports with the Ekman index have shown good agreement between offshore flow and the index, in this and other upwelling regions. Poorer agreement is found between offshore and onshore transport, which often appear to be locally out of balance.

Satellite imagery of sea surface temperature fields shows that the boundary between upwelled waters and open ocean is often strongly contorted into long filaments of cooler water stretching hundreds of kilometers out to sea. Filaments identified off the Iberian and African coasts appear to develop from instabilities of the along-frontal flow, often triggered by coastal or topographic irregularities such as capes or ridges. Some appear associated with offshore eddies, which may themselves be topographically anchored. Filaments tend to recur in the same positions year after year, as, for example, the one off north-west Spain at 42°N in **Figure 7**. They are associated with localized areas of net offshore flow (therefore local imbalance in the cross-shelf transport) that take the form of a narrow jet where the alongshore flow is diverted seaward. The figure shows a surface drifter following the filament offshore at a mean rate of $0.3\,\mathrm{m\,s^{-1}}$. This offshore transport can help effect exchange of water properties between shelf and deep ocean because of mixing along the filament boundary and interaction with the deeper waters beneath. As upwelled waters move offshore in the filaments, they gradually warm and become indistinguishable from surrounding waters so that any return flow to the coast is not easily identified in satellite images.

Poleward Undercurrent

Beneath the near-surface equatorward flow of the Canary Current, a subsurface current flows poleward, counter to the general circulation and tightly bound to the continental slope. It has been documented along the entire continental margin between the Gulf of Guinea and north-west Spain, and is a common feature of all eastern boundaries. Despite many direct and indirect observations of the flow, there are remarkably few systematic observations. The structure of the undercurrent is shown by measurements made close to 20°N (**Figure 8**). Its maximum speed is close to $0.1\,\mathrm{m\,s^{-1}}$ at about 150 m below the surface. The core extends about 300–400 m vertically and apparently less than 50 km horizontally (although its offshore limit was not directly observed). Above the undercurrent, shallow

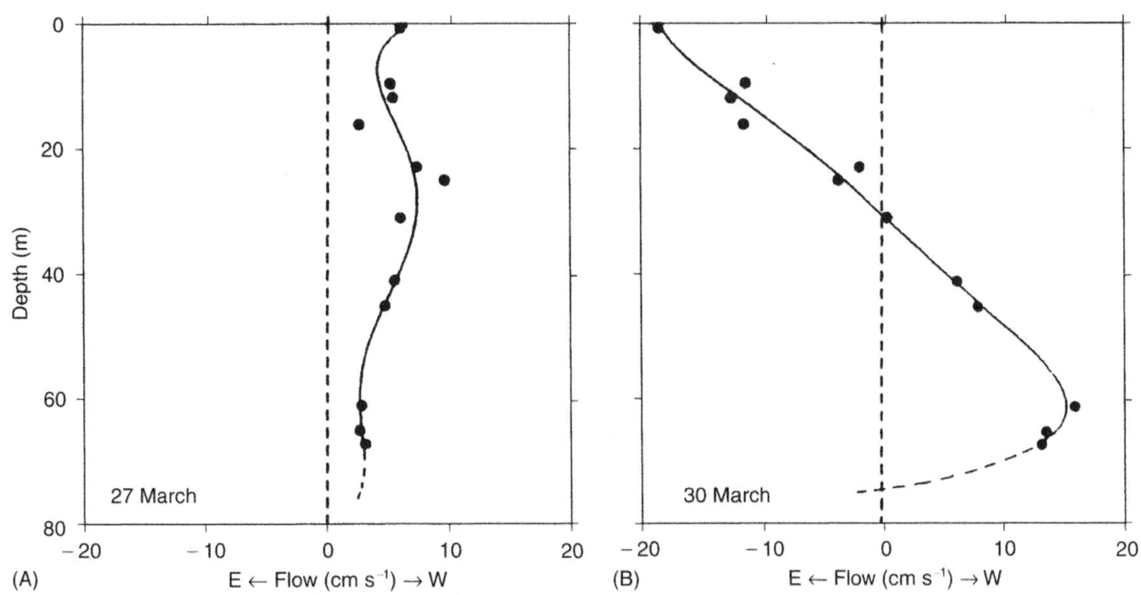

Figure 6 Profiles of cross-shelf flow measured by current meters (solid dots) in 76 m of water on the African continental shelf during (A) near-zero winds and (B) upwelling favorable wind.

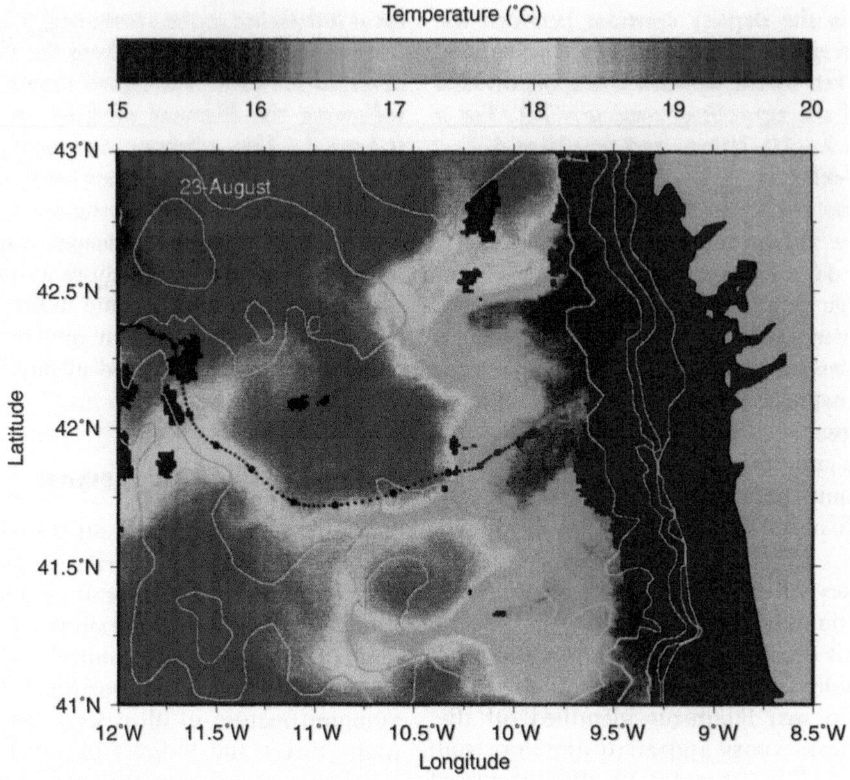

Figure 7 Satellite sea surface temperature image in August showing an upwelling filament extending 200 m offshore. Clouds obscure the image near to shore. A surface layer drift buoy traces the current along the filament. The dots mark daily positions starting on 14 August near shore. White curves mark the 50, 100, 200, 500, and 1000 m isobaths.

equatorward flow predominates, while in layers deeper than 500 m, Antarctic Intermediate Water is carried northward at depths around 900 m. Note also the weak undercurrent in **Figure 5**.

Along most of the eastern subtropical gyre the poleward flow is restricted to the subsurface layers, though it may surface when the Trade Winds weaken or turn northward. Off Iberia and south of 20°N it appears to extend to the sea surface for more of the annual cycle. In the latter area it forms the inshore loop of the cyclonic recirculation. Where it meets the Canary Current separating from the coast, some of the poleward flow continues northward as the undercurrent, carrying with it the typically warmer, fresher, and higher nutrient content South Atlantic Central Water of this region. The anomalous water can be traced as far as the Canary Islands (28°N) before mixing with the surrounding cooler and saltier North Atlantic Central Water dilutes it beyond recognition. The seasonal analysis showed that during autumn the poleward flow may occur at the sea surface, again reaching the Canary Islands, and the few available direct observations appear to corroborate this.

Off Iberia, most of the water column flows poleward, although the surface layer is flowing equatorward above 200 m depth in the long-term mean. During winter, all of the water column moves northward over the continental slope, but in summer, when the equatorward Trade Winds are present, the currents in the upper few hundred meters are driven equatorward. The undercurrent is known to extend deeply off Iberia, where it carries Mediterranean Intermediate Water at levels between 600 and 1500 m depth. This is water that has escaped through the Strait of Gibraltar and is constrained by the Earth's rotation to flow northward, hugging the continental slope. As it travels northward it tends to separate intermittently from the coast in various locations to form subsurface eddies known as Meddies. Three preferred paths are reported to carry the Mediterranean Water away from the Iberian continental slope: northward where the undercurrent extends beyond Cape Finisterre, north-westward west of the Galicia Bank, and south-westward off the Gorringe Bank. In each case the topographic feature seems to trigger the formation of the Meddies, which then migrate away from the boundary and can maintain their identity for up to 4 years.

One unresolved question is whether the poleward flow off NW Africa is in any sense continuous with that off Iberia. There are few observations of the

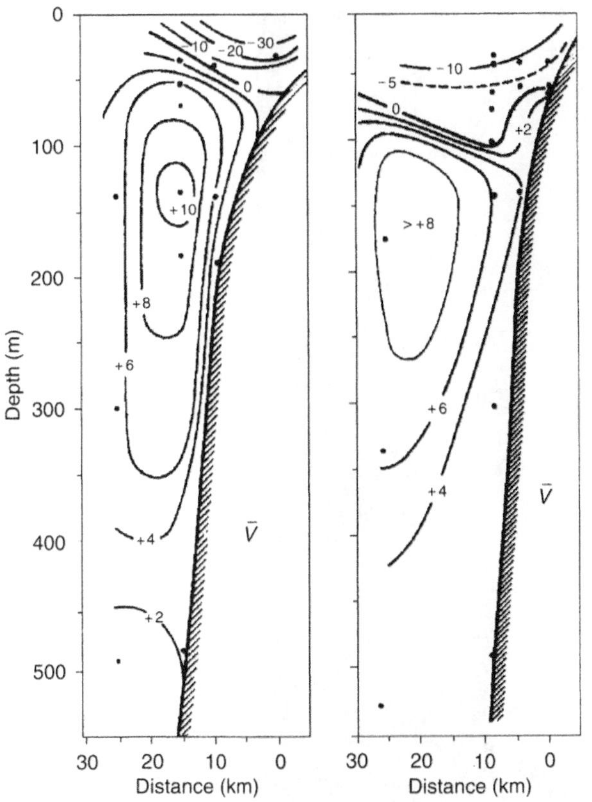

Figure 8 Two sections of alongshore flow measured by current meters (solid dots) near 20 N off north-west Africa, showing structure of the poleward undercurrent. Speeds are given in cm s^{-1}, northward positive.

undercurrent off northern Morocco and none that might indicate how the flow interacts with the deepening Mediterranean Intermediate Water. The latter is dense and sinks from shallow levels on leaving the Strait of Gibraltar to its equilibrium level around 1200 m. It therefore must pass through any undercurrent continuing north from Morocco to the Iberian slope in the Gulf of Cadiz.

Spatial Variability

Textbook pictures like **Figure 1** tend to show broad currents of weak unidirectional flows. However, measurements almost always indicate currents highly variable in both strength and direction. **Figure 9** shows a near-synoptic view of the currents, derived from the combined TOPEX and ERS-1 altimetry on 14 August 1993. The small sea surface height slope anomalies measured by satellites have been added to the mean summer surface elevations calculated with respect to a 400 m reference level from hydrography to compensate in part for the lack of the mean signal in the altimetry. The Azores Current meanders along latitude 32°N, gradually turning southward into the Canary Current. Little flow seems to come from the weak Portugal Current southwards. Near the African coast from 30°N the flow is southward as far as 20°N, where it turns abruptly offshore on meeting poleward flow from farther south. A large number of eddies are seen throughout the region, especially associated with the Azores Current and farther south. The resolution of the altimetry is limited by the ground track separation ($\sim$50 km or more) and the repeat interval ($>$15 days).

A survey of currents just south of the Canary Islands made near the same time (10–18 August 1993) shows patterns of flow on shorter scales as complex as in the satellite view (**Figure 10**). One might ask where the Canary Current is in this complexity. It lies, of course, in the average flow over the area of the survey, about 0.05 m s^{-1} toward the south west as expected. However, the instantaneous current in any location can have almost any direction and speed. The flow field is composed of narrow, strong jets of current, which may meander through the area changing their position with time, and eddy circulations that drift with the background flow. Here the alongshore, equatorward flow appears diverted around a 100 km diameter cyclonic eddy generated in the trough of bottom topography between Gran Canaria, Fuerteventura, and Africa. This eddy was only just resolved in the satellite analysis. This level of detail of field observation has only become available in the last two decades with the introduction of acoustic Doppler methods of determining upper-level currents from a moving research vessel. In this way rapid surveys can be made to reveal the intricate patterns of ocean currents, though still in an area limited in size by ship speed.

Numerical Models

Numerical modeling techniques for ocean circulation are improving rapidly as computer power allows more detailed calculations on finer grids. Recent results for the Canary Current region, representing the near-surface flow in early September, are shown in **Figure 11**. The model realistically depicts a meandering Azores Current at 35°N, which separates into branches entering the Gulf of Cadiz and flowing south-east into the Canary Current. Note the southward flow near-shore off NW Africa, typical of late summer upwelling conditions, and the prevalence of meanders and eddies throughout the field. Also noteworthy is the near-shore poleward current off Iberia, similar to conditions observed there in September with the cessation of upwelling and onset of the winter season.

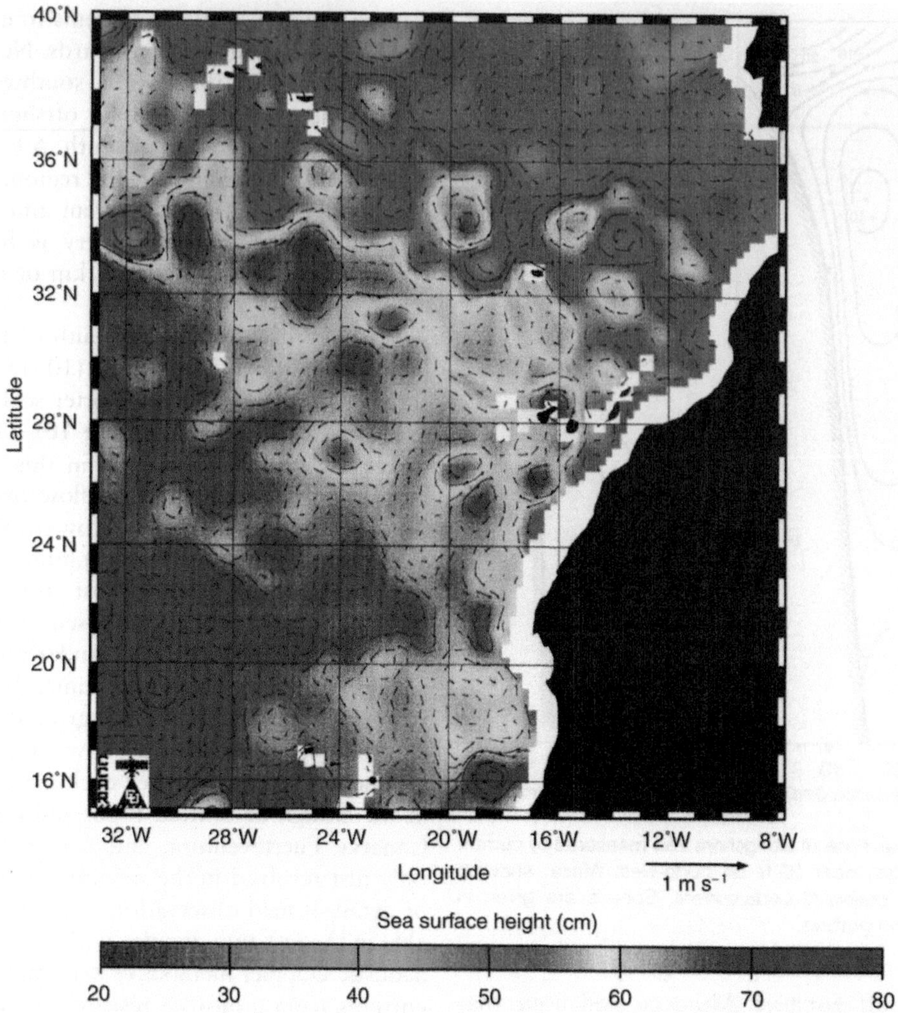

Figure 9 Surface geostrophic currents superimposed on sea surface height fields from TOPEX/ERS-1 altimeter (14 August 1993). The August mean sea surface height calculated from density fields has been added to compensate for lack of altimetric mean.

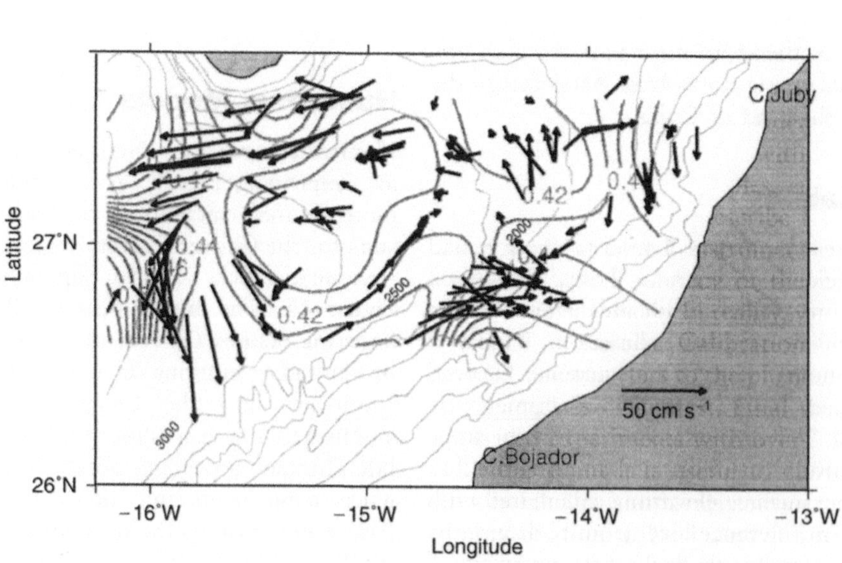

Figure 10 Field of current observed by acoustic Doppler current profiler near the time of **Figure 9**. Sea surface height contours are shown. Note the southward shelf flow turning offshore to describe a cyclonic eddy ~100 km diameter.

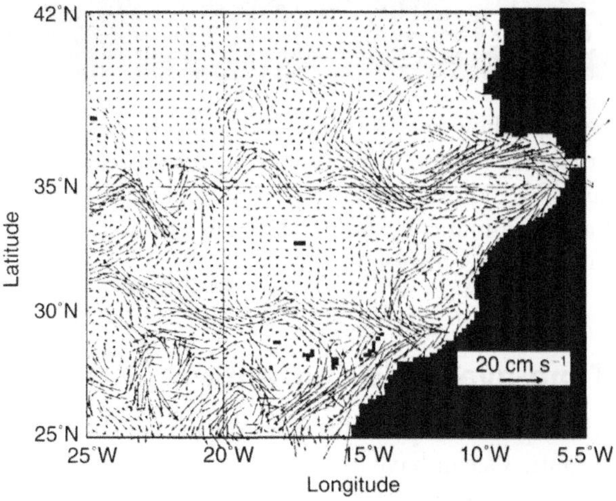

Figure 11 Numerical model results of calculated surface currents for September. (Johnson J and Stevens I 2000 *Deep Sea Research I*, 47(5): 875–900.)

This 'state of the art' numerical model is run using climatic winds, i.e., monthly averaged winds that vary smoothly through the year, and has a limited number of layers in the vertical. Nevertheless, the fine horizontal resolution allows the realistic reproduction of oceanic features often only partially sampled because of ship time and equipment constraints. Even this model is on a coarse scale compared to the size of many important features like islands or coastal capes. Entire islands are represented by one or two model grid points so the level of detail does not yet reproduce features on the scales seen in **Figure 10**. In the not too distant future, models driven by actual or forecast winds and incorporating ocean observations from monitoring networks will provide ocean forecasts that will rival in accuracy present-day meteorological models.

Conclusions

The overall features of the Canary and Portugal currents, such as the mean pattern and seasonal variation, are well established despite relatively few systematic observations. The Canary Current is fed from the Azores Current to a lesser extent from the weak Portugal Current. To north and south of the Canary Current proper, regions of predominantly northward flow persist through most of the year. These are apparently connected by a narrow undercurrent trapped to the continental slope near 300 m depth. Variability of the system is dominated by the Trade Wind, which varies on timescales of weeks, seasons, decades, and longer. The Trades directly force coastal upwelling and continental shelf currents throughout the region. Unresolved questions include the reality of the intense equatorward flow alongshore between 30° and 20°N suggested by the seasonal analyses, and the continuity of the undercurrent along the continental margin.

See also

Benguela Current. California and Alaska Currents. Canary and Portugal Currents. Ekman Transport and Pumping. Meddies and Subsurface Eddies. Mesoscale Eddies. Wind Driven Circulation.

Further Reading

Barton ED (1989) The poleward undercurrent on the eastern boundary of the Subtropical North Atlantic. In: Neshyba S, Smith RL, and Mooers CNK (eds.) *Poleward Flows Along Eastern Ocean Boundaries*, pp. 82–95. (Springer Lecture Note Series). Berlin: Springer-Verlag.

Barton ED (1998) Eastern boundary of the North Atlantic: Northwest Africa and Iberia. In: Brink KH and Robinson AR (eds.) *The Sea*, vol. 11: The Global Coastal Ocean: Regional Studies and Syntheses, ch. 22. New York: Wiley.

Brink KH (1997) Wind driven currents over the continental shelf. In: Brink KH and Robinson AR (eds.) *The Sea*, vol. 10: *The Global Coastal Ocean: Processes and Methods, ch. 1*. New York: Wiley.

Johnson J and Stevens I (2000) A fine resolution model of the eastern North Atlantic between the Azores, the Canary Islands and the Gibraltar Strait. *Deep-Sea Research I* 47(5): 875–900.

Krauss W (ed.) (1996) *The Warmwatersphere of the North Atlantic Ocean* chs 10–12. Berlin: Borntraeger.

Mann KH and Lazier JRN (eds.) (1991) *Dynamics of Marine Ecosystems*. Boston: Blackwell.

Mittelstaedt E (1983) The upwelling area off Northwest Africa – a description of phenomena related to coastal upwelling. *Progress in Oceanography* 12: 307–331.

Stramma L and Siedler G (1988) Seasonal changes in the North Atlantic Subtropical Gyre. *Journal of Geophysical Research* 93(C7): 8111–8118.

Tomczak M and Godfrey JS (1994) *Regional Oceanography: An Introduction*. Oxford: Elsevier.

ATLANTIC OCEAN EQUATORIAL CURRENTS

S. G. Philander, Princeton University, Princeton, NJ, USA

Introduction

The circulations of the tropical Atlantic and Pacific Oceans have much in common because similar trade winds, with similar seasonal fluctuations, prevail over both oceans. The salient features of these circulations are alternating bands of eastward- and westward-flowing currents in the surface layers (see **Figure 1**). Fluctuations of the currents in the two oceans have similarities not only on seasonal but even on interannual timescales; the Atlantic has a phenomenon that is the counterpart of El Niño in the Pacific. The two oceans also have significant differences. The Atlantic, but not the Pacific, has a net transport of heat from the southern into the northern hemisphere, mainly because of an intense, cross-equatorial coastal current in the Atlantic, the North Brazil Current. The similarities and differences between the tropical Atlantic and Pacific (and also the Indian Ocean) are of enormous interest to modelers because they provide invaluable checks on the theories and models that explain and simulate oceanic currents. Those currents play a central role in the Earth's climate, by influencing sea surface temperature patterns for example.

Time-averaged Currents

Although the trade winds that prevail over the tropical Atlantic Ocean have a westward component, the currents driven by those winds include the eastward North Equatorial Countercurrent, between the latitudes 3° and 10°N approximately. Sverdrup, in one of the early triumphs of dynamical oceanography, first pointed out that this current is attributable to the curl of the wind. Flanking this eastward current are westward currents to its north, the North Equatorial Current, and to its south, the South Equatorial Current. The latter current is particularly intense at the equator, where it can attain speeds in excess of $1\ \mathrm{m\,s^{-1}}$. **Figure 1**, a schematic map of the various currents, actually depicts conditions between July and September when the south-east trades are particularly intense and penetrate into the northern hemisphere.

Centered on the equator, and below the westward surface flow, is an intense eastward jet known as the

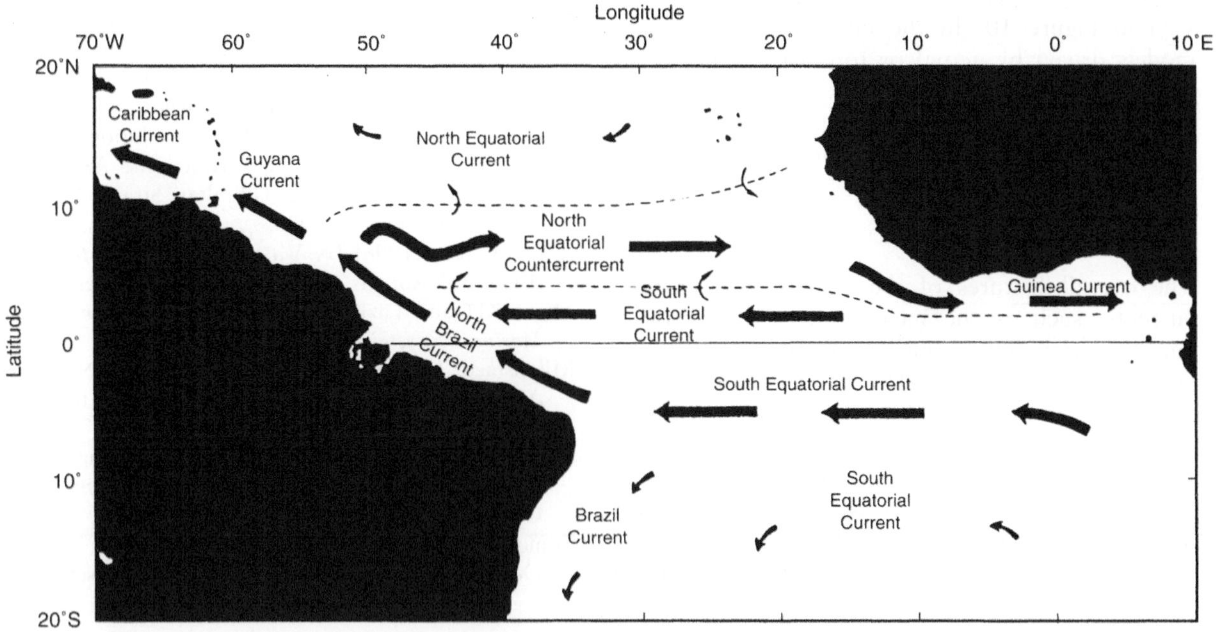

Figure 1 Schematic map showing the major surface currents of the tropical Atlantic Ocean between July and September when the North Equatorial Countercurrent (NECC) flows eastward into the Guinea Current in the Gulf of Guinea. From January to May the North Equatorial Countercurrent disappears and the surface flow is westward everywhere in the western tropical Atlantic.

Equatorial Undercurrent which amounts to a narrow ribbon that precisely marks the location of the equator. The undercurrent attains speeds on the order of 1 $m s^{-1}$ has a half-width of approximately 100 km; its core, in the thermocline, is at a depth of approximately 100 m in the west, and shoals towards the east. The current exists because the westward trade winds, in addition to driving divergent westward surface flow (upwelling is most intense at the equator), also maintain an eastward pressure force by piling up the warm surface waters in the western side of the ocean basin. That pressure force is associated with equatorward flow in the thermocline because of the Coriolis force. At the equator, where the Coriolis force vanishes, the pressure force is the source of momentum for the eastward Equatorial Undercurrent which, in a downstream direction, continually loses water because of intense equatorial upwelling which sustains the divergent, poleward Ekman flow in the surface layers.

Along the African coast, cold equatorward coastal currents, the Canary Current off north-west Africa, and the Benguela Current off south-west Africa, are driven by the components of the winds parallel to the coast. These currents, which are associated with intense coastal upwelling and low sea surface temperatures, feed the westward North and South Equatorial Currents respectively.

Along the coast of South America, the most prominent current is the North Brazil Current, which carries very warm water from about 5°N across the equator. Some of that water feeds the Equatorial Undercurrent, but much of it continues into the northern hemisphere. Further south along the coast of Brazil, the flow is southward.

The net north–south circulation associated with the various currents is a northward flow of warm surface waters, and a southward return flow of cold water at depth, resulting in a transport of heat from the southern into the northern Atlantic. The southward flow below the thermocline is part of the global thermohaline circulation, which involves the sinking of cold, saline waters in the northern Atlantic. The absence of such formation of deep water in the northern Pacific – that ocean is less saline than the northern Atlantic – is part of the reason why there is a northward transport of heat across the equator in the Atlantic but not the Pacific.

Seasonal Variations of the Currents

The seasonal variations of the winds are associated with the north–south movements of the Intertropical Convergence Zone (ITCZ), the band of cloudiness and heavy rains where the south-east and north-east trades meet. The south-east trades are most intense and penetrate into the northern hemisphere during the northern summer when the ITCZ is between 10° and 15°N. During those months the surface currents are particularly strong. The North Brazil Current, after crossing the equator, veers sharply eastward to feed the North Equatorial Countercurrent. The Equatorial Undercurrent is also strongest during this season when the east–west slope of the equatorial thermocline is at a maximum.

During the summer of the southern hemisphere, the zone where the north-east and south-east trades meet (the ITCZ) shifts equatorward so that the winds are relaxed along the equator. The North Brazil Current no longer veers offshore after crossing the equator, but continues to flow along the coast into the Gulf of Mexico. It is fed by surface flow that is westward at practically all latitudes in the tropics because, during this season, the eastward North Equatorial Countercurrent disappears from the surface layers, as is evident in **Figure 2**. At this time, the northward heat transport across 10°N is huge – on the order of a peta-watt; during the northern summer it is practically zero.

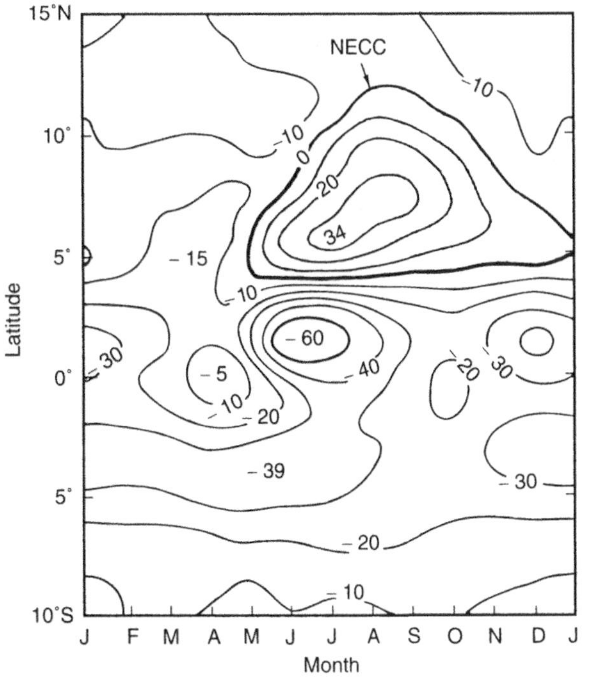

Figure 2 The seasonal disappearance of the North Equatorial Countercurrent from the western tropical Atlantic. The eastward velocity in cm s^{-1} (negative values correspond to westward flow) is shown as a function of latitude and month, starting in January. The data, which have been averaged over a band of longitudes in the western equatorial Atlantic from 23°W to 33°W, are from shipdrift records.

The upwelling along the west African coast, and the coastal currents too, are subject to large seasonal fluctuations in response to the variations in the local winds. Thus upwelling is most intense off south-western Africa, and surface temperatures there are at a minimum, during the late northern summer when the local alongshore winds are most intense. Off north-western Africa the season for such conditions is the late northern winter. The northern coast of the Gulf of Guinea (along 5°N approximately) also has seasonal upwelling, with lowest temperatures during the northern summer, even though the local winds along that coast have almost no seasonal cycle. In that region, changes in the depth of the thermocline (which separates warm surface waters from the cold water at depth) depend on winds everywhere in the equatorial Atlantic, even the winds off Brazil which are most intense during the northern summer when they cause a shoaling of the thermocline throughout the Gulf of Guinea.

If the winds over the ocean were suddenly to stop blowing, how long would it be before the currents in **Figure 1** disappear? The answer to this question (which is the same as asking how long it would take for the currents to be generated from a state of rest) is of central importance in climate studies because, associated with the currents, are sea surface temperature patterns that profoundly affect climate. (From a strictly atmospheric perspective, the cause of El Niño is a change in the surface temperature pattern of the tropical Pacific.) The Indian Ocean is ideal for studying these matters because there the abrupt onset of the south-west monsoons in May quickly generates the intense Somali Current along the eastern coast of Africa. Theoretical studies indicate that the generation of such currents, and more generally the adjustment of the ocean to a change in the winds, depend critically on waves (known as Rossby waves) that propagate across the ocean basin along the thermocline. The speed of those waves increases with decreasing latitude, reaching a maximum at the equator, which serves as a guide for the fastest waves – there they travel westward at about 50 cm s^{-1}. The equator is also a guide for a very rapid eastward traveling wave, a Kelvin wave with a speed on the order of 150 cm s^{-1}. The Somali Current near the equator can therefore be generated far more rapidly than can the Gulf Stream in mid-latitudes. The time it takes for the ocean to adjust (for the currents to be generated) depends not only on the speed of certain oceanic waves, but also on the width of the ocean basin. Hence it takes longer to generate the Kuroshio Current in the very wide Pacific, than the Gulf Stream in the smaller Atlantic.

If the winds change gradually rather than abruptly, then the timescale of the gradual changes relative to the time it takes the ocean to adjust determines the nature of the oceanic response. Thus winds that fluctuate on a timescale much longer than the adjustment time of the ocean will force an equilibrium response in which the ocean, at any given time, is in equilibrium with the winds at that time. (The currents and winds fluctuate essentially in phase.) From results such as these it can be inferred that the seasonally varying trade winds over the tropical Atlantic and Pacific Oceans should force an equilibrium response near the equator in the case of the small ocean basin, the Atlantic, but not in the case of the much larger Pacific. The measurements confirm this theoretical result: the seasonal variations of the currents and of the thermocline slope are in phase with the variations of the winds in the equatorial Atlantic, but not in the equatorial Pacific.

Interannual Variations

Given the similarities between the climates of the tropical Atlantic and Pacific – arid, cool conditions on the eastern sides, along the shores of Peru and south-western Africa, and warm moist conditions on the western sides – it should come as no surprise that the climate fluctuation known as El Niño has an Atlantic counterpart. As in the Pacific, such events involve a relaxation of the trades so that the warm waters that are usually confined to the western side of the basin flow eastward, causing a rise in sea surface temperatures off the south-west African coast where rainfall can increase significantly. To attribute this phenomenon to a relaxation of the trades is of course an oceanographic perspective. From a meteorological point of view, the warming of the eastern tropical Atlantic is the reason for the weakening of the winds and for several other changes in atmospheric conditions. This circular argument – changes in sea surface temperature are both the cause and consequence of changes in the winds – implies that interactions between the ocean and atmosphere are at the heart of the matter. Those interactions give rise to a variety of natural modes of oscillation which, in the Pacific, appear to be neutrally stable so that random atmospheric disturbances are able to sustain a continual oscillation, the Southern Oscillation, with a distinctive timescale on the order of 4 years. In the Atlantic the possible natural modes appear to be strongly damped and hence are far more sporadic than in the Pacific; there is no distinctive timescale for interannual fluctuations in the Atlantic. The main reason for this difference is the modest dimensions of the Atlantic relative to those of the Pacific. Some of the natural modes attributable to ocean–atmosphere interactions depend on the delayed response of the ocean to changes in the winds. If that delay is small, that is the case in an ocean basin of modest size – then the natural modes tend to be damped. Another factor that can inhibit interannual fluctuations is a particularly strong seasonal cycle. That cycle has a larger amplitude in the equatorial Atlantic than Pacific, because the influence of continents on the seasonal changes in the winds can exceed those of ocean–atmosphere interactions in a basin of small dimensions.

For a damped mode of oscillation to appear, a suitable perturbation is necessary. The occurrence of El Niño in the Pacific provides such a perturbation in the Atlantic by causing an intensification of the trade winds, and unusually low surface temperatures, in the Atlantic. (This is the impact of the presence of deep atmospheric convection over the eastern tropical Pacific during El Niño.) Apparently El Niño in the Pacific can amount to a preconditioning of the Atlantic because, on several occasions, El Niño in the Pacific was followed a year later by a similar phenomenon in the Atlantic. The amplitude of El Niño is generally much larger in the Pacific than Atlantic – the reason why the Pacific but not the Atlantic phenomenon is capable of a global impact.

El Niño, in the Atlantic and Pacific, has a structure that is essentially symmetrical about the equator. The Atlantic has an additional climate fluctuation that is anti-symmetrical relative to the equator, with sea surface temperatures that are high on one side of that line, low on the other side. The cross-equatorial winds then blow towards the warm side with exceptional intensity. If the higher temperatures are to the north, then the zonal band of heavy rains, the ITCZ, persists in a northerly position, bringing drought to north-eastern Brazil, and good rains to the Sahel, the region to the south of the Sahara desert in west Africa. The reverse happens when the ocean temperatures are high south of the equator, cool to the north.

Stability of the Currents

During the northern summer, the currents in the western equatorial Atlantic are so intense that they become unstable. One important factor is the enormous latitudinal shear between the eastward North Equatorial Countercurrent and the adjacent westward South Equatorial Current. The instabilities result in meanders that drift westward at a speed near 50 cm s^{-1}, that have a wavelength on the order of a 1000 km, and a period of approximately 1 month. The unstable conditions are confined to the western equatorial region where there is room for two or three waves at most – they sometimes appear in satellite photographs of sea surface temperature. The waves persist for a few months at most so that approximately three oscillations appear during the summer. The counterparts of these unstable waves in the eastern equatorial Pacific have a shorter period (close to 3 weeks) than in the Atlantic, cover a much larger region, and persist far longer. In the Pacific it is possible to observe very long wave trains – they can extend from the Galapagos Islands in the east to the dateline – that persist for many months.

See also

Brazil and Falklands (Malvinas) Currents. Current Systems in the Atlantic Ocean. Florida Current, Gulf Stream and Labrador Current. Rossby Waves.

Further Reading

The *Journal of Geophysical Research* Volume 103 (1998) is devoted to a series of excellent and detailed review articles on tropical ocean–atmosphere interactions, including an article on oceanic currents.

Carton J and Huang B (1994) Warm events in the tropical Atlantic. *Journal of Physical Oceanography* 24: 888–903.

Chang P, Ji L, and Li H (1997) A decadal climate variation in the tropical Atlantic ocean from thermodynamic air–sea interaction. *Nature* 385: 516–518.

Merle J, Fieux M, and Hisard P (1980) Annual signal and interannual anomalies of sea surface temperature in the eastern equatorial Atlantic. *Deep Sea Research* 26: 77–101.

Philander SGH (1990) *El Niño, La Niña and the Southern Oscillation*. New York: Academic Press.

Richardson PL and Walsh DW (1986) Mapping climatological seasonal variations of surface currents in the tropical Atlantic using ship drifts. *Journal of Geophysical Research* 91: 10537–10550.

THE OCEANS: THE PACIFIC OCEAN

KUROSHIO AND OYASHIO CURRENTS

B. Qiu, University of Hawaii at Manoa,
Hawaii, USA

Introduction

The Kuroshio and Oyashio Currents are the western boundary currents in the wind-driven, subtropical and subarctic circulations of the North Pacific Ocean. Translated from Japanese, Kuroshio literally means black ('kuro') stream ('shio') owing to the blackish – ultramarine to cobalt blue – color of its water. The 'blackness' of the Kuroshio Current stems from the fact that the downwelling-dominant subtropical North Pacific Ocean is low in biological productivity and is devoid of detritus and other organic material in the surface water. The subarctic North Pacific Ocean, on the other hand, is dominated by upwelling. The upwelled, nutrient-rich water feeds the Oyashio from the north and leads to its nomenclature, parent ('oya') stream ('shio').

The existence of a western boundary current to compensate for the interior Sverdrup flow is well understood from modern wind-driven ocean circulation theories. Individual western boundary currents, however, can differ greatly in their mean flow and variability characteristics due to different bottom topography, coastline geometry, and surface wind patterns that are involved. For example, the bimodal oscillation of the Kuroshio path south of Japan is a unique phenomenon detected in no other western boundary current of the world oceans. Similarly, interaction with the semi-enclosed and often ice-covered marginal seas and excessive precipitation over evaporation in the subarctic North Pacific Ocean make the Oyashio Current considerably different from its counterpart in the subarctic North Atlantic Ocean, the Labrador Current.

Because the Kuroshio and Oyashio Current sexert a great influence on the fisheries, hydrography, and meteorology of countries surrounding the western North Pacific Ocean, they have been the focus of a great amount of observation and research in the past. This article will provide a brief review of the dynamic aspects of the observed Kuroshio and Oyashio Currents: their origins, their mean flow patterns, and their variability on seasonal-to-interannual timescales. The article consists of two sections, the first focusing on the Kuroshio Current and the second on the Oyashio Current. Due to the vast geographical areas passed by the Kuroshio Current (**Figure 1**), the first section is divided into three subsections: the region upstream of the Tokara Strait, the region south of Japan, and the Kuroshio Extension region east of the Izu Ridge. As will become clear, the Kuroshio Current exhibits distinct characteristics in each of these geographical locations owing to the differing governing physics.

The Kuroshio Current

Region Upstream of the Tokara Strait

The Kuroshio Current originates east of the Philippine coast where the westward flowing North Equatorial Current (NEC) bifurcates into the northward-flowing Kuroshio Current and the southward-flowing Mindanao Current. At the sea surface, the NEC bifurcates nominally at 12°N–13°N, although this bifurcation latitude can change interannually from 11°N to 14.5°N. The NEC's bifurcation tends to migrate to the north during El Niño years and to the south during La Niña years. Below the sea surface, the NEC's bifurcation tends to shift northward with increasing depth. This tendency is due to the fact that the southern limb of the wind-driven subtropical gyre in the North Pacific shifts to the north with increasing depth.

Branching northward from the NEC, the Kuroshio Current east of the Philippine coast has a mean geostrophic volume transport, referenced to 1250 dbar, of 25 Sv (1 Sverdrup $= 10^6\,m^3\,s^{-1}$). Seasonally, the Kuroshio transport at this upstream location has a maximum ($\sim$30 Sv) in spring and a minimum ($\sim$19 Sv) in fall. Similar seasonal cycles are also found in the Kuroshio's transports in the East China Sea and across the Tokara Strait.

As the Kuroshio Current flows northward passing the Philippine coast, it encounters the Luzon Strait that connects the South China Sea with the open Pacific Ocean (**Figure 2**). The Luzon Strait has a width of 350 km and is 2500 m deep at its deepest point. In winter, part of the Kuroshio water has been observed to intrude into the Luzon Strait and form a loop current in the northern South China Sea (see the dashed line in **Figure 2**). The loop current can reach as far west as 117°E, where it is blocked by the presence of the shallow shelf break off the south-east coast of China. The formation of the loop current is

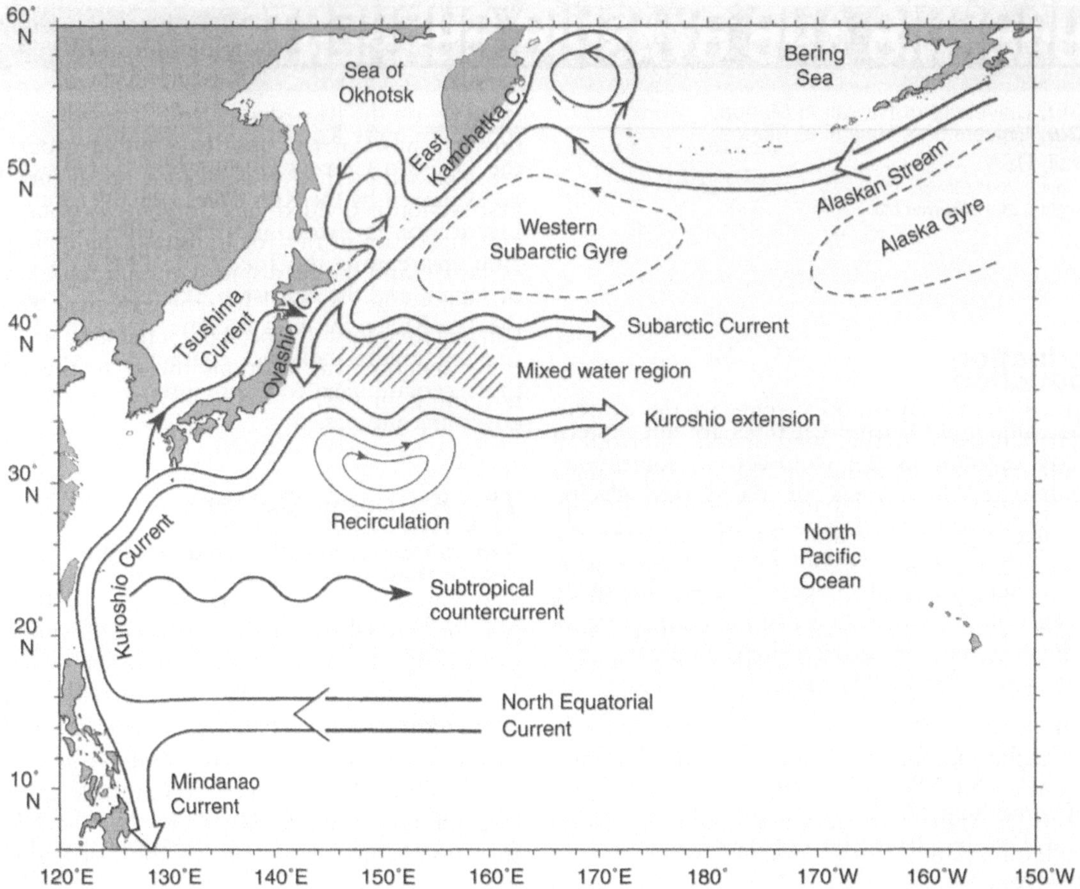

Figure 1 Schematic current patterns associated with the subtropical and subarctic gyres in the western North Pacific Ocean.

probably due to the north-east monsoon, prevailing from November to March, which deflects the surface Kuroshio water into the northern South China Sea. During the summer months from May to September when the south-west monsoon prevails, the Kuroshio Current passes the Luzon Strait without intrusion.

In the latitudinal band east of Taiwan (22°N–25°N), the northward-flowing Kuroshio Current has been observed to be highly variable in recent years. Repeat hydrographic and moored current meter measurements between Taiwan and the southernmost Ryukyu island of Iriomote show that the variability of the Kuroshio path and transport here are dominated by fluctuations with a period of 100 days. These observed fluctuations are caused by impinging energetic cyclonic and anticyclonic eddies migrating from the east. The Subtropical Counter current (STCC) is found in the latitudinal band of 22°N–25°N in the western North Pacific. The STCC, a shallow eastward-flowing current, is highly unstable due to its velocity shear with the underlying, westward-flowing NEC. The unstable waves generated by the instability of the STCC-NEC system tend to move westward while growing in amplitude. The cyclonic and anticyclonic eddies that impinge upon the Kuroshio east of Taiwan are results of these large-amplitude unstable waves. Indeed, satellite measurements of the sea level (**Figure 3**) show that the Kuroshio east of Taiwan has higher eddy variability than either its upstream counterpart along the Philippine coast or its downstream continuation in the East China Sea.

The Kuroshio Current enters the East China Sea through the passage between Taiwan and Iriomote Island. In the East China Sea, the Kuroshio path follows closely along the steep continental slope. Across the PN-line in the East China Sea (see **Figure 2** for its location), repeat hydrographic surveys have been conducted on a quarterly basis by the Japan Meteorological Agency since the mid-1950s. Based on the measurements from 1955 to 1998, the volume transport of the Kuroshio across this section has a mean of 24.6 Sv and a seasonal cycle with 24.7 Sv in winter, 25.4 Sv in spring, 25.2 Sv in summer, and 22.8 Sv in fall, respectively. In addition to this seasonal signal, large transport changes on

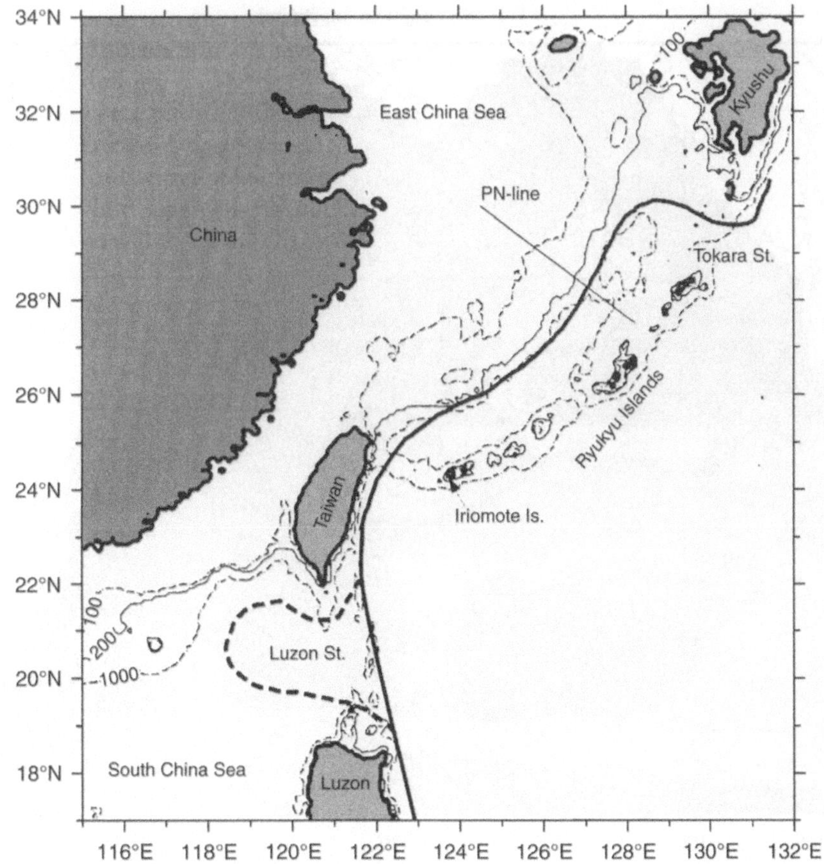

Figure 2 Schematic representation of the mean Kuroshio path (solid thick line) along the North Pacific western boundary. The thick dashed line south of Taiwan denotes the wintertime branching of the Kuroshio water into the Luzon Strait in the form of a loop current. PN-line denotes the repeat hydrographic section across which long-term Kuroshio volume transport is monitored (see **Figure 4**). Selective isobaths of 100 m, 200 m, and 1000 m are depicted.

longer timescales are also detected across this section (see **Figure 4**). One signal that stands out in the time-series of **Figure 4** is the one with the decadal timescale. Specifically, the Kuroshio transport prior to 1975 was low on average (22.5 Sv), whereas the mean transport value increased to 27.0 Sv after 1975. This decadal signal in the Kuroshio's volume transport is associated with the decadal Sverdrup transport change in the subtropical North Pacific Ocean.

Although the main body of the Kuroshio Current in the East China Sea is relatively stable due to the topographic constraint, large-amplitude meanders are frequently observed along the density front of the Kuroshio Current. The density front marks the shoreward edge of the Kuroshio Current and is located nominally along the 200 m isobath in the East China Sea. The frontal meanders commonly originate along the upstream Kuroshio front north east of Taiwan and they evolve rapidly while propagating downstreamward. The frontal meanders have typical wavelengths of 200–350 km, wave periods of 10–20 days, and downstreamward phase speeds of 10–25 cm s^{-1}. When reaching the Tokara Strait, the fully developed frontal meanders can shift the path of the Kuroshio Current in the strait by as much as 100 km.

Around 128°E–129°E and 30°N, the Kuroshio Current detaches from the continental slope and veers to the east toward the Tokara Strait. Notice that this area is also where part of the Kuroshio water is observed to intermittently penetrate northward onto the continental shelf to feed the Tsushima Current. The frontal meanders of the Kuroshio described above are important for the mixing and water mass exchanges between the cold, fresh continental shelf water and the warm, saline Kuroshio water along the shelf break of the East China Sea. It is this mixture of the water that forms the origin of the Tsushima Current. The volume transport of the Tsushima Current is estimated at 2 Sv.

Region South of Japan

The Kuroshio Current enters the deep Shikoku Basin through the Tokara Strait. Combined surface current

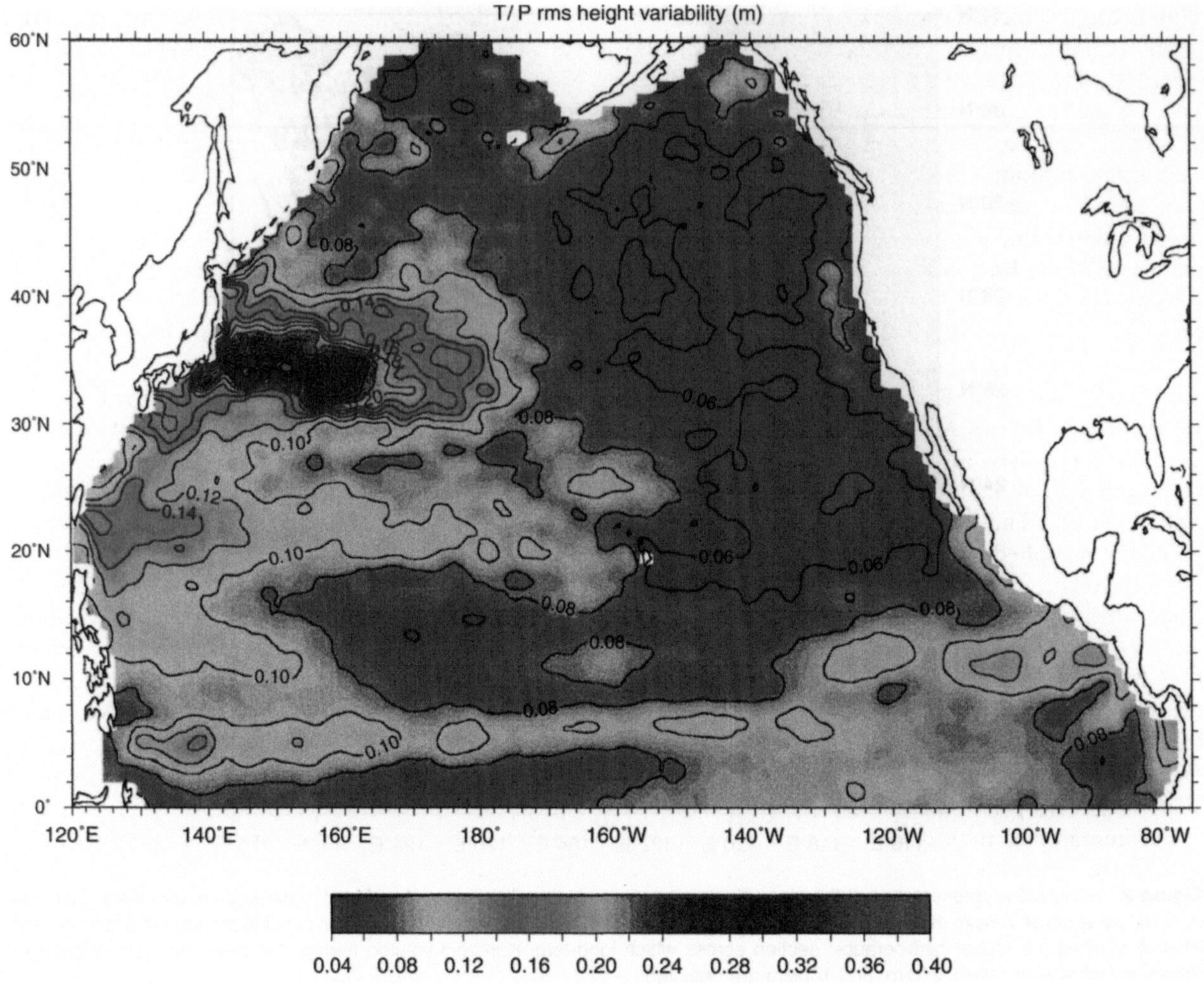

Figure 3 Map of the root-mean-square (rms) sea surface height variability in the North Pacific Ocean, based on the TOPEX/POSEIDON satellite altimetric measurements from October 1992 to December 1997. Maximum rms values of >0.4 m are found in the upstream Kuroshio Extension region south east of Japan. Sea surface height variability is also high in the latitudinal band east of Taiwan. (Adapted with permission from Qiu B (1999) Seasonal eddy Field modulation of the North Pacific Subtropical Countercurrent: TOPEX/Poseidom observations and theory. *Journal of Physical Oceanography* 29: 2471–2486.)

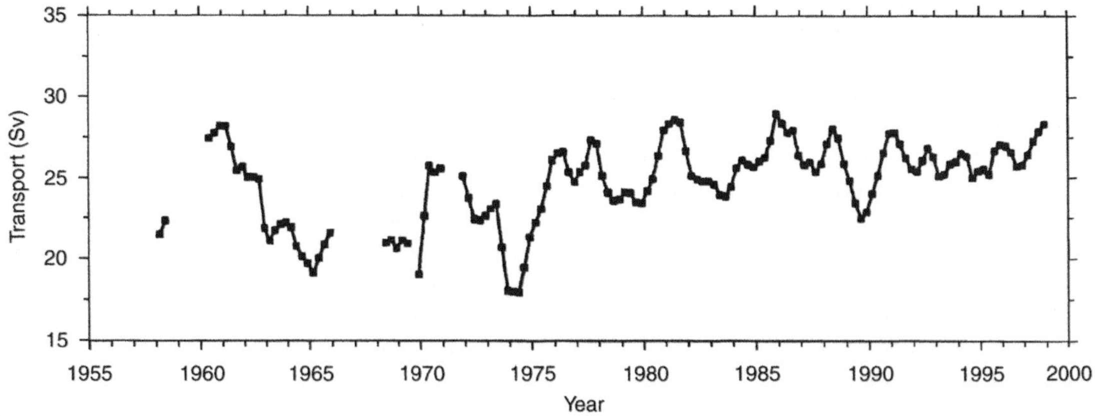

Figure 4 Time-series of the geostrophic volume transport of the Kuroshio across the PN-line in the East China Sea (see **Figure 2** for its location). Reference level is at 700 dbar. Quarterly available transport values have been low-pass filtered by the 1-year running mean averaging. (Data courtesy of Dr M. Kawabe of the University of Tokyo.)

and hydrographic observations show that the Kuroshio's volume transport through the Tokara Strait is about 30 Sv. Inference of transport from the sea level measurements suggests that the Kuroshio's transport across the Tokara Strait is maximum in spring/summer and minimum in fall, a seasonal cycle similar to that found in the upstream Kuroshio Current. Further downstream, offshore of Shikoku, the volume transport of the Kuroshio has a mean value of 55 Sv. This transport increase of the Kuroshio in the deep Shikoku Basin is in part due to the presence of an anticyclonic recirculation gyre south of the Kuroshio. Subtracting the contribution from this recirculation reduces the mean eastward transport to 42 Sv. Notice that this 'net' eastward transport of the Kuroshio is still larger than its inflow transport through the Tokara Strait. This increased transport, ~12 Sv, is probably supplied by the north-eastward-flowing current that has been occasionally observed along the eastern flank of the Ryukyu Islands. Near 139°E, the Kuroshio Current encounters the Izu Ridge. Due to the shallow northern section of the ridge, the Kuroshio Current exiting the Shikoku Basin is restricted to passing the Izu Ridge at either around 34°N where there is a deep passage, or south of 33°N where the ridge height drops.

On interannual timescales, the Kuroshio Current south of Japan is known for its bimodal path fluctuations. The 'straight path', shown schematically by path A in **Figure 5**, denotes when the Kuroshio flows closely along the Japan coast. The 'large-meander path', shown by path B in **Figure 5**, signifies when the Kuroshio takes a detouring offshore path. In addition to these two stable paths, the Kuroshio may take a third, relatively stable path that loops southward over the Izu Ridge. This path, depicted as path C in **Figure 5**, is commonly observed during transitions from a meandering state to a straight-path state. As the meander path of the Kuroshio can migrate spatially, a useful way of indexing the Kuroshio path is to use the mean distance of the Kuroshio axis from the Japan coast from 132°E to 140°E. South of Japan, the Kuroshio axis is well represented by the 16°C isotherm. Based on this representation and seasonal water temperature measurements, **Figure 6** shows the time-series of the Kuroshio path index from 1955 to 1998. A low index in **Figure 6** denotes a straight path, and a high index denotes an offshore meandering path of the Kuroshio. From 1955 to 1998, the Kuroshio large meanders occurred in 1959–62, 1975–79, 1982–88, and 1990. Clearly, the large-meanders occurrence is aperiodic. Once formed, the meander state can persist over a period ranging from a year to a decade. In contrast, transitions between the meander and straight-path states are rapid, often completed over a period of several months. It is worth noting that development of the large meanders is often preceded by the appearance of a small meander south of Kyushu, which migrates eastward and becomes stationary after reaching 136°E.

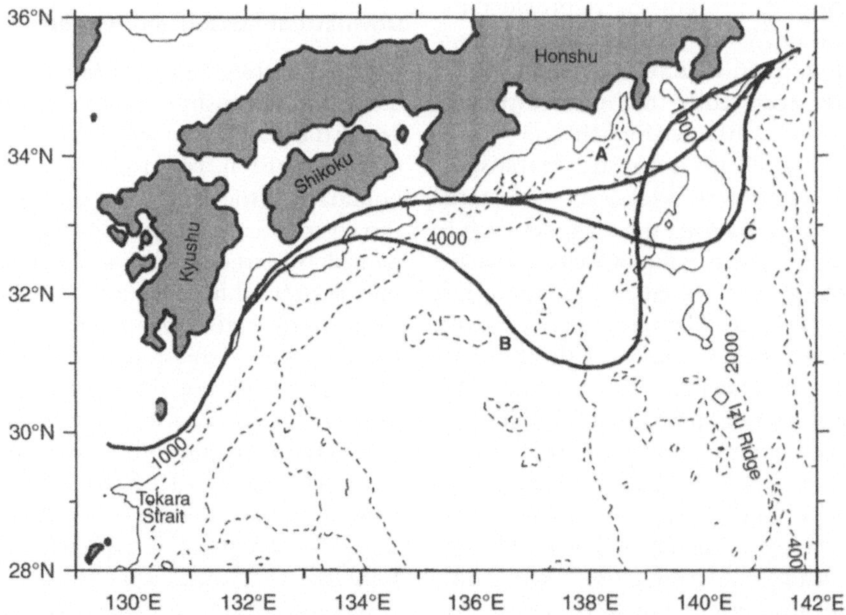

Figure 5 Schematic stable paths of the Kuroshio Current south of Japan. (Adapted with permission from Kawabe M (1985) Sea level variations at the Izu Islands and typical stable paths of the Kuroshio. *Journal of the Oceanography Society of Japan* 41: 307–326.) Selective isobaths of 1000 m, 2000 m, 4000 m, 6000 m, and 8000 m are included.

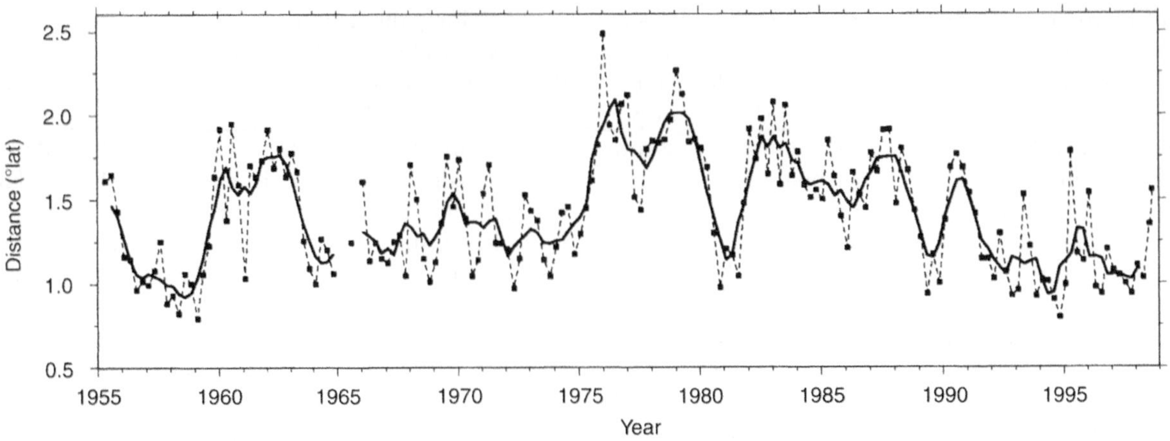

Figure 6 Time-series of the Kuroshio path index from 1955 to 1998, where the Kuroshio path index is defined as the offshore distance of the Kuroshio axis (inferred from the 16 C isotherm at the 200 m depth) averaged from 132 to 140 E. Solid dots denote the seasonal index values and the solid line indicates the annual average. (Adapted with permission from Qiu B and Miao W (2000) Kuroshio path variations south of Japan: Bimodality as a self-sustained internal oscillation. *Journal of Physical Oceanography* 30: 2124–2137.)

Several mechanisms have been proposed to explain the bimodal path variability of the Kuroshio south of Japan. Most studies have examined the relationship between the Kuroshio's path pattern and the changes in magnitude of the Kuroshio's upstream transport. Earlier studies of the Kuroshio path bimodality interpreted the meandering path as stationary Rossby lee wave generated by the protruding coastline of Kyushu. With this interpretation, the Kuroshio takes a meander path when the upstream transport is small and a straight path when it is large. By taking into account the realistic inclination of the Japan coast from due east, more recent studies have provided the following explanation. When the upstream transport is small, the straight path is stable as a result of the planetary vorticity acquired by the north-eastward-flowing Kuroshio being balanced by the eddy dissipation along the coast. When the upstream transport is large, positive vorticity is excessively generated along the Japan coast, inducing the meander path to develop downstream. In the intermediate transport range, the Kuroshio is in a multiple equilibrium state in which the meandering and straight paths coexist. Transitions between the two paths in this case are determined by changes in the upstream transport (e.g. the transition from a straight path to a meander path requires an increase in upstream transport).

A comparison between the Kuroshio path variation (**Figure 6**) and the Kuroshio's transport in the upstream East China Sea (**Figure 4**) shows that the 1959–62 large-meander event does correspond to a large upstream transport. However, this correspondence becomes less obvious after 1975, as there were times when the upstream transport was large, but no large meander was present. Assuming that the upstream Kuroshio transport after 1975 is in the multiple equilibrium regime, the correspondence between the path transition and the temporal change in the upstream transport (e.g. the required transport increase for the transition from a straight path to a meander path) is also inconclusive from the time-series presented in **Figure 4** and **6**. Given the low frequency and irregular nature of the Kuroshio path changes, future studies based on longer transport measurements are needed to further clarify the physics underlying the Kuroshio path bimodality.

Downstream Extension Region

After separating from the Japan coast at 140 E and 35 N, the Kuroshio enters the open basin of the North Pacific Ocean where it is renamed the Kuroshio Extension. Free from the constraint of coastal boundaries, the Kuroshio Extension has been observed to be an eastward-flowing inertial jet accompanied by large-amplitude meanders and energetic pinched-off eddies. **Figure 7** shows the mean temperature map at 300 m depth, in which the axis of the Kuroshio Extension is well represented by the 12 C isotherm. An interesting feature of the Kuroshio Extension east of Japan is the existence of two quasi-stationary meanders with their ridges located at 144 E and 150 E, respectively. The presence of these meanders along the mean path of the Kuroshio Extension has been interpreted as standing Rossby lee waves generated by the presence of the Izu Ridge. A competing theory also exists that regards the quasi-stationary meanders as being steered by the eddy-driven abyssal mean flows resulting from instability of the Kuroshio Extension jet.

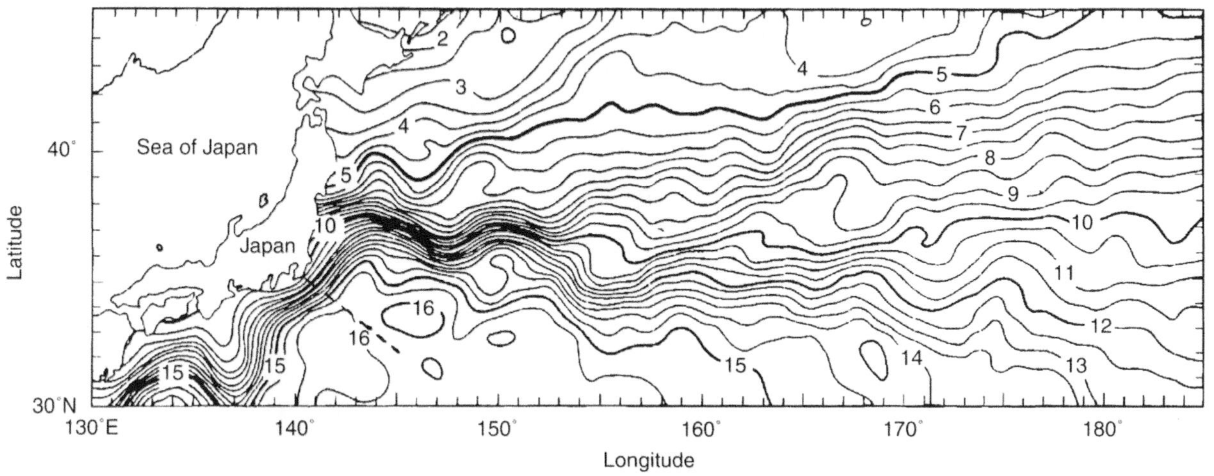

Figure 7 Mean temperature map (°C) at the 300 m depth from 1976 to 1980. (Adapted with permission from Mizuno K and White WB (1983) Annual and interannual variability in the Kuroshio Current system. *Journal of Physical Oceanography* 13: 1847–1867.)

Near 159°E, the Kuroshio Extension encounters the Shatsky Rise where it often bifurcates. The main body of the Kuroshio Extension continues eastward, and a secondary branch tends to extend northeastward to 40°N, where it joins the eastward-moving Subarctic Current. After overriding the Emperor Seamounts along 170°E, the mean path of the Kuroshio Extension becomes broadened and instantaneous flow patterns often show a multiple-jet structure associated with the eastward-flowing Kuroshio Extension. East of the dateline, the distinction between the Kuroshio Extension and the Subarctic Current is no longer clear, and together they form the broad, eastward-moving North Pacific Current.

As demonstrated in **Figure 3**, the Kuroshio Extension region has the highest level of eddy variability in the North Pacific Ocean. From the viewpoint of wind-driven ocean circulation, this high eddy variability is to be expected. Being a return flow compensating for the wind-driven subtropical interior circulation, the Kuroshio originates at a southern latitude where the ambient potential vorticity (PV) is relatively low. For the Kuroshio to smoothly rejoin the Sverdrup interior flow at the higher latitude, the low PV acquired by the Kuroshio in the south has to be removed by either dissipative or nonlinear forces along its western boundary path. For the narrow and swift Kuroshio Current, the dissipative force is insufficient to remove the low PV anomalies. The consequence of the Kuroshio's inability to effectively diffuse the PV anomalies along its path results in the accumulation of low PV water in its extension region, which generates an anticyclonic recirculation gyre and provides an energy source for flow instability. Due to the presence of the recirculation gyre (**Figure 8**), the eastward volume transport of the Kuroshio Extension can reach as high as 130 Sv south east of Japan. This is more than twice the maximum Sverdrup transport of about 50 Sv in the subtropical North Pacific. The inflated eastward transport is due to the presence of the recirculating flow to the south of the Kuroshio Extension. Although weak in surface velocity, **Figure 8** shows that the recirculating flow has a strong barotropic (i.e. depth-independent) component. As a consequence, the volume transport of the recirculation gyre in this case is as large as 80 Sv.

In addition to the high meso-scale eddy variability, the Kuroshio Extension also exhibits large-scale changes on interannual timescales. **Figure 9A** and **B** compares the sea surface height field in the Kuroshio Extension region in November 1992 with that in November 1995. In 1992, the Kuroshio Extension had a coherent zonal-jet structure extending beyond the dateline. The zonal mean axis position of the Kuroshio Extension from 141°E to 180°E in this case was located north of 35°N. In contrast, the jet-like structure in 1995 was no longer obvious near 160°E and the zonal mean axis position shifted to 34°N. Note that the changes in the zonal mean axis position of the Kuroshio Extension have interannual timescales (**Figure 9C**) and are associated with the changes in the strength of the southern recirculation gyre. As the recirculation gyre intensifies (as in 1992), it elongates zonally, increasing the zonal mean eastward transport of the Kuroshio Extension and shifting its mean position northward. When the recirculation gyre weakens (as in 1995), it decreases the eastward transport of the Kuroshio Extension and shifts its zonal mean position southward. At

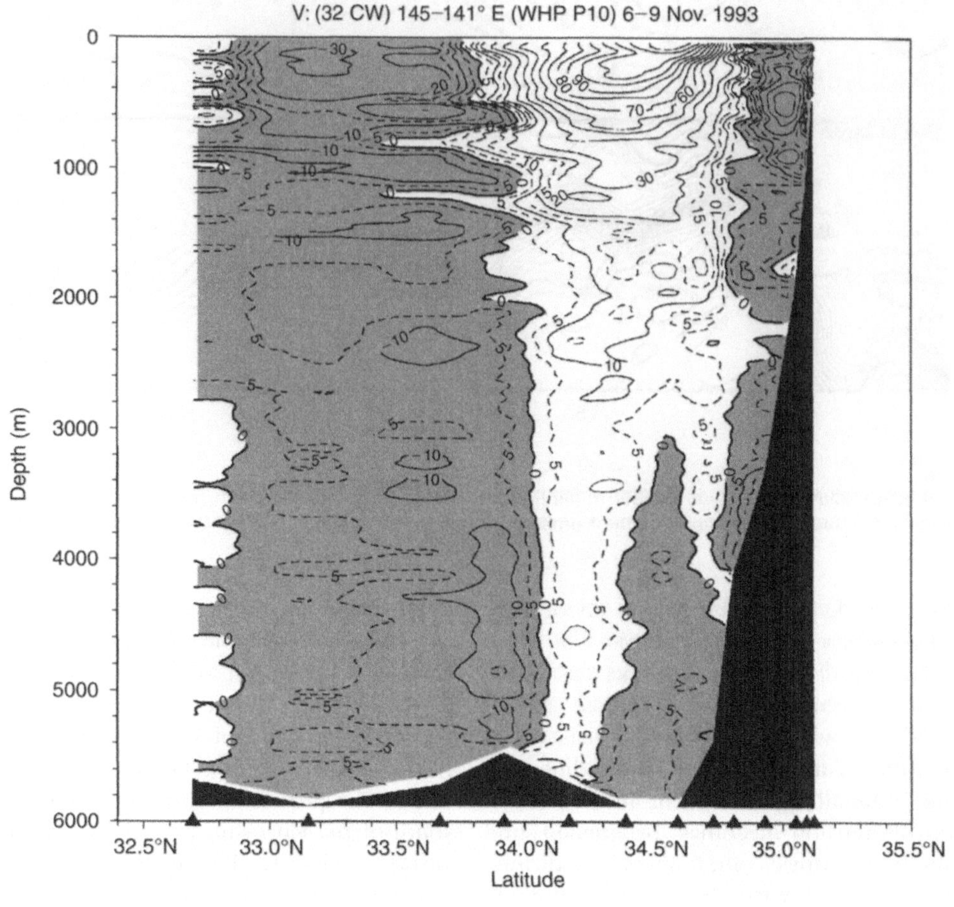

Figure 8 North-eastward velocity profile from lowered acoustic Doppler current meter profiler (ADCP) measurements along the WOCE P10 line south east of Japan in November 1993 (see the dashed line in **Figure 7** for its location). Units are cm s^{-1} and south-westward flow is shaded. (Figure courtesy of Drs E. Firing and P. Hacker of the University of Hawaii.)

present, the cause of the low-frequency changes of the recirculation gyre is unclear.

The Oyashio Current

Due to the southward protrusion of the Aleutian Islands, the wind-driven subarctic circulation in the North Pacific Ocean can be largely divided into two cyclonic subgyres: the Alaska Gyre to the east of the dateline and the Western Subarctic Gyre to the west (**Figure 1**). To the north, these two subgyres are connected by the Alaskan Stream, which flows south-westward along the Aleutian Islands as the western boundary current of the Alaska Gyre. Near the dateline, the baroclinic volume transport of the Alaskan Stream in the upper 3000 m layer is estimated at about15–20 Sv. As the Alaskan Stream flows further westward, the deep passages between 168°E and 172°E along the western Aleutian Islands allow part of the Alaskan Stream to enter the Bering Sea. In the deep part of the Bering Sea, the intruding Alaskan Stream circulates anticlockwise and forms the Bering Sea Gyre. The western limb of the Bering Sea Gyre becomes the East Kamchatka Current, which flows south-westward along the east coast of the Kamchatka Peninsula. The remaining part of the Alaskan Stream continues westward along the southern side of the Aleutian Islands and upon reaching the Kamchatka Peninsula, it joins the East Kamchatka Current as the latter exits the Bering Sea.

As the East Kamchatka Current continues south-westward and passes along the northern Kuril Islands, some of its water permeates into the Sea of Okhotsk. Inside the deep Kuril Basin in the Sea of Okhotsk, the intruding East Kamchatka Current water circulates in a cyclonic gyre. Much of this intruding water moves out of the Sea of Okhotsk through the Bussol Strait (46.5°N, 151.5°E), where it joins the rest of the south-westward-flowing East Kamchatka Current. The East Kamchatka Current is renamed the Oyashio Current south of the Bussol Strait. Because of the intrusion in the Sea of Okhotsk, the water properties of the Oyashio Current are different from those in the upstream East

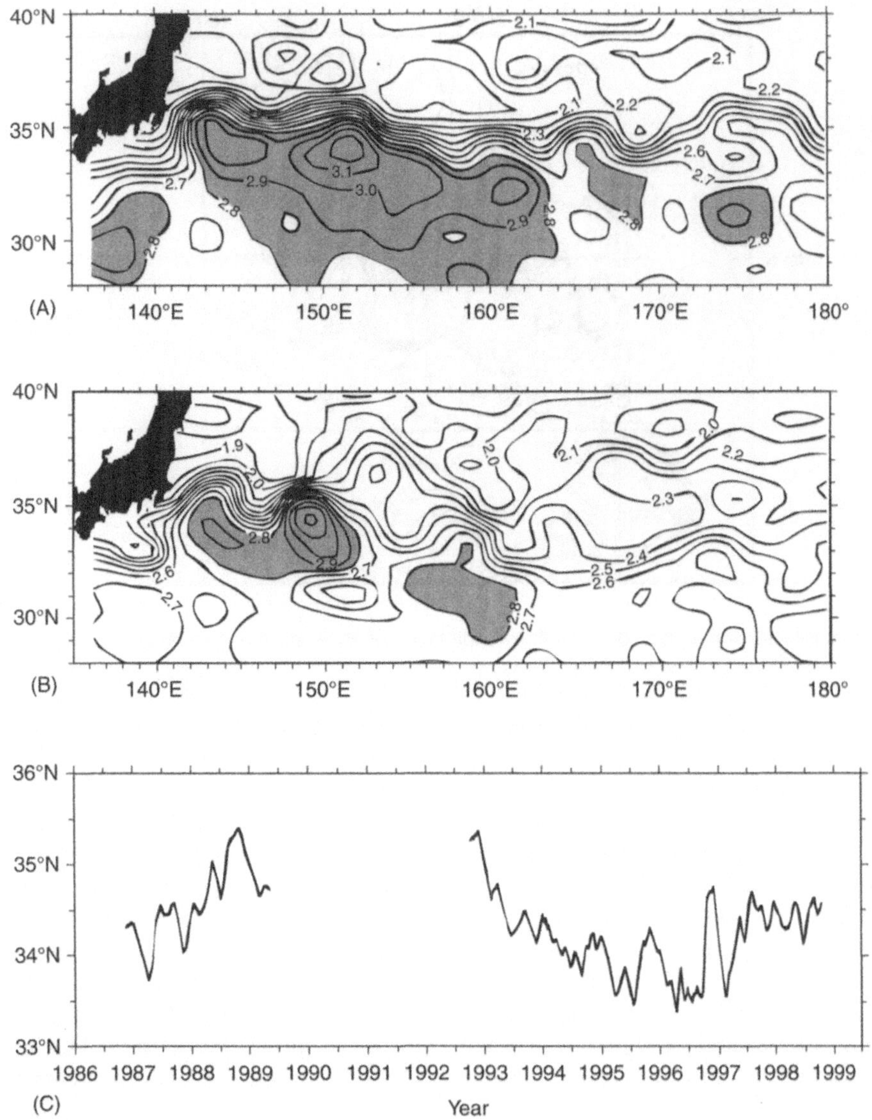

Figure 9 Sea surface height maps on (A) 20 November 1992 and (B) 15 November 1995 from the TOPEX/POSEIDON altimeter measurements. (C) Time-series of the mean axis position of the Kuroshio Extension from 141°E to 180°. (Adapted with permission from Qiu B (2000) Interannual variability of the Kuroshio Extension system and its impact on the wintertime SST field. *Journal of Physical Oceanography* 30: 1486–1502.)

Kamchatka Current. For example, the mesothermal water present in the East Kamchatka Current (i.e. the subsurface maximum temperature water appearing in the halocline at a depth of 150–200 m) is no longer observable in the Oyashio. While high dissolved oxygen content is confined to above the halocline in the upstream East Kamchatka Current, elevated dissolved oxygen values can be found throughout the upper 700 m depth of the Oyashio water.

The baroclinic volume transport of the Oyashio Current along the southern Kuril Islands and off Hokkaido has been estimated at 5–10 Sv from the geostrophic calculation with a reference level of no-motion at 1000 or 1500 m. Combining moored current meter and CTD (conductivity-temperature-depth) measurements, more recent observations along the continental slope south east of Hokkaido show that the Oyashio Current has a well-defined annual cycle: the flow tends to be strong, reaching from surface to bottom, in winter/spring, and it is weaker and confined to the layer shallower than 2000 m in summer and fall. The total (baroclinic + barotropic) volume transport reaches 20–30 Sv in winter and spring, whereas it is only 3–4 Sv in summer and fall. This annual signal in the Oyashio's total transport is in agreement with the nnual signal in the Sverdrup transport of the wind-driven North Pacific subArctic gyre.

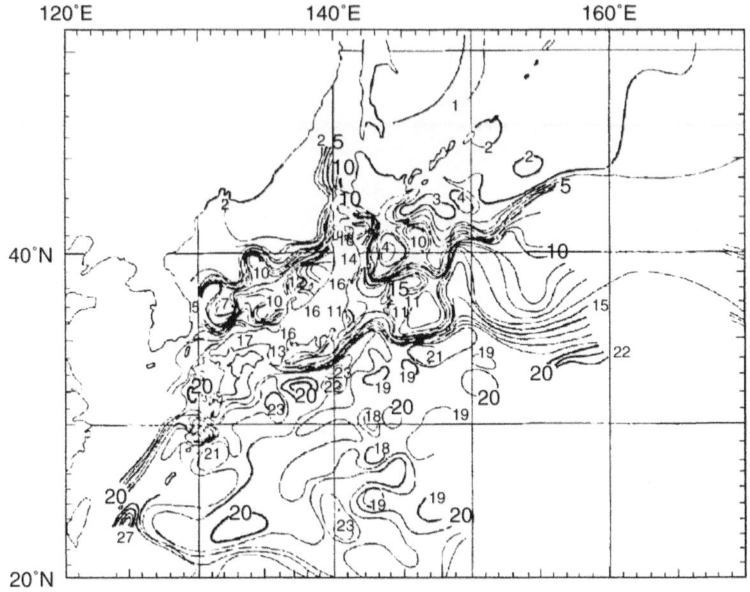

Figure 10 Water temperature map at the 100 m depth in September 1989 compiled by the Japan Meteorological Agency. Contour interval is 1 C.

After flowing south-westward along the coast of Hokkaido, the Oyashio Current splits into two paths. One path veers offshoreward and contributes to the east-north-eastward-flowing SubArctic Current. This path can be recognized in **Figure 10** by the eastward-veering isotherms along 42°N south east of Hokkaido. Because the Oyashio Current brings water of subarctic origin southward, the SubArctic Current is accompanied by a distinct temperature-salinity front between cold, fresher water to the north and warm, saltier water of subtropical origin to the south. This water mass front, referred to as the Oyashio Front or the Subarctic Front, has indicative temperature and salinity values of 5°C and 33.8 PSU at the 100 m depth. Across 165°E, combined moored current meter and CTD measurements show that the SubArctic Current around 41°N has a volume transport of 22 Sv in the upper 1000 m layer.

The second path of the Oyashio Current continues southward along the east coast of Honshu and is commonly known as the first Oyashio intrusion. As shown in **Figure 10**, an addition to this primary intrusion along the coast of Honshu, the southerly Oyashio intrusion is also frequently observed further

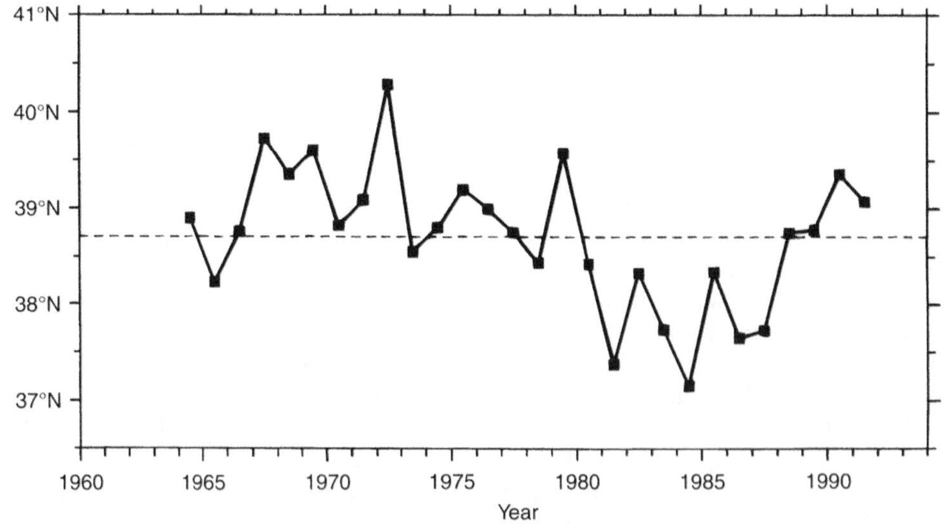

Figure 11 Time-series of the annually averaged southernmost latitude of the first Oyashio intrusion east of Honshu. The dashed line shows the mean latitude (38.7 N) over the period from 1964 to 1991. (Data courtesy of Dr K. Hanawa of Tohoku University.)

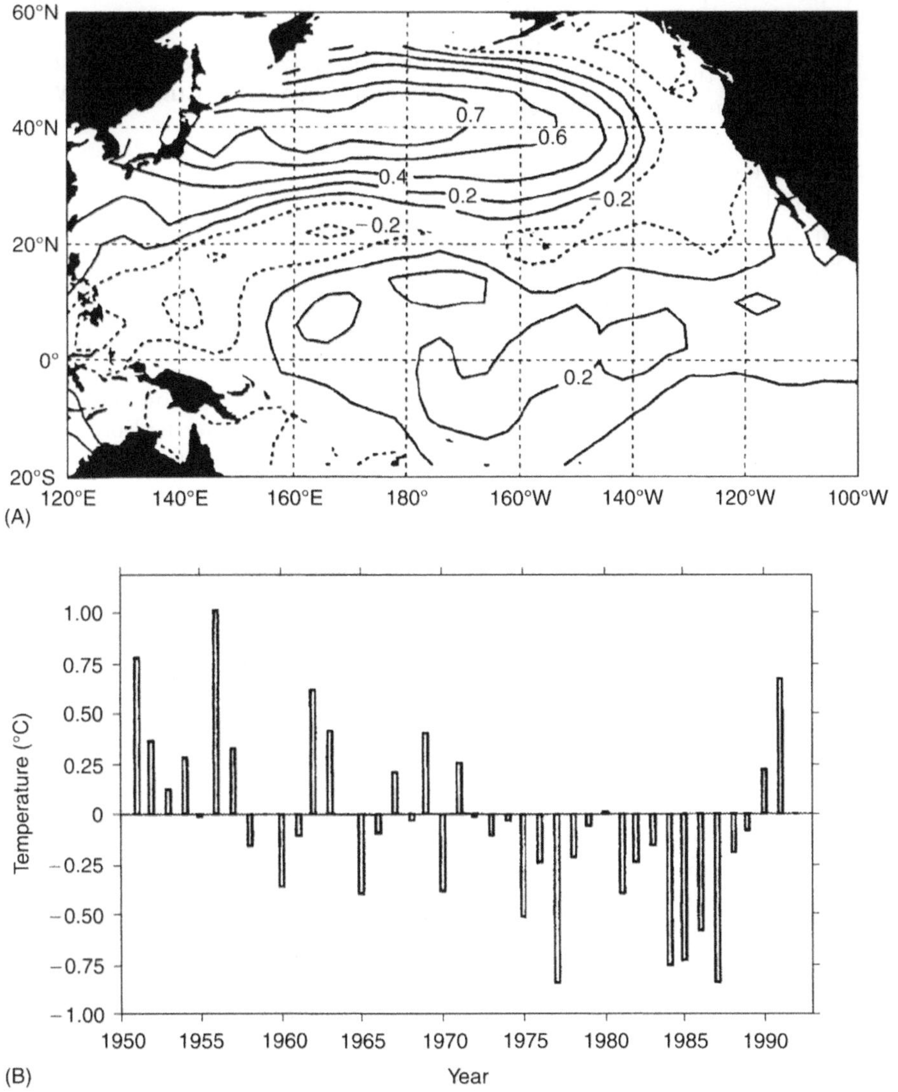

Figure 12 (A) Spatial pattern of the second empirical orthogonal function (EOF) mode of the wintertime sea surface temperature anomalies (1950–1992) in the Pacific Ocean. This mode explains 11% of the variance over the domain. (B) Time-series of thewintertime sea surface temperature anomalies averaged in the Kuroshio-Oyashio outflow region (32°N–46°N,136°E–176°W). (Adapted with permission from Deser C and Blackmon ML (1995) On the relationship between tropical and North Pacific sea surface variations. *Journal of Climate* 8:1677–1680.)

offshore along 147°E. This offshore branch is commonly known as the second Oyashio intrusion. The annual mean first Oyashio intrusion east of Honshu reaches on average the latitude 38.7°N, although in some years it can penetrate as far south as 37°N (see **Figure 11**). In addition to the year-to-year fluctuations, **Figure 11** shows that there is a trend for the Oyashio Current to penetrate farther southward after the mid-1970s. Both this long-term trend and the interannual changes in the Oyashio's intrusions seem to be related to the changes in the intensity of the Aleutian low atmospheric pressure system and the southward shift in the position of the mid-latitude westerlies. It is worth noting that the anomalous southward intrusion of the Oyashio Current not only influences the hydrographic conditions east of Honshu, it also affects the environmental conditions in the fishing ground and the regional climate (e.g. an anomalous southward intrusion tends to decrease the air temperature over eastern Japan).

Concluding Remarks

Because the Kuroshio and Oyashio Currents transport large amounts of water and heat efficiently in the meridional direction, there has been heightened interest in recent years in understanding the dynamic roles played by the time-varying Kuroshio and Oyashio Currents in influencing the climate through sea surface temperature (SST) anomalies. Indeed, outside the eastern equatorial Pacific Ocean, the largest SST variability on the interannual-to-decadal time-scale in the North Pacific Ocean resides in the Kuroshio Extension and the Oyashio outflow regions (**Figure 12**). Large-scale changes in the Kuroshio and Oyashio current systems can affect the SST anomaly field through warm/cold water advection, upwelling through the base of the mixed layer, and changes in the current paths and the level of the meso-scale eddy variability. At present, the relative roles played by these various physical processes are not clear.

This article summarizes many observed aspects of the Kuroshio and Oyashio Current systems, although due to the constraints of space, important subjects such as the water mass transformation processes in regions surrounding the Kuroshio and Oyashio and the impact of the Kuroshio and Oyashio variability upon the oceanographic conditions in coastal and marginal sea areas have not been addressed. It is worth emphasizing that our knowledge of the Kuroshio and Oyashio Currents has increased significantly due to the recent World Ocean Circulation Experiment (WOCE) program (observational phase: 1990–1997). Fortunately, many of the observational programs initiated under the WOCE program are being continued. With results from these new observations, we can expect an improved description of the Kuroshio and Oyashio Current systems in the near future, especially of the variability with timescales longer than those described in this article.

See also

Abyssal Currents. Okhotsk Sea Circulation. Pacific Ocean Equatorial Currents. Wind Driven Circulation.

Further Reading

Dodimead AJ, Favorite JF, and Hirano T (1963) Review of oceanography of the subarctic Pacific region. *Bulletin of International North Pacific Fisheries Commission* 13: 1–195.

Kawabe M (1995) Variations of current path, velocity, and volume transport of the Kuroshio in relation with the large meander. *Journal of Physical Oceanography* 25: 3103–3117.

Kawai H (1972) Hydrography of the Kuroshio Extension. In: Stommel H and Yoshida K (eds.) *Kuroshio – Its Physical Aspects*, pp. 235–354. Tokyo: University of Tokyo Press.

Mizuno K and White WB (1983) Annual and interannual variability in the Kuroshio Current system. *Journal of Physical Oceanography* 13: 1847–1867.

Nitani H (1972) Beginning of the Kuroshio. In: Stommel H and Yoshida K (eds.) *Kuroshio – Its Physical Aspects*, pp. 129–163. Tokyo: University of Tokyo Press.

Pickard GL and Emery WJ (eds.) (1990) *Descriptive Physical Oceanography: An Introduction 5th edn.* Oxford: Pergamon Press.

Shoji D (1972) Time variation of the Kuroshio south of Japan. In: Stommel H and Yoshida K (eds.) *Kuroshio – Its Physical Aspects*, pp. 217–234. Tokyo: University of Tokyo Press.

Taft BA (1972) Characteristics of the flow of the Kuroshio south of Japan. In: Stommel H and Yoshida K (eds.) *Kuroshio – Its Physical Aspects*, pp. 165–216. Tokyo: University of Tokyo Press.

Tomczak M and Godfrey JS (1994) *Regional Oceanography: An Introduction*. Oxford: Pergamon Press.

CALIFORNIA AND ALASKA CURRENTS

B. M. Hickey, University of Washington, Seattle, WA, USA
T. C. Royer, Old Dominion University, Norfolk, VA, USA

Introduction

The clockwise North Pacific Subtropical Gyre and the counterclockwise Subarctic or Alaska Gyre, the two principal current gyres of the North Pacific, have dimensions similar to those of the basins, i.e., several thousand kilometers. The West Wind Drift and Subarctic Current flow approximately zonally across the Pacific basin with origins in the Kuroshio and Oyashio Currents, respectively, in the western Pacific basin (**Figure 1**). As this broad eastward flow nears the west coast of North America, it bifurcates, splitting into the northward flowing Alaska Current, the eastern limb of the Alaska Gyre, and the southward flowing California Current, the eastern limb of the North Pacific Subtropical Gyre. This bifurcation takes place several hundred kilometers offshore and depends both on the ocean currents and the wind fields over the North Pacific. The water type is primarily Subarctic, relatively cool and fresh surface water. The volume transport within each of the currents is about 10–15 Sv. Both the California Current and the Alaska Current are coupled to current systems and processes along the adjacent continental margins, where the majority of the seasonal variability occurs. On both seasonal and ENSO (El Niño) timescales, the strengths of the two current systems vary out of phase. For each boundary current system (the Alaska system and the California system) both the large-scale gyres and the coastal current systems, as well as the interactions between them, are described in the text below.

The Alaska Current System

The Alaska Current, the eastern limb of the Subarctic Gyre, flows northward along the west coast of North America beginning at about 48–50°N (**Figure 2**). A companion coastal current flows parallel to the Alaska Current but closer to the coast. This current has several local names – the Vancouver Island Coastal Current, the Haida Current, and the Alaska Coastal Current. These flows move in a counterclockwise sense around the Gulf of Alaska, bringing relatively warmer water northward along the eastern boundary of the Pacific Ocean. As they follow the topography in the northern gulf, they are diverted westward.

Early explorations of the Subarctic North Pacific, initiated by the Russian czar, Peter the Great, illustrate this general circulation of the northern North Pacific region. In June 1741, two ships set out from the Kamchatka Peninsula to investigate the eastern

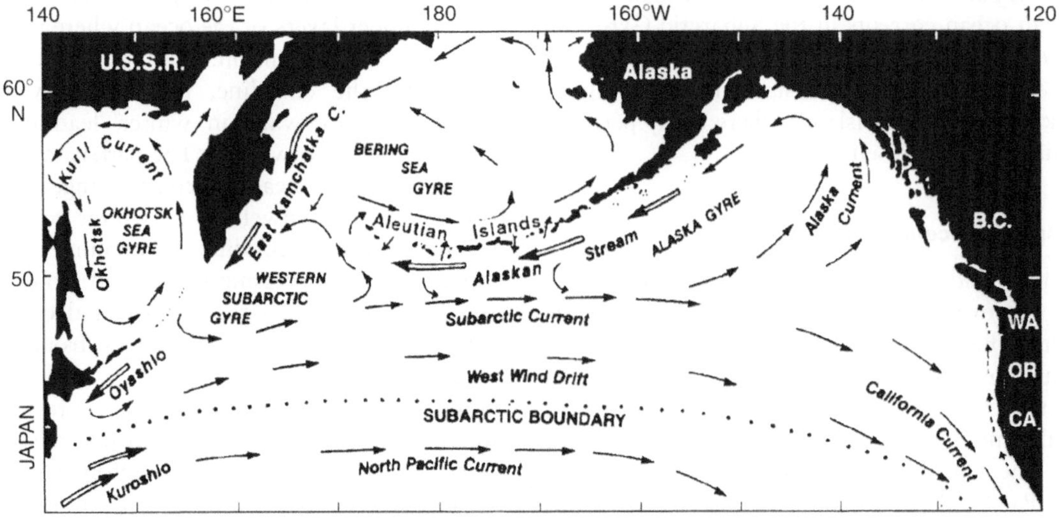

Figure 1 Schematic surface circulation of the North Pacific relative to 1000 db (i.e., assuming no flow near a depth of 1000 m) showing the Alaska and California Current systems. (Adapted with permission from Thomson, 1981.)

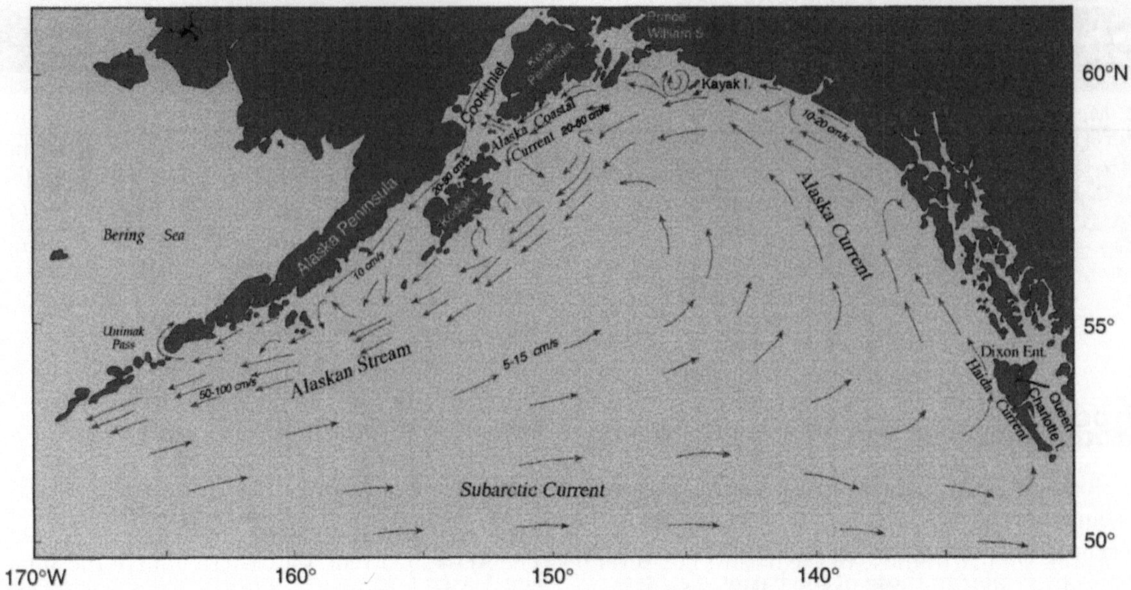

Figure 2 Schematic of North Pacific Subarctic Gyre with Alaska Current, Alaskan Stream and Alaska and Haida Coastal Currents. (Adapted with permission from Reed and Schumacher, 1986.)

side of this unexplored basin. They sailed following the currents and winds south-eastward and then eastward across the North Pacific at about 48–50°N. Near the date line, the two vessels were separated under foggy conditions, never to see one another again. Nevertheless, explorers on both ships eventually observed North America: from the *St. Peter* commanded by Vitus Bering, land was first seen off south-east Alaska on 15 July and from the *St. Paul* commanded by Aleksei Chirikov, land was seen near Kayak Island on 16 July. Both ships traveled back toward the west along the southern side of the Aleutian Islands, completing a counterclockwise path. This path is approximately the configuration of the mean ocean currents in the Subarctic Gyre. The *St. Paul* returned to Kamchatka in fall 1741, but the *St. Peter* ran aground on an island in the Bering Sea (now known as Bering Island) where many perished, including Bering himself.

Atmosphere–Ocean Interactions

In winter, cold, dry Siberian air masses frequently sweep out over the northern North Pacific, rapidly extracting heat and moisture from the ocean. The introduction of this heat into the atmosphere intensifies the atmospheric circulation. These storms generally move from west to east across the Pacific basin. The path of an individual storm across the North Pacific depends on the global scale atmospheric circulation which changes both seasonally and interannually. In winter, storm tracks often cross the Gulf of Alaska, resulting in a large zone of low atmospheric pressure known as the Aleutian Low. In summer, the North Pacific High strengthens and pushes northward into the gulf. Winter over the North Pacific Ocean is dominated by strong counterclockwise wind systems; in summer weak clockwise winds occur. Since the earth's rotational force ('Coriolis') tends to move near surface water to the right of the wind direction in the northern hemisphere, the counterclockwise winter winds over the Gulf of Alaska transport upper layer water shoreward, away from the central deep ocean. The subsurface waters move upward to replace these surface waters. The rising of these subsurface waters together with wind mixing brings nutrient-rich waters into the upper layers of the ocean where they can be utilized by phytoplankton. Over the continental shelf and along the coastline, the surface waters are transported shoreward in winter, leading to convergence and downwelling. The pattern of upwelling and downwelling near the coast changes seasonally from intense downwelling in winter to weak upwelling or neutral conditions in summer (**Figure 3**, upper panel). Progressing southward along the west coast of North America, the seasonal cycle of upwelling and downwelling changes from a downwelling-dominated wind system (northward winds along the coast) in the north to a more upwelling-dominated wind system (southward winds along the coast) farther south off Washington, Oregon, and California in the California Current system (see below).

The seasonal progression of storm tracks across the Gulf of Alaska also affects precipitation rates,

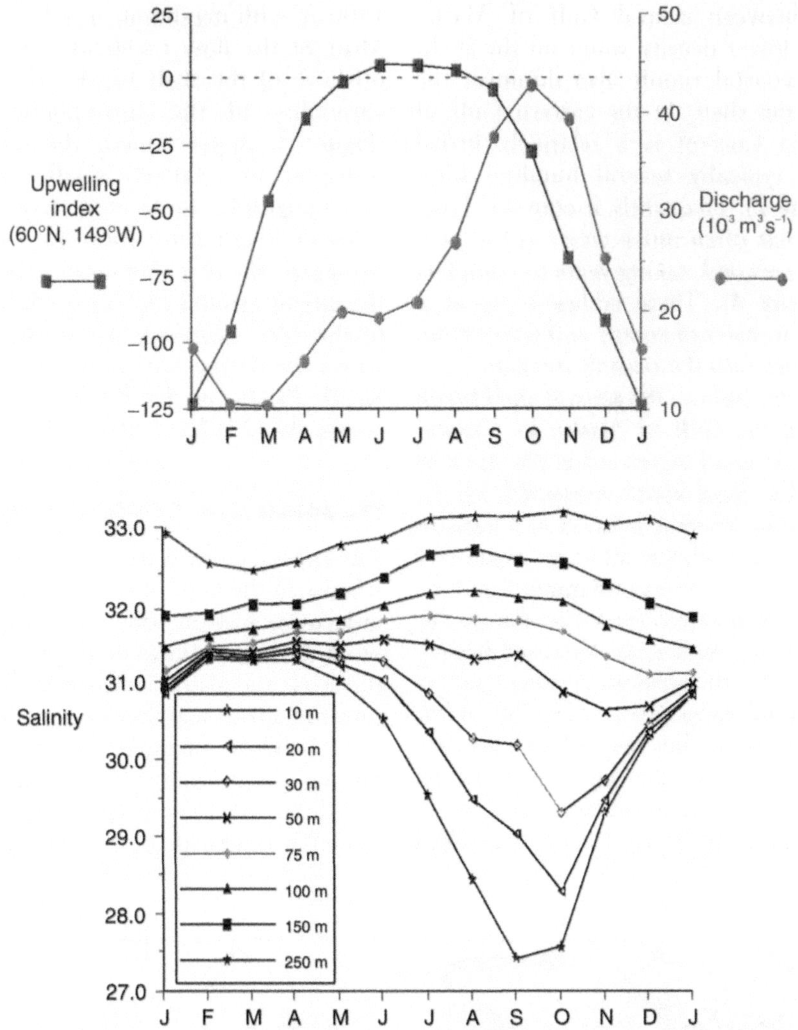

Figure 3 Seasonal changes in strength of upwelling/downwelling in the northern Gulf of Alaska and coastal freshwater discharge (upper panel). Upwelling (positive) or downwelling (negative) is indicated via the 'upwelling index', a measure of cross-shore transport by alongshelf winds (units are $m^3 s^{-1}$ per 100 m of coastline). Seasonal changes in salinity versus depth in the Alaska Coastal Current (60°N, 149°W) (lower panel).

especially along the coast. As these storms encounter the coastal mountain ranges that rim the eastern boundary of the northern North Pacific, heavy precipitation occurs. This provides vast quantities of fresh water in the coastal region and adds heat to the atmosphere. On average, about 2.4 m of rain and snow fall in a relatively narrow ($\sim$100 km) coastal drainage area and more than 8 m of precipitation have been reported for a single year. The majority of this precipitation runs directly into the ocean via coastal rivers when the air temperatures are above freezing. Otherwise, it is stored as snow and ice with about 20% of the region being glacial. The average annual coastal runoff is estimated to be 24 000 $m^3 s^{-1}$, about one third greater than the outflow of the Mississippi River. However, unlike the Mississippi, the runoff here is distributed along the coast in a number of smaller rivers rather than through a single major river. Fresh water is continually added to the coastal currents as they progress around the Gulf of Alaska. The coastal discharge (**Figure 3**, upper panel) is least in winter when most of the moisture is contained in snow and ice. A small peak occurs in spring corresponding to seasonal heating at lower elevations. However, maximum discharge occurs in September–November prior to annual freeze-up, when both precipitation caused by storms and runoff from snowmelt contribute to the coastal runoff.

The Alaska Current

The Alaska Current is affected by both winds and precipitation. Winds, which are usually downwelling-favorable along the coast, maintain the density contrast between central Gulf of Alaska water and fresher, lower density water on the shelf. Precipitation and coastal runoff also diminish the water density on the shelf. In the eastern Gulf of Alaska, the Alaska Current is a relatively broad meandering flow, typically several hundred kilometers wide (**Figure 2**). Frequently it contains large mesoscale eddies that often move westward at several centimeters per second, taking years to complete their journey (**Figure 4**). These eddies serve as a mechanism for the transfer of energy and water from the ocean boundaries into the ocean's interior.

The Alaska Current follows the general shelf break topography around the Gulf of Alaska in a counterclockwise sense. It turns westward at the apex of the Gulf of Alaska, then south-westward in the western Gulf of Alaska where it behaves as a western boundary current, intensifying into an organized flow known as the Alaskan Stream (**Figures 1** and **2**). This current is highly sheared vertically due to the density field (baroclinic), with a cross-current density gradient manifested by the offshore gradient in salinity. Surface salinities range from <30 PSU at the coast, to >31 PSU on the shelf, to >32.5 PSU over the central Gulf of Alaska. The baroclinic current transport near Kodiak Island in the western gulf (see **Figure 2**) is approximately 10 Sv in the upper 1500 m, with maximum speeds exceeding 1.0 m s^{-1}. Most of the flow (>80%) is found within 60 km offshore of the shelf break. Although the seasonal variability of the atmospheric forcing is large (**Figure 3**, upper panel), the seasonal changes in transport are relatively small – less than 10%. As the topography turns more westward, some of the Alaskan Stream turns southward where it rejoins the eastward flowing Subarctic Current, completing the circuit around the Alaska Gyre. However, most of the flow continues eastward and some fraction enters the Bering Sea, subsequently returning to the North Pacific in the Kamchatka Current that becomes the Oyashio Current (**Figure 1**).

The Alaska Gyre Coastal Currents

The coastal fresh water runoff affects the nearshore salinity in the Gulf of Alaska (**Figure 3**, lower panel) and causes a strong offshore gradient in density. The winds driving downwelling confine the fresh water to the coast and also drive currents along the coast. The resulting cross-shelf density gradient, through a balance between pressure forces and the earth's rotational forces, causes the coastal current in the Gulf of Alaska to be strongly baroclinic. The currents driven by these processes have been given specific

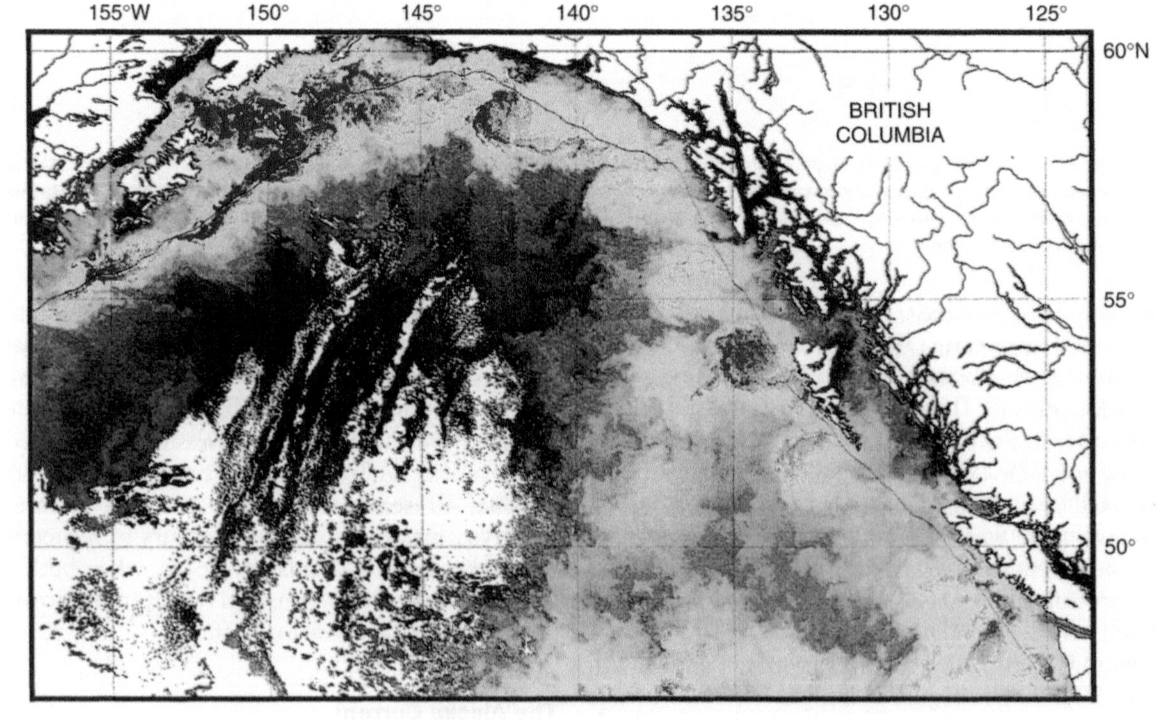

Figure 4 A composite sea surface temperature image for the Gulf of Alaska from 1, 2, 3 and 10 March 1995. The mean north–south temperature gradient has been removed to better elucidate eddy structure. Oceanic temperatures differed from the mean by −3°C (dark blue) to +3°C (red). Clouds and land are white. The dotted curve is the 1000 m depth contour. (Reproduced with permission from Thompson and Gower, 1998; copyright by the American Geophysical Union.)

names locally (the Vancouver Island Coastal Current off Vancouver Island, the Haida Current off central British Columbia, and the Alaska Coastal Current off the Alaskan coast). It is presently unknown whether the coastal current is continuous throughout the Gulf of Alaska at a given time. The coastal current likely begins with the Davidson Current, a wintertime current in the California Current system (see below) that links the two major current systems, at least in winter. The northward tending Alaska Coastal Current is strongest in winter and weakest in summer, exactly out of phase with the southward tending coastal currents in the California Current (see later sections).

The buoyancy and wind-driven coastal current flows along the coast with the coast on its right around the Gulf of Alaska, eventually flowing through Unimak Pass into the Bering Sea (see **Figure** 2). The flow has a typical width of about 30 km, a depth of 100–200 m and speeds $> 1.0\,\mathrm{m\,s^{-1}}$ have been reported. The mean transport is about 0.6 Sv with a seasonal variation of about 0.2 Sv. Transports of more than 2 Sv have been reported.

The coastal current is usually constrained along the coast by the downwelling winds. As it passes coastal openings at the entrance to Prince William Sound and Cook Inlet, some of the current enters these enclosures. Near Kayak Island (about 144 W) the coastal current is diverted across the shelf and some of it merges with the Alaska Current. In the western Gulf of Alaska, downwelling winds are less dominant and precipitation rates are lower. Here the current tends to diverge from the coast and spread across the shelf.

The coastal current is believed to be important for marine mammals and fish including salmon. Salmon might use chemical tracers carried by this current to navigate back to their original spawning streams. Similarly, the current is capable of carrying pollutants alongshore. In March 1989, more than 242 000 barrels of crude oil were spilled into Prince William Sound from T/V *Exxon Valdez*. The oil was carried westward in the Alaska Coastal Current more than 800 km in the span of about 2 months (about $0.15\,\mathrm{m\,s^{-1}}$).

The reduction in North Pacific sea surface salinity by precipitation and runoff (**Figure** 3, lower panel) creates a surface low-density lens that restricts vertical mixing in the Alaska Gyre and tends to prevent deep water formation. This is in sharp contrast to the deep circulation of other high latitude regions of the world's oceans such as the North Atlantic or Weddell Sea.

The California Current System

The California Current system (**Figure** 5) includes the southward California Current, the wintertime northward Davidson Current, the northward California Undercurrent (which flows over the continental slope beneath the southward flowing upper layers), the Southern California Countercurrent (or Eddy) as well as 'nameless' shelf and slope currents with primarily shorter-than-seasonal time scales. The California Current system includes one major river plume (the Columbia), several smaller estuaries, and (primarily in the north) numerous submarine canyons. The dominant scales and dynamics of the circulation over much of the California Current system are set by several characteristics of the physical environment; namely, strong winds, large alongshore scales for both the winds and the bottom topography, and a relatively narrow and deep continental shelf. Because of these characteristics, coastal-trapped waves (disturbances that travel northward along the shelf and slope) are efficiently generated and travel long distances toward the North Pole along the continental margins of much of western North America. Typical speeds are about $300–500\,\mathrm{km\,d^{-1}}$. These waves are integral to the current patterns observed in the California Current system and to their variability.

The California Current flows southward year-round offshore of the US and Mexican west coast from the shelf break to a distance of several hundred kilometers from the coast (**Figure** 5). The current is strongest at the sea surface, and generally extends over the upper 500 m of the water column. Seasonal mean speeds are $\sim 0.1\,\mathrm{m\,s^{-1}}$. The California Current in summer carries relatively colder, fresher water southward along the coast. South of Point Conception (the major indentation in the coastline near 35 N), a portion of the California Current turns southeastward and then shoreward and northward. This feature is known either as the 'Southern California Countercurrent' during periods when the flow successfully rounds Point Conception or the 'Southern California Eddy' when the flow recirculates within the southern California Bight (the indented region south of Point Conception; see inset map in **Figure** 5). The California Undercurrent is a relatively narrow feature ($\sim$ 10–40 km) flowing northward over the continental slope of the California Current system at depths of about 100–400 m as a nearly continuous feature, transporting warmer, saltier water of more southern origin northward along the coast. The Undercurrent has a jet-like structure, with the core of the jet usually located just seaward of the shelf break and peak speeds $\sim 0.3–0.5\,\mathrm{m\,s^{-1}}$. The Undercurrent divides into two components within the Southern California Bight, one flowing northwestward through the Santa Barbara Channel, the other flowing westward south of the island chain that

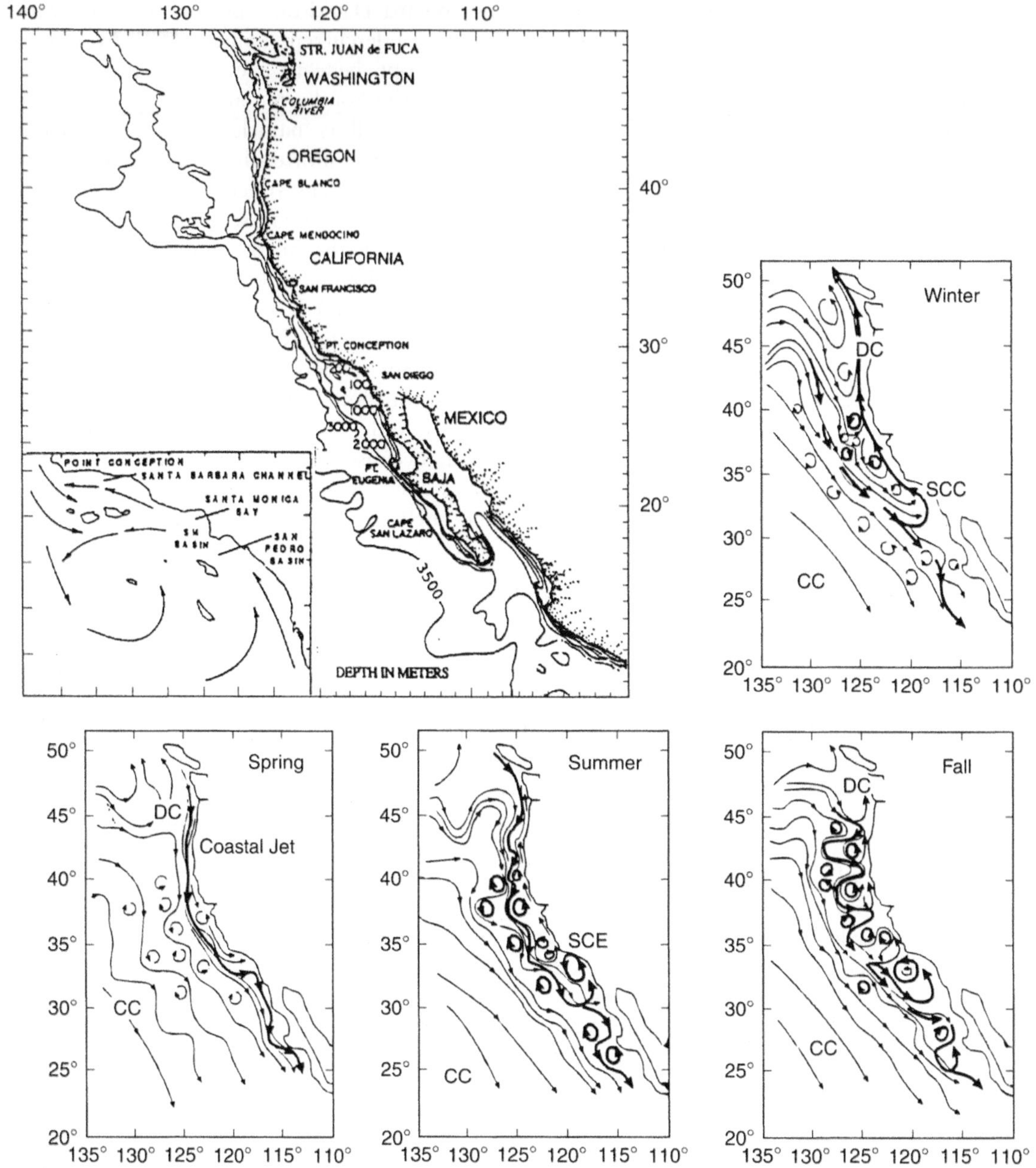

Figure 5 Schematic illustrating seasonal variation of large-scale boundary currents and coastal currents off the west coast of North America as well as important landmarks and bottom topography. (Adapted with permission from Strub and James, 2000.) An enlargement of the Southern California Bight, including the offshore islands, is given in the upper left panel. CC, California Current; DC, Davidson Current; SCC, Southern California Countercurrent; SCE, Southern California Eddy.

forms the southern side of the Santa Barbara Channel. A southward undercurrent occurs over the continental slope in winter at some latitudes. This undercurrent occurs at deeper depths than the northward undercurrent (~300–500 m). The existence of this southward undercurrent, like that of the northward undercurrent, likely depends on the co-occurrence of opposing local alongshore winds and alongshore sea level slope. The Davidson Current flows northward in fall and winter from Point Conception (~35°N) to at least Vancouver Island (~50°N) and may connect with the coastal currents that flow around the Alaska Gyre. This northward flow is generally broader (~100 km in width) and sometimes stronger than the corresponding subsurface northward flow in other seasons (the 'Undercurrent') and extends seaward of the slope. Poleward shelf flow, in the sense of a monthly mean phenomenon, is sometimes described as an expression of 'the Davidson Current'.

Winds in the California Current system are governed by atmospheric pressure systems – on average a low pressure system in winter and a high pressure system in summer, producing northward winds in winter and southward winds in summer over much of the California Current system. Because of the dramatic seasonal reversal in winds along much of the coast, currents and water properties of the California Current system also undergo large seasonal fluctuations. The southward flowing California Current and the northward flowing California Undercurrent are strongest in summer to early fall and weakest in winter. The majority of the seasonal variability occurs along the coastal boundaries rather than in the central basin. The seasonal signal in transport near the coast migrates northward and offshore in both the California and Alaska Current systems. The North Pacific Current, which feeds both the California Current and the Alaska Current (see **Figure 1**) has little seasonal variability in either strength or position. Much of the variability in the California Current is related to coastal wind variability. However, remote forcing (disturbances that are caused farther south and travel along the coastal margins as waves) is also likely important, particularly off California. The northward flowing Davidson Current is strongest in winter, as is the Southern California Countercurrent. Shelf currents along the coast from Point Conception to the Strait of Juan de Fuca are generally southward in the upper water column from early spring to summer and northward the rest of the year. The seasonal duration of southward flow usually increases with distance offshore and with proximity to the sea surface. A northward undercurrent is commonly observed on shelves during the summer and early fall. A strong tendency for northward flow throughout the water column exists over the inner shelf in all but the spring season.

The transition of currents and water properties over the shelf and slope between winter and spring, the 'spring transition', is a sudden and dramatic event in the California Current system. Along much of the coast sea level drops at least 10 cm during the transition, currents reverse from predominantly northward to predominantly southward within a period of several days and isopycnals tilt upward towards the coast. The transition is driven by changes in the large-scale wind field and these changes are a result of changes in the large-scale atmospheric pressure field over the Northeast Pacific and the California Current system. The transition includes both a local and a remotely forced response to the change in wind conditions. The fall transition is not as rapid or as dramatic as the spring transition.

Variability on Shorter-than-Seasonal Timescales

Seasonal conditions are often reversed for shorter periods of time in the California Current system. Changes in currents, water properties, and sea level over the shelf at most locations are dominated by wind forcing, with typical timescales of 3–10 days. Regions seaward of the shelf are dominated by jets, eddies, and in some locations, wave-like propagating disturbances, with typical timescales of 10–40 days. Along-shelf flow on the inner to mid shelf is primarily wind-driven and can be predicted with numerical models. However, the amplitude of current fluctuations is generally underpredicted, and the amount of variability predicted decreases offshore toward the shelf break (to $<20\%$ over the upper slope off northern California). Predictive capability for both temperature fluctuations and cross-shelf flow is very poor at the present time. The alongshelf currents include a response to both local wind forcing and wind forcing all along the coast south of a particular latitude ('remote' forcing). At any given time and location, the ratio of remote and local forcing varies. In winter, local wind forcing of currents dominates in regions where winter storms are accompanied by strong northward winds that increase northward (in the direction of propagating coastal-trapped waves). In summer, when winds increase southward, freely propagating coastal-trapped waves often contribute to current variability in the coastal regions of the Pacific Northwest. Off northern California, where wind stress is generally strongest in summer, both local and remote forcing are almost always important. Wind-driven currents in the Southern California Bight typically have much smaller alongshelf scales than in the region north of the Bight (20 km versus 500 km), weaker amplitudes, and weaker seasonal variation; likely a result of the much reduced winds and the narrow, more irregular shelves in this region.

The eddy/meander field in the regions seaward of the shelves in the California Current system has a seasonal variation, with maximum energy in summer, when upwelling is also at a maximum (**Figure 6**). Meandering jets extend from the sea surface to depths of over 200 m and separate fresher, warmer, chlorophyll-depleted water from colder, saltier, chlorophyll-rich, recently upwelled water. Jets are characterized by core speeds exceeding $0.5\ \mathrm{m\ s^{-1}}$ at the surface, widths of 50–75 km and total baroclinic transports of about 4 Sv. These filaments are particularly evident in regions which contain

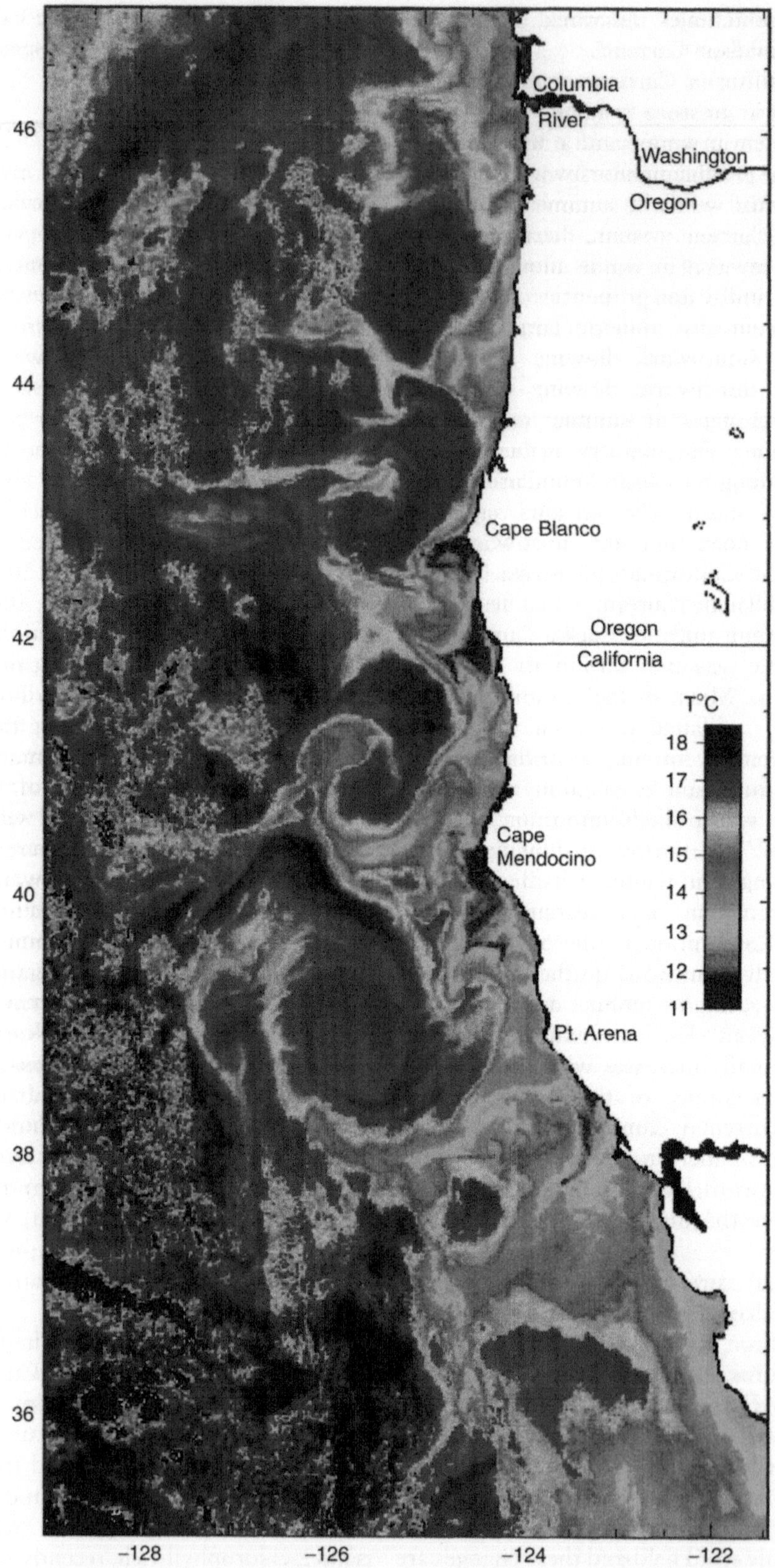

Figure 6 Satellite-derived image of sea surface temperature in the California Current system. The figure illustrates the jets, eddies, and meanders, many of which originate near coastal promontories. (AVHRR data collected and processed at Ocean Imaging, Inc., archived and made available at COAS/Oregon State University with funding from NSF and NASA in the US GLOBEC Northeast Pacific program.)

coastline irregularities, such as southern Oregon, northern California, and the Baja peninsula, and appear to be tied to these irregularities. One meandering jet can be traced continuously from southern Oregon, where it separates from the shelf, to southern California. Such separated coastal jets account for much of the energy, as well as seasonal and interannual variability in the California Current system. The jets and meanders are a dominant feature in satellite-derived images of sea surface temperature in the summer season (**Figure 6**).

Changes in currents with timescales similar to those of the eddy/meander field (15–40 days) dominate current variance over the slope within the southern California Bight. The majority of this variability is the signature of freely propagating coastal-trapped waves or disturbances, with an as yet unknown source along the coast of Baja California or even farther south. Such waves are likely to be important in other slope regions, but to date have not been separated from other sources of variability.

Water Properties and Upwelling

The California Current system contains waters of three primary types: Pacific Subarctic, North Pacific Central and Southern (sometimes termed 'Equatorial'). Pacific Subarctic water, characterized by relatively low salinity and temperature and high oxygen and nutrients, is advected southward in the outer edges of the California Current system and into the southern California Bight. Colder, fresher water from British Columbia and the US Pacific Northwest coastal region is also advected southward near the boundaries. North Pacific Central water, characterized by high salinity and temperature and low oxygen and nutrients, enters the California Current system from the west. Southern water, characterized by high salinity, temperature and nutrients, and low oxygen, enters the California Current system from the south with the northward flowing California Undercurrent. In general, salinity and temperature increase southward in the California Current system and also with depth.

Upwelling along the coast brings colder, saltier and more nutrient-rich water to the surface adjacent to the coast. In general, maximum upwelling (as seen at the sea surface) occurs off northern California, consistent with the alongshore maximum in alongshelf wind stress and hence mass transport away from the coast that causes deeper water to 'upwell' to fill the void (Ekman transport) (**Figure 7**). Maximum upwelling occurs in spring or summer in the California Current system. Stratification in the California Current system is remarkably similar at most locations and is largely controlled by the large-scale advection and upwelling of the water masses as described above.

With the possible exception of catastrophic storms (a '10-year storm') most river plumes on the coast between the tip of Baja California and the Strait of Juan de Fuca are relatively small, and their effects are likely confined to within one tidal excursion of the mouth of the river or estuary. The only large river plume in the California Current system is that from the Columbia River. The Columbia provides over 77% of the drainage between the Strait of Juan de Fuca and San Francisco Bay. As mentioned above, water from the Strait of Juan de Fuca, with origins in the Fraser River, also contributes to fresh water in the nearshore California Current in spring and summer.

The plume from the Columbia River is a dominant feature in near surface salinity (and, in some seasons, temperature) off the US west coast. On a seasonal basis, the plume flows northward over the shelf and slope in fall and winter, and southward well offshore of the shelf in spring and summer. In winter the Columbia plume provides a substantial fraction of fresh water to the Davidson Current, and, ultimately, the coastal currents of the Alaska Gyre. In summer, the Columbia provides fresh water to the California Current, giving rise to the low salinity signal and associated front used to trace the meandering coastal jet that separates from the shelf at Cape Blanco. Most other smaller rivers on the Pacific coast have significant river plumes only during major floods. In winter and spring, the Columbia plume has a dramatic effect on the Washington coast, producing time-variable currents as large as the wind-driven currents and flooding local estuaries with fresh water (fresher than that in much of the estuary). Short-term variability in both fresh water volume and plume location can have significant effects on shelf and slope currents as well as water properties.

Topographic Effects in the Alaska/California Current Systems

The outer edge of the continental shelf along the US west coast from California to the Aleutian Islands is intersected at many locations by submarine canyons. Canyons enhance the transfer of particles and water between the shelf and the deeper ocean. Upwelling of colder, nutrient-rich water can be enhanced by more than a factor of 10 over regions with a straight slope rather than a canyon indentation. For example, upwelling from a spur of Juan de Fuca canyon is responsible for enhanced seasonal productivity in that region. In the Alaska Current system, Hinchinbrook

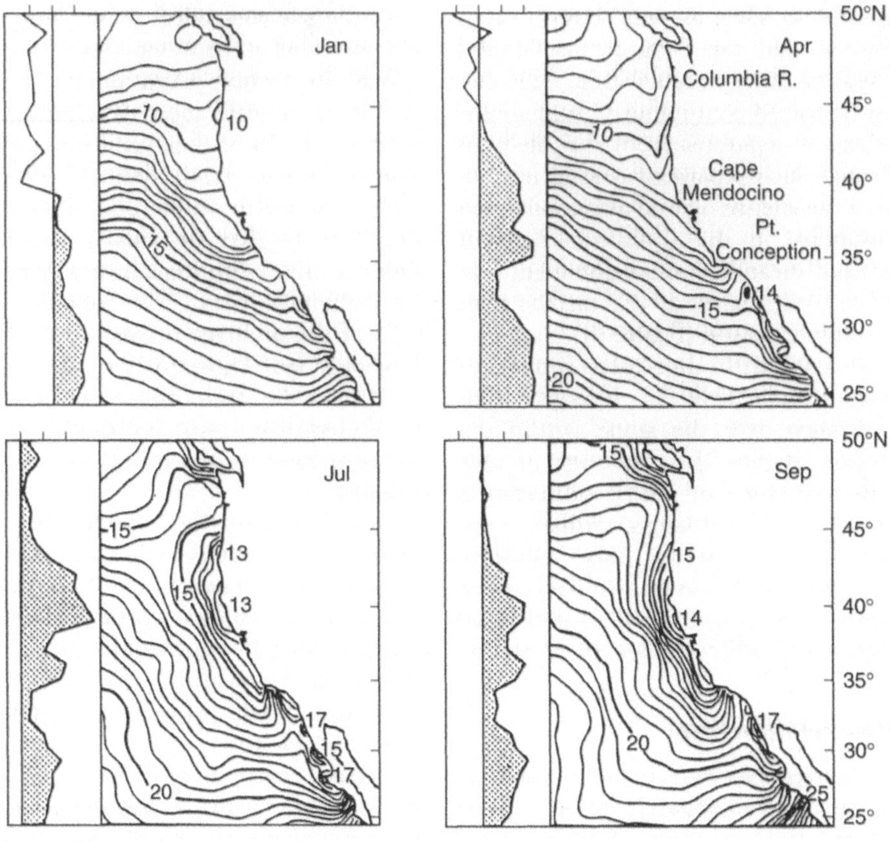

Figure 7 Maps of monthly mean sea surface temperature in the California Current System. The corresponding profile of cross-shore transport due to local winds is shown next to each map. Each tick represents a mass transport of $50 \times 10^3\,\text{kg s}^{-1}$ per 100 m of coastline. Offshore transport, indicative of upwelling conditions, is shaded. (Adapted with permission from Huyer, 1983.)

Canyon, which crosses the shelf adjacent to Prince William Sound, is a potential conduit for the exchange of deep waters with this coastal feature. Moreover, counterclockwise circulation patterns are generally observed both within and over submarine canyons (although not necessarily extending to the sea surface). Such eddies provide an effective mechanism for trapping particles such as suspended sediment or food for the biomass. Several major promontories and one major ridge (near Cape Mendocino) also occur along the coast adjacent to the California Current system (see **Figure 5**). Flow in the vicinity of such features is highly three-dimensional, generally producing offshore tending jets as well as counterclockwise eddies in their lee.

Interannual and Interdecadal Variability in the Alaska/California Current Systems

Year-to-year variability in the California and Alaska Current systems is significant in both physical and biological parameters. Surface transport in the two systems varies out of phase on El Niño timescales: when the Alaska Gyre strengthens, as during an El Niño, the California Current system weakens. As with seasonal changes, the majority of the changes in transport occur along the ocean boundaries in both systems. On interannual scales, some change in the North Pacific Current also occurs, but these changes follow those along the boundary. For the California Current system property anomalies of ~2–4°C and ~0.3 PSU have been observed at depths of 50–200 m from the sea surface, and these anomalies extend several hundred kilometers from the coast. Much of this variability is related to the El Niño (ENSO) phenomenon, occurring at periods of 3–7 years (**Figure 8**, right panels). In the Alaska Current system, subsurface temperature anomalies of >1.5°C have been observed at the coast 7–8 months after an El Niño has occurred at the equator. El Niño has been shown to affect the west coast both via an atmospheric route (i.e., by changes in the local winds which then cause changes in the local ocean currents and water properties) and by an oceanic route (i.e., by transmission of signals from the equator along the continental margin as propagating coastal-trapped

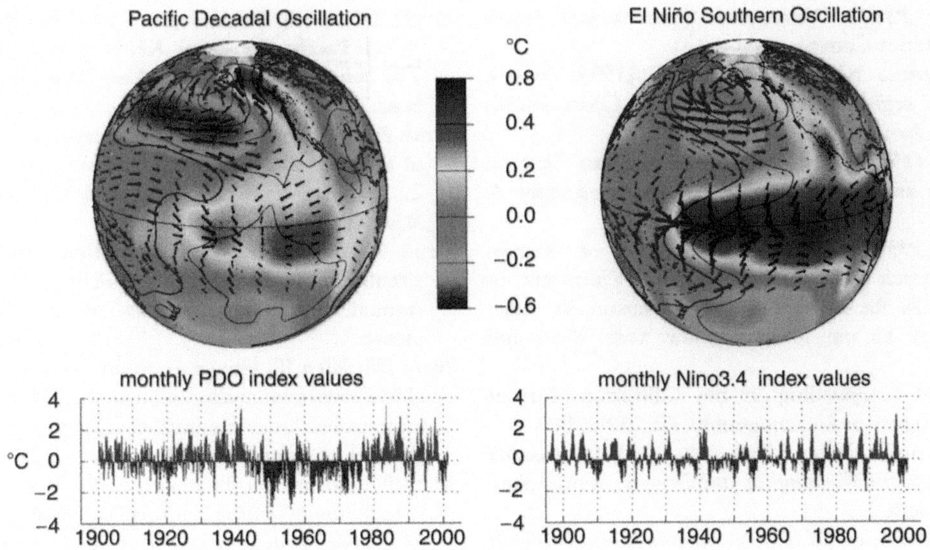

Figure 8 October-to-March average surface climate anomalies associated with a +1 standard deviation value in the Pacific Decadal Oscillation (PDO) index (left panel) and El Niño (cold-tongue) Index (right panel). Sea surface temperature anomalies are depicted by the color shading in degrees Celsius; surface wind stress anomalies are shown with vectors, with the largest vectors representing 10 $m^2 s^{-2}$ anomalies. Thin solid contours depict positive sea level pressure (SLP) anomalies; dashed contour lines depict negative SLP anomalies, with contour intervals at ±0.5, 1, 2 and 3 mb. The heavy solid contour depicts the SLP anomaly zero-line. (Courtesy of Steven Hare at the International Pacific Halibut Commission and Nathan Mantua at the Joint Institute for the Study of the Atmosphere and Oceans, University of Washington.)

waves). Local wind effects are more dominant in the Pacific Northwest and Alaska; remote forcing via waves is more dominant from central America to California. There is also evidence of increased formation of eddies along the shelf break in the eastern Gulf of Alaska during El Niño conditions.

The Northeast Pacific also has lower frequency fluctuations – periods of about 22 and 52 years, associated with the Pacific Decadal Oscillation (PDO) (**Figure 8**, left panels). PDO has oceanic and atmospheric patterns similar to those of El Niño, but has much longer duration. PDO is a pattern of low sea surface temperatures in the central North Pacific and high sea surface temperatures along the eastern Pacific boundary. The major pattern reverses at 25–30 year intervals. PDO was mostly positive from 1922 to 1942 and mostly negative (warm in the central gulf, cold nearshore) from 1942 to 1976 and has been mostly positive again through 1998. From fall 1998 until summer 2001 the PDO has been negative. The coastal precipitation follows a similar pattern and could reinforce the PDO locally. High rates of coastal precipitation would increase cross-shelf pressure differences, enhancing northward flow in the coastal currents as well as the Alaska Current through the balance between pressure and rotational forces. Relatively warm water from more southern latitudes would thus be advected northward, reinforcing the positive PDO sea surface temperature pattern. The ecosystem seems to respond to the PDO – salmon production in the northern Gulf of Alaska is positively correlated with PDO; Washington/Oregon salmon production is negatively correlated with PDO.

See also

Ocean Circulation. River Inputs.

Further Reading

Alverson DL and Pruter AL (eds.) (1972) *Bioenvironmental Studies of the Columbia River Estuary and Adjacent Ocean Regions*. Seattle: University of Washington Press.

Chavez FP and Collins CA (eds.) (1998) Studies of the California Current System Part 1. *Deep-Sea Research* 45(8–9): 1407–1904.

Chavez FP and Collins CA (eds.) (2000) Studies of the California Current System Part 2. *Deep-Sea Research* 47(5–6): 761–1176.

Divin VA (1993) *The Great Russian Navigator, A.I. Chirikov*. Fairbanks: University of Alaska Press.

Dodimead AJ, Favorite F, and Hirano T (1963) Review of the oceanography of the Subarctic Pacific. In: *Salmon of the North Pacific*. Vancouver: International North Pacific Fisheries Commission.

Hare SR, Mantua NJ, and Francis RC (1999) Inverse production regimes: Alaska and West Coast Pacific Salmon. *Fisheries* 24: 6–14.

Hickey BM (1979) The California Current System: hypotheses and facts. *Progress in Oceanography* 8: 191–279.

Hickey BM (1998) Coastal Oceanography of Western North America from the tip of Baja California to Vancouver Is. In: Brink KH and Robinson AR (eds.) *The Sea*, vol. 11, pp. 345–393. New York: Wiley and Sons.

Huyer A (1983) Upwelling in the California Current System. *Progress in Oceanography* 12: 259–284.

Landry MR and Hickey BM (eds.) (1989) *Coastal Oceanography of Washington and Oregon*. Amsterdam: Elsevier Science.

Lentz SJ and Beardsley RC (1991) *Introduction to CODE (Coastal Ocean Dynamics Experiment) A: Collection of Reprints*. Woods Hole, MA: Woods Hole Oceanographic Institution.

Lynn RS and Simpson JJ (1987) The California Current System: the seasonal variability of its physical characteristics. *Journal of Geophysical Research* 92(C12): 12 947–12 966.

Reed RK and Schumacher JD (1986) Physical oceanography. In: Hood DW and Zimmerman ST (eds.) *The Gulf of Alaska: Physical Environment and Biological Resources*. NOAA OCS Minerals Management Service MMS-86-0995. Springfield, VA.

Royer TC (1983) Observations of the Alaska Coastal Current. In: Gade H, Edwards A, and Svendsen H (eds.) *Coastal Oceanography*, pp. 9–30. New York: Plenum.

Royer TC (1998) Coastal ocean processes in the northern North Pacific. In: Brink KH and Robinson AR (eds.) *The Sea*, vol. 11, pp. 395–414. New York: John Wiley Sons.

Strub PT and James C (2000) Altimeter-derived variability of surface velocities in the California Current Systems: 2. Seasonal circulation and eddy statistics. *Deep-Sea Research II* 47(56): 831–870.

Strub PT and James C (2000) Altimeter-derived surface circulation in the large scale NE Pacific Gyres: Part 1. Annual Variability. *Progress in Oceanography* (in press).

Strub PT, Allen JS, Huyer A, Smith RL, and Beardsley RC (1987) Seasonal cycles of currents, temperatures, winds and sea level over the northeast Pacific continental shelf. *Journal of Geophysical Research* 92(C2): 1507–1526.

Strub PT, Kosro PM, Huyer A, *et al.* (1991) The nature of cold filaments in the California Current system. *Journal of Geophysical Research* 96(C8): 14 743–14 768.

Tabata S (1991) Annual and interannual variability of baroclinic transports across Line P in the northeast Pacific. *Deep-Sea Research* 38(supplement 1): S221–S245.

Thomson RE (1981) Oceanography of the British Columbia Coast. Canadian Spec. Pub. Fish. and Aquatic Sciences 86: 291 pp.

Thomson RE and Gower JFR (1998) A basin-scale oceanic instability in the Gulf of Alaska. *Journal of Geophysical Research* 103(C2): 3033–3040.

Wilson JG and Overland JE (1986) Meteorology. In: Hood DW and Zimmerman ST (eds.) *The Gulf of Alaska: Physical Environment and Biological Resources*, pp. 31–54. Springfield, VA: NOAA OCS Minerals Management Service MMS-86-0995.

PERU–CHILE CURRENT SYSTEM

J. Karstensen, Universität Kiel (IFM-GEOMAR), Kiel, Germany
O. Ulloa, Universidad de Concepción, Concepción, Chile

Introduction

The Peru–Chile Current System (PCCS) comprises the surface and subsurface flows in the eastern boundary current system along the Chilean and Peruvian coasts (**Figure 1**). The PCCS hosts four major currents: two flowing equatorward at the surface and two flowing poleward, one at the surface and one at subsurface. For individual currents of the PCCS a variety of names exist: the offshore equatorward flow has been called *Peru Current* (the name that relates the current into a regional geographical context is indicated in *italic* for purposes of clarity here), Oceanic Peru Current, Mentor Current, Humboldt Current (in honor of the German naturalist Alexander von Humboldt), or Peru-Humboldt Current. The equatorward flow near the coast has been called *Peru Coastal Current*, *Chile Coastal Current*, and sometimes the Inshore Peru Current. The poleward surface flow has been called the *Peru–Chile Countercurrent* while the poleward subsurface flow has been called *Peru–Chile Undercurrent* or Gunther Current (in honor of the British investigator Eustace Rolfe Gunther).

The PCCS has the largest meridional extent of all eastern boundary currents stretching from about 5° S to south of 42° S. The PCCS is similar to the other eastern boundary current regions in a number of aspects, for example, wind forcing, strong upwelling, and enhanced productivity. The PCCS has a rather tight connection to the equatorial Pacific and the globally strongest mode of interannual variability; the 'El Niño/Southern Oscillation (ENSO)' propagates via atmospheric and oceanic pathways into the PCCS and provokes specific physical and ecological responses. In fact, ENSO has become well known due to its severe socioeconomic consequences on the fishery activity off Peru and Chile.

The Currents

The primary driver of the currents in the PCCS is the frictional force of the wind on the ocean's surface, the wind stress. As the wind field has a pronounced seasonal signal it is useful to contrast the situation for austral summer (January to March) and austral winter (July to September) (**Figure 2**). On the large scale, the wind variability over the PCCS originates

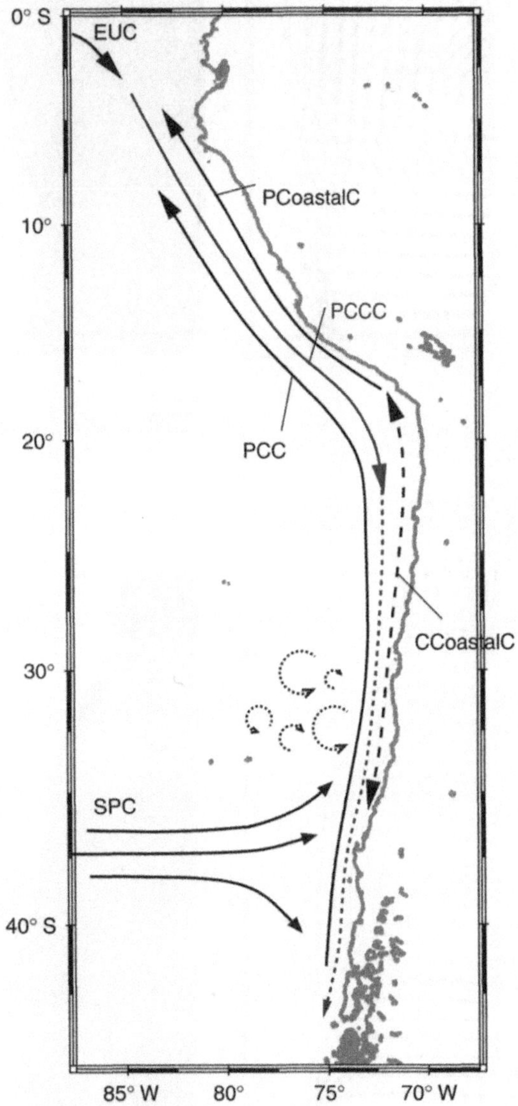

Figure 1 The principal surface flows path in the PCCS and associated temperature anomaly (red indicates warm current, blue indicates cold): Chile Coastal Current (CCoastalC), Peru Coastal Current (PCoastalC), Peru–Chile Countercurrent (PCCC), and Peru–Chile Current (PCC). On the large scale, supplies stem from the South Pacific Current (SPC) and the Equatorial Undercurrent (EUC). The circles at about 30° S symbolize intensive eddy formation. The subsurface poleward Peru–Chile Undercurrent (PCUC) is not drawn but approximately follows the CCoastalC and PCosatalC.

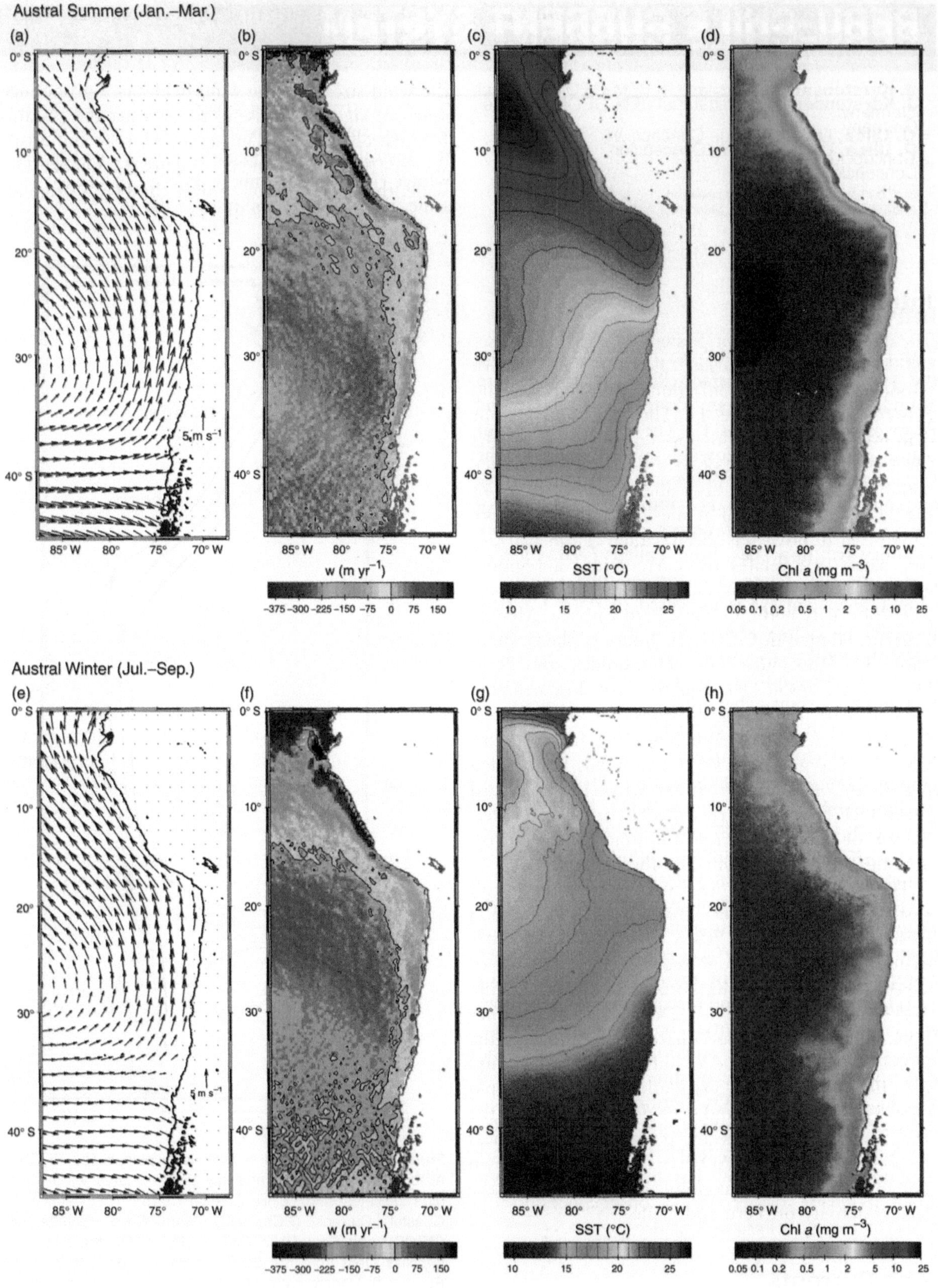

Figure 2 Wind field vectors (QuikSCAT) (a, e), Ekman pumping velocity (QuikSCAT) (b, f; upwelling negative), sea-surface temperature (AVHRR) (c, g), and surface chlorophyll *a* concentration (SeaWiFS/MODIS) (d, h) for austral summer (upper) and austral winter (lower).

from the meridional movement of the eastern Pacific Subtropical Anticyclone from approx. 32° S in austral summer to 28° S in austral winter (**Figures 2(a)** and **2(e)**). In the northern part of the region, off Peru, the movement of the Intertropical Convergence Zone (ITCZ) from approx. 5° to 10° N in austral winter is important. For the southern part of the region, in particular south of about 30° S, the alongshore propagation of atmospheric low-pressure systems is of importance. They originate from the west wind zone and upon impinging the coast propagate northward or southward steered by the Andes mountain range.

Over the PCCS the wind field has its dominant component parallel to the coast. One of the reasons for this is the configuration of the continent and the ocean which maintains through differential heating a large-scale pressure difference between the South American continent and the Southeast Pacific Ocean. The pressure difference leads to equatorward winds along the South American west coast. In addition, the Andes Mountain range, which closely follows the coastline, steers the low-level winds parallel to the coast. The along-shore equatorward wind stress generates an offshore Ekman transport and therefore intensive upwelling occurs along the coast. For the region off Peru, upwelling is strongest in austral winter (**Figure 2(f)**) while for the region between 30° and 40° S, off Chile, almost continuous upwelling is strongest in austral summer (**Figure 2(b)**). There is weaker but almost continuous upwelling for the region in between. South of about 42° S, winds are predominately poleward and associated with coastal downwelling.

The offshore transport of surface water causes a trough in the sea surface height (SSH) along the coast (**Figure 3**, left). Such a trough has dynamical consequences as it causes a zonal (west to east) pressure gradient force. On a rotating Earth and in cases where ocean bottom friction can be neglected, the pressure gradient force is balanced by the Coriolis force acting perpendicular (in the Southern Hemisphere to the left) to the movement. Oceanographers (and meteorologists) refer to the equilibrium of pressure gradient force and Coriolis force as geostrophy. To bring the zonal pressure gradient force along the coast in the PCCS region into a geostrophic balance an equatorward flow is required. This flow, which is in the same direction as the wind, is here associated with the Peru Coastal Current (PCoastC) and the Chile Coastal Current (CCoastC) (**Figure 1**). The coastal currents exist over the shelf area and reach only to shallow depths (less than 80 m). They have typical flow speeds of about $0.1\,\mathrm{m\,s^{-1}}$ in the northern part of the PCCS, decreasing to $0.02\,\mathrm{m\,s^{-1}}$ in the southern part. In austral winter, when atmospheric low-pressure systems impinge on the coast at about 30° S, episodic flow reversals of the CCoastC have been observed. The coastal currents are identified through a minimum in sea surface temperature. This is because they carry colder water from the south northward but more important is the entrainment of the cold upwelling waters into the flow.

Further offshore, at the boundary between the upwelling and downwelling favorable regions (**Figures 2(b)** and **2(f)**), the Peru–Chile Current (PCC) is found (**Figure 1**). This flow is independent of the coastal current and is sometimes considered to be the eastern branch of the subtropical gyre circulation of the South Pacific. Here the contrasting Ekman regimes modify the internal density field such that an equatorward flow is required. Again the PCC can be recognized at the surface from lower-than-average sea surface temperatures associated with the equatorward advection of colder waters.

About 100–300 km offshore, and between the equatorward flowing PCC and the coastal currents (CCoastalC and PCoastalC) is a poleward flow, the Peru–Chile Countercurrent (PCCC). The PCCC is most pronounced off the Peruvian Coast and transports warm and saline water of equatorial origin poleward. The driving mechanisms for the PCCC are not fully understood. It has been suggested that the flow is driven by Sverdrup dynamics as a result of the cyclonic wind stress curl. Other mechanisms, such as eddy/mean flow interaction, buoyancy forcing, or the multiple pattern of Ekman-driven up- and down welling, in particular off the Peruvian Coast, may contribute to the forcing of the current.

Upwelling in the PCCS coastal region brings cold (**Figures 2(c)** and **2(g)**) and nutrient-rich water from depths below 150 m to the surface. Local upwelling rates are higher than $600\,\mathrm{m\,yr^{-1}}$ (**Figures 2(b)** and **2(f)**), but it is clear that the transport is not only vertical but also 'horizontal' or, better to say, lateral, along surfaces of constant density, minimizing the required energy for the flow. The most intense upwelling in the PCCS occurs along the coast and the lateral transport at depths has a pronounced component toward the coast. This onshore transport must be in geostrophic balance and hence a north/south (meridional) pressure gradient along the coast is required for balancing the transport, with higher pressure toward the equator. This equilibrium of onshore transport and meridional pressure gradient is disturbed at the shelf break, acting as a barrier for the onshore flow and the flow stops. When the flow comes to a halt, the pressure gradient force is no longer balanced by the Coriolis force but drives a poleward meridional flow near and above the shelf.

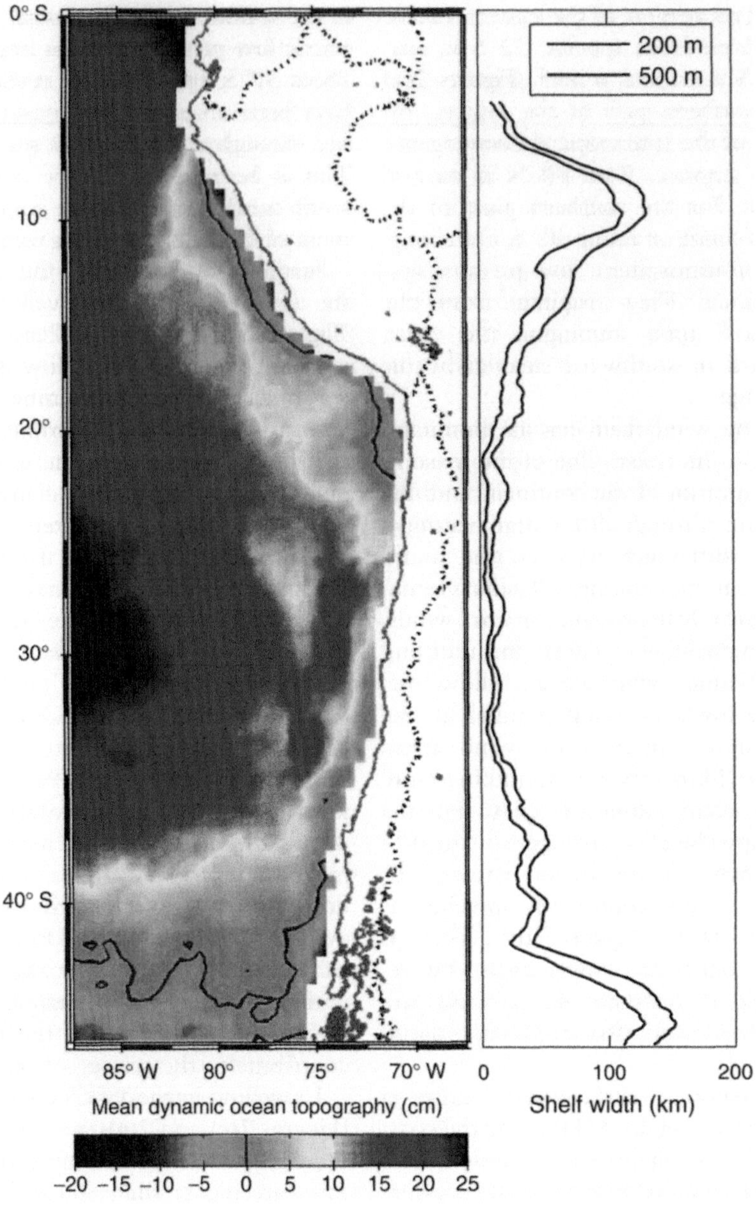

Figure 3 (Left) Absolute sea surface height (based on joint analysis of drifter data, satellite altimeter data, wind data, and a model geoid) and (right) width of the shelf with depth less than 200 m (blue) and less than 500 m (black). Black dotted line indicates political borders.

Now bottom-friction forces balance the pressure gradient force. This subsurface flow along the shelf in the PCCS is associated with the Peru–Chile Undercurrent (PCUC). The PCUC is typically found between 50- and >400-m depth, but it may sporadically even outcrop at the sea surface. The PCUC has flow speeds between 0.1 and 0.7 m s^{-1}. It is ultimately fed by the Pacific Equatorial Undercurrent (EUC) and thus transports warm and saline water, which is low in oxygen and high in nutrients content, southward. These unique characteristics of the water in the PCUC allow us to trace the flow to south of 42° S all along the coast.

The divergence of the eastward-flowing South Pacific Current (SPC), when it approaches the coast at about 40° S, also drives upwelling that may reach several hundreds of kilometers offshore. This upwelling is weaker than the coastal one and its location and intensity move southward in the Southern Hemisphere summer following the movement and intensification of the South Pacific Current.

The Ecosystem

The rich biological productivity of the PCCS depends primarily on the wind-driven coastal upwelling that brings colder and nutrient-rich subsurface waters into the euphotic zone. The upwelling is, in part, fed by water from the oxygen-minimum zone, located below the PCCS and the water has particularly high nutrient levels due to the remineralization of organic matter. When this nutrient-rich water is upwelled into the surface layer it is utilized by phytoplankton along with dissolved CO_2 (carbon dioxide) and light energy from the Sun. The productivity can be seen from the high chlorophyll *a* (Chl *a*) concentrations at the surface (**Figures 2(d)** and **2(h)**). The phytoplankton is, in turn, available for other components of the marine food web, including zooplankton and fish, making the region exceptionally productive.

The high phytoplankton biomass resulting from the fertilizing effect of the upwelling is evident in the coastal band off Peru and off southern Chile (south of about 30° S). For both regions, the temporal variability in near-shore sea surface temperature and Chl *a* concentrations are in phase with each other, suggesting that upwelling-favorable winds are the main driver for productivity here. However, it is not only the local upwelling intensity but also the supply of nutrient-rich waters to the region via the PCUC that maintains the productivity. The shelf is particularly wide in these regions (**Figure 3**, right) and the PCUC broadens and, in turn, facilitates the entrainment of the nutrient-rich waters into the euphotic zone over a wider area. In contrast, upwelling is weaker in the northern Chile region and there is virtually no shelf which makes the nutrient supply to the euphotic zone less effective and the productivity lower.

Offshore Chl *a* variability is not in phase with the coastal variability (cf. **Figures 2(d)** and **2(h)**). The reason for this is not yet clear, but possible mechanisms include differences in the timing of the maximum in nutrient supply and the effect of photoadaptation to the seasonal changes in the light field.

During the upwelling season, waters over the broad continental shelves off northern and central Peru as well as south of 30° S off Chile have extremely high Chl *a* concentrations ($>20\,\mathrm{mg\,m^{-3}}$), with large diatoms dominating the phytoplankton community. In contrast, over the narrow shelves in southern Peru and northern Chile, the chlorophyll concentrations are lower ($<2\,\mathrm{mg\,m^{-3}}$) and smaller phytoplankton species, including prymnesiophytes and cyanobacteria, dominate. Measurements of dissolved iron and experiments with iron addition off Peru have shown that primary productivity in the low-chlorophyll areas of the PCCS are limited by iron. This situation is similar to that found in the so-called high-nutrient low-chlorophyll (HNLC) regions, such as the equatorial eastern Pacific, the North Pacific, and the Southern Ocean. Thus, variability in the productivity of the PCCS cannot be due to changes in the intensity of the upwelling winds alone, but also due to changes in the availability of iron.

Zooplankton is considered the next trophic level in the food chain, going from phytoplankton to fish and top predators. In the PCCS, zooplankton is usually dominated by copepods and euphasids, but other groups, including gelatinous organisms, can become important in certain periods of the year. Nearly 60 species of copepods have been identified in the PCCS. Some of them are endemic, like the abundant *Calanus chilensis*, which is found to be associated with the areas of intensive upwelling. Within the euphasids, the most abundant one is the endemic species *Euphausia mucronata*, which has been shown to migrate vertically through the oxygen-minimum

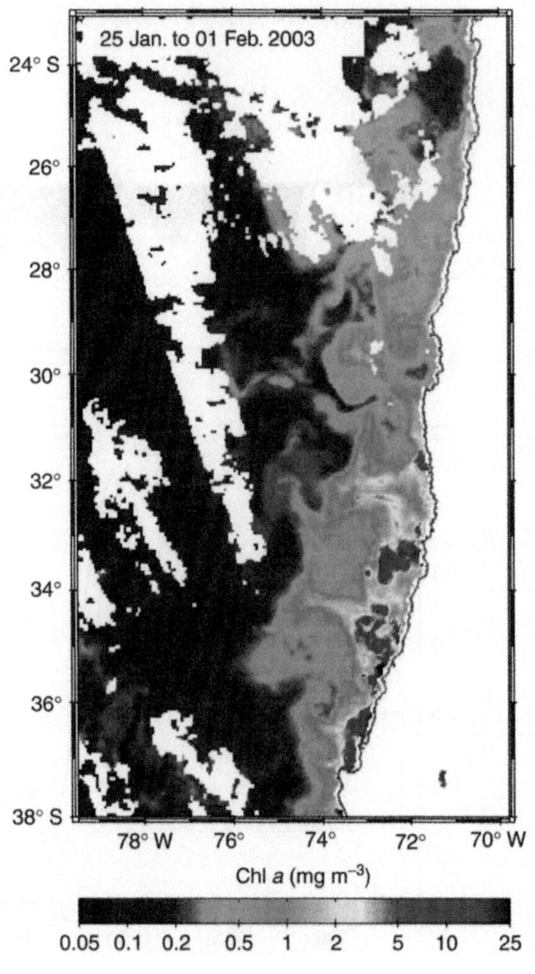

Figure 4 Weekly composite of *MODIS* satellite-derived Chl *a* concentrations for the period 25 Jan. to 1 Feb. 2003. White areas indicate data gaps (e.g., cloud cover).

zone. Another important one, appearing especially during El Niño years, is *Euphausia eximia*.

In order to avoid being permanently advected offshore and away from the food-rich upwelling regions along the coast, zooplankton often make use of their vertical migratory capacities to remain at the coast and 'travel back' on shore with the aid of the onshore subsurface flow. However, this mechanism does not appear to be widely 'used' by zooplankton species in the PCCS, presumably due to the presence of the intense and shallow oxygen-minimum zone that prevents many of them from vertical migration. Similarly, in other coastal systems, copepods show seasonal reproductive cycles with a few well-defined generations per year. In the PCCS, in contrast, they show continuous reproduction and multiple generations throughout the year. As there are virtually no food limitations in the upwelling areas along the coast, even during El Niño years, the near-coastal zooplankton growth rates appear to be mainly temperature-controlled. Recent evidence suggests that food quality can also be important for zooplankton reproduction and recruitment.

In the benthic environment of the continental shelf and upper slope with oxygen-deficient conditions, extensive mats of the giant, sulfide-oxidizing, nitrate-reducing bacteria *Thioploca* spp. can develop. This contrasts with the low abundance and diversity of the macrofauna found there. However, the benthic fauna that inhabits this particular environment, like certain polychaetes, presents particular adaptations to cope with the suboxia (dissolved oxygen content below $\sim 10\,\mu\text{mol kg}^{-1}$) or even anoxia (no oxygen available), and the building up of toxic hydrogen sulfide in the sediments.

The Variability

Snapshots of the ocean's surface property fields (**Figure 4**) reveal the existence of swirls and

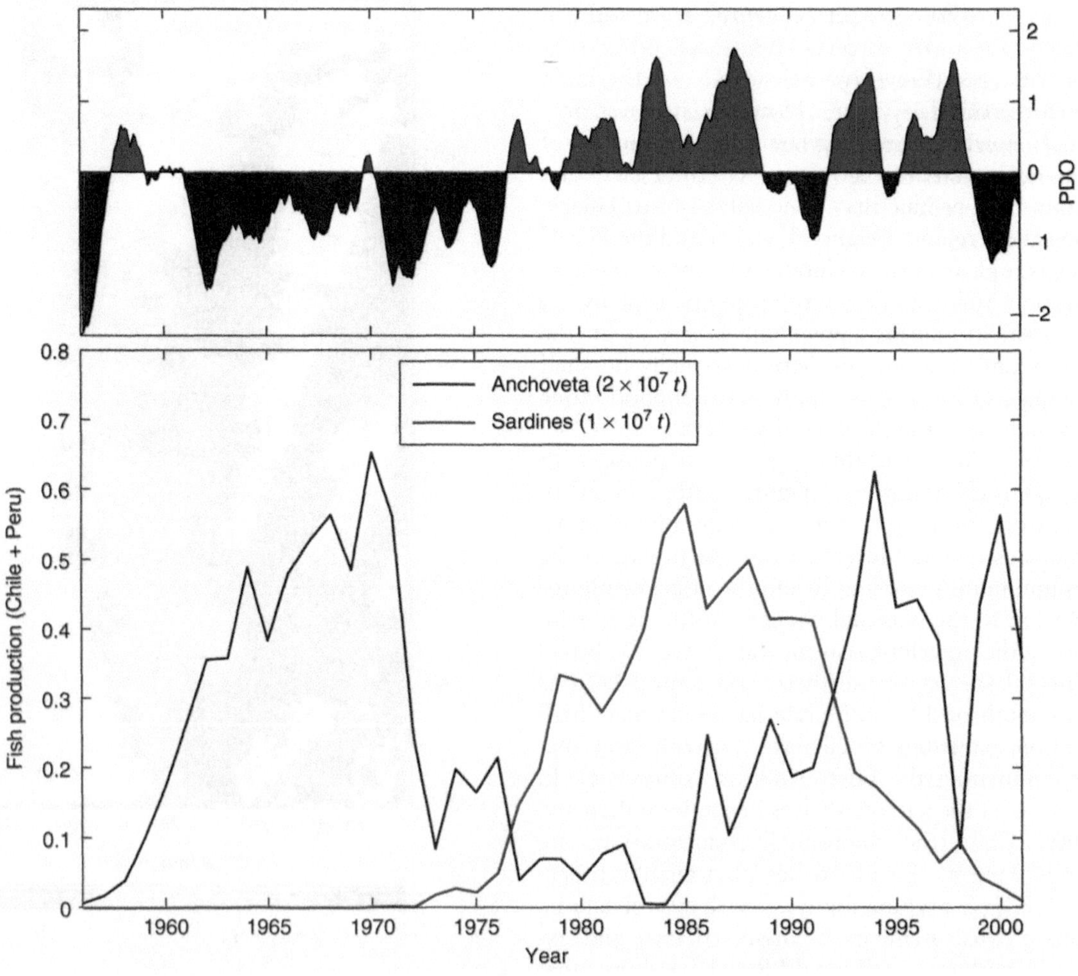

Figure 5 Time series of (upper) PDO index based on North Pacific sea surface temperatures and (lower) anchoveta (Peruvian anchovy; blue) and sardines (red) fish production as given by the Food and Agriculture Organization of the United Nations (FAO).

filaments, manifested in sharp gradients. They are associated with the generation and propagation of mesoscale eddies. Eddies may form from flow instabilities of the coastal currents, from westward propagating Rossby waves and from other coastal trapped waves (CTWs) propagating along the coast. The eddies act as an efficient distributor of nutrients and biomass from the coast to the open ocean into the so-called 'coastal transition zone'. This zone is particularly large between 25° and 40° S and may reach as far as 600–800 km off shore.

One important reason for the flow instability is the intermittent character of the upwelling intensity, with timescales from 3 to 10 days. This is true for the whole coast, but in the zone from 25° to 40° S a strong density gradient is formed between the coast and the open waters, in particular in summer through the increase in coastal upwelling. During intermittent upwelling events, instabilities are generated which, in turn, release the filaments of cold, nutrient-rich, and potentially productive waters far from the coast. After relaxation of the wind, these filaments are dissipated.

Interannual and decadal variability also plays an important role in the region. In particular, the effects of ENSO are evident. This influences the large-scale atmospheric as well as oceanic circulation. During an ENSO event, the wind field changes at the equator and warm waters of the western equatorial Pacific 'warm pool' region propagate to the eastern Pacific by means of a Kelvin wave and radiates in CTW poleward along the eastern boundary. As a consequence, the temperature (as well as the dissolved oxygen contents) of the surface ocean in the PCCS are significantly higher with a simultaneous deepening of the main thermocline. This deepening causes a reduction of the nutrient supply to the euphotic zone and thus a decrease in primary productivity, which has a negative impact all the way through the food chain.

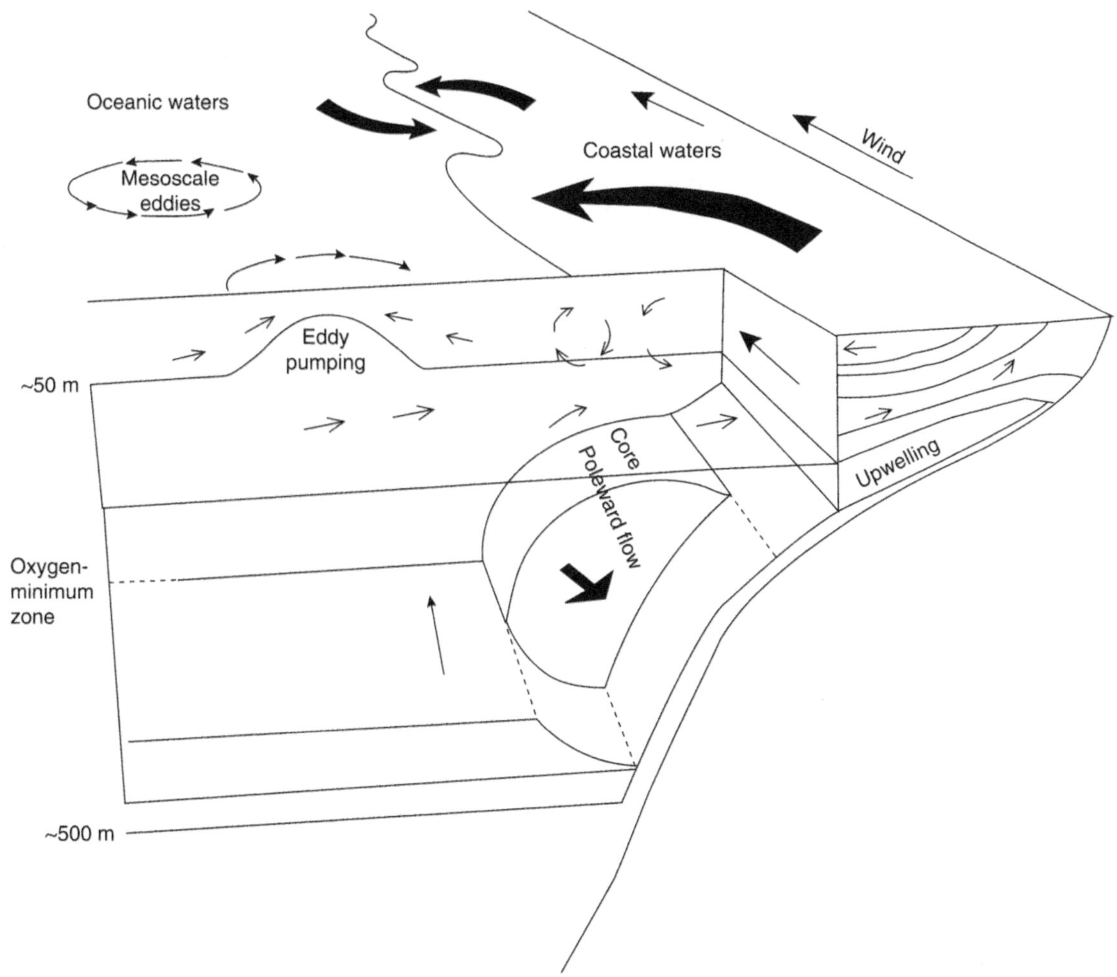

Figure 6 Schematic of physical processes and peculiarities of the hydrographic stratification in the PCCS. Printed with permission of Samuel Hormazábal.

It has been documented that the offshore extension of the high-productivity band reduces during El Niño years and a deepening of the oxycline (the transition between the surface water region with high oxygen content and the oxygen-minimum zone below) occurs. Species that are dependent on the oxygenated conditions of the surface waters then occupy a broader depth range or concentrate in deeper depths. Changes in migratory routes and nursery distributions have also been observed in some species such as jack mackerel.

The marked differences between the ecological effects of particularly strong El Niño events, like the 1982–83 versus the 1997–98, suggests that longer variability (e.g., Pacific Decadal Oscillation (PDO)) plays an important role in regulating ecological responses in the PCCS (**Figure 5**). As a first step, and associated with the PDO, two environmental stages have been identified: a warmer stage, termed 'El Viejo', where sardine abundance is promoted and a colder stage, termed 'La Vieja', where anchovy abundance is promoted.

It is the interaction of different time- and space scales which makes it difficult to fully understand the processes that control the physical and ecological functioning of the PCCS, as well as those of other eastern boundary current systems. A schematic of important physical processes for the PCCS is shown in **Figure 6**. The spectrum of variability is very wide. It covers spatial variability of a few kilometers, associated with filaments and mesoscale eddies, but the variability of basin-scale currents or large-scale atmospheric circulation is also of importance. Likewise, a wide range of temporal scales from the diurnal cycle to decadal times are important.

See also

Benguela Current. California and Alaska Currents. Canary and Portugal Currents. Ekman Transport and Pumping. Mesoscale Eddies. Rossby Waves.

Further Reading

Bakun A and Nelson CS (1991) The seasonal cycle of wind-stress curl in subtropical eastern boundary current regions. *Journal of Physical Oceanography* 21: 1815–1834.

Chavez FP, Ryan J, Lluch-Cota SE, and Niquen M (2003) From anchovies to sardines and back: Multidecadal change in the Pacific Ocean. *Science* 299: 217–221.

Hutchins DA, Hare CE, Weaver RC, *et al.* (2002) Phytoplankton iron limitation in the Humboldt Current and Peru upwelling. *Limnology and Oceanography* 47: 997–1011.

Montecino V, Strub PT, Chavez F, Thomas AC, Tarazona J, and Baumgartner TR (2006) Chapter 10: Bio-physical interactions off Western South America. In: Robinson AR and Brink KH (eds.) *The Sea, Vol. 14A: The Global Coastal Ocean – Interdisciplinary Regional Studies and Syntheses* (*Part 2. Regional Interdisciplinary Oceanography*). Cambridge, MA: Harvard University Press.

Philander SGH (1990) *El Nino, La Nina and the Southern Oscillation*, 289pp. San Diego, CA: Academic Press.

Strub PT, Mesias JM, Montecino V, Rutllant J, and Salinas S (1998) Coastal circulation off western South America. In: Robinson AR and Brink KH (eds.) *The Sea, Vol. 11: The Global Coastal Ocean – Regional Studies and Syntheses*, pp. 273–313. New York: Wiley.

EAST AUSTRALIAN CURRENT

G. Cresswell, CSIRO Marine Research,
Tasmania, Australia

The East Australian Current

The East Australian Current (EAC) is a strong western boundary current akin to the Gulf Stream, the Brazil and the Agulhas Currents, and the Kuroshio. Its maximum speed exceeds $2\,\mathrm{m\,s^{-1}}$ near the surface and its effects extend down several thousand meters. It has a distinct surface core of warm water with a trough-shaped cross-section that is about 100 km wide and 100 m deep. There is a northward undercurrent beneath the EAC that extends from the upper continental slope down to the abyssal plane at 4500 m. The EAC plays a pivotal role in the life cycles of marine fauna and flora of eastern Australia and weather systems respond to the patterns of warm water that it creates.

The strength of the EAC came as a surprise to explorer Captain James Cook: "Winds southerly, a fresh gale", he wrote in his log at sunset on 15 May 1770 as he neared Cape Byron. Seeking more sea-room for the night, he headed offshore until, "having increased our soundings to 78 fathoms, we wore and lay with her head in shore… At daylight we were surprised by finding ourselves farther to the southward than we were in the evening, and yet it had blown strong all night". The EAC had carried his ship, *Endeavour*, southward into the strong winds.

The EAC has two main sources (**Figure 1**): One is the Pacific Ocean South Equatorial Current that flows westward into the Coral Sea between the Solomon Islands at 11°S and northern New Caledonia at 19°S. As this current nears Australia's Great Barrier Reef it splits at 14°–18°S, with part going NW to the Gulf of Papua and part going SE as the first 'tributary' of the East Australian Current. The other source is the near-surface layer of salty waters of the central Tasman Sea that are linked to the central Pacific Ocean. North of 25°S the salty layer slips beneath the fresher and warmer waters of the tropics. Near Australia, as we will see, the salty waters are taken south to Tasmania by the EAC.

What is it that drives the EAC? The curl of the wind stress over much of the South Pacific Ocean moves its waters northward. The southward return flow in a narrow band against the western boundary is consequently very strong. At the same time, the westward propagation of Rossby waves accumulates energy against eastern Australia. As a result there is a band several hundred kilometers wide off central eastern Australia where the subsurface water

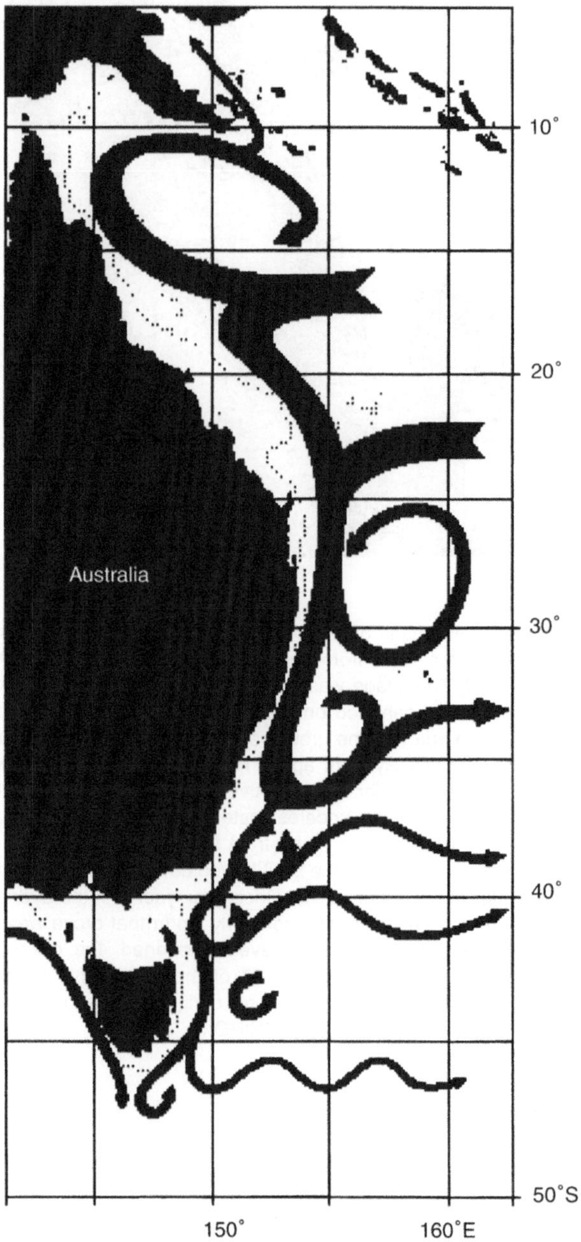

Figure 1 A schematic diagram of the East Australian Current showing its inflows and the eddies that are associated with it. The current off western Tasmania has been named the Zeehan Current.

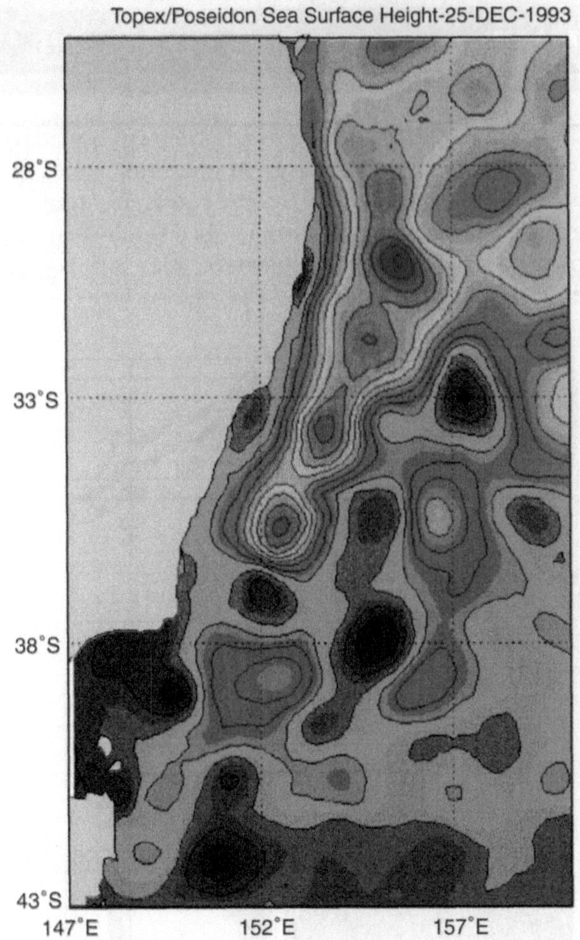

Figure 2 An image of sea surface topography prepared from radar altimeter data from the French-US satellite TOPEX/POSEIDON. The image was formed from satellite passes over a 17-day period centered on 25 December 1993. The passes were 300 km apart on the ground and the measurements along them were effectively 6 km apart. The data are contoured, but unless a pass goes, for example, over the center of an eddy its peak height will be an underestimate. The contour spacing is 10 cm; it is a little over a meter from the highest to the lowest parts of the sea surface. The EAC flowed down the western side of the ridge in this image. On 28 December, yachts in the Sydney to Hobart race encountered a southerly storm that drove waves into the opposing EAC. The waves steepened, the distance between crests decreased, and 67 of the 104 yachts sought safety in nearby ports owing to failures of rigging, hulls, and crew. Image provided by K. Ridgway, CSIRO.

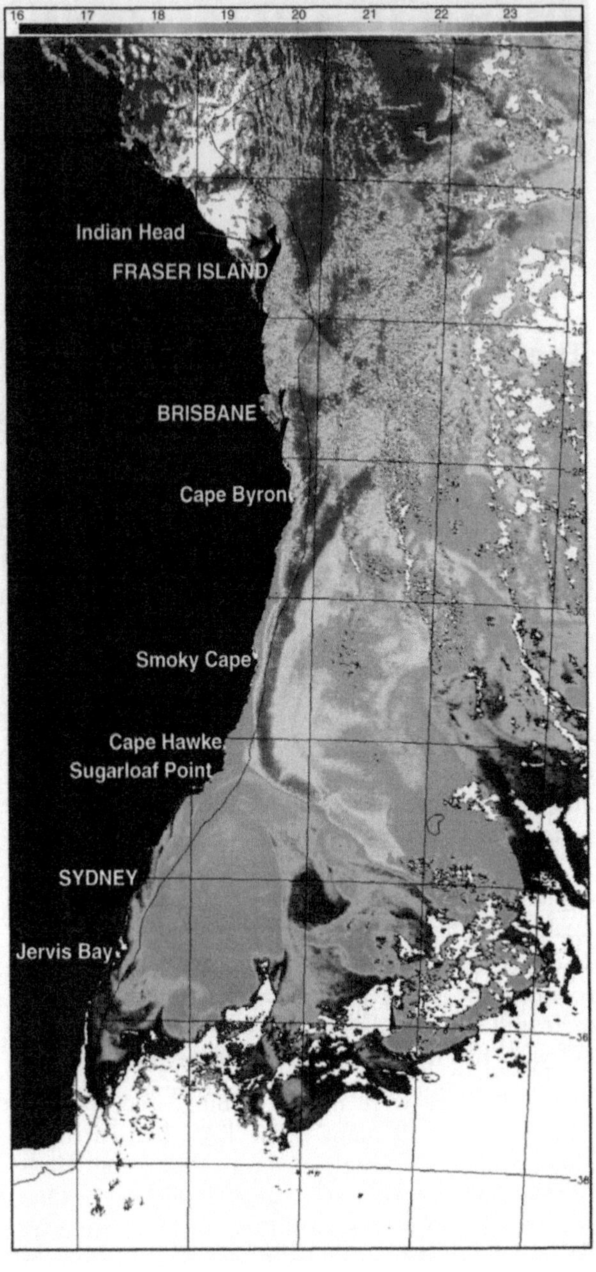

Figure 3 A NOAA satellite sea surface temperature image of the East Australian Current on 18 November 1991. The temperature scale is at the top of the image. There is a 2° × 2° latitude–longitude grid and the 200 m isobath (roughly the edge of the continental shelf edge) is marked as a black line. The white areas are cloud. Note the two streams converging at 30°S. A concurrent ship section (**Figure 4** suggests that the eastern stream was of high salinity while the stream from the Coral Sea was low salinity.

structure of the upper kilometer has been depressed by up to 300 meters. This means that the temperature at any depth in the band is higher, by as much 5°C, than that at the same depth in the neighbouring Tasman Sea. It follows that these waters, down to a 'depth of no motion', have lower density and, since a low-density water column will be taller than a high density one, the sea surface in the band stands as a ridge about one meter higher than its surroundings (**Figure 2**). In other words, it has a greater steric height. Surface currents are controlled by the shape and slope of the sea surface topography in the same way as winds are controlled by atmospheric pressure patterns. The currents follow contours of elevation

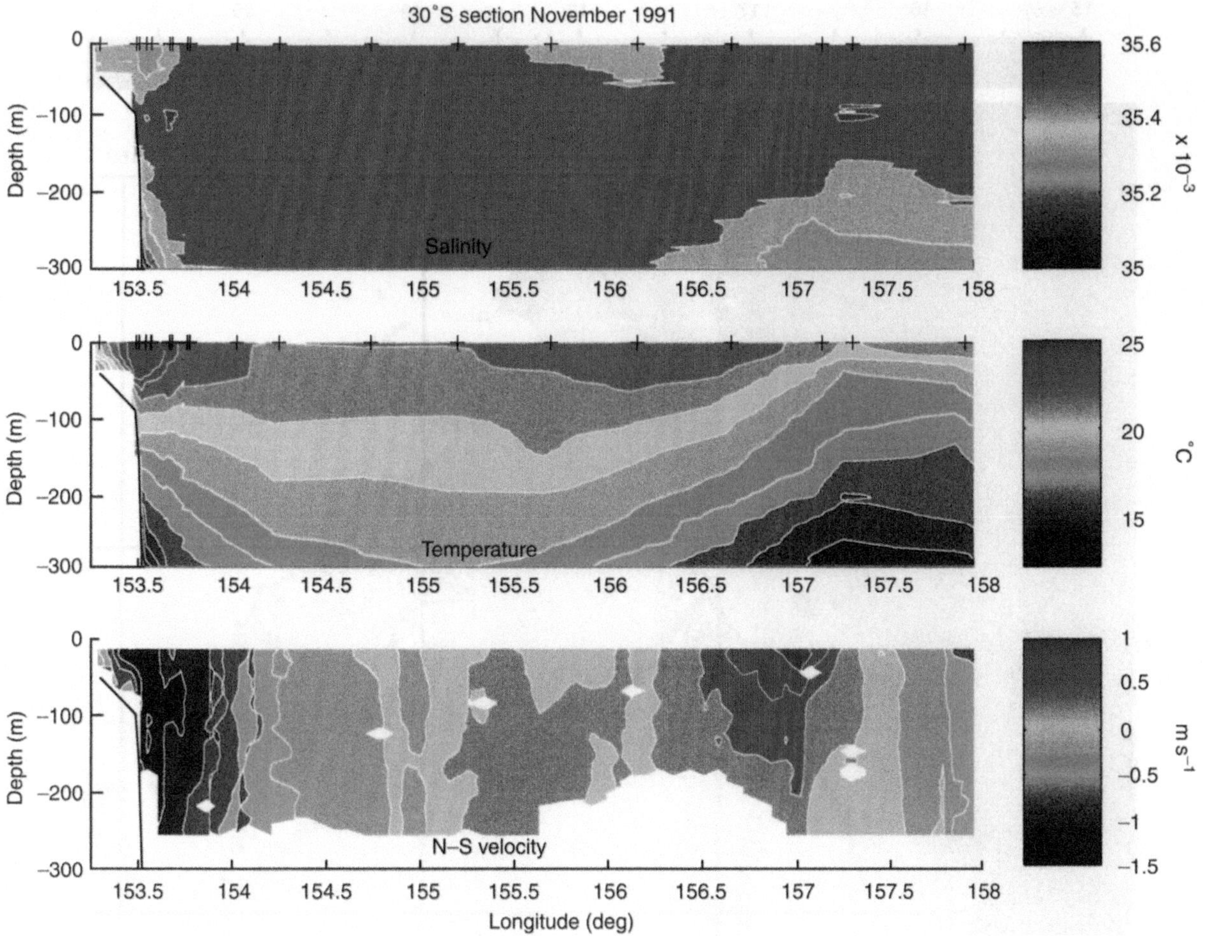

Figure 4 A ship section along 30°S out from Australia of salinity, temperature, and north–south velocity component (measured with an acoustic Doppler current profiler, ADCP). Note the warm, low-salinity feature with the trough-shaped section at the edge of the continental shelf. This is the fastest part of the EAC, but the ADCP data show this current to be much broader and deeper and to be carrying mainly higher-salinity west central Pacific water. The current measurements show the reverse (northward) flow of the EAC meander where the subsurface temperature structure has its greatest slope. On this section the maximum southward speed of the EAC was 1.2 m s^{-1} and the maximum reverse speed was 1.16 m s^{-1}. The innermost station on this section was at the 50 m isobath and suggested that a slope intrusion had upwelled to the surface because the surface and bottom water temperatures were 20°C and 17°C, respectively. Near and down from the shelf edge these temperatures were encountered at 100 m and 190 m, respectively.

and are strongest where the slopes of the sea surface are steepest. Similarly, deeper currents respond to patterns of subsurface steric height.

The low-salinity tributary from the Coral Sea, which is strongest in late summer, flows toward the western side of the ridge at about 25°S, where it joins high-salinity inflow from the east (**Figures 3** and **4**). The resulting EAC flows southward along the western side of the ridge, with higher sea surface elevation on its left (eastern) side. Peaks and saddles along the ridge force the current to meander, accelerate, and decelerate. The western edge of the EAC can spread in across the narrow continental shelf to influence the nearshore waters, while at the same time, as we will discuss later, driving intrusions of continental slope water onto the shelf. When the EAC reaches the southern end of the ridge, commonly near 33°S, it turns anticyclonically (anticlockwise in the southern hemisphere) until it has reversed direction and is running northward several hundred kilometers offshore. Part of it completes a circuit, rejoining the parent EAC, while the remainder meanders eastward along the Tasman Front toward New Zealand.

Various estimates have been made of the volume transport in the various components of the EAC system. From the surface to 2000 m depth the South Equatorial Current carries 50 Sverdrups into the Coral Sea. Of this, half flows southward into the EAC, which, at 30°S carries 55 Sv. After meandering

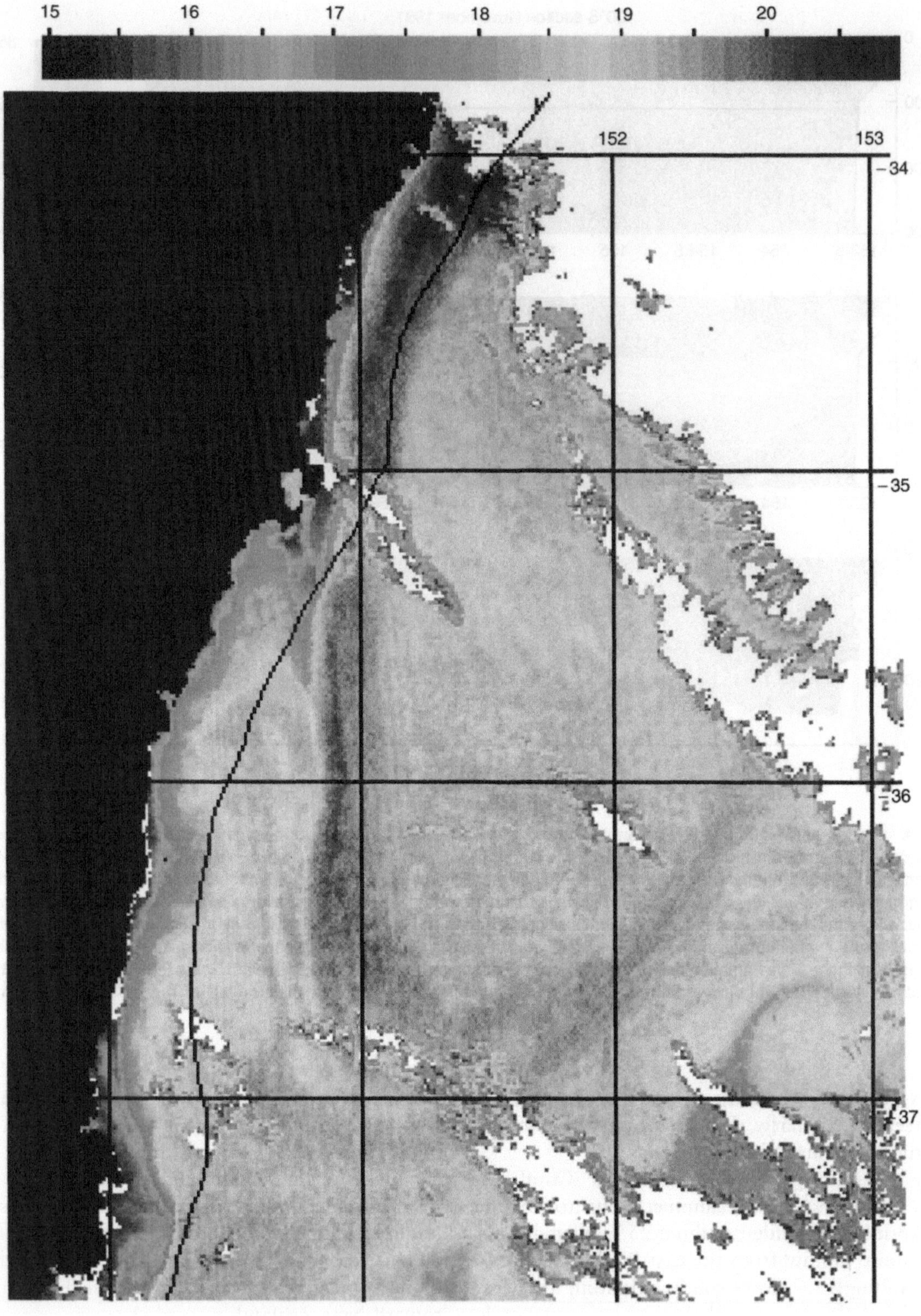

Figure 5 A NOAA sea surface temperature image of an eddy near Jervis Bay (35°S) in early December 1989. There is a 1° × 1° latitude–longitude grid and the 200 m isobath (roughly the edge of the continental shelf edge) is marked as a black line. The white areas are cloud. Note the warm band going around the elliptical eddy, the coastal upwelling, and a small cyclonic eddy centered near the intersection of 36°S and the shelf edge. (NOAA 11 Tm 45 S 2 Dec. 1989 15172. © 1998 CSIRO.)

to the north, 30 Sv recirculates back into the EAC and 25 Sv moves along the Tasman Front. The transport around a large anticyclonic eddy (next section) is about 55 Sv. Recent work suggests that up to 40 Sv can flow northward just south of the separation point of the EAC, perhaps on the western side of a cyclonic eddy.

Eddies

South of the ridge are 'warm-core' anticyclonic eddies that are 250 km in diameter with edge speeds of over 1 m s^{-1} (**Figures 5** and **6**). Just as Captain Cook had been surprised by the EAC, so was Commander J. Lort Stokes on HMS *Beagle* in July 1838 surprised by the currents in an eddy: "...from Hobarton we carried a strong fair wind to 40 miles east of Jervis Bay when we experienced a current that set us 40 miles S.E. in 24 hours; this was the more extraordinary as we did not feel it before, and scarcely afterwards".

The subsurface structure in the eddies is depressed by several hundred meters, such that at 400 m their temperature can be 8°C warmer than at the same

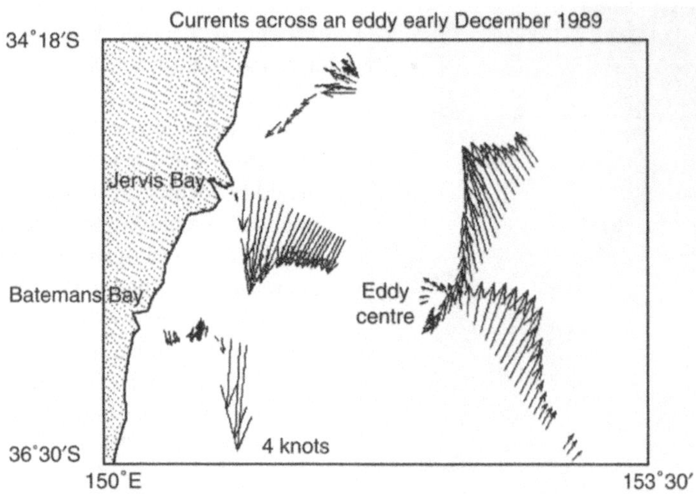

Figure 6 Current measurements with a shipboard ADCP around the time of the satellite image in **Figure 5**. Note that the current speed reaches almost 2 m s^{-1} on the western side of the elongated eddy.

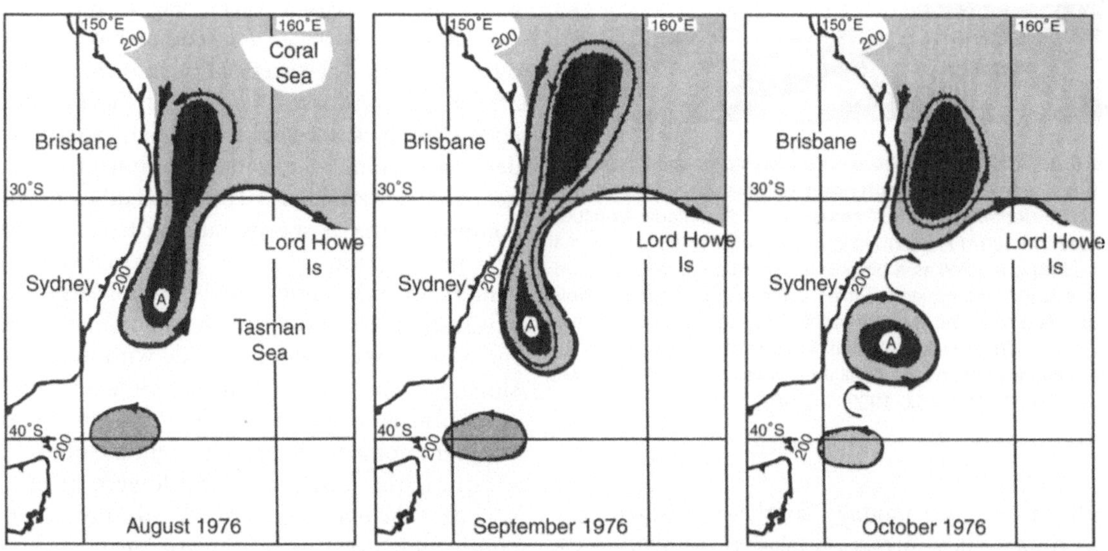

Figure 7 A cartoon showing how the westward propagation of an undulation on the Tasman Front causes the ridge of high sea surface elevation, around which flows the EAC, to pinch off to form a new anticyclonic eddy. (From Nilsson and Cresswell, 1986.)

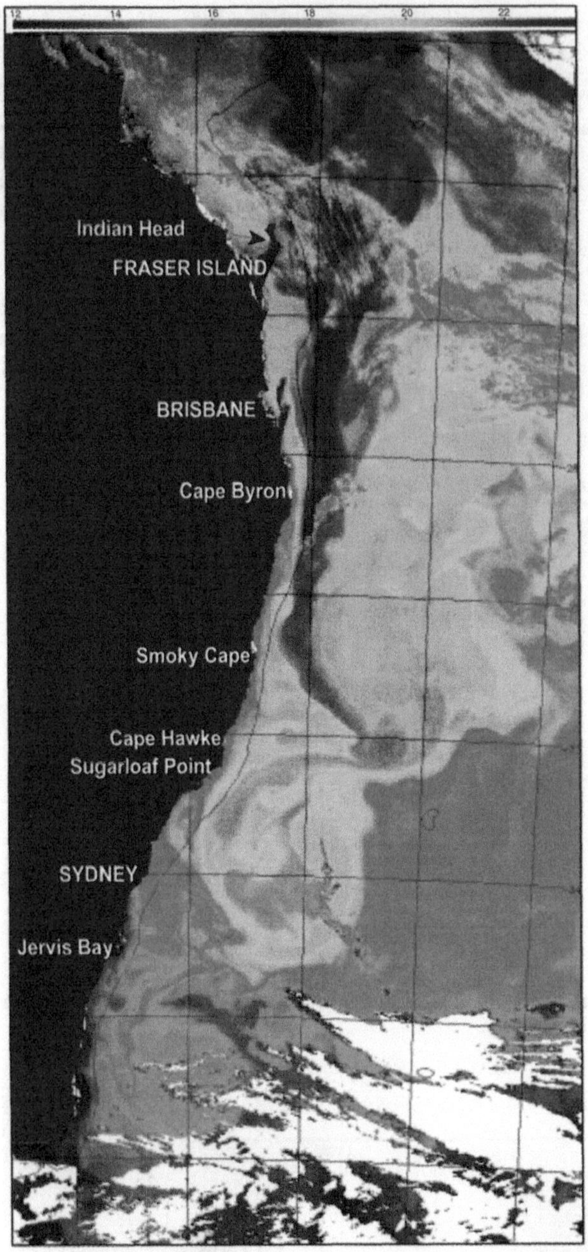

Figure 8 A NOAA satellite sea surface temperature image of the East Australian Current running southward and then out to sea on 29 September 1991. There is a 2° × 2° latitude–longitude grid and the 200 m isobath (roughly the edge of the continental shelf edge) is marked as a black line. The white areas are cloud. There are two anti-cyclonic eddies south of this meander. Note the cascade of warm water from the meander to one eddy and then to the next, as well as the small cyclonic eddies produced by current shear at the edges of all three features. (NOAA 11 TMS 45S 29 Sep. 1991 1615z. 1999 © CSIRO.)

depth in the surrounding southern Tasman Sea. The primary eddy formation process is linked to the westward propagation of Rossby Waves along the Tasman Front. These move its meanders and associated thermocline undulations towards Australia, even though the net flow along the Front is eastward. Several times each year a Rossby Wave reaches the ridge in the sea surface against Australia, initially constricting it and then causing it to pinch off a new eddy (**Figure 7**). These drift slowly southward. The eddies usually become isolated peaks in the sea surface topography that are not encircled by the surface core of the EAC. However, a transient saddle can form between an eddy and the ridge and this allows the warm EAC to 'reach' across to the eddy and encircle it (**Figure 8**). Part of this warm water may spread inward to cover the surface of the eddy, while part may escape the eddy's influence after encircling it several times. In summer, when the EAC is strongest, the ridge steps its way southward, successively coalescing with eddies, until it reaches the SE corner of the Australian continent at 38°S. These eddies retain their identities, with the ridge then consisting of a chain of eddy peaks and saddles. The elongated ridge opens a pathway for the EAC to cascade southward. The strength of this cascade progressively decreases because part of the EAC that encircles each eddy is lost to the northeast. Remnants of the EAC reach southern Tasmania at 43°S via smaller eddies and then overshoot by 200 km into the Southern Ocean (**Figure 9**). Each day in the fishing season fleets of 20–30 Japanese vessels follow parallel paths several kilometers apart across such eddies as they stream their 100 km long lines to catch tuna.

When the ridge retracts to the north it can spawn two or three eddies that are either new or rejuvenated in that the introduction of the new warm EAC water has increased their height above the surrounding sea surface. The eddies have lifetimes of over one year. The speeds in them increase from zero at the center to over $1\,\mathrm{m\,s^{-1}}$ at the perimeter. An eddy disk does not rotate stiffly: the rotation period decreases from 5 days at the perimeter to 1–2 days near the center (**Figure 10**). The eddies move along complex paths at speeds ranging from near-stationary up to $30\,\mathrm{km\,d^{-1}}$. The paths followed by their centers include anticlockwise loops about 200 km across that are described in about one month and these can cause eddies to collide with the continental slope. Such a collision distorts a circular eddy into an ellipse, around which the flow appears to conserve angular momentum, giving highest speeds ($>1.5\,\mathrm{m\,s^{-1}}$) near the minor axes and lowest speeds ($\sim 1\,\mathrm{m\,s^{-1}}$) near the major axes. While an eddy is against the continental slope its southward currents can extend in across the shelf. Eddies regain their near-circular shapes once they move out to sea.

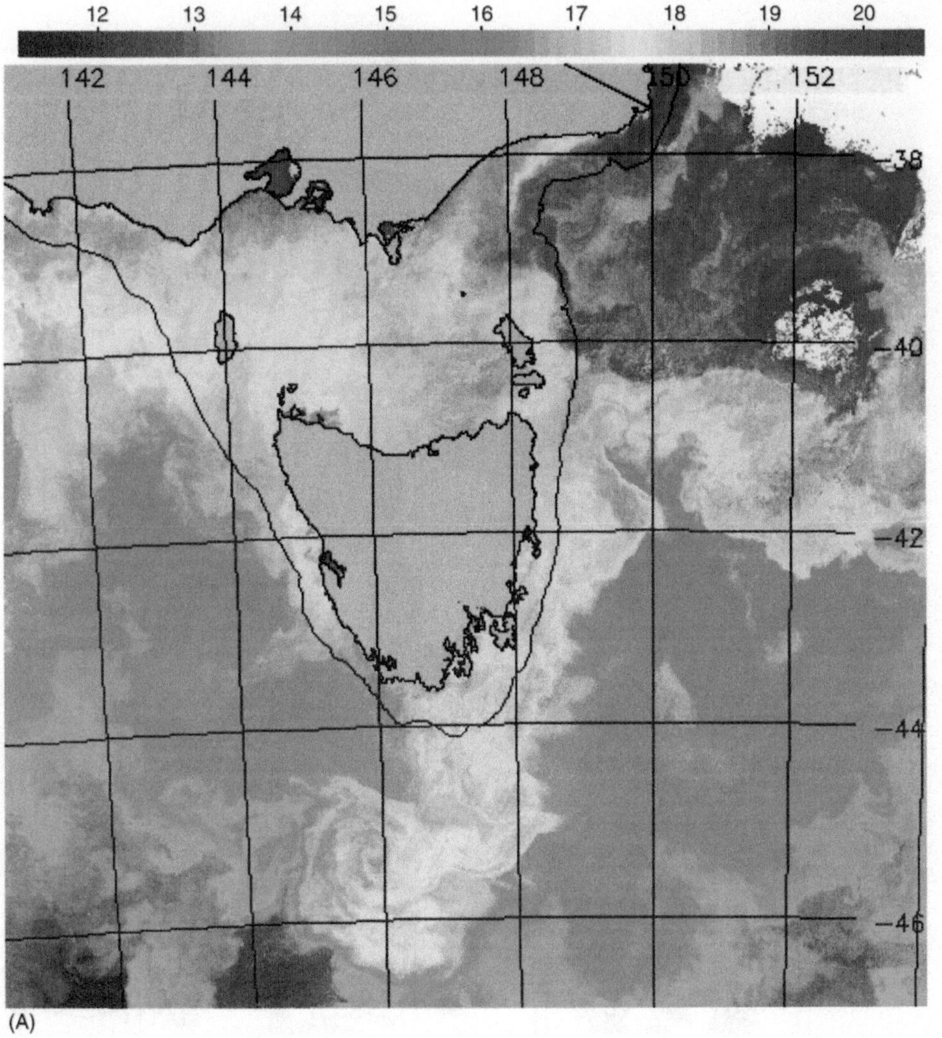

(A)

Figure 9 Winter and summer satellite sea surface temperature images from the NOAA11 satellite. The temperature scale is across the top of the images. Clouds are white. The shelf edge (200 m isobath) is marked as a thin black line. The winter image (A) shows the EAC apparently arrested off eastern Tasmania, while warm water of the Zeehan Current comes down the west coast and part way up the east coast of Tasmania. The summer image (B) shows the EAC to overshoot Tasmania by about 200 km. It entrains Zeehan Current water as it does this.

Not all eddies rotate anticyclonically. Between the ridge and the nearest warm-core eddy to the south – and between warm-core eddies themselves – can be depressions in the sea surface that are cold-core cyclonic eddies. On their inshore (western) sides these drive northward currents of up to 1 m s^{-1} near the shelf edge. Also, as the EAC separates from the shelf and slope at the southern end of the ridge it often develops instabilities that are carried along its edge. Each starts as a small meander to the west from which a warm plume then reaches back and around a growing cyclonic eddy. The instabilities can form every five days and are spaced at intervals of about 100 km. Similar small eddies form from the shear at the edges of large anticyclonic eddies. The continual formation of the small cyclonic eddies, each with doming of the water structure in its interior, lifts richer water into the photic zone where it can photosynthesize. Satellite color measurements reveal the widespread effects of this (**Figure 11**).

Occasionally the southward migration of the ridge will force two anti-cyclonic eddies together so that both are affected: they move several times anti-clockwise around one another until they coalesce into a large eddy. The process takes several weeks, during which the current speeds in the pair reach 2 m s^{-1}. The product highlights an interesting property of the anticyclonic eddies: Because their waters, like those of the ridge, are warmer down to several hundred meters than the surrounding waters, in

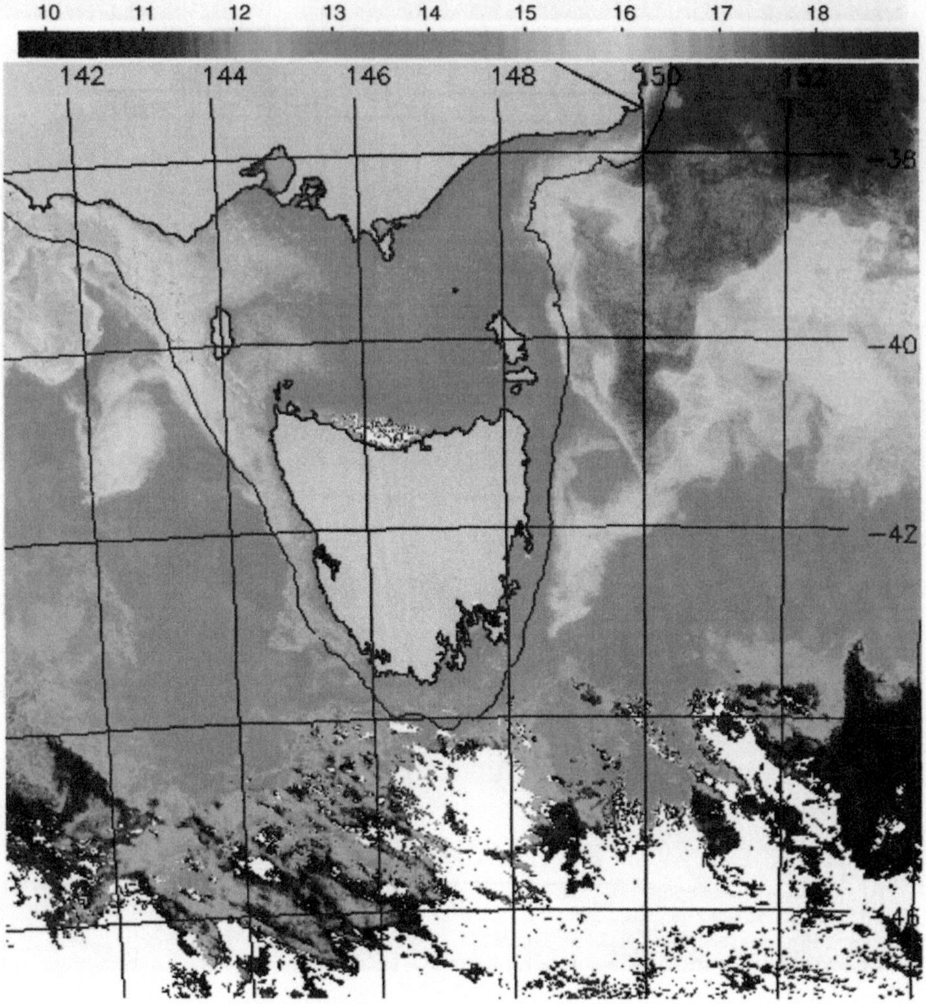

(B)

Figure 9 *Continued*

winter they lose heat at the surface more rapidly and they mix down to more than 400 m depth. The near-constant salinity and temperature in the mixed layer together serve as a unique signature for an eddy once it acquires a summer 'cap' owing to insolation and flooding by the EAC. When eddies coalesce, the new eddy has parts of their signature layers one above the other, according to their relative densities.

Effects on the Continental Shelf

We have already mentioned that the EAC (and its eddies) can influence the currents on the continental shelf of eastern Australia. The shelf is narrow, with a width of about 25 km, and its depth at mid-shelf is 60–80 m. The edge of the EAC can reach in to drive southward currents in excess of 0.5 m s^{-1} at promontories like Indian Head on Fraser Island, Cape Byron, Smoky Cape, Cape Hawke, Sugarloaf Point, and Jervis Bay. This means that it may be more of an influence on the circulation of the Australian shelf than is the Gulf Stream on the 70–120 km wide US eastern shelf with a typical depth of 30 m. Incursions of the EAC onto the shelf quickly overwhelm existing current patterns, replace large parts of the shelf waters, and appear to be a mechanism for driving cold, nutrient-rich intrusions of slope water from 200–300 m depth in towards the coast. The stress by the EAC on the bottom sets up a bottom boundary layer that moves in across the shelf at a slight angle. Near the coast, and notably downstream of

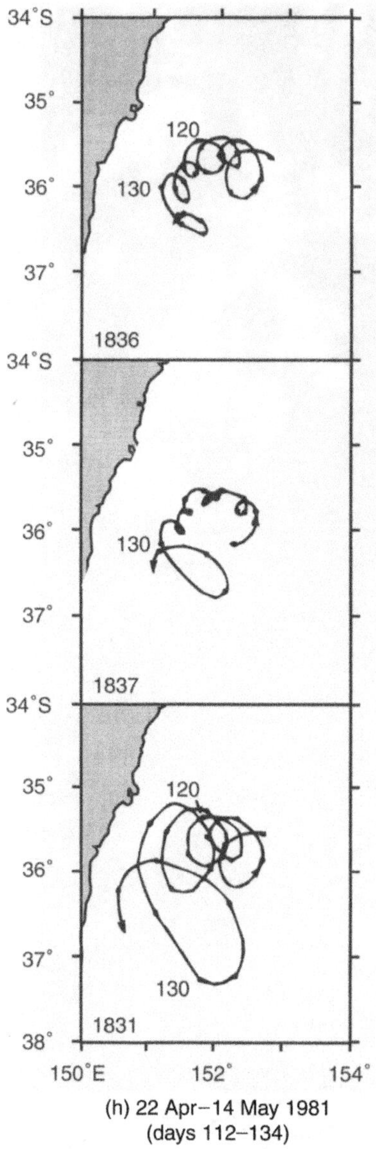

Figure 10 The paths followed by three satellite-tracked drifters in an eddy that became distorted into an ellipse after it collided with the continental slope of SE Australia. The dots on the tracks indicate daily intervals. Note the short rotation periods for drifter nearest the eddy center. (From Cresswell and Legeckis (1986) *Deep-Sea Research*.)

headlands (perhaps because of cyclonic motion that they induce), the intrusions may upwell to the surface. This process can be greatly assisted by northerly winds that drive the surface waters offshore in an Ekman layer, to be replaced by the upwelling of the slope intrusions near the shore. The upwelled waters photosynthesize, producing a peak in productivity at about 20–50 m that is a balance between light, nutrients, and grazing by zooplankton. The green waters are known to mariners and their spectacular patterns are clearly evident in color satellite imagery, contrasting with the transparent deep blue of the nutrient-poor EAC (**Figure 11**). The edge of the EAC can be seen in the images to carry the chlorophyll-rich waters along the shelf and well out to sea.

Summary

The near-surface core of the East Australian Current draws water both from the South Equatorial Current via the Coral Sea and from the central Tasman Sea. It flows southward to about 33°S, where it separates from the continent and executes a meander to the north. Part recirculates and part proceeds along the Tasman Front toward New Zealand. The meander spawns 250 km diameter anticyclonic eddies several times each year and these migrate southward. Smaller cyclonic eddies are found throughout the region. The separation point occurs near the southern end of a ridge in the sea surface topography. This ridge moves south and north, coalescing with the anticyclonic eddies that are also highs in the surface topography. The East Australian Current cascades southward along and around these structures, ultimately entering the Southern Ocean south of Tasmania each summer. The waters and currents of the East Australian Current and its eddies can extend in across the narrow continental shelf to the shore. Bottom friction can establish a bottom boundary layer that lifts continental slope water onto the shelf to upwell near the coast, particularly when the winds are northerly.

Glossary

Ekman layer The wind-driven component of transport in the Ekman or surface boundary layer is directed to the left (right) of the mean wind stress in the southern (northern) hemisphere.

Rossby waves Rossby waves occur in the atmosphere and the ocean. In the atmosphere they are high and low pressure cells that, while being carried eastward by strong westerly winds, they in fact propagate westward relative to the mean air flow. In the ocean the mean flow is weaker and undulations in the steric height propagate westward as Rossby waves.

Steric height and depth of no motion Measurements of temperature, salinity and pressure are needed to calculate steric height, which is the depth difference in the ocean between two surfaces of constant pressure. The deeper of these is often chosen to be a 'depth of no motion' which lies on a constant pressure surface and thus the currents are zero.

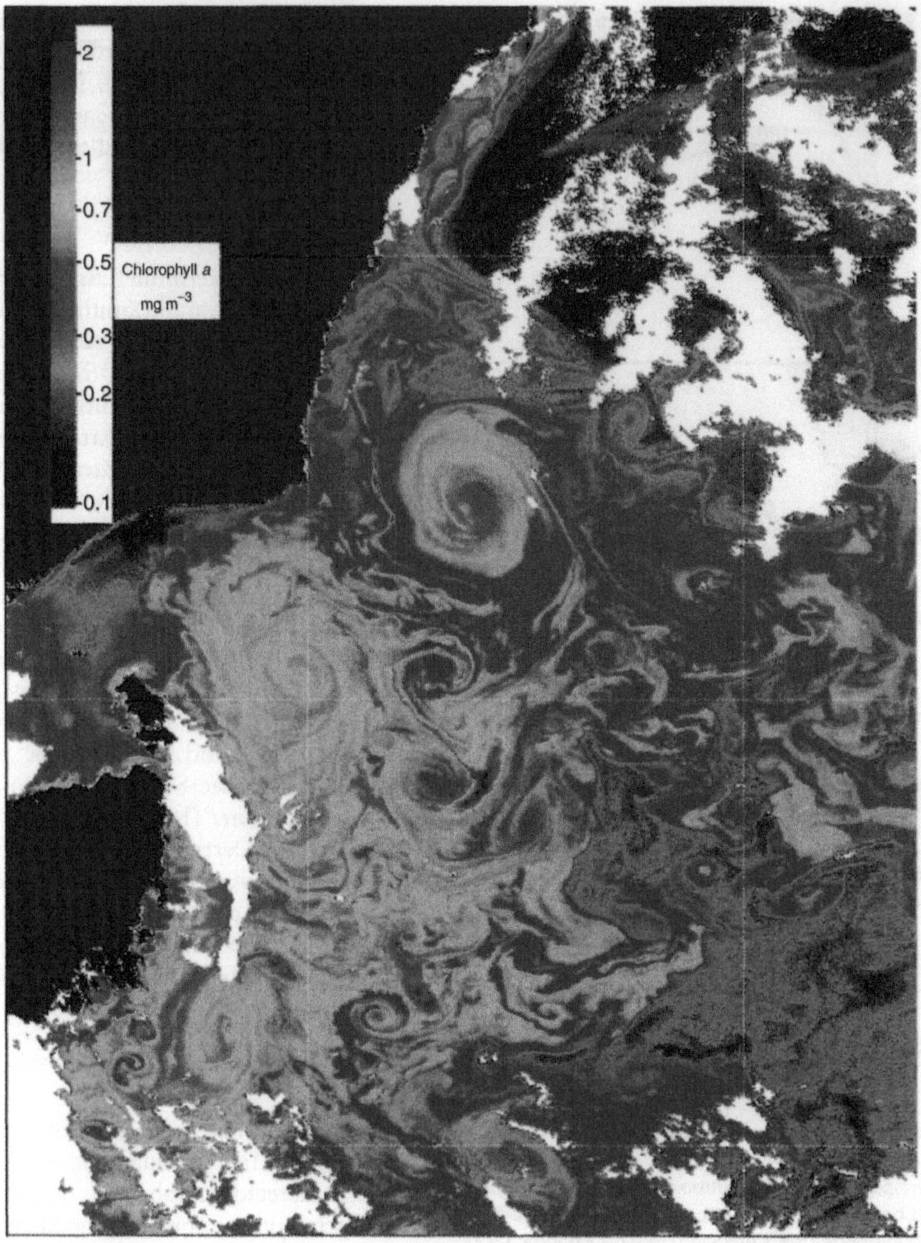

Figure 11 An image of chlorophyll concentration in the SW Tasman Sea inferred from color measurements from the SeaWIFS satellite on 22 November 1997. Clouds are white. In the top right the blue color marks the unproductive waters of the EAC. Inshore of those is a coastal upwelling from which two slugs of chlorophyll-rich waters are carried ~200 km southward along the continental shelf. There is an anticyclonic eddy at 38°S that has high chlorophyll concentrations, which is hard to understand without knowing the eddy's recent history. Southward from the mainland to southern Tasmania the chlorophyll concentrations are high, perhaps owing to the confluence and mixing of subantarctic and subtropical waters.

Steric height differences at the ocean surface across the East Australian Current are about 1 m.

See also

Ekman Transport and Pumping. Mesoscale Eddies. Pacific Ocean Equatorial Currents. Rossby Waves.

Further Reading

Nilsson CS and Cresswell GR (1980) The formation and evolution of East Australian Current warm-core eddies. *Progress in Oceanography* 9: 133–183.

Tomczak M and Godfrey JS (1994) *Regional Oceanography: An Introduction*. Oxford: Pergamon.

PACIFIC OCEAN EQUATORIAL CURRENTS

R. Lukas, University of Hawaii at Manoa, Hawaii, USA

Introduction

An essential characteristic of Pacific equatorial ocean currents is that they span the width of the Pacific basin (15 000 km at the Equator), linking to eastern and western boundary flows (**Figures 1** and **2**). While they have long zonal scales, the relatively strong near-equatorial flows have complex vertical and meridional structures. They exhibit energetic variability on timescales from days to years. In particular, the currents of the equatorial Pacific are considerably altered during El Niño/Southern Oscillation (ENSO) events.

These flows are subject to the distinctive physics associated with the equatorward decrease of the vertical component of the earth's rotation vector. The associated vanishing of the horizontal Coriolis force at the Equator results in relatively strong currents for a given wind stress or pressure gradient. Rapid adjustment of the currents to changing forcing is associated with a special class of internal wave motions termed linear equatorially trapped waves. There are several different types of waves with rich meridional and vertical structure, governed by dispersion relationships that tie zonal wavelength, meridional structure, and wave period together. The fastest waves cross the Pacific in only 2–3 months. With greater distance from the Equator, zonal propagation speeds become slower. The key feature is that these waves can transmit the signals of wind forcing to and from remote locations.

The Pacific equatorial surface currents are primarily wind-driven. Local forcing of the equatorial currents is dominated by surface wind stress (as opposed to heat and/or fresh water fluxes) and its variability. Variable wind forcing results in vertical pumping of the thermocline and subsequent dynamic adjustment, including radiation of equatorially trapped waves. The currents are not forced solely by local winds, however, because equatorially trapped waves carry wind-forcing signals across the entire basin, and similar boundary-trapped waves transmit information about forcing between the equator and higher latitudes. An important portion of the

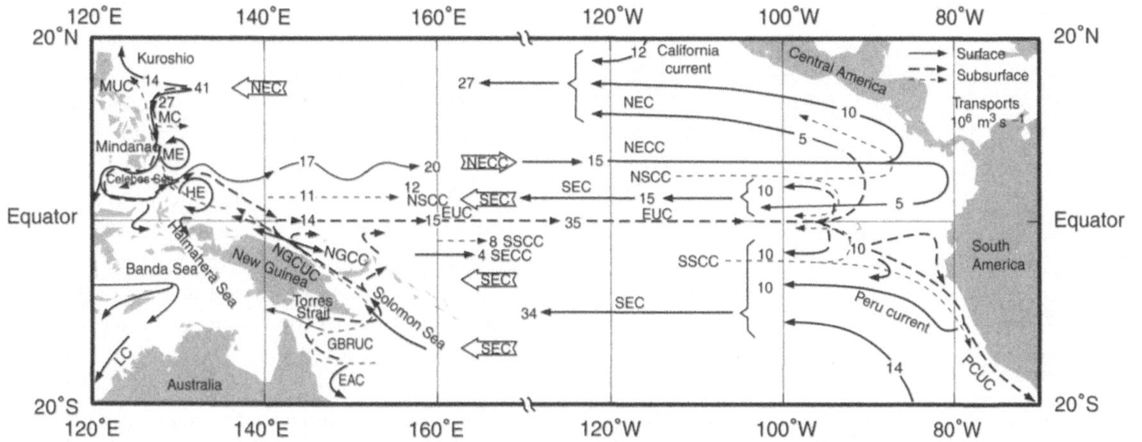

Figure 1 Schematic illustration of the major equatorial currents in the Pacific Ocean and their connections to eastern and western boundary currents. (Note the break in the longitude axis.) Surface currents are indicated by solid lines; subsurface currents are indicated by dashed lines, with deeper currents having lighter weight. Approximate average individual current transports (10^6 m^3 s^{-1}) are provided where known with some confidence. The equatorial surface currents are the North Equatorial Current (NEC), the South Equatorial Current (SEC), the North Equatorial Countercurrent (NECC), and the South Equatorial Countercurrent (SECC). The subsurface equatorial currents discussed here are the Equatorial Undercurrent (EUC), the Northern Subsurface Countercurrent (NSCC), and the Southern Subsurface Countercurrent (SSCC). Eastern boundary currents are the Peru Current and the Peru–Chile Undercurrent (PCUC). Western boundary surface currents are the Kuroshio, the Mindanao Current (MC), the New Guinea Coastal Current (NGCC) and the East Australia Current (EAC). Subsurface flows along the western boundary are the Mindanao Undercurrent (MUC), the New Guinea Coastal Undercurrent (NGCUC), and the Great Barrier Reef Undercurrent (GBRUC). The Mindanao Eddy (ME) and Halmahera Eddy (HE) are indicated.

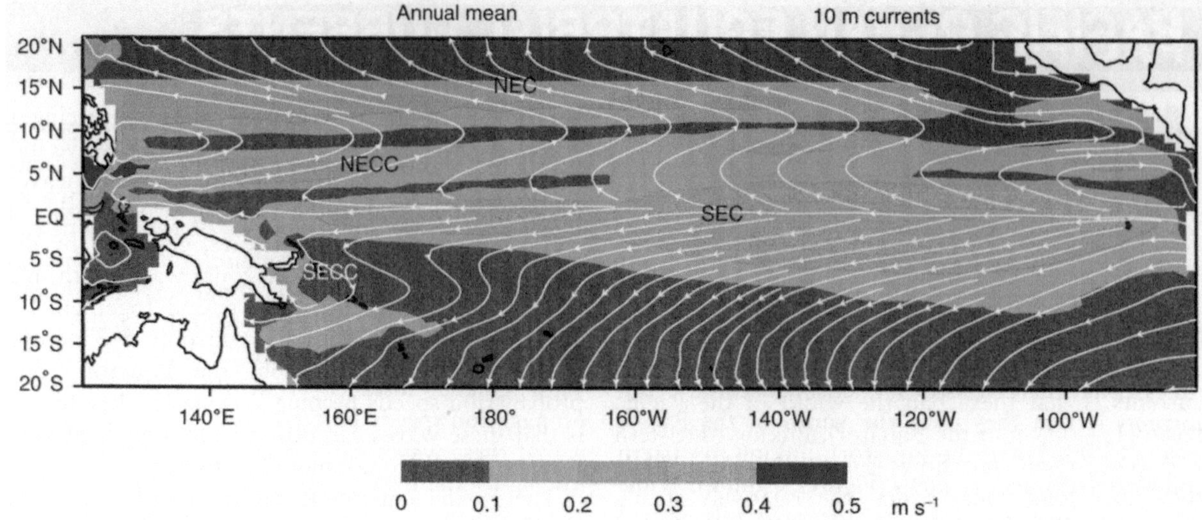

Figure 2 Map of long-term mean surface flow in the tropical Pacific. White lines and arrows indicate the direction of flow, while the colors indicate the speed of flow as given by the color bar. Current names are abbreviated as in **Figure 1**.

equatorial circulation is forced by winds and buoyancy forces (surface heat and fresh water fluxes) far from the equator.

The basic spatial structures and temporal variation of Pacific Ocean equatorial currents are presented here. Because systematic current measurements are available at only a few, widely separated locations, spatial variability is addressed primarily with the ocean assimilation/reanalysis from the US National Oceanic and Atmospheric Administration (NOAA) National Centers for Environmental Prediction, which combines a general circulation model of the ocean with atmospheric and ocean data to estimate the state of the tropical Pacific Ocean every week. Direct current measurements from several sites along the Equator maintained by the NOAA Pacific Marine Environmental Laboratory are used primarily to address temporal variability.

Mean Flow

Interior Flows

Zonal geostrophic flow, where meridional pressure gradients are balanced by Coriolis forces (due to flow on the rotating earth), dominates over meridional flow in the long-term annual-average Pacific equatorial circulation. In the surface layer (upper 50 m or so), currents directly driven by the generally westward Trade Winds typically flow poleward, superimposed on these strong zonal flows (**Figure 2**). The divergence of poleward-flowing surface currents is most pronounced along the Equator, leading to depth-dependent pressure gradients and strong vertical flow (called equatorial upwelling).

Surface currents The time-averaged surface flows (**Figure 2**) are dominated by the westward South Equatorial Current (SEC; between about 3°N and 20°S) and the North Equatorial Current (NEC; between about 10°N and 20°N). A persistent North Equatorial Countercurrent (NECC) flows eastward across the basin in the narrow band between about 5°N and 10°N. A weaker eastward-flowing South Equatorial Countercurrent (SECC) extends eastward from the region of the western boundary, but this flow only intermittently reaches the central and eastern Pacific.

North–south profiles of surface currents at three different longitudes across the Pacific basin clearly show the structure of these major zonal flows (**Figure 3**). The NEC, NECC, and SEC are strongest in the central Pacific where the Trade Winds are strongest. The SEC and NECC are considerably weaker in the west than in the east, reflecting the greater variability of the winds in the western Pacific. Very near the Equator in the central and eastern Pacific, there is a minimum in the speed of the SEC, and the relatively narrow filament of SEC north of the equator is stronger than the flow south of the Equator, except in the west. The SECC occurs between about 3°S and 10°S, but generally dissipates west of the dateline.

The meridional flows are considerably weaker than the zonal flow (**Figure 3**). The average currents have a northward component north of the Equator, and southward to the south of the Equator. The transition occurs rapidly very close to the Equator at all three longitudes, due to east-to-west trade winds and the change in sign of the Coriolis force at the

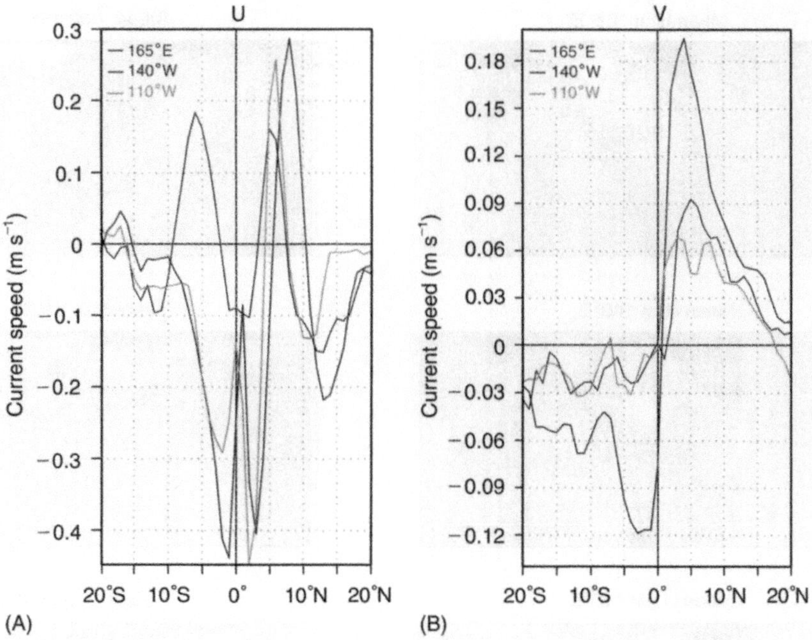

Figure 3 North–south profiles of zonal (A) and meridional (B) current in the western (165°E), central (140°W) and eastern (110°W) Pacific. Positive current is eastward and northward. Note the difference in the velocity scales between the two panels.

Equator. This divergence causes equatorial upwelling, which is responsible for colder surface temperatures along the Equator than just to the north or south.

Subsurface structure The zonal flow of the NECC is unusual among the surface currents in that it has a subsurface maximum near 50 m (**Figure 4**). This is due to its flowing eastward against the prevailing Trade winds. The directly wind-driven flow vanishes below the mixed layer (usually shallower than 100 m), and currents below are zonally-oriented except near the eastern and western boundaries.

The time-averaged subsurface flows are dominated by the eastward Equatorial Undercurrent (EUC; **Figures 4–6**), which is the strongest equatorial Pacific current, reaching speeds of about 1 m s^{-1} between 120°W and 140°W (**Figures 5** and **6**). The EUC is found within the very strong equatorial thermocline just below the westward SEC (**Figures 4** and **5**). The strong mean shear above the core of the EUC gives rise to strong vertical mixing. Note also the strong meridional shear of the zonal flow on either side of the EUC, and between the SEC and NECC.

Much weaker Northern and Southern Subsurface Countercurrents (NSCC and SSCC) flow eastward below the poleward flanks of the EUC (**Figure 4**). The westward Equatorial Intermediate Current (EIC) is found directly below the EUC across the Pacific. Both the EUC and EIC slope upward toward the east (**Figure 5**), tending to follow shoaling isopycnal surfaces. On the Equator, alternating deep equatorial jets (not shown) are found below the EIC.

The poleward wind-driven meridional flows are mostly confined to the upper 50 m (**Figure 4**). The central Pacific section (**Figure 4E**) shows meridional flow nearly symmetric with respect to the equator, with divergent poleward flow in the near-surface, and convergent equatorward flow near the core of the EUC. This classical picture is not seen in the eastern and western sections, due to a cross-equatorial component of the surface winds, especially in the east.

Boundary Flows

Surface The westward flow of the NEC impinges on the Philippines where it splits near 14°N into a northward-flowing Kuroshio Current and southward Mindanao Current (MC) along the western boundary (**Figures 1** and **2**). The MC has surface speeds exceeding 1 m s^{-1}, and the flow reaches to depths of 300–600 m. The upper 100 m flow splits at the south end of Mindanao Island, with a significant portion entering the Sulawesi Sea, and most of the rest retroflecting into the NECC. A small portion recirculates around the persistent Mindanao Eddy. A portion of the flow in the Sulawesi Sea transits through the Makassar Strait and ultimately through the Indonesian Seas and into the Indian Ocean, while the rest returns to the

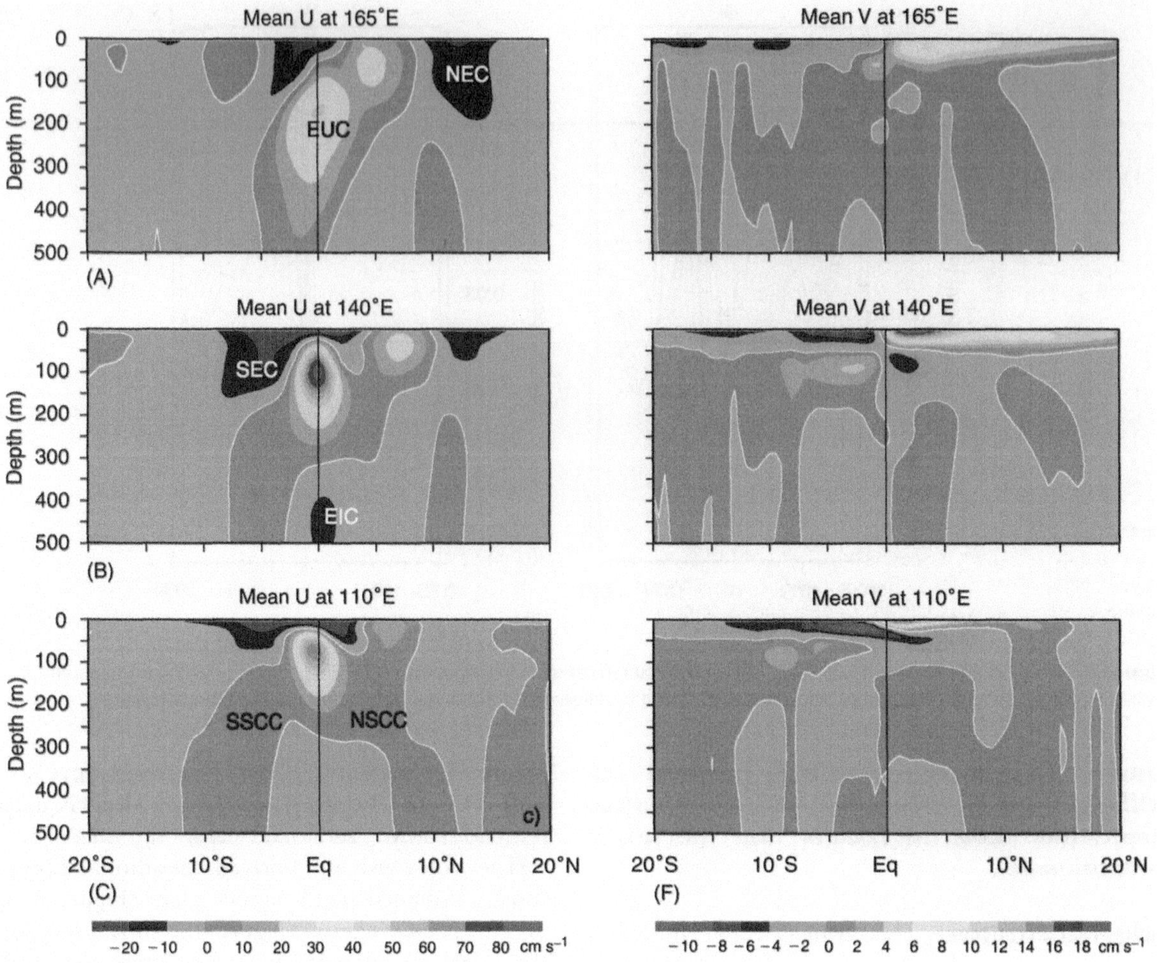

Figure 4 Vertical section showing the zonal (A–C) and meridional (D–F) components of mean flow in the upper 500 m of the western (A, D), central (B, E) and eastern (C, F) Pacific Ocean. Color bars give magnitude of flow (note that zonal and meridional scales are not the same); positive values are eastward and northward. Zero values are indicated by white contours.

Pacific to join the NECC (**Figure 6**). Deeper portions of the MC are also split between the Indonesian Throughflow and the Pacific equatorial circulation, with the latter flowing into the EUC and NSCC.

The SEC impinges on the north-eastern coast of Australia and the complex of islands including New Guinea. Similarly to the NEC, it splits into poleward and equatorward boundary flows near 14°S (**Figure 1**). The poleward branch is the East Australia Current, and the northward branch flows under a shallow southward surface flow as the Great Barrier Reef Undercurrent, eventually becoming the New Guinea Coastal Current (NGCC) and New Guinea Coastal Undercurrent (NGCUC), which follow a convoluted path around topographic features ending up with westward flow along the north coast of New Guinea. In this region, the surface flow of the NGCC reverses seasonally with the Asian winter monsoon westerly winds.

Low-latitude boundary currents in the eastern Pacific are not nearly as strong as along the western boundary, but they play a significant role in closing the circulation of the Pacific Ocean (**Figure 1**). The Peru Current flows northward along the west coast of South America, ultimately turning offshore into the SEC. North of the Equator, the NECC flows into the Gulf of Panama and retroflects around the Costa Rica Dome into the NEC, joining southward flow from the California Current.

Subsurface Along the western Pacific boundary (**Figure 1**), the NGCUC flows westward along the north coast of New Guinea with a maximum near 200 m, but extending to at least 800 m. The upper thermocline waters contribute a small fraction to the Indonesian Throughflow, with the rest retroflecting around the persistent Halmahera Eddy to form (with contributions from the MC) the

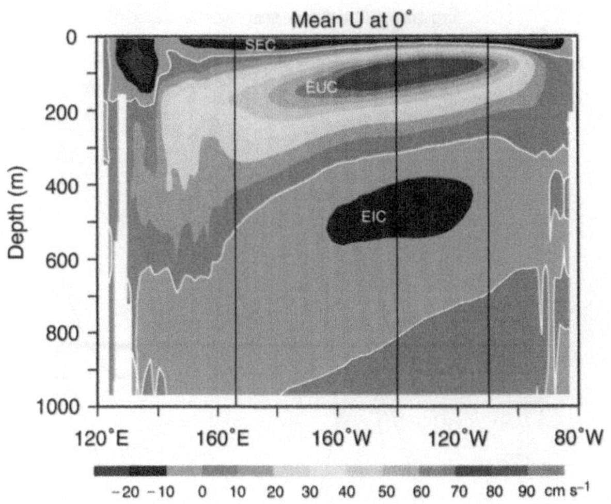

Figure 5 Zonal flow speed in the upper 1000 m in a vertical section along the Equator. Speeds are given in the color bar, with positive values eastward. The vertical lines indicate the longitudes of long-term current meter measurements (see **Figure 7**).

eastward-flowing EUC (**Figure 6**). The deeper portions of the NGCUC flow across the Equator and are traced into the weak Mindanao Undercurrent (MUC) which flows northward below the MC, carrying Antarctic Intermediate Water into the North Pacific.

In the east, the Peru–Chile Undercurrent flows poleward along the west coast of Ecuador, Peru and Chile, basically an extension of the EUC past the Galapagos Islands that then turns southward after converging at the coast of Ecuador. Some of these waters join the westward flows of the Peru Current and SEC through upwelling. The fate of waters flowing eastward in the NSCC and SSCC is not well known.

Variability

The variability of equatorial currents is complex, spanning a broad range of time and space scales. This variability is largely forced by changing winds; an important fraction of these wind changes are due to sea surface temperature changes in the equatorial zone, these being associated with changes in the currents and winds. These coupled variations include the well-known El Niño phenomenon, but also include the annual cycle.

Annual cycle

Although the annual cycle of wind forcing near the equator is not as extreme as in mid latitudes, the Pacific Trade Winds vary enough to force significant changes to the currents discussed above. Because of equatorial wave dynamics and coupling with the atmosphere, the relationship of the current variability to the winds is quite complex.

Figure 7 shows the long-term mean plus annual cycle of zonal and meridional current in the near-surface layer and at depth within the EUC for locations in the western, central and eastern Pacific. (Here, the annual cycle is the sum of annual and semiannual harmonics analyzed from direct current measurements.) The range of zonal current variation is about one order of magnitude larger than the meridional variations. The annual harmonic dominates the annual cycle of surface flow, except in the western Pacific where monsoon winds cross the

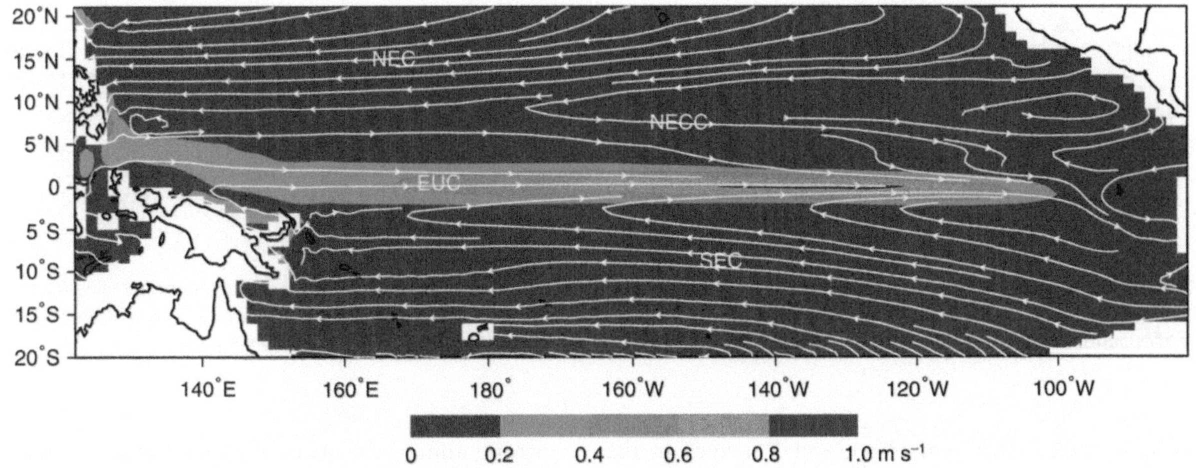

Figure 6 Map of long-term mean flow in the tropical Pacific on the isopycnal surface $\sigma_\Theta = 24.5\ \mathrm{kgm^{-3}}$, which lies within the high-speed core of the Equatorial Undercurrent. White lines and arrows indicate the direction of flow, while the colors indicate the speed of flow as given by the color bar. Current names are abbreviated as in **Figure 1**.

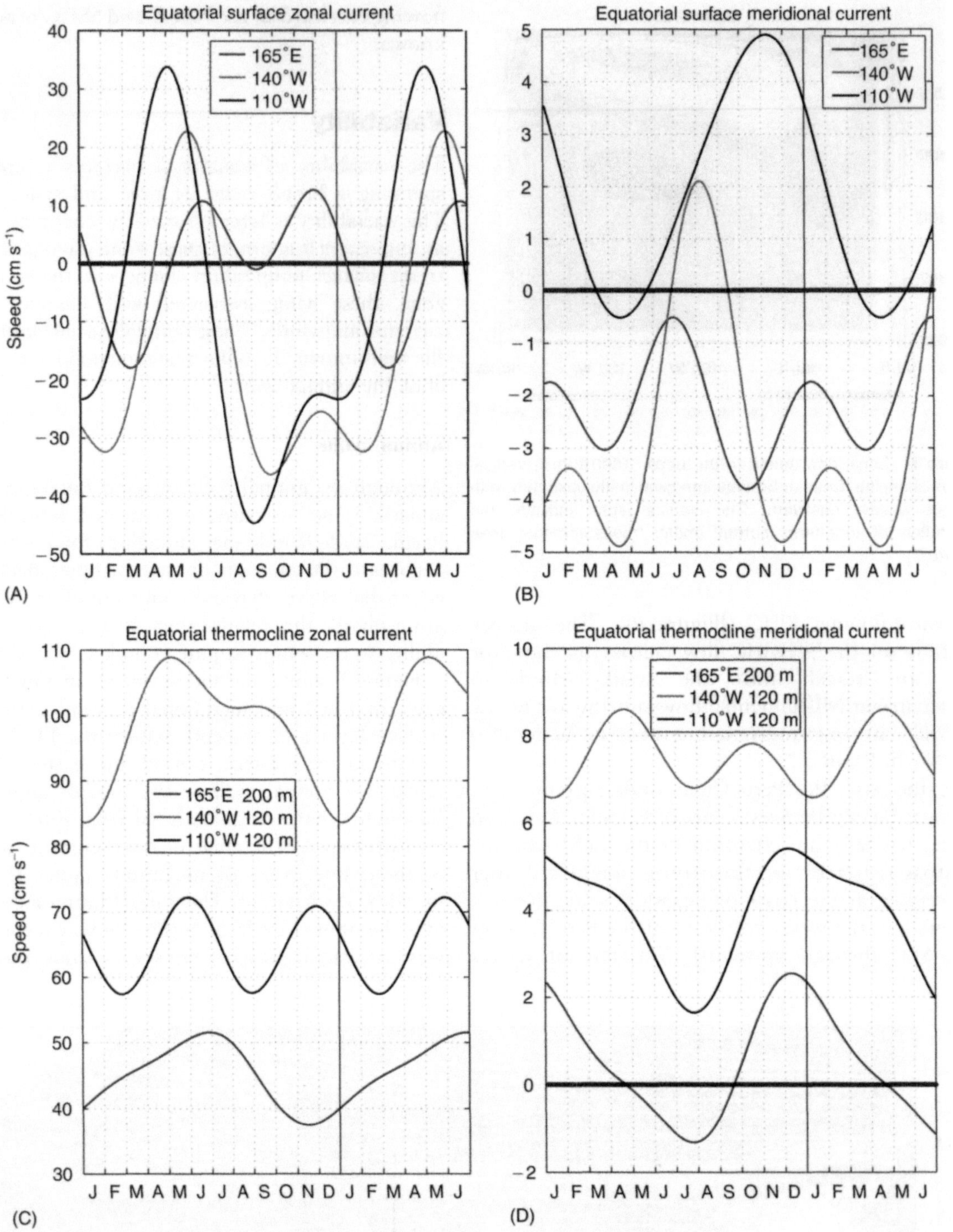

Figure 7 Mean annual variation of zonal (A, C) and meridional (B, D) currents at 10 m depth (A, B) and in the thermocline (C, D) on the Equator at three locations across the Pacific Ocean indicated in **Figure 5**. Because the thermocline is deeper in the western Pacific, currents are presented for a corresponding depth there. Note the different speed scales for each panel.

equator twice each year. Also, zonal current in the eastern equatorial thermocline and meridional current in the central equatorial thermocline show strong semiannual signals.

Strong annual variation of zonal surface current is sufficient to reverse the direction of the flow on the equator, especially during the Northern Hemisphere spring (**Figure 7A**). In the east, this feature occurs nearly every year, and has been erroneously described as a 'surfacing' of the EUC. In the central Pacific, such reversals are mainly observed during strong El Niño events, and its appearance in the annual cycle here may be due to the occurrence of several El Niño events during the record that was analyzed (1984–1998). It is noteworthy that the maximum eastward deviation of the annual cycle appears progressively later toward the west, thought to be coupled with westward propagation of the annual cycle of zonal wind and sea surface temperature.

The annual cycle in the strong eastward flow of the EUC within the thermocline (**Figure 7C**) is not large enough to reverse the current direction. (On interannual timescales, however, the flow may reverse—see below.)

El Niño/Southern Oscillation (ENSO)

The strongest variability of the zonal equatorial currents is associated with El Niño episodes that occurred in 1986–87, 1990–91, 1993, and 1997–98, and with La Niña episodes of 1984, 1988, and 1996. Current meter records that have had their mean and annual cycles removed are presented in **Figure 8**, showing that El Niño variations are large enough to reverse the westward flow of the SEC, especially in the western equatorial Pacific. Strong eastward surface flow in the warm water pool of the western equatorial Pacific (e.g., 1997) has been implicated in the warming of the sea surface in the central and eastern equatorial Pacific. Also, the eastward flow of the EUC is reversed during some of these events (e.g., in **Figure 8B** at 140°W during 1997). This disappearance of the EUC was first observed during the strong 1982–83 El Niño event.

The current systems off the Equator are also affected by ENSO, again through a combination of local wind forcing and remotely-forced baroclinic waves. The NECC strengthens in the early phases of El Niño. The off-equatorial portion of the SEC weakens during El Niño, and strengthens during La Niño. The interannual variations of the NEC show a

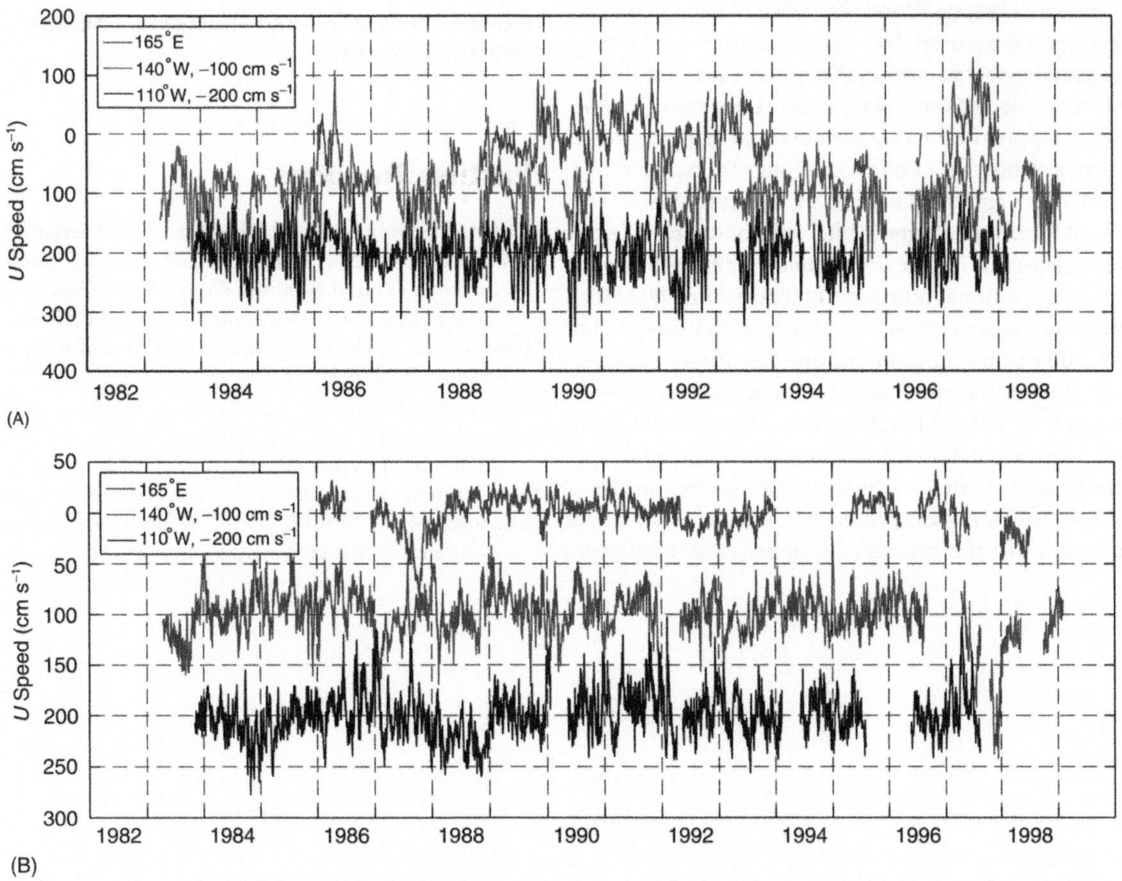

Figure 8 Time series of observed zonal currents, with mean and annual cycle removed, at three locations (indicated in **Figure 5**) along the Equator in the surface layer (A) and in the thermocline (B). Warm El Niño events are indicated by red shading along the time axis; blue shading indicates cold La Niño events. Missing observations are indicated by gaps. Note that time series have been offset from each other for clarity.

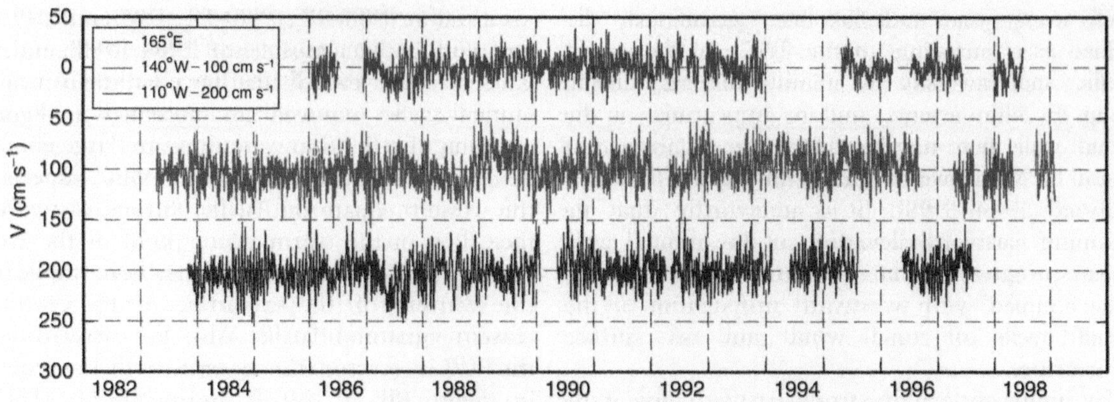

Figure 9 Time series of observed surface layer meridional currents, with mean and annual cycle removed, at three locations along the Equator. Missing observations are indicated by gaps. Note that time series have been offset from each other for clarity.

correlation with the NECC, but there are also strong variations with longer timescales than ENSO.

High-frequency current variability

Equatorially trapped waves with periods of a few days to a couple of months are quite energetic and ubiquitous (**Figures 8** and **9**). Zonal current fluctuations are dominated by intraseasonal fluctuations associated with the atmospheric intraseasonal (30–60 days) oscillation; eastward-propagating equatorially trapped Kelvin waves play an important role in transmitting this variability from the western Pacific to the eastern Pacific (e.g., **Figure 8A** during 1997). Meridional current fluctuations tend to have more energy at higher frequencies than the zonal current variations (compare **Figures 8** and **9**). Here, dominant periods are in the range 20–30 days. The amplitude of meridional current variability is largest at the surface in the east, and becomes somewhat smaller in the central Pacific, and much smaller in the west. The dominant mechanism is the tropical instability wave, which arises from the strong shears of the zonal flows discussed earlier. Modulation of the amplitudes of these waves occurs seasonally (largest amplitude in the northern fall) and interannually (small amplitude during El Niño) as the shears are affected by the annual cycle and ENSO.

See also

Ekman Transport and Pumping. Kuroshio and Oyashio Currents.

Further Reading

Godfrey JS, Johnson GC, McPhaden MJ, Reverdin G, and Wijffels S (2001) The tropical ocean circulation. In: Siedler G and Church J (eds.) *Ocean Circulation and Climate*, pp. 215–246. London: Academic Press.

Hastenrath S (1985) *Climate and Circulation of the Tropics*. Dordrecht: D. Reidel.

Lukas R (1986) The termination of the Equatorial Undercurrent in the Eastern Pacific. *Progress in Oceanography* 16: 63–90.

Neumann G (1968) *Ocean Currents*. Amsterdam: Elsevier.

Philander SGH (1990) *El Niño, La Niña, and the Southern Oscillation*. San Diego: Academic Press.

SEAS OF SOUTHEAST ASIA

J. T. Potemra and T. Qu, SOEST/IPRC, University of Hawai'i, Honolulu, HI, USA

Introduction

The Southeast Asian Seas lie between the western Pacific and eastern Indian Oceans (**Figure 1**). To the southeast, they are bounded by the Lesser Sunda Islands of Indonesia, but there are some gaps between these islands allowing for exchange of ocean waters between the SE Asian Seas and the Indian Ocean. Similarly to the east, these marginal seas meet the Pacific.

The main basins of the SE Asian seas are the South China Sea, the Celebes Sea (sometimes referred to as the Sulawesi Sea), and the Banda Sea. These basins all have depths exceeding a few thousand meters. The Sulu Sea, between the South China Sea and the Celebes Sea, is also deep, but the connections with the two adjacent seas are limited by islands and shallow sills and thus exchange of ocean water in and out of the Sulu Sea is probably limited. The Flores Sea, South of Sulawesi, also has a deep center, but in the case of the Flores Sea the connection to the Banda Sea is not very constrained. The Sulu and Flores Seas, as well as the numerous other seas in the region, all play a vital role in the region in terms of fisheries and commerce, but a detailed description of these seas is beyond the scope of this article.

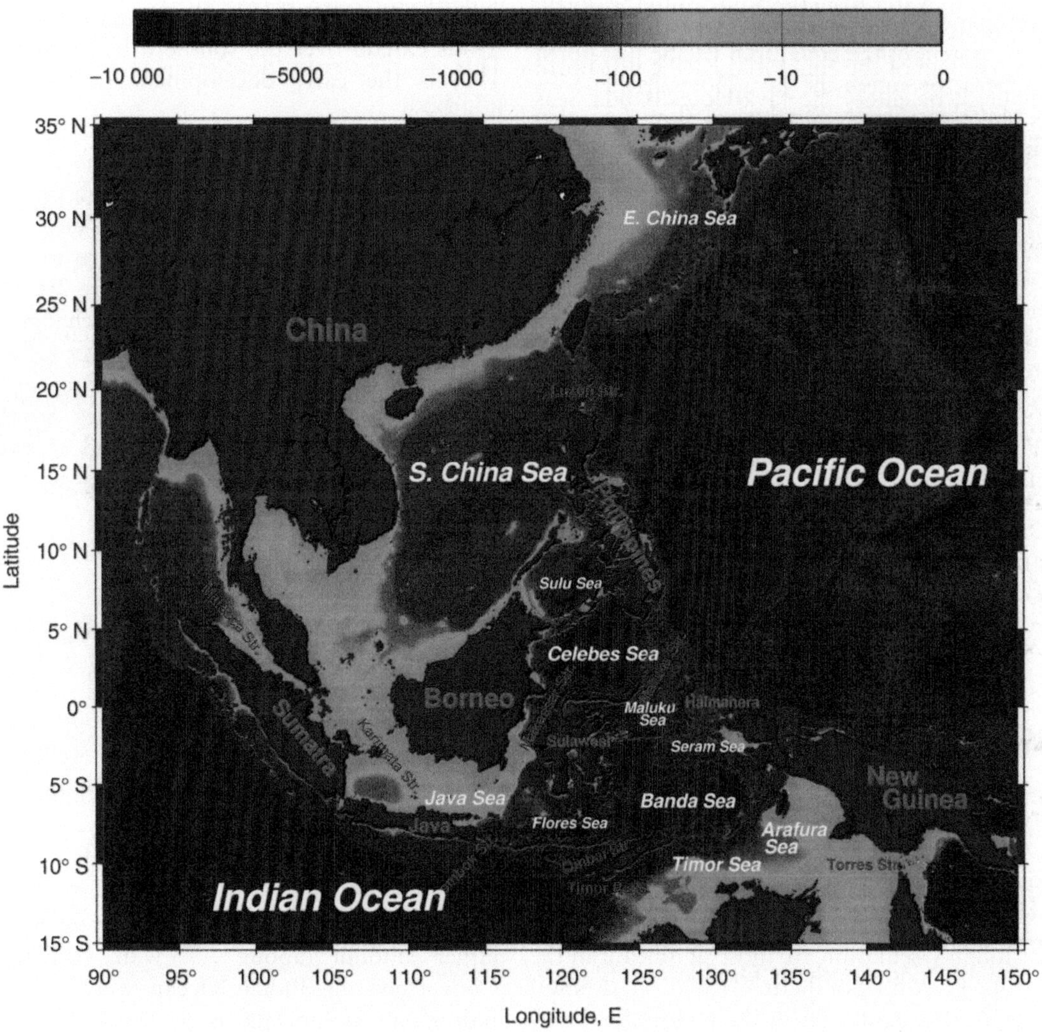

Figure 1 The Southeast Asian Seas. The color shading indicates ocean depth in meters, with scale given at the top.

The South China Sea is the largest of the SE Asian seas. The southern part is the shallow Sunda Shelf; south of approximately 5° N and west of about 110° E the water depths are less than 100 m. The central and northern parts of the basin have depths greater than 4000 m. The sea connects to the Pacific Ocean through the Luzon Strait between the Philippine Islands and Taiwan. To the north the South China Sea connects to the East China Sea via the Taiwan Strait, while to the south it connects to the Java Sea via the Karimata Strait and to the Sulu Sea via the Mindoro Strait. The Malacca Strait, between Sumatra and Malaysia, provides a connection to the Indian Ocean. This strait, however, is extremely narrow and net interbasin flow is probably small.

The most direct pathway of Pacific waters in the SE Asian seas is through the Celebes Sea. This sea is open to the Pacific on its eastern side, with a string of islands between Mindanao (Philippines) and Halmahera prohibiting the exchange. The Celebes Sea also receives water from the Sulu Sea in the north. The water that exits the Celebes Sea either passes into the current system of the equatorial Pacific just north of Halmahera, or enters the Java/Flores/Banda Seas via the Makassar Strait. Southward flow has also been observed through the Maluku Strait (east of Sulawesi), but this water probably originates in the Pacific and does not enter the Celebes Sea.

The Banda Sea, situated between New Guinea and Sulawesi, is the main region where uniquely identifiable Indonesian Sea Water (ISW, also call Banda Sea Water, BSW) is formed. This sea is extremely deep in the eastern part (called the Weber Deep), and connects to the Indian Ocean via the Ombai Strait and Timor Passage. While these two passages are narrow, their sills are quite deep, thus allowing for an exchange of deep waters between the Banda Sea and the Indian Ocean. The final strait through which significant exchange with the Indian Ocean (about 25% of the total) takes place is the Lombok Strait, between Bali and Lombok.

Surface Forcing

The SE Asian Seas are located in an extremely dynamic region that is under the direct influence of forcing on a variety of timescales. The main variability occurs due to the annual change of the monsoons. However, intraseasonal variations from the Madden Julian Oscillation are most significant in the region. The El Niño–Southern Oscillation (ENSO) variability is also significant in the region due to its proximity to the western Pacific warm pool. Here we will focus the discussion on the mean seasonal cycle.

Winds

As with most regions in the world oceans, surface wind forcing provides the greatest influence on ocean circulation. In January the northeast monsoon is fully developed. A high pressure is formed over Asia, bringing strong winds from the northeast over the South China Sea (**Figure 2(a)**). At 5° N the winds take a more northerly turn. Winds blow almost due south over the Karimata Strait. Further east, winds are weak, but still blow toward the south across the equator. South of the equator the winds turn to the east, forming the northwest monsoon over Indonesia. The equatorial trough (line of low pressure) lies along 10° S, south of which are the southeast trades.

Later in the year, the monsoon winds relax, and the equatorial trough moves north. By March (**Figure 2(b)**), the southeast trades of the Indian Ocean extend north and east, while northeast winds prevail over the Timor and Arafura Seas. In April the equatorial trough is over the equator and southeast winds extend over the southern Indonesian islands. This is the early development of the southeast monsoon.

By the northern summer months, the southeast monsoon strengthens over the southern Indonesian seas. Low pressure over Asia and high pressure over Australia become maximum. Winds continue southerly across the equator and turn to the northeast over the South China Sea (**Figure 2(c)**).

In September, the winds begin to change dramatically, returning to northeasterlies over the northern South China Sea. Winds near the equator become weak as they return to easterly trades. South of the equator, winds become easterly and southeasterly, leading to a return to the winter monsoon conditions (**Figure 2(d)**).

Surface Heat

The northern spring and fall months show the largest extremes in net surface heat fluxes in the region. During March through May, for example, the northern SE Asian Seas receive about $100\,\mathrm{W\,m^{-2}}$ (**Figure 3(b)**). In northern fall, the southern Indonesian seas are heated by a similar amount (**Figure 3(d)**).

In December through February, the South China Sea is slightly heated in the southern regions near Kalimantan, but the northern part is cooled by a larger amount (about $-75\,\mathrm{W\,m^{-2}}$; **Figure 3(a)**). Conversely, the Indian Ocean water south of the Indonesian archipelago is cooled during June, July, and August, while the seas to the north receive minimal warming (**Figure 3(c)**).

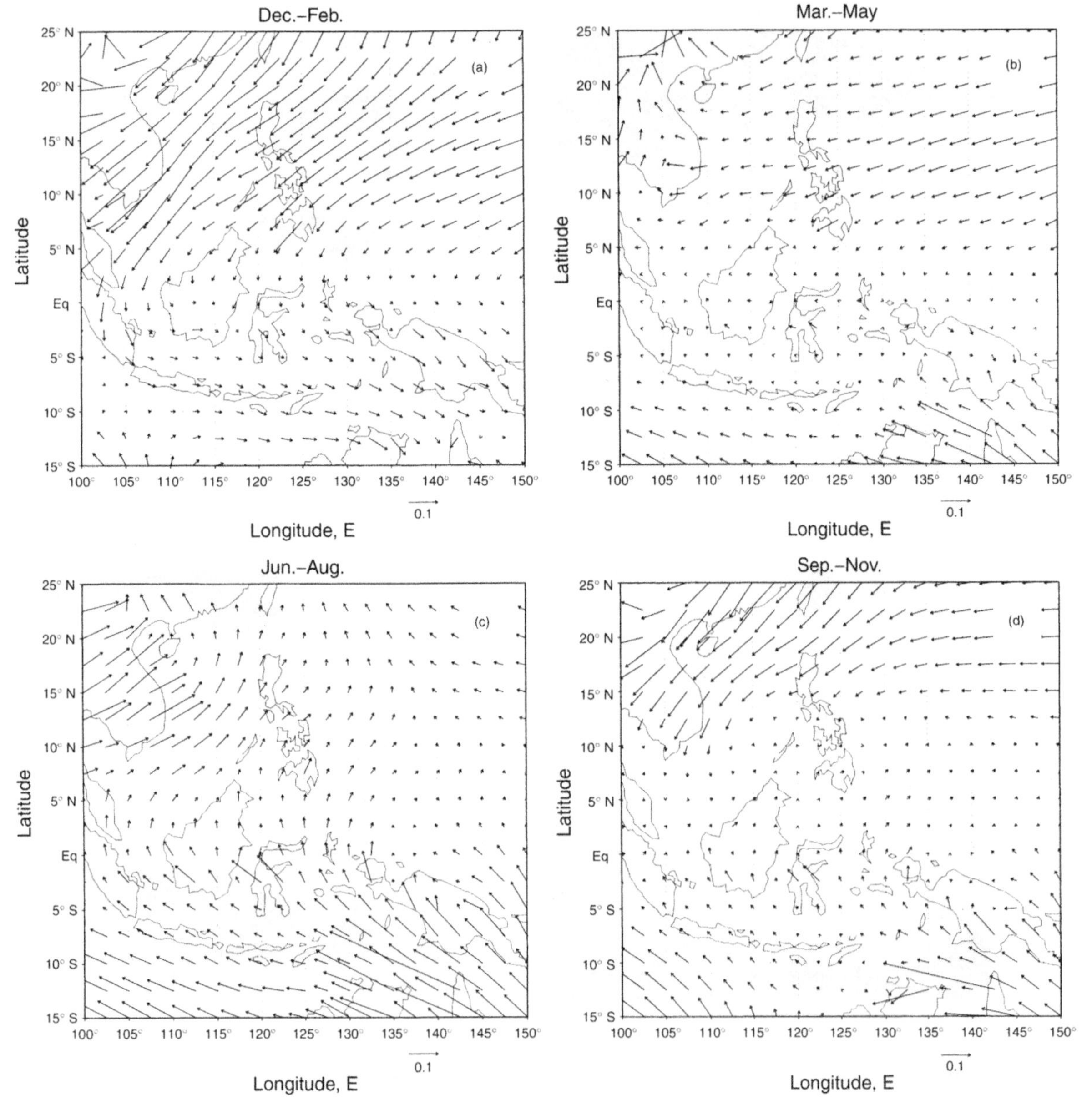

Figure 2 Seasonally averaged wind stress in N m $^{-2}$ from NOAA's National Center for Environmental Prediction (NCEP).

Surface Fresh Water

As with the winds, the precipitation over the SE Asian Seas migrates north and south with the monsoons. Early in the calendar year (**Figure 4(a)**) the center of maximum convection and rainfall is over the southern part of the region, providing over 10 mm d^{-1} (as a monthly mean) over the Arafura, Java, and Flores Seas. In contrast, the northern South China Sea receives very little rainfall during this time.

In March through May, the precipitation is weak over most of the region, with local areas of Papua New Guinea and Kalimantan receiving ~6–8 mm d^{-1}. At the height of the (northern) summer monsoon season, rainfall is maximum over the western Pacific warm pool, extending into the South China Sea (**Figure 4(c)**). The northern South China Sea receives maximum precipitation the following season (**Figure 4(d)**), and the cycle repeats thereafter.

While precipitation amounts are critical for land-based communities, perhaps more important for the ocean circulation and climate is the net freshwater flux, or evaporation minus precipitation ($E-P$). **Figure 5** shows the climatological $E-P$ flux, with negative values indicating an influx of fresh water.

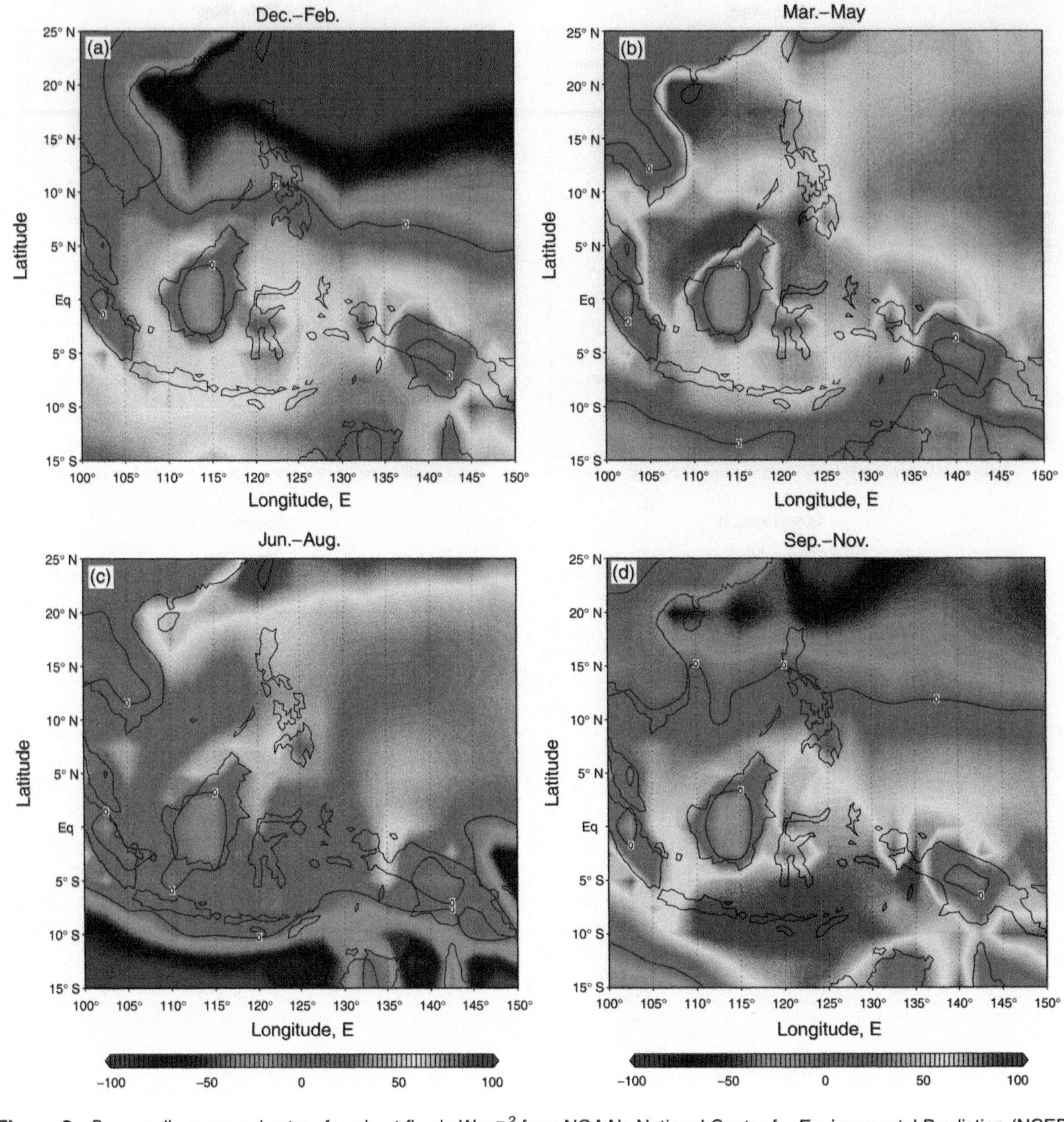

Figure 3 Seasonally averaged net surface heat flux in W m^{-2} from NOAA's National Center for Environmental Prediction (NCEP). Note that red colors indicate heating of the ocean surface.

The South Pacific is a region where evaporation exceeds precipitation throughout the year, and this is a region of high salinity. The opposite is true for the North Pacific, where sea surface salinities are lower. The SE Asian Seas have seasonally modulated $E - P$. As might be expected from **Figure 4**, the South China Sea has more evaporation in September through November, which may lead to higher surface salinities at this time. In the following season, however, the South China Sea has a net fresh water input (**Figure 5(a)**). The Banda Sea gets more fresh water in June through November during the SE monsoon, while in December it has more evaporation. These surface fluxes, therefore, lead to changes in surface salinities (discussed in the next section). When compared to actual surface salinities, one can infer upper ocean circulation patterns.

Oceanic Response to Surface Forcing

Not surprisingly, the sea surface temperature (SST) and sea surface salinity (SSS) are directly related to surface heat and freshwater fluxes. Nevertheless, it is useful to see how these properties change throughout the year. More importantly, these surface values are

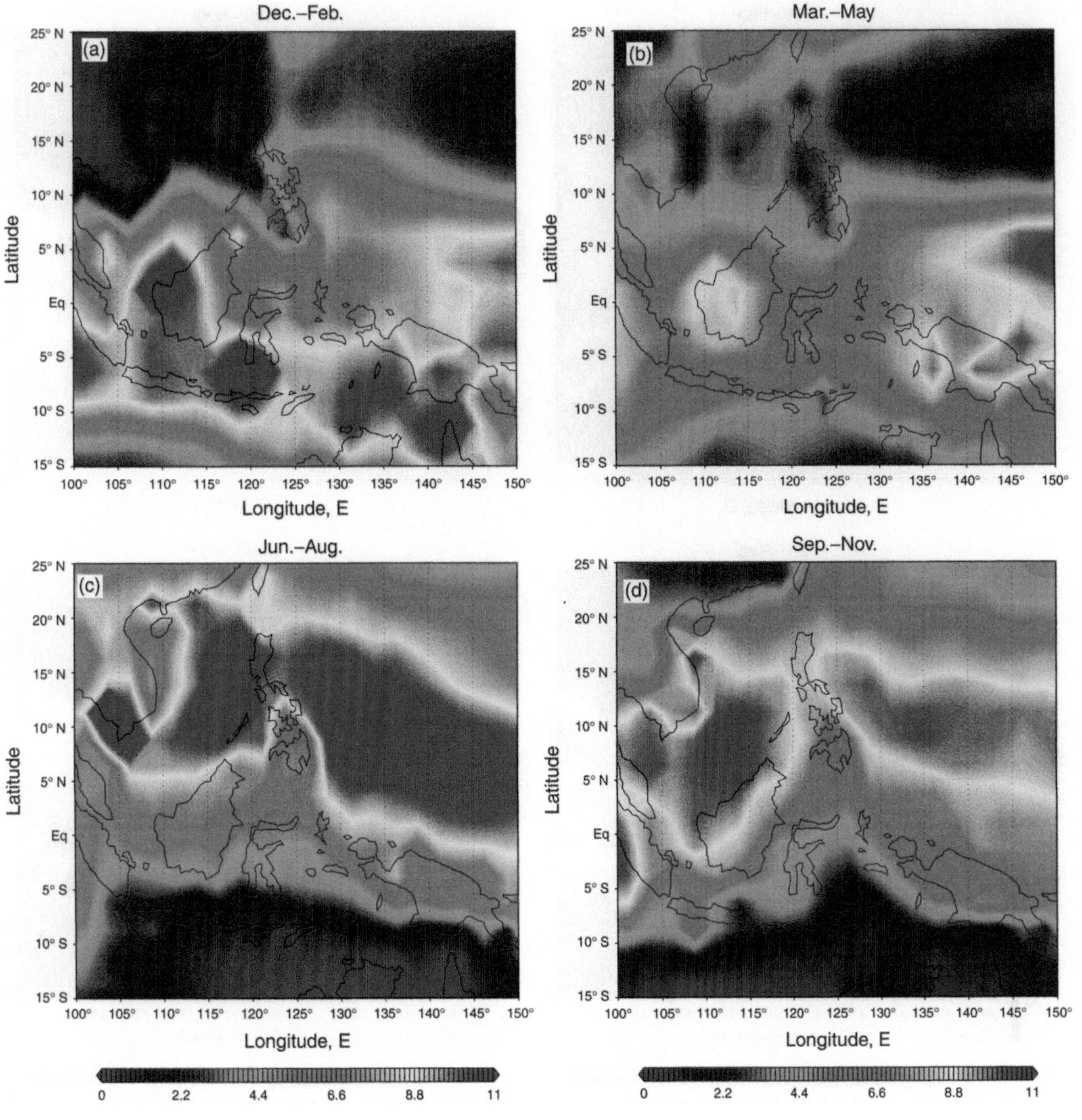

Figure 4 Seasonally averaged rainfall from the CPC merged analysis of precipitation (CMAP) in mm d^{-1}.

linked to atmospheric circulation, and their interannual variability, described later, could have implications for larger-scale climate. Likewise, upper ocean circulation can be proportional to surface winds. However, in this region the numerous straits and islands disrupt the flow, and remotely generated waves and circulation can add to the discrepancy between winds and oceanic flows. These surface quantities are described in more detail in this section.

Surface Temperature and Salinity

The SE Asian Seas are just to the west of the western Pacific warm pool and have the highest mean SSTs in the world oceans (>30° C). Within some of the deeper seas, vigorous upwelling and tidal mixing reduce the SSTs to a certain degree. **Figure 6(a)** shows the mean Dec.–Feb. SSTs in the region. The warm pool is at its southern extreme in this season, and the 28° C isotherm has a northern extent to about 10° N. Highest temperatures, reaching 30° C, are observed in the eastern Arafura Sea. SSTs in the South China Sea exhibit a east–west gradient, with temperatures on the northwestern side between 25 and 26° C while those on the southeastern side are closer to 28° C. This is due to a swift western boundary current (described later) as well as the enhanced northeasterly wind in the region that brings cold water south.

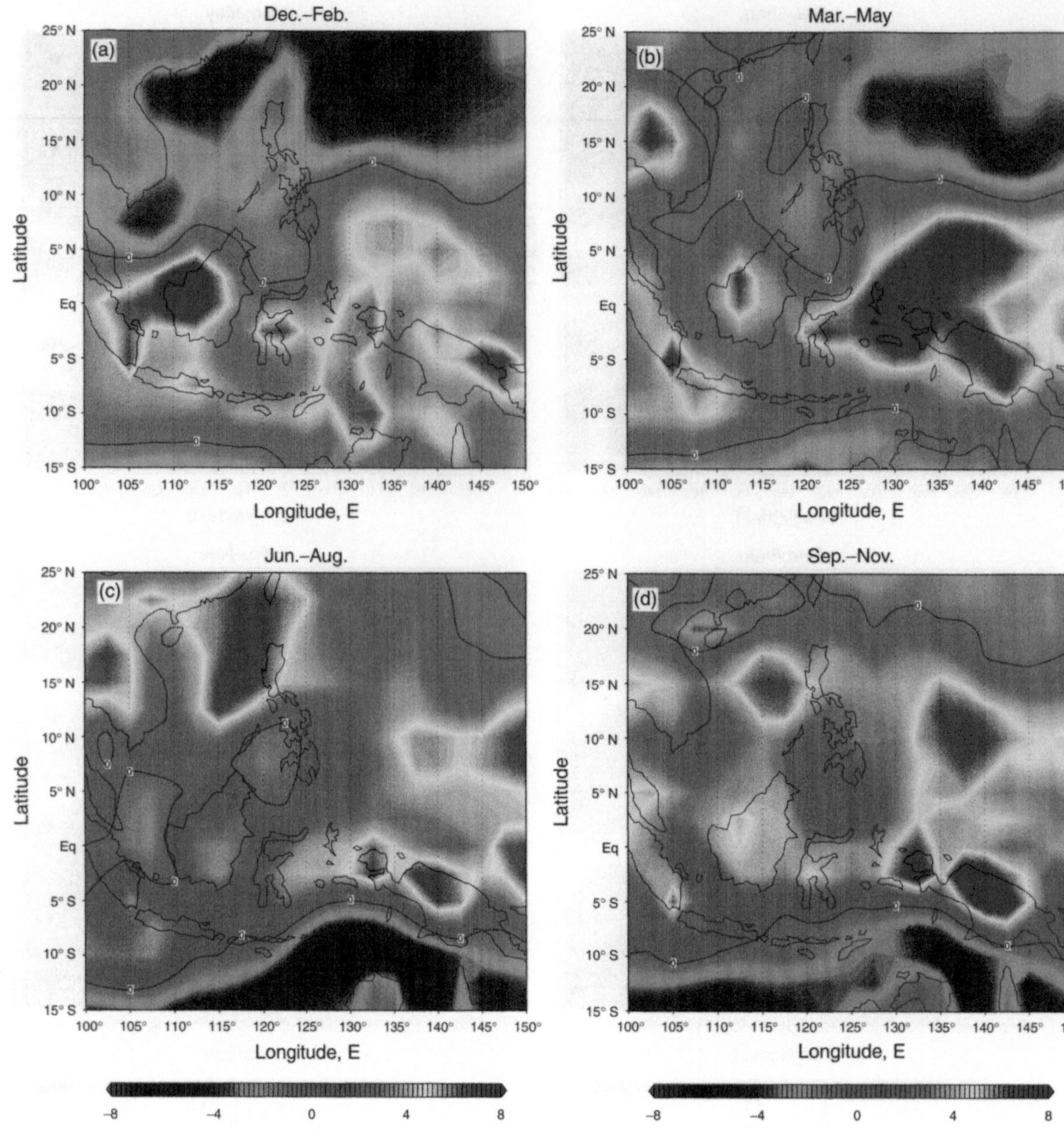

Figure 5 Seasonally averaged evaporation minus precipitation from NCEP reanalysis.

In the following season (**Figure 6(b)**) the cold tongue of water along the western boundary of the South China Sea retreats, and the 28° C isotherm extends from 10° to 15° N in the basin. Also at this time, SSTs in the warm pool and Makassar Strait begin to warm. The western South China Sea cold tongue remains apparent in June through August (**Figure 6(c)**), where temperatures along the coast are 28° C and below, while those in the center of the basin exceed 29° C. This is also the season when the warm pool reaches its northern extreme, and the 28° C isotherm is along 25° N. SSTs south of the equator, for example in the Arafura Sea and south of Java, are minimum at around 26° C. This is also a time when the SST gradients are greatest in the Banda Sea. The warm pool retreats to the equator in September, October, and November (**Figure 6(d)**). The cold western Southern China Sea gets intensified again with the development of the northwest monsoon, and SSTs in the Arafura Sea drop to 27° C.

Just as SSTs are closely linked to the surface heat fluxes, so SSS is linked to the surface freshwater flux. The seasonal changes are not as obvious as the SST (**Figure 7**). Perhaps most dramatic is the signal of low

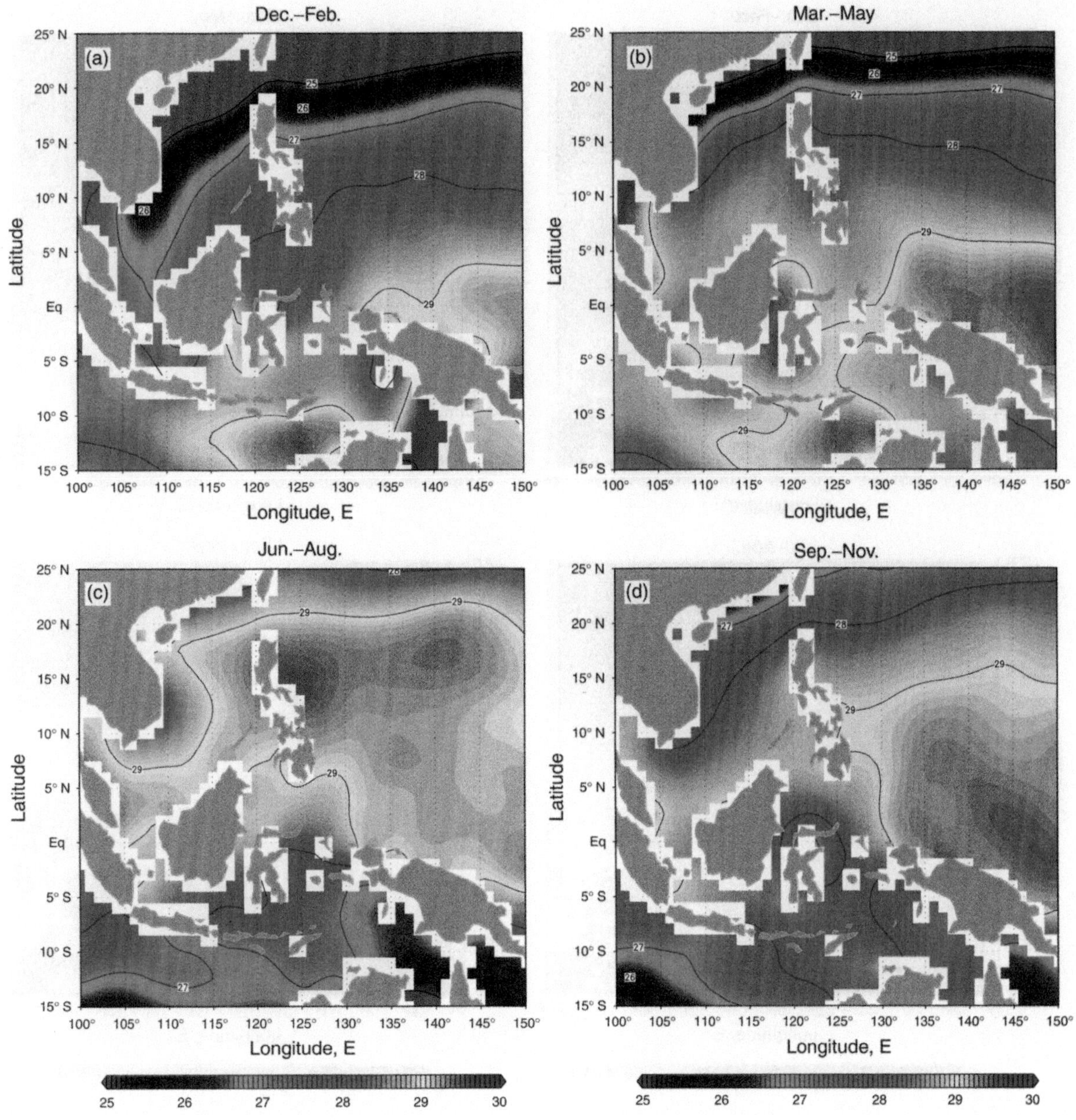

Figure 6 Seasonally averaged sea surface temperature in degrees Celsius from an NOAA climatology.

salinity in the SE Asian Seas compared to the South Pacific. This is primarily due to a large excess in precipitation over evaporation in the SE Asian Seas (Figure 5).

Surface Currents

Measuring the complex oceanic flows in the region is extremely difficult for several reasons. Most existing observations focused on flows in particular locations, and in particular seasons. A complete understanding of the entire current system in the region based on observations is not available. Instead, numerical models can be used to simulate the flows in the SE Asian Seas. The complex geometry of the region, with shallow sills, deep trenches, and intricate coastlines, presents a challenge for numerical models. Mixing, due to tides and interactions with topography, is parametrized in these models, and the results are typically not very accurate. Currents, however, are mainly wind driven (at least at the monthly timescale), and models can do much better

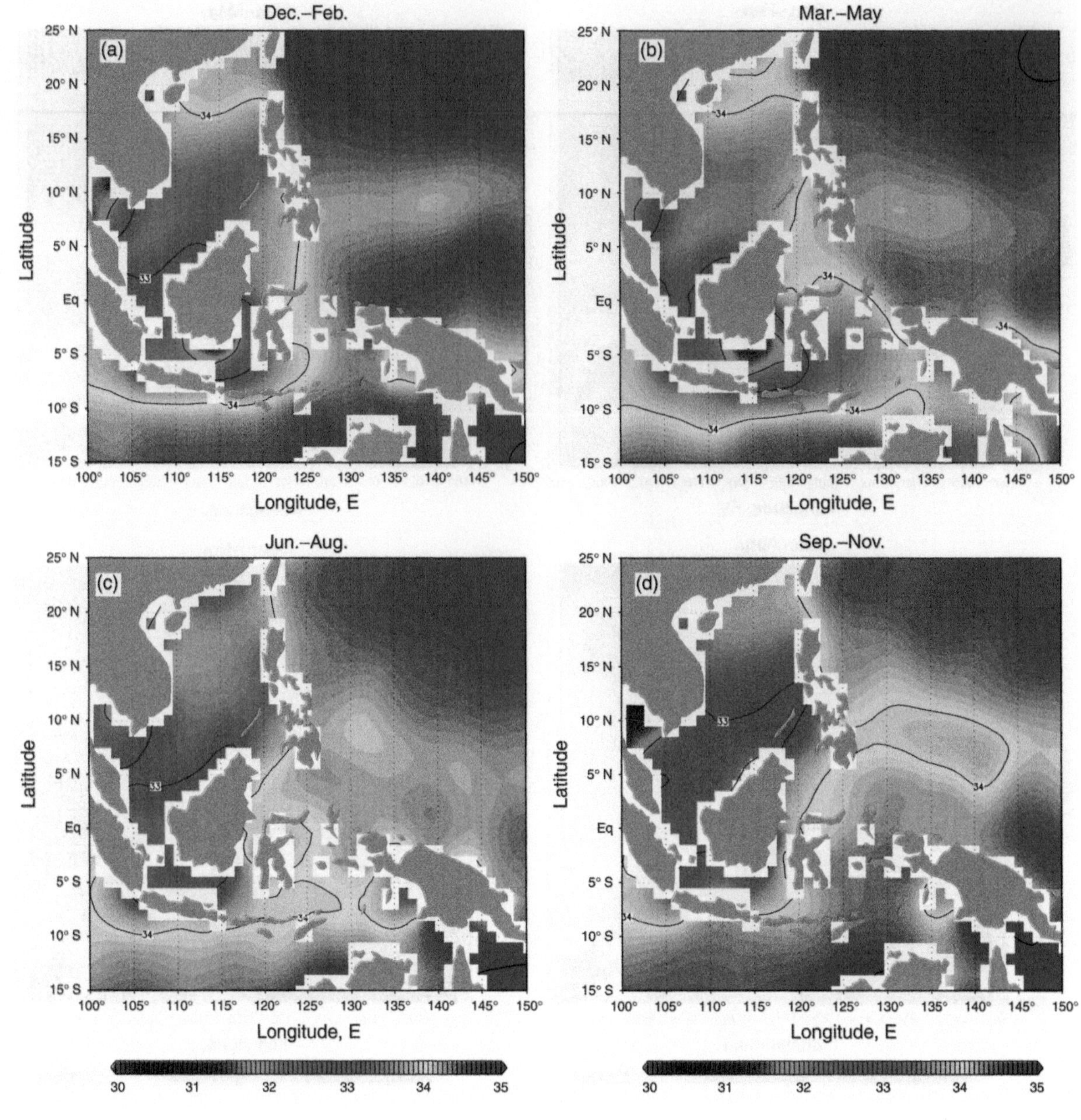

Figure 7 Seasonally averaged SSS from an NOAA climatology.

in simulating these. In the future, long-term, high-vertical-resolution observations will provide the necessary constraints to improve these models.

The mean, large-scale upper ocean circulation is from the Pacific into the South China, Celebes, and Halmahera Seas. From there, water flows into the Java, Flores, and Banda Seas where it gets mixed by tides and other processes. Ultimately the water exits to the Indian Ocean via the Lombok Strait, Ombai Strait, and Timor Passage. Numerical model results typically show this; however, the magnitude, vertical variability, and temporal variability, as well as the exact pathways, are model-dependent. A high-resolution, 1/16° model gives one interpretation (**Figure 8**).

The circulation patterns, as one might expect, are complicated and depend on seasons. The main, upper ocean current system in the western Pacific consists of broad, zonal flows in the interior and intense, narrow western boundary currents. Water from the central Pacific reaches the Philippine coast via the North Equatorial Current (NEC) that flows west between about 8° and 20° N. This current splits at the coast of the Philippines and forms the northward Kuroshio

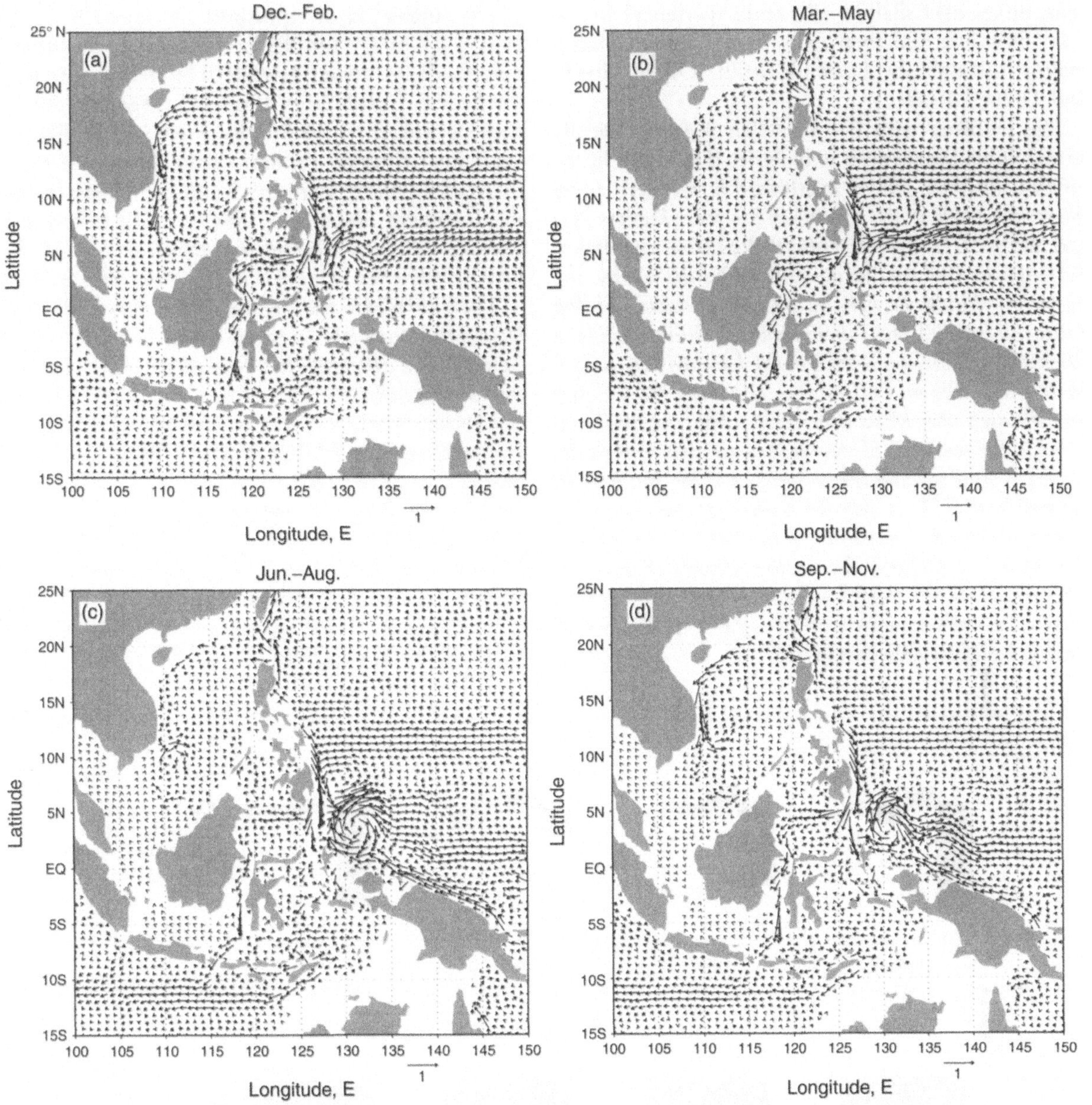

Figure 8 Seasonally averaged upper 100-m ocean currents from the high-resolution Navy Layered Ocean Model.

and the southward Mindanao Current (MC). As the MC flows to the south, part of it veers into the Celebes Sea, and the remainder turns east into the Pacific to form the North Equatorial Counter Current (NECC). As the MC turns east, the quasi-permanent counterclockwise Mindanao Eddy is formed. In the south, a similar situation arises. The South Equatorial Current (SEC) impinges on the east coast of Australia and Papua New Guinea. The western boundary currents formed by the SEC are the southward-flowing East Australia Current (EAC) and the northward-flowing North Queensland Current (NQC). Near the equator, the surface flow from the NQC reverses direction due to the monsoons, but there is a constant northward flow at depth (called the New Guinea Coastal Undercurrent (NGCUC)). The NQC/NGCUC system, like the MC, turns back to the east and supplies the North Equatorial Counter Current (NECC) and Equatorial Undercurrent (EUC), and the clockwise Halmahera Eddy is formed.

The main sources of inflow to the SE Asian Seas are therefore the MC and NGCC that provide Pacific waters into the Celebes Sea. The South China Sea also receives Pacific water from the Kuroshio via the Luzon Strait; from there part of the water continues southward into the Java Sea through the Karimata Strait and into the Sulu Sea via the Mindoro Strait. The Pacific water then passes within the Indonesian

Seas, mixes over shallow sills and experiences strong tidal motions, and then exits to the Indian Ocean mainly through the Lombok Strait, Ombai Strait, and Timor Passage.

The pathway with the SE Asian Seas is typically strong along the western boundaries (**Figure 8**). In the South China Sea, intense southward flow is seen along the western side of the sea late in the year (**Figures 8(a)** and **8(d)**). A large-scale counterclockwise circulation develops in the sea during this time, driven by the monsoon winds as well as by the intrusion of the Kuroshio through the Luzon Strait. A similar counterclockwise circulation is seen in the Celebes Sea, driven by the MC inflow. Part of the MC exits the Celebes Sea through the Makassar Strait. The upper ocean flows in the southern SE Asian Seas are strongest in the northern summer and fall months, and the Indonesian throughflow (ITF) is also the highest at this time.

The temporal variability is probably best seen in the transport through the various straits (**Figure 9**). The upper layer model velocities were integrated across the inflow and outflow straits in the South China, Celebes, and Banda Seas.

The upper layer transport is typically 3–4 Sv ($1\,\mathrm{Sv} = 10^6\,\mathrm{m}^3\,\mathrm{S}^{-1}$) in these straits, about an order of magnitude less than the western boundary currents in the Pacific. Luzon Strait transport (negative indicates water supplied to the South China Sea) is higher in September through February than during the rest of the year. Outflow from the South China Sea through the Karimata Strait and Taiwan Strait inflow peak later in the year, December through May, and flow into the Sulu Sea is highest in July through December.

The inflow into the Celebes Sea is also high during December through May, and the flow out of the Celebes Sea through the Makassar Strait becomes maximum prior to that, beginning in August when the southeasterly monsoons become strongest. The flow through the Makassar Strait either exits to the Indian Ocean through the Lombok Strait, or enters the Banda Sea. Transport from the Flores and Java Seas into the Banda Sea is highest in December through February, when the ITF transport is at a minimum. The ITF, a key component of the global heat transport in the oceans, is described in more detail in the next section.

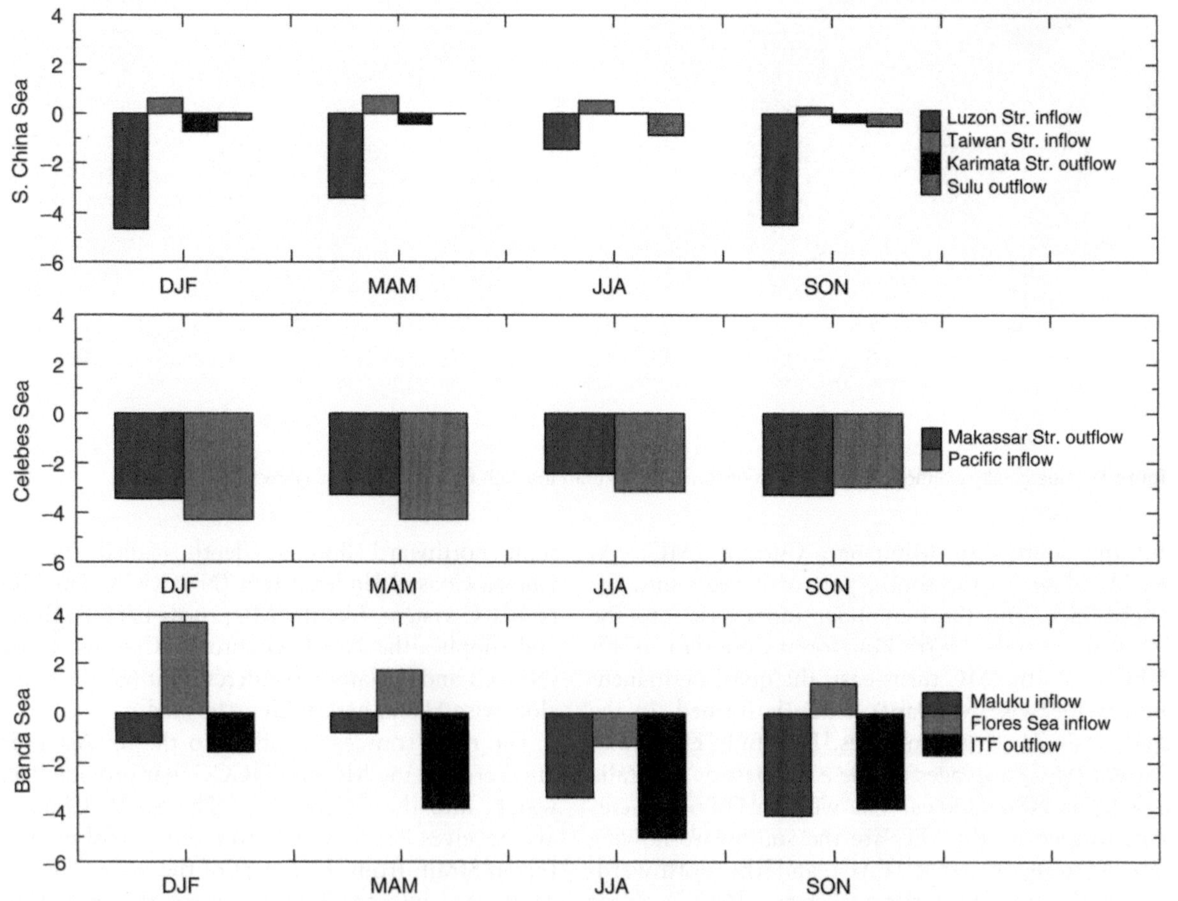

Figure 9 The model upper layer velocity was integrated across the straits connecting the South China, Celebes, and Banda Seas for each 3-month season. Negative values indicate southward or westward transport.

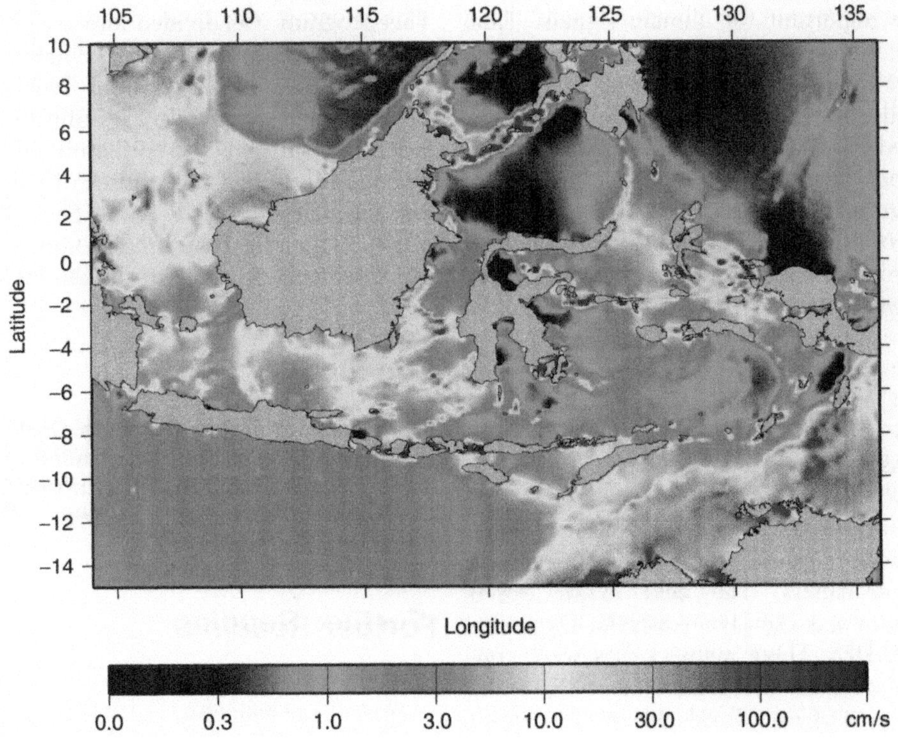

Figure 10 Maximum barotropic current velocity of the M_2 tide (equivalent to length of semimajor axis of tidal current ellipses). Note the nonlinear color bar (units of cm s^{-1}). Courtesy of Richard Ray.

SE Asian Seas' Connection to the Indian Ocean

The southern SE Asian Seas connect to the Indian Ocean through gaps in the Indonesian archipelago. The SE Asian Seas, and the restricted flows in the ITF, are an important factor for water mass formation in the region.

The ITF is thought to be forced by large-scale processes in both the Pacific and Indian Oceans. The Pacific trade winds allow a pressure difference to exist between the two oceans, which drives a mean flow from the Pacific to the Indian Ocean. Current estimates are that the mean ITF is about 10 Sv. Intraseasonal winds, seasonal monsoons and large-scale oceanic waves, as well as interannual forcing all adjust the ITF transport. The mean seasonal cycle, driven by local winds and remotely forced waves, ranges from 1 or 2 Sv to 16 or 17 Sv. The peak ITF occurs in July through September, when the SE monsoon is strongest. The minimum transport occurs during the opposite monsoon season. Semiannual variability is introduced from Indian Ocean forcing. This forcing causes coastal waves (called Kelvin waves) that propagate eastward along the south Sumatra/Java coastline. These waves inhibit the ITF during the spring and fall seasons.

Similarly, intraseasonal and interannual winds can adjust the ITF. For example, westerly wind bursts in the Indian Ocean associated with the MJO can cause Kelvin waves that reduce the ITF. On interannual timescales, the ITF is generally thought to be tied to ENSO events. During warm El Niño events, the Pacific easterly trades become reduced, and thus the Pacific to Indian Ocean pressure difference is reduced, and the ITF is lower than normal. However, again large-scale waves, local winds, and vertical effects tend to mask the ENSO–ITF relationship. Similarly, the Indian Ocean dipole (IOD), a large scale east–west phenomenon in the Indian Ocean, is also believed to affect ITF transport. During positive IOD, events lower than normal SSTs along south Sumatra/Java may enhance the pressure difference between the Pacific and Indian Ocean, therefore leading to a stronger than normal ITF. In this sense the ITF interannual variations are primarily a result of interplay between ENSO and IOD.

Other Considerations

This article provides a brief overview of the ocean state in the SE Asian Seas. There are several additional components to this system that cannot be described in detail here. The deep basins in the seas allow for a change in water mass properties, and could also be important for climate signals. Tidal flows are also critical for water mass formation, and the region sees some of the most strongest tides near the shallow sills. In some places, the tidal flows meet or even exceed the mean flows. **Figure 10** shows a numerical model approximation of the maximum barotropic current velocity of the M_2 tide.

All these contribute to the difficulty of making measurements in the region. Nevertheless, a few key observing campaigns have been carried out in the region.

The earliest large-scale effort was carried out by the Snellius I expedition in 1929–30 that made measurements of temperature and salinity throughout the SE Asian Seas. A Snellius II project was undertaken in mid-1984 through mid-1985, and these cruises measured the deep basins and flows, along with a wide range of biological, chemical, and geophysical parameters. The next major observational program was the Java–Australia Dynamics Experiment (JADE). Three main cruises were conducted in the southern ITF region in 1989, 1992, and 1995. As part of the World Ocean Circulation Experiment (WOCE), several hydrographic and repeat cruises were made in the SE Asian Seas. These include I10, a one-time cruise from Australia to Bali in 1995; IR6, a repeat hydrographic line from Australia to Bali from (1989 to 1995); IX1, an ongoing repeat hydrographic line between Australia and Java; PX2, an ongoing line from Java to New Guinea (through the Banda Sea); and IX22, an ongoing line from Australia to Timor. A large campaign called Arus Lintas Indonen (ARLINDO) was done in the Makassar Strait and Banda Sea regions in 1993–98. This program was divided into a mixing component and a circulation component. Finally, more recently, there is the INSTANT project. INSTANT stands for International Nusantara Stratification and Transport, and through this program numerous, simultaneous moorings and cruisework is being carried out throughout the SE Asian Seas. INSTANT will provide, for the first time, comprehensive *in situ* measurements of many of the SE Asian Seas.

See also

California and Alaska Currents. Current Systems in the Indian Ocean. East Australian Current. Indian Ocean Equatorial Currents. Kuroshio and Oyashio Currents. Mesoscale Eddies. Pacific Ocean Equatorial Currents.

Further Reading

Gordon AL (ed.) (2005) *Special Issue on the Indonesian Seas. Oceanography* 18: 14–127.

Lukas R, Yamagata T, and McCreary JP (eds.) (1996) *Special Issue on Pacific Low-Latitude Western Boundary Currents and the Indonesian Throughflow. Journal of Geophysical Research, Oceans* 101: 12209–12488.

Tomczak M and Godfrey JS (1994) *Regional Oceanography: An Introduction*. London: Pergamon.

Wyrtki K (1961) Scientific results of marine investigations of the South China Sea and the Gulf of Thailand 1959–1961. *NAGA Report*, vol. 2. La Jolla, CA: Scripps Institution of Oceanography.

THE OCEANS: THE INDONESIAN THROUGHFLOW

INDONESIAN THROUGHFLOW

J. Sprintall, University of California San Diego, La Jolla, CA, USA

Introduction

The Indonesian seas play a unique role in providing the only open pathway that connects two major ocean basins at tropical latitudes. On average, the sea level is higher on the western Pacific side of the Indonesian archipelago compared to the eastern Indian side. This pressure gradient generates a transport of water and their properties from the Pacific toward the Indian Ocean. This flow from the Pacific Ocean into the Indian Ocean is known as the Indonesian Throughflow. The warmer water that enters from the Pacific can be traced throughout the Indonesian seas, and then followed within the surface to intermediate depths as a distinct low-salinity tongue in the Indian Ocean. As such, the Throughflow forms the 'warm' water route for the global thermohaline circulation and therefore impacts the regional and global climate system. Because of its proximity to Asia and Australia, the circulation and transport in the Indonesian seas has a large seasonal variation due to the influence of the reversing annual wind patterns associated with the Asian–Australian monsoon system. During the different monsoon seasons, waters of different sources from both the Indian and Pacific Oceans flow into the Indonesian archipelago and cause variability in temperature, salinity, and other properties. Local processes within the Indonesian seas related to the regional monsoon winds such as upwelling and downwelling, along with the tides, air–sea heat fluxes, and voluminous precipitation and associated river runoff, also act to change the Pacific temperature and salinity stratification into the distinctly fresh Indonesian seas profile. These changes in the physical properties of the water are linked to the behavior, migration pattern, and the seasonal distribution of the phytoplankton and pelagic fish species that live within the Indonesian seas. Thus a knowledge and understanding of the pathways and variability of the Indonesian Throughflow and its properties is important for the region's people, who depend on the sea for their very food and livelihood, and also to help develop management plans to sustain these valuable and limited maritime resources.

Pathways and Water Masses

The combination of numerous islands, with narrow straits that connect a series of large, deep ocean basins within the Indonesian seas, provides a winding route for the Indonesian Throughflow (**Figure 1**). The surface to upper thermocline waters in the Indonesian seas are primarily drawn from the North Pacific subtropical waters and North Pacific Intermediate Water that flows southward in the Mindanao Current, east of the Philippines. These waters mostly take the 'western' Throughflow route through the Sulawesi Sea (sometimes known as the Celebes Sea) into the Makassar Strait. Within Makassar Strait, the 650-m-deep Dewakang sill permits only the upper thermocline waters to enter the Flores and Banda Seas, or to directly exit into the Indian Ocean via the shallow (350 m) Lombok Strait. Smaller contributions of North Pacific surface water may also take the 'eastern' Throughflow route, through the Maluku Sea and over the deeper (1940 m) sill of Lifamatola Strait. Lower thermocline and deeper water masses – Antarctic Intermediate Water (AAIW) and Circumpolar Deep Water – that originate in the South Pacific in the New Guinea Coastal Undercurrent, also enter the Indonesian seas via the eastern route through the Lifamatola Strait. These deeper water masses are saltier than the upper waters that originate from the North Pacific. Upper waters of South Pacific origin can also flow directly over the Halmahera Sea (blocked below 700-m depth) and into the internal Seram and Banda Seas.

Most of the transformation of the Pacific water masses into the distinctive Indonesian water profile of relatively isohaline water from the thermocline to near bottom occurs within the Banda Sea. As the different water masses flow toward the Banda Sea over the various Indonesian sills and straits, the abrupt changes in bottom topography cause intense diapycnal mixing. The region-average vertical diffusivities at the sills are greater than $10^{-3}\,m^2\,s^{-1}$, which is an order of magnitude larger than both that typically estimated for the interior ocean, and also for that which occurs within the interior of the Banda Sea itself. At least some of this diapycnal mixing is attributable to the strong tides within the Indonesian seas. At the sills, mixing can also occur via entrainment and shear-induced turbulence, including internal wave generation. It is also likely that hydraulic effects can restrict the flow and affect the circulation pattern in some straits due to the shoaling

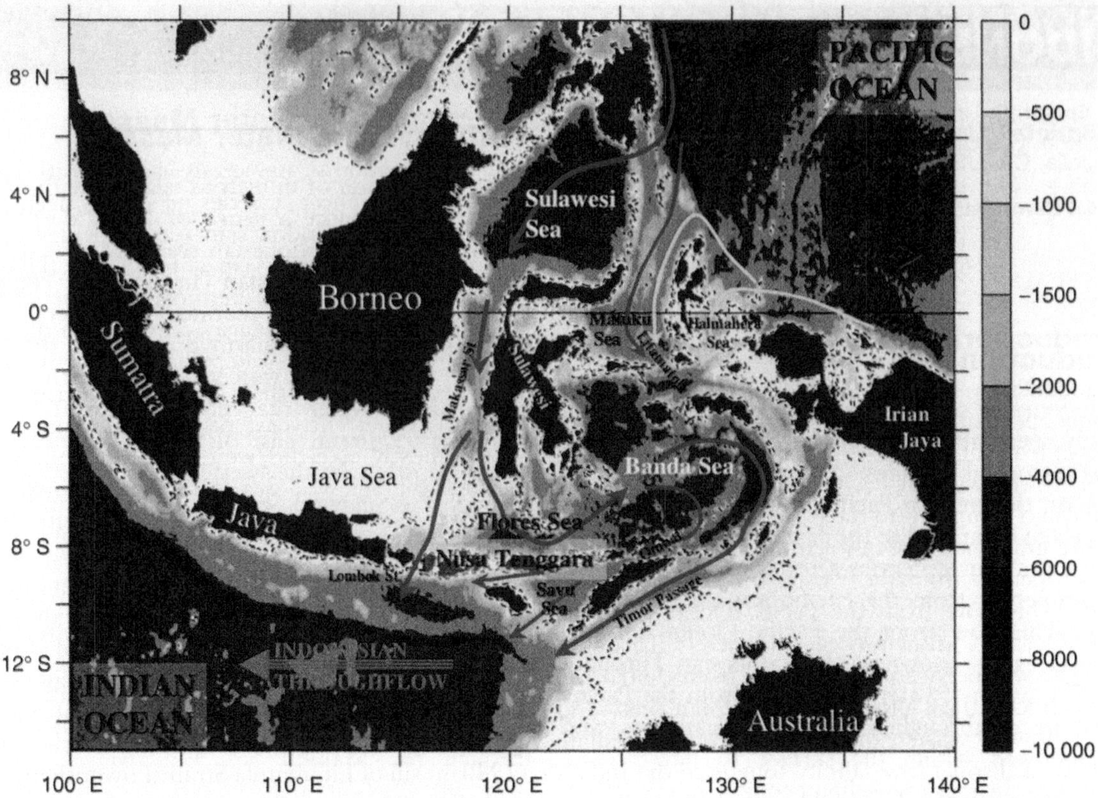

Figure 1 Map of the Indonesian region showing the bathymetry and pathways of the mean Indonesian Throughflow. Shallow and upper thermocline pathways from the North Pacific are shown by the red arrows, and lower thermocline to intermediate depth pathways are shown by the yellow arrows. Islands, seas, and straits referred to in the text are indicated. The 200-m isobath is shown by the dashed line.

and contraction of sills, although as yet there have been no suitable direct observations or process studies undertaken that may resolve this response. Density-driven (or isopycnal) flow dominates the deeper layers of South Pacific origin that enter through Lifamatola Strait into the Banda Sea, driving deep circulation patterns that keep the internal basins well ventilated. In the surface to upper thermocline layer, the diapycnal mixing combines with the tidal forcing, wind-driven upwelling, air–sea heat flux, and surface precipitation and river runoff to form the low-salinity Indonesian Throughflow Water (ITW: 34–34.5 psu). Below this sits the Indonesian Intermediate Water (IIW), which has a silica maximum and is slightly saltier (36 psu) as a result of diapycnal mixing of the intermediate Pacific water masses with a larger contribution from the saltier water masses of the South Pacific found at depth in the Banda basin.

From the Banda Sea, the Indonesian Throughflow exits into the southeast Indian Ocean via the Timor Passage (1250 m eastern sill depth at Leti Strait, and 1890 m at the western end), or through Ombai Strait (sill depth 3250 m) and then through the Savu Sea to Sumba Strait (900 m) and Savu Strait (1150 m). The Indonesian Throughflow waters are apparent as a freshwater jet in the westward-flowing South Equatorial Current (SEC) across the entire tropical Indian Ocean between 8° and 17° S. In the Indian Ocean, the total Throughflow is recognizable as the low-salinity ITW in the upper 100–200 m of surface water and the deeper-salinity minimum of the IIW which is found at ~600–1200-m depth, and hence above the eastern sill depth of Timor Passage. Although the ITW and IIW are separate cores even directly west of Timor Passage as they exit the Indonesian seas, they gradually increase in salinity and become even more distinct entities as the high-salinity maximum subtropical and equatorial waters of the open Indian Ocean encroach between them. The ITW and IIW lose their low-salinity signature more slowly than the intervening salinity maximum during westward advection across the Indian Ocean because the adjacent water masses on their isopycnals (surface layer and AAIW) are considerably fresher than other Indian Ocean water masses. Nonetheless, the slight increase in salt results in an increase in density and thus the Throughflow waters become deeper and erode toward the western Indian Ocean, although the latitude of both the core ITW and IIW remains remarkably constant within the SEC because of the nearly exact zonality of potential vorticity in the Tropics. At the western edge of the Indian Ocean, it is likely that much of the Throughflow enters the Agulhas Current system, although because of mixing with other water masses in this region and beyond, it becomes more difficult to directly trace their origin from the Indonesian seas. Nonetheless, numerous Agulhas eddies in the Atlantic have been identified as containing relatively fresh Indonesian thermocline water, providing observational evidence to support the numerical models that show the Indonesian Throughflow playing an important role in the global circulation.

Mean Throughflow Mass and Heat Flux Estimates

The magnitude and variability of the Indonesian Throughflow are still sources of major uncertainty for both the modeling and observational oceanography communities. They are the dominant sources of error in the basin-wide heat and freshwater budgets for the Pacific and Indian Oceans. Though general circulation models are gradually improving, they are unable at present to reproduce the narrow passages and convoluted bottom topography of the internal Indonesian seas in order to adequately resolve the structure and variability of the ITF transport. Thus the models still largely rely on the scantly available observations to provide guidance in their initialization and validation for simulations of the ocean circulation and climate.

Historical estimates of the mean mass transport of the Indonesian Throughflow are wide-ranging, from near zero to 30 Sv ($1\,Sv = 10^6\,m^3\,s^{-1}$). In part, this sizeable difference is due to the lack of direct measurements, but also because of the real variation that can severely alias mean estimates if survey periods are not sufficiently long. Recent mooring measurements of velocity in Makassar Strait in 1986–87 suggest a volume transport of around 8 Sv for the surface to upper thermocline waters that pass through this inflow channel. Long-term direct measurements of the lower thermocline to intermediate water masses have yet to be obtained, although water mass considerations and synoptic survey measurements of the Throughflow at the eastern edge of the Indian Ocean suggest a volume transport of 2–3 Sv. A 3-month current meter record at Lifamatola Strait in early 1985 determined a 1.5-Sv contribution of deep water to the Throughflow. Direct long-term measurements of the Throughflow within the exit passages have also been obtained at different times over the past few decades. The transport through Timor Passage (1988–89) and Ombai Strait (1995) appears to be roughly equally partitioned, at *c.* 4–6 Sv through each strait, whereas the transport through the shallower exit passage of Lombok Strait (1985) is ~2 Sv. Hence the total volume transport associated with the full-depth Throughflow is probably around 10–14 Sv, although because of the different sampling years of the different direct measurement programs this total volume transport estimate should be treated with caution. As will be discussed below, the volume and properties of the Throughflow are known to change with various phases of the El Niño–Southern Oscillation (ENSO) cycle, and hence this may have impact on the transport estimates from a given passage in a given year. Only with multiyear, simultaneous measurements of the full-depth velocity structure in all the major Throughflow passages can the volume transport be accurately determined. Direct measurements that meet these criteria to determine transport will become available from an array of moorings that were deployed as part of the International Nusantara Stratification and Transport (INSTANT) program. The 11 moorings were deployed in the major inflow and outflow passages in December 2003, turned around in June 2005, and are to be recovered in December 2006.

The heat and fresh water carried by the Indonesian Throughflow potentially impact the basin budgets of both the Pacific and Indian Oceans. Earlier estimates of the heat transport ranged from 0.5 to 1.0 PW (petawatt, with $1\,PW = 10^{15}\,W$) as the Throughflow was assumed to be surface-intensified, and reflect the relatively warm temperatures found in its western Pacific source waters. However, the recent observations from the Makassar moorings suggest that the strongest Throughflow velocities (southward in this strait) are found in the thermocline, and hence the cool transport-weighted temperature of ~15 °C leads to a mean heat transport of ~0.4–0.5 pW (relative to 0 °C). Including an as yet unresolved colder component from the IIW would only further reduce the Throughflow impact on the heat transport of the Southern Hemisphere. Because of the lack of subsurface salinity measurements, the contribution of the freshwater flux from the Indonesian Throughflow to the Indian Ocean and beyond is even less well constrained. However, since the salinity of the Pacific waters entering the Indonesian seas is fresher than the inflow from the Southern Ocean northward into the Pacific, the Throughflow probably represents a net transport of fresh water out of the Pacific and into the Indian Ocean where the salinity increases before being exported southward in the Agulhas Current.

Variability in the Throughflow Properties and Transport

Pacific, Indian, and regional Indonesian wind forcing modulates the Indonesian Throughflow mass and property transport over a wide range of timescales that combine to produce variability of the same order of magnitude as the long-term mean estimates cited above. In this section, we will discuss the cause of the different timescales of variability and their impact on the Throughflow. Because it is the main driving force of the Throughflow we begin with a discussion of the seasonal variability related to the Asian–Australasian monsoon regime. Forcing at other timescales from remote Pacific and Indian Ocean sources generally tends to modulate this fundamental seasonal response.

Monsoonal-Related Variability: Annual Timescales

The large-scale sea level difference between the easterly trade winds in the western Pacific and the reversing monsoonal winds over the southern Indonesian seas drives the annual change in the Throughflow. During the northwest monsoon from December to February, south of the equator the winds over Indonesia are to the southeast (**Figure 2(a)**). Through Ekman dynamics, the southeastward winds cause a drop in sea level along the northern Nusa Tenggara islands and warm surface waters accumulate in the Banda Sea, acting to reduce the Throughflow transport into the Indian Ocean. The downwelling south of Nusa Tenggara leads to warmer surface waters here. The convective rainfall associated with the northwest monsoon spans the Indonesian archipelago to produce heavy rainfall and subsequently high river runoff that together combine to freshen the surface waters, particularly in the Java Sea region. The flow from the Banda Sea toward the Indian Ocean is strongest during the southeast monsoon, from June through September, when the winds are to the northwest and more intense (**Figure 2(b)**). The strong winds during this monsoon result in a lower sea level to the south of Nusa Tenggara, and hence the flow through the exit passages is thought to be enhanced by this local Ekman response. Surface waters are cooler in the Banda Sea, and also south of Nusa Tenggara due to the wind-driven upwelling. The convective activities move northward of the equator so conditions in Indonesia are drier and surface salinity higher compared to the northwest monsoon. The wind-driven regime also leads to changes in the chlorophyll concentrations, which is elevated in regions of upwelling as nutrients from the deep are bought to the surface, and reduced where there is downwelling. Chlorophyll indicates phytoplankton abundance and therefore gives a broad view of the biological activity and associated fishery distribution during the different monsoon seasons.

While the surface response to the reversing monsoonal winds is clearly evident in **Figure 2**, the monsoonal effect on the Throughflow velocity and transport is much more complicated. While all the observations presently available in the exit passages do indeed show the expected maximum Throughflow occurring during the southeast monsoon period, the Makassar Strait mooring data showed the maximum southward flow occurred during the 1996–97 northwest monsoon period. The large differences in peak transport timing between the inflow and outflow straits are most likely due to storage of waters in the Banda Sea on seasonal timescales. It appears that the Banda Sea acts as a reservoir, filling up and deepening the thermocline during the northwest monsoon. During the more intense southeast monsoon, Ekman divergence in the Banda Sea combined with the lower sea level off the south coast of the Nusa Tenggara are more conducive to drawing waters into the Indian Ocean. The maximum Ekman convergence (or gain) of 1.7 Sv causes the thermocline to heave almost 40 m during October–November, although there is matching divergence (or loss) occurring in April–May of each year. Convergences and divergences of this magnitude can have a substantial impact on thermocline stratification, and thus may affect the sea surface temperature by changing the temperature of the water being entrained into the mixed layer. In addition, since the Banda Sea is also the primary site for conversion of the Pacific waters into the distinct Indonesian stratification profile, it is likely that the composition and magnitude of the stored waters could have a significant impact on the Indian Ocean heat, fresh water, and mass budgets.

The response of the surface properties to the shifting monsoonal winds has recently been suggested to play another important role in modifying the Throughflow velocity and transport. The buoyant low-salinity water present in the Java Sea and southern Makassar Strait during the northwest monsoon (**Figure 2(a)**) creates a northward pressure gradient that inhibits the warm surface water in Makassar Strait from flowing southward into the Indian Ocean, even though the predominant winds are southward during this monsoon. During the

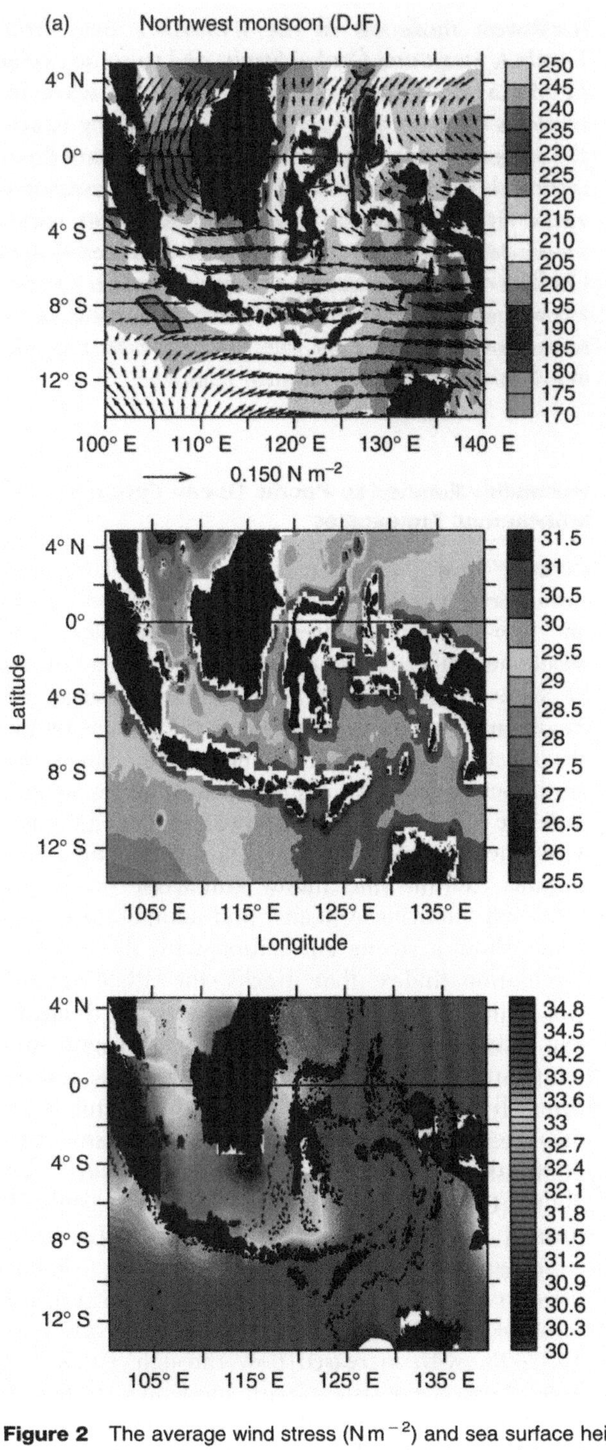

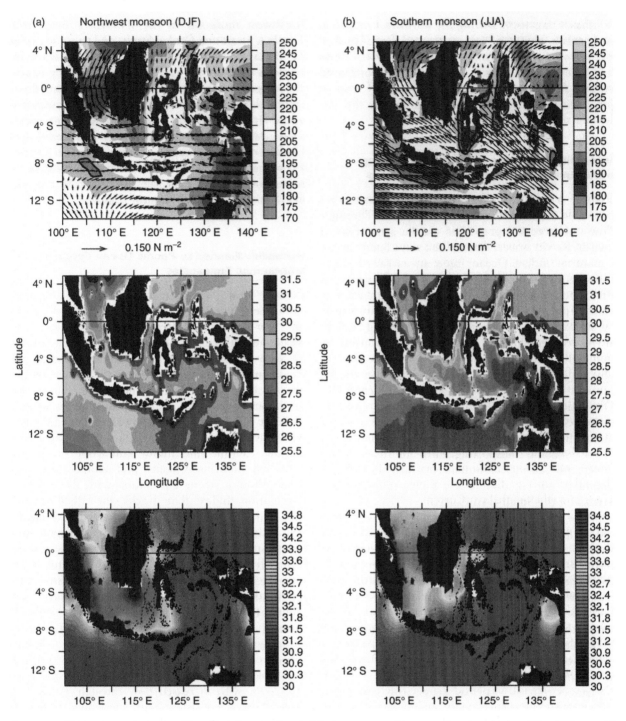

Figure 2 The average wind stress (N m^{-2}) and sea surface height (cm; top panels), sea surface temperature (°C; middle panels) and sea surface salinity (psu; bottom panels) during the (a) northwest monsoon (Dec.–Feb.) and (b) southeast monsoon (Jun.–Aug.). The 200-m isobath is shown by the dotted lines in the lower panel.

southeast monsoon, the surface waters are more saline (**Figure 2(b)**), the northward pressure gradient is eliminated, and the northward winds tend to constrain the surface Throughflow. Thus, it appears that during both monsoon seasons, the strongest southward Throughflow occurs at thermocline depths (100–200 m) in Makassar Strait, resulting in the recently determined cooler ocean heat transport cited in the above section. Although the data are still preliminary, the INSTANT moorings in the shallower Lombok Strait also show a subsurface southward velocity maximum at ~50 m during the southeast monsoon. During the northwest monsoon in Lombok Strait the main southward flow is pushed even deeper to ~100 m, as strong northward flows are found at the surface. Whether these northward surface flows are related to remote Indian Ocean or local wind forcing is further discussed in the following section.

Variability Related to Indian Ocean Forcing: Intraseasonal and Semiannual Timescales

Remote wind forcing from the Indian Ocean is responsible for variability in the Indonesian Throughflow transport and properties via the generation of coastal Kelvin waves. Anomalous wind bursts in the equatorial Indian Ocean force an eastward equatorial jet associated with an equatorial Kelvin wave that, upon impinging the west coast of Sumatra, results in poleward propagating coastal Kelvin waves. Westerly wind bursts force downwelling Kelvin waves semiannually during the monsoon transitions, nominally in May and November, and intraseasonally (30–90 days) most likely in response to the intraseasonal wind forcing of the Madden–Julian Oscillation. The poleward propagation of the downwelling coastal Kelvin wave raises the sea level along the Indonesian wave guide of Sumatra and Nusa Tenggara, transporting warm and fresh surface water and causing maximum eastward flow of the boundary current that flows along southern Nusa Tenggara, the South Java Current.

While there is some discrepancy between theoretical models as to whether the Indian Ocean-forced Kelvin wave energy can penetrate northward through Lombok Strait, the available observational data clearly show the northward flow of warm water that indicates the presence of Kelvin waves in both Lombok Strait and Makassar Strait, particularly during the May transition period. Furthermore, some of the Kelvin wave energy is also readily apparent in observations from Ombai Strait, further east along the coastal waveguide, during the May and November transition periods, suggesting that Indian Ocean water can penetrate through the Savu Sea and beyond. Commensurate with vertical ray-tracing theory, the semiannual energy from remotely forced Kelvin waves extends deeper in the water column with distance along the waveguide, and so the reversals in flow are evident well below the thermocline in all three straits. What proportion of the Kelvin wave energy is 'lost' through Lombok Strait versus that remaining in the coastal waveguide, and how this may be modulated by the annual cycle, is the focus of active research. Similarly, whether the northward surface flows observed during the northwest monsoon in the INSTANT data from Lombok Strait and Ombai Strait, and to some extent in Makassar Strait, are related to Kelvin wave intrusions or to locally generated wind forcing is also the subject of ongoing research. The northward flows during this monsoon phase have a clear intraseasonal variability that may be related to the remote winds associated with the intraseasonal Madden–Julian Oscillation in the eastern equatorial Indian Ocean. However, given the relatively shallow vertical penetration of these reversals, they could also be a locally wind-driven Ekman dynamical response.

Variability Related to Pacific Ocean Forcing: Interannual Timescales

Remote winds in the Pacific Ocean drive large-scale circulation that impacts the Indonesian Throughflow on low-frequency, interannual timescales. The weakening of Pacific trade winds that occurs during El Niño reduces the sea level gradient between the Pacific and Indian Oceans that is thought to be the driving force for the Throughflow. The changes are of the sense that mass and heat transport are smaller and the thermocline is shallower during El Niño, with the converse being true during La Niña time periods. Within the inflow Makassar Strait, the 1996–98 mooring velocity and temperature time series show a strong correlation with the Southern Oscillation Index that tracks the ENSO-related variability in the Pacific. Along the Nusa Tenggara exit passages, the relationship of transport and temperature to ENSO variability is much less certain, primarily because we lack the multiyear, full-depth measurements that are required to establish such a relationship. Surface geostrophic transport variability (0–100-m depth) over the period 1996–99 through the exit passages, inferred from the cross-strait pressure gradient measured by shallow pressure gauges, suggest diminished flow through Lombok and Ombai Straits during the El Niño years 1997–98, with increased flow through Timor Passage. This is consistent with modeling studies, in which there is stronger flow through Timor Passage and weaker flow through Ombai Strait during El Niño events. Interestingly, the surface geostrophic transport through Lombok Strait was strongly northward during the northwest monsoon of the 1997–98 El Niño event, amplified over the expected northward flows in this strait during this monsoon phase. It is possible, as discussed above, that the main southward Throughflow occurred at depth during this time period and hence was not measured by the shallow pressure gauges.

Interannual variability in sea level is also evident within the Banda Sea and along the Northwest Australian shelf region down to around 20° S. Both models and observations suggest that the remote Pacific tropical winds associated with ENSO drive both equatorial and off-equatorial Rossby waves that are responsible for the observed sea level and associated thermocline response. The Rossby wave signal is radiated through the Banda Sea, as well as entering the coastal waveguide at the western tip of New Guinea, and propagates poleward through Timor Strait and along the Australian shelf as coastally trapped waves. In turn, these trapped waves excite free westward propagating Rossby waves at the Australian coast that can be detected several hundreds of kilometers offshore in the southeast tropical Indian Ocean. Interannual Kelvin waves forced remotely by equatorial interannual zonal winds in the Indian Ocean have also been documented as impacting the thermocline and sea level variability along the Nusa Tenggara coastal waveguide, similar to those observed on intraseasonal and semiannual timescales. Again, both models and observations suggest these Kelvin waves penetrate through Lombok Strait and into the internal Indonesian seas. Thus, as suggested by Wijffels and Meyers, the Indonesian seas comprise the intersection of two oceanic wave guides from remotely generated wind forcing in both the Pacific and Indian Oceans on interannual timescales.

Conclusions

The mass and property characteristics of the Indonesian Throughflow vary over the full range of tidal to interannual timescales. Within the internal Indonesian seas, the Pacific temperature and salinity stratification, as well as the local sea surface temperature, are modified by the strong air–sea heat and freshwater fluxes, seasonal wind-induced upwelling, and large tidal forces. These ocean circulation and processes in the Indonesian seas influence not only the local climate but also the global climate through connections with the Pacific and Indian Ocean. For example, ruinous drought conditions in Australia have been linked to cool sea surface temperature anomalies in the Indonesian region. Similarly basin-wide anomalies in sea surface temperature, wind, and precipitation that spanned the Indonesian region during 1997–98 resulted in severe drought in Indonesia and catastrophic floods in eastern Africa. Thus, undoubtedly the oceanic heat and freshwater flux of the Indonesian Throughflow into the Indian Ocean affect the atmospheric–ocean coupling and are strongly linked to the evolution of interannual climate anomalies such as occur through the ENSO and monsoon systems. Understanding the variation in the Indonesian Throughflow is therefore crucial for understanding the coupled air–sea climate system, and the storage of the heat and fresh water that are ultimately redistributed throughout the world oceans by thermohaline circulation.

See also

Current Systems in the Indian Ocean. Ekman Transport and Pumping. Ocean Circulation: Meridional Overturning Circulation. Rossby Waves. Wind Driven Circulation.

Further Reading

Gordon AL (2001) Interocean exchange. In: Seidler G, Church J, and Gould J (eds.) *Ocean Circulation and Climate*, ch. 4.7, pp. 303–314. New York: Academic Press.

Talley LD and Sprintall J (2005) Deep expression of the Indonesian Throughflow: Indonesian intermediate water in the South Equatorial Current. *Journal of Geophysical Research* 110: C10009 (doi:10.1029/2004JC002826).

The Oceanography Society. (2005). *Special Issue: The Indonesian Seas. Oceanography* 18(4), 144pp.

Wijffels S and Meyers GA (2004) An intersection of oceanic wave guides: Variability in the Indonesian Through flow region. *Journal of Physical Oceanography* 34: 1232–1253.

Wyrtki K (1961) *Physical oceanography of the Southeast Asian waters. Scripps Institution of Oceanography NAGA Report 2*. San Diego, CA: Scripps Institution of Oceanography.

THE OCEANS: THE INDIAN OCEAN

CURRENT SYSTEMS IN THE INDIAN OCEAN

M. Fieux, Université Pierre et Marie Curie, Paris, France
G. Reverdin, LEGOS, Toulouse Cedex, France

Introduction

The Indian Ocean is the smallest of all the oceans and is in several respects quite different from the others. In particular, it is bounded by the Asian continent to the north. This meridional land–sea contrast has a strong influence on the winds, resulting in a complete seasonal reversal of the winds known as the monsoon system. The characteristics of the basin and of the wind regime are determinant for the currents, and will be described first in this article. The description of the currents has been separated into two main sections: the first for the southern part of the Indian Ocean which is not affected by the monsoons and is more akin to the other subtropical oceans; and the second for the northern part which undergoes forcing through the reversal of the monsoon winds. Some information on the deep circulation and a short conclusion are then provided.

Characteristics of the Indian Ocean Basin

The Indian Ocean basin is the smallest of the five great subdivisions of the world ocean with 49.10^6 km^2 out of the 361.10^6 km^2 of the global ocean (**Figure 1**). It is closed to the north around the latitude of the Tropic of Cancer by the Asian continent, which has important consequences on the ocean circulation. South of the equator, its western boundary is modified by the presence of the island of Madagascar. In the east, the basin is connected with the equatorial Pacific Ocean through the deep passages of the Indonesian Seas. The north of the Indian Ocean is made up of the large basins on either side of the Indian peninsula, the Arabian Sea in the west and the Bay of Bengal in the east which drains most of the river runoff from the Himalayas and the Indian subcontinent. The Arabian Sea is connected directly to the shallow Persian Gulf, and through the sill of Bab-el-Mandeb (110 m) to the deep Red Sea basin where high salinity waters are formed. In the south, the basin is largely open to the Antarctic Ocean between South Africa and Australia. The Indian Ocean limit to the south is the Subtropical Convergence, a hydrological limit where the meridional surface temperature gradient is maximum. At depth, the complicated system of ridges separates the Indian Ocean in many deep basins (**Figure 1**).

The Overlying Atmosphere

Due to the presence of the Asiatic continent to the north, the atmospheric circulation is quite different from the Pacific Ocean and the Atlantic Ocean, particularly north of 10°S. Seasonal heating and cooling of the atmosphere over Asia induces a seasonally varying monsoon circulation (**Figure 2**). For centuries it has been known that the winds north of around 10°S reverse with the seasons. A long time ago the Arabic traders along the east African coast made use of the fair currents and winds during their voyages. The word ‘monsoon’ comes from the Arabic word ‘mawsin’ meaning season. As the winds are the main driver of the currents, in particular near the surface, the main characteristics of the wind seasonal variability will be described below.

The wind seasonal variability over the ocean can be separated in four periods: the winter monsoon period, the summer monsoon period, and the two transition periods between the two monsoons.

Between December and March–April, north of the equator, the winter (NE) monsoon blows from the north east with a moderate strength. At the equator the winds are weak and usually from the north. Between the equator and the Intertropical Convergence Zone (ITCZ) which stretches zonally near 10°S between north Madagascar and south Sumatra, it blows from the north west. During that season (southern summer), the atmospheric pressure decreases over Australia and South Africa and the subtropical high pressure over the ocean, around 35°S, is weaker – as are the south-east trade winds during that season.

In April–May, during the transition period between the end of the NE monsoon and the beginning of the SW monsoon, the winds north of the equator calm down. At the equator moderate eastward winds blow, which contrasts with the westward winds over the equatorial Pacific and Atlantic Oceans.

From June to September–October during the SW monsoon, the winds reverse completely and north of the equator the summer monsoon blows steadily from the south west. The SW summer monsoon is much

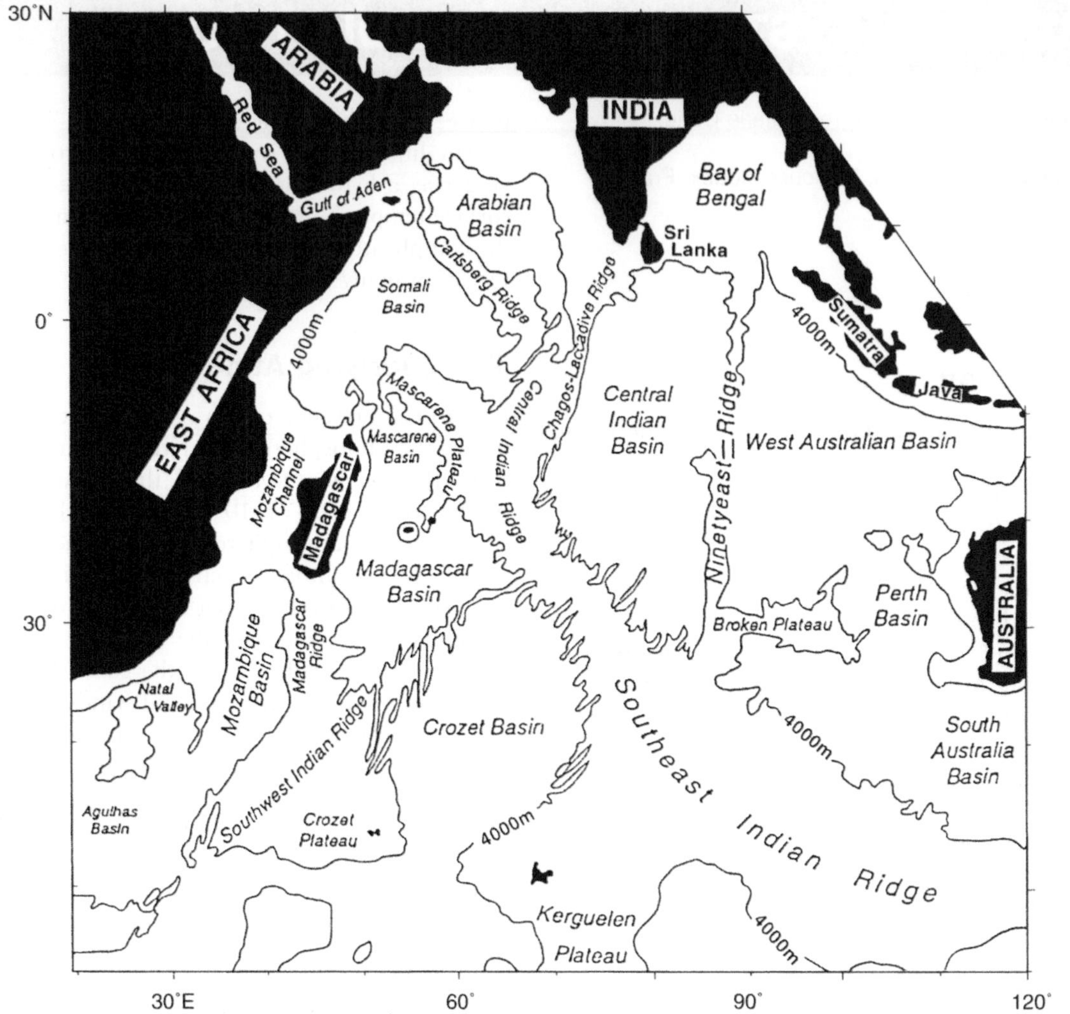

Figure 1 Map of the Indian Ocean with the different basins.

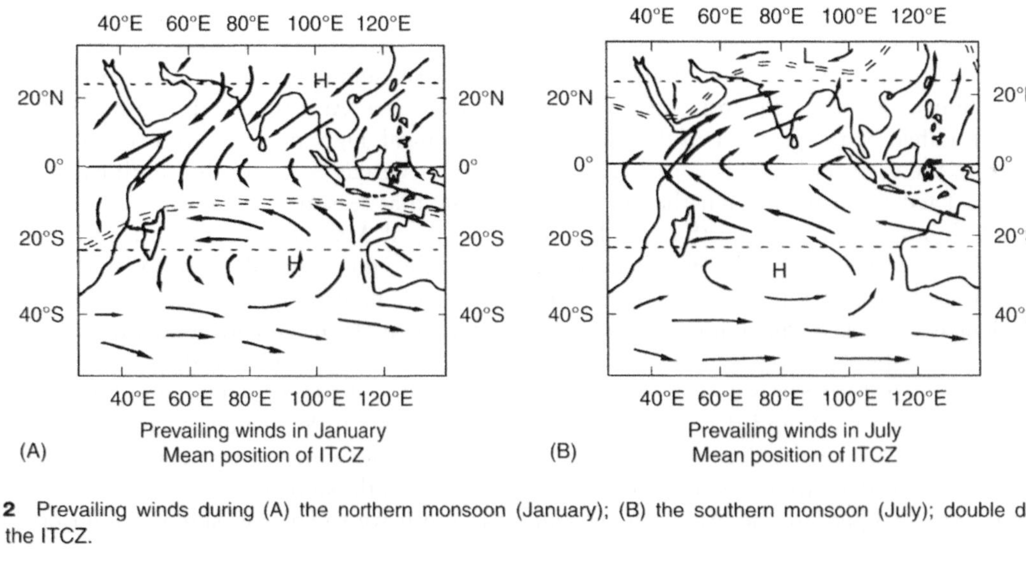

Figure 2 Prevailing winds during (A) the northern monsoon (January); (B) the southern monsoon (July); double dashed lines indicate the ITCZ.

stronger than the NE winter monsoon. A wind jet, also called the Finlater jet, develops along the high orography of the east African coast. As a consequence, the winds are the strongest on the western side of the Indian Ocean along the Somali coast towards the Arabian Sea, particularly north-east of Cape Guardafui (the horn of Africa) where the mean July wind speed is $12\,m\,s^{-1}$ with peaks exceeding $20\,m\,s^{-1}$. They are the strongest and the steadiest wind flow in the world. At the equator, the winds are moderate from the south and decrease eastward. In the southern Indian Ocean, the subtropical high pressure center intensifies and covers the whole width of the southern Indian Ocean during the southern winter (July) and the SE trades penetrate farther north than during the southern summer (January); they reach the equator in the western part of the ocean and are the strongest among the three oceans. During that season the air masses transported by the SE trade winds cross the equator in the west and continue, loaded with moisture, towards the Asian continent where they bring the awaited monsoon rainfall.

October–November corresponds to the second transition period between the end of the SW monsoon and the beginning of the NE monsoon. North of the Equator, the winds vanish and the sea surface temperature can exceed 30°C. At the equator moderate eastward winds blow again as during the first transition period, although they are usually slightly stronger.

This particular wind regime implies that at the equator, the zonal wind is dominated by a semi-annual period associated with the westerly winds of the transition periods, while the meridional wind presents a strongly annual period associated with the monsoon reversals. The winds off the equator also present a strong annual period.

Along the western Australian coast the northward winds, favorable for upwelling, are much weaker than in the eastern Pacific and Atlantic Oceans. They even drop down during the SW monsoon season. This is due to the different land–ocean distribution.

The seasonal changes of the winds south of latitude 10°S are smaller than to the north, and therefore the variability in the ocean circulation will also be smaller there. The next section will describe the currents in this area, before presenting the currents in the northern area.

The Currents in the Southern Part of the Indian Ocean

The strong anticyclonic subtropical gyre of the southern Indian Ocean is the result of the large wind stress curl between the Antarctic westerlies and the SE trade winds. Its northern branch is the westward-flowing South Equatorial Current (SEC) centered between 12°S and 20°S, fed in its northern part by the throughflow waters originating from the Indonesian Seas and corresponding to lower salinity waters (**Figure 3**). The SEC is the limit of the influence of the monsoon system. Its mean transport relative to the 1000 dbar level varies from 39 Sv in July–August to 33 Sv in January–February. Its latitudinal range varies between 8°S and 22°S in July–August and 10°S–20°S in January–February.

The SEC impinges on both the east coast of Madagascar and on the east African coast, resulting in several intensified boundary currents along these coasts. The SEC splits into a northward flow and a southward flow east of Madagascar near 17°S. The southern branch continues as the East Madagascar Current (EMC) carrying of the order of 20 Sv (0–1100 m), which ultimately joins the Mozambique Current and the Agulhas Current (AC) to the south west, besides some recirculation to the east into the subtropical gyre. The northern branch of the SEC splits again east of the African coast near cape Delgado (11°S) into the southward-flowing Mozambique Current (MC) and the northward-flowing East African Coastal Current (EACC). The Mozambique Current presents intense recirculation in the northern Mozambique Channel, but ultimately feeds roughly 20 Sv into the Agulhas Current which is the strongest western boundary current in the south Indian Ocean, transporting nearly 70 Sv.

The eastward-flowing south branch of the anticyclonic subtropical gyre of the southern Indian Ocean is part of the Antarctic Circumpolar Current (ACC). Nevertheless, north of the ACC, a South Indian Ocean Current (SIC) can be differentiated from the different cores of the ACC. The SIC comprises the eastward flow recirculating part of the Agulhas Current off South Africa and at depth transports North Atlantic Deep Water. The ACC transports Antarctic circumpolar waters with lower salinity than in the South Indian Ocean Current.

The eastern Indian Ocean is connected with the Pacific Ocean through channels in the Indonesian Archipelago. This has large consequences on the eastern Indian Ocean circulation and is expected to be one of the causes for the southward flow west of Australia, the Leeuwin Current, which flows opposite to the wind. The mean sea level, higher on the Pacific side than on the Indian Ocean side of the Indonesian Seas, drives a throughflow towards the Indian Ocean transporting warmer and fresher waters contributing to a high dynamic height in the north of the western Australian coast. This induces

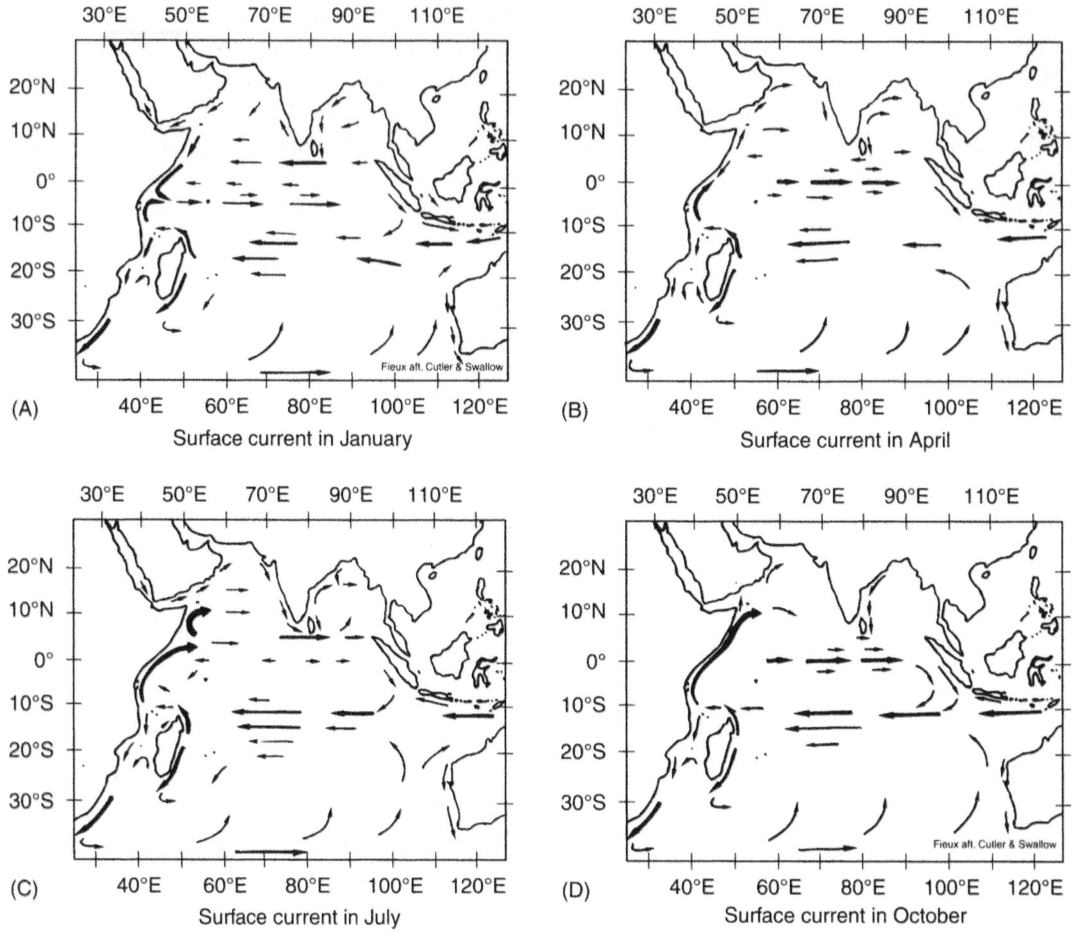

Figure 3 General surface circulation in the Indian Ocean: (A) during the NE monsoon; (B) during the transition period in April; (C) during the SW monsoon; (D) during the transition period in October. (Adapted from Cutler AN and Swallow JC (1984) *Surface Currents of the Indian Ocean* (to 25°S, 100°E): compiled from historical data archived by the Meteorological Office, Bracknell, UK. Institute of Oceanographic Sciences, Wormley, UK Rep. 187, 8pp and 36 charts.)

an alongshore pressure gradient off western Australia which drives the Leeuwin Current to the south and can even overwhelm the counteracting effect of the coastal upwelling induced by the weak southerly winds. At 22°S, the Leeuwin current is a 30–50 km wide and shallow (150–200 m) poleward jet close to the coast (max in May–June) with large intraseasonal interannual variability which is associated to an equatorward undercurrent.

Response of the Indian Ocean Circulation to the Wind Variability in the Northern Part

North of 10°S, the ocean circulation responds to the seasonally varying monsoon winds and as a consequence presents well defined seasonal characteristics. It is necessary to distinguish the circulations near the western boundaries, near the equator, in the northern ocean interior, and the eastern boundary current systems, which have different dynamics. They will be described successively, on the basis of observations as well as numerical or analytical modeling studies.

The Western Boundary Current System

As in the other oceans, the strongest currents are close to the western shores of the ocean as a result of the direction in which the earth rotates and the variation of the Coriolis parameter with latitude. North of 10°S (which is the limit of the monsoon influence), there are two western boundary currents: the East African Coastal Current (EACC) which always flows northward and the Somali Current which is the most intense and the most variable.

The EACC flows in continuity with the branch of the SEC which passes north of Madagascar and splits around 11°S. It runs northward throughout the year between latitudes 11°S and 3°S. Its surface speed can exceed 1 m s^{-1} during northern summer and its transport amounts to 20 Sv in the upper 500 m. Its northern end depends on the season. In northern winter, the EACC converges around 3°S–4°S with the south-going Somali Current to form the eastward South Equatorial Countercurrent. During the northern summer, the EACC merges into the north-going Somali Current.

The Somali Current is the most intense, but unlike the other western boundary currents it is highly variable due to the complete seasonal reversal of the winds. It was first studied during the International Indian Ocean Experiment in the 1960s, during INDEX (INDian ocean EXperiment which started in the 1970s) and during SINODE in the 1980s. Recently (1990–96), during WOCE (World Ocean Circulation Experiment) considerable amounts of new data were collected over the whole Indian Ocean. The Somali current develops in different phases in response to the winds.

During the transition period after the NE monsoon, in April, the Somali current flows south-westward along the coast south of 5°N, merging near the equator with the northward-flowing EACC. This feeds a south-eastward flow towards the ocean interior.

In early May, the Somali current responds rapidly to the onset of the local southerly winds and reverses northward in continuity with the northward EACC near the equator. By mid-May, the SW wind onset propagates northward and the current turns offshore towards the east at 4°N. North of that branch an upwelling wedge spreads out, bringing cold and enriched waters at the surface. Further to the north, the current flows northward from March onwards. When the onset of the strong summer monsoon winds occurs at these latitudes in June, the southern branch increases in strength and a strong anticyclonic gyre, called 'the great whirl' develops between 5°N and 10°N. Between the Somali coast and the northern branch of the great whirl a second upwelling wedge forms. Numerical models have shown that the location and motion of these structures are influenced by the distribution and strength of the wind forcing.

In August–September, when the winds decrease, the southern cold wedge propagates northward along the coast and meets with the northern one (although the latter probably also moves). It is only at that time that the Somali current is continuous from the equator up to 10°N and brings fresher waters into the Arabian Sea.

During the transition period in October–November, the northward Somali circulation decreases.

In December–February, during the NE winter monsoon, the Somali Current reverses southward from 10°N to 5°S where it converges with the northward EACC to form the South Equatorial Counter Current (SECC) flowing eastward. This countercurrent exists only during the NE winter monsoon and could be compared to the other Equatorial Counter Currents in the Atlantic Ocean and in the Pacific Ocean. It develops just north of the intertropical convergence zone (ITCZ) where the winds have an eastward component.

At the equator near Africa, the reversal affects only a thin surface layer below which (between 120 m and 400 m) there is a northward undercurrent, remnant of the SW monsoon season, followed again by a southward current below 400 m.

There are also western boundary currents in the Gulf of Bengal off the coasts of Sri Lanka and India.

From September to January, the currents are southward along the whole eastern coast of India and Sri Lanka, bringing fresh Bengal Bay water to low latitudes (6°N). In February–March the currents reverse to flow to the north along these coasts with a separation from the coast in the northern Bay of Bengal (19°N) in March. From April to August, the current reverses along the eastern coast of Sri Lanka where it flows to the south. Further north, from May to July, the separation from the coast of the northward current takes place around 16°N instead of 19°N. In July, north of 16°N, reversal to the south takes place. This seasonal cycle is markedly different from the one off the Somali coast. Modeling studies show that it is a response both to local winds, to the curl of the wind stress over the Bay of Bengal, with other contributions propagating along the coast from further east, and from the vicinity of the equator.

Equatorial Currents System

The winds at the equator are profoundly different in the Indian Ocean from the mostly westward winds in the Atlantic and the Pacific tropical oceans. Instead, in the Indian Ocean, there is a strong semi-annual cycle in the zonal winds and the mean zonal wind is westerly.

During the two transition periods between the monsoons, a strong eastward jet (called 'Wyrtki jet') occurs in a narrow band, trapped within 2°–3° of the equator, mostly in the central and eastern parts, driven by the equatorial westerly winds. Due to the efficiency with which zonal winds can accelerate zonal currents at the equator where the Coriolis force vanishes, the current speeds can rapidly reach >1 m s^{-1}. The jet usually peaks in November with velocities which can reach 1.5 m s^{-1}; it could also reach these values in May as there is large intraseasonal and interannual variability.

At the equator, in the middle of the Indian Ocean near Gan Island (73°E), measurements show currents throughout the upper 100 m in phase with local winds, which reverse four times a year. The associated change of current direction produces semi-annual variations in the thermocline depth and sea level. During periods of eastward flow the thermocline rises off Africa and falls off Sumatra corresponding to opposite displacements of the sea level. The set up of the jet is apparently triggered by the westerly winds during the transition periods. This forces a local response as well as waves propagating the response further to the east (Kelvin waves) and west (Rossby waves). The stopping of the jet seems to happen progressively from east to west, as it has been observed with drifting buoys. This is interpreted dynamically as westward propagating decelerating Rossby waves which are generated when the eastward jet reaches the coast of Sumatra.

During the fully developed SW monsoon in July–August, along the equator – aside from the extreme west where the strong north-eastward Somali Current occurs – the winds are southerly and light, and currents at the equator are weak and variable.

An eastward equatorial undercurrent embedded in the thermocline along the Indian Ocean equator exists only during January–June and is strongest in March at the end of the NE monsoon. It is confined between 2°30N and 2°30S and is weak east of 80°E. During the NE monsoon, it flows under a weak westward current until the eastward Wyrtki jet starts, then the whole upper layer flows eastward.

The Northern Interior Current System

During the NE monsoon in December–February, the northern Indian Ocean presents a current structure similar to those found in the other oceans. In the northern Arabian Sea, the circulation is not well defined during this season. There is a general westward flow south of 10°N, the North-east Monsoon Current (NMC), extending south to about 2°S, with speeds between 0.3 and 0.8 m s^{-1}. South of Sri Lanka, the current splits into a branch continuing westward and a branch which tends to follow the western coast of India, possibly after meandering in eddies off south-west India.

Further south between 2°S and 8°S the eastward South Equatorial Counter Current (SECC) flows eastward starting at the convergence of the southward Somali Current with the northward EACC (see above).

During the SW monsoon in the Arabian Sea, there is a general eastward flow in the South-west Monsoon Current (SMC), with more intense veins near 15°N as well as near 9–10°N (although the latter might be more developed in April–June) and near 5°N. In the Arabian Sea, there is some anticyclonic recirculation to the south of the northern eastward vein, very likely forced by the curl of the wind stress, and some indication of cyclonic eddies to the north of this circulation. The flow along the western coast of India is southward during the SW monsoon associated with a poleward undercurrent along the shelf which sometimes could reach the surface. In the northern Bay of Bengal, north of 15°N, the eastward currents are already set in April–May, which last until August (in particular near 15°N–17°N). Eastward currents also exist south of 8°N in the Bay of Bengal from April to September.

The currents are intensified south of Sri Lanka during both monsoons, resulting in particularly large eastward SMC and westward NMC. Numerous eddies are also present, especially in the western parts of the Arabian Sea and Bay of Bengal, associated with upwellings that are particularly strong off the Arabian coast during the SW monsoon.

The Eastern Boundary Current System Affected by the Monsoons

Along the eastern boundary the currents are also seasonally variable, except along Sumatra, south of the equator, where it is always south-eastward and flows against the winds in June–September. During the transition periods the Kelvin waves associated with the Wyrtki jet continue north-westward and south-eastward as coastal trapped Kelvin waves along the Indonesian islands. South of Java they reinforce the NE monsoon-driven south-eastward Java current, but work against the response of the Java boundary current to the SE monsoon onset in May–June. So when the equatorial Kelvin waves arrive along the Java coast in October–November, the reversal is faster than in May–June. In July–September when the Java current is north-westward, there is a convergence with the south-eastward Sumatra current south of Sumatra. It is also during that season that large-scale wind-driven upwelling occurs along the coast of Java.

The open eastern boundary of the Indian Ocean allows exchanges with the Pacific Ocean through the channels of the Indonesian Archipelago. This flow is called the Indonesian throughflow. It transports waters from the surface down to 1300–1400 m which are principally drawn from the northern Pacific and modified in the Indonesian archipelago under both effects of exchanges with the atmosphere and dynamical mixing. The resulting water entering the Indian Ocean is a well characterized Indonesian Water. The principal route of the throughflow goes through Makassar Strait then through Lombok Strait (350 m deep between Lombok and Bali), and north and south of Timor Island (mean sills depth around 1350 m). The few partial direct measurements show strong interannual, seasonal, and intra-seasonal variability. This throughflow is entrained westward into the SEC bringing fresh and warm water to the Indian Ocean. It is also part of the source waters of the southward Leeuwin current which brings fresh and warm waters to the south along the west Australian coast.

The surface flow along the western coast of India is usually south-eastward with a north-westward subsurface undercurrent, in particular during the SW monsoon with upwelling favorable winds. However, the currents reverse at the end of the year bringing a pulse of fresher Bay of Bengal water along the coast.

Deep Circulation

Most of the circulation described here concerns the upper part of the ocean. The deep circulation is relatively unknown. The Indian Ocean is separated into numerous deep basins connected through narrow passages (**Figure 1**). As a consequence, the Antarctic Bottom Water cannot reach the northern ocean basins. Recent long-term direct measurements were carried in the Crozet-Kerguelen Gap of the South-west Indian Ridge, one of the major channels through which Antarctic Bottom Water can move equatorwards. The annual northward transport of Antarctic Water at depth greater than 1600 m amounts to 11.5 Sv which, because it has undergone large dilution through mixing, corresponds to an initial volume of Antarctic Bottom Water of 2.5–3 Sv deduced from CFC distribution. By contrast, further north, in the Amirante Passage connecting at depth the Mascarene Basin and the Somali Basin, the flow of bottom water flowing northward has been estimated to be 2.5–3.8 Sv.

From the characteristics of the water masses, an intensification of the deep flow is found as deep western boundary currents against the eastern flanks of each meridional ridge separating the numerous deep basins. Some of this deep, intermediate, and subsurface water flowing northward in the Indian Ocean is upwelled and contributed to the cooling of the surface waters.

Conclusion

The northern Indian Ocean is a natural laboratory to study the effect of the wind on the oceanic circulation, as regularly twice a year the winds change direction rapidly and are particularly strong in the western boundary. The highest variability as well as the highest current speed of the world ocean are found in the Somali current.

At the equator, particularly between the two monsoon seasons, westerly wind bursts entrain a strong eastward equatorial jet twice a year which could have an effect on the strength of the transport coming from the Pacific Ocean. Some of the variability of these wind bursts seems to be related to the El Niño–La Niña climatic variability. Comparing the width and the external forcing of the Pacific and Indian oceans, the Indian Ocean at semi-annual frequency should behave dynamically like the Pacific at annual frequency.

See also

Somali Current.

Further Reading

Open University, Oceanography Course Team (1993) *Ocean Circulation*. Oxford: Pergamon Press.

Tomczak M and Godfrey S (1994) *Regional Oceanography: An Introduction*. Pergamon: Elsevier.

Fein JS and Stephens PL (eds.) (1987) *Monsoons*. Washington: John Wiley Sons.

INDIAN OCEAN EQUATORIAL CURRENTS

M. Fieux, Université Pierre et Marie Curie,
Paris Cedex, France

Introduction

Dynamically the equatorial area is a singular region on the earth because the Coriolis force is small, vanishing exactly at the equator. This results in a current structure that differs from that at other latitudes. Moreover, the equatorial current system of the Indian Ocean is entirely different from the current system found near the equator in the Pacific and Atlantic. This is principally due to its different wind forcing, which is described in the first section below. The systems of strictly equatorial currents at surface and at depth are reviewed in the second section. The third and fourth sections describe the North-East and South-West Monsoon Currents, north of the Equator, and the South Equatorial Countercurrent and the South Equatorial Current, south of the Equator.

The Atmospheric Circulation over the Equatorial Indian Ocean

The winds over the tropical Indian Ocean are quite different from the winds over the Atlantic and the Pacific tropical oceans, where the NE and SE trade winds blow always in the same direction. Instead, the Indian Ocean (**Figure 1**), north of 10°S, is under the influence of a monsoonal circulation, with complete reversal of the winds twice a year. The winter monsoon (December–March) blows from the NE in the Northern Hemisphere and from the NW south of the Equator toward the intertropical convergence zone (ITCZ) located near 10°S. The change in direction at the Equator comes from the change of sign of the Coriolis force. The summer monsoon (June–September) blows from the SW in the Northern Hemisphere in continuity with the SE trade winds of the Southern Hemisphere, particularly in the western part of the equatorial ocean. The winds are stronger during the summer monsoon season. At the equator, during the monsoons, the winds have a preponderant meridional component, southward during the winter monsoon and northward during the summer monsoon, particularly near the western boundary. The result is a strong annual cycle in the meridional component of the winds corresponding to the reversals between the NE and the SW monsoons, particularly in the western region along the Somali coast where the winds are the strongest.

Between the monsoons, during the two transition periods, at the Equator, moderate eastward winds blow in spring (April–May) and in fall (October–November), with maxima between 70°E and 90°E (**Figure 2**). They could blow into one or several eastward bursts with a large seasonal and interannual variability. During the NE monsoon, the mean zonal component of the wind is weakly westward west of 80°E, increasing in strength near the Somali coast. During the SW monsoon, the zonal components of the winds is weakly westward between 60°E and 70°E; west and east of that region, they are weakly eastward

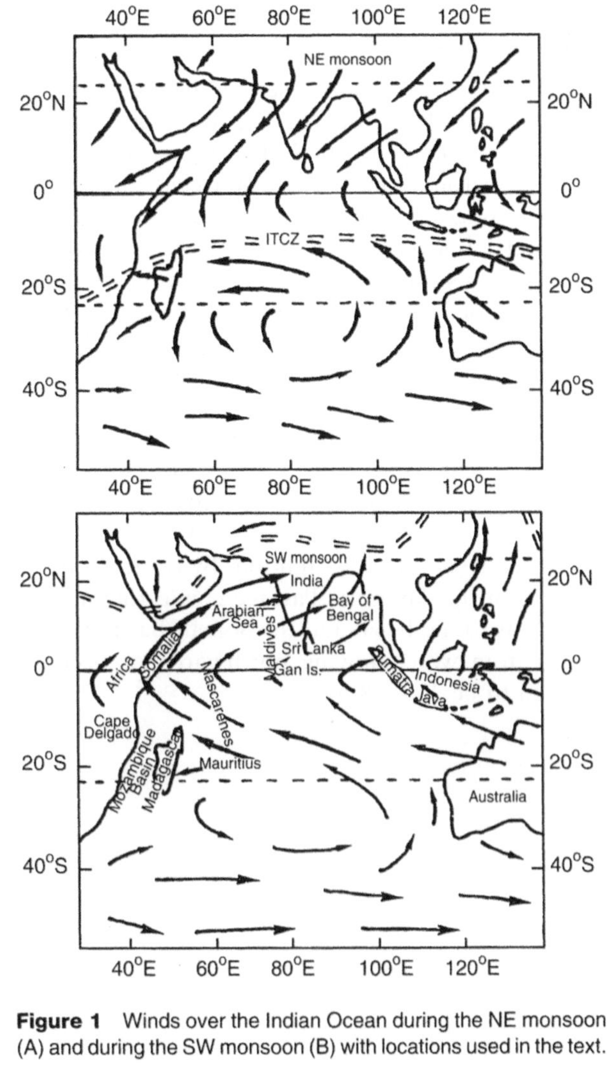

Figure 1 Winds over the Indian Ocean during the NE monsoon (A) and during the SW monsoon (B) with locations used in the text.

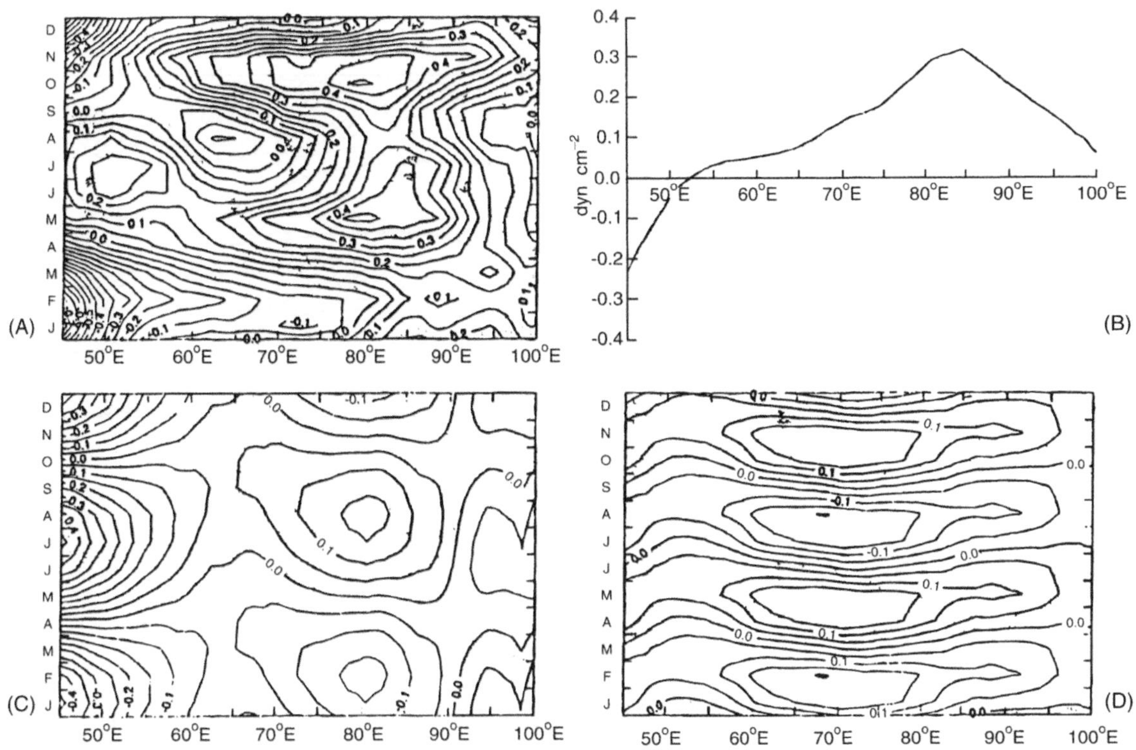

Figure 2 Longitude time plots of equatorial zonal wind stress averaged from 1°S to 1°N, determined from the FSU pseudostress climatology for the period 1970–1996: (A) the total wind in dyn cm^{-2}; (B) time-averaged mean zonal wind stress; (C) annual zonal wind stress component; (D) semiannual zonal wind stress component. Contour interval is 0.05 dyn cm^{-2}, eastward wind stress regions are shaded. (From W Han, JP McCreary, DLT Anderson, AJ Mariano (1999) Dynamics of the eastern surface jets in the equatorial Indian Ocean. *Journal of Physical Oceanography* 29: 2191–2209, (1999) copyright by the American Meteorological Society.)

with a maximum between 80°E and 90°E. As a result, the annual mean zonal wind is eastward, the opposite of what is found in the other equatorial oceans, and is maximum around 80°E to 85°E.

This particular wind regime is dominated by an annual and a semiannual cycle with similar amplitudes for the zonal components. The amplitude of the annual component is maximum near the western boundary associated with the stronger monsoon winds off Somalia. Relative annual component maxima are also found near 82°E and near the eastern boundary due to the southward monsoon winds extension south of Sri Lanka and along the Indonesian coast.

In contrast, the semiannual component of the equatorial zonal winds has a simple structure with a single maximum in the central ocean between 60°E and 80°E during spring and fall.

Currents at the Equator

Currents in the Extreme West, along the Somali Coast

Compared to the other major oceans, there are very few direct current measurements in the Indian Ocean and most of the surface currents information comes from the ship drifts and recently from satellite-tracked drifting buoys.

Along the Somali Coast, as seen above, the annual period is the dominant variability in the wind forcing. It is only relatively recently (1984–1986) that long-term direct current measurements were carried out at the equator within 200 km off the Somali coast. Above 150 m they show a seasonally reversing flow in phase with the NE and the SW monsoons, but a striking asymmetry between both monsoon seasons. While during the stronger SW summer monsoon the north-eastward Somali Current decays monotonically in the vertical, during the weaker NE winter monsoon the south-westward surface current is limited to the surface layer and a north-eastward countercurrent exists between about 150 m and 400 m, remnant of the SW monsoon current, followed again by weak south-westward current underneath down to about 1000 m (**Figure 3**). This sliced structure is confined to the equatorial region corresponding to the equatorial waveguide.

Surface Currents at the Equator, East of 52°E

Ship-drift climatology indicates that, at the equator, the surface currents reverse direction four times a

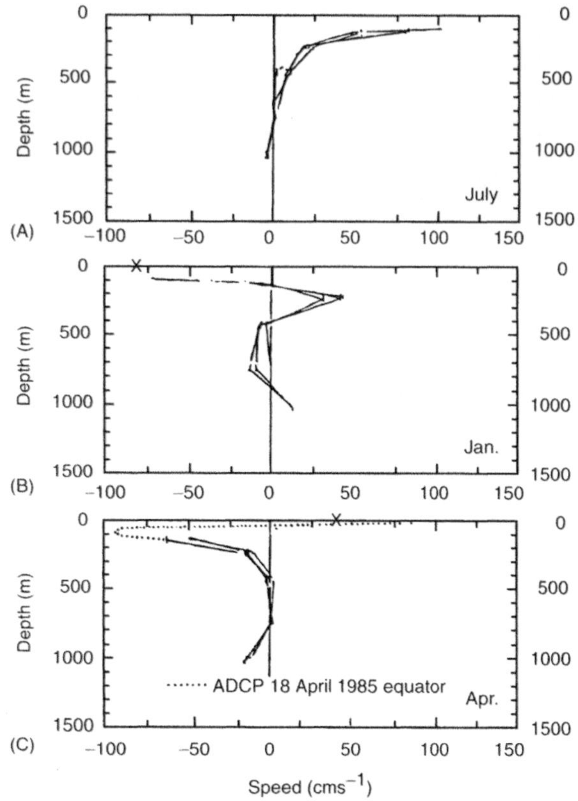

Figure 3 Monthly mean current profiles observed on three moorings in the Somali current at the equator during the NE monsoon (January 1985), the SW monsoon (July 1985), and during the transition period (April 1985); monthly mean of component parallel to coast. The crosses at the surface are historical ship drift data from Cutler and Swallow (1984). The dotted near-surface profile in April is from the ship mounted acoustic Doppler current profiler. (From F Schott (1986). Seasonal variation of cross-equatorial flow in the Somali current. *Journal of Geophysical Research* 91(C9): 10581–10584, copyright by the American Geophysical Union.)

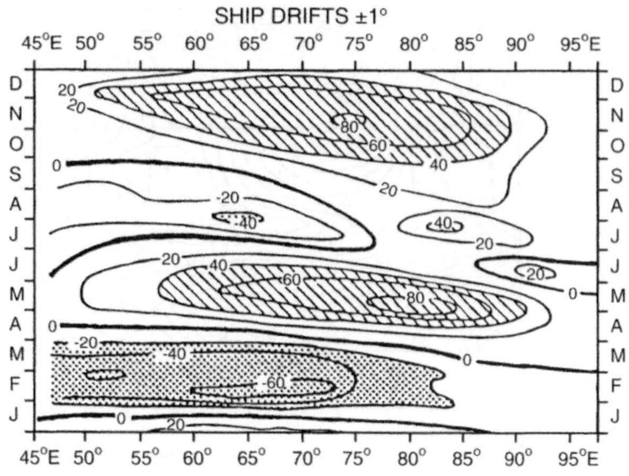

Figure 4 Mean ship drift estimates of the current between 1 N and 1 S in cm s^{-1}. Dots correspond to westward currents and stripes to eastward currents. (From G. Reverdin (1987). The upper equatorial Indian Ocean: The climatological seasonal cycle. *Journal of Physical Oceanography* 17: 903–927, copyright by the American Meteorological Society.)

year. During the two transition periods, under the influence of the equatorial eastward winds, a strong surface eastward jet (known as the 'Wyrtki jet' because Klaus Wyrtki first identified it by looking at different shipdrift atlases in the 1960s) sets up in a narrow band, trapped in the equatorial wave guide within 2–3° of the equator (**Figure 4**). The Wyrtki jet exists mostly in the central and eastern regions and disappears in the western part where the currents have a strong meridional component as shown above. Owing to the efficiency with which zonal winds can accelerate zonal currents at the equator where the Coriolis force disappears, the current speed can reach 1 m s^{-1}. The ship drift climatology indicates roughly the same strength for the two jets, with the October–November one (1 m s^{-1}) slightly stronger than the April–May one (0.90 m s^{-1}). As there is a large seasonal and interannual variability in the eastward winds, strong currents could be found in April–May instead of October–November during some years.

At the equator, in the middle of the Indian Ocean, the first long-term current measurements (1973–1975), near Gan Island (73°10E–0°41S), show energetic eastward currents throughout the upper

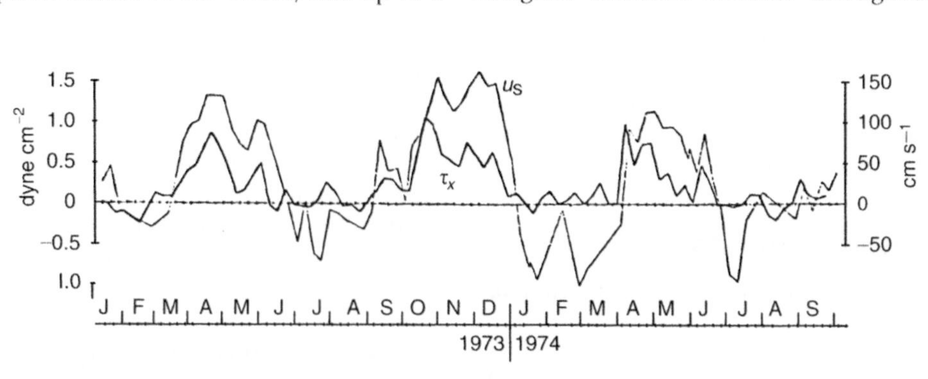

Figure 5 Comparison of weekly average zonal wind stress (τ_x) in dyne s^{-2} and surface zonal currents (u_s) in cm s^{-1} at Gan Island (73 10′E–0 41′S) in 1973–1975 (Reprinted from RA Knox. On a long series of measurements of Indian Ocean equatorial currents near Addu Atoll, copyright 1976, *Deep-Sea Research* 23: 211–221, with permission from Elsevier Science.)

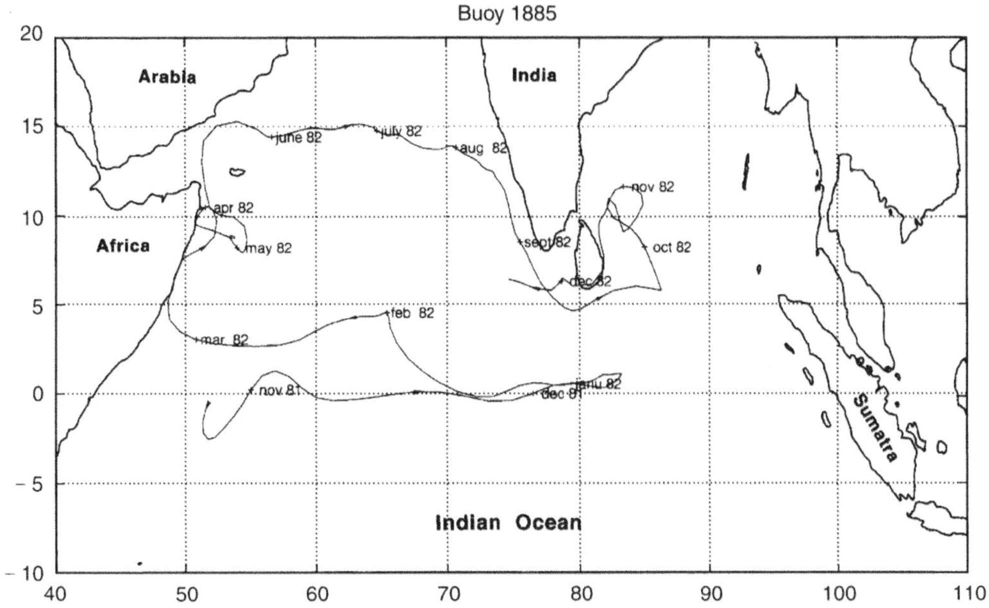

Figure 6 Trajectory of one drifting buoy (drogued at 20 m) entrained eastward into the equatorial Wyrtki jet in November–December 1981, then in the reversal in December–January, crossing the Arabian Sea, and undergoing the seasonal reversal of the current south of Sri Lanka. (From M. Fieux (1987) Océan Indien et Mousson *'Unesco Anton Bruun Lecture'* 15 pages, unpublished manuscript.)

100 m mixed layer in phase with the zonal eastward component of local winds during the two transition periods between the monsoons (**Figures 5, 6**). The establishment of this eastward jet could be explained through eastward-propagating Kelvin waves triggered by the eastward-going winds during the transition periods. The stopping of the jet is thought to be the result of the reflection of the eastward-propagating Kelvin waves on the eastern coast into westward-propagating equatorially trapped Rossby waves that progressively impede the jet from east to west. This has been observed with drifting buoys (**Figure 7**). The eastward currents at the equator are convergent (contrary to the westward currents). Consequently, drifting buoys launched in the jet stay in it and are a good means of observing its variability. In **Figure 8** the drifting buoys were entrained into the May jet. They then stopped during the SW monsoon drifting slowly south of the equator and were taken up again eastwards into the November jet.

Between the strong eastward flow periods, the equatorial surface currents are westward and much weaker (**Figures 4** and **6**). The associated change of current direction during each transition period produces semiannual variations in the thermocline depth and sea level. During periods of eastward flow when warm water is carried toward the east, the thermocline deepens off Sumatra and rises off Africa, corresponding to opposite displacements of the sea surface. Strong eastward flow at the equator entrains a surface convergence and as a result a downwelling in the upper layer, contrary to what is found in the

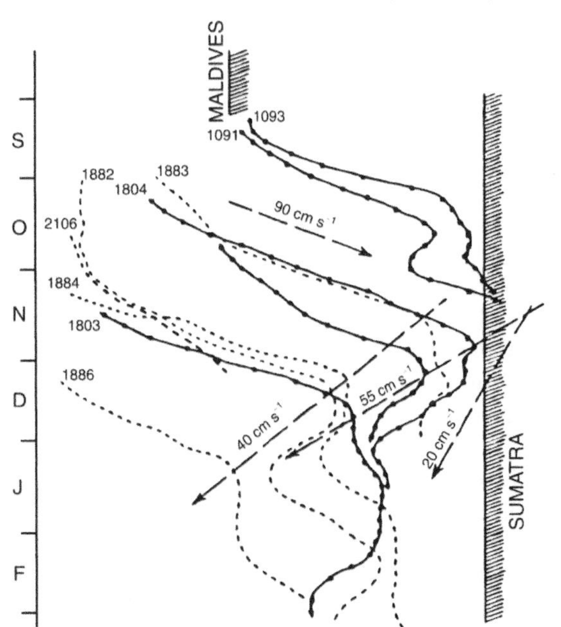

Figure 7 Drifting buoy trajectories between 60 E and Sumatra during the period of September–February. Full lines are for 1979–1980 and dashed lines are for 1980–1981. Dashed arrows indicate different speeds on this time–longitude diagram. During the first year the reversal propagation speed toward the west was around 55 cm s $^{-1}$, and around 40 cm s $^{-1}$ during the second year. (From G Reverdin, M Fieux, J Gonella and J Luyten (1983) Free drifting buoy measurements in the Indian Ocean equatorial jet. In: *Nihoul JCJ (ed.) Hydrodynamics of the Equatorial Ocean.* Amsterdam: Elsevier Science, 99–120.)

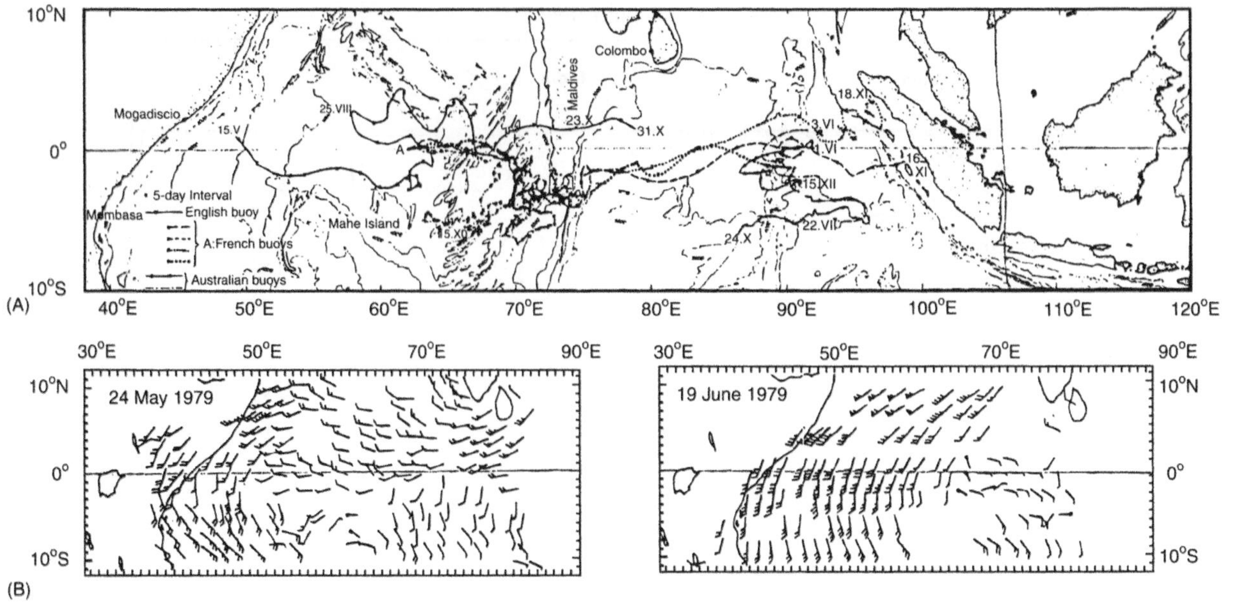

Figure 8 (A) Satellite-tracked drifting drogued (at 20 m) buoys trajectories in 1979; dots on the trajectories are spaced at 15-day intervals. (B) Wind vectors estimated from cloud trajectories on satellite images during the equatorial eastward wind period (24 May 1979) and during the fully developed SW monsoon (19 June 1979). (The short tick on the arrow corresponds to 5 knots, the longer one to 10 knots, and the triangle to 50 knots.) (Reprinted with permission from JR Luyten, M Fieux and J Gonella. Equatorial currents in the Western Indian Ocean, *Science* 209: 600–603, copyright 1980 American Association for the Advancement of Science.)

other oceans. It is only during the weak westward flow that there is a weak upwelling.

Long-term direct current measurements carried between Sri Lanka and 0°45′S, in 1993–1994, reveal a large seasonal asymmetry that year in the semi-annual eastward jet transport, with 35 Sv in November 1993 (surface velocities exceed 1.3 m s^{-1}) and only 5 Sv in May 1994. These transports were calculated with a lower boundary sets at 200 m as, sometimes, the eastward currents are not confined to one core but extend into the ray-like structures of the equatorial waves that continue to greater depths as in November 1993 (**Figure 9**). This large variability is due to a large seasonal and interannual variability of the zonal winds partly related to the Southern Oscillation. For example, during the El Niño of the century in 1997, the equatorial eastward winds in the Indian Ocean completely disappeared in October–November and were replaced by westward winds.

Numerical model results show that direct wind forcing is the dominant forcing mechanism of the equatorial surface jets. The semiannual response of the current to the wind is nearly three times as strong as the annual one, despite similar amplitudes in the corresponding wind components. This is due to the simpler zonal structure of the semiannual wind and to the resonance of the basin to the semiannual component. The mixed layer shear flow seems also to enhance the semiannual response. The jet is strengthened when the mixed layer is reduced, particularly when the precipitation during the northern hemisphere summer and fall thins the mixed layer in the eastern ocean and so could strengthen the November jet in the east. The reflection of equatorial Kelvin waves into westward going Rossby waves at the eastern boundary is also important in weakening and even canceling the directly forced eastward jet about 2 months after the wind onset. The presence of the Maldives islands blocks part of the equatorially trapped waves. The effect on the semiannual waves is to weaken both jets, and the effect on the annual waves tends to weaken the May jet and to strengthen the November jet.

Currents at Depth at the Equator

It is only during the NE monsoon, the season when the large-scale wind structure resembles the Pacific and the Atlantic ones, that a similar eastward equatorial undercurrent (EUC) embedded in the thermocline exists. Observations of the equatorial undercurrent are scarce. It was first observed during the International Indian Ocean Expedition (IIOE) in March–April 1963, between 53°E and 92°E, with speeds up to 0.8 m s^{-1}, then at Gan Island (73°E) in March 1973 with velocities of up to 1 m s^{-1}, but not the following year. In 1975 and 1976, it was found

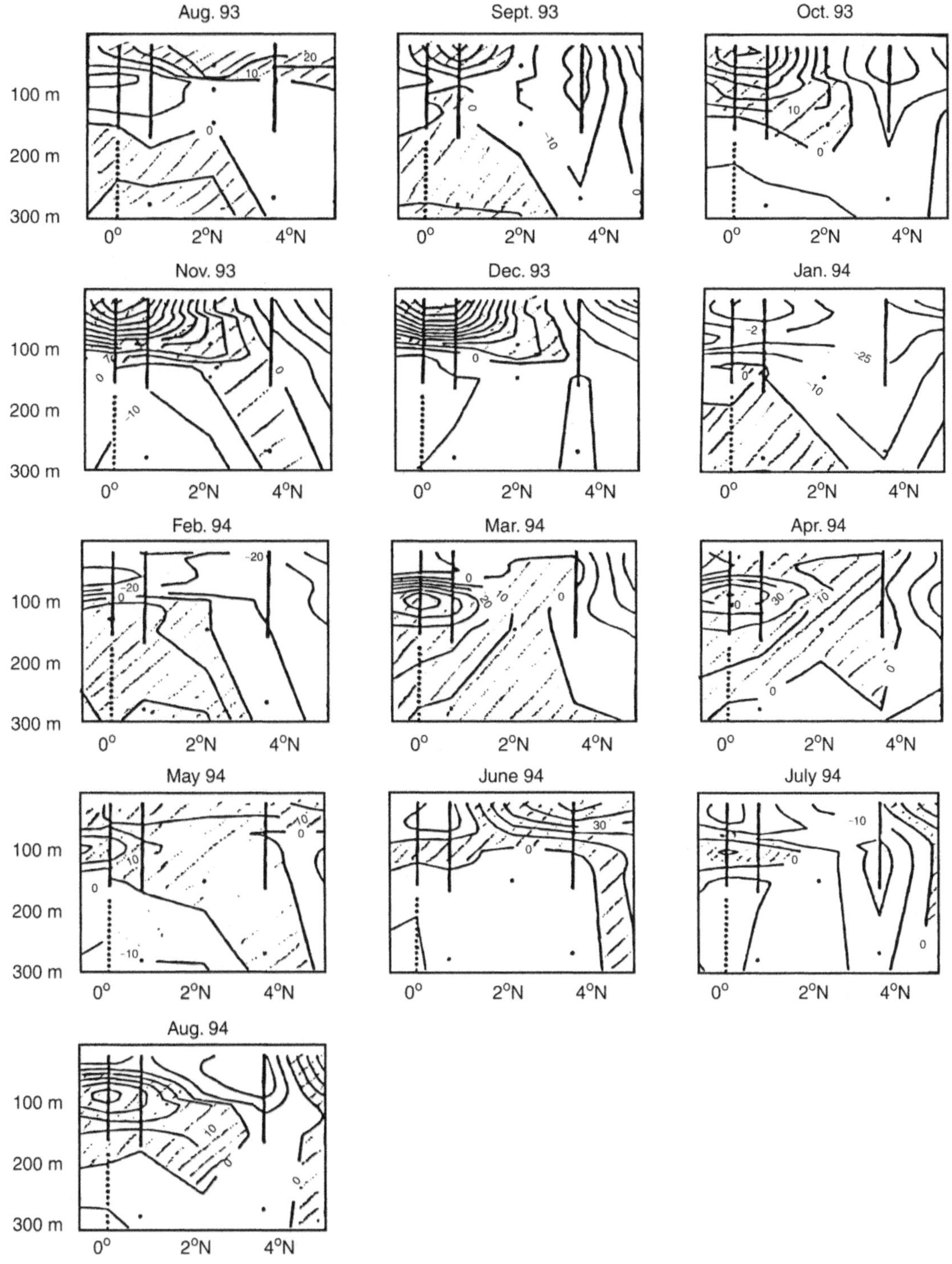

Figure 9 Monthly mean zonal velocities in cm s^{-1} for the upper 300 m, south of Sri Lanka, along 80°30′E, between August 1993 and August 1994. Contour interval is 10 cm s^{-1} and shaded areas indicate eastward currents. Note the unusual reappearance of the EUC in August 1994. (From J Reppin, F Schott and J Fischer (1999). Equatorial currents and transports in the upper central Indian Ocean: Annual cycle and interannual variability, *Journal of Geophysical Research* 104(C7): 15495–15514, copyright by the American Geophysical Union.)

from January extending into May–June at 55°30′E with observed speeds reaching 0.8 m s^{-1} in February and March at the end of the NE monsoon (**Figure 10**). Direct measurements give a maximum transport of 17 Sv, in March–April 1994, at the longitude of Sri Lanka (80°30′E) (**Figure 9**). It is confined between 2°30′N–2°30′S with slight meandering and is weak east of 80°E. Its core is around 100 m in the upper equatorial thermocline. During the NE monsoon, it flows under a weak westward current until April–May when the eastward Wyrtki jet starts, then the whole upper layer flows eastward. It stops in May–June but surprisingly, in 1994, it reappeared in August.

The existence of the EUC is related to equatorial westward winds, which force a westward surface current that builds up a zonal pressure gradient below the mixed layer that maintains the eastward undercurrent. The slope of the sea surface is opposite to the slope of the thermocline. The reappearance of the undercurrent in August 1994 is effectively related to anomalous onset of westward winds in the eastern part of the ocean, during May and June 1994. They force a westward surface current, again building a subsurface zonal pressure gradient, and thus an eastward undercurrent reappeared with some delay in August 1994.

In 1976, current profiles show that the vertical velocity structure in the vicinity of the equator is characterized by small vertical scales of order of 50–100 m in the upper layer increasing with depth throughout the water column. This deep jetlike structure is trapped to within 1° of the Equator and has a timescale of the order of several months.

One-year current-meter measurements made at 200 m, 500 m, and 750 m at the Equator, between 48°E and 62°E, and recently at 80°E, present a dominant semiannual reversal of the zonal component at all depths, much deeper than could be explained by direct wind forcing (**Figure 11**). Furthermore, these reversals do not happen at the same time at different depths. This seasonal cycle, with larger vertical scales, has been shown to penetrate vertically. The measurements suggest a mixture of equatorial Kelvin waves and long equatorial Rossby waves propagating downward from the surface where they are forced by the winds. The zonal velocity shows upward phase propagation, and downward energy propagation away from the surface. The behavior of the zonal currents is characteristic of an eastward-propagating equatorial Kelvin wave and a westward-propagating long equatorial Rossby wave. The ratio of the semiannual energy in the east current component to that in the north component is 40 to 1. This shows how much the equatorial ocean is a barrier to the meridional motions.

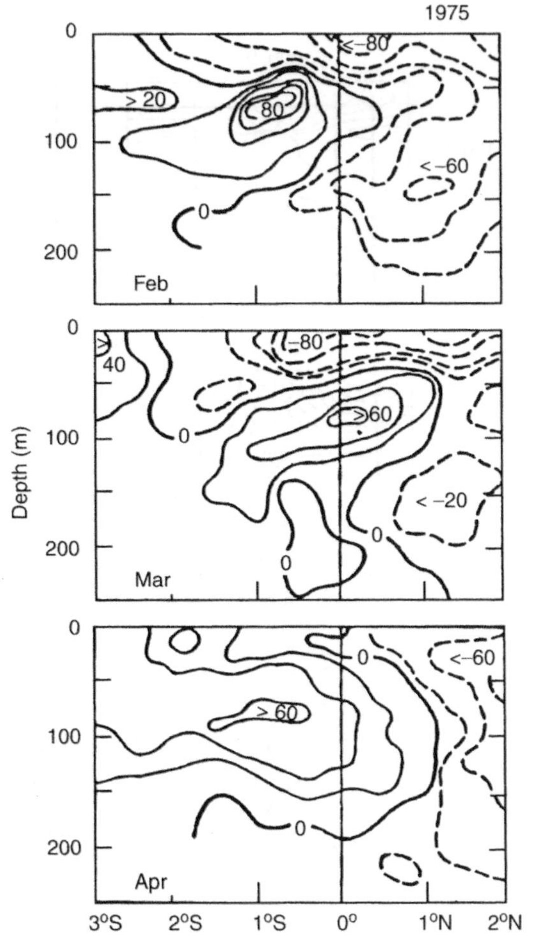

Figure 10 Meridional section along 53°E of zonal velocities in February, March, and April 1975 showing the equatorial undercurrent. Contours interval is 20 cm s^{-1}. Plain contours correspond to eastward currents and dashed contours to westward currents. (From A Leetmaa and H Stommel (1980) Equatorial current observations in the Western Indian Ocean in 1975 and 1976, *Journal of Physical Oceanography*, 10: 258–269, copyright by the American Meteorological Society.)

Currents North of the Equator

South of Sri Lanka

North of the Equator, the monsoon forcing drives a general eastward flow, the South-west Monsoon Current (SMC), during the fully developed SW monsoon, and a general westward flow, the North-east Monsoon Current (NMC), during the NE monsoon, extending south to about 2°S in January–February. The exchanges between the Arabian Sea and the Bay of Bengal are restricted to the south of

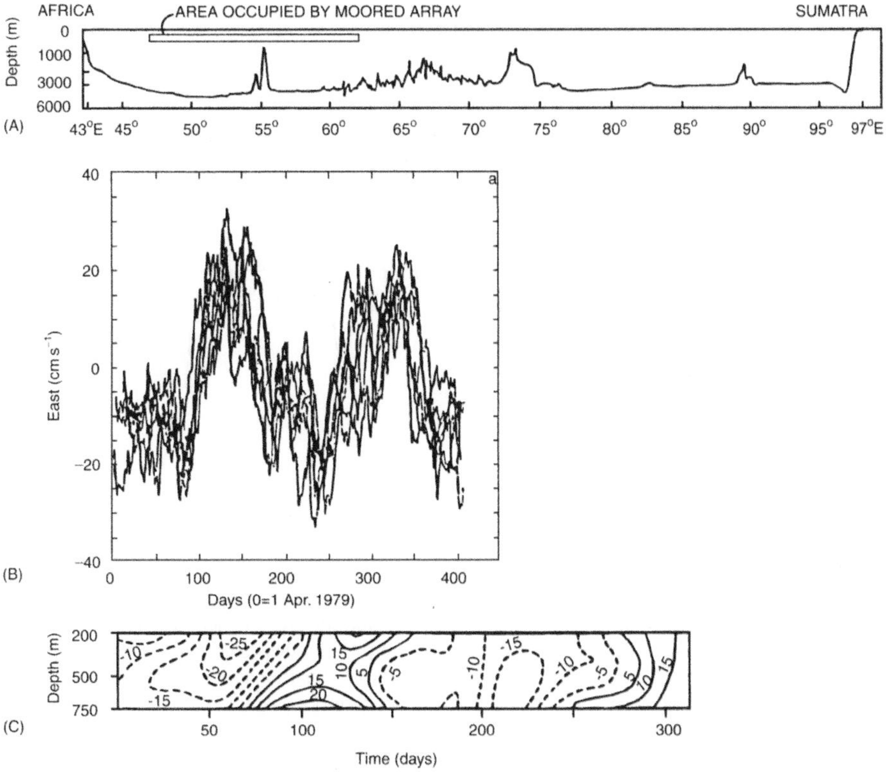

Figure 11 (A) Vertical section of the Indian Ocean along the Equator with the location of the current-meter array between 47 E and 62 E. (B) Time series of east velocity from seven current-meters at 750 m depth starting on April 1 1979. (C) Contour plot of low-frequency east velocity as a function of depth and time (day 01 April 1979). At each nominal depth (200, 500, 750 m), four records are filtered by a 30-day running mean and averaged together. The contour interval is 5 cm s^{-1}. (From JR Luyten and DH Roemmich (1982) Equatorial currents at semiannual period in the Indian Ocean, *Journal of Physical Oceanography* 12: 406–413, copyright by the American Meteorological Society.)

India and Sri Lanka. Drifting-buoy trajectories show the seasonal current reversal in that restricted region (**Figure 6**).

Direct observations carried out in 1991–1994 south of Sri Lanka show that the monsoon currents are mostly confined in the upper 100 m. In August 1993, the eastward SMC extended to the equator and retracted north of 4°30′N in September. It was replaced in October by the westward NMC, extending south to about 2°N, with speeds between 0.3 and 0.8 m s^{-1} (**Figure 9**). Shipdrifts show that, around the Maldives, the NMC splits into a branch that bends south-westward and a branch that follows the western coast of India, bringing low-salinity Bay of Bengal water into the Arabian Sea. The NMC maximum flow lies north of 4°N with a mean transport of 10–12 Sv (**Figure 12**). In May the current reverses eastwards into the SMC again. The SMC transport reaches about 8–10 Sv with speeds up to 0.75 m s^{-1} in July. The SMC is sometimes separated from the coast of Sri Lanka by a coastal westward countercurrent bringing Bay of Bengal water. The annual mean flow past Sri Lanka was 2–3 Sv westward in 1993–1994. Numerical models show that these monsoon currents are driven by the large-scale tropical wind field. Contrary to the equatorial circulation where the semiannual period prevails, the annual component dominates the upper layer flow north of 4°N.

Currents South of the Equator

Apart from shipdrifts and satellite-tracked drifting buoys, there are no direct long-term current measurements in the South Equatorial Countercurrent and only two in the South Equatorial Current.

The South Equatorial Countercurrent

In contrast to the Pacific and the Atlantic oceans, where a countercurrent is found year-long north of the Equator between the North Equatorial Current and the South Equatorial Current, in the Indian Ocean the countercurrent is found south of the Equator and has a large extent only during a short period of the year.

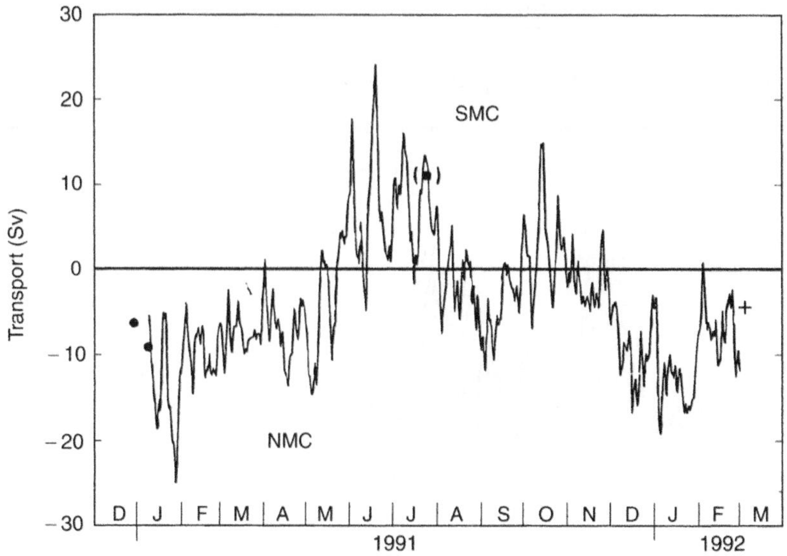

Figure 12 Seasonal variability of the transport in the upper 300 m from moorings south of Sri Lanka, between 3°45′N and 5°52′N, in 1991–1992. (The solid circles are from ship-mounted acoustic Doppler current profiler measurements in December 1990–January 1991; the solid circle in parentheses is for July 1991, and the cross is from Pegasus profiler measurements in March 1992.) (From F Schott, J Reppin, J Fisher and D Quadfasel, Currents and transports of the monsoon current south of Sri Lanka, *Journal of Geophysical Research* 99(C12): 25127–25141, copyright by the American Geophysical Union.)

During the winter monsoon, the Somali current flows southward, crosses the Equator, and merges with the northward flowing East Africa Coastal Current (EACC), at about 2–4°S, to form the eastward South Equatorial Countercurrent (SECC). It is a region of high eddy activity. The SECC is found between 2°S and 8°S during January–March. During that season the winds have an eastward component at those latitudes just north of the ITCZ. The speeds vary between 0.5 and 0.8 m s^{-1} in the west, getting weaker in the east. In March 1995, at 80°E, observed surface velocities exceed 0.7 m s^{-1} and the current extended to about 1100 m, with an eastward geostrophically deduced transport of about 55 Sv. In the east, part of it continues into the Java Current and part of it recirculates southward into the South Equatorial Current (SEC). It is detected in the meridional slope of the thermocline which slopes downward toward the Equator, with an opposite upward slope of the sea surface (**Figure 13**). Together with the SEC and the EACC, the SECC forms an elongated cyclonic gyre.

In the latitude range about 10–13°S and between 20°S and 25°S, recent measurements of the sea level variability through satellite altimetry show westward propagation of sea level anomalies corresponding to semiannual and annual Rossby wave characteristics. The wind-driven model results show westward propagation of Rossby waves in the shear zone between the SECC and the SEC that are obstructed and partially reflected by the Mascarenes banks (55–60°E). In the west, at the end of the winter season, in March–April, when the eastward winds start on the equator, the outflow from the EACC into the SECC begins to move northward toward the Equator and the eastward flow at that time is mostly equatorial.

The South Equatorial Current

The westward South Equatorial Current forced by the south-east trade winds extends south of the SECC. It represents the northern branch of the South Indian subtropical gyre. It is seen in the meridional downward slope of the thermocline towards the south (**Figure 13**). It is partly fed, in its northern part, between 10°S and 14°S, by the low-salinity throughflow jet originating from the western Pacific Ocean through the Indonesian Seas carrying about 4 to 12 Sv with larger extremes depending on the year. The low salinity extends down to about 1200 m. In March 1995, at 80°E, measured surface velocities reached 0.7 m s^{-1}. The SEC is the limit of the influence of the monsoon system and separates the northern and southern Indian Ocean. It is stronger during July–August when the SE trade winds are stronger. Its indirectly computed mean transport relative to the 1000 m level varies from 39 Sv in July–August to 33 Sv in January–February with large uncertainty. Its mean transport increases from east to west. Its latitudinal range varies between 8–22°S in July–August and 10–20°S in January–February. Its velocities range between 0.3 m s^{-1} and 0.7 m s^{-1}.

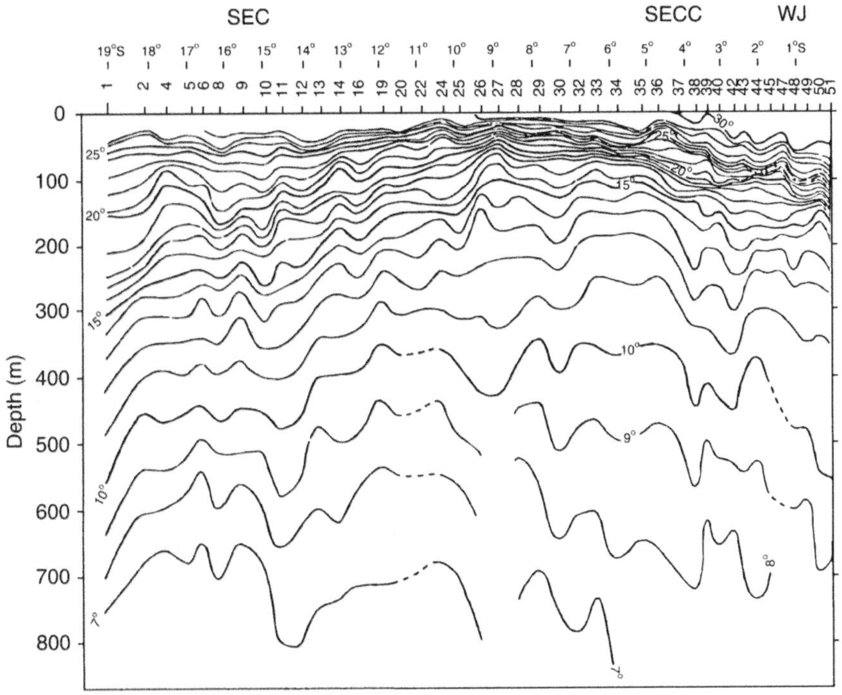

Figure 13 Meridional temperature section along 54–55 E in May 1981 with 1 C isotherm interval. The downward slope of the thermocline form 9 S toward the south corresponds to the SEC, and the downward slope toward the Equator corresponds to the SECC. Close to the Equator, the accentuated slope corresponds to the May Wyrtki jet. (From M Fieux and C Levy (1983) Seasonal observations in the western Indian Ocean. In: *Nihoul JCJ (ed.) Hydrodynamics of the Equatorial Ocean*. Amsterdam: Elsevier Science, 17–29.)

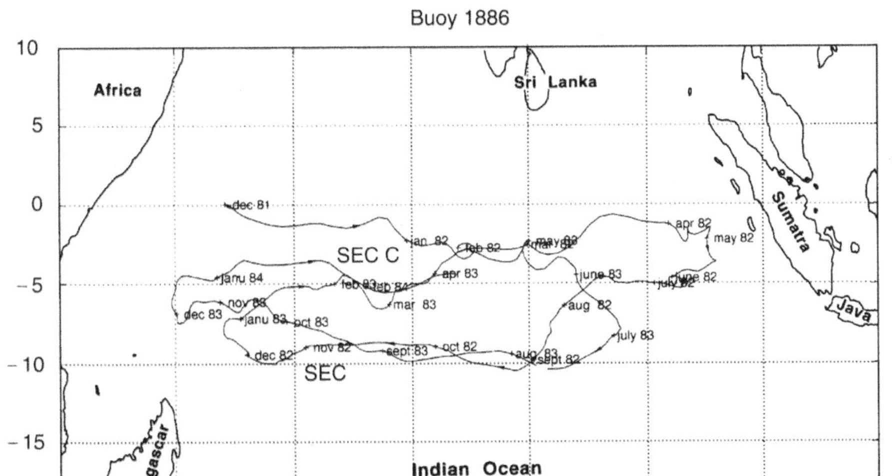

Figure 14 Satellite-tracked drifting buoy trajectory between November 1981 and February 1984. The buoy, drouged at 20 m, was entrained eastward in the November jet then in the SECC and back westward into the SEC. One year later it was again entained in the same elongated gyre. (From M Fieux (1987) Océan Indien et Mousson, *'Unesco Anton Bruun Lecture'*, unpublished manuscript.)

The SEC impinges both on the east coast of Madagascar and on the east African coast, resulting in several intensified boundary currents along these coasts. East of Madagascar, the SEC splits near 17°S into a northward flow, carrying about 27 Sv near 12°S, and a southward flow, transporting 20 Sv between the surface and 1100 m near 23°S. At 12°S energetic boundary current transport variations occur at the 40–55-day period, contributing to about 40% to the total transport variance, while at 23°S the 40–55-day period fluctuations contribute only 15% to the total transport variance. The northern branch of the SEC splits again east of the African coast near Cape Delgado (11°S) into the southward flowing Mozambique Current (MC) and the northward flowing East African Coastal Current (EACC).

Drifting-buoy trajectories describe an elongated cyclonic gyre of the equatorial current system composed of the equatorial eastward jet or the SECC, depending on the season, the off Sumatra SE current, the westward SEC, and the EACC. Some drifting buoys, launched during a transition period at the Equator, carried into the eastward Wyrtki jet and in the SECC, crossed the basin, then were driven southeastward into the Sumatra current, then crossed the basin westward into the SEC, and flowed back to the Equator in the western region exactly one year later when they were again carried into the SECC (**Figure 14**).

Conclusion

Owing to stronger winds in the tropical Indian ocean, the currents are stronger than in the Pacific and Atlantic Oceans but seasonally are highly variable. Away from the western boundary the equatorially trapped long waves (Kelvin and Rossby) explain most of the observed seasonal variations of the equatorial currents.

In contrast to the Atlantic and Pacific Oceans, equatorial upwelling is weak in the Indian Ocean. During the period of strong current, the surface flow is eastward, associated with a strong convergence at the surface inducing an equatorial downwelling. The upwelling regions are found instead north of the equator, along the Somalia, the Arabian, and the Indian coasts, and are seasonally depending.

The equatorial current structure in the Indian ocean is complex and further long-term observations as well as modeling efforts are needed to better understand its seasonal and interannual variability and its role in the large-scale meridional exchanges between the northern and the southern Indian Ocean.

See also

Agulhas Current. Current Systems in the Indian Ocean. Rossby Waves. Somali Current. Wind Driven Circulation.

Further Reading

Cutler AN and Swallow JC (1984) *Surface Currents of the Indian Ocean (to 25°S, 100°E): Compiled from Historical Data Archived by the Meteorological Of Tce, Brack-nell, UK*. Wormley, UK: Institute of Oceanographic Science (unpublished report) 187, 8pp. and 36 charts.

Fein JS and Stephens PL (eds.) (1987) *Monsoons*. Washington DC: NSF and Wiley.

Open University Oceanography Course Team (1993) *Ocean Circulation*. Oxford: Pergamon Press.

Tomczak M and Godfrey S (1994) *Regional Oceanography: An Introduction*. Oxford: Pergamon Elsevier.

SOMALI CURRENT

M. Fieux, Université-Pierre et Marie Curie, Paris, France

Introduction

The western Indian Ocean is the only region of the world where a large boundary current, as strong as the Gulf Stream, reverses twice a year in response to the wind reversals during the north-east winter monsoon and the south-west summer monsoon. This region of the Somali current is known to undergo the highest variability of the world ocean circulation.

Along the Somali coast, the reversals of winds and currents, known for many centuries, have been used by the Arabic traders for their navigation along the African coast and towards India. The term 'monsoon' comes from the Arabic word 'mawsin', which means seasonal.

Far from the large oceanographic research centers, the Indian Ocean used to be relatively poorly observed. The first large-scale international experiment was set up during the 1960s: the International Indian Ocean Experiment (IIOE), whose results were gathered into the Oceanographic IIOE Atlas. Until recently, the Somali current was known to reverse just twice a year with the direction of the monsoon, flowing south-westward during the NE monsoon and north-eastward during the SW monsoon. It is only with the international 'INDEX' experiment in 1979, under the initiative of Henry Stommel, that a careful survey of the response of the Somali current to the onset of the SW monsoon has been carried out.

Winds

The winds are the main driver of currents, in particular near the surface; therefore, we first recall their main characteristics. Their seasonal variability over the western boundary of the Indian Ocean can be described in four periods: the winter monsoon period, the summer monsoon period, and the two transition periods between the two monsoons (*see* Figure 2 of Current Systems in the Indian Ocean).

North of the Equator, the winter (NE) monsoon blows from the north east, with moderate strength, between December to March–April. At the Equator the winds are weak and usually from the north.

During the transition period between the end of the NE monsoon and the beginning of the SW monsoon, in April–May, the winds north of the Equator calm down. At the Equator moderate eastward winds blow, which contrast with the westward winds over the equatorial Pacific and Atlantic Oceans. As early as the end of March or early April, the SE monsoon starts south of the Equator, between the ITCZ (Inter-Tropical Convergence Zone) about 10°S and the Equator, with southerly winds along the East African Coast.

In most years, north of the Equator, the onset of the SW monsoon develops in two phases. The onset of the SW monsoon involves a reversal of the winds, which reach the Equator in early May as weak winds and progress northward along the Somali coast. Then a strong increase in wind occurs typically in late May to mid-June. In some years the onset can be gradual with no intermediate decrease between the first phase and the full-strength winds. They reach their full strength over the Arabian Sea usually in June–July. The summer (SW) monsoon blows steadily from the south west from June to September–October and is much stronger than the winter monsoon. Along the high orography of the east African coast, a low-level wind jet, called the Findlater jet (a kind of atmospheric western boundary flow) develops, bringing the strongest winds along the Somali coast toward the Arabian Sea, particularly north-east of Cape Guardafui (the horn of Africa). These are the strongest steady surface wind flows in the world, with mean July wind speed of $12\,m\,s^{-1}$ and peaks exceeding $20\,m\,s^{-1}$. At the Equator, the winds are moderate from the south and decrease eastward. In the southern Indian Ocean, during the southern winter (July), the SE Trades intensify and penetrate farther north than during the southern summer (January); they reach the Equator in the western part of the ocean and are the strongest in the three oceans. During that season the air masses transported by the SE Trade Winds cross the Equator in the west and flow, loaded with moisture, toward the Asian continent where they bring the awaited monsoon rainfall (for the Indian subcontinent, 'monsoon' means the wet monsoon, i.e., the SW monsoon).

October–November corresponds to the second transition period between the end of the SW monsoon and the beginning of the NE monsoon. North of the Equator, the winds vanish and the sea surface temperature can exceed 30°C. At the Equator moderate eastward winds blow again as during the first transition period.

This particular wind regime is dominated off the Equator by a strong annual period. At the Equator, the zonal wind component is dominated by a semi-annual period associated with the transition westerly winds, while the meridional wind component has a strongly annual periodicity associated with the monsoon reversals (*see* Figure 1 of Indian Ocean Equatorial Currents).

The Western Boundary Current System

As in the other oceans, the strongest currents are found close to the western shores of the ocean and are called the western boundary current system. In the western boundary region influenced by the monsoonal winds, i.e., north of 10°S, there are two western boundary currents: the East African Coastal Current (EACC), also called the Zanzibar current, which always flows north-eastward; and the Somali Current (SC), which is highly variable in contrast to other western boundary currents (*see* Figure 3 of Current Systems in the Indian Ocean). The Somali Current is the more intense, and reverses twice a year owing to the complete seasonal reversal of the winds. It has been particularly studied during the Indian Ocean Experiment (INDEX) that started in the 1970s and the SINODE (Surface Indian Ocean Dynamic Experiment) in the 1980s. In 1990–1996, during the WOCE (World Ocean Circulation Experiment), a large number of new data were collected over the whole Indian Ocean and particularly in the Somali Basin.

The East African Coastal Current (Also Called the Zanzibar Current)

The EACC is fed by the branch of the South Equatorial Current (SEC) that passes north of Madagascar and splits northward around 11°S. It runs northward throughout the year between latitudes 11°S and 4°S. The location of its northern end depends on the season. In the northern winter, during the NE monsoon, the EACC converges around 3°–4°S with the south-going Somali Current to form the eastward south equatorial counter current (SEC). It flows against light winds and is then the weakest (**Figure 1A**).

Direct current measurements at 2°S during March–April 1970 and 1971 indicated that the current reversed to the north at least one month before the onset of the SW monsoon over the interior of the north Indian Ocean, immediately after the southerly winds began along the East African coast at the beginning of April (15–20 knots) and the onset of the SE monsoon to the south. At that time, the current is very sensitive to small variations in the local wind direction. The boundary between the northward (EACC) and the southward (SC) flow was distinctly marked by changes in fauna and water properties with lower salinities in the EACC and higher salinities in the south-westward Somali Current. At that time the EACC is strengthened by the winds. By the end of April, at 2°S, the current was about 100 nautical miles wide with peak speeds of 2 m s^{-1} within few miles offshore.

In April 1985, of the 10 Sv passing northward at 5°S between 0 and 100 m, 4.5 Sv continued across the Equator and 5.5 Sv join the SECC from the south (**Figure 1A**). The total transport down to 300 m of the SECC was 23 Sv of which 17 Sv came from the EACC. Most of the subsurface transport of the SECC moved eastward between 2.5°S and 6°S across 45°E and most of the northward near surface boundary current crossing the Equator was turning eastward south of 1.5°N. At the Equator, within the layer 0–100 m, below the surface northward transport, 2.5 Sv still flowed south across the Equator (**Figure 1B**).

Under the onset of the SW monsoon, the EACC merges into the reversing Somali Current, which progresses northward. Its surface speed can then exceed 2 m s^{-1} and its transport amounts to 20 Sv in the upper 500 m with 14 Sv in the upper 100 m at 1°S. Below the surface, the deeper EACC current flows northward across the Equator at all seasons. During the NE monsoon, it becomes an undercurrent under the southward-flowing Somali current.

The Somali Current

North of the Equator, the Somali current develops in different phases in response to the onset of the monsoon winds. **Figure 2** shows the evolution of the circulation from the 1979 SW monsoon observations. **Figure 3** gives a schematic representation of the western boundary current system for the different seasons for the surface layer.

During the transition period at the end of the NE monsoon, in April–early May, the Somali current flows south-westward along the coast from 4°–5°N to the Equator, whereas south of the Equator the EACC flows north-eastward (see above). At that time, the two currents converge and turn offshore to the south east to form the SECC. North of 4°–5°N, the current is already north-eastward, fed by the NE monsoon current, which brings waters from the interior Arabian Sea driven by the wind stress curl, splitting into a northward boundary surface current between 5°N and 10°N associated with a southward subsurface current underneath, and a southward surface current towards the Equator.

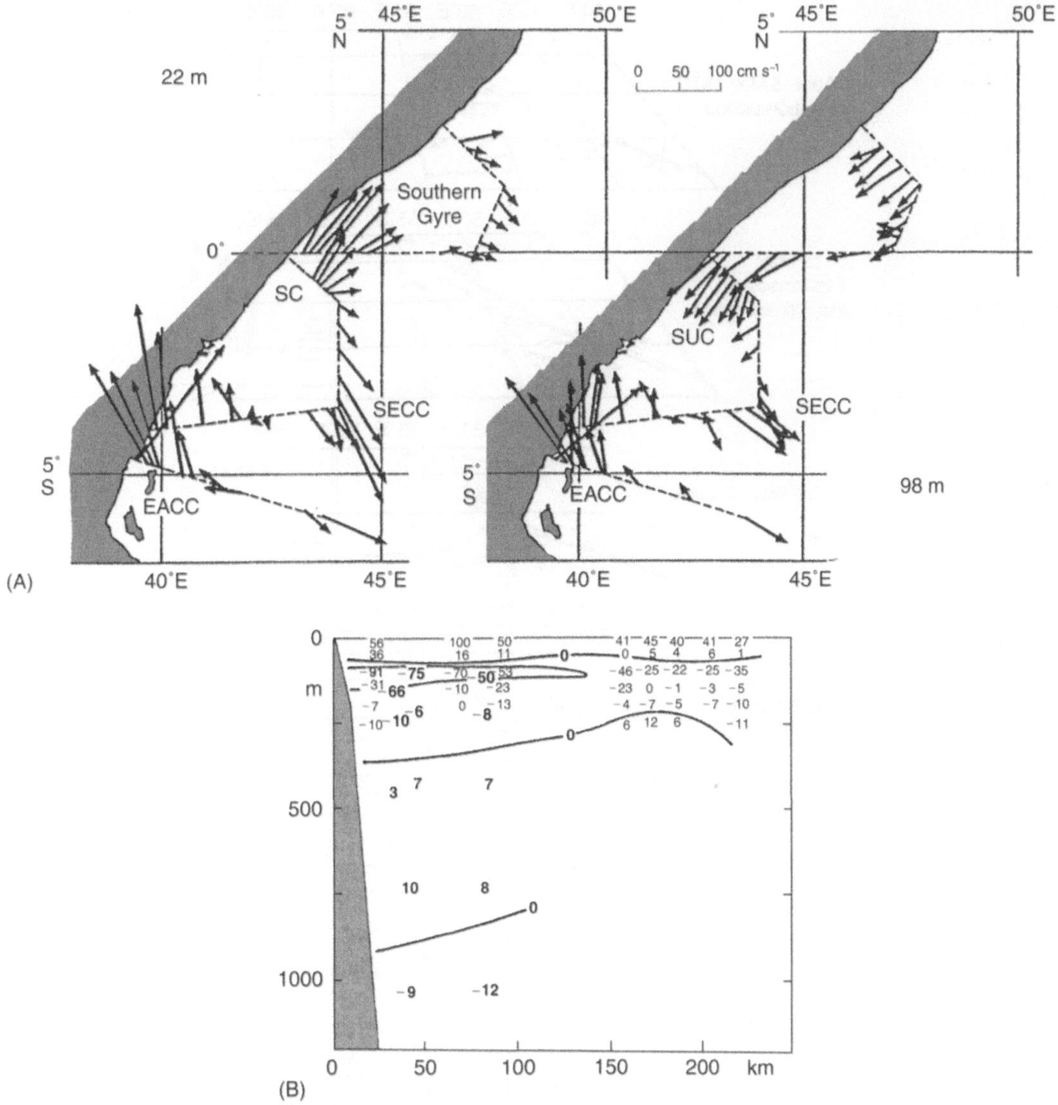

Figure 1 (A) Circulation along the East African Coast in April 1985 at 22 m and at 98 m depth, measured by shipboard Acoustic Doppler Current Profiler (ADCP), showing the EACC, the SECC, the reversal to the north at the equator of the Somali Current (SC), the Southern Gyre (SG) and the southward undercurrent (SUC) at the Equator. (B) Northward component of current along the equator in April 1985, from shipboard ADCP, in cm s^{-1}. Bold figures are mean northward components from moored currentmeters for the same period. (From Swallow et al., 1991, Structure and transport of the East African Coastal Current, *Journal of Geophysical Research* 96(C12) 22245-22257, 1991, copyright by the American Geophysical Union.)

During the early phase of the SW monsoon, in early May, the Somali current responds rapidly to the onset of the southerly winds at the Equator and reverses northward in continuity with the northward EACC, which crosses the Equator. It develops as a shallow cross-equatorial inertial current, turning offshore at about 3°N, where a cold upwelling wedge develops north of the turnoff latitude near the coast. As part of it recirculates southward across the Equator, it forms the anticyclonic Southern Gyre (**Figure 2**).

By mid-May, the SW wind onset propagates northward along the coast and the southern offshore-flowing branch, at 1°N to 3°N is strongly developed, with westward equatorial flow across 50°E indicating the recirculation of the Southern Gyre. North of that branch along the coast, the upwelling wedge spreads out bringing cold and enriched waters at the surface about 4°–5°N. Further to the north, the current is already northward from March. With southerly winds blowing parallel to the coast, a typical upwelling regime develops with northward surface flow, an undercurrent below and cold water along the coast.

When the onset of the strong summer monsoon winds occurs at these latitudes in June, the southern branch increases in strength and extends farther

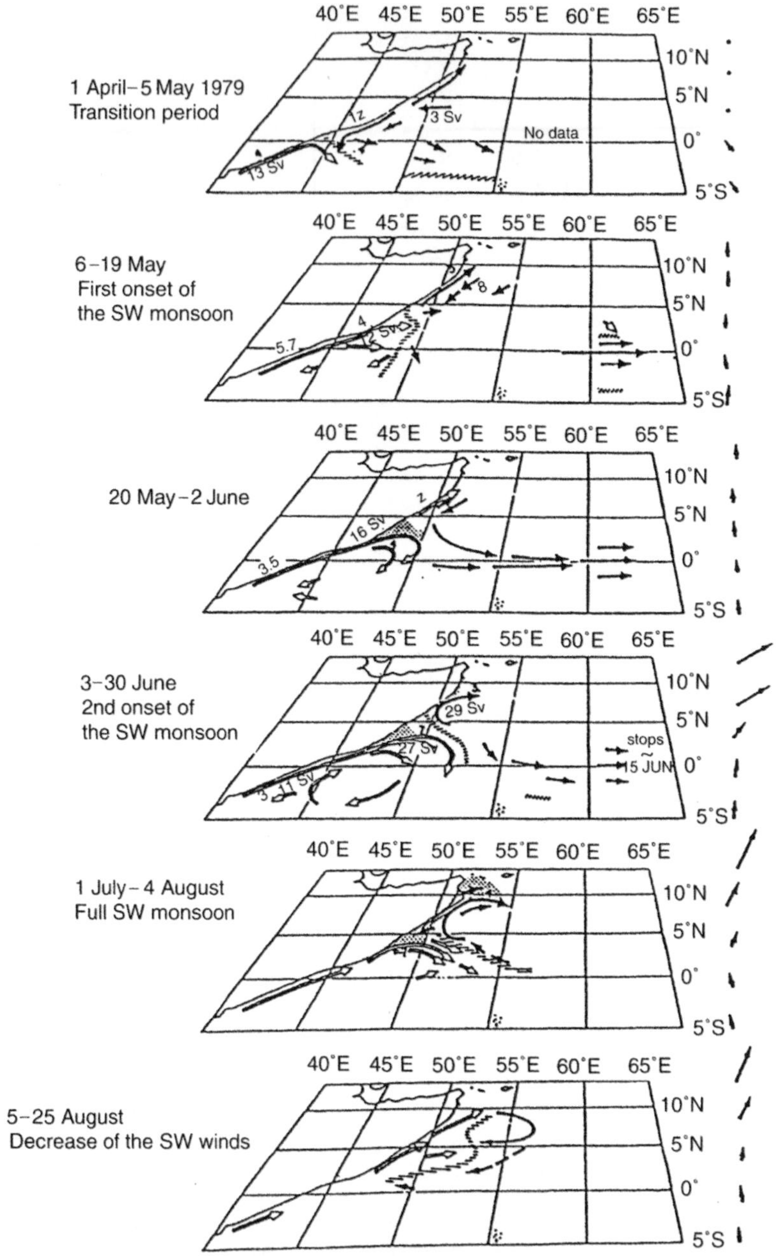

Figure 2 Evolution of the circulation during the onset of the SW monsoon from the 1979 observations during INDEX (speed in knots; transport in Sv = $10^6\,m^3\,s^{-1}$; open arrow = low salinity, solid arrow = high salinity, front, upwelling). The arrows on the right of the figure represent the wind stress observed at different latitudes along the coast; the full strength of the SW monsoon is reached in June (From M. Fieux, 1987, Circulation dans l'océan H an Indien occidental, Actes Colloque sur la Recherche Française dans les Terres Australes, Strasbourg, unpublished manuscript).

north (5°N) and a strong anticyclonic gyre, called the 'Great Whirl' develops between 5°N and 9°N with velocities at the surface higher than $2.5\,m\,s^{-1}$ and transports around 90 Sv between the surface and 1000 m, where currents exceeding $0.1\,m\,s^{-1}$ have been observed (**Figures 2** and **3**). Between the Somali coast and the northern branch of the Great Whirl a second strong upwelling wedge forms at its north-western flank where the flow turns offshore (**Figure 3**).

In August–September–October, depending on the year, when the winds decrease, it has been observed that the southern cold wedge propagates northward along the coast and coalesces with the northern one, which moves slightly northward (**Figure 4**). It is only at that time that the Somali current is continuous from the Equator up to the horn of Africa and brings fresher waters from the Southern Hemisphere into the Arabian Sea (**Figures 2, 5A–C, 6A,B**). The

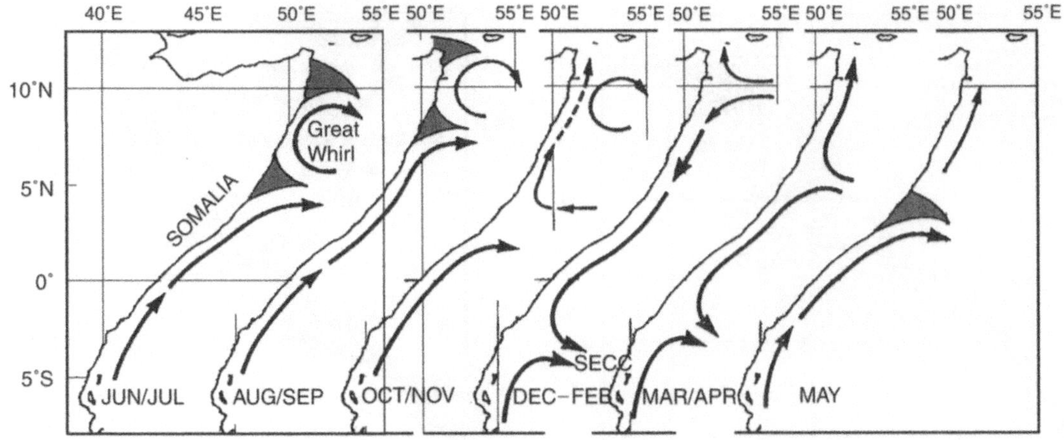

Figure 3 Somali Current flow patterns for the layer 0–100 m for different seasons with upwelling in grey (Reprinted from Deep Sea Research, 37(12), F. Schott et al., 1990, The Somali Current at the Equator: annual cycle of currents and transports in the upper 1000 m and connection to neighbouring latitudes, 1825–1848, copyright 1990, with permission from Elsevier Science).

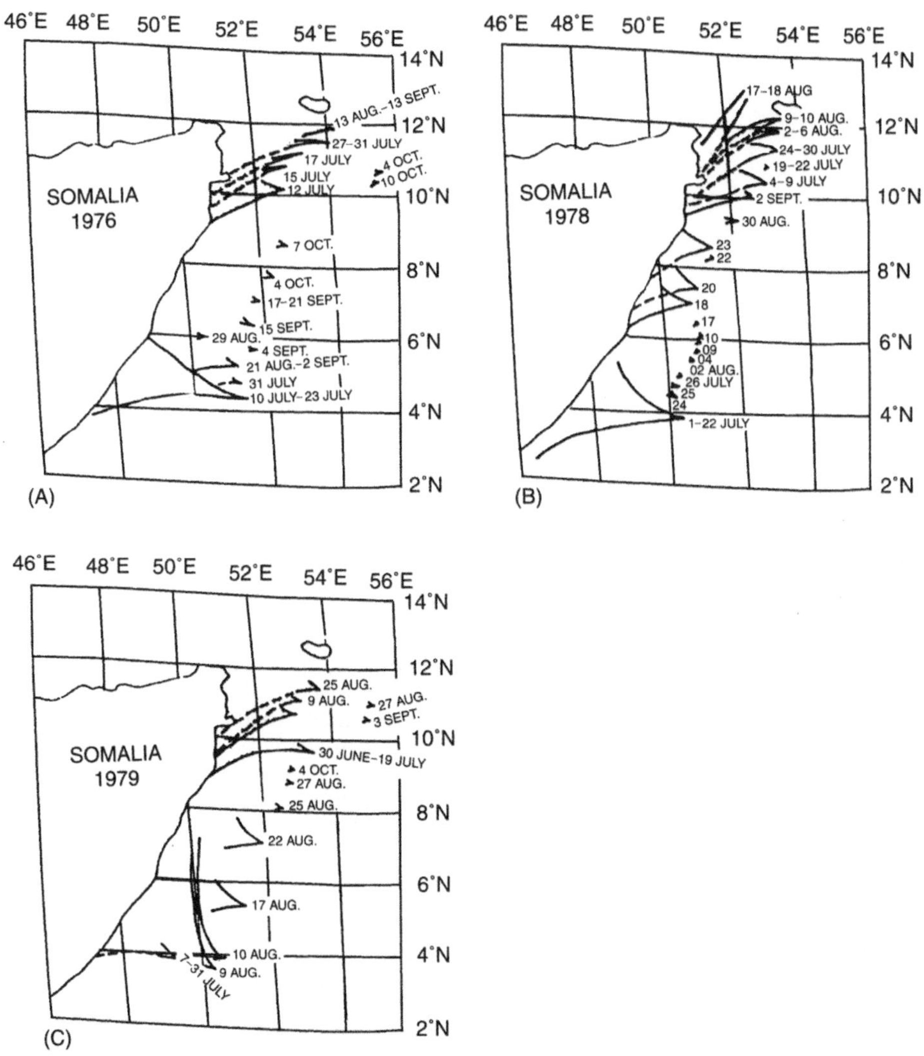

Figure 4 Propagation of upwelling wedges in 1976, 1978, and 1979, as seen in satellite infrared imagery (the northern one is in gray). (Reprinted from *Progress in Oceanography* 12, F. Schott, Monsoon response of the Somali Current and associated upwelling, 357–381, copyright 1983, with permission from Elsevier Science).

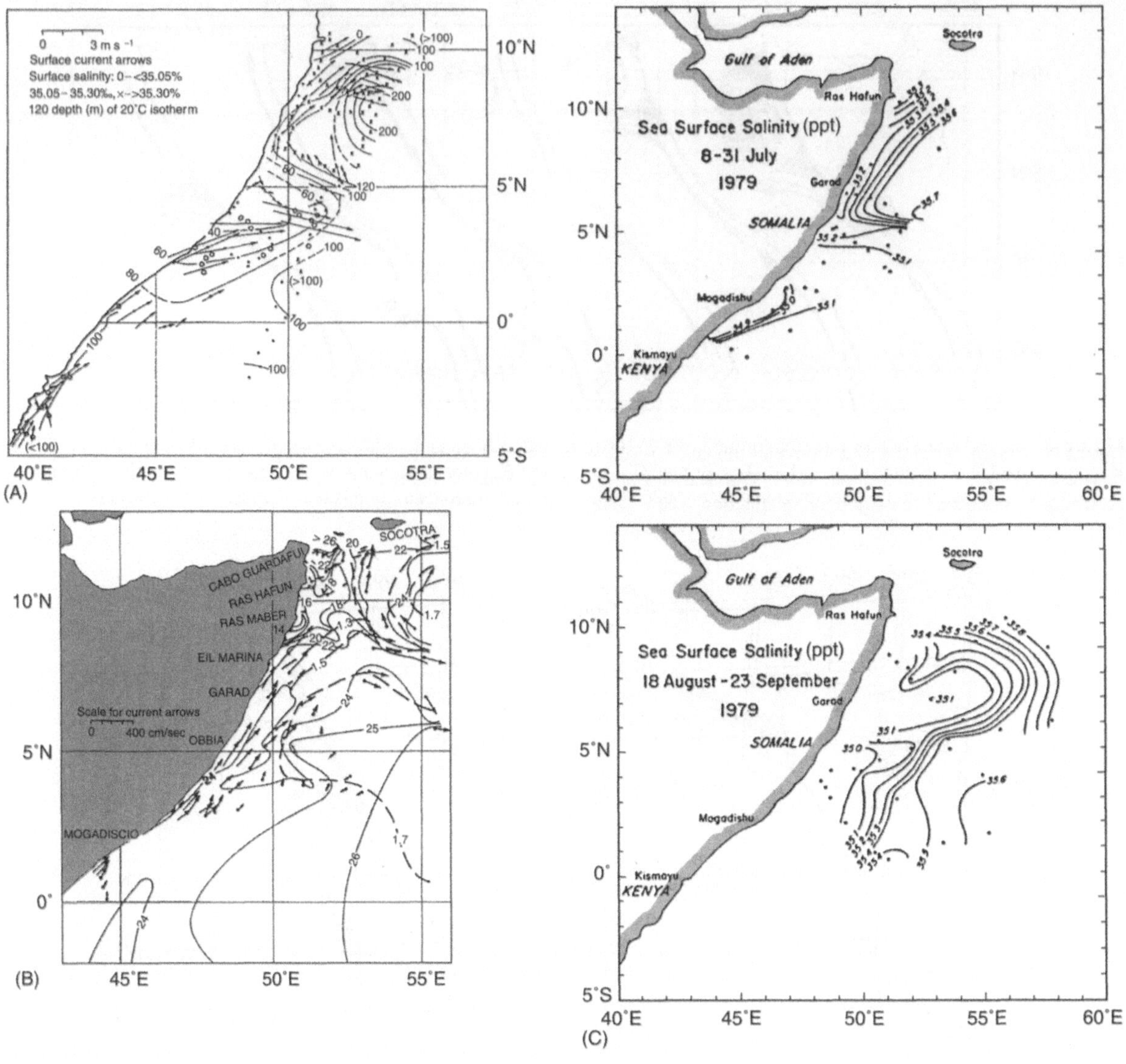

Figure 5 Variability of the circulation and salinity along the Somali coast. (A) 1 July–4 August 1979: surface currents (arrows), depth of the 20°C isotherm and surface salinity range (Reprinted from Swallow et al., 1983, Development of near-surface flow pattern and water mass distribution in the Somali Basin in response to the southwest monsoon of 1979, *Journal of Physical Oceanography*, 13, 1398–1415, with permission from American Meteorological Society). (B) 4 August–4 September 1964: currents at 10 m depth (arrows), dynamic heights of the sea surface relative to 1000 dbars in dyn. meters (heavy dashed) and sea surface temperatures in °C (solid) (Reprinted from *Progress in Oceanography*, 12, F. Schott, Monsoon response of the Somali Current and associated upwelling, 12, 357–381, copyright 1983, with permission from Elsevier Science (redrawn from J.C. Swallow and J.G. Bruce, 1966 and Warren *et al.*, 1966)). (C) Surface salinities during the existence of the two-gyre system 8–31 July 1979, and during and after the northward propagation of the southern cold wedge, 18 August–23 September 1979 (Reprinted from Progress in Oceanography, 12, F. Schott, Monsoon response of the Somali Current and associated upwelling, 357–381, copyright 1983, with permission from Elsevier Science). At that time the Somali Current is continuous along the coast and brings fresher water from the south at the end of the SW monsoon season.

breakdown of the two-gyre system can occur at speeds of up to 1 m s^{-1}, replacing a100 m thick and 100 km wide band of high-salinity water with lower-salinity water from south of the Equator, which represents a transport of 10 Sv.

At the end of the summer monsoon and during the transition to the winter monsoon, in October, the continuous Somali Current no longer exists; instead the cross-equatorial flow, characterized by low surface salinities (35–35.2) with a transport of 12 Sv in the upper 100 m, turns offshore south of 2.5°N. The northward current component through the equatorial section has a subsurface maximum of more than 1.50 m s^{-1} near the coast at 40 m depth and velocities of more than 0.5 m s^{-1} at 200 m depth. The cross-equatorial transport in the upper 100 m

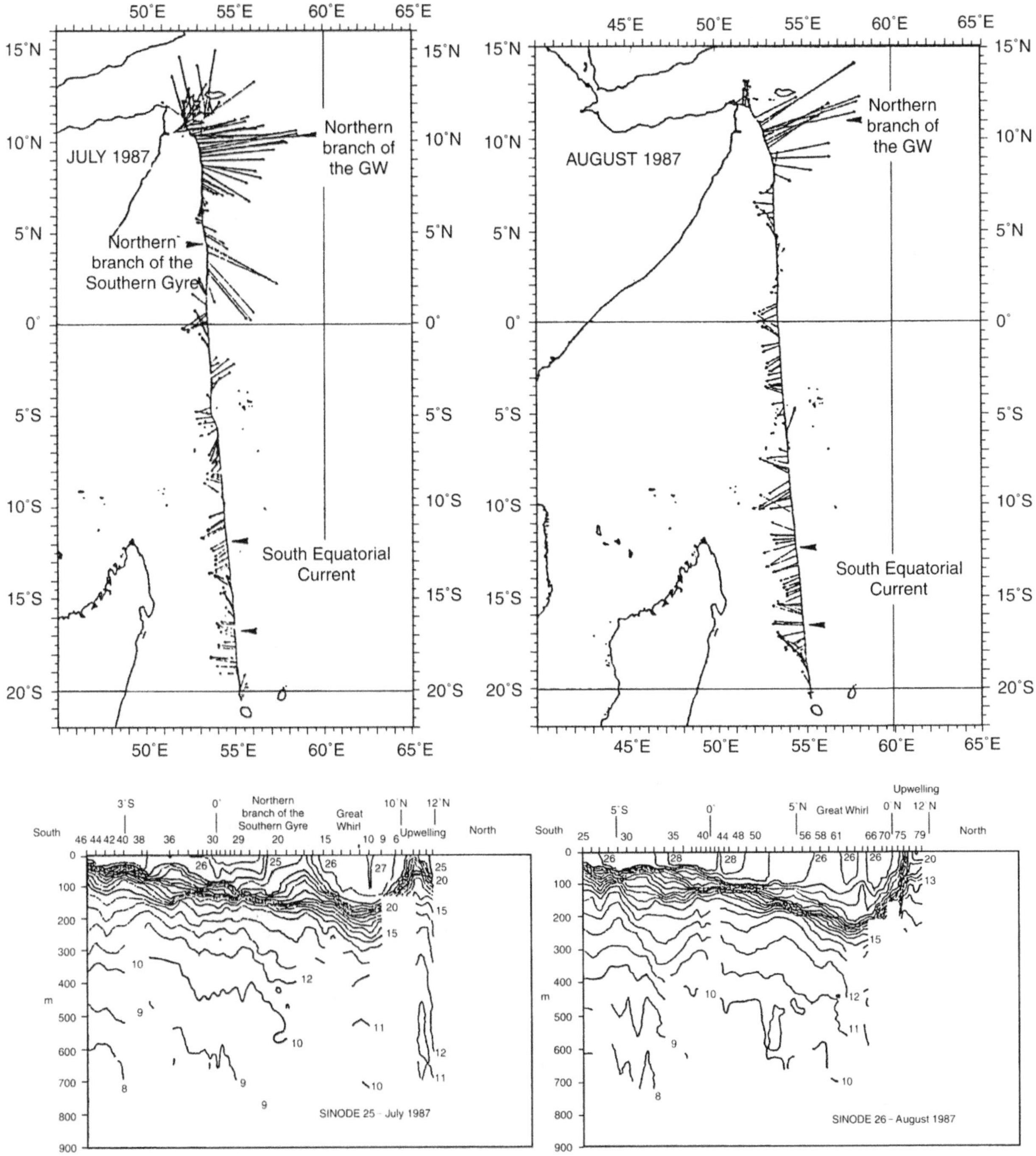

Figure 6 (A) Location of the section in July and August 1987 across the Great Whirl (GW) with corresponding surface drifts. (B) Corresponding temperature sections showing the strengthening of the northern front of the Great Whirl in August and the disappearance of the southern front in August compared to July at that longitude. (From M. Fieux, 1987, Circulation dans l'océan Indien occidental, Actes Colloque sur la Recherche Française dans les Terres Australes, Strasbourg, unpublished manuscript.)

was comparable to the 14 Sv transport at 1°S in the period May–June 1979. This means that the cross-equatorial transport of the Somali current in late autumn is very similar to that during the onset of the SW monsoon. The local winds are quite different during the two periods, whereas the Trade Winds over the subtropical south Indian ocean are similar, which suggests that, in October, the cross-equatorial flow is driven primarily by remote forcing through the inflow from the South Equatorial Current. Between 6°N and 11°N, the anticyclonic Great Whirl, marked by relatively high surface salinities (35.6–35.8), persists with transport of 33 Sv westward in the upper 250 m between 6°N and 8.5°N, and 32 Sv eastward between 8.5°N and 11.5°N. Then a southward undercurrent is established during the transition period while the Great Whirl weakens.

The northern gyre is not symmetrical, as can be seen from the depth and slope of the 20°C isotherm, which is much steeper on the northern flank of the gyre (**Figure 6B**). This can be explained by the strong nonlinearity of that system, causing a shift of high currents into the northern corner.

The northern Somali Current was found to be disconnected from the interior of the Arabian Sea in the latitude range 4°–12°N in terms of both water mass properties and current fields. Communication predominantly occurs through the passages between Socotra and the horn of Africa.

During the late phase of the SW monsoon, a third anticyclonic gyre appears north-east of the island of Socotra that is called the Socotra Gyre (**Figure** 7).

In summer 1993, a significant northward flow of 13 Sv was observed through the passage between the island of Abd al Kuri (west of Socotra) and Cape Guardafui (the horn of Africa). East of the Great Whirl, a band of northward warm water flow that provided low-latitude waters to the Socotra Gyre and the Socotra Passage separated the Great Whirl and the interior of the Arabian Sea. The net transport through the Socotra Passage is northward throughout most of the year.

During the transition period in October–November, the northward Somali circulation decreases, with a branch turning offshore south of 2.5°N.

In December–February, during the NE winter monsoon, the Somali current reverses southward from 10°N to 4°–5°S, where it converges with the northward EACC to form the SECC flowing eastward (see above).

Nonlinear reduced gravity numerical models driven by observed monthly mean winds are very successful in simulating the observed features of the circulation in this region, such as the formation and decay of the two-gyre system of the Somali Current during the SW monsoon. With interannually varying winds they also simulate a large interannual variability in the circulation.

Currents at Depth

Direct measurements made in 1984–1986 at the Equator, near Africa, show that the southward reversal implies only a thin surface layer below which, between 120 m and 400 m, there is a northward undercurrent, remnant of the SW monsoon season circulation (**Figures** 8 and 9), followed again by a southward current below 400 m. It results in a large variability of the cross-equatorial transport, which amounts to 21 Sv for the upper 500 m during the summer monsoon season and is close to zero for the winter monsoon mean transport. The annual mean cross-equatorial transport in the upper 500 m is 10 Sv northward, with very little transport in either season in the depth range 500–1000 m (**Figure 9**). Comparison of current profiles at 4°S in the EACC, at the Equator, and at 5°N in the Somali Current in both seasons shows that at 4°S the subsurface profile stays fairly constant while at 5°N drastic changes occur between the seasons as well as at the Equator (**Figure 10**). North of 5°N, there is less variability at subsurface, with the presence of a southward coastal undercurrent during most of the year except during the full strength of the deep-reaching Great Whirl in July–August to more than 1000 m, involving large deep transports. The Somali Current flow patterns

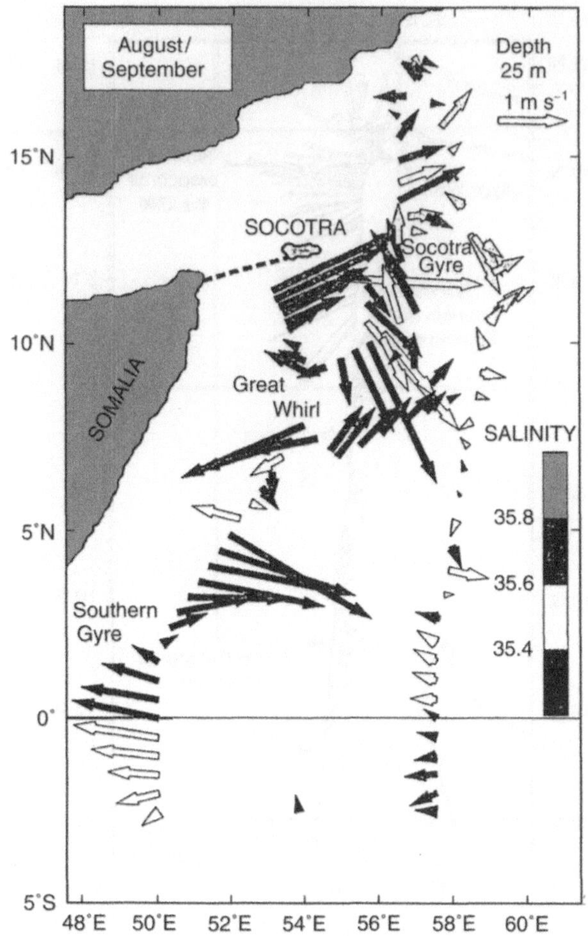

Figure 7 Near surface circulation at 25 m in August–September 1995, showing the Southern Gyre (SG), the Great Whirl (GW) and the Socotra Gyre (SG) together with surface salinities. (Reprinted from F. Schott et al., Summer monsoon response of the northern Somali Current, 1995, *Geophysical Research Letters*, 24(21), 2565–2568, 1997, copyright by the American Geophysical Union).

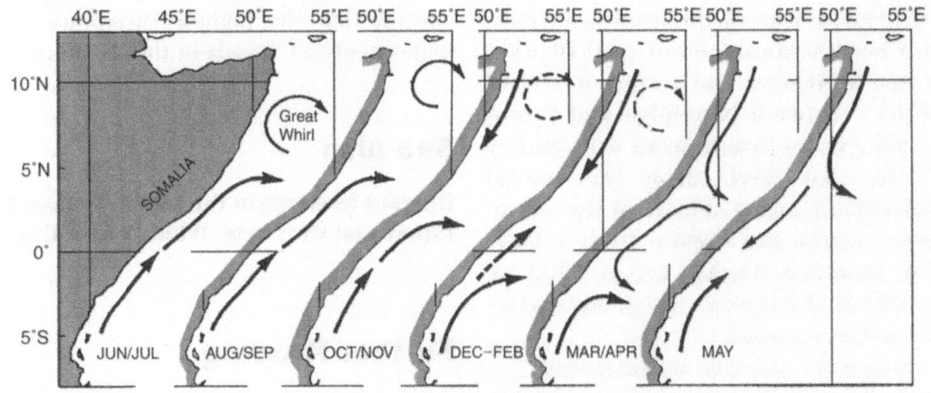

Figure 8 Schematic representation of Somali Current circulation patterns in the upper layer 100–400 m for different seasons. (Reprinted from *Deep Sea Research*, 37(12), F. Schott *et al.*, 1990, The Somali Current at the equator: annual cycle of currents and transports in the upper 1000 m and connection to neighbouring latitudes, 1825–1848, copyright 1990, with permission from Elsevier Science).

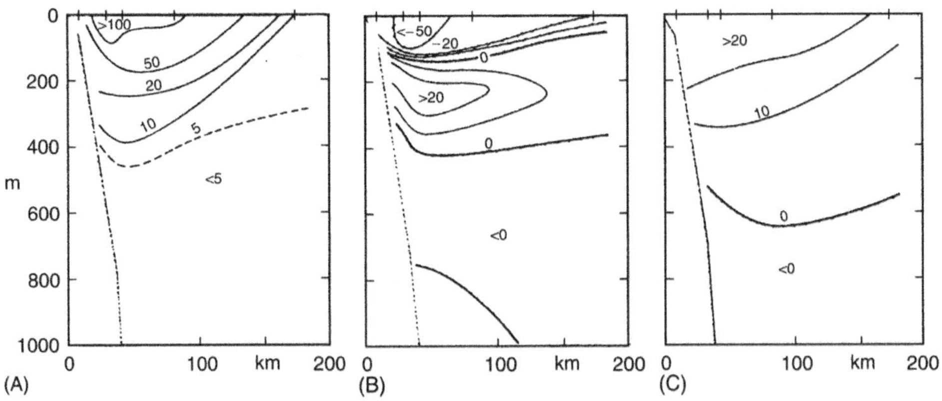

Figure 9 Mean sections of northward current component at the Equator (positive northward): (A) for summer monsoon (1 June–13 September); (B) for winter monsoon (15 December–15 February); (C) for overall mean, in cm s^{-1}. (Reprinted from *Deep Sea Research*, 37(12), F. Schott *et al.*, 1990, The Somali Current at the equator: annual cycle of currents and transports in the upper 1000 m and connection to neighbouring latitudes, 1825–1848, copyright 1990, with permission from Elsevier Science).

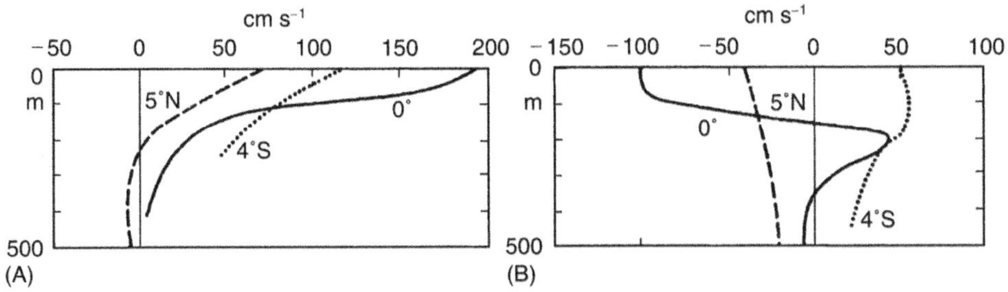

Figure 10 Current profiles for different seasons at 4°S, on the Equator, and at 5°N: (A) for summer monsoon; (B) for winter monsoon. (Reprinted from *Deep Sea Research*, 37(12), F. Schott *et al.*, 1990, The Somali Current at the equator: annual cycle of currents and transports in the upper 1000 m and connection to neighbouring latitudes, 1825–11848, copyright 1990, with permission from Elsevier Science).

for the different seasons in the 100–400 m layers are shown in **Figure 8**.

Deeper, below the Somali current, at the Equator during October 1984 to October 1986 at 1000 m, 1500 m, 2000 m, and 3000 m depth, the measured mean currents were very small. However, the only clear seasonal signal was observed at the 3000 m level, with a seasonal current parallel to the coast approximately in phase with the local surface winds. This reached a north-eastward mean of 0.10 m s^{-1} between June and September, and a south-westward mean of 0.06 m s^{-1} between November and February. This variability seems in agreement with salinity distribution near that level along the coastal boundary, with slightly higher salinity at the end of the NE monsoon season and slightly lower salinity during the SW monsoon. Higher up, at 2000 m, 1500 m, and 1000 m, the currents are dominated by events of 1–2 months duration.

From the long-term current measurements, it seems that at the Equator the semiannual variability is stronger than the annual variability even near the coast. Off the equator, the annual component dominates.

Numerical models have shown that the location and motion of eddies are influenced by the distribution and strength of the wind forcing; an increase in the winds leads to a southward displacement of the offshore turning of the southern branch; the northward motion of eddies is very dependent upon the coastal geometry; and the onset of the northward Somali current depends on local winds forcing and on the wind forcing far out at sea. Baroclinic Rossby waves generated by the strong offshore anticyclonic windstress curl have been found to be the generation mechanism of the Great Whirl.

Conclusion

The western boundary of the northern Indian Ocean is a remarkable natural laboratory for studying the effect of the wind on the oceanic circulation, as regularly twice a year the winds change direction rapidly and are particularly strong. It is in the Somali current that the highest variability as well as the highest current speeds in the world ocean are found.

See also

Current Systems in the Indian Ocean. Indian Ocean Equatorial Currents. Wind Driven Circulation.

Further Reading

Fein JS and Stephens PL (eds.) (1987) *Monsoons*. Washington, DC: Wiley Interscience.

Monsoon (1987) Fein JS and Stephens PL (ed.) NSF. A Wiley-Interscience Publication. Washington, USA: John Wiley Sons.

Open University Oceanography Course Team (1993) *Ocean Circulation*. Oxford: Pergamon Press.

Scott F (1983) Monsoon response of the Somali Current and associated upwelling. *Progress in Oceanography* 12: 357–381.

Schott F, Swallow JC, and Fieux M (1990) The Somali Current at the equator: annual cycle of currents and transports in the upper 1000 m and connection to neighbouring latitudes. *Deep Sea Research* 37(12): 1825–1848.

Schott F, Fischer J, Garternicht U, and Quadfasel D (1997) Summer monsoon response of the northern Somali Current, 1995. *Geophysical Research Letters* 24(21): 2565–2568.

Swallow JC, Molinari RL, Bruce JG, Brown OB, and Evans RH (1983) Development of near-surface flow pattern and water mass distribution in the Somali Basin in response to the southwest monsoon of 1979. *Journal of Physical Oceanography* 13: 1398–1415.

Tomczak M and Godfrey S (1994) *Regional Oceanography: An Introduction*. Oxford: Pergamon Press.

LEEUWIN CURRENT

G. Cresswell and C. M. Domingues, CSIRO Marine and Atmospheric Research, Hobart, TAS, Australia

Introduction

The Leeuwin Current (LC) is unusual among subtropical eastern boundary currents because it flows poleward instead of equatorward. Its annual cycle starts as far north as 17° S off northwestern Australia in March when sea level there increases due to the arrival of an annual sea level pulse from the western tropical Pacific via the Indonesian seas. The LC, marked at the surface by a shallow, warm, low-salinity core, flows down the sea level slope into a seasonally weakening equatorward wind stress to Cape Leeuwin at 34° S (**Figure 1**). There it turns eastward and flows toward the Great Australian Bight, which extends from about 125° to 135° E. As was pointed out recently by Ridgway and Condie, the arrival of the LC at the pivot point of Cape Leeuwin in April/May is fortuitous because it is then driven eastward by winter westerly wind forcing. By the time it reaches Tasmania (**Figure 2**), in a much diluted form, and with some name changes, it has traveled 5500 km from northwestern Australia. In addition to its tropical source, it receives contributions from the subtropical Indian Ocean to the west of Australia and from the subantarctic waters south of Australia. Off Perth, 32° S, its maximum poleward geostrophic transport is 5 Sv ($5 \times 10^6\,m^3\,s^{-1}$) during June–July. It is strongest during La Niña years.

The LC is an important factor in the life cycles of many commercially important marine creatures

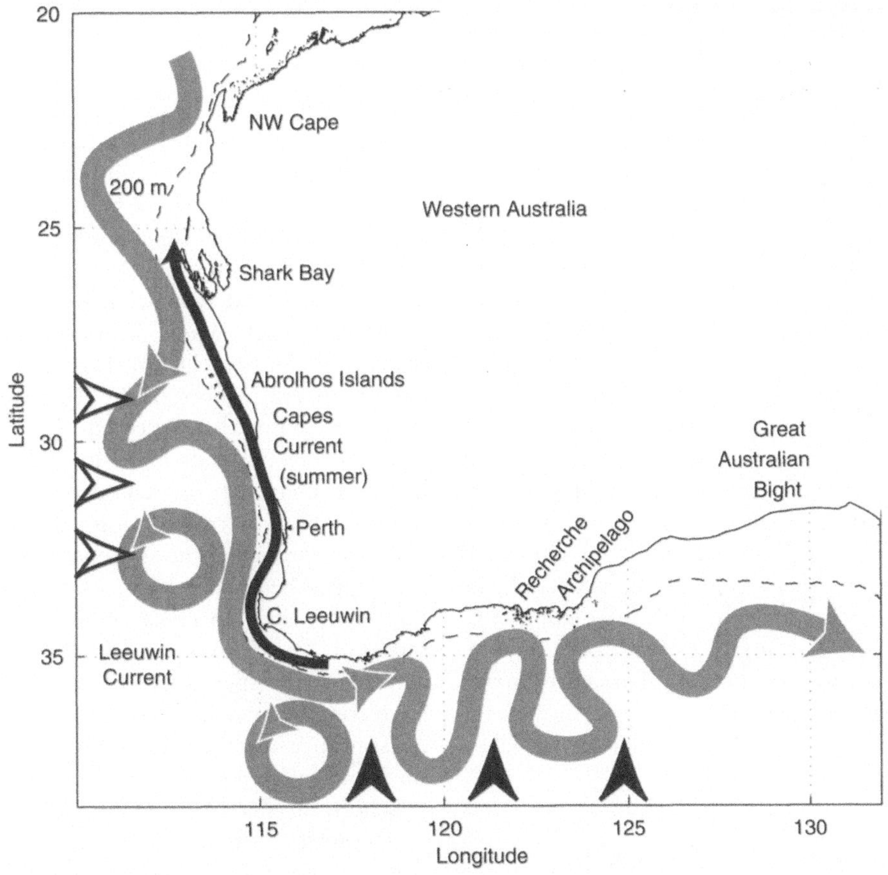

Figure 1 A schematic picture of the surface currents off southwestern Australia. The LC is marked by the broad gray arrow and flows year-round, being strongest in winter. The Capes Current (long black arrow on the continental shelf) is driven by southerly wind-forcing in summer. Subtropical water (open arrows) is entrained into the LC system from the west, as is subantarctic water (black arrows) from the south. Two eddies that have separated from the main current are shown.

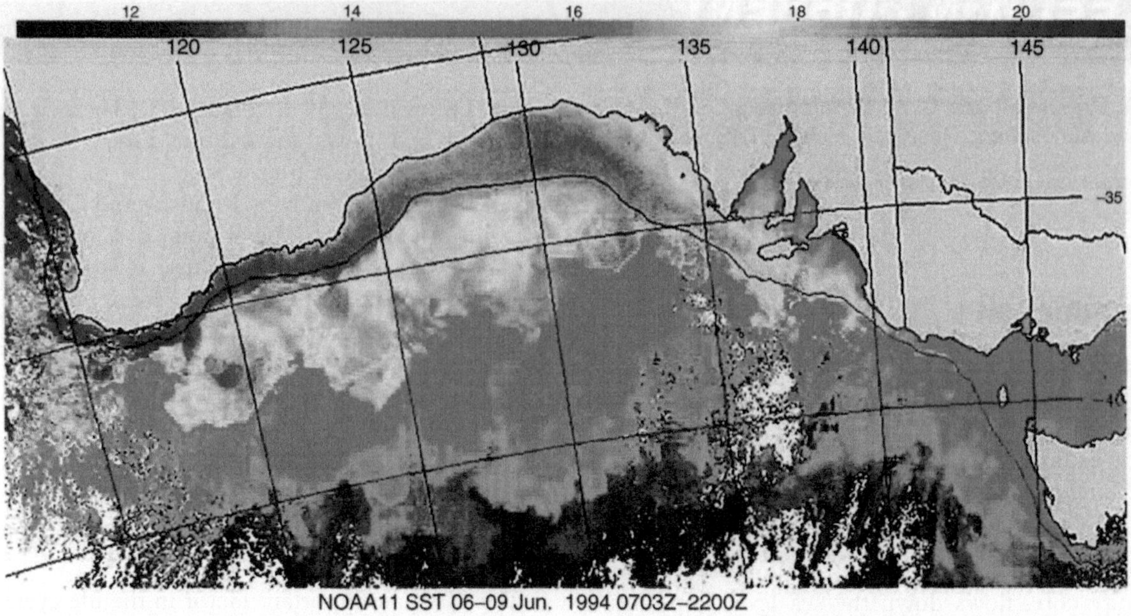

Figure 2 A composite satellite sea surface image from several days in Jun. 1994. There is a 5° × 5° latitude–longitude grid; land is gray; the shelf edge is marked by a thin black line; clouds are white; and the temperature scale (°C) appears at the top of the image. The warm LC (false color red) flows southward to the SW corner of Australia (Cape Leeuwin) and then eastward. There are eddies south of the LC. Copyright 2001, CSIRO Marine Research, Hobart.

along the Western Australian seaboard. Annual fluctuations in the numbers of western rock lobsters (Australia's most important single species fishery) recruited to the coastal reefs can be linked to the strength of the LC, which in turn can be related to El Niño/Southern Oscillation (ENSO). *Arripus trutta*, Australian salmon, migrate westward across the Great Australian Bight to spawn north of Cape Leeuwin in autumn; the seasonally strengthening LC then carries their eggs and young toward the Great Australian Bight. Young southern bluefin tuna *Thunnus maccoyii* that hatch between NW Australia and Indonesia commence their Southern Hemisphere odyssey with partial assistance from the LC; they are found south of WA as 1-year-olds. Corals are found farther south than otherwise might be expected; the southernmost hatchery for the red-tailed tropic bird *Phaeton rubricauda* is found near Cape Leeuwin; and other tropical marine flora and fauna are carried to the Great Australian Bight.

History

The evidence for an LC flowing from NW Australia south to Cape Leeuwin and then across toward Tasmania came piece by piece over almost two centuries. In May 1803, Matthew Flinders' vessel, *Investigator*, was set to the east between Cape Leeuwin (115° E) and several degrees of longitude eastward at a little over $0.5\,\mathrm{m\,s^{-1}}$. From there to the Recherche Archipelago (123° E) he found the current to increase with distance from the coast. Late in the nineteenth century, Saville-Kent suggested that tropical waters reached the Abrolhos Islands (~29° S) because of the warm waters and tropical marine flora and fauna, including corals, which he observed. Early in the twentieth century, Dakin learnt of a southward winter current inshore of the Abrolhos Islands from fishermen who also reported that it strengthened when northerly winds blew. In 1921, Halligan published a chart of the currents around Australia, showing how a cold and heavy Southern Ocean current approached the southwest coast of Australia and dipped beneath a warm southerly drift down the Western Australian coast. From Cape Leeuwin, he reported "a warm southerly and easterly surface current" with speeds of $0.2\,\mathrm{m\,s^{-1}}$. This fact must have come from ships that were close enough to the coast to measure their drifts with any precision – and therefore inshore of the strongest part of the LC.

Schott, in his book in 1935, showed charts of temperatures and currents for the world, with warm water rounding Cape Leeuwin and flowing eastward, more or less as we know the LC today. A decade later, Rochord started the continental shelf sampling station at the 60-m isobath off Rottnest Island (32° S); the station continues to be occupied every few weeks.

The time series of salinity, when combined with inferences from drift bottle releases and recoveries, suggested that low-salinity tropical water flowed southward to arrive at the station in autumn and winter and that there was a northward flow of high-salinity water in summer. In hindsight we recognize these as the Leeuwin and Capes Currents, respectively – in winter the LC spreads onto the continental shelf, while in summer it is seaward of the Rottnest station.

Phytoplankton also provided clues: among them, Wood in 1954 postulated a current from Cape Leeuwin to Tasmania based on the continuity of findings of certain species of dinoflagellates. In 1969, Hamon reported that there was an eastward flow toward the shelf north and south of the Abrolhos Islands. Gentilli in 1972 was the first to see the importance of the throughflow of tropical Pacific Ocean waters into the NE Indian Ocean: he reported that these waters flowed southward against WA down to the latitude of Rottnest Island (32° S). Andrews in 1977 confirmed Hamon's eastward inflow and found that it turned southward near the continent, spawned eddies, and turned eastward at Cape Leeuwin (cf. **Figure 1**). We now recognize this as the salty, subtropical input to the LC. In the same decade, Kitani reported a southward-flowing warm, low-salinity current between the Abrolhos Island and Fremantle as well as cyclonic and anticyclonic eddies.

When satellite-tracked drifters were first released off WA in the 1970s (**Figure 3**), they revealed that the LC not only flowed southward along the continental slope to the Cape Leeuwin vicinity, but that it also rounded the cape, on one occasion quickly accelerating to $1.5\,\mathrm{m\,s^{-1}}$, and then flowed eastward at

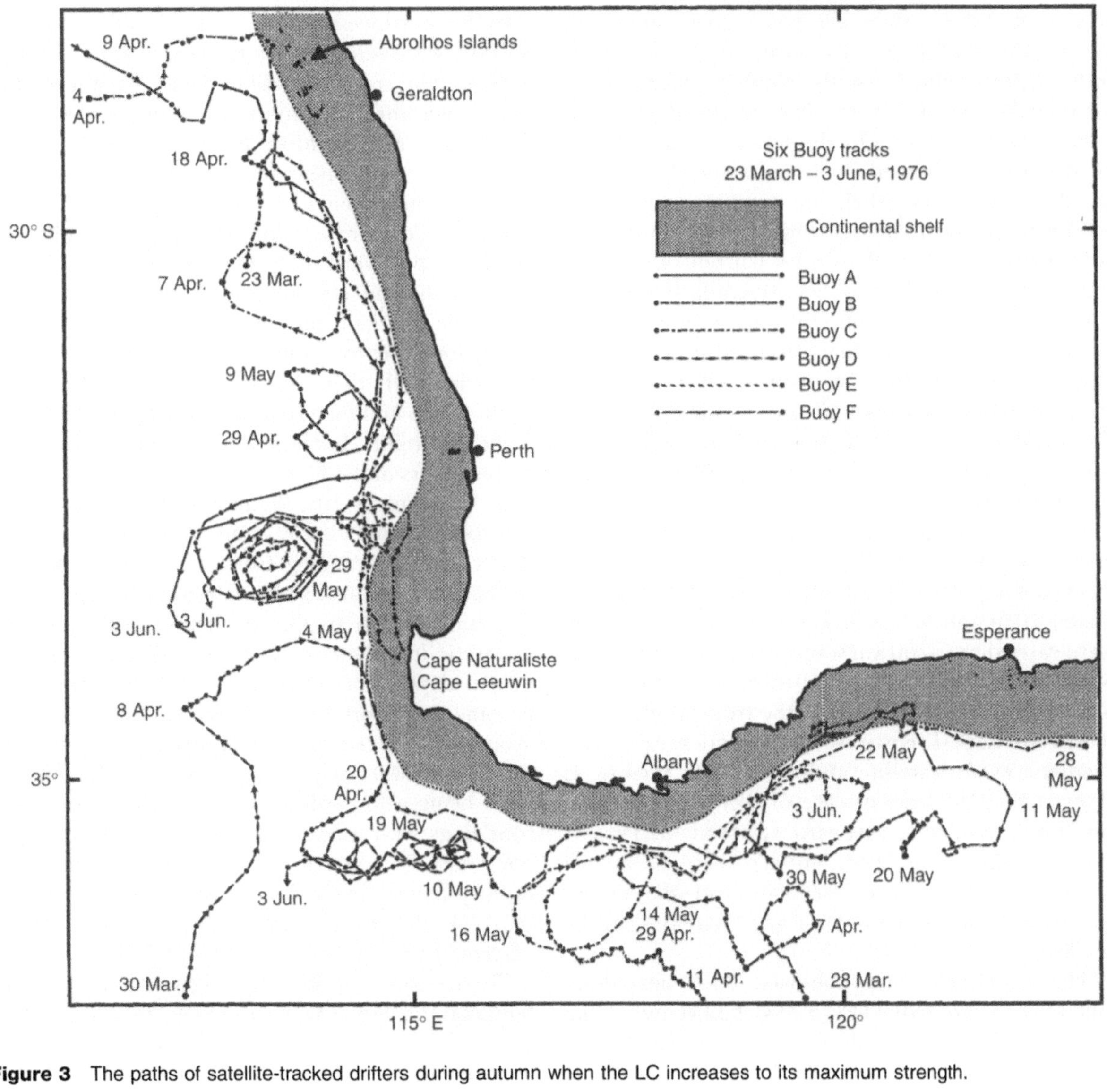

Figure 3 The paths of satellite-tracked drifters during autumn when the LC increases to its maximum strength.

speeds up to 1.8 m s^{-1} toward the western side of the Great Australian Bight. The drifters showed that the LC meandered and formed eddies; for reasons that are not clear, the drifters most frequently went into cyclonic eddies, but there was one example of two of them moving from near the shelf edge off Perth out into an anticyclonic eddy (no. 7 in the figure) that appeared to be in the process of forming.

The current south of WA, as indicated by the drifters, had small (~30 km) southward (anticyclonic) meanders separated by ~200 km. At the locations of the southward meanders there could be transient offshoots out to sea that took the drifters into cyclonic eddies, where they remained trapped for some weeks. In some instances, the drifter motions showed the cyclonic eddies to drift westward at about 7 km d^{-1}. The drifters decelerated when they entered the cooler cyclonic eddies and accelerated when they reentered the warmer LC. The groundings of several drifters along the south coast suggested that the LC could spread across the continental shelf. Some drifters came from the south to suggest that subantarctic waters were carried northward to interact with the LC. Off the western coast of WA the drifters also showed northward wind-driven flow on the continental shelf in summer, when the northward wind stress is strongest. This northward flow has been called the Capes Current and it includes waters upwelled off SW Australia (**Figure 4**).

Because of the significance of Cape Leeuwin as the pivot point for the change from southward flow down WA to eastward flow to the GAB, the LC was given its name by Cresswell and Golding in 1980. The *Leeuwin* was a VOC (Dutch East India Company) vessel that explored the coastline in 1622 en route to Batavia.

The first satellite sea surface temperature (SST) images in the late 1979s of the LC south of WA showed a number of small offshoots, with one being linked to an anticyclonic eddy so that the LC waters were carried over 200 km southward. The eddies and offshoots are now recognized to be a common feature of the LC. The arrival of the tropical waters of the LC at Cape Leeuwin in the austral autumn is a dramatic event: satellite images in 1987 (**Figure 4**) showed that the leading edge of the LC progressed southward and then eastward at about 0.25 m s^{-1} between March and April. In step with the arrival of the LC, there were 'decreases' in the northward winds off the west coast and 'increases' in the sea level.

The 6 March image showed cool near-shore waters westward from 118° E to Cape Leeuwin from upwelling driven by southeasterly winds. These were carried northward around Cape Leeuwin to contribute to the wind-driven Capes Current on the continental shelf (the green plume that crosses 33° S). The image for 15 April showed the northward intrusion of cold subantarctic water near 118° E. The intrusion turned to the east to run alongside the LC. The existence of such intrusions could, in retrospect, also be inferred from the tracks and temperatures of the 1970s drifters.

Water Masses off Western Australia

The water masses can largely be identified by their salinities, shown in the schematic diagram (**Figure 5**) drawn from several ship surveys. The diagram excludes continental shelf waters and currents such as the Capes Current. We note that the various water masses are found in layers variously atop, below, and alongside each other; that they move independently; and that mixing occurs across their boundaries.

In the northwest (top left) is the inflow of near-surface, low-salinity, warm tropical water (TW; blue; <34.5). This tropical input to the LC becomes saltier and cooler through mixing as it progresses southward (note that salinities lower than 35.5, that is, yellow, are found near the continent almost as far south as Cape Leeuwin). Air–sea fluxes play a small part in the temperature and salinity changes observed along the poleward path of the mean jet of the LC. Most of the changes seem to be due to eddy fluxes. We have marked the LC as a light blue semi-transparent arrow running southward to Cape Leeuwin and then eastward to the Recherche Archipelago and then beyond to the Great Australian Bight (top right). In reality, the LC flow is broken up by meanders and eddies.

In the west is the Hamon–Andrews inflow of subtropical water (STW) and South Indian Central Water (SICW), ranging from >35.5 (red) near the surface (due to an excess of evaporation over precipitation) down to the low-salinity (blue; <34.5) Antarctic Intermediate Water (AAIW) at more than 600-m depth. The salty STW (red) is entrained to become an important part of the LC and it is carried around Cape Leeuwin both near the surface and near the continental shelf edge, so that the LC changes from being a low-salinity current relative to its surroundings west of Australia to being a high-salinity one south of Australia. Salty STW and the lower-salinity central water sink beneath the LC to become part of the northward-flowing Leeuwin Undercurrent (LUC) between *c.* 300 and 700 m.

To the south of WA the green arrow marks the Subantarctic Mode Water (SAMW) that results from deep winter mixing at about 45° S. This can be traced as it moves northward off WA by its relatively high

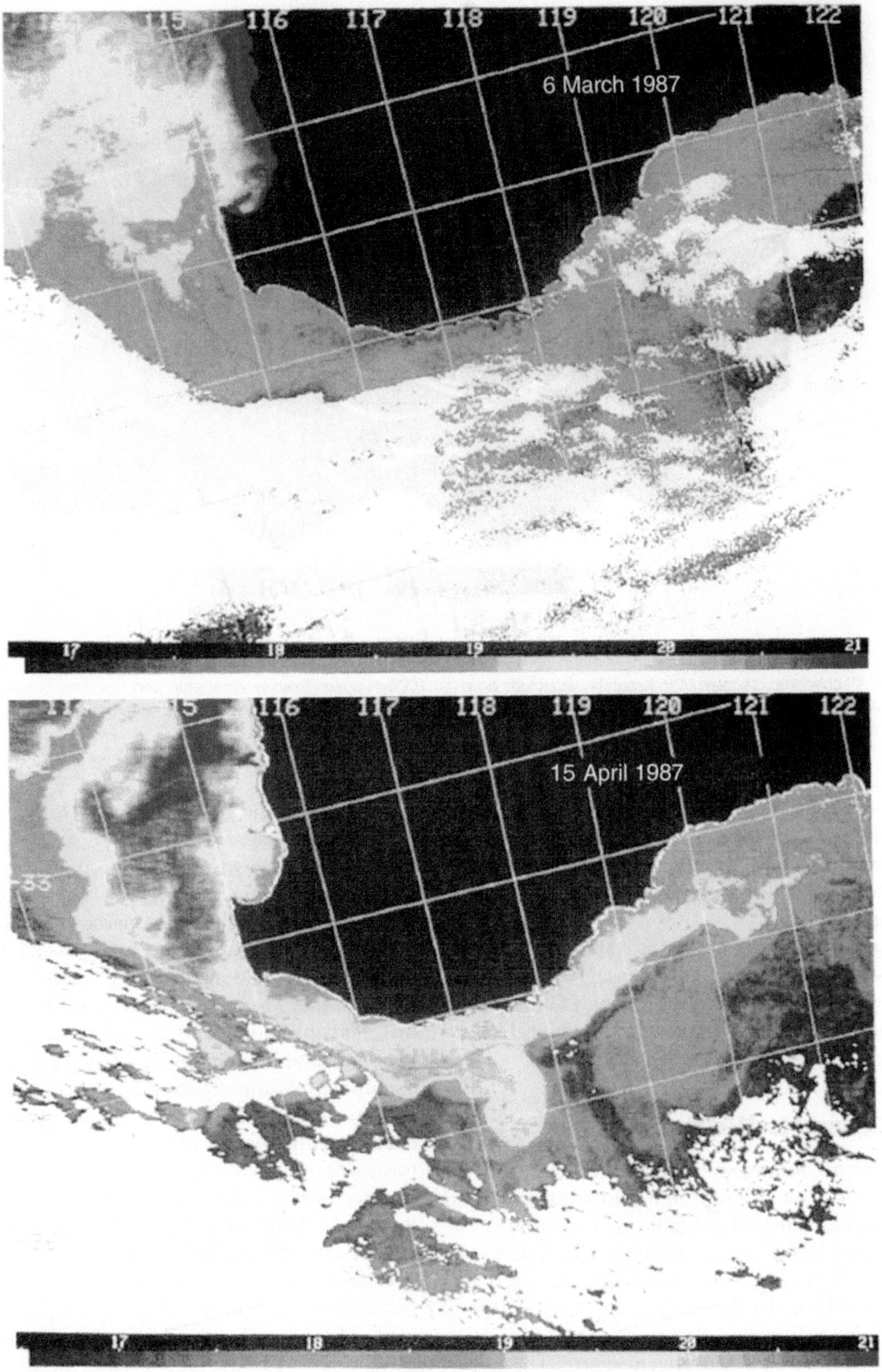

Figure 4 Satellite images 6 weeks apart showing the energetic nature of the arrival of the warm waters of the LC. There is a 1° × 1° latitude–longitude grid; land is black; the shelf edge is marked by a thin black line; clouds are white; and the temperature scale (°C) is at the bottom.

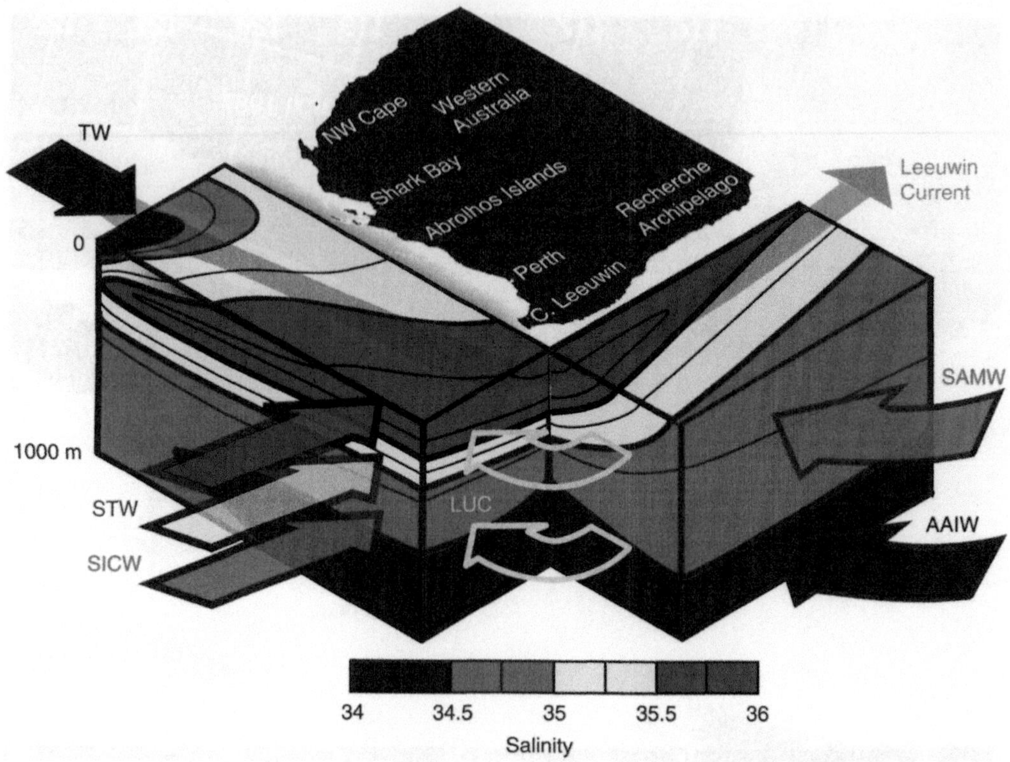

Figure 5 A schematic diagram of the salinity structure down to 1000 m drawn from several ship surveys. The boxes are 1000-m deep and about 200 km wide. The view is from above the southern Indian Ocean and the map is to orient the reader.

oxygen values as compared with the deeper component of the SICW from the west that has the same temperature and salinity values. The blue arrow indicates the AAIW originating at the surface much farther south.

The white arrows attempt to show the undercurrent in the south that flows westward to Cape Leeuwin beneath the eastward-flowing LC and then joins – or becomes – the northward-flowing LUC.

The LC and Its Eddies

The LC path southward from NW Cape to Cape Leeuwin is a meandering one and it spawns and interacts with anticyclonic and cyclonic eddies. The anticyclonic eddies can be 200 km across and they are more akin to eddies in western boundary systems, such as the EAC, having edge speeds of as much as $1\,\mathrm{m\,s^{-1}}$, long lifetimes (>1 year), and mixing to several hundred meters depth in winter. The eddies can be followed with satellite altimetry measurements as they drift into the central Indian Ocean. Both cyclonic and anticyclonic eddies can be tracked, with the former having a slight poleward bias to their drifts and the latter an equatorward one.

In **Figure 6**, the LC is seen near the upper continental slope where the salinity and temperature structure dip downward. The current has a subsurface maximum, probably because of strong southerly winds at the time. The structure farther out to sea reveals the upward doming associated with a cyclonic eddy. The high-salinity water is subtropical surface water from farther westward, although on the shelf insolation and evaporation have warmed the waters and made them more saline.

South of Western Australia the LC can be particularly strong (**Figure 7**) and the situation with eddies is different: weak anticyclonic eddies, that form as far east as Tasmania, drift westward to encounter the LC, and the continental slope near the Recherche Archipelago. There they steer the warmer LC waters offshore and around them (**Figure 8**) and over their tops, with the result that they strengthen significantly. The two eddies in the image drifted westward out into the Indian Ocean. The warm waters of the LC spread inshore across the shelf, although the current would be strongest just seaward of the shelf edge.

Ship measurements across an offshoot joining the LC to an anticyclonic eddy revealed a trough-shaped structure with southward flow of $0.5\,\mathrm{m\,s^{-1}}$ on the

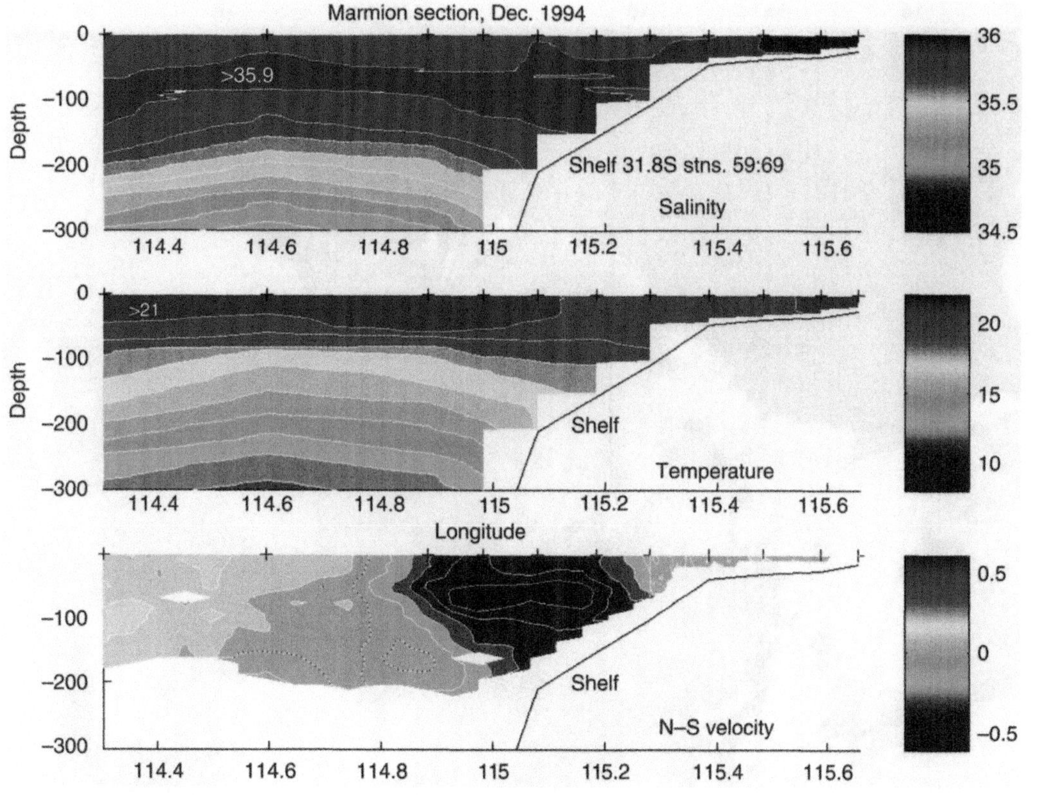

Figure 6 Ship measurements of salinity (practical salinity units) and temperature (°C) with a CTD, and north–south current with an acoustic Doppler current profiler (ADCP) west of Perth (32° S) in Dec. 1994.

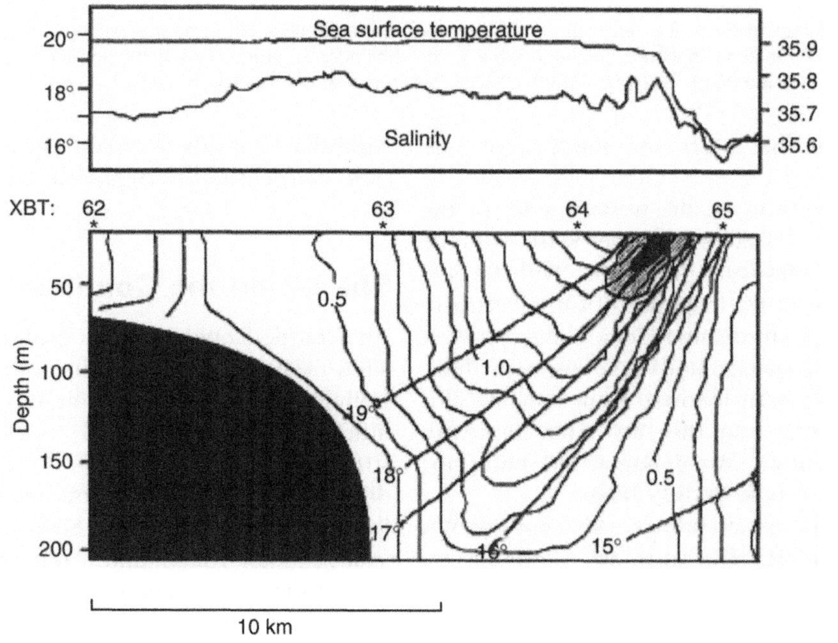

Figure 7 A north–south transect across the LC south of Western Australia near 117° E in June 1987. Upper panel: surface temperature and salinity across the Leeuwin Current. Lower panel: current speed from acoustic Doppler current profiler (ADCP) measurements, the temperature structure from expendable bathythermograph (XBT) casts, and Richardson numbers calculated from the current and temperature structure. Dark shading, <0.25 (the region of most active overturning); light shading, <0.5. The current speed reached 1.5 m s^{-1}. Reproduced from Cresswell GR and Peterson JL (1993) The Leeuwin Current south of Western Australia. *Australian Journal of Marine Freshwater Research* 44: 285–303.

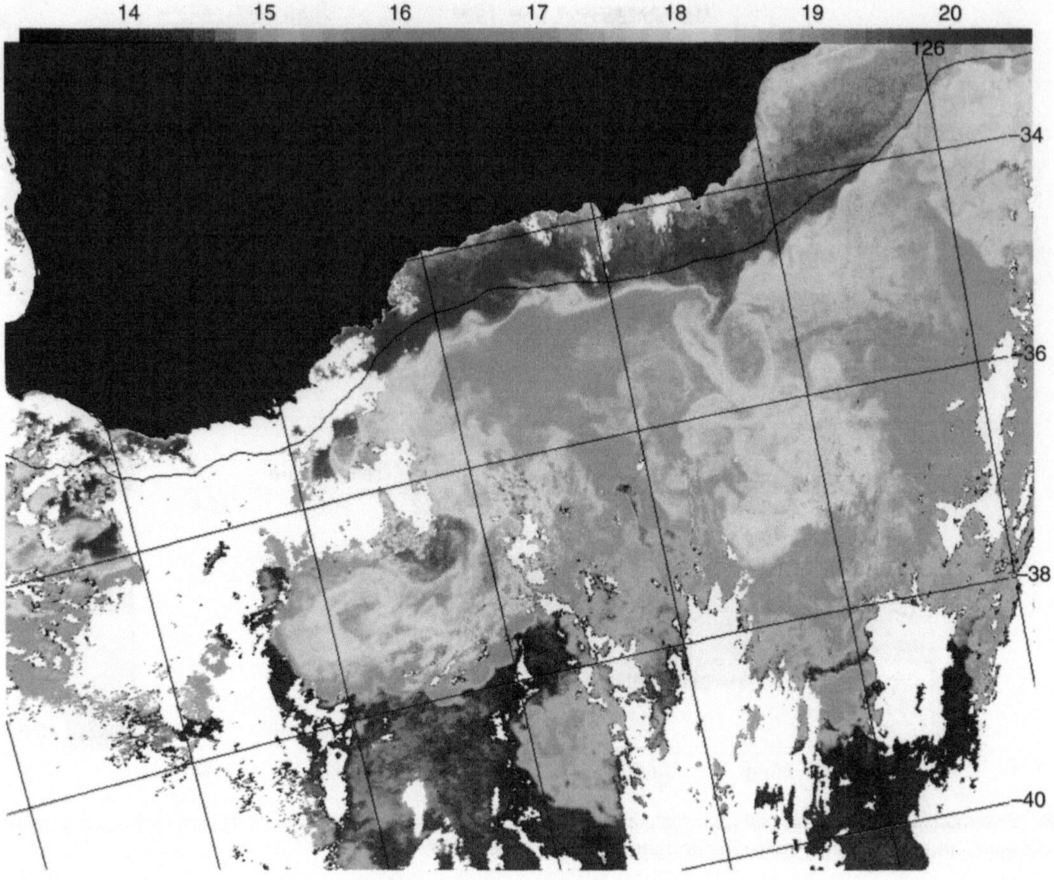

NOAA 1 1 SST 07 Jun 1994 0832Z

Figure 8 An SST image showing the sea south of WA between Cape Leeuwin (left) and the western Great Australian Bight (top right). There is a 1 ° × 1 ° latitude–longitude grid; land is black; the shelf edge is marked by a thin black line; clouds are white; and the temperature scale (°C) is at the top. Copyright 2000, CSIRO Division of Marine Research, Hobart.

western side and the reverse on the eastern side (**Figure 9**).

Cyclonic eddies form on the western sides of the seaward offshoots. The eddies drift westward, with the anticyclonic ones repeatedly interacting with the LC until they are west of the longitude of Cape Leeuwin. At times there can be three anticyclonic eddies between the Recherche Archipelago and Cape Leeuwin. Individual eddies have been followed using satellite altimetry for over a year from their first interaction with the LC at the Recherche Archipelago to beyond Cape Leeuwin and out into the Indian Ocean.

The structure of an anticyclonic eddy south of WA is shown in **Figure 10**. The north and south current speed maxima were $0.5\,\mathrm{m\,s^{-1}}$. The eddy vertical structure suggested that it had mixed to over 300-m depth during the winter and then, with the onset of summer, had been warmed near the surface. The depression of the water structure extended at least to 1000 m. The eddy was low in oxygen and nutrients as compared with the surrounding waters. This contrasts with the anticyclonic eddies formed by the LC west of Australia: there the formation process entrains relatively nutrient-rich waters from the continental shelf.

The LC on the Continental Shelf

An acoustic Doppler current profiler (ADCP) moored a few meters above the bottom at the 70-m isobath out from Perth gave hourly data at 4-m depth intervals for little over an year (**Figure 11**). In December, there was strong northward alongshore flow due to the wind-driven Capes Current that lowered the temperature at the instrument by 2°C. The onset of the LC was quite sudden in late March and it was present in the lower half of the water column until the end of the record. Nearer the surface the LC flow was often weaker, perhaps due to those waters being more susceptible to northward wind forcing from passing storms. The lower shelf waters, particularly from April to August, were drained out across the topography with the offshore flow peaking at over $0.1\,\mathrm{m\,s^{-1}}$ in May. Whether this was due to an Ekman bottom boundary layer

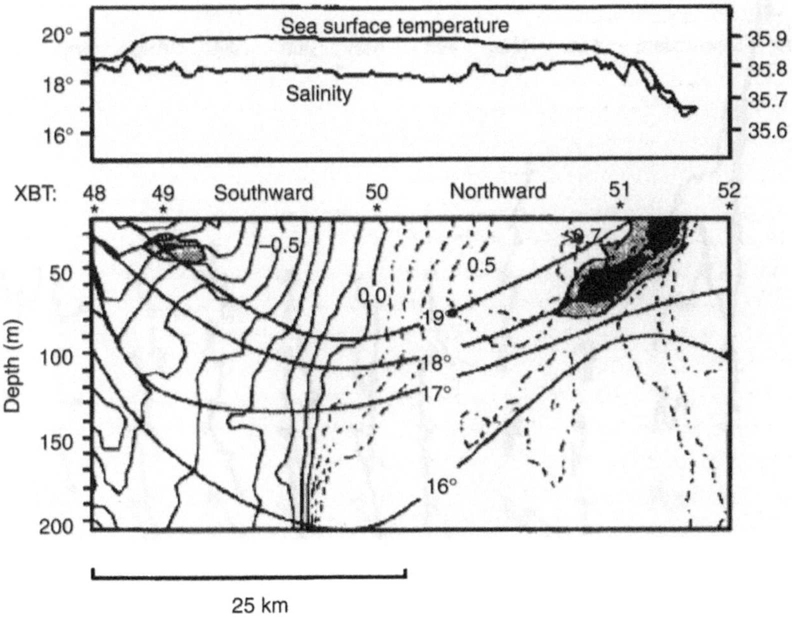

Figure 9 An east–west transect south of Western Australia at 117° E in Jun. 1987 across an offshoot connecting the LC to an anticyclonic eddy. Upper panel: surface temperature and salinity. Lower panel: the north–south current components from acoustic Doppler current profiler (ADCP) measurements, the temperature structure from expendable bathythermograph (XBT) casts, and Richardson numbers calculated from the current and temperature structure. Dark shading, <0.25; light shading, <0.5. Reproduced from Cresswell GR and Peterson JL (1993) The Leeuwin Current south of Western Australia. *Australian Journal of Marine Freshwater Research* 44: 285–303.

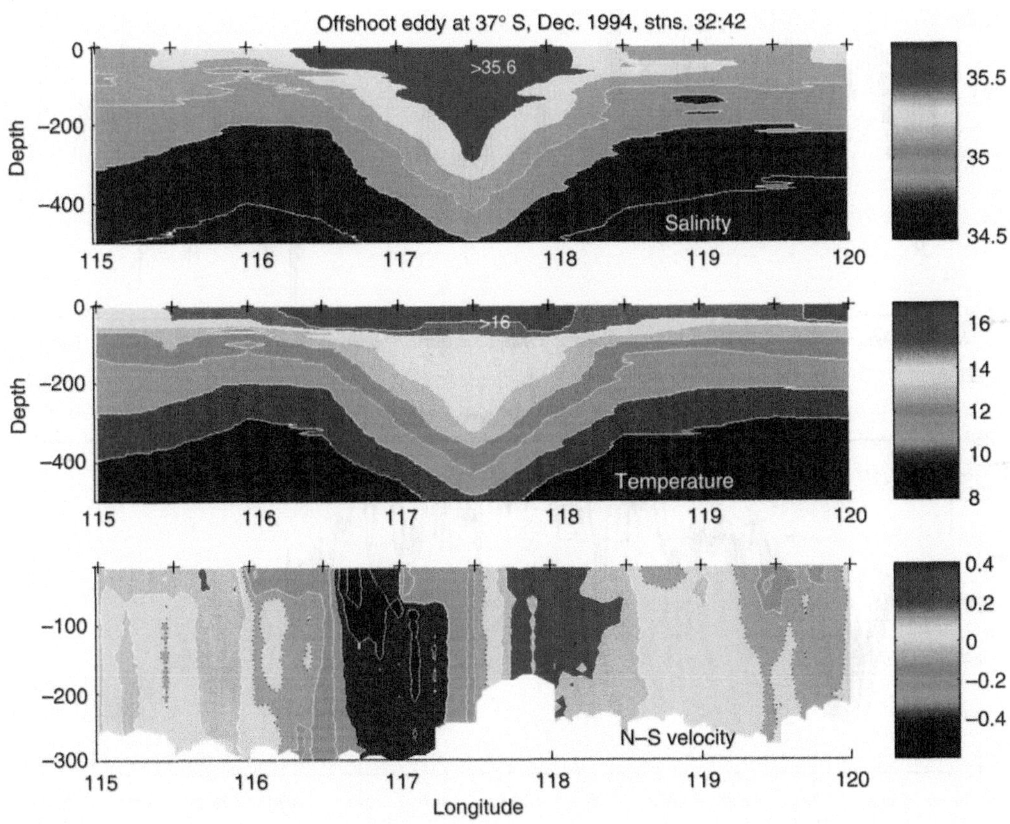

Figure 10 A transect of salinity (practical salinity units), temperature (°C), and north–south velocity component (m s^{-1}) from 38° S, 115° E to 37° S, 120° E across an anticyclonic eddy south of WA in Dec. 1994. The velocity component plot has been limited to the 300-m depth reached by the acoustic instrument taking the measurements.

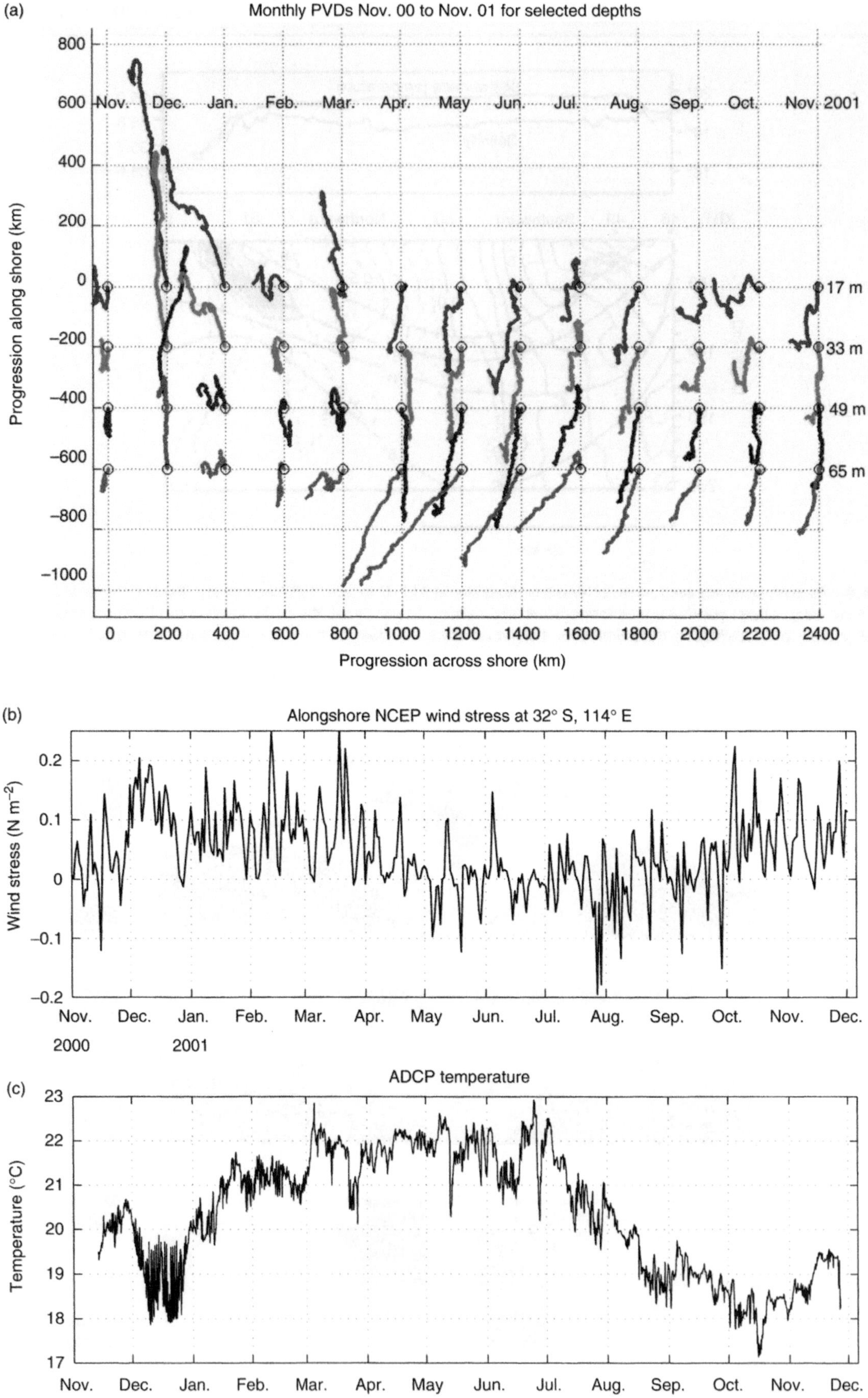

Figure 11 (a) The monthly progressive vector diagrams, or inferred water trajectories, for selected depths (every fourth depth bin) from Nov. 2000 to Nov. 2001. Only data from every fourth depth bin are shown. The trajectories are arranged in columns, with successive 200-km offsets, and by row, for the different depths. The axes are across and along the bottom topography. (b, c) The alongshore wind stress near the mooring site and the temperature recorded by the near-bottom instrument moored at the 70-m isobath.

beneath the LC, or to near-shore waters cooling and becoming denser and moving out across the shelf, or both, is not clear. Note that the near-bottom temperaturve (**Figure 11(c)**) was highest in the austral autumn/winter (March/July) due to the influence of the warm LC waters from the north.

See also

Current Systems in the Indian Ocean. Current Systems in the Southern Ocean. East Australian Current. Ekman Transport and Pumping. Meddies and Subsurface Eddies. Mesoscale Eddies.

Further Reading

Andrews JC (1977) Eddy structure and the West Australian Current. *Deep-Sea Research* 24: 1133–1148.

Cresswell GR and Golding TJ (1980) Observations of a south-flowing current in the southeastern Indian Ocean. *Deep-Sea Research* 27A: 449–466.

Cresswell GR and Griffin DA (2004) The Leeuwin Current, eddies and sub-Antarctic waters off south-western Australia. *Marine and Freshwater Research* 55: 267–276.

Cresswell GR and Peterson JL (1993) The Leeuwin Current south of Western Australia. *Australian Journal of Marine Freshwater Research* 44: 285–303.

Domingues CM, Wijffels SE, Maltrud ME, Church JA, and Tomczak M (2006) Role of eddies in cooling the Leeuwin current. *Geophysical Research Letters* 33: L05603 (doi:10.1029/2005GL025216).

Feng M, Meyers G, Pearce A, and Wijffels S (2003) Annual and interannual variations of the Leeuwin Current at 32°S. *Journal of Geophysical Research* 108(C11): 3355 (doi:10.1029/2002JC001763).

Fieux M, Molcard R, and Morrow R (2005) Water properties and transport of the Leeuwin Current and eddies off Western Australia. *Deep-Sea Research I* 52: 1617–1635.

Godfrey JS and Ridgway KR (1985) The large-scale environment of the poleward-flowing Leeuwin Current, Western Australia: Longshore steric height patterns, wind stresses and geostrophic flow. *Journal of Physical Oceanography* 15: 481–495.

Holloway PE and Nye HC (1985) Leeuwin Current and wind distributions on the southern part of the Australian North West Shelf between January 1982 and July 1983. *Australian Journal of Marine and Freshwater Research* 36: 123–137.

Middleton JF and Platov G (2003) The mean summertime circulation along Australia's southern shelves: A numerical study. *Journal of Physical Oceanography* 33: 2270–2287.

Morrow R, Birol F, Griffin D, and Sudre J (2004) Divergent pathways of cyclonic and anti-cyclonic ocean eddies. *Geophysical Research Letters* 31: L24311 (doi:10.1029/2004GL020974).

Morrow R, Fang FX, Fieux M, and Molcard R (2003) Anatomy of three warm-core Leeuwin Current eddies. *Deep-Sea Research II* 50: 2229–2243.

Pearce A and Pattiaratchi C (1999) The Capes Current: A summer countercurrent flowing past Cape Leeuwin and Cape Naturaliste, Western Australia. *Continental Shelf Research* 19: 401–420.

Pearce AF and Phillips BF (1988) ENSO events, the Leeuwin Current, and larval recruitment of the western rock lobster. *Journal Du Conseil* 45(1): 13–21.

Ridgway KR and Condie SA (2004) The 5500-km long boundary flow off western and southern Australia. *Journal of Geophysical Research* 109: C04017 (doi:10.1029/2003JC001921).

Schodlok MP, Tomczak M, and White N (1997) Deep sections through the South Australian Basin and across the Australian–Antarctic discordance. *Geophysical Research Letters* 22: 2785–2788.

Smith RL, Huyer A, Godfrey JS, and Church JA (1991) The Leeuwin Current off Western Australia, 1986–1987. *Journal of Physical Oceanography* 21: 323–345.

Waite AM, Thompson PA, Pesant S, *et al.* (2007) The Leeuwin Current and its eddies: An introductory overview. *Deep-Sea Research II* 54: 789–796.

AGULHAS CURRENT

J. R. E. Lutjeharms, University of Cape Town,
Rondebosch, South Africa

Introduction

The greater Agulhas Current forms the western boundary system of the circulation in the South Indian Ocean. Contrary to the flow of comparable subtropical gyres in other ocean basins, the sources of the Agulhas Current are interrupted by a substantial barrier, the island of Madagascar. This leads to the formation of two minor western boundary flows, the East Madagascar Current and the Mozambique drift. Once fully constituted off the coast of south-eastern Africa, the Agulhas Current proper can be considered to consist of two distinct parts: the northern and the southern current. The northern part flows along a steep continental shelf and its trajectory is extremely stable. The southern part flows along the wide shelf expanse of the Agulhas Bank and by contrast meanders widely. South of the African continent the Agulhas Current retroflects in a tight loop, with most of its waters subsequently flowing eastward as the Agulhas Return Current. This loop configuration is unstable and at irregular intervals it is pinched off to form a detached Agulhas ring. These rings, carrying warm and salty Indian Ocean water, drift into the South Atlantic Ocean. Some cross the full width of this ocean in the next 2–3 years, whereas many are dissipated within 5 months of being spawned. The Agulhas Return Current flows back into the South Indian Ocean along the Subtropical Convergence. This juxtaposition generates considerable mesoscale turbulence in the form of meanders and an assortment of eddies. Water from the Agulhas Return Current leaks northward, back into the subtropical gyre, along its full length. By about 70°E all Agulhas water has been lost to the eastward flow that subsequently continues as the South Indian Ocean Current.

Importance

Historically the Agulhas Current was one of the first ocean currents to receive a great deal of scientific attention. It was described in some detail as early as 1766 by Major James Rennell, preeminent British geographer at the time. This was followed by wide-ranging investigations by Dutch mariners such as Van Gogh and Andrau in the 1850s. This early interest was motivated purely by nautical concerns, the Agulhas Current constituting a formidable impediment to vessels sailing to India and to the East. Fundamental studies by German investigators dominated research on the Agulhas Current region in the 1930s, but this endeavour was terminated by the Second World War. During the past few decades a renaissance in interest in this current has occurred for totally different reasons.

It has been demonstrated that the greater Agulhas Current system (**Figure 1**) has a marked influence on the climate variability over the southern African subcontinent. It has also been shown that this current is a key link in the exchanges of water between ocean basins and thus probably has a special role in the oceans' influence on global climate. This renewed interest has stimulated a number of research cruises, the placement of current meter moorings, investigations by satellite remote sensing, as well as theoretical and modeling studies, all leading to an enormous increase in knowledge of the Agulhas Current.

Large-scale Circulation

The Agulhas Current forms part of the overall circulation of surface waters in the South Indian Ocean that is anticyclonic, i.e., anticlockwise in the Southern Hemisphere (**Figure 2**). On its eastern side, the equatorward flow is weak and dispersed, whereas the recirculation in the South West Indian Ocean is particularly strongly developed, penetrating to 1000 m depth at its centre. The southern border to the circulation is the Subtropical Convergence. This strong thermohaline front at roughly 41°S separates the characteristic flows and water masses of the subtropical gyre and those of the Antarctic Circumpolar Current that lies to the south. Along the Subtropical Convergence, the Agulhas Return Current and the South Indian Ocean Current carry their respective water masses eastward. To the north the gyral circulation is closed by the South Equatorial Current that is found between about 10° and 25°S and carries water from east to west.

At the eastern shores of Madagascar, the South Equatorial Current splits into a northern and a southern limb of the East Madagascar Current; about 70% going north along this shoreline, 30% heading south. Most of that heading north eventually

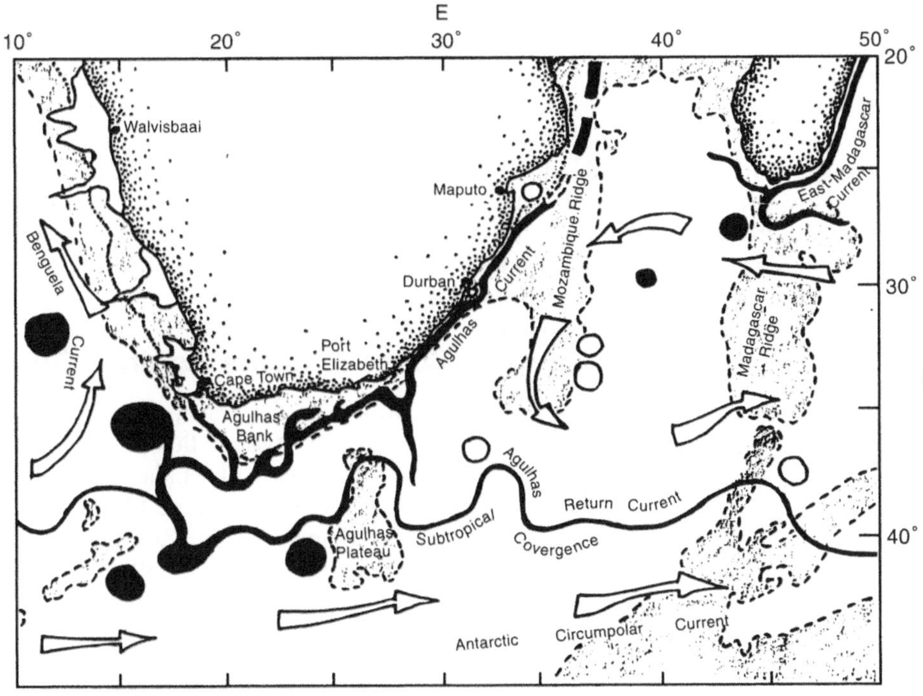

Figure 1 A conceptual portrayal of the greater Agulhas Current. Ocean regions shallower than 3000 m have been shaded. Intense currents are black, whereas the general background circulation is shown by open arrows. Cyclonic eddies are open; anticyclonic rings and eddies are black. Note the stability of the northern Agulhas Current, the meanders of the southern Agulhas Current, the tight retroflection loop, and the continuously weakening eastward flow of the Agulhas Return Current along the Subtropical Convergence. Agulhas rings (black) are advected by the Benguela Current past the extensive coastal upwelling off south-western Africa.

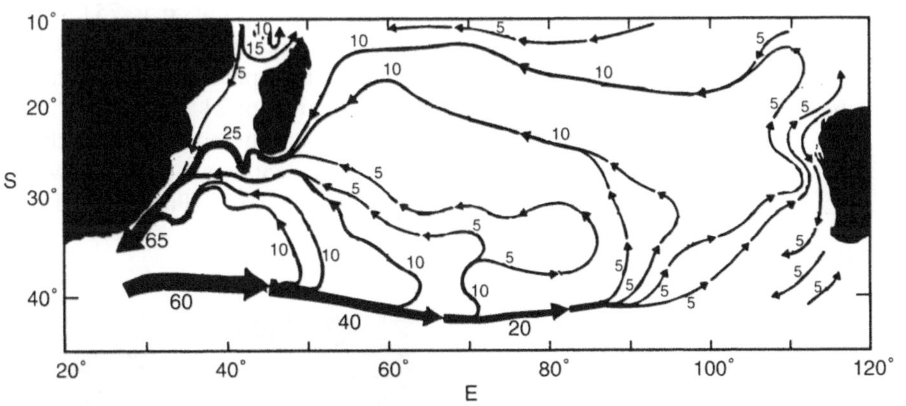

Figure 2 The baroclinic volume transport for the upper 1000 m of the South Indian Ocean. Values are in units of 10^6 m^3 s^{-1}. Note the very small contribution coming from the Mozambique Channel and the concentration of the recirculation in the South West Indian Ocean west of 70°E. This transport pattern, averaged over a long period, is not to be confused with the depictions of instantaneous currents in **Figures 1** and **4**. (After Stramma and Lutjeharms (1997) *Journal of Geophysical Research* 102(C3): 5513–5530. © American Geophysical Union.)

reaches the east coast of the African continent, where it forms the East African Coastal Current. There is some leakage from the Mozambique Channel and from the southern limb of the East Madagascar Current into the Agulhas Current, but most of the Agulhas Current's waters come from the subgyre of the South West Indian Ocean (see **Figure 2**). The Agulhas Current itself is narrow, deep, and fast; a typical western boundary current. All these surface currents are driven largely by the reigning wind systems.

The wind systems over the South West Indian Ocean fall largely outside the influence of the seasonally varying monsoonal winds of the North

Indian Ocean. The average air motion over the South Indian Ocean is dominated by a large-scale, anticyclonic circulation around a high-pressure system centered south-east of the island of Madagascar. This flow generally is stronger in austral summer than in winter. Strongest winds ($24\,m\,s^{-1}$) are found at 50°S latitude in summer, and weakest at 35°S ($2\,m\,s^{-1}$). A band of minimum wind stress extends across this ocean at 35°S. Next to the continental land masses the winds are usually aligned with the coasts. Apart from driving the surface currents, the atmosphere also has a considerable effect on the formation of certain water masses (**Figure 3**) that are typical for the region.

This temperature–salinity portrayal indicates the characteristic for each specific water mass in this ocean region. The fresher Tropical Surface Water is formed north of 20°S where there is an excess of precipitation over evaporation; Subtropical Surface Water is formed between 28° and 38°S, where this ratio is reversed. Subtropical Surface Water is found as a shallow subsurface salinity maximum in regions to the north and south of its region of formation. Intermediate waters lie at depths between 1000 and 2000 m and consist of Antarctic Intermediate Water and North Indian Intermediate Water (also called Red Sea Water). The former subducts between the Antarctic Polar Front and the Subtropical Convergence; the latter is formed owing to very high rates of evaporation in the Red Sea, the Arabian Sea and the Persian Gulf. Central Water, lying between the surface and the intermediate waters, is a mixture of these two. Below the intermediate waters are Indian Deep Water and North Atlantic Deep Water, each formed by subduction in the respective ocean regions after which they are named. All these respective water masses are involved in some way or other in the source currents of the Agulhas Current.

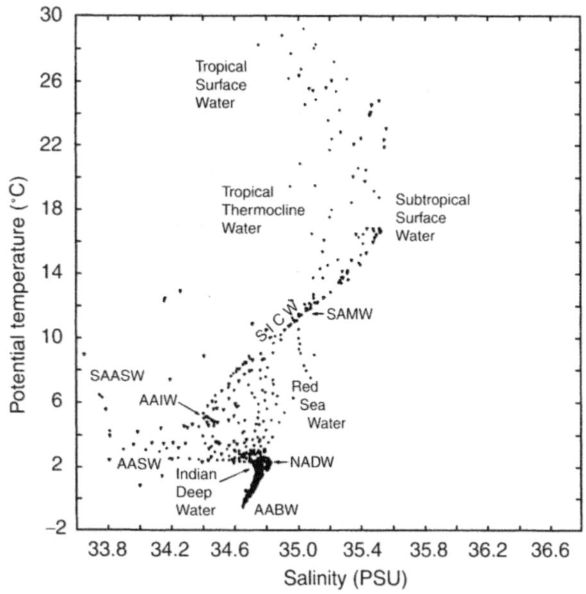

Figure 3 The relationship of potential temperature and salinity for waters in the western Indian Ocean. Some of the characteristic water masses to be found here are SICW, South Indian Central Water; SAMW, Subantarctic Mode Water; NADW, North Atlantic Deep Water; AABW, Antarctic Bottom Water; AASW, Antarctic Surface Water; AAIW, Antarctic Intermediate Water; and SAASW, Subantarctic Surface Water. (After Gordon *et al.* (1987) *Deep-Sea Research* 34(4): 565–599. © Elsevier Science.)

Sources of the Agulhas Current

According to the transport portrayed in **Figure 2**, 30% of the volume flux of the Agulhas Current derives from east of Madagascar, only 13% comes through the Mozambique Channel, and 67% is recirculated in a South West Indian Ocean subgyre. Note, however, that the inflow from east of Madagascar does not necessarily come from the East Madagascar Current (see **Figure 1**).

The southern limb of the East Madagascar Current starts at the bifurcation point of the South Equatorial Current at about 17°S along the east coast of Madagascar. Its surface speed here is roughly $1\,m\,s^{-1}$, increasing downstream to about $1.5\,m\,s^{-1}$. The current is very stable in both flux and trajectory and exhibits no clear seasonality in any of its characteristics. Using the $0.5\,m\,s^{-1}$ isotach as the outer limits of the current, it is 75 km wide, 200 m deep, and its core lies 50 km offshore. It carries Tropical as well as Subtropical Surface Water with a total volume transport of $21 \times 10^6\,m^3\,s^{-1}$ (i.e., 21 Sv). Where it overshoots the end of the shelf of Madagascar it retroflects, with most of its waters subsequently heading eastwards (**Figure 4**). There may be some

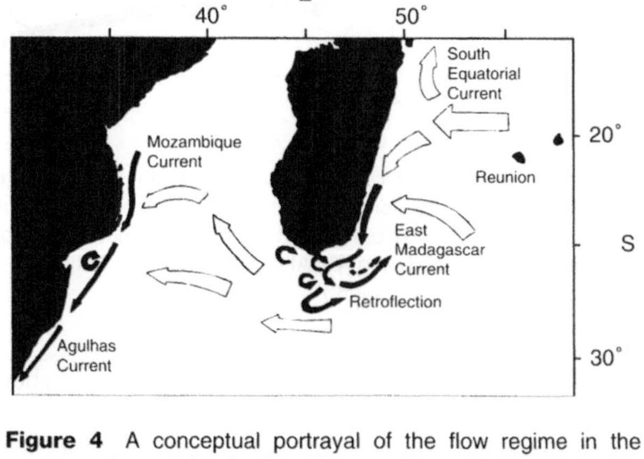

Figure 4 A conceptual portrayal of the flow regime in the source regions of the Agulhas Current. Narrow, intense currents are shown by black arrows. The East Madagascar Current is a miniature western boundary current that retroflects south of Madagascar. (After Lutjeharms *et al.* (1981) *Deep-Sea Research* **28**(9): 879–899. © Elsevier Science.)

leakage of East Madagascar Water into the Agulhas Current by way of rings and filaments, but this constitutes an insignificant contribution. The wider flow east of Madagascar, up to 240 km offshore, is $41 \times 10^6\,m^3\,s^{-1}$. A substantial part of this more extensive flow may eventually make its way into the Agulhas Current (**Figures 2** and **4**).

The flow through the Mozambique Channel was once thought to be the major contributor to the flux of the Agulhas Current. Now it is considered to be minor (**Figure 2**). The entire existence of a consistent, continuous Mozambique Current, flowing along the African coastline, has in fact been called into question.

The northern mouth of the Mozambique Channel is largely closed to subsurface flow by bottom ridges shallower than 2000 m, except at its western side. The surface circulation in the northern part of the channel is anticyclonic to an estimated depth of 1000 m. The eastern side of this flow, which might be the start of a Mozambique Current, is 250 km wide, has a surface speed of $0.3\,m\,s^{-1}$ and a volume flux of $6 \times 10^6\,m^3\,s^{-1}$. The water masses here are characteristic of the monsoonal regime to the north with no Antarctic Intermediate Water. There is some evidence for the presence of this water mass in the central part of the channel, but there still is no North Atlantic Deep Water. The rest is essentially Subtropical Surface Water of the South Indian variety. Occasional strong flows of $2\,m\,s^{-1}$ have been observed off the African coastline in this central part of the channel, but this is extremely variable. The southern third of the channel has all the thermohaline characteristics of the South West Indian Ocean, including the presence of North Atlantic Deep Water. Net volume flux through the southern mouth of the channel seems to be very changeble. Calculations have varied from $26 \times 10^6\,m^3\,s^{-1}$ southward to $5 \times 10^6\,m^3\,s^{-1}$ northward, above 1000 m.

The best-substantiated characteristic of the circulation in the Mozambique Channel is therefore its very high mesoscale variability. This argues against a persistent western boundary current and instead suggests a series of eddies moving southward down the channel. This scenario is consistent with most available observations and also some numerical models.

Northern Agulhas Current

Water from the South West Indian Ocean subgyre feeds into the Agulhas Current along its full length, but at a latitude of 27°S this current is nonetheless thought to be fully constituted (**Figure 5**).

This northern part of the Agulhas Current is characterized in particular by an extremely stable trajectory; its core meanders less than 15 km to either side. This stability is thought to be due to the strong slope of the continental shelf along which it flows. The current has a well-developed, inshore thermal front that meanders somewhat more extensively. The surface characteristics of the current are given in **Table 1**.

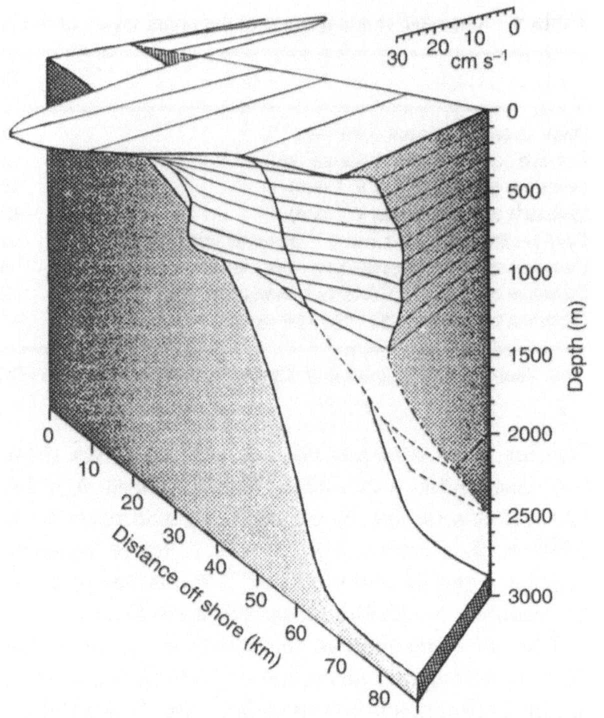

Figure 5 The spatial velocity structure of the northern Agulhas Current at Durban based on direct current measurements during a research cruise. Speeds below 1000 m have been estimated using a geostrophic calculation and are less certain. (After Duncan (1970) PhD dissertation, University of Hawaii.)

Surface temperatures decrease by about 2°C downstream. In the north they are at a maximum of 28°C in February and a minimum of 23°C in July. At Port Elizabeth, where the southern Agulhas Current starts, the maximum temperature is 25°C in January, with a minimum of 21°C in August. Surface salinities decrease from 35.5 PSU in the north to 35.3 PSU in the south.

At Durban the core of the current usually lies 20 km offshore and penetrates to a depth of 2500 m (see **Figure 5**). Between the $0.5\,m\,s^{-1}$ isotachs it is 90 km wide; its offshore termination being more disperse than its strong inshore edge. Its core slopes so that at 900 mdepth it lies 65 km offshore. Its total volume flux is $73 \times 10^6\,m^3\,s^{-1}$ and this increases by an estimated $6 \times 10^6\,m^3\,s^{-1}$ for every 100 km distance downstream in the northern Agulhas Current. Surface speeds at its core usually lie between 1.4 and $1.6\,m\,s^{-1}$, with occasional peaks of up to $2.6\,m\,s^{-1}$.

Table 1 Kinematic characteristics of the upper layers of the northern Agulhas Current

	Mean	*SD*	*Minimum*	*Maximum*
Peak speed in current core ($m\ s^{-1}$)	1.36	0.30		2.45
Current core offshore distance (km)	52	14	30	>100
Distance offshore $0.5\,m\ s^{-1}$ (km)	35	14	10	70
Distance offshore $1.0\,m\ s^{-1}$ (km)	42	14	25	95
Core width, between $1.0\,m\ s^{-1}$ isotachs (km)	34	15	10	>60
Distance offshore temperature max. (km)	58	20	35	>100
Distance offshore 15°C/200 m intersection (km)	50	15	25	90
Distance offshore 35.35 PSU salinity/200 m (km)	47	13	23	>100

After Pearce (1997), *Journal of Marine Research* 35(4): 731–753.

Neither these velocities nor the volume fluxes show any discernible seasonality. The northern Agulhas Current is underlain by an opposing undercurrent at 1200 m that carries $6 \times 10^6\,m^3\,s^{-1}$ water equatorward at a rate of about $0.3\,m\,s^{-1}$. It consists partially of modified North Indian Intermediate Water.

The invariant path of the northern Agulhas Current is interrupted during about 20% of the time by an intermittent, solitary meander – the Natal Pulse – that originates at the Natal Bight, an offset in the coast north of 30°S (see **Figure 6**). It translates downstream at a very steady 20 km per day, continuously growing in its lateral dimensions. On its landward side it encloses a cyclonic eddy that creates a strong coastal countercurrent as the Natal Pulse passes. This meander is triggered at the Natal Bight whenever the current intensity there exceeds a certain threshold, allowing baroclinic instability to develop away from the constraining shelf slope. The Natal Pulse is an important component of the current system since it may cause upstream retroflection at the Agulhas Plateau (see **Figure 1**) and may precipitate ring shedding at the Agulhas retroflection, far downstream.

Flow over the shelf adjacent to the northern Agulhas Current is dependent on the shelf morphology.

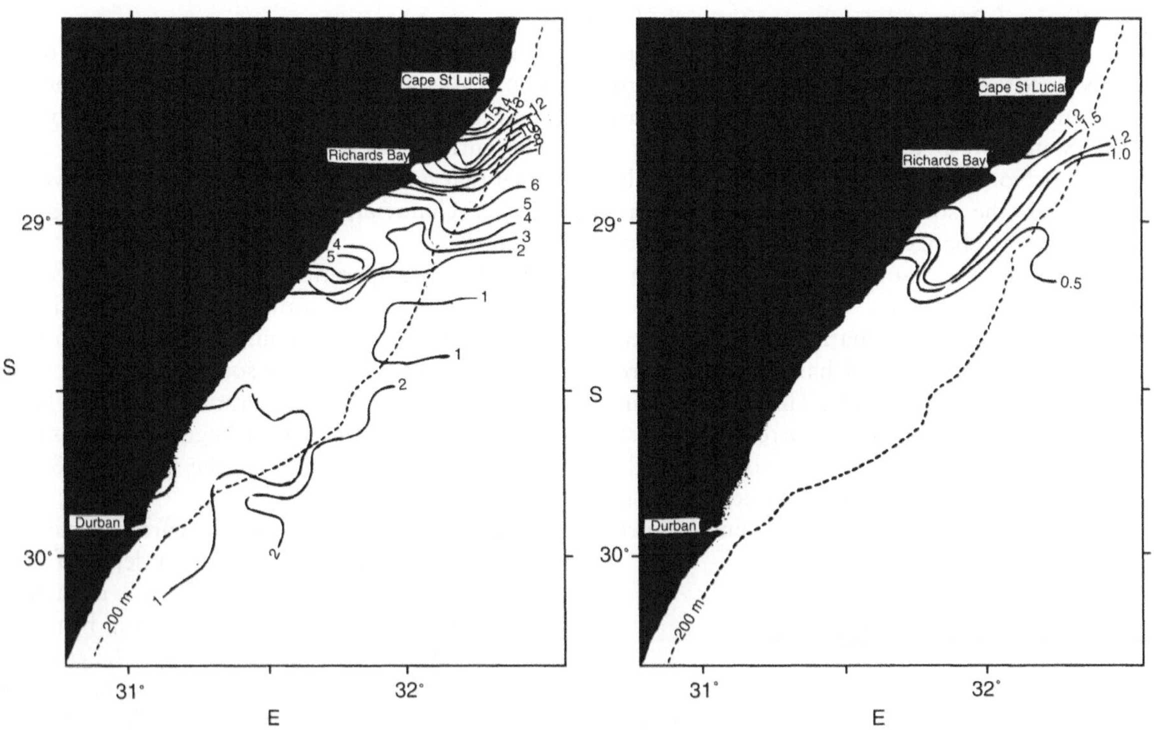

Figure 6 Current-induced upwelling in the Natal Bight off south-eastern Africa. The left panel gives the distribution of nitrate (in $\mu mol\,l^{-1}$) at 10 m depth and the right hand panel the simultaneous distribution of chlorophyll *a*. The active upwelling cell off Cape St Lucia with high nitrates and chlorophyll *a* is well circumscribed. (After Lutjeharms *et al.* (2000) *Continental Shelf Research* 20(14): 1907–1939. © Elsevier Science.)

Where the shelf is narrow, the flow is mostly parallel to the current. Over the Natal Bight, where the shelf is wider, the flow consist of cyclonic eddies. At the northern end of the Natal Bight, the current forces inshore upwelling (**Figure 6**).

The water in this upwelling cell may be 5°C colder than the adjacent current, have a high nutrient content, and exhibit enhanced biological primary productivity. The cold water thus upwelled flows over the bottom of the whole Natal Bight, strengthening the vertical layering over this shelf region. Off Durban a recurrent lee eddy is often observed.

Seaward of the Agulhas Current, off the Mozambique Ridge (see **Figure 1**), a large number of very intense deep-sea eddies have been observed. They may be at least 2000 m deep, 100 km in diameter, have surface speeds of $1\,m\,s^{-1}$ and circular transports of between 6×10^6 and $18 \times 10^6\,m^3\,s^{-1}$. Their lifetimes are estimated to be 1–3 years. Most of those observed are cyclonic, although a few anticyclonic ones have been seen. They seem to come from both the Mozambique Channel and from east of Madagascar, but their true origins remain unknown.

Southern Agulhas Current

In contrast to the northern Agulhas Current, the southern Agulhas Current is characterized by wide meanders as it flows past the Agulhas Bank south of Africa (see **Figure 1**). Meanders are present along this shelf edge at least 65% of the time. They have an average wavelength of 300 km and a phase speed that varies from 5 to 23 km per day.

Meanders usually have a trailing plume and an embedded, cyclonic lee eddy (**Figure 7**). There is evidence that these eddies are preferentially clustered in the eastern bight of the Agulhas Bank (see **Figure 1**). The dimensions of all these shear edge features change markedly with distance downstream (see **Table 2**).

Sea surface temperatures also change more readily with distance downstream here than they do in the northern Agulhas Current. Sea surface temperatures in the southern Agulhas Current reach a maximum of 26°C at Algoa Bay in February; 23°C off the southern tip of the Agulhas Bank. In August these temperatures are 21°C and 17°C, respectively.

The volume flux of the Agulhas Current off the southern tip of the Agulhas Bank has been estimated at $70 \times 10^6\,m^3\,s^{-1}$ down to 1500 m, i.e., about the same as that of the current to its full depth at Durban. Even though there is this increase, it seems that the

Table 2 Dimensions of shear edge features along the landward edge of the southern Agulhas Current

	Port Elizabeth	*Tip of Agulhas Bank*
Plume lengths at surface (km)	100	162
Plume widths at surface (km)	27	37
Diameter of enclosed eddy (km)	27	51
Plume dispersion from current (km)	50	150

After Lutjeharms *et al.* (1989) *Continental Shelf Research* 9(7): 597–616.

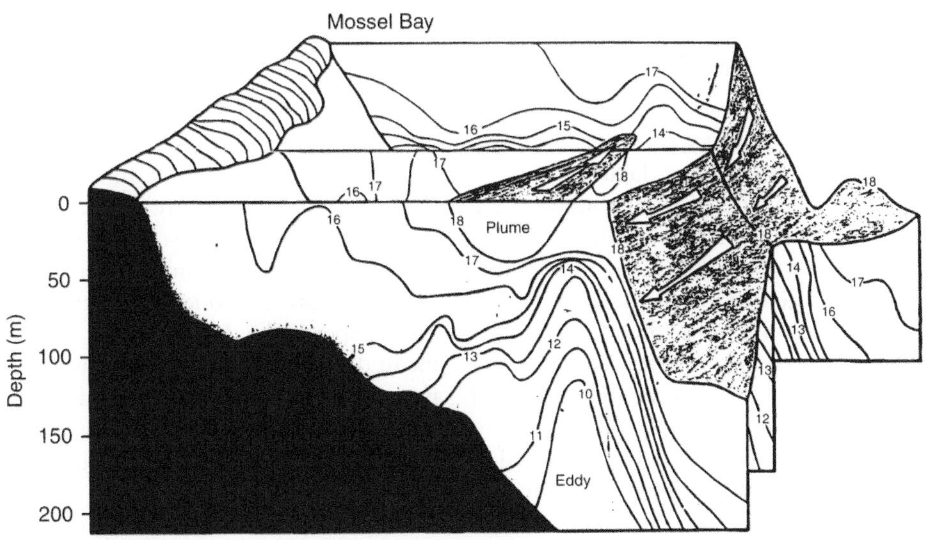

Figure 7 Three vertical temperature sections across the Agulhas Bank picture the thermal composition of the southern Agulhas Current, a shear-edge eddy, and its associated plume. The Agulhas Current lies outside the 18°C envelope. The plume has water warmer than 18°C while the core of the eddy has water colder than 10°C. (After Lutjeharms *et al.* (1989) *Continental Shelf Research* **9**(7): 1570–1583. © Elsevier Science.)

increase per unit distance downstream found in the northern Agulhas Current is not maintained in its southern part. Water masses in the current are generally the same in the northern and the southern part. The presence of Tropical Surface Water is maintained in the southern part, as are remnants of North Indian Intermediate Water (or Red Sea Water). Tropical Surface Water is mostly found at the inshore side of the current and derives from the Mozambique Current. Its presence and volume seem to be intermittent.

Some of the surface plumes generated by meanders in the far southern reaches of the Agulhas Current are advected past the western edge of the Agulhas Bank as Agulhas filaments (see **Figure 1**). They are present about 60% of the time and carry substantial amounts of heat that are rapidly lost to the much colder atmosphere. They also carry about $3–9 \times 10^{12}$ kg of salt per year into the South Atlantic; salt in excess to that of the waters already present there. On average they are 50 km wide and 50 m deep.

The southern Agulhas Current influences the water masses over the adjacent Agulhas Bank in three ways. First, plumes of warm Agulhas surface water may extend over the bank, heating the top layers (see **Figure 7**). Second, current-driven upwelling takes place off the far eastern side of the Agulhas Bank (see **Figure 1**) and this water flows westward and covers the greater part of the bottom of the shelf. This process cools the water column from below, leading to intense seasonal thermoclines over the Agulhas Bank. Third, most of the mean flow over the eastern part of this shelf is parallel to the current. At the southern tip of the Agulhas Bank, the current detaches from the shelf edge.

The Agulhas Retroflection

The region where the Agulhas Current then terminates south of Africa is characterized by its extremely high levels of mesoscale variability (**Figure 8**).

The measured eddy kinetic energy is higher here than in any comparable western boundary current such as the Kuroshio or the Gulf Stream. This is due to a number of dynamical traits of the current retroflection. First, the continuous progradation, or westward penetration, of the Agulhas retroflection loop into the South Atlantic Ocean causes substantial levels of variability. The loop has an average diameter of 340 km (±70 km) and progrades westward at about 10 km per day. The outer limits to its movement are 10° and 21°E. At its furthest extent a ring is shed by loop occlusion. This happens between 4 and 9 times per year. This ring spawning activity adds considerably to the general variability. Ring shedding events are usually preceded by the arrival of a Natal

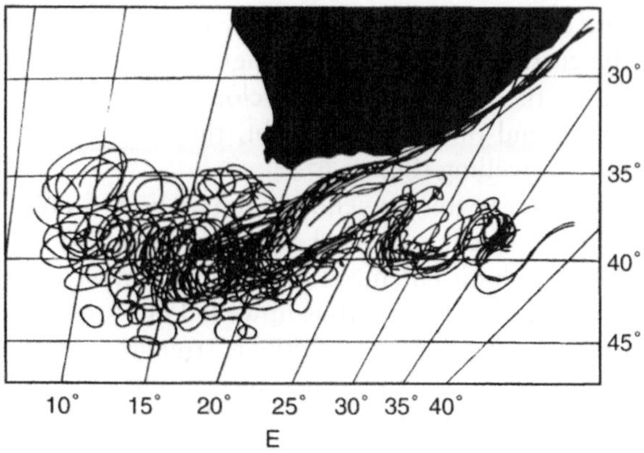

Figure 8 The high levels of mesoscale variability that are characteristic of the Agulhas Current retroflection and the Agulhas Return Current are here portrayed by the superimposed thermal borders at the sea surface as observed for a period of one year. The inshore border of the northern Agulhas Current is particularly stable, that of the southern Agulhas Current less so, whereas the Agulhas Return Current exhibits a tendency to prefer certain meanders. The Agulhas retroflection has a range of locations and is attended by a host of rings (to the north) and eddies (to the south). (After Lutjeharms and van Ballegooyen (1988) *Journal of Physical Oceanography* **18**(11): 1570–1580.)

Pulse on the Agulhas Current. The average lag time between initiation of a Natal Pulse off the Natal Bight and the shedding of a ring at the retroflection is 165 days. Between a newly formed ring and the reconstituted retroflection loop, a wedge of Subtropical Surface Water usually penetrates northward. Its water has a temperature of 17°C and salinity lower than 34.9 over the top 100 m, also adding to the variability of the region.

Newly formed rings retain the hydrographic and kinematic characteristics of the southern Agulhas Current (**Table 3**). Since the heat loss from such a ring may be between 80 and 160 W m^{-2} and since there is substantial evaporation in the region, the temperature and salinity of the upper layers of the features are considerably modified near the retroflection region. Rings are about 320 km (±100 km) in diameter at the sea surface. Estimated by the location of their maximum azimuthal speeds, the diameters are a reduced 240 km (±40 km). These radial speeds lie between 0.3 and 0.9 m s^{-1}. The mean depth of the 10°C isotherm in Agulhas rings, a proxy for the geostrophic speed of their water masses, is 650 m (±130 m). Further properties, as estimated by a number of investigators, are given in **Table 4**.

Rings move off into the South Atlantic Ocean at speeds of 4–8 km per day. Maximum translation rates of up to 16 km per day have been observed. They lose 50% of their sea surface height – and therefore by inference of their energy – during the first 4 months of their lifetime. A full 40% of rings never seem to leave the Cape Basin, off the south-eastern coast of Africa, at all but totally disintegrate here. This decay may well be enhanced by the splitting of rings. This process seems to be largely induced by rings passing over prominent features of the bottom topography such as seamounts. This rapid dissipation of these features means that a considerable part of all the excess salt, heat, energy, and vorticity carried by the rings is deposited exclusively in this corner of the South Atlantic. The remaining rings seem to have lifetimes between 2 and 3 years. They move westward across the full width of the South Atlantic Ocean, slightly to the left of the general background flow. A few of them interact with the upwelling front off the south-eastern coast of Africa, with upwelling filaments occasionally being wrapped around passing rings. These Agulhas rings play a crucial role in the interbasin exchange of waters between the South Indian and South Atlantic Oceans. This is partially quantified in **Table 5**.

Agulhas Return Current

That part of the Agulhas Current not involved in ring production flows back in an easterly direction on having successfully negotiated the retroflection. Here also there are very high levels of mesoscale variability with substantial meandering and eddies being shed to both sides of the Agulhas Return Current/ Subtropical Convergence. Much of this meandering

Table 3 Thermohaline characteristics of the principal water masses found at the Agulhas Current retroflection and vicinity

	Temperature range (°C)	*Salinity range (PSU)*
Surface Water	16.0 to 26.0	>35.50
Central Water		
South East Atlantic Ocean	6.0 to 16.0	34.50 to 35.50
South West Indian Ocean	8.0 to 15.0	34.60 to 35.50
Antarctic Intermediate Water		
South East Atlantic Ocean	2.0 to 6.0	33.80 to 34.80
South West Indian Ocean	2.0 to 10.0	33.80 to 34.80
Deep Water		
North Atlantic Deep Water (SE Atlantic)	1.5 to 4.0	34.80 to 35.00
Circumpolar Deep Water (SW Indian)	0.1 to 2.0	34.63 to 34.73
Antarctic Bottom Water	−0.9 to 1.7	34.63 to 34.72

After Valentine *et al.* (1993) *Deep-Sea Research* 40(6): 1285–1305. © Elsevier Science.

Table 5 Estimates of interbasin volume transport between the South Indian and the South Atlantic Oceans caused by ring shedding. The values were calculated by the investigators named here. For full references see De Ruijter *et al.* (1999)

Investigators	*Volume transport per ring ($10^6 m^3 s^{-1}$)*	*Referenced to*
Olson and Evans (1986)	0.5–0.6	$T > 10°C$
Duncombe Rae *et al.* (1989)	1.2	Total
Gordon and Haxby (1990)	1.0–1.5	$T > 10°C$
	2.0–3.0	Total
McCartney and Woodgate-Jones (1991)	0.4–1.1	Total
Van Ballegooyen *et al.* (1994)	1.1	$T > 10°C$
Byrne *et al.* (1995)	0.8–1.7	1000 db
Clement and Gordon (1995)	0.45–0.90	1500 db
Duncombe Rae *et al.* (1996)	0.65	Total
Goni *et al.* (1997)	1.0	$T > 10°C$

After De Ruijter WPM *et al.* (1999) Indian–Atlantic inter-ocean exchange: dynamics, estimation and impact. *Journal of Geophysical Research* 104(C9): 20885–20911. © American Geophysical Union.

Table 4 Physical properties of Agulhas rings as furnished by a number of independent investigators, calculated with respect to the characteristics of water in the South East Atlantic Ocean

Investigators	*Heat flux (10^{-3} PW)*	*Salt flux (10^5 kg/s)*	*Available potential energy (10^{15} J)*	*Kinetic energy (10^{15} J)*
Olson and Evans (1986)			30.5–51.4	6.2–8.7
Duncombe Rae *et al.* (1989)	25	6.3		
Duncombe Rae *et al.* (1992)			38.8	2.3
Van Ballegooyen *et al.* (1994)	7.5	4.2		
Byrne *et al.* (1995)			18	4.5
Clement and Gordon (1995)			7.0	7.0
Duncombe Rae *et al.* (1996)	1.7	1.1	11.3	2.0
Goni *et al.* (1997)			24	
Garzoli *et al.* (1996)	1.0–1.6	0.7–1.0	2.8–3.8	

After De Ruijter WPM *et al.* (1999) Indian–Altantic inter-ocean exchange: dynamics, estimation and impact. *Journal of Geophysical Research* 104(C9): 20885–29911. © American Geophysical Union, where full references can be found.

is brought about by the variable bathymetry over which the current has to pass.

The first obstacle to a purely zonal flow for the Agulhas Return Current is the Agulhas Plateau (see **Figure 1**). Here it carries out a northward meander of 290 ± 65 km. Cold eddies are frequently formed here with diameters of 280 ± 50 km, but rapidly warm to become indistinguishable from ambient surface waters. Warm eddies may in turn be shed to the south. One such eddy that has been observed closely remained in roughly the same position for 2 months, rotated every 3 days, and had a volume flux of $32 \times 10^6\,m^3\,s^{-1}$ to a depth of 1500 m. Downstream of the Agulhas Plateau the next meander lies at a distance of 450 ± 110 km. These meanders move upstream at about half a wavelength per season. Upstream of the Plateau there are also westward-propagating Rossby waves on the Agulhas Return Current/Subtropical Convergence. However, the direct correlation between the Agulhas Return Current and the Subtropical Convergence is not always straightforward. Sometimes they are in close juxtaposition, sometimes not. When they are not, two separate fronts, or even multiple fronts, may be formed (**Table 6**).

The Agulhas Current has a considerable effect on the Subtropical Convergence, forcing it to lie 5° of latitude farther south than in the South Atlantic Ocean and increasing its surface gradients so that a meridional gradient of 5°C in 35 km is not unknown.

The Agulhas Return Current starts off with characteristics nearly identical to those of the Agulhas Current. South of Africa it may exhibit a surface speed of $1.3\,m\,s^{-1}$ and a volume transport of $40 \times 10^6\,m^3\,s^{-1}$ to 1000 m. The parallel flow along the Subtropical Convergence would be $16\text{–}20 \times 10^6\,m^3\,s^{-1}$ at the same time. All these flow characteristics decrease rapidly as the current progresses eastward (**Figure 9**). By about 55°E, the surface velocity has reduced to $0.4\,m\,s^{-1}$ and the volume flux to about $19 \times 10^6\,m^3\,s^{-1}$. By a longitude of 70°E, at the Kerguelen Plateau, little of the Agulhas characteristics remain along the Subtropical Convergence, all of the Agulhas water having leaked off into the South West Indian Ocean subgyre. The speed, volume transport, as well as mesoscale variability associated with the Agulhas Return Current will all have declined here to values observed in the south-eastern Atlantic Ocean, upstream of any influence from the Agulhas Current. The Agulhas Return Current can therefore be considered to have terminated here. The continuing flow along the Subtropical Convergence east of here is known as the South Indian Ocean Current.

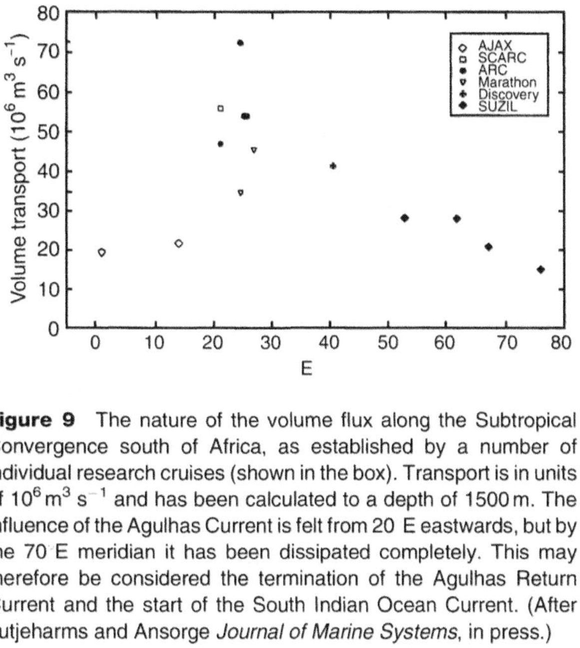

Figure 9 The nature of the volume flux along the Subtropical Convergence south of Africa, as established by a number of individual research cruises (shown in the box). Transport is in units of $10^6\,m^3\,s^{-1}$ and has been calculated to a depth of 1500 m. The influence of the Agulhas Current is felt from 20 E eastwards, but by the 70 E meridian it has been dissipated completely. This may therefore be considered the termination of the Agulhas Return Current and the start of the South Indian Ocean Current. (After Lutjeharms and Ansorge *Journal of Marine Systems*, in press.)

Table 6 The geographic location and the thermal characteristics of the surface expressions of the Agulhas Front as well as the Subtropical Convergence south of Africa. Values for the Agulhas Front were based on 24 crossings, that of the Subtropical Convergence on 70. Values in parentheses denote standard deviations for the calculated averages

	Latitudinal position				*Temperature*				
	From	*To*	*Middle*	*Width (km)*	*From*	*To*	*Middle*	*Range*	*Gradient (C km $^{-1}$)*
Agulhas Front	39 09′	40 01′	39 37′	96.3	21.0	15.7	18.4	5.4	0.102
	(01 16′)	(01 06′)	(01 14′)	69.1	(1.6)	(1.5)	(1.2)	(1.6)	0.106
Subtropical	40 35′	42 36′	41 40′	225.1	17.9	10.6	14.2	7.3	0.047
Convergence	(01 23′)	(01 32′)	(01 19′)	140.6	(2.1)	(1.8)	(1.7)	(1.9)	(0.043)

After Lutjeharms and Valentine (1984) *Deep-Sea Research*, 31(12): 1461–1476. © Elsevier Science.

Conclusion

The Agulhas Current is unusual as a western boundary current for a number of reasons. First, because the African continent terminates at relatively low latitudes, the current penetrates freely into the adjacent ocean basin and a substantial leakage between basins is feasible. Second, through the process of ring and filament shedding, an interaction between a western boundary current and an extensive coastal upwelling regime is brought about that is geographically not possible elsewhere. Third, the very stable nature of the northern Agulhas Current and its characteristic Natal Pulse creates a dynamic environment in which mesoscale disturbances can have profound circulatory effects downstream. The contemporary ignorance about the East Madagascar Current, about the circulation of the Mozambique Channel, and about the origin of midocean eddies in the South West Indian Ocean needs to be eliminated. Only then will a more realistic concept of the interactions between elements of the greater Agulhas Current system become possible.

See also

Mesoscale Eddies. Arctic Ocean Circulation.

Further Reading

De Ruijter WPM, Biastoch A, and Drijfhout SS *et al.* (1999) Indian–Atlantic inter-ocean exchange: dynamics, estimation and impact. *Journal of Geophysical Research* 104: 20885–20911.

Lutjeharms JRE (1996) The exchange of water between the South Indian and the South Atlantic. In: Wefer G, Berger WH, Siedler G, and Webb D (eds.) *The South Atlantic: Present and Past Circulation*, pp. 125–162. Berlin: Springer-Verlag.

Lutjeharms JRE (2001) *The Agulhas Current.* Berlin: Springer-Verlag.

Shannon LV (1985) The Benguela Ecosystem. 1. Evolution of the Benguela, physical features and processes. *Oceanography and Marine Biology, An Annual Review* 23: 105–182.

Shannon LV and Nelson G (1996) The Benguela: large scale features and processes and system variability. In: Wefer G, Berger WH, Siedler G, and Webb D (eds.) *The South Atlantic: Present and Past Circulation*, pp. 163–210. Berlin: Springer-Verlag.

THE OCEANS: THE SOUTHERN OCEAN

CURRENT SYSTEMS IN THE SOUTHERN OCEAN

A. L. Gordon, Lamont-Doherty Earth Observatory of Columbia University, Palisades, NY, USA

Summary

The Southern Ocean, encircling Antarctica, plays a major role in shaping the characteristics of the global ocean. It provides the most significant inter-ocean conduit by which waters of the three major oceans, the Atlantic, Pacific and Indian, can intermingle, acting to diminish their differences in temperature, salinity, and chemical properties. The most prominent current is the Antarctic Circumpolar Current, which transfers about 134 million $m^3 s^{-1}$ of sea water from west to east within a latitudinal range from 50° to 60°S. North of the Antarctic Circumpolar Current are the poleward limbs of the large subtropical gyres of the southern hemisphere, referred to as the South Atlantic, South Indian, and South Pacific Currents. Within large embayments of Antarctica, notably the Weddell and Ross Seas, south of the Antarctic Circumpolar Current are large clockwise flowing gyres. The Weddell Gyre carries about 30–50 million $m^3 s^{-1}$ of water. Along the continental margin of Antarctica is the coastal current that advects water from east to west. The coastal current is directed towards the north along the east coast of Antarctic Peninsula, forming the western boundary of the Weddell Gyre.

At the northern tip of the Antarctic Peninsula the coastal current is directed into the open ocean. The coastal waters injected into the open ocean separate the Antarctica Circumpolar Current from the Weddell Gyre, in what is called the Weddell-Scotia Confluence. Besides ocean currents flowing on nearly horizontal planes, the Southern Ocean experiences major overturning of ocean water. Overturning is forced by the production of dense surface water along the margins of Antarctica, leading to the formation of Antarctic Bottom Water. Within the

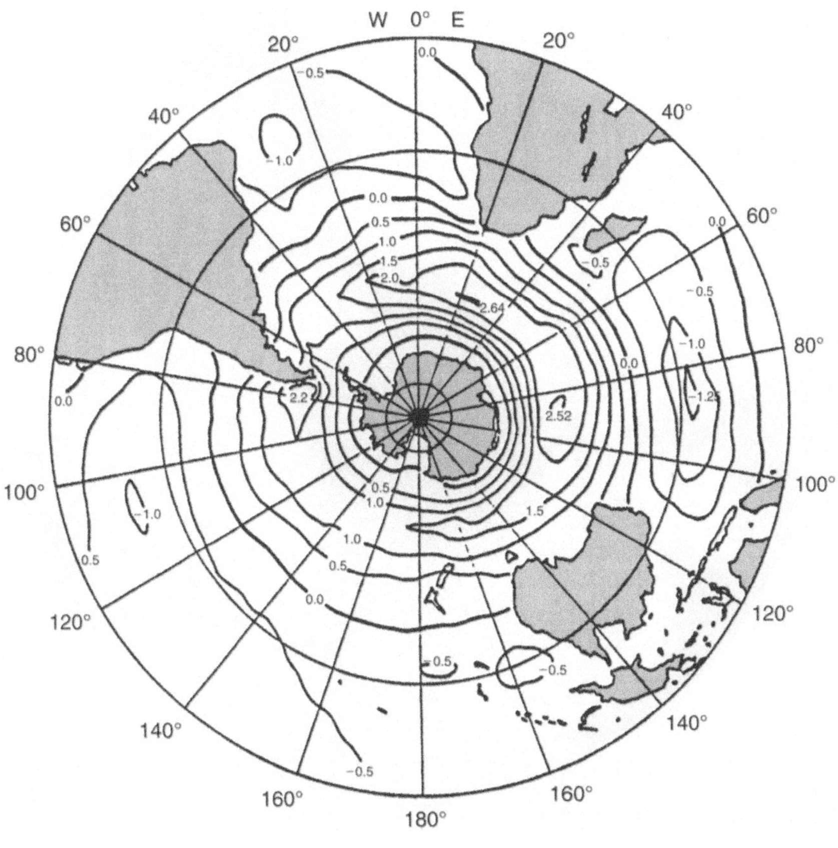

Figure 1 Wind stress in N m^{-2}. (Reproduced with permission from Nowlin and Klinck, 1986.)

Antarctic Circumpolar Current, at the Polar Front, surface waters sink under the more buoyant surface water to the north, forming Antarctic Intermediate Water, a low salinity intrusion spreading at a depth of nearly 1000 m under the main thermocline of subtropical ocean. The horizontal and vertical circulation influences the distribution of sea ice, which in turn modifies the heat, freshwater, and gas exchange between the Southern Ocean and polar atmosphere.

Introduction

Ocean currents are the product for the most part of the stress exerted on the sea surface by the wind. The winds are strong over the Southern Ocean, particular within the Indian Ocean and Australian sectors (**Figure 1**) and therefore drive a vigorous circulation (**Figure 2**). Strong westerlies (wind directed from west to east) extend from the subtropical high atmospheric pressure near 30°S, a latitude often used to define the northern limits of the Southern Ocean, to a belt of low atmospheric pressure at 65°S. South of 65°S the winds are easterlies, marking the northern edges of the polar high pressure over Antarctica.

Density changes of surface water induced by sea–air fluxes of heat and fresh water, often involving sea ice within the Southern Ocean, also produce circulation. However, buoyancy (sometimes referred to as thermohaline), circulation is sluggish and mainly occurs within the meridional vertical plane, as dense water sinking forces slow, compensatory upwelling of less dense resident water. Sinking of dense waters along the continental margins of Antarctica results in Antarctic Bottom Water.

Ocean currents are for the most part in equilibrium with the distribution of density within the ocean (**Figure 3**) satisfying approximately the so-called 'thermal wind equation' (*see* Arctic Ocean Circulation.) Along the Greenwich meridian the surface of equal density, or isopycnals, rise up towards the sea surface as latitude increases to the south. Strongly sloped isopycnals are coupled to strong ocean current, more precisely to strong geostrophic ocean currents, relative to the seafloor. As a rule of thumb, in the southern hemisphere higher

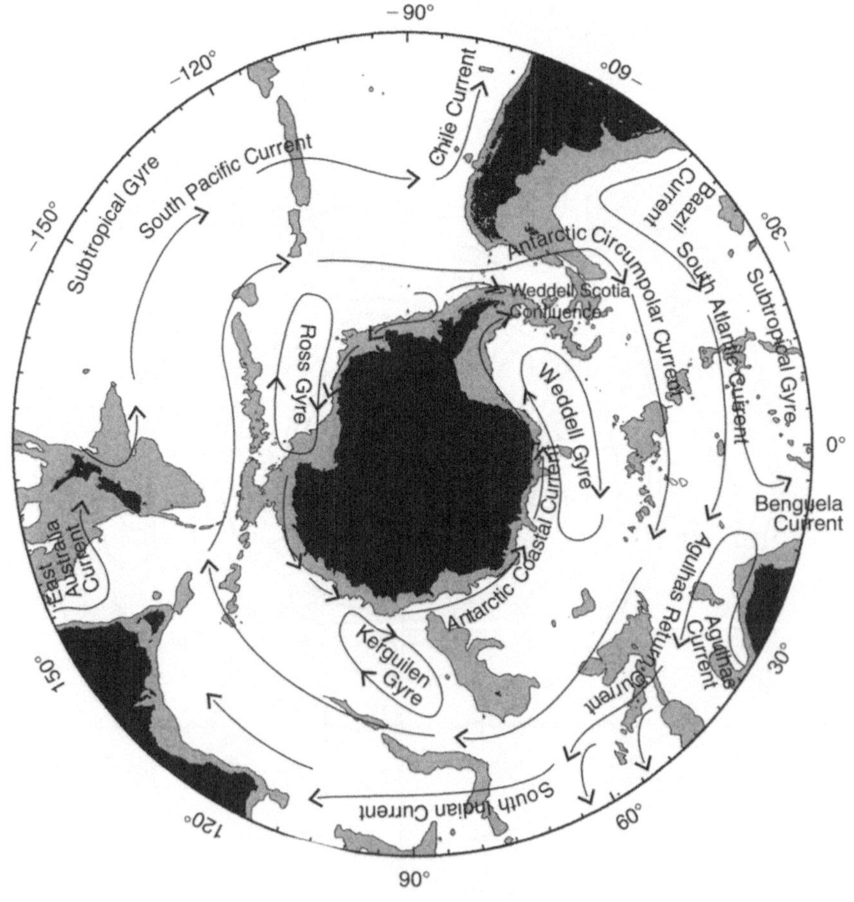

Figure 2 Schematic of general circulation of the Southern Ocean. Land is black; the shaded area marks ocean depths of <3000 m.

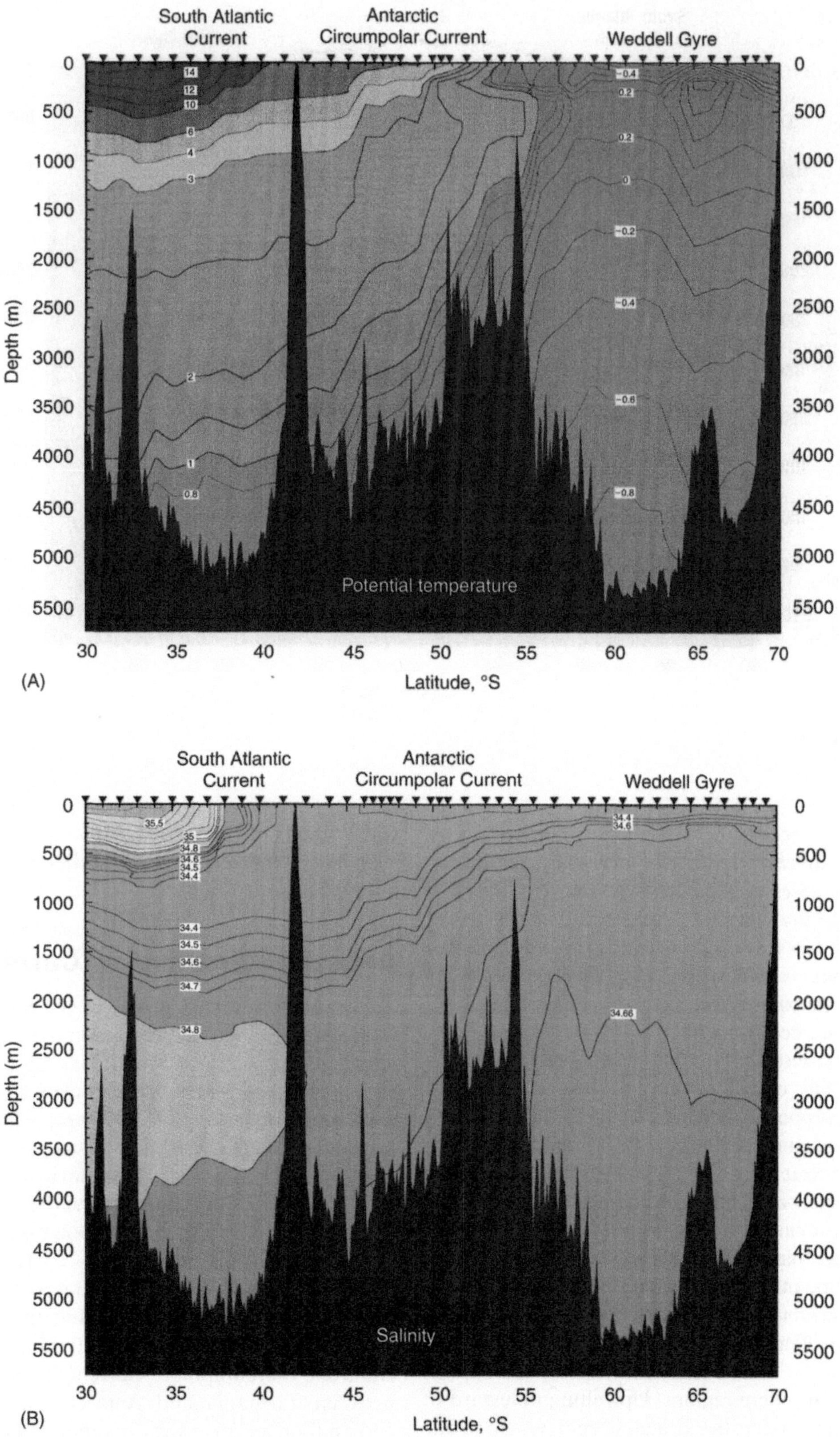

Figure 3 Potential (A) temperature, (B) salinity and (C) density (sigma-0) along the Greenwich meridian.

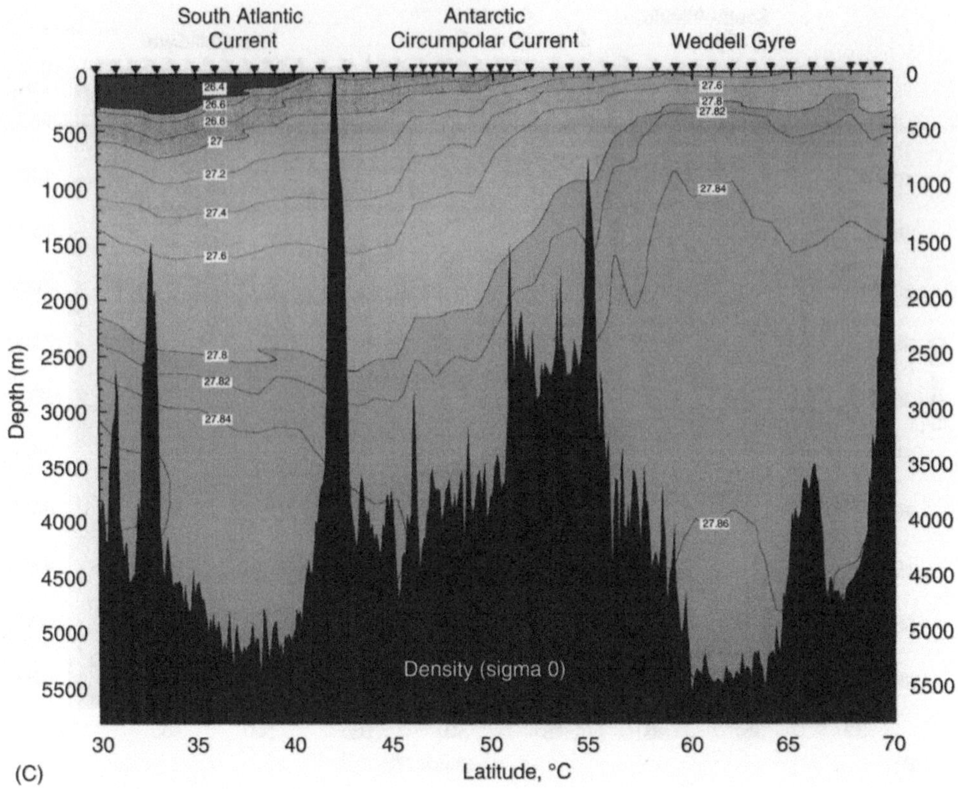

Figure 3 *Continued*

density water occurs to the right of the direction of the ocean current. Hence the increasing density as latitude increases is linked to west to east flow of water. Along the margin of Antarctica the descent of isopycnals marks a flow towards the east. Regions of rapid changes in temperature, salinity or density mark the positions of ocean fronts, often coinciding with a strong ocean current.

Maximum westerlies in the wind occur near 55°S, which roughly coincides with the axis of the Antarctic Circumpolar Current. To the south of this latitude, wind-induced northward Ekman transport of surface water results in a wide region of upwelling. North of 55°S the Ekman transport diminishes. This causes a region of surface water convergence. Near 55°S surface water sinks under the more buoyant surface water to the north, producing Antarctic Intermediate Water. Antarctic Intermediate Water forms a low salinity layer found at the base of the thermocline of the subtropical southern hemisphere regions. Upwelling poleward of the maximum westerlies brings deeper water to the sea surface to compensate for the sinking of Antarctic Bottom Water and Antarctic Intermediate Water. The upwelling also drives two large clockwise-flowing, cyclonic Gyres within the large embayments of Antarctica, marking the Weddell and Ross Seas (**Figure 2**) and a smaller one east of Kerguelen Plateau.

Antarctic Circumpolar Current

The most prominent current of the Southern Ocean is the west to east flowing Antarctic Circumpolar Current lying within a latitudinal range from 50° to 60°S. It is the greatest ocean current on the Earth, covering a distance of 21 000 km, with an average transport through the Drake Passage (between South America and Antarctic Peninsula) of 134 million $m^3 s^{-1}$ (134 Sv). The transport varies with time mirroring variations in the circumpolar wind field, from about 100 to 150 million $m^3 s^{-1}$. Transport is enhanced south of Australia by return of water to the Pacific Ocean lost to the Indian Ocean within the Indonesian Seas, by about 10 million $m^3 s^{-1}$. The Antarctic Circumpolar Current transport passing between Tasmania and Antarctica is estimated as 143 million $m^3 s^{-1}$, with a range from 131 to 158 million $m^3 s^{-1}$.

The Antarctic Circumpolar Current is a deep reaching or barotropic current, meaning that it extends to the seafloor. Because of this it is said that the

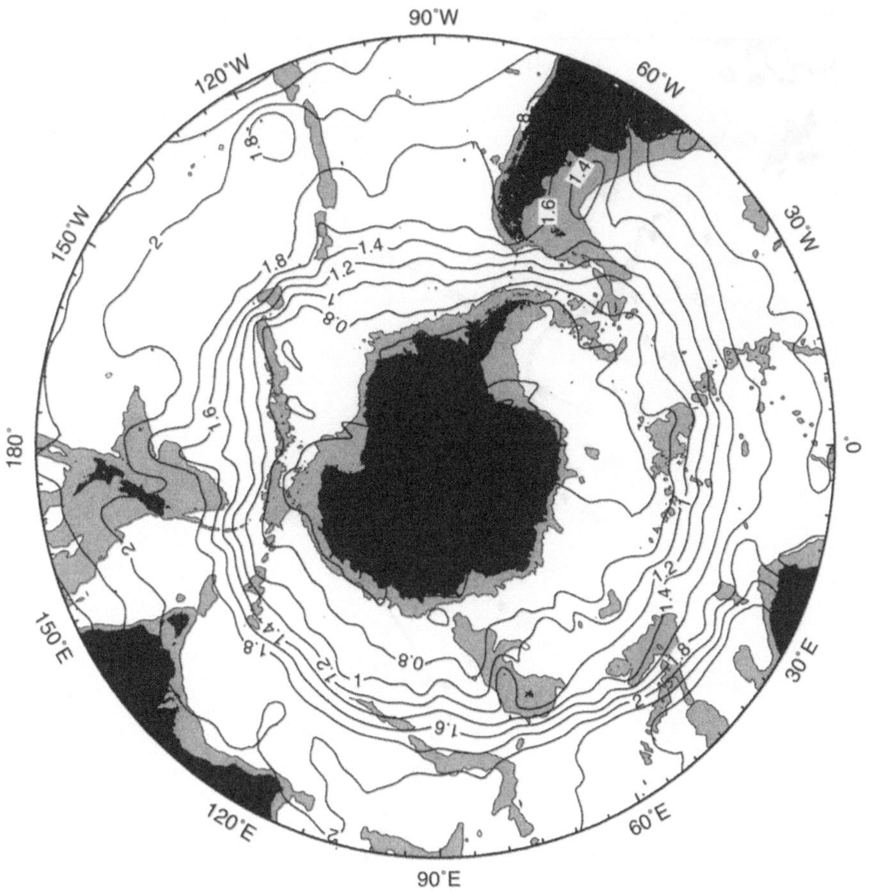

Figure 4 Anomaly of Southern Ocean sea level height relative to the 2×10^7 Pa pressure surface. The values are differences in dynamic meters (approximately equivalent to geometric meters) from that of a standard ocean (0°C, 35 PSU salinity). Land is black; the shaded area marks ocean depths of <3000 m. Geostrophic ocean currents in the southern hemisphere are directed so that lower sea level is to the right of the flow direction, which is aligned along lines of equal sea level height. The rise of sea level towards the north defines the eastward flowing Antarctic circumpolar current.

Antarctic Circumpolar Current 'feels' the shape of the sea floor and hence its path is steered by the seafloor topography (**Figure 4**). The flow following the southern deflection in the mid-ocean ridge reaches its southern-most position in the southwest Pacific Ocean, near 60°S. Upon passing through Drake Passage it turns sharply to the north, transversing the Atlantic Ocean near 50°S. As the ocean surface temperature pattern responds to the circulation pattern the surprising result is that the bottom topography is 'projected' in the sea surface temperature pattern.

Rather than a broad diffuse flow, the Antarctic Circumpolar Current is composed of a number of high speed filaments, separated by zones of low flow, or even reversed flow (towards the west). The jets are typically 40–50 km wide. Surface currents average about 30–40 $cm\,s^{-1}$ within the axes, and speeds of over 100 $cm\,s^{-1}$ are common. The high speed filaments are marked by ocean fronts, where the temperature and salinity stratification changes rapidly with latitude (**Figure 5**). Between these fronts are zones of similar stratification. The primary axis occurs at the polar front (**Figure 6**). Meanders of the flow axes and associated fronts displace these features by at least 100 km to either side of their mean position. Meanders occasionally produce detached eddies, in which pools of water ringed by a high speed current from one zone invade an adjacent zone. A characteristic of the Antarctic Circumpolar Current is its high degree of eddy activity (**Figure 7**). The most active eddy fields are observed where the Antarctic Circumpolar Current crosses submarine ridges or plateaus, as south of Australia, in the southwest Atlantic and south of Africa.

The correlation of the eddy currents and of the ocean temperature leads to significant poleward flux of ocean heat by the eddy processes. The poleward heat flux measured in the Drake Passage, if extrapolated all around Antarctica, is 0.3 PW, which

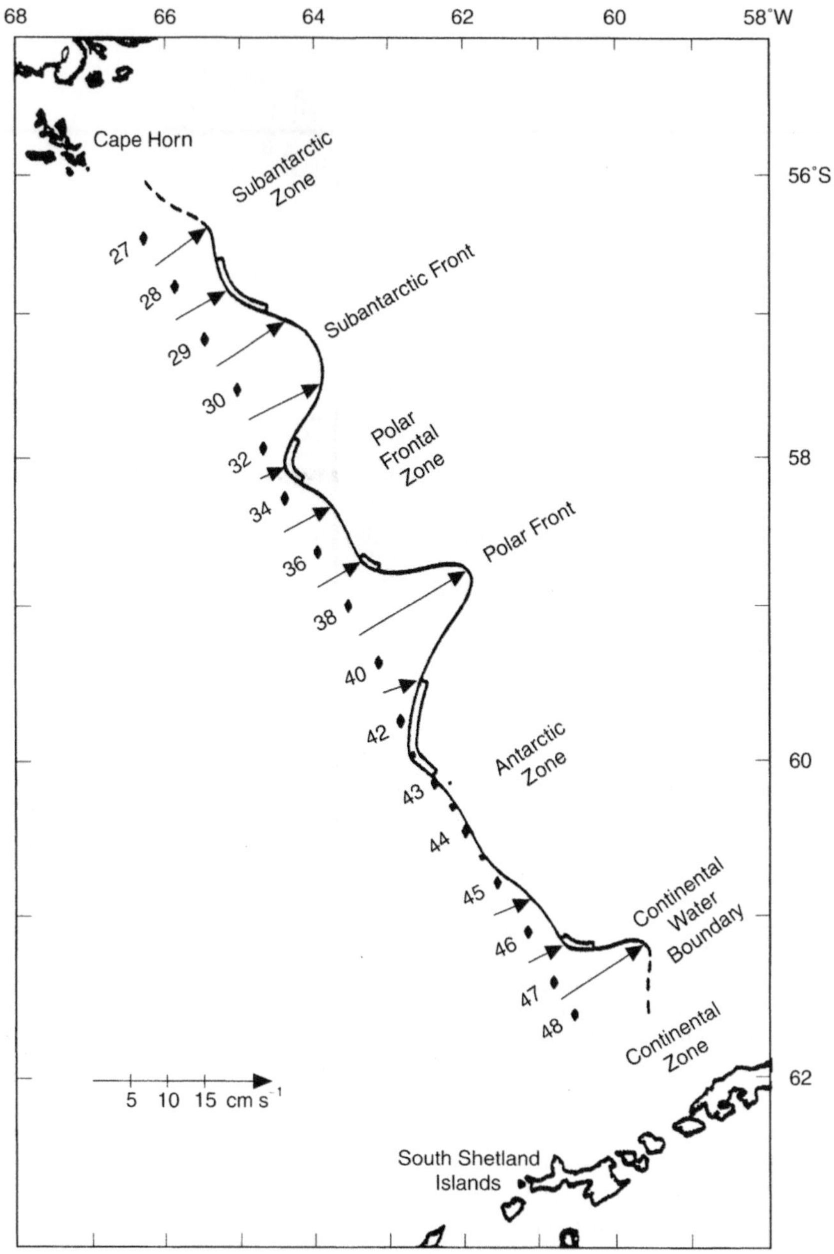

Figure 5 Vertical averaged geostrophic speeds of the upper 2500 m directed normal to a line across the Drake Passage. The positions of the front and stratification zones are shown. The highest flows defining the axes of the Antarctic Circumpolar Current coincide with the position of the ocean fronts. (Reproduced with permission from Clifford, 1983.)

can account for most of the meridional heat flux across 60°S. However caution is suggested as the Drake Passage eddy field may not be typical of the full circumpolar belt. Meanders and eddies of the Antarctic Circumpolar Current also act to carry wind-delivered momentum downward to the seafloor, where pressure forces (often referred to as form drag) acting on the slopes of bottom topographic features act to compensate the force of the wind. The downward transfer of momentum is integrally linked to the meridional fluxes of heat and fresh water by the eddy field, by baroclinic instability.

Weddell Gyre

The Weddell Gyre is the largest of the cyclonic Gyres occupying the region between the Antarctic Circumpolar Current and Antarctica, stretching from the Antarctic Peninsula to 30°E. The clockwise flow pattern is linked to doming of isopycnals and

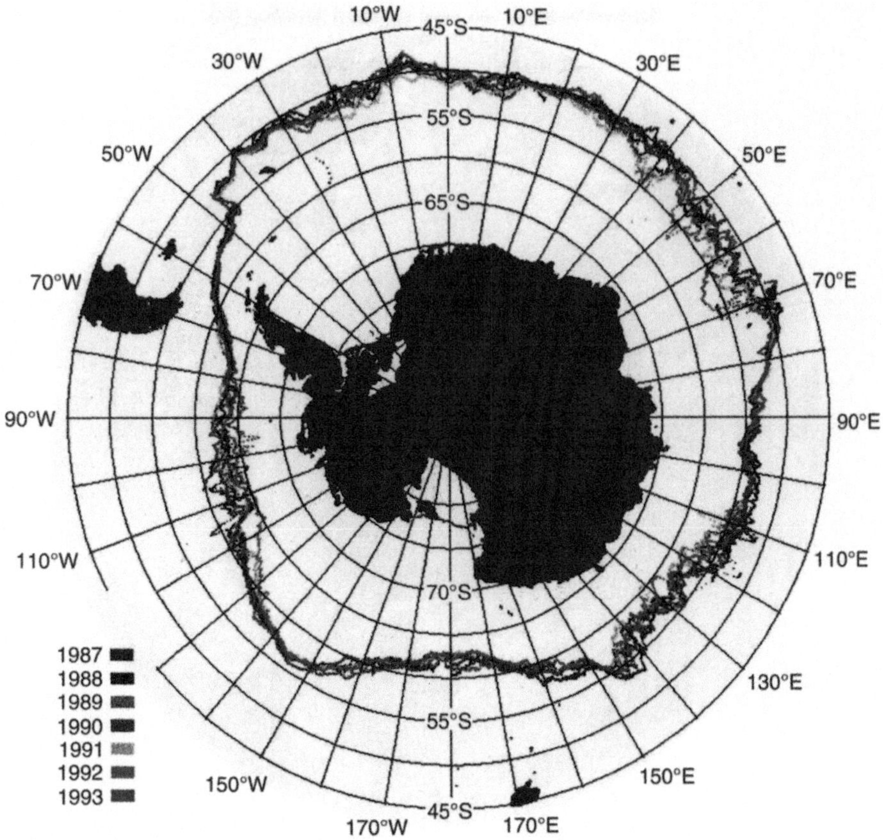

Figure 6 Position of the Antarctic polar front for the years 1987–1993 as revealed by satellite images of sea surface temperature. (Reproduced with permission from Moore *et al.*, 1999.)

upwelling of deep water within its central axis (**Figure 3**). As the deep water is warmer than the surface layer the Gyre injects heat into the surface layer, which limits the winter sea ice cover to a thickness of only 0.5 m. Along the Antarctic margins intense cooling of surface water lead to the formation of the coldest, densest ocean water masses of global importance: Antarctic Bottom Water. A branch of the Antarctic Circumpolar Current turns southward near 30°E forming the eastern limb of the Weddell Gyre. Some of this water turns westward along the coast of Antarctica (the rest continues to flow eastward, but at a more southern latitude than the axis of the Antarctic Circumpolar Current). Westward flowing coastal current is characteristic all around Antarctica, with a surface speed of about 10 cm s^{-1} as detected by satellite tracking of the drift of icebergs calved from Antarctica.

Within the Weddell Gyre the coastal current westward flow is blocked by the Antarctic Peninsula. Upon encountering the southern base of the peninsula, the coastal current turns northward, forming the western boundary current of the Weddell Gyre. At the northern tip of the Antarctic Peninsula, the western boundary current composed of the cold, low salinity stratification characteristic of Antarctic continental margin, is injected into the open ocean. This feature, called the Weddell-Scotia Confluence (**Figure 2**) separates the Antarctic Circumpolar Current from the interior of the Weddell Gyre. It can be traced as a low salinity band to the Greenwich Meridian. Along the sea floor Antarctic Bottom Water escapes from the Gyre, flowing northward within deep crevices in the seafloor morphology, into the Scotia Sea, South Sandwich Trench and south of Africa. Export of Bottom Water is compensated by import of circumpolar water along the eastern boundary.

Surface currents of the Weddell Gyre are weak, usually 10 cm s^{-1}, but the flow extends to the seafloor, as a strongly barotropic current. There is some evidence that the current increases along the seafloor of the continental slope, with speeds of up to 20 cm s^{-1}, associated with plumes of dense shelf water descending into the deep ocean as Antarctic Bottom Water. Observations of ocean currents during the period from 1989 to 1992 across the mouth of the Weddell Sea, stretching from Kapp Norvegia (71°20′S; 11°40′W) to the northern tip of the Antarctic Peninsula, find Gyre transport of about

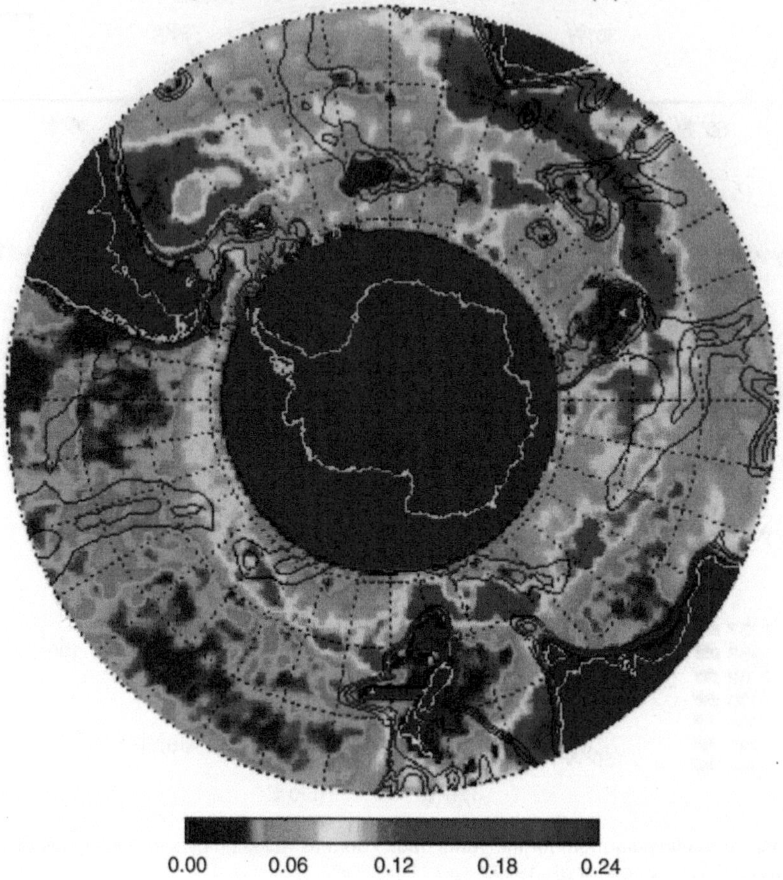

Figure 7 Variability of sea surface height from mean sea level, as revealed by Topex Poseidon satellite altimetric measurements of sea level for the period 1992–1999. Values are in meters. Variability of sea level is caused by meanders and eddies of the geostrophic flow field, or changes in its strength. (Provided by Donna Witter, Associated Research Scientist at Lamont-Doherty Earth Observatory.)

30 million $m^3 s^{-1}$ (30 Sv), most of which is contained in narrow jets following along the continental slope. An additional 10 million $m^3 s^{-1}$ (10 Sv) of transport is likely around the central axis of the Gyre, making a total recirculation transport around the Gyre of 40 million $m^3 s^{-1}$ (40 Sv). Export from the Gyre is not known exactly, but can be estimated as 5 million $m^3 s^{-1}$ (5 Sv) within the bottom layer.

See also

Agulhas Current. Antarctic Circumpolar Current. Atlantic Ocean Equatorial Currents. Indian Ocean Equatorial Currents.

Further Reading

Belkin IM and Gordon AL (1996) Southern ocean fronts from the Greenwich Meridian to Tasmania. *Journal of Geophysical Research* 101(C2): 3675–3696.

Clifford MA (1983) *A Descriptive Study of the Zonation of the Antarctic Circumpolar Current and its Relation to Wind Stress and Ice Cover.* MS thesis, Texas A & M University.

Gille S (1994) Mean sea surface height of the Antarctic Circumpolar Current from Geosat data: methods and application. *Journal of Geophysical Research* 99: 18 255–18 273.

Gille S and Kelly K (1996) Scales of spatial and temporal variability in the Southern Ocean. *Journal of Geophysical Research* 101: 8759–8773.

Gordon AL, Molinelli E, and Baker T (1978) Large-scale relative dynamic topography of the Southern Ocean. *Journal of Geophysical Research* 83: 3023–3032.

Hofmann EE (1985) The large-scale horizontal structure of the Antarctic Circumpolar Current from FGGE drifters. *Journal of Geophysical Research* 90: 7087–7097.

Nowlin W and Klinck J (1986) The physics of the Antarctic Circumpolar Current. *Reviews of Geophysics* 24(3): 469–491.

Moore JK, Abbott M, and Richman J (1999) Location and dynamics of the Antarctic Polar Front from satellite sea surface temperature data. *Journal of Geophysical Research* 104(C2): 3059–3073.

Orsi AH, Nowlin WD, and Whitworth T (1993) On the circulation and stratification of the Weddell Gyre. *Deep-Sea Research* 40: 169–203.

Orsi AH, Whitworth T, and Nowlin WD (1995) On the meridional extent and fronts of the Antarctic Circumpolar Current. *Deep-Sea Research* 42(5): 641–673.

Whitworth T (1980) Zonation and geostrophic flow of the Antarctic Circumpolar Current at Drake Passage. *Deep-Sea Research* 27: 497–507.

Whitworth T and Nowlin WD (1987) Water masses and currents of the Southern Ocean at the Greenwich Meridian. *Journal of Geophysical Research* 92: 6462–6476.

ANTARCTIC CIRCUMPOLAR CURRENT

S. R. Rintoul, CSIRO Antarctic Climate and Ecosystems Cooperative Research Centre, Hobart, TAS, Australia

Introduction

The Drake Passage between the South American and Antarctic continents is the only band of latitudes where the ocean circles the Earth, unblocked by land. The existence of this oceanic channel has profound implications for the global ocean circulation and climate. The Drake Passage gap permits the Antarctic Circumpolar Current (ACC) to exist, a system of ocean currents which flows from west to east along a roughly 25 000-km-long path circling Antarctica. In terms of transport, the ACC is the largest current in the world ocean, carrying about $137 \pm 8 \times 10^6\,m^3\,s^{-1}$ through the Drake Passage. The wind-driven ocean circulation theories that explain much of the ocean current patterns observed at other latitudes do not apply in an unbounded channel and the unique dynamics of the ACC have long been a puzzle for oceanographers. The three-dimensional circulation in the ACC belt is now understood to reflect the interplay of wind and buoyancy exchange with the atmosphere, water mass modification, eddy fluxes of heat and momentum, and strong interactions between the flow and bathymetry.

The strong eastward flow of the ACC has several important implications for the global ocean circulation and its influence on regional and global climate. By transporting water between the major ocean basins, the ACC tends to smooth out differences in water properties between the basins. The interbasin connection allows a global-scale pattern of ocean currents to be established, known as the thermohaline circulation (*see* Ocean Circulation: Meridional Overturning Circulation), which transports heat, moisture, and carbon dioxide around the globe and strongly influences the Earth's climate. The strong flow of the ACC is associated with steeply sloping density surfaces, which shoal to the south across the current and bring dense waters to the surface in the high-latitude Southern Ocean. Where the dense waters are exposed at the sea surface, they exchange heat, moisture, and gases like oxygen and carbon dioxide with the atmosphere. In this sense, the Southern Ocean provides a window to the deep sea. The ecology and biogeography of the Southern Ocean are influenced strongly by the ACC and the overturning circulation plays an important role in the marine carbon and nutrient cycles. The west-to-east flow of the ACC inhibits north–south exchanges across the current and isolates Antarctica from the warm waters to the north; the present glacial climate of Antarctica was not established until the South American and Antarctic continents began to drift apart about 30 Ma, opening a circumpolar channel.

Structure of the ACC

A schematic view of the major currents of the Southern Hemisphere oceans south of 20° S is shown in **Figure 1**. The flow of the ACC is focused in several jets, associated with sharp cross-stream gradients (or fronts) in temperature, salinity, and other properties. The three main fronts of the ACC – the Subantarctic Front, Polar Front, and southern ACC front – are indicated by the arrows circling Antarctica. To the south of the circumpolar flow of the ACC, clockwise gyres are found in the Weddell Sea, Ross Sea, and the Australian–Antarctic Basin (*see* Current Systems in the Southern Ocean). A westward flow associated with the Antarctic Slope Front and Antarctic Coastal Current is found near the continental shelf break around much of Antarctica. To the north of the ACC, water flows to the east in the southern limb of the large anticlockwise subtropical gyres in each basin. Exchanges of water masses between the ACC and the gyre circulations to the north and south are important components of the global circulation.

The distribution of water properties on a transect crossing the ACC is illustrated in **Figure 2**. The temperature and salinity of water masses are largely set at the sea surface, where there is active exchange with the atmosphere; nutrient and oxygen concentrations are also influenced by biological processes. Water masses carry these characteristics with them as they sink from the surface into the ocean interior. The major water masses of the Southern Ocean are associated with various property extrema that reflect their circulation and formation history. For example, the Subantarctic Mode Water is formed by deep convection in winter on the northern flank of the ACC, producing deep well-mixed layers that are rich in oxygen. The Antarctic Intermediate Water is the name given to the prominent salinity minimum layer north of the ACC. The Circumpolar Deep Water

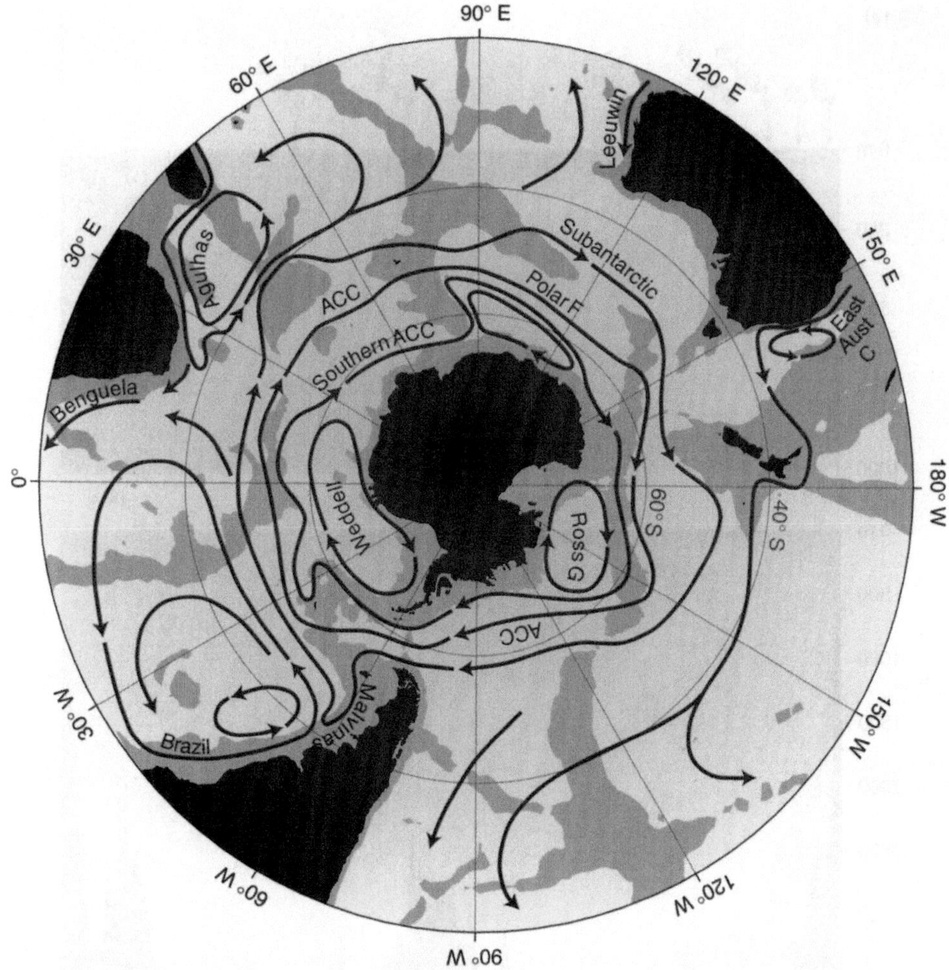

Figure 1 A schematic view of the major ocean currents of the Southern Hemisphere oceans south of 20° S. Depths shallower than 3500 m are shaded. C, current; G, gyre; F, front; ACC, Antarctic Circumpolar Current.

(CDW) is often divided into two layers: Upper CDW corresponds to the oxygen minimum layer, and Lower CDW corresponds to the deep salinity maximum layer. The relatively fresh layer near the seafloor is Antarctic Bottom Water, which sinks near Antarctica and carries water rich in oxygen and chlorofluorocarbons into the deep ocean (*see* Bottom Water Formation).

Water properties at a given depth change dramatically as the Southern Ocean is crossed from north to south (**Figure 2**). Surfaces of constant temperature, salinity, density, and other properties slope upward to the south. As a result, density layers found at 3000-m depth in subtropical latitudes approach the sea surface near Antarctica. The shoaling of density surfaces to the south is associated with the strong eastward flow of the ACC. Tilted density surfaces create pressure forces which drive ocean currents and an accompanying Coriolis force to balance the pressure force. (This balance of forces, known as geostrophy, describes the dynamics of all large-scale ocean currents (*see* Ocean Circulation).) In the Southern Hemisphere, an increase in density to the south supports an eastward flow (relative to the seafloor), as observed in the ACC.

The rise of temperature, salinity, and density surfaces to the south occurs in a series of steps, or rapid transitions, rather than as a uniform slope across the Southern Ocean (**Figure 2**). These rapid transitions are known as fronts. Because the strength of an ocean current is proportional to the magnitude of the horizontal density gradient, each of the fronts is associated with a maximum in velocity. Most of the flow of the ACC is concentrated in the fronts, with smaller transports observed between the fronts (**Figure 2(f)**). The zones between the fronts also coincide with regions of relatively uniform water properties at each depth. Unlike many fronts at lower latitudes, the ACC fronts extend from the sea surface to the seafloor. The current jets associated

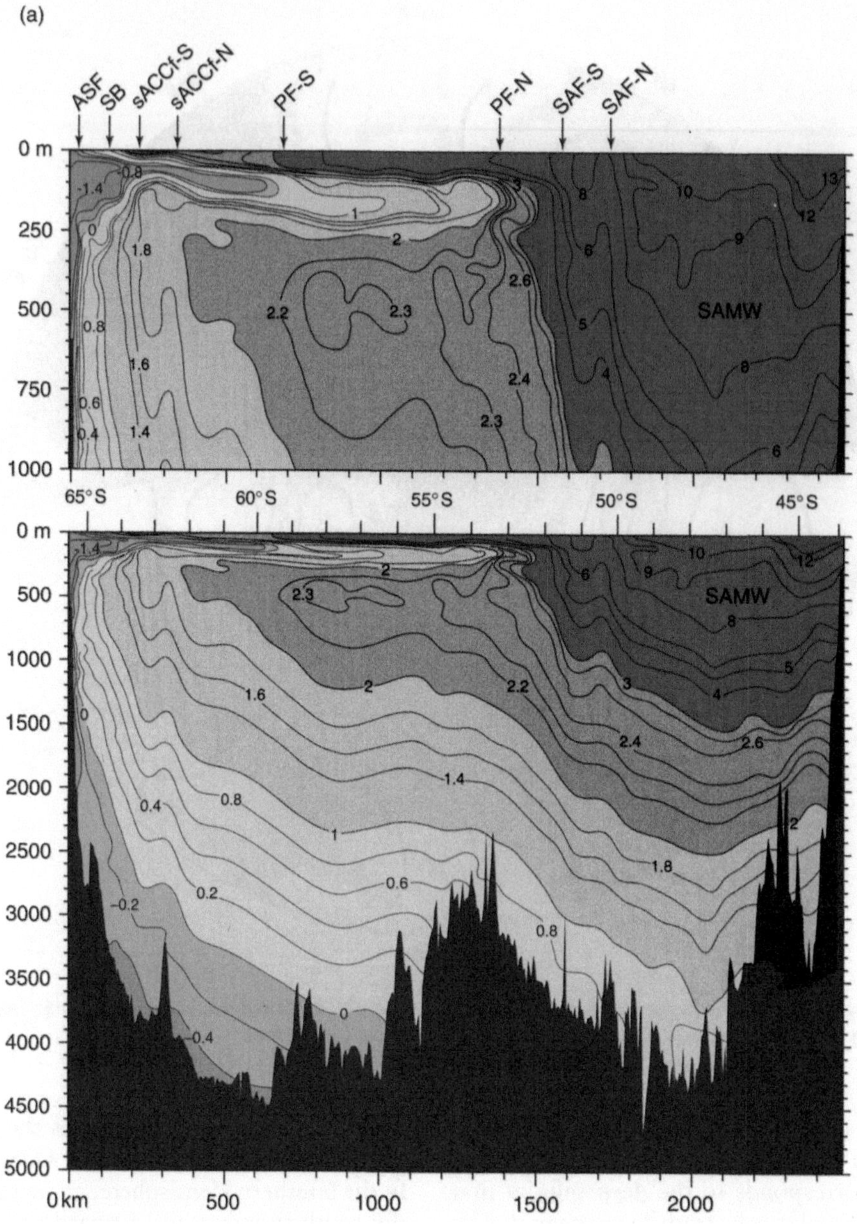

Figure 2 Property distributions along a roughly north–south section across the Southern Ocean at 140° E south of Australia (World Ocean Circulation Experiment section SR3): (a) potential temperature (°C), (b) salinity (PSS78), (c) neutral density (kg m^{-3}), (d) oxygen (μmol kg^{-1}), (e) chlorofluorocarbon 11 (CFC-11) (pM kg^{-1}), and (f) transport at each station pair (solid line, left axis) and cumulative transport from south to north (dashed line, right axis). Transport is in units of sverdrups (1 Sv = 10^6 m^3 s^{-1}). Contours slope upward from north to south; regions where the slope of the contours is steep correspond to the ACC fronts and to transport maxima. SAMW, Subantarctic Mode Water; LCDW, Lower Circumpolar Deep Water; AAIW, Antarctic Intermediate Water; UCDW, Upper Circumpolar Deep Water; CDW, Circumpolar Deep Water CDW; AABW, Antarctic Bottom Water. SAF, Subantarctic Front; PF, Polar Front; sACCf, southern ACC front; SB, southern boundary; ASF, Antarctic Slope Front; N and S indicate northern and southern branches of the primary fronts. The positive and negative peaks in transport labeled SAZ indicate a strong recirculation in the Subantarctic Zone north of the ACC at the time this section was occupied. Sections are adapted from Orsi AH and Whitworth T, III (2005) In: Sparrow M, Chapman P, and Gould J (eds.) *Hydrographic Atlas of the World Ocean Circulation Experiment (WOCE), Vol. 1: Southern Ocean.* Southampton: International WOCE Project Office (ISBN 0-904175-49-90), with permission (http://www.soc.soton.ac.uk).

with the fronts therefore also extend throughout the water column. The deep-reaching nature of the ACC fronts reflects the weak stratification of the Southern Ocean, where the change in density from surface waters to deep waters is small compared to lower latitudes. The mean current speeds of the ACC jets are relatively modest, typically less than 0.5 m s^{-1} (about 1 knot), with much weaker flow between the

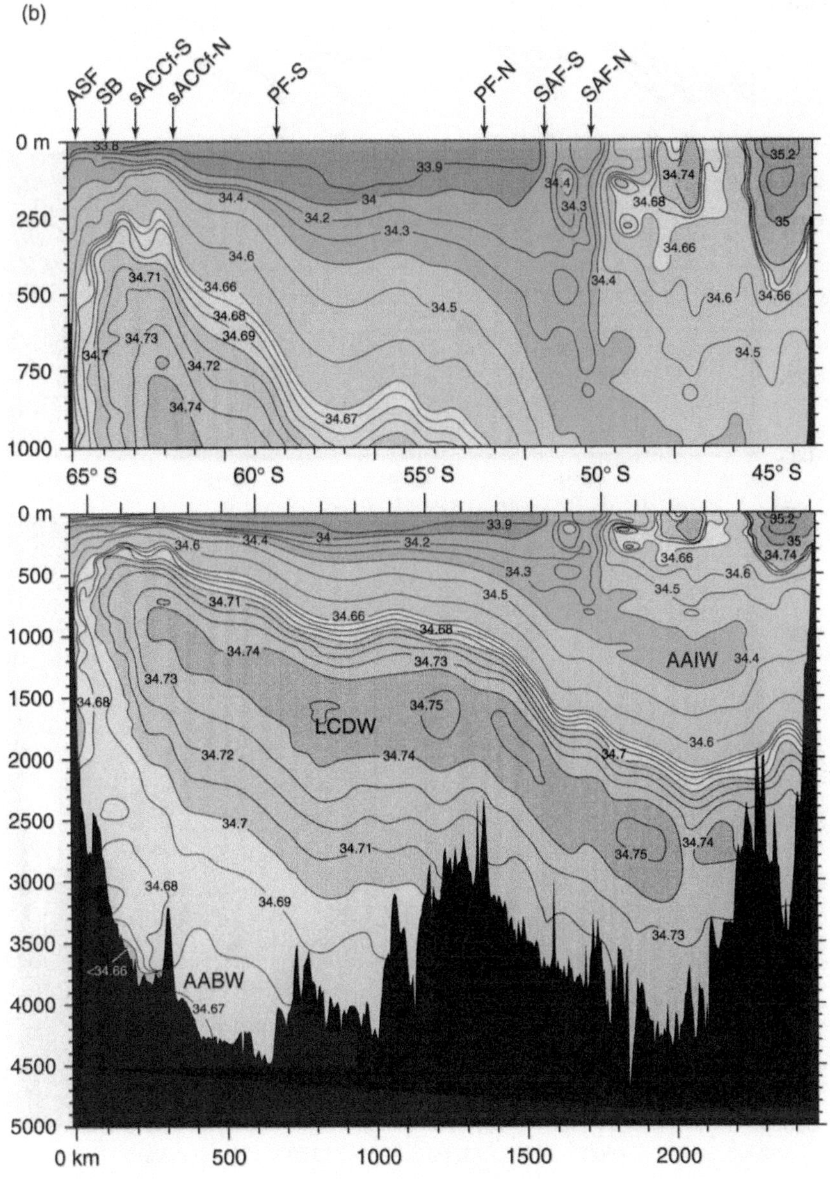

Figure 2 *Continued.*

jets. However, because the current is broad and deep, it carries a large transport.

Most studies of the ACC now recognize three main fronts. From north to south, these fronts are the Subantarctic Front, the Polar Front, and the southern ACC front. The southern limit of the ACC domain is defined by the 'southern boundary of the ACC', the southernmost streamline to pass through Drake Passage, which is also often associated with a weak front and current core. These fronts of the ACC are circumpolar in extent and can be found on any north–south transect across the Southern Ocean. Because the fronts extend to the seafloor and the stratification in the Southern Ocean is relatively weak, the position of the fronts is strongly influenced by bathymetry.

Between the fronts lie zones with more or less uniform physical and chemical characteristics: the Subantarctic Zone between the Subantarctic and Subtropical Fronts, the Polar Frontal Zone between the Subantarctic and the Polar Fronts, and the Antarctic Zone south of the Polar Front. Within each zone, the water properties tend to be similar at each depth and follow a similar seasonal cycle. For example, the Subantarctic Zone is characterized by very deep surface mixed layers in winter and low

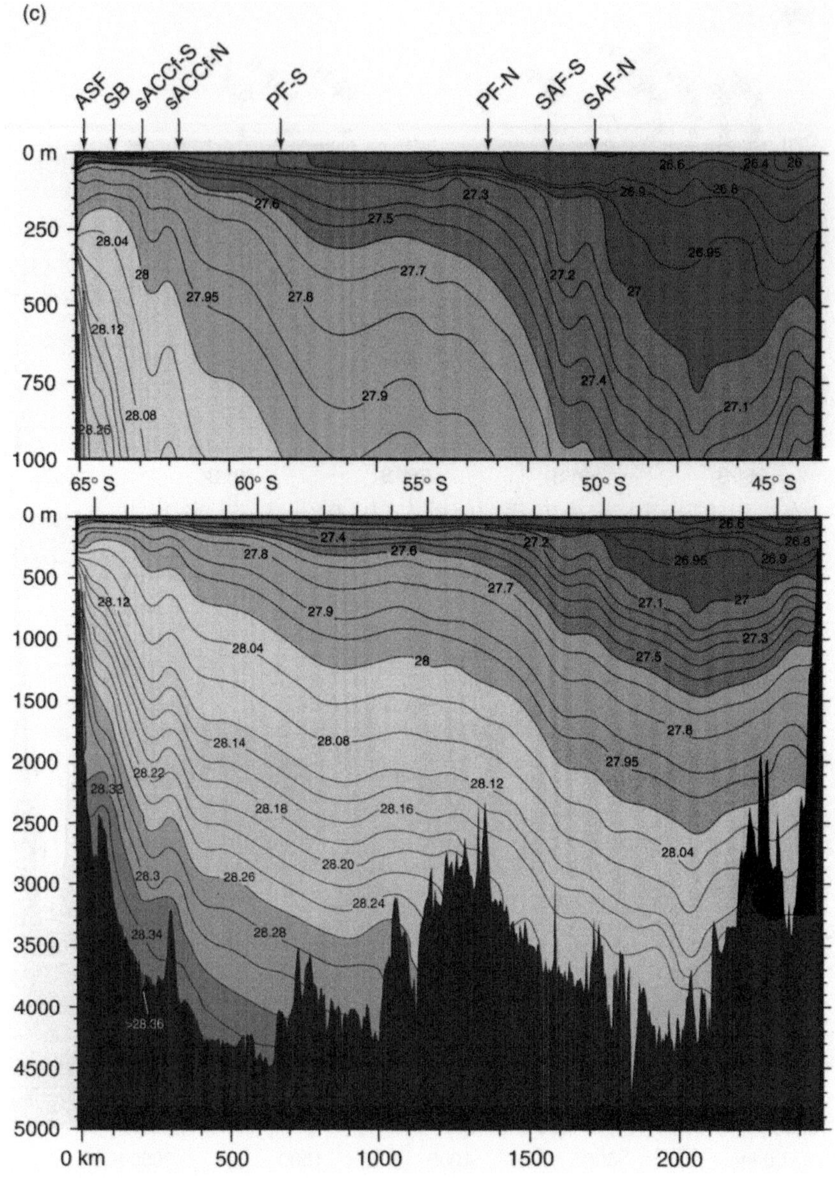

Figure 2 *Continued.*

nutrient concentrations (close to zero for silicic acid and somewhat higher concentrations of nitrate and phosphate). The Antarctic Zone is characterized by fresh surface waters, shallow summer mixed layers, and high concentrations of major nutrients like nitrate and silicic acid, but low concentrations of micronutrients like iron.

The zones delimited by the fronts of the ACC also define biogeographic zones populated by distinct species assemblages. For example, waters south of the Polar Front tend to be dominated by large phytoplankton such as diatoms (who need silicic acid) and large zooplankton, while coccolithophores and small zooplankton dominate north of the Subantarctic Front. The distribution and foraging patterns of larger animals (e.g., fish, seabirds, and marine mammals) are also influenced by the frontal structure of the ACC. In some cases, the fronts themselves tend to be associated with higher primary productivity. The higher productivity near fronts can be caused by advection of micronutrients by the current or by upwelling caused by eddies or by topographic interactions. The currents of the ACC can also play a direct role in ecosystem dynamics, for example, by carrying krill and larvae from the Antarctic Peninsula to South Georgia.

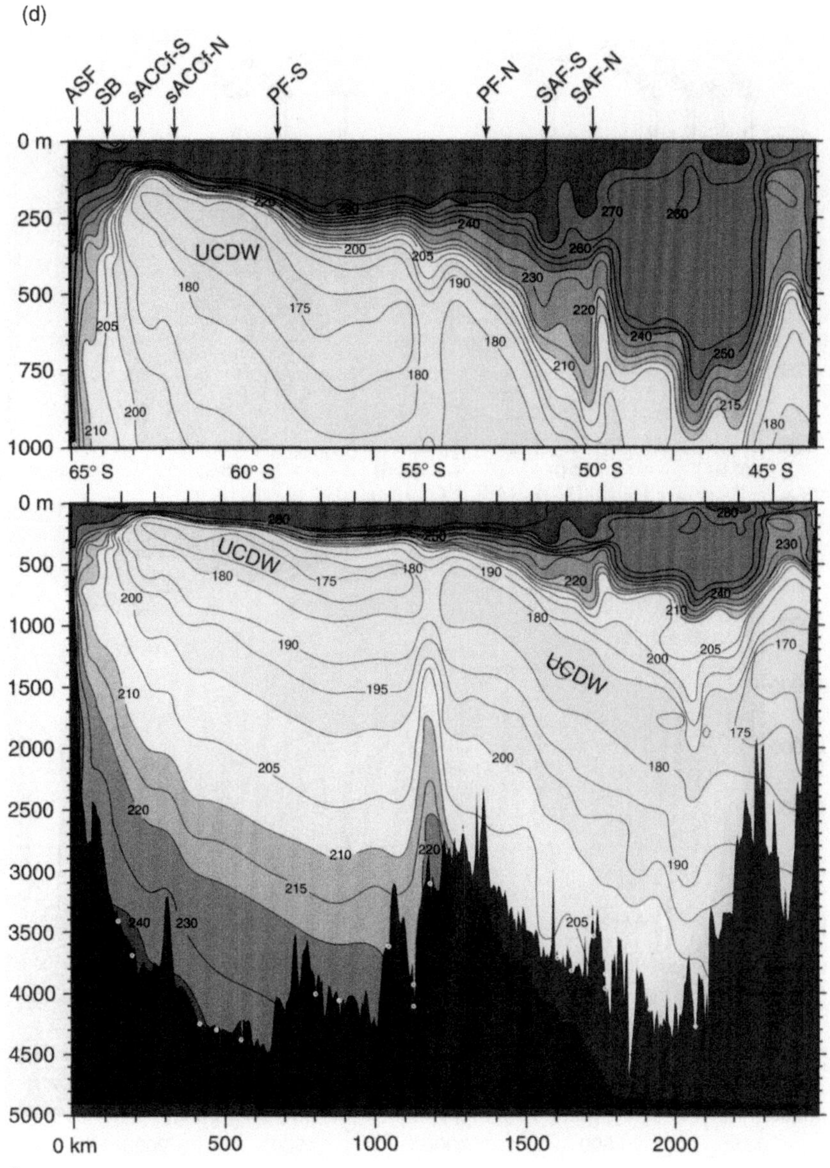

Figure 2 *Continued.*

Circulation of the ACC

A map of the elevation of the sea surface (dynamic height) shows how the position and intensity of the current varies along its circumpolar path (**Figure 3**). Contours of dynamic height are approximate streamlines for the flow, so the current flow is rapid in regions where the contours are closely spaced and weak in regions where the contours are widely separated. The steering of the fronts by large bathymetric features is clearly illustrated in **Figure 3**.

Recent studies using high-resolution sampling from ships and satellites have revealed a more complex structure to the ACC than previously appreciated. The ACC fronts consist of multiple branches, which merge and diverge in different regions and at different times along the circumpolar path of the current system (**Figures 3** and **4**). The multiple jets in the ACC reflect the tendency for geophysical flows on a sphere to self-organize into narrow, elongated, persistent zonal flows; similar circulation patterns form in the Earth's atmosphere and on other planets. The position of the fronts varies with time, but generally over a relatively small latitude range at any given longitude (typically $\pm 1°$ of latitude). The variability is larger downstream of major bathymetric features and in regions where the ACC fronts interact with the strong boundary currents of the subtropical gyres to the north (e.g., south of Africa).

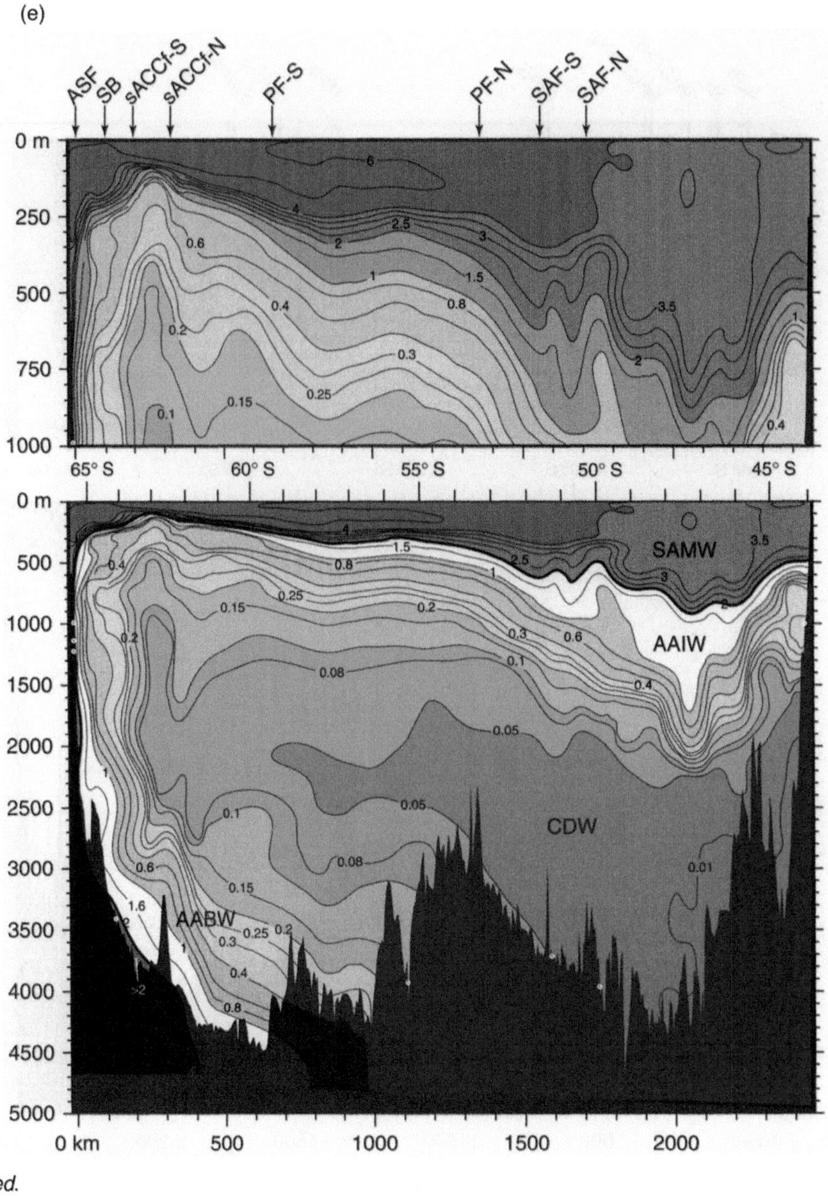

Figure 2 *Continued.*

Transport of the ACC

During the World Ocean Circulation Experiment (WOCE) in the 1990s, the transport through Drake Passage was estimated to be $137 \pm 8 \times 10^6\,m^3\,s^{-1}$, similar to the estimates made in the late 1970s. Because the Atlantic basin is nearly closed to the north of the Southern Ocean, the net transport between Africa and Antarctica must be very close to the Drake Passage transport (to within *c.* $1 \times 10^6\,m^3\,s^{-1}$). The transport between Australia and Antarctica must be somewhat greater, to compensate for the flow from the Pacific to the Indian Oceans through the Indonesian archipelago. Repeat transects during WOCE showed that the baroclinic transport south of Australia is $147 \pm 10 \times 10^6\,m^3\,s^{-1}$, consistent with estimates that about $10\text{–}15 \times 10^6\,m^3\,s^{-1}$ flows through the Indonesian passages.

The fronts, in particular the Subantarctic and Polar Fronts, carry most of the ACC transport. The relative contribution of these two fronts to the total transport varies around the circumpolar path. For example, south of Australia the Subantarctic Front carries 4 times more water to the east than the Polar Front, while in Drake Passage the transport carried by the two fronts is roughly equal in magnitude.

The transport of water masses by the ACC also changes with longitude. For example, the ACC carries an excess of intermediate density water

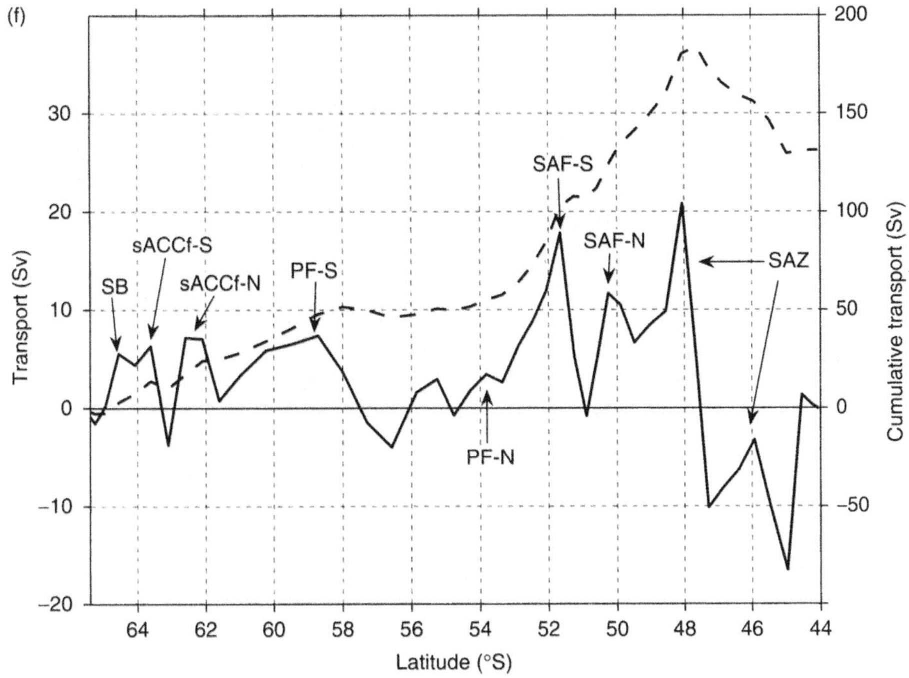

Figure 2 *Continued.*

into the Atlantic through Drake Passage, which is compensated by an excess of deep water leaving the basin south of Africa. These changes in water mass transports by the ACC reflect water mass transformations in the Atlantic basin, where relatively light Antarctic Intermediate Water is converted to denser North Atlantic Deep Water. The ACC also transports vast amounts of heat, fresh water, nutrients, carbon, and other properties between the ocean basins.

The transport of the ACC varies over a range of timescales. Multiyear deployments of bottom pressure recorders in Drake Passage during the late 1970s and 1990s suggest a standard deviation in net transport of about $8–10 \times 10^6 \, m^3 \, s^{-1}$. For periods shorter than about 6 months, most of the variability is due to changes in sea level (i.e., changes in the barotropic, or depth-independent, flow). Models and sea level measurements suggest these barotropic motions are highly correlated with changes in wind stress and tend to follow bathymetric contours (more precisely, the flow is along lines of constant planetary vorticity, where planetary vorticity is given by the Coriolis parameter (equal to twice the rotation rate of the Earth multiplied by the sine of the latitude) divided by the ocean depth). For longer periods, variations in the density field (and hence the baroclinic, or depth-varying, flow) also become important.

Dynamics of the ACC

The absence of continental barriers in the latitude band of Drake Passage makes the dynamics of the ACC distinctly different in character from those of currents at other latitudes. Simple wind-driven ocean circulation theory (the Sverdrup balance), which generally does a good job of describing the circulation of the upper ocean in basins bounded by continents, cannot be applied in the usual way in a continuous ocean channel. The dynamical balance of the ACC has therefore been a topic of great interest for many years.

The strong westerly winds over the Southern Ocean have long been recognized to help drive the ACC. The winds drive surface waters to the left of the wind (*see* Ekman Transport and Pumping), causing upwelling to the south of the wind stress maximum and downwelling to the north. This pattern of upwelling and downwelling helps to establish the tilt of density surfaces across the ACC and therefore its geostrophic flow.

Despite the absence of continental barriers, the Sverdrup theory of wind-driven currents has been applied in the Southern Ocean by assuming that relatively shallow bathymetric features act as 'effective continents'. However, such calculations assume no interaction between the current and the bathymetry, which we know to be a poor

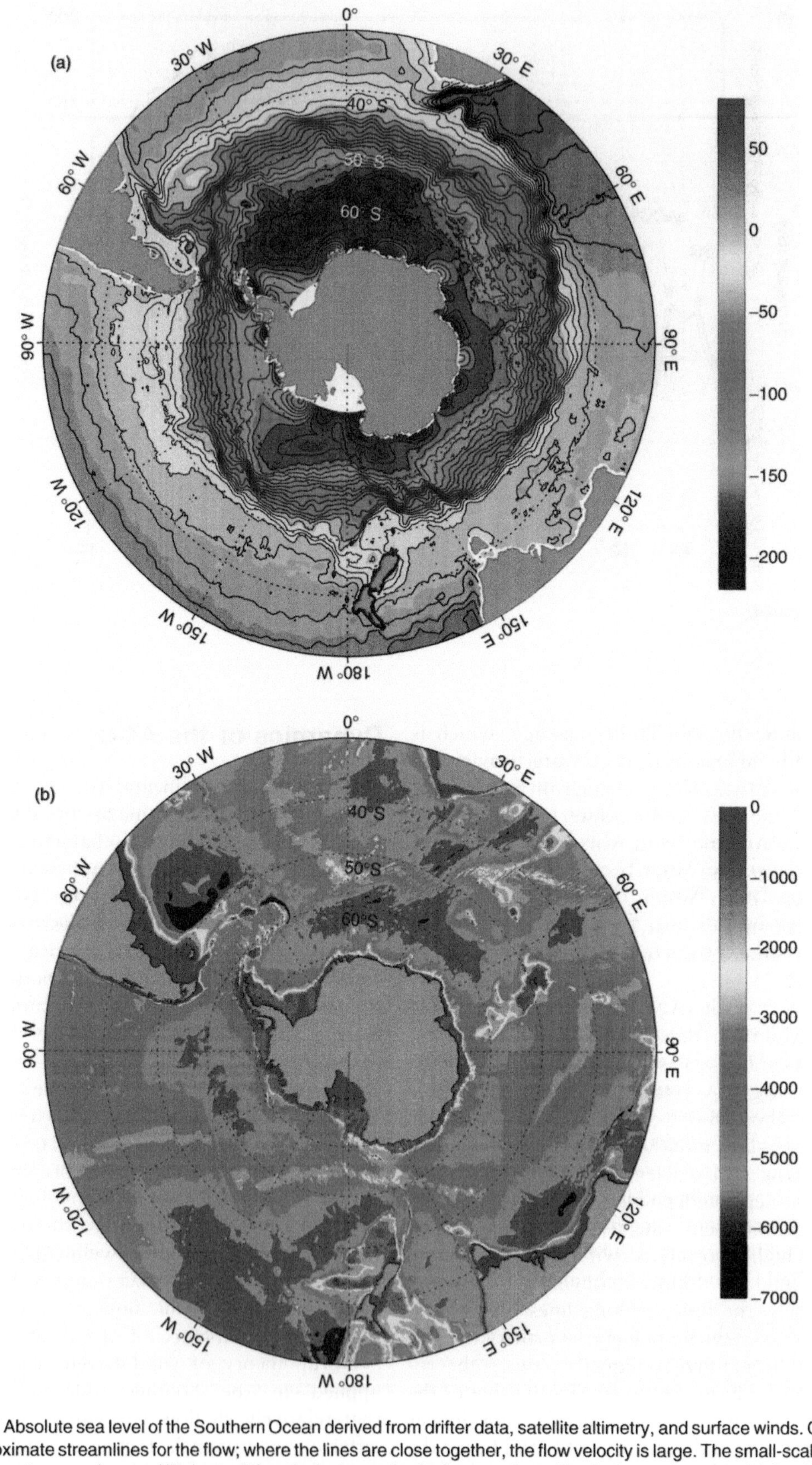

Figure 3 (a) Absolute sea level of the Southern Ocean derived from drifter data, satellite altimetry, and surface winds. Contours of sea level are approximate streamlines for the flow; where the lines are close together, the flow velocity is large. The small-scale features near the Antarctic coast are spurious and likely result from lack of sampling by floats and satellites in the sea ice zone. (b) Sea floor bathymetry (contour interval 500 m). (a) Courtesy Serguei Sokolov; data are from Niiler PP, Maximenko NA, and McWilliams JC (2003) Dynamically balanced absolute sea level of the global ocean derived from near-surface velocity observations. *Geophysical Research Letters* 30: 2164. (b) Courtesy Serguei Sokolov; bathymetry data are from ETOPO5, NOAA, National Geophysical Data Centre.

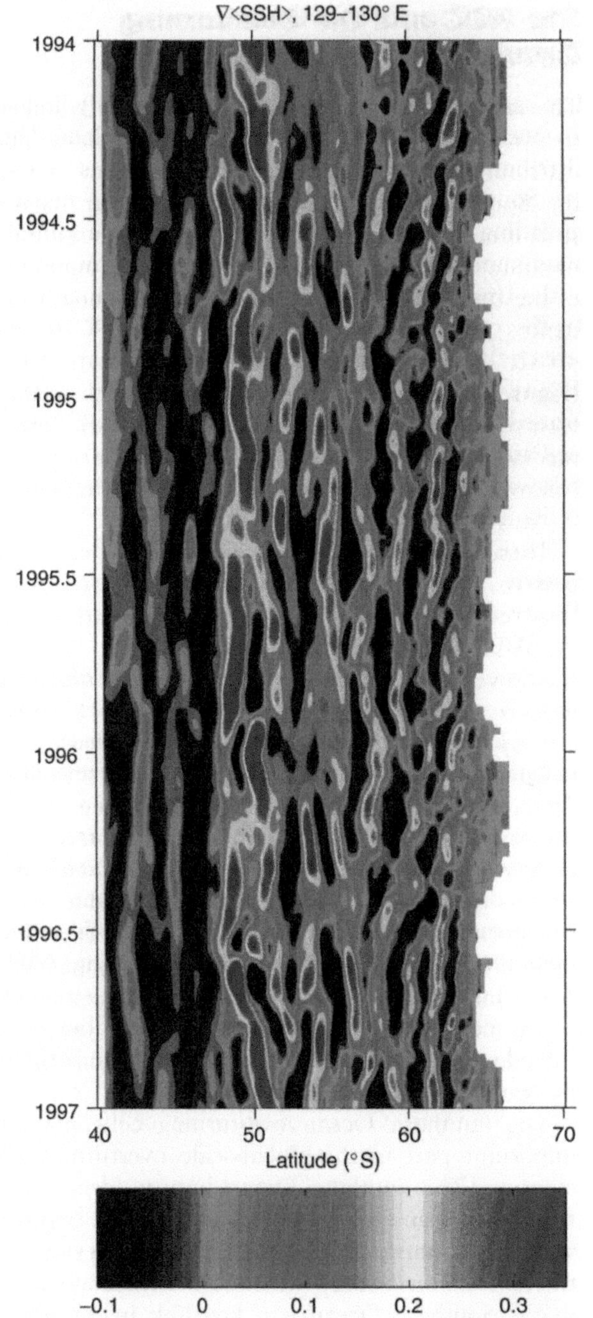

Figure 4 The north–south gradient of sea surface height (in m per 100 km) at 130° E, south of Australia. Large height gradients indicate strong currents. The multiple bands of large gradient correspond to the jets of the ACC. Note that the jets merge and diverge, and change in intensity with time, and also persist for many months at the same latitude. Reproduced with permission from Sokolov S and Rintoul SR (2007) Multiple jets of the Antarctic Circumpolar Current. *Journal of Physical Oceanography* 37: 1394–1412. © Copyright [2007] AMS.

assumption for the deep-reaching ACC. In addition, wind is not the only factor driving the ACC. The atmosphere also drives ocean currents by exchanging heat and fresh water with the ocean, causing the density of seawater to change. Exchange of fresh water can result from precipitation, evaporation, and the freezing and melting of ice, both sea ice and glacial ice in the form of icebergs and ice shelves. Because the speed of ocean currents is proportional to the horizontal gradient of density, any process that produces horizontal density gradients will drive ocean currents. In the case of the ACC, both the strong westerly winds and the air–sea exchange of buoyancy play a part in driving the current.

The momentum supplied to the Southern Ocean by the wind needs to be compensated in some way. The question of what balances the wind forcing has been a topic of debate for many decades. Recent studies have confirmed the early hypothesis by W. Munk and E. Palmén that interaction of the ACC with seafloor topography provides a force to balance the wind. This force, known as the bottom form stress, results when the ocean currents are organized such that there is higher pressure on one side of a ridge on the seafloor than is found on the other side. In the case of the ACC, higher pressure is generally found on the west side of topographic ridges or hills, providing a force from the solid Earth to the ocean that balances the wind stress at the sea surface. While these pressure differences are too small to observe directly, realistic numerical simulations clearly show this force balance in action.

Eddies produced by dynamical instabilities of the ACC fronts play a crucial role in establishing the momentum and heat balance of the Southern Ocean. The ACC has some of the most vigorous eddy activity observed in the ocean (**Figure** 5). Eddies are produced when dynamical processes release some of the energy stored in the sloping of density surfaces across the ACC, converting some of the energy in the mean flow into motions that vary with time, or eddies. This process is called baroclinic instability. The eddies transfer momentum vertically from the sea surface to the deep ocean, helping to set up the system of deep currents that interact with bathymetry to provide the bottom form stress. Both transient eddies (motions that vary with time) and standing eddies (deviations of the flow from the east–west average) contribute to the momentum transport. At the same time, the eddies carry heat poleward across the ACC, to compensate the heat lost to the cold atmosphere near Antarctica.

Attempts to relate ACC transport variability to variations in wind have been inconclusive. It is now understood that the dynamical balance of the ACC depends on a number of factors, including wind and buoyancy forcing, eddy–mean flow interaction, topographic form stress and the ocean stratification,

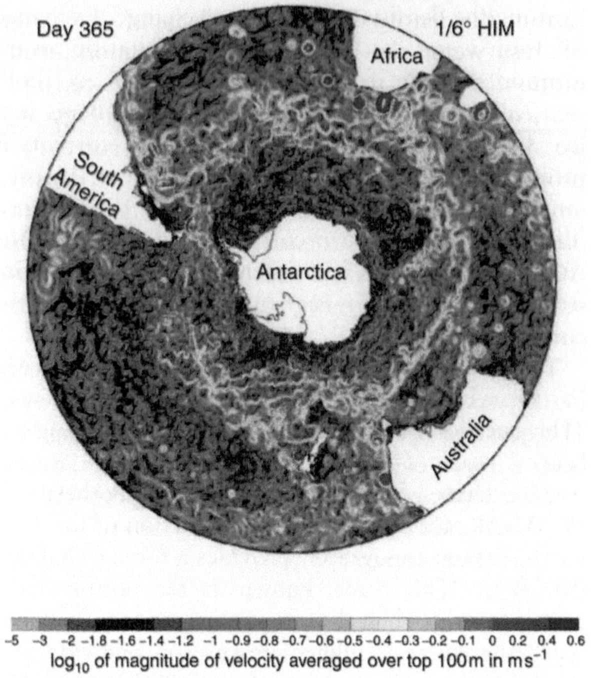

Figure 5 A snapshot of surface speed from a high-resolution numerical simulation of the ACC. The filamented, eddy-rich structure of the ACC stands out clearly. Adapted with permission from Hallberg R and Gnanadesikan A (2006) The role of eddies in determining the structure and response of the Southern Ocean overturning: Results from the Modelling Eddies in the Southern Ocean project. *Journal of Physical Oceanography* 36: 2232–2252. © Copyright [2007] AMS.

so a simple relationship between transport and wind should not be expected.

Recent improvements in ocean observing systems have allowed changes in the ACC to be assessed for the first time. Comparison of temperature profiles collected since the 1950s suggests that much of the Southern Ocean has warmed. The warming is largest in the ACC belt and is consistent with a southward shift of the current, allowing warm water to move south into areas previously occupied by cooler water. The southward movement of the ACC has been linked to a southward shift and strengthening of the westerly winds. The change in the winds, in turn, has been linked both to loss of ozone in the polar stratosphere and to enhanced greenhouse warming. The transport of the ACC has apparently not changed much over this time period, despite the change in wind forcing. Recent studies suggest the ACC may be in a regime in which the transport is insensitive to changes in wind: as the wind forcing increases, the ACC starts to speed up, but this causes more vigorous eddy fluxes that dissipate the extra energy in the mean flow.

The ACC and the Overturning Circulation

The eastward flow of the ACC is dynamically linked to a weaker circulation in the north–south plane. The distribution of water properties on transects across the Southern Ocean clearly reveals water masses spreading across the ACC. For example, the salinity maximum of the Lower CDW and oxygen minimum of the Upper CDW can be traced as they shoal from depths of 2000–3500 m north of the ACC to approach the sea surface south of the Polar Front (**Figure 2**). The high-oxygen, low-salinity waters formed in the Southern Ocean (Antarctic Intermediate Water and Antarctic Bottom Water) can be followed as they cross the ACC and enter the basins to the north.

These distributions reflect an ocean circulation pattern known as the overturning circulation (**Figure 6**). Deep water spreads to the south across the ACC and upwells at the sea surface. Some of the upwelled deep water is driven north beneath the westerly winds, gains heat and fresh water from the atmosphere, and therefore becomes less dense, and ultimately sinks to form Antarctic Intermediate Water and Subantarctic Mode Water. Deep water that upwells further south and closer to Antarctica is converted to denser Antarctic Bottom Water and returns to the north. The result of these water mass transformations is a circulation in the north–south plane that consists of two counter-rotating cells. According to the residual mean theory, the strength of the net overturning circulation (mean flow plus the eddy contribution) is determined by the surface buoyancy forcing.

The Southern Ocean overturning cells play an important part in the global-scale overturning circulation. The Southern Ocean imports deep water from the basins to the north, and exports bottom water and intermediate water. Recent studies suggest the conversion of deep water to intermediate water in the Southern Ocean is a key link in the global overturning circulation. For decades it has been assumed that the sinking of dense water in the polar regions was balanced by widespread upwelling at lower latitudes. However, measurements of mixing rates and large-scale tracer budgets suggest that mixing in the interior of the ocean is an order of magnitude too weak to support the upwelling required. The transformation of deep water to intermediate water by air–sea buoyancy exchange in the Southern Ocean provides an alternative means of connecting the upper and lower limbs of the global overturning circulation. Mixing likely also makes a contribution in regions of rough bathymetry,

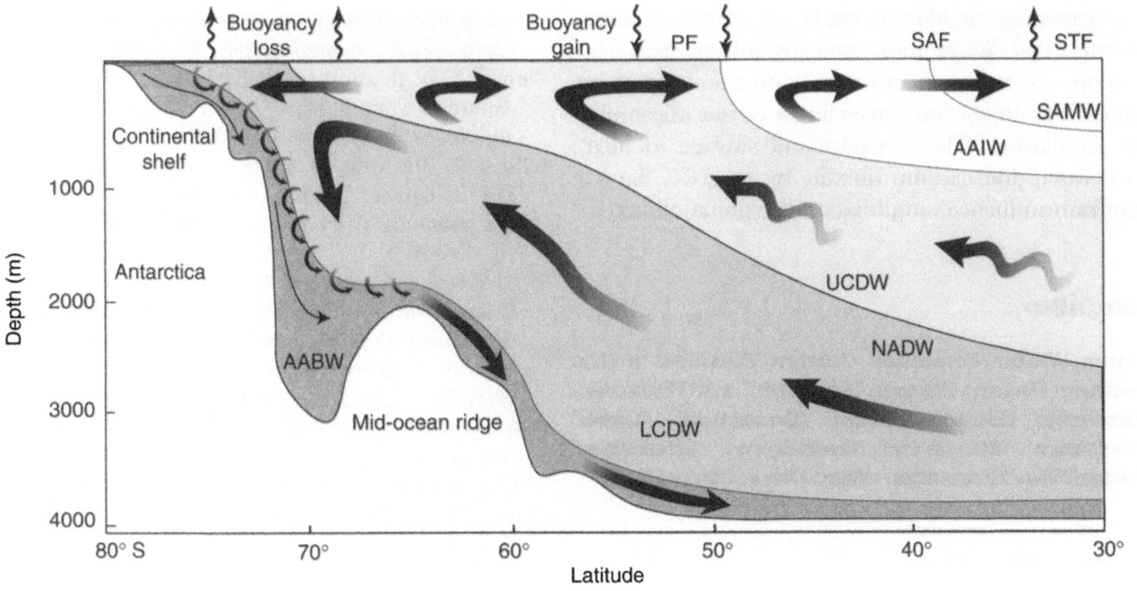

Figure 6 A schematic view of the Southern Ocean overturning circulation. Upper Circumpolar Deep Water (UCDW) upwells at high latitude and gains buoyancy (from heating, precipitation, and ice melt) as it is driven north, to ultimately sink as Antarctic Intermediate Water (AAIW) or Subantarctic Mode Water (SAMW). Lower Circumpolar Deep Water (LCDW) and North Atlantic Deep Water (NADW) upwell closer to the Antarctic continent, are made more dense (by cooling and salt rejected during sea ice formation), and sink to form Antarctic Bottom Water (AABW). The major southern ocean fronts are indicated. STF, Subtropical Front; SAF, Subantarctic Front; PF, Polar Front. Reproduced with permission from Speer K, Rintoul SR, and Sloyan B (2000) The diabatic Deacon cell. *Journal of Physical Oceanography* 30: 3212–3222. © Copyright [2007] AMS.

including the Southern Ocean, where elevated mixing rates have been measured.

Eddies spawned by the ACC make an important contribution to the overturning circulation. The eddies transfer mass across the Drake Passage gap, where the absence of land barriers means that there can be no net east–west pressure gradient and therefore no net north–south flow. Furthermore, the same forces that drive the overturning circulation (wind and buoyancy exchange) also drive the ACC. The west-to-east flow of the ACC and the overturning circulations cannot be understood in isolation. The two are intimately linked, and eddy fluxes, topographic interactions, and wind and buoyancy forcing are all important ingredients of the dynamical coupling between them.

The Southern Ocean overturning also has a large influence on global biogeochemical cycles. Upwelling of nutrient-rich deep water south of the ACC returns nutrients to the surface layer. The nutrients are exported from the Southern Ocean to lower latitudes by the overturning circulation, ultimately supporting a large fraction of global primary production. Water masses at the surface of the Southern Ocean exchange oxygen and carbon dioxide with the atmosphere and carry 'ventilated' water to the interior of the ocean when the water masses sink. Where deep water upwells south of the ACC, carbon dioxide is released to the atmosphere; where water sinks from the sea surface, carbon dioxide is carried from the atmosphere into the ocean. As a result of the overturning circulation, more of the carbon released by human activities is accumulating just north of the ACC than in any other latitude band of the ocean.

Summary

The ACC is the largest current in the world ocean, carrying about $137 \pm 8 \times 10^6\,m^3\,s^{-1}$ from west to east around Antarctica. By connecting the ocean basins, the ACC allows water masses and climate anomalies to propagate between the basins. The current flow is concentrated in a number of circumpolar fronts, which extend from the sea surface to the seafloor. The fronts also mark the boundaries between zones with distinct physical, chemical, and ecological characteristics. Eddies produced by dynamical instabilities of the fronts play an important part in the dynamics of the ACC by transporting momentum vertically and heat and mass poleward. Both wind and buoyancy forcing contribute to driving the ACC. Interaction between the deep-reaching flow and the bottom topography establish bottom form stresses to balance the wind forcing. The strong eastward flow of the ACC is intimately connected to an overturning circulation made up of two counter-rotating cells. Water mass transformations driven by exchange of heat and moisture with the atmosphere connect the upper and lower limbs of the thermohaline circulation. The transport and storage of heat, fresh water, and carbon dioxide by the ACC have a significant influence on global and regional climate.

See also

Bottom Water Formation. Current Systems in the Southern Ocean. Ekman Transport and Pumping. Mesoscale Eddies. Ocean Circulation. Ocean Circulation: Meridional Overturning Circulation. Weddell Sea Circulation. Wind Driven Circulation.

Further Reading

Cunningham S, Alderson SG, King BA, and Brandon MA (2003) Transport and variability of the Antarctic Circumpolar Current in Drake Passage. *Journal of Geophysical Research* 108: 8084 (doi:10.1029/2001 JC001147).

Deacon G (1984) *The Antarctic Circumpolar Ocean*. London: Cambridge University Press.

Gille ST (2002) Warming of the Southern Ocean since the 1950s. *Science* 295: 1275–1277.

Gordon AL and Molinelli E (1986) *Southern Ocean Atlas*. Washington, DC and New Delhi: National Science Foundation and Amerind Publishing.

Hallberg R and Gnanadesikan A (2006) The role of eddies in determining the structure and response of the Southern Ocean overturning: Results from the Modelling Eddies in the Southern Ocean project. *Journal of Physical Oceanography* 36: 2232–2252.

Munk WH and Palmén E (1951) Note on the dynamics of the Antarctic Circumpolar Current. *Tellus* 3: 53–55.

Niiler PP, Maximenko NA, and McWilliams JC (2003) Dynamically balanced absolute sea level of the global ocean derived from near-surface velocity observations. *Geophysical Research Letters* 30: 2164.

Nowlin WD Jr. and Klinck JM (1986) The physics of the Antarctic Circumpolar Current. *Review of Geophysics and Space Physics* 24: 469–491.

Olbers D, Borowski D, Volker C, and Wolff J-O (2004) The dynamical balance, transpor't and circulation of the Antarctic Circumpolar Current. *Antarctic Science* 16: 439–470.

Orsi AH and Whitworth T, III (2005) In: Sparrow M, Chapman P, and Gould J (eds.) *Hydrographic Atlas of the World Ocean Circulation Experiment (WOCE), Vol. 1: Southern Ocean*. Southampton: International WOCE Project Office (ISBN 0-904175-49-9).

Orsi AH, Whitworth T, III, and Nowlin WD (1995) On the meridional extent and fronts of the Antarctic Circumpolar Current. *Deep-Sea Research I* 42: 641–673.

Rintoul SR, Hughes C, and Olbers D (2001) The Antarctic Circumpolar Current system. In: Siedler G, Church J, and Gould J (eds.) *Ocean Circulation and Climate*, pp. 271–302. London: Academic Press.

Sokolov S and Rintoul SR (2007) Multiple jets of the Antarctic Circumpolar Current. *Journal of Physical Oceanography* 37: 1394–1412.

Speer K, Rintoul SR, and Sloyan B (2000) The diabatic Deacon cell. *Journal of Physical Oceanography* 30: 3212–3222.

Whitworth T, III (1980) Zonation and geostrophic flow of the Antarctic Circumpolar Current at Drake Passage. *Deep-Sea Research* 27: 497–507.

Whitworth T, III and Nowlin WD (1987) Water masses and currents of the Southern Ocean at the Greenwich Meridian. *Journal of Geophysical Research* 92: 6462–6476.

Relevant Website

http://www.soc.soton.ac.uk
– Electronic Atlas of WOCE Data.

SUB ICE-SHELF CIRCULATION AND PROCESSES

K. W. Nicholls, British Antarctic Survey,
Cambridge, UK

Introduction

Ice shelves are the floating extension of ice sheets. They extend from the grounding line, where the ice sheet first goes afloat, to the ice front, which usually takes the form of an ice cliff dropping down to the sea. Although there are several examples on the north coast of Greenland, the largest ice shelves are found in the Antarctic where they cover 40% of the continental shelf. Ice shelves can be up to 2 km thick and have horizontal extents of several hundreds of kilometers. The base of an ice shelf provides an intimate link between ocean and cryosphere. Three factors control the oceanographic regime beneath ice shelves: the geometry of the sub-ice shelf cavity, the oceanographic conditions beyond the ice front, and tidal activity. These factors combine with the thermodynamics of the interaction between sea water and the ice shelf base to yield various glaciological and oceanographic phenomena: intense basal melting near deep grounding lines and near ice fronts; deposition of ice crystals at the base of some ice shelves, resulting in the accretion of hundreds of meters of marine ice; production of sea water at temperatures below the surface freezing point, which may then contribute to the formation of Antarctic Bottom Water (*see* Bottom Water Formation); and the upwelling of relatively warm Circumpolar Deep Water.

Although the presence of the ice shelf itself makes measurement of the sub-ice shelf environment difficult, various field techniques have been used to study the processes and circulation within sub-ice shelf cavities. Rates of basal melting and freezing affect the flow of the ice and the nature of the ice–ocean interface, and so glaciological measurements can be used to infer the ice shelf's basal mass balance. Another indirect approach is to make ship-based oceanographic measurements along ice fronts. The properties of in-flowing and out-flowing water masses give clues as to the processes needed to transform the water masses. Direct measurements of oceanographic conditions beneath ice shelves have been made through natural access holes such as rifts, and via access holes created using thermal (mainly hot-water) drills. Numerical models of the sub-ice shelf regime have been developed to complement the field measurements. These range from simple one-dimensional models following a plume of water from the grounding line along the ice shelf base, to full three-dimensional models coupled with sea ice models, extending out to the continental shelf-break and beyond.

The close relationship between the geometry of the sub-ice shelf cavity and the interaction between the ice shelf and the ocean implies a strong dependence of the ice shelf/ocean system on the state of the ice sheet. During glacial climatic periods the geometry of ice shelves would have been radically different to their geometry today, and ice shelves probably played a different role in the climate system.

Geographical Setting

By far the majority of the world's ice shelves are found fringing the Antarctic coastline (**Figure 1**). Horizontal extents vary from a few tens to several hundreds of kilometers, and maximum thickness at the grounding line varies from a few tens of meters to 2 km. By area, the Ross Ice Shelf is the largest at around 500 000 km^2. The most massive, however, is the very much thicker Filchner-Ronne Ice Shelf in the southern Weddell Sea. Ice from the Antarctic Ice Sheet flows into ice shelves via fast-moving ice streams (**Figure 2**). As the ice moves seaward, further nourishment comes from snowfall, and, in some cases, from accretion of ice crystals at the ice shelf base. Ice is lost by melting at the ice shelf base and by calving of icebergs at the ice front. Current estimates suggest that basal melting is responsible for around 25% of the ice loss from Antarctic ice shelves; most of the remainder calves from the ice fronts as icebergs.

Over central Antarctica the weight of the ice sheet depresses the lithosphere such that the seafloor beneath many ice shelves deepens towards the grounding line. The effect of the lithospheric depression has probably been augmented during glacial periods by the scouring action of ice on the seafloor: at the glacial maxima the grounding line would have been much closer to the continental shelf-break. Since ice shelves become thinner towards the ice front and float freely in the ocean, a typical sub-ice shelf cavity has the shape of a cavern that dips downwards towards the grounding line (**Figure 2**).

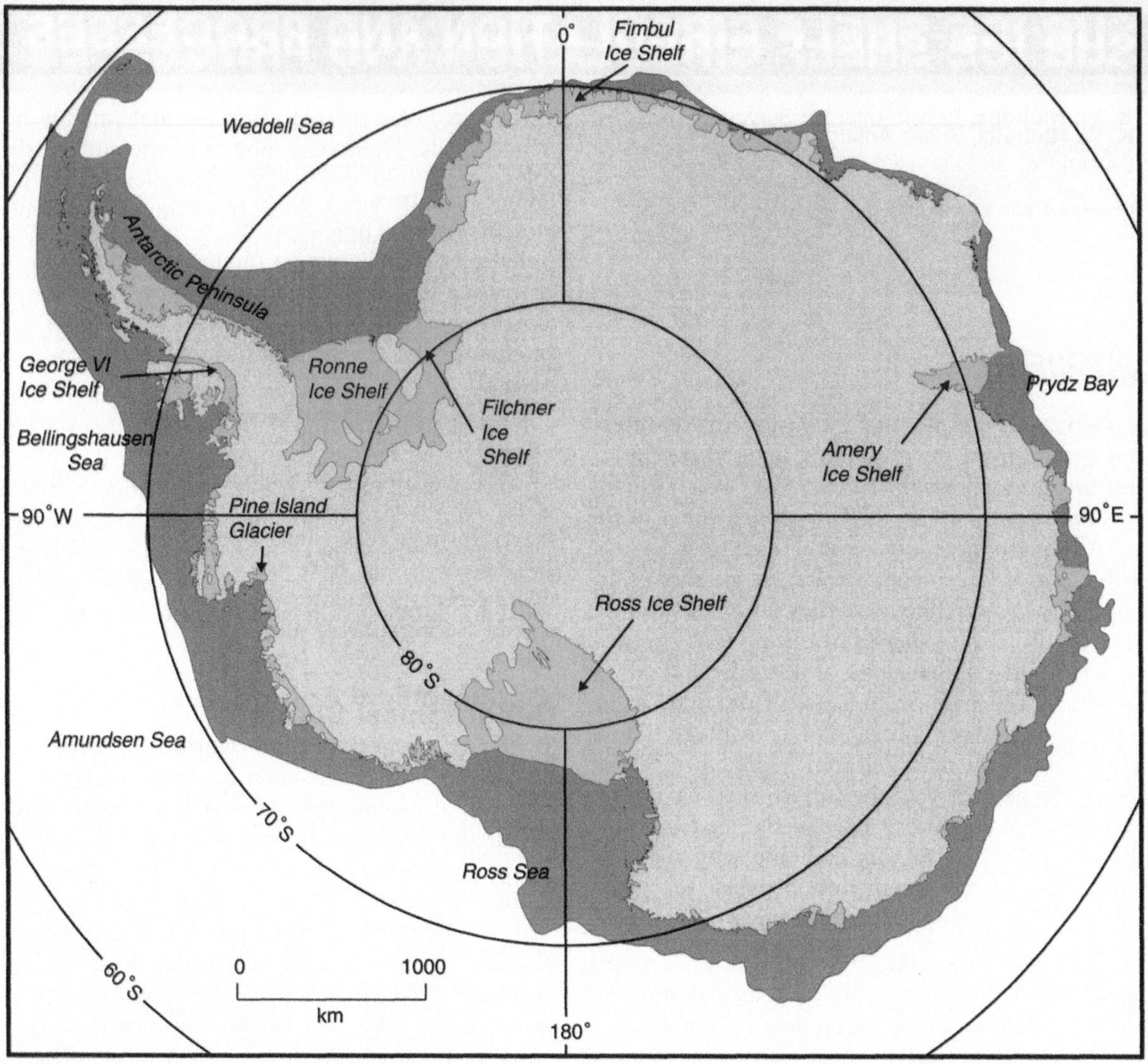

Figure 1 Map showing ice shelves (blue) covering about 40% of the continental shelf (dark gray) of Antarctica.

This geometry has important consequences for the ocean circulation within the cavity.

Oceanographic Setting

The oceanographic conditions over the Antarctic continental shelf depend on whether relatively warm, off-shelf water masses are able to cross the continental shelf-break.

For much of Antarctica a dynamic barrier at the shelf-break prevents advection of circumpolar deep water (CDW) onto the continental shelf itself. In these regions the principal process determining the oceanographic conditions is production of sea ice in coastal polynyas and leads, and the water column is largely dominated by high salinity shelf water (HSSW). Long residence times over some of the broader continental shelves, for example in the Ross and southern Weddell seas, enable HSSW to attain salinities of over 34.8 PSU. HSSW has a temperature at or near the surface freezing point (about -1.9°C), and is the densest water mass in Antarctic waters. Conditions over the continental shelves of the Bellingshausen and Amundsen seas (**Figure 1**) represent the other extreme. There, the barrier at the shelf-break appears to be either weak or absent. At a temperature of about 1°C, CDW floods the continental shelf.

Between these two extremes there are regions of continental shelf where tongues of modified warm deep water (MWDW) are able to penetrate the shelf-break barrier (**Figure 3**), in some cases reaching as far as ice fronts. MWDW comes from above the warm core of CDW: the continental shelf effectively skims

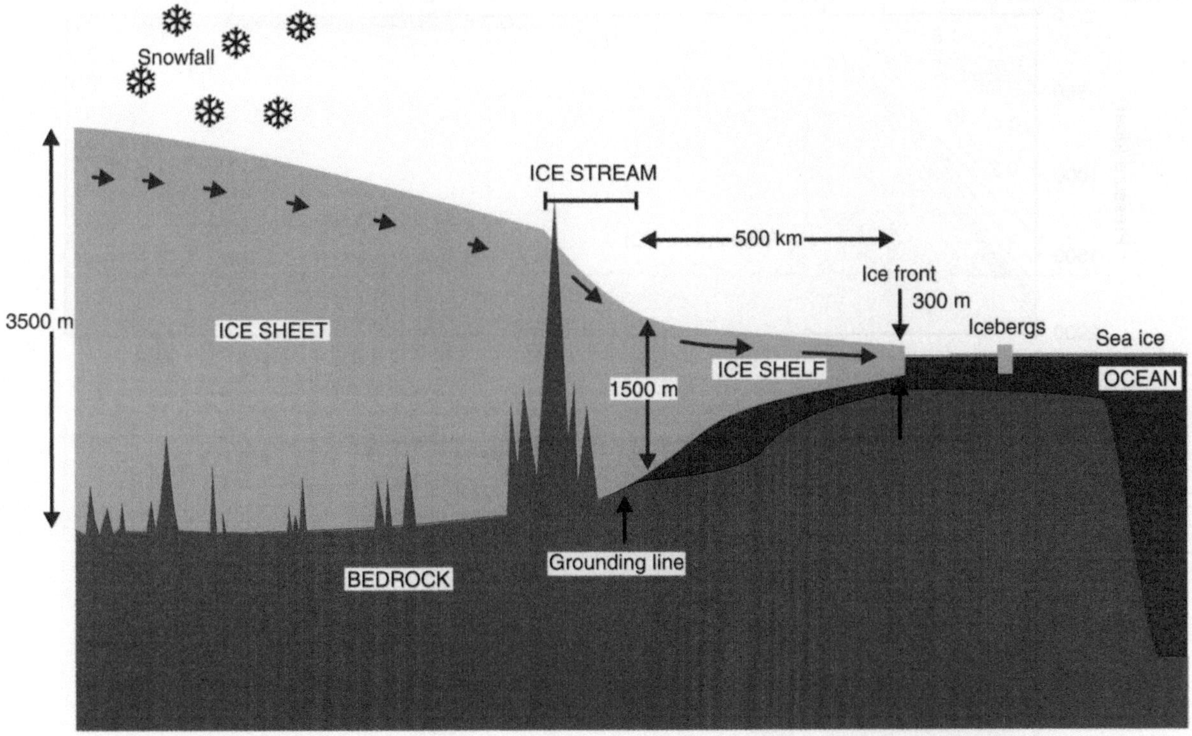

Figure 2 Schematic cross-section of the Antarctic ice sheet showing the transition from ice sheet to ice stream to ice shelf. Also shown is the depression of the lithosphere that results in the deepening of the seabed towards the continental interior.

off the shallower and cooler part of the water column.

What Happens When Ice Shelves Melt into Sea Water?

The freezing point of fresh water is 0°C at atmospheric pressure. When the water contains dissolved salts, the freezing point is depressed: at a salinity of around 34.7 PSU the freezing point is −1.9°C. Sea water at a temperature above −1.9°C is therefore capable of melting ice. The freezing point of water is also pressure dependent. Unlike most materials, the pressure dependence for water is negative: increasing the pressure decreases the freezing point. The freezing point T_f of sea water is approximated by:

$$T_f = aS + bS^{3/2} - cS^2 - dp$$

where $a = -5.75 \times 10^{-2}$°C PSU^{-1}, $b = 1.710523 \times 10^{-3}$°C PSU$^{-3/2}$, $c = -2.154996 \times 10^{-4}$°C PSU^{-2} and $d = -7.53 \times 10^{-4}$°C dbar^{-1}. S is the salinity in PSU, and p is the pressure in dbar. Every decibar increase in pressure therefore depresses the freezing point by 0.75 m°C. The depression of the freezing point with pressure has important consequences for the interaction between ice shelves and the ocean. Even though HSSW is already at the surface freezing point, if it can be brought into contact with an ice shelf base, melting will take place. As the freezing point at the base of deep ice shelves can be as much as 1.5°C lower than the surface freezing point, the melt rates can be high.

When ice melts into sea water the effect is to cool and freshen. Consider unit mass of water at temperature T_0, and salinity S_0 coming into contact with the base of an ice shelf where the *in situ* freezing point is T_f. The water first warms m kg of ice to the freezing point, and then supplies the latent heat necessary for melting. The resulting mixture of melt and sea water has temperature T and salinity S. If the initial temperature of the ice is T_i, the latent heat of melting is L, the specific heat capacity of sea water and ice, c_w and c_i, then heat and salt conservation requires that:

$$(T - T_f)(1 + m)c_w + m(c_i(T_f - T_i) + L) = (T_o - T_f)c_w$$

$$S(1 + m) = S_o$$

Eliminating m, and then expressing T as a function of S reveals the trajectory of the mixture in T–S space as a straight line passing through (S_0, T_0), with a gradient given by:

$$\frac{dT}{dS} = \frac{L}{S_o c_w} + \frac{(T_f - T_i)c_i}{S_o c_w} + \frac{(T_o - T_f)}{S_o}$$

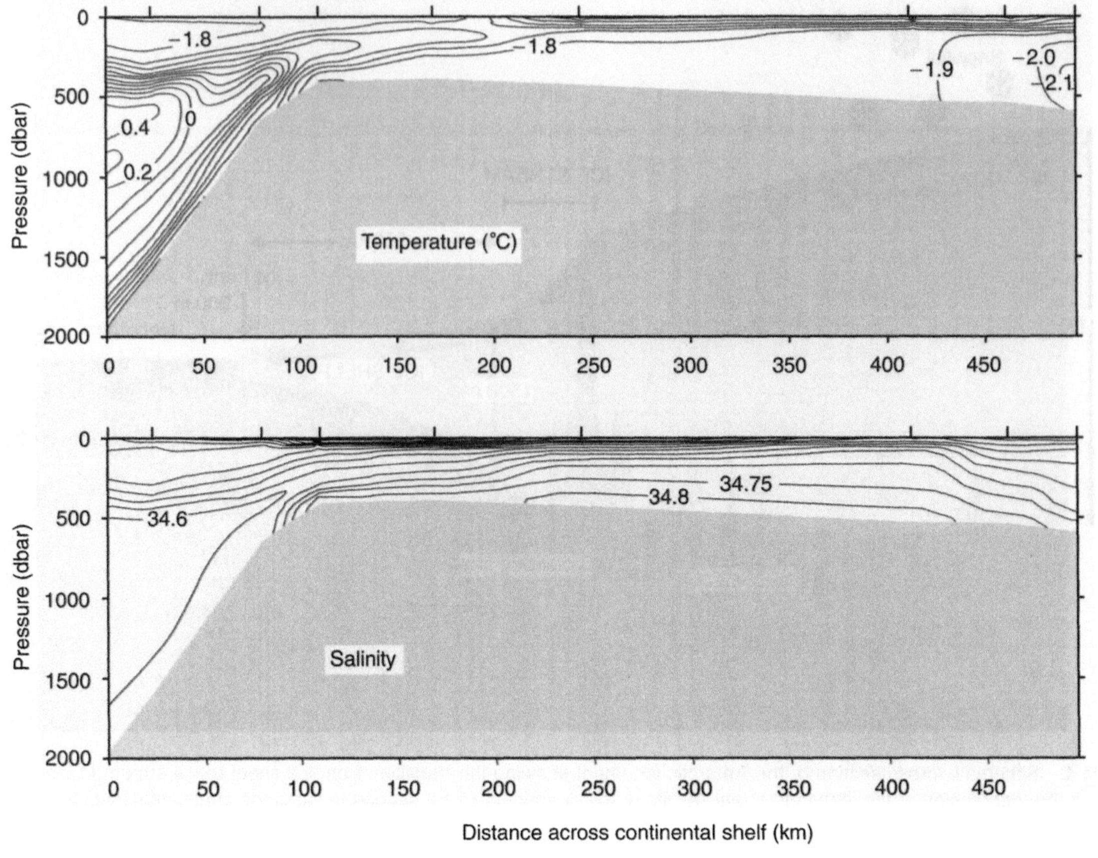

Figure 3 Hydrographic section over the continental slope and across the open continental shelf in the southern Weddell Sea, as far as the Ronne Ice Front. Water below the surface freezing point (−1.9°C) can be seen emerging from beneath the ice shelf. The majority of the continental shelf is dominated by HSSW, although in this location a tongue of warmer MWDW penetrates across the shelf-break. The station locations are shown by the heavy tick marks along the upper axes.

The gradient is dominated by the first term, which evaluates to about 2.4°C PSU^{-1}. In polar waters the third term is two orders of magnitude lower than the first; the second term results from the heat needed to warm the ice, and, at about a tenth the size of the first term, makes a measurable contribution to the gradient. This relationship allows the source water for sub-ice shelf processes to be found by inspection of the T–S properties of the resultant water masses. Examples of T–S plots from beneath ice shelves in warm and cold regimes are shown in **Figure 4**.

Two important passive tracers are introduced into sea water when glacial ice melts. When water evaporates from the ocean, molecules containing the lighter isotope of oxygen, ^{16}O, evaporate preferentially. Compared with sea water the snow that makes up the ice shelves is therefore low in ^{18}O. By comparing the $^{18}O/^{16}O$ ratios of the outflowing and inflowing water it is possible to calculate the concentration of melt water, provided the ratio is known for the glacial ice. Helium contained in the air bubbles in the ice is also introduced into the sea water when the ice melts. As helium's solubility in water increases with increasing water pressure, the concentration of dissolved helium in the melt water can be an order of magnitude greater than in ambient sea water, which has equilibrated with the atmosphere at surface pressure.

Modes of Sub-ice Shelf Circulation

Various distinguishable modes of circulation appear to be possible within a sub-ice shelf cavity. Which mode is active depends primarily on the oceanographic forcing from seaward of the ice front, but also on the geometry of the sub-ice shelf cavity. Thermohaline forcing drives three modes of circulation, although the tidal activity is thought to play an important role by supplying energy for vertical mixing. Another mode results from tidal residual currents.

Thermohaline Modes

Cold regime external ventilation Over the parts of the Antarctic continental shelf dominated by the

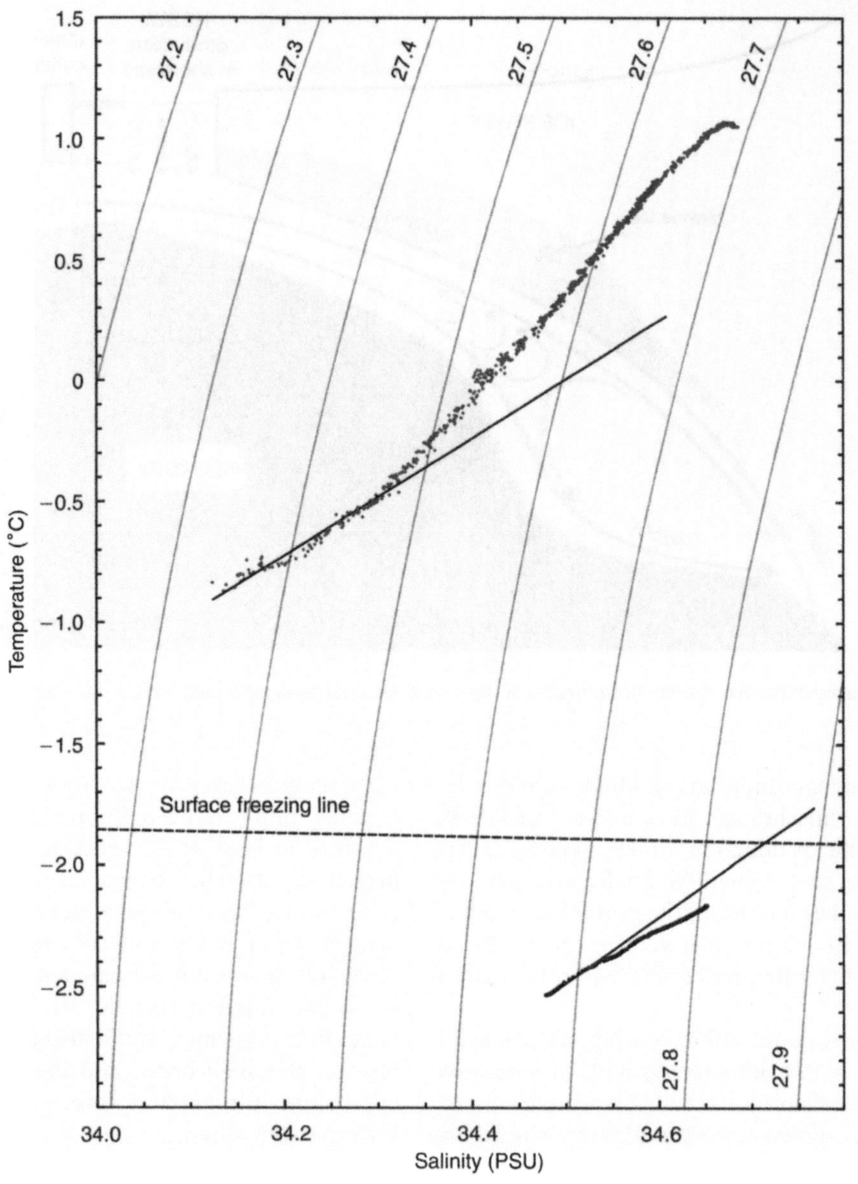

Figure 4 Temperature and salinity trajectories from CTD stations through the George VI Ice Shelf (red) and Ronne Ice Shelf (blue). The cold end of each trajectory corresponds to the base of the ice shelf. The straight lines are at the characteristic gradient for ice melting into sea water. For the Ronne data, as the source water will be HSSW at the surface freezing point, the intersection of the characteristic with the broken line gives the temperature and salinity of the source water. The isopycnals (gray lines) are referenced to sea level.

production of HSSW, such as in the southern Weddell Sea, the Ross Sea, and Prydz Bay, the circulation beneath large ice shelves is driven by the drainage of HSSW into the sub-ice shelf cavities. The schematic in **Figure 5** illustrates the circulation mode. HSSW drains down to the grounding line where tidal mixing brings it into contact with ice at depths of up to 2000 m. At such depths HSSW is up to 1.5°C warmer than the freezing point, and relatively rapid melting ensues (up to several meters of ice per year). The HSSW is cooled and diluted, converting it into ice shelf water (ISW), which is defined as water with a temperature below the surface freezing point.

ISW is relatively buoyant and rises up the inclined base of the ice shelf. As it loses depth the *in situ* freezing point rises also. If the ISW is not entraining sufficient HSSW, which is comparatively warm, the reduction in pressure will result in the water becoming *in situ* supercooled. Ice crystals are then able to form in the water column and possibly rise up and accrete at the base of the ice shelf. This 'snowfall' at

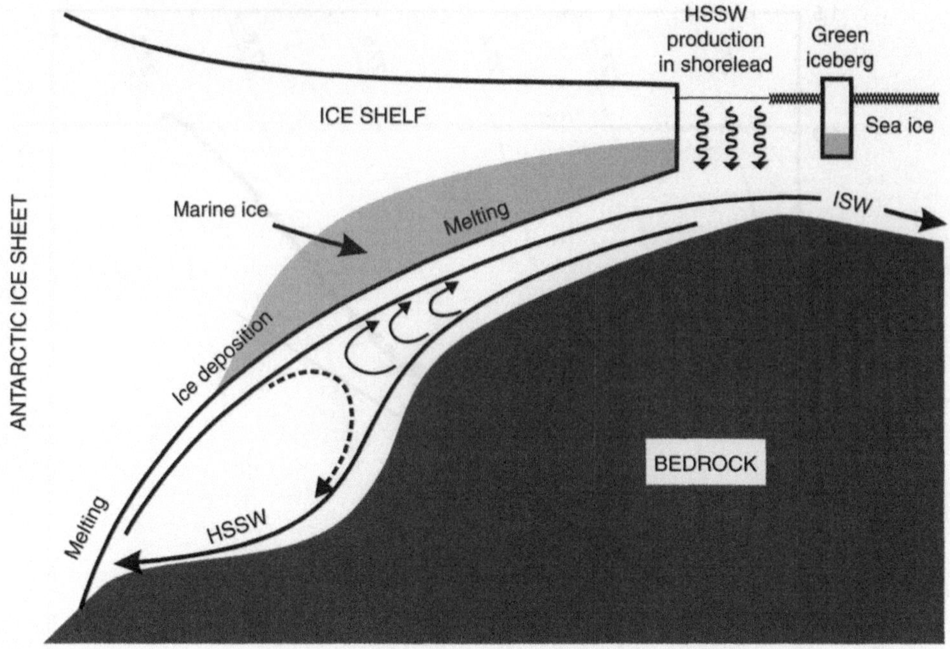

Figure 5 Schematic of the two thermohaline modes of sub-ice shelf circulation for a cold regime ice shelf.

the ice shelf base can build up hundreds of meters of what is termed 'marine ice'. Entrainment of HSSW, and the possible production of ice crystals, often result in the density of the ISW finally matching the ambient water density before the plume has reached the ice front. The plume then detaches from the ice shelf base, finally emerging at the ice front at mid-water depths.

The internal Rossby radius beneath ice shelves is typically only a few kilometers, and so rotational effects must be taken into account when considering the flow in three dimensions. HSSW flows beneath the ice shelf as a gravity current and is therefore gathered to the left (in the Southern Hemisphere) by the Coriolis force. As an organized flow, it then follows bathymetric contours. Once converted into ISW, the flow is again gathered to the left, following either the coast, or topography in the ice base. If the ISW plume fills the cavity, conservation of potential vorticity would demand that it follow contours of constant water column thickness. The step in water column thickness caused by the ice front then presents a topographic obstacle for the outflow of the ISW. However, the discontinuity can be reduced by the presence of trenches in the seafloor running across the ice front. This has been proposed as the mechanism that allows ISW to flow out from beneath the Filchner Ice Shelf, in the southern Weddell Sea (**Figure 1**).

Initial evidence for this mode of circulation came from ship-based oceanographic observations along the ice front of several of the larger ice shelves. Water with temperatures up to 0.3°C below the surface freezing point indicated interaction with ice at a depth of at least 400 m, and the $^{18}O/^{16}O$ ratio confirmed the presence of glacial melt water at a concentration of several parts per thousand. Nets cast near ice fronts for biological specimens occasionally recovered masses of ice platelets, again from depths of several hundred meters. The ISW flowing from beneath the Filchner Ice Shelf has been traced overflowing the shelf-break and descending the continental slope, ultimately to mix with deep waters and form bottom water.

Evidence from the ice shelf itself comes in the form of glaciological measurements. By assuming a steady state (the ice shelf neither thickening nor thinning with time at any given point) conservation arguments can be used to derive the basal mass balance at individual locations. The calculation needs measurements of the local ice thickness variation, the horizontal spreading rate of the ice as it flows under its own weight, the horizontal speed of the ice, and the surface accumulation rate. This technique has been applied to several ice shelves, but is time-consuming, and has rarely been used to provide a good areal coverage of basal mass balance. However, it has demonstrated that high basal melt rates do indeed exist near deep grounding lines; that the melt rates reduce away from the grounding line; that further still from the grounding line, melting frequently switches to freezing; and that the balance usually returns to melting as the ice front is approached.

One-dimensional models have been to study the development of ISW plumes from the grounding line to where they detach from the ice shelf base. The most sophisticated includes frazil ice dynamics, and suggests that the deposition of ice at the base depends not only on its formation in the water column, but also on the flow regime being quiet enough to allow the ice to settle at the ice base. As the flow regime usually depends on the basal topography, the deposition is often highly localized. For example, a reduction in basal slope reduces the forcing on the buoyant plume, thereby slowing it down and possibly allowing any ice platelets to be deposited.

Deposits of marine ice become part of the ice shelf itself, flowing with the overlying meteoric ice. This means that, although the marine ice is deposited in well-defined locations, it moves towards the ice front with the flow of the ice and may or may not all be melted off by the time it reaches the ice front. Icebergs that have calved from Amery Ice Front frequently roll over and reveal a thick layer of marine ice. Impurities in marine ice result in different optical properties, and these bergs are often termed 'green icebergs'.

Ice cores obtained from the central parts of the Amery and Ronne ice shelves have provided other direct evidence of the production of marine ice. The interface between the meteoric and marine ice is clearly visible – the ice changes from being white and bubbly, to clear and bubble-free. Unlike normal sea ice, which typically has a salinity of a few PSU, the salinity of marine ice was found to be below 0.1 PSU. The salinity in the cores is highest at the interface itself, decreasing with increasing depth. A different type of marine ice was found at the base of the Ross Ice Shelf. There, a core from near the base showed 6 m of congelation ice with a salinity of between 2 and 4 PSU. Congelation ice differs from marine ice in its formation mechanism, growing at the interface directly rather than being created as an accumulation of ice crystals that were originally formed in the water column.

Airborne downward-looking radar campaigns have mapped regions of ice shelf that are underlain by marine ice. The meteoric (freshwater) ice/marine ice interface returns a characteristically weak echo, but the return from marine ice/ocean boundary is generally not visible. By comparing the thickness of meteoric ice found using the radar with the surface elevation of the freely floating ice shelf, it is possible to calculate the thickness of marine ice accreted at the base. In some parts of the Ronne Ice Shelf basal accumulation rates of around $1\,\mathrm{m\,a^{-1}}$ result in a marine ice layer over 300 m thick, out of a total ice column depth of 500 m. Accumulation rates of that magnitude would be expected to be associated with high ISW fluxes. However, cruises along the Ronne Ice Front have been unsuccessful in finding commensurate ISW outflows.

Internal recirculation Three-dimensional models of the circulation beneath the Ronne Ice Shelf have revealed the possibility of an internal recirculation of ISW. This mode of circulation is driven by the difference in melting point between the deep ice at the grounding line, and the shallower ice in the central region of the ice shelf. The possibility of such a recirculation is indicated in **Figure 5** by the broken line. Intense deposition of ice in the freezing region salinifies the water column sufficiently to allow it to drain back towards the grounding line. In three dimensions, the recirculation consists of a gyre occupying a basin in the topography of water column thickness. The model predicts a gyre strength of around one Sverdrup ($10^6\,\mathrm{m^3\,s^{-1}}$).

This mode of circulation is effectively an 'ice pump' transporting ice from the deep grounding line regions to the central Ronne Ice Shelf. The mechanism does not result in a loss or gain of ice overall. The heat used to melt the ice at the grounding line is later recovered in the freezing region. The external heat needed to maintain the recirculation is therefore only the heat to warm the ice to the freezing point before it is melted. Ice leaves the continent at a temperature of around $-30^\circ\mathrm{C}$, and has a specific heat capacity of around $2010\,\mathrm{J\,kg^{-1}\,{}^\circ C^{-1}}$. As the latent heat of ice is $335\,\mathrm{kJ\,kg^{-1}}$, the heat required for warming is less than 20% of that required for melting. To support an internal redistribution of ice therefore requires a small fraction of the external heat that would be needed to melt and remove the ice from the system entirely. A corollary is that a recirculation of ISW effectively decouples much of the ice shelf base from external forcings that might be imposed, for example, by climate change.

Apart from the lack of a sizable ISW outflow from beneath the Ronne Ice Front, evidence in support of an ISW recirculation deep beneath the ice shelf is scarce, as it would require observations beneath the ice. Direct measurements of conditions beneath ice shelves are limited to a small number of sites. Fissures through George VI and Fimbul ice shelves (**Figure 1**) have allowed instruments to be deployed with varying degrees of success. The more important ice shelves, such as the Ross, Amery and Filchner-Ronne system have no naturally occurring access points. Instead, access holes have to be created using hot water, or other thermal-type drills. In the late 1970s researchers used various drilling techniques to gain access to the cavity at one location beneath the

Ross Ice Shelf before deploying various instruments. During the 1990s several access holes were made through the Ronne Ice Shelf, and data from these have lent support both to the external mode of circulation, and most recently, to the internal recirculation mode first predicted by numerical models.

Warm regime external ventilation The flooding of the Bellingshausen and Amundsen seas' continental shelf by barely modified CDW results in very high basal melt rates for the ice shelves in that sector. The floating portion of Pine Island Glacier (**Figure 1**) has a mean basal melt rate estimated to be around 12 m a^{-1}, compared with estimates of a few tens of centimeters per year for the Ross and Filchner-Ronne ice shelves. Basal melt rates for Pine Island Glacier are high even compared with other ice shelves in the region. George VI Ice Shelf on the west coast of the Antarctic Peninsula, for example, has an estimated mean basal melt rate of 2 m a^{-1}. The explanation for the intense melting beneath Pine Island Glacier can be found in the great depth at the grounding line. At over 1100 m, the ice shelf is 700 m thicker than George VI Ice Shelf, and this results in not only a lower freezing point, but also steeper basal slopes. The steep slope provides a stronger buoyancy forcing, and therefore greater turbulent heat transfer between the water and the ice.

The pattern of circulation in the cavities beneath warm regime ice shelves is significantly different to its cold regime counterpart. Measurements from ice front cruises show an inflow of warm CDW (+1.0°C), and an outflow of CDW mixed with glacial melt water. **Figure 6** shows a two-dimensional schematic of this mode of circulation. Over the open continental shelf the ambient water column consists of CDW overlain by colder, fresher water left over from sea ice production during the previous winter. Although the melt water-laden outflow is colder, fresher, and of lower density than the inflow, it is typically warmer and saltier than the overlying water, but of similar density. Somewhat counter-intuitively, therefore, the products of sub-glacial melt are often detected over the open continental shelf as relatively warm and salty intrusions in the upper layers. Again, measurements of oxygen isotope ratio, and also helium, provide the necessary confirmation that the upwelled CDW contains melt water from the base of ice shelves. In the case of warm regime ice shelves, melt water concentrations can be as high as a few percent.

Tidal Forcing

Except for within a few ice thicknesses of grounding lines, ice shelves float freely in the ocean, rising and falling with the tides. Tidal waves therefore propagate through the ice shelf-covered region, but are modified by three effects of the ice cover: the ice shelf base provides a second frictional surface, the draft of the ice shelf effectively reduces the water column thickness, and the step change in water column thickness at the ice front presents a topographic feature that has significant consequences for the generation of residual tidal currents and the propagation of topographic waves along the ice front.

Conversely, tides modify the oceanographic regime of sub-ice shelf cavities. Tidal motion helps transfer heat and salt beneath the ice front. This is a result

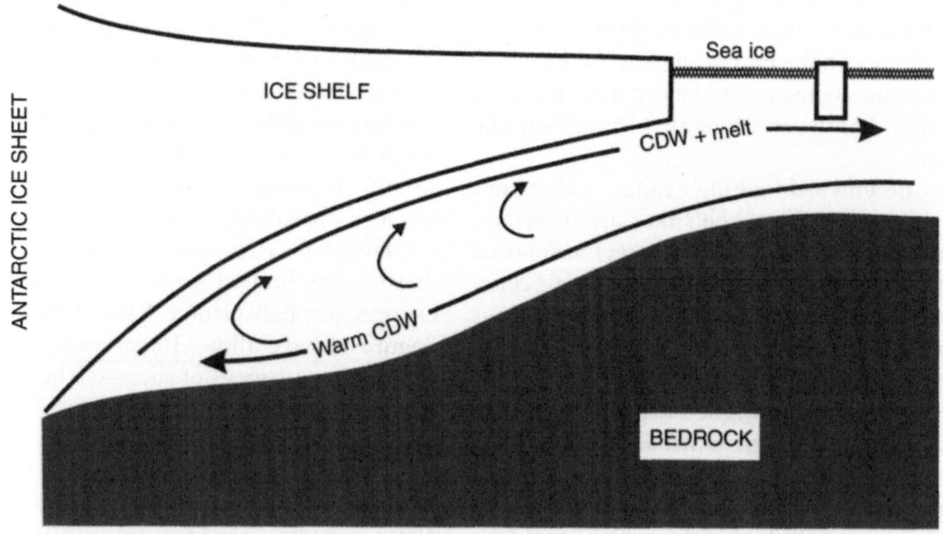

Figure 6 Schematic of the thermohaline mode of sub-ice shelf circulation for a warm regime ice shelf.

both of the regular tidal excursions, which take water a few kilometers into the cavity, and of residual tidal currents which, in the case of the Filchner-Ronne Ice Shelf, help ventilate the cavity far from the ice front. The effect of the regular advection of potentially seasonally warmed water from seaward of the ice shelf is to cause a dramatic increase in basal melt rates in the vicinity of the ice front. Deep beneath the ice shelf, tides and buoyancy provide the only forcing on the regime. Tidal activity contributes energy for vertical mixing, which brings the warmer, deeper waters into contact with the base of the ice shelf. **Figure 7A** shows modeled tidal ellipses for the M_2 semidiurnal tidal constituent for the southern Weddell Sea, including the sub-ice shelf domain. A map of the modeled residual currents for the area of the ice shelf is shown in **Figure 7B**. Apart from the activity near the ice front itself, a residual flow runs along the west coast of Berkner Island, deep under the ice shelf. However, this flow probably makes only a minor contribution to the ventilation of the cavity.

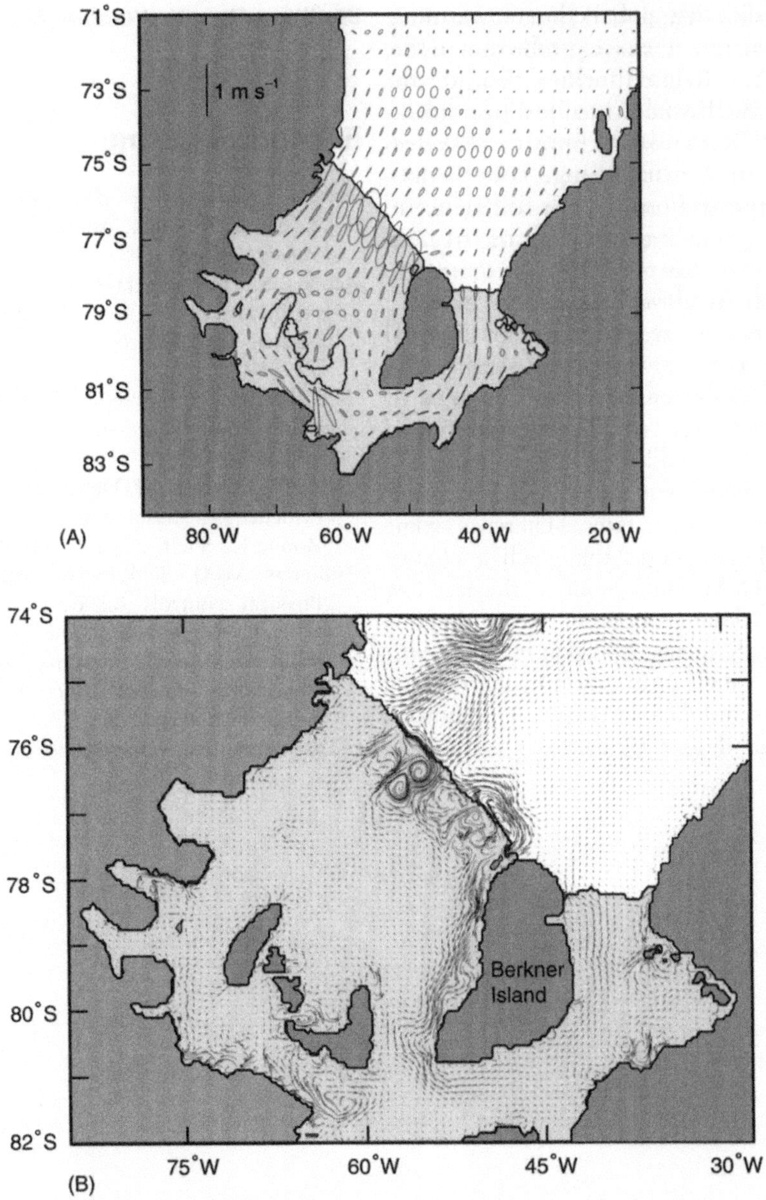

Figure 7 Results from a tidal model of the southern Weddell Sea, in the vicinity of the Ronne Ice Shelf. (A) The tidal ellipses for the dominant M_2 species. (B) Tidally induced residual currents.

How Does the Interaction between Ice Shelves and the Ocean Depend on Climate?

The response to climatic changes of sub-ice shelf circulation depends on the response of the oceanographic conditions over the open continental shelf. In the case of cold regime continental shelves, a reduction in sea ice would lead to a reduction in HSSW production. Model results, together with the implications of seasonality observed in the circulation beneath the Ronne Ice Shelf, suggest that drainage of HSSW beneath local ice shelves would then reduce, and that the net melting beneath those ice shelves would decrease as a consequence. Some general circulation models predict that global climatic warming would lead to a reduction in sea ice production in the southern Weddell Sea. Reduced melting beneath the Filchner-Ronne Ice Shelf would then lead to a thickening of the ice shelf. Recirculation beneath ice shelves is highly insensitive to climatic change. The thermohaline driving is dependent only on the difference in depths between the grounding lines and the freezing areas. A relatively small flux of HSSW is required to warm the ice in order to allow this mode to operate.

The largest ice shelves are in a cold continental shelf regime. If intrusions of warmer off-shelf water were to become more dominant in these areas, or if the shelf-break barrier were to collapse entirely and the regime switch from cold to warm, then the response of the ice shelves would be a dramatic increase in their basal melt rates. There is some evidence from sediment cores that such a change might have occurred at some point in the last few thousand years in what is now the warm regime Bellingshausen Sea. Evidence also points to the possibility that one ice shelf in that sector, the floating extension of Pine Island Glacier (**Figure 1**), might be a remnant of a much larger ice shelf.

During glacial maxima the Antarctic ice sheet thickens and the ice shelves become grounded. In many cases they ground as far as the shelf-break. There are two effects. The continental shelf becomes very limited in extent, and so there is little possibility for the production of HSSW; and where the ice shelves overhang the continental shelf-break, the only possible mode of circulation will be the warm regime mode. Substantial production of ISW during glacial conditions is therefore unlikely.

See also

Bottom Water Formation. Current Systems in the Southern Ocean. Weddell Sea Circulation.

Further Reading

Jenkins A and Doake CSM (1991) Ice–ocean interactions on Ronne Ice Shelf, Antarctica. *Journal of Geophysical Research* 96: 791–813.

Jacobs SS, Hellmer HH, Doake CSM, Jenkins A, and Frolich RM (1992) Melting of ice shelves and the mass balance of Antarctica. *Journal of Glaciology* 38: 375–387.

Nicholls KW (1997) Predicted reduction in basal melt rates of an Antarctic ice shelf in a warmer climate. *Nature* 388: 460–462.

Oerter H, Kipfstuhl J, Determann J, *et al.* (1992) Ice-core evidence for basal marine shelf ice in the Filchner-Ronne Ice Shelf. *Nature* 358: 399–401.

Williams MJM, Jenkins A, and Determann J (1998) Physical controls on ocean circulation beneath ice shelves revealed by numerical models. In: Jacobs SS and Weiss RF (eds.) *Ocean, Ice, and Atmosphere: Interactions at the Antarctic Continental Margin, Antarctic Research Series* 75, pp. 285–299. Washington DC: American Geophysical Union.

THE OCEANS: THE ARCTIC OCEAN

ARCTIC OCEAN CIRCULATION

B. Rudels, Finnish Institute of Marine Research, Helsinki, Finland

Introduction

The Arctic Ocean is the northernmost part of the Arctic Mediterranean Sea, which also comprises the Greenland Sea, the Iceland Sea, and the Norwegian Sea (the Nordic Seas) and is separated from the North Atlantic by the 500–850-m-deep Greenland–Scotland Ridge. The Arctic Ocean is a small, $9.4 \times 10^6\,km^2$, enclosed ocean. Its boundaries to the south are the Eurasian continent, Bering Strait, North America, Greenland, Fram Strait, and Svalbard. The shelf break from Svalbard southward to Norway closes the boundary. The Arctic Ocean lies almost entirely north of and occupies most of the region north of the polar circle. More than half of its area, 53%, consists of large, shallow shelves, the broad Eurasian marginal seas: the Barents Sea (200–300 m), the Kara Sea (50–100 m), the Laptev Sea (<50 m), the East Siberian Sea (<50 m) and the Chukchi Sea (50–100 m), and the narrower shelves north of North America and Greenland. The deep Arctic Ocean comprises two major basins, the Eurasian Basin and the Canadian (also called Amerasian) Basin, separated by the approximately 1600-m-deep Lomonosov Ridge. The Eurasian Basin is further divided into the Nansen and Amundsen basins by a mid-ocean ridge (the Gakkel Ridge), while the Canadian Basin is separated by the Alpha Ridge and the Mendeleyev Ridge into the Makarov and the Canada Basins. The Amundsen Basin is the deepest (~4500 m), while the maximum depths of the Makarov and the Nansen Basins are ~4000 m. The Canada Basin is slightly shallower (~3800 m) but by far the largest (**Figure 1**).

The Arctic Ocean water masses are primarily of Atlantic origin. Atlantic waters (AWs) enter the Arctic Ocean from the Nordic Seas through the 2600-m-deep Fram Strait and over the ~200-m-deep sills in the Barents Sea. The Arctic Ocean also receives low-salinity Pacific water through the shallow (45 m) and narrow (50 km) Bering Strait. The outflows occur through Fram Strait and through the shallow (150–230 m) and narrow channels in the Canadian Arctic Archipelago.

The physical oceanography of the Arctic Ocean is shaped by the severe high-latitude climate and the large freshwater input from runoff, ~0.1 Sv ($1\,Sv = 1 \times 10^6 m^3 s^{-1}$), and net precipitation, ~0.07 Sv. The Arctic Ocean is a strongly salinity stratified ocean that allows the surface water to cool to freezing temperature and ice to form in winter and to remain throughout the year in the central, deep part of the ocean.

The water masses in the Arctic Ocean can, because of the stratification, be identified by different vertical layers. Here five separate layers will be distinguished:

1. The ~50-m-thick upper, low-salinity polar mixed layer (PML) is homogenized in winter by freezing and haline convection, while in summer the upper 10–20 m become diluted by sea ice meltwater. The salinity, S, ranges from 30 to 32.5 in the Canadian Basin and between 32 and 34 in the Eurasian Basin.
2. The 100–250-m-thick halocline, with salinity increasing with depth, while the temperature remains close to freezing, $32.5 < S < 34.5$ (**Figure 2**).
3. The 400–700-m-thick Atlantic layer historically defined as subsurface water with potential temperature, θ, above 0 °C, $34.5 < S < 35$.
4. The intermediate water below the Atlantic layer that communicates freely across the Lomonosov Ridge, $-0.5\,°C < \theta < 0\,°C$, $34.87 < S < 34.92$.
5. The deep and bottom waters in the different basins, $-0.55\,°C < \theta < -0.5\,°C$, $34.92 < S < 34.96$ (Canadian Basin), $-0.97\,°C < \theta < -0.5\,°C$, $34.92 < S < 34.945$ (Eurasian Basin).

There are large lateral variations of the characteristics in these layers that depend upon the circulation and upon the mixing processes in the Arctic Ocean (**Figures 2** and **3**). A more detailed classification, especially for the deeper water masses, is given in **Table 1**. It should be kept in mind that this classification is not unique and that several others exist in the literature.

Circulation

The circulation of the uppermost layers of the Arctic Ocean has mainly been inferred from the ice drift, determined from satellites and from drifting buoys, while at deeper levels primarily the distributions of temperature and salinity and more recently of other tracers have been used to deduce the movements of the different water masses, often with

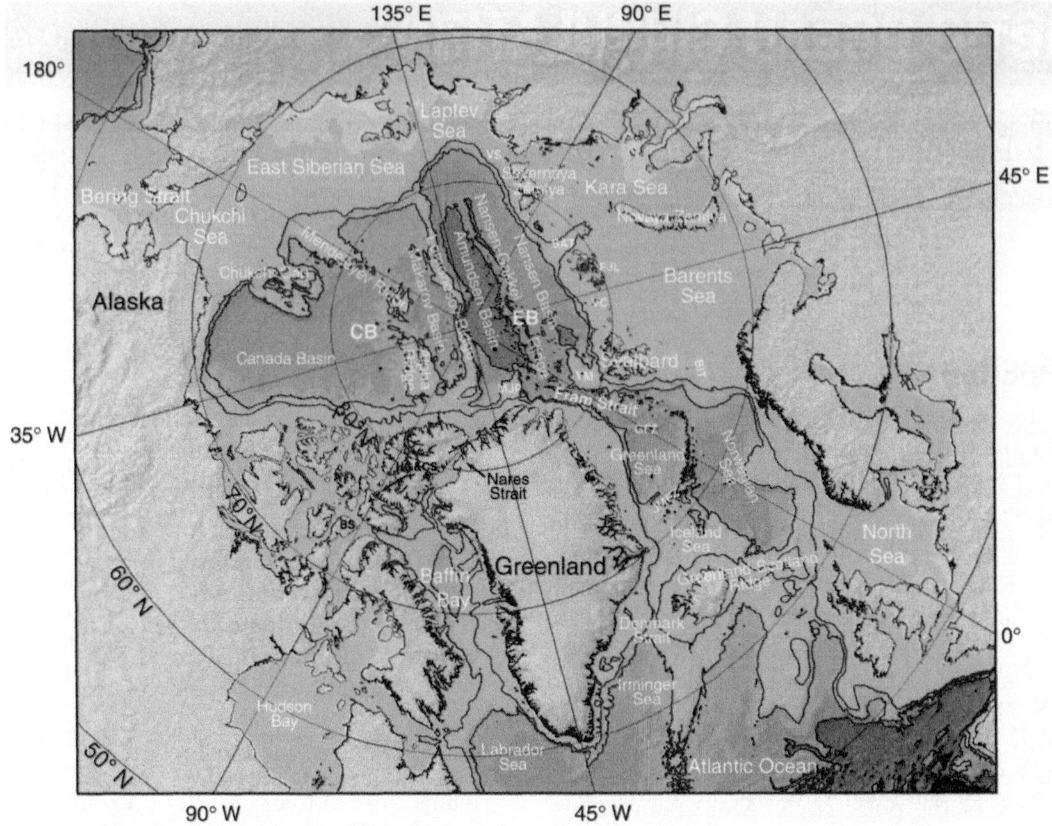

Figure 1 Map of the Arctic Mediterranean Sea showing geographical and bathymetric features. The bathymetry is from IBCAO (the International Bathymetric Chart of the Arctic Ocean; Jakobsson *et al.*, 2000; 2008) and the projection is Lambert equal area. The 500 and 2000 m isobaths are shown. All maps used here are made by Martin Jakobsson (personal communication). BIT, Bear Island Trough; CB, Canadian Basin; EB, Eurasian Basin; GFZ, Greenland Fracture Zone; MJP, Morris Jessup Plateau; JMFZ, Jan Mayen Fracture Zone; SAT, St. Anna Trough; YM, Yermak Plateau; VC, Victoria Channel; VS, Vilkiltskij Strait; FJL, Franz Josef Land; BS, Barrow Strait; HG & CS, Hell Gate and Cardigan Sound.

the assumption that the circulation is largely controlled by the bathymetry. Direct, moored current measurements have been scarce and mostly confined to the continental slope and to the Lomonosov Ridge. These measurements have confirmed the importance of the bathymetry for the circulation. In the deep basins current measurements have been made from drifting ice camps and more recently also from autonomous ice-tethered platforms, relaying the observations via satellite to shore. Subsurface drifters are just beginning to be used and observational efforts during the International Polar Year (IPY) 2007–09 are likely to significantly increase the knowledge of the circulation in the Arctic Ocean.

The motions of the ice cover and the surface water are predominantly forced by the wind, and the atmospheric high-pressure cell over the Arctic creates the anticyclonic Beaufort gyre in the Canada Basin. Ice leaks from the offshore side of the gyre and joins the Transpolar Drift (TPD) that brings ice from the Canada Basin toward Fram Strait. A second branch originating from the Siberian shelves, mainly from the Laptev Sea, carries ice across the Eurasian Basin. About 90% of the ice export (0.09 Sv) passes through Fram Strait (**Figure 4**).

The PML and the halocline are maintained by river runoff, ice melt, and the inflow of low-salinity water through Bering Strait. The lowest surface salinities and the thickest halocline are therefore observed in the Canada Basin. The inflow through Bering Strait, although affected by local winds, is in the last instance driven by a higher sea level in the North Pacific as compared to the Arctic Ocean. This creates a pressure gradient that forces the Pacific water northward into the Arctic Ocean. The Bering Strait inflow continues across the Chukchi Sea in four branches. One branch enters the East Siberian Sea, while one of the central inflow branches enters the Arctic Ocean along the Herald Canyon west of the Chukchi Plateau, and the other passes via the Central Gap east of the Chukchi Plateau into the Canada Basin. The easternmost branch reaches the Canada Basin along the Barrow Canyon close to Alaska. River runoff, mainly from the Mackenzie

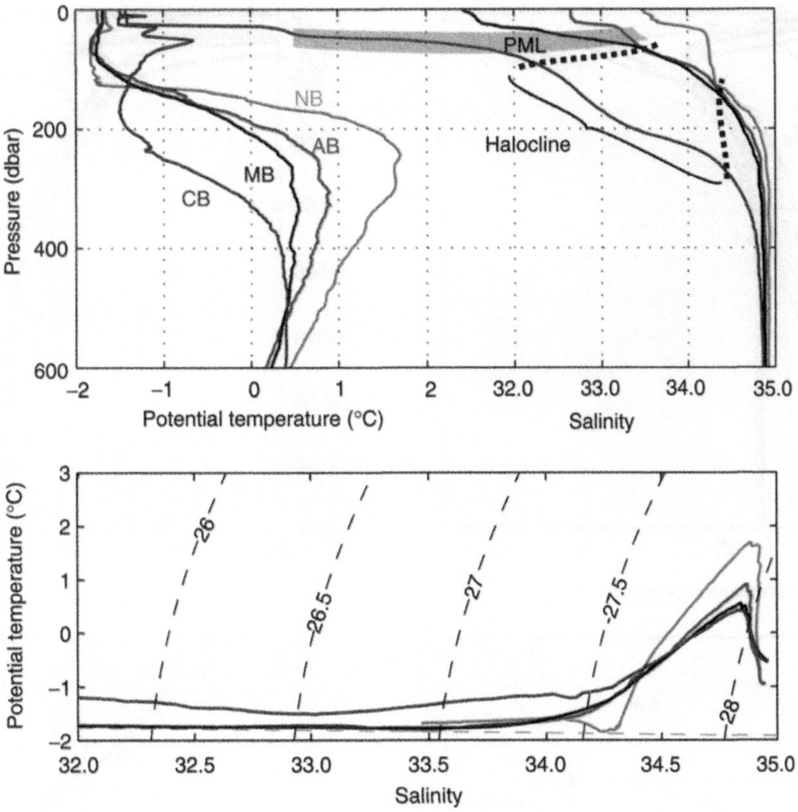

Figure 2 Potential temperature and salinity profiles and θS curves from the upper layers of the Nansen Basin (NB, dark yellow), Amundsen Basin (AB, green), Makarov Basin (MB, magenta), and Canada Basin (CB, blue). The PML and the halocline are indicated in the salinity profiles. Above the PML, the low-salinity layer due to seasonal ice melt is seen. The temperature maximum in the Canada Basin is due to the presence of Bering Strait Summer Water (BSSW) and the temperature minimum below indicates the upper halocline with $S \sim 33.1$, deriving from the colder, more saline Bering Strait Winter Water (BSWW) and from brine release and haline convection in the Chukchi Sea. No halocline is present in the Nansen Basin, only a deep winter mixed layer between the thermocline and the seasonal ice melt layer with temperature close to freezing. The curved shape of the Nansen Basin thermocline as seen in the θS diagram suggests wintertime haline convection with dense, saline parcels penetrating into the thermocline. Adapted from Rudels B, Jones EP, Schauer U, and Eriksson P (2004) Atlantic sources of the Arctic Ocean surface and halocline waters. *Polar Research* 23: 181–208.

River, adds freshwater to the Canada basin. The runoff peaks in early summer (June). The river runoff as well as most of the Pacific water becomes trapped in the anticyclonic Beaufort gyre, forming an oceanic high-pressure cell in the southern Canada Basin. The Pacific water leaves the Beaufort gyre and the Arctic Ocean mainly through the Canadian Arctic Archipelago, but a smaller fraction also exits, at least intermittently, through Fram Strait.

Warm AW crosses the Greenland–Scotland Ridge and continues toward the Arctic Ocean in the Norwegian Atlantic Current. The Norwegian Atlantic Current splits as it reaches the Bear Island Trough. One part enters the Barents Sea together with the Norwegian Coastal Current, which carries low-salinity water from the Baltic Sea and runoff from the Norwegian coast. The remaining part continues as the West Spitsbergen Current to Fram Strait. There the current again splits. Some AW recirculates westward in the strait to join the southward-flowing East Greenland Current, and the rest enters the Arctic Ocean in two streams. One stream flows over the Svalbard shelf and slope, the other passes west and north around the Yermak Plateau and then continues eastward, eventually joining the inner stream at the continental slope east of Svalbard. The deeper outer stream also transports intermediate and deep waters from the Nordic Seas into the Arctic Ocean.

As the AW enters the Arctic Ocean, it encounters, and melts, sea ice, and the upper part becomes colder and less saline. In winter this upper layer is homogenized by convection and mechanical stirring and cooled to freezing temperature. The salinity of the upper layer is 34.2–34.4, and it is advected with the warm AW core eastward along the Eurasian continental slope. In summer, seasonal ice melt creates a low-salinity upper layer, which is removed in

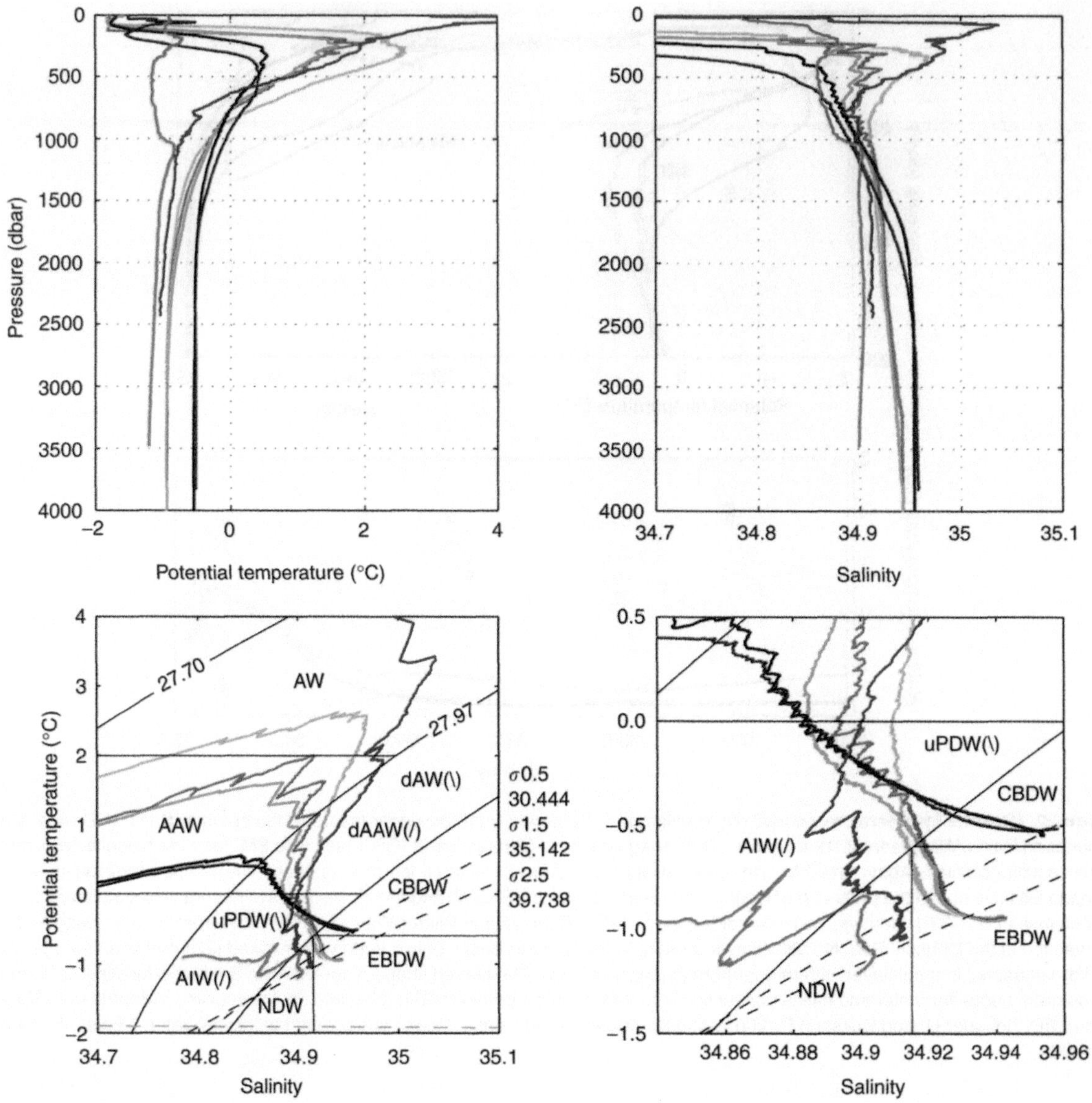

Figure 3 Characteristics of the water columns in different parts of the Arctic Mediterranean. Upper row: potential temperature and salinity profiles; lower row: θS curves. Green: The Greenland Sea, the ultimate source of the Arctic Intermediate Water (AIW) and the Nordic Seas Deep Water (NDW) entering the Arctic Ocean. Red: The West Spitsbergen Current in Fram Strait carrying warm Atlantic Water (AW), AIW, and NDW into the Arctic Ocean. Dark yellow: The Fram Strait branch at the continental slope of the Nansen Basin. Magenta: The interior Nansen Basin. Cyan: The Amundsen Basin. Black: The Makarov Basin. Blue: The Canada Basin. The shift in depth of the temperature maximum between the Makarov Basin and the Canada Basin is due to the stronger presence of Pacific water (PW) in the Canada Basin, displacing the deeper part of the water column. Note that the Canadian Basin deep waters becomes warmer than the Eurasian Basin deep waters below 1000 m and more saline below 1500 m, above the sill depth of the Lomonosov Ridge. In the θS diagrams, (/) indicates a stratification unstable in temperature or salinity, (\) indicates a stratification stable in both components. AAW, Arctic Atlantic Water; dAW, dense Atlantic Water; dAAW, dense Arctic Atlantic Water; uPDW, upper Polar Deep Water; CBDW, Canadian Basin Deep Water; EBDW, Eurasian Basin Deep Water. The $\sigma_{1.5}$ isopycnal shows the density at the sill depth of the Lomonosov Ridge and the $\sigma_{2.5}$ isopycnal the density at sill depth in Fram Strait.

winter by freezing and the upper layer is homogenized down to the thermocline. No cold halocline is present between the mixed layer and the AW in the Nansen Basin (see **Figure 2**). The Arctic Intermediate Water (AIW) and the Nordic Seas Deep Water (NDW) that enter the Arctic Ocean in the West Spitsbergen Current can be identified west and north of the Yermak Plateau as less saline and colder anomalies. Farther to the east these signals in temperature and salinity gradually disappear, but signs of the Nordic Seas waters can still be detected by other tracers, for example, CFCs.

Table 1 Simplified water mass classification for the Arctic Ocean

Water masses	*Abbreviation*	*Definition*	*Main source or origin*
Upper waters ($\sigma_\theta < 27.70$)			
Polar mixed layer	PML	$32 < S < 34$	Arctic Ocean
(Upper) halocline		$32.5 < S < 33.5$	Chukchi Sea
(Lower) halocline		$33.5 < S < 34.5$	Nansen Basin, Barents Sea
Intermediate waters I ($27.70 < \sigma_\theta < 27.97$)			
Atlantic Water	AW	$2 < \theta$	West Spitsbergen Current
Arctic Atlantic Water	AAW	$0 < \theta < 2$	Arctic Ocean (transformed)
Intermediate waters II ($27.97 < \sigma_\theta$, $\sigma_{0.5} < 30.444$)			
Dense Atlantic Water	DAW	$0 < \theta$, unstable in S (/)	West Spitsbergen Current
Dense Arctic Atlantic Water	DAAW	$0 < \theta$, stable in θ and S (\)	Arctic Ocean (transformed)
Arctic Intermediate Water	AIW	$\theta < 0$, unstable in S or θ (/)	Greenland Sea
Upper Polar Deep Water	UPDW	$\theta < 0$, stable in S and θ (\)	Arctic Ocean
Deep waters ($30.444 < \sigma_{0.5}$)			
Nordic Seas Deep Water	NDW	$S < 34.915$	Greenland Sea
Canadian Basin Deep Water	CBDW	$-0.6 < \theta$, $34.915 < S$	Canadian Basin
Eurasian Basin Deep Water	EBDW	$\theta < -0.6$, $34.915 < S$	Eurasian Basin

In the Barents Sea, the AW remains in contact with the atmosphere during most of its transit. It is cooled significantly and becomes freshened by net precipitation and by the melting of sea ice. Some of the water entering the Barents Sea returns as colder, denser water to the Norwegian Sea in the Bear Island Channel, but the major part reaches the eastern Barents Sea. In the Barents Sea, the AW becomes separated into three different water masses. (1) The bulk of the inflow is cooled and freshened by air–sea interaction and becomes denser. (2) Some of the AW reaches the shallow areas west of Novaya Zemlya and becomes transformed into saline, dense bottom water by the ice formation and brine rejection. (3) The upper part of the AW interacts with sea ice and a less saline, upper layer is formed by ice melt, which in winter becomes homogenized down to the thermocline. These waters all enter the Arctic Ocean, mainly by passing between Novaya Zemlya and Franz Josef Land into the Kara Sea and then continuing in the St. Anna Trough to the Arctic Ocean. However, some water reaches the Arctic Ocean directly from the Barents Sea along the Victoria Channel.

The Barents Sea branch follows the eastern side of the St. Anna Trough and then continues along the continental slope as an almost 1000-m-thick, cold, and weakly stratified water column, displacing the warmer, more saline Fram Strait inflow branch from the slope and deflecting the denser basin waters toward larger depths. Strong isopycnal mixing between the two branches takes place and inversions and irregular intrusive layers are formed (**Figure 5**, right column).

North of the Laptev Sea the two inflow branches have largely merged and bottom temperatures above 0 °C indicate that Fram Strait branch water again is present at the slope. The boundary current splits at the Lomonosov Ridge with one part continuing into the Canadian Basin, the rest returning along the ridge and the Amundsen Basin toward Fram Strait. The AW temperature in the boundary current is significantly reduced already north of the Laptev Sea. This could partly be due to mixing between the two branches and partly be the result of heat loss to the upper layers and the ice. However, it may also be an indication that some of the Fram Strait branch recirculates already in the Nansen Basin. The returning Fram Strait branch then becomes more prominent in the northern Nansen Basin, while the Barents Sea branch dominates in the Amundsen Basin and along the Lomonosov Ridge (**Figure 5**, left column).

The bulk of the Barents Sea branch in the St. Anna Trough is denser, colder, and less saline than the Atlantic layer and forms a distinct salinity minimum beneath the AW in the Amundsen Basin and above the Gakkel Ridge. This minimum identifies the upper Polar Deep Water (uPDW) in the Eurasian Basin and is located higher in the water column and is more distinct than the AIW minimum in the Nansen Basin deriving from the Nordic Seas. This is about the only area of the Arctic Ocean where a part of the water column is unstably stratified in salinity (**Figures 3** and **5**).

One part of the Barents Sea inflow, mostly comprising water from the Norwegian Coastal Current, enters the Kara Sea through the Kara Strait south of Novaya Zemlya. It mixes with the runoff from Ob

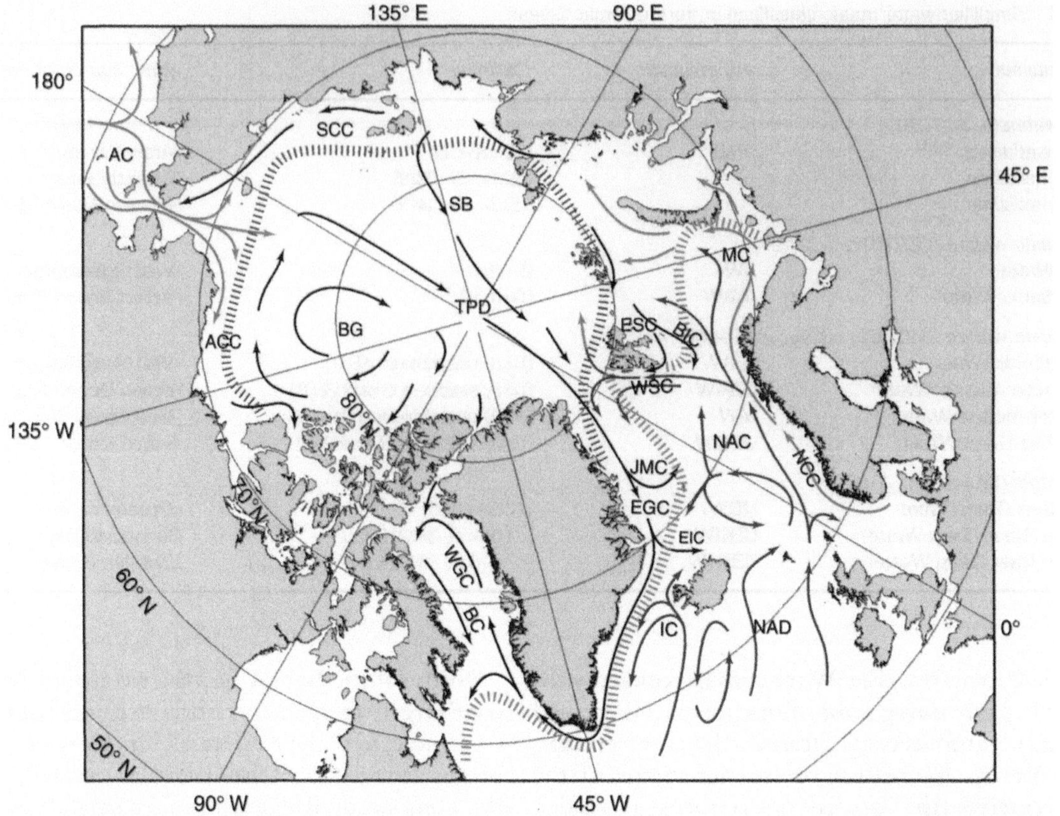

Figure 4 The circulation of the upper layers of the Arctic Mediterranean Sea. Warm Atlantic currents are indicated by red arrows, cold, less-saline polar and arctic currents by blue arrows. Low-salinity transformed currents are shown by green arrows. The maximum ice extent is shown by a blue and the minimum ice extent by red striped line. AC, Anadyr Current; ACC, Alaskan Coastal Current; BC, Baffin Current; BIC, Bear Island Current; BG, Beaufort gyre; EGS, East Greenland Current; EIC, East Iceland Current; ESC, East Spitsbergen Current; IC, Irminger Current; JMC, Jan Mayen Current; MC, Murman Current; NAD, North Atlantic Drift; NAC, Norwegian Atlantic Current; NCC, Norwegian Coastal Current; SB, Siberian branch (of the Transpolar Drift); SCC, Siberian Coastal Current; TPD, Transpolar Drift; WGC, West Greenland Current; WSC, West Spitsbergen Current.

and Yenisey, forming the low-salinity water present on the Kara Sea shelf. Most of this shelf water flows through the Vilkiltskij Strait into the Laptev Sea, where it receives additional freshwater, mostly from the Lena River. The main export of shelf water from the Eurasian shelves to the deep basins occurs across the Laptev Sea shelf break into the Amundsen Basin, and the low-salinity shelf water overruns the boundary current and the two inflow branches. The upper parts of the Fram Strait and the Barents Sea branches become isolated from the ice, the sea surface, and the atmosphere and evolve into halocline waters.

The Fram Strait branch supplies the halocline in the Amundsen Basin and the Makarov Basin and the lower halocline, beneath the Pacific water, in the northern part of the Canada Basin. The Barents Sea branch halocline, which remains at the continental slope, is affected by the stronger mixing occurring over the steep topography and becomes warmer by mixing with AW from below. It eventually supplies the lower halocline in the southern Canada Basin. The shelf water outflow supplies the PML in the Amundsen and Makarov Basins but it has high enough salinity also to contribute to the halocline waters in the Canada Basin, beneath most of the Pacific waters (**Figure 6**).

The properties of the Atlantic and intermediate waters in the Canada Basin are distinctly different from those in the Eurasian Basin, indicating that the waters of the boundary current become transformed along their paths in the Canadian Basin. The temperature maximum in the Atlantic layer becomes colder and less saline, while in the intermediate water range the temperature and the salinity increase and the intermediate uPDW salinity minimum disappears. The θS curves below the temperature maximum approach a straight line toward higher salinities and lower temperatures with increasing depth (**Figure 3**).

The boundary current circulates cyclonically around the Canada Basin and splits at the different

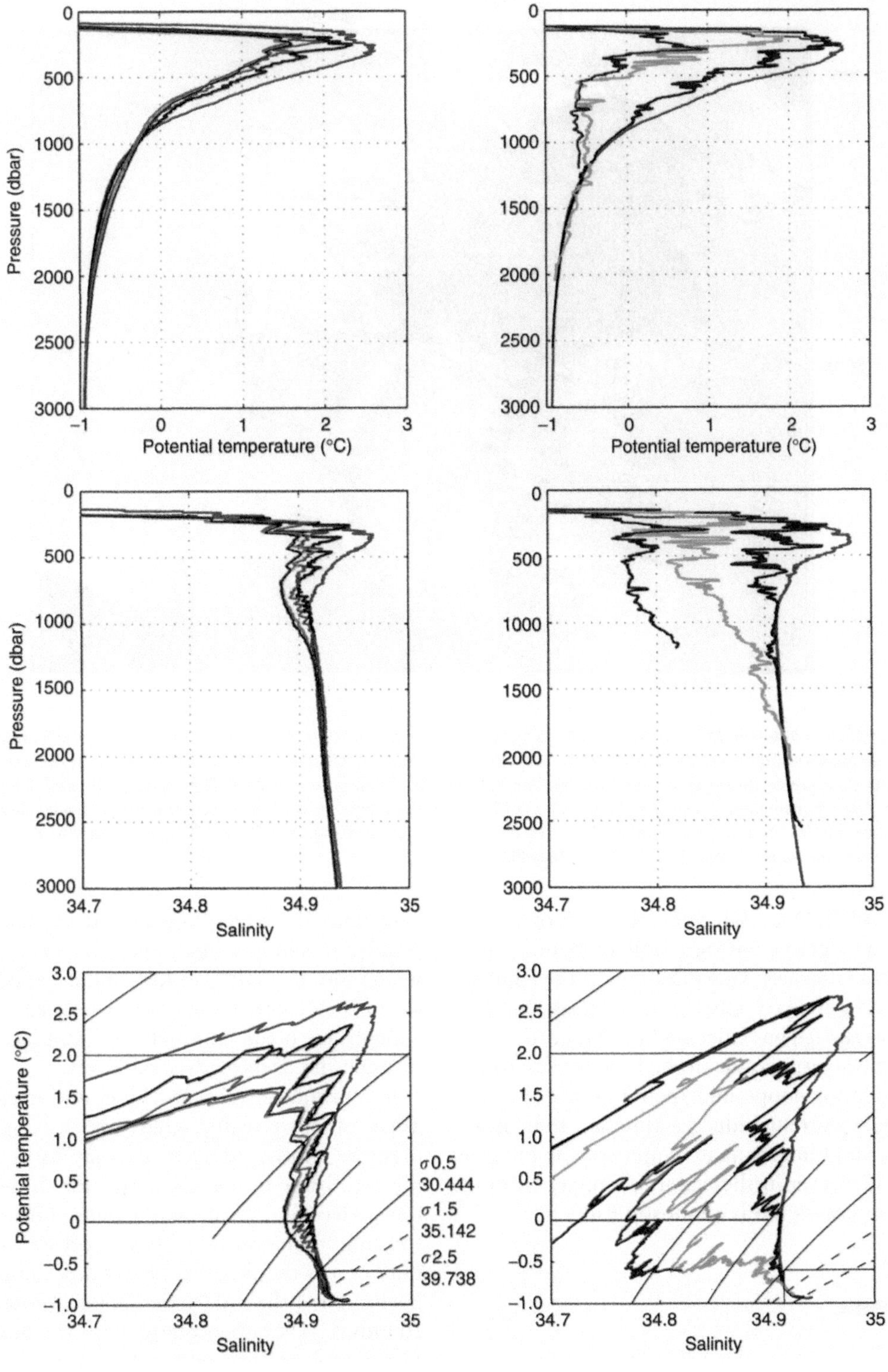

Figure 5 Potential temperature and salinity profiles and θS curves showing the interaction and interleaving between the Fram Strait branch and the Barents Sea branch north of Severnaya Zemlya (right column) and the water properties of the Nansen and Amundsen basins offshore of the Fram Strait branch (left column). Red stations: Fram Strait branch, blue station: the Barents Sea branch, black and cyan stations: active mixing between the branches. Interleaving is present not only in the AW but also in the deeper layers. Offshore of the Fram Strait branch, the warm, saline intrusions in the Nansen Basin (black and magenta stations) suggest a close recirculation of the Fram Strait branch in the Nansen Basin, while the colder, less saline intrusions in the Amundsen Basin (green and blue stations) indicate that the intermediate part of the water column here is dominated by Barents Sea branch waters.

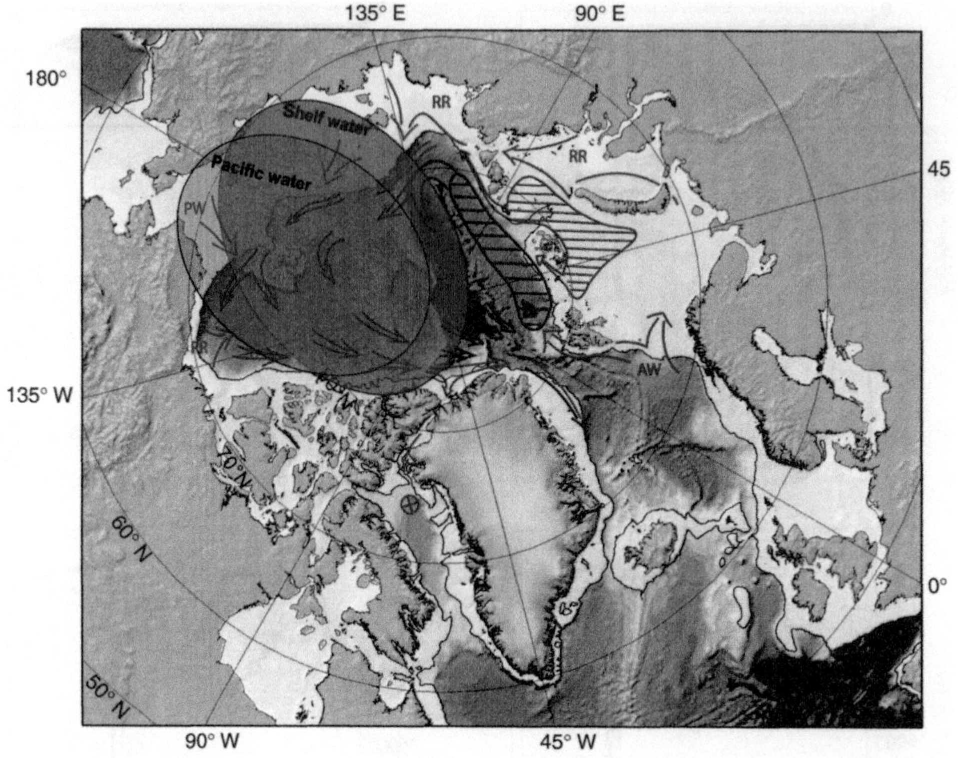

Figure 6 Circulation of the Atlantic-derived halocline waters and the distribution of Pacific water (orange) and Eurasian river runoff (green). The proposed source areas for the Fram Strait branch lower halocline water (black) and the Barents Sea branch lower halocline water (blue), and the circulation of these waters in the Arctic Ocean are indicated. RR, river runoff; PW, Pacific water; AW, Atlantic water. The cross indicates possible contribution of Barents Sea branch lower halocline water to the Baffin Bay bottom water. The isoline shown is 500 m. Based on Rudels B, Jones EP, Schauer U, and Eriksson P (2004) Atlantic sources of the Arctic Ocean surface and halocline waters. *Polar Research* 23: 181–208.

ridges (**Figure** 7). One loop enters the Makarov Basin along the Mendeleyev Ridge, and one loop penetrates into the northern Canada Basin at the Chukchi Plateau. The boundary current also enters the Canada Basin at the Alpha Ridge and the Makarov Basin at the North American side of the Lomonosov Ridge. These circulation loops and their interactions with the boundary current induce a time lag that makes the Atlantic and intermediate waters of the different basins distinct, even if the waters originate from the same inflow across the Lomonosov Ridge.

Processes

The climatic forcing on the Arctic Ocean is strong; the large variation of incoming solar radiation, the severe cooling during the polar night, and the intense weather systems, either locally formed polar lows or cyclones advected from lower latitudes, all contribute in forming the Arctic Ocean characteristics. The strong stability of the deep Arctic Ocean basins confines the effects of these forcing fields to the ice-ocean surface and to the PML, but the shallow Arctic shelves, which experience the largest impact of the forcing, the strong runoff in summer and excessive ice formation in winter, do not only add low-salinity upper water and ice to the central basins but also create waters dense enough to break through the stratification, transforming the deeper advected layers.

Ice formation releases brine, and in lee polynyas, areas of open water where the ice is removed by offshore winds, sufficient ice may form to allow the released brine to overcome the initial low salinity on the shelves and create saline dense bottom waters at freezing temperature. These bottom waters eventually cross the shelf break and descend into the deep basins until their densities match those of the surrounding. They then merge with the ambient water. Less-dense plumes add colder water to the upper layers, supplying water to the halocline.

This occurs in the Chukchi Sea, where the water of the Canada Basin upper halocline with salinity 33.1 is formed. In the Nansen Basin no halocline is present and the density range between the upper mixed layer and the AW is small. The shelf outflows then either enter the mixed layer or sink into and cool the thermocline and the Atlantic layer. The formation of

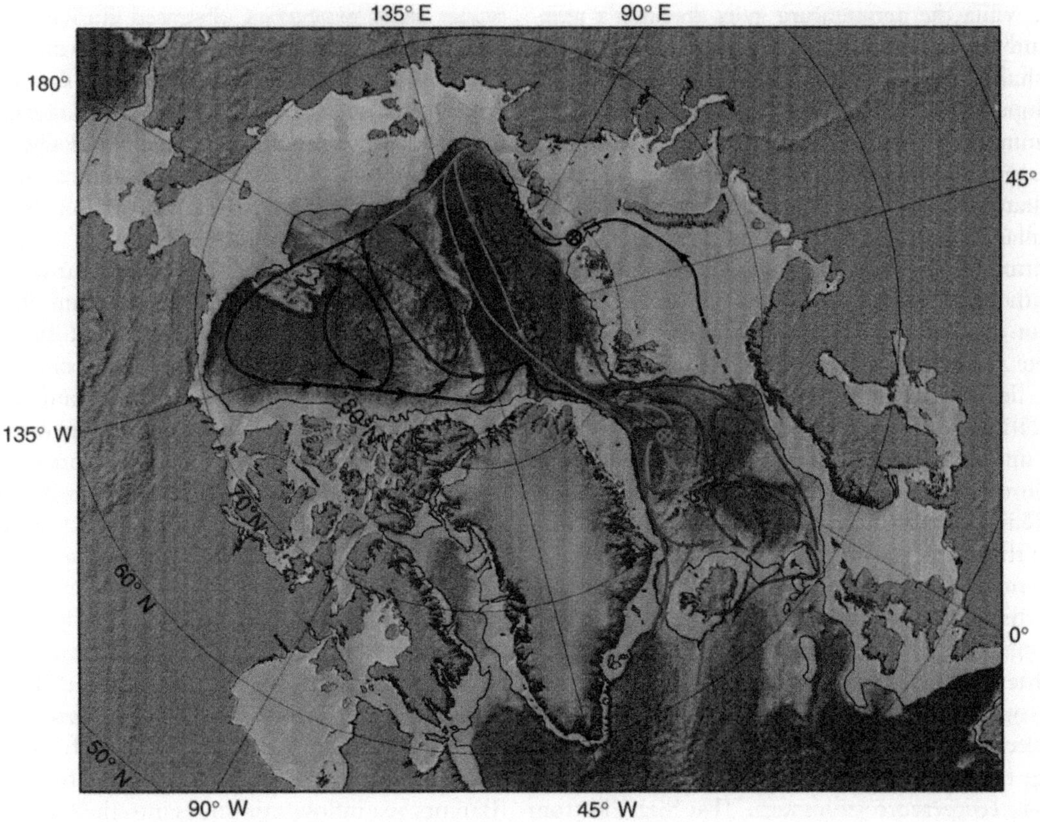

Figure 7 Schematics showing the circulation in the subsurface Atlantic and intermediate layers in the Arctic Mediterranean Sea. The interactions between the Barents Sea and Fram Strait inflow branches north of the Kara Sea as well as the recirculation and different inflow streams in Fram Strait and the overflows across the Greenland–Scotland Ridge are indicated. The isoline shown is 500 m. Based on Rudels B, Jones EP, Anderson LG, and Kattner G (1994) On the intermediate depth waters of the Arctic Ocean. In: Johannesen OM, Muench RD, and Overland JE (eds.) *AGU Geophysical Monographs 85*: *The Polar Oceans and Their Role in Shaping the Global Environment*, pp. 33–46. Washington, DC: American Geophysical Union.

halocline waters in the Eurasian Basin occurs, when the boundary current is overrun by the less saline shelf water, and the initial mixed layers become transformed into halocline waters.

Comparison between the intermediate and deep waters in the Nordic Seas and in the Arctic Ocean shows that the shelf-slope convection not only brings salt but also heat into the deeper layers (**Figures 3** and 5). This implies that the denser, more saline plumes are entraining warmer water, as they pass through the intermediate Atlantic layer. Observations of outflowing, cold and dense water from Storfjorden in southern Svalbard have shown that as the outflow reaches deeper than 1500 m, it has become warmer than the ambient water, supporting the idea of entraining boundary plumes. No plumes have yet been followed from their sources down the slope in the Arctic Ocean, but saline, warmer, and denser bottom layers have been observed deeper than 2000 m at the continental slope north of Severnaya Zemlya.

The deep waters of the Canadian and Eurasian Basins differ significantly, the Canadian Basin Deep Water (CBDW) being warmer and more saline than the Eurasian Basin Deep Water (EBDW). The difference in temperature was first believed due to the presence of the Lomonosov Ridge, blocking the cold deep water from the Nordic Seas, but the temperature difference starts well above the sill depth of the Lomonosov Ridge, showing that heat as well as salt is added to the uPDW and the CBDW in the Canadian Basin.

The differences between the Amundsen and Nansen Basins are small but clear. The Nansen Basin is slightly less saline than the Amundsen Basin between 1600 and 2600 m due to a stronger presence of NDW and AIW. In the Amundsen Basin a mid-depth (1700 m) salinity maximum or at least a sharp bend in the θS curves is observed. It is strongest closer to Greenland but is present over most of the basin. This maximum derives from the CBDW that crosses the Lomonosov Ridge and then penetrates into the central Amundsen Basin from the Greenland continental slope.

In the Nansen, Amundsen, and Canada Basins, the salinity of the deep water increases toward the bottom, while the temperature goes through a minimum and then increases slightly before an isothermal and isohaline bottom layer is reached (**Figure 3**). Shelf-slope convection cannot explain the temperature minimum in the deep waters, but it could be due to advection between the basins. The deep water in the Makarov Basin has at the sill depth (2400 m) of the Alpha Ridge similar θS characteristics as the temperature minimum in the Canada Basin and could supply the minimum. In the Eurasian Basin this explanation does not work. A minimum is present in both the Amundsen and Nansen Basins and these minima lie too deep to derive from the inflow of colder NDW. One possibility is that the St. Anna Trough intermittently conduits colder, denser bottom water, formed in the Barents Sea, into the Nansen Basin. This cold water would enter below the warm layers of the Fram Strait branch and by entraining less heat it could contribute colder water to the deeper part of the water column than the slope convection occurring farther east around Severnaya Zemlya.

Geothermal heat flux has been proposed as an energy source for heating and stirring the bottom layers, keeping them homogenous and gradually increasing their temperature, thereby creating the overlying temperature minimum. The high bottom salinity could then be a remnant from convection events that took place several hundred years ago and brought dense, saline, and cold water to the bottom, where it now is isolated and gradually becomes warmer, thicker, and less dense. Whether the bottom layers are kept homogenous by geothermal heating and convection from below, or if the stirring is due to mechanical mixing as the boundary plumes enter the bottom layer is an open question. However, geothermal heating and ventilation by shelf-slope convection are not two mutually exclusive processes and could operate simultaneously.

The Makarov Basin is different. The salinity becomes constant with depth, while the temperature still decreases, creating a temperature-stratified layer above the isothermal and isohaline bottom water (**Figure 3**). This rules out the gradual heating of an old, fossil bottom layer but requires that colder, rather than more saline, water renews the bottom layer. This then excludes shelf-slope convection as a dominant process. One possibility is that colder water from the Amundsen Basin is brought across the Lomonosov Ridge, either intermittently, forced by, for example, topographically trapped waves, or flowing through yet undetected passages in the ridge. The central part of the Lomonosov Ridge, where such an exchange was believed to occur, has recently been surveyed. A gap was found but not as deep as expected, barely 1900 m, and more critically, the water mass properties observed during the survey indicated that the flow of the densest water was from the Makarov Basin to the Amundsen Basin – in the wrong direction. This inflow of Makarov Basin water followed the Lomonosov Ridge toward Greenland and is likely to be one source, perhaps the most important one, for the mid-depth salinity maximum in the Amundsen Basin.

A flow onto the shelves is needed to compensate for the supply to the PML in summer and the outflow of dense water to the deeper layers and the export of ice in winter. This flow could come from the Arctic Ocean, along the bottom in summer and at the surface in winter, or along horizontally separated inflow and outflow paths across the shelf break. It could also be supplied from behind, like the Bering Strait inflow and the inflow to the Barents Sea. These inflow shelves differ from the inner shelves of the Kara, Laptev, and East Siberian Seas, where a two-way exchange across the shelf break might be more likely, even necessary. However, the difference is not as large as it might appear. Most of the Barents Sea inflow actually passes the Kara Sea before it enters the deep basins and much of the runoff from Ob and Yenisey, together with a substantial fraction of the Barents Sea inflow, continues into the Laptev Sea and then enters either the Amundsen Basin or the East Siberian Sea.

Such eastward flow is the usual fate of river plumes, which tend to follow the coastline eastward. This also holds for the East Siberian Coastal Current, which carries the runoff from perhaps the most 'inner' shelf of the Arctic Ocean, that of the East Siberian Sea, into the Chukchi Sea. The East Siberian Coastal Current occasionally, but rarely, passes south through Bering Strait into the Bering Sea.

The circulation pattern in the Arctic Ocean outlined above gives the impression that water masses are advected, with little mixing, in loops through the different basin, replacing one vintage water mass with another. This is not entirely the case. The structure of the salinity and temperature profiles shows sign of intrusive mixing and the presence of lenses or eddies of anomalous water masses in the water column. Eddies were first reported in the Canadian Basin halocline and were there considered related to the outflow of dense water from the Chukchi Sea. These eddies are mostly anticyclonic, 10–20 km in diameter and highly energetic with maximum velocities above $0.3\,\mathrm{m\,s^{-1}}$. Eddies have subsequently been observed in all water masses of the Arctic Ocean, although the velocities associated with the deeper-lying eddies have so far not been determined. Their water characteristics imply that the eddies have traveled considerable distances as coherent water bodies without their anomalous properties being removed by smaller-scale mixing, for example double-diffusive convection, merging the eddies with the Arctic Ocean water columns in different basins (**Figure 8**).

The intrusive layers are particularly intriguing (**Figures** 5 and 8). The largest property amplitudes are found at the frontal zones but the layers appear to reach over entire basins, making their extent perhaps the largest one observed in the World Ocean. The inversions allow for the release of the potential energy stored in the unstably stratified component by double-diffusive convection, which can drive the interleaving layers across the basins, and this has been suggested as a mechanism for transfer of heat from the boundary current into the deep basins. For the layers to expand, along-layer gradients in heat and salt are required. In the interior of the basins such gradients are often absent, and an alternative explanation for the wide extent of the intrusive layers is that they are formed in the frontal zones, expand initially, driven by double-diffusive fluxes, until the potential energy, available in the stratification, is removed. After this the layers remain as fossil structures that are advected around the main circulation gyres. The intrusions between the Fram Strait and the Barents Sea branch water observed in the Amundsen Basin and over the Gakkel Ridge have been interpreted as fossil structures, initially formed by the interactions between the Fram Strait and the Barents Sea branches north of the Kara Sea and now being advected toward Fram Strait (**Figure** 5).

Interleaving layers are observed in all background stratifications: saltfinger unstable, diffusively unstable, as well as when both components are stably stratified. In the stable–stable situation, disturbances extending across the front are necessary to create the initial inversions. Differential diffusion, taking into account the more rapid diffusion of heat in weakly turbulent surroundings, has been proposed as a mechanism for creating interleaving in a stable–stable stratification. The time required to create layers would then be in the order of years. This appears long, since interleaving layers are found very close to the area where the parent water masses first meet also when both components are stably stratified.

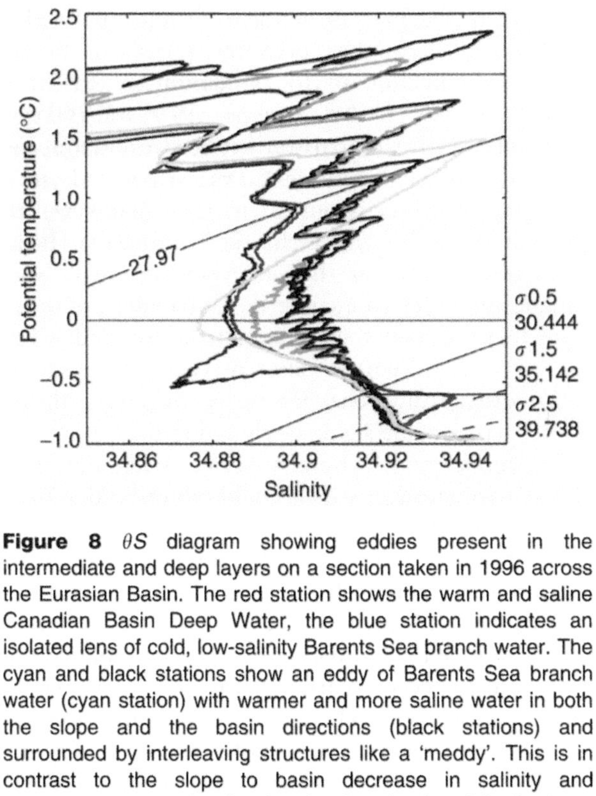

Figure 8 θS diagram showing eddies present in the intermediate and deep layers on a section taken in 1996 across the Eurasian Basin. The red station shows the warm and saline Canadian Basin Deep Water, the blue station indicates an isolated lens of cold, low-salinity Barents Sea branch water. The cyan and black stations show an eddy of Barents Sea branch water (cyan station) with warmer and more saline water in both the slope and the basin directions (black stations) and surrounded by interleaving structures like a 'meddy'. This is in contrast to the slope to basin decrease in salinity and temperature seen in the interleaving in the Atlantic layer (although here also an eddy (not shown) was detected). Finally an eddy of warm, saline Fram Strait branch water, yellow station, was present in the colder, Barents Sea branch dominated Atlantic layer in the Amundsen Basin.

Variability

Until 1990 an underlying assumption has been that the Arctic Ocean, at least in its deeper parts, is reasonably quiet and unchanging and that observations made during a longer period, 10–20 years, could be merged and used to describe the basic hydrographic conditions. The observations in the 1990s proved the Arctic Ocean to be as variable and changing as any other ocean.

An inflow of anomalously warm AW was reported in 1990 and has been observed propagating in the boundary current and into the different basins, changing the characteristics of the Atlantic layer in the Arctic Ocean. This warm inflow event persisted for almost a decade. Colder water then entered the Arctic Ocean for a short period after which the inflowing AW again became warmer. These inflow events can be traced upstream and originate from the input of warmer AW across the Greenland–Scotland Ridge. The pulses have also been followed in the Arctic Ocean, tracing several of the suggested circulation loops in the basins, giving timescales for the movements along the different loops. Model work has reproduced many of these events and their pathways around the Arctic Ocean.

The first pulse has now left the Eurasian Basin and partly exited the Arctic Ocean through Fram Strait. In the Amundsen Basin it has been replaced by colder water, while the part that entered the Makarov Basin at the Mendeleev Ridge has circulated around the

Makarov Basin and is now found at the Makarov Basin side of the Lomonosov Ridge, practically removing the temperature front previously present along the ridge. The pulse has also entered the northern Canada Basin at the Chukchi Plateau. Its movements around and south of the Chukchi Plateau have taken comparably long, and it has been proposed that the AW here enters not mainly in the boundary current but directly into the basin as intrusive, double-diffusively driven layers. Older water, previously found in the southern Canada Basin, has shifted northward along the slope and is seen penetrating along the Alpha Ridge into the northern Canada Basin.

The 1990 inflow event coincided with a strong, positive state of the North Atlantic Oscillation (NAO) and of the Arctic Oscillation (AO), which also affected the distribution of the runoff from the Siberian rivers. Instead of entering the Amundsen Basin from the Laptev Sea it continued eastward to the Makarov Basin and the northern Canada Basin. The upper Pacific water lens, which in the 1970s extended over the entire Canadian Basin to the Lomonosov Ridge, then contracted to the Canada Basin. Similar scenarios have been reproduced in model studies.

The shifting of the Pacific/Atlantic surface front as well as the river water (shelf water) front counterclockwise toward the Canada Basin elevated the effects of the warmer AW. In the Amundsen Basin, and partly in the Makarov Basin, it approached closer to the sea surface, into the levels previously occupied by the halocline, thus magnifying the increase of both temperature and salinity at these levels. The shifting of the river water front also caused an increase in salinity of the surface layers of the Amundsen and Makarov Basins. The area with a deep winter mixed layer, previously confined to the Nansen Basin, expanded into the Amundsen basin, almost reaching the Lomonosov Ridge. The winter convection reached down to 120–130 m and actually caused a temperature decrease immediately above the Atlantic layer. The river water front has during the 2000s moved back into the Amundsen Basin, almost as far as the Gakkel Ridge, indicating that the shelf water outflow from the Laptev Sea again primarily enters the Amundsen Basin, recreating the PML–halocline structure in that basin.

The inflow in the Barents Sea branch also appears to vary. The intermediate salinity minimum in the Eurasian Basin has become more pronounced and less saline uPDW has crossed the Lomonosov Ridge in the boundary current along the continental slope and entered the Canadian Basin. This has made the uPDW characteristics in the Makarov Basin more similar to those in the Eurasian Basin, with a curved rather than a straight θS curve below the temperature maximum. A question is, if these anomalies in the Canadian Basin will stay long enough to be removed by shelf-slope convection and interior mixing processes, recreating the older, smooth, θS characteristics, or if this inflow will create entirely new θS structures? If so, does this mean that the gyre circulation has changed and the communication between the Eurasian and Canadian Basins has increased, leading to a more rapid ventilation of the Canadian Basin? Historical data from the Russian archives indicate that sudden changes have occurred before, and the situation now observed may not be unique. The use of CTDs instead of Nansen bottles also reveals structures in the water column previously not resolved.

The circulation pattern in the Arctic Ocean responds to long periodic (decadal) variations in the atmospheric circulation, the NAO and the AO. The positive AO state increases the inflow of AW and weakens the Beaufort gyre, while the negative state leads to a well-developed Beaufort gyre and a smaller inflow of AW. The negative state then retains the fresh water and the ice, while the positive state acts to reduce the storage of ice and low-salinity upper waters, which, together with a larger inflow of AW, increases the mean salinity of the Arctic Ocean waters.

Perhaps the most prominent change observed in the Arctic Ocean has been the retreat of the ice cover. A reduction of the minimum ice extent of $1.5–2 \times 10^6\,km^2$ ($>20\%$) between 1979 and 2005 has been observed. Comparison between submarine observations of ice thicknesses 20 years apart indicates a thinning of the ice, from 3.1 to 1.8 m. However, the magnitude of this change has been contested. Thickness observations at the same position years apart might not be relevant, since the distribution of the ice will depend upon the forcing of the ice field, which will vary between the different years. Nevertheless, even disregarding changes in ice thickness, the ice cover has become significantly reduced.

A thinner and less extensive ice cover indicates a loss of ice storage, which could either be due to reduced formation of ice, or to an increased export of ice out of the Arctic Ocean. The amount of freshwater released by the reduction of the ice cover could have contributed to perhaps the largest fraction of the freshening of the Nordic Seas and to the subpolar gyre reported in recent years, but this does not give any information about where the phase change occurred, in the Arctic Ocean or south of Fram Strait. The time series of the ice export are not long enough to provide an answer and the knowledge of the liquid freshwater export is uncertain, if not completely absent.

The presence of the halocline between the PML and the Atlantic layer makes the stirring in the PML entrain cold water from below, and the ice cover is isolated from the heat stored in the Atlantic layer. The situation in the mid-1990s with no river (shelf) water present in the Amundsen Basin could lead to the entrainment of warm AW into the uppermost layer and thus bring heat to the sea/ice surface. This sensible heat rather than latent heat of ice formation could then be supplied to the atmosphere and less ice would form in winter. The heat might also directly melt the ice, but if the stirring of the mixed layer, driving the entrainment of heat from below, is due to haline convection, a change from freezing to melting would stop the convection, and the entrainment as well as the heat transport from below would cease.

Transports

The last 10 years have seen large programs focusing on measuring the transports through the key passages of the Arctic Ocean. Much of this activity has been coordinated by the ASOF (Arctic and Subarctic Ocean Fluxes) program but other projects have also been involved.

The observations of the Bering Strait inflow have largely confirmed the transport estimates proposed by Russian researchers 50 years ago. A mean inflow of 0.8 Sv of low-salinity ($S \leq 32$) water takes place. The seasonal variations are large; the inflow is 1.2 Sv in summer and 0.4 Sv in winter. The recent observations have shown that the freshwater transport through Bering Strait might be substantially (20%) higher than previously estimated, making it two-thirds of the river runoff. This reevaluation is due to the inclusion of the transport in the low-salinity Alaskan Coastal Current.

The transports through the Canadian Arctic Archipelago, notoriously difficult to measure due to the remoteness, the severe climate, and the proximity to the magnetic north pole that makes direction determinations extremely difficult, have in recent years been measured in the Hell Gate and Cardigan Strait (the Jones Sound) and in Barrow Strait (Lancaster Sound). This gives observed transports through two of the three main passages through the Archipelago. Transport measurements have also been made in Nares Strait, but year-long transport estimates from Nares Strait are not yet available.

The fluxes through the narrow Hell Gate are directed out of the Arctic Ocean, barotropic, and almost constant, while in the neighboring Cardigan Strait the flow is weaker and reversals are observed. The combined average transport is estimated to 0.3 Sv. The transports in the wider Barrow Strait are largely barotropic on the southern side of the channel and directed eastward, toward Lancaster Sound and Baffin Bay. A weak, baroclinic westward flow is observed on the northern side, indicating transport of runoff and penetration of water from Baffin Bay and Lancaster Sound. The flow is highly variable with an estimated mean transport of 0.75 Sv from the Arctic Ocean to Baffin Bay.

Models indicate that the Barrow Strait (Lancaster Sound) might contribute one-third to one-half of the total outflow through the Canadian Arctic Archipelago. Should this be correct the total transport would be 1.5–2 Sv. It is low-salinity, primarily Pacific, water that passes through the archipelago, but the bottom water in Baffin Bay, $-0.5\,^{\circ}C$ and 34.5, likely derives from the Arctic Ocean through Nares Strait and would then be supplied by lower halocline waters.

The transport of the AW to the Barents Sea between Bjørnøya and Norway has been measured continuously for 10 years. The mean net transport is into the Barents Sea, but there is a return flow of transformed, colder, and denser water. The mean net transport of AW to the Barents Sea has been estimated to 1.5 Sv. However, short periodic variations are large and in spring, due to changing wind conditions, a whole month of small net inflow, occasionally even outflow, has been observed. There are also indications of variability on longer timescales, 3–4 years, but no trend has been detected. The AW is warm and saline at the entrance, but it loses much of its heat during transit and does not contribute heat to the central Arctic Ocean. To the inflow of AW should be added a transport of 0.7 Sv of less saline (34.4) water from the Norwegian Coastal Current, increasing the net inflow to ~ 2.2 Sv.

Fram Strait, which also has been monitored regularly since 1997, has a two-directional flow, and not only polar surface water (PSW), comprising the PML and halocline waters, and AW but also intermediate and deep water masses pass through the strait. The flow is largely barotropic and highly variable in space and time. Eddies are present, both barotropic and baroclinic, which complicates the transport estimates. Most of the steady flow occurs in the northward-flowing West Spitsbergen Current to the east and the southward-flowing East Greenland Current to the west. The inflow comprises warm AW and colder, less-saline AIW and NDW, while the outflow carries sea ice, low-salinity PSW, cool Arctic Atlantic Water (AAW), uPDW with temperature between 0 and $-0.5\,^{\circ}C$ and CBDW, seen as a salinity maximum at 1700 m, and the colder, but also saline EBDW close to the bottom.

The observed total northward and southward transports are large, 10–15 Sv. The mean net transport is much smaller and lies between 1.5 and 2.5 Sv southward, out of the Arctic Ocean. A large recirculation is present in the strait. This appears partly associated with the barotropic eddies that drift westward along the sill of the strait and carry water from the West Spitsbergen Current to the East Greenland Current.

The transports through Fram Strait of largest climate importance are the export of ice and the heat carried by the AW. There is a large interannual variability but no trend has been found. The estimated mean ice export is ~0.9 Sv, while the outflow of PSW is ~1 Sv. The inflow of AW and heat also varies strongly from year to year but larger transports and higher temperatures have been observed in the last years, giving a flux ~50 TW. The average heat transport since 1997 is 35–40 TW. This is estimated relative to an assumed mean temperature of −0.1 °C in the Arctic Ocean. However, since the mass budget is not closed, the heat transport will depend upon the choice of reference temperature. The long-term mean inflow of AW, >2 °C, is *c.* 3 Sv.

Whether an increased heat transport through Fram Strait, connected with warmer AW and perhaps a stronger flow, also leads to more heat being available for the Arctic, for ice melt and for the atmosphere, is not yet clear. The fact that pulses of warm AW can be traced around the Arctic Ocean could imply that most of the heat is not lost but only stored and will eventually leave the Arctic Ocean through Fram Strait. The heat advected into the Arctic Ocean through Bering Strait is located closer to the sea surface and could have a larger impact on the heat flux to the ice cover from below and on the thickness of the ice cover. The large retreat in ice extent reported in the last years has mostly occurred in the southern Canada Basin, close to the Chukchi Sea.

Significance for Climate

The influence of the Arctic Ocean on the circulation at lower latitudes is mainly through the export of freshwater as ice and as low-salinity PSW, and through the export of dense intermediate and deep waters that contribute to the Greenland–Scotland overflow and to the North Atlantic Deep Water, enforcing the thermohaline circulation and the Atlantic meridional overturning circulation (AMOC). Of the around 6-Sv overflow water supplied to the North Atlantic Deep Water by the Arctic Mediterranean about 3 Sv have passed through the Arctic Ocean.

The outflow of ice and less-dense surface water could increase the stability of the water column in the convection areas to the south, in the Nordic Seas, and in the Labrador Sea. However, the fresh water largely remains in the East Greenland Current and mostly bypasses the Greenland Sea and Iceland Sea gyres. In recent years, ice has not been formed in the central Greenland Sea and the Greenland Sea has been dominated by thermal convection.

The AAW and the uPDW contribute to the Denmark Strait overflow, while the CBDW and the EBDW mainly enter the Greenland Sea, where they supply the mid-depth (1800 m) temperature maximum and the deep salinity maximum. In recent years, the local convection in the Greenland Sea has not penetrated through the temperature maximum and only less-dense AIW has been formed. The production of AIW has lead to a more direct contribution of the Greenland Sea to the AMOC, especially to the East Greenland Current and the Denmark Strait overflow. The denser Arctic Ocean deep water masses, previously continuing in the East Greenland Current to Denmark Strait are now entering the Greenland Sea and gradually replacing the old, colder, and less saline Greenland Sea deep and bottom waters, making the Greenland Sea water column more 'Arctic' in character.

See also

Bottom Water Formation. Meddies and Subsurface Eddies. Ocean Circulation. Ocean Circulation: Meridional Overturning Circulation. Overflows and Cascades. Rotating Gravity Currents.

Further Reading

Björk G, Jakobsson M, Rudels B, *et al.* (2007) Bathymetry and deep-water exchange across the central Lomonosov Ridge at 88–89° N. *Deep-Sea Research I* 54: 1197–1208 (doi:10.1016/j.dsr.2007.05.010).

Björk G, Söderqvist J, Winsor P, Nikolopoulos A, and Steele M (2002) Return of the cold halocline to the Amundsen Basin of the Arctic Ocean: Implications for the sea ice mass balance. *Geophysical Research Letters* 29(11): 1513 (doi:10.1029/2001GL014157).

Coachman LK and Aagaard K (1974) Physical oceanography of the Arctic and sub-Arctic seas. In: Herman Y (ed.) *Marine Geology and Oceanography of the Arctic Ocean*, pp. 1–72. New York: Springer.

Dickson B, Meincke J, and Rhines P (eds.) (2008) *Arctic–Subarctic Ocean Fluxes: Defining the Role of the Northern Seas in Climate*. New York: Springer.

Dickson RR, Rudels B, Dye S, Karcher M, Meincke J, and Yashayaev I (2007) Current estimates of freshwater Arctic and subarctic seas. *Progress in Oceanography* 73: 210–230 (doi:10.1016/j.pocean.2006.12.003).

Fahrbach E, Meincke J, Østerhus S, *et al.* (2001) Direct measurements of volume transports through Fram Strait. *Polar Research* 20: 217–224.

Jakobsson M, Cherkis NZ, Woodward J, Macnab R, and Coakley B (2000) New grid of Arctic bathymetry aids scientists and mapmakers. *EOS, Transactions of American Geophysical Union* 81(9): 89–96.

Jakobsson M, Macnab R, Mayer L, *et al.* (2008) An improved bathymetric portrayal of the Arctic Ocean: Implications for ocean modelling and geological, geophysical and oceanographic analyses. *Geophysical Research Letters* 35: L07602 (doi:10.1029/2008 GL0335220).

Johannesen OM, Muench RD, and Overland JE (eds.) (1994) *AGU Geophysical Monographs 85: The Polar Oceans and Their Role in Shaping the Global Environment*. Washington, DC: American Geophysical Union.

Jones EP, Rudels B, and Anderson LG (1995) Deep water of the Arctic Ocean: Origin and circulation. *Deep-Sea Research* 42: 737–760.

Leppäranta M (ed.) (1998) *Physics of Ice-Covered Seas*. Helsinki: Helsinki University Press.

Lewis EL, Jones EP, Lemke P, Prowse T, and Wadhams P (eds.) (2000) *The Freshwater Budget of the Arctic Ocean*. Dordrecht: Kluwer Academic Publishers.

Merryfield WJ (2002) Intrusions in double-diffusively stable Arctic waters: Evidence for differential mixing? *Journal of Physical Oceanography* 32: 1452–1459.

Nansen F (1902) Oceanography of the North Polar Basin. *Scientific Results III(9): The Norwegian North Polar Expedition 1893–96*. Oslo: Jacob Dybwad.

Peterson BJ, McClelland J, Curry R, Holmes RM, Walsh JE, and Aagaard K (2006) Trajectory shifts in the Arctic and subarctic freshwater cycle. *Science* 313: 1061–1066.

Prinsenberg SJ and Hamilton J (2005) Monitoring the volume, freshwater and heat fluxes passing through Lancaster Sound in the Canadian Arctic Archipelago. *Atmosphere-Ocean* 43: 1–22.

Quadfasel D, Sy A, Wells D, and Tunik A (1991) Warming of the Arctic. *Nature* 350: 385.

Rudels B, Jones EP, Anderson LG, and Kattner G (1994) On the intermediate depth waters of the Arctic Ocean. In: Johannesen OM, Muench RD, and Overland JE (eds.) *AGU Geophysical Monographs 85: The Polar Oceans and Their Role in Shaping the Global Environment*, pp. 33–46. Washington, DC: American Geophysical Union.

Rudels B, Jones EP, Schauer U, and Eriksson P (2004) Atlantic sources of the Arctic Ocean surface and halocline waters. *Polar Research* 23: 181–208.

Rudels B, Muench RD, Gunn J, Schauer U, and Friedrich HJ (2000) Evolution of the Arctic Ocean boundary current north of the Siberian shelves. *Journal of Marine Systems* 25: 77–99.

Schauer U, Fahrbach E, Østerhus S, and Rohardt G (2004) Arctic warming through Fram Strait: Oceanic heat transports from 3 years of measurements. *Journal of Geophysical Research* 109: C06026 (doi:10.1029/2003 JC001823).

Serreze MC, Barrett A, Slater AJ, *et al.* (2006) The large-scale freshwater cycle in the Arctic. *Journal of Geophysical Research* 111: C11010 (doi:10.1029/2005 JC003424).

Smith WO, Jr. (ed.) (1990) *Polar Oceanography, Part A: Physical Sciences*. San Diego: Academic Press.

Smith WO, Jr. and Grebmeier JM (eds.) (1995) *Coastal and Estuarine Studies: Arctic Oceanography, Marginal Ice Zones and Continental Shelves*. Washington, DC: American Geophysical Union.

Steele M and Boyd T (1998) Retreat of the cold halocline layer in the Arctic Ocean. *Journal of Geophysical Research* 100: 881–994.

Stein R and MacDonald RW (eds.) (2004) *The Organic Carbon Cycle in the Arctic Ocean*, 362pp. Berlin: Springer.

Swift JH, Aagaard K, Timokhov L, and Nikiforov EG (2005) Long-term variability of Arctic Ocean waters: Evidence from a reanalysis of the EWG data set. *Journal of Geophysical Research* 110: C03012 (doi:10.1029/ 2004JC002312).

Timmermanns M-L, Garrett C, and Carmack E (2003) The thermohaline structure and evolution of the deep water in the Canada Basin, Arctic Ocean. *Deep-Sea Research I* 50: 1305–1321 (doi:10.1016/S0967-0637(03)00125-0).

Untersteiner N (ed.) (1986) *The Geophysics of Sea Ice*. New York: Plenum.

Wadhams P, Gascard J-C, and Miller L (1990) Topical studies in oceanography: The European Subpolar Ocean Programme: ESOP. *Deep-Sea Research II* 46: 1011–1530.

Walsh D and Carmack EC (2002) A note on evanescent behavior of Arctic thermohaline intrusions. *Journal of Marine Research* 60: 281–310.

Wassmann P (ed.) (2006) Special Issue: Structure and Function of Contemporary Food Webs on Arctic Shelves: A Pan-Arctic Comparison. *Progress in Oceanography* 71(2–4): 123–477.

Wheeler PA (1997) Topical studies in oceanography: 1994 Arctic Ocean section. *Deep-Sea Research II* 44: 1483–1758.

Woodgate RA and Aagaard K (2005) Revising the Bering Strait freshwater flux into the Arctic Ocean. *Geophysical Research Letters* 32: L02602 (doi:1029/2004 GL021747).

Woodgate RA, Aagaard K, Muench RD, *et al.* (2001) The Arctic boundary current along the Eurasian slope and the adjacent Lomonosov Ridge. *Deep-Sea Research I* 48: 1757–1792.

THE ABYSSAL CIRCULATION

ABYSSAL CURRENTS

W. Zenk, Universität Kiel, Kiel, Germany

Introduction

Historically the term 'abyss' characterizes the dark, apparently bottomless ocean under extreme static pressure far beyond coastal and shelf areas. Today this ancient definition remains still rather unfocused in earth sciences. Geographers, marine biologists, and geologists use abyss for deep-sea regions with water depths exceeding 1000 or 4000 m. In physical oceanography a widely accepted definition of the abyss denotes the water column that ranges from the base of the main thermocline down to the seabed.

The main thermocline itself – occasionally also called warm-water sphere – extends laterally between the polar frontal zones of both hemispheres. In contrast to deep strata the surface exposition of the thermocline allows the direct exchange of heat, substances, and kinetic energy with the atmosphere. This wind-driven part of the water column and its fluctuations are consequently an immediate subject of weather and climatic conditions. The base of the thermocline at about 1000–1200 m represents the lower boundary of the warm-water sphere with temperatures well >5°C.

The abyss or cold-water sphere below, is clearly colder. Below 2000 m potential temperatures <4°C are found virtually everywhere. Below 4000 m values of 0–2°C are more characteristic. Until the advent of modern self-recording instrumentation abyssal currents were believed to be very slow ($<2\,\mathrm{cm\,s^{-1}}$) and negligible in comparison with rather vigorous and variable surface currents (sometimes $>100\,\mathrm{cm\,s^{-1}}$). Only subsurface passages in rises and ridges that subdivide ocean basins (**Figure 1**) seemed to allow for more energetic deep interior currents funneling through gaps and channels.

Until a few decades ago practically all knowledge about abyssal currents had to be inferred from the perspective of property fields like potential temperature, salinity, or potential density. Since these fields seem to be much more stable than velocity fields this method still yields reasonable general

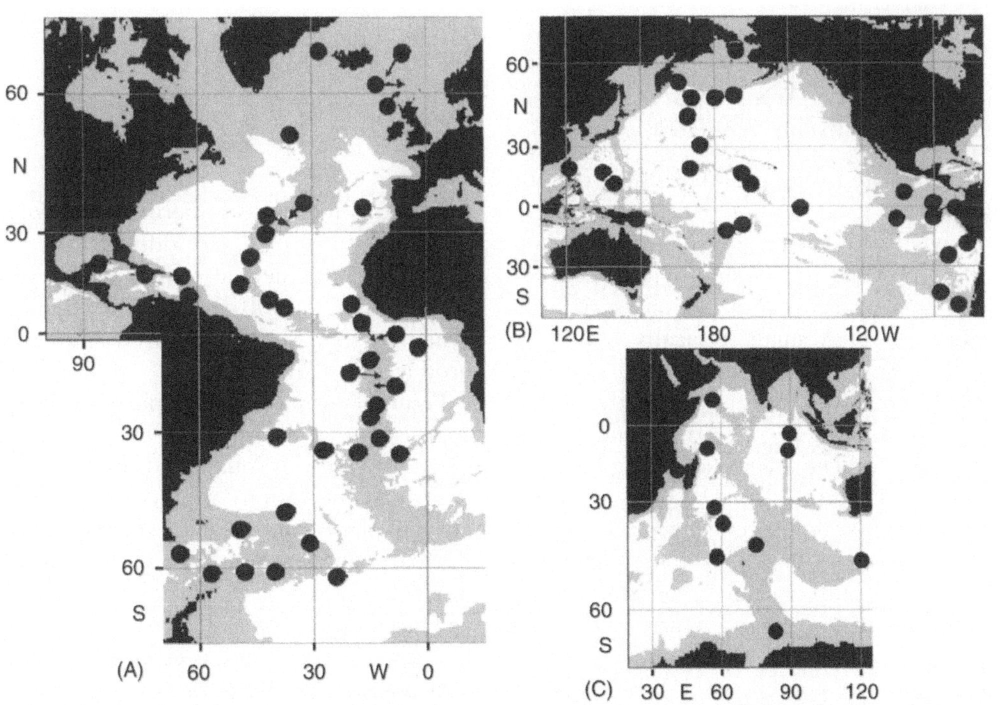

Figure 1 Major subsurface fracture zones, passages, and sills of the world's ocean. (A) Atlantic, (B) Pacific, (C) Indian Ocean. These topographic features represent important constraints for deep and bottom water spreading paths (after http://www.soc.soton.ac.uk/JRD/OCCAM/sills.html).

large-scale flow patterns even with scarce and non-synoptic data sets. However, for any resolution beyond the low-frequency band of the current spectrum direct current observations by floating objects, moored instruments, or remotely measuring methods are essential.

In low latitudes of all oceans the permanent presence of the enormous, almost invariant temperature gradient between the surface at >25°C and the floor at <2°C requires a deep advection of cold water. The latter circulates freshly ventilated cold water from polar latitudes towards the Equator. For reasons of continuity a compensating poleward flow of water heated from the surface through the thermocline must occur above the abyssal layer of the tropics and the subtropics. Hence, the cold water drift propelled by sinking of cooled water masses (convection in selected areas) accelerates an endless global circulation cell known as the meridional overturning circulation (MOC), occasionally called the global conveyor belt (**Figure 2**). Its bottom-nearest limb, and sometimes also one or two layers in motion above it are characterized by the abyssal circulation.

The Stommel-Arons Concept and Diffusivity in the Interior

The concept that rising water from a flat deep-sea bottom without density stratification has to be replaced by convectively formed water is a key element of the modern Stommel-Arons theory of abyssal circulation. While the freshly ventilated water sinks in only a few selected semi-enclosed polar regions, the rising process itself occurs over broad lateral scales everywhere at lower latitudes. As a consequence it is possible to distinguish between two different dynamical regimes of the abyssal circulation (**Figure 3**). (1) A small (typically only 100–200 km wide) corridor at the continental rises and near the slopes on the western margins of the ocean's basins is occupied by deep western boundary currents (DWBCs). (2) The huge remaining interiors of the basins, fed laterally by DWBCs, are ruled by a uniform broad-scale upwelling regime. The inherent vertical velocities imply vortex stretching on top of the abyss. Conservation of potential vorticity then requires the poleward return drift in the interior.

Not long after the concept of DWBC was hypothesized in 1958, the first observations confirming its existence were made beneath the Gulf Stream in the north-western Atlantic using neutrally buoyant Swallow floats. Shortly afterwards, the newly detected deep equatorward drift led to a supplementary experiment to find the slow poleward countercurrent of the interior flow. The test failed at the time (1962) in so far as no flow with a slow, persistently northward velocity component could be proved. Instead, the mesoscale eddy phenomenon at great depths was discovered.

In steady state the vertical flux of heat at the base of the thermocline in lower latitudes can be formulated as the balance of temperature advection and its vertical diffusion:

$$\vec{u}\cdot\nabla T + w\frac{\partial T}{\partial z} = k\frac{\partial^2 T}{\partial z^2} \qquad [1]$$

where $\vec{u}$ is the lateral current vector, T temperature, w vertical velocity component, z vertical coordinate, and k eddy diffusivity. This temperature equation implies for the interface between the base of the thermocline and the top of the abyss the generation of a vertical velocity to balance the upwelled cold water and the downward diffusion of heat. An assumed sinking rate of $20 \times 10^6\ \mathrm{m^3\,s^{-1}}$ in high latitudes and an active lower thermocline interface of $3 \times 10^8\,\mathrm{km^6}$ yields a global integral upwelling speed of $O(0.1\ \mathrm{dm\ d^{-1}})$. An adjoined downward diffusion with a diffusivity of $O(1\,\mathrm{cm^2\ s^{-1}})$ can be estimated under the assumption of a vertical scale of 1 km within the abyssal upwelling regime. **Figure 4** depicts a sketch of the integrated form of [1] under the assumption of negligible lateral currents along isotherms.

$$w \cdot T = k_d\frac{\partial T}{\partial z} \qquad [2]$$

where k_d is the cross-isothermal diffusivity.

Observations of k_d are difficult to conduct. On a global scale numerical values of k_d fluctuate in a wide range ($1\text{–}500\,\mathrm{cm^2\,s^{-1}}$). A summary of the available sparse estimates from the Atlantic is reproduced in **Figure 5**.

Figure 6 gives a rare example of the heterogeneity of cross-isopycnal diffusivity in the interior Brazil Basin. The diffusivity is quantified from uniform microstructure observations at all depths above the smooth abyssal plains and west of the Mid-Atlantic Ridge. Estimates range from very weak ($<0.1\,\mathrm{cm^2\,s^{-1}}$) to moderate ($>5\,\mathrm{cm^2\,s^{-1}}$) turbulent signals. The latter enhancement is observed above the rough flanks of the ridge. The response is especially prominent within 500 m above the bottom.

The abyss contains recirculation cells. Significant amounts of cold water do not participate directly in the global meridional overturning cell. They circulate horizontally in response to the upwelling process.

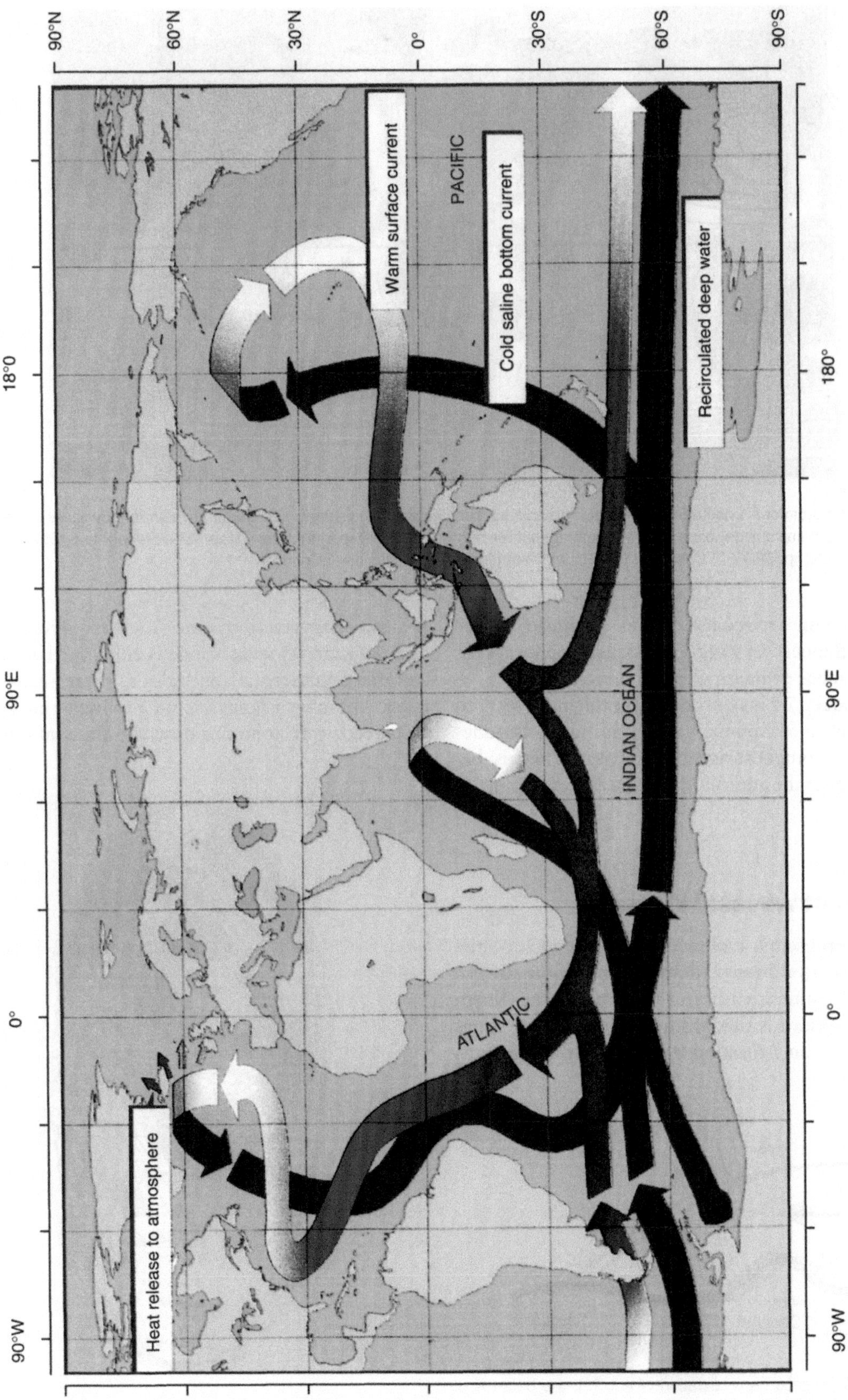

Figure 2 The Atlantic thermocline circulation as a key element of the global oceanic circulation. (After Broecker (1991), modified by Meier-Reimer.)

Figure 3 The Stommel-Arons (1958) concept of the abyssal circulation in the Atlantic. A system of western boundary currents feeds a slow broad-scale upwelling regime in the remaining interior of the basins. (Reprinted from *Deep-Sea Research* 5, Stommel H, The abyssal circulation, pp. 80–82, Copyright (1958), with permission from Elsevier Science.)

Such basin-wide recirculation cells (**Figure 7**) are distinctly influenced by the local bottom topography. Their persistence remains widely unexplored.

In summary, it is of considerable interest to quantify rates of sinking waters in high latitudes because the compensating abyssal upwelling is believed to drive the internal horizontal circulation of the oceans.

Sources of Abyssal Waters

Figure 8 represents a refined global update of the highly schematized deep Atlantic circulation pattern in **Figure 3**. Antarctica lies in the center. The North Pole cap is split; it is situated towards the outer edges of the Pacific and Atlantic blocks. All three oceans are interconnected by the Southern Ocean as the circum-Antarctic water ring is called. In this strongly simplified pattern of the global water mass circulation the abyss ranges from the lower third of the displayed water column down to the seafloor.

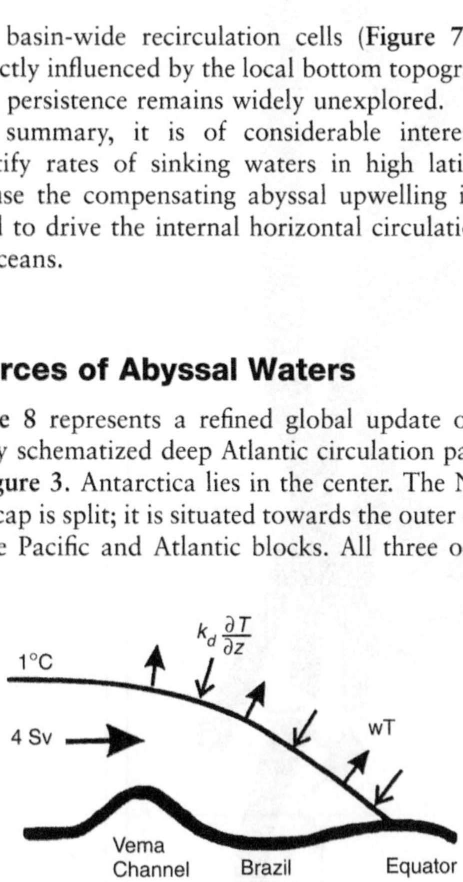

Figure 4 Schematic representation of eqn [2], the balance of downward diffusivity and upwelling across the 1°C isotherm level of the western South Atlantic. (After Hogg *et al.*, 1994 © Springer Verlag.)

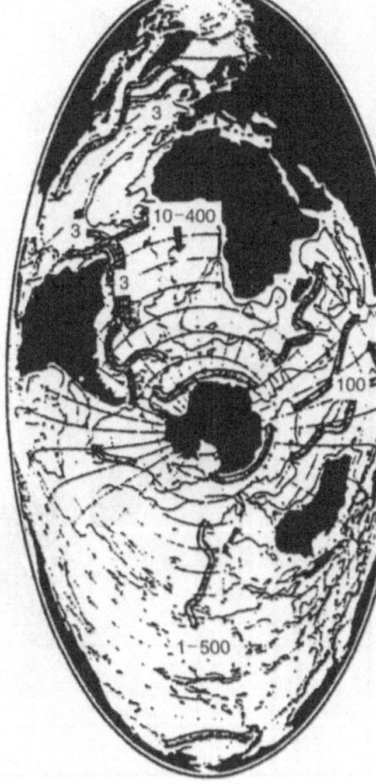

Figure 5 Some estimates of cross-isotherm diffusivity in $cm^2\ s^{-1}$ in the abyss of the world's oceans. (After Hogg, 2001, © Academic Press.)

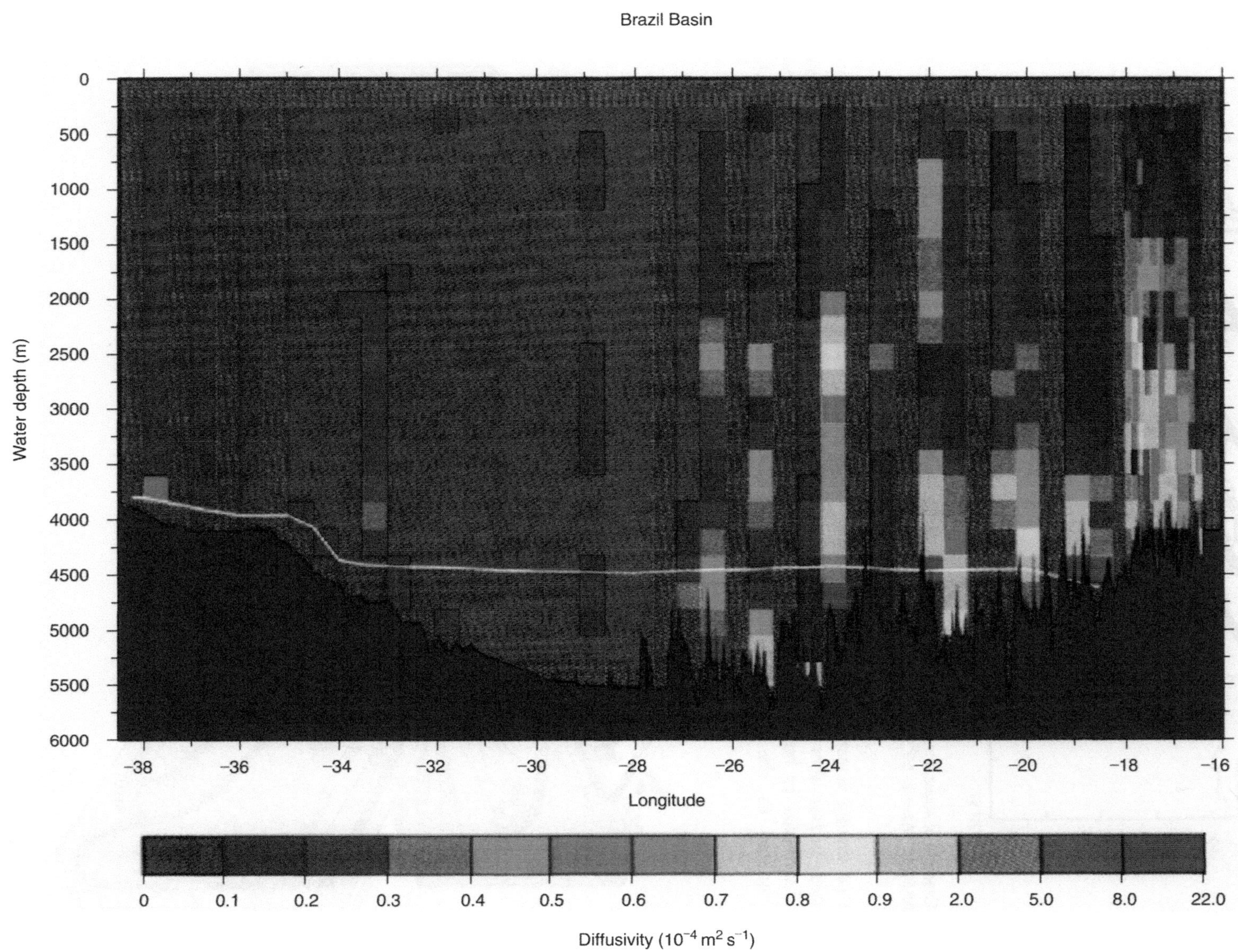

Figure 6 Directly observed distribution of diffusivity in the South Atlantic. The section covers the range between the continental rise off Brazil and the western flanks of the Middle Atlantic Ridge. Highest diffusivity values are correlated with the rough topography on the slopes of the ridge. (After Polzin *et al.*, 1997 © *Science*.)

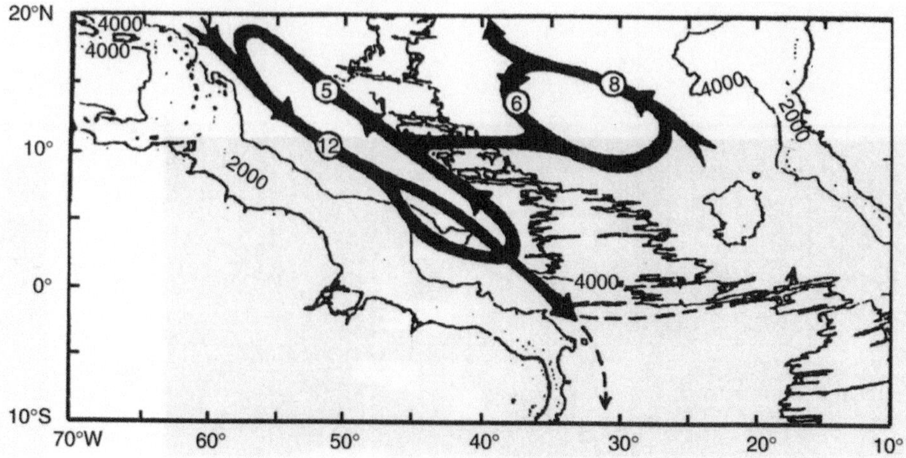

Figure 7 Example of abyssal recirculation cells in the tropical North Atlantic Ocean. Numbers indicate volume transports of lower North Atlantic Deep Water in $10^6\ m^3\ s^{-1}$. (After Friedrichs and Hall, 1993. Courtesy of the *Journal of Marine Research*.)

Regions of sinking waters are symbolized in **Figure 8** by near-surface downward arrows. They are unevenly distributed in both hemispheres. In fact, they are limited to the Antarctic Circumpolar Current System (ACCS) and to the high latitudes of the North Atlantic. In addition to bottom and deep water sources there are contributions from outflows of marginal seas (not shown in **Figure 8**). Although

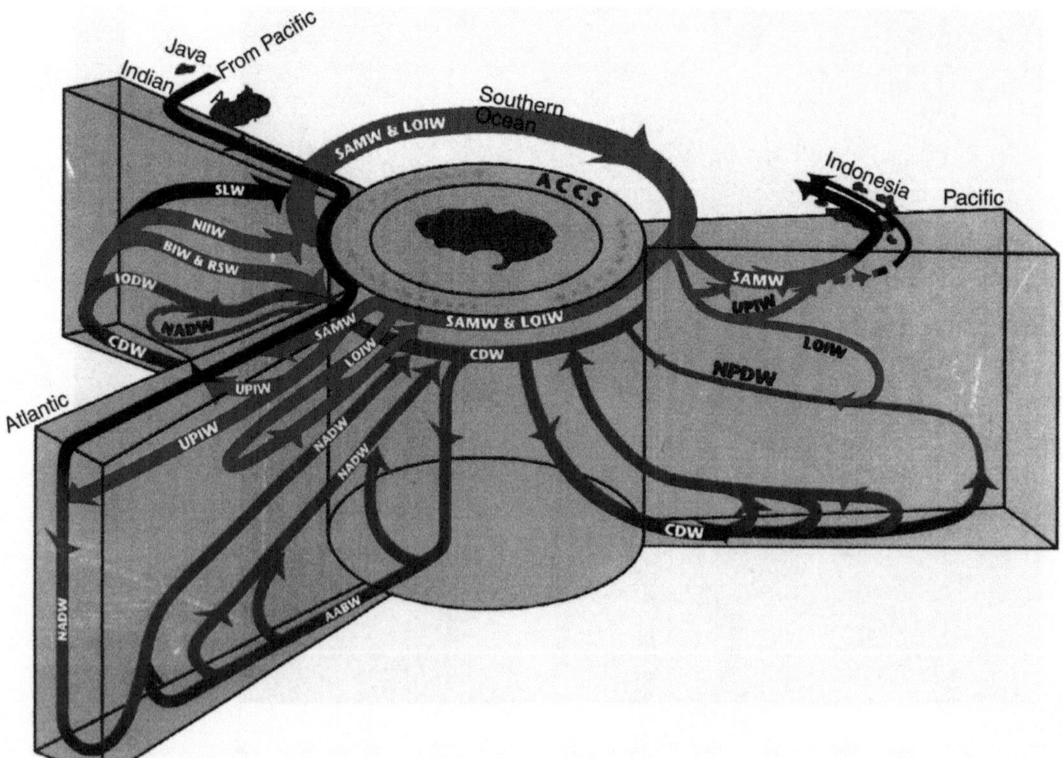

Figure 8 The complex global circulation. Colors: red and purple, thermocline waters; green and blue, abyssal waters. For details see text. Abbreviations: SAMW, SubAntarctic Mode Water; LOIW, Lower Intermediate Water; SLW, Surface Layer Water; UPIW, Upper Intermediate Water; NIIW, North-west Indian Intermediate Water; BIW, Banda Intermediate Water; RSW, Red Sea Water; CDW, Circumpolar Deep Water; NADW, North Atlantic Deep Water; NPDW, North Pacific Deep Water; IODW, Indian Ocean Deep Water; AABW, Antarctic Bottom Water. (Reproduced with permission from the Woods Hole Oceanographic Institution, Schmitz 1996a, 1996b.)

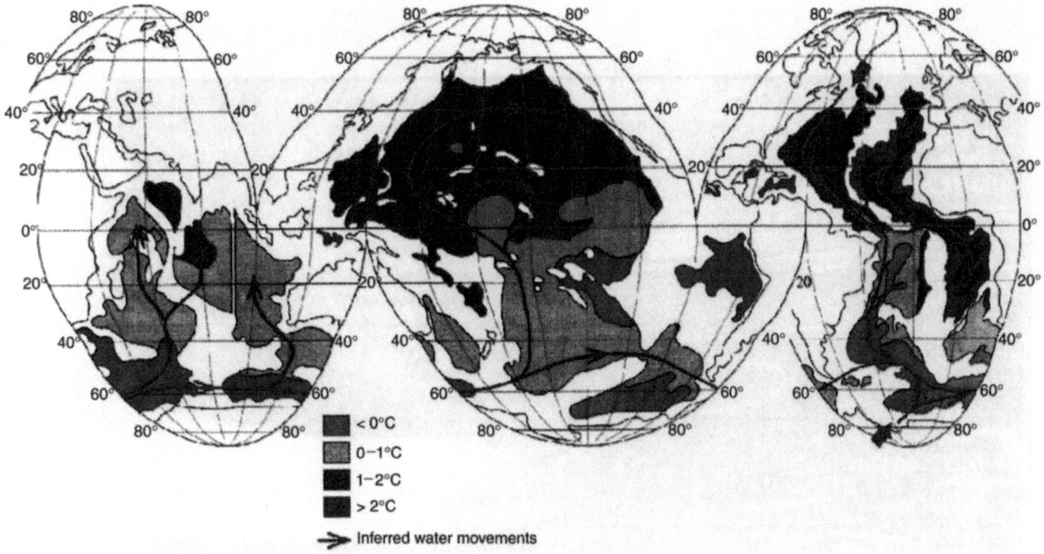

Figure 9 Global near-bottom potential temperature distribution and inferred flow paths of abyssal waters at 4000 m depth. (Reproduced with permission from an article by Charnock, The Atmosphere and the Ocean; in Summerhayes CP and Thorpe SA, *Oceanography, An Illustrated Guide* © 1996 Manson Publishing.)

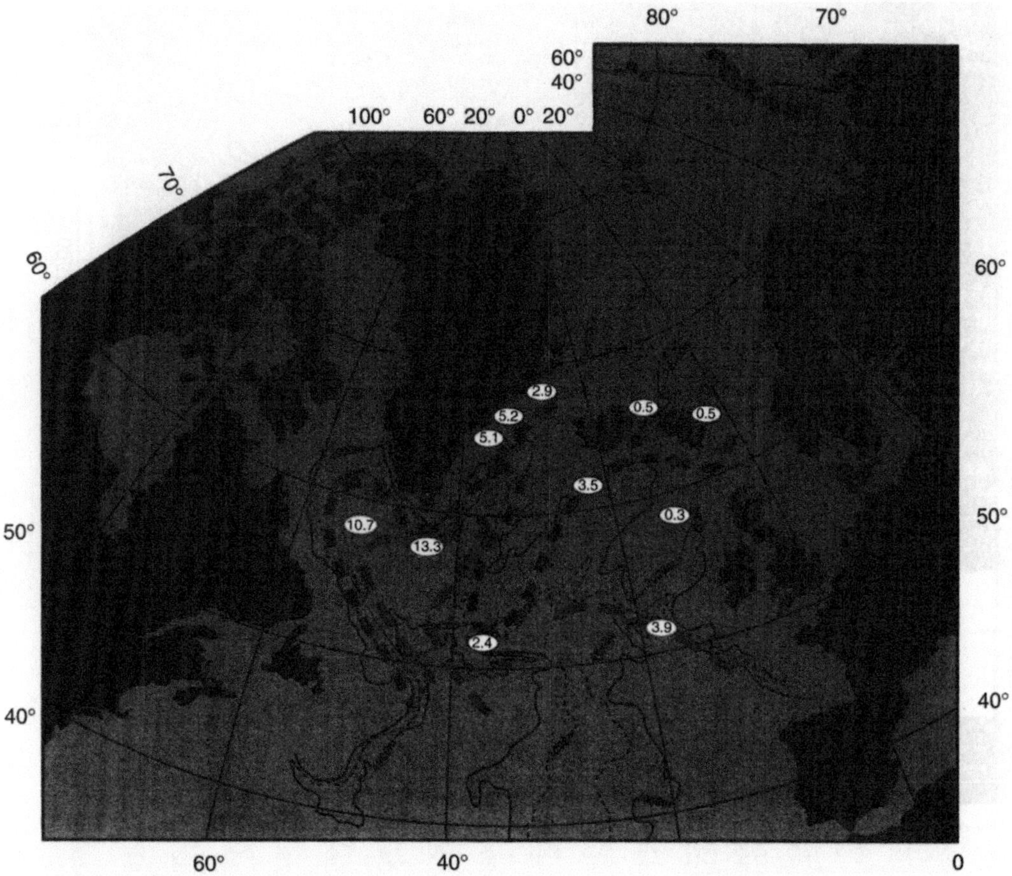

Figure 10 Examples of deep entrainment from the subpolar North Atlantic. Numbers represent volume transports in 10^6 m^3 s^{-1}. Note the observed significant increase of transport along the east Greenland side which is caused by Denmark Strait Overflow Water. The Denmark Strait between Iceland and Greenland is also called Greenland Strait. (Reproduced with permission from Dickson RR and Brown J, *Journal of Geophysical Research* 99, 12319–12341 (1994) © American Geophysical Union.)

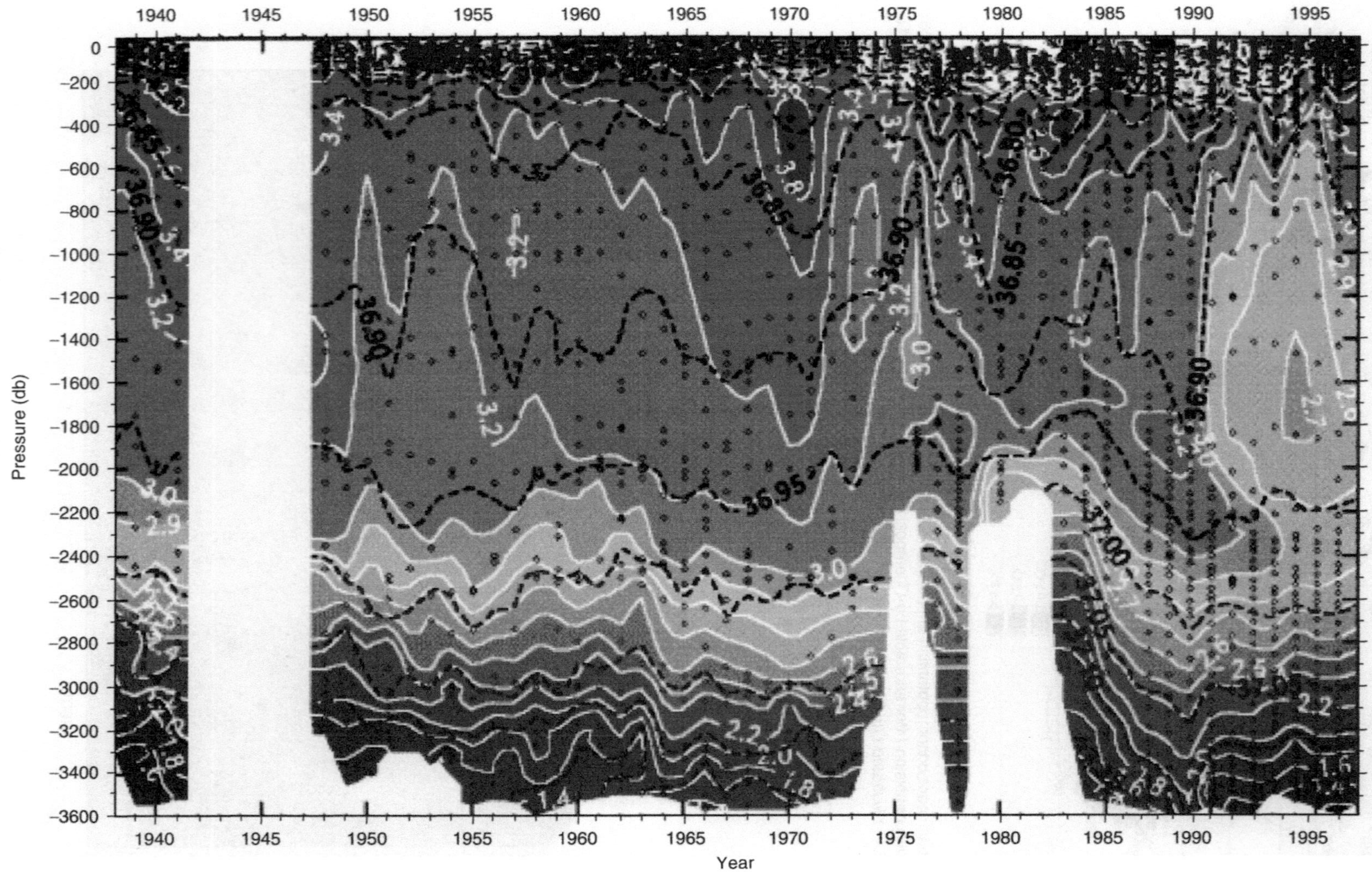

Figure 11 Demonstration of interannual variability (1938–97) of potential temperature in °C (colors and white isotherms) and potential density $\delta_{1.5}$ (kg m^{-3}) 1.5 (dashed black lines) referenced to 1500 dbar pressure level. (After I Yashayaev *et al.*, www.mar.dfo-mpo.gc.ca/science/ocean/woce/labsea/labsea_poster.html.)

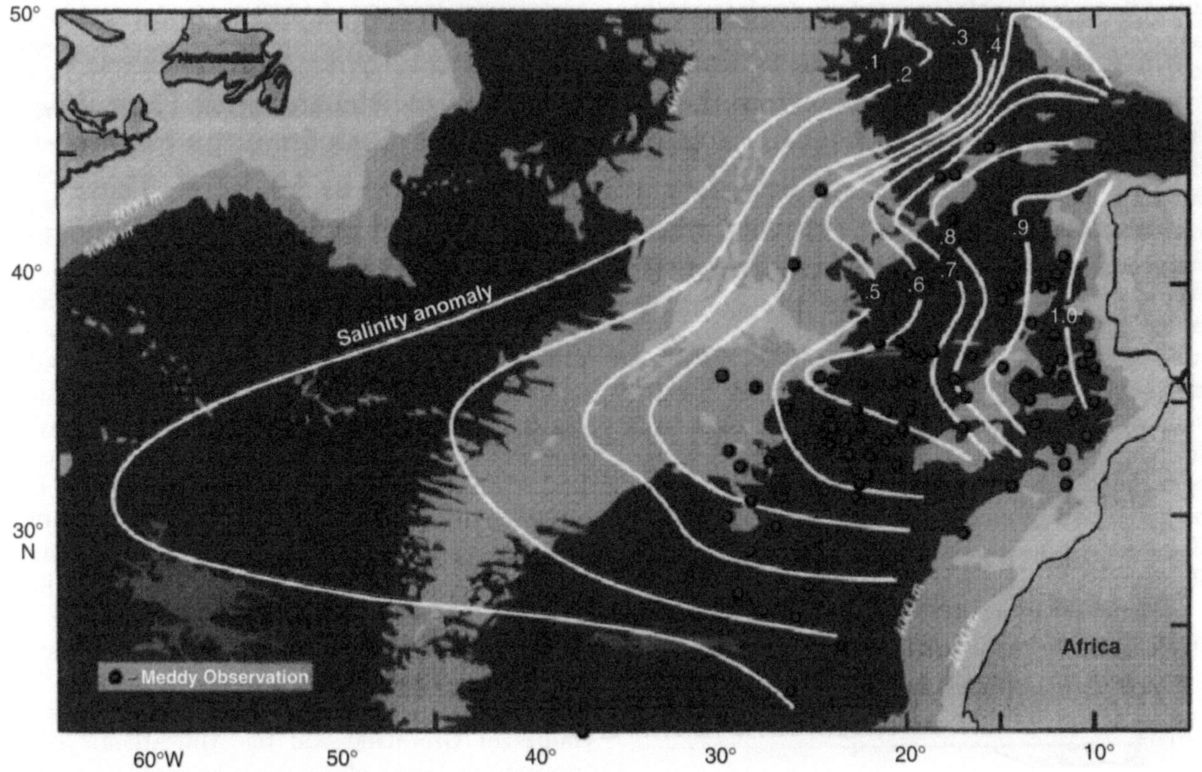

Figure 12 Influence of the Mediterranean Water on the salinity distribution at 1100 m of the North Atlantic. Isohalines represent positive anomalies relative to 35.01. Circles depict observed high salinity eddies of Mediterranean Water (Meddies). (Reprinted from *Progress in Oceanography* 4, Richardson PL, Bower AS and Zenk W, A census of meddies tracked by floats, pp. 209–250 © (2000) with permission from Elsevier Science and the American Meteorological Society.)

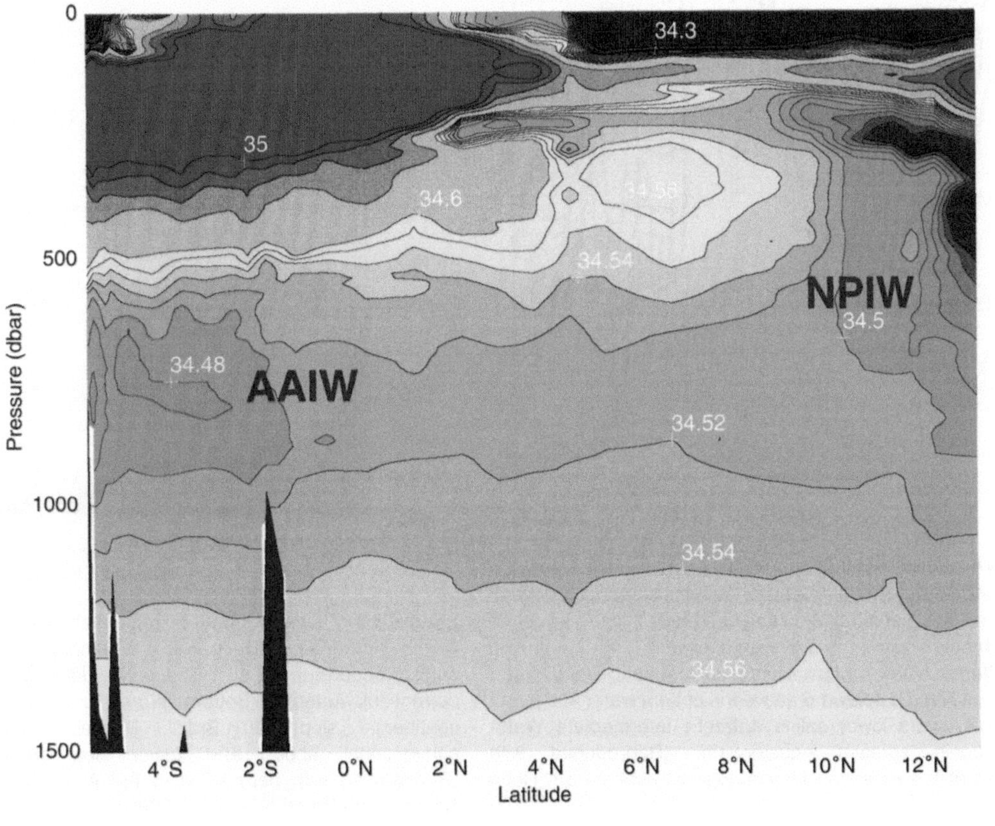

Figure 13 Salinity distribution (PSU) of Antarctic and North Pacific Intermediate Waters in the tropical western Pacific along 150°E. (After Holfort and Zenk, 1999.)

these flows reach only intermediate depth levels they affect characteristic water masses of the abyss beneath their own spreading level.

Several discrete locations for water sinks have been identified around Antarctica. Completely homogenizing wintertime convection down to the bottom of this semi-enclosed basin was observed only in Bransfield Strait south of the South Shetland Islands. However, in all other locations around Antarctica cooled and freshly ventilated water sinks to the bottom where it behaves like a contour current on the rotating earth. These densest waters in thin layers on the bottom mix with surrounding slightly warmer water masses and finally are transformed into Antarctic Bottom Water (AABW) spreading equatorward in all oceans. Known sinking areas are the Weddell Sea, Ross Sea, Wilkes-Adélie Coast, and some locations off Enderby Land and the Prydz Bay.

Because the generation of AABW is characterized by highest densities (due to the freezing temperatures in the source regions), AABW can be detected by low (potential) temperature signals close to the bottom everywhere in the Southern Hemisphere and in lower latitudes of the Northern Hemisphere. **Figure 9** demonstrates the global deep flow at 4000 m inferred from temperature distribution in the world ocean. The main production regions of sinking waters in the Weddell Sea (and near Iceland in the Northern Hemisphere) are labeled by arrows.

The northern counterpart of AABW is called North Atlantic Deep Water (NADW). Since its potential density is slightly different than that of AABW both water masses create an abyssal stratification where they encounter. Formation areas of NADW are located in the Nordic seas, i.e. the Norwegian and Greenland Seas. Convectively formed deep water from polar regions spills through outflow channels over sills to the north west and south east of Iceland. The outflow products are called Denmark Strait and Iceland Scotland Overflow Waters. On their way into the subpolar North Atlantic, the swift overflow plumes with speeds exceeding 50 cm s^{-1} are subject to strong topographic control along the slopes of Greenland and the Mid-Atlantic Ridge, respectively. The high propagation speed of overflow waters favors further transformations by entrainment from intermediate waters increasing its transport rate in selected regions by a factor of 2 or more (**Figure 10**).

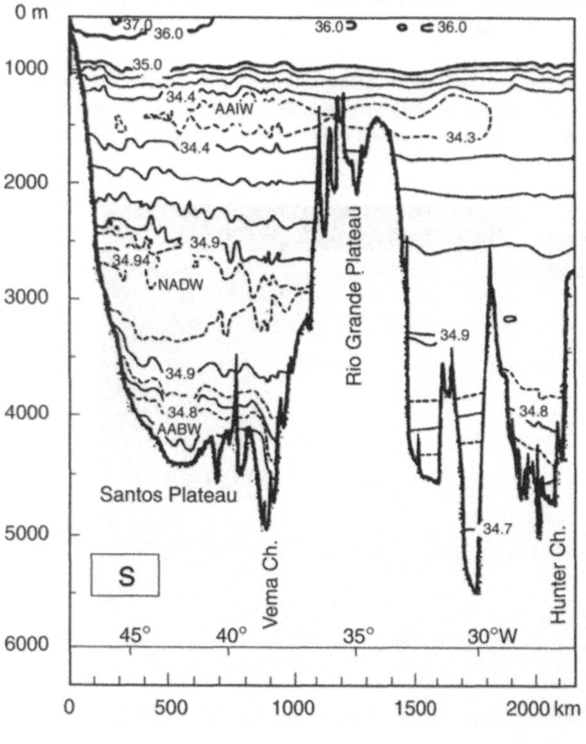

Figure 14 Salinity distribution (PSU) on a section across the Rio Grande Rise, South Atlantic, at nominally 30°S between the slope off Brazil and the Middle Atlantic Ridge. The Vema and Hunter Channels allow an active equatorward flux of Antarctic Bottom Water (AABW). More saline North Atlantic Deep Water (NADW) is advected poleward above the bottom water. On top of the abyssal layers lower saline Antarctic Intermediate Water (AAIW) again spreads equatorward. (Reproduced with permission of the American Meteorological Society from Hogg *et al.*, 1999.)

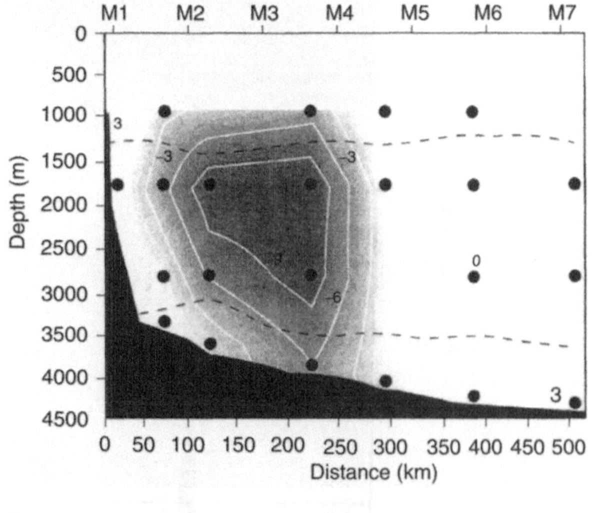

Figure 15 Eulerian long-term observation of the southward-flowing North Atlantic Deep Water at nominally 18°S. Isoclines of current speeds in cm s^{-1}. The prime deep western boundary current of Antarctic Bottom Water is pressed against the continental slope off Brazil. Dashed lines denote the conventional upper and lower boundaries of NADW. (Reproduced with permission of the American Meteorological Society from Weatherley *et al.* 2000.)

Additional mixing occurs with seasonally generated Labrador Sea Water. The latter also is formed by deep wintertime convection; its seasonal generation rate fluctuates substantially on interannual scales (**Figure 11**).

Local formation regions are not restricted to the western and central Labrador Sea itself. The western Irminger Sea seems to contribute to NADW formation as well.

Formation and sinking rates due to deep convection of AABW around Antarctica and of NADW in the North Atlantic are about equal in volume. Each hemisphere contributes $\sim 10 \times 10^6\ m^3\ s^{-1}$. In the Southern Ocean generation sites are less focused compared with the Nordic Seas of the Atlantic. In the Pacific and Indian Oceans northern sources of abyssal waters are unknown (or at least are insignificant in the Pacific).

Overlying waters from two additional sources influence the abyssal circulation. (1) The excess evaporation over precipitation in the Mediterranean and Red Seas increases salinities and densities of these marginal seas substantially in comparison with the adjacent oceans. These density differences force exchange currents in the connecting straits (Strait of Gibraltar and Bab-el-Mandeb). The outflows sink to their equilibrium levels where they form pronounced intermediate high salinity layers in the North Atlantic (**Figure 12**) and northern Indic, respectively. (2) In selected regions of the southern polar frontal zone notable amounts of low salinity Antarctic Intermediate Water (AAIW) downwell and become

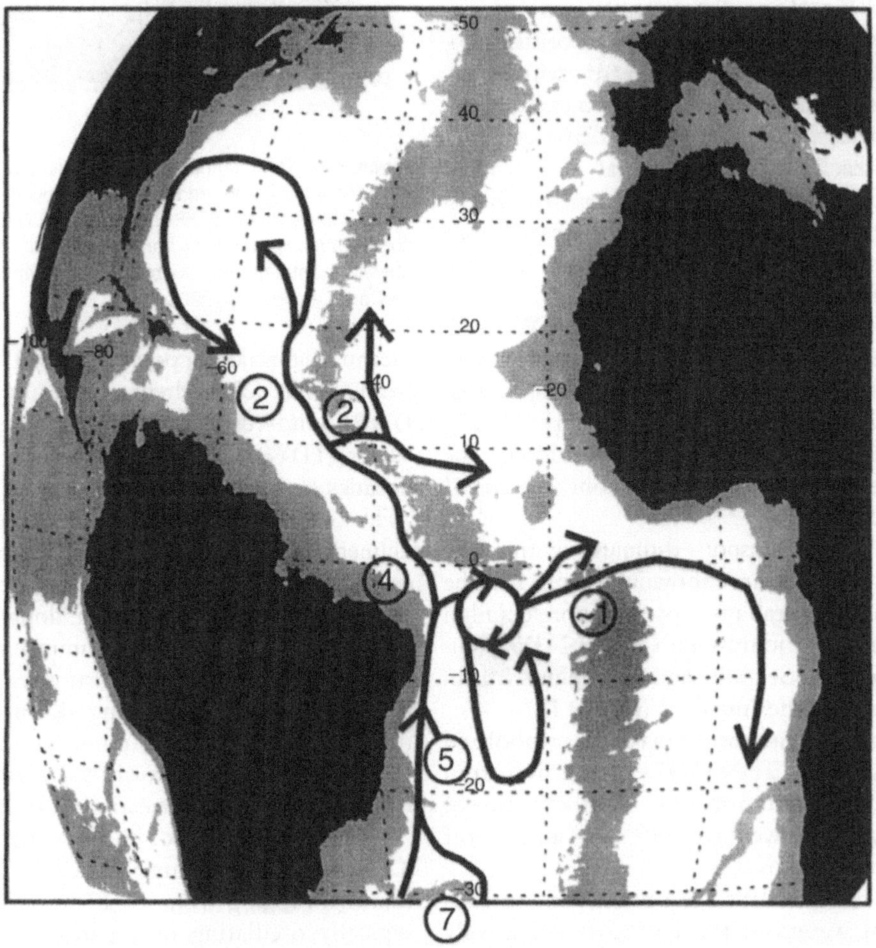

Figure 16 Spreading paths of Antarctic Bottom Water in the Atlantic after miscellaneous observations. Numbers indicate volume fluxes in $10^6\ m^3\ s^{-1}$. Originally the AABW spreading path is restricted to the western side of the South Atlantic. It enters the subtropical region through the Vema and Hunter Channels at 30°S. The eastern basin is only accessible via the Vema Fracture Zone (11°S) and through the Romanche and Chain Fracture Zones near the equator. (Reproduced with permission of the American Meteorological Society from Stephens and Marshall, 2000.)

part of the intermediate circulation. AAIW is found northward of subpolar regions in all three oceans. In the Pacific there is also a northern mode of intermediate water which is called North Pacific Intermediate Water (**Figure 13**).

Deep Western Boundary Currents (DWBC)

In contrast to the slow interior drift in ocean basins DWBC are much easier to access by direct observations. They can be revealed by property characteristics of parameters like temperature, salinity, nutrients or other chemical tracers, or be quantified by moored instrumentation. Following the schematic circulation diagram (**Figure 8**) we expect three stacked opposite DWBC cores at intermediate and abyssal levels in the South Atlantic: northward propagating AABW with mixing components of Circumpolar Deep Water and of the Weddell Sea Deep Water, southward-flowing NADW at least reaching the southern rim of the subtropics, and AAIW with a northward drift again. Water mass distributions and their dynamical imprints are exemplarily displayed in form of a hydrographic section across the Rio Grande Rise at ~30°S. This ridge separates the abyssal Argentine from the Brazil Basins (**Figure 14**). Long-term current observations with moored instruments farther north at 18°S are reproduced in **Figure 15**.

The core of the southward flowing NADW with averaged speeds of >10 cm s^{-1} is centered at 2200 m about 200 km offshore. The integrated transport of the NADW plume amounts to 39×10^6 m^3 s^{-1}. The deepest northward AABW component along the continental rise with long-slope speeds of <3 cm s^{-1} is less developed.

The most recent transport estimates within the AABW from the Atlantic are shown in **Figure 16**. The northern part of this graph is shown in more detail in a schematic diagram jointly with lower NADW invading the Sargasso Sea and adjacent regions from the northern end of the Americas (**Figure 17**).

Mixing of both abyssal water masses is symbolized by stars in **Figure 17**. The NADW drift grows from $\sim 13 \times 10$ m^6 s^{-1} south of the southern tip of Greenland (Cape Farewell) to $\sim 40 \times 10^6$ m^6 s^{-1} off the Caribbean island arc.

In contrast to the Atlantic the basin-wide spreading of abyssal waters in the Pacific is exclusively controlled by the Antarctic Circumpolar Current System (ACCS). The lowest stratum is filled with Circumpolar Deep Water (CDW), a mixture of converted AABW and transformed NADW, possibly homogenized by repeat circulation in the Southern Ocean. Potential temperatures of CDW at 32°S range from 0.6 to 1.2°C. It enters the South Pacific as a DWBC and returns to the ACCS as Pacific Deep Water (PDW) after internal mixing and upwelling decades or centuries later.

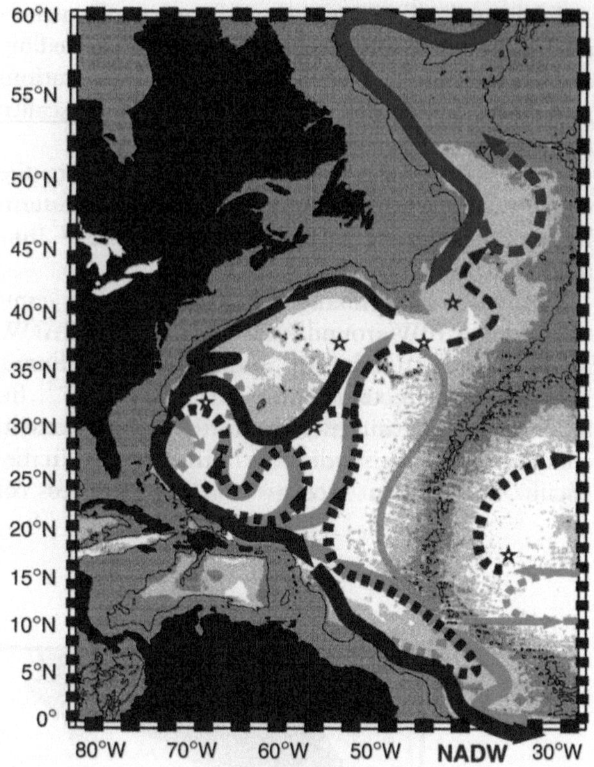

Figure 17 The complex abyssal circulation in the western North Atlantic. Colors: red and brown, flow of North Atlantic Deep Water; blue, Antarctic Bottom Water. Dashed lines indicate conceivable recirculation branches. Stars denote regions of abyssal entrainment. (Reproduced with permission from McCartney and Curry, 2001.)

The region north east of northern New Zealand delineates the gateway for CDW into the southeastPacific Basin (**Figure 18**). Its boundary current system was found from 2-year long observations to be ~700 km wide. The maximum mean velocity of the applied current meter array was 9.6 cm s^{-1} on the eastern flank of the Tonga-Kermadec Ridge. The time-averaged transport amounts to 15.8×10^6 m^6 s^{-1} with a standard error of 9.2×10^6 m^6 s^{-1} for the focused northward advection of the CDW core.

A striking aspect of the overall observations at 32°S are the total transport (CDW plus PDW) fluctuations ranging from $(-17$ to $+51) \times 10^6 m^6 s^{-1}$, typically oscillating over periods of a few months with amplitudes of 1–2 times the mean.

About 20° farther north the northward flow of CDW into the main basins of the Pacific is topographically controlled by the Samoan Passage. Recent observations with moored current meters have

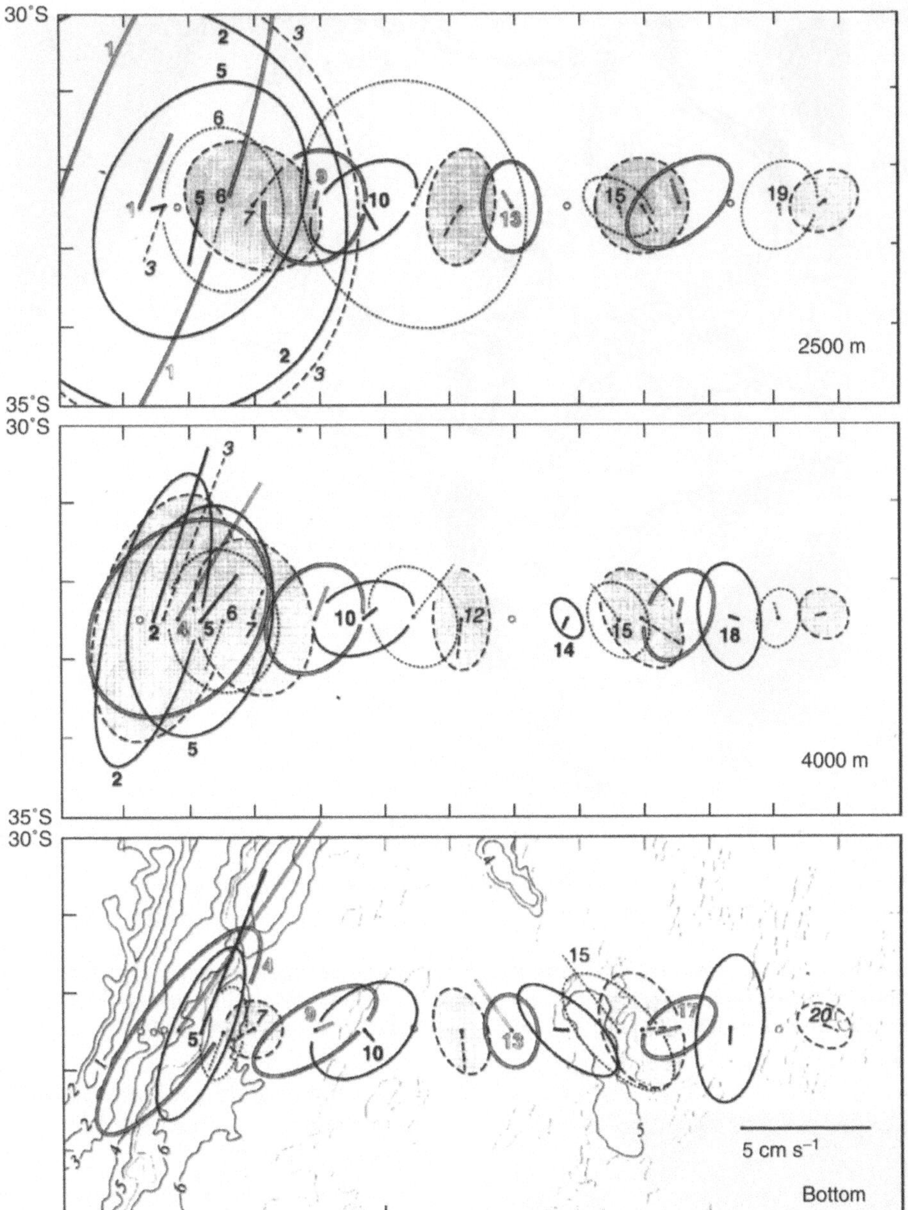

Figure 18 Long-term observations of the deep western boundary current system at 32°30′S at three instrumented abyssal layers (2500, 4000 m, near-bottom) of the Pacific. Mooring locations at dots are shown by the origin of the presented mean velocity vectors. Ellipses inform about the current's stability in form of the root mean square amplitudes. Small italics in the lower panel indicate depths in km. Bolder numbers are mooring identifiers. Note the clear bottom-intensified flow of Antarctic Bottom Water. The returning North Pacific Deep Water flow is less confined and highly variable. (Reprinted from *Progress in Oceanography* 43, Whitworth *et al.*, on the deep western-boundary current in the Southwest Pacific Basin, 1–54 © (1999) Elsevier Science and the American Meteorological Society.)

yielded northward transports of $10.6 \pm 1.7 \times 10^6\,\mathrm{m}^6\,\mathrm{s}^{-1}$ from an 18 month measuring program.

In accordance with the Stommel-Arons concept further decreases of DWBC transports were estimated at 10°N × (9.6×10^6 $\mathrm{m}^6\,\mathrm{s}^{-1}$) and 24°N ((4.9 and 9.1) × 10^6 $\mathrm{m}^6\,\mathrm{s}^{-1}$) from snapshot hydrographic surveys (**Figure 19**).

To date estimates of the returning PDW transports are rare and quantitatively inconsistent.

The abyssal conditions of the Indic resemble those of the Pacific: no deep convective sources in the north are available (**Figure** 8). Instead, the deep Indian Ocean is controlled by deep and bottom waters from the ACCS (CDW) with a particular influence from

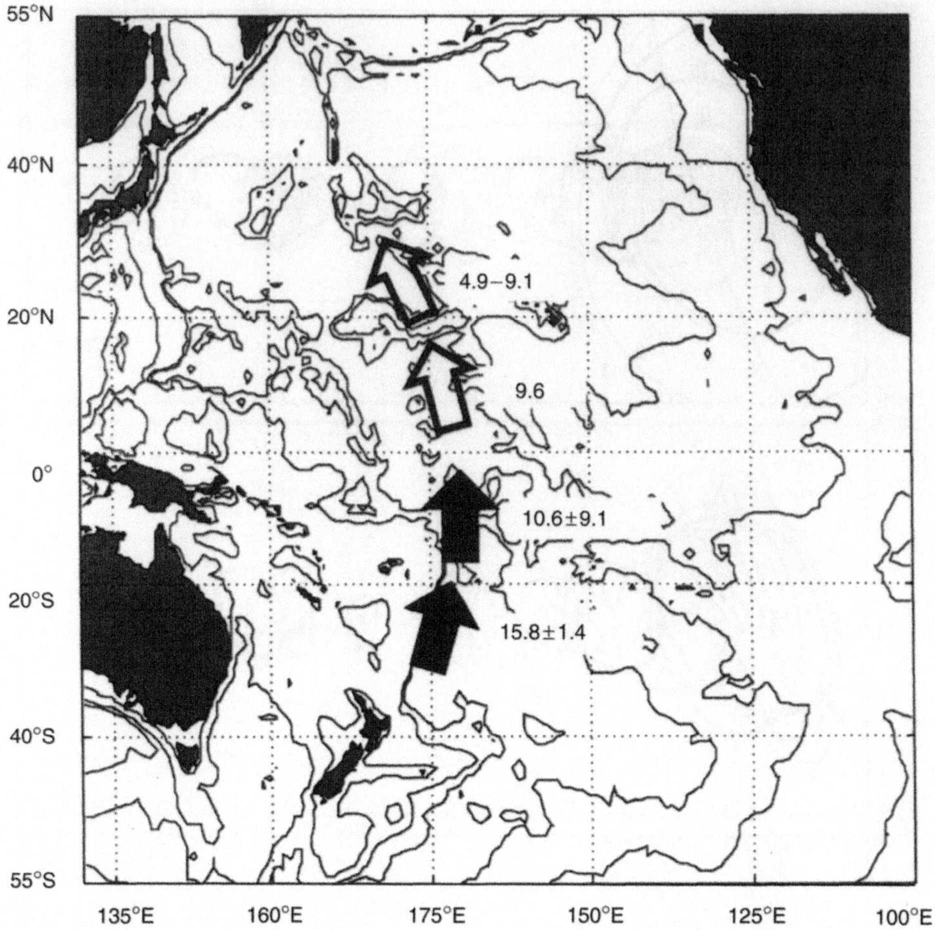

Figure 19 Inflow of Antarctic Bottom Water into the western basins of the subtropical/tropical Pacific. Heavy arrows were inferred from long-term current meter observations, open arrows represent single realizations from hydrography. Numbers indicate volumes of the deep western boundary current transports in $10^6\ m^3\ s^{-1}$. (After Hogg, 2001 © Academic Press.)

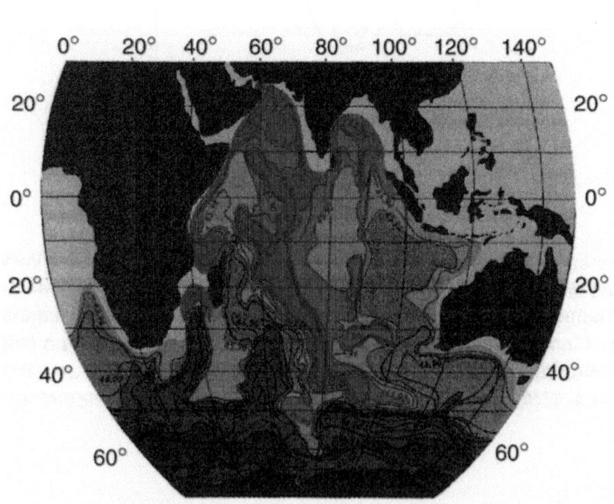

Figure 20 The distribution of potential density σ_4 in kg m^{-3} referenced to 4000 dbar is used as a tracer for the spreading of Antarctic Bottom Water in the Indian Ocean. Note the three distinct tongues of dense water invading the Indian Ocean from the Southern Ocean. (Adapted from Mantyla AW and Reid JL, *Journal of Geophysical Research* 100, pp. 2417–2439 (1995) © by the American Geophysical Union and from Schmitz, 1996b.)

the southern South Atlantic (NADW). The Indonesian passages are too shallow to affect the Indic abyss immediately.

The topography of the Indic is different to that of the other oceans. Several meridionally aligned ridges divide this ocean into sub-basins. The pathways of deep flows are therefore more complex then assumed in the simple approach by the Stommel-Arons concept. **Figure 20** displays the distribution of water density, referenced to 4000 dbar at the bottom of the Indic. The graph enables a view of large-scale bottom water spreading on the base of hydrographic observations.

Three major inlets for deep water (CDW or modified NADW) are obvious at the northern tips of the Mozambique (38°E) and the Crozet (60°E) Basins. The spreading into the Australian Basin (125°E) is constrained by the Australian–Antarctic Discordance at 50°S, 123°E. The near-Equator Somali and Australian Basins are only accessible for deep flows through the Amirante and the Diamartina

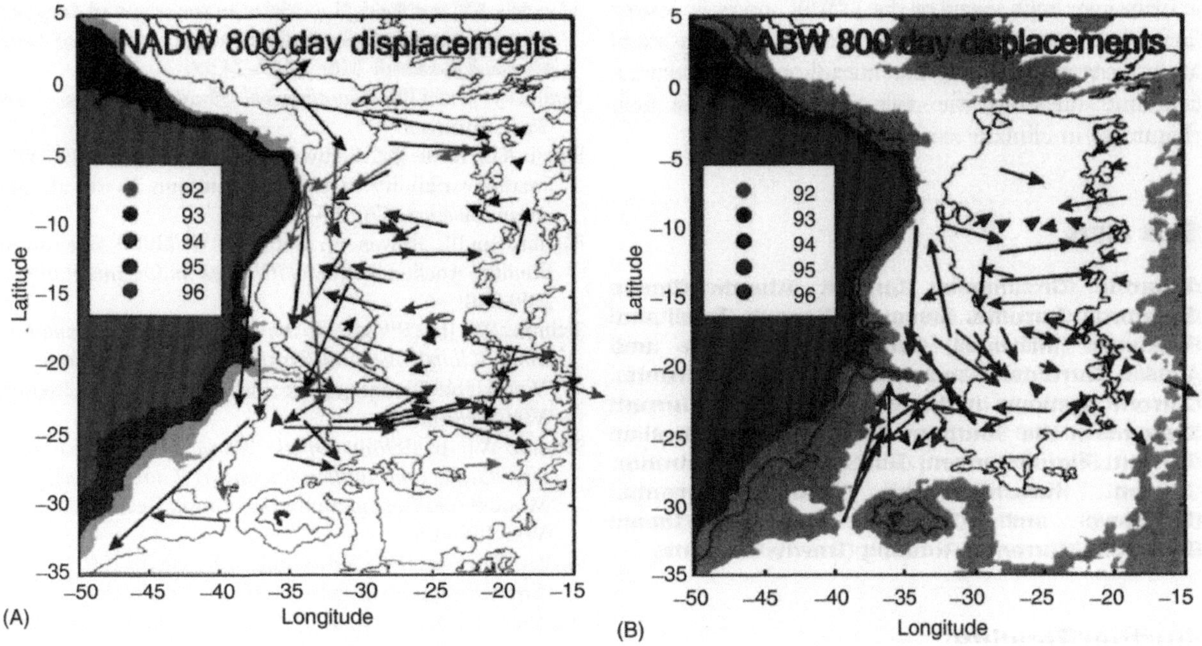

Figure 21 Displacement vectors from 800 days of the drift of North Atlantic Deep (A) and Antarctic Bottom Waters (B) in the Brazil Basin. Nominal observation levels were 2500 m (A) and 4000 m (B). Note the clear signal of boundary currents and the predominant zonal structure of the flow in the central basins. Numbers + 1900 indicate the years in which the Lagrangian current observations by neutrally buoyant floats were started. (From Hogg, 2001 © Academic Press.)

Passages, respectively. Long-term transport estimates of CDW spreading are rare. They lie clearly below the $10 \times 10^6\,m^6\,s^{-1}$ range.

Conclusions

During the past experimental period of the World Ocean Circulation Experiment (1990–99) significant progress in understanding of the abyssal circulation has been achieved. Deep Western Boundary Currents have been quantified on virtually all rises along continents and parallel to meridionally aligned ocean ridges. Pathways and transports of Deep Western Boundary Currents (DWBCs) were identified and quantified mostly with moored current meter arrays. Today the concept of a quasi-steady mean deep circulation has been outdated. In view of the omnipresent eddy kinetic energy with surprisingly large fluctuations more observations on a wide scale of temporal and spatial scales from the whole water column are essential. The need for long-term time-series in 'ocean observatories' has been recognized in international programs like the Climate Variability and Predictability Programme (CLIVAR) or the Global Ocean Observation System (GOOS).

The Stommel-Arons concept seems to be confirmed in respect to the balance of abyssal upwelling and downward heat flux through the base of the main thermocline. However, although the first Lagrangian vector time-series from the abyss of the Brazil Basin are now available, according to the theory, the slow poleward return drift remains inaccessible to direct observations, as was found some 30 years ago in the deep Sargasso Sea. In contrast to expectations, net flow displacements in the Brazil Basin over 800 days indicate preferred zonal advection of bottom water partly in opposing directions (**Figure 21**).

Besides current meter moorings improved instrumental approaches with autonomous arrays of density recorders ('moored geostrophy') or acoustic ocean tomography at a basin-scale lie at the frontier of new technologies for future integral observations. Such methods attenuate eddy noise before data are recorded, and hence reveal motion characteristics that may be closer to robust circulation patterns.

Progress has also been made with respect to the experimental determination of abyssal diffusivities. Heat transports through deep passages appears to be significantly higher and steady compared with unconstrained areas. This implies enhanced mixing above passages converting them into important stirring agents of the abyss.

Finally, recirculation cells delineate a significant limb between the swift DWBC regime and the interior slow drift regions. These deep powerful current loops on a sub-basin-wide scale may explain the enormously high speed of the DWBC cores in lower latitudes, particularly in the Atlantic. Their variability strongly affects the meridional overturning cell and questions the role of the conveyor belt (**Figure 2**) in climate variability.

See also

Antarctic Circumpolar Current. Atlantic Ocean Equatorial Currents. Benguela Current. Brazil and Falklands (Malvinas) Currents. California and Alaska Currents. Canary and Portugal Currents. Current Systems in the Atlantic Ocean. Current Systems in the Southern Ocean. East Australian Current. Florida Current, Gulf Stream and Labrador Current. Kuroshio and Oyashio Currents. Overflows and Cascades. Pacific Ocean Equatorial Currents. Rotating Gravity Currents.

Further Reading

Broecker W (1991) The great ocean conveyor. *Oceanography* 4: 79–89.

Dickson RR and Brown J (1994) The production of North Atlantic Deep Water: Sources, rates and pathwalks. *Journal of Geophysical Research* 99: 12319–12341.

Friedrichs MAM and Hall MM (1993) Deep circulation in the tropical North Atlantic. *Journal of Marine Research* 51: 697–736.

Hogg N (2001) Deep circulation. In: Siedler G, Church J, and Gould JW (eds.) *Ocean Circulation and Climate*. London: Academic Press.

Hogg N, Siedler G, and Zenk W (1999) Circulation and variability at the southern boundary of the Brazil Basin. *Journal of Physical Oceanography* 29: 145–157.

McCartney M and Curry R (2001) Abyssal potential vorticity in the western North Atlantic and the formation of Lower North Atlantic Deep Water. *Progress in Oceanography* (in preparation).

Mantyla AW and Reid JL (1995) On the origin of deep and bottom waters of the Indian Ocean. *Journal of Geophysical Research* 100: 2417–2439.

Pedlosky J (1979) *Geophysical Fluid Dynamics*. New York: Springer.

Polzin KL, Toole JM, Ledwell JR, and Schmitt RW (1997) Spatial variability of turbulent mixing in the abyssal ocean. *Science* 276: 93–96.

Richardson PL, Bower AS, and Zenk W (2000) A census of Meddies tracked by floats. *Progress in Oceanography* 4: 209–250.

Schmitz WJ Jr (1996a) *On the World Ocean Circulation: Some Global Features/North Atlantic Circulation*. 1, Woods Hole Oceanographic Institution, Technical Report, WHOI-96-03.

Schmitz WJ Jr (1996b) *On the World Ocean Circulation: The Pacific and Indian Oceans/A Global Update*. 2, Woods Hole Oceanographic Institution, Technical Report, WHOI-96-08.

Siedler G, Church J, and Gould JW (eds.) (2001) *Ocean Circulation and Climate*. London: Academic Press.

Stephens JC and Marshall DP (2000) Dynamical pathways of Antarctic Bottom Water in the Atlantic. *Journal of Physical Oceanography* 30: 622–640.

Stommel H (1958) The abyssal circulation. *Deep-Sea Research* 5: 80–82.

Summerhayes CP and Thorpe SA (1996) *Oceanography. An Illustrated Guide*. London: Manson Publishing.

Warren BA and Wunsch C (1981) *Evolution in Physical Oceanography*.

Weatherly GL, Kim YY, and Kontar EA (2000) Eulerian measurements of the North Atlantic Deep Water Western Boundary Current at 18°S. *Journal of Physical Oceanography* 30: 971–986.

Wefer G, Berger WH, Siedler G, and Webb DJ (eds.) (1996) *The South Atlantic: Present and Past Circulation*. Berlin: Springer-Verlag.

Whitworth T III, Warren BA, Nowlin WD Jr, *et al.* (1999) On the deep western-boundary current in the Southwest Pacific Basin. *Progress in Oceanography* 43: 1–54.

FLOW THROUGH DEEP OCEAN PASSAGES

N. G. Hogg, Woods Hole Oceanographic Institution, Woods Hole, MA, USA

Introduction

It is commonly stated that the relief of the ocean bottom is more extreme than that of the Earth's surface, the depth of the deepest ocean trenches being greater than the height of the highest mountain peaks. The ocean floor, in many ways, does resemble the familiar surface of the Earth – there are long mountain ranges such as the Mid-Atlantic Ridge that run many thousands of kilometers and there are quite flat abyssal plains, such as the Sohm Abyssal Plain south of Newfoundland, that cover many thousands of square kilometers. This topography of the ocean floor plays a very important role in governing the circulation of the bottom three-fourths of the ocean that lies beneath the warm surface waters.

The circulation of the abyss is largely governed by forces acting on the ocean surface: wind, evaporation, precipitation, heating, and cooling. In addition, instabilities of the upper ocean circulation lead to motions whose effects penetrate to the deep either in the form of eddies or larger-scale flows forced by those eddies. In the context of this article, the currents produced by heating, cooling, and evaporation, otherwise known as buoyancy forcing, are the most important as they lead to a global-scale convective circulation. During polar winters, surface waters are made dense by cooling, either directly, or through removal of fresh water by ice formation, and evaporation. These waters are then heavier than those below them and this causes convection and gradual descent to intermediate and bottom depths. They are replaced by less-dense, generally warmer, waters, thus producing large-scale convection connecting all the great ocean basins, sometimes referred to as the 'great conveyor belt' or the meridional overturning circulation. A remarkable feature of this phenomenon is that the sinking regions are few in number and occur at isolated sites only in the northern reaches of the North Atlantic and around Antarctica.

The bottom limb of this circulation forms up into what are known as deep western boundary currents: relatively swift and narrow currents generally found along the western deep continental rises and slopes of the ocean basins. The reasons for this remarkable asymmetry are beyond the scope of this article but result from the Earth's spherical shape and its rotation. The continental margins and ocean basins are broken by various ridges cut by deep passages which these bottom currents must navigate in a manner consistent with the laws of physics (see **Figure 1**). The way in which the deep and bottom waters deal with this challenge, and its larger implications, are the focus of this article.

Some Physics

In the context of hydroelectric power generation and flood control, engineers have understood for some time how to control the flow of water by adjusting the 'head' above the dam using spillways. The basic physics are quite well understood. Consider the flow of water along a straight channel with rectangular cross section that has a bump across the bottom whose height is adjustable. As the height of this bump is slowly increased, the water flowing over it must accelerate in order that the same amount of water flows through the diminished cross section, as is demanded by the law of conservation of mass. Bernoulli's law implies that pressure must decrease, just as it does when air accelerates over an airplane's wing, thereby inducing lift. In a fluid with a free surface this decreased pressure is accommodated by a lowering of the surface, as the pressure at any depth, to a good approximation, is just the weight of the overlying water. As the bump rises further, the water continues to accelerate over it and the water surface continues to fall until a point is reached at which this cannot continue: the upstream–downstream symmetry breaks and the water then cascades down over the bump. At this point, the bump has effectively become a dam and now 'controls' the flow of water over it. Further increases in the height of the dam now demand an increase in water depth upstream.

The mathematical solution for the problem described above can be found by combining Bernoulli's law with the equation describing the conservation of mass. This produces a cubic equation with three roots, only one of which is physical and stable to small perturbations. A critical bump height can be found above which no solutions exist for the specified set of upstream conditions: at this point either the upstream flow rate or height must increase in order that a solution be found. This situation is known as hydraulic control. It can also be shown that conditions at the control point, that is, the top of

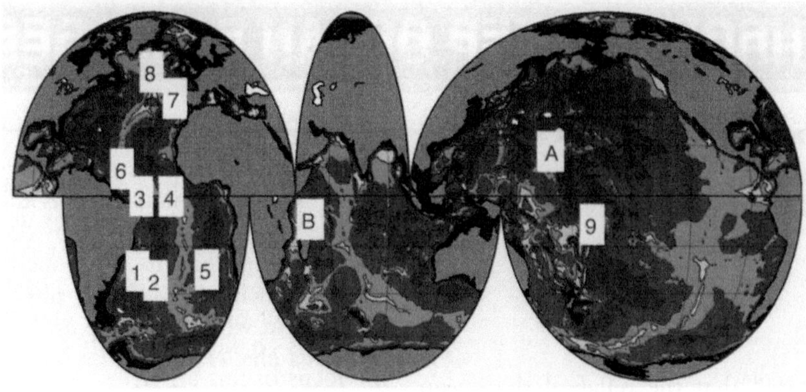

Figure 1 The bottom relief of the world's oceans. Deep passages that are mentioned in this article are identified by number as: (1) Vema Channel, (2) Hunter Channel, (3) an unnamed equatorial passage from the Brazil Basin to the Ceara Abyssal Plain, (4) Romanche and Chain Fracture Zones, (5) Walvis Passage, (6) the Vema Gap, (7) Discovery Gap, (8) Charlie-Gibbs Fracture Zone, (9) the Samoan Passage, (A) Wake Island Passage, and (B) Amirante Passage.

the bump, are such that the phase speed of long gravity waves matches the flow speed, and information about downstream conditions can no longer be transmitted upstream. The critical point is reached when the volume flux, Q, is related to the upstream height, h, of the free surface above the dam by the relation:

$$Q = \left(\frac{2}{3}\right)^{2/3} \sqrt{gh}L$$

where g is the acceleration due to gravity and L is the width of the channel.

The above relation has been known for some time and applies well to swiftly flowing rivers in which the water moves an appreciable distance, relative to its width, in the course of a local pendulum day. The deep ocean situation, on the other hand, is distinguished by much broader, more slowly moving currents which do not cover such great distances in this time and, consequently, feel the influence of the Earth's rotation. A nondimensional quantity known as the Rossby number, which measures the relative importance of the Earth's rotation, is the ratio of the fluid speed to the product of the current width, L, and the Coriolis frequency, f – twice the Earth's rotation speed times the sine of the latitude (or 2π divided by the length of the pendulum day). A large river that might be 1 km wide and travel at 2 kt gives a value of about 40 for the Rossby number at mid-latitudes and supports the notion that the Earth's rotation can be ignored. For the deep ocean 'rivers', on the other hand, a typical width is 100 times larger and flow speeds are a fraction of a knot, say 0.5 kt, yielding 0.03 for the Rossby number and indicating that the Coriolis force must also be accounted for. In this situation, the formula relating volume flux to the free surface height is somewhat more complicated:

$$Q = \frac{gh^2}{2f} \quad \text{if} \quad L > \left(\frac{2gh}{f^2}\right)^2$$

$$Q = \left(\frac{2}{3}\right)^{2/3} \sqrt{gh}\left(1 - \frac{f^2L^2}{8gh}\right), \quad \text{otherwise}$$

The first formula applies to a channel that is so wide that the acceleration-induced tilt of the surface (see below) is great enough to intersect the bottom on the left-hand side, looking downstream (Northern Hemisphere). The second formula approaches the nonrotating one quoted earlier as the Coriolis frequency or the current width approaches zero. Important assumptions include that the channel cross section is rectangular and that the water is derived from an upstream reservoir that is quiescent and very deep. These formulas also apply to situations in which a single layer of fluid is moving so that the full force of gravity is felt across the surface. For deep, dense currents, the effective gravitational force is reduced by a factor, typically 10^{-3}, related to the vertical density gradient because it is overlain by a slightly lighter fluid.

The Coriolis force acts to the right of any motion, in the Northern Hemisphere, and must be balanced by opposing pressure forces if a current is to flow steadily in any particular direction. These pressure forces arise from tilts of the ocean surface (e.g., there is a rise of *c.* 1 m across currents such as the Gulf Stream) and horizontal density differences within the ocean. For deep ocean boundary currents this balance between the Coriolis and pressure forces results in constant density surfaces which slope downward to the right several hundreds of meters over the width of the current for a current which gets more intense as the bottom is approached (again, in the Northern Hemisphere). As flow accelerates over a sill or on being squeezed through a narrow passage, we will expect that these density surface tilts will become magnified.

The formulas above give a prediction for the volume flux of fluid through a deep passage, given the height of the fluid interface above the sill and the dimensions of the passage itself. It has been assumed that a single deep layer is moving under a less-dense, still layer above, so a choice must be made as to which density surface should approximate the separation into two layers. Typically, this is done by inspection of upstream and downstream density profiles. The height above the bottom at which they diverge is taken as the point at which the upstream–downstream asymmetry occurs, thus indicative of hydraulic control.

Next, we shall review the important deep passages where measurements have been made and compare the estimated volume flux with that given by the second relation above.

Important Deep Passages

Figure 1 illustrates the complexity of the ocean floor, divided into basins by rugged ridges and numerous seamount chains. This necessarily produces countless deep passages through which water can flow from one basin to another. Only a few are considered significant enough to have been investigated. These are generally the deepest along the paths of important deep currents, interrupting their otherwise ponderous flow away from their source regions. Away from these deep-convection sites, the densest water in the ocean is known as Antarctic Bottom Water (AABW) or Circumpolar Deep Water, and, consequently, the water flowing through the major deep passages is almost exclusively this water mass. A brief survey, by ocean basin, follows. The time-averaged transport of water through a passage that is quoted below is, perhaps, the most important quantity being sought by oceanographers, but it should be emphasized that the flow is never steady and can, in fact, reverse for brief periods.

The Atlantic Ocean

AABW flows northward along the east coast of South America. After crossing the Scotian Arc north of the Drake Passage, it then travels along the western margin of the Argentine Basin at depths greater than about 3500 m. The northern end of the Argentine Basin is bounded by the Santos Plateau and the Rio Grande Rise which extend to the Mid-Atlantic Ridge but are fissured by two important deep channels: the Vema and Hunter with the former being the best defined and studied. **Figure 2** shows the topography of the Vema Channel and an expanded view near the sill, as determined by modern echo-sounding gear which is capable of mapping a swath of the bottom under a moving ship.

The Vema Channel extends *c.* 400 km and constricts the bottom water both because it is narrow and because it shallows, somewhat like a mountain pass. Moored arrays in the late 1970s and the early 1990s both gave estimates of close to $4 \times 10^6\,m^3\,s^{-1}$ for the transport of AABW through the channel. The topographic constriction accelerates the flow and results in large slopes to the isopycnals, as mentioned above. In the Vema Channel, the highest velocities are found above the bottom, and the balancing of pressure gradients and the Coriolis force then produce a fan-like cross-channel density distribution with the heaviest and coldest water found hugging the eastern side.

The Hunter Channel to the east is considerably more complex. Measurements there indicate a smaller transport of nearly $3 \times 10^6\,m^3\,s^{-1}$, although the sparseness of the array relative to the challenging topography indicates a substantial uncertainty in this estimate.

Continuing northward along the western boundary of the Brazil Basin after traversing these passages, the AABW next is confronted with two constrictions at the equator: an equatorial passage leading to the Ceara Abyssal Plain and the western North Atlantic, and the Romanche and Chain Fracture Zones leading through the Mid-Atlantic Ridge to the eastern Atlantic. Both passages have been well instrumented with current meters and estimates for the throughflow of AABW are $1.2 \times 10^6\,m^3\,s^{-1}$ for the Romanche and Chain and $2 \times 10^6\,m^3\,s^{-1}$ for the equatorial passage into the Cearra Abyssal Plain. The Romanche–Chain complex has been particularly well surveyed using swath technology: a three-dimensional view of this region is reproduced in **Figure 3**. The profound influence of this deep passage on the distribution of water properties is made clear in **Figure 4**, a potential temperature section along the axis, where we see the rapid fall of isotherms just downstream of the main sill.

From the above estimates of transport it is noted that the amount of AABW entering the Brazil Basin exceeds that leaving by about a factor of 2, a fact which has permitted the estimation of the rate of mixing of deep water masses (see the 'Broader implications' section).

A small supply of AABW also enters the eastern South Atlantic through the Walvis Passage in the Walvis Ridge – a not very well-defined depression in

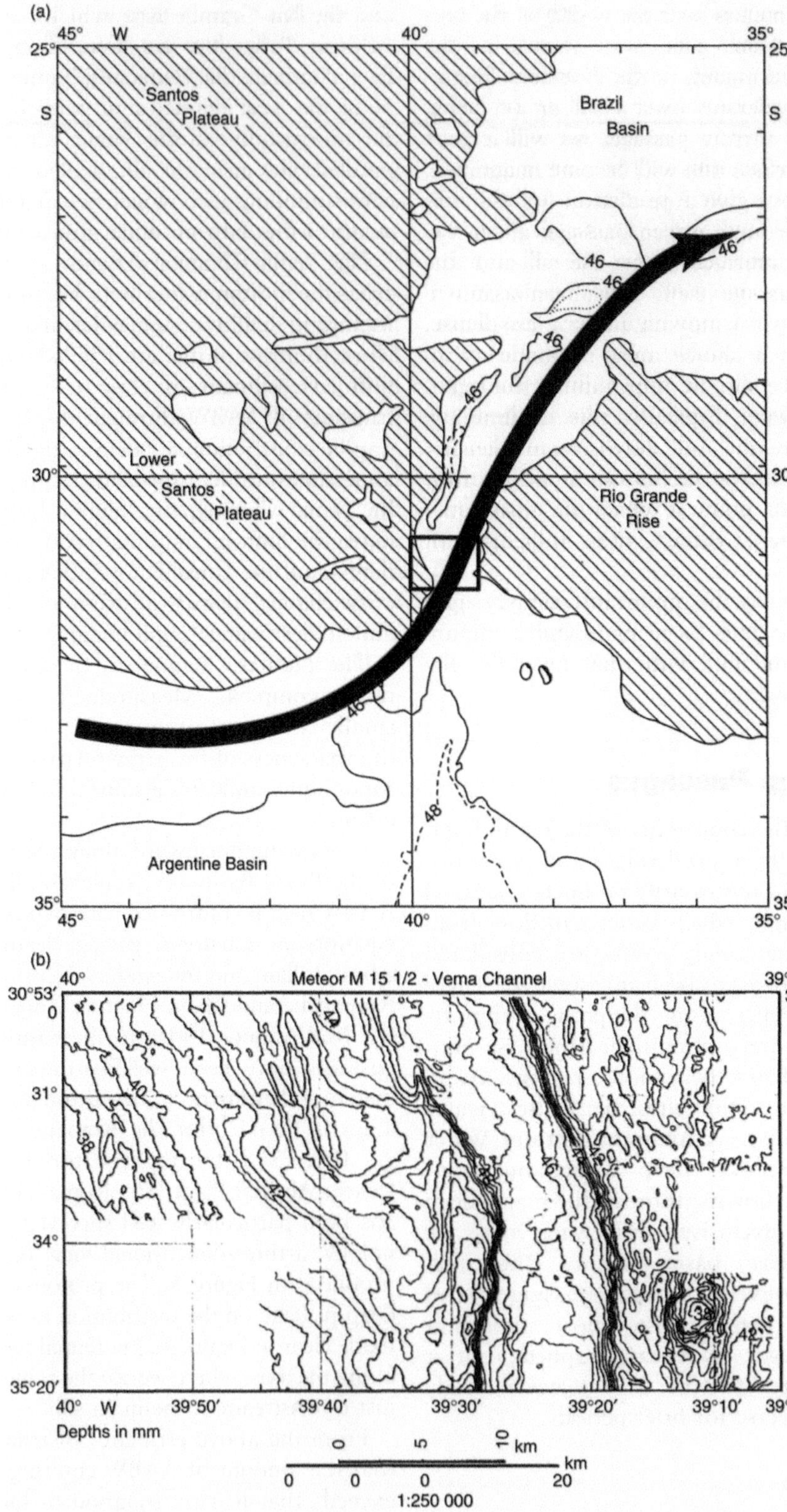

Figure 2 The topography of the Vema Channel which allows bottom water to flow between the Argentine Basin to the south and the Brazil Basin to the north. (a) The full length of the channel is shown. Hatching shows areas that are less than 4600 m deep. Units on contours are in hundreds of meters. (b) A detailed survey surrounding the channel's sill – the shallowest point along the axis of the channel – and delineated by the box in (a). This was measured using modern depth-sounding instrumentation, which permits a swath of the bottom to be observed as the ship moves forward.

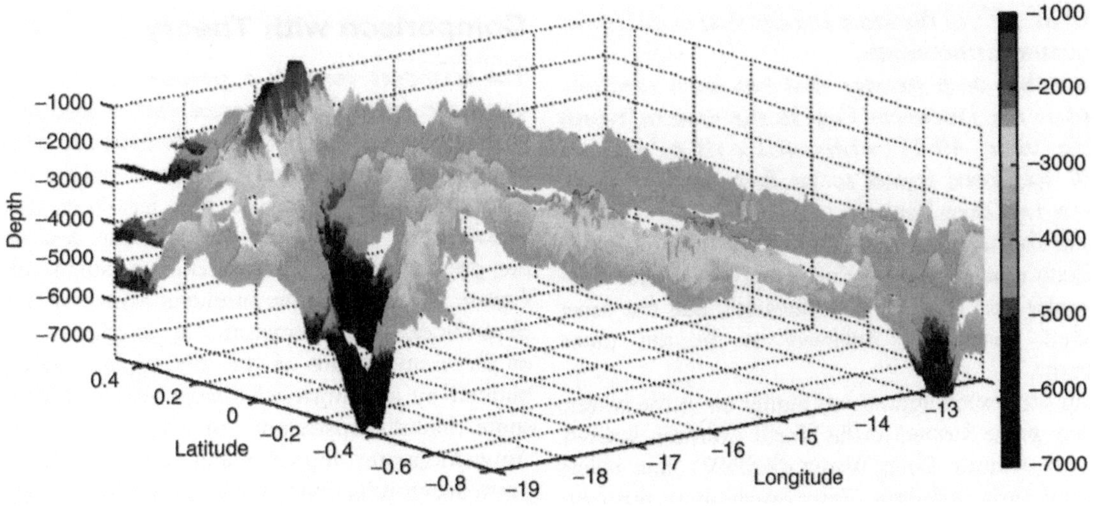

Figure 3 A three-dimensional view of the topography of the Romanche and Chain Fracture Zones which permit bottom water to flow through the Mid-Atlantic Ridge and thereby connect the western basin with the eastern one. Again, this was observed using a swath depth sounder. The Romanche Fracture Zone is the deep fissure seen near 0.4° S latitude while the Chain can be seen near 13° S.

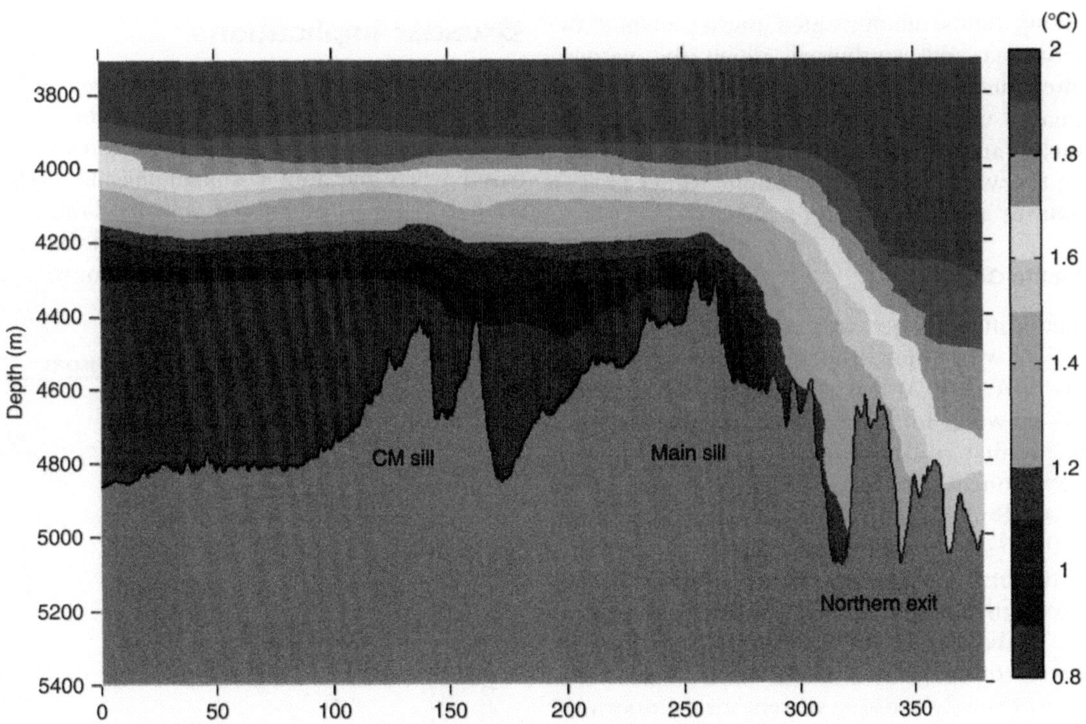

Figure 4 The variation of potential temperature along the axis of the Romanche Fracture Zone revealing the blocking caused by the main sill and the resemblance to water flowing over a dam.

a ridge system which connects southern Africa with the Mid-Atlantic Ridge and separates the Angola and Cape Basins.

From the Ceara Abyssal Plain, AABW enters the western North Atlantic but instead of continuing as a Deep Western Boundary Current, it flows northward along the western flanks of the Mid-Atlantic Ridge until it reaches about 11° N where another deep fracture zone, known as the Vema Gap (also known as the Vema Fracture Zone), cuts through the ridge to the east. Various investigations put the transport of bottom water through this gap at about

$2 \times 10^6\, m^3\, s^{-1}$, or the same amount that came across the equator farther south.

One other deep passage that has been carefully studied is the Discovery Gap in the eastern North Atlantic near 40° N where $0.2 \times 10^6\, m^3\, s^{-1}$ of AABW has been found to be flowing northward between two deep basins.

Given the rugged nature of the Mid-Atlantic Ridge it is likely that there are other deep passes that bottom water can find its way through, but the ones described above are believed to be the most important.

Being somewhat lighter and higher up in the water, the deep water formed in the North Atlantic, known as North Atlantic Deep Water (NADW), has a less torturous path to follow. After overflowing into the eastern North Atlantic through the Faroe Channel, this water mass passes through the Mid-Atlantic Ridge at the Charlie-Gibbs Fracture Zone near 53° N. It then turns north and meets up with another tributary at the Denmark Straits. From there it begins a long, almost uninterrupted journey around the Labrador Sea and southward along the western margin of the North Atlantic, eventually arriving at the equator where a part heads east through the Romanche and Chain Fracture Zones. In fact, though, the whole of the Mid-Atlantic Ridge crest is quite porous at the depth of the NADW.

The Pacific Ocean

Although a much larger ocean, the Pacific seems to present less well defined topographic constraints to the flow of AABW. At least, few passages have been studied. Flowing northward from the circumpolar region, the first real constriction is at the Samoan Passage through the Robie Ridge at about 10° S where an 18-month current meter array measured *c.* $6 \times 10^6\, m^3\, s^{-1}$ flowing northward.

The bottom topography farther north is more complex with no well defined constrictions except for the Wake Island Passage at 18° N where a northward transport of $3.6 \times 10^6\, m^3\, s^{-1}$ has been estimated from 2-year-long current meter moorings.

The Indian Ocean

The Indian Ocean has several meridional ridges but few gaps in them have been discovered to be important for the northward progression of AABW. The Amirante Passage connecting the Mascarene Basin with the Somali Basin in the Northwest Indian Ocean is one. No long-term measurements with current meters have been made but a transport of bottom water northward of about $1.5 \times 10^6\, m^3\, s^{-1}$ has been estimated from hydrographic measurements.

Comparison with Theory

The transport through a passage can be predicted using the formulas presented earlier and sufficient hydrographic data to be able to estimate an appropriate value for the upstream head and the effective gravitational acceleration. This has been done for a subset of the passages that were described in the previous section and a comparison is given in **Figure 5**. Generally the predicted transport is higher than the observed, sometimes by as much as a factor of 2. A small part of the discrepancy can be explained by an improved theory which does not assume that the upstream reservoir is infinitely deep. An even greater improvement can be made by using more accurate geometry. For example, if a parabolic cross section is used for the Vema Channel, the predicted transport is reduced to $4.5 \times 10^6\, m^3\, s^{-1}$, not far from the observed $4 \times 10^6\, m^3\, s^{-1}$.

Broader Implications

The understanding of flow through deep passages is more than just of intellectual interest for several reasons. These are important connections between ocean basins where the flow is topographically constrained, so that it can be monitored and the flow of bottom water quantified as has been detailed above. Determining the amount of water flowing through a passage

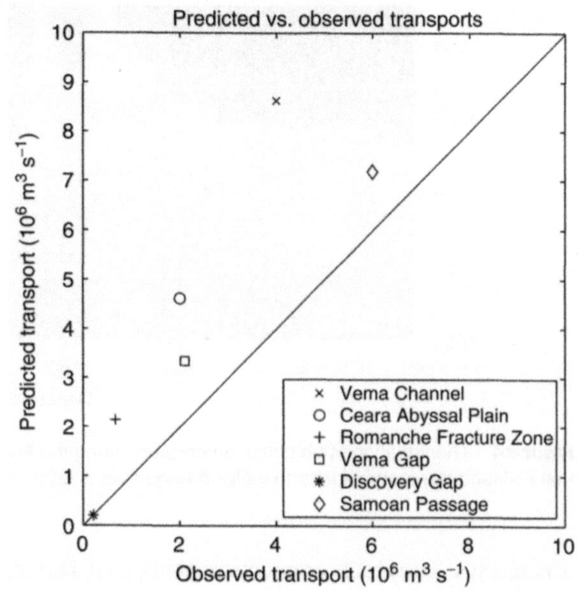

Figure 5 A comparison of measured bottom water fluxes through passages with that predicted using hydraulic control theory and simplified rectangular cross sections for the various passages. Improved predictions can be made using more accurate topography.

is an expensive and laborious task usually involving arrays of current meters deployed over long-enough periods that turbulent fluctuations can be averaged out to obtain accurate estimates. Modern-day mooring technology puts practical limits on these deployments on the order of 2 years, so few of these deep passages have been investigated for long enough to observe signals that might be related to climate change. One that has is the Vema Channel connecting the Argentine and Brazil Basins where hydrographic measurements have been made regularly since the early 1970s and moored measurements since about 1990. Although the flux of AABW through the passage has remained reasonably constant at $4 \times 10^6\,m^3\,s^{-1}$, the minimum temperature of the water coming through the passage increased abruptly in the early 1990s and the lowest temperatures have generally increased ever since, as shown in **Figure 6**.

Anyone who has watched water flowing over a dam will have noticed the dramatic change from the swiftly moving but smooth surface at the top of the dam to the violent turbulence and eddying at the bottom. One might expect there to be an analogous process occurring in deep passages but evidence is scant except in the Romanche Fracture Zone, where measurements with sensitive instruments from which mixing can be inferred indicate elevated rates some 1000 times higher than found upstream away from the fracture zones. To be fair, similar measurements have not been made elsewhere but this is the only passage studied for which the Earth's rotation is an unimportant factor: hence the water can tumble downstream from the sill without being steered along the bathymetry by the Coriolis force as happens in the Vema Channel, for example. The fact that the properties of the most dense water in the Vema Channel change little over its 400-km length suggest that mixing is not an important process there.

This being said, though, estimates of mixing across density surfaces in the deep ocean can be made using knowledge of the fluxes through deep passages. Taking the Vema Channel as an example, of the $4 \times 10^6\,m^3\,s^{-1}$ of AABW that fluxes northward into the Brazil Basin, *c.* $2 \times 10^6\,m^3\,s^{-1}$ is colder than $0\,°C$. No water this cold is found within the Romanche or Chain Fracture Zone and none is observed in the equatorial passage into the western North Atlantic, either. In fact, information from hydrographic sections within the Brazil Basin shows that this cold water penetrates no farther than about halfway up the basin. The conclusion is that either the deep water is not in a steady state and is continuing to increase its volume within the basin, or the cold water is migrating upward across isotherms.

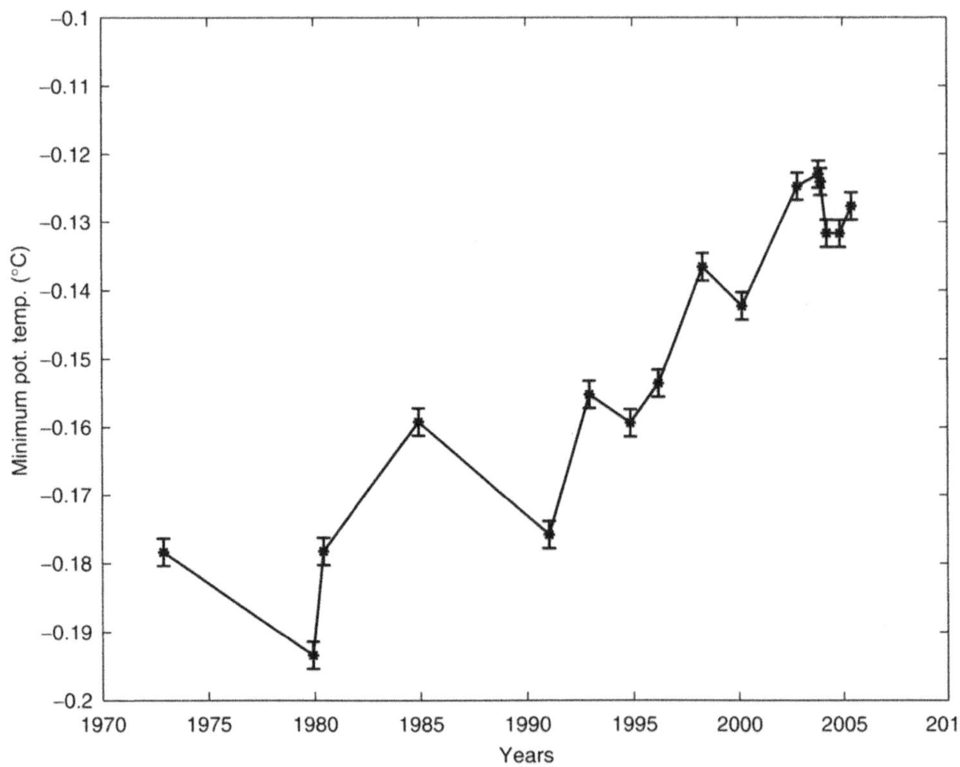

Figure 6 The temperature of the coldest water flowing through the Vema Channel as measured at various times since 1970.

The measurements shown in **Figure 5** might support the former explanation but it can be easily shown that the expansion into the basin should be observable over the period of time that accurate enough measurements be made: essentially from the German *Meteor* Expedition of the 1920s until today. Such a change has not been observed. Taking the second explanation and using the principle of conservation of mass allows one to calculate the rate that fluid must, on average, cross a particular isotherm. This upward migration of cold water must be warmed by a downward flux of heat through diffusion. Conservation of heat then permits an estimation of the thermal diffusion coefficient. This is not the molecular value: small-scale processes such as breaking internal waves give elevated mixing rates in the ocean. One of the great quests on oceanography has been to estimate and map out mixing rates in the ocean. Deep basins, such as the Brazil Basin, have permitted calculation of regionally averaged mixing rates. These calculations have given numbers varying from around $10^{-4}\,m^2\,s^{-1}$ (the molecular value is $10^{-7}\,m^2\,s^{-1}$) to well over $10^{-2}\,m^2\,s^{-1}$.

Such mixing rates are much higher than those found higher in the water column where $10^{-5}\,m^2\,s^{-1}$ is more typical. Careful surveys have discovered that these higher rates are associated with topographic roughness such as that of the Mid-Atlantic Ridge, most likely as a result of the barotropic tide flowing over the corrugations, radiating internal waves that then break and mix the surrounding fluid.

Final Thoughts

Clearly these deep passages put important constraints on the flow of deep and bottom water in the ocean both through their control on the actual fluxes and through potential mixing related to the local acceleration of the flow and the possibility of hydraulic jumps downstream of the control point. Therefore, they are an important part of the global overturning circulation and, consequently, play a possible role in long-term climate change. However, they are typically no more than a few tens of kilometers wide and present a real challenge to current numerical modeling efforts aimed at both reproducing today's circulation and predicting climate change over the next few hundred years. This is an area of active research at present.

See also

Abyssal Currents.

Further Reading

Gill AE (1977) The hydraulics of rotating-channel flow. *Journal of Fluid Mechanics* 80: 641–671.

Hall M, McCartney M, and Whitehead JA (1997) Antarctic Bottom Water flux in the equatorial western Atlantic. *Journal of Physical Oceanography* 27: 1903–1926.

Hogg N, Biscaye P, Gardner W, and Schmitz WJ, Jr. (1982) On the transport and modification of Antarctic bottom water in the Vema Channel. *Journal of Marine Research* 40(supplement): 231–263.

Mercier H and Speer KG (1998) Transport of bottom water in the Romanche Fracture Zone and the Chain Fracture Zone. *Journal of Physical Oceanography* 28: 779–790.

Morris MY, Hall MM, St. Laurent LC, and Hogg NG (2001) Abyssal mixing in the Brazil Basin. *Journal of Physical Oceanography* 31: 3331–3348.

Polzin KL, Toole JM, Ledwell JR, and Schmitt RW (1997) Spatial variability of turbulent mixing in the abyssal ocean. *Science* 276: 93–96.

Pratt LJ and Whitehead JA (2008) *Rotating Hydraulics: Nonlinear Topographic Effects in the Ocean and Atmosphere*. New York: Springer/Kluwer.

Rudnick DL (1997) Direct velocity measurements in the Samoan Passage. *Journal of Geophysical Research* 102: 3293–3302.

Whitehead JA (1998) Topographic control of oceanic flows in deep passages and straits. *Reviews of Geophysics* 36: 423–440.

BOTTOM WATER FORMATION

A. L. Gordon, Columbia University,
Palisades, NY, USA

Introduction

Meridional sections of temperature and salinity through the Pacific and Atlantic Oceans (**Figure 1**) reveal that in the Pacific below 2000 m, more than half of the ocean depth, the water is colder than 2°C. The Atlantic is somewhat warmer, but there too the lower 1000 m of the ocean is well below 2°C. Only within the surface layer, generally less than the upper 500 m of the ocean is the water warmer than 10°C, amounting to only 10% of the total ocean volume. The coldness of the deep ocean is due to interaction of the ocean with the polar atmosphere. There, surface water reaches the freezing point of sea water. Streams of very cold water can be traced spreading primarily from the Antarctic along the sea floor, warming en route by mixing with overlying water, into the world's oceans (**Figure 2**).

The coldest bottom water, Antarctic Bottom Water (AABW), is derived from the shores of Antarctica. There, freezing point, high oxygen concentration, water is produced during the winter over the continental shelf. At a few sites the shelf water salinity is

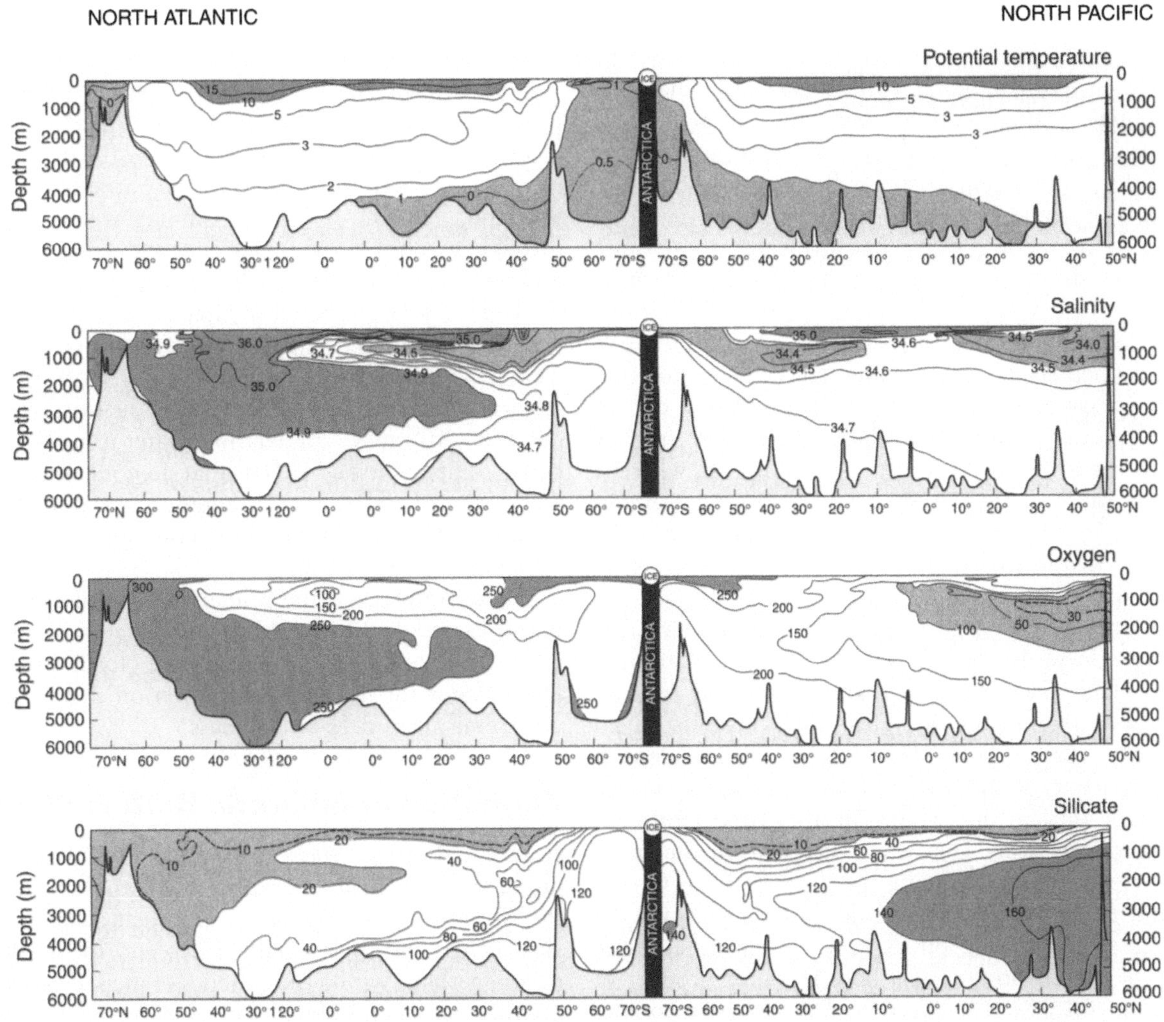

Figure 1 Meridional sections of temperature and salinity through the Pacific and Atlantic Oceans. Antarctica is at the center of the figure.

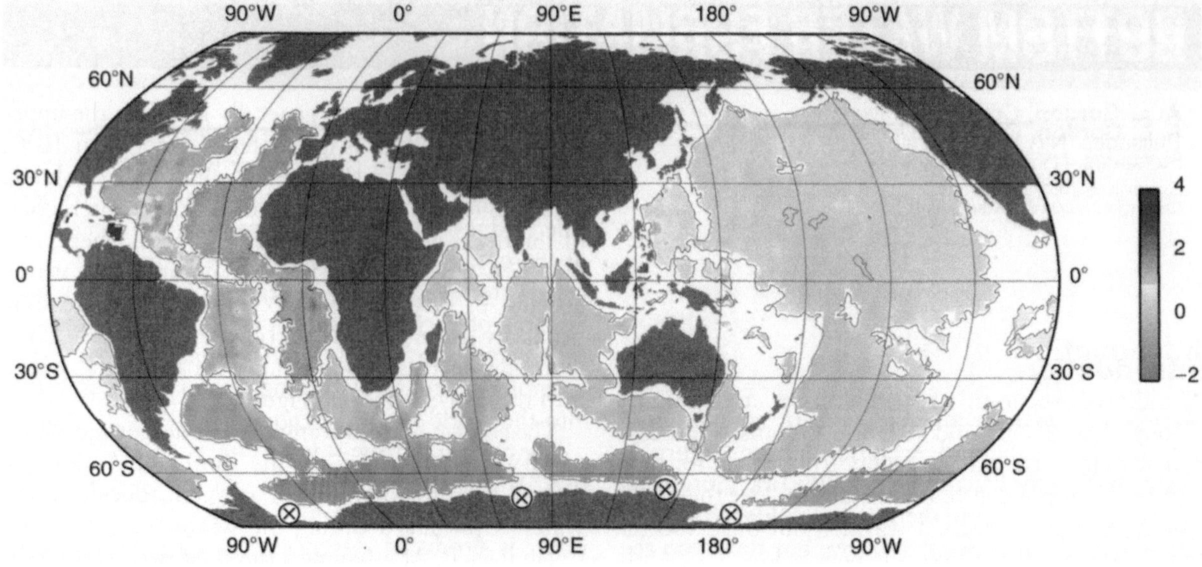

Figure 2 Bottom potential temperature along the seafloor, for oceanic areas deeper than 4000 m. The four symbols along the coast of Antarctica mark the places where Antarctic Bottom water forms.

sufficiently high, greater than 34.61‰, that, on cooling to the freezing point, the surface water density is sufficiently high to allow it to sink to great depths of the ocean. As the shelf water descends over the continental slope into the deep ocean it mixes with adjacent deep water, but this water is also quite cold so the final product arriving at the seafloor at the foot of Antarctica is about $-1.0°C$. Definitions used by different authors vary, but generally AABW is defined as having a potential temperature (the temperature corrected for adiabatic heating due to hydrostatic pressure) less than 0°C. AABW spreads into the lower 1000 m of the world ocean, where it cools and renews oxygen concentrations drawn down by oxidation of organic material within the deep ocean. AABW is said to ventilate the deep ocean.

In the Atlantic Ocean the 2°C isotherm marks the base of a wedge of relatively salty water, associated with high dissolved oxygen and low silicate concentrations (see **Figure 1**). This water mass is called North Atlantic Deep Water (NADW). The densest component of NADW is formed as cold surface waters during the winter in the Greenland and Norwegian Seas. This water sinks to fill the basin north of a ridge spanning the distance from Greenland to Scotland. Excess cold water overflows the ridge crest, mixing on descent with warmer more saline water, producing a bottom water product of about $+1.0°C$. The overflow water stays in contact with the sea floor to near 40°N in the Atlantic Ocean, where on spreading southward it is lifted over the remnants of denser AABW.

Export of Greenland and Norwegian Sea bottom water has been estimated from a series of current measurements. Transports of about $2 \times 10^6\ m^3\ s^{-1}$ of near 0.4°C water occur between the Faroe Bank and Scotland, $1 \times 10^6\ m^3\ s^{-1}$ of similar water passes through notches between Iceland and Faroe Bank, and $3 \times 10^6\ m^3\ s^{-1}$ of near 0°C is exported through the Denmark Strait, between Greenland and Iceland. The overflow plumes rapidly entrain warmer waters, producing bottom water of near $+1.0°C$. With entrainment of other deep water, a production rate of about $8 \times 10^6\ m^3\ s^{-1}$ of overflow water is likely. Less dense components of NADW, that do not contact the seafloor are formed in the Labrador Sea and Mediterranean Sea. The total production of NADW is estimated as $15 \times 10^6\ m^3\ s^{-1}$.

As the Antarctic is the primary source of the cold bottom waters of the world ocean, Antarctic Bottom Water is discussed in this article. See **North Atlantic Deep Water** for further information on that Northern Hemisphere deep water mass.

Formation of Antarctic Bottom Water

Antarctic Bottom Water is formed at a few sites along the continental margin of Antarctica during the winter months (**Figure 2**). The shelf dense water slips across the shelf break to descend to the deep ocean, perhaps within the confines of incised canyons on the continental slope. The shelf water is made dense (**Figure 3**) by the very cold winds from Antarctica, which spur the formation of sea ice. Normally sea ice acts to insulate the ocean from further

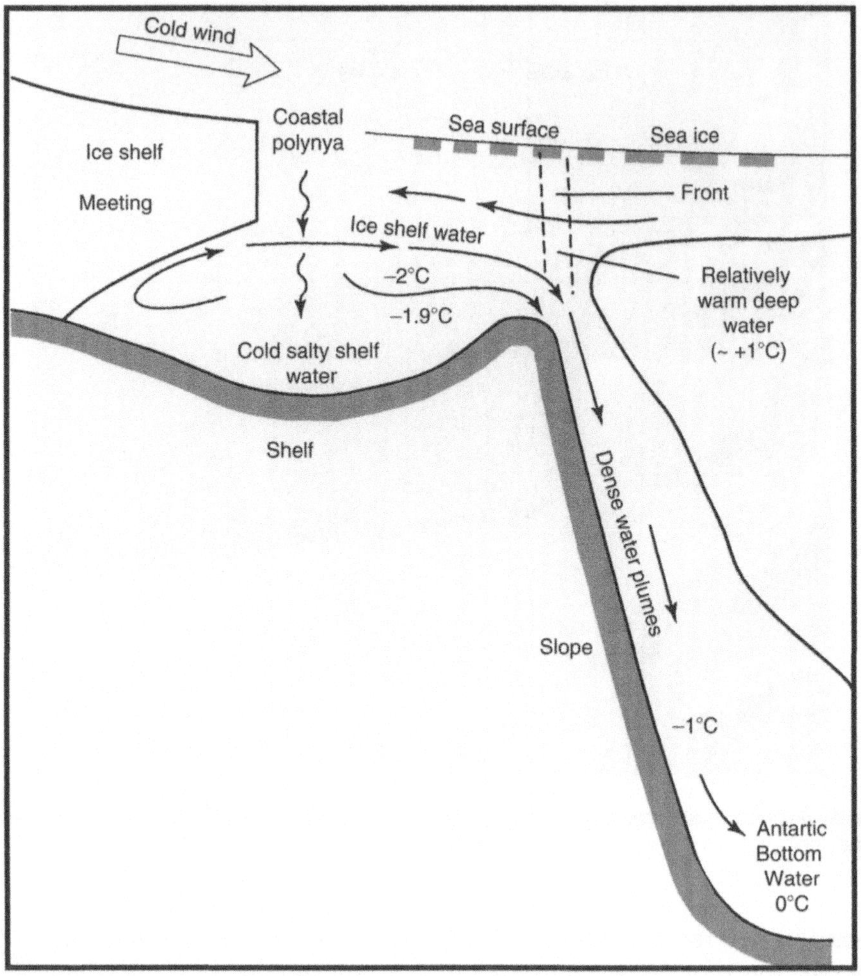

Figure 3 Schematic of circulation and water masses in the Antarctic continental shelf.

heat loss and thus attenuates the continued formation of sea ice. But along the shores of Antarctica sea ice is blown northward by the strong winds descending over the cold glacial ice sheet of Antarctica. The removal of the sea ice exposes the ocean water to the full blast of the cold air, forming coastal polynyas (persistent bands of ice-free ocean adjacent to Antarctica; **Figure 4**). Production and removal of yet more sea ice continues within the coastal polynyas, which act as 'sea ice factories'. As sea ice has a lower salinity than the sea water from which it formed, approximately 5‰ versus 34.5‰ of the sea water, salt rejected during ice formation, concentrates in the remaining freezing point sea water making it saltier, and hence denser. The exposure of the ocean to the atmosphere also raises the dissolved oxygen concentration within the shelf water. As the ability of sea water to hold oxygen increases with lowering temperature, the oxygen concentration of the shelf water is very high, about 8 ml l^{-1}.

The shelf water at some sites, such as the Weddell and Ross seas, is made even colder on contact with floating ice shelves that are composed of glacial (freshwater) ice. Ocean contact with glacial ice occurs not only at the northern face of the ice sheet, but also at hundreds of meters depth along the bases of floating ice shelves. As the freezing point of sea water is lowered with increasing pressure (-0.07°C per 100 m of depth), the shelf water in contact with the base of the ice shelves often at a depth of many hundreds of meters, attains temperatures well below -2.0°C. The cooling of shelf water in contact with the glacial ice is linked to melting of the glacial ice, hence this very cold water is slightly less saline than the remaining shelf water. Ocean–glacial ice interaction is believed to be a major factor in controlling Antarctica's glacial ice mass balance and stability. The resultant water, called ice shelf water, can be identified as streams within the main mass of shelf water by its low potential temperatures. Ice shelf

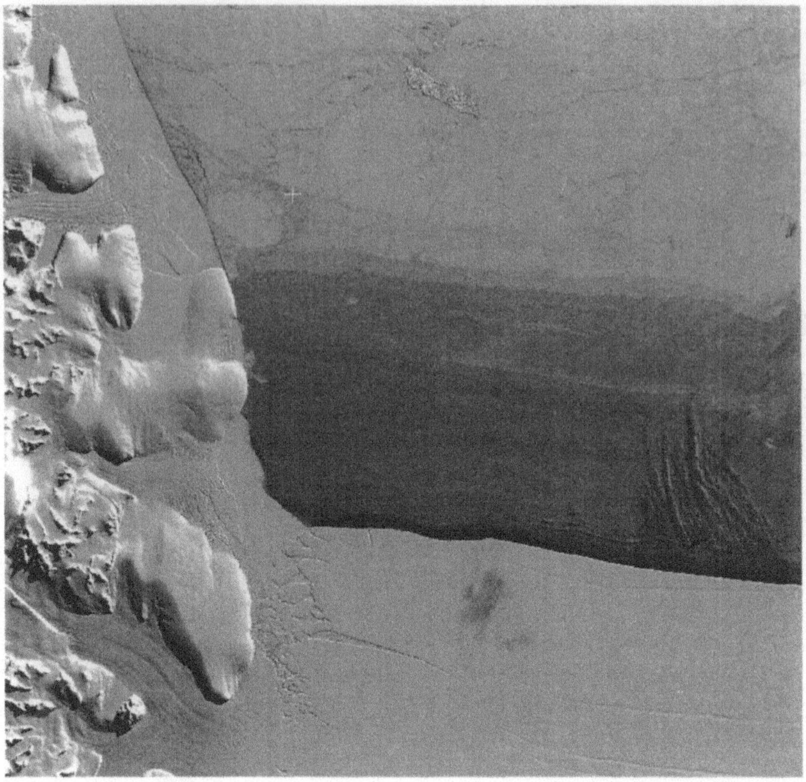

Figure 4 Satellite image of a coastal polynya shown as the dark region extending from the ice-covered shore and sea ice on the left. (*See also* **Polynyas**.)

water with potential temperatures as low as -2.2°C have been measured. This very cold water may act to encourage formation of AABW, as the seawater compressibility increases with lowered temperature. Thus as the shelf water begins its descent to the deep ocean, compressibility of the ice shelf water induces water of greater density, which accelerates the descent, limiting mixing with adjacent water.

As shelf water escapes the shelf environs, offshore water must flow onto the shelf to compensate for the shelf water loss. The offshore water is warmer than the dense shelf water, with temperatures closer to $+1.0^{\circ}$C. Once on the shelf this water cools to renew the reservoir of freezing point shelf water. The onshore flow is drawn from deep water of the world ocean that slowly flows southward and upward, crossing the Antarctic Circumpolar Current. It may be viewed as 'old' AABW returning from its northern sojourn. In this way the overturning thermohaline circulation cell forced by AABW formation is closed; the escape of very cold shelf water, spreading northward, mixing en route with warmer overlying water, eventually results in upwelling to return to the Antarctic. The whole process takes some hundreds of years. Only now can the chlorofluorocarbons (CFC) added to the ocean surface layer in the last 70 or so years, be detected reaching along the seafloor into the midlatitudes of the Southern Hemisphere.

Formation Rate of Antarctic Bottom Water

Measurement of the formation and escape of Antarctic shelf water and the subsequent formation of Antarctic Bottom Water is difficult because the formation regions are geographically remote, covered by sea ice year-round, and distributed along a shelf break frontal region that extends more than 18 000 km around Antarctica. In addition, the thermohaline properties of source waters vary spatially, which can lead to quite different ideas regarding the mixing 'recipes' and specific processes leading to the ultimate cold products. It can be argued that for every 1×10^6 m^3 of shelf water escaping 2×10^6 m^3 of AABW is formed.

The best observed bottom water formation is in the Weddell Sea (the extreme southern Atlantic Ocean). There, deep-reaching convective plumes of very cold surface water descend over the continental slope into the deep ocean (**Figure** 5), producing Weddell Sea Bottom Water. This is a particularly cold form of AABW, having initial potential temperatures

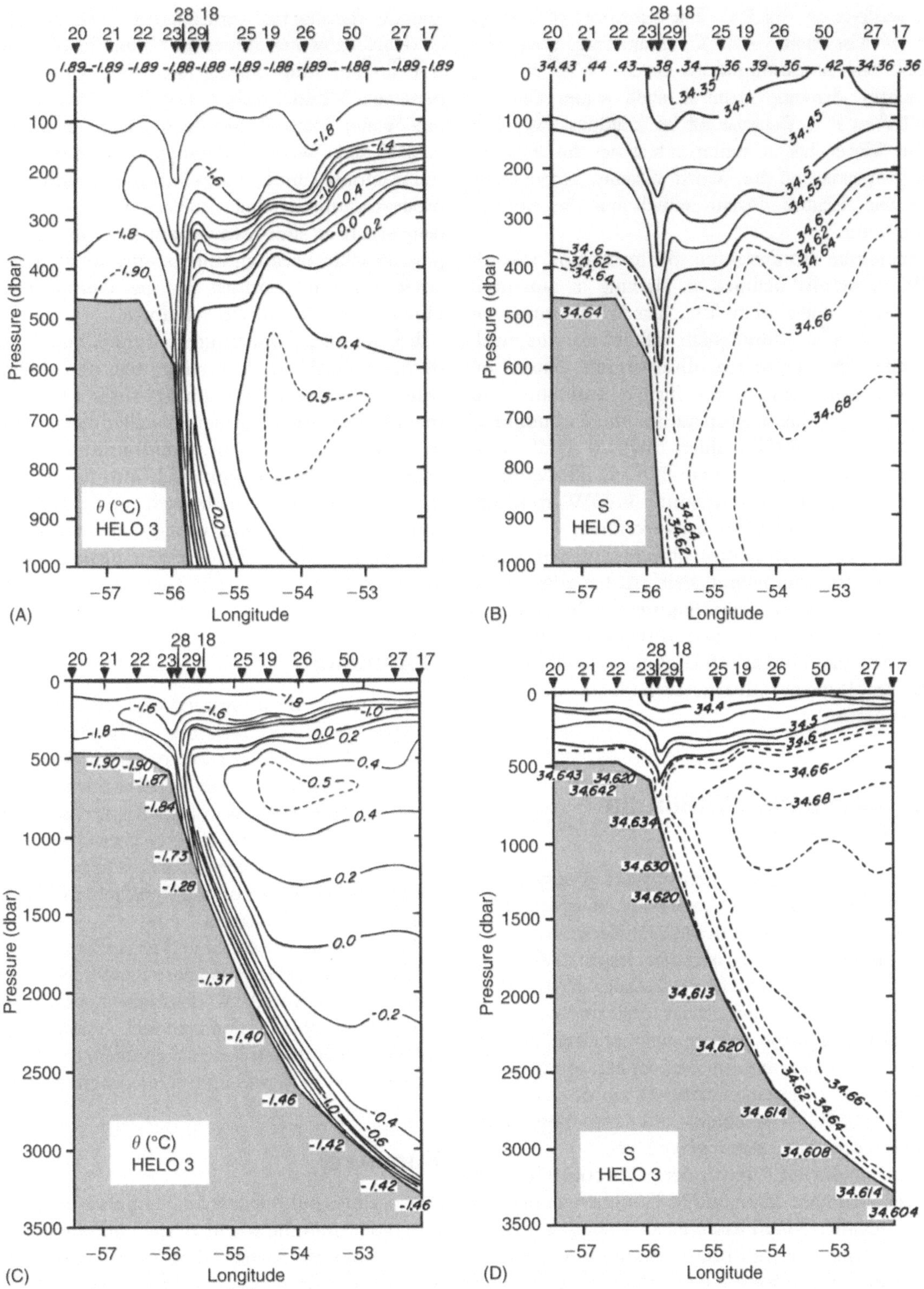

Figure 5 Potential temperature (A,C) and salinity (B,D) along a section at 67°40′S in the western Weddell Sea, showing the varied stratification over the continental shelf and slope. The sea floor is stippled. The values inserted along the seafloor are the bottom temperature and salinity within the descending plume of dense water. The sharp change in water properties on the upper 500 m at 56°W marks the shelf/slope front. (Reproduced from Gordon AL (1998) Western Weddell Sea thermohaline stratification. In: Jacobs SS and Weiss R (eds) *Ocean, Ice and Atmosphere*, Antarctic Research Series, vol. 75, pp. 215–240. Washington, DC: American Geophysical Union.)

at the seafloor of $-1.5°C$. The transport of bottom water of less than $-0.7°C$ emanating from the Weddell Sea is estimated to be $2–5 \times 10^6 m^3 s^{-1}$, presumably drawing from a shelf water flux of $1–3 \times 10^6 m^3 s^{-1}$. Because of its coldness Weddell Bottom Water has a major effect on the bottom water properties of the World Ocean, particularly within the Atlantic Ocean which has the coldest bottom water.

Circumpolar estimates for the formation rate of AABW, generally defined as having a potential temperature of less than 0°C, are in the range of $10–15 \times 10^6 m^3 s^{-1}$, but such values are not well constrained by the sparse observations. Based on CFC measurements a firm estimate of $9.4 \times 10^6 m^3 s^{-1}$ has been made for the circumpolar production of AABW colder than $-1.0°C$ descending to depths greater than 2500 m. This value is similar to estimates of descending NADW from the Greenland and Norwegian Seas overflow.

It is not clear what controls the rate of shelf water export from the continental shelf. At the edge of the shelf is a strong ocean front (**Figure 5**). Movement of this front may be associated with escape of dense shelf water. The presence of canyons incised into the continental slope may also act as paths for descent to the deep ocean.

Deep Convection Within the Southern Ocean

In addition to descending dense water plumes over the continental slope of Antarctica, deep reaching convection over the deep ocean may occur. In winter, a thin veneer of sea ice stretches from Antarctica northward, reaching half the distance to the Antarctic Circumpolar Current. A delicate balance is achieved between the cold atmosphere and upward flux of oceanic heat into the atmosphere, resulting in the formation of approximately 0.6 m of sea ice. In this balance cold, low-salinity water sits stably over a warmer, more saline deep water layer.

During the austral winters of 1974 to 1976, near the Greenwich Meridian and 66°S in the vicinity of a seamount called Maud Rise, the ice displayed strange behavior, which has not been repeated since. A large region normally covered by sea ice in winter remained ice free throughout the winter, though it was surrounded by sea ice. This remarkable anomaly, is referred to as the Weddell Polynya. Though a full Weddell Polynya has not been observed since the mid-1970s, short lived polynyas (lasting roughly one week) are frequently observed by satellite imaging in the Maud Rise region. During the Weddell Polynya episode the normal stratification was disturbed as cold surface water convected to depths of 3000 m. It is estimated that during the three winters of the persistent Weddell Polynya, $1–3 \times 10^6 m^3 s^{-1}$ of surface water entered the deep ocean, as a form of AABW. The Weddell Polynya convection may represent another mode of operation of Southern Ocean processes, one in which surface water sinks into the deep ocean at sites away from the continental margins, in what oceanographers refer to as open ocean convection, in contrast to the continental slope plume convection discussed above.

It is not clear what triggered the Weddell Polynya of the mid-1970s. It is clear that if enough deep water can be brought to the sea surface to melt all of the ice cover, then further upwelled deep water, not to be diluted by ice melt, would produce cold saline water that can then sink back into the deep ocean. What would cause enhanced upwelling of deep water? This is not clear, though interaction of ocean circulation with the Maud rise is suspected to play a key role.

Conclusion

Bottom water of the western Weddell Sea in the early 1990s is colder and fresher than observed in previous decades. The same is true for the Southern Ocean south of Australia. These observations suggest an increased role of the low salinity, ice shelf water in recent decades. However, the period over which observations have been taken within the hostile Southern Ocean is not long enough to place much importance on this trend.

The presence of sea ice and the rather small spatial and temporal scales associated with the convective plumes, makes AABW formation processes very difficult to observe and model. Advancement of AABW research represents a significant technological challenge to field and computer oceanographers.

Summary

The ocean is cold. Its average temperature of 3.5°C is far colder than the warm veneer capping much of the ocean. Waters warmer than 10°C amount to only 10% of the total ocean volume; about 75% of the ocean is colder than 4°C. Along the seafloor the ocean temperature is near 0°C. The cold bottom water is derived from Southern Ocean, the ocean belt surrounding Antarctica. Sea water at its freezing point of $-1.9°C$ is formed in winter over the continental shelf of Antarctica. Where the salt content of shelf water is high enough, roughly 34.61‰ (parts per thousand) the water is sufficiently dense to descend as convective plumes over the continental slope into the adjacent deep ocean. In so doing Antarctic Bottom Water is formed. It is estimated that on average between 10 and $15 \times 10^6\,m^3$ of Antarctic bottom water forms every second! Antarctic Bottom Water spreads away from Antarctica into the world oceans, chilling the deep ocean to temperatures near 0 C. Bottom water warms en route on mixing with warmer overlying waters. Cold winter water also forms in the Greenland and Norwegian Seas of the northern North Atlantic. This water ponds up behind a submarine ridge spanning the distance from Greenland to Scotland. This water overflows the ridge crest into the ocean to the south. As the overflow water mixes with warmer saltier water during descent into the deep ocean, it results in a warmer, more saline water mass than Antarctic Bottom water. The Greenland and Norwegian Sea overflow water forms the densest component of the water mass called North Atlantic Deep Water and is estimated to form at a rate of around $8 \times 10^6\,m^3\,s^{-1}$. The overflow water stays in contact with the seafloor until the northern fringes of Antarctic Bottom Water encountered in the North Atlantic near 40°N, lifts it to shallower levels.

See also

Antarctic Circumpolar Current. Rotating Gravity Currents. Sub Ice-Shelf Circulation and Processes. Weddell Sea Circulation.

Further Reading

Fahrbach E, Rohardt G, Scheele N, *et al.* (1995) Formation and discharge of deep and bottom water in the northwestern Weddell Sea. *Journal of Marine Research* 53(4): 515–538.

Foster TD and Carmack EC (1976) Frontal zone mixing and Antarctic Bottom Water formation in the southern Weddell Sea. *Deep-Sea Research* 23: 301–317.

Gordon AL and Tchernia P (1972) Waters off Adelie Coast. *Antarctic Research Series*, vol. 19, pp. 59–69. Washington, DC: American Geophysical Union.

Jacobs SS, Fairbanks R, and Horibe Y (1985) Origin and evolution of water masses near the Antarctic continental marginE: vidence from $H_2{}^{18}O/H_2{}^{16}O$ ratio in seawater. In: Jacobs SS (ed.) *Oceanography of Antarctic Continental Margin*, Antarctic Research Series, vol. 43, pp. 59–85. Washington, DC: American Geophysical Union.

Jacobs SS and Weiss R (eds.) (1998) *Ocean. Ice and Atmosphere: Interactions at the Antarctic Continental Margin*, Antarctic Research Series, vol. 75. Washington DC: American Geophysical Union.

Nunes RA and Lennon GW (1996) Physical oceanography of the Prydz Bay region of Antarctic waters. *Deep-Sea Research* 43(5): 603–641.

Orsi AH, Johnson GC, and Bullister JL (1999) Circulation, mixing, and production of Antarctic Bottom Water. *Progress in Oceanography* 43: 55–109.

Tomczak M and Godfrey JS (1994) *Regional Oceanography: An Introduction*. London: Pergamon Press.

DEEP-SEA SEDIMENT DRIFTS

D. A. V. Stow, University of Southampton,
Southampton, UK

Introduction

The recognition that sediment flux in the deep ocean basins might be influenced by bottom currents driven by thermohaline circulation was first proposed by the German physical oceanographer George Wust in 1936. His, however, was a lone voice, decried by other physical oceanographers and unheard by most geologists. It was not until the 1960s, following pioneering work by the American team of Bruce Heezen and Charlie Hollister, that the concept once more came before a critical scientific community, but this time with combined geological and oceanographic evidence that was irrefutable.

A seminal paper of 1966 demonstrated the very significant effects of contour-following bottom currents (also known as contour currents) in shaping sedimentation on the deep continental rise off eastern North America. The deposits of these currents soon became known as contourites, and the very large, elongate sediment bodies made up largely of contourites were termed sediment drifts. Both were the result of semipermanent alongslope processes rather than downslope event processes. The ensuing decade saw a profusion of research on contourites and bottom currents in and beneath the present-day oceans, coupled with their inaccurate identification in ancient rocks exposed on land.

By the late 1970s and early 1980s, the present author had helped establish the standard facies models for contourites, and demonstrated the direct link between bottom current strength and nature of the contourite facies, especially grain size. Discrimination was made between contourites and other deep-sea facies, such as turbidites deposited by catastrophic downslope flows and hemipelagites that result from continuous vertical settling in the open ocean. Since then, much progress has been made on the types and distribution of sediment drifts, the nature and variability of bottom currents, and the correct identification of fossil contourites.

Of particular importance has been the work at Cambridge University in decoding the often very subtle signatures captured in contourites in terms of variation in deep-sea paleocirculation. As this is closely linked to climate, the drift successions of ocean basins hold one of the best records of past climate change. This clear environmental significance, together with the recognition that sandy contourites are potential reservoirs for deep-sea oil and gas, has spurred much current research in the field.

Bottom Currents

At the present day, deep-ocean bottom water is formed by the cooling and sinking of surface water at high latitudes and the deep slow thermohaline circulation of these polar water masses throughout the world's ocean (**Figure 1** and **2**). Antarctic Bottom Water (AABW), the coldest, densest, and hence deepest water in the oceans, forms close to and beneath floating ice shelves around Antarctica, with localized areas of major generation such as the Weddell Sea. Once formed at the surface, partly by cooling and partly as freezing sea water leaves behind water of greater salinity, AABW rapidly descends the continental slope, circulates eastwards around the continent and then flows northwards through deep-ocean gateways into the Pacific, Atlantic and Indian Oceans.

Arctic Bottom Water (ABW) forms in the vicinity of the subpolar surface water gyre in the Norwegian and Greenland Seas and then overflows intermittently to the south through narrow gateways across the Scotland–Iceland–Greenland topographic barrier. It mixes with cold deep Labrador Sea water as it flows south along the Greenland–North American continental margin. Above these bottom waters, the ocean basins are compartmentalized into water masses with different temperature, salinity, and density characteristics.

Bottom waters generally move very slowly (1–2 cm s^{-1}) throughout the ocean basins, but are significantly affected by the Coriolis Force, which results from the Earth's spin, and by topography. The Coriolis effect is to constrain water masses against the continental slopes on the western margins of basins, where they become restricted and intensified forming distinct Western Boundary Undercurrents that commonly attain velocities of 10–20 cm s^{-1} and exceed 100 cm s^{-1} where the flow is particularly restricted. Topographic flow constriction is greater on steeper slopes as well as through narrow passages or gateways on the deep seafloor.

Bottom currents are a semipermanent part of the thermohaline circulation pattern, and sufficiently

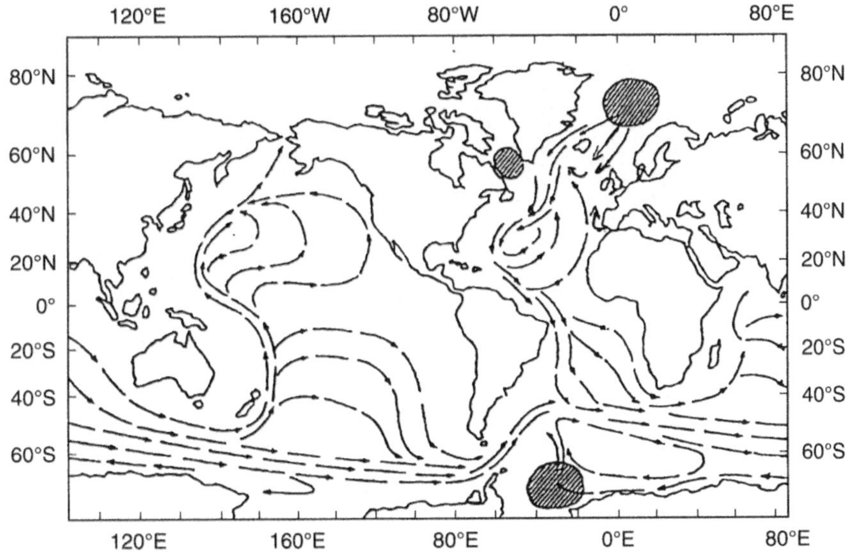

Figure 1 Global pattern of abyssal circulation. Shaded areas are regions of production of bottom waters. (After Stow*et al.*, 1996).

competent in parts to erode, transport and deposit sediment. They are also highly variable in velocity, direction, and location. Mean flow velocity generally decreases from the core to the margins of the current, where large eddies peel off and move at high angles or in a reverse direction to the main flow. Tidal, seasonal, and less regular periodicities have been recorded during long-term measurements, and complete flow reversals are common. Variation in kinetic energy at the seafloor results in the alternation of short (days to weeks) episodes of high velocity known as benthic storms, and longer periods (weeks to months) of lower velocity. Benthic storms lead to sediment erosion and the resuspension of large volumes of sediment into the bottom nepheloid layer. They appear to correspond to episodes of high surface kinetic energy due to local storms.

Deep and intermediate depth water is also formed from relatively warm surface waters that are subject to excessive evaporation at low latitudes, and hence to an increase in relative density. This process is generally most effective in semi-enclosed marginal seas and basins. The Mediterranean Sea is currently the principal source of warm, highly saline, intermediate water, that flows out through the Strait of Gibraltar and then northwards along the Iberian and north European margin. At different periods of Earth history warm saline bottom waters will have been equally or more important than cold water masses.

Contourite Drifts

Contourite accumulations can be grouped into five main classes on the basis of their overall morphology: (I) contourite sheet drifts; (II) elongate mounded drifts; (III) channel-related drifts; (IV) confined drifts; and (V) modified drift–turbidite systems (**Table 1**, **Figure 3**). It is important to note, however, that these distinctive morphologies are simply type members within a continuous spectrum, so that all hybrid types may also occur. They are also found at all depths within the oceans, including all deep-water (>2000 m) and mid-water (300–2000 m) settings. Those current-controlled sediment bodies that occur in shallower water (50–300 m) on the outer shelf or uppermost slope are not considered contourite drifts *sensu stricto*. The occurrence and geometry of these different types is controlled principally by five interrelated factors: the morphological context or bathymetric framework; the current velocity and variability; the amount and type of sediment available; the length of time over which the bottom current processes have operated; and modification by interaction with downslope processes and their deposits.

Contourite Sheet Drifts

These form extensive very low-relief accumulations, either as part of the fill of basin plains or plastered against the continental margin. They comprise a layer of more or less constant thickness (up to a few hundred meters) that covers a large area, but that demonstrates a very slight decrease in thickness towards its margins, i.e., having a very broad low-mounded geometry. The internal seismofacies is typically one of low amplitude, discontinuous reflectors or, in some parts, is more or less transparent. They may be covered by large fields of sediment waves, as in the case of the South Brazilian and

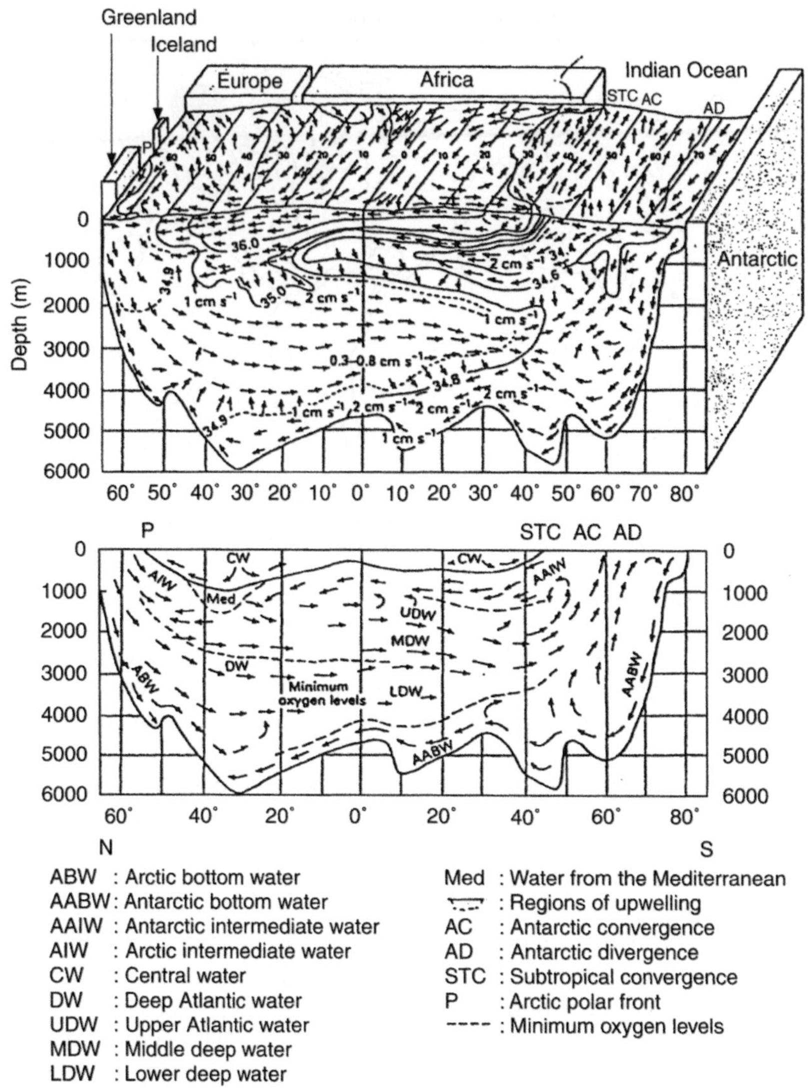

Figure 2 Bottom water masses in the North Atlantic Ocean (Reproduced from Stow *et al.*, 1996).

Argentinian basins where they are also capped in the central region by giant elongate bifurcated drifts.

The different hydrological and morphological contexts define either abyssal sheets or slope sheets (also known as plastered drifts). The former carpet the floors of abyssal plains and other deep-water basins including those of the South Atlantic and the central Rockall trough in the north-east Atlantic. The basin margin relief partially traps the bottom currents and determines a very complex gyratory circulation. Slope sheets occur near the foot of slopes where outwelling or downwelling bottom currents exist, such as in the Gulf of Cadiz as a result of the deep Mediterranean Sea Water outwelling at an intermediate water level into the Atlantic, or around the Antarctic margins as a result of the formation and downwelling of cold AABW. They are also found plastered against the slope at any level, particularly where gentle relief and smooth topography favors a broad nonfocused bottom current, such as along the Hebrides margin and Scotian margin.

Abyssal sheet drifts typically comprise fine-grained contourite facies, including silts and muds, biogenic-rich pelagic material, or manganiferous red clay, interbedded with other basin plain facies. Accumulation rates are generally low – around 2–4 cm ky^{-1}. Slope sheets are more varied in grain size, composition and rates of accumulation. Thick sandy contourites have been recovered from base-of-slope sheets in the Gulf of Cadiz, and rates of over 20 cm ky^{-1} (1000 years) are found in sandy–muddy contourite sheets on the Hebridean slope.

Elongate Mounded Drifts

This type of contourite accumulation is distinctly mounded and elongate in shape with variable

Table 1 Drift morphology, classification and dimensions

Drift type	*Subdivisions*	*Size (km^2)*	*Examples*
Contourite sheet drift	Abyssal sheet	10^5–10^6	Argentine basin; Gloria Drift
	Slope (plastered sheet)	10^3–10^4	Gulf of Cadiz; Campos margin
	Slope (patch) sheets	10^3	
Elongated mounded drift	Detached drift	10^3–10^5	Eirek drift; Blake drift
	Separated drift	10^3–10^4	Feni drift; Faro drift
Channel-related drift	Patch-drift	10–10^3	North-east Rockall trough
	Contourite-fan	10^3–10^5	Vema Channel exit
Confined drift		10^3–10^5	Sumba drift; East Chatham rise
Modified drift–turbidite systems	Extended turbidite bodies	10^3–10^4	Columbia levee South Brazil Basin; Hikurangi fandrift
	Sculptured turbidite bodies	10^3–10^4	South-east Weddell Sea
	Intercalated turbidite–contourite bodies	Can be very extensive	Hatteras rise

dimensions: lengths from a few tens of kilometers to over 1000 km, length to width ratios of 2 : 1 to 10 : 1, and thicknesses up to 2 km. They may occur anywhere from the outer shelf/upper slope, such as those east of New Zealand to the abyssal plains, depending on the depth at which the bottom current flows. They are very common throughout the North Atlantic, but also occur in all the other ocean basins and some marginal seas. One or both lateral margins are generally flanked by distinct moats along which the flow axis occurs and which experience intermittent erosion and nondeposition. Elongate drifts associated with channels or confined basins are classified separately.

Both the elongation trend and direction of progradation are dependent on an interaction between the local topography, the current system and intensity, and the Coriolis Force. Elongation is generally parallel or subparallel to the margin, with both detached and separated types recognized, but progradation can lead to parts of the drift being elongated almost perpendicular to the margin. Internal seismic character reflects the individual style of progradation, typically with lenticular, convex-upward depositional units overlying a major erosional discontinuity. Fields of migrating sediment waves are common.

Sedimentation rates depend very much on the amount and supply of material to the bottom currents. On average, rates are greater than for sheet drifts, being between 2 and 10 cm ky^{-1}, but may range from <2 cm ky^{-1} for open ocean pelagic biogenic-rich drifts, to >60 cm ky^{-1}, for some marginal drifts (e.g., along the Hebridean margin). The sediment type also varies according to input, including biogenic, volcaniclastic, and terrigenous types. Grain size varies from muddy to sandy as a result of long-term fluctuations in bottom current strength.

Channel-related Drifts

This type of contourite deposit is related to deep channels, passageways or gateways through which the bottom circulation is constrained so that flow velocities are markedly increased (e.g., Vema Channel, Kane Gap, Samoan Passage, Almirante Passage, Faroe-Shetland Channel etc.). Gateways are very important narrow conduits that cut across the sills between ocean basins and thereby allow the exchange of deep and intermediate water masses. In addition to significant erosion and scouring of the passage floor, irregular discontinuous sediment bodies are deposited on the floor and flanks of the channel, as axial and lateral patch drifts, and at the downcurrent exit of the channel, as a contourite fan.

Patch drifts are typically small (a few tens of square kilometers in area, 10–150 m thick) and either irregular in shape or elongate in the direction of flow. They can be reflector-free or with a more chaotic seismic facies, and may have either a sheet or mounded geometry. Contourite fans are much larger cone-shaped deposits, up to 100 km or more in width and radius and 300 m in thickness (e.g., the Vema contourite fan).

Channel floor deposits include patches of coarse-grained (sand and gravel) lag contourites, mud–clast contourites and associated hiatuses that result from substrate erosion, as well as patch drifts of finer-grained muddy and silty contourites where current velocities are locally reduced. Manganiferous mud contourites and nodules are also typical in places. Accumulation rates range from very low, due to nondeposition and erosion, to as much as 10 cm ky^{-1} in some patch drifts and contourite fans.

Confined Drifts

Relatively few examples are currently known of drifts confined within small basins. These typically

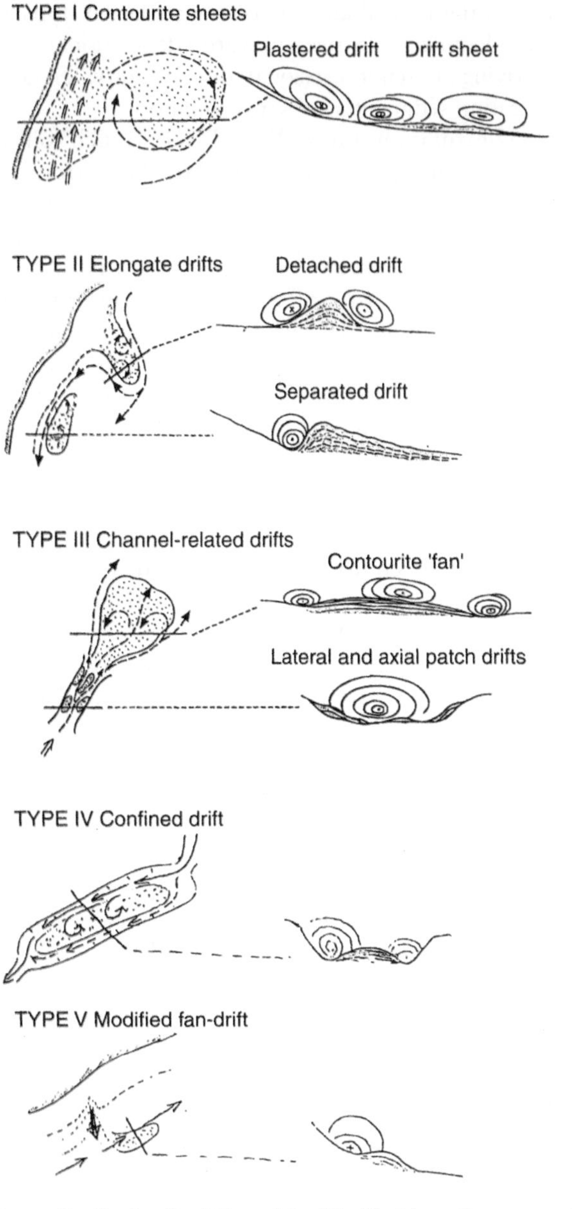

Figure 3 Contourite drift models. (Modified from Faugeres *et al.*, 1999.)

occur in tectonically active areas, such as the Sumba drift in the Sumba forearc basin of the Indonesian arc system, the Meiji drift in the Aleutian trench and an unnamed drift in the Falkland Trough. Apart from their topographic confinement, the gross seismic character appears similar to mounded elongate drifts with distinct moats along both margins. Sediment type and grain size depend very much on the nature of input to the bottom current system.

Modified Drift-turbidite Systems: Process Interaction

The interaction of downslope and alongslope processes and deposits at all scales is the normal condition on the margins as well as within the central parts of present ocean basins. Interaction with slow pelagic and hemipelagic accumulation is also the norm, but these deposits do not substantially affect the drift type or morphology. Over a relatively long timescale, there has been an alternation of periods during which either downslope or alongslope processes have dominated as a result of variations in climate, sealevel and bottom circulation coupled with basin morphology and margin topography. This has been particularly true since the late Eocene onset of the current period of intense thermohaline circulation, and with the marked alternation of depositional style reflecting glacial–interglacial episodes during the past 2 My (million years).

At the scale of the drift deposit, this interaction can have different expressions as exemplified in the following examples.

1. Scotian Margin: regular interbedding of thin muddy contourite sheets deposited during interglacial periods and fine-grained turbidites dominant during glacials; marked asymmetry of channel levees on the Laurentian Fan, with the larger levees and extended tail in the direction of the dominant bottom current flow.
2. Cape Hatteras Margin: complex imbrication of downslope and alongslope deposits on the lower continental rise, that has been referred to as a companion drift-fan.
3. The Chatham–Kermadec Margin: the deep western boundary current in this region scours and erodes the Bounty Fan south of the Chatham Rise and directly incorporates fine-grained material from turbidity currents that have traveled down the Hikurangi Channel. This material, together with hemipelagic material, is swept north from the downstream end of the turbidity current channel to form a fan-drift deposit.
4. West Antarctic Peninsula Margin: eight large sediment mounds, elongated perpendicular to the margin and separated by turbidity current channels, have an asymmetry that indicates construction by entrainment of the suspended load of down-channel turbidity currents within the ambient south-westerly directed bottom currents and their deposition downcurrent.
5. Hebridean Margin: complex pattern of intercalation of downslope (slides, debrites, and turbidites), alongslope contourites and glaciomarine hemipelagites in both time and space; the alongslope distribution of these mixed facies types by the northward-directed slope current has led to the term composite slope-front fan for the Barra Fan.

Erosional Discontinuities

The architecture of deposits within a drift is complex, stressing variations of the processes and accumulation rates linked to changes in current activity. In many cases, the history of contourite drift construction is marked by an alternation of periods of sedimentation and erosion or nondeposition, the latter corresponding to a greater instability of and/or a drastic change in current regime. The result is the superposition of depositional units whose general geometry is lenticular and whose limits correspond to major discontinuities, that are more or less strongly erosive. These discontinuities can be traced at the scale of the accumulation as a whole and are marked by a strong-amplitude continuous reflector, commonly marking a change in seisomofacies linked to variation in current strength. Such extensive and synchronous discontinuities are typical of most drifts. The principal characteristics of drifts evident in seismic records are shown in **Figure 4**.

Contourite Sediment Facies

Several different contourite facies can be recognized on the basis of variations in grain size and composition. These are listed and briefly described below and illustrated in **Figure 5** and **6**.

- Siliciclastic contourites (muddy, silty, sandy and gravel-rich variation)
- Shale-clast/shale-chip contourites (all compositions possible)
- Volcaniclastic contourites (muddy–silty–sandy variations)
- Calcareous biogenic contourites (calcilutite, -siltite, -arenite variations)
- Siliceous biogenic contourites (mainly sand grade)
- Manganiferous muddy contourites (+ manganiferous nodules/pavements)

Muddy contourites

These are homogeneous, poorly bedded and highly biouturbated, with rare primary lamination (partly destroyed by bioturbation), and irregular winnowed concentrations of coarser material. They have a silty-clay grain size, poor sorting, and a mixed terrigenous (or volcaniclastic)–biogenic composition. The components are in part local, including a pelagic contribution, and in part far-traveled.

Silty contourites

These, which are also referred to as mottled silty contourites commonly show bioturbational mottling to indistinct discontinuous lamination, and are gradationally interbedded with both muddy and sandy contourite facies. Sharp to irregular tops and bases of silty layers are common, together with thin lenses of coarser material. They have a poorly sorted clayey-sandy silt size and a mixed composition.

Sandy contourites

These occur as both thin irregular layers and as much thicker units within the finer-grained facies and are generally thoroughly bioturbated throughout. In some cases, rare primary horizontal and cross-lamination is preserved (though partially destroyed by bioturbation), together with irregular erosional contacts and coarser concentrations or lags. The mean grain size is normally no greater than fine sand, and sorting is mostly poor due to bioturbational mixing, but more rarely clean and well-sorted sands occur. Both positive and negative grading may be present. A mixed terrigenous–biogenic composition is typical, with evidence of abrasion, fragmented bioclasts and iron oxide staining.

Gravel-rich contourites

These are common in drifts at high latitudes as a result of input from ice-rafted material. Under relatively low-velocity currents, the gravel and coarse sandy material remains as a passive input into the contourite sequence and is not subsequently reworked to any great extent by bottom currents. Gravel lags indicative of more extensive winnowing have been noted from both glacigenic contourites and from shallow straits, narrow moats, and passageways, where gravel pavements are

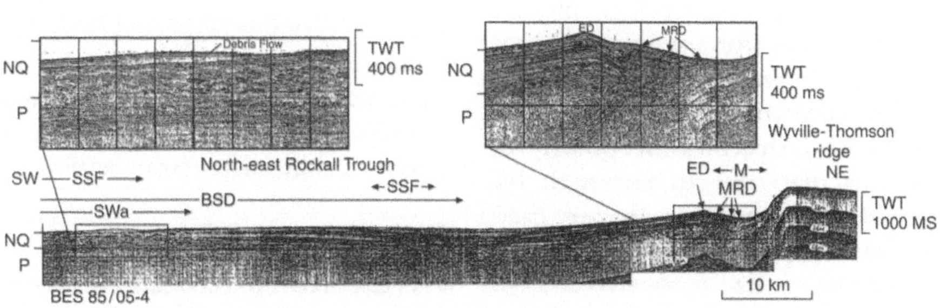

Figure 4 Seismic profiles of actual drift systems.

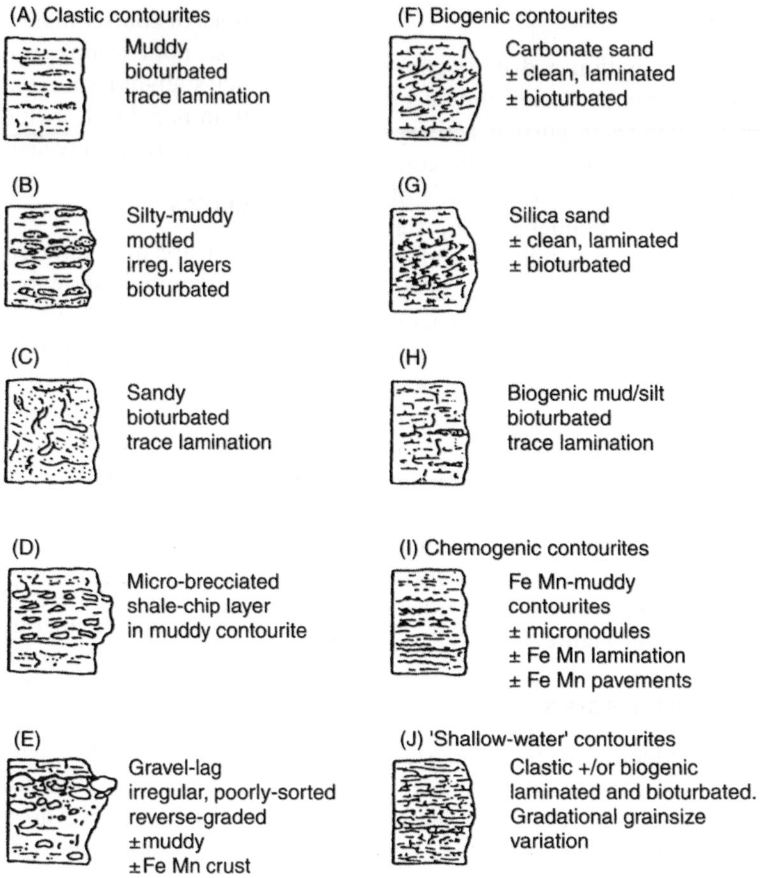

Figure 5 Contourite facies models for clastic, biogenic, chemogenic, and 'shallow-water' contourites. (Reproduced from Stow *et al.*, 1996.)

locally developed in response to high-velocity bottom current activity

Shale-clast or shale-chip layers

These have been recognized in both muddy and sandy contourites from relatively few locations. They result from substrate erosion under relatively strong bottom currents, where erosion has led to a firmer substrate and in some cases burrowing on the omission surface has helped to break up the semi-firm muds.

Calcareous and siliceous biogenic contourites

These occur in regions of dominant pelagic biogenic input, including open ocean sites and beneath areas of upwelling. In most cases bedding is indistinct, but may be enhanced by cyclic variations in composition, and primary sedimentary structures are poorly developed or absent, in part due to thorough bioturbation as in siliciclastic contourites. In rare cases, the primary lamination appears to have been well preserved. The mean grain size is most commonly silty clay, clayey silt or muddy-sandy, poorly sorted and with a distinct sand-size fraction representing the coarser biogenic particles that have not been too fragmented during transport. The composition is typically pelagic to hemipelagic, with nannofossils and foraminifera as dominant elements in the calcareous contourites and radiolaria or diatoms dominant in the siliceous facies. Many of the biogenic particles are fragmented and stained with either iron oxides or manganese dioxde. There is a variable admixture of terrigenous or volcaniclastic material.

Manganiferous contourites

These manganiferous or ferromanganiferous-rich horizons are common. This metal enrichment may occur as very fine dispersed particles, as a coating on individual particles of the background sediment, as fine encrusted horizons or laminae, or as micronodules. It has been observed in both muddy and biogenic contourites from several drifts.

Bottom-current influence

It is important to recognize that bottom currents will influence, to a greater or lesser extent, other deep-water sediments, particularly pelagic, hemipelagic,

Figure 6 Photographs of contourite facies from cores drilled through existing drift systems. Vertical scales labelled in cm.

turbiditic, and glacigenic, both during and after deposition. Where the influence is marked and deposition occurs in a drift, then the sediment is termed contourite. Where the influence is less severe, so that features of the original deposit type remain dominant, then the sediment is said to have been influenced by bottom currents, as in bottom-current reworked turbidites. Some more-laminated facies, as well as the thin, clean, cross-laminated sands originally described from the north-east American margin, are most likely of this type.

Contourite Sequences and Current Velocity

Muddy, silty, and sandy contourites, of siliciclastic, volcaniclastic, or mixed composition, commonly occur in composite sequences or partial sequences a few decimeters in thickness. The ideal or complete sequence shows overall negative grading from muddy through silty to sandy contourites and then positive grading back through silty to muddy contourite facies (**Figure** 7). Such sequences of grain size and facies variation are now widely recognized, although not always fully developed, and are most probably related to long-term fluctuations in the mean current velocity. Not enough data exist to be certain of the timescale of these cycles, though some evidence points towards 5000–20 000 cycles for certain marginal drifts.

The occurrence of widespread hiatuses in the deep-ocean sediment record is best related to episodes of particularly intense bottom currents. More locally,

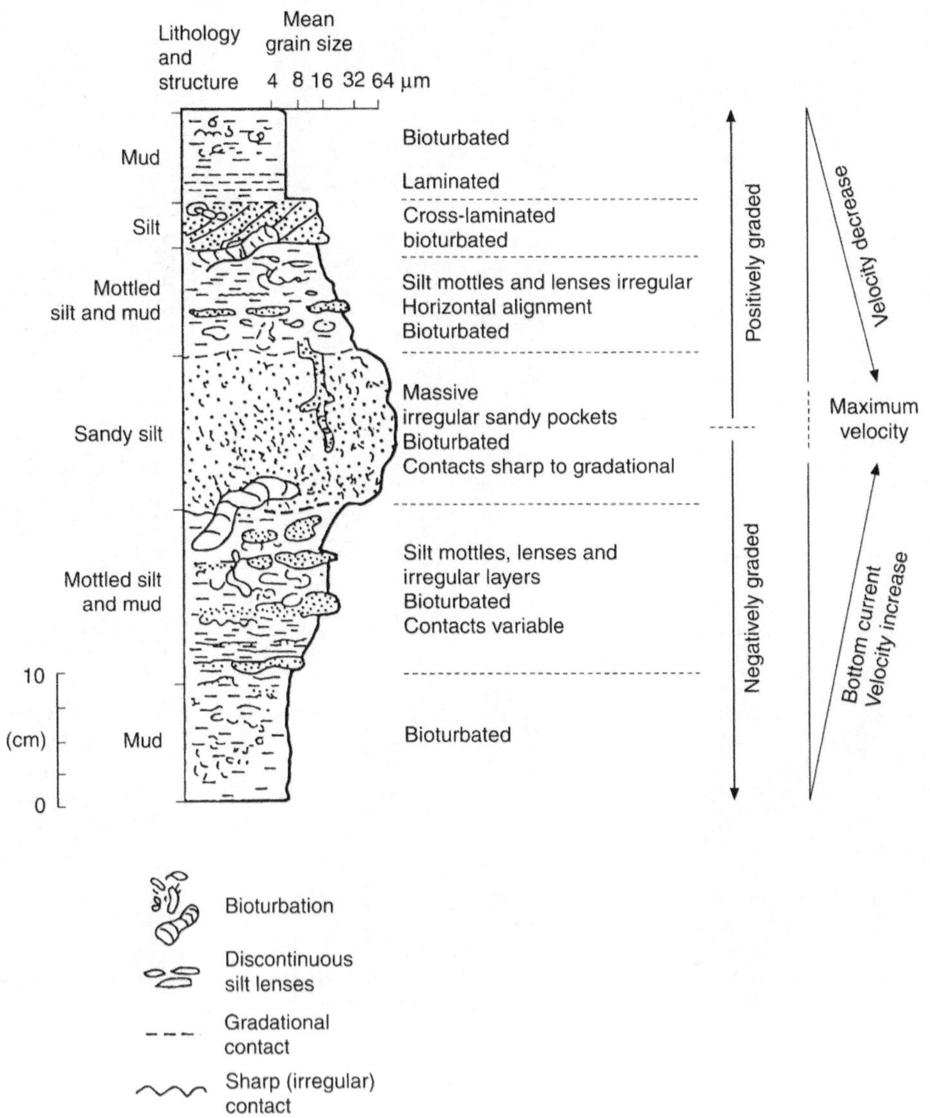

Figure 7 Composite contourite facies model showing grain size variation through a mud–silt–sand contourite sequence. (Modified from Stow *et al.*, 1996.)

such strong currents result in significant sediment winnowing and the accumulation of sand, gravel, and shale-clast contourites. Thick units of sandy contourites together with sandy turbidites reworked by the bottom current are potentially important as hydrocarbon reservoirs where suitably buried in association with source rocks.

Biogenic contourites typically occur in similar sequences of a decimetric scale that show distinct variation in biogenic/terrigenous ratio, generally linked to the grain size variation. This cyclic facies pattern has a longer timescale, in the few examples from which there is good dating, and is closely analogous to the Milankovitch cyclicity recognized in many pelagic and hemipelagic successions. It is, therefore, believed to be driven by the same mechanism of orbital forcing superimposed on changes in bottom-current velocity.

The link between contourite sequences and changes in paleoclimate and paleocirculation is an extremely important one. Where such sequences can be correctly decoded then a more accurate understanding of the paleo-ocean and its environment can be built up.

See also

Bottom Water Formation.

Further Reading

Faugeres JC, Stow DAV, Imbert P, Viana A, and Wynn RB (1999) Seismic features diagnostic of contourite drifts. *Marine Geology*, 162: 1–38.

Heezen BC, Hollister CD, and Ruddiman WF (1966) Shaping the continental rise by deep geostrophic contour currents. *Science* 152: 502–508.

McCave IN, Manighetti B, and Robinson SG (1995) Sortable silt and fine sediment size/composition slicing: parameters for paleocurrent speed and paleoceanography. *Paleoceanography* 10: 593–610.

Nowell ARM and Hollister CD (eds.) (1985) Deep ocean sediment transport – preliminary results of the high energy benthic boundary layer experiment. *Marine Geology* 66.

Pickering KT, Hiscott RN, and Hein FJ (1989) *Deep-Marine Environments: Clastic Sedimentation and Tectonics*. London: Unwin Hyman.

Stow DAV and Faugeres JC (eds.) (1993) Contourites and Bottom Currents, *Sedimentary Geology*, Special Volume 82, 1–310.

Stow DAV, Reading HG, Collinson J (1996) Deep seas. In: Reading HG (ed.) *Sedimentary Environments and Facies*. 3rd edn, pp. 380–442. Blackwell Science Publishers.

Stow DAV and Faugeres JC (eds.) (1998) Contourites, turbidites and process interaction. *Sedimentary Geology Special Issue* 115.

Stow DAV and Mayall M (eds.) (2000) Deep-water sedimentary systems: new models for the 21st century. *Marine and Petroleum Geology*, Special Volume 17.

ENCLOSED OR SEMI-ENCLOSED SEAS, FJORDS, ESTUARIES AND RIVERS

MEDITERRANEAN SEA CIRCULATION

A. R. Robinson and W. G. Leslie, Harvard University, Cambridge, MA, USA
A. Theocharis, National Centre for Marine Research (NCMR), Hellinikon, Athens, Greece
A. Lascaratos, University of Athens, Athens, Greece

Introduction

The Mediterranean Sea is a mid-latitude semi-enclosed sea, or almost isolated oceanic system. Many processes which are fundamental to the general circulation of the world ocean also occur within the Mediterranean, either identically or analogously. The Mediterranean Sea exchanges water, salt, heat, and other properties with the North Atlantic Ocean. The North Atlantic is known to play an important role in the global thermohaline circulation, as the major site of deep- and bottom-water formation for the global thermohaline cell (conveyor belt) which encompasses the Atlantic, Southern, Indian, and Pacific Oceans. The salty water of Mediterranean origin may affect water formation processes and variabilities and even the stability of the global thermohaline equilibrium state.

The geography of the esntire Mediterranean is shown in **Figure 1A** and the distribution of deep-sea topography and the complex arrangement of coasts and islands in **Figure 1B**. The Mediterranean Sea is composed of two nearly equal size basins, connected by the Strait of Sicily. The Adriatic extends northward between Italy and the Balkans, communicating with the eastern Mediterranean basin through the Strait of Otranto. The Aegean lies between Greece and Turkey, connected to the eastern basin through the several straits of the Grecian Island Arc. The Mediterranean circulation is forced by water exchange through the various straits, by wind stress, and by buoyancy flux at the surface due to freshwater and heat fluxes. Evaporation 1.27 m/year, Precipitation 0.59 m/year, Mediterranean outflow (through the Gibraltar) $\sim$1.0 Sv, the inflow exceeds outflow by 5% (0.05 Sv) to compensate the water deficit of the Mediterranean, fresh water input 0.67 m/year, which comprises precipitation, river runoff and the Black Sea input, Net salt flux towards the Atlantic $\sim 2 \times 10^6$ kg/s.

Research on Mediterranean Sea general circulation and thermohaline circulations and their variabilities, and the identification and quantification of critical processes relevant to ocean and climate dynamics involves several issues. Conceptual, methodological, technical, and scientific issues include, for example, the formulation of multiscale (e.g., basin, sub-basin, mesoscale) interactive nonlinear dynamical models; the parametrization of air–sea interactions and fluxes; the determination of specific regional processes of water formation and transformations; the representation of convection and boundary conditions in general circulation models. A three-component nonlinear ocean system is involved whose components are: (1) air–sea interactions, (2) water mass formations and transformations, and (3) circulation elements and structures. The focus here is on the circulation elements and their variabilities. However, in order to describe the circulation, water masses must be identified and described.

Multiscale Circulation and Variabilities

The new picture of the general circulation in the Mediterranean Sea which is emerging is complex, and composed of three predominant and interacting spatial scales: basin scale (including the thermohaline (vertical) circulation), sub-basin scale, and mesoscale. Complexity and scales arise from the multiple driving forces, from strong topographic and coastal influences, and from internal dynamical processes. There exist: free and boundary currents and jets which bifurcate, meander and grow and shed ring vortices; permanent and recurrent sub-basin scale cyclonic and anticyclonic gyres; and small but energetic mesoscale eddies. As the scales are interacting, aspects of all are necessarily discussed when discussing any individual scale. The path for spreading of Levantine Intermediate Water (LIW) from the region of formation to adjacent seas together with the thermohaline circulations are shown in **Figure 2**; where the entire Mediterranean is schematically shown as two connected basins (western and eastern). The internal thermohaline cells existing in the western and eastern Mediterranean have interesting analogies and differences to each other and to the global thermohaline circulation. In the western basin (**Figure 3A**) the basin-scale thermohaline cell is driven by deep water formed in the Gulf of Lions and

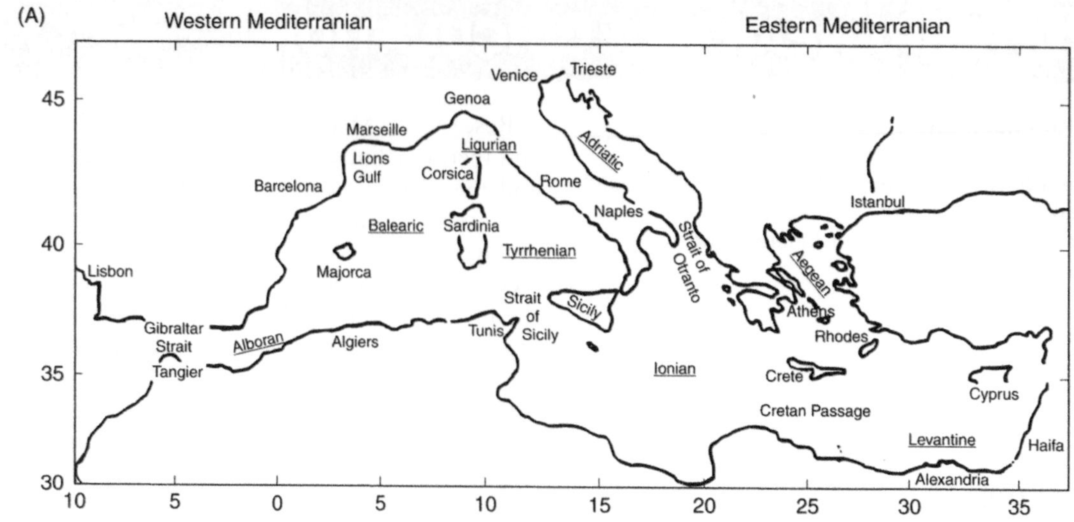

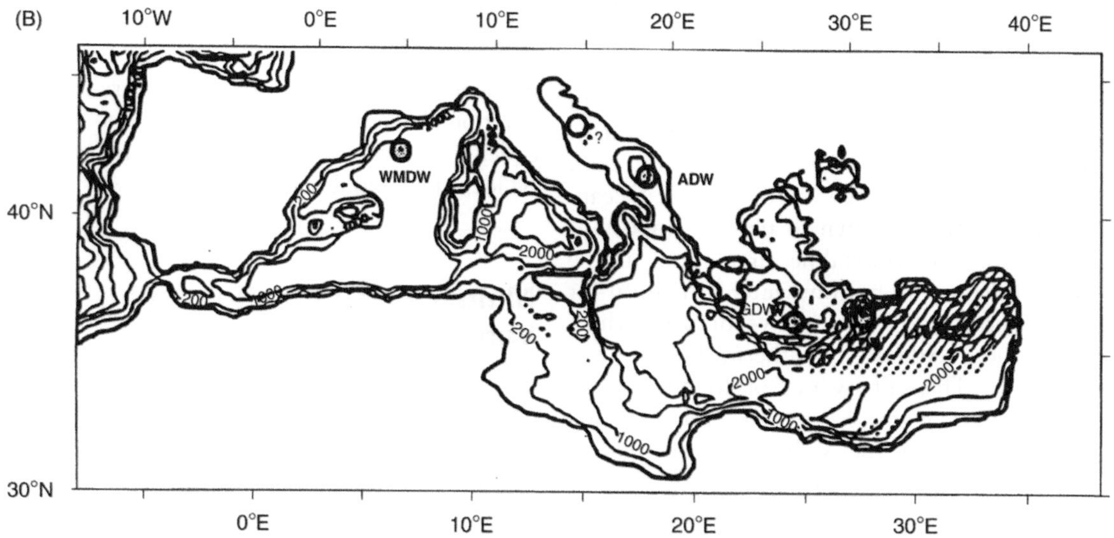

Figure 1 (A) The Mediterranean Sea geography and nomenclature of the major sub-basins and straits. (B) The bottom topography of the Mediterranean Sea (contour interval is 1000 m) and the locations of the different water mass formations.

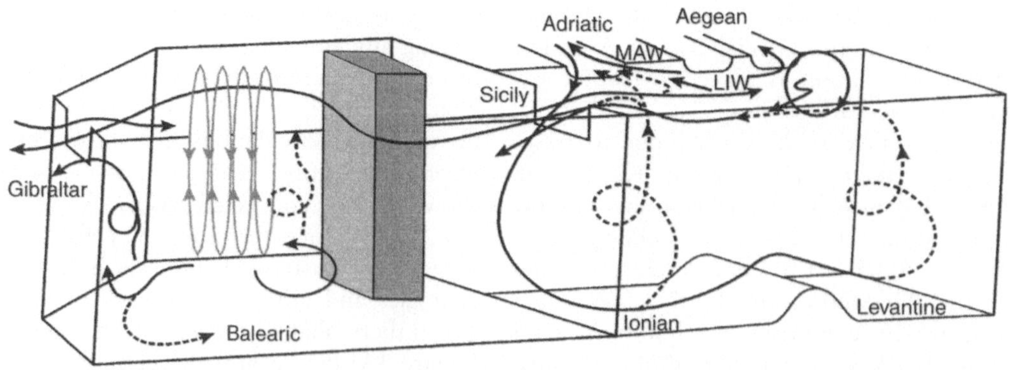

Figure 2 Schematic of thermohaline cells and path of Levantine Intermediate Water (LIW) in the entire Mediterranean.

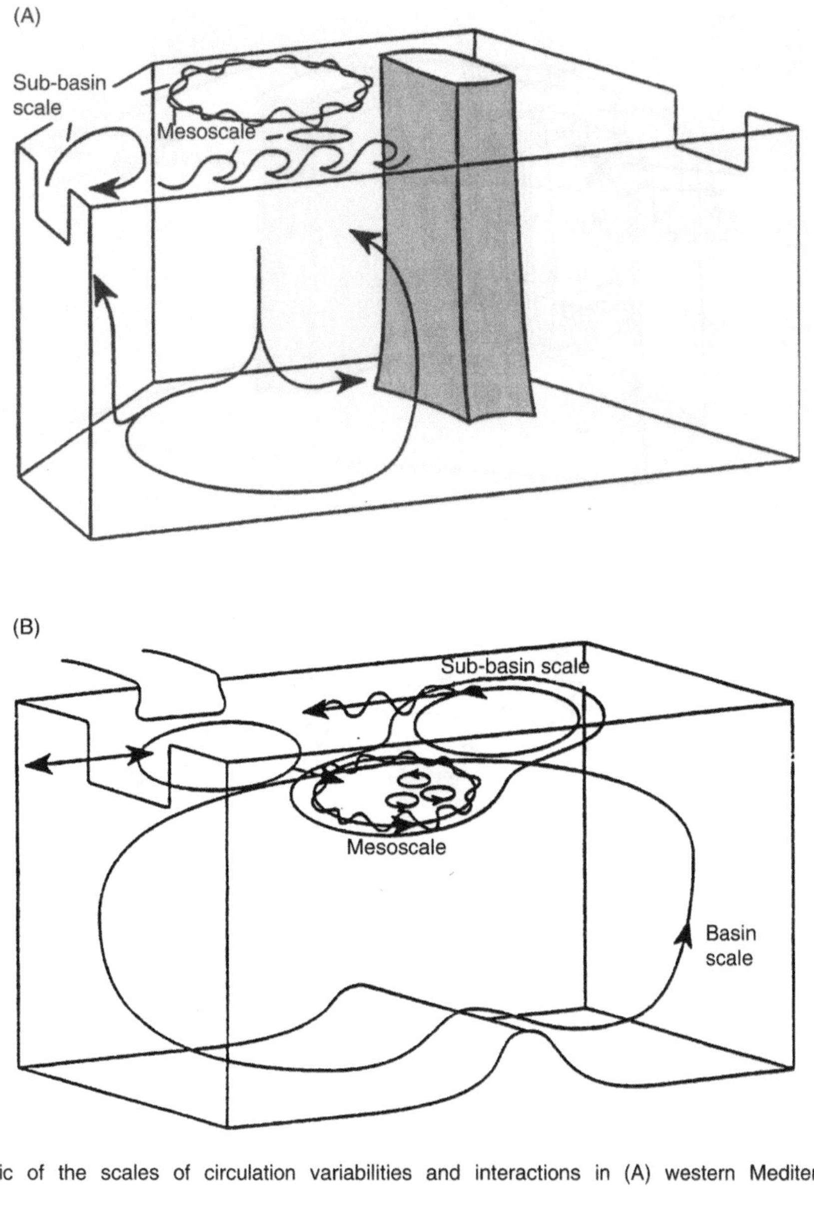

Figure 3 Schematic of the scales of circulation variabilities and interactions in (A) western Mediterranean, (B) eastern Mediterranean.

spreading from there. Important sub-basin scale gyres in the main thermocline in the Alboran and Balearic Seas have been identified. Intense mesoscale activity exists and is shown by instabilities along the coastal current, mid-sea eddies and along the outer rim swirl flow of a sub-basin scale gyre. The basin scale thermohaline cell of the eastern basin is depicted generically in **Figure 3B** and discussed in more detail in the next section. The basin scale general circulation of the main thermocline is composed of dominantly energetic sub-basin scale gyres linked by sub-basin scale jets. The active mesoscale is shown by a field of internal eddies, meanders along the border swirl flow of a sub-basin scale gyre, and as meandering jet segments. The Atlantic Water jet with its instabilities, bifurcations, and multiple pathways, which travels from Gibraltar to the Levantine is a basin scale feature not depicted in **Figure 3**; this also pertains to the intermediate water return flow.

Large-scale Circulation

Processes of global relevance for ocean climate dynamics include thermohaline circulation, water mass formation and transformation, dispersion, and mixing. These processes are schematically shown in **Figure 4A** and **B** for the western and the eastern basins. The Mediterranean basins are evaporation basins (lagoons), with freshwater flux from the

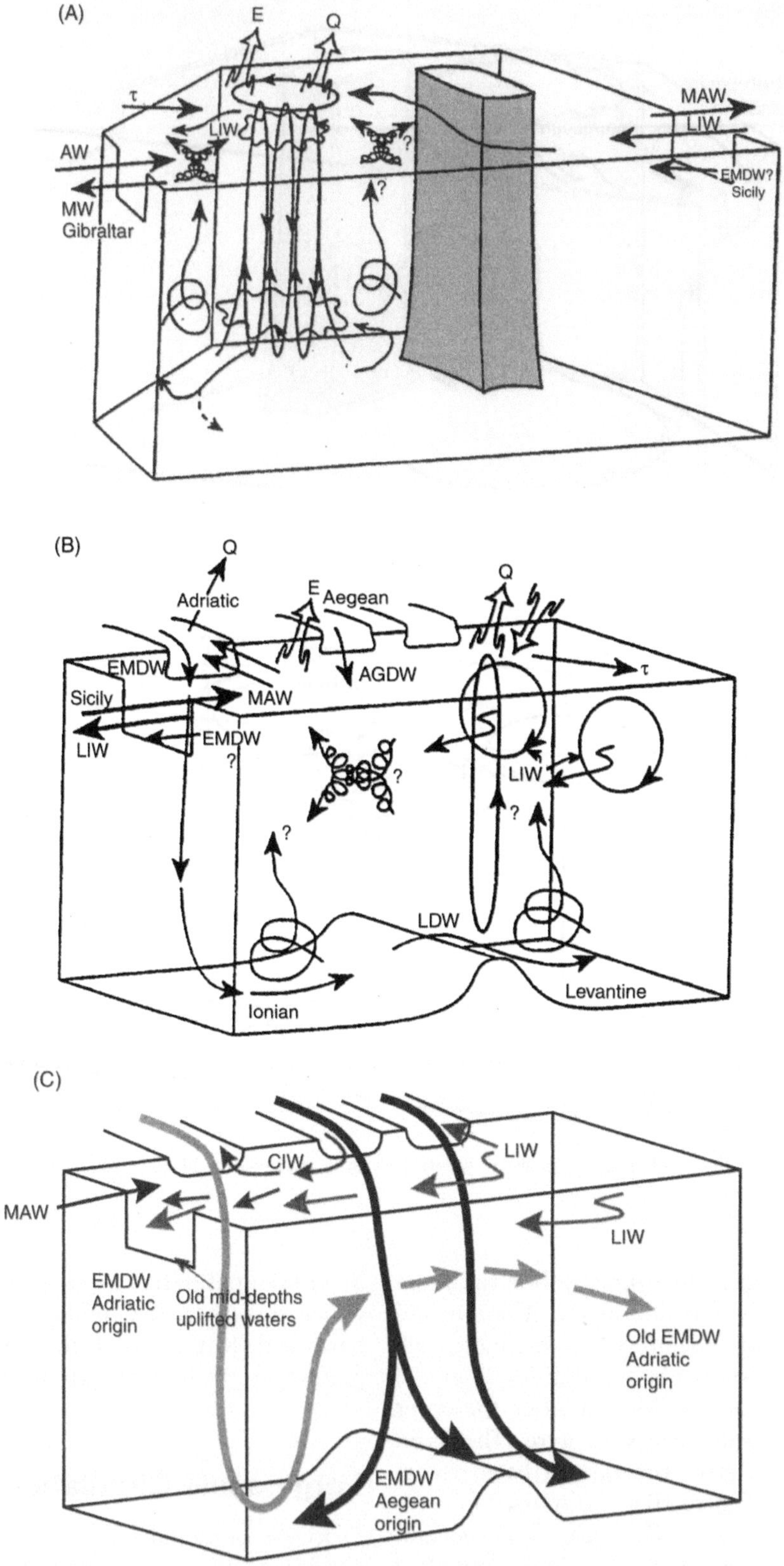

Figure 4 Processes of air–sea interaction, water mass formation, dispersion, and transformation. (A) western Mediterranean, (B) eastern Mediterranean, (C) eastern Mediterranean (post-eastern Mediterranean Transient).

Atlantic through the Gibraltar Straits and into the eastern Mediterranean through the Sicily Straits. Relatively fresh waters of Atlantic origin circulating in the Mediterranean increase in density because evaporation (E) exceeds precipitation (advective salinity preconditioning), and then form new water masses via convection events driven by intense local cooling (Q) from winter storms. Bottom water is produced: for the western basin (WMDW) in the Gulf of Lions (**Figure 4A**) and for the eastern basin (**Figure 4B**) in the southern Adriatic (EMDW, which plunges down through the Otranto Straits). Recent observations also indicate deep water (LDW) formation in the north-eastern Levantine basin during exceptionally cold winters, where intermediate water (LIW) is regularly formed seasonally. Evidence now shows that LIW formation occurs over much of the Levantine basin, but preferentially in the north, probably due to meteorological factors. The LIW is an important water mass which circulates through both the eastern and western basins and contributes predominantly to the efflux from Gibraltar to the Atlantic, mixed with some EMDW and together with WMDW. Additionally, intermediate and deep (but not bottom) waters formed in the Aegean (AGDW) are provided to the eastern basin through its straits. As will be seen below, that water formerly known as AGDW, is now identified as Cretan Intermediate Water (CIW) and Cretan Deep Water (CDW). Important research questions relate to the preconditioning, formation, spreading, dispersion, and mixing of these water masses. These include: sources of forced and internal variabilities; the spectrum and relative amounts of water types formed, recirculating within the Mediterranean basins, and fluxing through the straits, and the actual locations of upwelling.

A basin-wide qualitative description of the thermohaline circulation in the western basin of the Mediterranean Sea has recently been provided by Millot (see Further Reading). Results based on cruises in December 1988 and August 1989 indicated that the deep layer in the western Mediterranean was 0.12°C warmer and about 0.33 PSU more saline than in 1959. Analysis of these data together with those from earlier cruises has shown a trend of continuously increasing temperatures in recent decades. Based on the consideration of the heat and water budget in the Mediterranean, the deep-water temperature trend was originally speculated to be the result of greenhouse gas-included local warming. A more recent argument considers the anthropomorphic reduction of river water flux into the eastern basin to be the main cause of this warming trend.

During the 18th and 19th centuries a number of observations of temperature and salinity were made mostly in the surface down to intermediate layers, in certain areas of the Mediterranean. Progressively, the investigators extended their measurements into deeper layers to understand the distribution of the parameters, both horizontally and vertically. Sometimes the values were surprisingly close to the correct values but in other cases they presented significant differences due mainly to the primitive instrumentation and methods used by this time. Initially, some believed that the deep waters are motionless. Later the researchers noted the variations of the parameters occurring on a daily, monthly and seasonal basis in the upper layers and began to think about the movements and renewal of the deep waters. Regarding the distribution of salinity extremely confused and erroneous views prevailed up to 1870. However, it was already known by this time that the fresh water input is much lower and does not compensate evaporation from the sea surface. Furthermore, they noted that the source of the salt of the deep water is the surface layer. Several theories on the mechanisms governing the renewal and oxygenation of the deep layers were formulated. Moreover, they succeeded to measure currents and structure primitive maps showing prevailing circulation patterns, as the Atlantic Water inflow and the Black Sea outflow towards the Northeast Aegean.

Since the beginning of the twentieth century, when the first investigations in the Mediterranean Sea took place (1908), up to the mid-1980s, both the intermediate and deep conveyor belts of the eastern basin presented rather constant characteristics. The Adriatic has been historically considered as the main contributor to the deep and bottom waters of the Ionian and Levantine basins, thus indicating an almost perfectly repeating cycle in both water mass characteristics and formation rates during this long period. Roether and Schlitzer found in 1991 that the thermohaline circulation in the eastern basin consists of a single coherent convective cell which connects the Levantine and Ionian basins and has a turnover time of 125 years below 1200 m. Their results indicated that the water formed in the Adriatic is a mixture of surface water (AW) and intermediate Levantine water (LIW) from the Mediterranean. The Aegean has also been reported as a possible secondary source, providing dense waters to the lower intermediate and/or deep layers, namely Cretan Intermediate Water (CIW), that affected mainly the adjacent to the Cretan Arc region of the eastern Mediterranean. Since 1946 increased densities were observed in the southern Aegean Sea in 1959–65 and 1970–73. These events occurred under extreme meteorological conditions. However, the quantities of the dense water produced were never enough to affect the whole eastern Mediterranean. The traditional historical picture of water properties is illustrated in **Figure 5A** by a west–east vertical section of salinity through the eastern Mediterranean.

After 1987, the most important changes in the thermohaline circulation and water properties basin-wide ever detected in the Mediterranean occurred. The Aegean, which had only been a minor contributor to the deep waters, became more effective than the Adriatic as a new source of deep and bottom waters of the eastern Mediterranean. This source gradually provided a warmer, more saline,

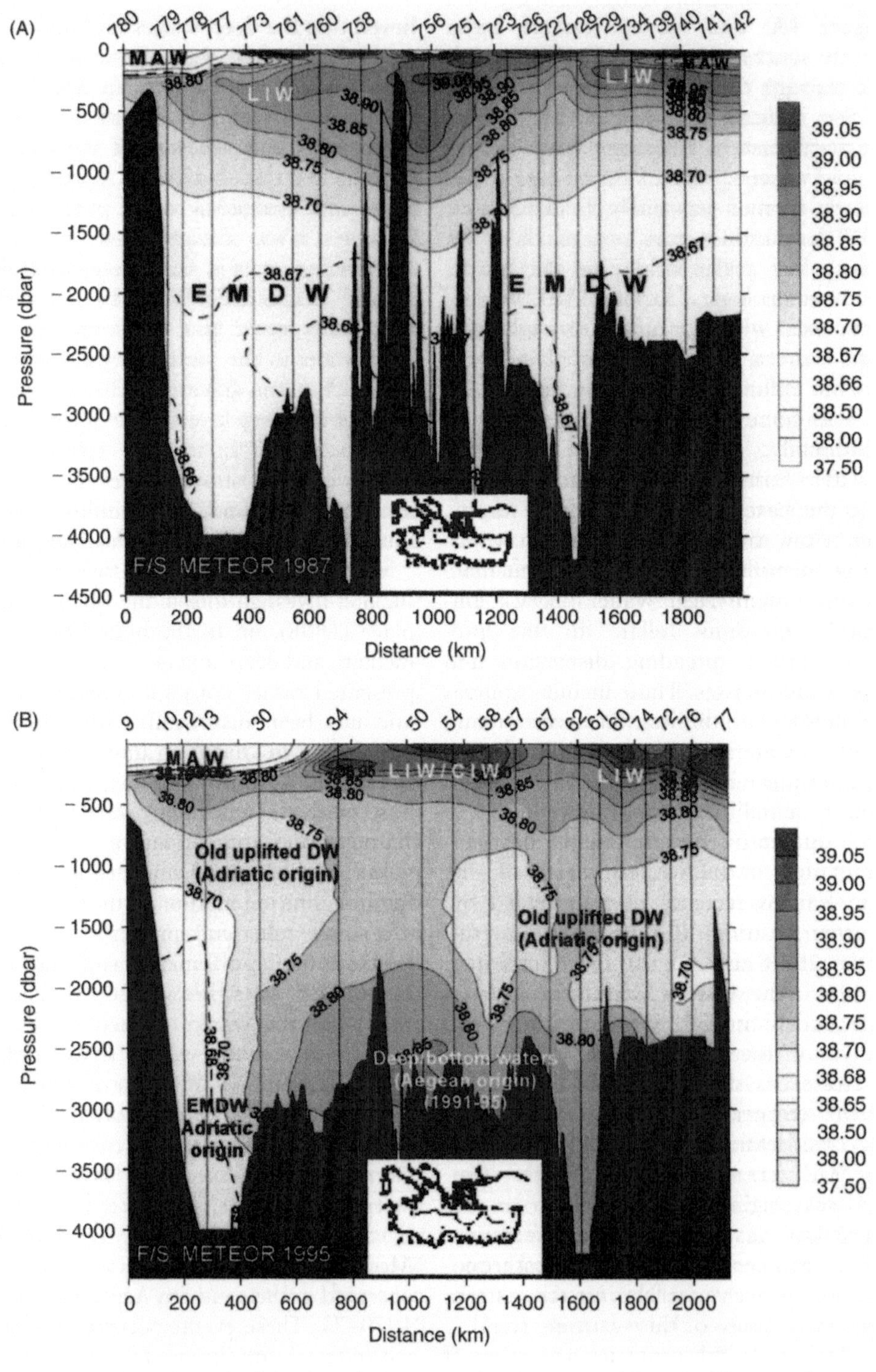

Figure 5 West–east vertical sections of salinity through the eastern Mediterranean: (A) 1987, (B) 1995, (C) 1999.

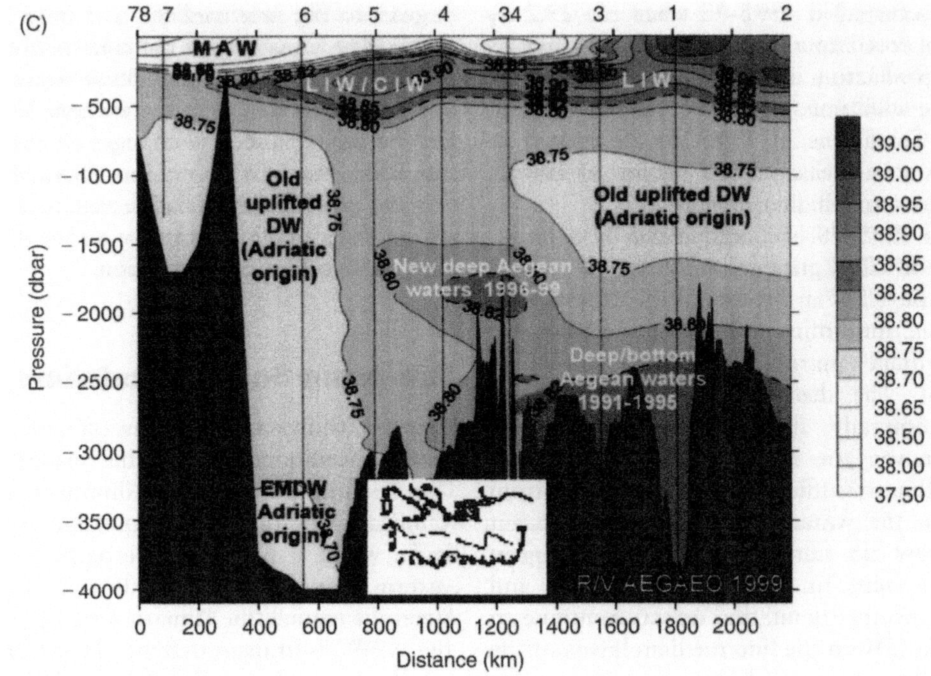

Figure 5 *Continued*

and denser deep-water mass than the previously existing Eastern Mediterranean Deep (and bottom) Water (EMDW) of Adriatic origin. Its overall production was estimated for the period 1989–95 at more than 7 Sv, which is three times higher than that of the Adriatic. After 1990, CIW appeared to be formed in the southern Aegean with modified characteristics. This warmer and more saline CIW (less dense than the older one) exits the Aegean mainly through the western Cretan Arc Straits and spreads in the intermediate layers, the so-called LIW horizons, in the major part of the Ionian Sea, blocking the westward route of the LIW.

These changes have altered the deep/internal and upper/open conveyor belts of the eastern Mediterranean. This abrupt shift in the Mediterranean 'ocean climate' has been named the Eastern Mediterranean Transient (EMT). Several hypotheses have been proposed concerning possible causes of this unique thermohaline event, including: (1) internal redistribution of salt, (2) changes in the local atmospheric forcing combined with long term salinity change, (3) changes in circulation patterns leading to blocking situations concerning the Modified Atlantic Water (MAW) and the LIW, and (4) variations in the fresher water of Black Sea origin input through the Strait of Dardanelles.

The production of denser than usual local deep water started in winter 1987, in the Kiklades plateau of the southern Aegean. The combination of continuous salinity increase in the southern Aegean during the period 1987–92, followed by significant temperature drop in 1992 and 1993 caused massive dense water formation. The overall salinity increase in the Cretan Sea was about 0.1 PSU, due to a persistent period of reduced precipitation over the Aegean and the eastern Mediterranean. This meteorological event might be attributed to larger scale atmospheric variability as the North Atlantic Oscillation. Moreover, the net upper layer (0–200 m) salt transport into the Aegean from the Levantine was increased one to four times within the period 1987–94 due not only to the dry period but also to significant changes of the characteristic water mass pathways. This was a secondary source of salt for the south Aegean that has further preconditioned dense water formation. The second period is characterized by cooling of the deep waters by about 0.35°C, related to the exceptionally cold winters of 1992 and 1993. The strongest winter heat loss since 1985 in the Adriatic and since 1979 in the Aegean was observed in 1992. During this winter an almost complete overturning of the water column occurred in the Cretan Sea. The density of the newly formed water, namely Cretan Deep Water (CDW), reached its maximum value in 1994–95 in the Cretan Sea of the southern Aegean. The massive dense water production caused a strong deep outflow through the Cretan Arc Straits towards the Ionian and Levantine basins. Interestingly, the peak of the production rate, about 3 Sv, occurred in 1991–92 when the 29.2 σ_T isopycnal was raised up to the surface layer. While its deep-water production in the Aegean is becoming more effective with time, that in the Adriatic stopped after 1992. Conditions in 1995 are illustrated in **Figure 5B** by a west–east vertical section of salinity through the eastern Mediterranean.

The period 1995–98 is characterized by continuous decrease of CDW production, from 1 to 0.3 Sv. The level of the CDW at the area of the deep Cretan Arc Straits (i.e. Antikithira in the West and Kassos in the East) is found approximately at the sill depths (800–1000 m). The deep outflow has also been weakened, especially from the western Cretan Straits. Moreover, the density of the outflowing water is no longer sufficient to sink to the bottom and therefore the water coming from the recent Aegean outflow has settled above the old Aegean bottom-water mass, in layers between 1500 and 2500 m. On the other hand, the Aegean continues to contribute the CIW to the intermediate layers of the eastern Mediterranean. Salinity in the eastern Mediterranean in 1999 is shown in **Figure 5C**.

The intrusion of the dense Aegean waters has initiated a series of modifications not only in the hydrology and the dynamics of the entire basin, but also in the chemical structure and some biological parameters of the ecosystem. The dense, highly oxygenated CDW has filled the deep and bottom parts of the eastern Mediterranean, replacing the old EMDW of Adriatic origin, which has been uplifted several hundred meters. This process brought the oxygen-poor, nutrient-rich waters closer to the surface, so that in some regions winter mixing might bring extra nutrients to the euphotic zone, enhancing the biological production. Since 1991, the above mentioned uplifted old EMDW of Adriatic origin has reached shallow enough depths outside the Aegean and especially in the vicinity of the Straits of the Cretan Arc, to intrude the Aegean (Cretan Sea) and compensate its deep outflow (CDW outflow). These waters, namely Transitional Mediterranean Water (TMW), gradually formed a distinct intermediate layer (150–500 m) in the south Aegean, characterized by temperature, salinity and oxygen minima, and nutrient maxima. This has enhanced the previously weak stratification and enriched with nutrients one of the most oligotrophic seas in the world. This new structure prevents winter convection deeper than 250 m. Finally, in 1998–99, the presence of the TMW was much reduced, mainly as a result of mixing.

The simultaneous changes in both the upper and deep conveyor belts of the eastern Mediterranean may affect the processes and the water characteristics of the neighboring seas. The contribution of the Aegean to the intermediate and deep layers is still active. The variability in the intermediate waters can alter the preconditioning of dense water formation in the Adriatic as well as in the western Mediterranean. On the other hand, the changes in the deep waters can affect the LIW formation characteristics. Whether the present thermohaline regime will eventually return to its previous state or arrive at a new equilibrium is still an open question.

Sub-basin Scale Circulation

Figure 6 shows the patterns of circulation in the western Mediterranean for the various water types. The Atlantic Water in the Alboran Sea flows anticyclonically in the western portion of the western basin, while a more variable pattern occurs in the eastern portion. The vein flowing from Spain to Algeria is named the Almeria-Oran Jet. Further east, the MAW is transported by the Algerian Current, which is relatively narrow (30–50 km) and deep (200–400 m) in the west, but it becomes wider and thinner while progressing eastwards along the Algerian slope till the Channel of Sardinia. Meanders of few tens of kilometers, often 'coastal eddies' (**Figure 7**), are generated due to the unstable character of the current. The cyclonic eddies are relatively superficial and short-lived, while the anticyclones last for weeks or months. The current and its associated mesoscale phenomena can be disturbed by the 'open sea eddies.' The buffer zone that is formed by the MAW reservoir in the Algerian Basin disconnects the inflow from the outflow at relatively short timescales typically.

Large mesoscale variability characterizes the Channel of Sicily. In the Tyrrhenian Sea both the flow along Sicily and the Italian peninsula and the mesoscale activity in the open sea are the dominant features. The flows of MAW west and east of Corsica join and form the so-called Liguro-Provenco-Catalan Current, which is the 'Northern Current' of the Basin along the south-west European coasts. Mesoscale activity is more intense in winter, when this current becomes thicker and narrower than in summer. There is also strong seasonal variability in the mesoscale in the Balearic Sea. Intense barotropic mesoscale eddy activity propagates seaward from the coastline around the sea from winter to spring, and induces a seasonal variability in the open sea.

There is evidence that the Winter Intermediate Water (WIW) formed in the Ligurian Sea and the Gulf of Lions can be in larger amounts than that of Western Mediterranean Deep Water (WMDW). Because of their appropriate or shallower depths,

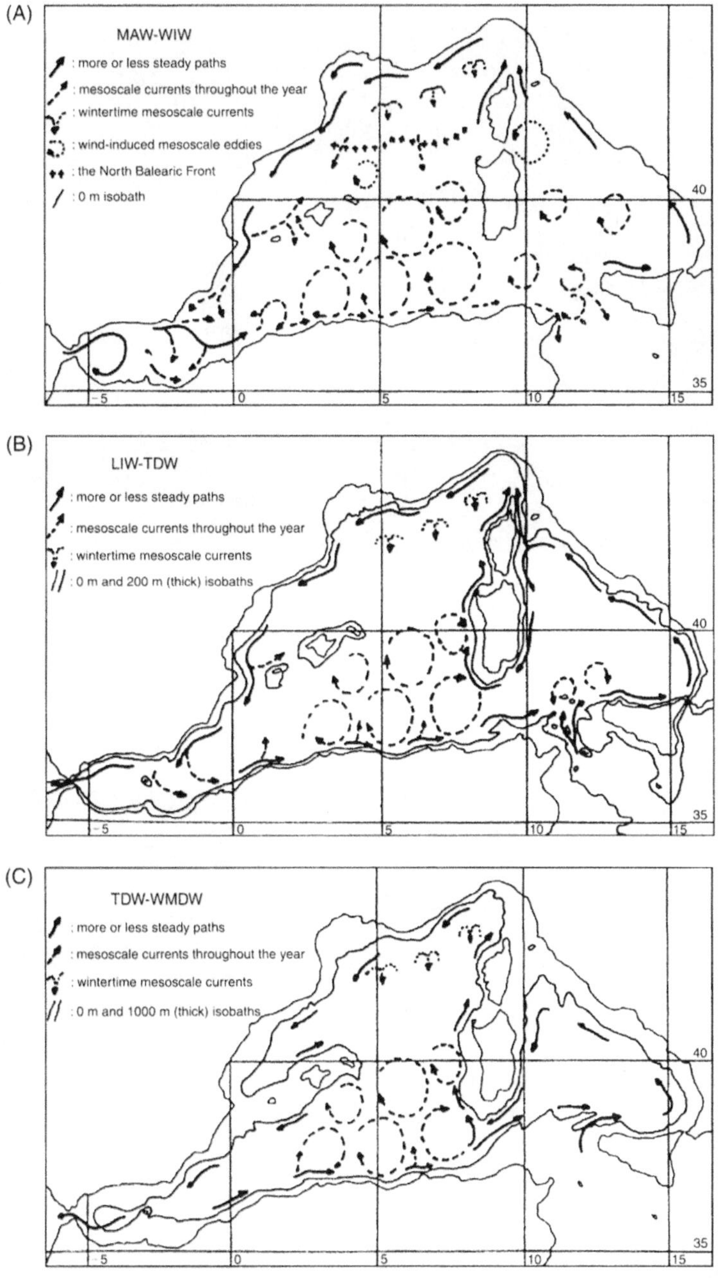

Figure 6 Schematics of the circulation of water masses in the western Mediterranean. (A) MAW–WIW; (B) LIW–TDW; (C) TDW–WMDW (Reproduced with permission from Millot, 1999).

these WIW can flow out at Gibraltar with LIW more easily than the WMDW. Furthermore, apart from the LIW there are also other intermediate waters of eastern Mediterranean origin that circulate and participate in the processes of the western topography. After filling the Algero–Provencal Basin up to depths ~2000 m, the WMDW intrudes into the deep Tyrrhenian (~3900 m). The amount of unmixed WMDW in the western Mediterranean and especially in the south Tyrrhenian Sea is automatically controlled by the density of the cascading flow from the Channel of Sicily and thus from the dense water formation processes in the eastern Mediterranean. The south Tyrrhenian is a key place for the mixing and transformation of the water masses; the processes within the eastern Mediterranean play a dominant role in the entire Mediterranean Basin.

In the eastern basin energetic sub-basin scale features (jets and gyres) are linked to construct the basin-wide circulation. Important variabilities exist and include: (1) shape, position, and strength of permanent sub-basin gyres and their unstable lobes,

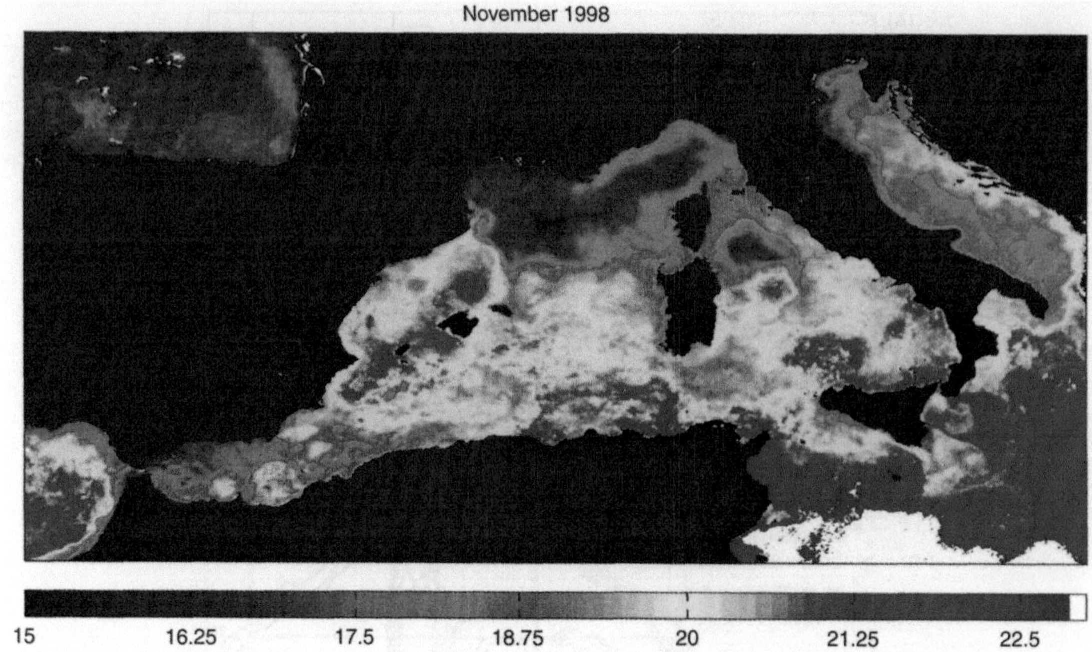

Figure 7 Satellite imagery of sea surface temperature during November 1998 in the Western Mediterranean.

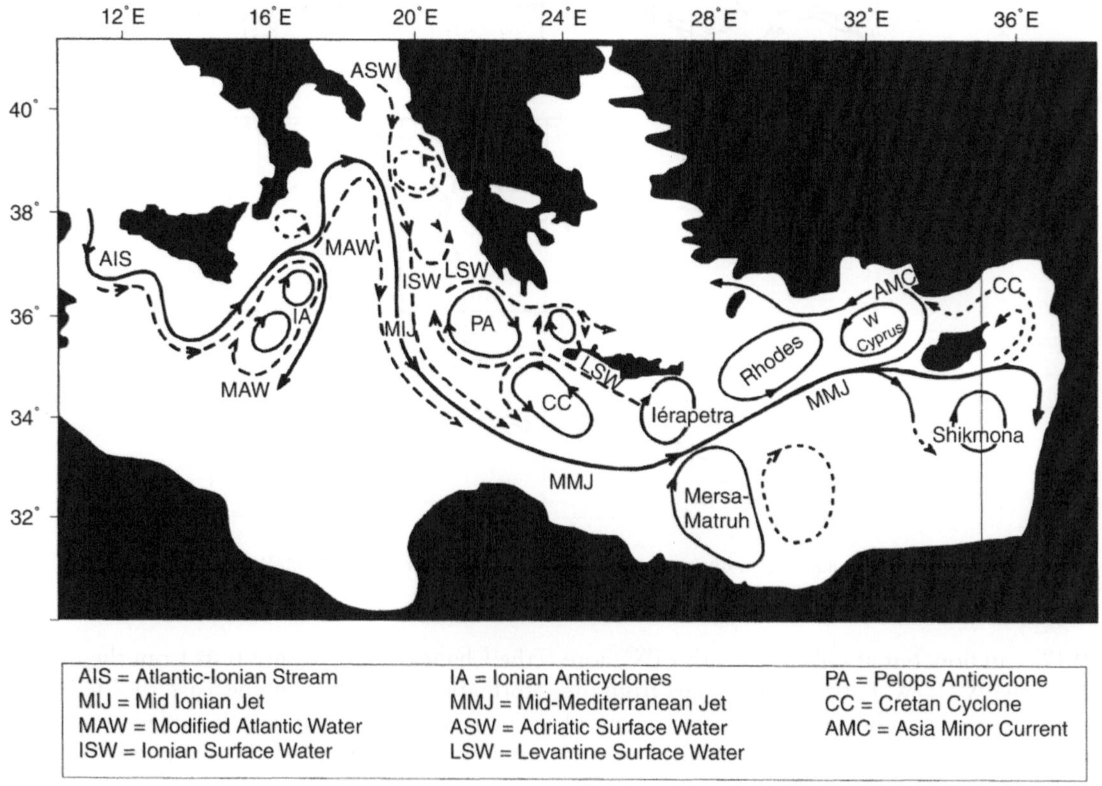

Figure 8 Sub-basin scale and mesoscale circulation features in the eastern Mediterranean (Reproduced with permission from Malanotte-Rizzoli et al., 1997 after Robinson and Golnaraghi, 1994).

multi-centers, mesoscale meanders, and swirls; (2) meander pattern, bifurcation structure, and strength of permanent jets; and (3) occurrence of transient eddies and aperiodic eddies, jets, and filaments. Figure 8 shows a conceptual model in which a jet of Atlantic Water enters the eastern basin through the

Table 1 Upper thermocline circulation features

Feature	*Type*	*ON85*	*MA86*	*MA87*	*AS87*	*SO91*	*JA95*	*S97*	*ON98*
AIS	P	–	–	Y	Y	Y	Y	Y	N
MMC	P	Y	Y	–	Y	Y	Y	–	Y
AMC	P	Y	Y	–	Y	Y	Y	–	Y
CC	R	Y	N	Y	N	–	–	–	–
Se Lev. Jets	T	Y	Y	–	Y	–	–	–	–
Rhodes C	P	Y	Y	–	Y	Y	Y	–	Y
West Cyprus C	P	Y	Y	–	Y	Y	Y	–	–
MMA	P	Y	Y	–	Y	Y	Y	–	Y
Cretan C	P	Y	?	–	Y	Y	–	Y	Y
Shikmona AC	R	Y	Y	–	Y	Y	–	–	–
Latakia C	R	Y	N	N	Y	–	–	–	–
Antalya AC	R	?	Y	–	N	–	–	–	–
Pelops AC	P	–	Y	Y	Y	Y	–	Y	Y
Ionian eddies AC	T	–	–	–	Y	Y	–	Y	N
Cretan Sea eddies	T	Y	Y	–	Y	–	Y	–	Y
Ierapetra	R	Y	N	Y	Y	Y	Y	–	Y

Straits of Sicily, meanders through the interior of the Ionian Sea, which is believed to feed the Mid-Mediterranean Jet, and continues to flow through the central Levantine all the way to the shores of Israel. In the Levantine basin, this Mid-Mediterranean Jet bifurcates, one branch flows towards Cyprus and then northward to feed the Asia Minor Current, and a second branch separates, flows eastward, and then turns southward. Important sub-basin features include: the Rhodes cyclonic gyre, the Mersa Matruh anticyclonic gyre, and the south-eastern Levantine system of anticyclonic eddies, among which is the recurrent Shikmona eddy south of Cyprus. The diameter of the gyres is generally between 200 and 350 km. Flow in the upper thermocline is in the order of 10–20 cm s^{-1}. A tabulation of circulation features in the eastern Mediterranean and their characteristics is presented in **Table 1**.

Figure 7 shows the upper-thermocline main circulation features and surface waters' pathways. **Figure 4** presents the thermohaline (intermediate and deep) circulation, which has a significant vertical component. Finally, **Figure 5** presents the vertical structure of the water masses in three different periods in order to follow/show the continuous transformation of the water mass structure and characteristics in the recent 13 years, that is the period of the Eastern Mediterranean Transient.

During the period 1991–95, a large three-lobe anticyclonic feature developed in the south-western Levantine (from the eastern end of the Cretan Passage, 26°E up to 31°E), blocking the free westward LIW flow, from the Levantine to the Ionian, and causing a recirculation of the LIW within the west Levantine Basin. Although, multiple, coherent anticyclonic eddies were also quite common in the area before 1991 (as the Ierapetra and Mersa-Matruh), the 1991–95 pattern differs significantly, with three anticyclones of relatively larger size covering the entire area. This feature seems to comprise the Mersa-Matruh and the Ierapetra Anticyclone. Moreover, the 1998–99 infrared SST images (**Figure 9**) indicated that the area was still occupied by large anticyclonic structures. The data sets collected in late 1998 and early 1999 indicated that this circulation pattern had been reversed to cyclonic, confirming the transient nature of these eddies. Consequently, the Atlantic Ionian Stream (AIS) was not flowing from Sicily towards the northern Ionian, but directly eastwards crossing the central Ionian towards the Cretan Passage (**Table 2**).

The seasonal variability of the circulation of the late 1980s in the south Aegean Sea has been replaced by a rather constant pattern in the period of the EMT (1991–98). Therefore, the Cretan Sea eddies were in a seasonal evolution in the 1980s (always present), while in the 1990s there was a constant succession of three main eddies (one cyclone in the west, one anticyclone in the central region and again one cyclone in the east) that presented spatial variability.

Mesoscale Circulation

The horizontal scale of mesoscale eddies is generally related to, but somewhat larger than, the Rossby radius of deformation. In the Mediterranean the internal radius is O(10–14) km or four times smaller than the typical values for much of the world ocean. The study of mesoscale instabilities, meandering, and eddying thus requires a very fine resolution sampling. For this reason, only recently different mesoscale features were found in both the western and eastern

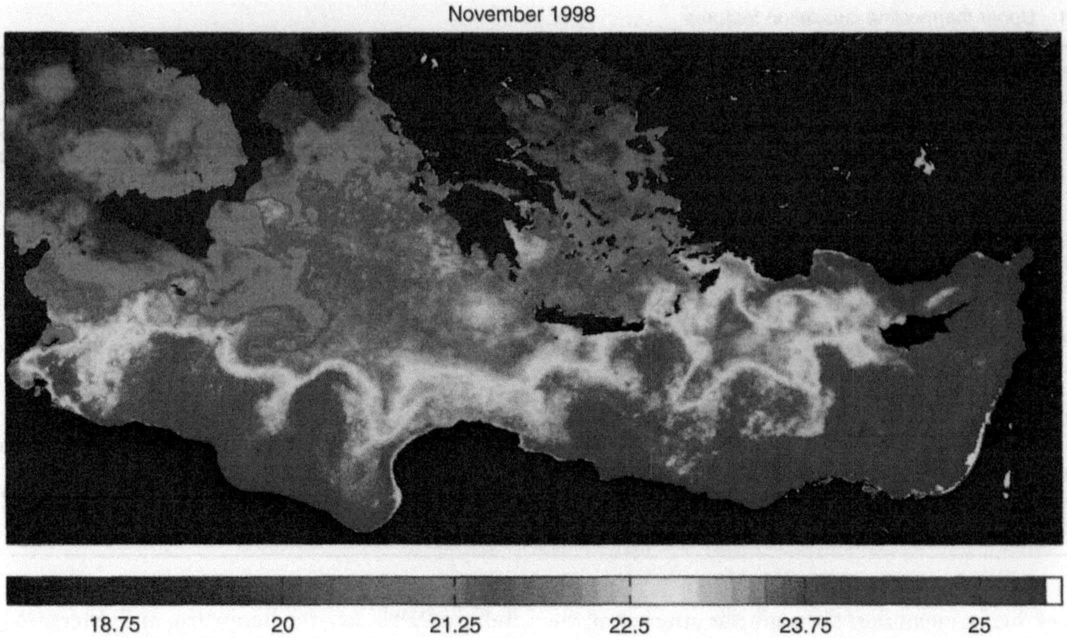

Figure 9 Satellite imagery of sea surface temperature during November 1998 in the Eastern Mediterranean.

Table 2 Mediterranean water masses

Water mass name	*Acronym*
Aegean Deep Water	AGDW
Adriatic Water	ASW
Cretan Deep Water	CDW
Cretan Intermediate Water	CIW
Eastern Mediterranean Deep Water	EMDW
Eastern Mediterranean Transient	EMT
Levantine Deep Water	LDW
Levantine Intermediate Water	LIW
Modified Atlantic Water	MAW
Transitional Mediterranean Water	TMW
Winter Intermediate Water	WIW
Western Mediterranean Deep Water	WMDW

basins including the mesoscale variabilities associated with the coastal currents in the western basin and open sea mesoscale energetic eddies in the Levantine basin.

In the western basin, intense mesoscale phenomena (**Figure 6**) have been detected using satellite information and current measurements. Mesoscale activity occurs as instabilities along the coastal currents (i.e., the Algerian Current) leading to the formation of mesoscale eddies which can eventually move across the basin or interact with the current itself. Along the Algerian Current cyclonic and anticyclonic eddies develop and evolve over several months as they slowly drift eastward (a few kilometers per day). The anticyclonic eddies generally increase in size and detach from the coast. Some may drift near the continental slope of Sardinia, where a well-defined flow of LIW exists. Here they are able to pull fragments of LIW seaward. Old offshore eddies extend deep in the water column and last from several months to as much as a year. They sometimes enter the coastal regions and interact with the Algerian Current. In the coastal zones the mesoscale currents appear to be strongly sheared in the vertical.

This clearly indicates that eddies can modify the circulation over a relatively wide area and for relatively long periods of time. The coastal eddies along the Algerian coast can be especially vigorous, inducing currents of 20–30 cm s^{-1} strength for periods of a few weeks. More complicated variations of the currents have also been measured at 300 m and sometimes at 1000 m.

Mesoscale activity has been observed in the northern basin (i.e., along the western and northern Corsican Currents). Coastal Corsican eddies are typically anticyclonic and located either offshore or along the coast of Corsica. A number of experiments were conducted to investigate the mesoscale phenomena in the Ligurian Sea. The results of a 1-year current meter array are shown for the southern coastal zone in the Corsican Channel (**Figure 10D**). Mesoscale currents are characterized by permanent occurrence and by a baroclinic structure with relatively large amplitude at the surface, moderate at the intermediate level and still noticeable at depth, thus indicating large vertical shear of the horizontal currents.

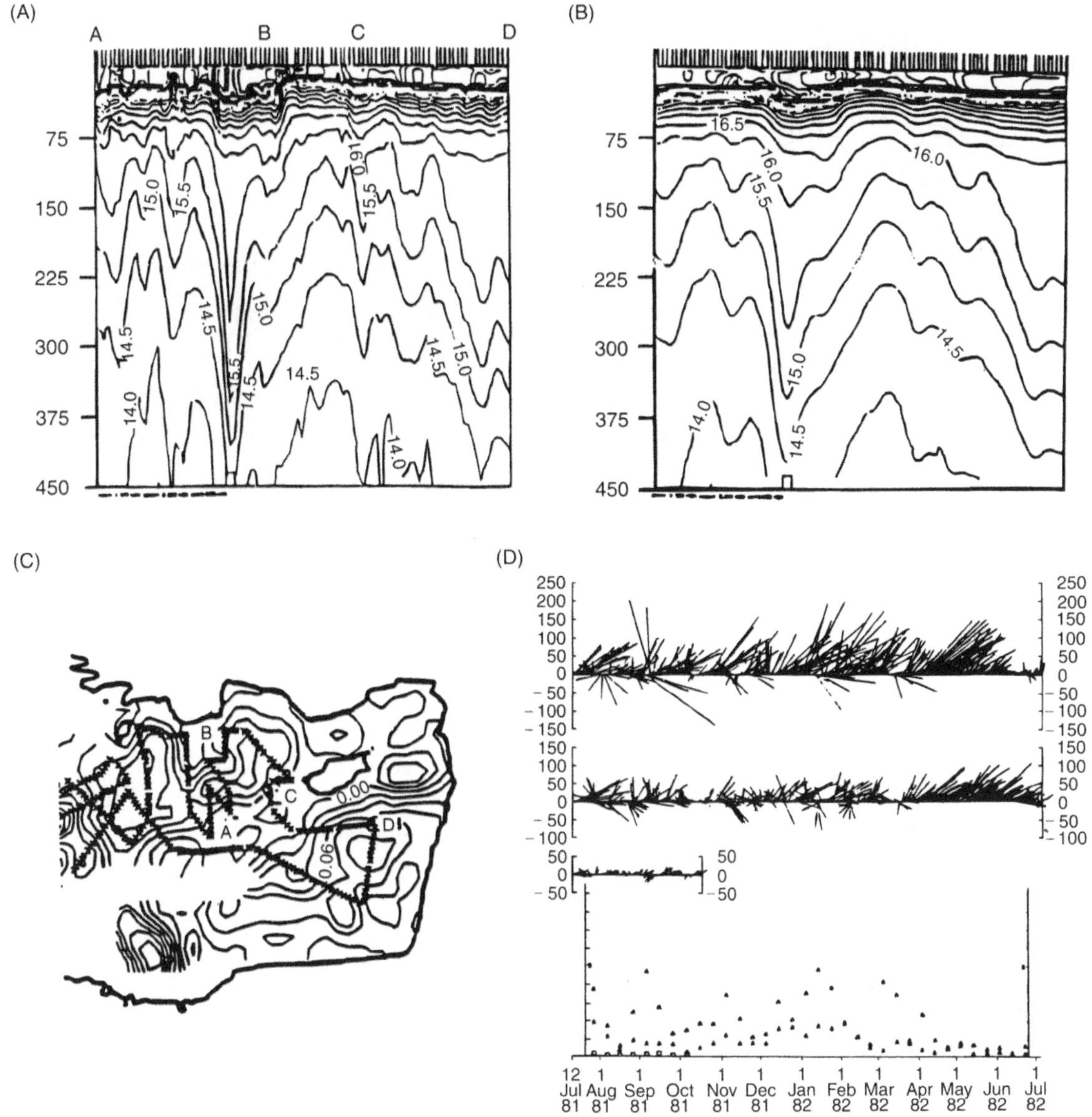

Figure 10 (A) Mesoscale temperature cross-section from XBTs in AS87 POEM cruise along section ABCD. (B) Filtered temperature cross-section using a pyramid filter with 50 km influential radius. (C) Location of the cross-section superimposed on the dynamic height anomaly from AS87 survey (excluding XBTs). (D) Velocities from a current meter array in the Corsican Channel.

Dedicated high-resolution sampling in the Levantine basin led to the discovery of open ocean mesoscale energetic eddies, as well as jets and filaments. This was confirmed by a mesoscale experiment in August–September 1987 in the eastern basin. Mesoscale eddies dynamically interacting with the general circulation occur with diameters in the order of 40–80 km. From this analysis a notable energetic sub-basin/mesoscale interaction in the Levantine basin and a remarkable thermostad in the Ionian have been revealed.

Figure 10A shows a temperature cross-section from XBT profiles collected in summer 1987 across the Mid-Mediterranean Jet, West Cyprus Gyre, MMJ, and the northern border of the Shikmona eddy (section ABCD shown in **Figure 10C**). **Figure 10B** shows the identical XBT section after a pyramid filter, with horizontal influential distance of 50 km. The filter has removed very small scale features while maintaining the mesoscale structure.

Modeling

Vigorous research in the 1980s and the developing picture of the multiscale Mediterranean circulation

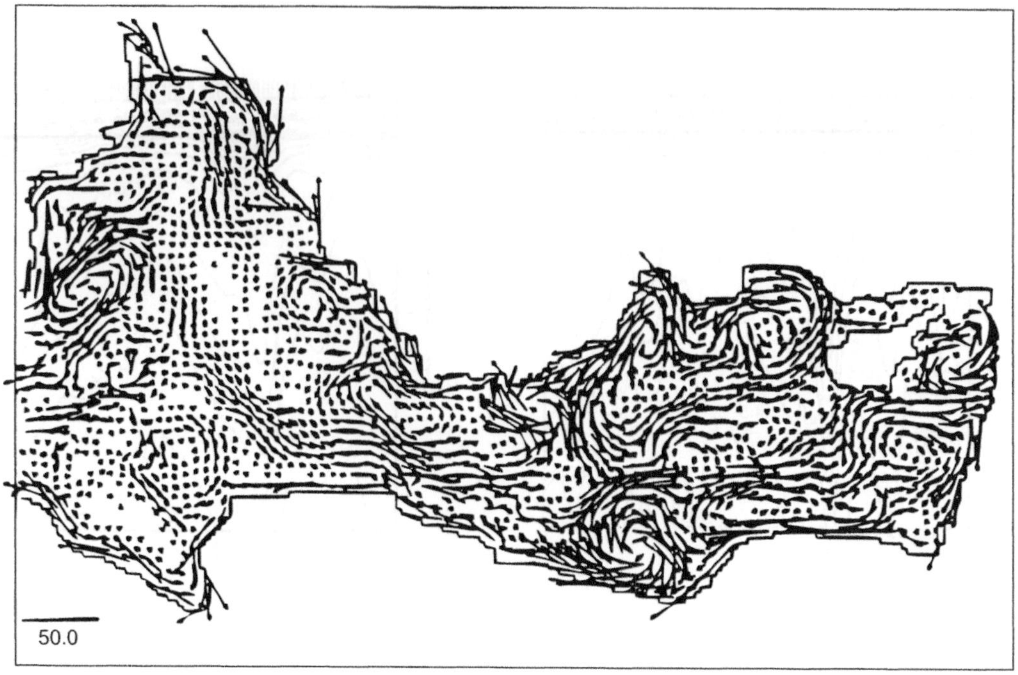

Figure 11 Velocity field for the eastern Mediterranean at 10 m depth from an eddy-resolving primitive equation dynamical model.

were accompanied by a new era of numerical modeling on all scales. Modeling efforts included: water mass models, general circulation models, and data assimilative models. Dynamics in the models include: primitive equations, non-hydrostatic formulations, and quasi-geostrophy. The assimilation of the cooperative eastern Mediterranean surveys of the 1980s and 1990s into dynamical models played a significant role in the identification of sub-basin scale features. The numerical model results shown in **Figure 11** depict the existence of numerous sub-basin scale features, as schematized in **Figure 7**.

In recent years numerical modeling of the general circulation of the Mediterranean Sea has advanced greatly. Increased computer power has allowed the design of eddy resolving models with grid spacing of one-eighth and one-sixteenth of a degree for the whole basin and higher for parts of it. An example output from such a model, forced with perpetual year atmospheric forcing, which includes the seasonal cycle but not interannual variability, is shown in **Figure 12**. Many of these models incorporate sophisticated atmospheric forcing parameterizations (e.g., interactive schemes) which successfully mimic existing feedback mechanisms between the atmosphere and the ocean. Studies have been carried out using perpetual year atmospheric forcing, mostly aimed at studying the seasonal cycle, as well as interannual atmospheric forcing. Studies have mainly focused on reproducing and understanding the seasonal cycle, the deep- and intermediate-water formation processes and the interannual variability of the Mediterranean. They have shown the existence of a strong response of the Mediterranean Sea to seasonal and interannual atmospheric forcing. Both seasonal and interannual variability of the Mediterranean seems to occur on the sub-basin gyre scale. The Ionian and eastern Levantine areas are found to be more prone to interannual changes than the rest of the Mediterranean. Sensitivity experiments to atmospheric forcing show that large anomalies in winter wind events can shift the time of occurrence of the seasonal cycle. This introduces the concept of a 'memory' of the system which 'preconditions' the sea at timescales of the order of one season to 1 year.

In the deep- and intermediate-water formation studies, the use of high frequency (6 h) atmospheric forcing (in contrast to previously used monthly forcing) in correctly reproducing the observed convection depths and formation rates was found to be crucial. This shows the intermittent and often violent nature of the convention process, which is linked to a series of specific storm events that occur during each winter rather than to a gradual and continuous cooling over winter. The use of high-resolution numerical models in both the western and the eastern Mediterranean allowed the study of the role of baroclinic eddies, which are formed at the periphery of the chimney within the cyclonic gyre by instabilities

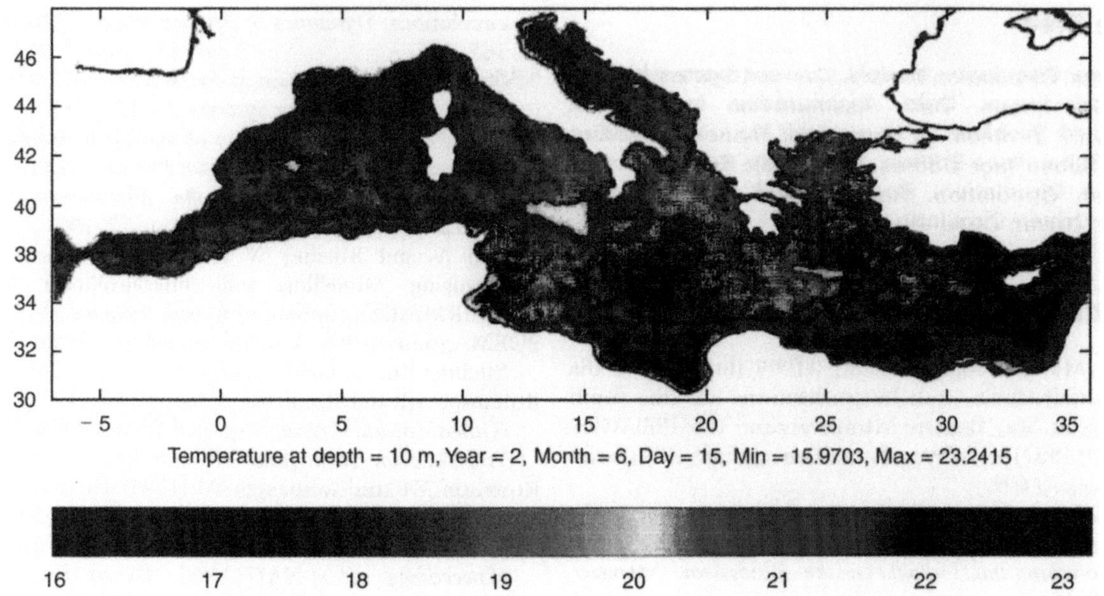

Figure 12 Temperature and superimposed velocity vectors at 10 m depth from a numerical simulation of the entire Mediterranean Sea.

of the meandering rim current, in open ocean convection. These eddies were shown to advect buoyancy horizontally towards the center of the chimney, thus reducing the effectiveness of the atmospheric cooling in producing a deep convected mixed layer. These results are in agreement with previous theoretical and laboratory work.

The LIW layer which extends over the whole Mediterranean was found to play an important role both in the western (Gulf of Lions) and the eastern (Adriatic) deep-water formation sites and more specifically in 'preconditioning' the formation process. It was shown that the existence of this layer greatly influences the depth of the winter convection penetration in these areas. This is related to the fact that the LIW layer with its high salt content decreases the density contrast at intermediate layers, thus allowing convection to penetrate deeper. This result shows the existence of teleconnections and inter-dependencies between sub-basins of the Mediterranean.

A number of numerical models have been developed to simulate and understand the origins and the evolution of the Eastern Mediterranean Transient. These models indicate that the observed changes can be at least partially explained as a response to variability in atmospheric forcing. Sensitivity experiments, in which the observed precipitation anomaly of 1989–90 and 1992–93 was not included, did not reproduce properly the EMT. This confirms that this factor was a significant contributor to the occurrence and evolution of the Eastern Mediterranean Transient, since it acted as a 'preconditioner' to the latter by importantly increasing the salinity in the area.

The enhanced deep water production in the Aegean has implied a deposition of salt in the deep and bottom layers with a simultaneous decrease higher up. As the turnover rate for waters below 1200 m has been estimated to exceed 100 years, this extra salt will take many decades to return into the upper waters. Its return, however, might well induce changes in the thermohaline circulation, considering the dependence of the two potential sources of deep water on the salinity preconditioning. It will therefore take many decades before the eastern Mediterranean returns to a new quasi-steady state. An interesting question in this connection is whether the system will recover its previous mode of operation with a single source of deep-water production in the Adriatic or evolve into an entirely different, perhaps even an unanticipated direction.

Conclusion

The Mediterranean Sea is now known to have a complex thermohaline, wind, and water flux-driven multi-scale circulation with interactive variabilities. Recent vigorous research, both experimental and modeling, has led to this interesting and complex picture. However, the complete story has not yet been told. We must wait to see the story unfold and see how many states of the circulation exist, what changes occur and whether or not conditions repeat.

See also

Coastal Circulation Models. Current Systems in the Mediterranean. Data Assimilation in Models. Forward Problem in Numerical Models. Meddies and Subsurface Eddies. Mesoscale Eddies. Arctic Ocean Circulation. Regional and Shelf Models. Wind Driven Circulation.

Further Reading

Angel MV and Smith R (eds.) (1999) Insights into the hydrodynamics and biogeochemistry of the South Aegean Sea, Eastern Mediterranean: The PELAGOS (EU) PROJECT. *Progress in Oceanography* 44 (special issue): 1-699.

Briand F (ed.) (2000) CIESM Workshop Series no.10, *The Eastern Mediterranean Climatic Transient: its Origin, Evolution and Impact on the Ecosystem*. Monaco: CIESM.

Chu PC and Gascard JC (eds.) (1991) *Elsevier Oceanography Series, Deep Convection and Deep Water Formation in the Oceans*. Elsevier.

Lascaratos A, Roether W, Nittis K, and Klein B (1999) Recent changes in deep water formation and spreading in the Eastern Mediterranean Sea. *Progress in Oceanography* 44: 5–36.

Malanotte-Rizzoli P (ed.) (1996) *Elsevier Oceanography Series, Modern Approaches to Data Assimilation in Ocean Modeling*.

Malanotte-Rizzoli P and Eremeev VN (eds.) (1999) *The Eastern Mediterranean as a Laboratory Basin for the Assessment of Contrasting Ecosystems, NATO Science Series – Environmental Sercurity* vol. 51, Dordrecht: Klumer Academic.

Malanotte-Rizzoli P, Manca BB, and Ribera d'Acala M (1999) The Eastern Mediterranean in the 80s and in the 90s: The big transition in the intermediate and deep circulations. *Dynamics of Atmosphere and Oceans* 29: 365–395.

Millot C (1999) Circulation in the Western Mediterranean Sea. *Journal of Marine Systems* 20: 423–442.

Nielsen JN (1912) Hydrography of the Mediterranean and Adjacent Waters. In: *Report of the Danish Oceanographic Expedition 1908–1910 to the Mediterranean and Adjacent Waters*, 1, Copenhagen, pp. 72–191.

Pinardi N and Roether W (eds) Mediterranean Eddy Resolving Modelling and InterDisplinary Studies (MERMAIDS). *Journal of Marine Systems* 18: 1–3.

POEM group (1992) General circulation of the Eastern Mediterranean. *Earth Sciences Review* 32: 285–308.

Robinson AR and Brink KH (eds.) (1998) *The Sea: The Global Coastal Ocean, Regional Studies and Syntheses*, Vol. 11. New York: John Wiley and Sons.

Robinson AR and Golnaraghi M (1994) The physical and dynamical oceanography of the Mediterranean Sea. In: Malanotee-Rizzoli P and Robinson AR (eds.) *Proceedings of a NATO-ASI, Ocean Processes in Climate Dynamics: Global and Mediterranean Examples*, pp. 255–306. Dordecht: Kluwer Academic.

Robinson AR and Malanott-Rizzoli P (eds.) *Physical Oceanography of the Eastern Mediterranean Sea, Deep Sea Research*, vol. 40(6) (Special Issue), Oxford: Pergamon Press.

Roether W, Manca B, and Klein B *et al.* (1996) Recent changes in the Eastern Mediterranean deep waters. *Science* 271: 333–335.

Theocharis A and Kontoyiannis H (1999) Interannual variability of the circulation and hydrography in the eastern Mediterranean (1986–1995). In: Malanotte-Rizzoli P and Eremeev VN (eds.) *NATO Science Series – Environmental Security* vol. 51, *The Eastern Mediterranean as a Laboratory Basin for the Assessment of Contrasting Ecosystems*, pp. 453–464. Dordrecht: Kluwer Academic.

CURRENT SYSTEMS IN THE MEDITERRANEAN SEA

P. Malanotte-Rizzoli, Massachusetts Institute of Technology, Cambridge, MA, USA

Introduction

In the last two decades the Mediterranean Sea has been the object of renewed interest in the oceanographic community thanks to the formulation and execution of international collaborative programs, such as the UNESCO/IOC sponsored Programme de Recherche Internationale en Méditerranée Occidentale (PRIMO) in the Western basin and Physical Oceanography of the Eastern Mediterranean (POEM) Programme in the Eastern Mediterranean, followed up by the recent effort undertaken by the European community under the Marine Science and Technology (MAST) banner.

Two main reasons form the basis of this scientific effort. The Mediterranean is a midlatitude semi-enclosed sea, exchanging water and other properties with the North Atlantic Ocean, which plays a crucial role in the global thermohaline circulation through the formation of North Atlantic Deep Water (NADW) in the polar Greenland and Labrador Seas. The upper intermediate layer of the North Atlantic is replenished with a very salty water mass that spreads out from the Mediterranean through the connecting narrow Straits of Gibraltar. This salty water is formed in the easternmost Mediterranean, the Levantine basin, as Levantine Intermediate Water (LIW). The Gibraltar Straits are thus a point source of heat and salt for the North Atlantic at all depths from 1000 m to more than 2500 m. Tongues of this warm, salty water extend through the Atlantic interior, both northward along the coast of Europe and westward towards America on all isopycnals from $\sigma_1 = 31.938\,\text{kg}\,\text{m}^{-3}$ to $\sigma_3 = 41.44\,\text{kg}\,\text{m}^{-3}$, thus crucially preconditioning the NADW convective cells in the polar seas.

The second reason is that many dynamical processes which are fundamental to the world ocean circulation also occur within the Mediterranean, such as deep convection cells completely analogous to the NADW cell, with convective sites in both the Western and Eastern basins. Both of these basins are moreover endowed with a deep thermohaline circulation, the equivalent of the global conveyor belt. Thus the Mediterranean provides a laboratory basin for general circulation studies. The two basins, Western and Eastern, can be studied quite independently, as they are connected through the shallow Sicily Straits, with the deepest threshold at ~250 m, which prevents direct communication between the subsurface layers.

This paper first provides a short review of the climatology characterizing the entire basin. The western and eastern basins are then discussed separately. Particular attention is devoted to the Eastern Mediterranean, where a major transient has occurred in the last decade that documents the existence of multiple states for the Eastern Mediterranean internal thermohaline circulation.

Morphology and Climatology

The Mediterranean Sea (**Figure 1A**) is an enclosed basin connected to the Atlantic Ocean by the narrow and shallow Strait of Gibraltar (width ~ 13 km; sill depth ~300 m). It is composed of two similar size basins, western and eastern, connected by the Strait of Sicily (width ~35 km; sill depth ~250 m). The Western Mediterranean has a triangular shape, with the Ligurian Sea at its apex and a large topographic plateau in the Balearic Sea (**Figure 1B**). The islands of Corsica and Sardinia separate the Balearic Sea from the Tyrrhenian Sea, where the bottom relief has the more complex shape of a deep, corrugated valley. The Balearic and Tyrrhenian Seas join in the south in a wide passage between Sardinia and Sicily that leads to the Sicily Straits and the eastern basin.

The Eastern Mediterranean has a more complicated structure than the western, with a much more irregular, complex topography constituted by a succession of deep valleys, ridges, and localized pits (**Figure 1B**). Four sub-basins can be defined in the Eastern Mediterranean (**Figure 1A**): the Ionian, the Levantine, the Adriatic, and the Aegean Seas. The Ionian Sea is the deepest in its central part, ending in the shallow Gulf of Sirte at its southernmost end. The Cretan passage leads from the Ionian into the Levantine basin, that reaches its maximum depth in a localized depression south-east of the island of Rhodes.

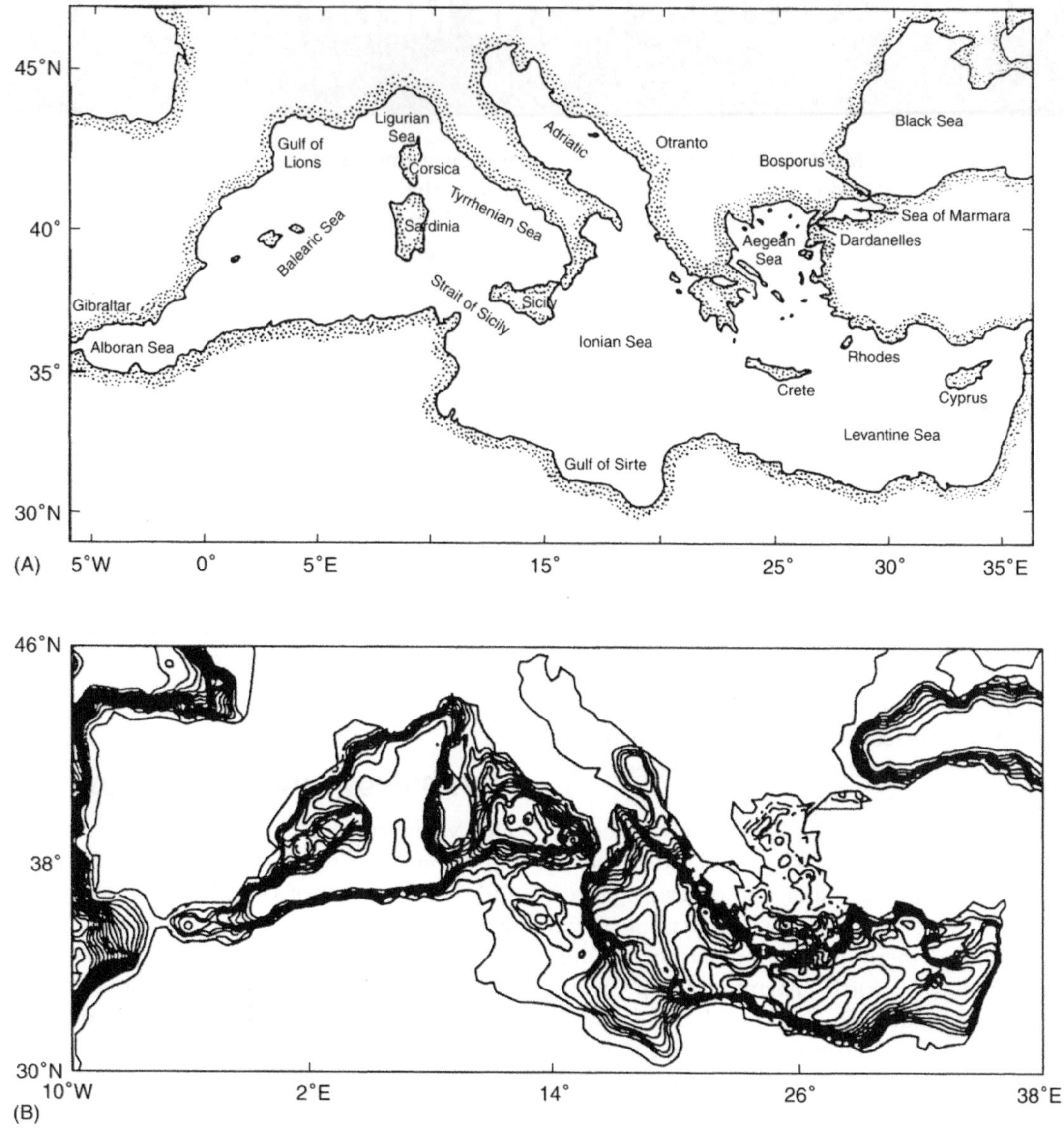

Figure 1 Morphology of the Mediterranean Sea: (A) geography of the basin; (B) bathymetry.

The hydrology and the circulation of the Mediterranean Sea have been known in overall generality for some time. For instance it is well known that the Mediterranean basins, both western and eastern, are evaporation basins (lagoons), with freshwater flux from the Atlantic through the Gibraltar Straits and into the Eastern Mediterranean through the Sicily Straits. The Atlantic Water (AW) mass entering through Gibraltar increases in density because evaporation exceeds precipitation and becomes Modified Atlantic Water (MAW) in its route to the Levantine basin. New water masses are formed here via convection events driven by intense local cooling and evaporation from winter storms. Bottom water is produced in localized convection sites, for the western basin in the Gulf of Lions (Western Mediterranean Deep Water, WMDW) and for the eastern basin in the southern Adriatic (Eastern Mediterranean Deep Water, EMDW). Recent observations also indicate Levantine Deep Water (LDW) formation in the north-eastern Levantine basin during exceptionally cold winters, where Levantine Intermediate Water (LIW) is regularly formed seasonally. Evidence has emerged that LIW formation occurs over much of the Levantine basin but preferentially in the north probably due to meteorological factors. The LIW is the important water mass which circulates westward through both the eastern and western basins and

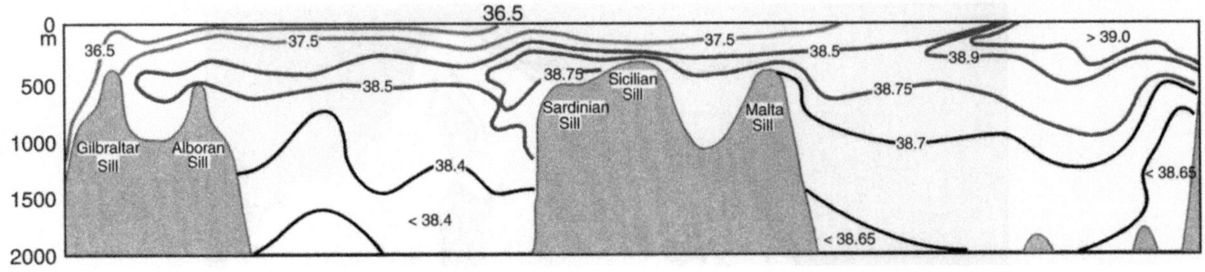

Figure 2 The open thermohaline cell of the Mediterranean upper layer.

contributes predominantly to the efflux from Gibraltar to the Atlantic, mixed with some EMDW and WMDW.

The western and eastern basins are connected in the upper layer, ~200 m thick, through the 'open thermohaline cell' of the basin, schematized in **Figure 2**.

While the entire Mediterranean is a 'lagoon' for the North Atlantic, the eastern basin itself is a 'lagoon' for the western one. The Atlantic Water (AW) mass enters the Mediterranean at Gibraltar with a typical salinity of $S = 36.15$ PSU and temperature $T = 15°C$. The AW becomes Modified Atlantic Water (MAW) through diffusive processes in its pathway eastward and can be identified as a subsurface salinity minimum below ~30 m depth. At the Sicily Straits, the saltier MAW has a salinity $S \leq 37.5$ PSU reaching a maximum of $S < 38.9$ PSU in the Levantine basin. Here in its northern part, winter episodes of cold, dry winds blowing from the mainland under surface cooling and evaporative fluxes lead to the formation of LIW, which has $39.0 \leq S \leq 39.2$ PSU and $15°C < \theta < 16°C$ at the formation sites, the most important of which is the well-known Rhodes gyre. The LIW return route is westward in the layer between 200 and 600 m depth. LIW becomes progressively colder and fresher. At the Sicily Straits typical LIW core values are $\theta = 14.3°C$ and $S \leq 38.8$ PSU. At the Gibraltar Straits LIW is diluted to $S \leq 38.5$ PSU, and spreads out in the North Atlantic, becoming Upper North Atlantic Intermediate Water. Secondary pathways of LIW will be discussed separately for the two basins.

The Western Mediterranean

The upper thermocline circulation (upper ~200 m) of the Western basin is schematized in **Figure 3A**.

The MAW in the Alboran Sea describes a quasi-permanent anticyclonic gyre in the west and a more variable circuit in the eastern Alboran. Further east, the MAW is entrained in the strong meandering Algerian current, whose instabilities lead to the formation of anticyclonic eddies (diameter ~50–100 km) all along the coast of Algeria. These eddies grow in size; some may detach from the coast and drift into the interior of the Balearic Sea. A quasi-stationary cyclonic path of MAW has been observed around the Balearic Sea leading to the formation of the Western Corsican Current west of Corsica. A steady cyclonic path of MAW is also present in the Tyrrhenian Sea, that intrudes into Northern Ligurian Sea, where it joins the Western Corsican Current producing a return south-westward flow along the Italian, French, and Spanish coasts, towards the Alboran Sea, that is called the Northern Current. The latter shows strong seasonal variability, becoming more intense and narrower in wintertime when it develops intense meanders, and splits into multiple branches in the southern Balearic sea.

Figure 3B shows the pathway of LIW emerging from the Sicily Straits into the Western Mediterranean in the intermediate layer, 200–600 in depth.

LIW follows a cyclonic route all around the Tyrrhenian Sea, and splits into two branches at the northern tip of Corsica. One branch enters directly into the Ligurian Sea, the second circulates around Sardinia and Corsica, merges with the previous branch and successively flows cyclonically around the Balearic Sea. This major LIW branch enters the Gulf of Lions, where it plays a crucial role in preconditioning the winter convective cell of WMDW located here. WMDW has been observed to form in the Gulf of Lions basically every year, under winter episodes of cold, dry Mistral wind blowing from France. Here the mixed, ventilating chimney (~100 km in diameter) can reach 2000 m depth.

It must be pointed out that the mean LIW pathway is still controversial. Numerical simulations as well as data analysis indicate a major direct route of LIW from the Sicily Straits to Gibraltar. On the other hand, strong observational evidence suggests the pattern presented in **Figure 3B**, with the LIW cyclonic circuit around the Tyrrhenian, the islands of

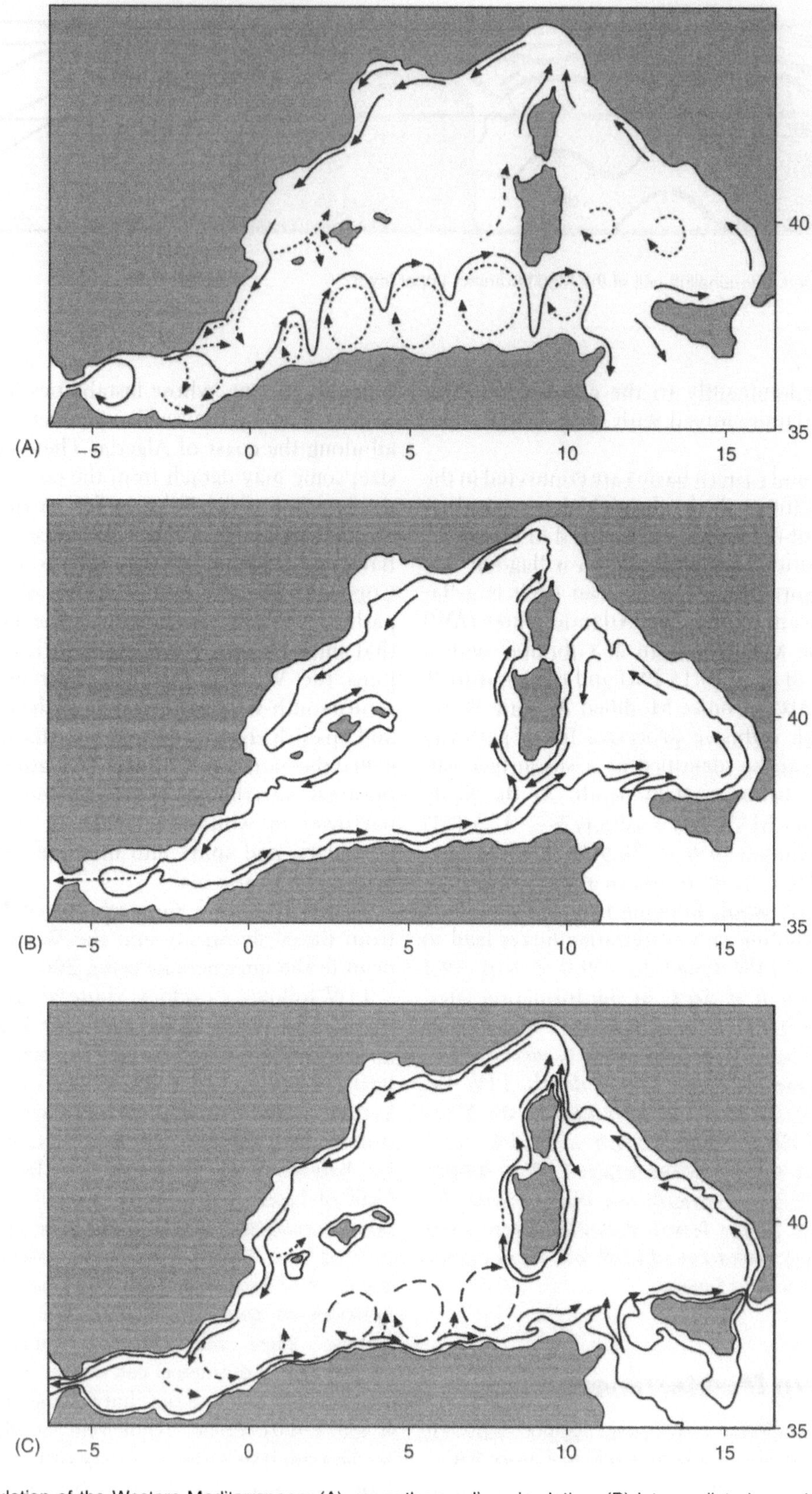

Figure 3 Circulation of the Western Mediterranean: (A) upper thermocline circulation; (B) intermediate layer circulation with LIW pathways; (C) deep thermohaline circulation with WMDW pathways.

Sardinia and Corsica and finally the Balearic Sea. Here, in the southern, eastern part, the LIW pathway bifurcates, with one branch proceeding towards and exiting from Gibraltar and a second returning eastward along the Algerian coast.

Figure 3C schematizes the thermohaline cell of the Western Mediterranean, driven by the winter convection in the Gulf of Lions.

Even though the exact routes of WMDW are also under debate, the pattern of **Figure 3C** is based on available observations and indicates spreading at depth of WMDW both towards the Sicily and the Gibraltar Straits, following a circuitous cyclonic route that leads it throughout the Balearic and Tyrrhenian Seas. The deep WMDW flow is obviously affected by topography. In the Tyrrhenian Sea, the WMDW joins the Tyrrhenian Dense Water present in the deep layers. At Gibraltar, upwelling of WMDW occurs, mixing with the overlying LIW, and contributing (what it is believed to be a small proportion) to the outflow from Gibraltar into the northern Atlantic.

The Eastern Mediterranean

The field work for the POEM program which was carried out in the period 1985–95, has definitively established the existence of three dominant scales interacting in the general circulation pattern; the basin-scale, i.e. the intermediate and deep thermohaline circulation; the sub-basin scale, characterizing the upper thermocline; and the mesoscale, defined by a ubiquitous and energetic eddy field. The POEM observational evidence has moreover shown that the Eastern Mediterranean has undergone a startling transition in the intermediate and deep basin circulations between the 1980s and the 1990s, the first documented example of the existence of multiple thermohaline states.

The upper thermocline circulation, on the other side, which is embedded in the Mediterranean open thermohaline cell, has remained very consistent throughout the two decades. The building blocks of the upper thermocline circulation are sub-basin-scale gyres and permanent, or quasi-permanent, cyclonic and anticyclonic structures interconnected by intense jets and meandering currents. The schematic representation of the upper thermocline circulation is given in **Figure 4**.

All the structures depicted in **Figure 4** are robust and persisted in the 1980s and the 1990s, albeit with modulations in strength and areal extension. Some differences were present but only in the Levantine Sea. At the Sicily Straits, the entering MAW is advected by the strong Atlantic Ionian Stream (AIS)

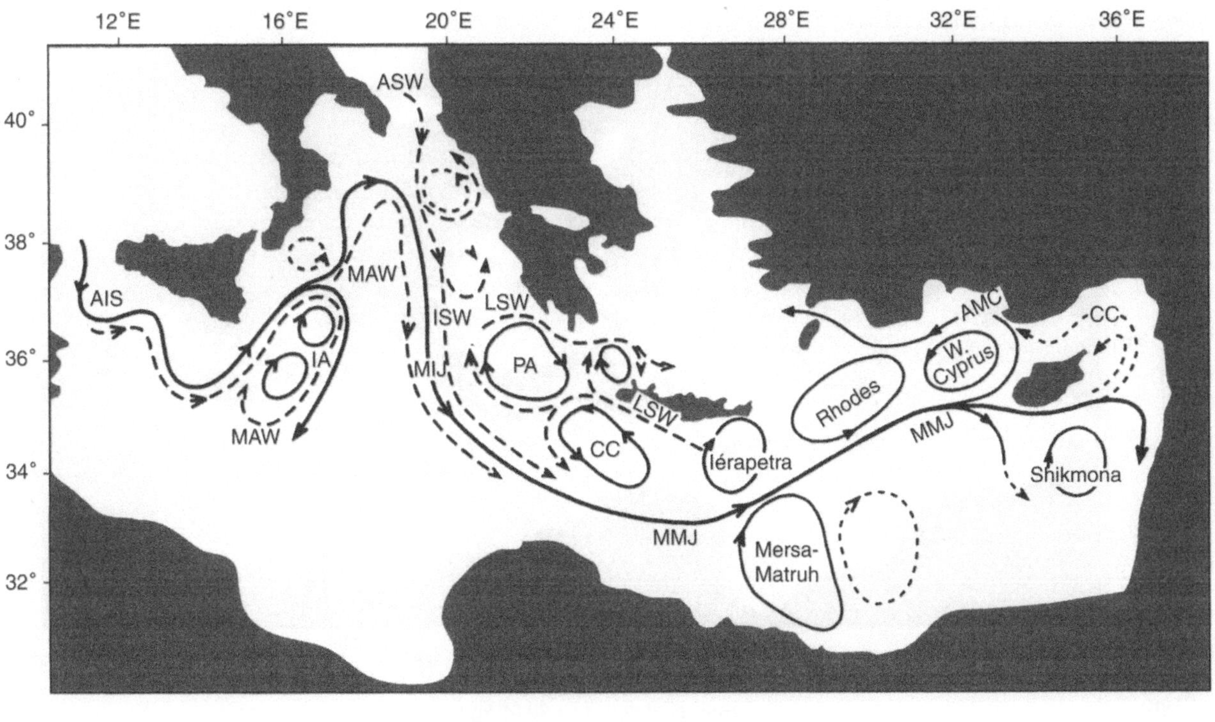

Figure 4 Eastern Mediterranean: Schematic representation of the upper thermocline circulation. AIS, Atlantic Ionian Stream; AMC, Asia Minor Current; ASW, Adriatic Surface Water; CC, Cretan Cyclone; IA; Ionian anticyclones; ISW, Ionian Surface Water; LSW, Levantine Surface Water; MAW, Modified Atlantic Water; MIJ, Mid-Ionian Jet; MMJ, Mid-Mediterranean Jet; PA, Pelops anticyclone.

jet, which forms a broad meander in the Ionian Sea, bifurcating into two main branches. One branch turns directly southward towards the African coast enclosing an intense anticyclonic area with multiple centers, the Ionian anticyclones (IA) which penetrate deeply into the intermediate layer. The second AIS branch protrudes into the north-eastern extremity, then turns southward forming the strong Mid-Ionian Jet (MIJ) that crosses the entire Ionian sea meridionally, thereafter veering eastward through the Cretan channel where it becomes the Mid-Mediterranean Jet (MMJ). Strong permanent features are

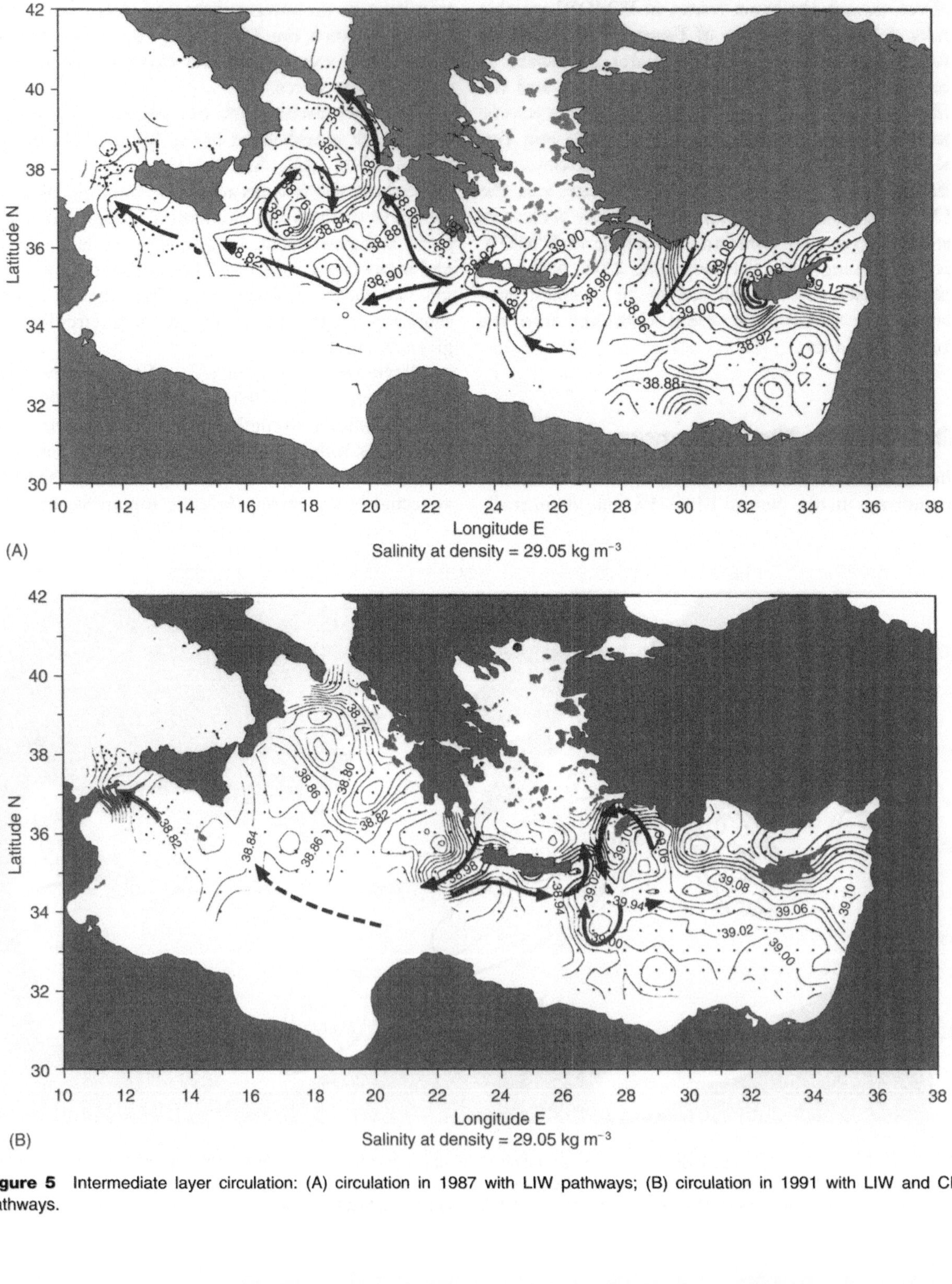

Figure 5 Intermediate layer circulation: (A) circulation in 1987 with LIW pathways; (B) circulation in 1991 with LIW and CIW pathways.

the Pelops anticyclone (PA) that has a strong barotropic component and penetrates to 800–1000 m depth. The Cretan cyclone, located south of Crete, is on the other side confined to the upper thermocline. A further permanent structure in the Cretan passage is the strong Ierapetra anticyclone, also South of Crete. The MMJ from the Cretan passage intrudes into the Eastern Levantine where it separates a northern overall cyclonic region from a southern anticyclonic region. The northern region comprises two well defined, permanent cyclones, the Rhodes gyre, site of LIW and LDW formation, and the western Cyprus cyclone. The southern anticyclonic area also comprises multiple centers, the strongest and most robust of which is the Mersa-Matruh anticyclone, located just south of the Rhodes gyre. A quasi-permanent structure, the Shikmona anticyclone, is present in the easternmost Levantine. In the 1990s, the only major difference was constituted by the appearance of a third anticyclonic center in the southern Levantine, of which the MMJ constituted the Northern [illegible] This anticyclone pushed westward the Mersa-Matruh

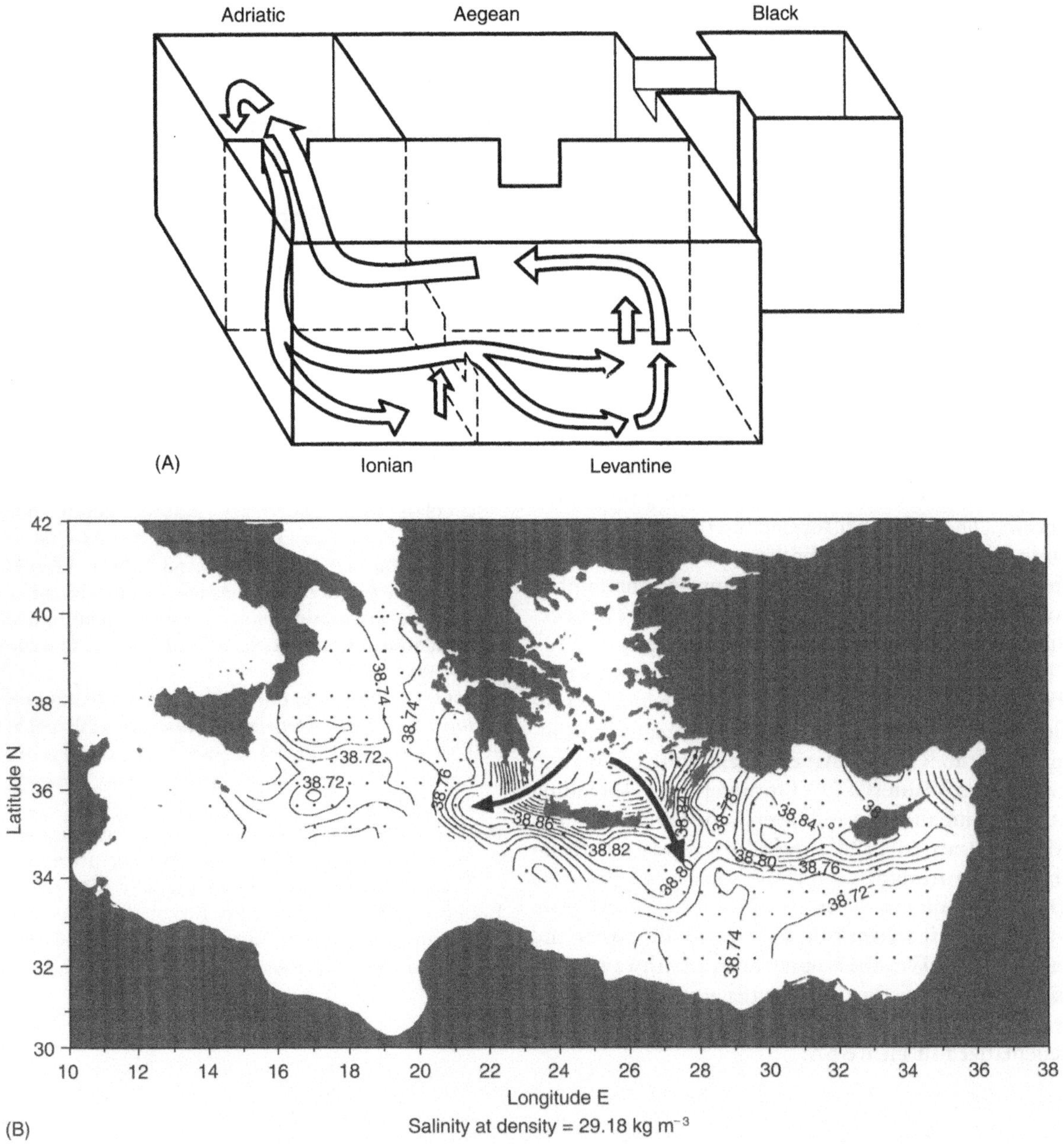

Figure 6 Deep thermohaline circulation: (A) schematic representation of the deep thermohaline cell in 1987; (B) deep pathways of CDW in 1991.

anticyclone, thus forming with the Ierapetra gyre a three-lobed intense anticyclonic area that induced a stronger meandering in the MMJ, confining it to its northern rim.

The dramatic transition occurred in the intermediate and deep layer circulations. Definitive observational evidence has been presented that this change was already present in 1991 and persisted through 1995–96, while the 1985–87 situation was completely different. In the intermediate layer, 250–600 dbar depth, characterized by LIW spreading on the isopycnal horizons $\sigma_0 = 29.00–29.05–29.10\,\mathrm{kg\,m^{-3}}$, the 1987 LIW circulation is depicted in **Figure 5A** for $\sigma_0 = 29.05\,\mathrm{kg\ m^{-3}}$.

LIW, formed in the northern Levantine in the Rhodes gyre, follows its 'classical' pathway, spreading towards the Sicily Straits through the Cretan channel and the Ionian interior. A second major route produced by the veering by the Pelops anticyclone is northward along the Greek coastline towards the Otranto Straits. Here LIW enters the southern Adriatic Sea to precondition the deep convective cell where Adriatic Deep Water (ADW) occurs. ADW spreads at depth out of the Otranto Strait to become EMDW.

The situation in 1991 (shown in **Figure 5B**, again the salinity distribution on $\sigma_0 = 29.05$ to $\mathrm{kg\ m^{-3}}$) is completely different. Now Cretan Intermediate Water (CIW) formed inside the Cretan/Aegean sea, substitutes for LIW, exiting from the western Cretan Arc Straits in a well defined tongue and filling the Ionian interior. The LIW is still formed in the northern Levantine, but its westbound pathway is blocked by the three-lobed anticyclonic region now present in the southern Levantine, which induces a local LIW cyclonic recirculation inside the Levantine itself.

This startling change is due to the fact that in the 1990s the 'driving engine' of the intermediate, transitional and deep layer circulations became the interior of the Cretan/Aegean Sea, with CIW and Cretan Deep Water (CDW) forming there, spreading out and filling the abyssal layers of the entire Eastern Mediterranean. In 1987, the EMDW was formed in the southern Adriatic as ADW, spread out from the Otranto Strait into the entire Eastern Mediterranean, upwelled to the transitional/intermediate layers and returned as LIW in the upper warm pathway to the southern Adriatic, thus closing the internal thermohaline cell. The 1987 'conveyor belt' of the basin is schematized in **Figure 6A**.

In 1991 the transitional and deep water masses were also formed in the Cretan/Aegean Sea, and they spread out from the western and eastern Cretan Arc Straits on all the horizons $\sigma_0 = \geq 29.15\,\mathrm{kg\ m^{-3}}$, as shown in **Figure 6B**

These denser isopycnals rose to much shallower depths in 1991 than in 1987, thus greatly increasing by advection the salt content of the intermediate layer. In the Ionian Sea, CDW pushes the old and slightly denser EMDW of southern Adriatic origin to the west and downward to the near bottom layer. However, the closing pathways of the Eastern Mediterranean deep thermohaline cell in the 1990s are not yet clearly identified.

See also

Mediterranean Sea Circulation.

Further Reading

Klein B, Roether W, and Manca BB (1999) The large deep water transient in the Eastern Mediterranean. *Deep-Sea Research I* 46: 371–414.

Malanotte-Rizzoli P and Robinson AR (eds.) (1994) *Ocean Processes in Climate Dynamics: Global and Mediterranean ExamplesN: ATO-ASI Series C*, vol. 419. Kluwer Academic Publisher

Malanotte-Rizzoli P, Manca BB, and Ribera d'Alcala M (1997) A synthesis of the Ionian Sea hydrography, circulation and water mass pathways during POEM-Phase I. *Progress in Oceanography* 39: 153–204.

Malanotte-Rizzoli P, Manca BB, and Ribera d'Alcala M (1999) The Eastern Mediterranean in the 80s and in the 90s: the big transition in the intermediate and deep circulations. *Dynamics of Atmospheres and Oceans* 29: 365–395.

Millot C (1999) Circulation in the Western Mediterranean. *Journal of Marine Systems Special volume* 20: 423–442.

POEM Group (1992) The general circulation of the Eastern Mediterranean. *Earth Science Review* 32: 285–309.

Reid JJ (1994) On the total geostrophic circulation of the North Atlantic oceanf: low patterns, tracers, and transports. *Progress in Oceanography* 33: 1–92.

Robinson AR and Malanotte-Rizzoli P (1993) Physical Oceanography of the Eastern Mediterranean. *Deep-Sea Research (Special Issue)* 40: 1073–1332.

RED SEA CIRCULATION

D. Quadfasel, Niels Bohr Institute, Copenhagen, Denmark

Introduction

The Red Sea is a fiord type marginal sea in the north-west of the Indian Ocean, exposed to the arid climate of northern Africa and Arabia. Strong evaporation leads to a buoyancy-driven overturning circulation and to the production of a dense and highly saline water mass, the Red Sea Water. Near the surface the circulation is substantially modified through the forcing by the monsoon winds. Red Sea water spills at a rate of $0.37 \times 10^6\ m^3\ s^{-1}$ through the Strait of Bab El Mandeb into the Gulf of Aden and spreads at intermediate depths in the Indian Ocean, where it can be traced as far south as the southern tip of Africa. The circulation of the Red Sea appears not to be controlled by the hydraulics in the strait. This article summarizes the circulation internal to the Red Sea, the formation and transformation of the deep water masses and the exchange of water between the Red Sea and the Gulf of Aden.

The Red Sea was created as part of the spreading rift system between the African and the Arabian plates. It stretches between 12° and 20°N from tropical to subtropical latitudes (**Figure 1**), is about 2000 km long, on average 220 km wide and reaches depths of up to 2900 m, but its mean depth is only 560 m. The only connection to the oceans is in the south to the Gulf of Aden–Indian Ocean via the narrow Strait of Bab El Mandeb. The shallow Hanish sill of 163 m depth in the north of the strait limits the free communication to the upper part of the water column. Northern appendices to the Red Sea are the Gulf of Suez, a shallow shelf basin, and the 1800 m deep Gulf of Aqaba, a smaller version of the Red Sea itself, bounded by a 260 m deep sill in the Strait of Tiran. Both regions are important for the production of the deep water of the Red Sea.

The circulation of the Red Sea is driven by local atmospheric forces, through buoyancy and momentum fluxes. The net evaporative freshwater flux in the Red Sea is among the highest in the world, reaching $2\ m\ year^{-1}$ with some seasonal variability associated with the monsoon wind system. During summer, values are highest in the north, but during winter in the south. Although the strong evaporation leads to sea surface salinities of up to 42 PSU (practical salinity units) in the Gulf of Suez, the buoyancy flux associated with the latent and sensible heat losses is an order of magnitude larger than that caused by the freshwater flux. In the north, where convection and water mass formation take place, the total buoyancy flux amounts to more than $20 \times 10^{-8}\ m^2\ s^{-3}$. Here surface temperatures may drop below 20°C in winter.

The surface wind field over the Red Sea is guided by the high mountains and plateaux on either side of the basin, and its dominant component is in the along-axis direction. In the southern part winds reverse twice per year. During the south-west monsoon from June to September they blow towards south south-east, whereas during the north-east monsoon and the transition times between the monsoon periods (October to May) the airflow is directed towards north north-west. North of about 20°N the wind is blowing towards south south-east throughout the year, being strongest during summer and weakest during winter. Laterally the mean wind stress

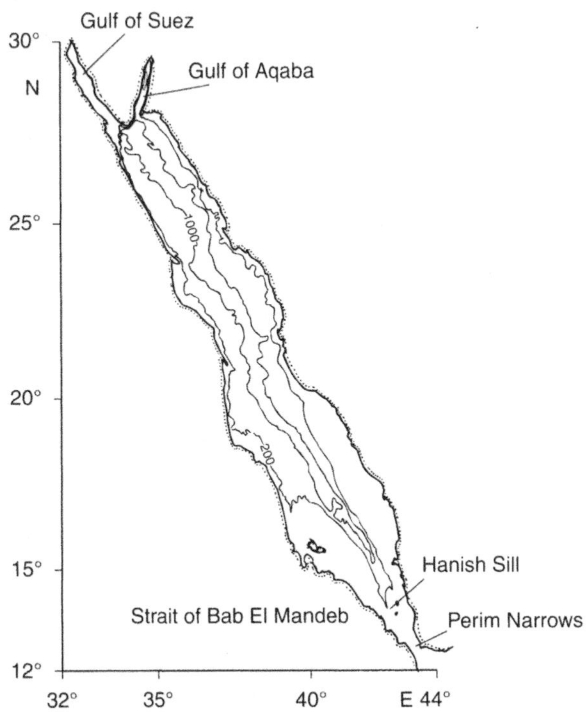

Figure 1 Bathymetry of the Red Sea with 200 m and 1000 m depth contours. Geographical names mentioned in the text are indicated.

decreases from the central axis towards the boundaries on either side, due to lateral friction effects.

The geographical setting of the Red Sea with a length to width ratio of about ten has frequently led to the use of a two-dimensional channel approach to model the circulation. Also most of the observational data have been collected along the main axis of the Red Sea, limiting the interpretation of physical features to the two channel dimensions. The more recent three-dimensional surveys and modeling studies have, however, shown that at least in the upper layers significant cross-axis flows exist. Mesoscale gyres and eddies are an important part of the Red Sea circulation.

Hydrographic Structure

In terms of the classical hydrographic parameters the Red Sea can be viewed as a two-layer system consisting of a deep layer below sill depth, which is very homogeneous in temperature and salinity, and an upper layer, comprising the surface mixed layer and the strongly stratified thermocline. This upper layer is about 200–300 m deep. It is composed of warm and relatively fresh surface water that enters from the Indian Ocean and is then modified due to air–sea interaction. Evaporation increases the salinity, from 36 PSU in the Strait of Bab El Mandeb to above 40 PSU in the north, and cools the surface water, although locally some heating may also take place due to radiative heat gain. Convection and vertical mixing renews the subsurface water which then returns southward in the lower part of the thermocline.

The temperature and salinity distribution of the upper layer changes substantially between the two monsoon seasons, due to the varying atmospheric forcing (**Figure 2**). In winter the vertical temperature stratification is stable throughout the Red Sea. Highest surface temperatures are found around 18°N, in the zone of wind convergence and thus low wind speeds. Toward the north, surface temperatures decrease to less than 20°C in the Gulf of Suez. In the thermocline below the mixed layer, temperatures decrease to the deep water value. This pattern is consistent with the above described overturning circulation, typical for concentration basins, with an inflow at the surface and an outflow of colder and more saline, and thus denser water in the lower thermocline. It is this water that constitutes the major part of the outflow at the bottom of the strait of Bab El Mandeb. In contrast during the summer the temperature stratification in the southern part of the Red Sea is unstable. A cold wedge penetrates below the shallow mixed layer from the Gulf of Aden into the Red Sea and outflows of warmer water exist above and below. This cold wedge has been observed as far north as 20°N. This three-layer structure in the southern Red Sea is caused by the northerly monsoon winds, which during the summer season between June and August drive the surface water out of the Red Sea.

The full depth hydrographic section shown in **Figure 3** was taken during October 1982. In the upper layer the structure is still that of the south-west monsoon, with an intrusion of cold and low salinity water at intermediate depths. Surface temperatures are highest just north of Bab El Mandeb (~32°C) decreasing to 27°C in the north, and surface salinities increase from 37.5 PSU in the strait to more than 40 PSU in the north. Compared to these changes in the upper layer the deep water temperatures and salinities are almost homogeneous with values of 21.6°C for temperature and 40.6 PSU for salinity. Horizontal differences between the northern and the southern part of the Red Sea do not exceed 0.2 K and 0.02 PSU, respectively. In contrast, the distribution of dissolved oxygen (and other chemical parameters) does show some structure in the deep water. There is a pronounced oxygen minimum around 300 m depth in the southern part of the Red Sea, whereas values are relatively high near the bottom and in the northern part. The distributions of chemical parameters have been used to evaluate the pattern and magnitude of the deep circulation. New deep water is formed in the north, it sinks to the bottom, spreads southward and upwells slowly. The intermediate depth oxygen minimum is thus a result of the balance between upwelling of Red Sea Deep Water and downward diffusion of thermocline water.

Circulation in the Upper Layer

The annual mean circulation above the pycnocline is largely driven by thermohaline forces. The excess of evaporation over precipitation results in a loss of 0. 03 × 10^6 m^3 s^{-1} and forces a near surface inflow through the Strait of Bab El Mandeb. Applying Knudsen's formula and balancing mass and salt fluxes through the strait and the sea surface results in an overall strength of the circulation and thus an outflow into the Gulf of Aden of 0.33 × 10^6 m^3 s^{-1}. Direct measurements of the exchanges in the strait have confirmed this number.

These mean conditions have been successfully modeled as a turbulent convective flow driven by a constant buoyancy flux B_0 at the sea surface. The surface currents scale with $(B_0^2x)^{1/3}$ where x is the distance from the inner boundary of the estuary, and the overturning circulation is confined to the layer above sill depth. Superimposed on this pattern is the

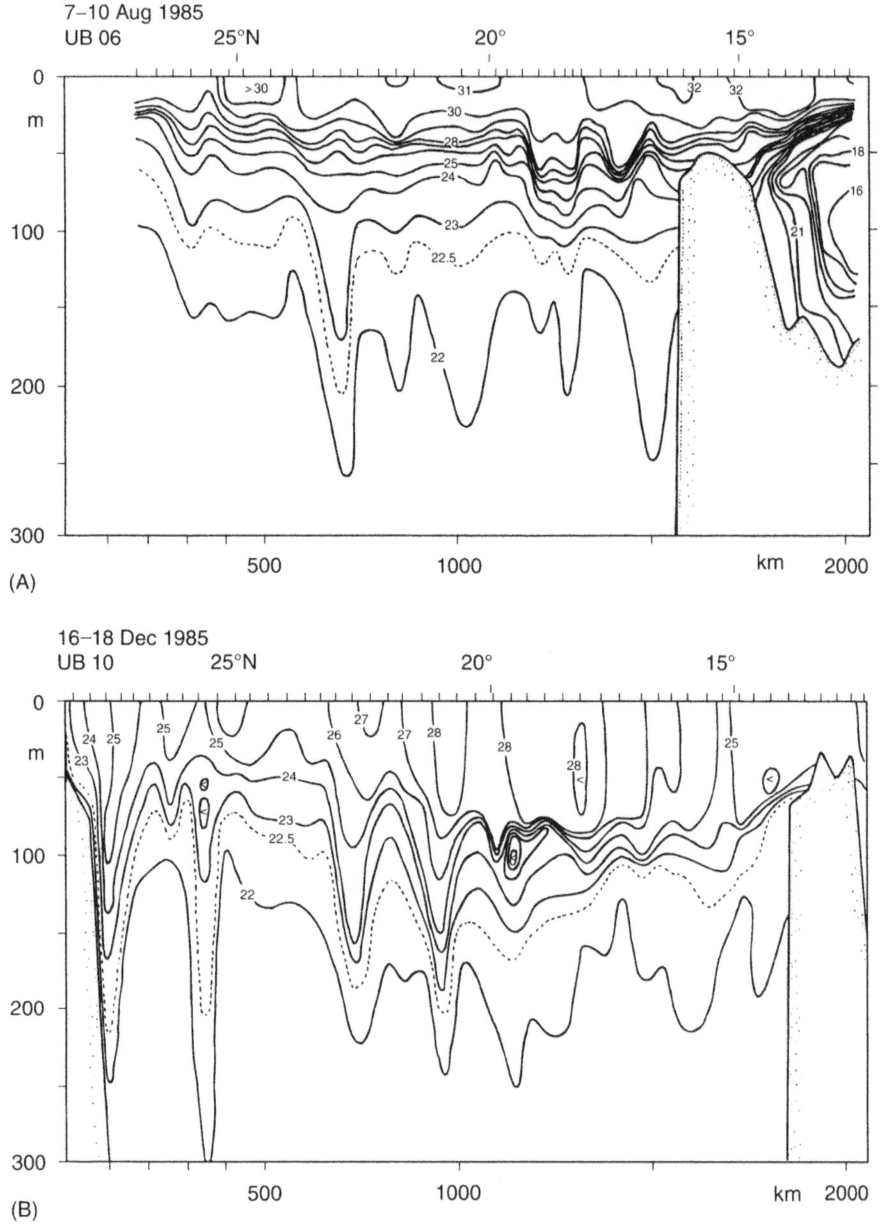

Figure 2 Distribution of temperature (°C) along the axis of the Red Sea in the upper 300 m during the two monsoon seasons in (A) August and (B) December, based on measurements with expendable bathythermographs.

wind-driven circulation forced by the seasonally changing monsoon winds. It leads to a southward surface flow in the southern part of the Red Sea during summer. The inflow compensating the two outflows at the surface and at the bottom then occurs at intermediate depths and a three-layer system develops in and north of the strait. During winter the wind enforces the thermohaline circulation, resulting in the classical two-layer system. Linear inverse box models applied to hydrographic and chemical data have in principle also been able to reproduce the overturning circulation and confirmed the dominant role of the buoyancy forcing over the wind forcing. However, due to the lack of geo-chemical data the details of the upper circulation have not been delineated. Only few direct current measurements have been made within the Red Sea proper, outside the strait regions, and the above described surface circulation patterns are mainly based on ship drift observations. In general, mean near surface current speeds do not exceed 0.1 m s^{-1} except in the southern narrows of the sill region and the strait of Bab El Mandeb.

This two-dimensional description of the upper layer circulation is, however, incomplete. Along-axis

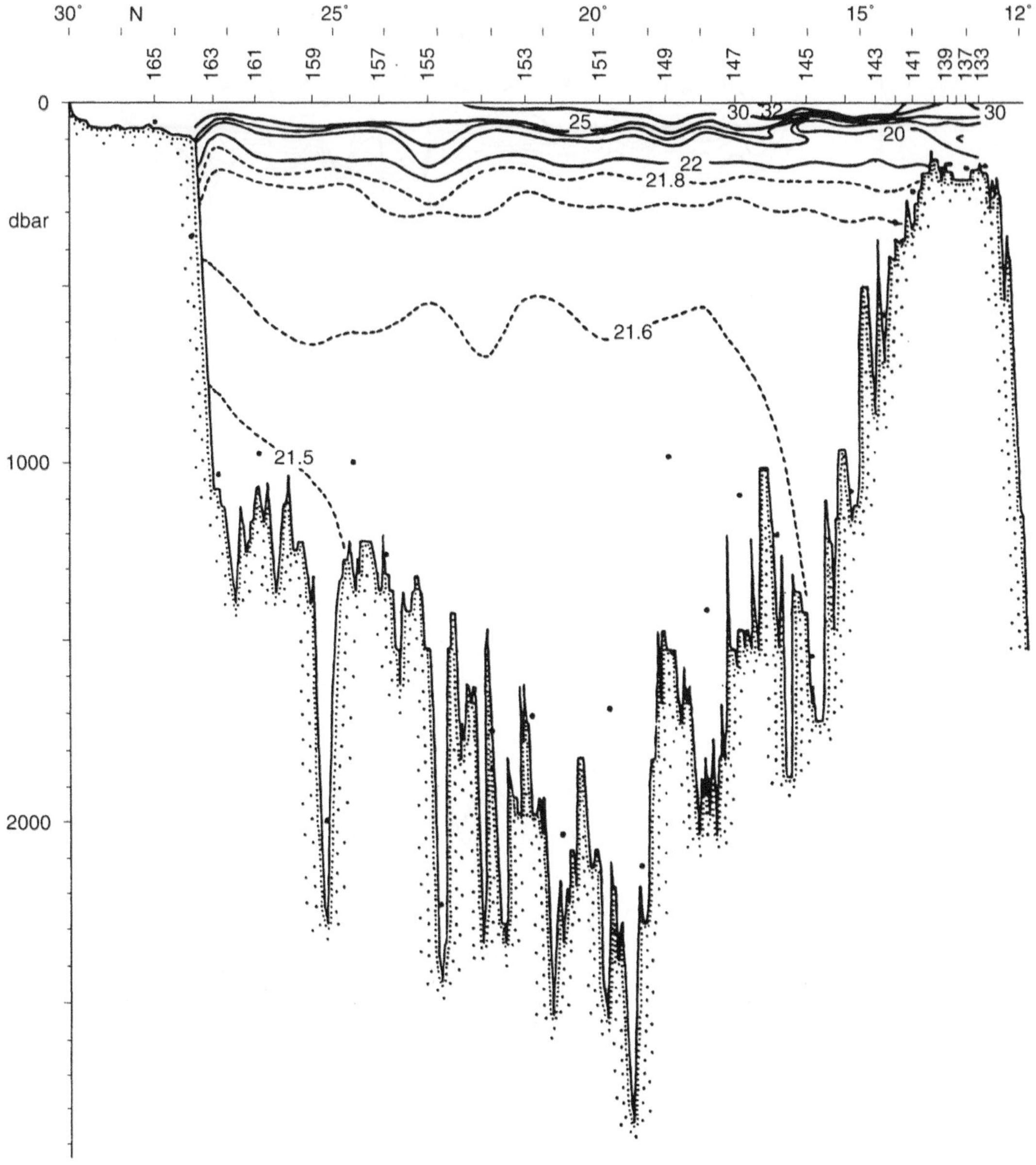

Figure 3 Vertical distribution of (A) potential temperature (°C), (B) salinity (PSU) (C), potential density and (D) dissolved oxygen content (ml l^{-1}) along the central axis of the Red Sea during October 1982.

hydrographic sections show strong depressions of the thermo- and haloclines with horizontal scales of the basin width, which are superimposed on the mean hydrographic structure (**Figure 2** and **3**). The depressions, indicative of anticyclonic eddies or gyres, are mostly found in regions where the Red Sea widens, i.e. around 18°N, 20°N and 23°N. The gyres have also been seen in the few direct current measurements available and are associated with transports of up to $3 \times 10^6\ m^3\ s^{-1}$, an order of magnitude larger than those of the mean circulation. It has been suggested that they are formed through the strong rotational wind field over the Red Sea and are locked in by the topography. Also high-resolution three-dimensional circulation models have recently been used to simulate the current patterns in the Red Sea in more detail. Probably the most important result of these studies is the prediction of narrow boundary currents, at both the western and eastern coasts, that partly detach from the boundary to form closed circulation cells in the interior. The large-scale meridional pressure gradient caused by evaporation is suggested as a driving force for the boundary currents.

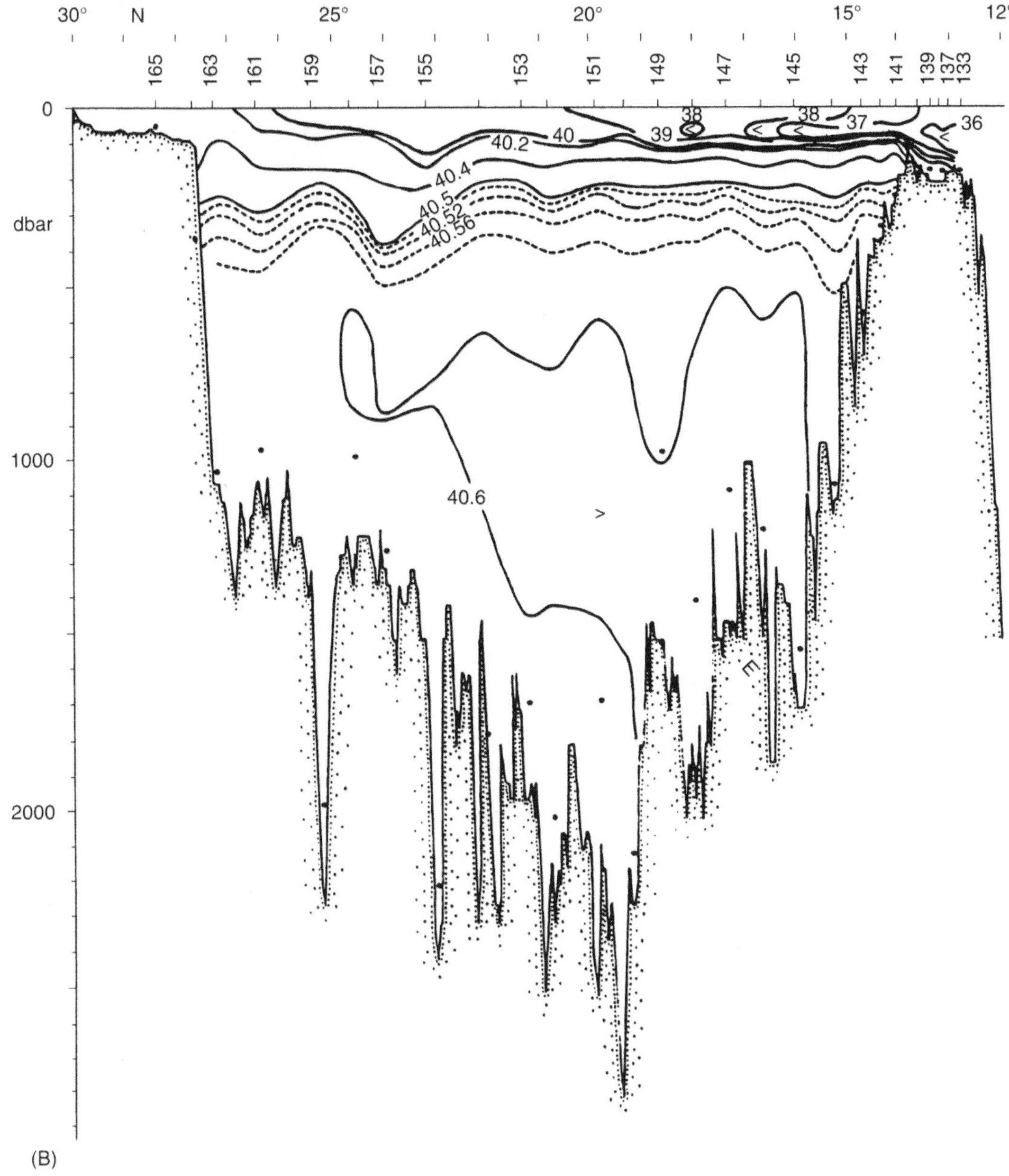

Figure 3 (Continued)

Deep Circulations

The renewal and circulation of the deep water of the Red Sea has attracted more scientific research than the surface circulation. Temperature and salinity in the deep layer of the Red Sea are almost homogeneous (**Figure 2** and **3A, B**), pointing towards a sluggish renewal, but the high degree of oxygen saturation of more than 50% near the bottom suggests just the opposite.

Three different sources for the deep water in the Red Sea were proposed based on observations made during the International Indian Ocean Expedition of the 1960s: outflow and plume convection of winter bottom water from the Gulf of Suez; overflow and plume convection of intermediate water from the Gulf of Aqaba; and open ocean convection and subduction south of the Sinai Peninsula. The individual contributions of these sources could not be established, but the overall deep water renewal was estimated to $0.06 \times 10^6\ m^3\,s^{-1}$ by fitting hydrographic data of the interior of the Red Sea to a simple one-dimensional vertical advection–diffusion model. This transport value corresponds to a residence time of the deep

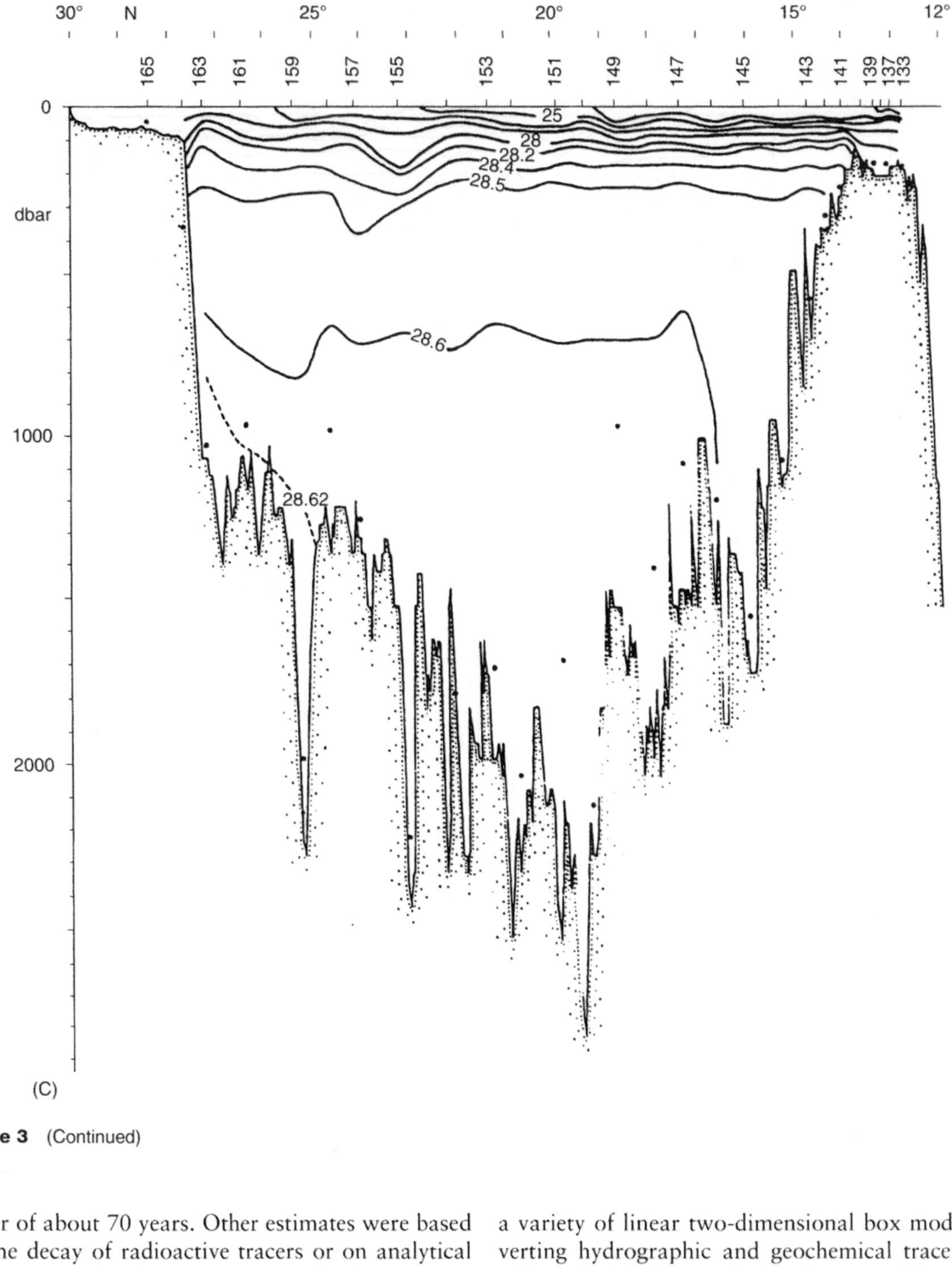

Figure 3 (Continued)

water of about 70 years. Other estimates were based on the decay of radioactive tracers or on analytical modeling studies. These gave somewhat longer timescales of up to 300 years for the deep water renewal. Finally, short-term direct current measurements in the dense winter outflows from the Gulf of Suez and from the Gulf of Aqaba showed transports of 0.08 and 0.03×10^6 m^3 s^{-1}, respectively. Adding a mixing contribution to the descending plumes led to renewal times of the deep water of the order 100 years.

More recently the ventilation of the deep water and the associated circulation has been studied using a variety of linear two-dimensional box models, inverting hydrographic and geochemical tracer data. Although differing in detail, these models estimate renewal timescales of only 30–40 years and suggest a ventilation of the upper deep water through open ocean convection and subduction and of the lower deep water through slope convection fed by the outflows from the two gulfs.

All these estimates of the deep water renewal are either based on mean climatological data or on sets of measurements that can be considered synoptic with respect to the calculated residence times. They

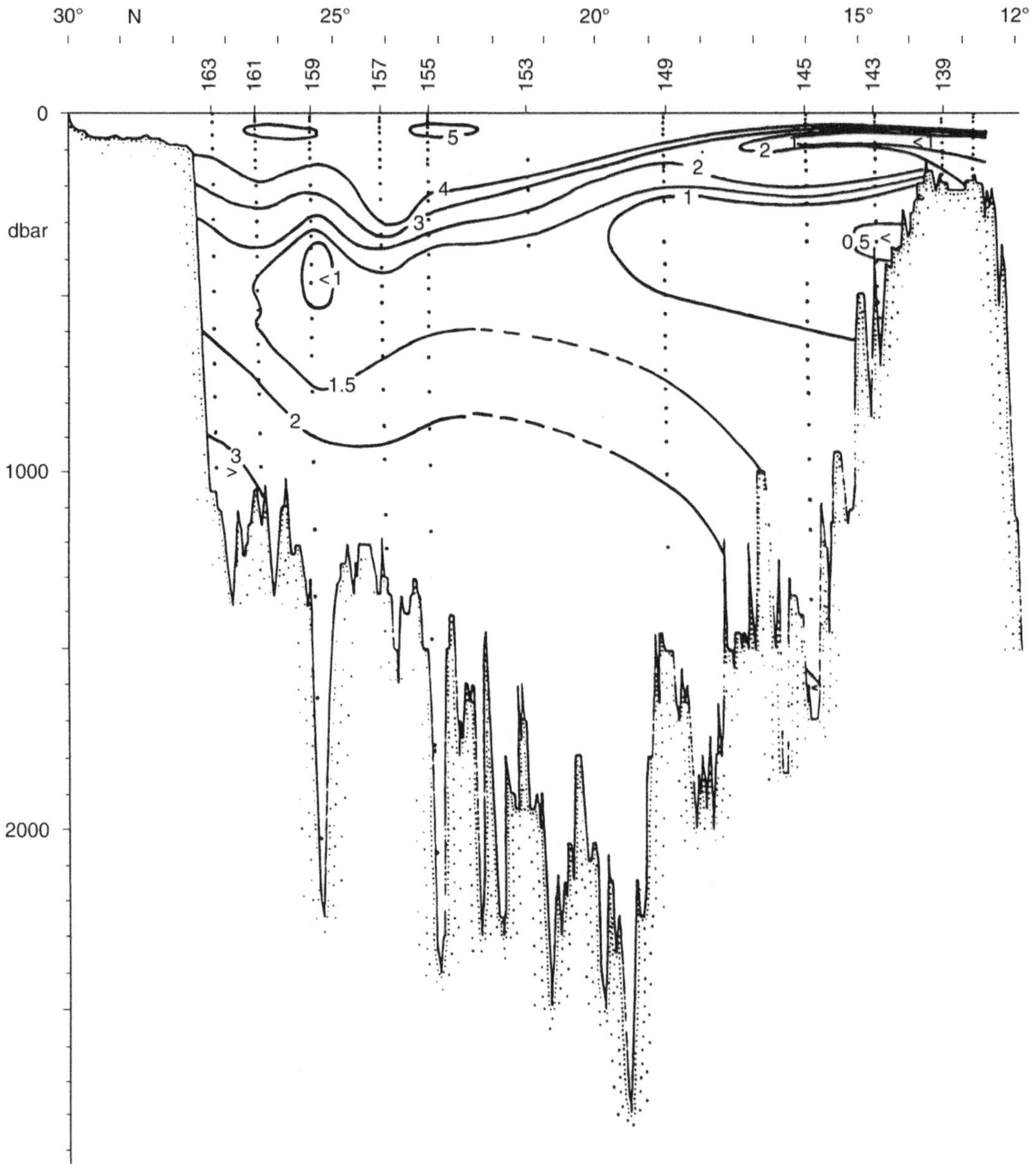

(D)

Figure 3 (Continued)

use information on tracer history and decay or short-term current measurements. Variability in the deep water formation is not considered.

Intermittency of the deep water renewal has been demonstrated based on a time series of hydrographic data separated by one or several winters. During a survey in May 1983, just seven months after the homogeneous structure shown in **Figure 3** was observed, the $\Theta - S$ characteristics in the northern part of the Red Sea had changed drastically whereas in the southern part only marginal differences were observed. Between the slope of the Gulf of Suez and about 25°N deep water temperatures had dropped by up to 0.3 K (**Figure 4**), the salinity distribution showed a parallel freshening of up to 0.06 PSU and the oxygen content had increased by $0.5\,\mathrm{ml\ l^{-1}}$. Cooler and fresher water has entered this regime and mixed with the older deep water of the previous year. Since the lowest temperatures were found near the bottom, the deep-water renewal took place in the form of plume convection caused by dense outflow of bottom water from the Gulf of Suez, plunging down the northern continental slope. Four years later, during summer 1987, this newly formed water had

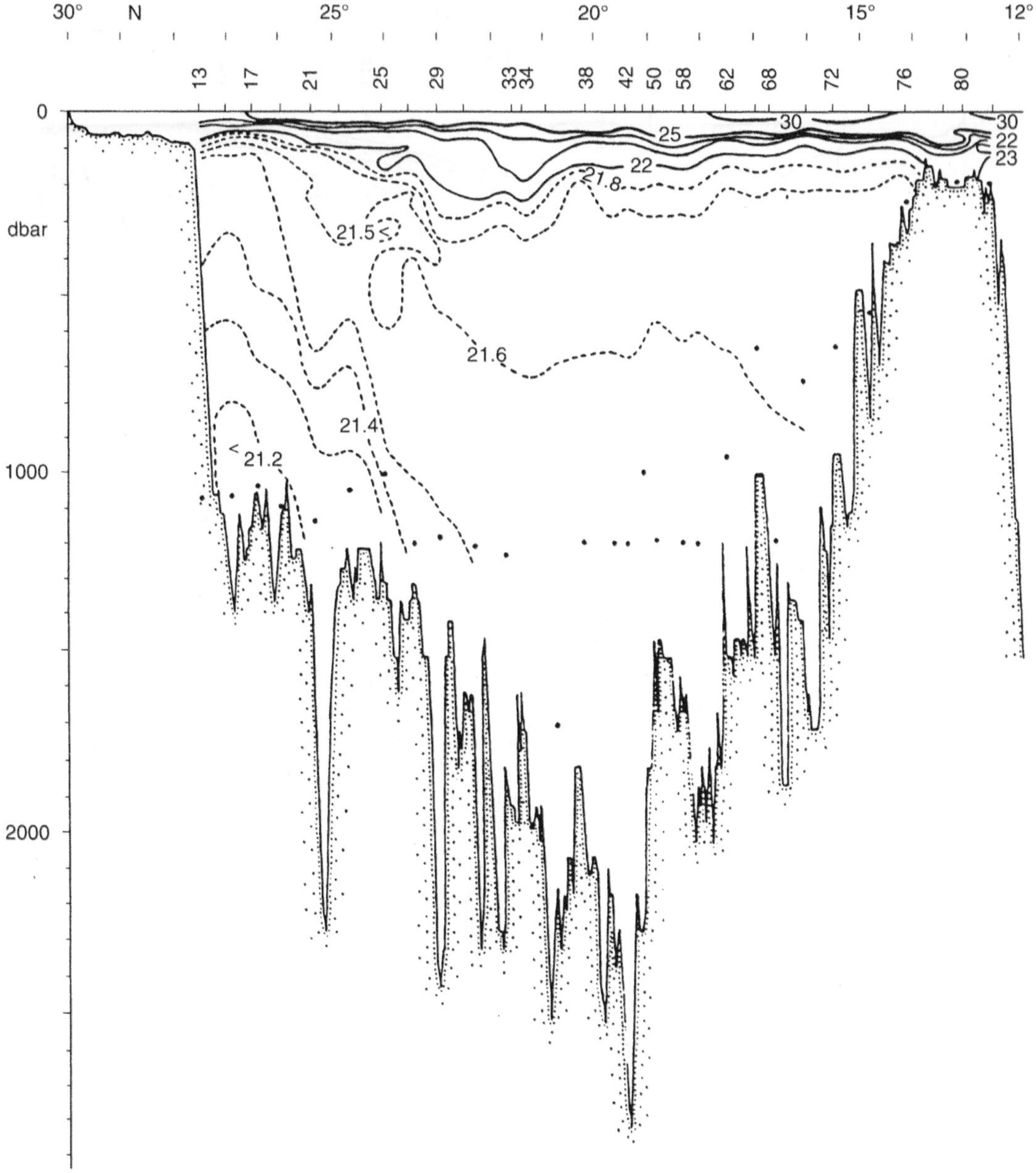

Figure 4 Vertical distribution of potential temperature (C) along the central axis of the Red Sea during May 1983. Dots mark the maximum depth of measurements at the respective station.

spread southward along the entire Red Sea, having ventilated the deep basin within less than a decade. This sets a lower limit on the near bottom currents of 1–2 cm s^{-1}.

The renewal of the Red Sea Deep Water and its circulation may thus be summarised as follows (**Figure 5**). The deeper part of the basin below about 600 m is ventilated by slope convection fed by the outflows from the Gulf of Suez and the Gulf of Aqaba. Due to the large reservoir of the Gulf of Aqaba the latter outflow is more steady and provides the more permanent source for the deep water. The density of the winter water in the Gulf of Suez shows strong interannual variability and its contribution to the deep basin of the Red Sea is thus intermittent. These two sources feed a southward bottom current which upwells and in part recirculates northward above the bottom flow. The upper deep water is ventilated through shallow plume convection, when the bottom waters of the adjacent shelves are not dense enough to sink to the bottom of the Red Sea, and through open ocean convection south of the Sinai peninsula. Both sources feed a southward flow that merges with the southward flow of the lower thermocline in the interior of the basin and make up the outflow of Red Sea Water through the Strait of Bab El Mandeb.

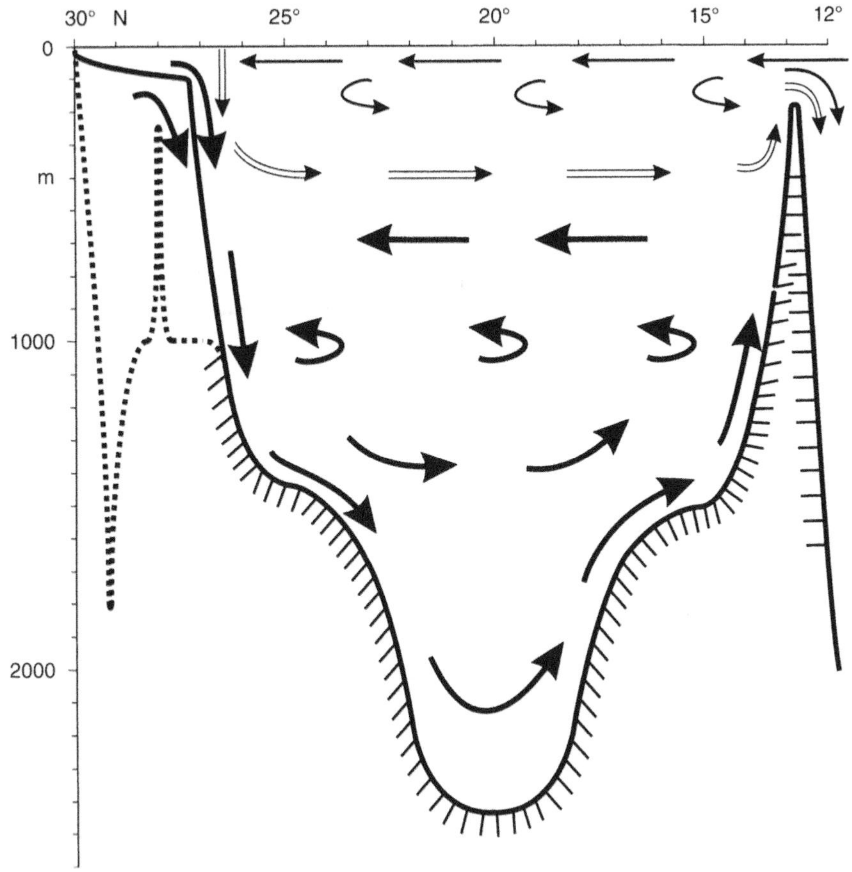

Figure 5 Schematic of the circulation in the upper and deep water layers in the Red Sea. Light arrows indicate the circulation cell in the upper layer, double arrows that of the upper deep water that is driven by open ocean convection and bold arrows the deep circulation driven by plume convection. The circulation pattern is based on interpretation of hydrographic data and few direct current observations. The dotted topography in the north is that of the Gulf of Aqaba.

Exchanges with the Gulf of Aden

The Strait of Bab El Mandeb is about 150 km long and connects the Red Sea with the Gulf of Aden and the Indian Ocean. With the 163 m deep sill near the Hanish Islands in the north and the 18 km wide Perim Narrows in the south it is a bottleneck for the exchange between both basins. Both locations are candidates for hydraulic control of the water exchange.

The buoyancy driven large-scale overturning circulation of the Red Sea supports a two-layer exchange through the strait, with an inflow at the surface and an outflow of dense water at the bottom. Simple budget calculations for the Red Sea point toward an inflow at a mean rate of $0.36 \times 10^6\,m^3\,s^{-1}$ and outflow of $0.33 \times 10^6\,m^3\,s^{-1}$ balancing the annual mean freshwater loss and keeping the overall salt content of the Red Sea constant.

The local wind field over the southern Red Sea modifies this picture and strengthens the exchange during winter, when winds blow towards the north. During the height of summer, between June and August, the southward blowing winds force a shallow outflow of surface water from the Red Sea and cause upwelling in the Gulf of Aden. This sets up a northward pressure gradient which drives the cold inflow at intermediate depths, sandwiched between the surface and the bottom outflows. These layered structures of the throughflow and their seasonal modulation has been documented during a number of hydrographic surveys and from some short-term current measurements with moored instrumentation, carried out since the 1960s. More recently an almost complete seasonal cycle has been documented with a ten months mooring array.

The total exchange through the strait varies between a high of $0.7 \times 10^6\,m^3\,s^{-1}$ during February and a low of $0.35 \times 10^6\,m^3\,s^{-1}$ during August, the deep outflow of high salinity Red Sea Water changes from 0.7 to less than $0.05 \times 10^6\,m^3\,s^{-1}$ between those seasons (**Figure 6**). This means that during the three to four month long summer season the fluxes are mainly associated with the wind-driven upper circulation cell rather than with the thermohaline

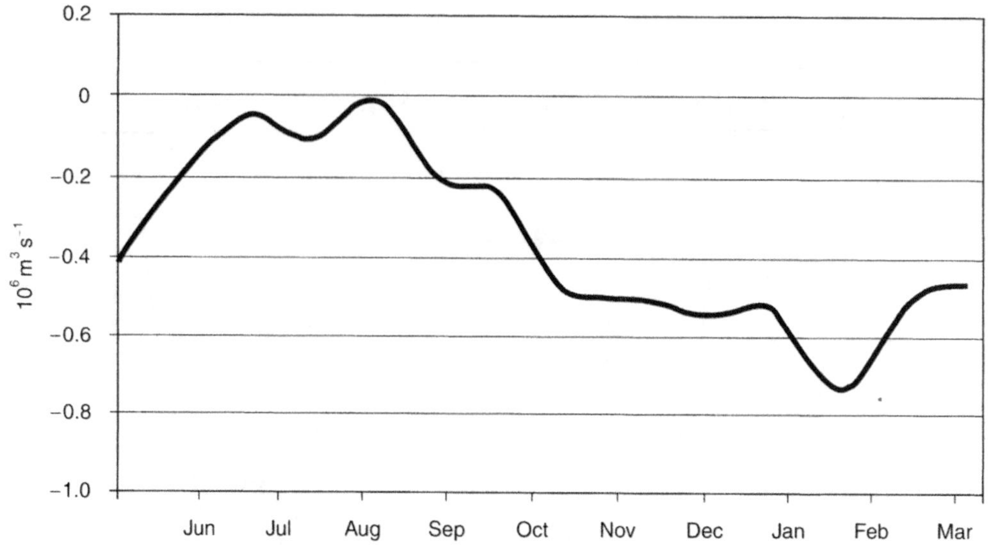

Figure 6 Time series of transport in the outflow of Red Sea Water in the Strait of Bab El Mandeb.

overturning driving the deep outflow. However, one should bear in mind, that from the observations the fluxes in the upper layers are not as well established as those in the deep outflow, due to the widening geometry of the strait and the poorer coverage with moored instruments. The mean outflow of dense Red Sea Water ($0.37 \times 10^6\,m^3\,s^{-1}$), however, corresponds closely to the prediction based on the evaporation budget.

Mean current speeds in the deep outflow vary from less than $0.2\,m\,s^{-1}$ to more than $1.0\,m\,s^{-1}$ between the seasons. Superimposed on this annual signal are shorter-term current fluctuations, from several days to weeks, that may be associated with the passage of coastal trapped waves or are caused directly by the local wind forcing. Transport variability associated with these fluctuations, are most pronounced in the winter season and can be as high as $0.4 \times 10^6\,m^3\,s^{-1}$. The strongest variability is caused by the semidiurnal tides, which reach amplitudes of $0.5\,m\,s^{-1}$ during spring conditions.

Hydraulic control does not appear to limit the exchanges through the strait on the mean seasonal cycle. In the long-term data from moored instrumentation critical or supercritical flow conditions were only seen for the second vertical mode, where internal waves propagating into the Red Sea may get arrested. This limits the outflow of upper deep Red Sea Water but does not hamper the overall exchange. The situation may, however, be different during extreme conditions associated with the passage of coastal trapped waves and/or tides.

The seasonally changing structure and strength of the flow in the strait has an impact on the water mass characteristics of the dense outflow that actually reaches the Gulf of Aden, sinks and spreads in the Indian Ocean. In the northern part of the strait, temperatures and salinities are fairly constant throughout the year, despite the changing transport rates of the outflow. In the strait towards the southern exit, vertical mixing with the overlying water reduces the salinity during both seasons, but more so in summer, and leads to a warming in winter and a cooling in summer (**Table 1**). The seasonal change in the density of the outflow is, however, small, as these two effects almost cancel each other. In the Gulf of Aden the dense Red Sea Water sinks down the continental slope to a depth of equilibrium density, primarily along two channels in the topography. During descent in the two channels mixing with the ambient water of the Gulf of Aden reduces the density further and dilutes the Red Sea Water by a factor of about three. The endproducts then spread at depths of around 700 m and 1100 m.

Table 1 Characteristics of the dense Red Sea outflow

	Summer	*Winter*
Temperature (C)	20.7	22.9
Salinity (PSU)	39.0	39.9
Density (kg m^{-3})	27.63	27.69
Transport ($10^6\,m^3\,s^{-1}$)	0.05	0.70

Conclusions

The basic dynamics of the circulation of the Red Sea, both for the upper stratified layer and in the nearly homogeneous deep trench, have been known since the 1960s. Heat and freshwater fluxes associated with the strong evaporation lead to the formation of the highly saline Red Sea Water and drive both the shallow overturning circulation above sill depth and the deep circulation. The latter is forced through deep convection in the far north of the Red Sea. More recently the ventilation time of the deep water was calculated by use of geochemical tracers fitted to box models. Estimates converge on around 30 years, but it has also been found that deep convection is intermittent and a strong convection event can renew the deep water within less than a decade. The exchanges between the Red Sea and the Indian Ocean, in particular its seasonal variability, have also recently been quantified. The outflow of dense Red Sea Water varies between $0.05 \times 10^6\ m^3\ s^{-1}$ in summer and $0.70 \times 10^6\ m^3\ s^{-1}$ in winter, with an annual mean of $0.37 \times 10^6\ m^3\ s^{-1}$. The exchanges are not limited by hydraulic control and thus the Gulf of Aden/Indian Ocean might influence the dynamics of the Red Sea. The possible remote forcing of the Red Sea circulation from the Indian Ocean and a better understanding of its variability remain a challenging aspect of future Red Sea research.

See also

Current Systems in the Indian Ocean. Flows in Straits and Channels. Wind Driven Circulation.

Further Reading

Cember RP (1988) On the sources, formation and circulation of Red Sea deep water. *Journal of Geophysical Research* 93: 8175–8191.

Pratt LJ, Johns W, Murray SP, and Katsumata K (1999) Hydraulic interpretation of direct velocity measurements in the Bab al Mandab. *Journal of Physical Oceanography* 29(11): 2769–2784.

BLACK SEA CIRCULATION

G. I. Shapiro, University of Plymouth, Plymouth, UK

Introduction

The Black Sea is a unique marine environment, representing the largest land-locked basin in the world. It is situated at the southeastern edge of Europe and is connected to the remote waters of the Atlantic Ocean via the Marmara, Aegean, and Mediterranean seas through a chain of narrow straits: Bosporus, Dardanelles, and Gibraltar. The Black Sea extends for 1167 km from the most westerly point in the Burgas Bay, Bulgaria (27° 27′ E), to the most easterly point near the town of Kobuleti, Georgia (41° 42′) and for 624 km from the most northerly point in Beresan Liman, Ukraine (46° 32′), to Giresun, Turkey (40° 55′), in the south. The longest extents of water are 1150 and 610 km in the zonal and meridional directions, respectively. The sea has a surface area of 413 490 km^2, maximum depth of 2245 m, a volume of *c.* 529 955 km^3, and a mean depth of 1197 m.

The apple-shaped Crimean Peninsula penetrates into the Black Sea from the north and separates it from its largest arm, the shallow Sea of Azov, which is linked to the Black Sea through the narrow Kerch Strait. Otherwise, the Black Sea coastline is fairly regular and has a length of 4445 km. The Black Sea contains only a few small islands, the largest being Zmeyiny in the west and Beresan in the northwest. The shores of the Black Sea are bounded by Bulgaria (whose length of shoreline is 300 km) and Romania (225 km) on the west, Ukraine (1540 km) and Russia (475 km) on the north, Georgia (310 km) on the east, and Turkey (1595 km) on the south (**Figure 1**). These data should be taken with caution as the length of coastline is a fractal measurement; if a more detailed map is used, the coastline will be longer.

The seabed is divided into the shelf, the continental slope, and the deep-sea depression. The shelf occupies a large area (*c.* 25% of the total) in the northwestern part of the Black Sea where it is over 200 km wide, has a depth ranging from 0 to 160 m, and also receives major freshwater discharges from Europe's large rivers. Near the Caucasian and Anatolian coasts, the shelf rarely exceeds 15 km in width and has numerous submarine canyons and channel extensions.

In the geological past, the connection of the Black Sea to the oceans opened and closed a few times. Around 5–7 Ma the ancient Tethys Ocean disintegrated, giving birth to the enclosed Sarmatian Sea-lake. Consequently the modern Black, Caspian, and Aral Seas originated from its parts. During the Ice Age of the Pleistocene epoch, the level of the Black Sea rose, and the sea was connected to the Mediterranean and Caspian Seas several times. In the postglacial period the Black Sea contracted, became

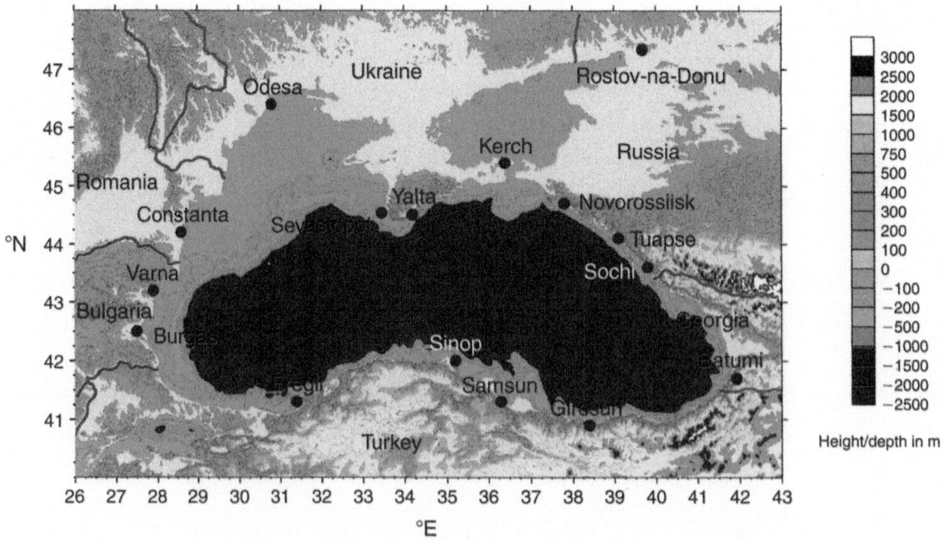

Figure 1 Bathymetric map of the Black Sea based on ETOPO2 data.

a brackish lake, and its water level fell below that of the ocean. Evidence suggests that *c.* 6000–8000 years ago the connection to the Mediterranean Basin was restored and the freshwater fauna of the Black Sea were replaced by the saltwater fauna of the Mediterranean Sea.

The Black Sea has been studied since the time of Phoenician sailors. The Greeks, who colonized the shores of the sea in sixth to eighth century BC, called it *Pontos Euxeinos* (meaning hospitable sea), and had a fairly good estimate of its width (*c.* 640 km in modern units); however, they highly exaggerated its length, as gathered from writings by the famous Greek historian Herodotus (fifth century BC). The Romans colonized its shores in the third to first century BC. In the tenth century AD, the sea was often called the Russian Sea, as it was used by the Russians as a trade route 'from Varyags to the Greeks' connecting the Baltic Sea and Byzantium, and from the fifteenth to the eighteenth century it was a Turkish 'lake'. A comprehensive atlas of the Black Sea was published in 1842, which was based on the 11-year-long observational campaign led by Captain E.P. Manganary and showed details of bottom topography of the coastal waters and general currents. Oceanographic measurements (temperature and water density) started in 1868 on board the Russian corvette *Lvitsa* ('Lioness'). A breakthrough in the multidisciplinary study of the Black Sea was made in the 1920s: more than 1000 deep-water stations were occupied in 1924–28, measuring an extensive set of physical and chemical parameters. Incidentally, the first general scheme of basin-scale surface currents was presented by biologist N.M. Knipovich in 1932, and then improved by oceanographer G. Neumann in 1942. A database of hydrographic measurements in the Black Sea from the beginning of the century until the 1980s exists through the efforts of the erstwhile USSR and other countries. After 1990, coordinated multinational surveys were carried out with much increased coverage, and improved resolution and quality of data.

Water Budget

Thermohaline properties of the Black Sea and hence the geostrophic circulation depend to a large extent upon the balance of inflowing and outflowing waters. There is no consensus about the specific figures, partly due to interannual and interdecadal variability of various components – river discharges, precipitation/evaporation, inflow of low-salinity (11 psu) water from the Sea of Azov and high-salinity (34.9 psu) Marmara Sea water with the Lower Bosporus Current, outflow of low-salinity (17 psu) Black Sea water with the Upper Bosporus Current into the Marmara Sea and through the Kerch Strait into the Sea of Azov. Current estimates of average values for the water inflow/outflow for the period 1923–40 and 1945–85 based on a large number of individual studies are reported in **Table 1**.

Table 1 Freshwater balance for the Black Sea (km^3 per year)

Freshwater supply by rivers	338
Precipitation	238
Inflow from Marmara Sea	176
Inflow from the Sea of Azov	50
Evaporation	−396
Outflow into Marmara Sea	−371
Outflow into the Sea of Azov	−33

Adapted from Simonov AI and Altman EN (eds.) (1991) *Hydrometeorology and Hydrochemistry of the Seas of the USSR: The Black Sea,* issue 1, 449pp. St. Petersburg: Gidrometizdat.

The Black Sea has a positive water balance, in which the inputs from freshwater sources (rivers and precipitation) exceed losses by evaporation by 180 km^3 per year; this causes freshening of the upper layer, which is balanced by mixing with lower, more-saline waters, and explains why the salinity in the upper layer of the Black Sea (17–18 psu) is only half that of the World Ocean (35 psu).

The Black Sea drainage area is *c.* 2 000 000 km^2, which is nearly 5 times larger than its surface area, and covers almost a third of Europe (**Figure 2**). This results in a disproportionally large freshwater input. Unlike the Mediterranean, the Black Sea is an estuarine-type basin (or dilution basin).

The principal source of freshwater is the Danube, which supplies on average 209 km^3 of freshwater per year, although the instantaneous fluxes range from 4×10^3 to $9\times10^3\,m^3\,s^{-1}$ owing to seasonal and interannual fluctuations. Other large feeders are the Dnieper, Southern Bug, Dniester, and Rioni, and also the Don and Kuban via the Sea of Azov. The rivers flowing into the northern and western parts of the Black Sea carry much silt and form deltas, sandbars, and lagoons.

An interesting two-layer current system (to and from the Black Sea) exists in the Bosporus Strait. These currents were first studied *in situ* and by laboratory experiments by Admiral S. Makarov in the nineteenth century. They are highly variable and dependent upon meteorological and hydrological forcing on both ends. The upper current (out of the Black sea) has a typical speed of $0.9\,m\,s^{-1}$ accelerating in places to $2\,m\,s^{-1}$ or more. The average speed of the lower current (below 20–40 m) exceeds $0.9\,m\,s^{-1}$.

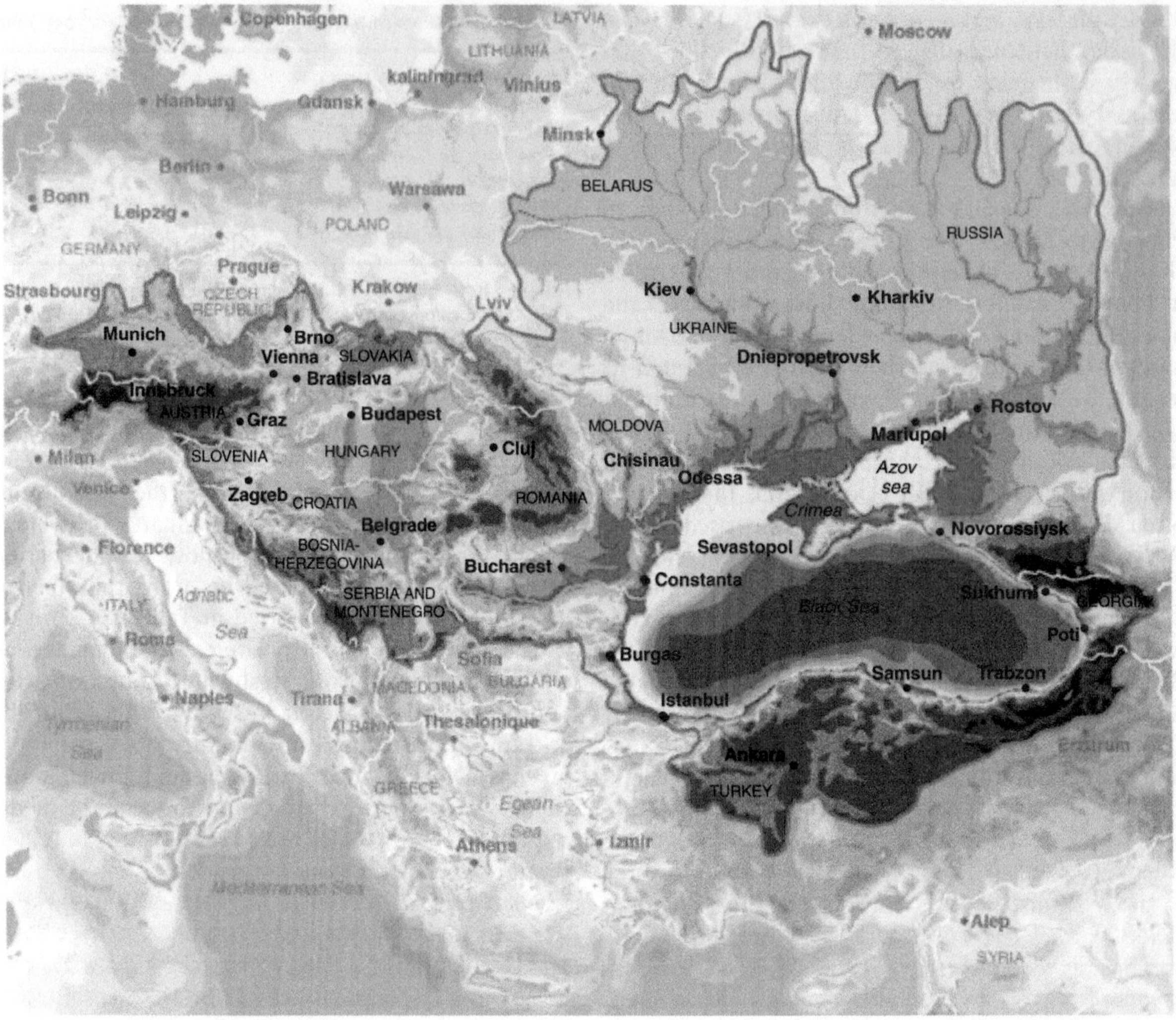

Figure 2 Drainage area of the Black Sea (shown in strong color). Graphic by Philippe Rekacewicz, UNIP/GRID-Arendal, http://maps.grida.no/go/graphic/drainage_in_the_black_sea_area.

Meteorological Forcing

Circulation in the Black Sea is a product of complex interaction of meteorological and thermohaline factors, influenced by detail of the coastline and bottom topography. The climate of the Black Sea is predominantly continental. Despite its modest size, the Black Sea experiences different climatic forcing in its different parts.

In winter, severe cold and dry northeastern winds prevail over the northern and northwestern parts of the sea, lowering the sea temperature and bringing frequent storms. Cold and violent bora winds reach up to $30\,\mathrm{m\,s^{-1}}$ in the coastal region of Novorossiysk, just to the east of the Kerch Strait, making the harbor unusable for up to 70 days in a year. The eastern and southeastern parts of the sea are protected by the Caucasus mountains from the cold winds and hence retain their relatively warm temperature. This causes basin-scale air temperature and air pressure gradients, which lead to the formation of cyclonic circulation patterns in the atmosphere above the sea. The western and southern areas are influenced by warm and humid Atlantic air, which enters the Black Sea with the Mediterranean cyclones. The monthly mean air pressure at sea level for the month of January clearly shows its cyclonic nature (**Figure 3(a)**).

The lowest monthly average air temperature in the northern part of the Black Sea is about −2.5 °C in January (near Ochakiv) and occasionally falls as low as −30 °C. The coastal waters freeze up to a month or more, while the shallow bays, river mouths, and limans freeze up to 2–3 months. During this season the southern coast of the Crimea and the sheltered eastern and southeastern coasts enjoy a temperature of 6–8 °C, well above the freezing point.

In the summer, the Azores high-pressure center extends far into Europe and influences much of the

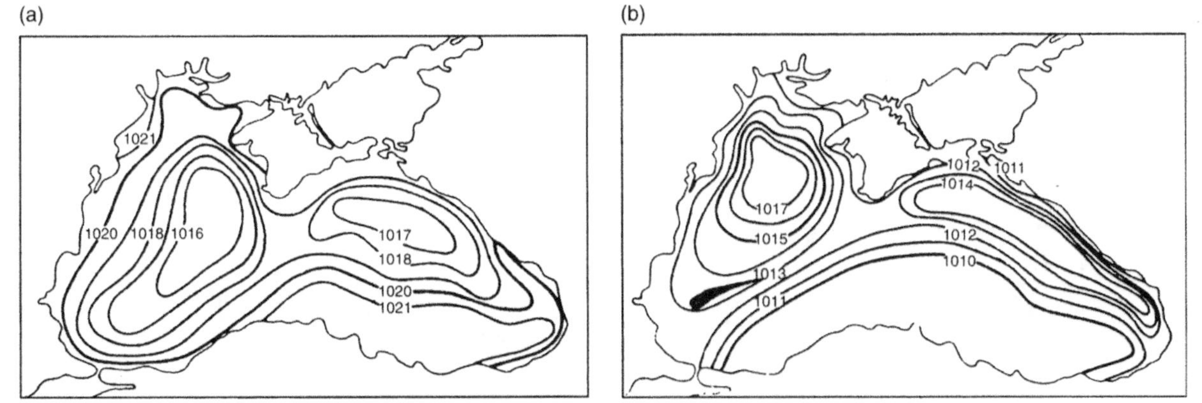

Figure 3 Mean atmospheric pressure at sea level in January (left panel) and July (right panel). Adapted from Leonov AK (1960) *Regional Oceanography, Part 1: The Black Sea*, pp. 623–765. Leningrad: Gidrometizdat.

Black Sea, causing anticyclonic air flow above the sea, and stable weather with clear skies from May to end of September; see **Figure 3(b)**. The average July temperature is more uniform across the sea: in the north it is 22–23 °C and in the south, 24 °C. The northwestern coast has the lowest annual precipitation (300 mm), and the Caucasian coast has the highest (up to 1800 mm). The air pressure field and associated winds in July lose their cyclonic character and in many areas the vorticity of the wind becomes negative (anticyclonic); see the map of mean air pressure at sea level in **Figure 3(b)**.

Water Mass Structure

A combination of salt and heat budgets results in the unique thermohaline structure of water masses in the Black Sea. The most important feature of the Black Sea is that oxygen is dissolved and a rich sea life is made possible only in the upper water levels – the Black Sea is the world's largest water body containing hydrogen sulfide. This is because it has a strong pycnocline, located at around 100–150-m depth, which separates the top layer (salinity 17.5–18.5 psu) influenced by fluvial inputs and the bottom layer (21–22 psu) fed by warm salty waters from the Mediterranean Sea through the Bosporus – so the sea is stratified and cannot be mixed even by strong winds or winter convection. This leads to oxygen depletion in layers below 80–150 m, which occupy up to 90% of the volume of the sea. In the anoxic deep waters, organic matter degradation uses oxygen bound in nitrates, and especially in sulfates; this generates hydrogen sulfide and leads to a gloomy, 'dead' zone populated only by adapted bacteria (**Figure 4**).

Due to the low salinity, the fauna and flora of the Black Sea are qualitatively poor as compared to the Mediterranean Sea. The Mediterranean has about 7000 plants and animal species, while the Black Sea has only about 1200. On the other hand, primary productivity based on nutrients coming from large rivers is relatively high.

A cold intermediate layer (CIL; usually defined by 8 °C temperature contour) sits at the upper boundary of the main pycnocline. In the winter this layer is fed by even colder surface waters at the centers of cyclonic gyres and through cascading from the northwest shelf; in summer the surface waters get warmer and CIW becomes the coldest water in the Black Sea. An additional seasonal summer pycnocline usually develops between 10- and 40-m depth (**Figure 5**).

Basin-Scale Circulation

Basin-scale circulation consists of a coherent, cyclonic boundary current often called the Rim Current or the Main Black Sea Current, and two sub-basin cyclonic gyres which are particularly prominent in winter and form a circulation pattern known as 'Knipovich spectacles'; see **Figure 6**. These gyres are related to the shape of the coastline, which divides the sea into two sub-basins. Occasionally this basic circulation encompasses a smaller anticyclonic gyre south of the Crimean peninsula, a number of cyclonic and anticyclonic vortices and a quasi-permanent anticyclonic gyre in the southeast (Batumi Eddy). Due to conservation of potential vorticity, the Rim Current tends to flow along the depth contours. Observations show that the location of the Rim Current generally coincides well with that of the continental slope: the current flows typically at a distance of 15–25 km from the shore; however, it can be as little as 3–5 km near the Georgian coast, or as great as 200 km in the northwestern part due to the

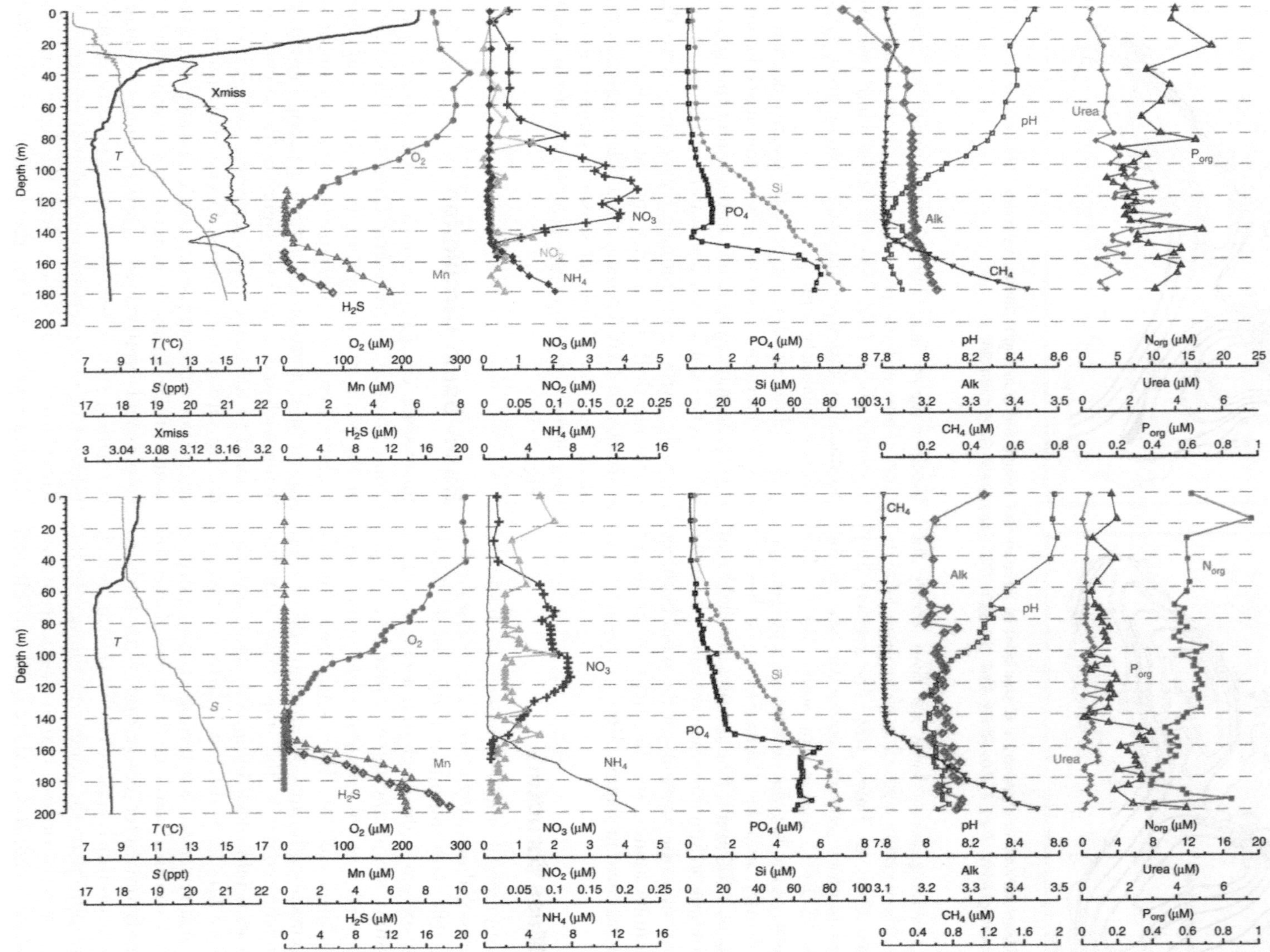

Figure 4 Vertical distribution of temperature (T), salinity (S), density (σ_θ), transmission (Trans), oxygen (O_2), hydrogen sulfide (H_2S), total manganese (Mn^{2+}), silicates (Si), nitrates (NO_3), nitrites (NO_2), ammonia (NH_4), urea (Urea), phosphates (PO_4), and organic phosphorus (P_{org}) at a summer station near Gelendzhik. Concentrations of chemical parameters are in μM. Upper panel shows summer profiles measured on 2 July 2002 at 44° 516N; 37° 872E; lower panel shows winter profiles measured on 26 January 2004 at 44° 415N; 37° 317E. From Yakushev EV, Podymov OI, and Chasovnikov VK (2005) The Black Sea seasonal changes in the hydrochemical structure of the redox zone. *Oceanography* 18(2): 48–55.

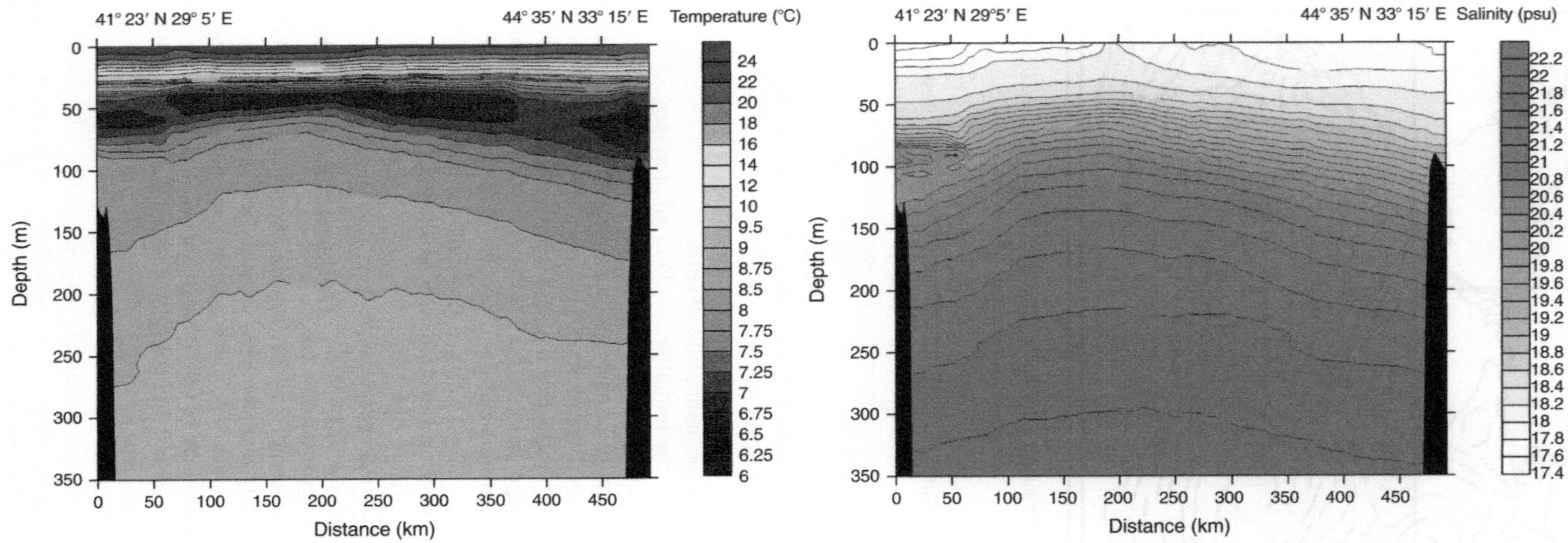

Figure 5 Composite temperature and salinity distributions across the western Black Sea gyre on a transect from Istanbul to Sevastopol. Based on data from 522 stations taken in the month of July over the period 1956–96 and collated by Suvorov AM, Eremeev VN, Belokopytov VN, *et al.* (2004) *Digital Atlas: Physical Oceanography of the Black Sea* (CD-ROM). Kiev: Environmental Services Data and Information Management Program, Marine Hydrophysical Institute of the National Academy of Sciences of Ukraine. Note the dome-shaped structure of the CIL, main thermocline, and the halocline due to cyclonic general circulation in the Black Sea.

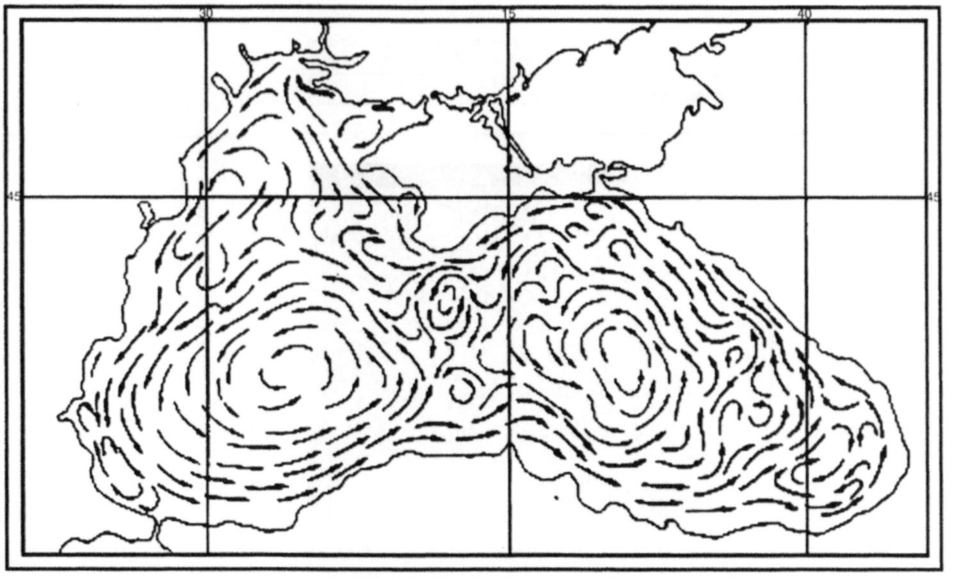

Figure 6 Mean basin-scale circulation – 'Knipovich spectacles'. Reproduced from *Nauka i Zhizn*, 2006, No. 2, http://www.nkj.ru/archive/articles/4019/.

extensive shelf. The Rim Current has a width of 40–60 km and a typical speed of 0.3–0.5 m s^{-1}, sometimes accelerating to 1 ms^{-1}. The Rim Current is subject to meandering both on- and offshore.

Specific physical processes that form the observed circulation pattern are complicated. Classically, the cyclonic wind pattern (positive curl of wind stress) has been recognized as the main forcing for the cyclonic surface circulation. On the other hand, numerical studies indicate a seasonal thermohaline circulation driven by nonuniform surface fluxes, and density distribution due to river runoff complementary to the wind-driven circulation. Thermohaline sources in the shelf regions generate density fronts and doming of density surfaces around the entire sea. This results in surface currents of comparable magnitude.

The low-salinity surface waters, which are formed over the northwest shelf due to intense river discharge, travel with the Rim Current and reach the Anatolian coast in a modified form due to mixing; the travel time from the Danube River mouth to the Bosporus Strait is 1–2 months. Currents in the coastal zone and on the shelf are highly variable both in time and space but rarely exceed 0.3 m s^{-1}. The Rim Current is clearly seen up to a depth of 300–400 m. There were early hypotheses that at depth there is a permanent counter-current. This was not confirmed later, however; occasionally weak counter-currents were recorded above the continental slope, probably induced by mesoscale eddies. Currents in the central parts of the sea are weak, typically 0.05–0.15 m s^{-1}, and extend to greater depths. The western and eastern cyclonic subgyres slightly change their location from year to year and are known to move north after cold winters.

The thermohaline component of the Black Sea circulation (caused by differences in density) can already be seen from an early calculation of geostrophic currents based on temperature and salinity observations of the 1920s and 1930s, as shown in **Figure 7**.

Diurnal and semi-diurnal tides are nearly nonexistent in the Black Sea – the tidal amplitude at Constanţa is 7 cm, and at Sevastopol only 1–3 cm. However, the water level of the Black Sea is subject to seasonal fluctuations averaging about 20 cm. The two-layer density structure of the Black Sea is reflected in the circulation pattern – the most energetic currents are concentrated in the upper 200 m, above the permanent pycnocline.

Mesoscale Dynamics

In the 1980s a strange behavior of the Rim Current was recorded near the town of Gelendzhik located on the northeastern coast. The along-shore current sometimes flowed in the opposite direction for a few days, producing a bimodal distribution in the direction probability diagram at the coastal station (3 miles offshore), while 20 nautical miles offshore the prevailing direction of currents was toward northwest as usual; see **Figure 8**.

This was later found to be caused by mesoscale (30–50 km in size) coastal anticyclonic eddies, which

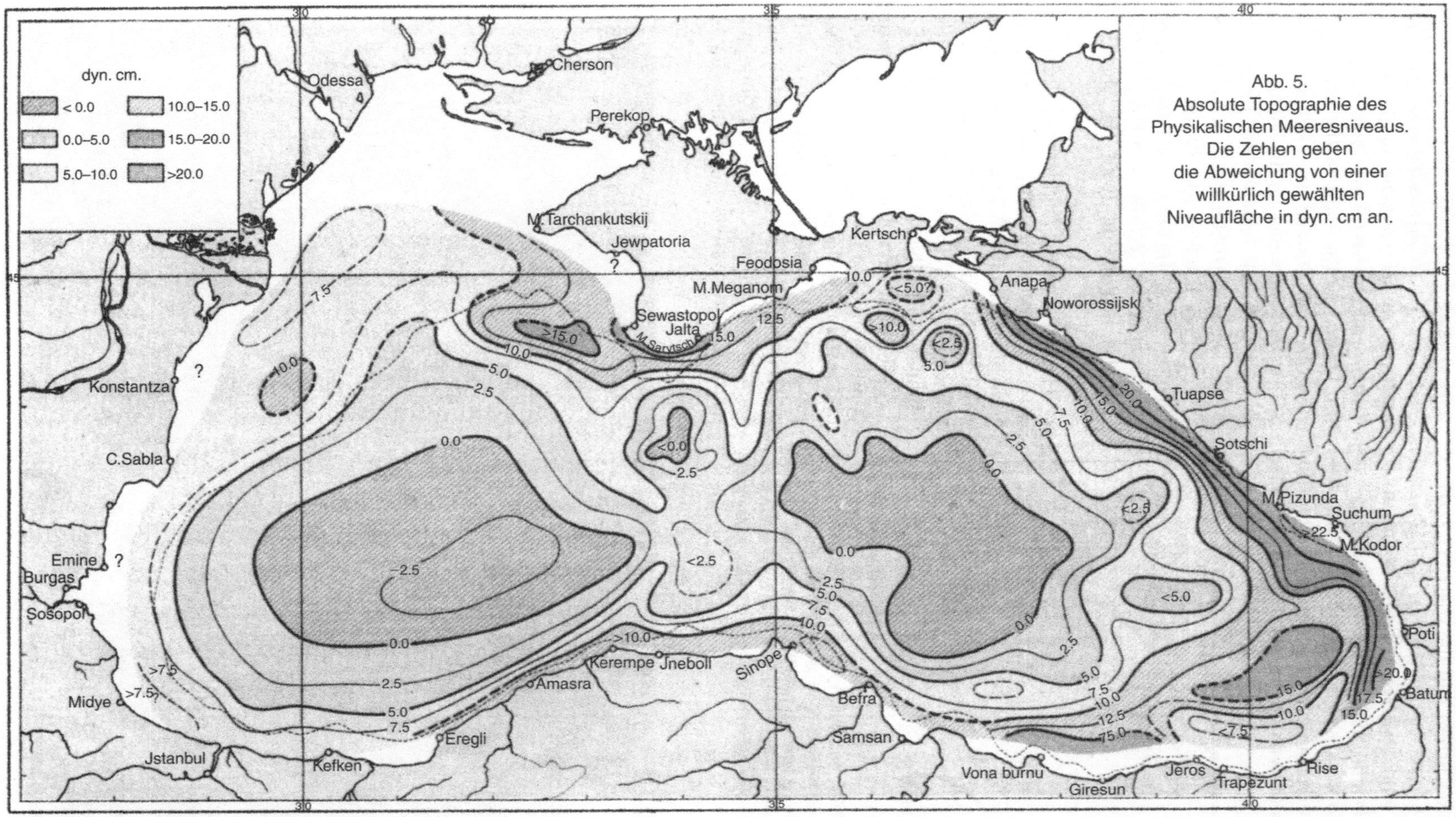

Figure 7 Dynamic topography and geostrophic currents in the Black Sea. From Neumann G (1942), adapted from Neumann G and Pierson WJ, Jr. (1966) *Principles of Physical Oceanography*. New York: Prentice-Hall.

are typically elongated, anticyclonic gyres wedged between the Rim Current and the coast. The coastal anticyclones are formed principally by fluid elements shedding from elongated recirculating wakes structures formed behind coastal headlands. They are often related to anticyclonic meanders of the Rim Current, which typically have length scales of 100–200 km. Charts of dynamic topography (in dyn mm) clearly show the horizontal structure of the coastal anticyclones.

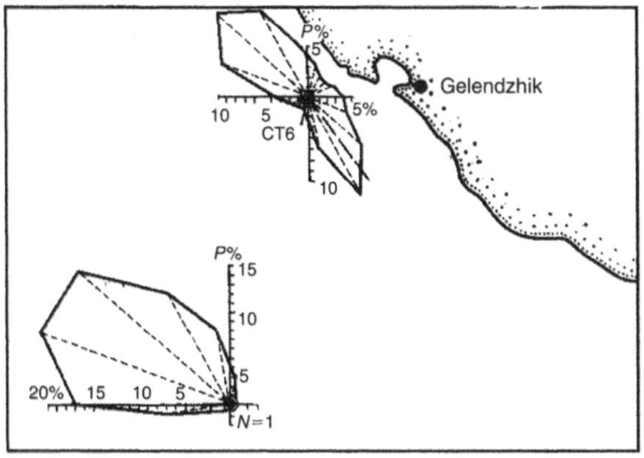

Figure 8 Probability distribution of current direction at two locations in the northeastern part of the Black Sea. Coastal station shows a strong bimodal pattern. Adapted from Titov VB (1992) About the vortex role in the formation of current regime on the Black. Sea shelf and in the coastal zone ecology. *Oceanology* 32(1): 39–48.

In contrast to mesoscale rings generated by meandering of jet currents in the ocean (i.e., Gulf Stream rings), coastal anticyclones in the Black Sea are located to the right of the main jet, and hence they are not formed by circling of an extended anticyclonic meander on the left-hand side of the jet, but rather by the instability of the Rim Current due to the high horizontal shear and horizontal friction.

The Rim Current can be identified on satellite images as it transports colored waters rich in phytoplankton, yellow substance, and suspended particles originating from the shelf areas, as in **Figure 9**.

The upper ocean circulation pattern also reveals a series of recurrent anticyclonic eddies (sometimes individually named Sevastopol, Kali-akra, Bosporus, Sakarya, Sinop, Kizilirmak, Caucasus, and Crimea) on the periphery between the Rim Current and the undulations of the coast.

Even in the presence of a strong stratification, the continental slope can affect the dynamics, structure, and stability of the near-shore current as well as the processes of generation, movement, and transformation of eddies, particularly over the shelf edge and the top part of the continental slope, where the sea depth is comparatively low. Since eddies are one of

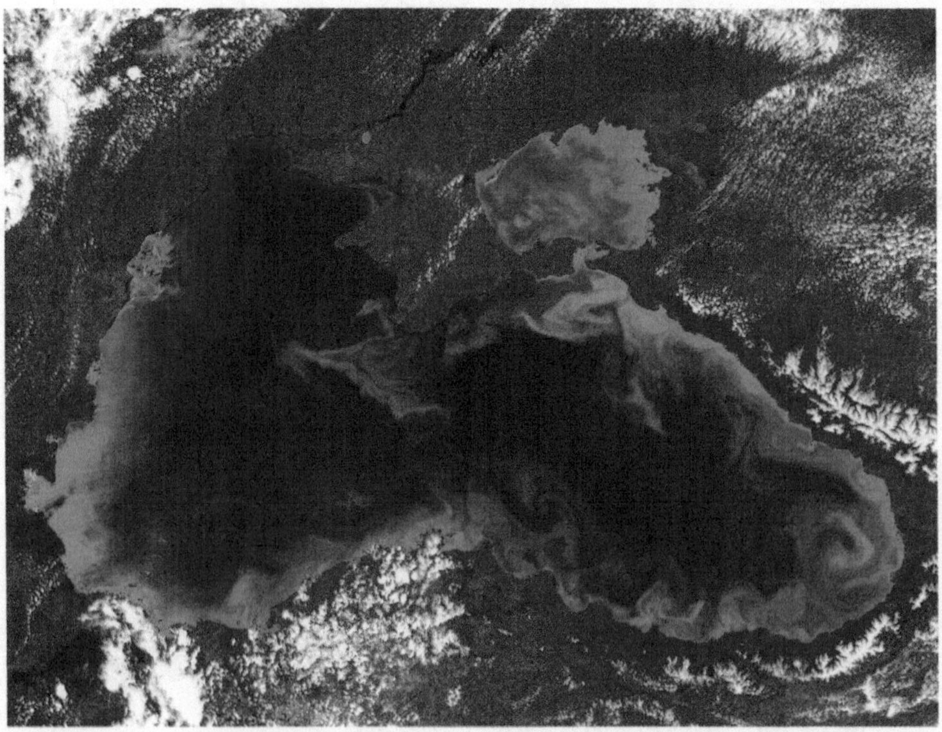

Figure 9 Satellite image of the Black Sea based on *MODIS* data from NASA, 22 May 2004. Colors are exaggerated to show the contrasting coastal and open sea water masses. Adapted from http://visibleearth.nasa.gov.

the main mechanisms of the interaction of the nearshore zone with the deep sea, the character and intensity of the water exchange over the slope must be affected by the topography of the continental slope.

In the central part of the Black Sea, larger anticyclonic eddies of 80–100 km in diameter have been observed both from *in situ* measurements and with satellite imagery; these are quite deep with currents penetrating the main pycnocline. Their life span ranges from 1 to 8 months and the orbital velocities can reach 0.5 m s^{-1} at the surface. The open-sea eddies can interact with the Rim Currrent, deflect it away from the coast, and cause intense horizontal exchanges and mixing between the shelf and deep sea.

Numerical models and satellite altimetry have shown that Black Sea eddies tend to form in the eastern basin and propagate westward as Rossby waves with a speed of *c.* 0.03 m s^{-1}. The narrow Black Sea section south of the Crimean peninsula strongly affects that eddy propagation. Dissipation increases in the western basin where eddies slow down and their scales become smaller.

The deep-sea anticyclones originate from those coastal anticyclonic eddies which manage to escape from their near-shore locations when the Rim Current weakens, becomes intermittent, and its main stream moves further offshore. Strong mesoscale activity takes place over the wide northwest shelf where both cyclones and anticyclones are generated. Shelf-born eddies are sometimes transported by the Rim Current counterclockwise, and disintegrate due to friction or collision with other current systems such as upwellings.

Temporal Variability

Due to the temporal variability of currents it is not easy to obtain a snapshot of the circulation system. In recent years, numerical modeling has become an important tool to study Black Sea circulation. An example of model simulation of sea surface elevation and related geostrophic currents based on climatic mean external forcing is given in **Figure 10**.

A second example is shown in **Figure 11**: when the models assimilate real-time observational data (mostly from satellite-borne altimeters) they are able to provide a snapshot of Black Sea circulation on a specific date. **Figure 11** shows enhanced mesoscale activity – eddies, filaments, and meanders are formed in all areas of the sea. Their size range is a few tens of kilometres, consistent with the fact that the strong thermocline makes the internal Rossby radius equal to 15–20 km. A set of marginal anticyclones is clearly defined both in climate-driven and operational velocity charts (**Figures 10** and **11**).

Model simulations have shown that, regardless of the forcing mechanism, the Rim Current is absent if the topography is not included. Changes in bottom slope and coastline orientation along the coasts generate meanders and instability features in the Rim

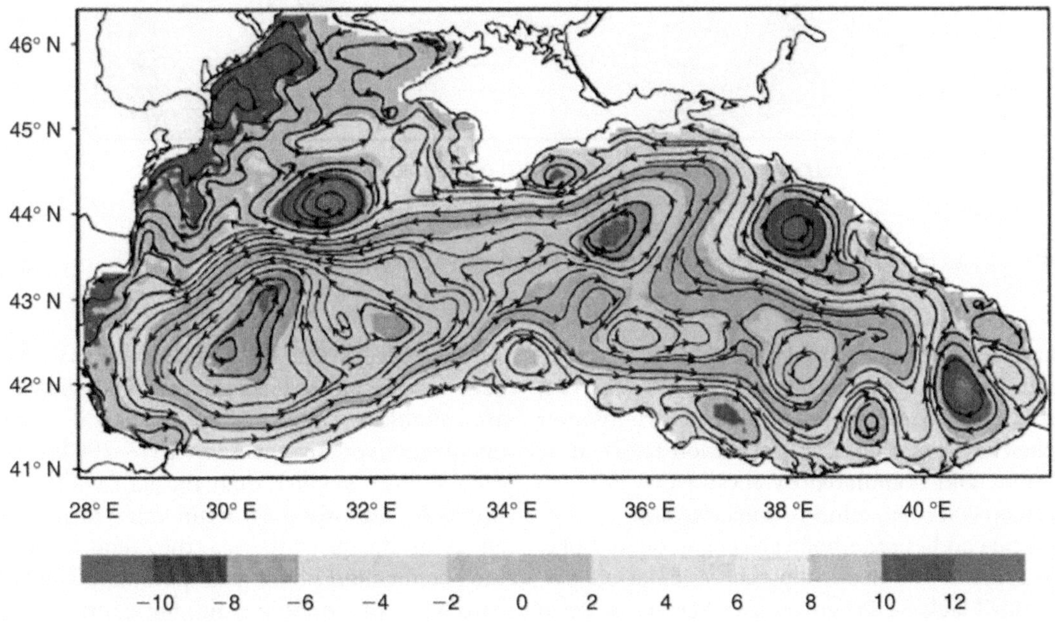

Figure 10 Snapshot of sea level elevation (cm) and surface currents from a numerical model. From Staneva JV, Dietrich DE, Stanev EV, and Bowman MJ (2001) Rim Current and coastal eddy mechanisms in an eddy-resolving Black Sea general circulation model. *Journal of Marine Systems* 31: 137–157.

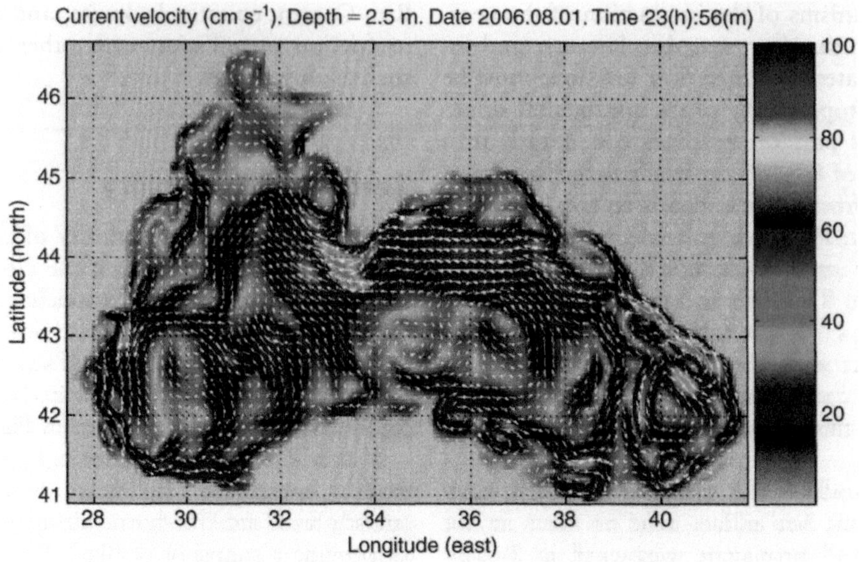

Figure 11 Current velocities obtained with a numerical model, which assimilates satellite altimetry data. From Korotaev G, *et al.* (2006) http://dvs.net.ua.

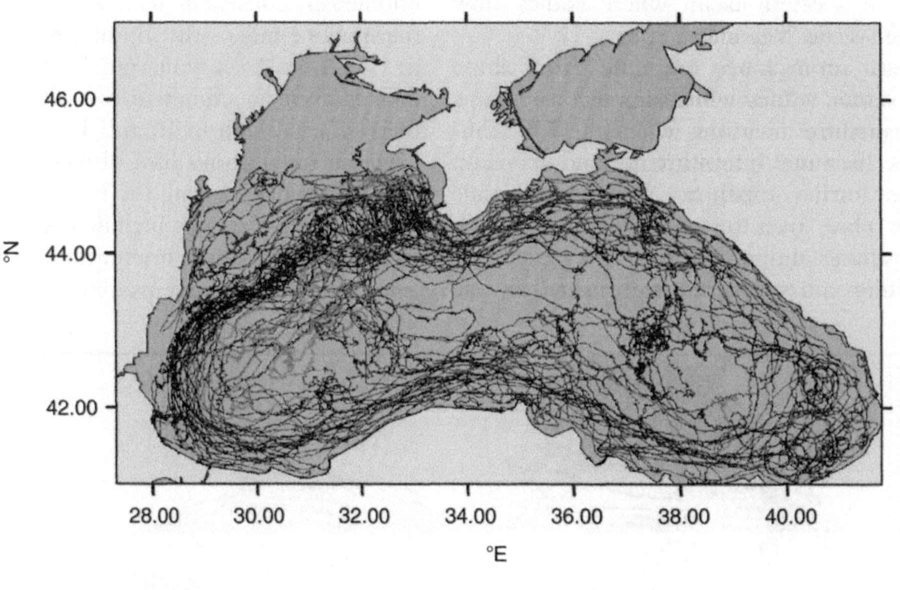

Figure 12 Tracks of 54 surface drifters in the Black Sea. Courtesy of S.V. Motyzhev, 2006.

Current on a wide range of space scales and timescales. The Rim Current is stronger, more coherent, and attached to the continental slope in the winter when the wind is strong; the circulation tends to be intermittent and dominated by mesoscale eddies in the summer when the wind is much weaker.

Mesoscale eddies are often very energetic and play a much greater role in forming the circulation pattern in the Black Sea than was earlier thought; in particular, anticyclonic meanders and eddies lying between the Rim Current and the coast provide an efficient exchange mechanism between the areas of different depth (i.e., shelf and deep sea). Mesoscale eddies and localized jets (filaments) of various origins are capable of transporting the coastal waters to the open sea over distances up to ~200 km, which is only slightly smaller than the half-width of the sea itself. As mesoscale structures are formed over the entire perimeter of the sea, they tend to homogenize the chemical and biological parameters over its area. This kind of horizontal mixing cannot be achieved by a large-scale current alone.

New information on basin- and mesoscale circulation was obtained from a recent international

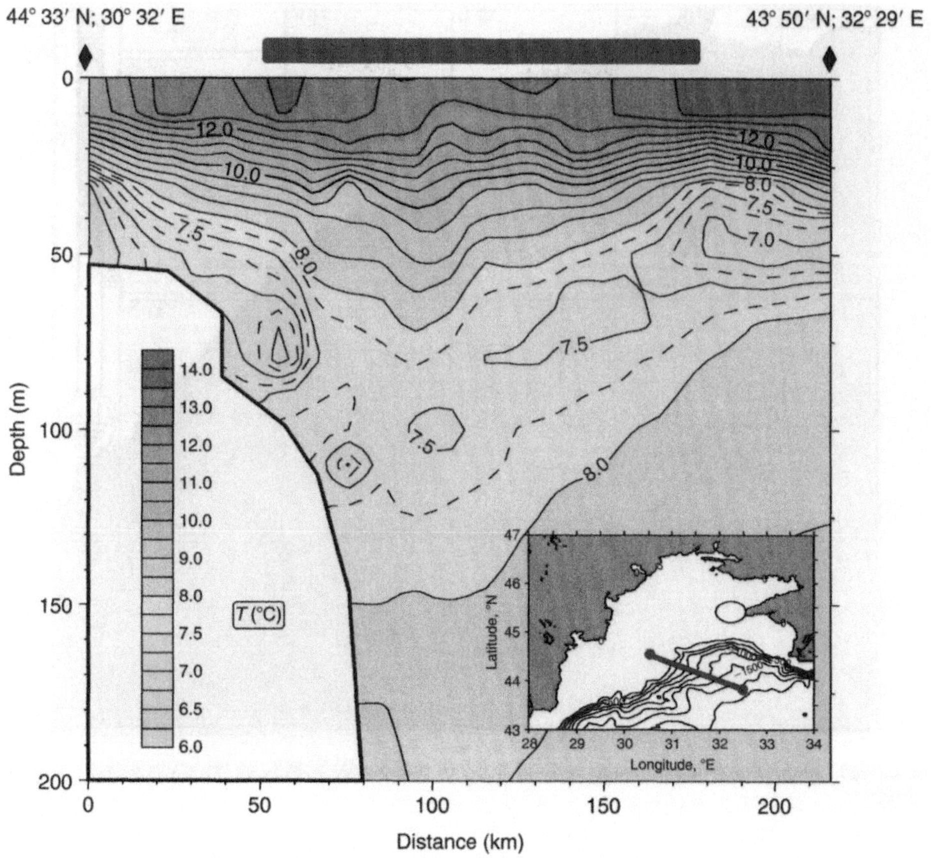

Figure 13 Temperature section through Sevastopol Eddy, May 2004. The bar at the top shows the horizontal extent of the eddy. The eddy deepens temperature contours below 15-m depth and lifts them up in the top layer. Cascading of cold winter water from the shelf is clearly seen. Reproduced from Shapiro GI *et al.* (2005) http://www.research.plymouth.ac.uk/shelf/projects/Black-sea/Black-sea.html.

drifter experiment (1999–2003), where 54 satellite-tracked drifters were deployed in the open part of the Black Sea (**Figure 12**). However, half of the drifters terminated on coastal shoals, driven there by mesoscale currents from the deep sea. Of the 54 drifters, only 11 executed at least one full circuit around the entire basin. Although the drifters showed the general cyclonic circulation in the Black Sea, they did not reveal the eastern and western sub-basin cyclonic gyres which are represented in traditional schemes of the Black Sea circulation.

Coherent Structures and Overturning Circulation

One interesting mesoscale feature is the Sevastopol Eddy located southwest of Crimean peninsula. Fine-scale anticyclonic eddies and vortex filaments that peel off the Crimean headlands merge into a particularly strong and coherent recirculation eddy during the period of weak stratification. Mesoscale processes near the Crimean peninsula and their

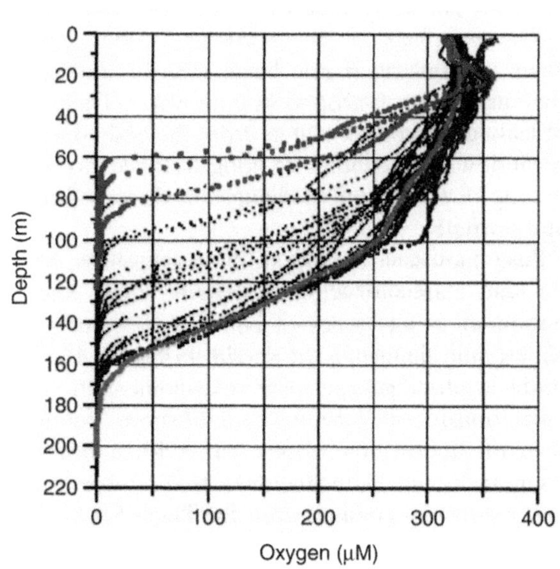

Figure 14 Anticyclonic eddy facilitates downward penetration of oxygen. The open sea stations outside the eddy (red dots) show a sharp decrease of oxygen concentrations below the upper mixed layer. Greatest deepening of oxygenated waters takes place in the region where the eddy is in contact with the continental slope (blue dots). From Romanov AS *et al.* (2006).

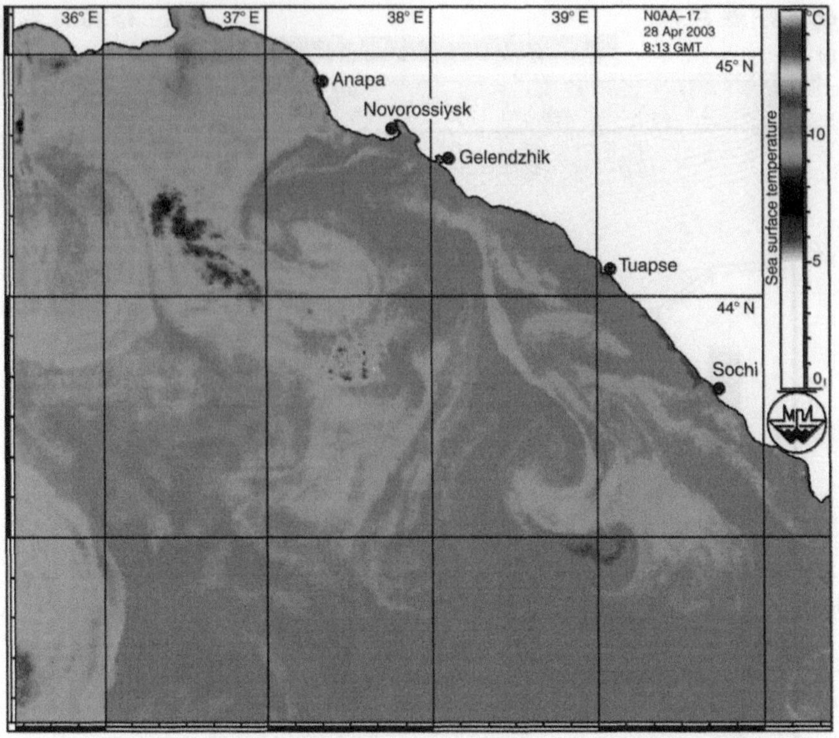

Figure 15 Sea surface temperature map showing various types of mesoscale activity in the northeastern Black Sea. Courtesy of S.V. Stanichny, 2003.

interaction with the shelf can substantially affect the vertical mixing of chemicals and biological organisms. The distortion of the thermocline and the deepening of oxygenated waters by the Sevastopol Eddy are shown in **Figures 13** and **14**, respectively.

The periphery of the Sevastopol Eddy entrains water originating in the basin interior and on the shelf and wraps it around its core water. The process of leakage of coastal waters from the eddy due to its breakdown contributes to long-distance horizontal mixing of water, and exchanges of dissolved matter and vorticity.

Busy mesoscale activity in the northeastern part of the sea, characterized by a narrow shelf, and formation of a few types of coherent structures (jets, eddies, and filaments) are shown in **Figure 15**. Some of the localized jets give birth to cyclonic–anticyclonic pairs (mushroom currents); an offshore example is seen off Tuapse, the vortex pair with a near-shore component eddy is located next to Novorossiysk.

Overturning circulation in the Black Sea is driven by winter convection. Cold dense water is produced more efficiently over the shelves and in the centers of sub-basin cyclonic eddies than elsewhere as the water layer involved in cooling is thinner there. This is due to limited water depth over the shelves and upward doming of the pycnocline in cyclones. Winter convection feeds the intermediate cold layer, which then spreads over the entire sea due to horizontal mixing enhanced by mesoscale eddies and basin-scale oscillations.

See also

Coastal Circulation Models. Estuarine Circulation. Meddies and Subsurface Eddies. Mediterranean Sea Circulation. Mesoscale Eddies. Ocean Circulation. Ocean Circulation: Meridional Overturning Circulation. Wind Driven Circulation.

Further Reading

Izdar E and Murray JW (eds.) (1991) *NATO Advanced Science Institute Series, Series C, Vol. 351: Black Sea Oceanography*. Dordrecht: Kluwer.

Leonov AK (1960) *Regional Oceanography, Part 1: The Black Sea*, pp. 623–765. Leningrad: Gidrometizdat.

Murray JW (2005) *Special Issue: The Black Sea. Oceanography* 18(2).

Neumann G and Pierson WJ, Jr. (1966) *Principles of Physical Oceanography*. New York: Prentice-Hall.

Oguz T, Latun VS, Latif MA, *et al.* (1993) Circulation in the surface and intermediate layers of the Black Sea. *Deep Sea Research* 40: 1597–1612.

Özsoy E and Ünlüata Ü (1997) Oceanography of the Black Sea: A review of some recent results. *Earth Science Review* 42: 231–272.

Simonov AI and Altman EN (eds.) (1991) *Hydrometeorology and Hydrochemistry of the Seas of the USSR: The Black Sea*, issue 1, 449pp. St. Petersburg: Gidrometizdat.

Sorokin Yu I (2002) *The Black Sea Ecology and Oceanography*, 875pp. Leiden: Backhuys.

Staneva JV, Dietrich DE, Stanev EV, and Bowman MJ (2001) Rim Current and coastal eddy mechanisms in an eddy-resolving Black Sea general circulation model. *Journal of Marine Systems* 31: 137–157.

Suvorov AM, Eremeev VN, Belokopytov VN, *et al.* (2004) *Digital Atlas: Physical Oceanography of the Black Sea* (CD-ROM). Kiev: Environmental Services Data and Information Management Program, Marine Hydrophysical Institute of the National Academy of Sciences of Ukraine.

Titov VB (1992) About the vortex role in the formation of current regime on the Black. Sea shelf and in the coastal zone ecology. *Oceanology* 32(1): 39–48.

UNEP/GRID-Arendal (2001) Drainage in the Black Sea Area. In: *UNEP/GRID-Arendal Maps and Graphics Library*. http://maps.grida.no/go/graphic/drainage_in_the_black_sea_area (accessed Feb. 2008).

Yakushev EV, Podymov OI, and Chasovnikov VK (2005) The Black Sea seasonal changes in the hydrochemical structure of the redox zone. *Oceanography* 18(2): 48–55.

Zatsepin AG and Flint MV (eds.) (2002) *Multidisciplinary Investigations of the Northeast Part of the Black Sea*, 475pp. Moscow: Nauka.

Relevant Websites

http://dvs.net.ua
– Marine Hydrophysical Institute National Academy of Sciences of Ukraine, Remote Sensing Department.

http://research.plymouth.ac.uk
– Shelf Sea Oceanography and Meteorology research group.

http://blacksea.orlyonok.ru
– The Living Black Sea Marine Environmental Education Program.

http://visibleearth.nasa.gov
– Visible Earth.

BALTIC SEA CIRCULATION

W. Krauss, Institut für Meereskunde an der Universität Kiel, Kiel, Germany

Introduction

The Baltic Sea consists of a number of sub-basins connected by straits and deep channels. They have an important influence on the current system. A map of the Baltic Sea with topographic subdivision and bottom topography is shown in **Figure 1**.

The North Sea, via Skagerrak and Kattegat, is connected to the Baltic Proper by three pathways, called the Danish Straits (**Figure 2**). The eastern route, the Sound, is only 2 km wide at its narrowest location, shallow (sill depth 8 m) and only about 55 km long. It is the shortest inflow route for saline water. The central pathway is about 180 km long and about 13 m deep on the average, 25–30 m along the axis. It consists of the Great Belt and the Fehmarnbelt and is terminated at the east by the Darss sill (18 m deep), the shallowest sill for the main inflow. The third connection is the Little Belt, having a cross section of only 16 000 m^2 compared to 255 000 m^2 of the Great Belt and 80 000 m^2 of the Sound. Therefore, about 70% of the in- and outflows occur through the Great Belt; the Little Belt is negligible.

Moving from west to east, the Arkona Basin (45 m) is connected to the Bornholm Basin (95 m) by the Bornholm Channel. The Stolpe Channel (20 km wide by the 60 m isobath) allows inflow into the Gdansk Basin (110 m) and the Eastern Gotland Basin with a maximum depth of 250 m. Further to the north the Farö Deep is followed by the Northern Basin (200 m) which extends to the east towards the Gulf of Finland as a deep channel with decreasing width and depth. By contrast, the Gulf of Bothnia is separated from the deeper layers of the Baltic Proper by the Aland Sea with its numerous islands. The Gulf of Bothnia consists of two basins, the Bothnian Sea and the Bothnian Bay.

The deep basins and the connecting channels are important for the water exchange between the North Sea and the Baltic Sea and determine the current structure. Due to the large river runoff, the Baltic Sea is the largest brackish water body of the world with a volume of 21 000 km^3. The mean annual total contribution of all rivers for the period 1950–1990 was 446 km^3 $year^{-1}$. This would correspond to a sea-level rise of 1.18 m $year^{-1}$ if the Baltic Sea were not connected to the North Sea. An additional surplus of fresh water results from the difference of precipitation minus evaporation, which amounts to about 60 km^3 $year^{-1}$.

The river runoff shows both a seasonal cycle and interannual variations. The lowest and highest annual values differ from the mean value by −27% and +22%, respectively. The seasonal runoff varies between 25 000 m^3 s^{-1} in spring and 12 000 m^3 s^{-1} in winter.

As a consequence of this freshwater supply the salinity in the eastern and northern parts of the Baltic Sea is reduced to about 4 PSU (practical salinity units), increasing to about 8 PSU in the central and western Baltic Sea. Due to the associated density difference saline water penetrates from the Kattegat (30–34 PSU) through the Danish Straits and the channels into the Baltic Sea and yields a strong halocline, which separates the water of the deeper layers from the brackish upper water masses (**Figure 3**). As a consequence vertical mixing is strongly reduced.

The Baltic Sea extends from about 54°N to 66°N thus ranging from mild and humid to a subarctic climate. Frequent and complex synoptic-scale cyclonic activity and subsynoptic-scale depressions are characteristic, leading to a high variability of the prevailing westerly winds. The wind fields produce a highly variable current system, especially in the upper layers, superimposed on the weak baroclinic flow field induced by the salinity differences. Beneath the wind-mixed layer bottom topography has a strong steering influence. The wind-induced currents and the associated sea-level variations may drastically change, when parts of the Baltic Sea are ice-covered. Sea ice occurs every year in the Baltic Sea. Under normal winter conditions about 45% of the Baltic Sea is covered with ice and the ice season lasts about 6 months in the northern parts. Severe winters may lead to almost total ice coverage. Ice first appears in the innermost parts of the Bothnian Bay during mid-November. In a normal winter, the entire Bothnian Sea, Aland Sea, Gulf of Finland and the northernmost part of the Baltic Proper are covered by ice. In severe ice-winters the Kattegat, the Belt Sea, the Sound and large parts of the Baltic Proper are also ice covered.

Wind and thermohaline forcing determine the currents of the Baltic Sea, strongly influenced by bottom topography and ice coverage. A large

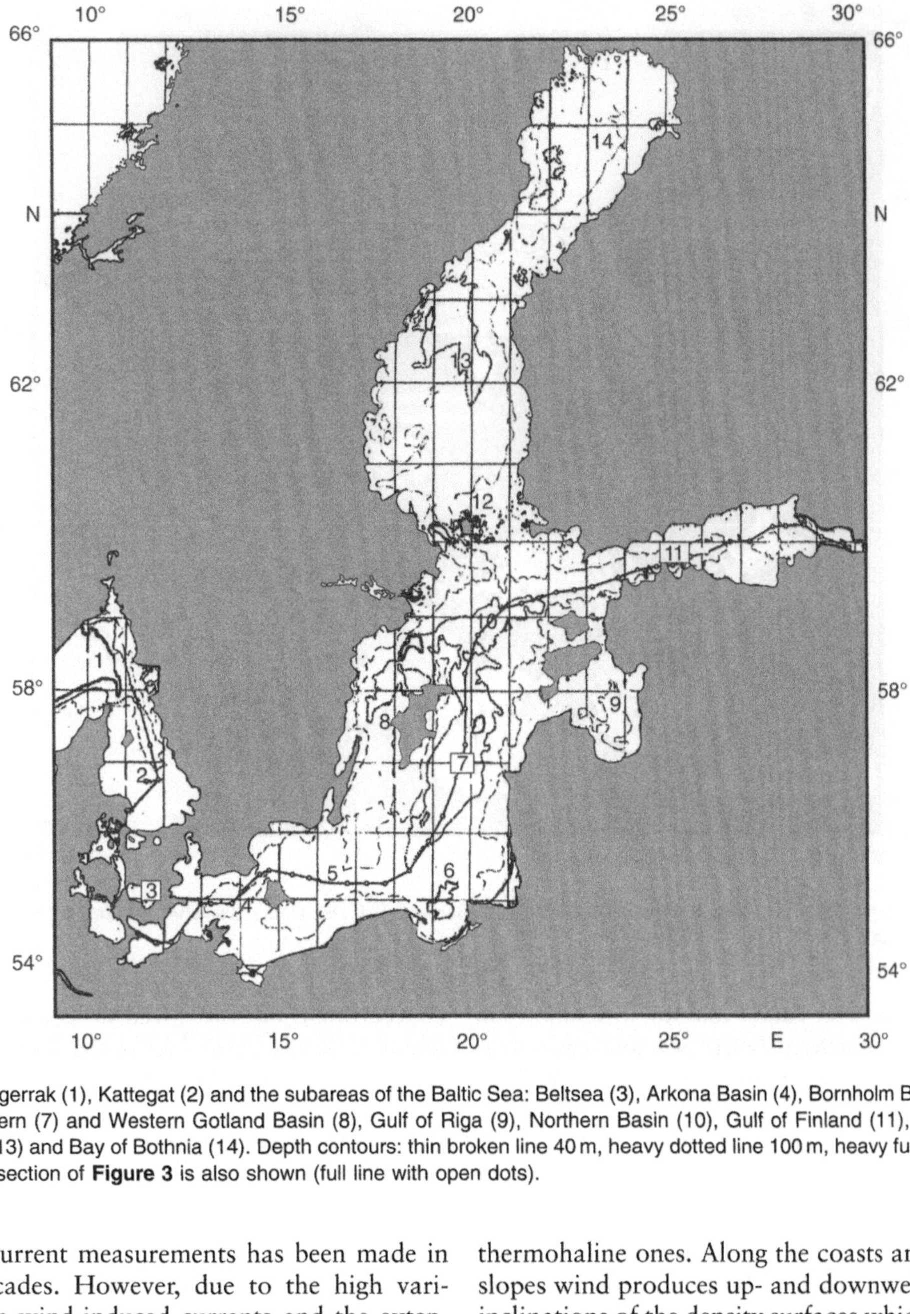

Figure 1 Skagerrak (1), Kattegat (2) and the subareas of the Baltic Sea: Beltsea (3), Arkona Basin (4), Bornholm Basin (5), Gdansk Basin (6), Eastern (7) and Western Gotland Basin (8), Gulf of Riga (9), Northern Basin (10), Gulf of Finland (11), Aland Sea (12), Bothnian Sea (13) and Bay of Bothnia (14). Depth contours: thin broken line 40 m, heavy dotted line 100 m, heavy full line 200 m. The location of the section of **Figure 3** is also shown (full line with open dots).

number of current measurements has been made in the past decades. However, due to the high variability of the wind-induced currents and the extensive fishing activities, which make it impossible to install observational systems in some areas, it was not possible in the past to derive a consistent circulation pattern from observations. In recent years, three-dimensional models, combined with data assimilation, have improved in such a way that both the mean circulation and its variability can now be described with sufficient accuracy.

The Estuarine Circulation

In a stratified sea it is generally not possible to separate wind-induced currents properly from thermohaline ones. Along the coasts and the bottom slopes wind produces up- and downwelling and thus inclinations of the density surfaces which may largely amplify the wind effects. However, some insight is gained by considering the horizontal pressure gradients, which result from the mean surface inclination and the horizontal density differences. They produce an estuarine type of circulation.

Figure 3 shows the mean salinity section from the Kattegat along the main axis of the Baltic Sea into the Gulf of Finland. Except at the level of the summer thermocline (about 25–30 m depth) the density distribution is similar. Near the bottom the density decreases from about 25 kg m^{-3} in the Kattegat to 9 kg m^{-3} in the Gotland Basin and less than 4 kg m^{-3} in the interior of the Gulf of Finland.

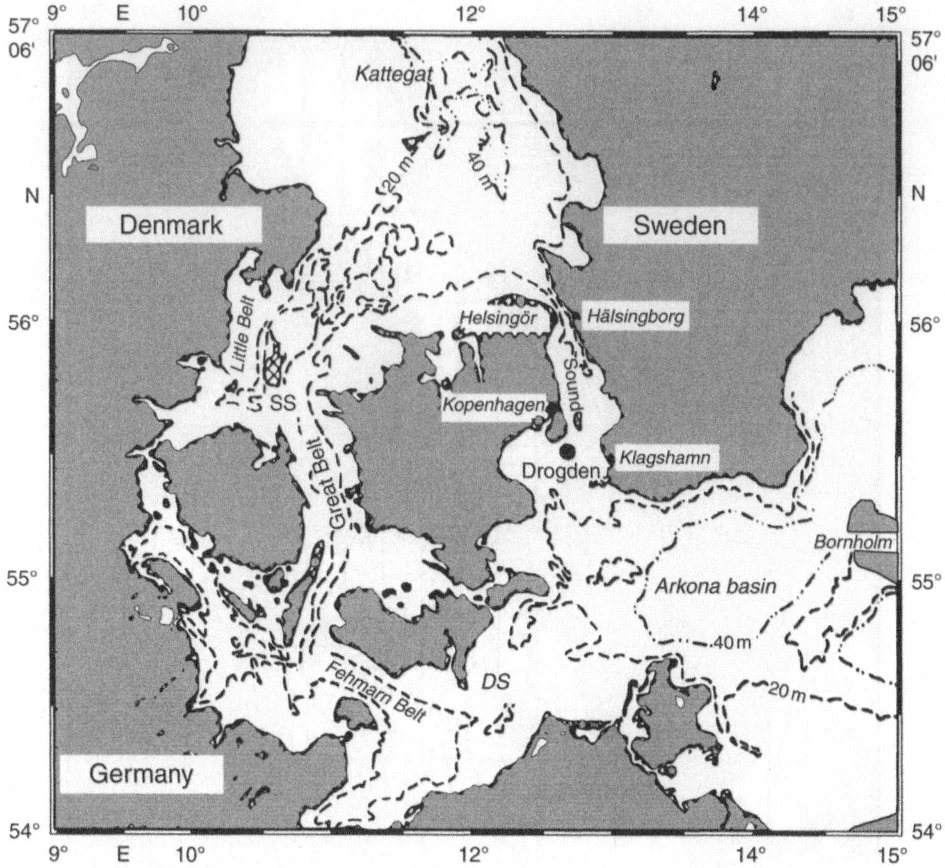

Figure 2 The Danish Straits, Kattegat and Arkona Basin. SS, Samsoe Sill, DS, Darss Sill. The Belt Sea is the area between the Kattegat and Arkona Basin.

Because of the river runoff and the low salinity in the eastern and northern Baltic Sea the sea level is higher by 0.3 m in the interior of the Gulf of Finland and the Gulf of Bothnia compared to that in the Belt Sea. This sea level inclination produces a pressure gradient which drives the upper layers out of the Baltic Sea. Due to the Coriolis force this outflow is concentrated on the Swedish coast. On average about 1250 km^3 $year^{-1}$ leave the Baltic Sea through the Danish Straits. As compensation about 740 km^3 $year^{-1}$ of saline, dense and oxygen-rich water penetrates in the deeper layers through the Danish Straits into the Arkona Basin and from there through the Bornholm Channel into the Bornholm Basin.

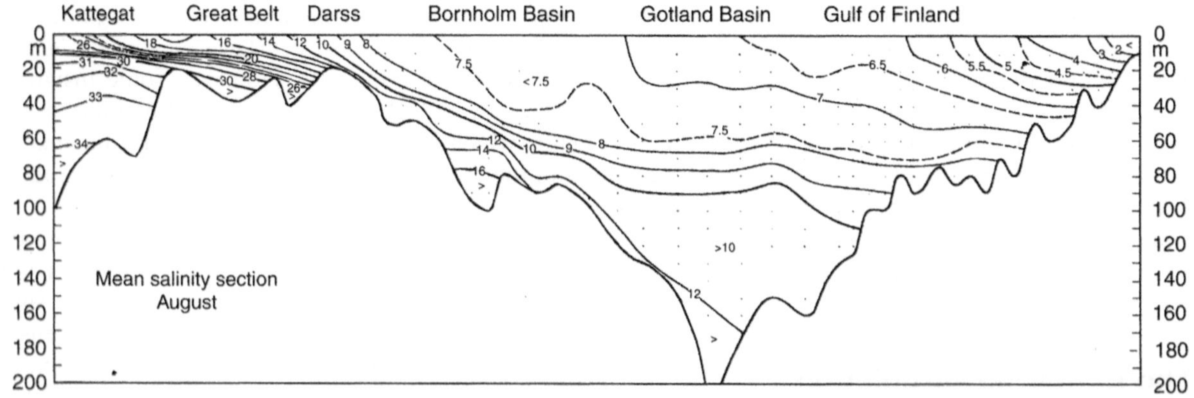

Figure 3 Mean salinity (practical salinity units) distribution (1902–1956) along a section from the Kattegat to the Gulf of Finland (for location see **Figure 1**).

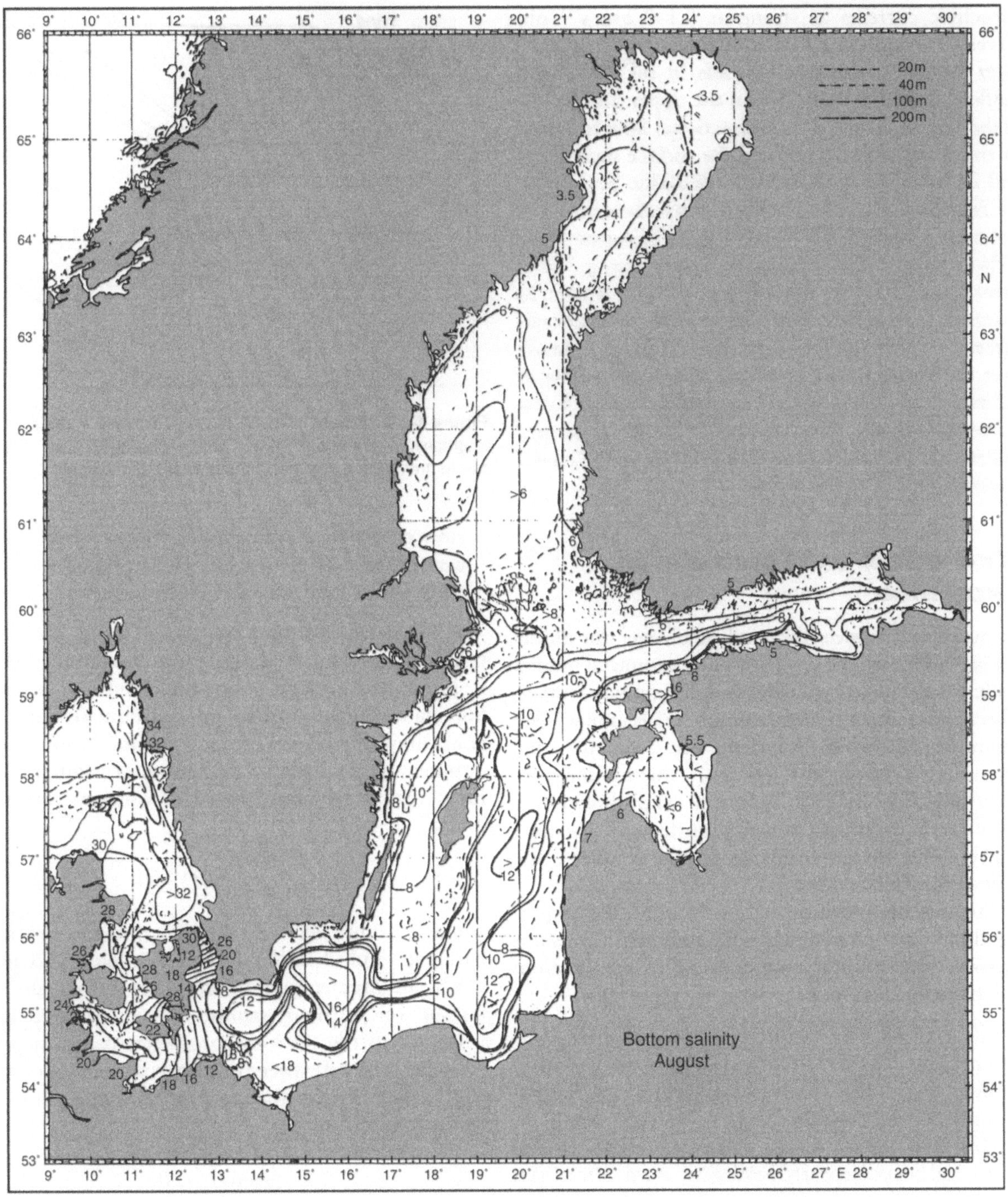

Figure 4 Mean salinity (practical salinity units; PSU) distribution at the bottom of the Baltic Sea (1902–1956).

Horizontal pressure gradients due to the density gradients, which overcompensate the sea level gradient, act as forcing term. The path of this inflow further to the north is shown in **Figure 4** by the salinity distribution close to the bottom. Water of 12 PSU flows through the Stolpe Channel into the Gdansk Basin and further into the Eastern Gotland Basin. Part of this water continues into the Northern Basin, and another part flows around Gotland and south into the Western Gotland Basin. However, these observations also represent the cyclonic wind-produced circulation in the Baltic Proper and are only partially due to thermohaline driving (see next section).

Only in the Danish Straits, where the salinity differences are large and the flow is channeled, are baroclinic currents as strong as 0.1–0.2 m s^{-1} observed. During light and moderate winds, a two-layer system of currents prevails in the Belt, with outflow from the Baltic Sea in the upper and inflow in the deeper layers. This bidirectional flow is driven by the contrast in density and the surface inclination and includes the outflow from the Baltic Sea due to the fresh water surplus. In the Fehmarnbelt this flow system is observed in 50% of all cases in summer and 29% in winter. In the [illegible] the [illegible] numbers are 95% in summer and 65% in winter. In contrast to the Belt with its typical bidirectional current system, outflow into the Kattegat extends over the entire Sound, reaching maximum values of 0.2 m s^{-1} near the surface. This flow forms the Baltic Current along the Swedish coast in the Kattegat and Skagerrak which becomes the Norwegian Coastal Current further to the north.

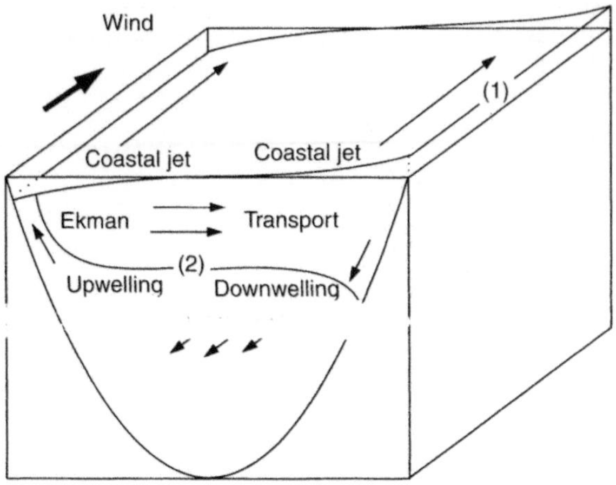

Figure 5 Schematic of wind-produced currents in an infinitely long basin. Sea surface (1) and density surfaces (2) are inclined due to the Ekman transport perpendicular to the coast.

Wind-driven and Thermohaline Circulation

As mentioned above the wind field over the Baltic Sea is highly variable with a well pronounced annual cycle. Maximum speed is reached during November–March with an absolute maximum in December. The minimum occurs in May/June. In December the maximum mean speed is 7–9 m s^{-1}, the minimum in May/June is 4–5 m s^{-1}. The mean wind direction is from south west/west towards north east/east, i.e., nearly along the line from the western Baltic to the Gulf of Finland.

The current system produced by wind blowing in the direction of an infinitely long basin with constant cross-section and horizontally uniform stratification is depicted in **Figure 5**.

1. In the surface layers there results an Ekman current which is deflected to the right of the wind direction by the Coriolis force, yielding an Ekman transport in cross direction.
2. This Ekman transport produces a sea-level rise on the right-hand coast (looking in wind direction) and a fall on the left-hand side. Furthermore, downwelling occurs on the right-hand side and upwelling on the left-hand side, resulting in baroclinicity of the same sign at both coasts. Note the different length scales of deflection of the sea surface and the isopycnals, which are governed by the external and internal Rossby radius, a measure which describes the adjustment of the deflection under gravity of a rotating fluid.
3. Consequently, geostrophically balanced coastal jets are produced by the inclination of the sea surface along both coasts in wind direction and a slow return flow compensates this transport in the central area of the basin. The inclination of the density surfaces enhances these currents.

If the depth contours are closed, as in the deep basins of the Baltic Sea, the return flow in the center is prohibited. Instead topographic effects become dominant. Outside of the up- and downwelling regions along the coasts the flow in the basins is determined by the sea level inclination of **Figure 5**. For these regions Rossby's potential vorticity theorem holds,

$$d/dt((\zeta + f)/H) = 0 \qquad [1]$$

where ζ is the relative vorticity, f is the Coriolis parameter and H is depth. In basins like the Baltic, which are large enough for rotation to play an important role in the dynamics, but not so large that variations of the Coriolis parameter must be taken into account, and where the mean currents are weak ($\xi \ll f$), the vorticity theorem for the mean flow, **v**, can be reduced to

$$\mathbf{v} \cdot \mathrm{grad}\ H = 0 \qquad [2]$$

i.e. the vertically averaged currents below the Ekman layer try to follow the depth contours.

The wind-driven currents of the Baltic Sea show much of this typical response besides the strong vertical stratification, e.g. coastal jets are often observed along the eastern slopes with current speeds up to 1 m s^{-1}. Numerical models reveal clear evidence of characteristic persistent circulation patterns which comprise mostly the sub-basins with less transport between the basins. **Figure 6** shows the annual average of the vertically integrated transport

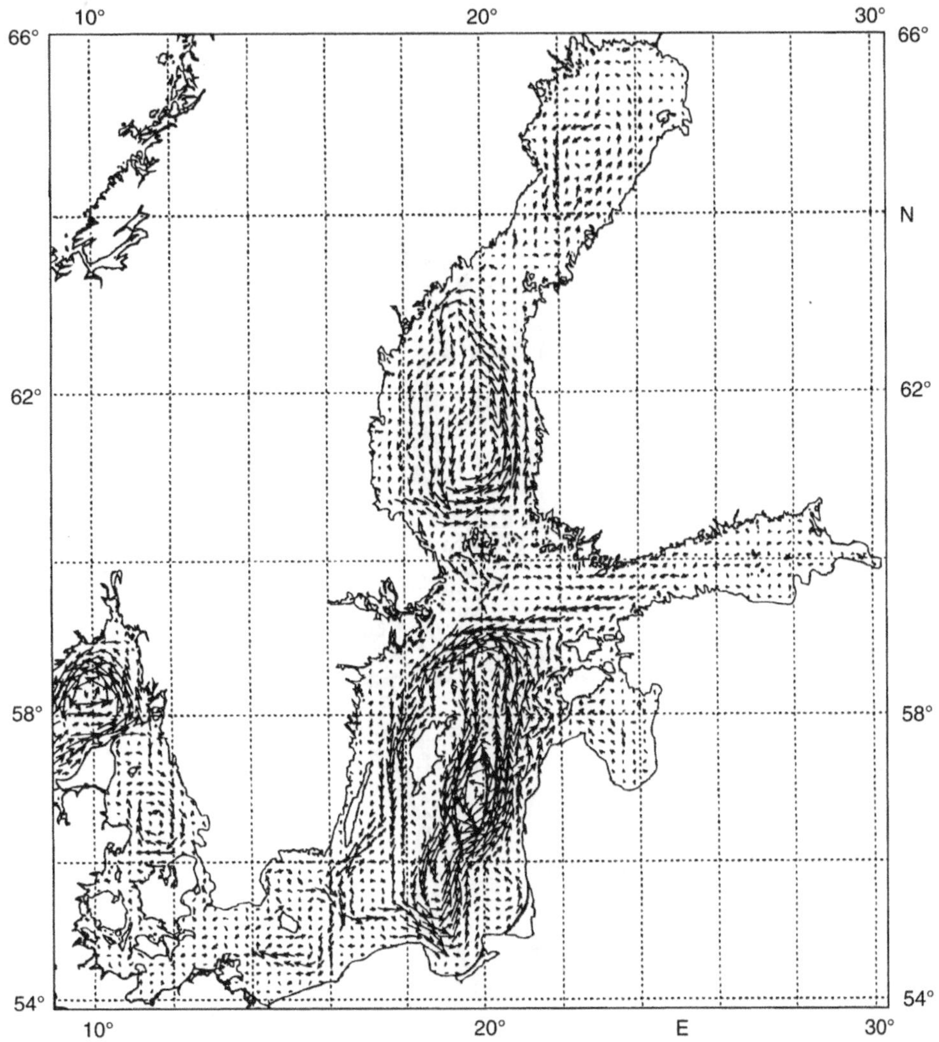

Figure 6 Annual average of the depth-integrated transports ($m^3\ s^{-1}$) for 1992 (model results). Maximum vector $0.934 \times 10^5\ m^3\ s^{-1}$.

for the year 1992. As can be seen from this figure the most pronounced structure with the highest stability is a cyclonic circulation cell comprising the Eastern Gotland basin. Most of the water is recirculating in the Eastern Gotland Basin, but part of it flows back along the western slope of the Western Gotland Basin, continues towards the Bornholm Basin and closes the return flow through the Stolpe Channel. On average the stability of this circulation pattern is higher than 50%. The northward flow along the eastern Baltic slope is much more pronounced than the southward flow along the Swedish slope. This can be interpreted according to **Figure 5**: the currents along the eastern slope are intensified by the coastal jet and correspondingly reduced along the Swedish coast. Similar patterns as in the Gotland Basin are observed in all other basins.

The transport associated with these wind-induced patterns is about an order of magnitude higher in the Stolpe Channel than the fresh water supply by the river runoff. Although highly variable between the years, this area shows the most intense transport rates of the entire Baltic Sea.

The bottom currents are only slightly different from the vertically averaged ones. Bottom currents towards the east are found in the Bornholm and in the Stolpe Channel with stabilities higher than 50%. The annual average of the bottom currents in the Bornholm Channel reaches about $0.1\ m\ s^{-1}$ and in the Stolpe Channel $0.05–0.08\ m\ s^{-1}$. This demonstrates the importance of these channels for the inflow of saline water into the eastern basins. However, the strongest bottom currents are in most cases not coincident with the highest stability. This indicates that these bottom currents are not as permanent as the vertically averaged currents.

As expected from **Figure 5** up- and downwelling occurs along the coasts. **Figure 7** shows the vertical

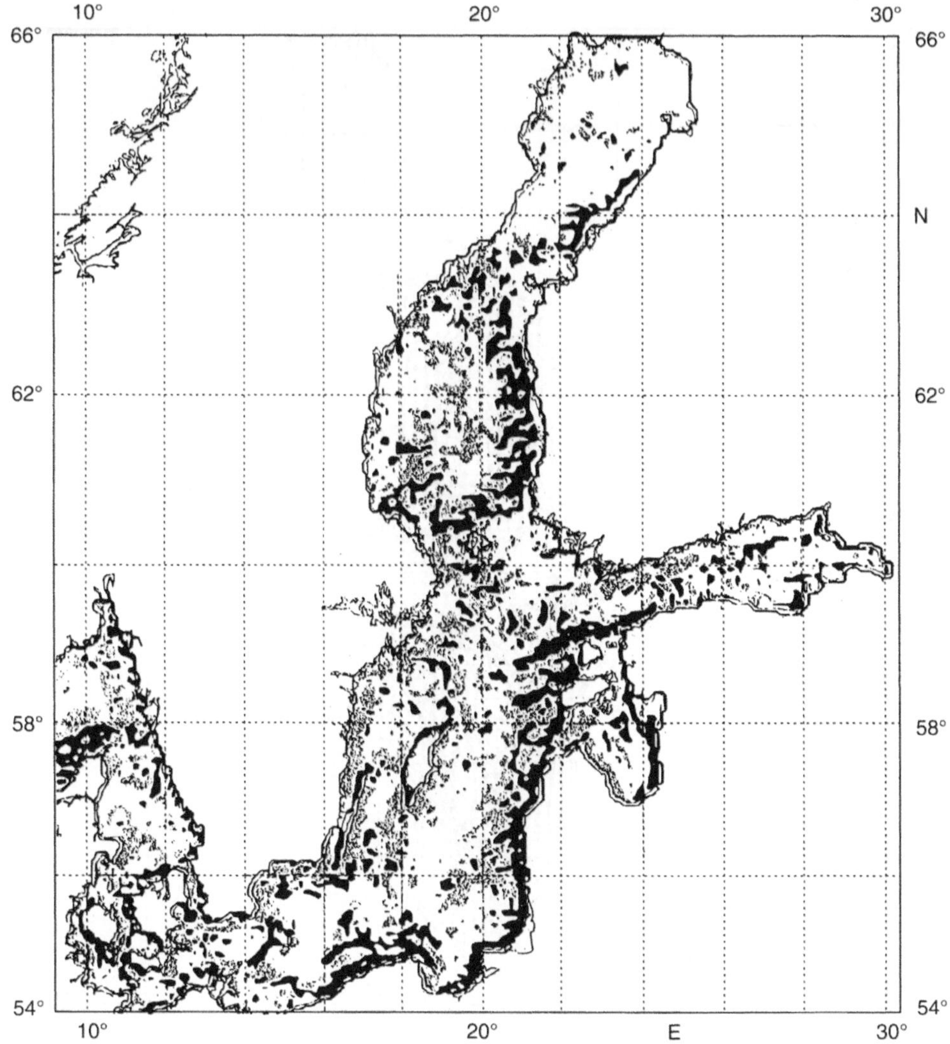

Figure 7 Upwelling and downwelling in the Baltic Sea based on four-year vertical averages of vertical velocities (model results). Grey areas, upwelling $> 4 \times 10^{-6}\,m\,s^{-1}$; black areas, downwelling $< 4 \times 10^{-6}\,m\,s^{-1}$.

average of the vertical velocities for a four-year modeling period. Along the western coast of the Baltic upwelling occurs at a distance of about 20–30 km offshore, and along the eastern coast a continuous band of downwelling can be seen. Larger islands and topographic features have similar effects. The areas can be regarded as the most active areas in the vertical exchange of water masses.

Currents in the Danish Straits and Major Inflow Events

The currents in the Danish Straits are highly variable, mainly determined by the sea-level differences between the Skagerrak and the Arkona Basin. Westerly winds, for example, pile up water in the Skagerrak–Kattegat area and simultaneously lower the sea level in the Arkona Basin. Large sea-level differences are therefore common along the Danish Straits. The flow in these straits responds like the flow in a manometer between two large basins.

Table 1 depicts the mean daily outflow rates and the standard deviations through the straits for some years (model results). Typically the standard deviation is 30–40 times larger than the mean flow. Actual mass transport across Darss Sill calculated from observations is depicted in **Figure 8** for 1994 and 1995, showing this high variability. In the Great Belt the speed reaches more than $\pm 1\,m\,s^{-1}$, and in the Öresund as much as $\pm 1.5\,m\,s^{-1}$. At Darss Sill, where the central channel widens, $\pm 0.5\,m\,s^{-1}$ is observed.

The inflow of saline and oxygen-rich water along the bottom of the Baltic Sea is essential for bottom-living animals in the Baltic Proper. Oxygen depletion in the deep basins causes oxygen deficiencies and the

Table 1 Mean daily outflow rates and standard deviation through the Danish Straits (model results) (1 km^3 day^{-1} = 11.6 × 10^{-3} Sv)

Year	*Mean outflow (km^{3}day^{-1})*	*Standard deviation (km^{3}day^{-1})*
1986	1.28	41.0
1988	1.39	54.4
1993	2.26	63.6
1994	1.02	59.1

production of hydrogen sulfide after long periods of stagnation.

Inflow of water of intermediate salinity occurs more or less continuously, as mentioned above, but this water is normally not dense enough to replace the bottom water in the deep basins. Renewal occurs only intermittently after periods of strong and persistent westerly winds. The greatest renewal during the last 50 years occurred in 1951, followed by a stagnation period until 1963. By then about one-third of the area of the Gotland Basin was covered with hydrogen sulfide-enriched water near the bottom. The inflows in 1963 and 1969 only partially replaced these water masses. The situation became worse during the next decades; the last major inflow occurred in 1993, which is the best documented and modeled inflow event.

The following conditions are most favorable for a major inflow, which usually occurs between November and January.

1. An extended period of easterly winds must drive water out of the Baltic and lower the sea level in the entire Baltic Sea. Maximum observed sea-level difference between Kattegat and Baltic Sea then amounts to about 0.7 m on average (maximum value 1.66 m). During this period the Kattegat front, which separates the saline Kattegat water from the outflowing Baltic Sea water, is pushed far to the north.
2. This period must be followed by a phase of strong westerly winds over the North Sea and the Baltic, such that the sea level and barocline gradients drive the water back into the Baltic. This period lasts typically 2–3 weeks, during which the brackish water of the Straits is flowing into the Arkona Basin and allows the saline water of the Kattegat to penetrate towards Darss and Drogden Sill. Due to this inflow, sea level rises in the Baltic up to normal or even above normal.

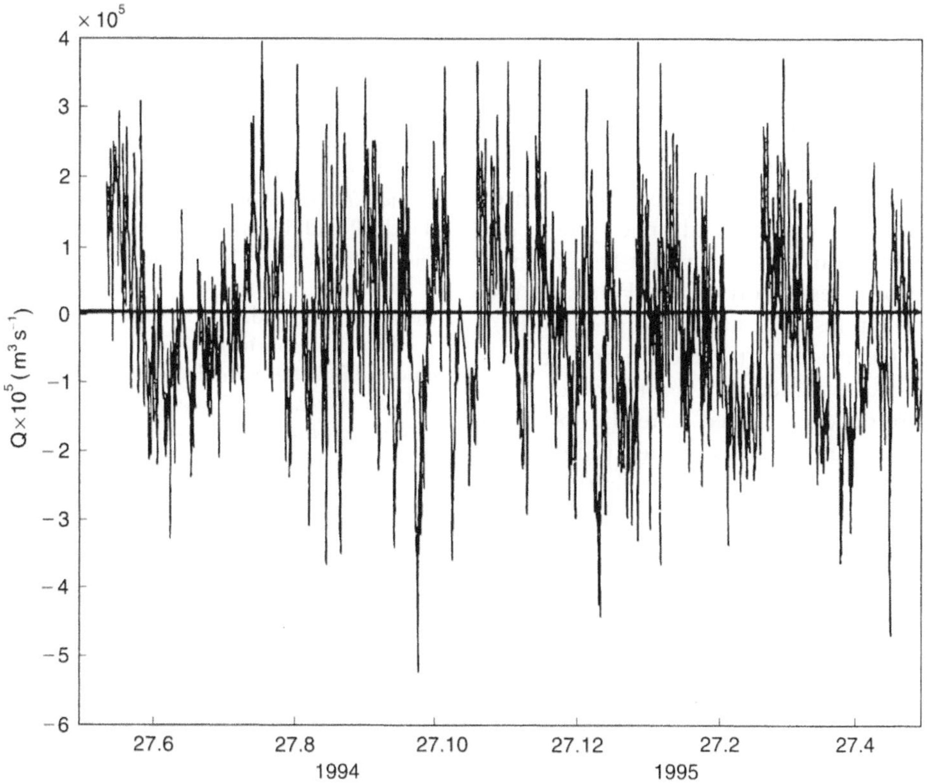

Figure 8 Mass transport over Darss Sill during June 1994–April 1995 based on two stations at the sill. '27.6' on the time axis means 27 June.

3. A major inflow of saline water starts when the Kattegat water, usually reduced in salinity to 22 PSU by vertical mixing, reaches Darss and Drogden Sill. It then penetrates along the bottom into the Arkona Basin. During major inflows the water transports 2×10^9–5×10^9 tonnes of salt into the Baltic Sea. The water takes about 22 days on average ($0.11\,\mathrm{m\,s^{-1}}$) to pass the Arkona Basin. The reason for this slow propagation is doming of the inflowing water. The heavy water forms a large cyclonic eddy (geostrophic flow around the dense water) which prevents the water from flowing rapidly through the basin. Thus the heavy water remains there as a pool over extended periods, its salinity may be further reduced by wind-induced mixing and the water in the upper layers may even flow back into the Danish Straits. After passing Bornholm Channel the same occurs in the Bornholm Basin, where the water takes about 40 days to reach the center. In many cases it does not fill the basin up to 60 m (the sill depth of Stolpe Channel) and its further penetration into the Gotland Basin depends on the wind conditions.

Oscillatory Currents

A pronounced phenomenon in the current variations is the appearance of inertial oscillations, a response of the Ekman currents to the varying wind conditions. They include about 30% of the observed kinetic energy within the wind-mixed layer. Their frequency is determined by the Coriolis parameter, slightly modified by stratification.

Furthermore, seiches (standing waves of the entire Baltic Sea, transformed to amphidromic systems due to the Earth's rotation) are a dominant feature in all tide gauge records, especially at the Finnish coast. These free oscillations were first recognized by F. A. Farel in Lake Geneva, and the local name 'seiches' has generally been adopted as a term for free oscillations of more or less enclosed water basins. They occur when the wind decreases after piling-up of water at one end of the basin. In the Baltic Sea the seiches are heavily damped due to the complicated configuration of the sea, the division into several subbasins and the open connection to the North Sea, which allow water exchange up to $4 \times 10^5\,\mathrm{m^3\,s^{-1}}$ (**Figure** 8). Typically, we find only two or three oscillations.

Observations indicate that the dominant oscillation belongs to the western Baltic–Gulf of Finland system, which represents the second mode of the entire Baltic. The period varies between 26 and 27 h (theoretically 26.4 h). The Gulf of Bothnia is little influenced by this mode, in contrast to the first mode (period 31 h). Both the first and second modes show amphidromic points north of Gotland, where the vertical displacement vanishes. This allows the tide gauge of Landsort (south of Stockholm) to be used as reference station for the mean sea level of the Baltic Sea.

Currents due to seiches are most pronounced in the shallow areas. Tidal currents are negligible in the Baltic Sea east of Darss Sill.

See also

Ekman Transport and Pumping. Rossby Waves.

Further Reading

Bock KH (1971) Monatskarten des Salzgehaltes der Ostsee, dargestelt für verschiedene Tiefenhorizonte. *Deutsche Hydrographische Zeitschrift, Ergänzungsheft Reihe B* 12: 1–147.

Lehman A and Hinrichsen H-H (2000) On the wind driven and thermohaline circulation of the Baltic Sea. *Physics and Chemistry of the Earth B* 25(2): 183–189.

Mälkki P and Tamsalu R (1985) *Physical Features of the Baltic Sea.* Finnish Marine Research No. 252, 110 pp. Helsinki.

Rheinheimer G (ed.) (1996) *Meereskunde der Ostsee* 2nd edn. Berlin: Springer.

Simons TJ (1980) *Circulation Models of Lakes and Inland Seas. Canadian Bulletin of Fisheries and Aquatic Sciences, Bulletin 203.* Ottawa: Department of Fisheries and Oceans.

Wübber Ch and Krauss W (1979) The two-dimensional seiches of the Baltic Sea. *Oceanologica Acta* 2: 435–446.

NORTH SEA CIRCULATION

M. J. Howarth, Proudman Oceanographic Laboratory, Wirral, UK

Introduction

Currents in the North Sea, as in any continental shelf sea, occur in response to forcing by tides, winds, density gradients (arising from freshwater input), and pressure gradients. For most of the North Sea the dominant motion is tidal, with the next most significant motion generated by wind forcing. The response is determined by the sea's topography and bathymetry and is modified by, and modifies, its density distribution. Currents occur over a range of scales – a practical minimum is indicated in each case, ignoring waves and turbulence which are outside the scope of this article. For time the range is from minutes to years, horizontally in space from <1 km to basin-wide (1000 km), and vertically from 1 m to the full water depth. Current amplitudes range generally from $0.01\,\mathrm{m\,s^{-1}}$ to $2\,\mathrm{m\,s^{-1}}$ (in a very few places to in excess of $5\,\mathrm{m\,s^{-1}}$). The North Sea has been extensively studied for more than 100 years and its physical oceanography has been widely reviewed (see Further Reading section).

Topography

The North Sea is a semi-enclosed, wide continental shelf sea (**Figure 1**). It stretches southward for about 1000 km from the Atlantic Ocean in the north, and is about 500 km broad, from west to east. The northern boundary is composed of two connections. Firstly, the main connection with the ocean at the shelf edge is between the Shetland Islands and Norway. Secondly, from the mainland of Scotland to the Orkney and Shetland Islands via the Pentland Firth and the Fair Isle Channel, the connection is with the continental shelf seas to the west and north of Scotland. The North Sea has two other, narrow, connections. One is at its southern end, via the Dover Strait to the relatively salty English Channel (and ultimately again Atlantic) water. The second, on its eastern side, is via the Skagerrak and Kattegat to fresh Baltic water.

Topographically the North Sea can be split into three (**Figure 2**). In the south, bounded by about 54°N and the shallow Dogger Bank, depths are <40 m. Secondly, to the north of this the water depth increases from 40 m to 200 m at the shelf edge. Thirdly, the deep Norwegian Trench is in the north east, penetrating from the Norwegian Sea southward along the coast of Norway. The deepest depths in the trench, about 750 m, occur in the Skagerrak, off southern Norway and western Sweden, whilst the minimum axial depth is about 225 m off western Norway. This has a significant impact on the sea's dynamics.

Meteorology

The North Sea lies at temperate latitudes, between 51°N and 62°N, and so experiences pronounced seasonal changes in meteorological conditions. From September to April, in particular, it is exposed to the effect of a series of storms. These usually track eastward to the north of the British Isles, taking about a day to pass by the North Sea, accompanied by strong winds from the south west, west, and north west. Although these are the dominant wind directions, strong winds can blow from any direction. To some extent the British Isles shelter the North Sea from the wind's full effect since the only wind direction uninterrupted by land is from the north. (This is more significant for waves than currents.) The size of the storms is generally of the same order as that of the North Sea, or larger, and they can recur every few days. On average gale force winds blow on 30 days per year. The extreme 'one-in-50-year' hourly mean wind speeds are estimated to decrease from $39\,\mathrm{m\,s^{-1}}$ in the north to $32\,\mathrm{m\,s^{-1}}$ in the south, slightly stronger in the center of the North Sea compared with near the coast.

River Discharges

The main input of fresh water into the North Sea is from rivers, including the Rhine and Elbe, discharging along its southern coast (Belgium, The Netherlands, Germany). On average this input is $4000\,\mathrm{m^3\,s^{-1}}$, but it is highly variable from day to day and also from year to year. This fresh water has a significant impact on the salinity and density distribution (lower salinities along the continental coast and in the German Bight) and ultimately on the long-term circulation. A further $1500\,\mathrm{m^3\,s^{-1}}$ of fresh water on average is discharged from Scottish and English rivers. More fresh water flows into the Kattegat from the Baltic, on average $15\,000\,\mathrm{m^3\,s^{-1}}$ in the form of a low salinity (8.7 PSU) flow of

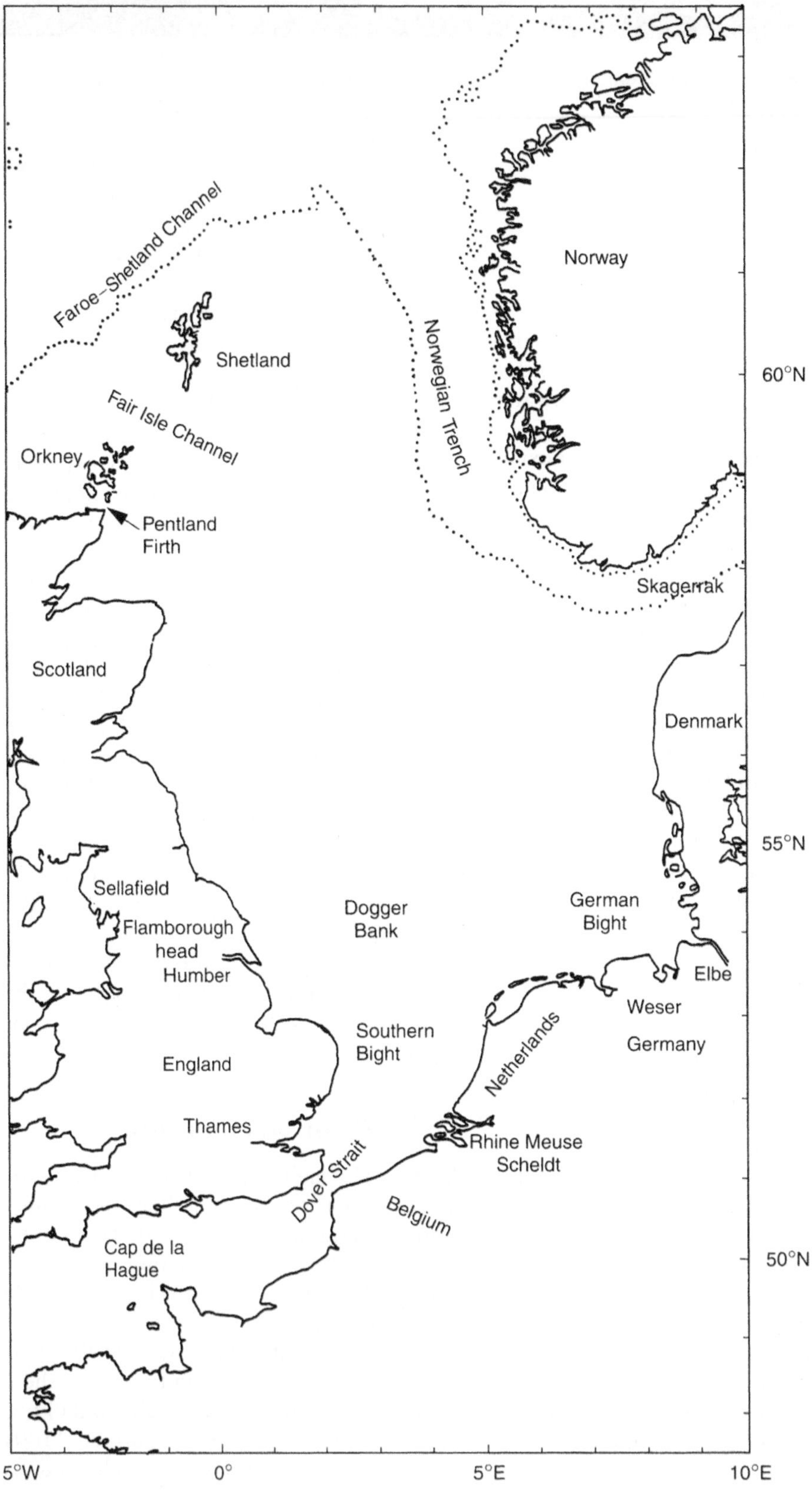

Figure 1 Map of the North Sea.

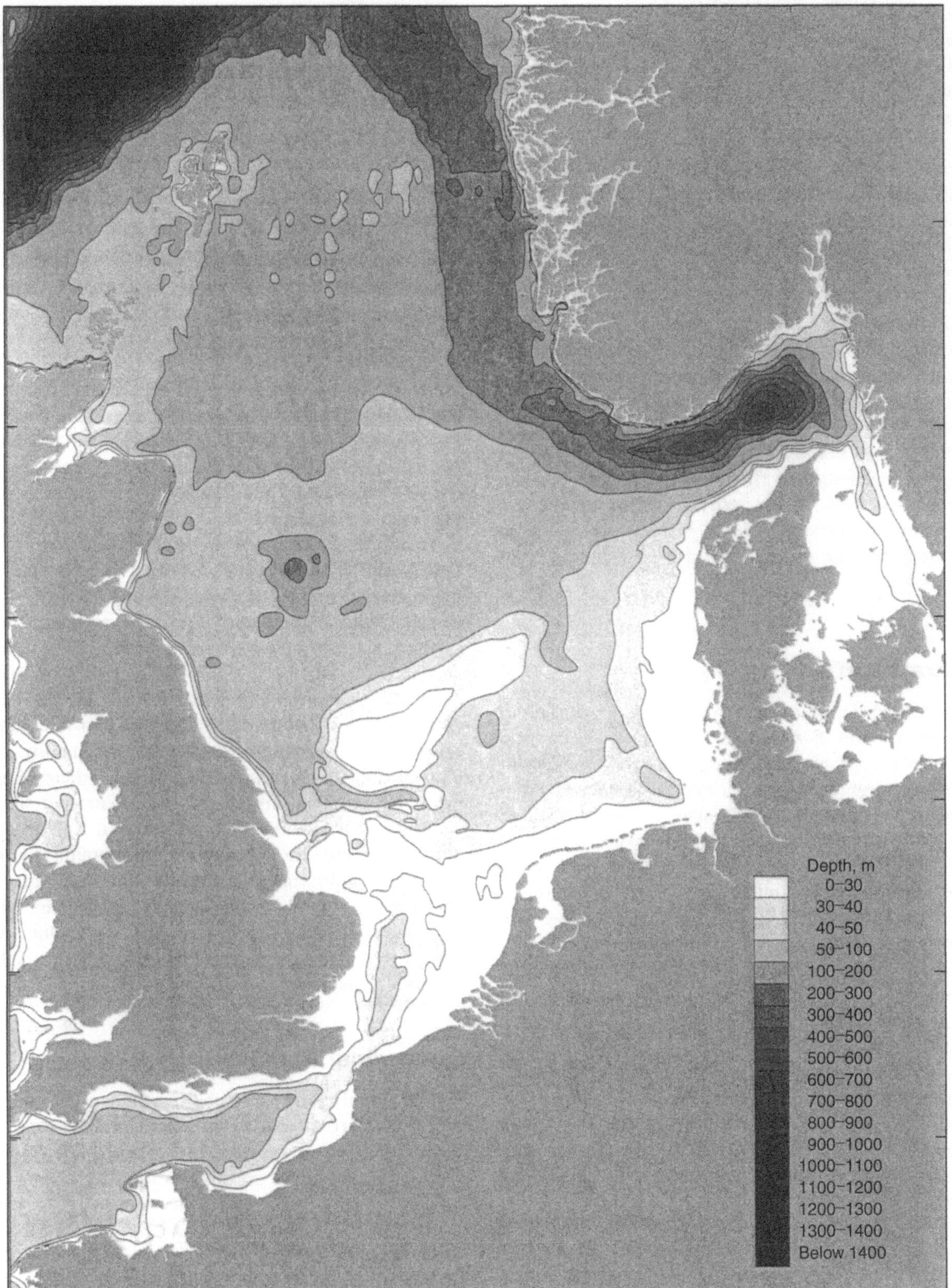

Figure 2 Bathymetry of the North Sea. (Reproduced with permission from North Sea Task Force, 1993.)

30 000 m^3 s^{-1} in a surface layer partially compensated by a flow of 15 000 m^3 s^{-1} of higher salinity (17.4 PSU) North Sea water into the Baltic in a near-bed layer. However, the instantaneous exchange between the Kattegat and the Baltic Sea is by no means a steady two-layer flow, but is predominantly barotropic (in response to wind forcing and the resulting sea level slopes) and can be up to 300 000 m^3 s^{-1} in either direction. The fresher water flows into the Norwegian Coastal Current, a surface current following the Norwegian coast, and exits the North Sea. Generally there is little exchange with water in the rest of the North Sea, although on occasion a surface layer of low salinity water can spread westward across much of the northern North Sea. The boundary between the Norwegian Coastal Current and Atlantic water in the northern North Sea (see **Figure** 3) is not smooth but is subject to meanders with a wavelength of 50–100 km which can break off into gyres and vortex pairs generally propagating northward. Orbital speeds within the gyres are asymmetric; up to 2 m s^{-1} has been reported in the upper layer.

Stratification

The behavior of currents depends on whether the water column is well mixed or stratified. For some processes, for instance the response to wind forcing, stratification leads to the surface layer becoming decoupled from the bed layer – the sharper the transition from surface to bed layers (called the thermocline, for temperature) the greater the degree of decoupling. Well-mixed and stratified regions are separated by sharp fronts. Stratification arises when mixing is insufficient to mix down lighter water at the sea surface – the water might be lighter either because of solar heat input during summer or because of freshwater river discharge. The energy available for mixing is primarily derived from the tides and its effect on the water column depends on the water depth. Hence the southern North Sea, where depths are shallow and tidal currents strong, tends to remain well mixed throughout the year, apart from regions close to the coast affected by river plumes.

Solar heat input causes most of the northern North Sea to stratify between April/May and October/December, with a well-mixed surface layer of about 30–40 m deep. In autumn heat loss at the surface leads to the surface mixed layer deepening and cooling until the bottom is reached.

Stratification caused by river discharge can occur at any time of the year, the fresher water tending to form a thin surface layer in a plume about 30 km wide which stays close to the coast. The nature of the plume is very variable, depending on freshwater discharge, wind, and tide (both during the semi-diurnal and the spring-neap cycles). Mean currents in the plume can be up to 0.2 m s^{-1}.

Measurements

The long-term circulation (**Figure** 3) has been investigated since before 1900 by studying the returns of surface and seabed drifters and the distribution of tracers, initially salinity. More recently dissolved radionuclides have been used as tracers, primarily cesium-137 and technetium-99 discharged into the sea at very low levels in waste from nuclear fuel reprocessing plants, mainly at Sellafield on the west coast of the UK and Cap de la Hague on the north coast of France. These have given unambiguous confirmation of circulation paths.

Since the 1960s variations in currents on short time scales have been measured at fixed points by deploying moorings, often for about a month, with current meters, fitted with rotors or propellers, recording data every few minutes. In the 1990s a new instrument, acoustic Doppler current profiler (ADCP), became available which can measure the current profile throughout the majority of the water column in continental shelf seas. Frequently deployed in a seabed frame the current is determined from the Doppler shift in the back-scattered signal transmitted at 20° or 30° to the vertical. Currents at different depths are determined by chopping the return signal into segments. Another new technique is shore-based hf radar, where again the Doppler shift in a back-scattered signal is used to determine surface currents over a region, typically in 1 km cells, out to a range of about 30 km.

Numerical Models

At any one time measurements can give only an estimate of currents with very limited spatial coverage compared with the extent of the North Sea. However, these can be complemented by numerical models to give fuller spatial coverage. The basic hydrodynamic equations are well known and have been solved on regular finite difference or irregular finite element grids. Currents vary in two horizontal directions and in the vertical, but the models are particularly quick if the equations are depth-averaged, which is especially useful for the estimation of tide and storm surge elevations. Such models are now run operationally in association with

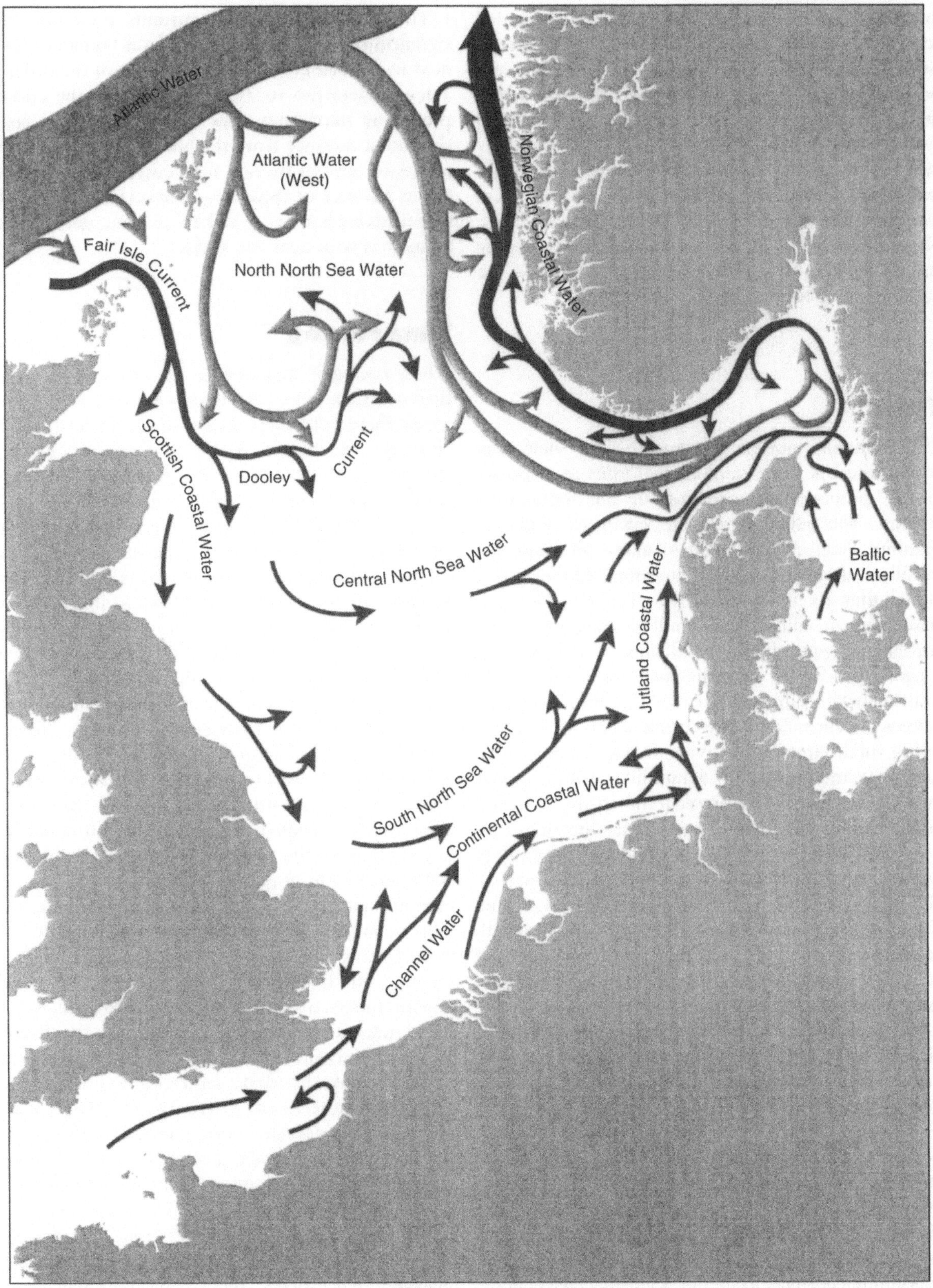

Figure 3 Schematic diagram of the circulation of the North Sea. The thickness of the lines is indicative of the magnitude of volume transport. (Reproduced with permission from North Sea Task Force, 1993.)

meteorological models (the only sufficiently detailed source of meteorological forcing data to drive realistic sea models). Three-dimensional models are being developed which predict the tidal and wind-driven current profile and the effects of horizontal and vertical density gradients. Given the difficulty and expense of making accurate current measurements the future for prediction is via numerical models tested against a few critical measurements. Numerical models are also an integral component of process studies, since the significance of terms in the equations can be isolated and determined.

Tides

The dominant motion for most of the shelf seas around the UK is the semi-diurnal tides in response to the gravitational attraction of the moon and the sun. The tides are regular, repetitive, and predictable. Tidal currents control many physical and nonphysical processes and determine mixing, even though they are predominantly cyclical (to and fro or describing an ellipse) and tend to have small net movement, typically 0.01–0.03 m s^{-1}, except near irregularities in the coastline such as headlands. In addition to the familiar basic M_2 12.4 hour cycle, the tides experience fortnightly spring–neap (on average spring tidal currents are twice as big as neap tidal currents), monthly, and 6 monthly cycles (larger spring tides usually occur near the spring and autumn equinoxes). Away from coasts the tidal currents can be accurately approximated (better than 0.1 m s^{-1}) by just four constituents, or frequencies – M_2, S_2, N_2, and K_2, with periods between 11.97 and 12.66 hours. Near the shelf edge there can be localized amplification of the diurnal constituents O_1 and K_1. Only close to the coasts, in shallow water and near headlands, and in estuaries are distortions to the tides (represented by higher harmonics) significant. Clearly this is a very important region for predicting sea levels for navigation, but represents a small proportion of the area of the North Sea.

The tides enter the North Sea from the Atlantic Ocean north of Scotland and sweep round it in anticlockwise sense as a progressive (Kelvin) wave gradually losing energy by working against bottom friction. By the time Norway is reached tidal currents are relatively weak, < 0.1 m s^{-1} at spring tides. Maximum currents occur in localized coastal constrictions, such as the Pentland Firth (in excess of 5 m s^{-1} at spring tides) and the Dover Strait (2 m s^{-1}). Elsewhere currents at spring tides rarely exceed 1 m s^{-1} (**Figure 4**).

The amplitude of tidal currents does not vary significantly with depth, except in a boundary layer near to the bed generated by friction, of the order of a few meters (up to 10 m) thick. Here the current profile at maximum flow is approximately logarithmic in distance from the bottom. Also due to the effects of bottom friction tidal currents near the bed are in advance of those near the surface, so that the tide turns earlier, by up to half an hour, near the bed compared with near the surface.

Wind Forcing

Wind forcing is responsible for the second largest currents after tides over most of the region, and in areas where tidal currents are weak (for instance off Norway), the largest. By the very nature of storms the forcing is intermittent and seasonal. The largest storms tend to occur in winter, when the water column is homogeneous. Two effects are important. Firstly, sea level gradients are set up against a coast counterbalancing the wind stress, since the North Sea is semi-enclosed and so has no long approximately straight coasts, in contrast with many other continental shelf seas. Secondly there is the direct action of the wind stress at the sea surface.

The sea level gradients lead to storm surges at the coast, with the threat of coastal flooding. Sizeable currents can be generated during the setting up and relaxing of the sea level gradients. Measurement of these currents at depth is scarce not only because the environment is harsh during storms, although less severe than at the surface, but also because long deployments are required – at least over a 6 month period including winter to ensure an adequate sampling of storms. Seabed-mounted ADCPs are helpful, both because the storm environment is less likely to impinge on data quality and also because longer measurements are possible. The only feasible way of predicting extreme currents, for instance the 'one-in-50-year' current for the design of offshore structures, is via numerical models, either run for long periods or forced by a series of extreme events. The accuracy of extreme sea level estimates is better than that for extreme currents, because models can be validated against long records (many years) of coastal elevations from tide gauges.

The direct current response at the sea surface to the action of the wind stress is very difficult to measure, since waves tend to be large when winds are strong, but it appears to be limited to at most the top 25 m of the water column. (A very useful remote sensing technique here is hf radar which measures surface currents.) The response is rapid, less than a

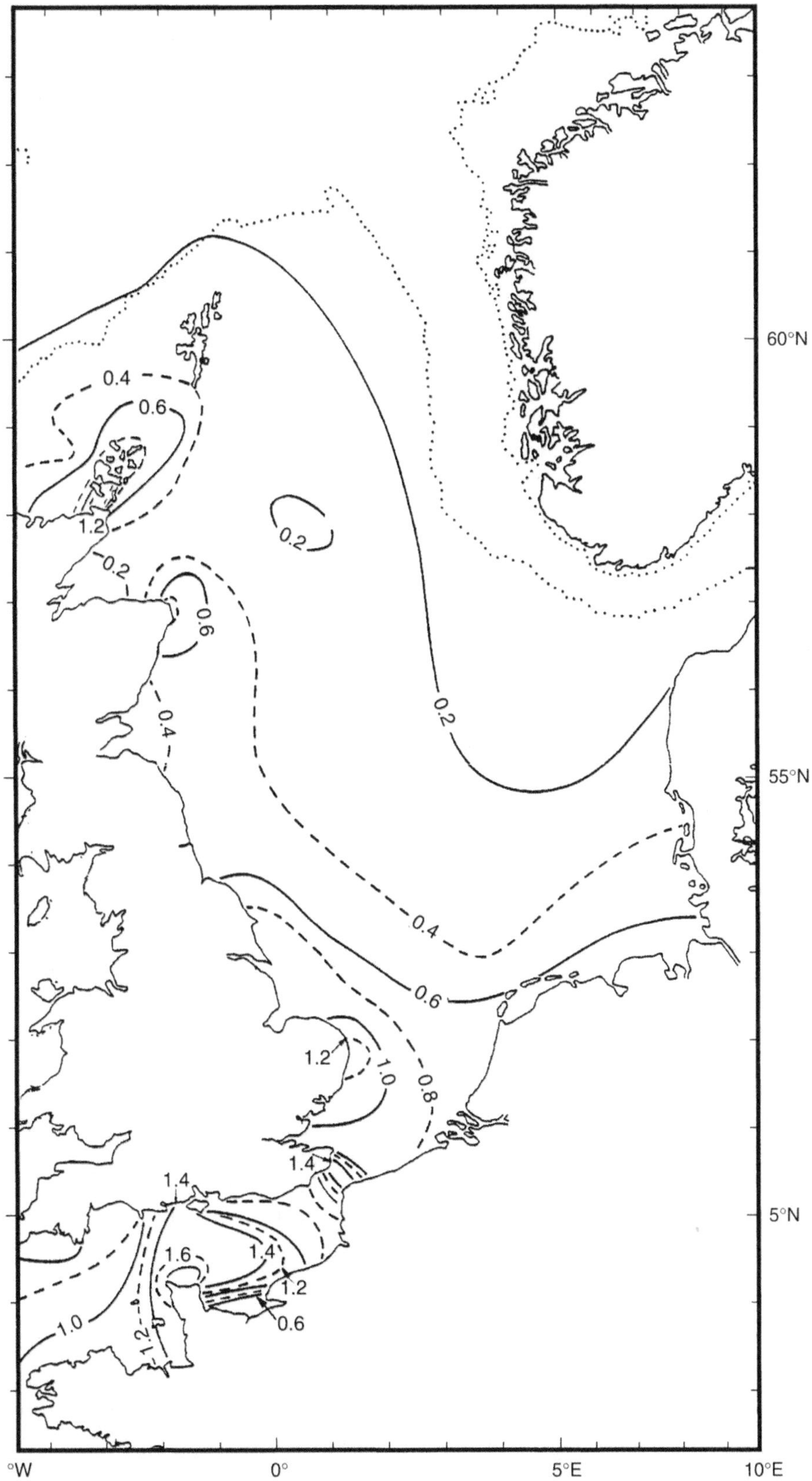

Figure 4 Estimate of the maximum depth-averaged currents for an average spring tide, in m s^{-1}.

few hours, and dependent on the roughness of the sea surface (i.e. on the waves and whitecapping). The surface current amplitude used, for instance in oil spill studies, is usually taken to be in the range 1–6% of the wind speed, with a typical value of 3%. The higher ratio appears to be more appropriate for developing seas. The direction of the wind drift is a few degrees to the right of the wind. Since there are very few accurate measurements to test models critically, the accuracy of models in this region is uncertain.

Circulation

Circulation is the long-term movement of water. Within the North Sea it is forced by the net tides (for instance, responsible for 50% of water transport in the western North Sea), the mean winds, and density gradients, principally caused by freshwater input from rivers and the Baltic. All these forces tend to act in the same sense, described below. There are seasonal variations, as storms mainly occur in winter, river discharge has an annual cycle, and in summer the water column stratifies in some regions. Over short periods the net movement of water is likely to be determined by the last storm or storms. The circulation pattern outlined below is enhanced by winds from the south west and can be reversed by strong winds from south and east. Although of considerable and long-standing interest to oceanographers – to physicists, fisheries biologists, and to those studying the movement of suspended sediments and pollution – it is difficult to measure and quantify since in most places its magnitude is $< 0.1\,\mathrm{m\,s^{-1}}$, much smaller than tidal or extreme wind-driven currents. It frequently has very short space scales, changing significantly in magnitude over a few kilometers and changing direction by 180° from the surface to the seabed.

There is, however, an overall long-term mean pattern (**Figure 3**), shown up clearly in the distributions of tracers whose major features in terms of direction have been known since the early 1900s. There are large uncertainties on estimates of amplitudes. The flow is broadly anticlockwise round the coasts of the North Sea, with weak and varied circulation in its center. The mean coastal flow is southward past Scotland and England into the Southern Bight, where there are inputs of salty water through the Dover Straits and of fresh water down the main rivers, all of which pass through large industrialized regions (Humber, Thames, Meuse, Scheldt, Rhine, Weser, Elbe) and on into the German Bight, flowing northward past Denmark in the Jutland current to join the Norwegian Coastal Current in the Skagerrak. The average input through the Dover Strait is about $0.1 \times 10^6\,\mathrm{m^3\,s^{-1}}$, but the flow there is strongly influenced by the wind and can reverse, moving south-westward out of the North Sea. There is also some flow from the coast of East Anglia to the north Dutch coast.

There are major inflows (in excess of $10^6\,\mathrm{m^3\,s^{-1}}$) of water of Atlantic origin across the northern boundary, but very little penetrates far into the North Sea. The larger portion flows in on the western side of the Norwegian Trench and recirculates in the Skagerrak, flowing out along the eastern side of the Norwegian Trench underneath the Norwegian Coastal Current. A smaller inflow of mixed Atlantic and shelf water flows in on either side of the Shetland Islands and turns eastward to follow the 100 m contour across the North Sea, eventually to flow out along the eastern side of the Norwegian trench. The majority of water passing through the North Sea goes through the Skagerrak. The only major outflow is along the eastern side of the Norwegian Trench, with magnitude of about $(1.3\text{–}1.8) \times 10^6\,\mathrm{m^3\,s^{-1}}$.

Effects of Stratification

Stratification profoundly affects the nature of currents in two ways. Firstly currents are driven by the density gradients associated with fronts – the sharp boundaries between stratified and well-mixed water. The currents tend to be along the fronts with transverse exchanges inhibited. Secondly, the structure of motion in the water column is affected by the region of vertical density change – either because the surface and bed layers are decoupled or because internal waves can propagate along the density interface. Theoretically there is no difference between stratification caused by heat input and that caused by fresh water; dynamically all that matters is the density difference. In practice the consequences of thermal stratification in summer can be easier to study because the extent of the surface mixed layer and of the water depth are more manageable and because regions of thermal stratification are predictable and dependable.

Fronts

The main front separating the thermally stratified water in summer to the north from the well-mixed water to the south starts from Flamborough Head, bifurcates around the Dogger Bank and passes to the north of the Frisian Islands. Tidal current speeds and water depths determine the front's position. Currents along the front in a west to east sense are typically about $0.05\,\mathrm{m\,s^{-1}}$ but can be up to $0.15\,\mathrm{m\,s^{-1}}$.

Inertial Currents

Inertial currents are generated by pulses of wind. The currents rotate clockwise in the Northern Hemisphere with a period in hours of 12/sine (latitude), which, in the North sea, is from 13.6 h at 62°N to 14.6 h at 55°N. In stratified water their vertical structure is primarily the first baroclinic mode; the currents are 180° out of phase above and below the thermocline and integrate to zero over the water column. Because of the phase reversal large shears can be generated across the thermocline – the largest in the water column. Their decay time, if not reinvigorated by another wind pulse, is of the order of days. ADCPs covering the majority of the water column are ideal instruments for measuring inertial currents, which can show all the characteristics indicated above. In the North Sea inertial current speeds up to 0.25 m s^{-1} have been observed in the surface well-mixed layer, which for most of the period of stratification is the shallower and hence the layer where inertial currents are stronger. During stratification some energy at or near inertial periods is present most of the time, but not always with a simple first baroclinic mode structure. The situation is different from the open ocean – in shelf seas the water depth is limited and tidal friction/mixing is ever present.

Internal waves

Internal waves are ubiquitous in stratified water. They propagate along the thermocline density interface as a cyclical vertical movement of the density interface, which can be quite large, over 10 m, with only a 0 (1 cm) movement of the sea surface. The restoring force is primarily the buoyancy force. Internal interfacial waves are analogous to surface waves but their phase speed is less. They can have any period between the local inertial (see above) and the Brunt-Väissälä or buoyancy period, which depends on the degree of stratification and can be as short as a few minutes for sharp stratification. The corresponding currents are out of phase above and below the density interface and, for small amplitude waves, sum to zero in the vertical. It is only possible to give generalities since each situation is unique. Because vertical accelerations are significant, non-hydrostatic numerical models are required to reproduce the dynamics. These are rare for shelf-wide models.

Internal tides are often generated as the tides cross sharp changes in bathymetry, such as the shelf break. The internal tide then propagates both into the shelf sea and into the ocean, usually in the form of a pulse of waves with periods of order 30 minutes, repeated with a tidal periodicity.

Shelf Edge

The main distinguishing feature of the shelf edge is the pronounced change in water depth over a short distance, from shelf depths (< 200 m) to oceanic (usually several thousand meters). At the shelf break between the North Sea and the Faroe–Shetland Channel water depths deepen from 200 m to 1000–1500 m over a distance of tens of kilometers. Compared with other shelf edges, even around the UK, this is relatively gentle and the gradients are relatively smooth. A mixture of dynamical processes can exist at the shelf edge involving a combination of stratification and steep bottom slope. Three relevant to the North Sea are internal tides (see above), the slope current flowing north-westward, and coastal trapped waves with super-inertial periods which can, for instance, lead to enhanced diurnal tides (see above).

Conclusions

The North Sea is a typical semi-enclosed continental shelf sea, with a wide range of tidal conditions, winds, circulation, and effects of stratification. To interpret and understand the dynamics it is not sufficient just to study the currents but also seabed pressures/coastal elevations, the (horizontal and vertical) density field which is not static, meteorological forcing (wind, atmospheric pressure, heat input at sea surface), and river discharges.

See also

Regional and Shelf Models.

Further Reading

Charnock H, Dyer KR, and Huthnance JM (eds.) (1994) *Understanding the North Sea System*. London: Chapman and Hall.

Eisma D (1987) The North Sea: an overview. *Philosophical Transactions of the Royal Society B* 316: 461–485.

Lee AJ (1980) North Seap: hysical oceanography. *Elsevier Oceanography Series* 24B: 467–493.

North Sea Task Force (1993) *North Sea Quality Status Report 1993. Oslo and Paris Commissions, London.* Fredensborg, Denmark: Olsen Olsen.

Otto L, Zimmerman JTF, and Furnes GK (1990) Review of the physical oceanography of the North Sea. *Netherlands Journal of Sea Research* 26: 161–238.

Rodhe J (1998) The Baltic and North Seasa: process-oriented review of the physical oceanography. In: Robinson AR and Brink KH (eds.) *The Sea*, 11, pp. 699–732. Chichester: John Wiley Sons.

INTRA-AMERICAS SEAS

G. A. Maul, Florida Institute of Technology,
Melbourne, FL, USA

Introduction

The Intra-Americas Sea (IAS) is a semi-enclosed salt-water body of the tropical and subtropical western Atlantic Ocean that comprises the Caribbean Sea, the Gulf of Mexico, the Straits of Florida, the Bahamas, the Guianas, and the adjacent waters. Biogeographically, the IAS includes the estuarine, coastal, shelf, and pelagic waters from the mouth of the Amazon River at the equator off Brazil, to Bermuda and westward to the shores of North, Central, and South America. Geographically, the boundaries may be set approximately as $\phi = 0°$ to 32°N latitude, and $\lambda = 50$–98°W longitude. **Figure 1** summarizes the geographical setting.

Early oceanographic explorations of the region were by European scientists who chose to name the IAS (the Caribbean Sea in particular) the 'American Mediterranean'. While superficially this terminology describes the IAS as a similar semi-enclosed sea where evaporation (E) exceeds precipitation (P) plus river runoff (R), $E > P + R$, the Mediterranean Sea is markedly different in character from its western Atlantic counterpart. Also, the IAS was broken into smaller components, and little attention was paid to the Caribbean Sea and by Gulf of Mexico oceanographers and *vice versa*. Conversely, the Straits of Florida and the water currents of the Gulf Stream system are perhaps the most widely studied oceanographic features on Earth.

With the coming of significant international cooperation between scientists from throughout all the Americas, the IAS began to be appreciated as a

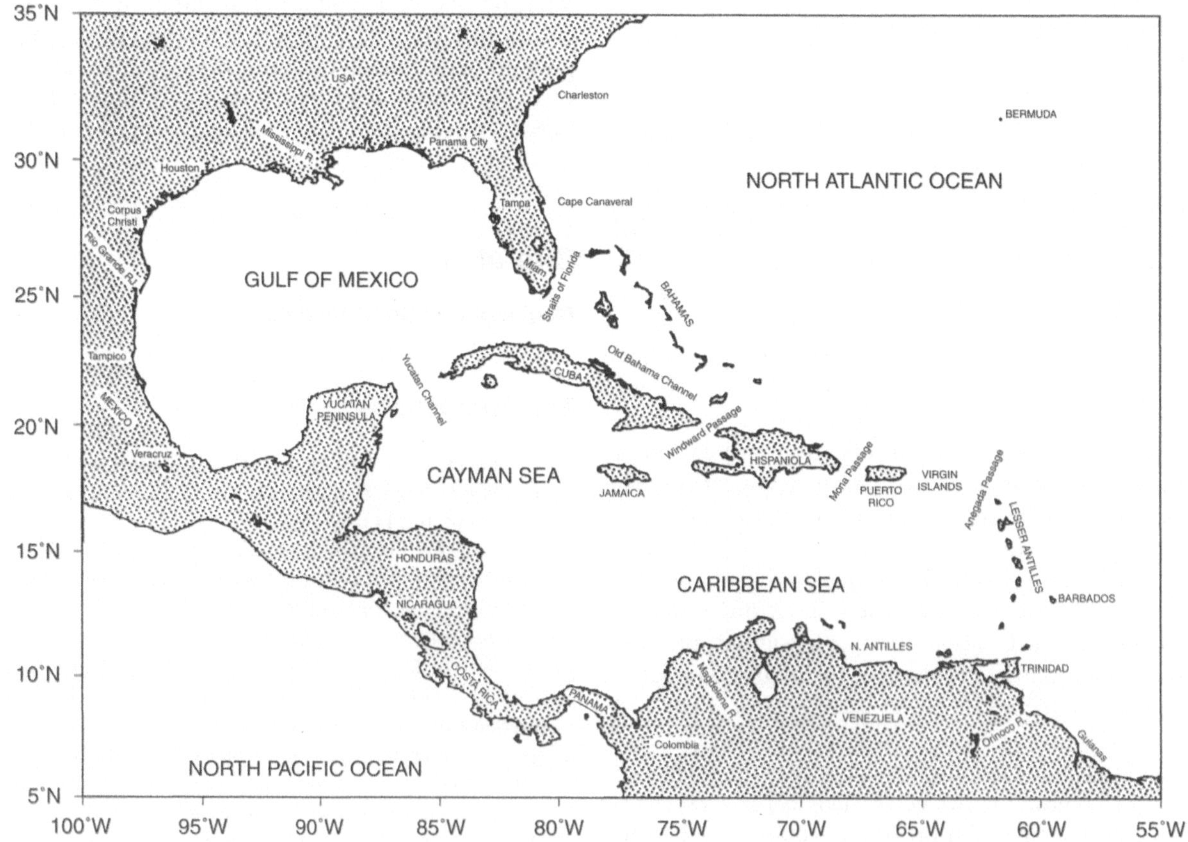

Figure 1 The Intra-Americas Sea, a semi-enclosed water body of the subtropical and tropical western North Atlantic Ocean. Place names in accord with the US Board on Geographic Names.

unified body of water distinctly different from the early and mid-twentieth century European perspective. An inclusive term was needed to integrate not only the oceanography of the region, but its meteorology and maritime socioeconomic connectivity as well. Thus the term Intra-Americas Sea was developed from the multilingual and multicultural heritage shared by its people.

Regional Overview

Pre-Columbian indigenous peoples of the Intra-Americas Sea certainly knew of the oceanic currents and atmospheric winds. Caribes from the south, Tainos in the middle Antilles, Arawaks from the north, Mayas and Olmecs to the west, all moved freely from island to island to continent, presumably with some knowledge of the currents and winds we have named Caribbean Current, Trade Winds, Gulf Stream, Hurricane, and Guianas Current. European explorers and conquistadors relearned this information, not from the IAS's inhabitants, but from hardship after experiencing what was so well known already. James A. Mitchner's 1989 novel *Caribbean* imagines so well what science could have learned directly.

As regards the geological setting, the IAS encompasses three tectonic plates: the North American Plate, the Caribbean Plate, and the South American Plate (the Cocos and Nazca Plates mark Pacific tectonic boundaries but are not significantly involved in the air–sea regime discussed herein). About 3 Ma the Caribbean Plate drifting from west to east closed the gap between North and South America, creating Panama and deflecting oceanic flow northward. Central America and the eastern Caribbean margin are volcanically active today. Associated with tectonics are earthquakes and seismic sea waves (tsunami) that are part of the circulation regime, as well as the complex bottom topography that channels water movement and perturbs the atmospheric circulation.

Geological forces then not only form the coasts and islands of the IAS; they are a central element in appreciating the flow of air and water. Geophysical fluids obey Newton's laws of motion, specifically $\sum F = \sum m \cdot a$ (where F is force, m is mass, and a is acceleration), the laws of thermodynamics, and continuity of mass. The air and water of the IAS are accordingly connected to all of Earth's ocean–atmosphere continuum, and have special complexities as they flow around and through the passages, channels, and straits within the region. Thus detailed knowledge of the water depths and land heights, and their attendant frictional characteristics, is essential to appreciating the flow patterns discussed below.

Sill depths control much of the oceanic flow patterns in the IAS. It has been reported that the deepest sill is in the Yucatan Channel (2040 m) between Mexico and Cuba, that the average sill-depth of the Antillean Arc is 1200 m, but that the Jungfern-Anegada Passage is 1815 m deep (between the Virgin Islands and the Lesser Antilles), and that the Windward Passage separating Cuba and Hispaniola is 1690 m deep. Within the IAS are many much deeper basins than these controlling sills, leading to the notion that the waters deeper than the sill depths are moved by convection rather than advection. This infers that the IAS has a two-flow regime: an upper layer mostly influenced by wind-driven advection, and a deep layer controlled by overflows. The controlling sill depth of the Straits of Florida is about 800 m, which means that the outflow of the IAS is topographically accelerated. **Figure 2** summarizes the IAS water depth information.

Lastly, to appreciate the ocean currents of the Intra-Americas Sea, the structure of atmospheric forcing needs to be mentioned. The southern portion of the IAS is under the influence of the Inter-Tropical Convergence Zone, the ITCZ. In the Northern Hemisphere summer, the ITCZ migrates northward and the easterly Trade Winds at the southern boundary of the Bermuda meteorological high-pressure zone dominate the IAS to its northern extent. As boreal winter approaches, the ITCZ migrates southward and the midlatitude frontal passages sweep across the IAS to south of Cuba and sometimes almost to South America. During early autumn, the atmosphere is characterized by a series of tropical cyclones, that from time to time reach intensities known as the West Indian Hurricane. These severe atmospheric disturbances are cyclonic circulation features that bring not only strong winds, storm surge, and the associated damage, but also much needed precipitation and flushing of shallow bays and estuaries.

Climatologically, the Köppen classification system would place the northern IAS in a *Cfa* (humid subtropical) category; coastal Central America, the Greater Antilles and the Bahamas as *Aw* (tropical wet and dry) with sections as *Af* (tropical rain forest) including the Lesser Antilles. The southern and southeastern coastal IAS is classified as *BSh* and *BW* (semi-arid or steppe, and arid desert, respectively), with *Am* (tropical monsoon) along the coasts of the Guianas and Brazil. These classifications are perturbed by interannual and decadal climate oscillations, in particular by El Niño–Southern Oscillation (ENSO) events, which tend to cause cold wet winters in the northern IAS (particularly Florida and Cuba) and warm dry autumn conditions in Panama, coastal Colombia and Venezuela, the Guianas, and northern Brazil.

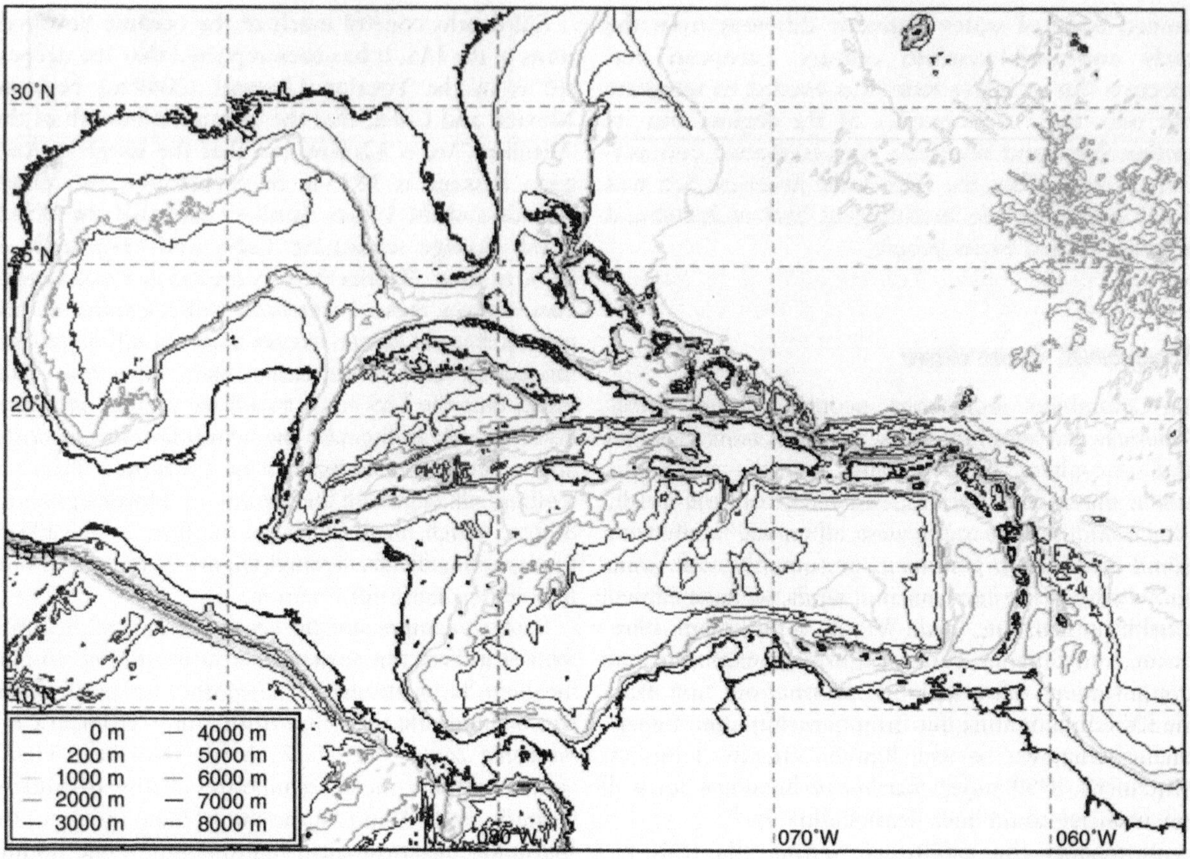

Figure 2 Bottom topography of the Intra-Americas Sea; water depths in meters.

Surface Throughflow Regime

Perhaps the best way to envision IAS surface currents is to take a Lagrangian perspective, and imagine floating on a northbound satellite-tracked buoy passing the equator off Brazil as part of the 'oceanic conveyor belt'. As the water parcel travels northward, perhaps entraining some Amazon River water, the dominant physics is conservation of potential vorticity, $\mathrm{d}/\mathrm{d}t[\varsigma + f/H] = 0$, where relative vorticity $\varsigma = \partial v/\partial x - \partial u/\partial y$, the Coriolis parameter $f = 2\Omega \sin \phi$ with latitude ϕ, and water depth H. For a given H, as the parcel flows northward, f increases and ζ must decrease, forcing an anticyclonic (clockwise) turning. The region where this occurs is called the North Brazil Current 'retroflection' and can be seen in satellite images as a distinct offshore turning of the current. Some of this water continues toward the east in the North Atlantic Equatorial Counter-Current, but some advects up the South American coast in the Guianas Current to the Lesser Antilles Arc, sometimes as an anticyclonic eddy. Much of this is evident in **Figure 3**.

The northward-flowing water parcel usually passes Barbados and often flows through the passages of the Lesser Antilles, carrying anticyclonic vorticity and perhaps Amazon riverine particles and biota into the Caribbean Sea. Under the influence of the Northeast Trade Winds, the Caribbean Current moves westward at a leisurely pace of perhaps $0.2\,\mathrm{m\,s^{-1}}$, but with notable meandering and eddying along the path. Sometimes under this same wind regime, water from the Orinoco River is seen to be carried completely across the eastern Caribbean Sea to Puerto Rico and Hispaniola, particularly in late summer.

Similarly, in the Panama-Colombia Bight, the wind-stress (τ) curl, $\partial\tau_y/\partial x - \partial\tau_x/\partial y$, causes an $r \approx 150\,\mathrm{km}$ radius eddy to spin-up and spin-down annually, the so-called Panama-Colombia Gyre (PCG). In the vicinity of the PCG a major South American river, Colombia's Magdalena, flows into the ocean where its waters and its flotsam mix with the sea, and are carried to distant shores by ocean currents. The PCG is but one feature of the IAS surface current variability now being simulated in numerical models, and being observed by systems such as satellite altimeters and radiometers.

As the meandering, eddying, Caribbean Current approaches the Central America coast, it is forced anticyclonically northward into the Yucatan

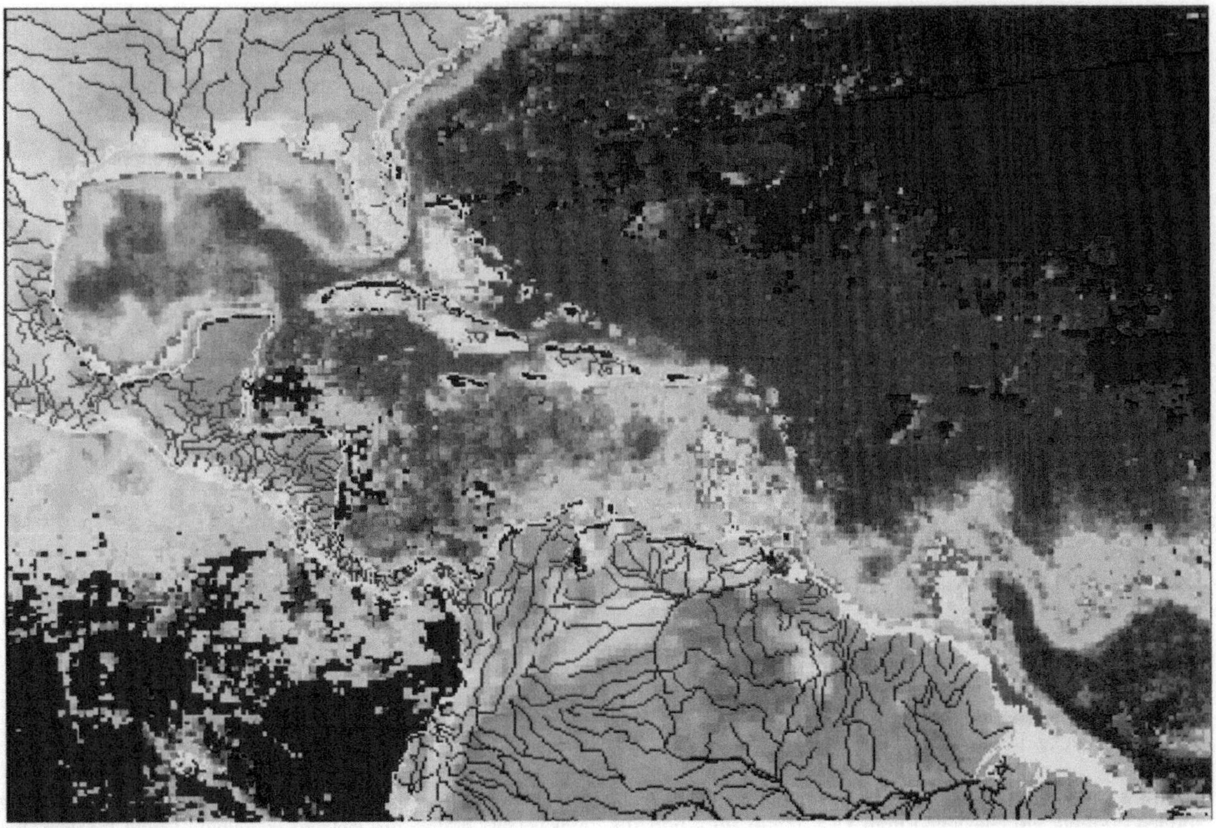

Figure 3 Ocean color composite of the Intra-Americas Sea showing concentration of chlorophyll + phaeophytins and suspended sediments. Image from observations of the Coastal Zone Color Scanner (NASA) compiled by Frank Muller-Karger for October 1979.

Channel. In the area off Belize and Yucatan Mexico, this IAS current takes on the characteristics of the Gulf Stream: a deep ($z \approx 1200$ m) western boundary current with swift surface flows of more than $\vec{v} = 1$ m s^{-1} and a distinct cyclonic horizontal velocity shear boundary, $\partial\vec{v}/\partial x$, along the western edge. Here the stream is known as the Yucatan Current, and it is a northward flow connecting the Caribbean Sea with the eastern Gulf of Mexico.

Gallegos (1996) has shown that the Yucatan Current is highly geostrophic, the balance of forces (per unit mass) being $f\vec{v} = 1/\rho \; \partial p/\partial\vec{n}$, where ρ is the density of sea water, and $\vec{v}$ is the current speed at right angles to the horizontal pressure gradient, $\partial p/\partial\vec{n}$. Gallegos also studied the temperature–salinity (T–S) structure of the Yucatan Current and has concluded that it has T–S properties similar to those in the offing of Cape Hatteras. As this branch of the Gulf Stream system flows northward, it forces a vigorous upwelling regime along the eastern Campeche Bank that supports one of the IAS's greatest fisheries.

Once in the Gulf of Mexico, the Yucatan Current is known as the Gulf Loop Current because of its characteristic anticyclonic looping from northward to eastward to southward to eastward again as it exits the Gulf of Mexico through the Straits of Florida. This clockwise turning of the Gulf Loop Current is part of a cycle of growth (penetrating into the Gulf of Mexico almost to the latitude of the Mississippi River Delta, $\phi \approx 30°$N), then turning eastward and southward to run along the west Florida escarpment. Near the latitude of Key West ($\phi \approx 24°$N), the current turns sharply cyclonically and begins to run eastward in the Straits of Florida, where it is now called the Florida Current. Once the Gulf Loop Current has reached its maximum latitudinal extent, a large anticyclonic current ring, 100–150 km radius, separates from the flow, and the main current reforms farther south near the latitude of the Florida Keys. In its southernmost position, the Gulf Loop Current flows into the Gulf of Mexico, and turns rather sharply in an anticyclonic turn to exit almost directly into the Straits of Florida (**Figure 4**).

Anticyclonic Gulf Loop Current rings have all the T–S and flow features of the Gulf Stream system, just as in the Yucatan Channel and in the Straits of Florida. The ageostrophic dynamic balance is $\pm\vec{v}^2/r + f\vec{v} - 1/\rho \; \partial p/\partial\vec{n} = 0$, where r is the radius of curvature, and where $\pm\vec{v}^2/r$ is positive in cyclonic

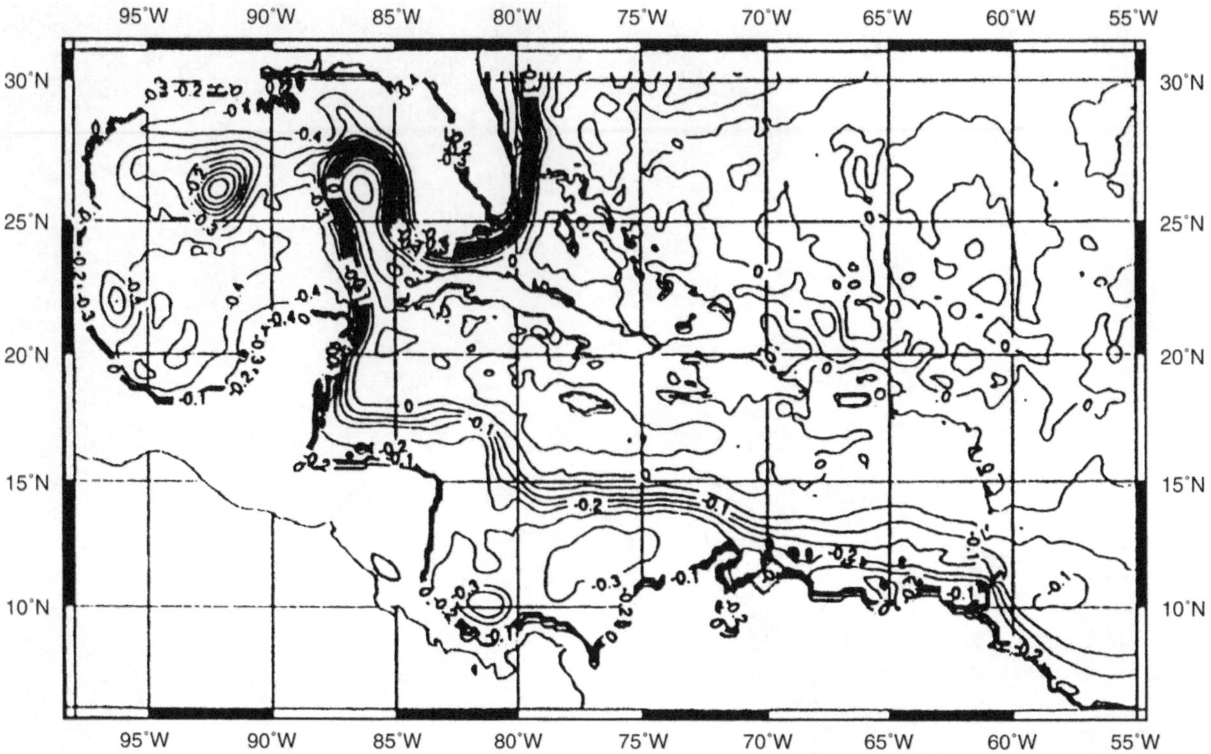

Figure 4 Sea surface height (h) in meters from numerical model calculations as reported in Mooers and Maul (1998). Geostrophic surface currents are calculated from $f\vec{v} = 1/\rho \partial p/\partial \vec{n} = g\partial h/\partial \vec{n}$. Anticyclonic eddies are isolated concentric height maxima, the largest of which is shown in the western Gulf of Mexico; cyclonic eddies have the opposite surface height field.

curvature and negative in anticyclonic. Accordingly, flow is super-geostrophic in anticyclonic turns, and sub-geostrophic in cyclonic turns. Identical dynamics describe midlatitude upper tropospheric flows in Earth's atmosphere, notably in the Jet Stream.

A separated Gulf Loop Current ring travels westward into the western Gulf of Mexico, most probably by a self-propulsion mechanism associated with the beta effect, $\beta = \partial f/\partial y$, of differing Coriolis parameter between the southern and northern ring edges. Using the hydrostatic equation, $\partial p = \rho g \partial z$, the horizontal pressure gradient term $1/\rho\ \partial p/\partial \vec{n}$ may be written as $g\partial h/\partial \vec{n}$, and for an anticyclonic Gulf Loop Current ring with a diameter of 300 km and $\partial h = 0.75$ m, it is calculated using $-\vec{v}^2/r + f\vec{v} - g\partial h/\partial \vec{n} = 0$ that eddies self-propagate at speeds of 5–10 cm s^{-1} (5–10 km d^{-1})($\approx 5 - 10$ km d^{-1}). Direct observations from satellite-tracked buoys and from satellite altimeter measurements of sea surface height (h) substantially agree with such calculations.

As these Gulf Loop Current rings travel to the west, they begin to spin down, losing their momentum per unit volume, $\rho\vec{v}$, to horizontal friction and eventually mixing in with the ambient Gulf of Mexico Common Water. The lifetime of the current rings is typically 6 months, and they carry with them the temperature, salinity, and other characteristics of their source region, the Caribbean Sea. Approximately 3×10^{13} kg y^{-1} of salt is injected into the western Gulf of Mexico by an average Gulf Loop Current ring. Thus the salt balance of the western Gulf of Mexico is decidedly nonMediterranean, $E + \delta = P + R$, because in the IAS the classical evaporation/precipitation/runoff equation requires an additional term δ to account for the infusion of high salinity water from the rings. Typical values for these terms are $E - P \approx 35$ cm y^{-1}, $R \approx 75$ cm y^{-1}, and the volume of fresh water to maintain the salt balance $\delta \approx 40$ cm y^{-1}.

The Gulf Loop Current interacts with the fourth great riverine system of the IAS, the Mississippi River. As the Gulf Loop Current nears the Mississippi Delta, it is observed to entrain or advect the river water to the east. Mississippi River water has been observed by its low surface salinity all along the eastern edge of the Current, into the Straits of Florida, and up the east coast of the USA at least to Georgia ($\phi \approx 32°$N), the northern boundary of the IAS. Thus the four great rivers of the IAS, the Amazon, the Orinoco, the Magdalena, and the Mississippi, and the many smaller tributaries, are all known to interact with the oceanic flows and to be carried great distances by them. This is an important transport mechanism in the IAS whereby riverine flotsam and jetsam is found on distant shores. This same flow-through regime is also responsible for the considerable impact of tars from maritime commerce and from oil drilling on the highly valuable tourist beaches of the area (*cf*. **Figure 3**).

Gulf Loop Current ring shedding seems to have a cycle of 10–11 months on average, with some rings being shed in as few as 6 months and others taking 17 months. This is a surprising frequency since it is not at the annual harmonic where many other oceanic features have a spectral peak. Gulf Loop Current ring shedding has been simulated by IAS numerical models, and super-annual periods are often calculated. The cycle of ring formation does not seem to be forced by the unmistakable annual cycle of volume transport, $\int\int v(x, z)\mathrm{d}x\,\mathrm{d}z$ in the Gulf Stream system, with its maximum in June and minimum in October, nor by variability in the Caribbean Current along 15°N, which has a spectral peak at about 75 days. Connectivity between flow variability in the Caribbean Sea and the Gulf of Mexico seems remarkably weak.

In the Straits of Florida, the Florida Current turns cyclonically as it passes between Cuba, Florida, and the Bahamas. The lesser passages of the Straits of Florida contribute small amounts to the total volume transport, which by now is at the level of 30 G s^{-1} ($30 \times 10^6\ \mathrm{m}^3\ \mathrm{s}^{-1}$) or about 30 Sv. At the latitude of Palm Beach, Florida ($\phi \approx 27°$N), the meridional oceanic heat transport, $\int\int \rho CpTv(x, z)\mathrm{d}x\,\mathrm{d}z$, is about 1.5 petawatts (1.5×10^{15}W), an amount comparable with the atmosphere at the same latitude. Transport is known to vary on timescales ranging from days to years. Fortnightly volume transport changes are observed to range from 20 Sv to 40 Sv, a value much larger than that of the annual cycle, which is more likely less than ± 5 Sv.

In the narrow confines of the northern Straits of Florida, lateral friction is significantly larger than in the open sea. Here, from extensive *in situ* studies, the Guldberg-Mohn friction coefficient J in quasi-geostrophic flow $fv = g\partial h/\partial x + Jv$ has been estimated at approximately $4 \times 10^{-5}\ \mathrm{s}^{-1}$, one to two orders of magnitude larger than away from continental boundaries. In addition, the Florida Current axis meanders to the west when transport is high and to the east when it is less. Such detailed transport variations in other regions of the IAS are not as well documented as in the Straits of Florida between Palm Beach and the Bahamas.

North of the Straits of Florida the current is called the Gulf Stream, a name it retains to the offing of Nova Scotia. The flow northward to Cape Hatteras, North Carolina ($\phi \approx 35°$N), seems to be topographically controlled, whereby H in the potential vorticity equation (above) directs onshore and offshore meandering. Near Charleston, South Carolina ($\phi \approx 32°$N), a notably shallow area decreases H significantly, and the current responds by turning anticyclonically ($\varsigma < 0$) into deeper water. Once offshore, the larger value of H causes the flow to turn cyclonically again ($\varsigma > 0$) in a series of meanders as it progresses downstream (*cf*. **Figure 3**).

Along the outer boundary of the IAS, a surface flow with a great deal of changeability is observed, called the Antilles Current. This intermittent current, carrying on average about 15 Sv, progresses northward up the margin of the Caribbean Sea, along the eastern outer banks of the Bahamas, and eventually joins the Gulf Stream north of Cape Canaveral, Florida ($\phi \approx 29°$N). Satellite-tracked buoys and subsurface floats both show the latitude of the Bahamas to be an area of great temporal and spatial variability, with many eddies of various sizes, and with inflows to the Straits of Florida through the Old Bahama Channel and the Northwest Providence Channel. The general sense is that of converging surface flows all feeding the Gulf Stream system.

Subsurface Flow Regime

The strong surface currents of the Gulf Stream system decrease with depth. Mathematically, this can be explored by applying Leibnitz' rule to the integral form of the hydrostatic equation $p = \int_h^z \rho g dz$. The geostrophic equation (above) can then be expressed as:

$$\rho f \vec{v} = \rho g \partial h/\partial \vec{n} + \int_h^z g \partial \rho/\partial \vec{n} dz$$

where the first term on the right-hand side is the barotropic term, and the integral term on the right-hand side is the baroclinic term. Facing downstream $\partial\rho/\partial\vec{n}$ is negative, and the surface current (at $z = h$) decreases with depth until the two terms on the right-hand side become equal and opposite. This depth is called the level-of-no-motion and $\vec{v} = 0$. In the Yucatan Channel, the level-of-no-motion is approximately 1200 m, but in the northern Straits of Florida ($\phi \approx 27°$N) northward-flowing currents as much as 0.3 m s^{-1} reach to the seafloor at 800 m.

Details of the flows into and out of the Intra-Americas Sea at depth in other passages are less well known than the surface flows. Numerical models and observations suggest a general inflow though the passages of the Lesser and Greater Antilles into the Caribbean Sea, an outflow through the Yucatan Channel into the Gulf of Mexico, and continuing flow into the North Atlantic Ocean north of the

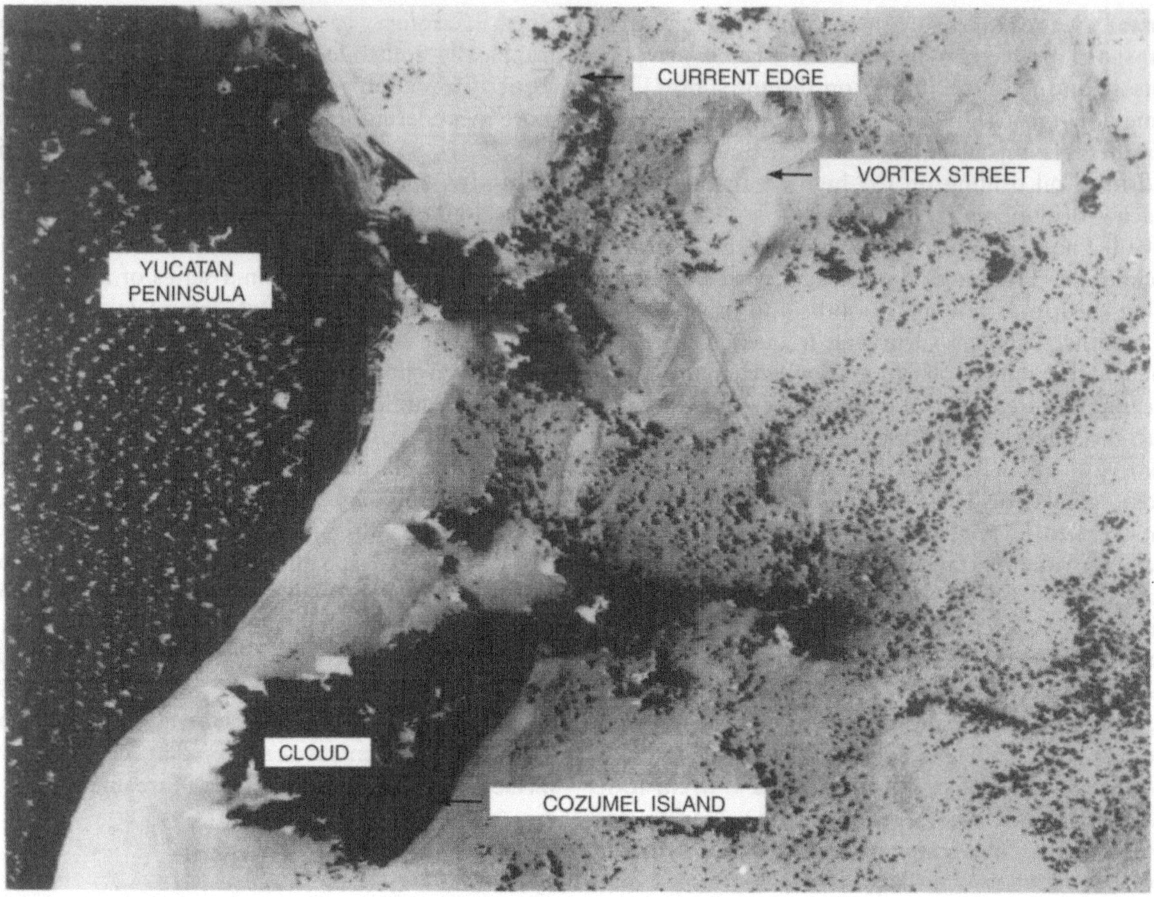

Figure 5 LANDSAT negative image of surface-wave glitter patterns showing von Karman vortices downstream of Cozumel Island in the Yucatan Channel. For scaling, Cozumel Island is approximately 50 km long.

Bahamas. Near the sill of the Yucatan Channel, approximately 100 m above the ocean floor, the flow is decidedly from the Gulf of Mexico into the Caribbean Sea. In the Windward Passage, there is also evidence of north-eastward outflow at depth, but it seems not to be as persistent as that in the Yucatan Channel.

A major characteristic of the deep waters of the IAS is their nearly isothermal and isohaline profiles below the depth of the major sills. The ocean, being a stratified fluid, tends to inhibit vertical mixing. Thus the sub-sill depth waters are characterized by near-zero vertical density gradients, $\partial\rho/\partial z \approx 0$, and are neutrally stable. Deep IAS waters have the T–S characteristics of the offshore waters of the juxtaposed North Atlantic Ocean, which seem to spill over the sills from time to time to replenish and ventilate the water interior to the IAS. Thus there must be a surging of sorts to bring into the Caribbean basin in particular, renewing mid-depth North Atlantic Common Water.

Along the eastern margin of the IAS at about 2000–3000 m water depth is a southward-flowing mid-level current called the Deep Western Boundary Current (DWBC). The DWBC has its genesis in the thermohaline circulation of the North Atlantic Ocean, north of the Denmark Strait, and is part of the global conveyor belt. The DWBC flows along the entire outer boundary of the IAS, and has a volume transport of approximately 15 Sv. This extremely important flow is climatically linked to the role of the ocean in Earth's heat budget and may participate in the complexities of the IAS's deeper flow patterns.

Other Currents in the IAS

Tidal currents in the IAS are generally weaker than in other semi-enclosed seas. The tides are typically semi-diurnal on the Atlantic Ocean margin of the IAS (M_2 and S_2 constituents usually), and progressively become diurnal in the Gulf of Mexico where the K_1 and O_1 constituents dominate. Estuaries such as the Mississippi Delta are of the salt-wedge category, mostly because the tidal currents and ranges are small and the river flows very large (average for the Mississippi River is about 10^3 km^3 y$^{-1} \approx 0.03$ Sv). Tidal currents around many IAS islands are similarly weak, with extremes rarely exceeding $\vec{v} = 1\,\mathrm{m\,s^{-1}}$ even in passes through the many bar-built barrier island lagoons.

Inertial currents are a ubiquitous feature of the ocean, and are characterized by periods $\vec{v} = 12^h/\sin\,\phi$. In the northern IAS, inertial currents often have periods equal to the dominant diurnal tidal currents, such as the K_1 or the O_1 because $\sin\phi \approx 0.5$. At these critical latitudes, the inertial currents are not separable from the diurnal currents in the tidal spectrum. Inertial currents, when they occur, are intermittent and have velocities typically below $\vec{v} = 0.2\,\mathrm{m\,s^{-1}}$.

Current flows past islands can induce complex patterns in their lee. Numerical models and observations suggest von Karman vortex streets downstream of many island land masses (**Figure 5**). Such complex currents can cause engineering design complexities, particularly regarding waste disposal and spills. Similarly, with the normally low wave heights so characteristic of the IAS, the longshore and littoral currents are also weak, although in certain areas, especially the east coast of Florida, dangerous wave-induced rip currents are very common.

Perhaps the most significant physical marine hazard in the Intra-Americas Sea is the combined storm surge and inverted barometer effect associated with hurricanes. Along linear coasts, the water level elevation from a major storm can exceed 7 m, on top of which may be 3 m or greater wind waves. Little is directly known of the currents associated with storm surges, but indirect evidence suggests that they can exceed $\vec{v} = 1\text{–}2\,\mathrm{m\,s^{-1}}$, especially in the vicinity of harbor entrances and inlets. Small islands are less at risk from large storm surge-driven currents than are long, low coasts with shallow offshore bottom topography. While storm surge currents are transient features of the IAS with timescales of approximately half a day, they can be costly to infrastructure, and very dangerous to human life.

Upwelling is a vertical current of note in the IAS, especially along the long east–west tending north coast of South America. Here the zonal wind stress τ_x can force an Ekman upwelling with mass transport per unit width $M_y = \int \rho v dz$ given by $M_y = \tau_x/f$.

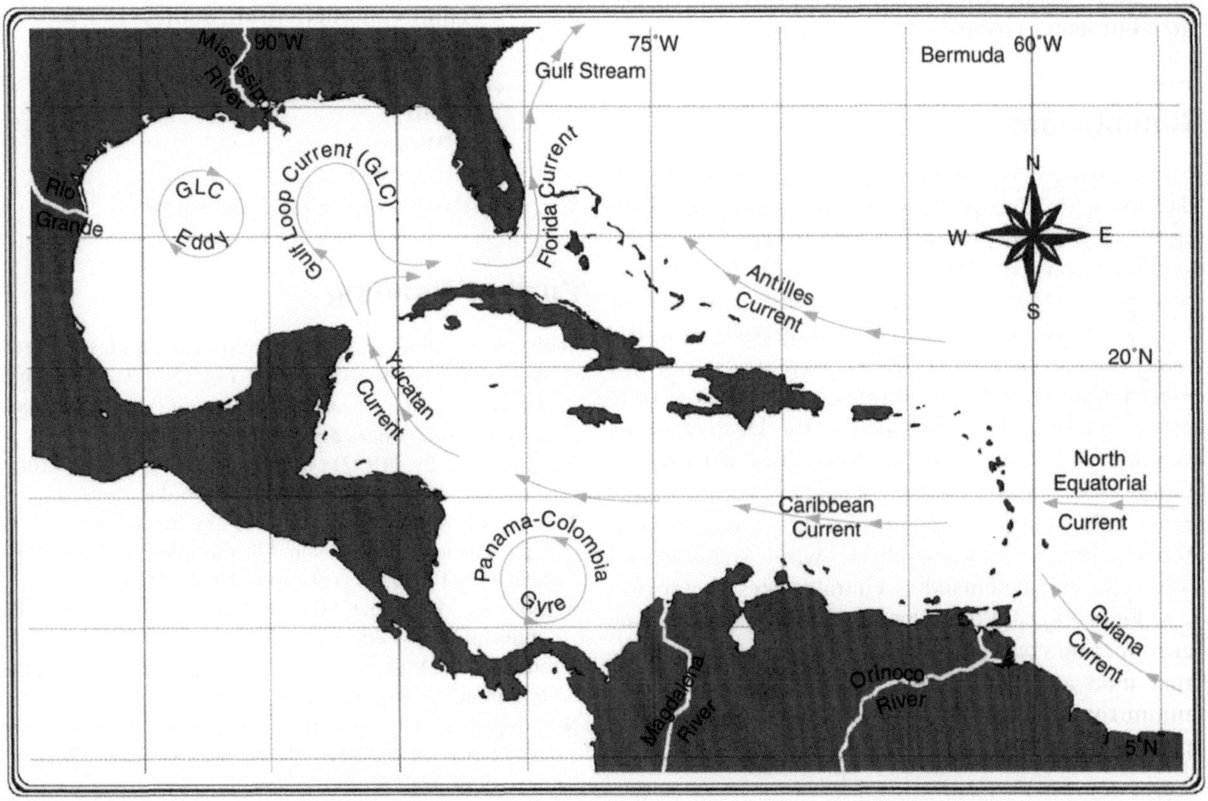

Figure 6 Summary of surface currents in the Intra-Americas Sea. Maximum IAS sea surface height variability $h = \pm 24$ cm is centered in the Gulf Loop Current (GLC) at $\phi = 26°$N, $\lambda = 88°$W; a second maximum in the Caribbean Current of $h = \pm 12$ cm is centered at $\phi = 15°$N, $\lambda = 77°$W.

Accordingly, the easterly Trade Winds force a northward mass transport along the coast, and the ocean responds with classical coastal upwelling as seen in the 3°C cooler sea surface temperatures along Venezuela. The same physical circumstances cause open-ocean Ekman pumping in the wake of hurricanes. The hurricane's large cyclonic wind stress, $\tau = \rho_{air} C_d v_{air}^2$, forces mass transport $M_{x,y}$ in all directions away from the storm center with attendant lifting of the thermocline and upwelling. Lower sea surface temperatures are often observed as a cool streak in the wake of these intense air–sea storm systems.

While the danger from seismic sea waves (tsunami) is recognized as another although largely unappreciated natural hazard of the IAS, it is the waves and wave particle motions that create currents of such great danger. Caribbean tsunami waves have been observed to exceed 9 m in height. Since a tsunami is a progressive shallow-water wave with celerity $c = \sqrt{gH}$, the maximum currents come at the wave crest and at the wave trough. These currents probably exceed $\vec{v} = 10\ \mathrm{m\,s^{-1}}$, and have timescales of several minutes. In that short amount of time however, even more danger exists than with storm surge currents, and the small islands are equally as vulnerable as are the continental coasts.

Conclusions

Intra-Americas Sea surface currents (**Figure 6**) are dominated by a single fact: the IAS is the formation region of the Gulf stream system. Except along the northern coast of the Gulf of Mexico, the volume transport of the interior thermohaline component of IAS currents is minuscule compared with the wind-driven component. While there are important external and peripheral currents associated with the global thermohaline flow such as the Deep Western Boundary Current, it is the North Brazil Current–Guianas Current–Caribbean Current–Yucatan Current–Gulf Loop Current–Florida Current–Gulf Stream family of atmospheric wind stress-forced advective movements that characterizes the region (*cf*. **Figure 4**). All these 'currents' are in reality one current that, coupled with air–sea heat and moisture fluxes and winds, integrate into a single continuum that connect the Intra-Americas Sea and its peoples.

Nomenclature

c	wave celerity
f	Coriolis parameter
g	gravity
h	sea surface height
$\vec{n}$	direction vector parallel to pressure gradient
p	pressure
r	radius of curvature
t	time
u,v,w	eastward, northward, upward speed
$\vec{v}$	velocity vector orthogonal to $\vec{n}$
x, y, z	east, north, vertical Cartesian coordinates
C_d	air-sea drag coefRcient
C_p	specific heat
E	evaporation
Gl	gigaliters
H	water depth
M	mass transport
P	precipitation
R	river runoff
S	salinity
Sv	Sverdrups
T	temperature
W	watts
δ	salinity anomaly from eddies
ϕ	latitude
λ	longitude
ρ	density
ζ	vorticity
τ	wind stress
Ω	Earth's rate of angular rotation

Further Reading

Gallegos A (1996) Descriptive physical oceanography of the Caribbean Sea. In: Maul GA (ed.) *Small Islands: Marine Science and Sustainable Development*. Washington: American Geophysical Union.

Maul GA (ed.) (1993) *Climatic Change in the Intra-Americas Sea*. London: Edward Arnold.

Mooers CNK and Maul GA (1998) Intra-Americas Sea Circulation. In: Robinson AR and Brink KH (eds.) *The Sea*, Vol. 11. New York: John Wiley & Son.

Murphy SJ, Hurlburt HH, and O'Brien JJ (1999) The connectivity of eddy variability in the Caribbean Sea, the Gulf of Mexico, and the Atlantic Ocean. *Journal of Geophysical Research* 104: 1431–1453.

Schmitz WJ Jr (1995) On the interbasin-scale thermohaline circulation. *Reviews in Geophysics* 33: 151–173.

OKHOTSK SEA CIRCULATION

L. D. Talley, Scripps Institution of Oceanography, La Jolla, CA, USA

Introduction

The Okhotsk Sea (**Figure 1**) is one of the marginal seas of the north-western North Pacific. The circulation in the Okhotsk Sea is mainly counterclockwise. The Okhotsk Sea is the formation region for the intermediate water layer of the North Pacific. Water entering the Okhotsk Sea from the North Pacific is transformed in temperature, salinity, oxygen, and other properties through ice processes, convection, and vigorous mixing before returning to the North Pacific. Relatively saline water from the Japan Sea assists in making Okhotsk Sea waters denser than those of the Bering Sea, which otherwise has similar processes but which does not produce intermediate water. Tides are exceptionally large within the Okhotsk which has broad, shallow continental shelves, providing a significant location for dissipation of tidal energy.

Geography

The Okhotsk Sea is enclosed by the Russian coastline to the north, Sakhalin and Hokkaido Islands to the west, the Kamchatka peninsula to the east and the Kuril Islands to the south east. Two other marginal

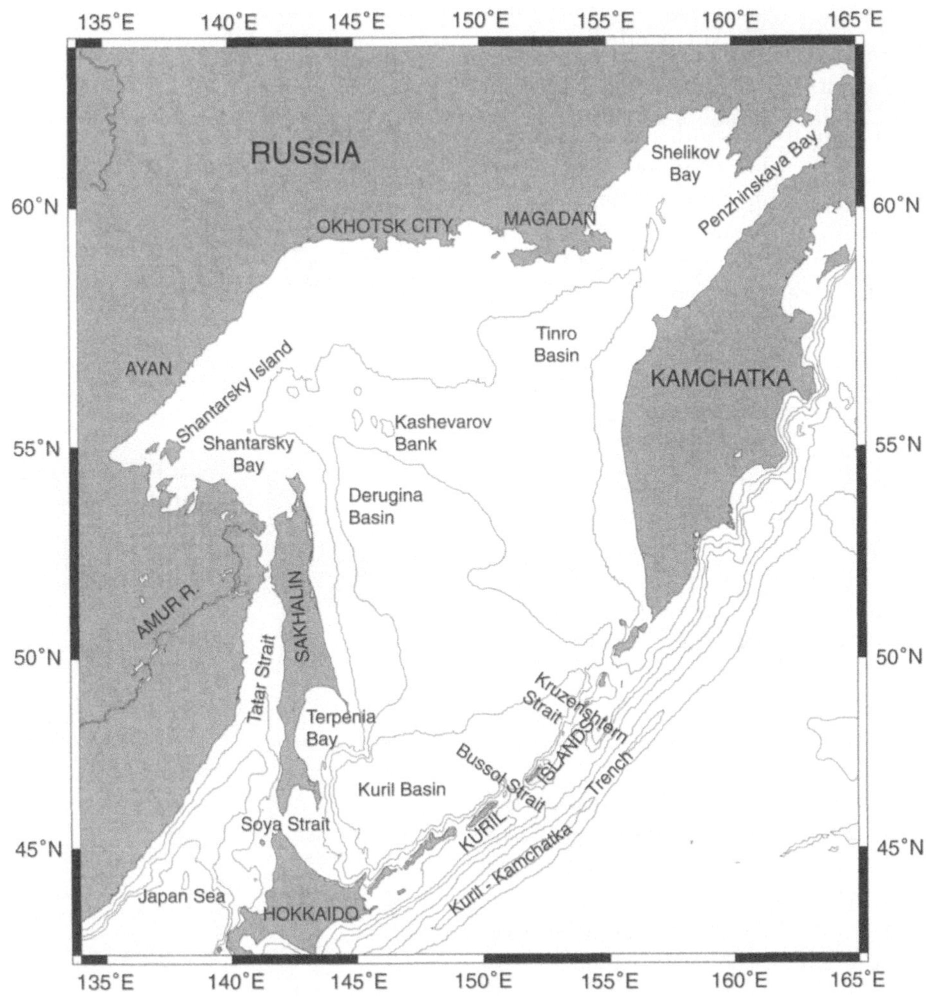

Figure 1 Okhotsk Sea geography.

seas are nearby: the Bering Sea east of Kamchatka, and the Japan (East) Sea west of Sakhalin and Hokkaido. These three marginal seas are characterized by limited connections to the North Pacific, relatively deep basins, and the presence of sea ice in winter. The Okhotsk Sea has a highly productive fishery, and is the site of explorative oil lease sites. The Okhotsk also occupies a unique role as the highest density surface source of waters for the North Pacific, feeding into the intermediate depth layer down to about 2000 m. All deeper water (and indeed, much of the water even in this intermediate layer) comes from the Southern Hemisphere.

The Okhotsk Sea is connected to the North Pacific through 13 straits between the numerous Kuril Islands. Because of the political history of this region, most straits and islands bear both Russian and Japanese names. The two deepest straits are Bussol' and Kruzenshtern, with sill depths of 2318 m and 1920 m, respectively, through which there is significant exchange of water with the North Pacific. Other straits of importance for exchange are Friza and Chetvertyy Straits, both with sill depths of about 600 m. The Okhotsk Sea is connected to the Japan Sea through two straits on either end of Sakhalin: Soya (La Perouse) Strait to the south and Tatar Strait to the north. Soya Strait, while shallow (55 m), is an important source of warm, saline water for the Okhotsk Sea. Tatar Strait is extremely shallow (5 m) and there is little exchange through it. The Okhotsk Sea is not connected directly to the Bering Sea, but much of the water flowing into the Okhotsk Sea originates in a current from the western Bering Sea.

Within the Okhotsk Sea there are three deep basins, with the greatest depth being 3390 m in the Kuril Basin. An important characteristic of the Okhotsk Sea is its very broad continental shelves in the north; these impact tidal energy dissipation and formation of dense waters in winter. The fairly shallow, isolated Kashevarov Bank is found in the north west. A major river, the Amur, drains into the Okhotsk Sea north-west of Sakhalin.

Tides

The Okhotsk Sea is one of the major tidal dissipation areas of the world ocean as a result of its very broad continental shelves. Tides in the North Pacific rotate counterclockwise. The tidal energy passing by the Kuril Islands enters the Okhotsk Sea. Tides around the Kuril Islands can have associated currents of 4–8 knots (20–40 cm s^{-1}). The currents through each of the Kuril Straits associated with the tides are northward at the eastern side of each strait and southward at the western side, leading to a net clockwise circulation of water around each of the Kuril Islands. Within the Okhotsk Sea, the maximum tidal currents and sea surface height displacements are in the northern bays and on Kashevarov Bank (**Figure 2**). Sea surface displacements can reach several meters in Penzhinskaya Bay in the north east. Maximum energy dissipation occurs in Shelikov Bay, with a somewhat less important site on Kashevarov Bank.

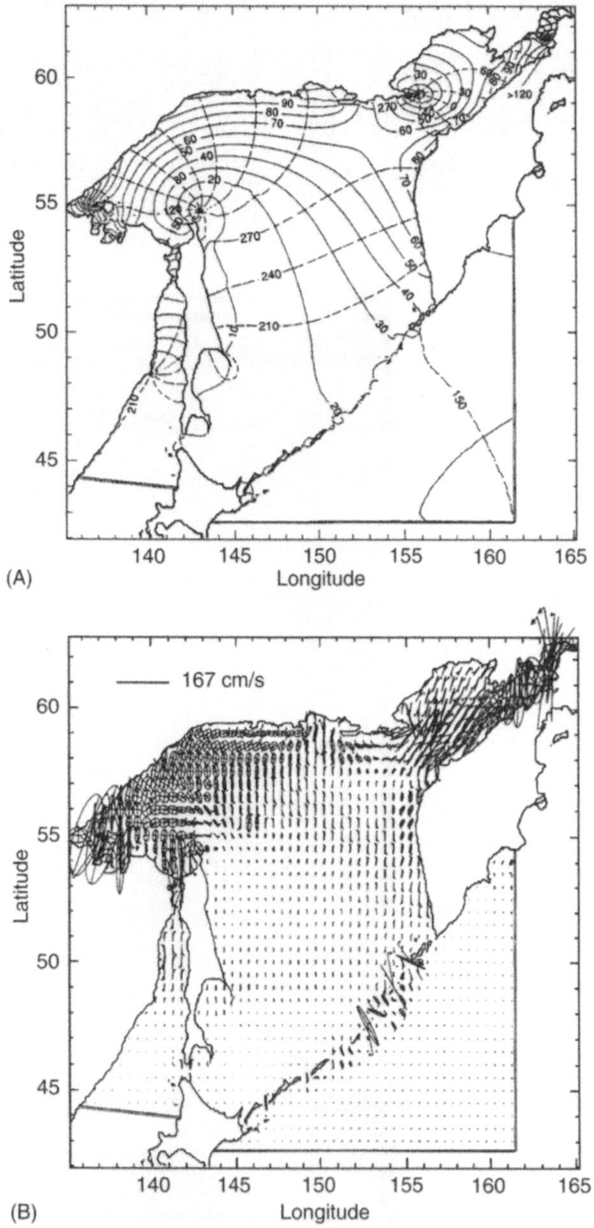

Figure 2 (A) The amplitude (cm) and phase and (B) current ellipses of the dominant semi-diurnal tide (M_2) in the Okhotsk Sea, from Kowalik and Polyakov (1998).

Circulation and Eddy Field

The mean circulation of the Okhotsk Sea is counterclockwise (**Figure 3**). Because ocean currents are in geostrophic balance, this means that there is low pressure in the center of the Okhotsk Sea. The flow in the Okhotsk Sea is driven by the same wind field that drives the cyclonic circulation of the adjacent subpolar North Pacific. The prevailing winds are the western side of the Aleutian Low. These winds create upwelling in the subpolar region and Okhotsk Sea. Upwelling over a broad region causes counterclockwise (cyclonic) flow in the Northern Hemisphere.

North Pacific water enters the Okhotsk Sea through the northernmost passages through the Kurils, primarily Kruzenshtern and Chetvertyy Straits. The North Pacific water comes from the East Kamchatka Current, which is a narrow southward boundary current along the eastern coast of Kamchatka, from the Bering Sea. The water that leaves the Okhotsk Sea through the southern Kuril Islands turns southward along the Kurils and joins the narrow, strong Oyashio. The Oyashio flows to the southern coast of Hokkaido, where it turns eastward and enters the North Pacific gyres. The net exchange between the Okhotsk Sea and the North Pacific is superimposed on clockwise flow around each of the Kuril Islands, driven by the strong tidal currents.

Within the Okhotsk Sea, the inflow from the Pacific feeds a broad northward flow in the east called the West Kamchatka Current. The West Kamchatka Current is fragmented and broad and often contains an inshore countercurrent. The warmth of the West Kamchatka Current region keeps this region ice-free throughout the year.

Along the northern shelves there is westward flow. The shelf flow rounds the northern tip of Sakhalin (Cape Elizabeth) and collects into a swift, narrow boundary current flowing southward along the east

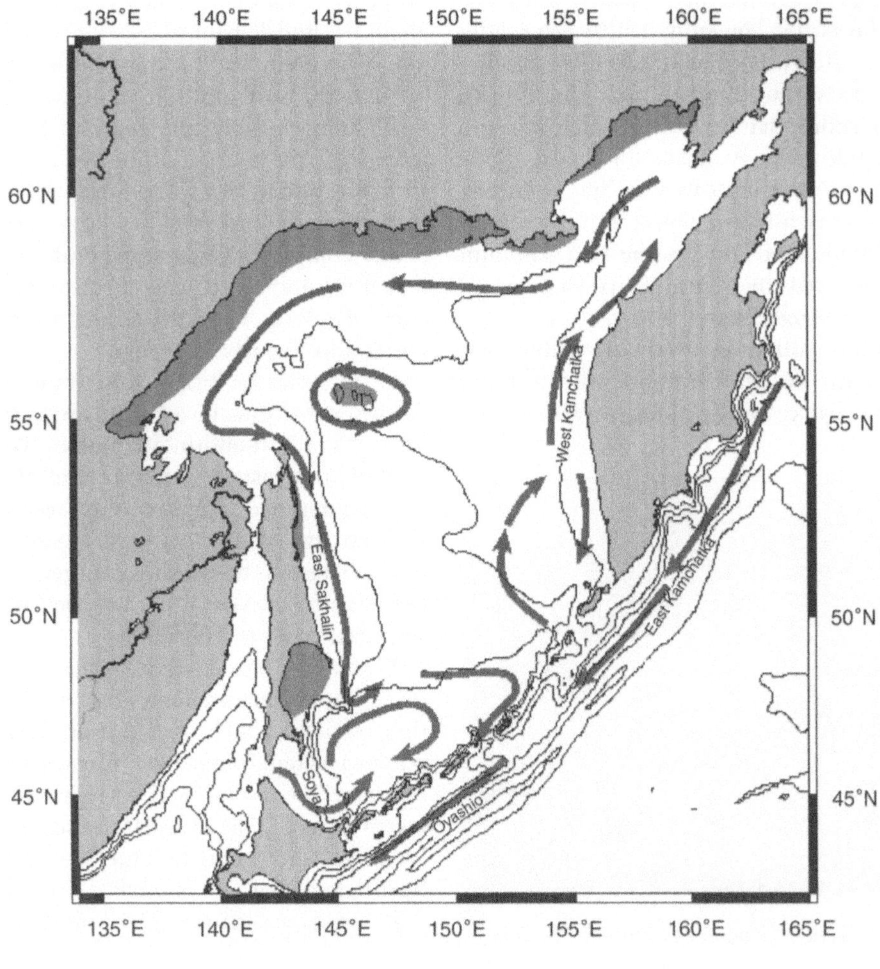

Figure 3 Mean circulation of the Okhotsk Sea, after numerous sources. (See Talley and Nagata, 1995, for collection of the many cartoons of the flow.)

side of Sakhalin, called the East Sakhalin Current. Where the East Sakhalin Current ends at Cape Terpeniya at the southern end of Sakhalin, it feeds into an eddy field with net transport toward Bussol' Strait, in the center of the Kuril Islands. Southward motion of the winter ice pack from Cape Terpeniya to Hokkaido suggests that some of the East Sakhalin Current also continues southward.

Water also enters the Okhotsk Sea from the Japan Sea, through Soya Strait. The narrow Soya Current carries this water along the northern coast of Hokkaido, moving towards the Kuril Islands. The typical speed of the Soya Current is 25–50 cm s^{-1}, with greater speeds in Soya Strait. The Soya Current peaks in summer and is submerged or very weak in winter, which allows ice to form along Hokkaido in the absence of this relatively warm water. The Soya Current water joins the other waters exiting the Okhotsk Sea through the southern Kuril Islands, including through Bussol' Strait.

The Okhotsk Sea has a vigorous eddy field, meaning that often it is difficult to discern the mean flow because of the presence of moving, transient eddies of about 100–150 km scale. The eddy field is especially pronounced in the Kuril Basin as tracked by satellite imagery of the sea surface temperature. The two to four Kuril Basin eddies that are formed each year are clockwise (anticyclonic). A mean clockwise flow, perhaps divided into smaller subgyres, has been discerned in this eddy-rich basin. This anticyclonic permanent flow or eddy field conveys the East Sakhalin Current waters towards the central Kuril Islands and hence to the exit through Bussol' Strait.

The Soya Current often has dramatic eddies that form as the current becomes unstable and rolls up horizontally into backward-breaking waves. Eddies of the Soya Current also form mushroom-shaped vortices. Soya Current eddies have been photographed and tracked using loose sea ice in winter (**Figure 4**).

Figure 4 Loose sea ice in the Soya Current, Showing a counterclockwise eddy off the Hokkaido coast from an aircraft at altitude 1000 ft. The eddy was about 20 km in diameter. From Wakatsuchi and Ohshima (1990).

Sea Ice in the Okhotsk Sea

Ice forms every winter in the Okhotsk Sea and melts away completely every summer. Thus all ice in the Okhotsk is first-year ice. A small amount of ice is usually present by the end of October in the coastal areas of Shelikof and Penzhinskaya Bays, with formation by mid-November in Shantarsky Bay and off the west coast of Kamchatka. Ice expansion is rapid and by December ice is found throughout much of the northern Okhotsk Sea and around Sakhalin. Ice forms off Hokkaido by early January. Maximum ice extent in the Okhotsk occurs in late March when ice can cover almost the entire Okhotsk Sea in a heavy-ice winter. The south-eastern Okhotsk Sea, where relatively warm North Pacific waters enter, remains ice-free in even the most extreme winters. Fast ice (attached to land) is found in late winter in Shantarsky, Shelikov, and Penzhinskaya Bays and around Sakhalin. Maximum ice thickness in the Okhotsk is about 1.5 m in the north and 1 m in the central Okhotsk Sea.

Circulation along the coast of Hokkaido is dominated by the warm, saline Soya Current entering from the Japan Sea. The Soya Current inhibits ice formation. It also exhibits large seasonality, being nearly completely submerged under fresh, cold water in winter and so ice can form along the Hokkaido coast. Pack ice from the Sakhalin area also reaches Hokkaido in early February.

Ice melt begins in late March and usually finishes by the end of June or early July, with the last vestiges of ice usually found in Shantarsky Bay. In heavy-ice winters, this last ice melts in late July.

In winter, pack ice often flows out of the Okhotsk Sea into the Pacific through the southern Kuril Islands. This and the fresh water generated by ice melt form the fresh coastal part of the Oyashio along the southern coast of Hokkaido.

Within the ice-covered zones, there are usually several areas that are either free of ice or contain only thin frazil ice. Such openings are called polynyas. A polynya often forms over Kashevarov Bank in the north, where strong tidal currents continually upwell warm waters to the sea surface and keep the region ice-free. The heat flux maintaining this polynya is provided by an upwelling of 0.3–0.6 m d^{-1} of water at 2°C.

Polynyas are also usually found in a narrow band along the north-western and northern coast between Shantarsky and Tauskaya Bays. The coastal polynyas are kept open by northerly or north-westerly winds that force the coastal ice offshore. Therefore ice forms continuously within these polynyas. Similar wind-forced polynyas are found in several spots along Sakhalin and in Terpeniya Bay at the southern end of Sakhalin.

Sea ice is always fresher than the sea water from which it forms since salt is rejected from the developing ice lattice. The rejected salt collects in pockets of brine within the ice and drips out at the bottom of the ice. This brine rejection process increases the salinity of the waters underneath the sea ice, which thus increases the density of these waters. This densification process is most effective in areas of active ice formation, such as the wind-created coastal polynyas, and where the water depth is not too great, so that the brine is less diluted as it mixes into the underlying water. In the Okhotsk Sea, the broad northern shelf with its coastal polynya is a site of active dense water formation.

Water Properties of the Okhotsk Sea

The main water source for the Okhotsk Sea is the North Pacific just east of the Kuril Islands and upstream of the Okhotsk Sea in the East Kamchatka Current. These North Pacific waters are characterized by a shallow temperature minimum below the sea surface, which is often called the 'dichothermal' layer (**Figure 5A**). Below this is a temperature maximum (the 'mesothermal' layer). The temperature minimum is supported by the existence of a low salinity surface layer (**Figure 5B**) since both temperature and salinity contribute to seawater density. The temperature minimum is a remnant of winter cooling, although it may reflect cooling farther upstream in the flow. The Okhotsk Sea waters also have this structure, although modified from the North Pacific's, as described below.

The second major source of water for the Okhotsk Sea is the Japan Sea, through Soya Strait. Japan Sea water is relatively saline, since it all originates as a branch of the Kuroshio at much lower latitude. Even though net precipitation in the Japan Sea reduces the salinity of the Kuroshio waters that enter it, the Soya Current waters feeding into the Okhotsk Sea are relatively saline. There is no similar source of relatively high salinity water for the Bering Sea. The resulting difference in overall salinity between the Okhotsk and Bering Seas is likely the main reason that the intermediate waters of the North Pacific are ventilated (originate at the sea surface) in the Okhotsk Sea rather than in the higher latitude Bering Sea.

Salinity within the Okhotsk Sea is reduced by flow from the Amur River and local precipitation. Sea ice

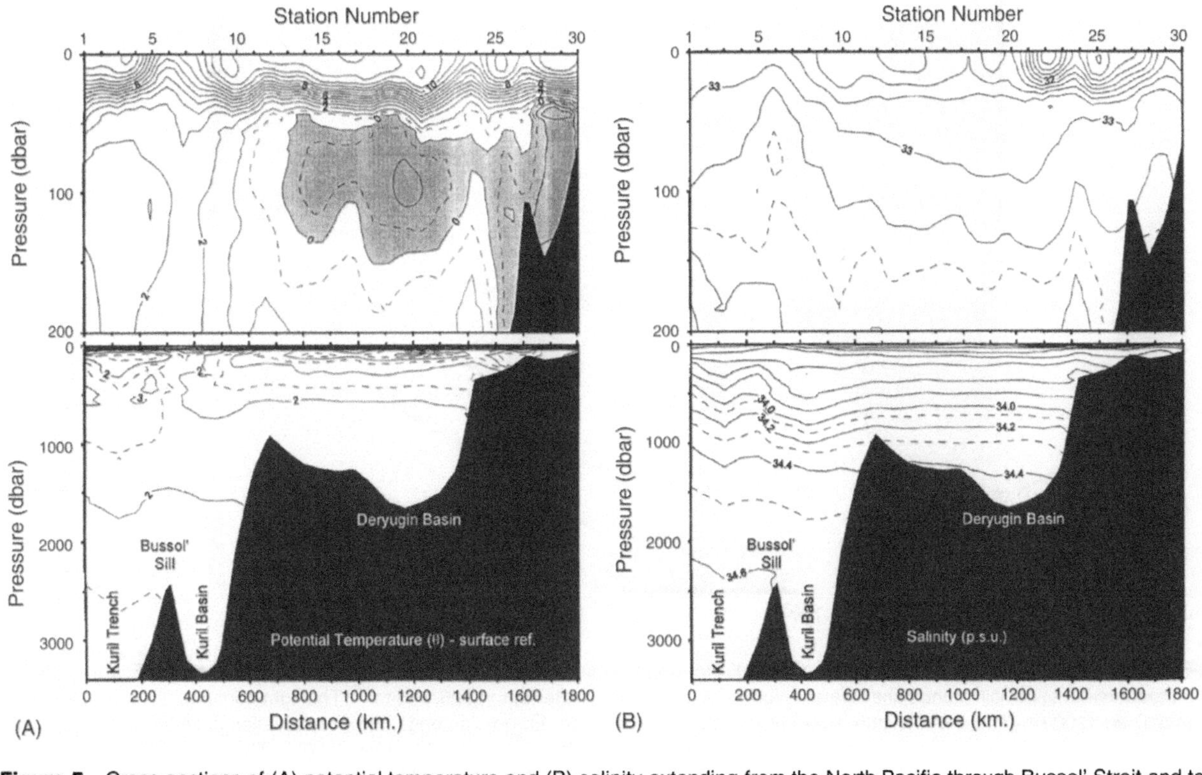

Figure 5 Cross-sections of (A) potential temperature and (B) salinity extending from the North Pacific through Bussol' Strait and to the northwest Okhotsk Sea, from Freeland *et al.* (1998).

production each winter creates higher density waters through brine rejection from the ice, and sea ice melt in spring and summer produces a freshened surface layer.

Within the Okhotsk Sea the dichothermal layer is colder than outside in the Pacific. Ice formation throughout most of the Okhotsk Sea depresses the temperature of the minimum. In the north-west Okhotsk Sea, the temperature minimum is close to the freezing point due to ice production in this shallow region in winter, causing the waters to the shelf bottom to be near freezing. Fresh, cold, oxygenated water penetrates much deeper with the Okhotsk Sea than in the East Kamchatka Current which is its primary source. The temperature maximum in the Okhotsk Sea is near 1000 m, considerably deeper than the mesothermal layer in the East Kamchatka Current which lies at about 300 m.

In summer, the surface salinity is very low, <32.8 PSU, and considerably lower near the Amur River outflow. The generally low salinity is due to ice melt and river discharge. Salinity at the temperature minimum is about 33.0 PSU and then increases gradually to the bottom, consonant with the North Pacific source of the deep waters.

Several important water transformation processes occur in the Okhotsk: densification resulting from ice formation, convection resulting from cooling,

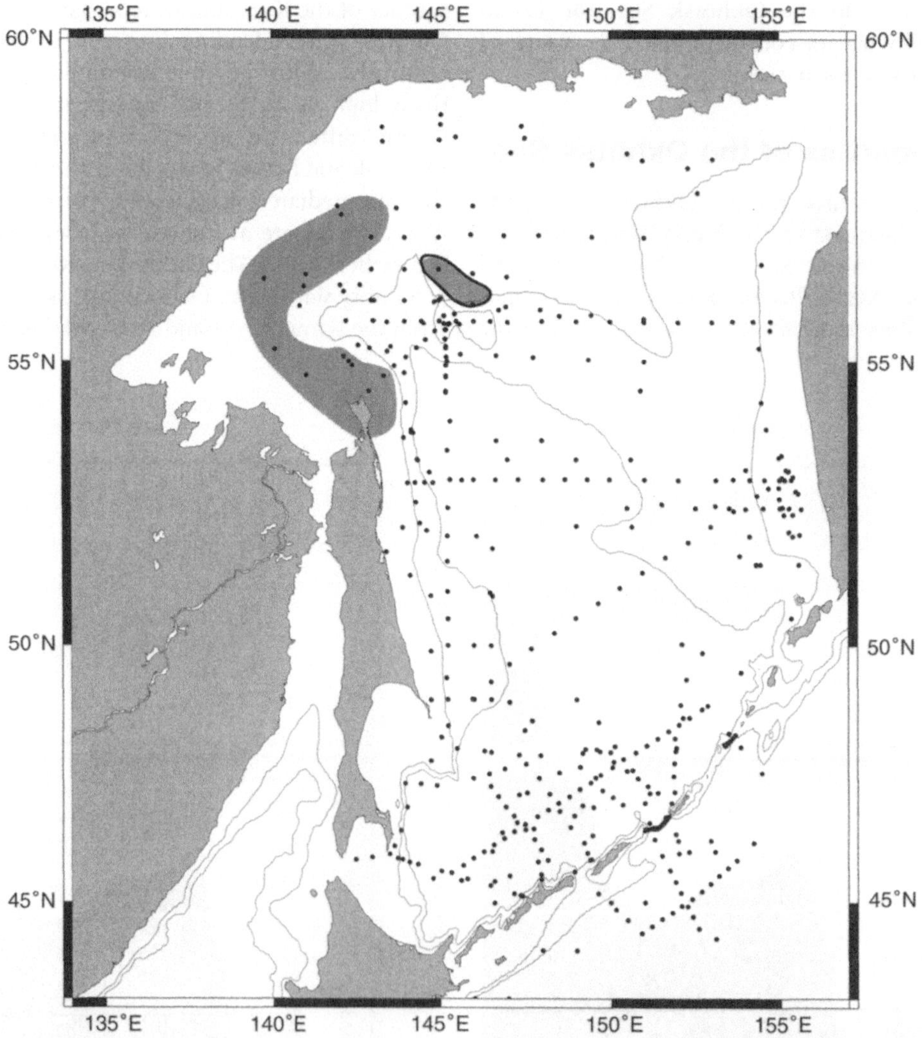

Figure 6 Bottom density (in units of kg/m^{-3}–1000) for depths less than 300 m is shown with the heavy contours. Bottom density is shaded where especially dense shelf water is found, with density greater than 1026.9 kg m^{-3}. The light contours show bottom depths of 300 and 1000 m. The dots indicate where observations were made. Large diamonds show observation positions where the water depth is less than 300 m. Large stars indicate that the temperature at the bottom is colder than -1.0°C. The data are from Kitani (1973) and from surveys of the Okhotsk Sea in 1994 and 1995, with the latter data provided by Rogachev (Pacific Oceanological Institute, Vladivostok, Russia) and Riser (University of Washington).

input of low salinity at the surface from rivers, precipitation and ice melt, and mixing which is greatly accentuated in the Okhotsk because of its large tidal amplitudes. Ice formation over the northern shelves, particularly in the coastal polynyas, creates a dense shelf water that moves cyclonically around to Shantarsky Bay and then out past the northern end of Sakhalin and down along the sloping side to feed the dichothermal layer. The density of the new shelf water (**Figure 6**) can be surmised from nonwinter observations. Based on data collected over many years, it appears that the maximum density of the shelf waters and their offshore mixture is 1027.2 kg m^{-3}.

Convection due to heat loss occurs in the Okhotsk Sea, notably in the Kuril Basin to a depth of about 500 m. The maximum density affected by convection is about 1026.85 kg m^{-3}, and so ice formation on the shelves creates denser water than does convection. This limit on the density created by convection is set by the salinity of surface waters in the Kuril Basin and the maximum density they can reach when cooled to freezing. (Densification through ice formation has negligible effect in deep water since the rejected salt mixes into a thick water column.) However, the convective mixing is important as a signature of Okhotsk Sea water transformation as the waters enter the Oyashio and move southward into the North Pacific, since the thickness of the newly convected layer is retained to some extent.

Mixing is a much more significant process in the Okhotsk than in the open North Pacific because of the large tidal amplitudes and topography within the sea. Two locations especially deserve mention – in the straits between the Kuril Islands and over Kashevarov Bank in the north west. Mixing over the latter moves relatively warm water to the sea surface, melting out the sea ice there. Mixing over the nearby shelves may also be an important factor in setting the maximum density of the winter shelf waters, since the mixing could bring higher salinity waters from offshore onto the shelves. In the Kuril Straits, tidal currents are large, and oppose each other on opposite sides of each strait. In particular, water properties in Bussol' Strait have long been observed to be strongly mixed, to the bottom of the strait. This mixing brings the high oxygen of the upper waters down to the sill depth and is the major source for the North Pacific down to this depth of oxygen and other atmospheric gases such as chlorofluorocarbons.

Connection of the Okhotsk Sea with North Pacific Processes

The Okhotsk Sea is an important factor in air–sea exchange and overturn in the North Pacific. The 'ventilation' processes of the Okhotsk directly affect densities higher than elsewhere in the North Pacific. Deep and bottom waters are not formed at the sea surface in the North Pacific – globally they are formed only in the North Atlantic and around Antarctica. However, the intermediate layer of the North Pacific, between about 500 and 2000 m, is ventilated through Okhotsk Sea processes, similar to the impact of intermediate water formation in the Labrador Sea of the North Atlantic and Southern Hemisphere intermediate water formation around southern South America.

The intermediate layer of the North Pacific lies between the salinity minimum found in the subtropical region, at a density of 1026.8 kg m^{-3}, and the deep waters found below 2000 m, or a density of about 1027.6 kg m^{-3}. The most recently ventilated water in the North Pacific, as marked by laterally high oxygen (**Figure 7**) and chlorofluorocarbons and other gases of atmospheric origin, is found in the north west, in the neighborhood of the Okhotsk Sea. The intermediate layer is also freshest in this area. The evidence described above indicates that the Okhotsk Sea is the source of the ventilation. The bottom of the North Pacific's ventilated intermediate water layer is set by the sill depth of Bussol' Strait, that is, by the maximum depth and density of the vertical mixing that moves ventilated waters downward in the Okhotsk Sea, with outflow into the North Pacific.

The water that leaves the Okhotsk Sea through Bussol' Strait turns southward and becomes the Oyashio. Its properties are significantly different from those of the East Kamchatka Current, that feeds water into the Okhotsk Sea farther north. The Oyashio continues southward to the southern end of Hokkaido and then turns offshore and to the east.

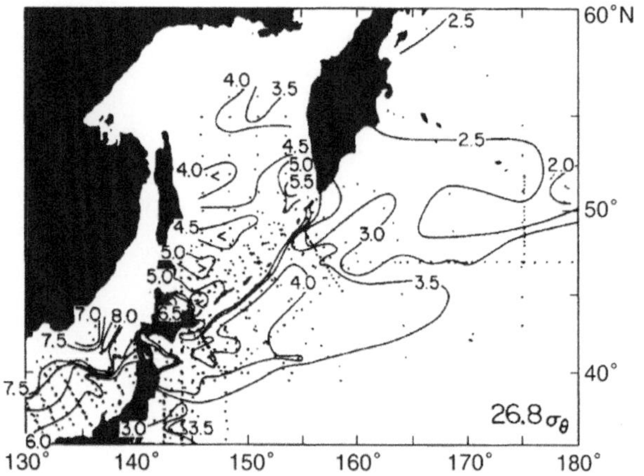

Figure 7 Oxygen on a constant density surface in the North Pacific Intermediate Water layer, from Talley (1991). This surface lies at about 300–400 m depth within the Okhotsk Sea.

The Oyashio waters, including the Okhotsk Sea products, then enter the interior of the North Pacific.

See also

Bottom Water Formation. Kuroshio and Oyashio Currents. Wind Driven Circulation.

Further Reading

Alfultis MA and Martin S (1987) Satellite passive microwave studies of the Sea of Okhotsk ice cover and its relation to oceanic processes, 1978–1982. *Journal of Geophysical Research* 92: 13 013–13 028.

Favorite F, Dodimead AJ, and Nasu K (1976) *Oceanography of the Subarctic Pacific region, 1960–71.* Vancouver: International North Pacific Fisheries Commission 33: 1–187.

Freeland HJ, Bychkov AS, Whitney F, *et al.* (1998) WOCE section P1W in the Sea of Okhotsk – 1. Oceanographic data description. *Journal of Geophysical Research* 103: 15 613–15 623.

Kitani K (1973) An oceanographic study of the Okhotsk Sea – particularly in regard to cold waters. *Bulletin of Far Seas Fisheries Research Laboratory* 9: 45–76.

Kowalik Z and Polyakov I (1998) Tides in the Sea of Okhotsk. *Journal of Physical Oceanography* 28: 1389–1409.

Moroshkin KV (1966) *Water Masses of the Okhotsk Sea.* Moscow: Nauka. (Translated from the Russian by National Technical Information Services, 1968.)

Preller RH and Hogan PJ (1998) Oceanography of the Sea of Okhotsk and the Japan/East Sea. In: Robinson AR and Brink KH (eds.) *The Sea*, vol. 11, New York: John Wiley and Sons.

Reid JL (1973) Northwest Pacific ocean waters in winter. *Johns Hopkins Oceanographic Studies*, Baltimore, MD: The Johns Hopkins Press 5: 1–96.

Talley LD (1991) An Okhotsk Sea water anomaly: implications for ventilation in the North Pacific. *Deep-Sea Research* 38 (Suppl): S171–S190.

Talley LD and Nagata Y (1995) *The Okhotsk Sea and Oyashio Region.* PICES Scientific Report No. 2. Sidney BC. Canada: North Pacific Marine Science Organization (PICES), Institute of Ocean Sciences

Wakatsuchi M and Martin S (1990) Satellite observations of the ice cover of the Kuril Basin Region of the Okhotsk Sea and its relation to the regional oceanography. *Journal of Geophysical Research* 95: 13 393–13 410.

Wakatsuchi M and Ohshima K (1990) Observations of ice-ocean eddy streets in the Sea of Okhotsk off the Hokkaido coast using radar images. *Journal of Physical Oceanography* 20: 585–594.

Warner MJ, Bullister JL, Wisegarver DP, *et al.* (1996) Basin-wide distributions of chlorofluorocarbons CFC-11 and CFC-12 in the North Pacific – 1985–1989. *Journal of Geophysical Research* 101: 20 525–20 542.

WEDDELL SEA CIRCULATION

E. Fahrbach and A. Beckmann, Alfred-Wegener-Institut für Polar- und Meeresforschung, Bremerhaven, Germany

Introduction

The Weddell Sea is an area of intense air–sea interaction and vertical exchange. The resulting cold water masses participate in the global thermohaline circulation as deep and bottom waters. The large-scale horizontal circulation in the Weddell Sea is dominated by a cyclonic (clockwise) gyre conforming to the coastline and topographic features like midocean ridges and submarine escarpments. The flow is driven by both wind and thermohaline forcing. The meridional component at the eastern edge of the gyre transports upper ocean water from the Antarctic Circumpolar Current to the Antarctic Continent, where its density increases by ocean–ice–atmosphere interaction processes. Part of this newly formed water leaves the inner Weddell Sea northward and escapes through gaps and passages and thus ventilates the deep world ocean. Upwelling in the Antarctic Divergence and sinking plumes along the continental slope cause a large-scale overturning motion. Intermittently, open ocean deep convection might also play a role in deep water renewal. The observational database of direct current measurements is rather weak. Consequently, the most reliable basin-wide estimates of ocean currents are obtained from adequately validated numerical models.

Limits of the Weddell Sea

Geographically, the Weddell Sea is a southern embayment of the Atlantic Ocean, bounded to the west by the Antarctic Peninsula up to Joinville Island, to the east by Coats Land on the Antarctic Continent with the north-eastern limit at 73°25′S, 20°00′W, and to the south by the Filchner–Ronne Ice Shelf. In these limits it covers an area of 2 800 000 km^2.

From an oceanographic point of view, however, it is more adequate to consider the closed cyclonic circulation cell as one dynamically connected regime; hence the Weddell Sea is often considered as the area of the Weddell Gyre which extends eastward as far as 30°E or 40°E (**Figure 1**). Then, the Weddell Sea is bounded to the north by the South Scotia, the North Weddell and the Southwest Indian Ridges where the Weddell Front marks the transition from the water masses of the Weddell Gyre to the Antarctic Circumpolar Current. At the eastern boundary no obvious front separates the Weddell and the Circumpolar water masses which leaves the eastern boundary diffuse. Obviously, limits based on water mass properties and current branches are time dependent.

The Effect of Ice–Ocean Interaction on the Currents

Ocean currents are wind-driven and thermohaline-driven; the particular situation in the Weddell Sea is even more complex due to the presence of the seasonally variable sea ice belt. During the winter months, sea ice covers almost all of the Weddell Sea (exceptions are the occurrences of coastal or, less frequently, open ocean polynyas); during summer, only a relatively small area in the southwestern Weddell Sea remains ice covered.

The presence of sea ice affects the ocean currents both through the modification of the momentum transfer from the wind to the ocean and through density changes caused by the salt gain during the freezing phase and the fresh water release during melting. In areas of stagnant sea ice the internal ice stresses balance the momentum supplied by the wind and the momentum input into the ocean is largely reduced; this situation is typical for the western Weddell Sea. The freeze and melt cycle of sea ice (and the corresponding freshwater fluxes) is an equally important component in the thermohaline forcing of the currents in the Weddell Sea; as freezing and melting regions are not identical, the sea ice drift with wind and ocean currents causes a net northward freshwater transport. This leads to a salt gain in the interior Weddell Sea and a freshwater gain at its northern boundary.

The Database on Ocean Currents in the Weddell Sea

The database to derive ocean currents in the Weddell Sea is small compared to other ocean areas. This is due to a number of reasons.

- The uninhabited Antarctic Continent does not require intensive shipping traffic and consequently

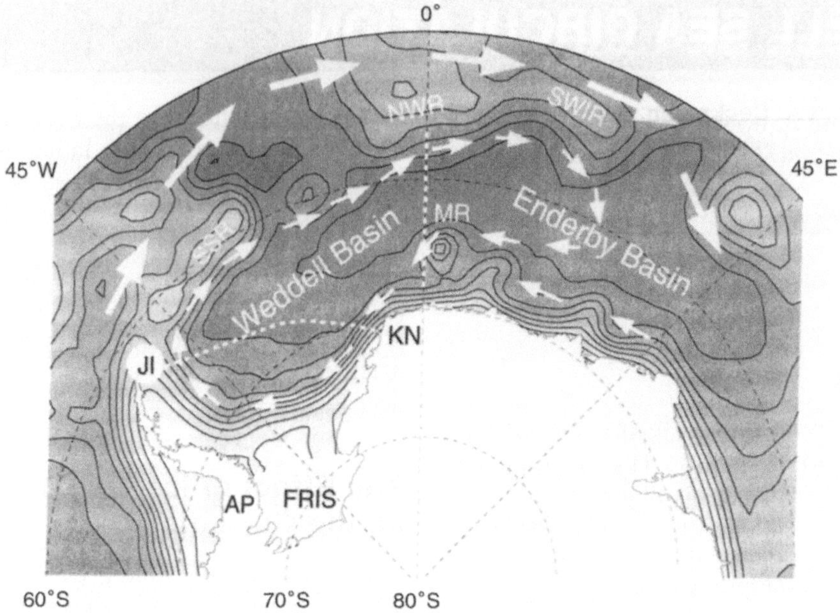

Figure 1 Schematic representation of the Weddell Gyre and the Antarctic Circumpolar Current in the Atlantic and Indian sectors of the Southern Ocean. The locations of the two transects with direct current measurements are indicated by lines. AP, Antarctic Peninsula; FRIS, Filchner–Ronne Ice Shelf; JI, Joinville Island; KN, Kapp Norvegia; MR, Maud Rise.

no long-term records of ship observations are available.

- The sea ice cover restricts shipping traffic to selected areas and part of the year (austral summer).
- The sea ice does not move exactly with the surface ocean currents; consequently ice buoys or satellite-derived ice motion do not represent surface ocean currents.
- The weak stratification in polar and subpolar oceans implies a strong barotropic component of the currents. This restricts the reliability of current estimates by use of geostropic currents with unknown reference velocities or the geopotential anomaly (dynamic topography) with a fixed reference level.
- In large areas of the Weddell Sea the time-mean currents are relatively weak. They are therefore masked by high frequency motion (tides and inertial motion), especially when the period of observations is short.

Most of the current patterns in the Weddell Sea are derived indirectly either by tracking water mass properties or from the geopotential anomaly (dynamic topography) measurements. Water mass properties (**Figure 2**) give good qualitative results because water masses of circumpolar origin enter the Weddell Gyre in the north and the major modification areas are in the southern parts of the Weddell Sea. The modified water leaves the Weddell Sea to the north. However, the regional variations in the water mass characteristics do not allow quantitative estimates of current velocities and volume transport. Because of the strong barotropic component, the current pattern derived from water mass properties holds for most of the water column except for the bottom water plumes and the boundary layers.

In certain areas information on currents is available from moored instruments. They are concentrated on transects between Joinville Island and Kapp Norvegia, along the Greenwich Meridian and off the Filchner-Ronne Ice Shelf (**Figure 3**). The data allow to determine the horizontal and vertical structure of the currents on the basis of records which are at least several months long. However, moored instruments can not be used in the upper ocean layer due to the effect of sea ice and icebergs which might damage the moorings. In the eastern Weddell Gyre current information is available from ALACE floats drifting at approximately 750 m depth (**Figure 3**).

Information on the surface currents (i.e., within the Ekman layer) can be obtained indirectly, if wind and sea ice motion is known. For example, weather center (e.g., European Centre for Medium-Range Weather Forecasts (ECMWF), National Centers for Environmental Prediction (NCEP)) surface winds are used to calculate sea ice drift. Differences in observed and calculated ice buoy drift are then attributed to the surface currents. Iceberg drift data can be included in this analysis.

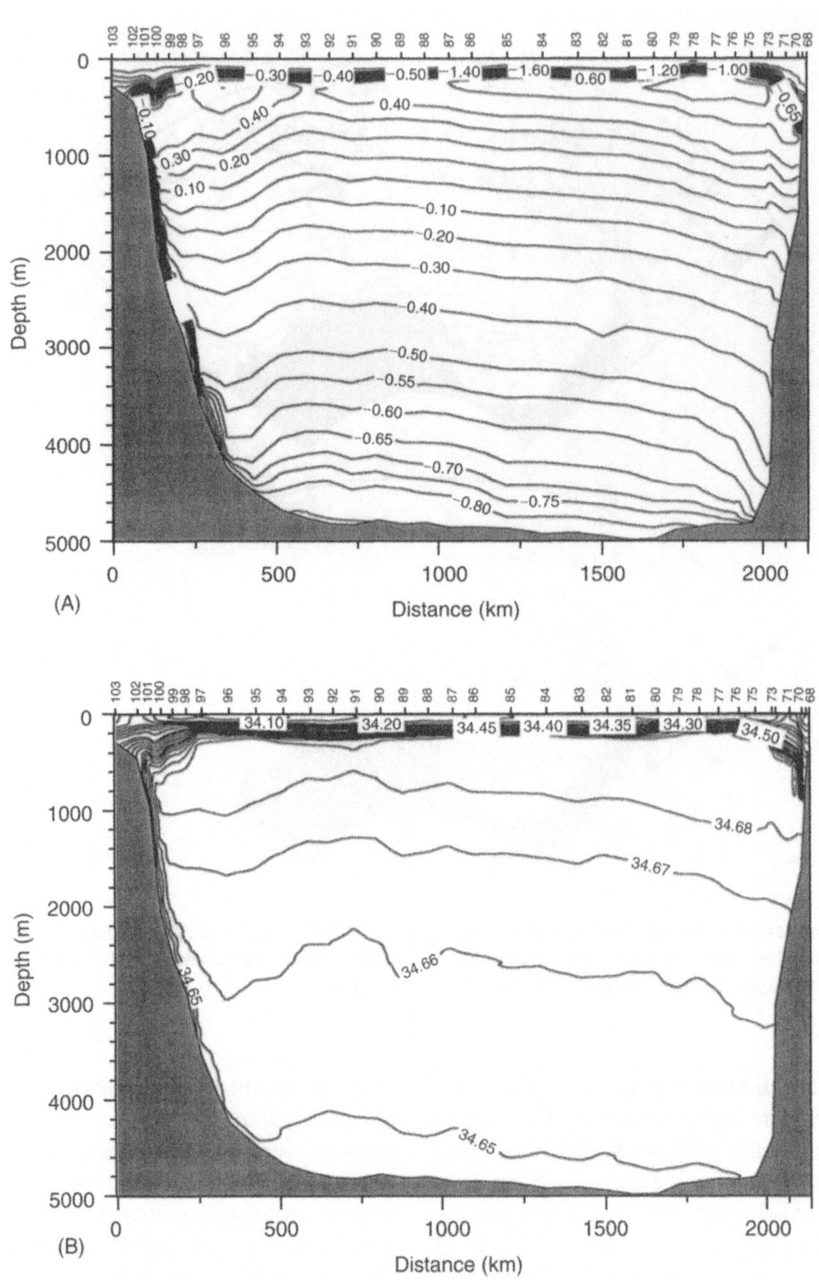

Figure 2 Distribution of the potential temperature (°C) (A) and salinity in PSU (B) on a vertical transect from the northern tip of the Antarctic Peninsula (left) to Kapp Norvegia (right) obtained with RV *Polarstern* in 1996.

Uncertainties in these estimates arise from the highly variable contribution of internal ice forces (ice pressure). Furthermore, buoy observations are rare and require extensive interpolation or extrapolation in both time and space. The resulting flow fields are rather smooth and do not contain much detail. Some improvement may be gained from satellite-derived sea ice drift.

As a direct consequence of the sparse observational data, the most consistent estimates of large-scale ocean currents in the Weddell Sea stem from numerical models, which have been validated rigorously against available observations. They extend into regions without observational data, include the upper ocean layers which are not accessible by moored instruments and cover time periods when no measurements are made. This is of particular interest because of the wide range of variability detected in the measurements.

Current state-of-the-art numerical tools to simulate the large-scale circulation and water mass structure in the ocean are coupled sea ice-ocean models, which are driven by atmospheric data sets from weather center analyses. They allow for ocean

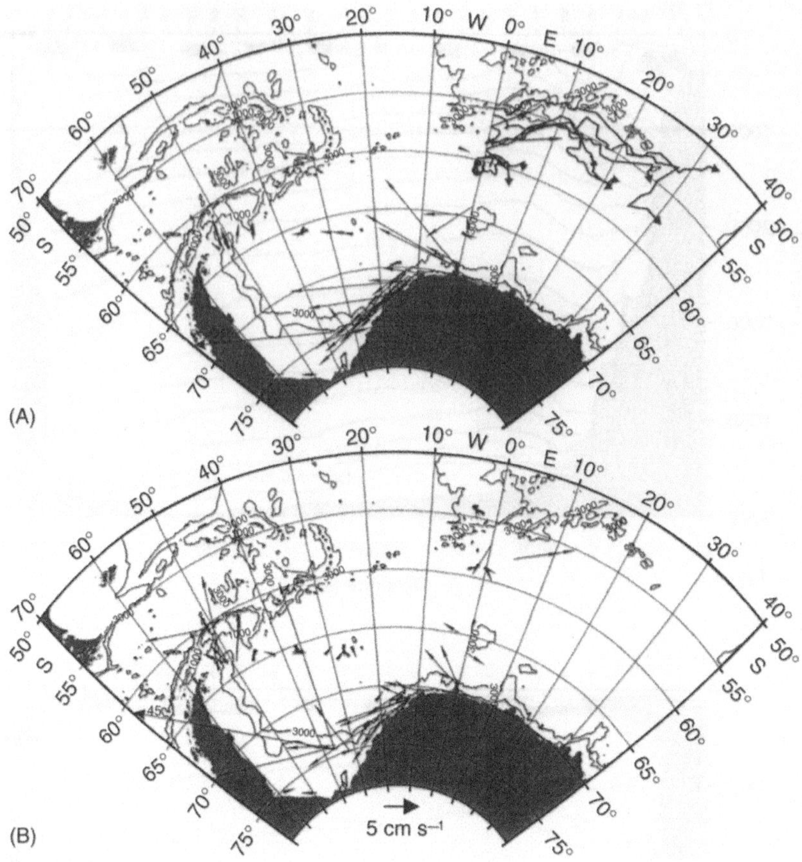

Figure 3 Ocean currents in the Weddell Sea obtained by direct measurements with recording periods between several months and two years. The currents in the upper ocean layer to a depth of 750 m (A) include data from moored instruments and subsurface floats. Those in the near bottom layer (B) include data up to 200 m from the bottom.

and ice dynamics and thermodynamics, as well as the feedback between both components of the climate system. Advanced models of the Weddell Sea also include the ice shelf cavities (Filchner–Ronne Ice Shelf, Larsen Ice Shelf and the ice shelves in the eastern Weddell Sea). The ocean component is based on the hydrostatic primitive equations. For an optimal representation of ocean dynamics in shallow shelf areas and over a sloping bottom a terrain-following vertical coordinate is useful. The sea ice component describes sea ice as a viscous-plastic medium. The model computes temperature, salinity, horizontal and vertical motion in the ocean as well as ice and snow thickness, and ice concentration. The horizontal resolution varies between 20 and 100 km horizontally and 10 and 400 m vertically. This excludes many details of coastline and topography but preserves the ability to capture the main features necessary to simulate the large-scale features. Multiyear integrations are carried out with these models and averaged in time to obtain a picture of the general circulation.

A validation of the coupled model system has to take into account oceanic transport estimates along selected sections (see **Figure 1**), satellite-based observations of the annual cycle of sea ice concentration, pointwise measurements of sea ice thickness and Lagrangian observations of sea ice drift. In all four categories, agreement within the limits of measurement uncertainty can be achieved.

A consistent picture of the three-dimensional ocean circulation in the Weddell Sea is obtained from multiyear simulations of the circumpolar ocean, using ECMWF atmospheric data for the late 1980s and early 1990s. The model indicates that the wind-driven and thermohaline components are of similar importance in forcing the barotropic flow.

The Structure of the Ocean Currents in the Weddell Sea

The dominant feature of the currents in the Weddell Sea is the cyclonic gyre. It appears clearly in the time-mean streamlines and barotropic currents (**Figure 4A**) of the numerical model. There is a pronounced double cell structure, caused by either the presence of Maud Rise, a seamount with its center at 65°S, 2°30′E or/and the inflow from the north of water of circumpolar origin at about 20°W. The volume transport across sections along the Greenwich Meridian amounts to 50 Sv. The transport across the section from the northern tip of the Antarctic Peninsula to Kapp Norvegia is 30 Sv (**Figure 1**). The velocities in the boundary currents are relatively high, up to $50\,\mathrm{cm\,s^{-1}}$ and in the interior almost stagnant conditions prevail (**Figure 3**).

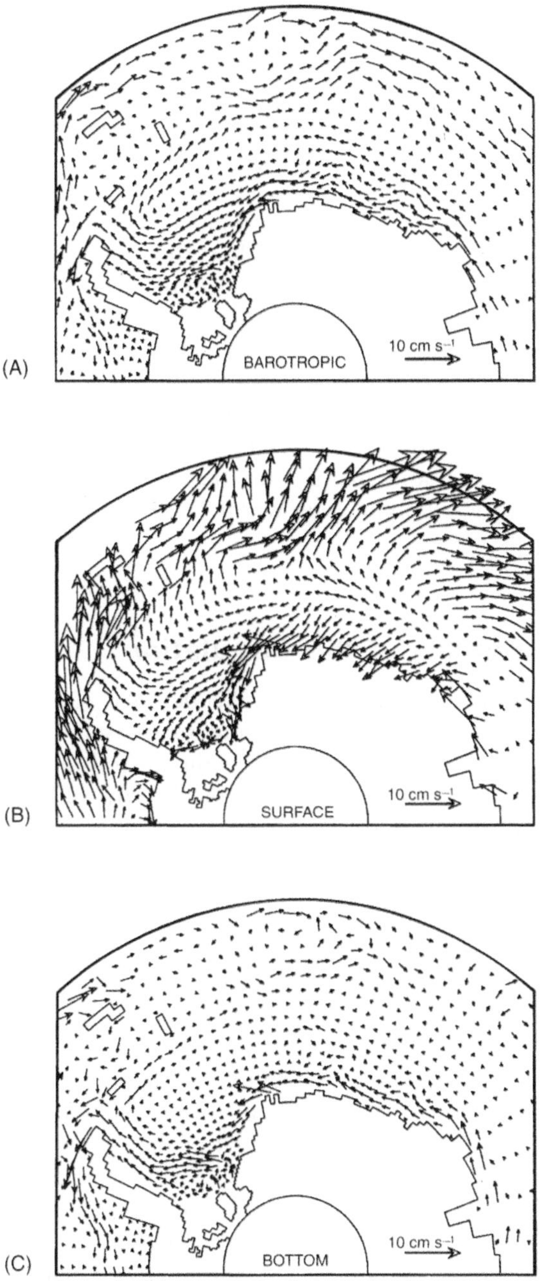

Figure 4 Annual mean current field in the Weddell Sea from a numerical simulation of the circumpolar ocean; (A) barotropic; (B) surface velocity vectors, and (C) bottom.

The surface circulation in the eastern Weddell Gyre is characterized by the Antarctic Divergence, which divides the onshore flow south of 63°S from the predominantly equatorward flow to the north (**Figure 4B**). This pattern is disrupted in the western Weddell Sea, where the presence of the Antarctic Peninsula causes a generally northward flow. The observed sea ice drift patterns reflect both this wind-driven surface velocity field, and the barotropic flow, through the sea surface inclination.

The wind-driven flow in the surface Ekman layer leads to coastal downwelling, with a corresponding offshore (downslope) component in the bottom boundary layer (**Figure 4C**). The near-bottom flow is clearly concentrated along the continental slope and numerical results suggest that part of it continues westward past the tip of the Antarctic Peninsula.

The best-known part of the ocean current system in the Weddell Sea is the Antarctic Coastal Current which follows the Antarctic coastline from the east to the west. It is partly driven by the persistent Antarctic east winds and partly by thermohaline forcing evidenced by the differences between shelf and open ocean water masses. The east winds force an onshore Ekman transport which is balanced by an upward sea level inclination towards the coast which leads to westward currents along the coast.

Its seaward extent is defined in various ways; either, it includes all westward flow between the coast and the center of the cyclonic gyre, or it is limited to the near-shore part including the 'shelf front' which is formed by differences between the water mass characteristics of the shelf waters and the open ocean surface layer. The near-shore water is seasonally highly variable and can be less or more saline than the open ocean water, depending on the relative importance of the melt water input from the continent and the salt release due to sea ice formation. Enhanced (vertical) mixing on the shelf also contributes to the horizontal water mass differences. The density gradient-related shelf front gives rise to a frontal jet. The frontal jet forms a local maximum of the coastal current.

The path of the coastal current roughly follows the depth contours, i.e., is mainly along-slope. It is mainly barotropic, but a baroclinic component is superimposed which causes a decrease of the flow with depth but in certain areas the flow is bottom-intensified.

The flow speed in the coastal current averaged over weekly or monthly periods ranges between 10 and 50 cm s^{-1}. The corresponding volume transport amounts to 10 to 15 Sv. In some areas, the current may separate into two (or more) quasiparallel branches. This is often triggered by the bottom topography, e.g., near Kapp Norvegia where a relatively flat plateau interrupts the continental slope at 1500–2500 m depth. Eventually, an undercurrent towards the east is found on the upper slope.

The core of the offshore branch of the coastal current (200–1500 m) transports warm deep water westward which originates from Circumpolar Deep Water (**Figure 2**). This warm and saline water mass can be used to trace the gyre flow along the coast. The westward-decreasing temperature and salinity anomaly reveals the exchange with the ambient water masses. In addition to the along-slope flow, cross-slope circulation cells transport modified warm deep water up the continental slope and on to the shelf, where it either loses its heat to the atmosphere or melts sea or shelf ice. The heat supply by warm deep water is the major heat source for shelf ice melt. In the open ocean it controls the sea ice thickness, because haline convection due to sea ice formation can bring the heat from the warm deep water into the surface mixed layer and control sea ice growth.

In the southern Weddell Sea where the shelf widens to several hundred kilometers the Antarctic Coastal Current splits into two branches; one follows the coastline onto the shelf and the other continues along the continental slope. Both branches join again at the Antarctic Peninsula.

The shelf areas in the southern and western Weddell Sea with a depth up to 500 m are the origin of downslope plumes of dense water. These water masses usually descend gradually down the slope as they follow the general along-slope path of the water masses of the coastal current. Alternatively, they may be guided directly downslope at topographical features like ridges or canyons. They form either Weddell Sea Deep Water by mixing with adjacent water masses or interleaving in intermediate depths, or Weddell Sea Bottom Water which mixes afterwards with the layers on top of it to form again Weddell Sea Deep Water, which fills most of the Weddell Basin.

The northward flow along the Antarctic Peninsula is relatively well studied. It ranges between 25 and 30 Sv whereas the transport of newly formed Weddell Sea Bottom Water amounts only to a few Sv.

At the tip of the Antarctic Peninsula, the Weddell Sea shelf waters are injected between those from the Antarctic Circumpolar Current and the Weddell Sea proper. By this process two fronts are formed: the Weddell Front in the south and the Scotia Front in the north which enclose the 'Weddell–Scotia Confluence' zone. Water masses from the Weddell–Scotia Confluence are dense enough to sink and to contribute to the renewal of the global oceans deep water. The location of the flow band is strongly controlled by bottom topography. Mixing and local atmosphere ocean exchange can cause further modifications.

The Weddell Front follows the North Weddell and the Southwest Indian Ridges to the east. The related northern current band follows those structures and is strongly affected by their irregularities which generate meridional perturbations of the zonal field. It is most likely that Circumpolar Deep Water enters the Weddell Gyre in such excursions and that Weddell Sea Deep Water leaves it. Horizontal structures in the Warm Deep Water core in the gyre suggest this exchange and affect the gyre structure. A separation into two adjacent subgyres, as indicated earlier, might be due to those intrusions.

The trapping of the gyre flow along the midocean ridge ends at the eastern edge of the Southwest Indian Ridge. There, the northern part of the gyre flow seems to split into an eddy field, consisting of cold eddies with water from the Weddell Gyre and warm eddies with a core of Circumpolar Deep Water. The eddies drift in the remnant flow field to the south-west and merge into the southern band of the gyre supplying the Warm Deep Water flow in the gyre.

Variability of the Circulation

Seasonal variations in the Weddell Sea circulation have been observed in the Antarctic Coastal Current, which reaches a maximum in the austral winter. A similar cycle is superimposed on the flow of bottom water in the north-western Weddell Sea. However, outside the boundary currents the seasonal cycle can only be derived from numerical model results and appears to be relatively small. The Weddell Gyre transport is larger in winter than in summer, due to stronger winds. Thermohaline effects on the large-scale circulation are not felt on a seasonal scale but on longer timescales.

This interannual variability is dominated by variations in sea ice cover and formation. Numerical studies indicate that the signal of the Antarctic Circumpolar Wave (with a typical period of four years) influences the whole Weddell Sea. A four-year periodicity can be detected, mainly in response to meridional wind stress anomalies; strong southerly winds in the western Weddell Sea lead to increased ice export causing more ice formation and deep water production during the following winter. With a time lag of about one year, this newly formed bottom water will begin to cross the South Scotia Ridge northward and spread into the global ocean.

Circulation fluctuations on longer timescales certainly exist, but they have not been investigated conclusively.

The Role of the Weddell Gyre in the Global Ocean Circulation

A significant part of the water mass transformation in the Southern Ocean occurs in the Weddell Sea. A census of the water colder than 0°C south of the Polar Front revealed that 66% of it was from Weddell Sea Bottom Water, 25% from Adélie Land Bottom Water and 7% from Ross Sea Bottom Water.

Whereas the quasi-zonal Antarctic Circumpolar Current system is a barrier for meridional exchange, the subpolar gyres in the Ross and Weddell Seas have sufficiently strong meridional flow components to allow for significant meridional heat and freshwater transports. The eastern branch of the cyclonic circulation of the Weddell Gyre advects water masses from the subantarctic water belt towards the Antarctic coast where intense atmosphere–ocean interaction will lead to a decrease in temperature and an increase in salinity.

This occurs mainly in coastal polynyas, induced by a strong offshore wind component with cold air from the continent. The irregular structure of the coastline forming capes and embayments leads regionally to offshore winds even if the large-scale directions of the winds is parallel to the coast. In the coastal polynyas the oceanic heat loss to the atmosphere can exceed 500 W m^{-2}. The salt gain due to sea ice formation has to compensate the fresh water gain by glacial melt water from the continent and precipitation which had desalinated the previously upwelled Circumpolar Deep Water. The relative importance of melting icebergs as a regional enhancement of the freshwater gain is still unclear. If the salt release is strong enough, the density increases until the water sinks and forms bottom water directly by plumes sinking down the continental slope or by further cooling during the circulation under the ice shelves. Both forms of dense shelf water form plumes on the continental slope which either reach the bottom of the Weddell Basin or enter the open ocean at a depth level according to their own density by interleaving. Due to the nonlinearity of the equation of state of sea water, descending plumes can be formed or enhanced by the thermobaric effect.

Eventually the regime of the deep water renewal by plumes along the continental slopes can switch to deep open ocean convection. This was most likely happening during the 1970s when a large open ocean polynya was observed west of Maud Rise. The polynya and open ocean convection are in intensive interaction, because the heat loss due to open water in winter cools the water column sufficiently to form deep water whilst on the other hand, the normal supply of Warm Deep Water from deeper layers can maintain the polynya. The polynya formation could be caused by advection of warmer water from the north or by changing atmospheric forcing.

Weddell Sea Deep Water leaves the Weddell Sea to the north and represents the major cold source of water for the globally spreading bottom water. The water mass formation in the Weddell Sea, therefore, represents a major part of the global thermohaline overturning circulation. The combined effects of wind-induced downwelling and thermohaline driven sinking of dense water masses from the shelves of the inner Weddell Sea generate a large-scale overturning motion in the Southern Ocean. Circumpolar ice–ocean model simulations indicate maximum overturning transports of about 20 Sv, half of which originate in the Weddell Sea sector. At the same time, estimates from observations show that only a relatively small part (2–3 Sv) seems to be directly in contact with the surface, thus ventilating the deep ocean.

See also

Antarctic Circumpolar Current. Bottom Water Formation. Ekman Transport and Pumping. Rotating Gravity Currents.

Further Reading

Beckmann A, Hellmer HH, and Timmermann R (1999) A numerical model of the Weddell Seal: arge-scale circulation and water mass distribution. *Journal of Geophysical Research* 104: 23 375–23 391.

Carmack EC (1986) Circulation and mixing in ice-covered waters. In: Untersteiner N (ed.) *The Geophysics of Sea Ice*, pp. 641–712. New York: Plenum.

Carmack EC (1990) Large-scale physical oceanography of polar oceans. In: Smith WO Jr (ed.) *Polar Oceanography*, Part A: *Physical Science*, pp. 171–222. San Diego: Academic Press.

Fahrbach E, Klepikov A and Schröder M (1998) Circulation and water masses in the Weddell Sea. *Physics of Ice-Covered Seas*, vol. 2, pp. 569–604. Helsinki: Helsinki University Press.

Gordon AL, Martinson DG, and Taylor HW (1981) The wind-driven circulation in the Weddell–Enderby Basin. *Deep-Sea Research* 28: 151–163.

Haidvogel DB and Beckmann A (1988) Numerical modeling of the coastal ocean. In: Brink KH and Robinson AR (eds.) *The Sea*, vol. 10, pp. 457–482. New York: John Wiley and Sons.

Hofmann EE and Klinck JM (1998) Hydrography and circulation of the Antarctic continental shelf: 150°E to the Greenwich meridian. In: Robinson AR and Brink KH (eds) *The Sea*, vol. 11, 997–1042. New York: John Wiley and Sons.

Jacobs SS (ed.) (1985) *Oceanology of the Antarctic Continental Shelf*, Antarctic Research Series, 43. Washington, DC: American Geophysical Union.

Jacobs SS and Weiss RF (eds.) (1998) *Ocean, Ice and Atmosphere: Interactions at the Antarctic Continental Margin*, Antarctic Research Series, 75. Washington DC: American Geophysical Union.

Muench RD and Gordon AL (1995) Circulation and transport of water along the western Weddell Sea margin. *Journal of Geophysical Research* 100: 18503–18515.

Orsi AH, Nowling Jr WD, and Whitworth III T (1993) On the circulation and stratification of the Weddell Gyre. *Deep-Sea Research* 40: 169–203.

Whitworth III T, Nowling Jr WD, Orsi AH, Locarnini RA, and Smith SG (1994) Weddell Sea Shelf Water in the Bransfield Strait and Weddell–Scotia Confluence. *Deep-Sea Research* 41: 629–641.

FIORD CIRCULATION

A. Stigebrandt, University of Gothenburg, Gothenburg, Sweden

Introduction

Fiords are glacially carved oceanic intrusions into land. They are often deep and narrow with a sill in the mouth. Waters from neighboring seas and locally supplied fresh water fill up the fiords, often leading to strong stratification. During transport into and stay in the fiord, mixing processes modify the properties of imported water masses. From the top downward, the fiord water is appropriately partitioned into surface water, intermediary water, and, beneath the sill level, basin water. Fiord circulation is forced both externally and internally. External forcing is provided by temporal variations of both sea level (e.g., tides) and density of the water column outside the fiord mouth. Internal forcing is provided by freshwater supply, winds, and tides in the fiord. The response of circulation and mixing in the different water masses to a certain forcing depends very much on characteristics of the fiord topography. The circulation of basin water is critically dependent on diapycnal mixing. Hence we focus on fiord circulation from a hydrodynamic point of view. Major hydrodynamic processes and simple quantitative models of the main types of circulation are presented.

Basic Concepts in Fiord Descriptions

Some key elements of the hydrography and dynamics of fiords are shown in **Figure 1**. The surface water may have reduced salinity due to freshwater supply. It is kept locally well mixed by the wind and the thickness is typically of the order of a few meters. The thickness of the intermediary layer, reaching down to the sill level, depends strongly on the sill depth. This layer may be thin and even missing in fiords with shallow sills. Surface and intermediary waters have free connection with the coastal area through the fiord's mouth. Basin water, the densest water in the fiord, is trapped behind the sill. It is vertically stratified but the density varies less than in the layers above. A typical vertical distribution of density, $\rho(z)$, in a strongly stratified fiord is shown in **Figure 1**. The area (water) outside the fiord is denoted here as coastal area (water).

In most fiords, temporal variations of the density of the coastal water are crucial for the water exchange, both above and below the sill level. Surface and intermediary waters in short fiords with relatively wide mouths may be exchanged quickly (i.e., in days). The vertical stratification in the intermediary layer in such a fiord is usually quite similar to the

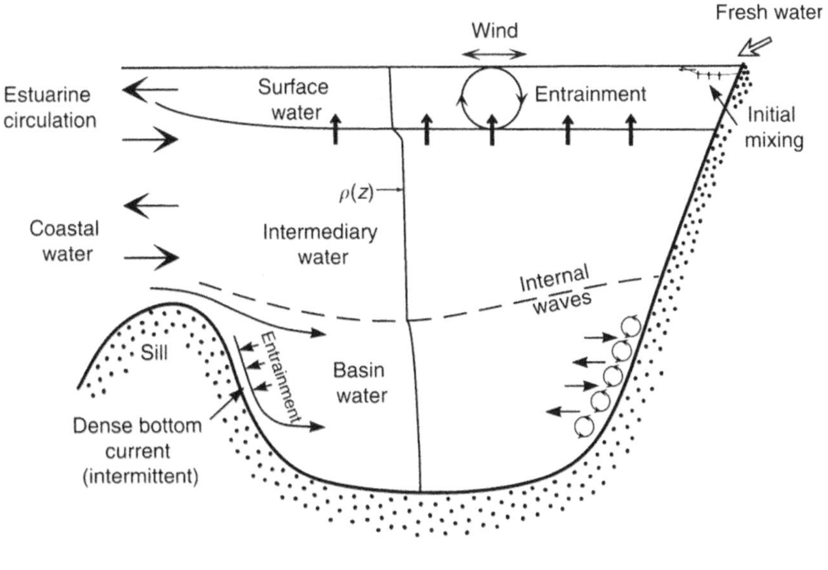

Figure 1 Basic features of fiord hydrography and circulation. Turbulent mixing in the basin water occurs mainly close to the bottom boundary, with increased intensity close to topographic features where internal tides are generated.

stratification outside the fiord although there is some phase lag, with the coastal stratification leading before that in the fiord. Short residence time for water above sill level means that the pelagic ecology may change rapidly due to advection. Denser coastal water, occasionally appearing above sill level, intrudes into the fiord and sinks down along the seabed. It then forms a turbulent dense bottom current that entrains ambient water, decreasing the density and increasing the volume flow of the current (**Figure 1**). The intruding water replaces residing basin water. During so-called stagnation periods, when the density of the coastal water above sill level is less than the density of the basin water, a pycnocline develops at or just below sill depth in the fiord. In particular, during extended stagnation periods, oxygen deficit and even anoxia may develop. The density of basin water decreases slowly due to turbulent mixing, transporting less-dense water from above into this layer. Tides are usually the main energy source for deep-water turbulence in fiords. Baroclinic wave drag, acting on barotropic tidal flow across sills separating stratified basins, is the process controlling this energy transfer.

Major Hydrodynamic Processes in Fiord Circulation

Understanding fiord circulation requires knowledge of some basic physical oceanographic processes such as diapycnal mixing and a variety of strait flows, regulating the flow through the mouth. Natural modes of oscillation (seiches) as well as sill-induced processes like baroclinic wave drag, tidal jets, and hydraulic jumps may constitute part of the fiord response to time-dependent forcing.

Flow through fiord mouths is driven mainly by barotropic and baroclinic longitudinal pressure gradients. The barotropic pressure gradient, constant from sea surface to seabed, is due to differing sea levels in the fiord and the coastal area. Baroclinic pressure gradients arise from differing vertical density distributions in the fiord and coastal area. The baroclinic pressure gradient varies with depth. Different types of flow resistance and mechanisms of hydraulic control may modify flow through a mouth. Barotropic forcing usually dominates in shallow fiord mouths while baroclinic forcing may dominate in deeper mouths.

Three main mechanisms cause resistance to barotropic flow in straits. First is the friction against the seabed. Second is the large-scale form drag, due to large-scale longitudinal variations of the vertical cross-sectional area of the strait causing contraction followed by expansion of the flow. And third is the baroclinic wave drag. This is due to generation of baroclinic (internal) waves in the adjacent stratified basins.

Barotropic flow Q through a straight, rectangular, narrow, and shallow strait of width B, depth D, and length L, connecting a wide fiord and the wide coastal area may be computed from

$$Q^2 = \frac{2g\Delta\eta B^2 D^2}{1 + 2C_D(L/D)} \qquad [1]$$

The sign of Q equals that of $\Delta\eta = h_o - h_i$, h_o (h_i) is the sea level in the coastal area (fiord). This equation contains both large-scale topographic drag and drag due to bottom friction (drag coefficient C_D). Large-scale topographic drag is the greater of the two in short straits (i.e., where $L/D < 2/C_D$).

To get first estimates of baroclinic flows and processes, it is often quite relevant to approximate a continuous stratification by a two-layer stratification, with two homogeneous layers of specified thickness and density difference $\Delta\rho$ on top of each other. The reduced gravity of the less-dense water is $g' = g\Delta\rho/\rho_0$, where g is the acceleration of gravity and ρ_0 a reference density.

In stratified waters, fluctuating barotropic flows (e.g., tides) over sills are subject to baroclinic wave drag. This is important in fiords because much of the power transferred to baroclinic motions apparently ends up in deep-water turbulence. Assuming a two-layer approximation of the fiord stratification, with the pycnocline at sill depth, the barotropic to baroclinic energy transfer E_j from the jth tidal component is given by

$$E_j = \frac{\rho_0}{2}\omega_j^2 a_j^2 \frac{A_f^2}{A_m} \frac{H_b}{H_b + H_t} c_i \qquad [2]$$

Here ω_j is the frequency and a_j the amplitude of the tidal component. A_f is the horizontal surface area of the fiord, A_m the vertical cross-sectional area of the mouth, H_t (H_b) sill depth (mean depth of the basin water), and

$$c_i = \sqrt{g' \frac{H_t H_b}{H_t + H_b}}$$

the speed of long internal waves in the fiord. Baroclinic wave drag occurs if the speed of the barotropic flow in the mouth is less than c_i. If the barotropic speed is higher, a tidal jet develops on the lee side of the sill together with a number of flow phenomena like internal hydraulic jumps and associated internal waves of supertidal frequencies.

Baroclinic flows in straits may be influenced by stationary internal waves imposing a baroclinic hydraulic control. For a two-layer approximation of the stratification in the mouth the flow is hydraulically controlled when the following condition, formulated by Stommel and Farmer in 1953, is fulfilled:

$$\frac{u_{1m}^2}{g'H_{1m}} + \frac{u_{2m}^2}{g'H_{2m}} = 1 \qquad [3]$$

Here u_{1m} (u_{2m}) and H_{1m} (H_{2m}) are speed and thickness, respectively, of the upper (lower) layer in the mouth. Equation [3] may serve as a dynamic boundary condition for fiord circulation as it does in the model of the surface layer presented later in this article. It has also been applied to very large fiords like those in the Bothnian Bay and the Black Sea. Experiments show that superposed barotropic currents just modulate the flow. However, if the barotropic speed is greater than the speed of internal waves in the mouth these are swept away and the baroclinic control cannot be established and the transport capacity of the strait with respect to the two water masses increases. In wide fiords, the rotation of the Earth may limit the width of baroclinic currents to the order of the internal Rossby radius $\sqrt{g'H_1}/f$. Here H_1 is the thickness of the upper layer in the fiord and f the Coriolis parameter. The outflow from the surface layer is then essentially geostrophically balanced and the transport is $Q_1 = g'H_1^2/2f$. This expression has been used as boundary condition for the outflow of surface water, for example, from Kattegat and the Arctic Ocean through Fram Strait.

Diapycnal (vertical) mixing processes may modify the water masses in fiords. In the surface layer, the wind creates turbulence that homogenizes the surface layer vertically and entrains seawater from below. The rate of entrainment may be described by a vertical velocity, w_e, defined by

$$w_e = \frac{m_0 u_*^3}{g'H_1} \qquad [4]$$

Here u_* is the friction velocity in the surface layer, linearly related to the wind speed, and m_0 (~ 0.8) is essentially an efficiency factor, well known from seasonal pycnocline models. H_1 is the thickness of the surface layer in the fiord and g' the buoyancy of surface water relative to the underlying water. Buoyancy fluxes through the sea surface modify the entrainment velocity.

Due to their sporadic and ephemeral character, there are only few direct observations of dense bottom currents in fiords. However, from both observations and modeling of dense bottom currents in the Baltic it appears that entrainment velocity may be described by eqn [4]. For this application, u_* is proportional to the current speed, H_1 is the current thickness, and g' the buoyancy of ambient water relative to the dense bottom current.

Observational evidence strongly supports the idea that diapycnal mixing in the basin water of most fiords is driven essentially by tidal energy, released by baroclinic wave drag at sills. The details of the energy cascade, from baroclinic wave drag to small-scale turbulence, are still not properly understood. **Figure 1** leaves the impression that energy transfer to small-scale turbulence and diapycnal mixing takes place in the inner reaches of a fiord. Here internal waves, for example, waves generated by baroclinic wave drag at the sill, are supposed to break against sloping bottoms. A tracer experiment in the Oslo Fiord suggests that mixing essentially occurs along the rim of the basin. Recently, the temporal and spatial distributions of turbulent mixing in the basin waters of a few fiords have been mapped. Measurements in the Gullmar Fiord suggest that much of the mixing takes place close to the sill where most of the barotropic to baroclinic energy transfer takes place.

In a column of the basin water the mean rate of work against the buoyancy forces is given by

$$W = W_0 + \frac{Rf \sum_{j=1}^{n} E_j}{A_t} \qquad [5]$$

Here W_0 is the nontidal energy supply, n the number of tidal components, and E_j may be obtained from eqn [2]. A_t is the horizontal surface area of the fiord at sill level and Rf the flux Richardson number, the efficiency of turbulence with respect to diapycnal mixing. Estimates from numerous fiords show that $Rf \sim 0.06$. Experimental evidence shows that in fiords with a tidal jet at the mouth, most of the released energy dissipates above sill level and only a small fraction contributes to mixing in the deep water.

Simple Quantitative Models of Fiord Circulation

The Surface Layer

The upper layers in fiords may be exchanged due to so-called estuarine circulation, caused by the combination of freshwater supply and vertical mixing. This is essentially a baroclinic circulation, driven by density differences between the upper layers in the fiord and coastal area, respectively. However, if the sill is very shallow, the water exchange tends to be performed by barotropic flow with alternating direction, forced by the fluctuating sea level outside the fiord.

The following equations describe the steady-state volume and salt conservation of the surface layer:

$$Q_1 = Q_2 + Q_f \quad [6]$$

$$Q_1 S_1 = Q_2 S_2 \quad [7]$$

Here Q_f is the freshwater supply, Q_1 (Q_2) the outflow (inflow) of surface (sea-) water, and S_1 (S_2) the salinity of the surface water (seawater). Equations [6] and [7] give:

$$Q_1 = Q_f \frac{S_2 - S_1}{S_2} \quad [8]$$

This equation has been used throughout the past century for diagnostic estimates of the magnitude of estuarine circulation from measurements of S_1, S_2, and Q_f.

Density varies with both salinity and temperature. In brackish waters, density (ρ) variations are often dominated by salinity (S) variations. For simplified analytical models, one may take advantage of this and use the equation of state for brackish water:

$$\rho = \rho_f(1 + \beta S) \quad [9]$$

Here ρ_f is the density of fresh water and the so-called salt contraction coefficient β equals 0.000 8 S^{-1}. The density difference $\Delta\rho$ between two homogeneous layers, with salinity difference ΔS, then equals $\rho_f \beta \Delta S$, and the buoyancy $g' = g\Delta\rho/\rho$ equals $g\beta\Delta S$. A continuous stratification in a salt-stratified system may be replaced by a dynamically equivalent two-layer stratification. This requires that the two-layer and observed stratification (1) contain the same amount of fresh water and (2) have the same potential energy.

Stationary estuarine circulation in fiords with deep sills To investigate how salinity S_1 and thickness H_1 of the surface layer depend on wind and freshwater supply, the following simple model may be illustrative. It is assumed that the fiord mouth is deep and even narrow compared to the fiord. Then the so-called compensation current into the fiord is deep and slow compared with the current of outflowing surface water. The hydraulic control condition in the mouth, eqn [3], is then simplified to $u_{1m}^2 = g' H_{1m}$. The thickness of the surface layer H_1 in the fiord is related to that in the mouth by $H_1 = \varphi H_{1m}$. Entrainment of seawater of salinity S_2 into the surface layer is described by eqn [4]. Under these assumptions, expressions for the thickness H_1 and salinity S_1 of the surface layer in the fiord may be derived:

$$H_1 = \frac{G}{2Q_f g'} + \varphi\left(\frac{Q_f^2}{g' B_m^2}\right)^{1/3} \quad [10]$$

$$S_1 = \frac{S_2 G}{G + 2\varphi\left[Q_f^5\left(\frac{g'}{B_m}\right)^2\right]^{1/3}} \quad [11]$$

Here B_m is the width of the control section in the mouth, $G = CW_s^3 A_f$, $C = 2.5 \times 10^{-9}$ is an empirical constant containing, among others, the drag coefficient for air flow over the sea surface, W_s the wind speed and $g' = g\beta S_2$. Theoretically, the value of φ is expected to be in the range 1.5–1.7 for fiords where $B_m/B_f \leq 1/4$ where B_f is the width of the fiord inside the mouth. The value of φ should be smaller for wider mouths. Observations in fiords with narrow mouths give φ values in the range 1.5–2.5.

The left term in the expression for H_1 is the so-called Monin–Obukhov length, known from the theory of geophysical turbulent boundary layers with vertical buoyancy fluxes, and the right term is the freshwater thickness H_{1f}, hydraulically controlled by the mouth. The salinity of the surface layer S_1 increases with increasing wind speed and decreasing freshwater supply. For a given freshwater supply, strong winds may apparently multiply the outflow as compared with Q_f (cf. eqn [8]).

The freshwater volume in the fiord is $V_f = H_{1f} A_f$. The residence time of fresh water in the fiord, $\tau_f = V_f/Q_f$ is given by

$$\tau_f = \varphi A_f \left(\frac{1}{g' B_m^2}\right)^{1/3} Q_f^{-1/3} \quad [12]$$

The residence time thus decreases with the freshwater supply. It should be noted that H_{1f}, τ_f, and V_f are independent of the rate of wind mixing.

Water exchange through very shallow and narrow mouths In very shallow and narrow fiord mouths, barotropic flow usually dominates. The instantaneous flow which can be estimated using eqn [1] is typically unidirectional and the direction depends on the sign of $\Delta\eta$, the sea level difference across the mouth. If the mouth is extremely shallow and narrow, tides and other sea level fluctuations in the fiord will have smaller amplitude than in the coastal area due to the choking effect of the mouth. A number of choked fiords and other semi-enclosed water bodies have been described in the literature, such as Framvaren and Nordaasvannet (Norway), Sechelt Inlet (Canada), the Baltic Sea, the Black Sea, and tropical lagoons.

The Intermediary Layer

Density variations in the coastal water above sill level give rise to water exchange across the mouth, termed intermediary water exchange. The stratification in the fiord strives toward that in the coastal water. Intermediary circulation increases in importance with increasing sill depth. It has been found that baroclinic intermediary circulation is the dominating circulation component in a majority of Scandinavian and Baltic fiords and bays. This is probably true also in other regions, although there have been few investigations quantifying intermediary circulation.

Despite its often-dominating contribution to water exchange in fiords, the intermediary circulation has remained astonishingly anonymous and in many studies of inshore waters even completely overlooked. One obvious reason for this is that a simple formula to quantify the mean rate of intermediary water exchange, Q_i, has been available only during the last decade (eqn [13]).

$$Q_i = \gamma\sqrt{B_m H_t A_f \frac{g\Delta M}{\rho}} \qquad [13]$$

Here the dimensionless empirical constant γ equals 17×10^{-4}, as estimated for Scandinavian conditions, and ΔM the standard deviation of the weight of the water column down to sill level ($\mathrm{kg\,m^{-2}}$) in the coastal water. The latter should be a hydrodynamically reasonable measure of the mean strength of the baroclinic forcing. Statistics of scattered historic hydrographic measurements may be used to compute ΔM. Equation [13] should be regarded as a precursor to a formula, which is yet to be developed, accommodating for the frequency dependence of ΔM.

A conservative estimate of the mean residence time for water above the sill is $\tau_i = V_i/Q_i$, where V_i is the volume of the fiord above sill level. The residence time may be shorter if other types of circulation contribute to the water exchange.

The Basin Water

The time between two consecutive exchanges of basin water, often called the residence time, can be partitioned into the stagnation time (when there is no inflow into the basin) and the filling time (the time it takes to fill the basin with new deep water). The filling time is determined by the flow rate of new deep water through the mouth and the fiord volume below sill level. In most fiords the filling time is much shorter than the stagnation time. The spectral distribution of the density variability in the coastal water determines the recurrence time of water of density higher than a certain value. Knowing the spectral distribution of the variability of the coastal density and the rate of density decrease of the basin water due to diapycnal mixing, one may estimate the recurrence time of water exchange. In fiords with short filling time, this should be inversely proportional to the rate of diapycnal mixing. A very long stagnation period may occur only if the rate of vertical mixing is very low and the density in the coastal water has a long-period component of appreciable amplitude.

A rough estimate of the mean rate of diapycnal mixing in the basin water may be obtained from eqn [14]:

$$\frac{\mathrm{d}\rho}{\mathrm{d}t} = -\frac{CW}{gH_b^2} \qquad [14]$$

Here the empirical constant C equals 2.0 and W may be obtained from eqn [5]. The vertical diffusivity κ at the level z in the basin water may be computed from the empirical expression in eqn [15]:

$$\kappa(z) = \frac{W/H_b}{\rho\bar{N}^2} c_\kappa \left(\frac{N(z)}{\bar{N}}\right)^{-1.5} \qquad [15]$$

Here $\bar{N}$ is the volume-weighted vertical average of the buoyancy frequency $N(z)$ and c_κ (~ 1) is an empirical constant.

If the filling time of the fiord basin is very long, the basin will be filled not only with the densest but also with less dense coastal water. The basin water in such a fiord will thus have lower density than the basin water in a neighboring fiord with similar conditions except for a much shorter filling time. If the filling time is sufficiently long, this will determine the residence time, which will not change if the rate of vertical mixing changes. The fiord basin is then said to be overmixed. The Baltic and the Black Seas are two overmixed systems for which the transport capacities of the mouths determine the residence time.

Conclusions

This article on fiord circulation demonstrates that several oceanographic processes of general occurrence are involved in fiord circulation. Being sheltered from winds and waves, fiords are excellent large-scale laboratories for studying these processes. The mechanics of water exchange of surface and intermediary layers, as described here, should apply equally well to narrow bays lacking sills.

Nomenclature

a_j	amplitude of the *j*th tidal component
A_f	horizontal surface area of the fiord
A_m	vertical cross-sectional area of the mouth
A_t	horizontal surface area of the fiord at sill depth
B	width of the mouth channel
B_f	width of the fiord inside the mouth
B_m	width of the mouth at the control section
c_κ	empirical constant (≈ 1)
C	empirical constant ($= 2.5 \times 10^{-9}$)
C_D	drag coefficient for flow over the seabed
D	depth of the mouth channel
E_j	barotropic to baroclinic energy transfer from the *j*th tidal component
g	acceleration of gravity
g'	($= g\Delta\rho/\rho_0$) buoyancy
G	wind factor
h_i	sea level in the fiord
h_o	sea level in the coastal area
$H_{1m}(H_{2m})$	thickness of upper (lower) layer in the mouth
H_b	mean depth of the basin water
H_t	sill depth
L	length of mouth channel
m_0	efficiency factor ($= 0.8$)
$N(z)$	buoyancy frequency at depth z
$\bar{N}$	volume-weighted buoyancy frequency in the deepwater
Q_1	flow out of the surface layer
Q_2	flow into the fiord beneath the surface layer
Q_f	freshwater supply
Rf	Richardson flux number, efficiency factor
S_1 (S_2)	salinity of upper (lower) layer
t	time
u_{1m} (u_{2m})	speed of upper (lower) layer in the mouth
u_*	friction velocity
V_f	volume of fresh water in the upper layer
W	rate of work against buoyancy forces in a column of basin water
Ws	wind speed
z	vertical coordinate
β	salt contraction of water ($= 0.000\,8$ S^{-1})
γ	empirical constant ($= 17 \times 10^{-4}$)
ΔM	standard deviation of weight of water column between sea surface and sill level
$\Delta\eta$	sea level difference across the mouth ($h_o - h_i$)
$\Delta\rho$	density difference
ρ_0	reference density
ρ_f	density of fresh water
$\rho(z)$	density as function of depth z
τ_f	residence time for fresh water in the upper layer
φ	contraction factor ($= 1.5$ for fiords with narrow mouths)
ω_j	frequency of the *j*th tidal component

See also

Flows in Straits and Channels.

Further Reading

Arneborg L, Janzen C, Liljebladh B, Rippeth TP, Simpson JH, and Stigebrandt A (2004) Spatial variability of diapycnal mixing and turbulent dissipation rates in a stagnant fiord basin. *Journal of Physical Oceanography* 34: 1679–1691.

Aure J, Molvær J, and Stigebrandt A (1997) Observations of inshore water exchange forced by fluctuating offshore density field. *Marine Pollution Bulletin* 33: 112–119.

Farmer DM and Freeland HJ (1983) The physical oceanography of fiords. *Progress in Oceanography* 12: 147–220.

Freeland HJ, Farmer DM and Levings DC (eds.) (1980) *Fiord Oceanography*, 715pp. New York: Plenum.

Stigebrandt A (1999) Resistance to barotropic tidal flow by baroclinic wave drag. *Journal of Physical Oceanography* 29: 191–197.

Stigebrandt A and Aure J (1989) Vertical mixing in the basin waters of fiords. *Journal of Physical Oceanography* 19: 917–926.

ESTUARINE CIRCULATION

K. Dyer, University of Plymouth, Plymouth, UK

Introduction

Estuaries are formed at the mouths of rivers where the fresh river water interacts and mixes with the salt water of the sea. Even though there is only about a 2% difference in density between the two water masses, the horizontal and vertical gradients in density causes the water circulation, and the mixing created by the tides, to be very variable in space and time, resulting in long residence times for pollutants and trapping of sedimentary particles. Both salinity and temperature affect density, but the salinity changes are normally of greatest influence.

Most estuaries are the result of a dramatic rise in sea level of about 100 m during the last 10 000 years, following the end of the Pleistocene glaciation. The river valleys were flooded by the sea, and the valleys infilled with sediment to varying extents. The degree of infilling provides a wide range of topographic forms for the estuaries. Those estuaries that have had large inputs of sediment have been filled, and may have been built out into deltas, where the sediment flux is extreme. These are typical of tropical and monsoon areas. Where the sediment discharge was less, the estuary may still have many of the morphological attributes of river valleys: a sinuous, meandering outline, a triangular cross-sectional form with a deep central channel, and wide shallow flood plains. These are termed drowned river valleys, and are typical of higher latitudes. Where the land mass was previously covered by the glaciers, the river valleys may have been drastically overdeepened, and the U-shaped valleys became fiords.

Because of the shallow water, the sheltered anchorages, and the ready access to the hinterland, estuaries have become the centers of habitation and of industrial development, and this has produced problems of pollution and environmental degradation. The solution to these problems requires an understanding of how the water flows and mixes, and how the sediment accumulates.

Definition

Estuaries can be defined and classified in many ways, depending on whether one is a geologist/geographer, a physicist, an engineer, or a biologist. The most comprehensive physical definition is that,

> An estuary is a semi-enclosed coastal body of water that has free connection to the open sea, extending into the river as far as the limit of tidal influence, and within which sea water is measurably diluted with fresh water derived from land drainage.

There are normally three zones in an estuary: an outer zone where the salinity is close to that of the open sea, and the horizontal gradients are low; a middle zone where there is rapid change in the horizontal gradients and where mixing occurs; and an upper or riverine zone, where the water may be fresh throughout the tide despite there being a tidal rise and fall in water level.

Tides in Estuaries

The tidal rise and fall of sea level is generated in the ocean and travels into the estuary, becoming modified by the shoreline and by shallow water. The tidal context is microtidal when the range is less than 2 m; mesotidal when the range is between 2 m and 4 m; macrotidal when it is between 4 m and 6 m; and hypertidal when it is greater than 6 m. A narrowing of the estuary will cause the range of the tide to increase landward. However, this is counteracted by the friction of the seabed on the water flow, and the fact that some of the tidal energy is reflected back toward the sea. The result is that the time of high and low water is later at the head of the estuary than at the mouth, and the currents turn some tens of minutes after the maximum and minimum water level. This phase difference means that the tidal wave is a standing wave with a progressive component. In high tidal range estuaries, the currents and the range of the tide generally increase toward the head, until in the riverine section the river flow becomes important. These estuaries are often funnel-shaped with rapidly converging sides. The volume of water between high and low tide, known as the tidal prism, is large compared with the volume of water in the estuary at low tide. Since the speed of progression of the tidal wave increases with water depth, the flooding tide rises more quickly than the ebbing tide, leading to an asymmetrical tidal curve and currents on the flood tide larger than those on the ebb tide. These are flood-dominant estuaries. The high tidal range leads to high current velocities, which can produce considerable sediment transport, also dominant in the flood direction.

When the estuary is relatively shallow compared with the tidal range, the wet cross-sectional area is greater at high water than at low water, and the discharge of water per unit of velocity is greater also. There is thus a greater discharge landward near high water than seaward around low water. This creates a Stokes Drift towards the head of the estuary that has to be compensated for by an extra increment of tidally averaged flow toward the mouth in the deeper areas. The Eulerian mean flow measured at a fixed point, the nontidal drift, is therefore greater than the flow due to the river discharge. Additionally, in the upper estuary where intertidal areas are extensive, the deeper channel changes from being flood-dominant to ebb dominance. This produces an obvious location for siltation and the need for dredging.

In microtidal estuaries, friction exceeds the effects of convergence and the tidal range diminishes toward the head. These estuaries are generally fan-shaped, with a narrow mouth and extensive shallow water areas. The currents at the mouth are larger than those inside, and the ebb current velocities are larger than the flood currents, i.e., the estuary would be ebb-dominant.

It has been observed that the ratio of certain estuarine dimensions frequently appears to be constant, leading to the concept of an equilibrium estuary. In particular, the variations of breadth and depth along the estuary are often exponential in form. The O'Brien relationship shows that the tidal prism volume is related to the cross-sectional area of the estuary. As the tidal prism increases, the increased currents through the cross-section cause erosion and the area increases until equilibrium is reached. This implies that there is ultimately a balance between the amount of sediment carried landward on the flood tide and that carried seaward on the ebb, leading to zero net accumulation. This is a widely used concept for the prediction of morphological change.

Circulation Types

From a morphological point of view, estuaries can be classified into many categories in terms of their shape and their geological development. These reflect the degree of infilling by sediment since the ice age. Present-day processes are also important in distinguishing those estuaries dominated by tidal currents and those whose mouths are drastically affected by the spits and bars created by wave-induced longshore drift of sediment. However,though important, these classifications do not tell us much about the water and how its movement affects the discharge of pollutants and of sediment. The classifications, originally proposed by Pritchard in 1952, describe the tidally averaged differences in vertical and longitudinal salinity structure, and the mean water velocity profiles. These have been extended to consider the processes during the tide that contribute to the averages.

Salt Wedge Estuaries

Where the river discharges into a microtidal estuary, the fresher river water tends to flow out on top of the slightly denser sea water that rests almost stationary on the seabed and forms a wedge of salt water penetrating toward the head of the estuary. The salt wedge has a very sharp salinity interface at the upper surface – a halocline. There is a certain amount of friction between the two layers and the velocity shear between the rapidly flowing surface layer and the almost stationary salt wedge produces small waves on the halocline; these can break when the velocity shear between the layers is large enough. The breaking waves inject some of the salty lower layer water into the upper layer, thus enhancing its salinity. This process, known as entrainment, is equivalent to an upward flow of salt water. It is not a very efficient process, but requires a small compensating inflow in the salt wedge. Salt wedge estuaries have almost fresh water on the surface throughout, and almost pure salt water near the bed. The slight amount of tidal movement of the water does not create much mixing, except in some of the shallower water areas, but does act to renew the salty water trapped in hollows on the bed. This process is shown in **Figure 1**. During the tide, the stratification remains high, but the halocline becomes eroded from its top surface during the ebb tide and rises during the flood tide because of new salt water intruding along the bottom. Fiords often have a shallow rock sill at their mouth that isolates the deeper interior basins. The renewal of bottom water is infrequent, often only every few years. Because fiords are so deep the lower layer is effectively stationary, and the mixing is dominated by entrainment.

Partially Mixed Estuaries

In mesotidal and macrotidal estuaries the tidal movement drives the whole water mass up and down the estuary each tide. The current velocities near to the seabed are large, producing turbulence, and

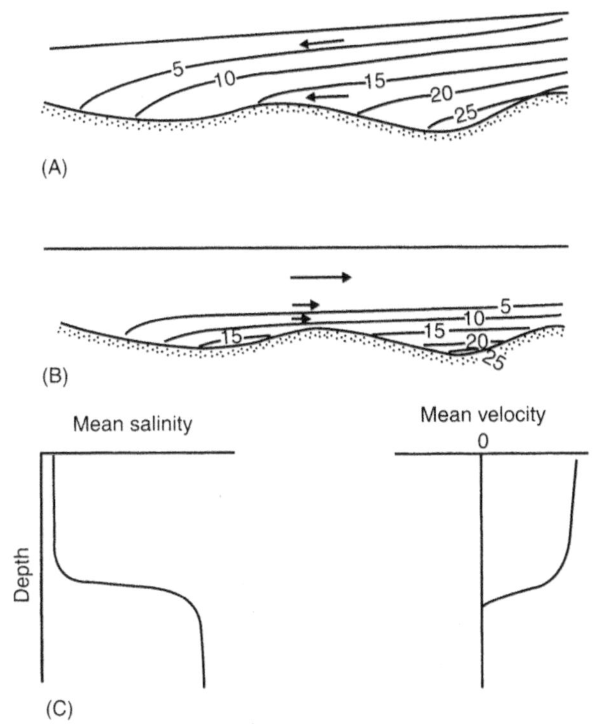

Figure 1 Diagrammatic longitudinal salinity and velocity distributions in a salt wedge estuary in PSU. (A) At mid flood tide. (B) At mid ebb tide. Contours show representative salinities; arrows show relative strength of currents. (C) Tidally averaged mean salinity and velocity profiles.

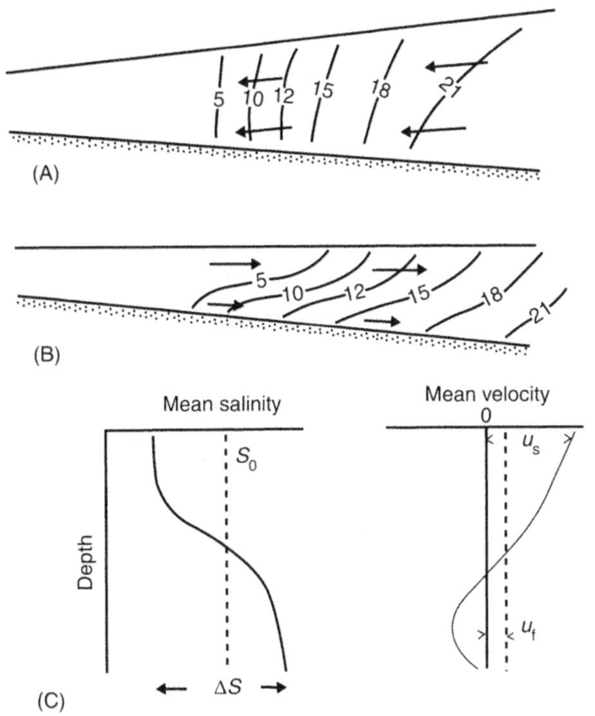

Figure 2 Diagrammatic longitudinal salinity and velocity distributions in a partially mixed estuary in PSU. (A) At mid flood tide. (B) At mid ebb tide. Contours show representative salinities; arrows show relative strength of currents. (C) Tidally averaged mean salinity and velocity profiles, including the definition of parameters for **Figure 4**: S, salinity; ΔS, salinity difference normalized by S_0 and circulation parameter u_s/u_f; u_s, surface mean flow; u_f, depth mean flow.

creating mixing of the water column. This is a much more efficient process for mixing than entrainment, but it is greatest near the bed, whereas entrainment is maximum in mid-water. However, both processes can be active together. Turbulent mixing is a two-way process, mixing fresher water downward and salty water upward. As a result, the mean vertical profile of salinity has a more gentle halocline (**Figure 2**). The salt intrusion is now a much more dynamic feature, changing its structure regularly with a tidal periodicity. During the flood tide, the surface water travels faster up the estuary than the water nearer the bed, and the salinity difference between surface and bed is minimized. Conversely, during the ebb, the fresher surface water is carried over the near-bottom salty water, and the stratification is enhanced. This process is known as tidal straining. It is to be expected that entrainment will be large during the ebb tide, and turbulent mixing will be dominant on the flood tide. Since each tide must discharge an amount of fresh water equivalent to an increment of the river inflow, and the water involved in this is now salty, there must be a significant mean advection of salt water up the estuary near to the bed to compensate for that discharged. There is thus a mean outflow of water and salt on the surface, and a mean inflow near the bed. The latter must diminish toward the head of the salt intrusion, and there is a level of no net motion near to the halocline where the sense of the mean flow velocity changes. At the landward tip of the salt intrusion there is a convergence, a null zone, where the tidally averaged near-bed flow diminishes to zero. The vertical turbulent mixing acting in combination with the mean horizontal advection is known as vertical gravitational circulation. Because of the meandering shape of the estuary, the flow is not straight but tends to spiral on the bends. This leads to the development of lateral differences in salinity and velocity, the shallower regions generally being better mixed and ebb-dominated.

Well-mixed Estuaries

When the tidal range is macrotidal or hypertidal, the turbulent mixing produced by the water flow across the bed is active throughout the water column, with the result that there is very little stratification. Nevertheless, there can be considerable lateral differences across the estuary in salinity and in mean flow velocity, as if the vertical circulation were turned on its side. This results in the currents on one side of the channel being flood-dominated while those on the other side are ebb-dominated, often separated by a mid-channel bank (**Figure 3**). As a consequence, during the tide the water tends to flow preferentially landward up one side of the bank, cross the channel at high water, and ebb down the other side. The bed sediment also is driven to circulate around the bank in the same way.

The above descriptive classification shows the general relationship between the structure of an estuary, the tidal range, and the river flow. Quantified classifications depend on producing numerical criteria that relate these variables. When the flow ratio - the ratio of river flow per tidal cycle to the tidal prism volume – is 1 or greater, the estuary is highly stratified. When it is about 0.25, the estuary is partially mixed; and for less than 0.1 it is well mixed. Alternatively, the ratio of the river discharge velocity (R/A, where R is the river volume discharge, and A is the cross-sectional area) divided by the root-mean-square tidal velocity gives values of less than 10^{-2} for well-mixed conditions and greater than 10^{-1} for stratified conditions. There are many alternative proposals that depend on describing the processes controlling the tidally averaged vertical profiles of salinity and of current velocity. One example uses a stratification parameter, which relates the surface to near-bed salinity difference (ΔS) normalized by the depth mean salinity (S_0), and a circulation parameter that expresses the ratio of the surface mean flow (u_s) to the depth mean flow (u_f) (**Figure 2**). Use of these classification schemes shows that estuaries plot as a series of points depending on position in the estuary, on river discharge, and on tidal range. They form part of a continuous sequence, rather than falling into distinct types, and an estuary can change in both space and time. Near the head of the estuary the river flow will be more important than nearer the mouth, and consequently the structure may be more stratified. Alternatively, if the tidal currents rise toward the head of the estuary, better-mixed conditions may prevail. **Figure 4** shows the classification scheme based on the stratification–circulation parameters, together with indication of the direction in which an estuary would plot for different circumstances.

Changes in river discharge of water force the estuarine circulation to respond. An increase will push the salt intrusion downstream and increase the stratification. Conversely, a decrease will allow the salt intrusion to creep further landward, and the stratification will decrease because the tidal motion

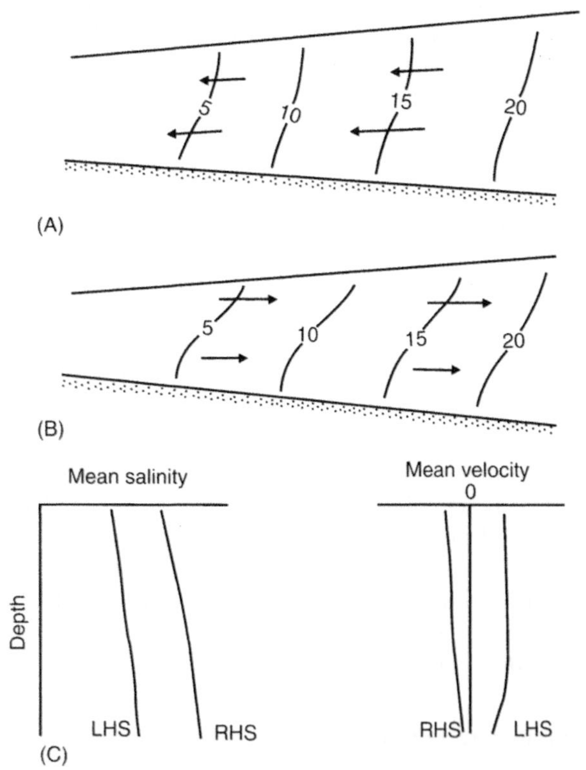

Figure 3 Diagrammatic plan views of the horizontal distribution of salinity in a well-mixed estuary (PSU). (A) At mid flood tide. (B) At mid ebb tide. Contours show representative salinities; arrows show relative strengths of currents. (C) Vertical mean profiles for left-hand side and right-hand side of the estuary looking down estuary.

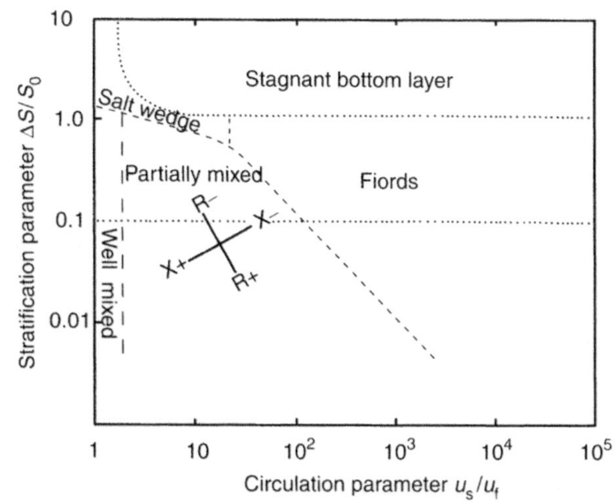

Figure 4 Quantified estuarine classification scheme based on stratification and circulation parameters. R − and R + show the direction of movement of the plot of an estuarine location with increase or decrease in river discharge; X − and X + with movement toward the estuarine head or mouth. Parameters are defined in **Figure 2**. From Dyer (1972) with permission. © John Wiley & Sons.

will become relatively more important than the stabilizing effect of the fresh water input. Because of the drastic variation of tidal elevation and currents between spring and neap tides, well-mixed conditions may occur at spring tides and partially mixed at neaps. This must occur by a relatively stronger mixing on the flood tide during the increasing tidal range, and by increased stabilization on the ebb during decreasing tidal range. The change between the two conditions may occur very rapidly at a critical tidal range as the effects of mixing rapidly break down the stratification.

Additionally, a wind blowing landward along the estuary will tend to restrict the surface outflow, and may even reverse the sense of the vertical gravitational circulation. A down-estuary wind would enhance the stratification. Also, atmospheric pressure changes will cause a total outflow or inflow of water, as the mean water level responds. Thus large variations in the mixing and water circulation are likely on the timescale of a few days, the progression rate of atmospheric depressions.

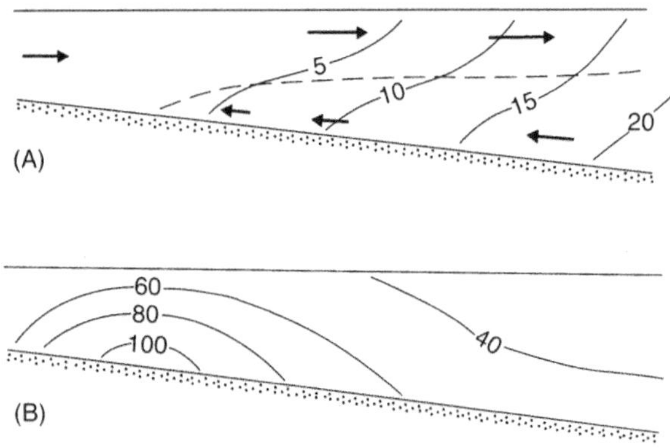

Figure 5 (A) Diagrammatic longitudinal distribution of tidally averaged salinity and current velocities in a partially mixed estuary. Contours show salinities; arrows show relative current velocities. Dashed line shows level of no net motion. (B) Diagrammatic distribution of mean suspended sediment concentration (in ppm), showing the turbidity maximum.

Flushing

Within the estuary there is a volume of fresh water that is continually being carried out to sea and replaced by new inputs of river water. At the mouth, not all of the brackish water discharged on the ebb tide returns on the flood, the proportion being very variable depending on the coastal circulation close to the mouth, and fresh water comprises a fraction of that lost. The flushing or residence time is the time taken to replace the existing fresh water in the estuary at a rate equal to the river discharge. From direct measurements of the salinity distribution it is possible to calculate the volume of accumulated fresh water and the flushing time for various river flows. The flushing time changes rapidly with discharge at low river discharge, but slowly at high river flow, being buffered to a certain extent by the consequent changes in stratification. Flushing times are generally of the order of days to tens of days, rising with the size of the estuary.

Turbidity Maximum

A distinctive feature of partially and well-mixed estuaries is the turbidity maximum. This is a zone of high concentrations of fine suspended sediment, higher than in the river or lower down the estuary, that depends upon the estuarine circulation for its existence. The maximum is located near the head of the salt intrusion (**Figure** 5), and the maximum concentrations tend to rise with the tidal range, from of the order of 100 ppm to 10 000 ppm, though there are variations depending on the availability of sediment. Often the total amount of suspended sediment exceeds several million tonnes. The location of the maximum coincides with a mud-reach of bed sediment, and with muddy intertidal areas. The continual movement of the turbidity maximum and the sorting of the sediment appears to create a gradient of sediment grain size along the estuary. Closer to the mouth, the bed is generally dominated by more sandy sediment. Because of the affinity of the fine particles for contaminants, prediction of the characteristics of the turbidity maximum is important. The position of the maximum changes with river discharge and with tidal range, but with a time lag behind the variation in the salt intrusion position. During the tide there are drastic changes in concentration. At high water the maximum is located well up the estuary and concentrations are relatively low because of settling. During the ebb the current reentrains the sediment from the bed, and advects it down the estuary. Settling again occurs at low water and there is further reentrainment and advection on the flood tide. Thus there is active cycling of sediment between the water column and the bed, and the whole mass becomes very well sorted in the process. The behavior of the sediment will depend on the settling velocity and the threshold for erosion. The settling velocity varies with concentration and with the intensity of the turbulence because flocculation of the particles occurs, forming large, loose but fast-settling floes. Settling causes high-concentration layers to build up on the bed, and at neap tides these may persist throughout the tide, forming layers often seen on echo-sounders as fluid mud.

The turbidity maximum is maintained by a number of trapping mechanisms. Vertical gravitational circulation carries the fine particles and flocs downstream on the surface, and they settle toward the bed because the turbulent mixing is reduced by the stratification. Near the bed the landward mean flow carries the suspended particles toward the head of the salt intrusion, together with new material coming in from the sea. There they meet other particles carried downstream by the river flow and are suspended by the high currents. Additionally, the asymmetry of the currents during the tide leads to tidal pumping of sediment, the flood-dominant currents ensuring that the sediment transport on the flood exceeds that on the ebb. The sediment is thus pumped landward until the asymmetry of the currents is reversed by the river flow – somewhere landward of the tip of the salt intrusion. The pumping process is affected by the settling of sediment to the bed and its reerosion, which introduces phase lags between the concentration and current velocity variations and enhances the asymmetry in the sediment transport rate during the tide. Measurements have shown that all of these processes are important, but tidal pumping is dominant in well-mixed estuaries. Both vertical gravitational circulation and tidal pumping are important in partially mixed estuaries. As a result the sediment in the turbidity maximum is derived from the coastal sea as much as from the rivers.

In microtidal salt wedge estuaries the surface fresh water carries sediment from the river seaward throughout the estuary, discharging into the sea at high river flow. The clearer water in the underlying salt wedge is often revealed in the wakes of ships.

Estuarine Modeling

The water flows transport salt, thus affecting the density distribution, and the densities affect the longitudinal pressure gradients that drive the water flow. Turbulent eddies produced by the flows cause exchanges of momentum as well as of salt, and produce frictional forces that help to resist the flow. Estuarine models attempt to predict the salt distribution, the water circulation and the sediment transport through application of the fundamental equations for mass, momentum, and sediment continuity, which formalize the above interactions. In principle, it is possible to measure directly most of the terms in these equations, apart from those that involve the turbulent exchanges. In practice, approximations are required to solve the equations. Either tidally averaged or within-tide conditions can be modeled. For sediment, the definition of the settling velocity and the threshold are required.

One-dimensional (1D) models assume that the estuarine characteristics are constant across the cross-section, and that the estuary is of straight prismatic shape. For tidally average conditions the assumption is that the seaward advection of salt on the mean flow is balanced by a landward dispersion of salt at a rate determined by the horizontal salinity gradient and a longitudinal eddy dispersion coefficient. This coefficient obviously incorporates the effects of turbulent mixing, the tidal oscillation and any actual nonuniformity of conditions.

Two dimensional (2D) models can either assume that conditions are constant with depth but vary across the estuary (2DH), or are constant across the estuary but vary with depth (2DV). The eddy dispersion and eddy viscosity coefficients will be different for the two situations. Using the 2DH model one would not be able to explore the effects of vertical gravitational circulation, whereas with the 2DV model the effects of the shallow water sides of the channel would be missing. In many cases a further approximation may be made by assuming the water is of constant density.

Three-dimensional models are obviously the ideal as they represent fully the vertical and horizontal profiles of velocity, density, and suspended sediment concentration. The only unknown terms are those relating to the turbulent mixing, the diffusion terms. Although computationally 3D models are becoming less costly to run, it is still difficult to obtain sufficient data to calibrate and validate them.

Once realistic flow models have been constructed, they can be used to explore the transport of sediment and estuarine water quality. However, the limitations on the field data restricts their use for prediction. For instance, the majority of the sediment travels very near to the bed, and it is difficult to measure the concentrations of layers only a few centimeters thick.

An active topic for research and modeling at present is the prediction of morphological change in estuaries resulting from sea level rise. This involves integrating the sediment transport over the tide, for many months or years, in which case the accumulated errors can become overriding. These results can then be compared with the simple empirical relationships stemming from considerations of estuarine equilibrium.

See also

Coastal Circulation Models. Fjord Circulation. Rotating Gravity Currents.

Further Reading

Dyer KR (1986) *Coastal and Estuarine Sediment Dynamics*. Chichester: Wiley.

Dyer KR (1997) *Estuaries: A Physical Introduction*, 2nd edn. Chichester: Wiley.

Kjerfve BJ (1988) *Hydrodynamics of Estuaries*, vol. I: *Estuarine Oceanography*; vol. II: *Hydrodynamics*. Boca Raton, FL: CRC Press.

Lewis R (1997) *Dispersion in Estuaries and Coastal Waters*. Chichester: Wiley.

Officer CB (1976) *Physical Oceanography of Estuaries (and Associated Coastal Waters)*. New York: Wiley.

Perillo GME (1995) *Geomorphology and Sedimentology of Estuaries*. Amsterdam: Elsevier.

RIVER INPUTS

J. D. Milliman, College of William and Mary,
Gloucester, VA, USA

Introduction

Rivers represent the major link between land and the ocean. Presently rivers annually discharge about 35 000 km^3 of freshwater and 20–22 × 10^9 tonnes of solid and dissolved sediment to the global ocean. The freshwater discharge compensates for most of the net evaporation loss over the ocean surface, groundwater and ice-melt discharge accounting for the remainder. As such, rivers can play a major role in defining the physical, chemical, and geological character of the estuaries and coastal areas into which they flow.

Historically many oceanographers have considered the land–ocean boundary to lie at the mouth of an estuary, and some view it as being at the head of an estuary (or, said another way, at the mouth of the river). A more holistic view of the land–ocean interface, however, might include the river basins that drain into the estuary. This rather unconventional view of the land–sea interface is particularly important when considering the impact of short- and medium-term changes in land use and climate, and how they may affect the coastal and global ocean.

Uneven Global Database

A major hurdle in assessing and quantifying fluvial discharge to the global ocean is the uneven quantity and quality of river data. Because they are more likely to be utilized for transportation, irrigation, and damming, large rivers are generally better documented than smaller rivers, even though, as will be seen below, the many thousands of small rivers collectively play an important role in the transfer of terrigenous sediment to the global ocean. Moreover, while the database for many North American and European rivers spans 50 years or longer, many rivers in Central and South America, Africa and Asia are poorly documented, despite the fact that many of these rivers have large water and sediment inputs and are particularly susceptible to natural and anthropogenic changes.

The problem of data quality is magnified by the fact that the available database spans the latter half of the twentieth century, some of the data having been collected >20–40 years ago, when flow patterns may have been considerably different to present-day patterns. In many cases, more recent data can reflect anthropogenically influenced conditions, often augmented by natural change. The Yellow River in northern China, for example, is considered to have one of the highest sediment loads in the world, 1.1 billion tonnes y^{-1}. In recent years, however, its load has averaged <100 million tons, in response to drought and increased human removal of river water. How, then, should the average sediment load of the Yellow River be reported – as a long-term average or in its presently reduced state?

Finally, mean discharge values for rivers cannot reflect short-term events, nor do they necessarily reflect the flux for a given year. Floods (often related to the El Ninõ/La Niña events) can have particularly large impacts on smaller and/or arid rivers, such that mean discharges or sediment loads may have little relevance to short-term values. Despite these caveats, mean values can offer sedimentologists and geochemists considerable insight into the fluxes (and fates) from land to the sea.

Because of the uneven database (and the often difficult access to many of the data), only in recent years has it been possible to gather a sufficient quantity and diversity of data to permit a quantitative understanding of the factors controlling fluvial fluxes to the ocean. Recent efforts by Meybeck and Ragu (1996) and Milliman and Farnsworth (2002) collectively have resulted in a database for about 1500 rivers, whose drainage basin areas collectively represent about 85% of the land area emptying into the global ocean. It is on this database that much of the following discussion is based.

Quantity and Quality of Fluvial Discharge

Freshwater Discharge

River discharge is a function of meteorological runoff (precipitation minus evaporation) and drainage basin area. River basins with high runoff but small drainage area (e.g., rivers draining Indonesia, the Philippines, and Taiwan) can have discharges as great as rivers with much larger basin areas but low runoff (**Table 1**). In contrast, some rivers with low

Table 1 Basin area, runoff, and discharge for various global rivers

River	*Basin area* ($\times 10^3 km^2$)	*Runoff* ($mm\ y^{-1}$)	*Discharge* ($km^3 y^{-1}$)
Amazon	6300	1000	6300
Congo	3800	360	1350
Ganges/Bramaputra	1650	680	1120
Orinoco	1100	1000	1100
Yangtze (Changjiang)	1800	510	910
Parana/Uruguay	2800	240	670
Yenisei	2600	240	620
Mekong	800	690	550
Lena	2500	210	520
Mississippi	3300	150	490
Choshui (Taiwan)	3.1	1970	6.1
James (USA)	20	310	6.2
Cunene (Angola)	100	68	6.8
Limpopo (Mozambique)	380	14	5.3

The first 10 rivers represent the highest discharge of all world rivers, the Amazon having discharge equal to the combined discharge of the next seven largest rivers. Basin areas of these rivers vary from 6300 (Amazon) to 800 (Mekong) $\times 10^3$ km^2, and runoffs vary from 1000 (Amazon) to 150 mm y^{-1} (Mississippi). The great variation in runoff can be seen in the example of the last four rivers, each of which has roughly the same mean annual discharge (5.3–6.8 km^3 y^{-1}), but whose basin areas and runoffs vary by roughly two orders of magnitude.

runoff (such as the Lena and Yenisei) have high discharges by virtue of their large drainage basin areas. The Amazon River has both a large basin (comprising 35% of South America) and a high runoff; as such its freshwater discharge equals the combined discharge of the next seven largest rivers (**Table 1**). Not only are the coastal waters along north-eastern South America affected by this enormous discharge, but the Amazon influence can be seen as far north as the Caribbean >2000 km away. The influence of basin area and runoff in controlling discharge is particularly evident in the bottom four rivers listed in **Table 1**, all of which have similar discharges (5.3–6.2 km^3 y^{-1}) even though their drainage basin areas and discharges can vary by two orders of magnitude.

Sediment Discharge

A river's sediment load consists of both suspended sediment and bed load. The latter, which moves by traction or saltation along the river bed, is generally assumed to represent 10% (or less) of the total sediment load. Suspended sediment includes both wash load (mostly clay and silt that is more or less continually in suspension) and bed material load (which is suspended only during higher flow); bed material includes coarse silt and sand that may move as bed load during lower flow. A sediment-rating curve is used to calculate suspended sediment concentration (or load), which relates measured concentrations (or loads) with river flow. Sediment load generally increases exponentially with flow, so that a two-order of magnitude increase in flow may result in three to four orders of magnitude greater suspended sediment concentration.

In contrast to water discharge, which is mainly a function of runoff and basin area, the quantity and character of a river's sediment load also depend on the topography and geology of the drainage basin, land use, and climate (which influences vegetation as well as the impact of episodic floods). Mountainous rivers tend to have higher loads than rivers draining lower elevations, and sediment loads in rivers eroding young, soft rock (e.g., siltstone) are greater than rivers flowing over old, hard rock (e.g., granite) (**Figure 1**). Areas with high rainfall generally have higher rates of erosion, although heavy floods in arid climates periodically can carry huge amounts of sediment.

The size of the drainage basin also plays an important role. Small rivers can have one to two orders of magnitude greater sediment load per unit basin area (commonly termed sediment yield) (**Figure 1**) than larger rivers because they have less flood plain area on which sediment can be stored, which means a greater possibility of eroded sediment being discharged directly downstream. Stated another way, a considerable amount of the sediment eroded in the headwaters of a large river may be stored along the river course, whereas most of the sediment eroded along a small river can be discharged directly to the sea, with little or no storage.

Small rivers are also more susceptible to floods, during which large volumes of sediment can be transported. Large river basins, in contrast, tend to be self-modulating, peak floods in one part of the basin are often offset by normal or dry conditions in another part of the basin. The impact of a flood on a small, arid river can be illustrated by two 1-day floods on the Santa Clara River (north of Los Angeles) in January and February 1969, during which more sediment was transported than the river's cumulative sediment load for the preceding 25 years! The combined effect of high sediment yields from small rivers can be seen in New Guinea, whose more than 250 rivers collectively discharge more sediment to the ocean than the Amazon River, whose basin area is seven times larger than the entire island.

Our expanded database allows us to group river basins on the basis of geology, climate, and basin area, thereby providing a better understanding of how these factors individually and collectively influence sediment load. Rivers draining the young,

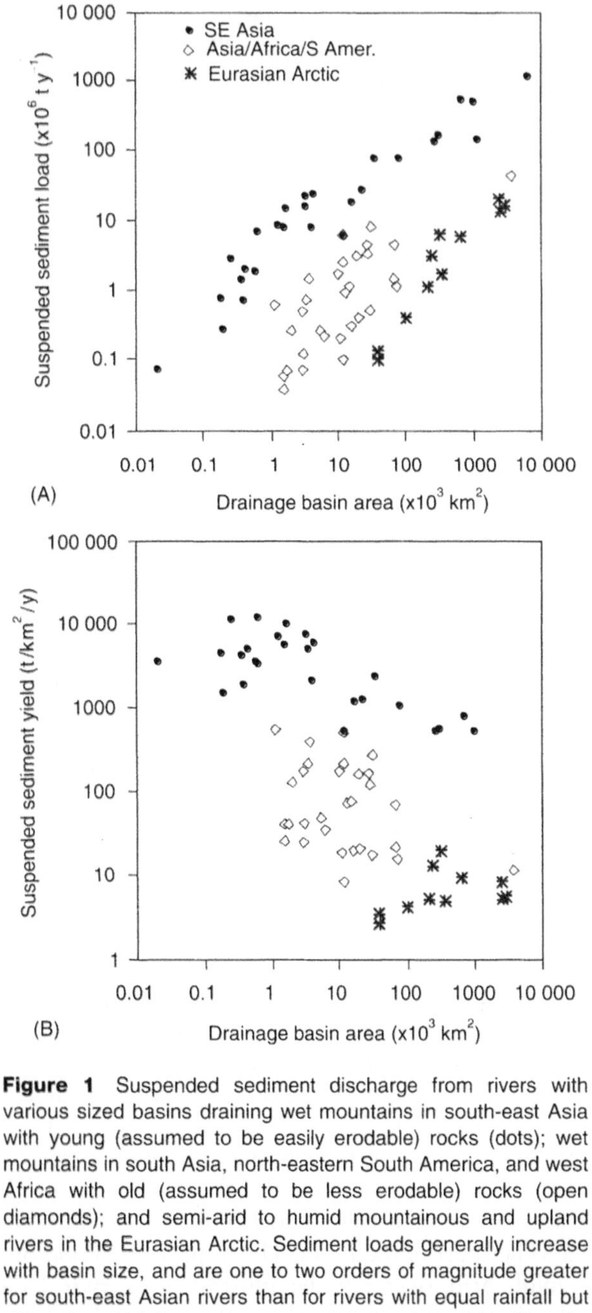

Figure 1 Suspended sediment discharge from rivers with various sized basins draining wet mountains in south-east Asia with young (assumed to be easily erodable) rocks (dots); wet mountains in south Asia, north-eastern South America, and west Africa with old (assumed to be less erodable) rocks (open diamonds); and semi-arid to humid mountainous and upland rivers in the Eurasian Arctic. Sediment loads generally increase with basin size, and are one to two orders of magnitude greater for south-east Asian rivers than for rivers with equal rainfall but old rocks, which in turn are greater than for rivers with old rocks but less precipitation (A). Sediment yields for smallest south-east Asian rivers are as much as two orders of magnitude greater than for the largest, whereas the yield for Eurasian Arctic rivers shows little change with basin size (B).

easily eroded rocks in the wet mountains of south-east Asia, for instance, have one to two orders of magnitude greater sediment loads (and sediment yields) than similar-sized rivers draining older rocks in wet Asian mountains (e.g., in India or Malaysia), which in turn have higher loads and yields than the rivers from the older, drier mountains in the Eurasian Arctic (**Figure 1**).

Because of the many variables that determine a river's sediment load, it is extremely difficult to calculate the cumulative sediment load discharged from a land mass without knowing the area, morphology, geology, and climate for every river draining that land mass. This problem can be seen in the last four rivers listed in **Table 2**. The Fitzroy-East drains a seasonally arid, low-lying, older terrain in north-eastern Australia, whereas the Mad River in northern California (with a much smaller drainage basin), drains rainy, young mountains. Although the mean annual loads of the two rivers are equal, the Mad has 112-fold greater sediment yield than the Fitzroy-East. The Santa Ynez, located just north of Santa Barbara, California, has a similar load and yield, but because it has much less runoff, its average suspended sediment concentration is 13 times greater than the Mad's.

Dissolved Solid Discharge

The amount of dissolved material discharged from a river depends on the concentration of dissolved ions and the quantity of water flow. Because dissolved concentrations often vary inversely with flow, a river's dissolved load often is more constant throughout the year than its suspended load.

Dissolved solid concentrations in river water reflect the nature of the rock over which the water flows, but the character of the dissolved ions is controlled by climate as well as lithology. Rivers with different climates but draining similar lithologies can have similar total dissolved loads compared with rivers with similar climates but draining different rock types. For example, high-latitude rivers, such as those draining the Eurasian Arctic, have similar dissolved solid concentrations to rivers draining older lithologies in southern Asia, north-eastern South America and west Africa, but lower concentrations than rivers draining young, wet mountains in south-east Asia (**Figure 2**).

The concentration of dissolved solids shows little variation with basin area, small rivers often having concentrations as high as large rivers (**Figure 2B**). At first this seems surprising, since it might be assumed that small rivers would have low dissolved concentrations, given the short residence time of flowing water. This suggests an important role of ground water in both the dissolution and storage of river water, allowing even rivers draining small basins to discharge relatively high concentrations of dissolved ions.

Four of the 10 rivers with highest dissolved solid discharge (Salween, MacKenzie, St Lawrence, and Rhine) in **Table 3** are not among the world leaders in

Table 2 Basin area, TSS (mg l^{-1}), annual sediment load, and sediment yield (t km^2 per year) for various global rivers

River	*Basin area* ($\times 10^3 km^2$)	*TSS* (gl^{-1})	*Sediment load* ($\times 10^6 t\ y^{-1}$)	*Sediment yield* ($t\ km^{-2} y^{-1}$)
Amazon	6300	0.19	1200	190
Yellow (Huanghe)	750	25	1100	1500
Ganges/Bramaputra	1650	0.95	1060	640
Yangtze (Changjiang)	1800	0.51	470	260
Mississippi	3300	0.82	400 (150)	120 (45)
Irrawaddy	430	0.6	260	600
Indus	980	2.8	250 (100)	250 (100)
Orinoco	1100	0.19	210	190
Copper	63	1.4	130(?)	2100
Magdalena	260	0.61	140	540
Fitzroy-East (Australia)	140	0.31	2.2	16
Arno (Italy)	8.2	0.69	2.2	270
Santa Ynez (USA)	2	23	2.3	1100
Mad (USA)	1.2	1.7	2.2	1800

Loads and yields in parentheses represent present-day values, the result of river damming and diversion. The first 10 rivers represent the highest sediment loads of all world rivers, the Amazon, Ganges/Brahmaputra and Yellow rivers all having approximately equal loads of about 1100×10^6 t y^{-1}. No other river has an average sediment load greater than 470×10^6 t y^{-1}). Discharges for these rivers vary from 6300 (Amazon) to $43 \times km^3\ y^{-1}$ (Yellow), and basin areas from 6300 (Amazon) to $63 \times 10^3\ km^2\ y^{-1}$ (Copper). Corresponding average suspended matter concentrations and yields vary from 0.19 to 25 and 190 to 2100, respectively. The great variation in values can be seen in the example of the last four rivers, which have similar annual sediment loads ($2.2–2.3 \times 10^6$ t y^{-1}), but whose average sediment concentrations and yields vary by several orders of magnitude.

terms of either water or sediment discharge (**Tables 1** and **2**). The prominence of the Rhine, which globally can be considered a second-order river in terms of basin area, discharge, and sediment load, stems from the very high dissolved ionic concentrations (19 times greater than the Amazon), largely the result of anthropogenic influence in its watershed (see below).

The last four rivers in **Table 3** reflect the diversity of rivers with similar dissolved solid discharges. The Cunene, in Angola, is an arid river that discharges about as much water as the Citandy in Indonesia, but drains more than 20 times the watershed area. As such, total dissolved solid (TDS) values for the two rivers are similar (51 vs 62 mg l^{-1}), but the dissolved yield (TDS divided by basin area) of the Cunene is <1% that of the Citandy. In contrast, the Ems River, in Germany, has a similar dissolved load to the Citandy, but concentrations are roughly three times greater.

Fluvial Discharge to the Global Ocean

Collectively, the world rivers annually discharge about 35 000 km^3 of fresh water to the ocean. More than half of this comes from the two areas with highest precipitation, south-east Asia/Oceania and north-eastern South America, even though these two areas collectively account for somewhat less than 20% of the total land area draining into the global ocean. Rivers from areas with little precipitation, such as the Canadian and Eurasian Arctic, have little discharge (4800 km^3) relative to the large land area that they drain ($>20 \times 10^6\ km^2$).

A total of about $20–22 \times 10^9$ t y^{-1} of solid and dissolved sediment is discharged annually. Our estimate for suspended sediment discharge (18×10^9 t y^{-1}) is less accurate than our estimate for dissolved sediment (3.9×10^9 t y^{-1}) because of the difficulty in factoring in both basin area (see above) and human impact (see below). Given the young, wet mountains and the large anthropogenic influence in south-east Asia, it is perhaps not surprising that this region accounts for 75% of the suspended sediment discharged to the global ocean. In fact, the six high-standing islands in the East Indies (Sumatra, Java, Celebes, Borneo, Timor, and New Guinea) collectively may discharge as much sediment (4.1×10^9 t y^{-1}) as all non-Asian rivers combined. Southeast Asia rivers are also the leading exporters of dissolved solids to the global ocean, 1.4×10^9 t y^{-1} (35% of the global total), but Europe and eastern North America are also important, accounting for another 25%.

Another way to view river fluxes is to consider the ocean basins into which they empty. While the Arctic Ocean occupies <5% of the global ocean basin area ($17 \times 10^6\ km^2$), the total watershed draining into the Arctic is $21 \times 10^6\ km^2$, meaning a land/ocean ratio of 1.2. In contrast, the South Pacific accounts for one-quarter of the global ocean area, but its land/ocean ratio is only 0.05 (**Table 4**). The greatest fresh water input occurs in the North Atlantic (largely

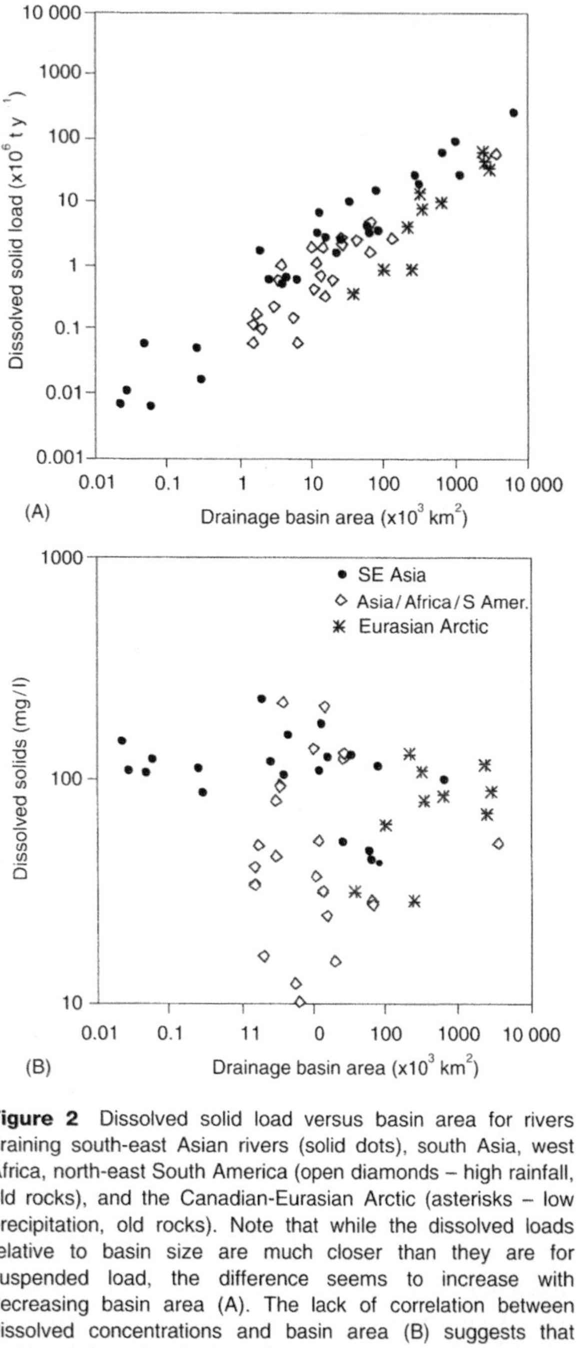

Figure 2 Dissolved solid load versus basin area for rivers draining south-east Asian rivers (solid dots), south Asia, west Africa, north-east South America (open diamonds – high rainfall, old rocks), and the Canadian-Eurasian Arctic (asterisks – low precipitation, old rocks). Note that while the dissolved loads relative to basin size are much closer than they are for suspended load, the difference seems to increase with decreasing basin area (A). The lack of correlation between dissolved concentrations and basin area (B) suggests that residence time of a river's surface water may play less of a role than ground water in determining the quantity and character of the dissolved solid fraction.

because of the Amazon and, to a lesser extent, the Orinoco and Mississippi), but the greatest input per unit basin area is in the Arctic (28 cm y^{-1} if evenly distributed over the entire basin). The least discharge per unit area of ocean basin is the South Pacific (4.5 cm y^{-1} distributed over the entire basin). In terms of suspended sediment load, the major sinks are the Pacific and Indian oceans (11.1 and 4×10^9 t y^{-1}, respectively), whereas the North Atlantic is the major sink for dissolved solids (1.35×10^9 t y^{-1}; **Table 4**), largely due to the high dissolved loads of European and eastern North American rivers.

This is not to say, of course, that the fresh water or its suspended and dissolved loads are evenly distributed throughout the ocean basins. In fact, most fluvial identity is lost soon after the river discharges into the ocean due to mixing, flocculation, and chemical uptake. In many rivers, much of the sediment is sequestered on deltas or in the coastal zone. In most broad, passive margins, in fact, little sediment presently escapes the inner shelf, and very little reaches the outer continental margin. During low stands of sea level, on the other hand, most fluvial sediment is discharged directly to the deep sea, where it can be redistributed far from its source(s) via mass wasting (e.g., slumping and turbidity currents). This contrast between sediment discharge from passive margins during high and low stands of sea level is the underlying basis for sequence stratigraphy. However, in narrow active margins, often bordered by young mountains, many rivers are relatively small (e.g., the western Americas, East Indies) and therefore often more responsive to episodic events. As such, sediment can escape the relatively narrow continental shelves, although the ultimate fate of this sediment is still not well documented.

Changes in Fluvial Processes and Fluxes – Natural and Anthropogenic

The preceding discussion is based mostly on data collected in the past 50 years. In most cases it reflects neither natural conditions nor long-term conditions representing the geological past. While the subject is still being actively debated, there seems little question that river discharge during the last glacial maximum (LGM), 15 000–20 000 years ago, differed greatly from present-day patterns. Northern rivers were either seasonal or did not flow except during periodic ice melts. Humid and sub-humid tropical watersheds, on the other hand, may have experienced far less precipitation than they do at present.

With the post-LGM climatic warming and ensuing ice melt, river flow increased. Scattered terrestrial and marine data suggest that during the latest Pleistocene and earliest Holocene, erosion rates and sediment delivery increased dramatically as glacially eroded debris was transported by increased river flow. As vegetation became more firmly established in the mid-Holocene, however, erosion rates apparently decreased, and perhaps would have remained relatively low except for human interference.

Table 3 Basin area, TDS (mg l^{-1}), annual dissolved load, and dissolved yield (t/km^2 per year) for various global rivers

River	*Basin area ($\times 10^3 km^2$)*	*TDS (mg l^{-1})*	*Dissolved load ($\times 10^6 t\ y^{-1}$)*	*Dissolved yield (t $km^{-2} y^{-1}$)*
Amazon	6300	43	270	43
Yangtze (Changjiang)	1800	200	180	100
Ganges/Bramaputra	1650	130	150	91
Mississippi	3300	280	140	42
Irrawaddy	430	230	98	230
Salween	320	310	65	200
MacKenzie	1800	210	64	35
Parana/Uruguay	2800	92	62	22
St Lawrence	1200	180	62	52
Rhine	220	810	60	270
Cunene (Angola)	110	51	0.35	3
Torne (Norway)	39	30	0.37	9
Ems (Germany)	8	180	0.34	42
Citandy (Indonesia)	4.8	62	0.38	79

The first 10 rivers represent the highest dissolved loads of all world rivers, only the Amazon, Yangtze, Ganges/Brahmaputra, and Mississippi rivers having annual loads >100 $\times 10^6$ t y^{-1}. Discharges vary from 6300 (Amazon) to 74 $\times$ km^3 y^{-1} (Rhine), and basin areas from 6300 (Amazon) to 220 $\times 10^3$ km^2 (Rhine). Corresponding average dissolved concentrations and yields vary from 43 to 810 and 43 to 270, respectively. The bottom rivers have similar annual dissolved loads (0.34–0.38 $\times 10^6$ t y^{-1}), but their average sediment concentrations and yields vary by factors of 6–26 (respectively), reflecting both natural and anthropogenic influences.

Few terrestrial environments have been as affected by man's activities as have river basins, which is not surprising considering the variety of uses that humans have for rivers and their drainage basins: agriculture and irrigation, navigation, hydroelectric power, flood control, industry, etc. Few, if any, modern rivers have escaped human impact, and with exception of the Amazon and a few northern rivers, it is difficult to imagine a river whose flow has not been strongly affected by anthropogenic activities. Natural ground cover helps the landscape resist erosion, whereas deforestation, road construction, agriculture, and urbanization all can result in channelized flow and increased erosion. Erosion rates and corresponding sediment discharge of rivers draining much of southern Asia and Oceania have increased substantially because of human activities, locally as much as 10-fold. While land erosion in some areas of the world recently has been decreasing (e.g., Italy, France, and Spain) due to decreased farming and increased reforestation; deforestation and poor land conservation practices elsewhere, particularly in the developing world, have led to accelerated increases in both land erosion and fluvial sediment loads.

While increased river basin management has led to very low suspended loads for most northern European rivers, mining and industrial activities have resulted in greatly elevated dissolved concentrations.

Table 4 Cumulative oceanic areas, drainage basin areas, and discharge of water, suspended and dissolved solids of rivers draining into various areas of the global ocean

Oceanic area	*Basin area ($\times 10^6$ km^2)*	*Drainage basin ($\times 10^6$ km^2)*	*Water discharge (km^3 y^{-1})*	*Sediment load ($\times 10^6$ t y^{-1})*	*Dissolved load ($\times 10^6$ t y^{-1})*
North Atlantic	44	30	12 800	2500	1350
South Atlantic	46	12	3300	400	240
North Pacific	83	15	6000	7200	660
South Pacific	94	5	4300	3900	650
Indian	74	14	4000	4000	520
Arctic	17	21	4800	350	480
Total	358	98	35 200	18 000	3900

For this compilation, it is assumed that Sumatra and Java empty into the Indian Ocean, and that the other high-standing islands in Indonesia discharge into the Pacific Ocean. Rivers discharging into the Black Sea and Mediterranean area are assumed to be part of the North Atlantic drainage system. (Data from Milliman and Farnsworth, 2001.)

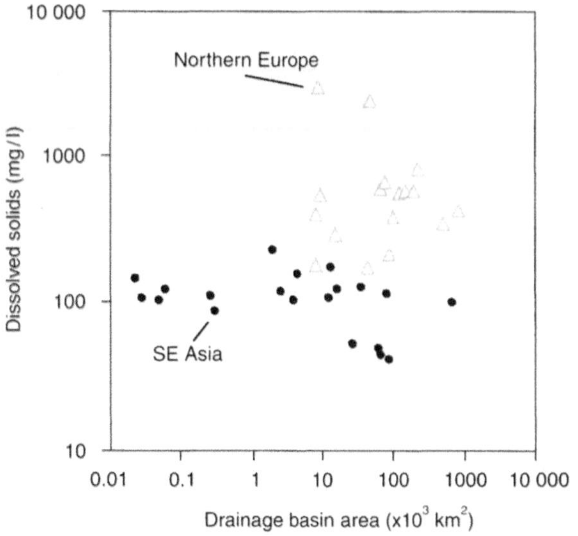

Figure 3 Dissolved solid concentration for south-east Asian rivers (also shown in **Figure 2**) compared with the much higher concentrations seen in northern European rivers (excluding those in Scandinavia). The markedly higher levels of dissolved solids in northern European rivers almost certainly reflects greater mining and industrial activity.

Compared to south-east Asian rivers, whose waters generally contain 100–200 mg l^{-1} dissolved solids, some European rivers, such as the Elbe, can have concentrations greater than 1000 mg l^{-1} (**Figure 3**), in large part due to the mining of salt deposits for potassium. Since the fall of the Soviet empire, attempts have been made to clean up many European rivers, but as of the late 1990s, they still had the highest dissolved concentrations in the world.

Interestingly, while sediment loads of rivers may be increasing, sediment flux to the ocean may be decreasing because of increased river diversion (e.g., irrigation and flood protection through levees) and stoppage (dams). As of the late 1990s there were >24 000 major dams in operation around the world. The nearly 200 dams along the Mississippi River, for example, have reduced sediment discharge to the Gulf of Mexico by >60%. In the Indus River, construction of irrigation barrages in the late 1940s led to an 80% reduction in the river's sediment load. Even more impressive is the almost complete cessation of sediment discharge from the Colorado and Nile rivers in response to dam construction. Not only have water discharge and sediment flux decreased, but high and low flows have been greatly modulated; the effect of this modulated flow, in contrast to strong seasonal signals, on the health of estuaries is still not clear.

Decreased freshwater discharge can affect the circulation of shelf waters. Many of the nutrients utilized in coastal primary production are derived from upwelled outer shelf and slope waters as they are advected landward to offset offshore flow of surface waters. Decreased river discharge from the Yangtze River resulting from construction of the Three Gorges Dam might decrease the shoreward flow of nutrient-rich bottom shelf waters in the East China Sea, which could decrease primary production in an area that is highly dependent on its rich fisheries.

Retaining river water within man-made reservoirs also can affect water quality. For example, reservoir retention of silicate-rich river water can lead to diatom blooms within the man-made lakes and thus depletion of silicate within the river water. One result is that increased ratios of dissolved nitrate and phosphate to dissolved silica may have helped change primary production in coastal areas from diatom-based to dinoflagellates and coccolithophorids. One result of this altered production may be increased hypoxia in coastal and shelf waters in the north-western Black Sea and other areas off large rivers.

Taken in total, negative human impact on river systems seems to be increasing, and these anthropogenic changes are occurring faster in the developing world than in Europe or North America. Increased land degradation (partly the result of increased population pressures) in northern Africa, for instance, contrasts strongly with decreased erosion in neighboring southern Europe. Dam construction in Europe and North America has slackened in recent years, both in response to environmental concerns and the lack of further sites of dam construction, but dams continue to be built in Africa and Asia.

Considering increased water management and usage, together with increased land degradation, the coastal ocean almost certainly will look different 100 years from now. One can only hope that the engineers and planners in the future have the foresight to understand potential impacts of drainage basin change and to minimize their effects.

See also

Estuarine Circulation. Flows in Straits and Channels. Arctic Ocean Circulation.

Further Reading

Berner RA and Berner EK (1997) Silicate weathering and climate. In: Ruddiman RF (ed.) *Tectonic Uplift and Climate Change*, pp. 353–365. New York: Plenum Press.

Chen CTA (2000) The Three Gorges Dam: reducing the upwelling and thus productivity in the East China Sea. *Geophysics Research Letters* 27: 381–383.

Douglas I (1996) The impact of land-use changes, especially logging, shifting cultivation, mining and urbanization on sediment yields in tropical Southeast Asia: a review with special reference to Borneo. *Int. Assoc. Hydrol. Sci. Publ.* 236: 463–472.

Edmond JM and Huh YS (1997) Chemical weathering yields from basement and orogenic terrains in hot and cold climates. In: Ruddiman RF (ed.) *Tectonic Uplift and Climate Change*, pp. 329–351. New York: Plenum Press.

Humborg C, Conley DJ, Rahm L, *et al.* (2000) Silicon retention in river basins: far-reaching effects on biogeochemistry and aquatic food webs in coastal marine environments. *Ambio*.

Lisitzin AP (1996) *Oceanic Sedimentation, Lithology and Geochemistry*. Washington, DC: American Geophysics Union.

Meybeck M (1994) Origin and variable composition of present day river-borne material. *Material Fluxes on the Surface of the Earth, National Research Council Studies in Geophysics*, pp. 61–73. Washington: National Academy Press.

Meybeck M and Ragu A (1996) *River Discharges to the Oceans. An Assessment of Suspended Solids, Major Ions and Nutrients*. GEMS-EAP Report.

Milliman JD (1995) Sediment discharge to the ocean from small mountainous rivers: the New Guinea example. *Geo-Mar Lett* 15: 127–133.

Milliman JD and Farnsworth KL (2002) *River Runoff, Erosion and Delivery to the Coastal Ocean: A Global Analysis*. Oxford University Press (in press).

Milliman JD and Meade RH (1983) World-wide delivery of river sediment to the oceans. *J Geol* 91: 1–21.

Milliman JD and Syvitski JPM (1992) Geomorphic/tectonic control of sediment discharge to the ocean: the importance of small mountainous rivers. *J Geol* 100: 525–544.

Milliman JD, Ren M-E, Qin YS, and Saito Y (1987) Man's influence on the erosion and transport of sediment by Asian rivers: the Yellow River (Huanghe) example. *J Geol* 95: 751–762.

Milliman JD, Farnworth KL, and Albertin CS (1999) Flux and fate of fluvial sediments leaving large islands in the East Indies. *Journal of Sea Research* 41: 97–107.

Thomas MF and Thorp MB (1995) Geomorphic response to rapid climatic and hydrologic change during the late Pleistocene and early Holocene in the humid and sub-humid tropics. *Quarterly Science Review* 14: 193–207.

Walling DE (1995) Suspended sediment yields in a changing environment. In: Gurnell A and Petts G (eds.) *Changing River Channels*, pp. 149–176. Chichester: John Wiley Sons.

GRAVITY AND TURBIDITY CURRENTS, AND FLOWS IN CHANNELS

OVERFLOWS AND CASCADES

G. F. Lane-Serff, University of Manchester, Manchester, UK

Introduction

When dense water enters an ocean basin it often does so at a shallow depth passing through a narrow strait or over a sill from a neighboring basin or marginal sea, or flowing down the continental slope from a shelf sea. The dense water flow is affected by the Earth's rotation and so does not flow directly downslope but instead turns to flow partly along the slope. Large-scale, continuous flows from one ocean basin or large sea down into another are referred to as 'overflows', while smaller-scale, intermittent flows of dense water from shelf seas down continental slopes are referred to as 'cascades'. In both cases, the flows can be summarized as gravity currents on slopes in a rotating system (*see* Rotating Gravity Currents).

The deep ocean is divided into many sub-basins by ridges, so that overflows play an important role in the global thermohaline circulation, providing the mechanism by which dense waters created in the polar regions flow from one ocean basin down into another. The flow speeds in overflows can be 10–100 times faster than most other deep-ocean flows, generating turbulence and entraining ambient seawater into the overflow. Thus the mixing of deep waters is often dominated by the mixing that happens at overflows.

The combination of density contrast, slope, and rotation produces complicated dynamics that are still not fully understood. The effects of the overflow are not limited to the waters immediately around the overflow, but can extend up through the water column producing, in some cases, eddies that can be observed at the sea surface, hundreds of meters above the overflow. These flows can be characterized in terms of the density contrast, flow rate, slope, water depth, and effect of the Earth's rotation.

Observations

Waters of increased density are formed at a wide range of locations throughout the globe. In low latitudes, waters of increased density are formed through evaporation increasing the salinity, such as in the Mediterranean Sea and the Red Sea. At mid-latitudes, wintertime cooling on continental shelves leads to the formation of cold, dense waters that flow down the continental slope into the deep ocean. In polar regions cooling again increases the density, but this can be further enhanced by brine rejection (and thus increased salinity) during ice formation.

The salinity of the Mediterranean is approximately 2 parts per thousand more than the Atlantic. The resulting density difference drives an exchange flow through the Strait of Gibraltar, with fresher Atlantic water entering the Mediterranean at the surface while the denser, more saline Mediterranean water flows into the Atlantic beneath. The flow of Mediterranean water through the strait is approximately 0.7 Sv ($1\,\mathrm{Sv} = 10^6\,\mathrm{m}^3\,\mathrm{s}^{-1}$). The dense water descends the slope in the eastern Gulf of Cadiz into the Atlantic, with the flow deflected to the right by Coriolis forces and ambient Atlantic water entrained into this fast-moving flow (**Figure 1**). The path of the Mediterranean water is strongly influenced by the local bathymetry, especially by canyons that cut the continental slope and provide 'shortcuts' into deeper water. With the strong entrainment of Atlantic water the overflow reaches a neutral level at a depth of approximately 1000 m, where the density of the overflow matches that of the ambient seawater. The Mediterranean waters then flow along the continental slope at this level, although rotating lenses of water are shed from the flow, perhaps through the strong changes in direction of the flow caused by the capes along the flow path. These relatively well-mixed lenses of high-salinity water (known as 'meddies') have been observed to propagate around the North Atlantic as coherent features for considerable periods of time.

On the northwest European shelf, wintertime cooling results in colder waters on the shelf than in the neighboring deeper waters. In the deeper waters, the cooling produces a surface mixed layer that is deeper than the waters on the shelf. For the same heat loss, the shallower layer of water on the shelf becomes colder than the deeper surface layer off-shelf. Furthermore, during the spring the off-shelf, deep waters are replaced more quickly by warmer water from the south while the flows of waters onto and off the shelf are relatively slow. As the surface waters on the shelf begin to warm and stratify, the deeper waters on the shelf retain their lower temperatures, becoming significantly colder than any nearby waters. These waters tend to flow off-shelf in intermittent cascades. While cascades are smaller in

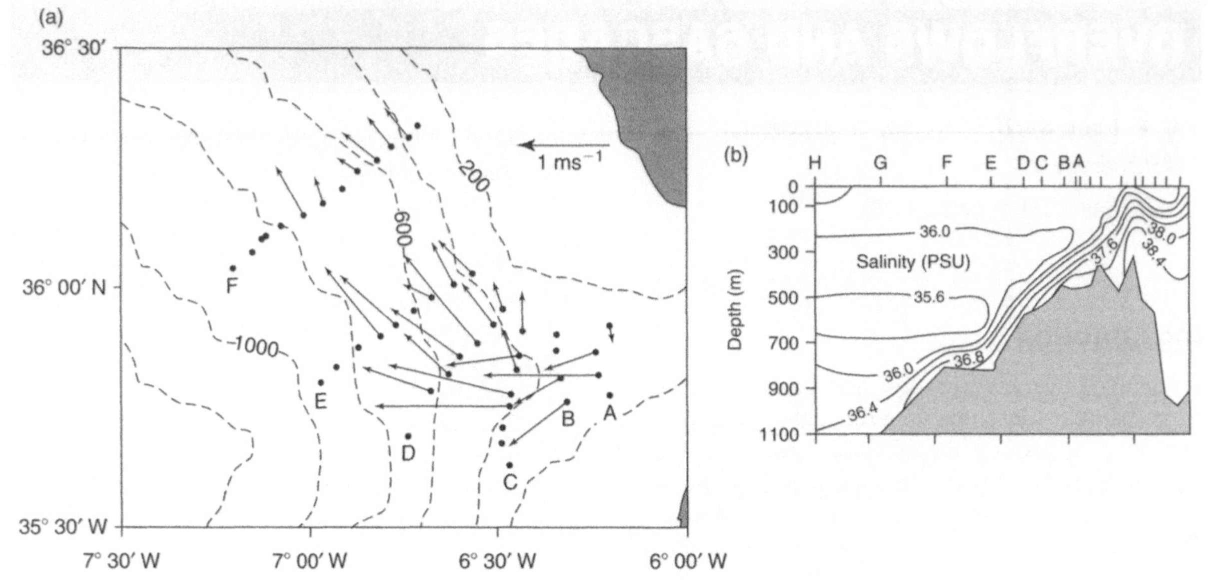

Figure 1 The Mediterranean outflow. (a) The peak outflow velocities in the region just to the west of the Strait of Gibraltar, with depths given in meters. Note how the flow is initially downslope before turning and flowing along the slope. (b) Salinity section along the axis of the outflow, with the letters A–F corresponding to the lines of stations in (a) (positions G and H lie beyond the regions covered by (a)). Reproduced from Price JF, Baringer MO, Lueck RG, *et al.* (1993) Mediterranean outflow mixing and dynamics. *Science* 259: 1277–1282.

magnitude and shorter in duration than overflows, they obey the same dynamics, turning under the influence of the Earth's rotation to flow along rather than directly downslope. Cascades are also observed flowing off the shelf to the east of Bass Strait, between Tasmania and the mainland of Australia. These cascades reach a flow rate of up to 3 Sv, and have a typical duration of 2–3 days. More generally, throughout the world's oceans, typical cascade fluxes (measured at the shelf edge) are *c.* 0.05–0.08 Sv per 100 km of shelf.

Dense waters formed in the Arctic Ocean flow south through the Denmark Strait, which lies between Iceland and Greenland, forming one of the largest overflows with a flux of 2.7 Sv. The sill depth at Denmark Strait is approximately 700 m and the overflow descends down into the North Atlantic reaching a depth of between 1000 and 1500 m before flowing along the slope. The overflow remains at approximately this depth for at least 500 km. It forms cyclonic eddies in the overlying water which have been tracked by deep-drogued buoys (**Figure 2**). Measurements from these, and from moored current meters, show that these eddies have a regular frequency and a strong vorticity.

Similar overflows are observed in the Southern Hemisphere. Dense ice shelf water (with a temperature below the surface freezing point) flows out of the Filchner Depression in the southeast corner of the Weddell Sea. This flow (with a flux of 0.5 Sv) mixes as it flows down into the Weddell Sea, eventually contributing to the formation of Antarctic Deep and Bottom Waters. A flow in the opposite direction can be found in the southwest corner of the Weddell Sea, where high-salinity shelf water, formed by brine rejection during sea ice formation in open water, flows southward underneath the Ronne Ice Shelf. In the case of flow under an ice shelf, the dynamics of the flow is complicated by presence of the ice shelf above.

Flow Characteristics

Parameters

Overflows and cascades are formed by sources of dense water on slopes in a rotating system. The source and the system into which the source is flowing can be characterized by five main parameters (**Figure 3**):

1. the source flux (Q), volume per unit time;
2. the density contrast between the source water and the ambient seawater ($\Delta\rho$);
3. the tangent of the slope angle (β);
4. the local value of the Coriolis parameter (f); and
5. the depth of ambient seawater above the source (H).

Although the overflow may initially be constrained by flowing through a narrow strait or channel, the width and depth of the overflow are not generally

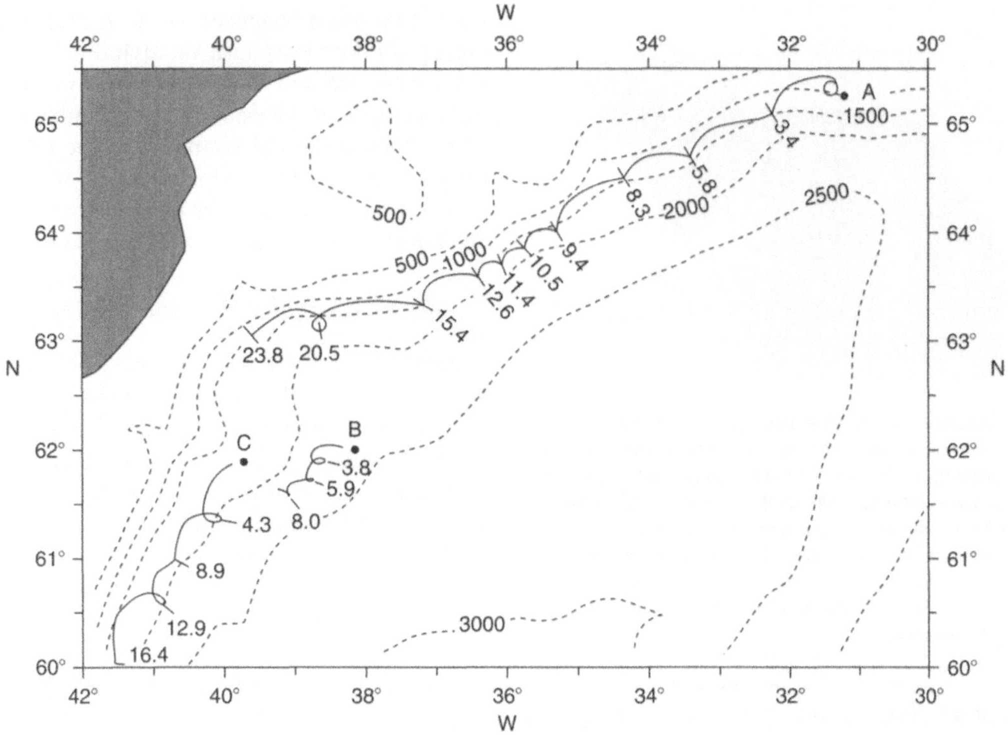

Figure 2 Trajectories of satellite-tracked buoys trapped in cyclonic eddies on the East Greenland continental slope, downstream of the Denmark Strait overflow. Tick marks give time in days, depths are given in meters. The flow here is mainly along the slope, with a superimposed cyclonic (anticlockwise) motion. Reproduced from Krauss W (1996) A note on overflow eddies. *Deep Sea Research Part I* 43(10): 1661–1667.

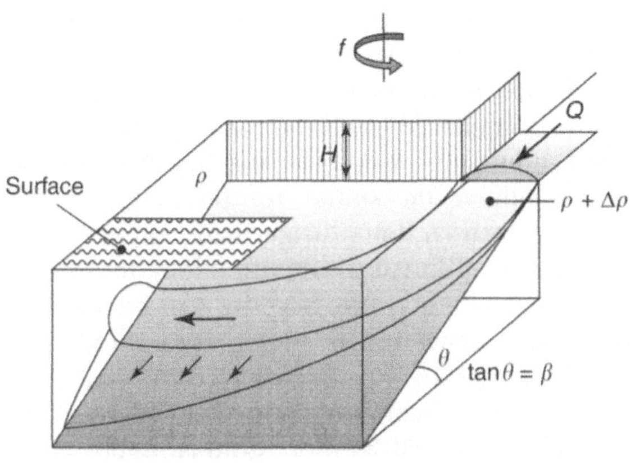

Figure 3 The basic parameters and behavior of overflows. A source (flow rate Q) of relatively dense fluid ($\rho + \Delta\rho$, where the ambient seawater has density ρ) flows onto a slope (of tangent β) in water of depth H. The local value of the Coriolis parameter (which varies with latitude) is f. Initially the fluid flows down the slope, turning to the right (in the Northern Hemisphere) under the influence of the Earth's rotation until the main flow is flowing along the slope. A thin viscous sublayer continuously drains the main flow, taking fluid at an angle down the slope.

independent parameters. The shape of the overflow (and thus its width and height) adjusts in response to the effects of gravity and rotation as it flows over the slope. Where the ambient seawater is stratified, the last parameter (the depth of ambient water above the source, H) may be replaced by a vertical length scale dependent on the strength and nature of the stratification (see below).

Strong currents in the ambient seawater into which the overflow is flowing have an effect on the overflow, as do irregularities in the slope. In particular, channels and canyons can direct the dense fluid downslope much more rapidly than a similar flow on a smooth slope.

Basic Behavior

A basic description of the behavior of overflows can be given based on the results of laboratory experiments, field observations, and numerical models. As the dense water leaves the channel or flows over the sill that marks the source at the top of the slope, it initially flows directly down the slope under the influence of gravity. The effects of the Earth's rotation deflect the current (to the left in the Southern Hemisphere, to the right in the Northern Hemisphere) so that the flow curves to eventually flow mainly along the slope, maintaining its depth. The distance over which the adjustment from downslope

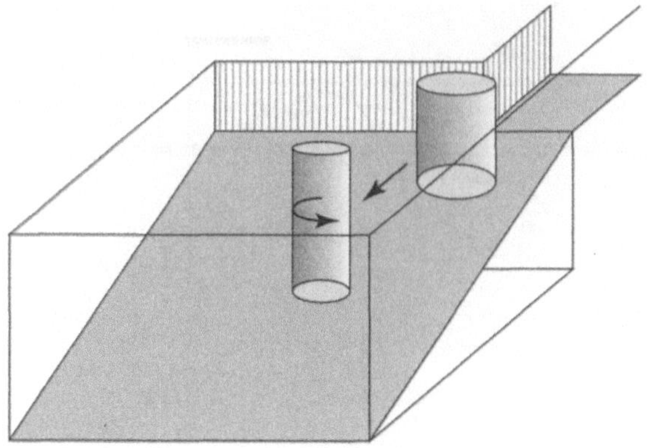

Figure 4 Columns of ambient fluid taken into deeper water stretch and become thinner. Viewed from a stationary observer outside the rotating Earth, columns that appear stationary on the Earth are actually rotating (with vertical vorticity f). When the column becomes thinner it must spin even faster to conserve angular momentum (as ice skaters spin faster when they draw their arms in), giving it cyclonic vorticity relative to the Earth. (Cyclonic is anticlockwise in the Northern Hemisphere, clockwise in the Southern Hemisphere.)

to along-slope flow takes place scales with the Rossby radius of deformation (discussed below). The along-slope flow maintains an inviscid geostrophic balance, but is continuously drained by a viscous Ekman layer at its base. The viscous draining flow takes fluid from the base of the current at an angle down the slope.

The inviscid along-slope flow is not always steady. The overflow carries columns of overlying ambient seawater out into deeper water so that these columns of ambient seawater will be stretched (**Figure 4**). This stretching produces cyclonic vorticity in the columns as they conserve their angular momentum. The vorticity in the overlying water breaks up the dense overflow into a series of domes. The cyclonic eddies in the ambient water lie above the domes of dense overflow water and the eddy-dome structures propagate along the slope (but the dense fluid is still continuously drained by the viscous sublayer). The downslope motion of columns of ambient fluid may occur in the initial downslope flow and adjustment near the source, or as a result of instabilities in the along-slope flow further from the source (see the section 'Instabilities and mixing').

Scalings

We make use of the parameters introduced above, but it is useful to express density contrast in terms of the reduced gravity $g' = (\Delta\rho/\rho)g$, where g is the acceleration due to gravity and ρ is the density of the ambient seawater. Two different horizontal length scales have been proposed as the appropriate Rossby radius, the first based on the dynamics of a rotating gravity current and the second based on the injection of fluid of the same density as the ambient seawater. These depend on the source flux, the local value of Coriolis parameter (which varies with latitude), and (for the first type) the density contrast between the dense current and the ambient seawater, with

$$R_1 = (2Qg')^{1/4} f^{-3/4} \quad \text{and} \quad R_2 = (Q/f)^{1/3}$$

Typically both these scalings give a Rossby radius of order 10 km for the cases described above. If a column is taken into deeper water by a distance R over a slope of tangent β, then the column will increase in height by an amount βR. This increase in height divided by the original height, H, gives the relative stretching of the water columns and thus gives useful nondimensional parameters (depending on the version of R used):

$$\Gamma = \beta R_1/H \quad \text{or} \quad \mathrm{G} = \beta R_2/H$$

In practice, these parameters are very similar in magnitude and it has yet to be established which is the most useful in describing the behavior of overflows. Comparisons with experiments show that an increase in these parameters (i.e., increased stretching) gives an increase in the frequency of eddy production. Where the ambient seawater is stratified (e.g., by a strong thermocline above the source) this can put an effective 'lid' on the vertical influence of the flow and the total depth H is replaced by a height scale derived from the stratification (this would be the height of the thermocline above the source for the simple example mentioned earlier, but a height scale can be obtained for more complicated stratification too). In Antarctica a more physical lid is provided by floating ice shelves, and eddies generated by dense high-salinity water flowing under the ice shelves are affected by the sudden changes in ambient water depth at the front of ice shelves, as well as more gradual depth changes beneath them. Where eddies are formed that extend up to the underside of an ice shelf this will have an effect on heat transfer and thus melting, although the scale of the contribution of these eddies to ice shelf melting has yet to be quantified.

The appropriate scales for the speed and thickness of the viscous draining layer can also be estimated. This makes it possible to estimate the along-slope distance over which the initial flux is entirely drained into the viscous Ekman layer:

$$Y = \frac{Qf^{3/2}}{g'\beta\sqrt{2v}}$$

where υ is the molecular viscosity for laboratory experiments (which have smooth, laminar boundary layers) or a vertical eddy viscosity for oceanographic flows (which have turbulent boundary layers).

Dividing Y by a Rossby radius gives a nondimensional parameter that can be used to describe the importance of the draining flow, for example:

$$Y/R_1 = \frac{Q^{3/4} f^{9/4}}{2^{3/4} g'^{5/4} \upsilon^{1/2} \beta}$$

The draining flow is expected to dominate when this parameter is small. In laboratory experiments it has been found that the speed at which the eddies propagate along the slope is determined by Y/R, with slower propagation speeds for low values of Y/R and no eddies at all for sufficiently small values ($Y/R < 1$, approximately).

Modeling and Interpretation

Streamtube Models

In developing a mathematical description of overflows there has been considerable use of 'streamtube' models. These are 'integral' models (meaning that the results can be obtained by integrating the equations of motion over planes perpendicular to the flow) that describe the flow in terms of mean properties as a function of the distance (e.g., s) along the flow centerline (**Figure 5**). Thus the overflow is described by a mean centerline speed, $U(s)$, density contrast, $\Delta\rho(s)$, and shape parameters (e.g., width and height $w(s)$ and $h(s)$). With prescribed parameterizations for the entrainment of ambient seawater into the flow and for the frictional drag at the base of the flow, the path of the overflow centerline $(x(s), y(s))$ can be calculated.

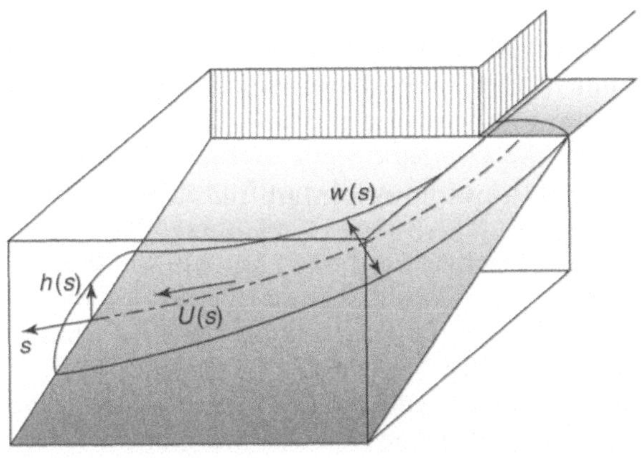

Figure 5 Streamtube models represent the overflow in terms of a mean centerline position with mean overflow properties described as functions of the distance (s) along the centerline.

Streamtube models give a simple mathematical formulation that can be used to analyze some of the aspects of overflows. However, the averaging of properties throughout the flow (especially momentum) leads to a somewhat misleading picture of the flow. While the continuously drained along-slope flow (described earlier) does have a mean center position that moves gradually downslope, thinking of the flow in that way does not help comparisons with observations. In many overflows the effect of the bottom friction is not distributed throughout the overflow, but confined to a lower boundary layer, giving the flow sketched in **Figure 3**. However, there are some strong, turbulent overflows (e.g., Mediterranean outflow) in which properties are well mixed and the streamtube approach provides a useful model for direct comparison with oceanographic observations and numerical models.

Laboratory Models

Laboratory experiments allow a wide range of parameters and flows to be examined in detail and much of the insight into the behavior of overflows has come from laboratory modeling. Experiments are conducted in tanks mounted on rotating tables to simulate the effects of the Earth's rotation (see **Figure 6** for an example). Typically the scale of the laboratory tanks is of the order of $\leqslant 1$ m, but rotating tables with diameters of up to 13 m are used. In order to apply the results from the laboratory experiments to oceanographic flows it is necessary to scale the parameters carefully. The inviscid aspects of the flow can be simulated with confidence in the laboratory (e.g., by matching the values of the nondimensional stretching parameters) so that the along-slope flow and eddy formation processes are likely to be well represented. However, in the laboratory experiments the viscous draining layer is smooth and usually laminar and driven by molecular viscosity, while in the corresponding oceanographic flows the Ekman layer is a turbulent boundary layer. It is therefore necessary to identify an effective vertical eddy viscosity in the oceanographic flows to allow comparison of the draining flow with the laboratory experiments. Reasonable choices of the turbulent eddy viscosity have allowed successful comparisons between the laboratory experiments and oceanographic measurements, but a better understanding of the turbulent oceanic bottom boundary layer would allow more reliable predictions.

In addition to flows where the ambient fluid is homogeneous, experiments have also been conducted where the flow is directed into a stratified ambient fluid. As mentioned earlier, the effect of

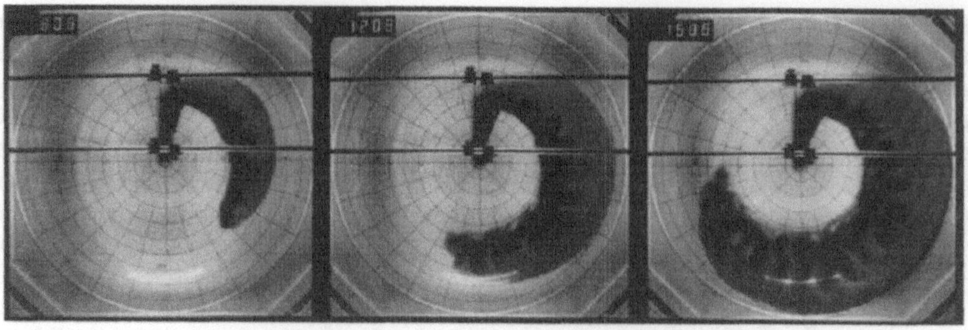

Figure 6 A sequence of photographs of a laboratory experiment (plan view). Dense fluid (dyed) is released on an axisymmetric hill. The inviscid core flows around the hill at constant depth (darkest fluid) while a thin layer of fluid drains down the slope (toward the edge of the image). The dense fluid breaks up into a series of domes with 'subplumes' extending downslope from the domes. Reproduced from Baines PG and Condie S (1998) Observations and modelling of Antarctic downslope flows: A review. In: Jacobs SS and Weiss RF (eds.) *Antarctic Research Series, Vol. 75: Ocean, Ice, and Atmosphere – Interactions at the Antarctic Continental Margin*, pp. 29–49. Washington, DC: American Geophysical Union.

stratification is to reduce the effective vertical height scale, thus replacing the fluid depth H with a smaller height. This increases the stretching parameters and enhances eddy production. There is some evidence that the propagation speed of the eddies along the slope is reduced by the presence of stratification, perhaps because of internal wave generation. In experiments with a homogeneous ambient fluid, any mixed fluid created at the interface between the overflow and the ambient fluid is still denser than the ambient fluid and eventually rejoins the current. Where the ambient fluid is stratified, mixed fluid may have the same density as ambient fluid above the final depth reached by the main overflow and thus may detrain from the overflow at shallower depths. This distributed injection of overflow water into a stratified ambient has been observed in detail in nonrotating experiments but has yet to be quantified in rotating experiments.

Numerical Models

In numerical ocean models the ocean is generally divided into a series of boxes, the horizontal dimensions of which are at least 10 km and often much larger (100 km or so for many climate simulations). These boxes are too large to represent overflows accurately since typical overflow widths are of the order of 10 km or less. There are also problems with the accurate representation of mixing at overflows, which often has a dramatic effect on the wider ocean circulation. For example, in a recent project to compare three different models of the North Atlantic, the different models showed very different behavior for meridional overturning and water mass formation. These differences were found to be strongly dependent on the way the mixing of the Denmark Strait overflow was treated. Of the numerical models in that comparison, the isopycnic model (which treats the ocean as made up of a series of layers of uniform density) seems the most promising, since it had too little mixing (when compared with oceanographic observations), whereas the standard level model (with 'step-like' bottom topography) and sigma-coordinate model (with terrain-following coordinates) had too much mixing.

Numerical models of idealized geometries (similar to the laboratory models described above) with high resolution have shown similar results to the laboratory models, with both the formation of eddies and draining downslope flow represented provided that the resolution is high enough. These simpler models give useful insights into how numerical ocean models might be improved. In addition to resolution and mixing, the way the bottom drag is treated is also important. The inclusion of special benthic boundary layer models in future ocean models may address this problem.

Instabilities and Mixing

One of the striking features of overflows is their ability to generate strong cyclonic vortices (or eddies) in the overlying water. These have been observed above the Denmark Strait overflow (**Figure 2**) and even above overflows beneath the Ronne Ice Shelf (Antarctica). These eddies fall into two main groups: strong eddies that form very close to the source (referred to as PV for potential vorticity eddies) and those that form after the along-slope flow has been established (BI for baroclinic instability eddies). In flows with PV eddies, much of the dense fluid is concentrated in the domes moving along the slope (although with a thin viscous draining layer), while the BI eddies are accompanied by more substantial 'subplumes' of dense fluid extending downslope from the flow. The BI eddies have a weaker effect on the overlying fluid and appear to have a weaker effect on mixing.

Both types of eddies are observed in laboratory experiments and numerical models, and the instabilities leading to BI eddies have been analyzed mathematically. From the mathematical studies an 'interaction' parameter has been defined:

$$\mu = d/\beta R_H$$

where d is the depth of the overflow current as it flows along the slope and R_H is a Rossby radius based on the total depth of the fluid:

$$R_H = \frac{\sqrt{g'H}}{f}$$

Thus βR_H is the vertical scale corresponding to horizontal motions of scale R_H on a slope of tangent β, and the interaction parameter compares this vertical scale to the depth of the current. Although the mathematical analysis is formally valid for small values of μ, laboratory experiments suggest that BI eddies occur for large μ, while PV eddies occur for small μ.

While these eddy types have been identified, and the conditions under which they occur in laboratory experiments have been approximately defined, the clear characterization of eddies in oceanographic flows has yet to be attempted. The viscous draining flows and mixing behavior of real overflows have also not been quantified in much detail, although shallow layers of relatively unmixed fluid extending downslope with a more mixed flow remaining at constant depth have been observed (e.g., downstream of the Faeroes–Shetland channel in the Northeast Atlantic).

Summary

The flow of dense fluid down slopes occurs at a range of scales in the oceans, from small, temporary cascades from shallow shelf seas to the large, continuous flows into major ocean basins. These flows play an important role in controlling the large-scale thermohaline circulation, but their representation in numerical ocean models (especially those used for climate prediction) is poor because of problems with resolution, mixing, and bottom drag. Overflows display a rich behavior and can have a strong effect on the overlying seawater. This behavior is being illuminated through laboratory experiments, numerical and mathematical models, and increasingly detailed and sophisticated field observations.

See also

Bottom Water Formation. Flows in Straits and Channels. Meddies and Subsurface Eddies. Rotating Gravity Currents. Ocean Circulation: Meridional Overturning Circulation. Topographic Eddies.

Further Reading

Baines PG and Condie S (1998) Observations and modelling of Antarctic downslope flows: A review. In: Jacobs SS and Weiss RF (eds.) *Antarctic Research Series, Vol. 75: Ocean, Ice, and Atmosphere – Interactions at the Antarctic Continental Margin*, pp. 29–49. Washington, DC: American Geophysical Union.

Griffiths RW (1986) Gravity currents in rotating systems. *Annual Review of Fluid Mechanics* 18: 59–89.

Hansen B and Osterhus S (2000) North Atlantic–Nordic Seas exchanges. *Progress in Oceanography* 45(2): 109–208.

Ivanov VV, Shapiro GI, Huthnance JM, Aleynik DL, and Golovin PN (2004) Cascades of dense water around the world ocean. *Progress in Oceanography* 60: 47–98.

Krauss W (1996) A note on overflow eddies. *Deep Sea Research Part I* 43(10): 1661–1667.

Lane-Serff GF and Baines PG (2000) Eddy formation by overflows in stratified water. *Journal of Physical Oceanography* 30(2): 327–337.

Price JF and Baringer MO (1994) Outflows and deep-water production by marginal seas. *Progress in Oceanography* 33(3): 161–200.

Price JF, Baringer MO, Lueck RG, *et al.* (1993) Mediterranean outflow mixing and dynamics. *Science* 259: 1277–1282.

Roberts MJ, Marsh R, New AL, and Wood RA (1996) An intercomparison of a Bryan-Cox-type ocean model and an isopycnic ocean model. Part 1: The subpolar gyre and high-latitude processes. *Journal of Physical Oceanography* 26(8): 1495–1527.

Simpson JE (1997) *Gravity Currents: In the Environment and the Laboratory*, 2nd edn. Cambridge, UK: Cambridge University Press.

ROTATING GRAVITY CURRENTS

J. A. Whitehead, Woods Hole Oceanographic Institution, Woods Hole, MA, USA

Introduction

The term gravity current refers to unidirectional flow of fluid with a free surface in a field of gravity. The free surface commonly separates that fluid from a second fluid of different density, although it is possible that the second fluid has negligible effect on the flow, as in the case of water flowing under air. Such currents are also called density currents or buoyancy currents. The distinguishing feature of a gravity current is that the force of gravity acts on the density variation to create a buoyancy force that produces the motion. Additional effects due to earth rotation are frequently found upon such currents in the ocean.

Much knowledge of buoyancy-driven currents in the ocean comes from studies that originated in engineering, and in natural sciences for which earth rotation is not important. When the door is opened between rooms with different air temperatures, a flow and counterflow quickly start to transport heat from the warm to the cold room. In mines, the presence of gas or differential heating can lead to buoyant forces and produce an intense outflow through narrow openings at higher elevations of the mine. During fires, hot gas can accumulate under ceilings and ascend stairwells as an inverted density current. In rivers, bores and fronts can become intense enough to produce very rapid changes in water elevations and velocity during floods or, in some cases, during extreme tide events. In the atmosphere, strong local effects produce katabatic winds, fronts, sea-breezes, and outflows of chilled air from thunderstorms. In some cases suspended particles cause density variation of the bulk fluid. Thus, such currents are found during dust storms, as avalanches, and as downflows of pyroclastic suspensions during volcanic eruptions. They can also be found as turbidity currents on mountains and within the ocean.

Gravity currents in the ocean come in many similar forms, with density differences arising from turbidity, salinity, or temperature variations. They usually originate in constricted regions that separate waters of different density. Thus they frequently originate near river mouths, at the openings of estuaries, at sea straits, and near saddle points separating deep ocean basins.

Roughly speaking, the most rapidly occurring currents have lifetimes of less than a day. Density difference between the density current and surrounding ocean water ranges from 1% to 10% and is usually produced by sediment or salinity differences. The thickness is limited by the fact that the water rapidly flows away from the source region. It is thought typically to range from ten to a few hundred meters. These large and energetic currents are usually formed in response to sudden triggering events such as floods, sudden shifts in high winds, earthquakes, landslides, or volcanic eruptions. Good measurements are often lacking owing to the extremely rapid nature of the events. The relation of such currents to surface weather or tectonic events means that they are usually located near land, from the coast to the bottom of the continental slope or even within bays and estuaries. Examples range from great sediment currents emerging from landslides or from the mouths of rivers and estuaries to fresh water layers in fjords and bays. Both are found more commonly during flood seasons.

The greatest documented rapid currents are the giant ocean turbidity currents that course down the continental shelf break into the abyssal ocean. Not only are these believed to erode the continental rise, but they also propagate hundreds of kilometers over flat terrain (principally the abyssal plains) and supply sediment to cover these sedimented regions. Large currents from volcanic ash settling in ocean water are also known indirectly from the sediment record. Although these may involve relatively immense amounts of mass flux, none has been directly measured. Unusually large amounts of fresh water may flow out of bays under certain meteorological conditions. The outflow of fresh water from the Baltic occasionally increases enormously during storms and propagates along the coast of Norway as a gravity current. The current has, at times, endangered oil platforms with its intensity. There is speculation that large outflows of fresh water may have occurred at the end of some ice ages.

Currents with intermediate timescales from one day to a few months have layer depths from 10 to 100 m. These are usually associated with salinity variations in the sea water with density variations ranging from 0.1% to 1%. These currents are often unsteady, ebbing and waning in periods ranging from days to months or even years. Many are found at the outlets of rivers, bays, or marginal seas. Surface currents have lower salinity than the ambient ocean. They tend to flow along coastlines owing to the influence of the earth's rotation. These currents often radiate waves along their outer edge, and are bounded offshore by a semipermanent front that separates the fresh water from the ocean water. Bottom currents have higher salinity than the ambient water. The dense salty water descends over the sloping continental shelf and slope, and occasionally selects pathways determined by submarine canyons.

The slowest to change are the largest density currents with widths of 20–1000 km and layer depths from 100 m to 1 km. They typically vary in intensity over periods of a few months to many years. These too can be composed of a layer of warmer or fresher water lying above deep denser water, or, alternatively, of a layer of colder or saltier dense water that sinks below the ambient warmer water into the deep ocean. Such currents are associated with temperature changes of one or two degrees, or salinity variations of 0.1 to 1 ppt or less.

Observations

A large turbidity current was triggered by an earthquake near the Grand Banks Newfoundland in 1929. Submarine cables were sequentially broken at progressively greater depths for a lateral extent of over 300 km during an interval of 13 hours. Later geological surveys yielded estimates of volumetric displacement over a thousand times greater than any recorded on land. Despite the importance of turbidity currents, their direct observation is limited to occasional reports because of their rapid dispersion time and the lack of forecasts of the earthquakes that start them. The fossil evidence of the sedimentary record indicates they are a major mechanism for dispersion of sediment on the deep ocean floor. Such rapid turbidity currents have also been found on a smaller scale in canyons on the continental rise and offshore of islands. They are a principal mechanism for erosion of young ocean islands. And finally, they are candidates for tsunami production after an earthquake, since their energy release and volume displacement can greatly exceed the energy of the earthquake itself.

(A)

Figure 1 (A) Atlantic, (B) Pacific, and (C) Indian Ocean locations of deep ocean sills. The arrows show the direction of flow. Light shading indicates depths between 4 and 5 km; dark shading is deeper. Some surface straits that produce gravity currents (bidirectional arrows, e.g, the Strait of Gibraltar) are also shown. The volume flux for these flows have estimates ranging from 0.01 to 10 Sv.

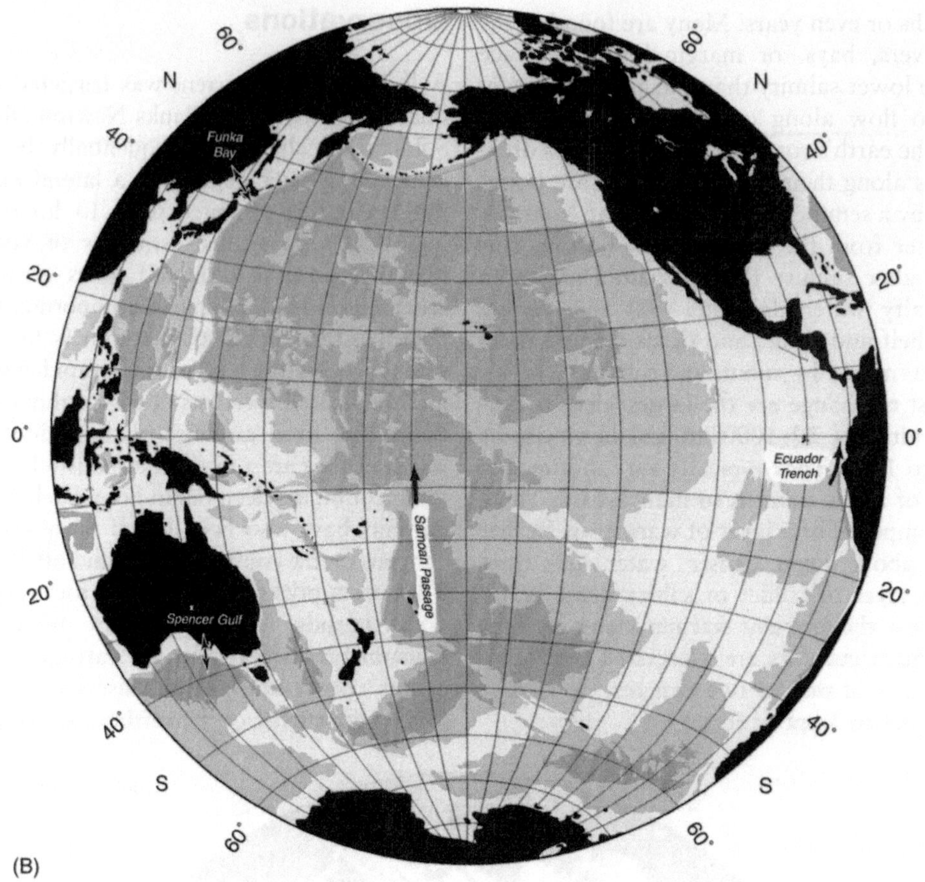

Figure 1 Continued

Low-salinity plumes flowing along coastlines are probably the best-studied of all ocean gravity currents. Their offshore edge is easily seen by airplane and satellite. Some freshwater density currents, for example the Norwegian Coastal Current shown in **Figure 3**, have been known by fishermen for centuries. A similar kind of current is found flowing eastward out of the Tsugaru Strait between Hokkaido and Honshu, Japan. The current veers to the south and forms a warm eddy during certain seasons. This eddy is well known to the fishermen.

If the layer is only slightly larger than 10 m in depth, wind events, with periods of a few days, may have an important effect. An example is the outflow of fresh water from Chesapeake Bay. The plume of fresh water frequently curves to the right as it leaves the mouth of the bay and flows along the coast of Virginia toward Cape Hatteras. A flow to the right is consistent with the direction that theory and laboratory models have indicated would be produced by earth rotation. At some stage wind is seen to blow the freshwater plume offshore and the current breaks up, probably owing to the lack of a coastline to support the pressure field of a unidirectional gravity current with the earth's rotation. Similar effects have been seen offshore of the mouth of Delaware Bay.

The fresh water outflow through straits from large interior seas is found to be great enough that the current persists for weeks or months even in the face of wind fluctuations. The Norwegian coastal current mentioned above is a good example. The Baltic current empties about $15\,000\,\mathrm{m}^3\,\mathrm{s}^{-1}$ of fresh water into the Skagerrak. The water mixes with salt water, turns northward in association with the Jutland current coming from the south, and flows along the coast of Norway. This coastal current moves northward for many hundreds of kilometers. Fluctuations take days or weeks to move from their source in the Skagerrak to the northern end of the coastal current.

Two well-known currents driven by salinity difference are produced by the salinity-driven exchange flow through the Strait of Gibraltar. The Mediterranean Sea is about 10% saltier than the surface Atlantic water owing to the aridity and consequent evaporation of the Mediterranean region. In the Strait, there is a surface inflow of fresher Atlantic

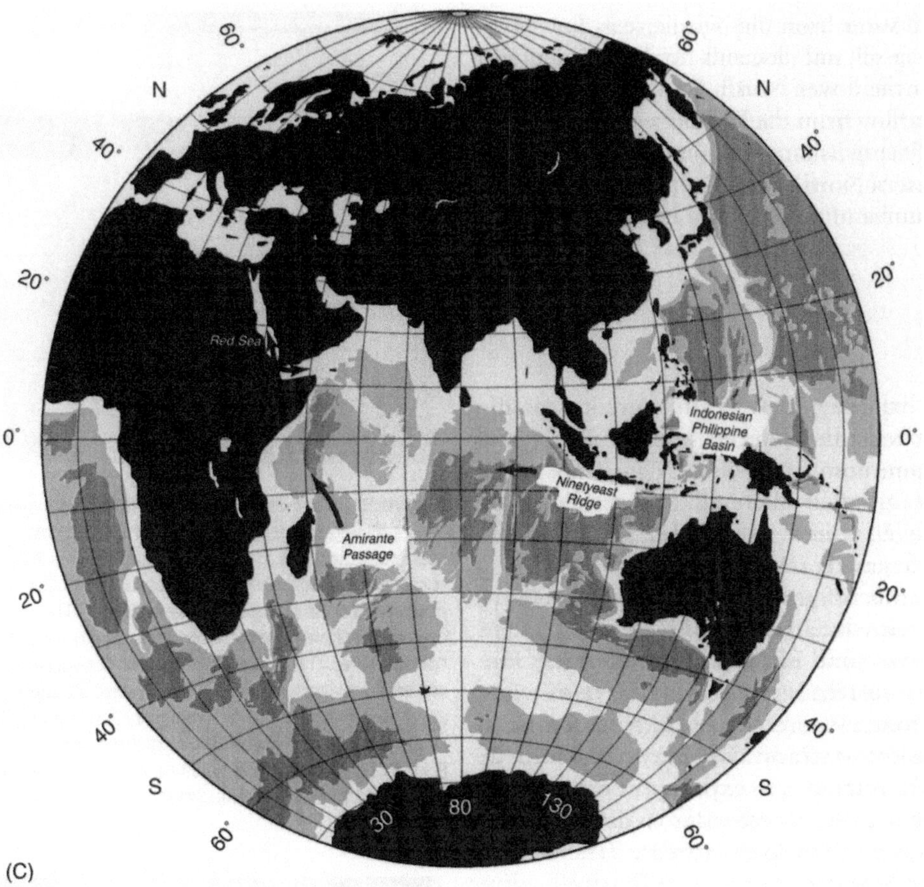

(C)

Figure 1 Continued

water (known even in ancient times) and an underlying outflow of saltier Mediterranean water that was detected in 1870, after more than two centuries of speculation that it existed. The eastward-flowing inflow encircles a clockwise (anticyclonic) gyre in the Alboran Sea by flowing around the northern edge of the gyre. The current then veers southward along the eastern portion of the gyre and encounters the coast of North Africa. It then curves eastward and proceeds along the coast of Africa as the Algerian current. This current is also found to meander offshore and back again with wavelengths lying between 25 and 100 km. The current has seasonal and monthly fluctuations and the gyres expand and contract accordingly. The westward flow of deeper salty water leaves the Mediterranean and descends roughly to 1 km depth in the Atlantic. Part of the Mediterranean water breaks up and forms internal eddies that spread from their origin throughout the eastern North Atlantic and contribute to a salinity maximum at that depth. The remainder continues to flow northward along the continental shelf break at roughly a depth of 1000 m along the coast of Portugal and Spain.

There are numerous other currents of lower salinity that flow along the coastlines of the continents. The Gaspe current flows seaward along the coast of the Gaspe Peninsula of Quebec, driven by the river runoff from the St. Lawrence estuary. The East Greenland current conveys lower salinity water southward along the shelf and shelf break of Eastern Greenland. This large current contributes to the salt/freshwater balance of the Arctic. The Labrador Current flows southward along the East Coast of Canada, bringing icebergs southward into the shipping lanes.

Submerged gravity currents are associated with important deep and bottom water circulation. Flows from one deep ocean basin to another produce intense localized flows at the sill between the two deep basins. This sill is thought to be located at the col, or deepest saddle point between basins, although some frictional effects may move the location of the most important regions downstream. Measurements of temperature, salinity, and velocity have been taken in the localized vicinity of about ten large oceanic sills in all the oceans. The most thoroughly studied is the Denmark Straits overflow between Greenland and

Iceland. Cold water from the Nordic seas flows over the 650 m deep sill and descends to about a depth of 3 km to form the lower North Atlantic Deep water. A second overflow from the Nordic seas crosses a sill south of the Faeroe Islands at about 800 m depth and enters the eastern North Atlantic. Overflows are also found at a number of other sills at locations shown in **Figure 1**.

Theory

The primary balance in any density current, rotating or not, is between the buoyant force of gravity and change in momentum of the fluid. This leads to the velocity scale given by Bernoulli's principle of the form $v = \sqrt{g'D}$, where $g' = g\delta\rho/\rho$, is commonly called reduced gravity, D is vertical layer thickness, g is the acceleration of gravity, and $\delta\rho/\rho$ is the relative density difference due to thermal or salinity content between a layer and surrounding sea water. Frequently, such a current is set up as the layer of water flows away from a source region. Such currents can be either transient or steady. One way to measure the effect of Earth rotation is to express a timescale that is given by the number of seconds in a day divided by 4π times the sine of the local latitude. The formula for this timescale is thus $\tau(\text{secs}) = 6876/\sin\theta$, a time duration that is about 2.7 h at 45°. For currents of longer duration than such a timescale, the flowlines will curve toward the right in the Northern hemisphere, and to the left in the Southern hemisphere from the Coriolis force of the earth's rotation. The only way for gravity currents to move large distances thereafter is to curve around and flow next to a coast or front. In such a case, the width of the current is limited to $v\tau$. Thus the volumetric flux is limited to size $v^2D\tau = g'D^2\tau$. Gravity currents flowing along a coastline like this have a balance between a pressure gradient at right angles to the flow and the Coriolis force. For example, a current produced by a salinity change of 1.4 ppt and a depth of 100 m has the velocity scale $1\ \text{m s}^{-1}$, lateral width of about 10 km, and a volume flux of about $10^6\ \text{m}^3\ \text{s}^{-1}$.

Inviscid 'rotating hydraulics' theory has been developed to enlarge the understanding of a variety of nonlinear rotating flows. If the flow is fed from a region in which the layer of water is elevated by height H, velocity along the wall scales with $v_w = \sqrt{2g'(H - D)}$ in the absence of friction. Many gravity currents have been monitored by satellite data or with fixed arrays of instruments in the ocean. Flows through eight of the deep ocean sills shown in **Figure 2** have been found to follow the above scaling within a range of about a factor of 2 or 3. The effects

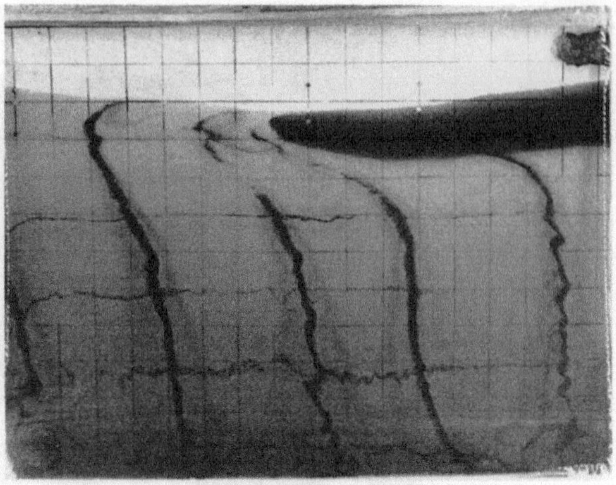

Figure 2 Top view of a dyed surface fresh water gravity current on a rotating turntable. The current flows over and around clear salt water above a flat tilted bottom with a slope of 30° from the horizontal. The coastline is located at the top of the picture; the deep region is located at the bottom. The turntable is rotating counterclockwise, which corresponds to a model of the Northern hemisphere. The fluid was fed with a source at the right. Density of the fluid and turntable rotation are adjusted to make a width scale $v\tau$ of about 0.1 m, which is the spacing of the grid lines. Distorted lines of dye show displacement around the current. Some of the displacement is made by shelf waves that propagate ahead of the nose of the gravity current.

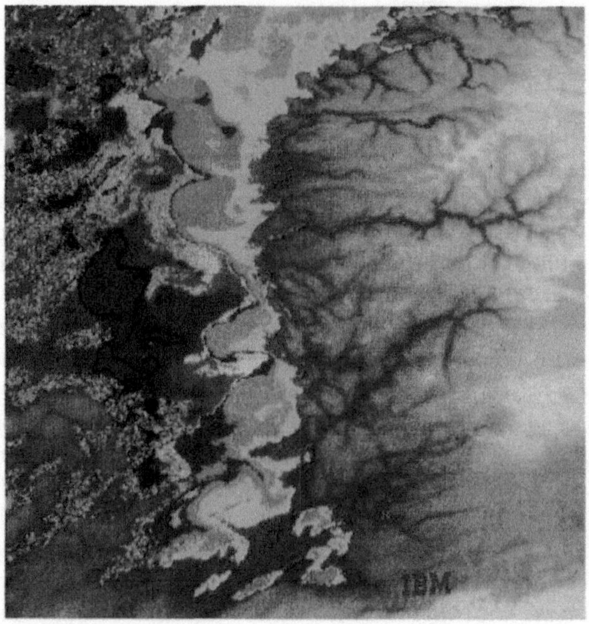

Figure 3 The Norwegian Coastal Current. The colors reveal variations of surface temperature with blue colder and yellow-red warmer. However the current itself has a density change from salinity variations that are much greater than those due to temperature. The current flows northward along the coast of Norway as a gravity current of lower-salinity water from fresh water outflow along Norway and the Baltic. Eddies along the outer edge are frequently visible. In some cases, an energized current has been seen as a northward-propagating nose.

of bottom friction, interfacial mixing, and complex ocean floor topography remain poorly understood at present.

Concluding Remarks

The velocity scale $v = \sqrt{g'D}$, timescale $\tau(\mathrm{secs}) = 6876/\sin\theta$, and width scale $v\tau$ reproduce gravity currents of oceanic flows in a variety of settings as long as the width, duration, and depth lie within certain ranges. As one gets to water shallower than roughly 100 m, the frictional effects of turbulent boundary layers are believed to modify the flows substantially. The detailed response of gravity currents to this drag is poorly understood and is the subject of substantial research. Currents wider than about 100 km begin to be influenced by the effects of a spherical earth or large variations in bottom shape. The radiation of shelf waves can attenuate gravity currents by leaking the momentum into shelf modes in front of the nose of the gravity current. The illustration in **Figure 2** shows dye displacement from a shelf wave current next to a fresh water density current. This current is viewed from above in an experiment performed on a rotating turntable over a sloping bottom. In contrast to the photograph of the Norwegian Coastal Current in **Figure 3**, where eddies are visible at the outer edge of the current, the bottom slope seems to stabilize this current but also allow smoother and more wavelike currents in the clear water surrounding the dark intrusion. In climate studies, equatorial Kelvin waves and internal Rossby waves are all important possible features of the gravity current component of tropical oscillations such as El Niño and ENSO (El Niño Southern Oscillation).

See also

Overflows and Cascades.

Further Reading

Griffiths RW (1986) Gravity currents in rotating systems. *Annual Review of Fluid Mechanics* 18: 59–89.

Pratt LJ (ed.) (1989) *The Physical Oceanography of Sea Straits*. Dordrecht: Kluwer Academic Publishers.

Pratt LJ and Lundberg PA (1991) Hydraulics of rotating straits and sill flow. *Annual Review of Fluid Mechanics* 23: 81–106.

Saetre R and Mork M (eds.) (1981) *The Norwegian Coastal Current*. Proceedings from the Norwegian Coastal Current Symposium, Bergen, vols. 1, 2.

Simpson JE (1997) *Gravity Currents*. 2nd edn. Cambridge: Cambridge University Press.

Whitehead JA (1998) Topographic control of ocean flows in deep passages and straits. *Reviews of Geophysics* 36: 423–440.

FLOWS IN STRAITS AND CHANNELS

D. M. Farmer, Institute of Ocean Sciences, Sidney, British Columbia, Canada

Introduction

The term sea strait is used to describe a channel connecting different water bodies. It may refer to narrow passes between an island and the mainland, such as the Strait of Messina, or to channels serving as the primary or sole connection between enclosed seas and the open ocean, such as the Strait of Gibraltar or the Bosphorus. A further distinction exists between straits that are wide relative to the internal Rossby radius of deformation such as the Denmark Strait and the Faroe Bank Channel, for which rotation plays a dominant role, and narrower straits where rotation is of lesser or negligible significance. An intermediate case is the Strait of Gibraltar, where rotation results in a tilted interface but does not dominate the exchange process. In many deep-water channels, such as the Samoan Gap, only the deep water is controlled by the topography. The upper layer in these cases is relatively passive and not dynamically linked to the deeper flow, except by virtue of the imposed density difference between the layers. However, the concept of flow control also applies to these cases and so is appropriately considered here.

Exchange through a strait may also involve movement of water in more than two distinct layers. For example, in Bab el Mandab during summer the surface layer reverses and colder, lower-salinity water intrudes at intermediate depths from the Gulf of Aden into the Red Sea, above the denser, deep-water overflow.

In general, confinement of the flow in straits and channels tends to amplify tidal and atmospherically forced currents. A crucial aspect of circulation in semienclosed seas is the exchange of water with the open ocean through the connecting strait. An important class of flows, referred to as maximal exchange flows, applies when the strait exercises control over the bidirectional transport. The distinction between maximal and submaximal flow conditions, which may change in any single strait on a seasonal or longer-term basis, provides a further characteristic with which to classify straits.

Environmental and strategic considerations have motivated widespread research of straits in recent years and there have been a number of detailed observational and theoretical studies leading to new insights on exchange dynamics, mixing, internal wave generation, implications for climate change, and other aspects. Many estuaries are connected to the ocean through narrow channels and are subject to similar exchange flow constraints. Flow characteristics in these smaller channels have served as excellent surrogates for the study of processes occurring at larger scales.

Topographic details can have a profound effect on the exchange. Many straits have a shallower sill in addition to being narrower than the adjacent water bodies. The location of the shallowest section relative to the narrowest portion of the strait can determine whether or not variability in the stratification of adjoining water bodies can propagate into the strait and thus affect the rate at which exchange takes place.

History

It has long been recognized that straits are often characterized by persistent flows in one direction or another, and these have motivated scientific investigation and analysis. Two examples for which scientific comment dates back many centuries are the Straits of Gibraltar and the Bosphorus or Strait of Istanbul. It is a striking fact that both of these straits exhibit a persistent flow inward to the Mediterranean. Attempts to explain this apparent violation of continuity included the proposed existence of underground channels through which the surplus drained into the interior of the earth. It was recognized that evaporation played a role, although evaporation alone removed insufficient water to account for the exchange. Towards the end of the seventeenth century, Luigi Marsigli carried out a beautiful laboratory experiment with salt and fresh water that exchanged as density currents past a barrier. This demonstrated the mechanism of bidirectional flow, consistent with Marsigli's own observations in the Bosphorus. In 1755 Waitz explained a similar exchange flow in the Strait of Gibraltar, subsequently confirmed by drogue measurements, thus accounting for the large surface flow into the Mediterranean.

The fact that water moves in opposite directions at different depths through the Strait raises the question of possible controls on the rate of exchange. This is a problem of quite general significance in fluid mechanics, having application, for example, to calculation of the temperature of thoroughly mixed air in a heated room communicating through an open door with outside air of lower temperature. Stommel was one of the first to develop a modern explanation of internal exchange controls and in the 1950s described the situation in which thoroughly mixed water in an estuary was exchanged with the open ocean through a strait. The estuary was said to be 'overmixed' in the sense that the bidirectional exchange had achieved a maximum rate for the given density difference and channel geometry owing to the presence of internal flow control. For a given influx of fresh water, the salinity of the estuary was then determined. This flow state is now recognized as an important special case of bidirectional exchange and may occur in any strait that is not so wide as to be dominated by Coriolis effects, provided a suitable set of bounding conditions exist at either end.

Two-layer Exchange Flows in Straits

In the simplest case, where rotation is negligible and the strait is short enough for mixing and friction to be unimportant, the two-layer exchange flow can be analyzed in terms of the frictionless theory of layered flow. We refer to the flow as being controlled when the combined Froude number G^2 is unity:

$$G^2 = F_1^2 + F_2^2 = 1 \qquad [1]$$

where

$$F_i^2 = \frac{u_i^2}{g' y_i} \qquad [2]$$

u_i is flow speed and y_i is the thickness of the upper ($i = 1$) and lower ($i = 2$) layer respectively, and

$$g' = g\,\frac{\rho_2 - \rho_1}{\rho_2} \qquad [3]$$

is the reduced gravity due to the differences in density ρ_i of each layer. F_{i^2} is referred to as the layer Froude number. When the flow is controlled, adjustments in the depth of the interface can travel in one direction, but not in the other. Thus an internal control acts as an information gate that blocks adjustments of interface depth, in the form of long internal waves, from propagating against the flow. The control condition [1] corresponds to the point at which such a wave is just arrested and it always separates subcritical from supercritical flow. Subcritical flow occurs when the combined Froude number is less than unity such that waves can travel in both directions. Supercritical flow occurs when the Froude number is greater than unity such that waves can only travel downstream. Steady-state solutions to the frictionless exchange flow can be found from the Bernoulli and continuity equations and have been discussed in detail in the literature.

Maximal exchange flow constitutes the steady-state limit when water of differing densities is free to move in opposite directions. In general, maximal flow requires two separate locations where eqn [1] is satisfied. Where there is both a sill and a contraction, the relative locations of these topographic features are important in determining whether or not the reservoir conditions, specifically the interface depths in each adjoining water body, can influence stratification within the controlling region, thus eliminating at least one control and preventing maximal exchange from taking place. If the narrowest section is located toward the end of the strait bounded by denser water and the shallowest section is toward the end bounded by less-dense water and if controls [1] occur at both of these locations, the requirements for maximal exchange are met. Additional controls may also occur, for example, in the form of a sequence of deeper sills downstream of the first, but these have no direct influence on the exchange rate.

An internal control separates supercritical from subcritical flow. The depth of the interface between the two layers is asymmetrical about a control position (i.e., the interface depth increases or decreases continuously as one moves through the control) and thus easily recognized in oceanographic data. Although the control acts on both layers, in the more general case of a strait containing both a sill and contraction, a sill acts primarily on the lower layer, whereas the contraction acts primarily on the upper layer. In the case of exchange flows through a strait, maximal exchange requires that the supercritical flow be directed away from the control and towards the nearest reservoir; that is, the two control locations for which eqn [1] is satisfied are separated by flow that is subcritical ($G^2 < 1$). The presence of supercritical conditions on either side of a subcritical interior portion, the 'control section' lying between the two controls, ensures that adjustments in the level of the interface in the adjoining water bodies cannot propagate into the strait and thereby influence the exchange, although adjustments in interface depths at one control can communicate through the subcritical part of the flow with the other control. The exchange within a strait is therefore determined entirely by the local geometry and the densities of the exchanging water masses and is therefore maximal. The bounding supercritical states isolate the control from the adjacent seas. Of course, the supercritical flow must always match the subcritical conditions in each reservoir far away from the strait and generally does so through an internal jump. In the special case of a simple contraction with no sill and in the absence of barotropic forcing, such as that due to the tide, fresh water discharge or atmospheric pressure differences, the two controls can be thought of as having coalesced, so that the subcritical portion vanishes and the location at which eqn [1] is satisfied separates two supercritical flows.

The above description covers maximal exchange. If the interface (or usually the thermocline) is sufficiently shallow in the less dense reservoir (i.e., the Atlantic Ocean in the case of the Strait of Gibraltar) or sufficiently deep in the denser reservoir (the Alboran Sea in the case of Gibraltar) to prevent formation of a supercritical flow, the nearest control location is said to be 'flooded' and flow in this portion of the strait is subcritical. The exchange is then subject only to a single control, for example, at the sill if the contraction is flooded, and is therefore submaximal. In this case the exchange rate is a function not only of the densities of the exchanging water masses but also of the stratification depth in the reservoir adjacent to the flooded contraction. Both maximal and submaximal conditions have been observed in the Strait of Gibraltar. However, the frequency with which transitions take place between one state and the other, which has important implications for the interpretation of longer-term variability in the Mediterranean, remains to be established.

Barotropic Forcing of Exchange Flows

Natural flows in straits are rarely steady: barotropic forcing by tides, meteorological effects and changes in the stratification at either end may modify the rate of exchange. Quasi-steady solutions of the exchange equations may still be valid, provided the influence of the forcing is properly accommodated. Steady solutions for maximal exchange in the presence of a sill and a contraction remain valid if interfacial depth adjustments at one control communicate through the subcritical portion of the strait with the other control, subject to a delay that is short relative to the time scale of forcing. In the Strait of Gibraltar this condition is met approximately for tidal forcing.

Barotropic forcing modifies the exchange within the maximal state, but may also be strong enough to arrest one of the layers, thus inhibiting the bidirectional flow altogether and temporarily overiding the internal control. This situation is common in many coastal environments with large tidal currents. In the quasi-steady case the effect of barotropic forcing can be investigated by appropriate modification of the exchange equations to accommodate relative differences in the transport of each layer. The resulting exchange falls naturally into three regimes (**Figure 1**). In the 'moderate' regime both layers continue to exchange, but with an adjustment of interface depth to maintain control at the sill and the contraction. If the forcing is from the less-dense towards the denser reservoir, the interface drops; for forcing in the other direction, it rises. If the forcing is great enough to cause the interface to intersect the seafloor or the surface, one of the layers is arrested.

'Strong' forcing occurs when the barotropic pressure gradient from one reservoir to the other is great enough to push the interface downstream of the sill. In this case control over the sill crest is lost, leaving just a single layer over the crest, which is therefore uncontrolled (see **Figure 1A** and **C**). The steeply inclined stratification intersects the surface in a front, a phenomenon commonly observed in coastal waters. The front is characterized by sharp changes in water properties and may entrain bubbles to considerable depth. In general, tidal forcing tends to increase the overall exchange through the strait beyond the maximal exchange rate in the absence of tides. This is particularly true in straits such as Gibraltar, where the dependence of flow rate on tidal modulation is strongly nonlinear.

The dimensional barotropic transport per unit width is defined as

$$Q = u_{1s}y_{1s} + u_{2s}y_{2s} \qquad [4]$$

where y_{is} is the dimensional thickness of each layer ($i = 1, 2$) and u_{1s}, u_{2s} are the corresponding flow speeds; the subscript 's' refers to values above the sill crest (see **Figure 1**). Considering the case illustrated in **Figure 1D** ($Q > 0$), the barotropic transport for which two-layer exchange is lost and the interface intersects the surface is then shown to be

$$Q = (2/3)^{3/2} h_s \sqrt{g' h_s} \qquad [5]$$

where $h_s = y_{1s} + y_{2s}$. The corresponding limit for flow in the other direction (in **Figure 1A**) is

$$Q = -h_s \sqrt{g' h_s} \qquad [6]$$

for which the interface intersects the sill crest and the dense water layer is arrested. At still stronger forcing in this direction, the point of intersection moves down the lee face of the sill (**Figure 1G**).

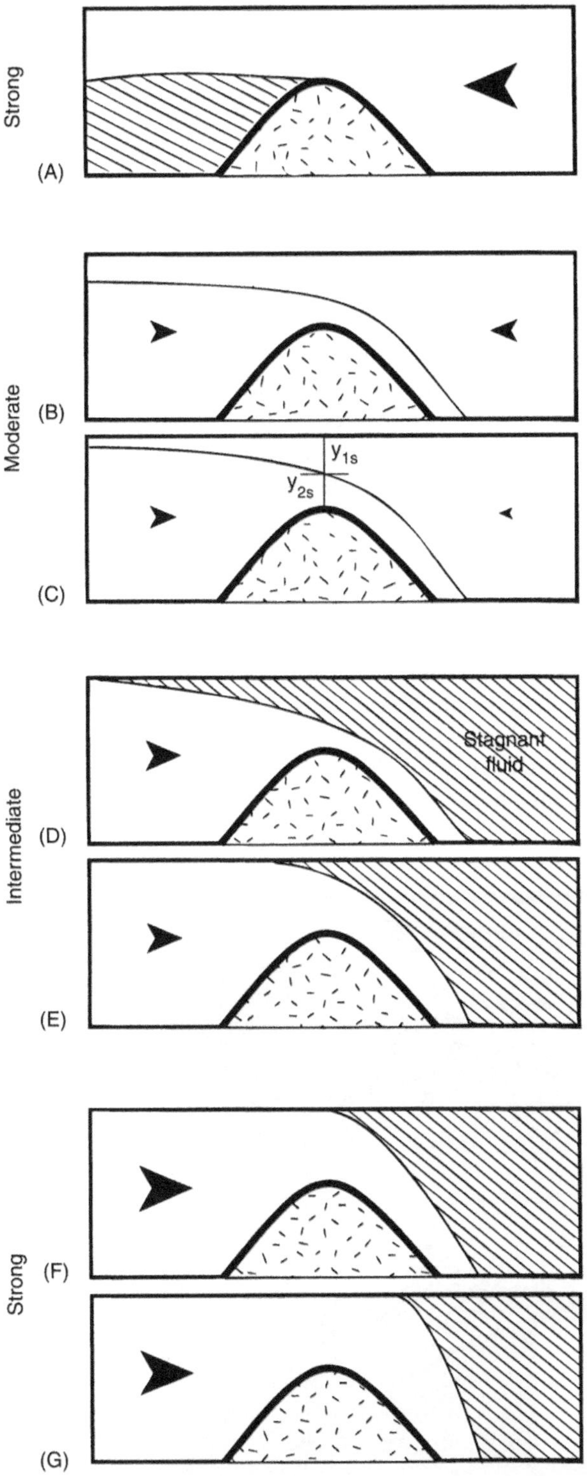

Figure 1 Quasi-steady response of a two-layer exchange flow through a strait subject to barotropic forcing. With strong enough forcing from the less dense reservoir on the right, the denser layer is arrested and the interface intersects the crest of the sill (A). A similar effect occurs with strong forcing from the dense reservoir on the left, in which case the interface intersects the surface (D) and may be pushed downstream of the sill (G). (Adapted from Farmer Armi, 1986.)

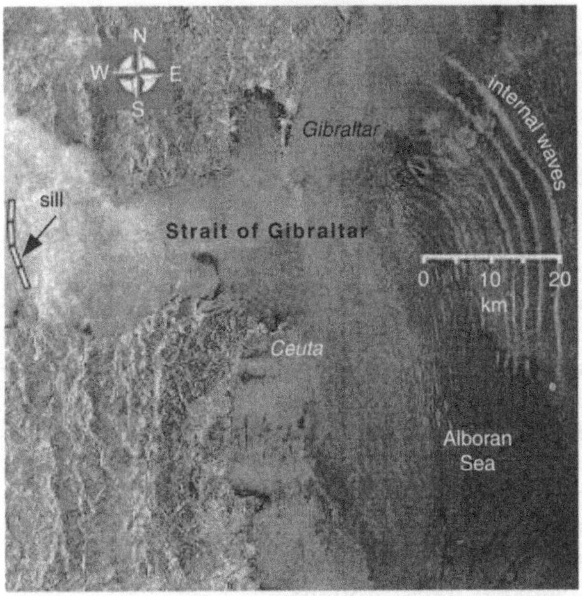

Figure 2 An ERS-1 synthetic aperture radar (SAR) image, showing internal waves at 22.39 UT, January 1, 1993, formed over the sill of the Strait of Gibraltar, radiating eastward into the Alboran Sea. (Photograph: European Space Agency.) The strait is approximately 27 km wide at the sill and 14 km wide in the narrowest section to the east of the sill.

The unsteady character of tidal forcing in straits where the flow is controlled can have a further effect leading to strong surface signatures in remotely sensed images. The release of potential energy stored in the deformation of the interface downstream of the control as the tide slackens can generate large-amplitude nonlinear internal waves. These propagate away from the sill in the form of an undular bore. In the Strait of Gibraltar, for example, they are generated toward the end of the ebb flow as the internal hydraulic control over the sill is lost. They are observed to travel east along the strait and into the Alboran Sea where they spread radially before dissipating (**Figure 2**).

Mixing in Straits and over Sills

Exchange flows in straits experience enhanced shear between the exchanging layers. Under certain circumstances this leads to instability and mixing. In longer straits, such as the Bosphorus, mixing can produce significant changes in the layer densities. Combined with frictional effects, this results in an internal response that differs from the frictionless results discussed above. In contrast to the frictionless prediction, there can be a marked slope within the subcritical portion of the flow and the exit control of the upper layer is displaced downstream with respect to the active layer.

Mixing can also result from tidal effects. Small-scale processes leading to mixing have been examined in detail over some sills where they are also seen to play a role in the establishment of controlled flow. A particularly well-studied example is Knight Inlet, British Columbia, where instabilities form on the interface over the sill crest (**Figure 3**).

Acceleration of the flow in the supercritical layer downstream of the crest creates an asymmetric instability that ejects fluid upward from the deeper layer. This in turn forms an intermediate layer of weakly mixed fluid that fills in as the downslope flow becomes established. When the tidal current slackens, the mixed layer can intrude upstream just beneath the fresher surface layer. While Knight Inlet is perhaps unique in the extent to which it has been studied, similar processes can be expected wherever stratified flow occurs over abrupt topography such as commonly found in straits.

Effects Due to Rotation

The hydraulics of flow with uniform potential vorticity are very similar to classical hydraulics, provided there is no separation from the sidewalls. Scale analysis based on the geostrophic relation suggests that separation occurs if the channel width is greater than the distance L,

$$L \sim \frac{2(g'\bar{Y})^{1/2} f^{-1}}{\bar{U}/(g'\bar{Y})^{1/2}} \qquad [7]$$

where f is the Coriolis parameter and $\bar{Y}$ and $\bar{U}$ are representative depth and velocity scales within the strait. In the absence of separation, maximal exchange in a rotating channel flow can still occur, with the control being exercised through a Kelvin wave. The situation becomes more complicated, however, when the width is sufficient for the flow to separate. Control is then exercised by a frontal wave with strong cross-stream velocities.

As with all hydraulic approaches, irrespective of rotation, the calculation involves integration over layers, for example, from the seafloor to the interface and from the interface to the surface in a two-layer flow. For rotational flows, once the potential vorticity is assumed, integration can be carried out and the cross-stream structure of the flow is fully determined. It has been shown that when the flow is critical with respect to the frontal wave, a stagnation point occurs on the right sidewall (in the northern hemisphere), independent of the potential-vorticity

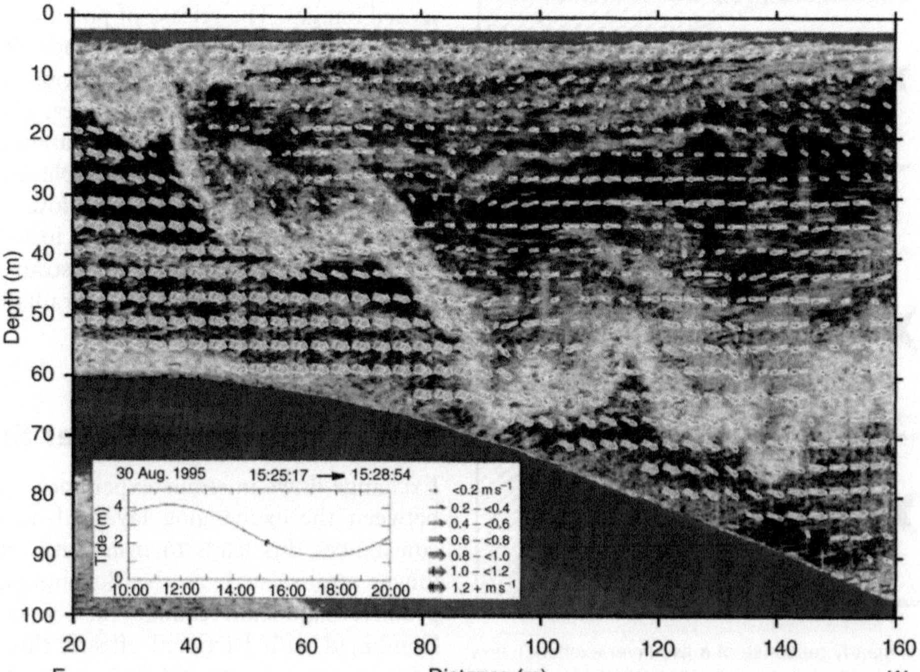

Figure 3 Instabilities on the sheared interface of controlled flow over the sill in Knight Inlet. The instabilities are asymmetrical, leading to injection of water from the supercritical lower layer into the slowly moving upper layer. The flow state corresponds to the bottom illustration in **Figure 1**. The tidal current is from left to right, with arrows indicating current vectors; a weak recirculation exists within the upper layer. The inset indicates the phase of the tide at the time of measurement. Adapted from Farmer DM and Armi L (1999) *Proceedings of the Royal Society, Series A* 445: 3221–3258.

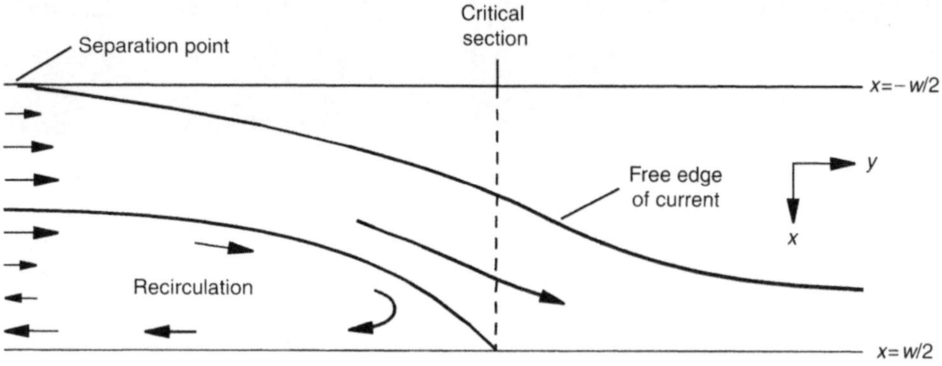

Figure 4 Plan view of exchange flow in a rotating strait, showing separated single-layer flow at the critical section of a strait, with a corresponding stagnation point on the opposite wall. Recirculation occurs upstream. Adapted with permission from Pratt and Lundberg (1991).

distribution. The two-layer portion of the flow crosses over the strait in a distance of order the internal Rossby radius; under certain conditions recirculation may occur as shown in **Figure 4**. Time dependent adjustment of strait flows where rotation is important gives rise to internal bores or shock waves that have a two-dimensional structure. Even where rotation is of minor importance, transverse variability and flow separation can also occur owing to abrupt changes in channel width or direction. This is observed, for example, in the Bosphorus, where the channel geometry leads to marked transverse variability.

See also

Estuarine Circulation. Overflows and Cascades.

Further Reading

Armi L (1986) The hydraulics of two flowing layers with different densities. *Journal of Fluid Mechanics* 163: 27–58.

Armi L and Farmer DM (1988) The flow of Atlantic water through the Strait of Gibraltar. *Progress in Oceanography* 21(1): 1–105.

Assaf G and Hecht A (1974) Sea straits: a dynamical model. *Deep-Sea Research* 21: 947–958.

Briand F (ed.) (1996) Dynamics of Mediterranean straits and channels. *Bulletin de l'Institut océanographicque, numéro spécial* 17. Monaco: Musee océanographique.

Deacon M (1985) An early theory of ocean circulation: J. S. Von Waitz and his exploration of the currents in the Strait of Gibraltar. *Progress in Oceanography* 14: 89–101.

Farmer DM and Armi L (1986) Maximal two-layer exchange over a sill and through the combination of a sill and contraction with barotropic flow. *Journal of Fluid Mechanics* 164: 53–76.

Farmer DM and Armi L (1988) The flow of Mediterranean water through the Strait of Gibraltar. *Progress in Oceanography* 21(1): 1–105.

Murray SP and Johns W (1997) Direct observations of seasonal exchange through the Bab el Mandab Strait. *Geophysics Research Letters* 24: 2557–2560.

Pratt LJ (ed.) (1990) *The Physical Oceanography of Sea Straits*, NATO ASI Series, Series C: Mathematical and Physical Sciences, vol. 318.

Pratt LJ and Lundberg PA (1991) Hydraulics of rotating strait and sill flow. *Annual Review of Fluid Mechanics* 23: 81–106.

Stommel H and Farmer HG (1953) Control of salinity in an estuary by a transition. *Journal of Marine Research* 12: 13–20.

SLIDES, SLUMPS, DEBRIS FLOW, AND TURBIDITY CURRENTS

G. Shanmugam, The University of Texas at Arlington, Arlington, TX, USA

Introduction

Since the birth of modern deep-sea exploration by the voyage of HMS *Challenger* (21 December 1872–24 May 1876), organized by the Royal Society of London and the Royal Navy, oceanographers have made considerable progress in understanding the world's oceans. However, the physical processes that are responsible for transporting sediment into the deep sea are still poorly understood. This is simply because the physics and hydrodynamics of these processes are difficult to observe and measure directly in deep-marine environments. This observational impediment has created a great challenge for communicating the mechanics of gravity processes with clarity. Thus a plethora of confusing concepts and related terminologies exist. The primary objective of this article is to bring some clarity to this issue by combining sound principles of fluid mechanics, laboratory experiments, study of modern deep-marine systems, and detailed examination of core and outcrop.

A clear understanding of deep-marine processes is critical because the petroleum industry is increasingly moving exploration into the deep-marine realm to meet the growing demand for oil and gas. Furthermore, mass movements on continental margins constitute major geohazards. Velocities of subaerial debris flows reached up to 500 km h^{-1}. Submarine landslides, triggered by the 1929 Grand Banks earthquake (offshore Newfoundland, Canada), traveled at a speed of 67 km h^{-1}. Such catastrophic events could destroy offshore oil-drilling platforms. Such events could result in oil spills and cost human lives. Therefore, numerous international scientific projects have been carried out to understand continental margins and related mass movements.

Terminology

The term 'deep marine' commonly refers to bathyal environments, occurring seaward of the continental shelf break (>200-m water depth), on the continental slope and the basin (**Figure 1**). The continental rise, which represents that part of the continental margin between continental slope and abyssal plain, is included here under the broad term 'basin'.

Two types of classifications and related terminologies exist for gravity-driven processes: (1) mechanics-based terms and (2) velocity-based terms.

Mechanics-Based Terms

Continental margins provide an ideal setting for slope failure, which is the collapse of slope sediment (**Figure 2**). Following a failure, the failed sediment moves downslope under the pull of gravity. Such gravity-driven processes exhibit extreme variability in mechanics of sediment transport, ranging from mobility of kilometer-size solid blocks on the seafloor to transport of millimeter-size particles in suspension of dilute turbulent flows. Gravity-driven processes are broadly classified into two types based on the physics and hydrodynamics of sediment mobility: (1) mass transport and (2) sediment flows (**Table 1**).

Mass transport is a general term used for the failure, dislodgement, and downslope movement of sediment under the influence of gravity. Mass transport is composed of both slides and slumps. Sediment flow is an abbreviated term for sediment-gravity flow. It is composed of both debris flows and turbidity currents (**Figure 3**). In rare cases, debris flow may be classified both as mass transport and as sediment flow. Thus the term mass transport is used for slide, slump, and debris flow, but not for turbidity current. Mass transport can operate in both subaerial and subaqueous environments, but turbidity currents can operate only in subaqueous environments. The advantage of this classification is that physical features preserved in a deposit directly represent the physics of sediment movement that existed at the final moments of deposition. The link between the deposit and the physics of the depositional process can be established by practicing the principle of process sedimentology, which is detailed bed-by-bed description of sedimentary rocks and their process interpretation.

The terms slide and slump are used for both a process and a deposit. The term debrite refers to the deposit of debris flows, and the term turbidite represents the deposit of turbidity currents. These

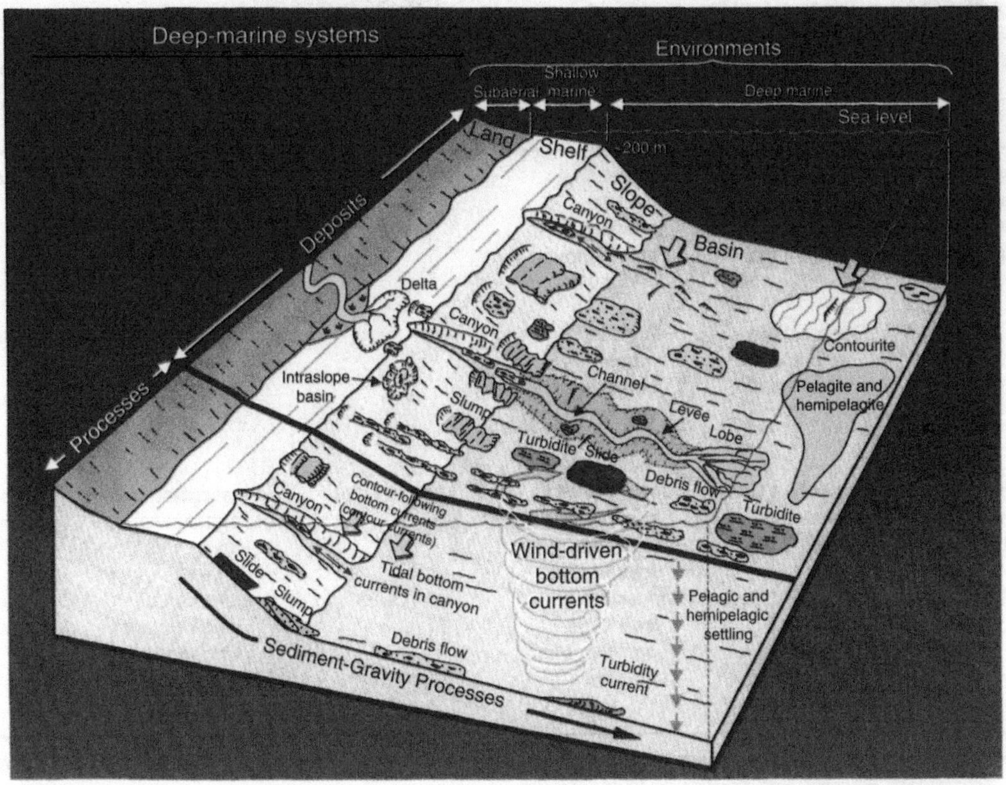

Figure 1 Schematic diagram showing complex deep-marine sedimentary environments occurring at water depths deeper than 200 m (shelf-slope break). In general, sediment transport in shallow-marine (shelf) environments is characterized by tides and waves, whereas sediment transport in deep-marine (slope and basin) environments is characterized by mass transport (i.e., slides, slumps, and debris flows), and sediment flows (i.e., debris flows and turbidity currents). Internal waves and up and down tidal bottom currents in submarine canyons (opposing arrows), and along-slope movement of contour-following bottom currents are important processes in deep-marine environments. Circular motion of wind-driven bottom currents can be explained by eddies induced by the Loop Current in the Gulf of Mexico or by the Gulf Stream Gyre in the North Atlantic. However, such bottom currents are not the focus of this article. From Shanmugam G (2003) Deep-marine tidal bottom currents and their reworked sands in modern and ancient submarine canyons. *Marine and Petroleum Geology* 20: 471–491.

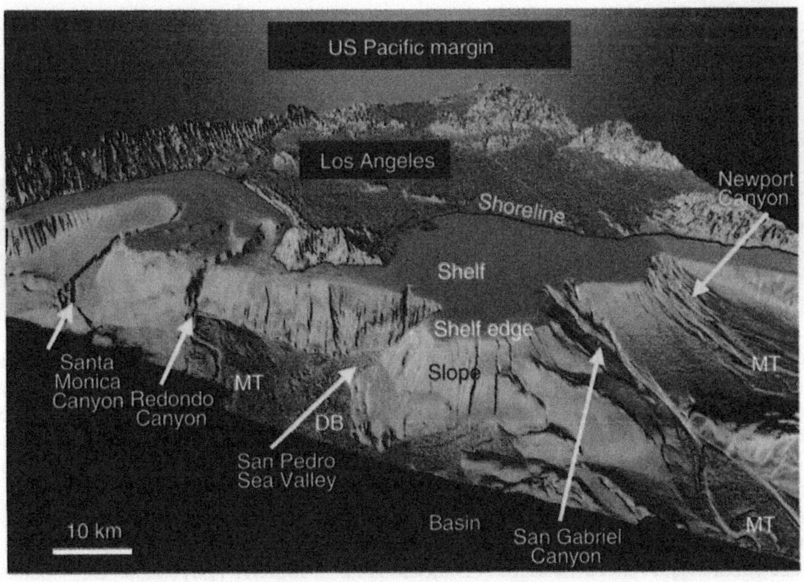

Figure 2 Multibeam bathymetric image of the US Pacific margin, offshore Los Angeles (California), showing well-developed shelf, slope, and basin settings. The canyon head of the Redondo Canyon is at a water depth of 10 m near the shoreline. DB, debris blocks of debris flows; MT, mass-transport deposits. Vertical exaggeration is 6 ×. Modified from USGS (2007) US Geological Survey: Perspective view of Los Angeles Margin. http://wrgis.wr.usgs.gov/dds/dds-55/pacmaps/la_pers2.htm (accessed 11 May 2008).

Table 1 A classification of subaqueous gravity-driven processes

Major type	*Nature of moving material*	*Nature of movement*	*Sediment concentration (vol. %)*	*Fluid rheology and flow state*	*Depositional process*
Mass transport (also known as mass movement, mass wasting, or landslide)	Coherent Mass without internal deformation	Translational motion between stable ground and moving mass	Not applicable	Not applicable	Slide
	Coherent Mass with internal deformation	Rotational motion between stable ground and moving mass			Slump
Sediment flow (in cases, mass transport)	Incoherent Body (sediment–water slurry)	Movement of sediment–water slurry *en masse*	High (25–95)	Plastic rheology and laminar state	Debris flow (mass flow)
Sediment flow	Incoherent Body (water-supported particles in suspension)	Movement of individual particles within the flow	Low (1–23)	Newtonian rheology and turbulent state	Turbidity current

Reproduced from Shanmugam G (2006). *Deep-Water Processes and Facies Models: Implications for Sandstone Petroleum Reservoirs.* Amsterdam: Elsevier, with permission from Elsevier.

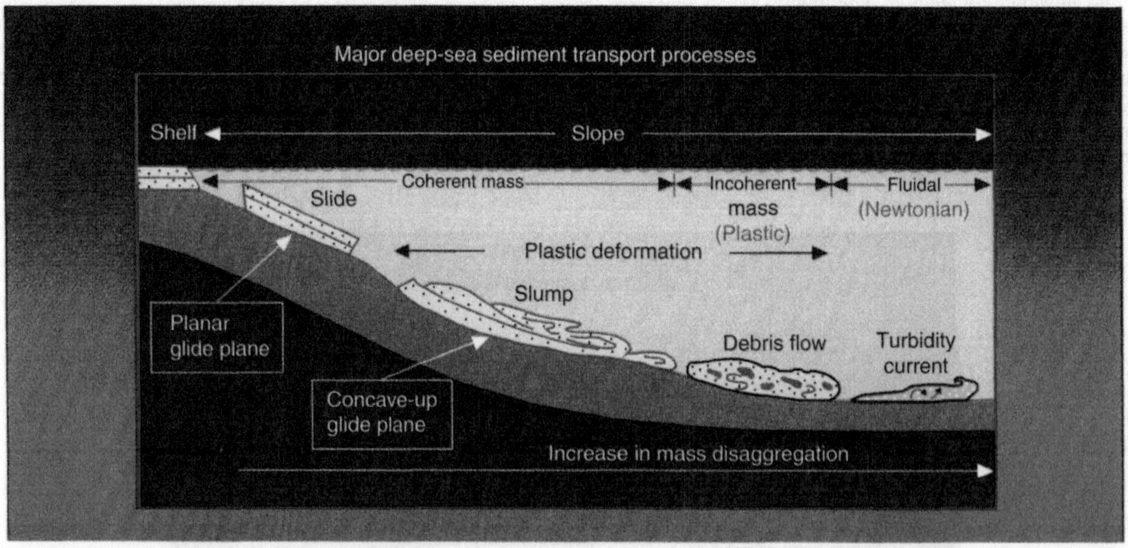

Figure 3 Schematic diagram showing four common types of gravity-driven downslope processes that transport sediment into deep-marine environments. A slide represents a coherent translational mass transport of a block or strata on a planar glide plane (shear surface) without internal deformation. A slide may be transformed into a slump, which represents a coherent rotational mass transport of a block or strata on a concave-up glide plane (shear surface) with internal deformation. Upon addition of fluid during downslope movement, slumped material may transform into a debris flow, which transports sediment as an incoherent body in which intergranular movements predominate over shear-surface movements. A debris flow behaves as a plastic laminar flow with strength. As fluid content increases in debris flow, the flow may evolve into Newtonian turbidity current. Not all turbidity currents, however, evolve from debris flows. Some turbidity currents may evolve directly from sediment failures. Turbidity currents can develop near the shelf edge, on the slope, or in distal basinal settings. From Shanmugam G, Lehtonen LR, Straume T, Syvertsen SE, Hodgkinson RJ, and Skibeli M (1994) Slump and debris flow dominated upper slope facies in the Cretaceous of the Norwegian and northern North Seas (61°–67° N): Implications for sand distribution. *American Association of Petroleum Geologists Bulletin* 78: 910–937.

process-product terms are also applicable to deep-water lakes. For example, turbidites have been recognized in the world's deepest lake (1637 m deep), Lake Baikal in south-central Siberia.

There are several synonymous terms in use:

1. rotational slide = slump;
2. mass transport = mass movement = mass wasting = landslide;
3. muddy debris flow = mud flow = slurry flow;
4. sandy debris flow = granular flow.

Velocity-Based Terms

Examples of velocity-based terms are:

1. rock fall: free falling of detached rock mass that increases in velocity with fall;
2. flow slide: fast-moving mass transport;
3. rock slide: fast-moving mass transport;
4. debris avalanche: fast-moving mass transport;
5. mud flow: fast-moving debris flow;
6. sturzstrom: fast-moving mass transport;
7. debris slide: slow-moving mass transport;
8. creep: slow-moving mass transport.

These velocity-based terms are subjective because the distinction between a fast-moving and a slow-moving mass has not been defined using a precise velocity value. It is also difficult to measure velocities because of common destruction of velocity meters by catastrophic mass transport events. Furthermore, there are no scientific criteria to determine the absolute velocities of sediment movement in the ancient rock record. This is because sedimentary features preserved in the deposit do not reflect absolute velocities. Therefore, velocity-based terms, although popular, are not practical.

Initiation of Movement

The initiation of mass movements and sediment flows is closely related to shelf-edge sediment failures. These failures along continental margins are triggered by one or more of the following causes:

1. *Earthquakes*. Earthquakes and related shaking can cause an increase in shear stress, which is the downslope component of the total stress. Submarine mass movements have been attributed to the 1929 Grand Banks earthquake off the US East Coast and Canada.
2. *Tectonic oversteepening*. Tectonic compression has elevated the northern flank of the Santa Barbara Basin and overturned the slope in southern California along the US Pacific margin. Uplift (thrusting) along these slopes has led to oversteepening and sediment failure.
3. *Depositional oversteepening*. Mass movements have been attributed to oversteepening near the mouth of the Magdalena River, Colombia.
4. *Depositional loading*. Delta-front mass movements have been attributed to rapid sedimentation in the Mississippi River Delta, Gulf of Mexico.
5. *Hydrostatic loading*. Slope-failure deposits originated during the late Quaternary sea level rise on the eastern Tyrrhenian margin. These slope failures were attributed to rapid drowning of unconsolidated sediment, which resulted in increased hydrostatic loading.
6. *Glacial loading*. Submarine mass movements have been attributed to glacial loading and unloading along the Scotian margin in the North Atlantic.
7. *Eustatic changes in sea level*. Shelf-edge sediment failures and related gravity-driven processes have been attributed to worldwide fall in sea level, close to the shelf edge.
8. *Tsunamis*. The advancing wave front from a tsunami is capable of generating large hydrodynamic pressures on the seafloor that would produce soil movements and slope instabilities.
9. *Tropical cyclones*. Category 5 hurricane Katrina (2005) induced sediment failures and mudflows near the shelf edge in the Gulf of Mexico.
10. *Submarine volcanic activity*. Submarine mass movements have been attributed to volcanic activity in Hawaiian Islands.
11. *Salt movements*. Submarine mass movements along the flanks of intraslope basins have been attributed to mobilization of underlying salt masses in the Gulf of Mexico.
12. *Biologic erosion*. Submarine mass movements have been associated with erosion of the walls and floors of submarine canyons by invertebrates and fishes and boring by animals in offshore California.
13. *Gas hydrate decomposition*. Submarine mass movements have been associated with seepage of methane hydrate and related collapse of unconsolidated sediment along the US Atlantic Margin.

Mass Transport

Various methods are used to recognize mass transport deposits (slides and slumps) in modern submarine environments and their deposits in the ancient geologic record. These methods are (1) direct observations; (2) indirect velocity calculations; (3) remote-sensing technology (e.g., seismic profiling); and (4) examination of the rock (see Glossary).

Slides

A slide is a coherent mass of sediment that moves along a planar glide plane and shows no internal deformation (**Figure 3**). Slides represent translational shear-surface movements. Submarine slides can travel hundreds of kilometers on continental slopes (**Figure 4**). Long-runout distances of up to 800 km for slides have been documented (**Table 2**). Submarine slides are common in fiords because the submerged sides of glacial valleys are steep and because the rate of sedimentation is high due to sediment-laden rivers that drain glaciers into fiords. Submarine canyon walls are also prone to generate slides because of their steep gradients.

Multibeam bathymetric data show that the northern flank of the Santa Barbara Channel (Southern California) has experienced massive slope failures that resulted in the large (130 km^2) Goleta landslide complex (**Figure 5(a)**). Approximately 1.75 km^3 has been displaced by this slide during the Holocene (approximately the last 10 000 years). This complex has an upslope zone of evacuation and a downslope zone of accumulation (**Figure 5(b)**). It measures 14.6 km long, extending from a depth of 90 m to nearly 574 m, and is 10.5 km wide. It contains both surfical slump blocks and muddy debris flows in three distinct segments (**Figure 5(b)**).

Slides are capable of transporting gravel and coarse-grained sand because of their inherent strength. General characteristics of slides are:

- gravel to mud lithofacies;
- upslope areas with tensional faults;
- area of evacuation near the shelf edge (**Figures 4** and **5(a)**);
- occur commonly on slopes of 1–4 °;
- long-runout distances of up to 800 km (**Table 2**);
- transported sandy slide blocks encased in deep-water muddy matrix (**Figure 6**);
- primary basal glide plane or décollement (core and outcrop) (**Figure 7(a)**);
- basal shear zone (core and outcrop) (**Figure 7(b)**);
- secondary internal glide planes (core and outcrop) (**Figure 7(a)**);
- associated slumps (core and outcrop) (**Figure 6**);
- transformation of slides into debris flows in frontal zone;
- associated clastic injections (core and outcrop) (**Figure 7(a)**);
- sheet-like geometry (seismic and outcrop) (**Figure 6**);
- common in areas of tectonic activity, earthquakes, steep gradients, salt movements, and rapid sedimentation.

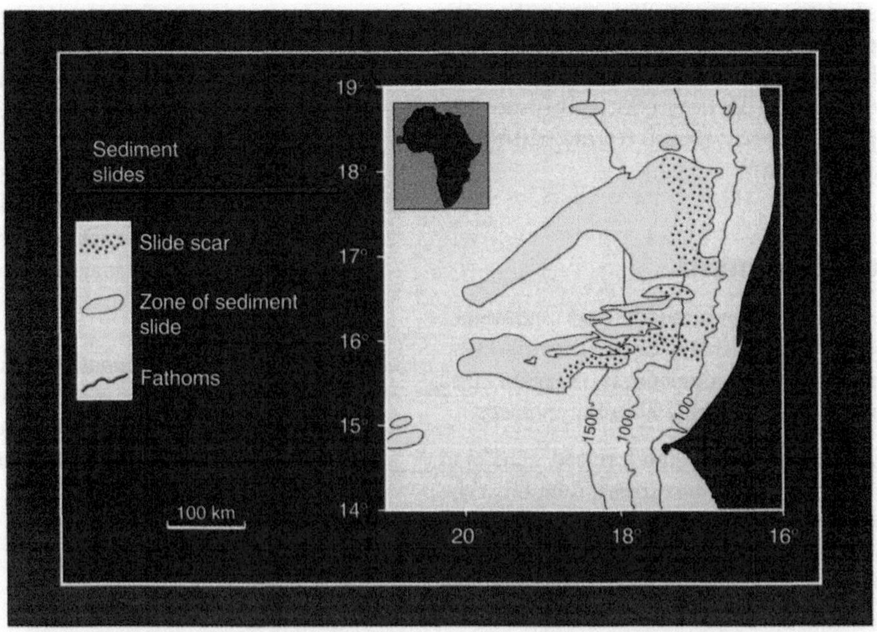

Figure 4 Long-distance transport of detached slide blocks from the shelf edge at about 100 fathom (i.e., 600 ft or 183 m) contour, offshore northwestern Africa. Note that two slide blocks near the 15° latitude marker have traveled nearly 300 km from the shelf edge (i.e., 100 fathom contour). Three contour lines (100, 1000, 1500 fathom) show increasing water depths to the west. From Jacobi RD (1976) Sediment slides on the northwestern continental margin of Africa. *Marine Geology* 22: 157–173.

Table 2 Long-runout distances of modern mass transport processes

Name and location	*Runout distance (km)*	*Data*	*Process*
Storegga, Norway	800	Seismic and GLORIA side-scan sonar images	Slide, slump, and debris flow
Agulhas, S Africa	750	Seismic	Slide and slump
Hatteras, US Atlantic margin	~500	Seismic and core	Slump and debris flow
Saharan NW African margin	>400	Seismic and core	Slump and debris flow
Mauritania–Senegal, NW African margin	~300	Seismic and core	Slump and debris flow
Nuuanu, NE Oahu (Hawaii)	235	GLORIA side-scan sonar images	Mass transport
Wailau, N Molakai (Hawaii)	<195	GLORIA side-scan sonar images	Mass transport
Rockall, NE Atlantic	160	Seismic	Mass transport
Clark, SW Maui (Hawaii)	150	GLORIA side-scan sonar images	Mass transport
N Kauai, N Kauai (Hawaii)	140	GLORIA side-scan sonar images	Mass transport
East Breaks (West), Gulf of Mexico	110	Seismic and core	Slump and debris flow
Grand Banks, Newfoundland	>100	Seismic and core	Mass transport
Ruatoria, New Zealand	100	Seismic	Mass transport
Alika-2, W Hawaii (Hawaii)	95	GLORIA side-scan sonar images	Mass transport
Kaena, NE Oahu (Hawaii)	80	GLORIA side-scan sonar images	Mass transport
Bassein, Bay of Bengal	55	Seismic	Slide and debris flow
Kidnappers, New Zealand	45	Seismic	Slump and slide
Munson–Nygren, New England	45	Seismic	Slump and debris flow
Ranger, Baja California	35	Seismic	Mass transport

On the modern Norwegian continental margin, large mass transport deposits occur on the northern and southern flanks of the Voring Plateau. The Storegga slide (offshore Norway), for example, has a maximum thickness of 430 m and a length of more than 800 km. The Storegga slide on the southern flank of the plateau exhibits mounded seismic patterns in sparker profiles. Although it is called a slide in publications, the core of this deposit is composed primarily of slumps and debrites. Unlike kilometers-wide modern slides that can be mapped using multibeam mapping systems and seismic reflection profiles, huge ancient slides are difficult to recognize in outcrops because of limited sizes of outcrops.

A summary of width:thickness ratio of modern and ancient slides is given in **Table 3**.

Slumps

A slump is a coherent mass of sediment that moves on a concave-up glide plane and undergoes rotational movements causing internal deformation (**Figure 3**). Slumps represent rotational shear-surface movements. In multibeam bathymetric data, distinguishing slides from slumps may be difficult because internal deformation cannot be resolved. In seismic profiles, however, slumps may be recognized because of their chaotic reflections. Therefore, a general term mass transport is preferred when interpreting bathymetric images. Slumps are capable of transporting gravel and coarse-grained sand because of their inherent strength. General characteristics of slumps are:

- gravel to mud lithofacies;
- basal zone of shearing (core and outcrop);
- upslope areas with tensional faults (**Figure 8**);
- downslope edges with compressional folding or thrusting (i.e., toe thrusts) (**Figure 8**);
- slump folds interbedded with undeformed layers (core and outcrop) (**Figure 9**);
- irregular upper contact (core and outcrop);
- chaotic bedding in heterolithic facies (core and outcrop);
- steeply dipping and truncated layers (core and outcrop) (**Figure 10**);
- associated slides (core and outcrop) (**Figure 6**);

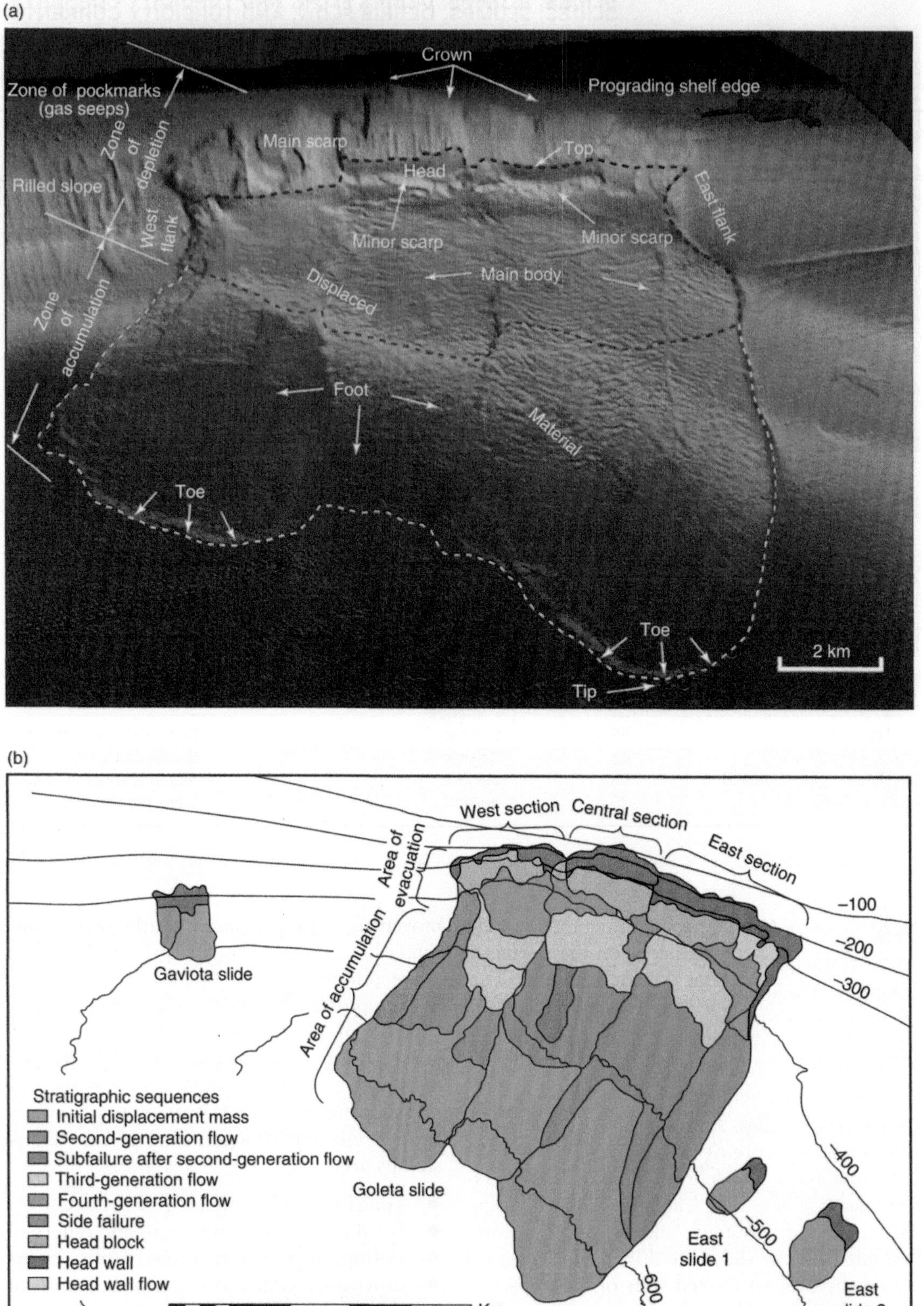

Figure 5 (a) Multibeam bathymetric image of the Goleta slide complex in the Santa Barbara Channel, Southern California. Note lobe-like (dashed line) distribution of displaced material that was apparently detached from the main scarp near the shelf edge. This mass transport complex is composed of multiple segments of failed material. (b) Sketch of the Goleta mass transport complex in the Santa Barbara Channel, Southern California (a) showing three distinct segments (i.e., west, central, and east). Contour intervals (−100, −200, −300, −400, −500, and −600) are in meters. (a) From Greene HG, Murai LY, Watts P, *et al.* (2006) Submarine landslides in the Santa Barbara Channel as potential tsunami sources. *Natural Hazards and Earth System Sciences* 6: 63–88. (b) From Greene HG, Murai LY, Watts P, *et al.* (2006) Submarine landslides in the Santa Barbara Channel as potential tsunami sources. *Natural Hazards and Earth System Sciences* 6: 63–88.

Figure 6 Outcrop photograph showing sheet-like geometry of an ancient sandy submarine slide (1000 m long and 50 m thick) encased in deep-water mudstone facies. Note the large sandstone sheet with rotated/slumped edge (left). Person (arrow): 1.8 m tall. Ablation Point Formation, Kimmeridgian (Jurassic), Alexander Island, Antarctica. Photo courtesy of D.J.M. Macdonald.

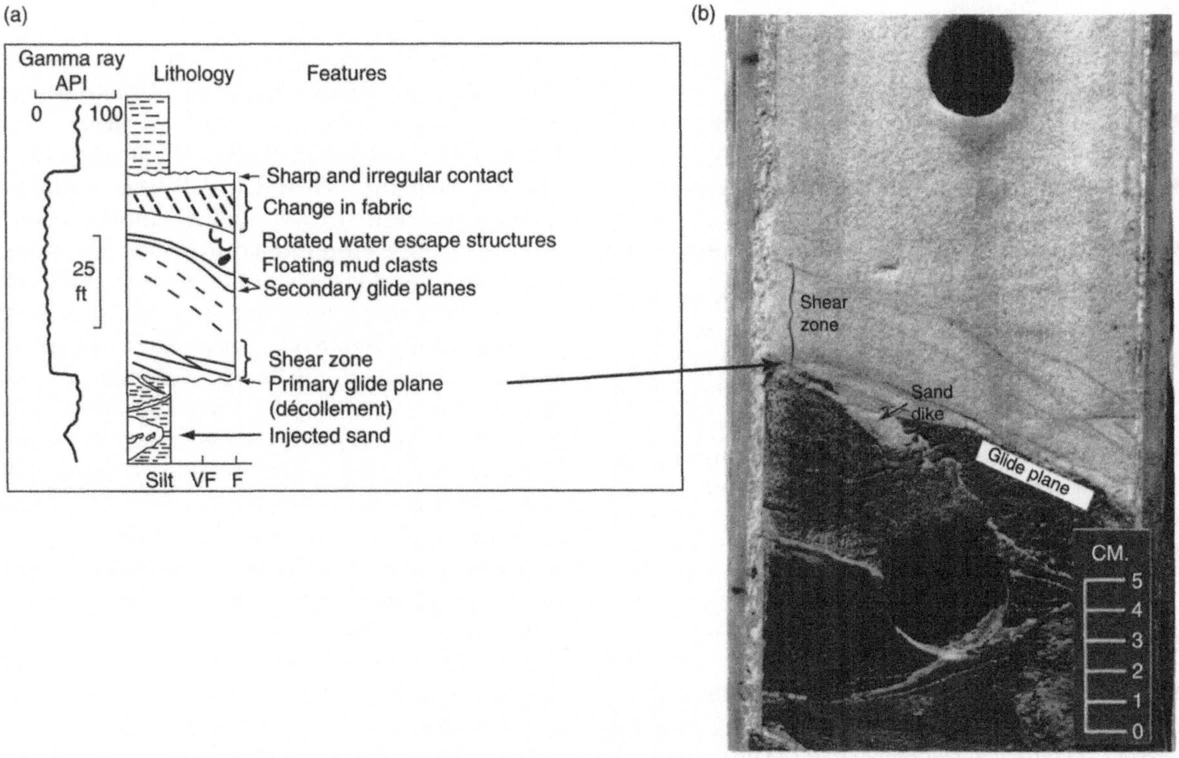

Figure 7 (a) Sketch of a cored interval of a sandy slide/slump unit showing blocky wireline log motif (left) and sedimentological details (right). (b) Core photograph showing the basal contact of the sandy slide (arrow). Note a sand dike (i.e., injectite) at the base of shear zone. Eocene, North Sea. Compare this small-scale slide (15 m thick) with a large-scale slide (50 m thick) in **Figure 6**. From Shanmugam G (2006) *Deep-Water Processes and Facies Models: Implications for Sandstone Petroleum Reservoirs.* Amsterdam: Elsevier.

Table 3 Dimensions of modern and ancient mass transport deposits

Example	*Width:thickness ratio (observed dimensions)*
Slide, Lower Carboniferous, England	7:1 (100 m wide/long, 15 m thick)
Slide, Cambrian–Ordovician, Nevada	30:1 (30 m wide/long, 1 m thick)
Slide, Jurassic, Antarctica	45:1 (20 km wide/long, 440 m thick)
Slide, Modern, US Atlantic margin	40–80:1 (2–4 km wide/long, 50 m thick)
Slide, Modern, Gulf of Alaska	130:1 (15 km wide/long, 115 m thick)
Slide, Middle Pliocene, Gulf of Mexico	250:1 (150 km wide/long, 600 m thick)
Slide/slump/debris flow/turbidite 5000–8000 BP, Norwegian continental margin	675:1 (290 km wide/long, 430 m thick)
Slump, Cambrian–Ordovician, Nevada	10:1 (100 m wide/long, 10 m thick)
Slump, Aptian–Albian, Antarctica	10:1 (3.5 km wide/long, 350 m thick)
Slump/slide/debris flow, Lower Eocene, Gryphon Field, UK	21:1 (2.6 km wide/long, 120 m thick)
Slump/slide/debris flow, Paleocene, Faeroe Basin, north of Shetland Islands	28:1 (7 km wide/long, 245 m thick)
Slump, Modern, SE Africa	171:1 (64 km wide/long, 374 m thick)
Slump, Carboniferous, England	500:1 (5 km wide/long, 10 m thick)
Slump, Lower Eocene, Spain	900–3600:1 (18 km wide/long, 5–20 m thick)
Debrite, Modern, British Columbia	12:1 (50 m wide/long, 4 m thick)
Debrite, Cambrian–Ordovician, Nevada	30:1 (300 m wide/long, 10 m thick)
Debrite, Modern, US Atlantic margin	500–5000:1 (10–100 km wide/long, 20 m thick)
Debrite, Quaternary, Baffin Bay	1250:1 (75 km wide/long, 60 m thick)
Turbidite (depositional lobe) Cretaceous, California	167:1 (10 km wide/long, 60 m thick)
Turbidite (depositional lobe) Lower Pliocene, Italy	1200:1 (30 km wide/long, 25 m thick)
Turbidite (basin plain) Miocene, Italy	11400:1 (57 km wide/long, 5 m thick)
Turbidite (basin plain) 16 000 BP Hatteras Abyssal Plain	125 000:1 (500 km wide/long, 4 m thick)

Reproduced from Shanmugam G (2006) *Deep-Water Processes and Facies Models: Implications for Sandstone Petroleum Reservoirs*. Amsterdam: Elsevier, with permission from Elsevier.

- associated sand injections (core and outcrop) (**Figure 10**);
- lenticular to sheet-like geometry with irregular thickness (seismic and outcrop);
- contorted bedding has been recognized in Formation MicroImager (FMI);
- chaotic facies on high-resolution seismic profiles.

In Hawaiian Islands, the Waianae Volcano comprises the western half of O'ahu Island. Several submersible dives and multibeam bathymetric imaging have confirmed the timing of seabed failure that formed the Waianae mass transport (slump?) complex. Multiple collapses and deformation events resulted in compound mass wasting features on the volcano's southwest flank. This complex is the largest in Hawaii, covering an area of about 5500 km^2.

Sediment Flows

Sediment flows (i.e., sediment-gravity flows) are composed of four types: (1) debris flow, (2) turbidity current, (3) fluidized sediment, and (4) grain flow. In this article, the focus is on debris flows and turbidity currents because of their importance. These two processes are distinguished from one another on the basis of fluid rheology and flow state. The rheology of fluids can be expressed as a relationship between applied shear stress and rate of shear strain (**Figure 11**). Newtonian fluids (i.e., fluids with no inherent strength), like water, will begin to deform the moment shear stress is applied, and the deformation is linearly proportional to stress. In contrast, some naturally occurring materials (i.e., fluids with strength) will not deform until their yield stress has been exceeded (**Figure 11**); once their yield stress is exceeded, deformation is linear. Such materials with strength (e.g., wet concrete) are considered to be Bingham plastics (**Figure 11**). For flows that exhibit plastic rheology, the term plastic flow is appropriate. Using rheology as the basis, deep-water sediment flows are divided into two broad groups, namely (1) Newtonian flows that represent turbidity currents and (2) plastic flows that represent debris flows.

In addition to fluid rheology, flow state is used in distinguishing laminar debris flows from turbulent turbidity currents. The difference between laminar and turbulent flows was demonstrated in 1883 by Osborne Reynolds, an Irish engineer, by injecting a thin stream of dye into the flow of water through a glass tube. At low rates of flow, the dye stream traveled in a straight path. This regular motion of

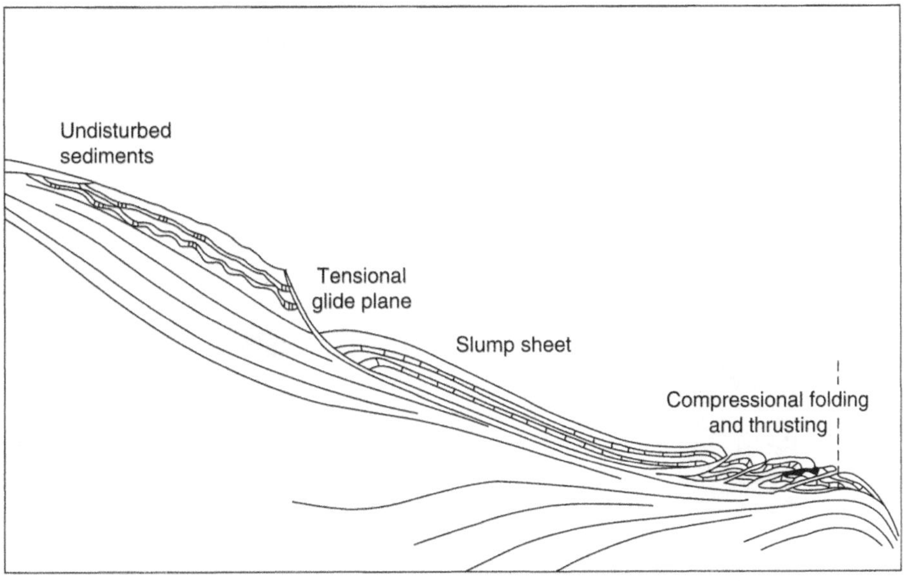

Figure 8 Sketch of a submarine slump sheet showing tensional glide plane in the updip detachment area and compressional folding and thrusting in the downdip frontal zone. From Lewis KB (1971) Slumping on a continental slope inclined at 1°–4°. *Sedimentology* 16: 97–110.

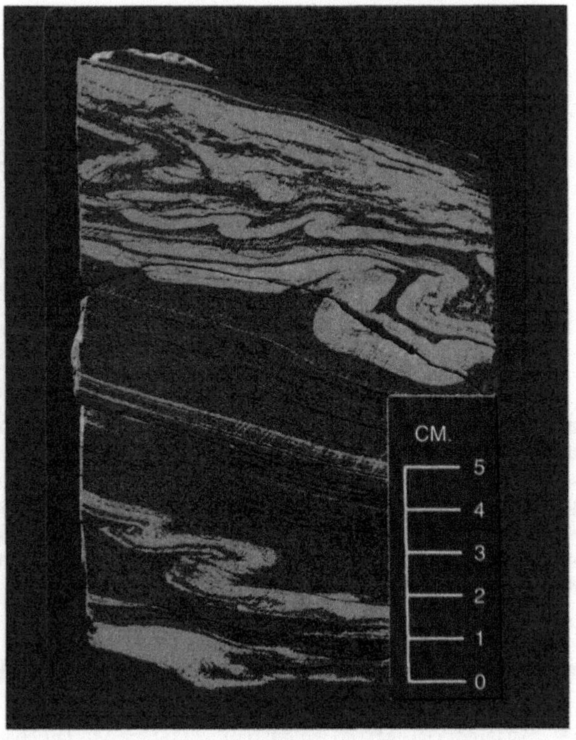

Figure 9 Core photograph showing alternation of contorted and uncontorted siltstone (light color) and claystone (dark color) layers of slump origin. This feature is called slump folding. Paleocene, North Sea. Reproduced from Shanmugam (2006). *Deep-Water Processes and Facies Models: Implications for Sandstone Petroleum Reservoirs*. Amsterdam: Elsevier, with permission from Elsevier.

fluid in parallel layers, without macroscopic mixing across the layers, is called a laminar flow. At higher flow rates, the dye stream broke up into chaotic eddies. Such an irregular fluid motion, with macroscopic mixing across the layers, is called a turbulent flow. The change from laminar to turbulent flow occurs at a critical Reynolds number (the ratio between inertia and viscous forces) of *c.* 2000 (**Figure 11**).

Debris Flows

A debris flow is a sediment flow with plastic rheology and laminar state from which deposition occurs through 'freezing' *en masse*. The terms debris flow and mass flow are used interchangeably because each exhibits plastic flow behavior with shear stress distributed throughout the mass. In debris flows, intergranular movements predominate over shear-surface movements. Although most debris flows move as incoherent material, some plastic flows may be transitional in behavior between coherent mass movements and incoherent sediment flows (**Table 1**). Debris flows may be mud-rich (i.e., muddy debris flows), sand-rich (i.e., sandy debris flows), or mixed types. Sandy debrites comprise important petroleum reservoirs in the North Sea, Norwegian Sea, Nigeria, Equatorial Guinea, Gabon, Gulf of Mexico, Brazil, and India. In multibeam bathymetric data, recognition of debrites is possible.

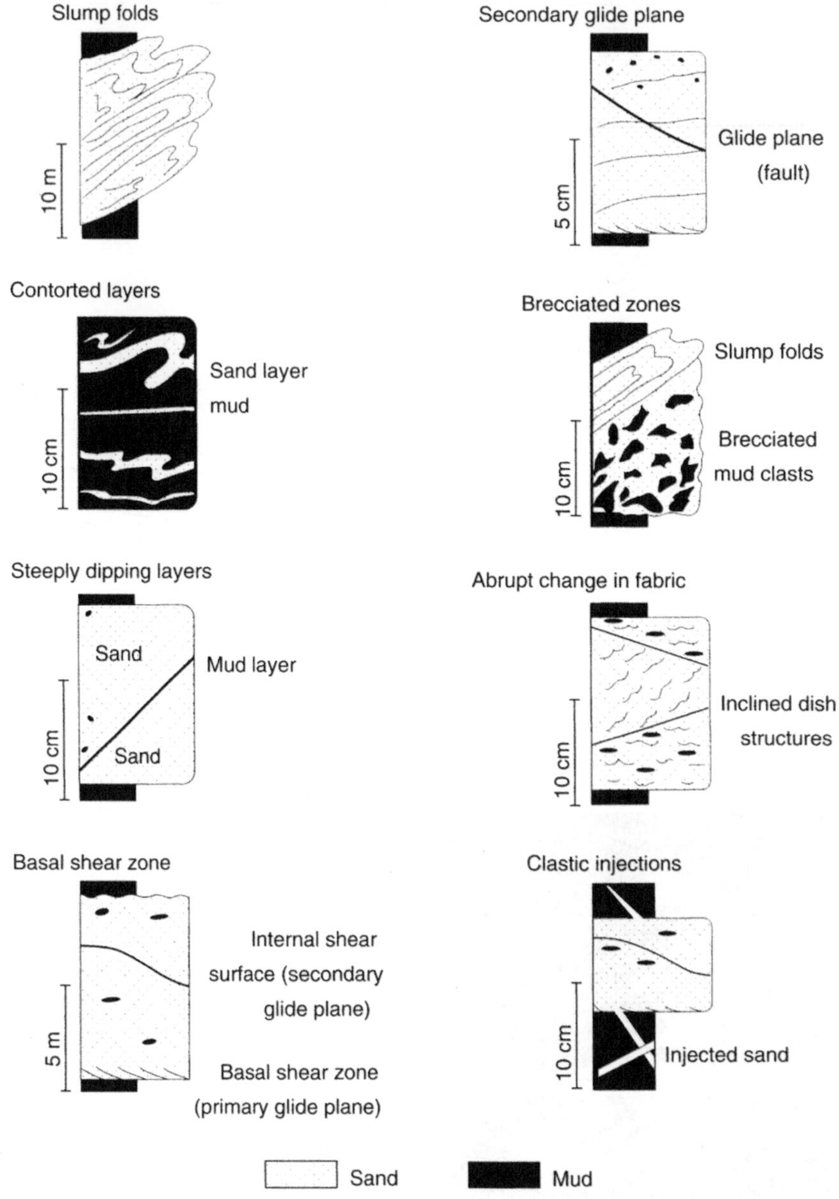

Figure 10 Summary of features associated with slump deposits observed in core and outcrop. Slump fold, an intraformational fold produced by deformation of soft sediment; contorted layer, deformed sediment layer; basal shear zone, the basal part of a rock unit that has been crushed and brecciated by many subparallel fractures due to shear strain; glide plane, slip surface along which major displacement occurs; brecciated zone, an interval that contains angular fragments caused by crushing of the rock; dish structures, concave – up (like a dish) structures caused by upward-escaping fluids in the sediment; clastic injections, natural injection of clastic (transported) sedimentary material (usually sand) into a host rock (usually mud). From Shanmugam G (2006) *Deep-Water Processes and Facies Models: Implications for Sandstone Petroleum Reservoirs*. Amsterdam: Elsevier.

Debris flows are capable of transporting gravel and coarse-grained sand because of their inherent strength. General characteristics of muddy and sandy debrites are:

- gravel to mud lithofacies;
- lobe-like distribution (map view) in the Gulf of Mexico (**Figure 12**);
- tongue-like distribution (map view) in the North Atlantic (**Figure 13**);
- floating or rafted mudstone clasts near the tops of sandy beds (core and outcrop) (**Figure 14**);
- projected clasts (core and outcrop);
- planar clast fabric (core and outcrop) (**Figure 14**);
- brecciated mudstone clasts in sandy matrix (core and outcrop);

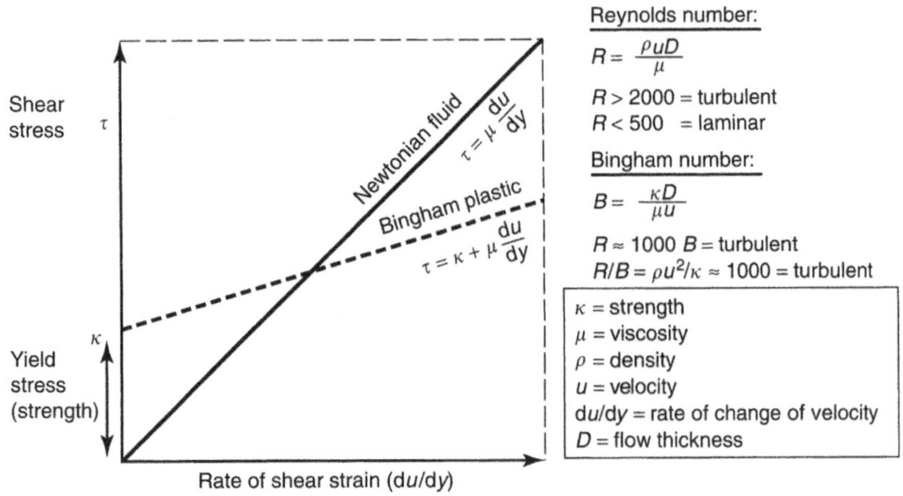

Figure 11 Graph showing rheology (stress–strain relationships) of Newtonian fluids and Bingham plastics. Note that the fundamental rheological difference between debris flows (Bingham plastics) and turbidity currents (Newtonian fluids) is that debris flows exhibit strength, whereas turbidity currents do not. Reynolds number is used for determining whether a flow is turbulent (turbidity current) or laminar (debris flow) in state. From Shanmugam G (1997) The Bouma sequence and the turbidite mind set. *Earth-Science Reviews* 42: 201–229.

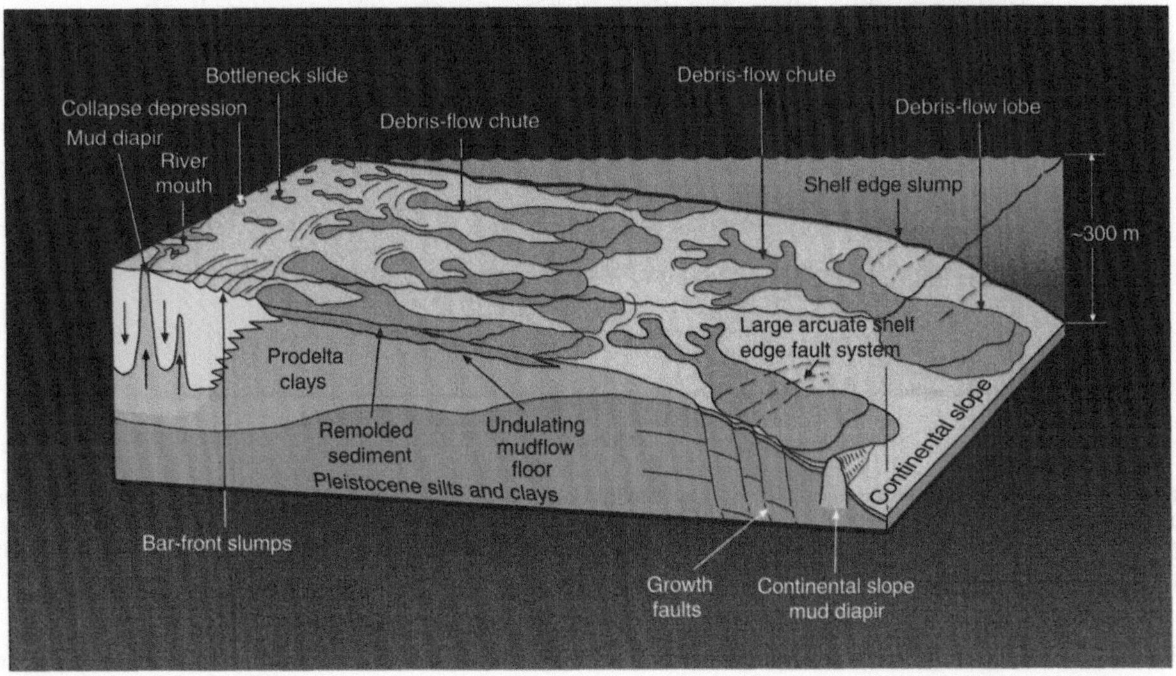

Figure 12 Schematic diagram showing lobe-like distribution of submarine debris flows in front of the Mississippi River Delta, Gulf of Mexico. This shelf-edge deltaic setting, associated with high sedimentation rate, is prone to develop ubiquitous mud diapers and contemporary faults. Mud diaper, intrusion of mud into overlying sediment causing dome-shaped structure; chute, channel. From Coleman JM and Prior DB (1982) Deltaic environments. In: Scholle PA and Spearing D (eds.) *American Association of Petroleum Geologists Memoir 31: Sandstone Depositional Environments*, pp. 139–178. Tulsa, OK: American Association of Petroleum Geologists.

- inverse grading of rock fragments (core and outcrop);
- inverse grading, normal grading, inverse to normal grading, and no grading of matrix (core and outcrop);
- floating quartz granules (core and outcrop);
- inverse grading of granules in sandy matrix (core and outcrop);
- pockets of gravels (core and outcrop);
- irregular, sharp upper contacts (core and outcrop);

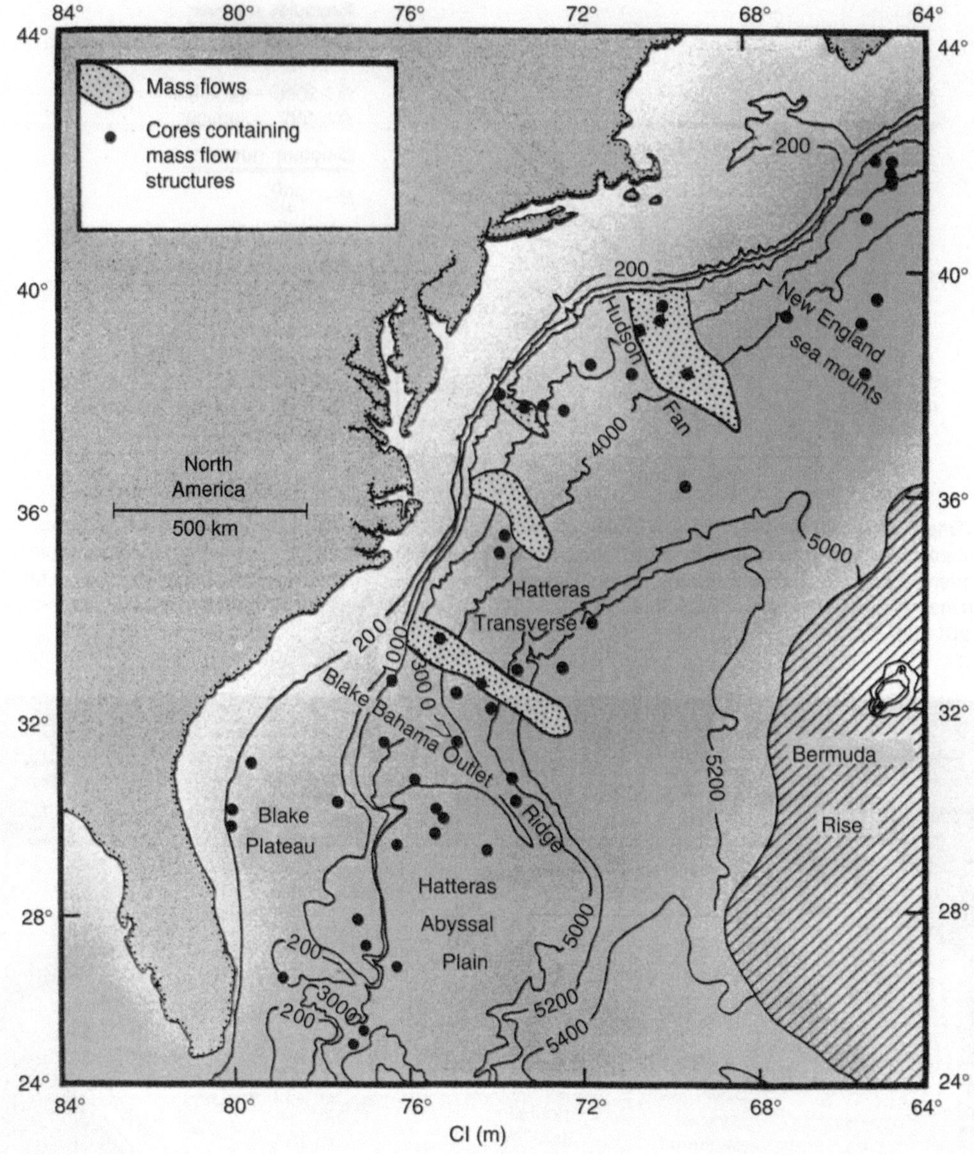

Figure 13 Tongue-like distribution of mass flows (i.e., debris flows) in the North Atlantic. Debrite units are about 500 km long, 10–100 km wide, and 20 m thick. Note a debrite tongue has traveled to a depth of about 5250 m water depth. CI, contour intervals. From Embley RW (1980) The role of mass transport in the distribution and character of deep-ocean sediments with special reference to the North Atlantic. *Marine Geology* 38: 23–50.

- side-by-side occurrence of garnet granules (density: 3.5–4.3) and quartz granules (density: 2.65) (core and outcrop);
- lenticular geometry (**Figure 15**).

The modern Amazon submarine channel has two major debrite deposits (east and west). The western debrite unit is about 250 km long, 100 km wide, and 125 m thick. In the US Atlantic margin, debrite units are about 500 km long, 10–100 km wide, and 20 m thick (**Figure 13**). In offshore northwest Africa, the Canary debrite is about 600 km long, 60–100 km wide, and 5–20 m thick. Submarine debris flows and their flow-transformation induced turbidity currents have been reported to travel over 1500 km from their triggering point on the northwest African margin.

Turbidity Currents

A turbidity current is a sediment flow with Newtonian rheology and turbulent state in which sediment is supported by turbulence and from which deposition occurs through suspension settling. Turbidity currents exhibit unsteady and nonuniform flow behavior (**Figure 16**). Turbidity currents are surge-type waning flows. As they flow downslope,

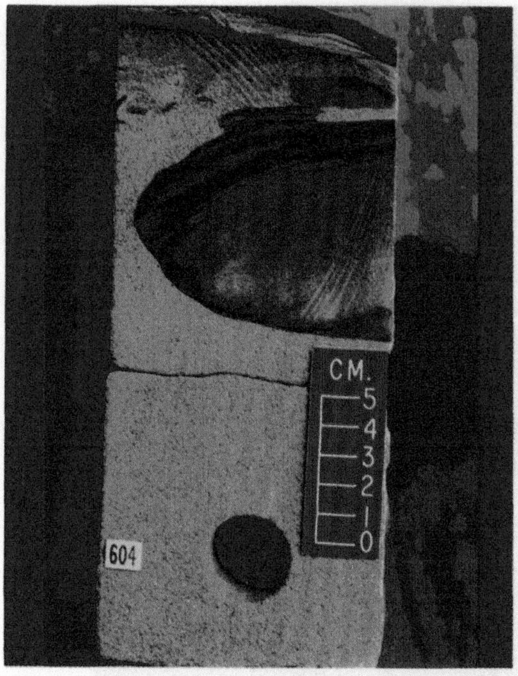

Figure 14 Core photograph of massive fine-grained sandstone showing floating mudstone clasts (above the scale) of different sizes. Note planar clast fabric (i.e., long axis of clast is aligned parallel to bedding surface). Note sharp and irregular upper bedding contact (top of photo). Paleocene, North Sea. Reproduced from Shanmugam G (2006). *Deep-Water Processes and Facies Models: Implications for Sandstone Petroleum Reservoirs*. Amsterdam: Elsevier, with permission from Elsevier.

turbidity currents invariably entrain ambient fluid (seawater) in their frontal head portion due to turbulent mixing (**Figure 16**).

With increasing fluid content, plastic debris flows may tend to become Newtonian turbidity currents (**Figure 3**). However, not all turbidity currents evolve from debris flows. Some turbidity currents may evolve directly from sediment failures. Although turbidity currents may constitute a distal end member in basinal areas, they can occur in any part of the system (i.e., shelf edge, slope, and basin). In seismic profiles and multibeam bathymetric images, it is impossible to recognize turbidites.

Turbidity currents cannot transport gravel and coarse-grained sand in suspension because they do not possess the strength. General characteristics of turbidites are:

- fine-grained sand to mud;
- normal grading (core and outcrop) (**Figure 17**);
- sharp or erosional basal contact (core and outcrop) (**Figure 17**);
- gradational upper contact (core and outcrop) (**Figure 17**);
- thin layers, commonly centimeters thick (core and outcrop) (**Figure 18**);
- sheet-like geometry in basinal settings (outcrop) (**Figure 18**);
- lenticular geometry that may develop in channel-fill settings.

Figure 15 Outcrop photograph showing lenticular geometry (dashed line) of debrite with floating clasts (arrow heads). Cretaceous, Tourmaline Beach, California. Reproduced from Shanmugam G (2006). *Deep-Water Processes and Facies Models: Implications for Sandstone Petroleum Reservoirs*. Amsterdam: Elsevier, with permission from Elsevier.

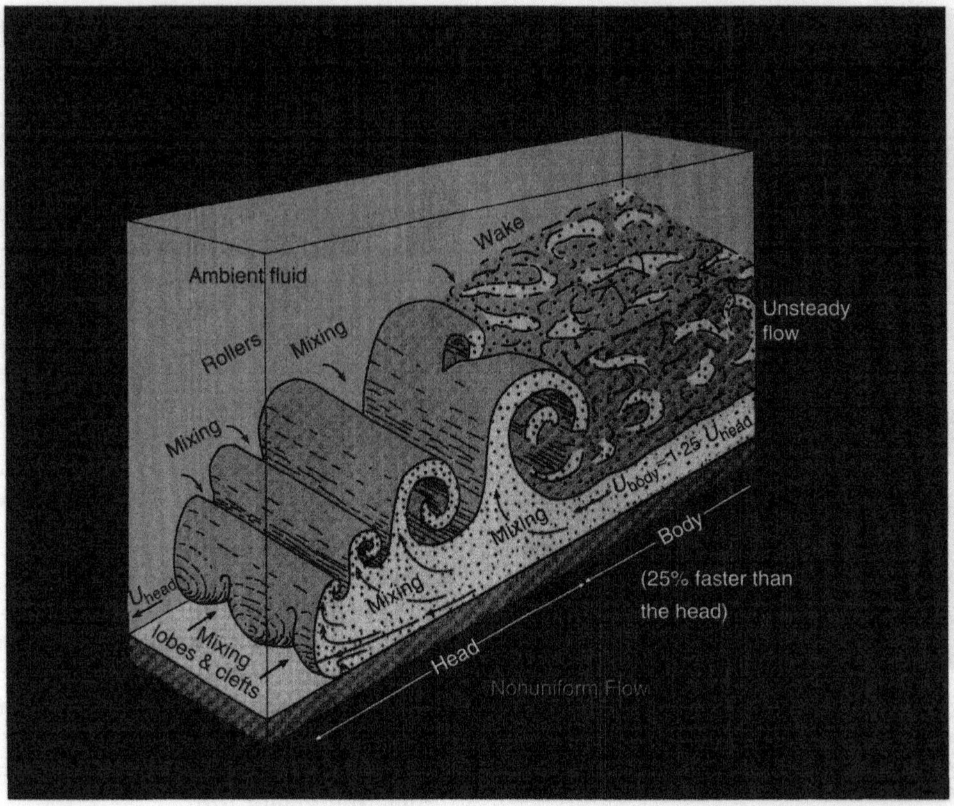

Figure 16 Schematic illustration showing the leading head portion of an unsteady, nonuniform, and turbulent turbidity current. Due to turbulent mixing, turbidity currents invariably entrain ambient fluid (seawater) at their head regions. Modified from Allen JRL (1985) Loose-boundary hydraulics and fluid mechanics: Selected advances since 1961. In: Brenchley PJ and Williams BPJ (eds.) *Sedimentology: Recent Developments and Applied Aspects*, pp. 7–28. Oxford, UK: Blackwell.

In the Hatteras Abyssal Plain (North Atlantic), turbidites have been estimated to be 500 km wide and up to 4 m thick (**Table 3**).

Selective emphasis of turbidity currents Turbidity currents have received a skewed emphasis in the literature. This may be attributed to existing myths about turbidity currents and turbidites. Myth no. 1: Some turbidity currents may be nonturbulent (i.e., laminar) flows. Reality: All turbidity currents are turbulent flows. Myth no. 2: Some turbidity currents may be waxing flows. Reality: All turbidity currents are waning flows. Myth no. 3: Some turbidity currents may be plastic in rheology. Reality: All turbidity currents are Newtonian in rheology. Myth no. 4: Some turbidity currents may be high in sediment concentration (25–95% by volume). Reality: All turbidity currents are low in sediment concentration (1–23% by volume). Because high sediment concentration damps turbulence, high-concentration (i.e., high-density) turbidity currents cannot exist. Myth no. 5: All turbidity currents are high-velocity flows and therefore they elude documentation. Reality:

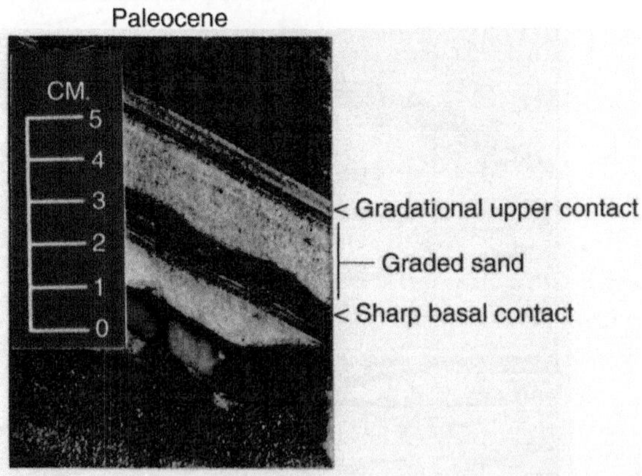

Figure 17 Core photograph showing a sandy unit with normal grading (i.e., grain size decreases upward), sharp basal contact, and gradational upper contact. These features are interpreted to be deposition from a waning turbidity current. Dark intervals are mudstone. Paleocene, North Sea.

Turbidity currents can operate under a wide range of velocity conditions. Myth no. 6: Turbidity currents are believed to operate in modern oceans. Reality: No one has ever documented turbidity

Figure 18 Outcrop photograph showing tilted thin-bedded turbidite sandstone beds with sheet-like geometry, Lower Eocene, Zumaya, northern Spain. Reproduced from Shanmugam G (2006). *Deep-Water Processes and Facies Models: Implications for Sandstone Petroleum Reservoirs.* Amsterdam: Elsevier, with permission from Elsevier.

currents in modern oceans based on the physics of the flow. Myth no. 7: Some turbidites are deposits of debris flows. Reality: All turbidites are the exclusive deposits of turbidity currents. Myth no. 8: Some turbidites exhibit inverse grading. Reality: All turbidites develop normal grading. Myth no. 9: Cross-bedding is a product of turbidity currents. Reality: Cross-bedding is a product of bottom currents. Myth no. 10: Turbidite beds can be recognized in seismic reflection profiles. Reality: Normally graded turbidite units, centimeters in thickness, cannot be resolved on seismic profiles. As a consequence, many debris flows and their deposits have been misclassified as turbidity currents and turbidites.

Sediment Transport via Submarine Canyons

Submarine canyons, which are steep-sided valleys incised into the continental shelf and slope, serve as major conduits for sediment transport from land and the shelf to the deep-sea environment. Although downslope sediment transport occurs both inside and outside of submarine canyons, canyons play a critical role because steep canyon walls are prone to slope failures. Submarine canyons are prominent erosional features along both the US Pacific (**Figure 2**) and the Atlantic margins. Many submarine canyons in the US Atlantic margin commence at a depth of about 200 m near the shelf edge, but heads of California canyons in the US Pacific margin begin at an average depth of about 35 m. The Redondo Canyon, for example, commences at a depth of 10 m near the shoreline (**Figure 2**). Such a scenario would allow for a quick transfer of sediment from shallow-marine into deep-marine environments. In the San Pedro Sea Valley, large debris blocks have been recognized as submarine landslides. Some researchers have proposed that these submarine landslides may have triggered local tsunamis. The significance of this relationship is that tsunamis can trigger submarine landslides, which in turn can trigger tsunamis. Such mutual triggering mechanisms can result in frequent sediment failures in deep-marine environments.

Tsunamis and tropical cyclones are important factors in transferring sediment into deep-marine environments via submarine canyons. For example, a rapid (190 cm s^{-1}) sediment flow was recorded in the Scripps Submarine Canyon (La Jolla, California) during the passage of a storm front on 24 November 1968 over La Jolla. In the Scripps Canyon, a large slump mass of about 105 m^3 in size was triggered by the May 1975 storm.

Hurricane Hugo, which passed over St. Croix in the US Virgin Islands on 17 September 1989, had generated winds in excess of 110 knots (204 km h^{-1}, category 3 in the Saffir–Simpson scale) and waves 6–7 m in height. In the Salt River submarine canyon (>100 m deep), offshore St. Croix, a current meter measured net downcanyon currents reaching velocities of 2 m s^{-1} and oscillatory flows up to 4 m s^{-1}. Hugo had caused erosion of 2 m of sand in the Salt River Canyon at a depth of about 30 m. A minimum of 2 million kg of sediment were flushed down the Salt River Canyon into deep water. The transport rate associated with hurricane Hugo was 11 orders of magnitude greater than the rate measured during a fair-weather period. In the Salt River Canyon, much of the soft reef cover (e.g., sponges) had been eroded away by the power of the hurricane. Debris composed of palm fronds, trash, and pieces of boats found in the canyon were the evidence for storm-generated debris flows. Storm-induced sediment flows have also been reported in a submarine canyon off Bangladesh, in the Capbreton Canyon, Bay of Biscay in SW France, and in the Eel Canyon, Northern California, among others. In short, sediment transport in submarine canyons is accelerated by tropical cyclones and tsunamis in the world's oceans.

Glossary

abyssal plain The deepest and flat part of the ocean floor that occupies depths between 2000 and 6000 m (6560 and 19 680 ft).

ancient The term refers to deep-marine systems that are older than the Quaternary period, which began approximately 1.8 Ma.

basal shear zone The basal part of a rock unit that has been crushed and brecciated by many subparallel fractures due to shear strain.

bathyal Ocean floor that occupies depths between 200 (shelf edge) and 4000 m (656 and 13 120 ft). Note that abyssal plains may occur at bathyal depths.

bathymetry The measurement of seafloor depth and the charting of seafloor topography.

brecciated clasts Angular mudstone clasts in a rock due to crushing or other deformation.

brecciated zone An interval that contains angular fragments caused by crushing or breakage of the rock.

clastic sediment Solid fragmental material (unconsolidated) that originates from weathering and is transported and deposited by air, water, ice, or other processes (e.g., mass movements).

continental margin The ocean floor that occupies between the shoreline and the abyssal plain. It consists of shelf, slope, and basin (Figure 2).

contorted bedding Extremely disorganized, crumpled, convoluted, twisted, or folded bedding. Synonym: chaotic bedding.

core A cylindrical sample of a rock type extracted from underground or seabed. It is obtained by drilling into the subsurface with a hollow steel tube called a corer. During the downward drilling and coring, the sample is pushed upward into the tube. After coring, the rock-filled tube is brought to the surface. In the laboratory, the core is slabbed perpendicular to bedding. Finally, the slabbed flat surface of the core is examined for geological bedding contacts, sedimentary structures, grain-size variations, deformation, fossil content, etc.

dish structures Concave-up (like a dish) structures caused by upward-escaping water in the sediment.

floating mud clasts Occurrence of mud clasts at some distance above the basal bedding contact of a rock unit.

flow Continuous, irreversible deformation of sediment–water mixture that occurs in response to applied stress (**Figure 11**).

fluid A material that flows.

fluid dynamics A branch of fluid mechanics that deals with the study of fluids (liquids and gases) in motion.

fluid mechanics Study of the properties and behaviors of fluids.

geohazards Natural disasters (hazards), such as earthquakes, landslides, tsunamis, tropical cyclones, rogue (freak) waves, floods, volcanic events, sea level rise, karst-related subsidence (sink holes), geomagnetic storms; coastal upwelling; deep-ocean currents, etc.

heterolithic facies Thinly interbedded (millimeter- to decimeter-scale) sandstones and mudstones.

hydrodynamics A branch of fluid dynamics that deals with the study of liquids in motion.

injectite Injected material (usually sand) into a host rock (usually mudstone). Injections are common in igneous rocks.

inverse grading Upward increase in average grain size from the basal contact to the upper contact within a single depositional unit.

lithofacies A rock unit that is distinguished from adjacent rock units based on its lithologic (i.e., physical, chemical, and biological) properties (see Rock).

lobe A rounded, protruded, wide frontal part of a deposit in map view.

methodology Four methods are in use for recognizing slides, slumps, debris flows, and turbidity currents and their deposits. *Method 1*: Direct observations – Deep-sea diving by a diver allows direct observations of submarine mass movements. The technique has limitations in terms of diving depth and diving time. These constraints can be overcome by using a remotely operated deep submergence vehicle, which would allow observations at greater depths and for longer time. Both remotely operated vehicles (ROVs) and manned submersibles are used for underwater photographic and video documentation of submarine processes. *Method 2*: Indirect velocity calculations – A standard practice has been to calculate velocity of catastrophic submarine events based on the timing of submarine cable breaks. The best example of this method is the 1929 Grand Banks earthquake (Canada) and related cable breaks. This method is not useful for recognizing individual type of mass movement (e.g., slide vs. slump). *Method 3*: Remote sensing technology – In the 1950s, conventional echo sounding was used to construct seafloor profiles. This was done by emitting sound pulses from a ship and by recording return echos from the sea bottom. Today, several types of seismic profiling techniques are available depending on the desired degree of resolutions. Although popular in the petroleum industry and academia, seismic profiles cannot resolve subtle sedimentological features that are required to distinguish turbidites from debrites. In the 1970s, the most significant progress in mapping the seafloor was made by adopting multibeam side-scan sonar survey. The Sea MARC 1 (Seafloor Mapping and Remote Characterization) system uses up to 5-km-broad swath of the seafloor. The GLORIA (Geological Long Range Inclined Asdic) system uses up to 45-km-broad swath of the seafloor. The advantage of GLORIA is that it can map an area of 27 700 $km^2 day^{-1}$. In the 1990s, multibeam mapping systems were adopted to map the seafloor. This system utilizes hull-mounted sonar arrays that collect bathymetric soundings. The ship's position is determined by Global Positioning System (GPS). Because the transducer arrays are hull mounted rather than towed in a vehicle behind the ship, the data are gathered with navigational accuracy of about 1 m and depth resolution of 50 cm. Two of the types of data collected are bathymetry (seafloor depth) and backscatter (data that can provide insight into the geologic makeup of the seafloor). An example is a bathymetric image of the US Pacific margin with mass transport deposits (**Figure 2**). The US National Geophysical Data Center (NGDC) maintains a website of bathymetric images of continental margins. Although morphological features seen on bathymetric images are useful for recognizing mass transport as a general mechanism, these images may not be useful for distinguishing slides from slumps. Such a distinction requires direct examination of the rock in detail. *Method 4*: Examination of the rock – Direct examination of core and outcrop is the most reliable method for recognizing individual deposits of slide, slump, debris flow, and turbidity current. This method known as process sedimentology, is the foundation for reconstructing ancient depositional environments and for understanding sandstone petroleum reservoirs.

modern The term refers to present-day deep-marine systems that are still active or that have been active since the Quaternary period that began approximately 1.8 Ma.

nonuniform flow Spatial changes in velocity at a moment in time.

normal grading Upward decrease in average grain size from the basal contact to the upper contact within a single depositional unit composed of a single rock type. It should not contain any floating mudstone clasts or outsized quartz granules. In turbidity currents, waning flows deposit successively finer and finer sediment, resulting in a normal grading (see waning flows).

outcrop A natural exposure of the bedrock without soil capping (e.g., along river-cut subaerial canyon walls or submarine canyon walls) or an artificial exposure of the bedrock due to excavation for roads, tunnels, or quarries.

planar clast fabric Alignment of long axis of clasts parallel to bedding (i.e., horizontal). This fabric implies laminar flow at the time of deposition.

primary basal glide plane (or décollement) The basal slip surface along which major displacement occurs.

projected clasts Upward projection of mudstone clasts above the bedding surface of host rock (e.g., sand). This feature implies freezing from a laminar flow at the time of deposition.

rock The term is used for (1) an aggregate of one or more minerals (e.g., sandstone); (2) a body of undifferentiated mineral matter (e.g., obsidian); and (3) solid organic matter (e.g., coal).

scarp A relatively straight, cliff-like face or slope of considerable linear extent, breaking the continuity of the land by failure or faulting. Scarp is an abbreviated form of the term escarpment.

secondary glide plane Internal slip surface within the rock unit along which minor displacement occurs.

sediment flows They represent sediment-gravity flows. They are classified into four types based on sediment-support mechanisms: (1) turbidity current with turbulence; (2) fluidized sediment flow with upward moving intergranular flow; (3) grain flow with grain interaction (i.e., dispersive pressure); and (4) debris flow with matrix strength. Although all turbidity currents are turbulent in state, not all turbulent flows are turbidity currents. For example, subaerial river currents are turbulent, but they are not turbidity currents. River currents are fluid-gravity flows in which fluid is directly driven by gravity. In sediment-gravity flows, however, the interstitial fluid is driven by the grains moving down slope under the influence of gravity. Thus turbidity currents cannot operate without their entrained sediment, whereas river currents can do so. River currents are subaerial flows, whereas turbidity currents are subaqueous flows.

sediment flux (1) A flowing sediment–water mixture. (2) Transfer of sediment.

sedimentology Scientific study of sediments (unconsolidated) and sedimentary rocks (consolidated) in terms of their description, classification, origin, and diagenesis. It is concerned with physical, chemical, and biological processes and products. This article deals with physical sedimentology and its branch, process sedimentology.

submarine canyon A steep-sided valley that incises into the continental shelf and slope. Canyons serve as major conduits for sediment transport from land and the shelf to the deep-sea environment. Smaller erosional features on the continental slope are commonly termed gullies; however, there are no standardized criteria to distinguish canyons from gullies. Similarly, the distinction between submarine canyons and submarine erosional channels is not straightforward. Thus, alternative terms, such as gullies, channels, troughs, trenches, fault valleys, and sea valleys, are in use for submarine canyons in the published literature.

tropical cyclone It is a meteorological phenomenon characterized by a closed circulation system around a center of low pressure, driven by heat energy released as moist air drawn in over warm ocean waters rises and condenses. Structurally, it is a large, rotating system of clouds, wind, and thunderstorms. The name underscores their origin in the Tropics and their cyclonic nature. Worldwide, formation of tropical cyclones peaks in late summer months when water temperatures are warmest. In the Bay of Bengal, tropical cyclone activity has double peaks; one in April and May before the onset of the monsoon, and another in October and November just after. Cyclone is a broader category that includes both storms and hurricanes as members. Cyclones in the Northern Hemisphere represent closed counterclockwise circulation. They are classified based on maximum sustained wind velocity as follows:

- tropical depression: 37–61 km h^{-1};
- tropical storm: 62–119 km h^{-1};
- tropical hurricane (Atlantic Ocean): $>$119 km h^{-1};
- tropical typhoon (Pacific or Indian Ocean): $>$119 km h^{-1}.The Saffir–Simpson hurricane scale:

- category 1: 119–153 km h^{-1};
- category 2: 154–177 km h^{-1};
- category 3: 178–209 km h^{-1};
- category 4: 210–249 km h^{-1};
- category 5: $>$249 km h^{-1}.

tsunami Oceanographic phenomena that are characterized by a water wave or series of waves with long wavelengths and long periods. They are caused by an impulsive vertical displacement of the body of water by earthquakes, landslides, volcanic explosions, or extraterrestrial (meteorite) impacts. The link between tsunamis and sediment flux in the world's oceans involves four stages: (1) triggering stage, (2) tsunami stage, (3) transformation stage, and (4) depositional stage. During the triggering stage, earthquakes, volcanic explosions, undersea landslides, and meteorite impacts can trigger displacement of the sea surface, causing tsunami waves. During the tsunami stage, tsunami waves carry energy traveling through the water, but these waves do not move the water. The incoming wave is depleted in entrained sediment. This stage is one of energy transfer, and it does not involve sediment transport. During the transformation stage, the incoming tsunami waves tend to erode and incorporate sediment into waves near the coast. This sediment-entrainment process transforms sediment-depleted waves into outgoing mass transport processes and sediment flows. During the depositional stage, deposition from slides, slumps, debris flows, and turbidity currents would occur.

unsteady flow Temporal changes in velocity through a fixed point in space.

waning flow Unsteady flow in which velocity becomes slower and slower at a fixed point through time. As a result, waning flows would deposit successively finer and finer sediment, resulting in a normal grading.

waxing flow Unsteady flow in which velocity becomes faster and faster at a fixed point through time.

Further Reading

Allen JRL (1985) Loose-boundary hydraulics and fluid mechanics: Selected advances since 1961. In: Brenchley PJ and Williams BPJ (eds.) *Sedimentology: Recent Developments and Applied Aspects*, pp. 7–28. Oxford, UK: Blackwell.

Coleman JM and Prior DB (1982) Deltaic environments. In: Scholle PA and Spearing D (eds.) *American Association of Petroleum Geologists Memoir 31: Sandstone Depositional Environments*, pp. 139–178. Tulsa, OK: American Association of Petroleum Geologists.

Dingle RV (1977) The anatomy of a large submarine slump on a sheared continental margin (SE Africa). *Journal of Geological Society of London* 134: 293–310.

Dott RH Jr. (1963) Dynamics of subaqueous gravity depositional processes. *American Association of Petroleum Geologists Bulletin* 47: 104–128.

Embley RW (1980) The role of mass transport in the distribution and character of deep-ocean sediments with special reference to the North Atlantic. *Marine Geology* 38: 23–50.

Greene HG, Murai LY, Watts P, *et al.* (2006) Submarine landslides in the Santa Barbara Channel as potential tsunami sources. *Natural Hazards and Earth System Sciences* 6: 63–88.

Hampton MA, Lee HJ, and Locat J (1996) Submarine landslides. *Reviews of Geophysics* 34: 33–59.

Hubbard DK (1992) Hurricane-induced sediment transport in open shelf tropical systems – an example from St. Croix, US Virgin Islands. *Journal of Sedimentary Petrology* 62: 946–960.

Jacobi RD (1976) Sediment slides on the northwestern continental margin of Africa. *Marine Geology* 22: 157–173.

Lewis KB (1971) Slumping on a continental slope inclined at 1 –4 . *Sedimentology* 16: 97–110.

Locat J and Mienert J (eds.) (2003) *Submarine Mass Movements and Their Consequences*. Dordrecht: Kluwer.

Middleton GV and Hampton MA (1973) Sediment gravity flows: Mechanics of flow and deposition. In: Middleton GV and Bouma AH (eds.) *Turbidites and Deep-Water Sedimentation*, pp. 1–38. Los Angeles, CA: Pacific Section Society of Economic Paleontologists and Mineralogists.

Sanders JE (1965) Primary sedimentary structures formed by turbidity currents and related resedimentation mechanisms. In: Middleton GV (ed.) *Society of Economic Paleontologists and Mineralogists Special Publiation 12: Primary Sedimentary Structures and Their Hydrodynamic Interpretation*, 192–219.

Schwab WC, Lee HJ, and Twichell DC, (eds.) (1993) Submarine Landslides: Selected Studies in the US Exclusive Economic Zone. *US Geological Survey Bulletin 2002*.

Shanmugam G (1996) High-density turbidity currents: Are they sandy debris flows? *Journal of Sedimentary Research* 66: 2–10.

Shanmugam G (1997) The Bouma sequence and the turbidite mind set. *Earth-Science Reviews* 42: 201–229.

Shanmugam G (2002) Ten turbidite myths. *Earth-Science Reviews* 58: 311–341.

Shanmugam G (2003) Deep-marine tidal bottom currents and their reworked sands in modern and ancient submarine canyons. *Marine and Petroleum Geology* 20: 471–491.

Shanmugam G (2006) *Deep-Water Processes and Facies Models: Implications for Sandstone Petroleum Reservoirs*. Amsterdam: Elsevier.

Shanmugam G (2006) The tsunamite problem. *Journal of Sedimentary Research* 76: 718–730.

Shanmugam G (2008) The constructive functions of tropical cyclones and tsunamis on deep-water sand deposition during sea level highstand: Implications for petroleum exploration. *American Association of Petroleum Geologists Bulletin* 92: 443–471.

Shanmugam G, Lehtonen LR, Straume T, Syvertsen SE, Hodgkinson RJ, and Skibeli M (1994) Slump and debris flow dominated upper slope facies in the Cretaceous of the Norwegian and northern North Seas (61 –67 N): Implications for sand distribution. *American Association of Petroleum Geologists Bulletin* 78: 910–937.

Shepard FP and Dill RF (1966) *Submarine Canyons and Other Sea Valleys*. Chicago: Rand McNally.

Talling PJ, Wynn RB, Masson DG, *et al.* (2007) Onset of submarine debris flow deposition far from original giant landslide. *Nature* 450: 541–544.

USGS (2007) US Geological Survey: Perspective view of Los Angeles Margin. http://wrgis.wr.usgs.gov/dds/dds-55/pacmaps/la_pers2.htm (accesed May 11, 2008).

Varnes DJ (1978) Slope movement types and processes. In: Schuster RL and Krizek RJ (eds.) *Transportation Research Board Special Report 176: Landslides: Analysis and Control*, pp. 11–33. Washington, DC: National Academy of Science.

Relevant Websites

http://www.ngdc.noaa.gov
– NGDC Coastal Relief Model (images and data), NGDC (National Geophysical Data Center).

http://www.mbari.org
– Submarine Volcanism: Hawaiian Landslides, MBARI (Monterey Bay Aquarium Research Institute)

http://www.nhc.noaa.gov
– The Saffir–Simpson Hurricane Scale, NOAA/National Weather Service.

EDDIES AND WAVES

MESOSCALE EDDIES

P. B. Rhines, University of Washington, School of Oceanography, Seattle, WA, USA

Introduction

Mesoscale eddies are energetic, swirling, time-dependent circulations about 100 km in width, found almost everywhere in the ocean. Several modern observational techniques will be used to profile these 'cells' of current, and to describe briefly their impact on the physical, chemical, biological, and geophysical aspects of the ocean.

The ocean is turbulent. Viewed either with a microscope or from an orbiting satellite, the movements of sea water shift and meander, and eddying motions are almost everywhere. These unsteady currents give the ocean a rich 'texture' (**Figure 1**). If you stir a bathtub filled with ordinary water, it will quickly be populated with eddies: whirling, unstable circulations that are chaotically unpredictable. There is also a circulation of the water with larger scale, that is, broader and deeper movements. The 'mission' of the eddies is to fragment and mix the flow, and to transport quantities like heat and trace chemicals across it. In a remarkably short time (considering the smallness of viscous friction in water) the energy in the swirling basin will have greatly diminished. The bath will also cool much more quickly than one would estimate, based on simple conduction of heat across the fluid into the air above.

One may think about the fineness of the pattern of fluid motion in analogy to the resolution of an image on a computer screen. In a bathtub, the fluid eddies have scales from about 1 mm to 1 m, hence spanning a thousandfold range of sizes. In the oceans, the smallest circulations are also a few millimeters in size, but the largest are of the order of 10 000 km in diameter: this represents a range in scale of about 10^{10} between the smallest and the largest. There is

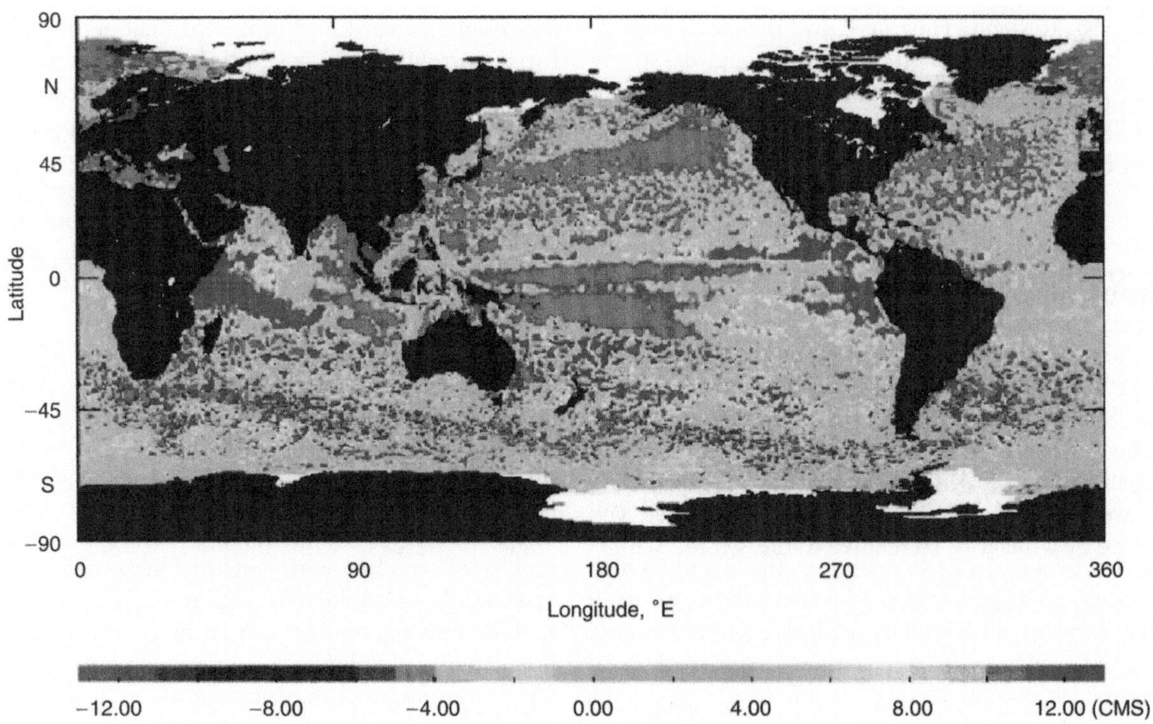

Figure 1 Sea surface elevation (cm) from the Topex-Poseidon and ERS satellites, 25 March 1998. The mean sea surface elevation for this time of year has been subtracted, so that only 'anomalies' from normal conditions are shown. The speckled pattern shows mesoscale eddies almost everywhere. In addition there are larger-scale patterns associated with El Niño (where the Equatorial Pacific has more level sea surface than normal) and large bands of high and low sea surface at middle latitudes. These may be associated with climate variability. The home web site for Topex-Poseidon satellites is http://topex-www.jpl.nasa.gov

thus room for many sizes of motion, each with a distinct dynamical nature: from tiny eddies that strongly feel viscosity, to 'mesoscale eddies' that strongly feel the Earth's rotation, to great 'gyres' of circulation filling entire oceans that feel also the curvature of the Earth. At scales in between are also numerous types of wave motion.

Mesoscale eddies are whirling and localized yet they densely populate the ocean. Typically 100 km across, their size varies with latitude and other factors of their environment: energy level, nearly bottom topography, and the nature of their generation. Eddies need not necessarily be round, with circular streamlines. They are often generated by unstable meandering of an intense current like the Gulf Stream. In this case the waving deflection of the Stream is itself a form of latent eddy, which may eventually grow and 'break' to form a circular eddy (as an ocean surface wave grows and 'breaks' at a beach).

Eddies are important because they have so much kinetic energy, and because they can transport momentum and trace water properties. They have deep 'roots' that often reach 5 km or more downward, carrying energy and momentum to the seafloor. They are responsible for the irreversible mixing of waters with different properties. Mesoscale eddies are typically as energetic as the concentrated currents that give birth to them. They may owe their existence to several sources other than meandering of strong currents: for example, direct generation by winds or cooling at the sea surface; flow over a rough seafloor or past islands and coastal promentories; or generation by mixing or waves of smaller scale.

'Geography' of Mesoscale Eddies

Before describing the 'physics' of mesoscale eddies, we should discuss their 'geography.' A satellite image of the surface of the global ocean can be assembled from many orbits, as the Earth turns below. A particularly basic measurement is that of the height of the sea surface. If ordinary waves are averaged out, we are left with a surface smooth to the eye yet varying by a meter or so relative to the 'geoid,' which determines the gravitational horizon (the geoid itself is permanently distorted by seafloor topography and, by itself, yields useful approximate maps of the seafloor elevation). Small variations in height of the sea surface correspond to small variations in pressure in the ocean below. Lines of constant pressure (isobars) are approximate lines of flow, or streamlines for horizontal circulation.

If one subtracts from this field the time-averaged sea surface height the result (**Figure 1**) is a dramatic display of time-varying mesoscale eddies: they are nearly everywhere. Additionally one sees in this image from the Topex-Poseidon and European Remote Sensing satellites the large-scale variation of the sea surface along the Equator in the Pacific Ocean. This anomalous state is characteristic of El Niño, when the Trade Winds fail to blow westward with normal intensity. Usually the winds pile up water at the west end of the Equator, but if they are absent the sea 'sloshes' back, one-quarter of the way round the Earth, toward South America. Animations of this field can be seen on the World Wide Web (for example, at http://topex-www.jpl.nasa.gov, and many of the features are seen to move westward.

Mesoscale eddies (as currents at the ocean surface) are particularly apparent in **Figure 1** along the paths of intense, major ocean currents. These delineate the Antarctic Circumpolar Current round Antarctica, which has a 'saw-tooth' form, flowing south-eastward across the South Indian and Pacific Oceans, and jogging northward where it encounters major seafloor ridges or gaps (at the Campbell Plateau south of New Zealand and the Drake Passage between South America and Antarctica, for example). In each subtropical ocean there are western boundary currents like the Gulf Stream and Kuroshio, which are marked by time-dependent energy after they leave the coasts and flow eastward and poleward. The jetlike equatorial currents show fine-scale energy that is more related to meandering than to separated, circular eddies. The westward flow in the low subtropical latitudes develops eddies in mid-ocean. Altimetry measurement has a large 'footprint' that misses eddies smaller than about 50 km in diameter. From direct measurements in the sea we know that the texture of the circulation includes mesoscale eddies smaller than this, particularly at high latitudes.

Orbiting satellites do more for us than produce images. Freely drifting instruments on the sea surface, and at great depth below the surface, tell us 'where the water goes.' These can be tracked by satellites and acoustic networks. Rather than delineating a smooth pattern of general circulation, drifters in the North Atlantic (**Figure 2**) show a tangle of tracks, with intense mesoscale eddies causing the gyrelike circulation to be nearly obscured. This region of the Atlantic involves the subpolar gyre, circulating counterclockwise north of 48° N latitude, and the subtropical gyre, circulation clockwise to the south. The Gulf Stream leaves the US coast at Cape Hatteras, in the southwest corner of the figure. It flows east-northeast and rounds the Grand Banks of Newfoundland, flowing north to about the latitude of Newfoundland, where it separates from the coast again, joining the subpolar

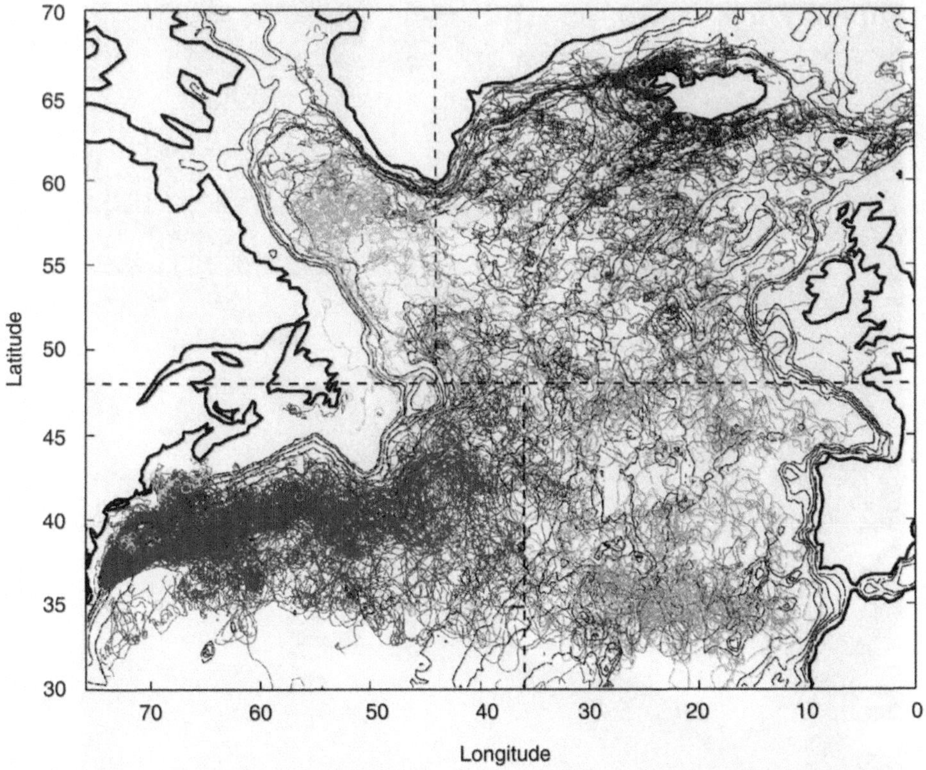

Figure 2 Tracks of drifting buoys ('drifters') on the sea surface, launched during 1996–1999 by Dr. P.P. Niiler, analyzed by J. Cuny. Tracks are colored corresponding to the 'box' (dashed lines) they were launched in. The tracks show intense eddy activity, superimposed on the general circulation. Typical duration of a track is 200 days. There is a mean movement of surface waters counterclockwise around this pattern; the Gulf Stream dominates the yellow tracks moving from west to east, and progressing into the other boxes. The purple tracks from the north-east move quickly westward, round the Labrador Sea (the north-west box) in strong boundary currents. This figure symbolizes the challenge of describing the ocean circulation in the presence of mesoscale eddies. There are several kinds of drifting floats, many of which also move vertically to record profiles of temperature, salinity and other properties; these involve some remarkable new technologies. Examples of web sites showing surface and deep-ocean currents using Lagrangian drifters include www.http://www.whoi.edu/science/PO/dept and http://flux.ocean.washington.edu/

gyre. The intense boundary current running westward around Greenland is clearly visible as the drifters invade from the east. The kinetic energy associated with eddies exceeds that in the time-averaged currents by factors ranging from 1.5 or so (in the jetlike current cores) to 50 or more (in the 'quiet' regions far from intense mean currents).

Radiometers on satellites record images at many different wavelengths; in the infrared (typically between 3.7 and 13 μm wavelength), and at even longer wavelengths of 'microwaves,' the radiation is strongly related to the temperature of the water at the sea surface (sea surface temperature, SST). Images in visible light show the texture of ocean color, which is strongly correlated with biological activity. These same images record 'sun glitter' patterns that are textured by ocean currents. Radiometers typically cannot resolve features less than a kilometer wide, though visible-light imaging can distinguish features down to tens of meters. Satellites actively transmitting beams of radiation can sense the sea surface elevation, slope, and roughness. Fine ripples and sharp surface wave crests give other sensors (as with synthetic aperature radar (SAR) satellites) resolution down to 20 m or so. These measurements tell us much about the surface currents and winds just above the sea surface.

Using SST sensors we now zoom in on a smaller region of ocean. SST patterns are shaped also by the movement of heat in ocean currents. The Gulf Stream (**Figure 3**) is visible as a warm, red band with sharp edges, carrying tropical heat northward on the west side of the North Atlantic. It shows a warm mesoscale eddy breaking off its northern edge. There are also many features evident of finer scale than was visible using the altimeter data (**Figure 1**). As with the global pictures of sea surface elevation, SST satellite images can be viewed as animations (e.g., www.nesdis.noaa.gov/). This involves removing the obstacle of clouds (though some sensors, like the radiometers in the TRMM (Tropical Rainfall Measuring Mission) satellite can see SST right through

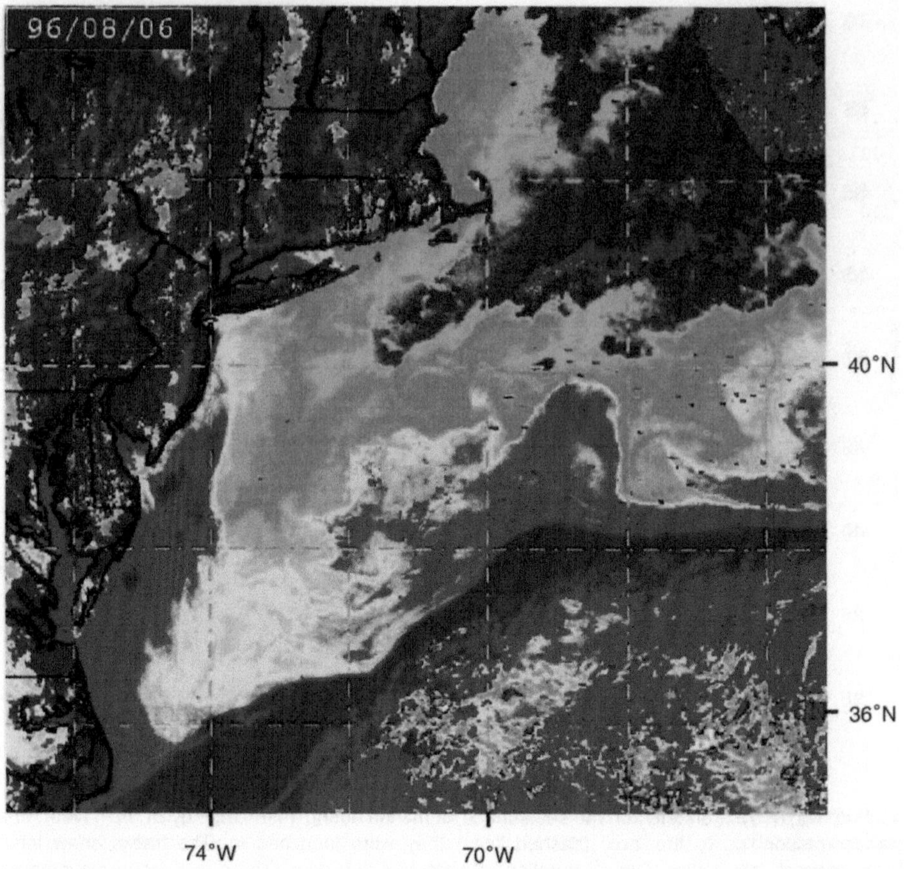

Figure 3 Sea surface temperature (SST) patterns in the Gulf Stream and adjacent warm waters of the Sargasso Sea (to the south) and cold, shallow water on the continental shelf. Color indicates temperature, ranging from purple, blue, green, yellow, orange to red as one moves from cold to warm. The Gulf Stream is the narrow, deep red feature flowing rapidly from south-west (left) to north-east. Its instability spawns mesoscale eddies. Each gridded box is one degree wide, and hence the north–south size of each box is 111 km.

clouds). Viewing these animations, the trained eye will see a wealth of phenomena, from the swing of the seasons, to boundary currents, tropical instability waves, upwelling of cold waters at the coasts and Equator, and ubiquitous mesoscale eddies.

Fritz Fuglister of Woods Hole Oceanographic Institution, once an artist during the Great Depression, pioneered the mapping of Gulf Stream eddies with painstaking ship surveys. It was a task befitting his training, and the 'false color' renditions used here to show temperature, are surely a high form of natural art. As well as being a warm current, the Gulf Stream is also a front separating the warm (red, orange) saline tropical waters of the Sargasso Sea to the south, from the fresher, colder (green, blue, purple) subpolar waters to the north. Despite the time of year (August), waters flowing south from the Labrador Current chill the coastal region as far south as Cape Hatteras. The Gulf Stream front was first mapped in 1768 by Benjamin Franklin, whose cousin Timothy Folger was familiar with it, as a site where whales could be found.

The roundish feature breaking off the north wall of the Gulf Stream in **Figure 3** is an example of an eddy formed by instability of a current and its associated temperature front. This instability can draw its energy from two sources: the kinetic energy of the current or the gravitational potential energy of the tilted stratification. As the instability grows, the Stream meanders wildly. Like an oxbow in a sinuous river, it can break off and become an isolated eddy. Here the eddies are sometimes called 'rings' because they are like rings of Gulf Stream water enclosing a trapped, foreign water mass. Meanders toward the north thus break off on the north side of the Stream and form warm eddies (relative to the cold waters around them). Conversely, southward meanders break off, encapsulating cold water to form 'cold' rings that wander south-westward and are often absorbed back into the Gulf Stream. The net effect is an exchange of water across the front: the Gulf Stream is a 'mixer.'

Biological communities are strongly affected by ocean circulation and eddies. In the East Australia

Current (**Figures 4** and **5**) the color of the sea surface can be used to estimate chlorophyll concentration (**Figure 4**) in green plants (phytoplankton). This current is, like the Gulf Stream, a western boundary current. The sea surface temperature for the same region at approximately the same time is shown in **Figure 5**. The two figures show the differing texture of the two properties, temperature and phytoplankton (plant growth). Temperature is strongly affected by the atmosphere, which erases the memory of SST patterns. Biological activity can persist for longer times, and hence the patterns show streakiness – longer persistence of fine details.

Baroclinic and Barotropic Eddies

Eddies produced by the shearing motion of a current or by its store of gravitational potential energy are part of a life cycle of energy transformation. There is a natural evolution of the eddies toward greater width, and toward greater vertical penetration. With the right circumstances the cycle can continue until the eddies reach to the seafloor with nearly identical horizontal currents at every depth. This is known as a 'barotropic' state, whereas currents that decrease or increase with depth are termed 'baroclinic.'

Baroclinic currents obey a balance of Coriolis forces and pressure forces in the horizontal, and gravity and pressure forces in the vertical: this is known as the 'thermal' wind balance.' It establishes a close relationship between horizontal variations in fluid density (as in an ocean front separating warm water from cold) and vertical variations in current velocity (as in a current whose velocity decreases as one moves downward from the sea surface). It is a key connection which, for more than a century, has allowed oceanographers to infer currents from observations of the temperature and salinity in the ocean (for temperature and salinity and pressure together determine the fluid density). Thus, for example, the Gulf Stream front, which in cross-section has tilted lines of constant density, is the site of strong vertical variations in current. These relationships are visible in **Figure 6**, showing a cross-section of potential temperature and salinity in the northern Atlantic. These high-resolution data show the upper layer of warm water that dominates the southern and eastern parts of the section, floating on a bed of much colder, denser water. The sloping surfaces of constant temperature and salinity are evidence of thermal wind velocities associated with the general circulation, and the smaller-scale wiggles show mesoscale eddies. The seafloor topography is dominated by the Mid-Atlantic Ridge.

Near Cape Farewell, Greenland (the left end of the section), the subpolar waters reach right to the surface. The salinity section shows the warm water to be saline (of subtropical origin), while the deeper and more northern waters are of much lower salinity, owing to the sources of fresh water at high latitude. (Plots like this can be seen, or made to order, using software available at http://odf.ucsd.edu/OceanAtlas (the Ocean Atlas system) or http://www.awibremerhaven.de/GEO/eWOCE (the Ocean Data View system).) The lower plots show the vertical profiles of potential temperature versus depth, and potential temperature versus salinity,

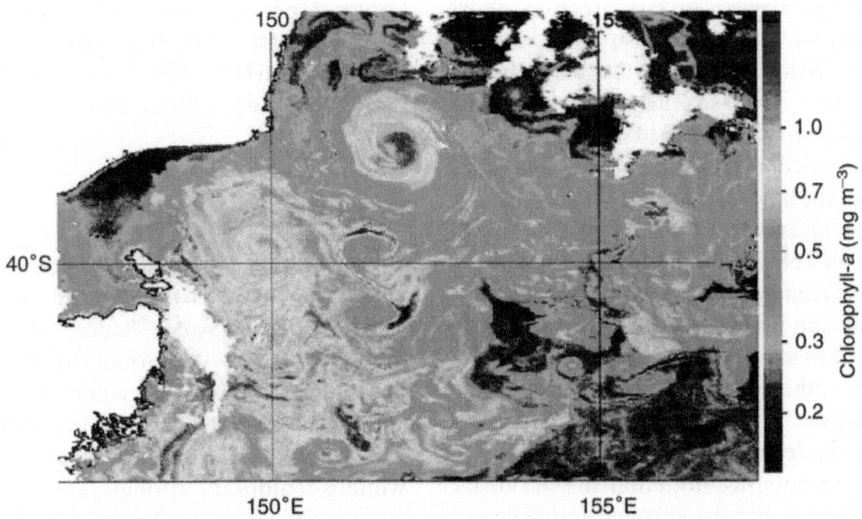

Figure 4 Chlorophyll-*a* concentration inferred from ocean color, SeaWIFS satellite. This is the East Australia Current along the coast of New South Wales. Note the richer content of finely textured eddies. Characteristically, ocean color and other 'tracers' can develop a more finely filamented structure than can temperature, whose patterns are erased by heat exchange with the atmosphere. Latitude and longitude (the parallels 40°S latitude, 150°E and 155°E longitude) are shown (http://www.marine.csiro.au/~lband/SEAWIFS/).

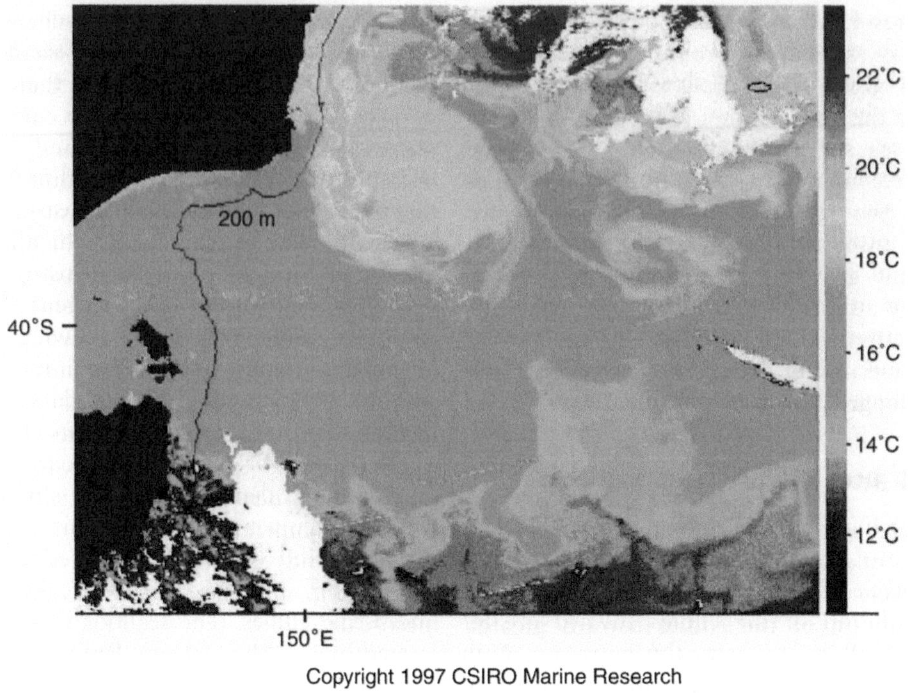

Figure 5 Sea surface temperature (°C) in the East Australia Current, showing a field of anticyclonic eddies. This is approximately the same region and time as in **Figure 4**, with warm waters in the north flowing from the tropics, meeting cold waters of the Southern Ocean. White regions are clouds. Temperature scale shown at right. (This image is from http://www.marine.csiro.au/~lband/SEAWIFS/; there are many web sites providing SST imagery, for example http://www.rsmas.miami.edu/groups/rrsl/,http://www.el~ino.noaa.gov/, and http://fermi.jhuapl.edu/avhrr/sst.html).

for the entire dataset (with colors indicating the very low values of dissolved silicate in this highly ventilated part of the world ocean).

When we see eddies in the surface temperature we can thus infer that there will be variations in the currents from one vertical level to the next. Typically, warm eddies appear in cross-section as depressions in surfaces of constant temperature, while cold eddies are 'domes' of deep water elevated toward the surface. The sea surface has upward deflection opposite to that of the underlying density layers (provided the currents diminish as one moves downward from the surface). Usually this is the case, and this fits the picture of warm anticyclonically rotating eddies and cold cyclonically rotating eddies. In the Northern Hemisphere, cyclonic means counterclockwise, and the reverse in the Southern Hemi sphere. There are exceptions to this rule, typically occurring when the eddies are generated deep beneath the surface.

Capping off this description of the vertical variation in ocean currents, we note that the sea surface elevation reveals not only the existence and shape of surface ocean current patterns but also their sense of rotation. The 'lows' in sea surface elevation are low-pressure cells beneath, and hence are cyclonic, while 'highs' in sea surface elevation are anticyclonic.

These difficult dynamical connections take on practical significance when one considers the biology of the ocean. Nutrients are richly abundant deep in the ocean, yet they need to be drawn up to the sunlit surface waters to produce chlorophyll-rich phytoplankton. Stable density layering of the oceans, however, provides a strong barrier to vertical movement of water. Anything that can lift deep water nearer the surface is likely to promote life, and this is just what cold, cyclonic eddies do.

Formation of Eddies

Eddies and thermal wind balance are also strongly in evidence in the coastal zones of the ocean. The long stretch of the eastern Pacific, shown in **Figure 7**, extending from California to Washington, shows cold waters upwelling where the north winds of summer blow surface waters offshore. The southward-flowing California Current, and narrower upwelling region are strongly unstable, and mesoscale eddies grow rapidly. Nutrient-rich cold waters promote growth right through the entire food chain, from plankton to whales and sea birds. Eddies act to exchange water between the shallow continental shelf and deeper ocean to the west.

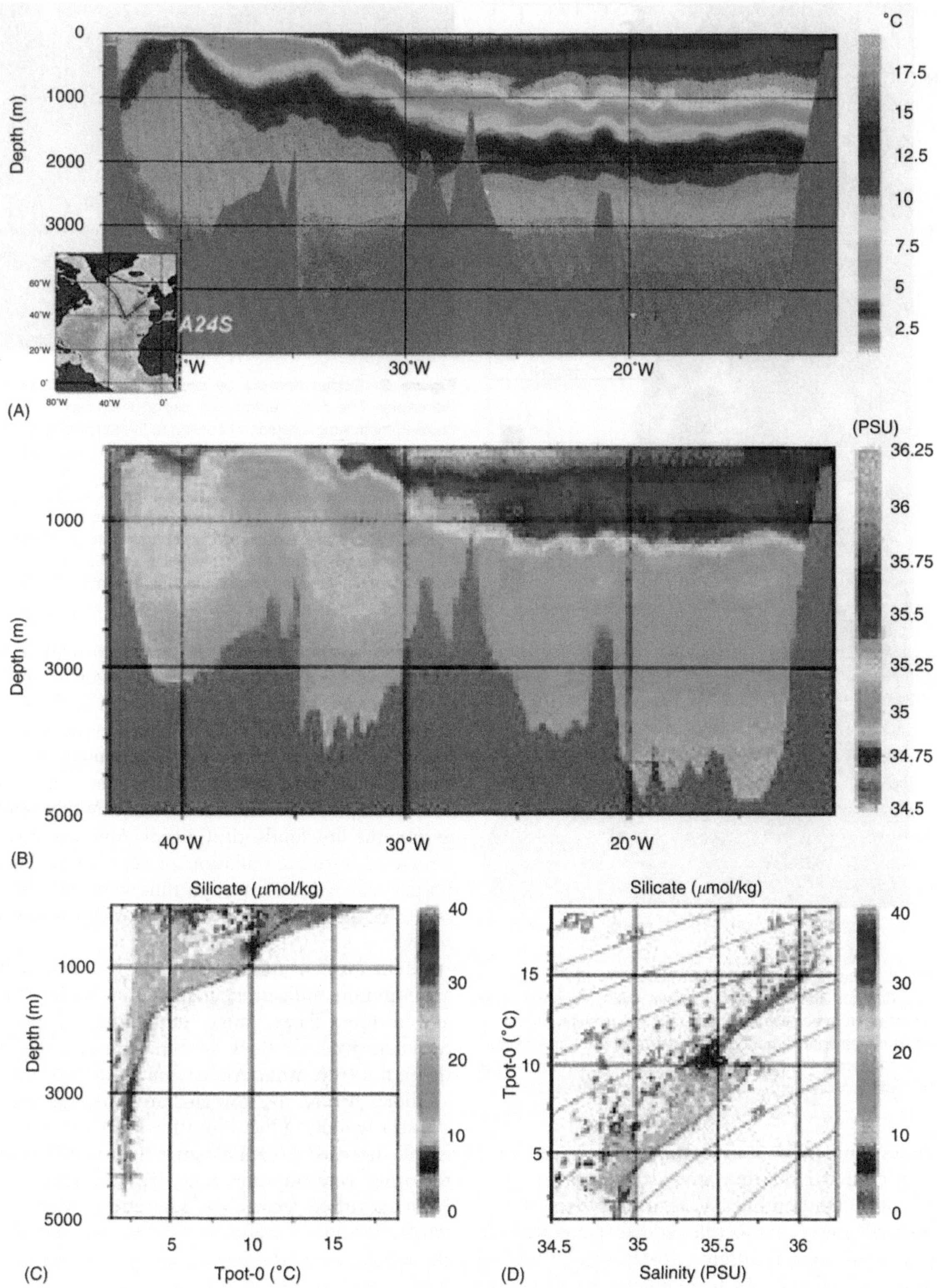

Figure 6 Section view of (A) temperature and (B) salinity along a ship-track in the northern Atlantic (the vee-shaped path shown in the map inset), from the WOCE hydrographic program. Surfaces of constant fluid density have a form broadly similar to the temperature and salinity surfaces. Also shown are salinity along the same section (lower left), vertical profiles of potential temperature (lower center) and plots of potential temperature against salinity (lower right). (Plot courtesy of Dr. Rainer Schlitzer.)

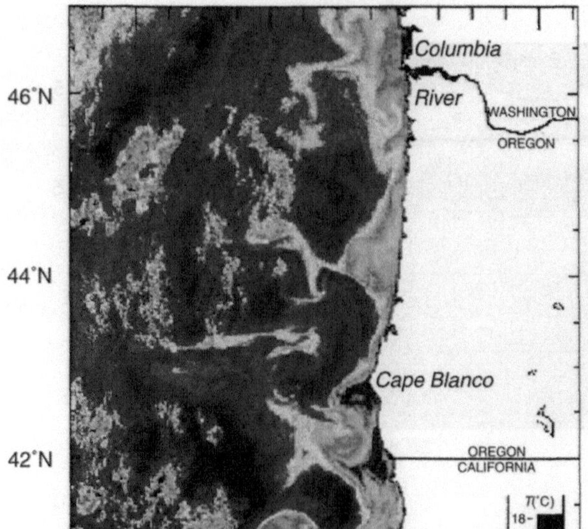

Figure 7 Coastal upwelling and eddies in the California Current. (Oregon State University.). Blue (cold) coastal waters are drawn up from below the surface, and are rich with nutrients. New instruments enable us to do 'cat-scans' of the upper ocean using instruments 'flying' behind a rapidly moving vessel, tethered with a cable (e.g., http://www.oce.orst.edu/research).

Eddies formed by convection can be seen over most of the Earth, but they are particularly energetic in the cold, high latitudes. A laboratory experiment (**Figure** 8) shows mesoscale eddies generated by cooling of the water surface. Rotation of the fluid organizes the eddies, which are much bigger than the convective plumes directly generated by cooling. In the Labrador Sea, cold winds from the Canadian Arctic sweep over the water and cool it intensively (at a rate exceeding 800 W per m^2 of sea surface, in a cold-air outbreak, and averaging $300\,W\,m^{-2}$ for an entire winter month).

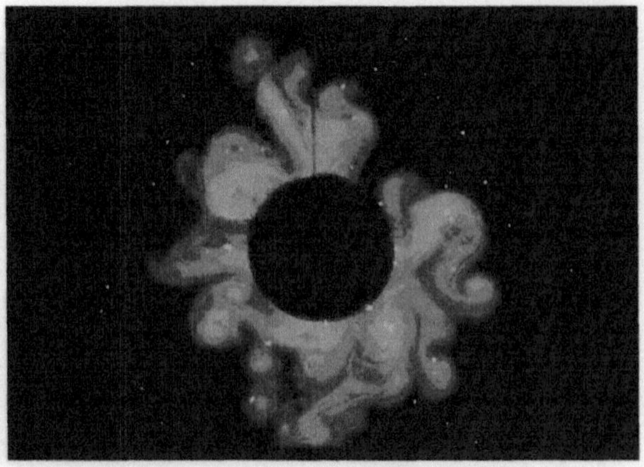

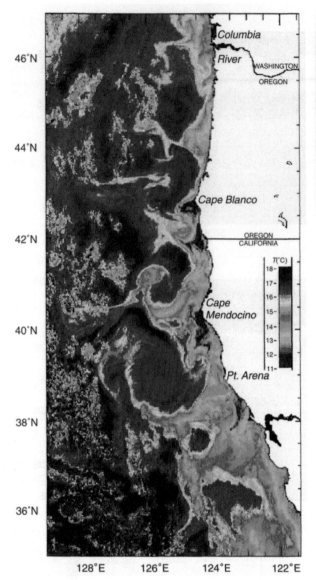

Figure 8 Eddies formed by cooling a rotating fluid in the laboratory. The dark central disk sits at the water surface and cools it, mimicking a region of cooling to the atmosphere. Coriolis forces give the thermal convection form, initially as small plumes (a few hundred meters across in the ocean), subsequently as mesoscale eddies with scale of 10–100 km, which dominate the scale model experiment here (oceanic flows and waves can be studied using scale models in the laboratory (e.g., Geophysical Fluid Dynamics Laboratory, University of Washington, http://www.ocean.washington.edu/research/gfd/gfd.html).

Eddies formed directly by winds blowing on the sea surface are thought to occur widely, and yet the large size of wind patterns is not well-matched to the small, roughly 50 km diameter of mesoscale eddies. However, near ocean boundaries, wind forcing can have demonstrable effect on eddies (for example, the westward Trade Winds spilling across the lowlands of Central America create a strong eddy-rich circulation in the eastern Pacific). Larger-scale eddies, more in tune with wind forcing take on the characteristics of Rossby waves (see below).

Eddies formed by flow over an irregular seafloor are common, and can be identified in tracks of floats and drifters. These range across the spectrum of turbulent sizes, all the way to the grand scale setting the path of the Antarctic Circumpolar Current.

Eddies formed by flow past an irregular coastline are seen widely. When fluid flows past a cylindrical island, it sheds a regular pattern of eddies with alternating rotation direction. This is known as a Karman vortex street. The interesting thing is that, when the same experiment is done in a laboratory, the regularity of the vortex street disappears as the flow is made stronger or the cylinder is made larger. At the much greater scales of oceanic flow it is at first surprising that the turbulence regime is not encountered. The likely reason is that fluid motions restricted to two dimensions cannot fragment their energy into a full state of turbulence as readily as can a fluid with full freedom to move in all three dimensions. Thus, your kitchen sink looks more turbulent than does a much larger ocean basin.

The Physical Properties of Eddies

Some basic physical effects If it were not for the Earth's rotation, its associated Coriolis forces, and its spherical shape and (or) complex bottom topography, the eddies shown in these figures would be much larger in scale. To discuss these effects we need to review some of the basic physics of the ocean.

On the great scale of the circulation of the oceans, there are several physical forces at work, particularly buoyancy forces and Coriolis forces. Buoyancy arises because both the water temperature and the concentration of dissolved salts (called the 'salinity' – kg of dissolved salt per kg of sea water) affect the density (expressed as the mass of 1 m^3 of sea water). Coriolis forces arise ultimately from the rotation of the Earth.

Buoyancy produces a layered ocean, with dense fluids beneath less dense fluid. A measure of its importance is the buoyancy frequency or Brunt–Vaisala frequency, N, measured in radians per second. If a region of sea water were lifted upward and then released, it would settle back to its original depth, bobbing about it with a frequency N. The bobbing period ($2\pi/N$) varies from a few minutes in the upper ocean to several hours at great depth. Stable stratification greatly limits vertical motion of the fluid, for tremendous energy is required to lift fluid against gravity. Yet the deep ocean is 'ventilated' at high latitude. Cold air from the continents and the Arctic is particularly effective at cooling the ocean, making the waters dense enough to sink.

Coriolis forces greatly restrict the motion of the fluid oceans and atmosphere. Their importance is measured simply by the rotational frequency, Ω, of the planet (2π per day). At locations other than the poles, this effect is diminished by the sine of the latitude (θ); hence the important frequency, say f, is $2\Omega \sin(\theta)$ which is just equal to the frequency of a Foucault pendulum.

Horizontal structure and size A number of factors are at work determining the diameters of mesoscale eddies. One central idea is that if the buoyancy forces and Coriolis forces are of similar strength, the width, call it λ, will be approximately given by $\lambda = NH/f$, where N and f are as defined above, and H is the vertical scale of the eddy. This is known as the Rossby deformation radius, after Carl Gustav Rossby, a pioneer in both oceanography and atmospheric sciences. For eddies with vertical scale comparable with the ocean depth, the size λ ranges from a few hundred kilometers in the tropics to 10 km or so at high latitudes. This great range of variation comes from the tendency for the high-latitude ocean to have weaker density stratification (small N) and larger Coriolis frequency f. λ also represents the horizontal distance traveled by a simple internal gravity wave in a half-pendulum day. The same dynamical eddies exist in the atmosphere, yet are much larger in horizontal scale. They are the basic high- and low-pressure cells seen on weather maps. Their 1000 km diameter (roughly) is also estimated by λ, which is much larger because the buoyancy frequency of the atmosphere is so much greater on average, than that of the ocean.

Vertical structure The oceans are full of three-dimensional structures. The general circulation involves 'arteries' of flow, often narrow horizontally (say, 50 km wide) and vertically (say, 1 km or less, thick). Cross-sections of velocity or trace properties marking the circulation illustrate this. Because eddies are often spawned as instabilities of major currents, they too may be three-dimensional, and of limited extent in the vertical. Such structures, which are termed 'baroclinic,' may have a range of vertical scales, H.

In addition, both the general circulation and eddies can exist in a form with no variation of horizontal current from the top of the ocean to the seafloor. These 'tall' currents and eddies, termed barotropic, are distinct and important. They disturb the density field only slightly, and hence are invisible in classic hydrographic sections. For this reason, they were not well understood or observed by early oceanographers. Barotropic flows have a signature at the sea surface, but otherwise have no gravitational potential energy. Tall, barotropic eddies evolve rapidly and are strongly associated with Rossby waves.

Rossby waves; potential vorticity Finally, the shape of the planet is important. Its nearly spherical form causes Coriolis effects to change with latitude, and this leads to a new, rather exotic phenomenon known as Rossby waves. Water on a spinning planet is endowed with a 'stiffness' along lines parallel with the planet's axis. This stiffness does many things to the circulation, tending to restrict motion to lie east and west. More generally, in the presence of valleys and ridges on the seafloor, currents can circulate freely along curves of constant $\sin(\theta)$ divided by depth. These are simply curves of constant ocean depth, if we measure the depth parallel to the Earth's axis rather than vertically. Such pathways of freely flowing water are known as 'geostrophic contours.'

The physics of mesoscale eddies is described well by an exotic property of the fluid: the potential vorticity. We are interested in many different things in fluids: their velocity, temperature, density, salinity, etc., but certain of these properties are particularly illuminating. The fluid density, for example, is active in determining buoyant forces in the fluid. After correcting for pressure effects, the density is also a marker of fluid motion, so long as mixing and diffusive effects can be ignored: it stays the same (after that correction), if one follows a moving parcel of fluid. Maps of this corrected 'potential' density both give us dynamical information about currents and show us something about the mass distribution of the oceans. Potential vorticity also has the property that it remains constant, as we follow a parcel of fluid, until mixing or external forces or heating is felt. The quantity describes the 'spin' of a fluid parcel, including the rotation of the Earth, and also including a measure of the thickness of the fluid layer. From knowledge of the field of potential vorticity, one can calculate much about the currents and displacement of the mass field of the ocean. As a more general definition, geostrophic contours become lines of constant potential vorticity, which cover surfaces of constant (potential) density.

This same stiffness imparted by the Earth's rotation produces wave motions if fluid is pushed across, rather than along, geostrophic contours. These Rossby waves are 'information carriers' that help to form the general circulation. They are themselves unsteady currents, whose patterns radiate principally horizontally from where they are generated. A laboratory experiment, **Figure 9**, shows Rossby waves in a basin centered on a virtual North Pole. All of the motion in this experiment is generated by a small, oscillating body in the lower left of the figure (beneath the black rectangle). The Rossby waves are seen as wavy deflections of the central band of dye. Constant–latitude circles become marked with colored dyes as east–west currents develop in response to the Rossby waves' shaping of the general circulation.

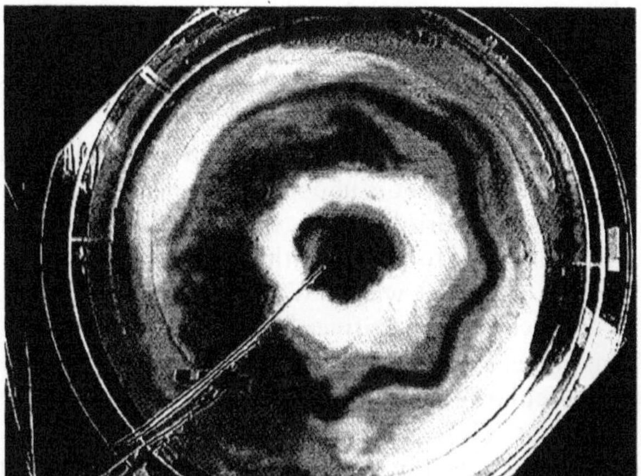

Figure 9 Rossby waves in a laboratory simulation of the circulation of an ocean centered on the North Pole. The waves are visible as undulations of the dye line, and are propagating eastward away from the wave source. Nevertheless, the wavecrests seem to move westward (clockwise). The source of the wave motion is a small oscillating cylinder at the lower left (black band). Geophysical Fluid Dynamics Laboratory, University of Washington.

These Rossby waves are 'weak' eddies. If they 'break,' that is, deform the basic fluid greatly and irreversibly, they fulfill our picture of turbulence: chaotic, with active stirring and mixing of trace properties, like the colored dyes here. The 'rotational stiffness' that makes the waves possible also greatly limits the north–south movement of fluid. Thus, the polar cap in this experiment is virtually unmixed, and is chemically isolated from the lower latitudes. This is just the physics at work in the atmosphere, in defining the polar ozone depletion zones ('ozone holes').

Rossby waves, and their more violent cousins the mesoscale eddies, help to set the fundamental force balances of the general circulation. They redistribute the momentum of ocean currents horizontally (as in **Figure 9**), establishing and reshaping currents in horizontal planes. But the ocean is three-dimensional, and these waves and eddies are also active in transferring momentum downward from the sea surface. In **Figure 1** the Antarctic Circumpolar Current is driven eastward by the strongest sustained winds on Earth. It thus becomes the greatest of ocean currents (in terms of transport and potential and kinetic energy). The eastward force of the winds is balanced by pressure forces exerted by the ridges and gaps of the seafloor topography. To connect these opposing forces, Rossby waves and eddies are active in transporting momentum downward. Elsewhere in the world ocean, eddies also provide essential communication of momentum downward from the surface. They drive deep gyres known as inertial recirculations, and establish the form of the deep roots of currents like the Gulf Stream. Mesoscale eddies also stir and mix the potential vorticity field. With weak ocean currents, the geostrophic contours, or free-flow pathways, tend to lie east and west. In order to develop the great gyres of circulation, with substantial north–south flow, the ocean has to reorganize its potential vorticity field accordingly. This is accomplished both by eddy activity and by the dynamical reshaping of the large-scale oceanic density field.

We have argued that long waves, involving much subsurface activity, are an important cousin of mesoscale eddies. Such waves are particularly visible in the tropical oceans. El Niño is an interaction between atmosphere and oceans. While it recurs somewhat unpredictably, in ways not yet fully understood, oceanic wave propagation along the Equator adds a 'delayed memory' to the process (such propagation is clearly visible in satellite altimeter and SST animations; see www.pmel.noaa.gov). Precursors to El Niño are recognizable, and give roughly six months of predictability at present. Sea surface temperature anomalies on 18 January 1999, during La Niña (the opposite phase to El Niño) involved an unusually cold eastern tropical Pacific and warm core in the subtropical North Pacific (**Figure 10**). The pattern is decorated with mesoscale eddies of much smaller scale and unstable waves on the equatorial westward jet. It is interesting that as one moves from high latitude toward the Equator, the Rossby scale, λ, increases markedly and mesoscale eddies become larger and more wavelike. Energy sources in the strong Equatorial current system give rise to tropical instability waves, which appear to play an important role in both dynamics and biology.

Modelling Techniques

Figures 8 and 9 showed laboratory simulations of mesoscale eddies and Rossby waves. Some, but not all, physical effects active in ocean circulation can be modeled in the fluids laboratory. A persistent problem is the exaggerated effect of friction and molecular diffusion of heat and salt in the small scale of a model. Beginning in the 1970s, computers were developed with enough speed and memory to solve adequately the physical equations of motion, using methods of numerical approximation. Fully turbulent flows that have eluded theoreticians for hundreds of years have suddenly become accessible to 'numerical experiments.' These experiments have problems analogous of those in the fluids laboratory: limited resolution of fine details. Yet analysis of the flow is far easier in a computer model than

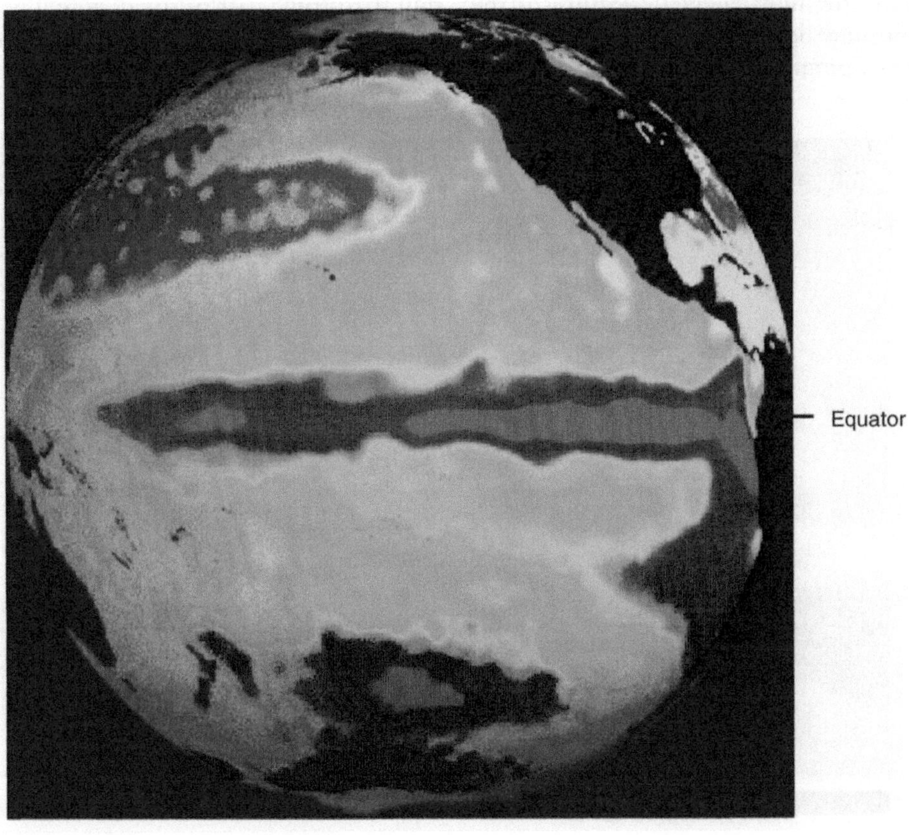

Figure 10 Sea surface temperature map (showing the temperature anomaly, or difference between the temperature and its seasonal mean at each point) in the eastern Pacific Ocean for 18 January 1999 (http://podaac.jpl.nasa.gov/sst/intro.html). The low (purple, blue) temperatures along the Equator represent the unusually strong easterly (that is, westward) winds associated with La Niña. Mesoscale eddies appear in middle latitudes, and also as tropical instability waves along the Equatorial circulation. Both satellite observations and *in situ* instruments, moored or drifting or shipborne, contribute to our understanding of the tropical oceans (e.g., http://www.pmel.noaa.gov).

a laboratory model: everything about the computer-modeled flow is measurable.

We have pointed out the range of length scales needed to describe oceanic motions, from 1 mm to 10 billion times that large. Computer models currently describe a range of scales, typically, of only a thousandfold; with these we can simulate a flow that is a few hundred eddies 'wide.' What does this mean in terms of representing the global ocean?

Computer experiments were originally developed for the atmospheric fluid, with weather prediction a principal goal. There are many similarities between atmospheric and oceanic circulations and eddies, but there are also striking differences. Particularly, Rossby's deformation scale, λ, is much smaller in the ocean than the atmosphere. λ is an estimate of the diameter of mesoscale eddies, and this means that the great high- and low-pressure patterns on a weather map, with cyclonic and anticyclonic systems typically having 1000 km diameter, are dynamically similar to 100 km wide ocean eddies. The texture in **Figure 1** is much more fine-grained than that of a weather map showing atmospheric pressure patterns. These eddies are the most energetic features of the circulation and must be resolved by the models for many purposes. Computer models of the ocean are thus global in extent yet need gridpoint spacing significantly less than λ. Ten-kilometer spacing of gridpoints is typical in a modern 'eddy resolving' ocean model (with vertical resolution of a few hundred meters), and even this is not enough to resolve fully eddies and boundary currents. Atmospheric models with grids having ten times greater spacing are considered to have 'high resolution.'

There is another striking, nearly devastating, obstacle to ocean modeling: evolution of the circulation is much slower than with comparable features of the atmospheric circulation. As we know from experience, the atmosphere adjusts to changes in solar heating after a period of a few months. Its weather features are rapidly destroyed by friction at the ground after just a few days. Thus a 10-year simulation of the atmosphere is very long indeed. However, the oceanic circulation responds much more slowly to changes in forcing, and this requires much longer simulation experiments. If the winds or solar heating at the sea surface are changed, the ocean will begin to respond quickly. Rossby waves will transmit the changes across major oceans in a few weeks. The tall, barotropic part of the flow will begin to adjust. Currents along the western boundaries and the Equator will quickly change. But the deeper, more

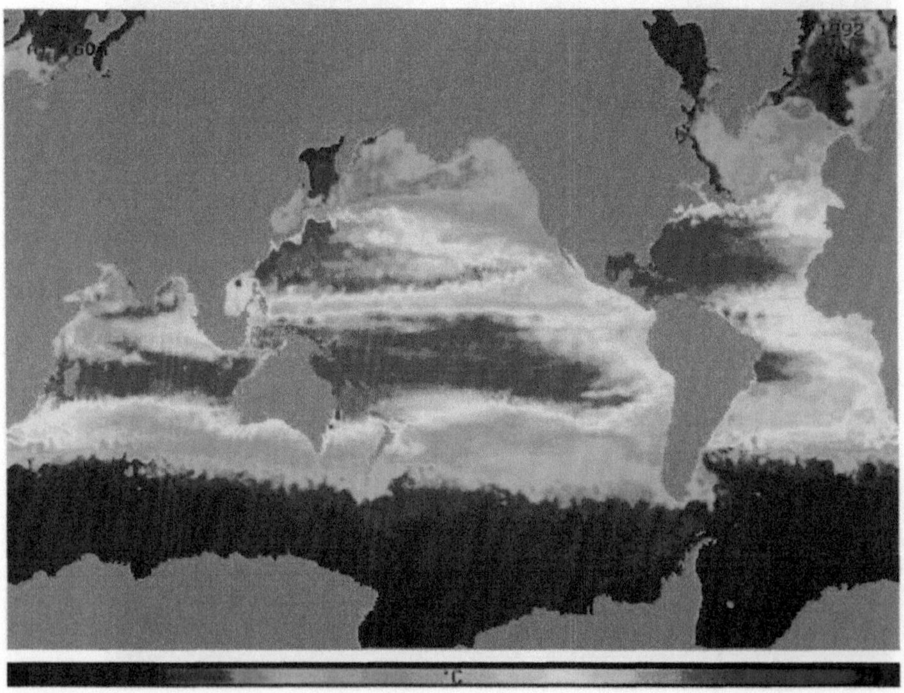

Figure 11 Temperature at 160 m depth, from the Parallel Ocean Processing model (http://vislab-www.nps.navy.mil/~braccio/) of the Naval Postgraduate School and Los Alamos National Laboratory. This is a snapshot during a single day in January 1992. The numerical model is driven by observed meterological winds and is also 'restored' back toward surface ocean observations. Notice the fine pattern of mesoscale eddies superimposed on the warm, (red) subtropical gyres and the cool (blue) high-latitude circulations. Grid points of this computer model are spaced 1/6° of latitude apart, giving reasonable coverage of mesoscale eddies at low and middle latitudes. This map of temperature differs from **Figure 10**, which shows only the anomaly field.

baroclinic flow will take more than 10 years to adjust, and the great, global meridional overturning circulation will not fully redistribute heat, salt, and trace chemicals and biological fields for several thousand years. These key features – the smallness of strong ocean currents and their slow evolution in time – can be seen in animations of computer model runs (for example at http://vislab-www.nps.navy.mil/~braccio/ for the Naval Postgraduate School and Los Alamos National Laboratory POP model; or www.http://panoramix.rsmas.miami.edu/micom/ for the University of Miami isopycnal ocean model). These simulations (**Figure 11**) use the full power of our largest computers, and are wonderful renditions of an entire world of ocean physics.

Computer simulations of weather and climate have to be run for many thousands of years if they are to encompass the full range of oceanic adjustment. In practice this cannot yet be done with 10 km grid spacing, and climate modelers instead use coarse resolution (typically 100–400 km grid spacing) and simulate the action of mesoscale eddies. They do this with exaggerated friction, and diffusion of heat and salinity that is much larger than in reality. These 'sticky, conductive' oceans may provide models of some of the important oceanic transport of heat and fresh water, and have much interesting structure, but they lack the full detail of both boundary currents and mesoscale eddies. The art of 'parametrizing' the effects of eddies so as to allow their neglect in detail is an active area of current research.

Conclusion

We have argued that mesoscale eddies contain large kinetic energy, comparable with that of the time-averaged ocean circulation. Eddies are crucial to the transport of heat, momentum, trace chemicals, biological communities, and the oxygen and nutrients relating to life in the sea. They are also active in air–sea interaction, both through response to weather and in shaping the patterns of warmth that drive the entire atmospheric circulation.

As a member of the huge family of turbulent motions, eddies contribute to the stirring and mixing of the oceans, to the creation of its basic, layered density field, and to its general circulation. The fundamental physics of eddies is expressed in terms of its potential vorticity, which is a tracerlike property that 'moves with the fluid.' The distribution of potential vorticity can be turned into knowledge of the currents and fluid density variations. The smallness and great energy of mesoscale eddies, the great thermal and chemical capacity of the oceans, and the slowness of the circulation conspire to challenge computer models, but rapidly increasing computer power is producing ever better representations of the ocean's fabric. At present, rather short-lived experiments (a few decades duration) can be carried out that resolve the global field of eddies, intense currents, and wind-driven gyres, whereas the slower features important to long-term climate change cannot be examined while also resolving mesoscale eddies. Nevertheless, several important physical processes like turbulent mixing, convection, upper mixed layer dynamics, and interaction with complex bottom topography are not yet well simulated by computer models. Many of the important applications of physical circulation in the oceans involve vertical motion: for biological communities, for transport of trace gases and their exchange with the atmosphere, for ocean/atmospheric climate interaction. This vertical motion of the fluid is particularly difficult to predict without fully resolving the detail of mesoscale – and smaller – features.

See also

Ocean Circulation. Rossby Waves.

Further Reading

Summerhayes CP and Thorpe SA (1996) *Oceanography, An Illustrated Guide*. New York: Wiley.

MEDDIES AND SUB-SURFACE EDDIES

H. T. Rossby, University of Rhode Island, Graduate School of Oceanography, Kingston, RI, USA

Introduction

Meddies and related types of circular motion in the ocean belong to a class of eddy activity characterized by a highly coherent, axisymmetric circulation in the horizontal plane. These stable features have a lifetime measured in years, during which time they may drift thousands of kilometers, carrying with them waters from where they were formed. They play an important, but as yet inadequately defined and quantified, role in the transport and exchange of waters between different regions. Understanding these processes is of fundamental importance for a correct characterization of subsurface eddy processes in the ocean and their representation or parametrization in ocean circulation models.

These subsurface eddies, shaped as very thin disks or lenses with an aspect ratio of $\sim 1{:}50$ to $\sim 1{:}100$, have a core body that rotates virtually as a solid disk, surrounded by a perimeter region of strong radial shear. Among the largest and most conspicuous of this type of eddy motion, the meddy (for Mediterranean eddy), can have diameters exceeding 100 km and life spans measured in years. We now know that this type of eddy motion, discovered in 1976, occurs in many regions of the world ocean, at shallow depths and deep, in tropical, subpolar, and arctic waters. Some eddies, such as the meddies, rotate anticyclonically, but, evidently just as likely, lenses may rotate in the other direction. From the growing observational database it now appears that the meddies do not merely drift with, but can in fact move through the surrounding waters. Their ubiquity, longevity, and mobility render them of potentially great importance in the transport, exchange, and mixing of waters between different water masses in the ocean. But there is much about these enigmatic features we have yet to understand.

Definitions

On scales of tens of kilometers and larger in the ocean, fluid motion is in geostrophic balance, meaning that the pressure gradient is balanced by the Coriolis force. These forces act at right angles to the direction of motion and thus do not tend to accelerate or alter the pattern of flow. For the circular motion of eddies discussed in this article, particle motion is curved rather than straight. This adds a radial acceleration, v^2/r, also perpendicular to the direction of motion, into the momentum balance, giving rise to a cyclogeostrophic balance or flow:

$$fv + \frac{v^2}{r} = -\frac{1}{\rho}\frac{\partial p}{\partial r} \qquad [1]$$

where r is the radius of the curved motion, v is the azimuthal velocity, p is pressure, $f(= 2\Omega \sin(\text{latitude}))$ is the Coriolis parameter, Ω is the angular velocity of the earth, and ρ is the density of the fluid. Again, because the forces act at right angles to the direction of motion, they do not alter the pattern of flow, i.e., the pattern is self-preserving. We call the clockwise motion of the (northern hemisphere) meddies anticyclonic, because they have a pressure maximum in the center. Low-pressure cyclonic eddies rotate in the opposite direction. Cyclones and anticyclones rotate in the opposite direction in the southern hemisphere.

Potential vorticity expresses the circulation per unit volume of fluid and for our purposes can be written as $(f + \zeta)/h =$ constant, where f is the Coriolis parameter and ζ represents the relative vorticity of a layer of fluid with thickness h. The conservation of this quantity, in the absence of forcing or dissipation, severely constrains the movement of fluids. For steady axisymmetric motion, a fluid parcel's potential vorticity is automatically conserved.

A measure of the intensity of rotation is given by the ratio of the relative vorticity of the core of the lens to the planetary vorticity, the Rossby number: $R = \zeta/f$. Typical R values for meddies range between -0.1 and -0.6, with the most extreme value reported $= -0.85$. Another number, the Burger number, expresses the ratio of strength of relative vorticity to the vortex stretching terms in the potential vorticity equation and is normally written as $N^2H^2/(f^2L^2)$. The Burger number can also be defined as the ratio of the available potential energy to kinetic energy of the lens.

SOFAR (sound fixing and ranging) and the related RAFOS floats reveal how fluid parcels drift, disperse, and mix in the ocean from their trajectories, which are determined by acoustic triangulation. SOFAR floats transmit signals to stationary hydrophones;

RAFOS floats listen to moored acoustic sound sources. The travel times multiplied by the speed of sound in the ocean give the distances to within a few kilometers accuracy. Isopycnal RAFOS floats can also drift with the waters in the vertical. Because isopycnals move up or down or as fluid slides up or down along isopycnals such as across the Gulf Stream, isopycnal floats will accurately follow that motion. These acoustically tracked floats have been major contributors to our knowledge of the subsurface eddy field.

History

During an oceanographic cruise in the Fall of 1976 to study ocean currents east of the Bahamas, a large body of very warm and salty water was observed at 1000 m depth. Shaped as a thin lens, it had a core diameter of nearly 150 km and a thickness of 500 m. Nothing like this had been observed before. Several SOFAR floats deployed in it to study the currents revealed a clockwise circular motion with a 10-day rotation period for the innermost float at 10 km radius. The proximity of this warm saline lens to the well-known Mediterranean Salt Tongue that stretches west across the ocean (centered near 30°N) clearly suggested a Mediterranean origin. This discovery stimulated a search for similar temperature–salinity anomalies in the eastern Atlantic, and before long it became clear that these lenses, coined meddies for Mediterranean eddies, have a widespread distribution in the eastern Atlantic west of Spain and Portugal. We now know that these lenses belong to a class of coherent motion most characterized as thin spinning disks with a thickness to diameter (or aspect) ratio of about 1 : 100. While meddies rotate only anticyclonically, other subsurface eddies or lenses may rotate in the other direction. Their overall diameters range from perhaps less than 10 km to as large as the 150 km of the original meddy, which remains one of the largest ever found. During lifetimes of months to years, they may travel thousands of kilometers. The focus here is first on the well-studied meddies, their structure, origin, patterns of drift, and decay. Other observations that illustrate the ubiquity of this class of eddy motion are then discussed.

The Meddy

Structure

The typical meddy has a very distinctive density and velocity field. A vertical cross-section will reveal a spreading of the isopycnals such that the core or center of the eddy has weak stratification bounded by layers above and below with very high stratification (**Figure 1**). This leads to a pressure field that is higher inside the lens than outside, which requires for equilibrium an anticyclonic rotation (clockwise in the northern hemisphere). The top panels show typical profiles of temperature, salinity, and density of meddy 'Sharon' in 1984 and in 1985, one year later. Averaged over the volume of the meddy the contributions of high temperature and salinity to density must cancel and equal the average density of the displaced waters (Archimedes' principle). Hence, the core waters at 12°C and 36.2 PSU, have the same density as the surrounding waters at 8°C and 35.6 PSU at 1000 m where the density profiles can be seen to cross. But the vertical spread of the isopycnals results in radial density gradients such that for the lower half of the meddy the density inside is less, and for the upper half is higher, than that of the surrounding waters. If we imagine that at great depth the pressure is the same everywhere, then as we ascend into the meddy, the hydrostatic pressure will decrease less rapidly than outside so that pressure in the center of the core exceeds that of the surrounding waters by about 500 Pa (5 mbar). It is this excess pressure that maintains the orbital or rotary motion of the eddy in cyclogeostrophic balance. If we continue up through the meddy, the greater density of the core waters leads to a more rapid pressure drop than outside such that topside of the meddy at the surface the radial pressure all but vanishes. (This does not always apply, some meddies, especially recently formed ones, may have a surface signature.)

Meddy 'Sharon,' by far the best-documented meddy, was visited four times over a 2-year period during which detailed surveys of the density, velocity, and microstructure were conducted. Orbiting SOFAR floats trapped in the meddy made it easy to relocate for subsequent visits. Vertical profiles of horizontal velocity show a maximum at about 1000 m depth and increasing linearly outward. Beyond a certain radius the velocity field decreases rapidly. **Figure 1D** shows the azimuthal velocity as a function of radius. The sharp transition from a linear increase to radial decay indicates the radial limit of solid body rotation. This rotation rate has a maximum at mid-depth (near 1000 m) and decreases slightly above and below. But at each depth the rotation appears to be that of a solid body; that is, the meddy can also be characterized as a stack of disks, each rotating at its own rate with the one in the center having the highest rotation frequency (**Figure 1E**). This is consistent with a density field that is nearly uniform in the horizontal yet stratified in the vertical.

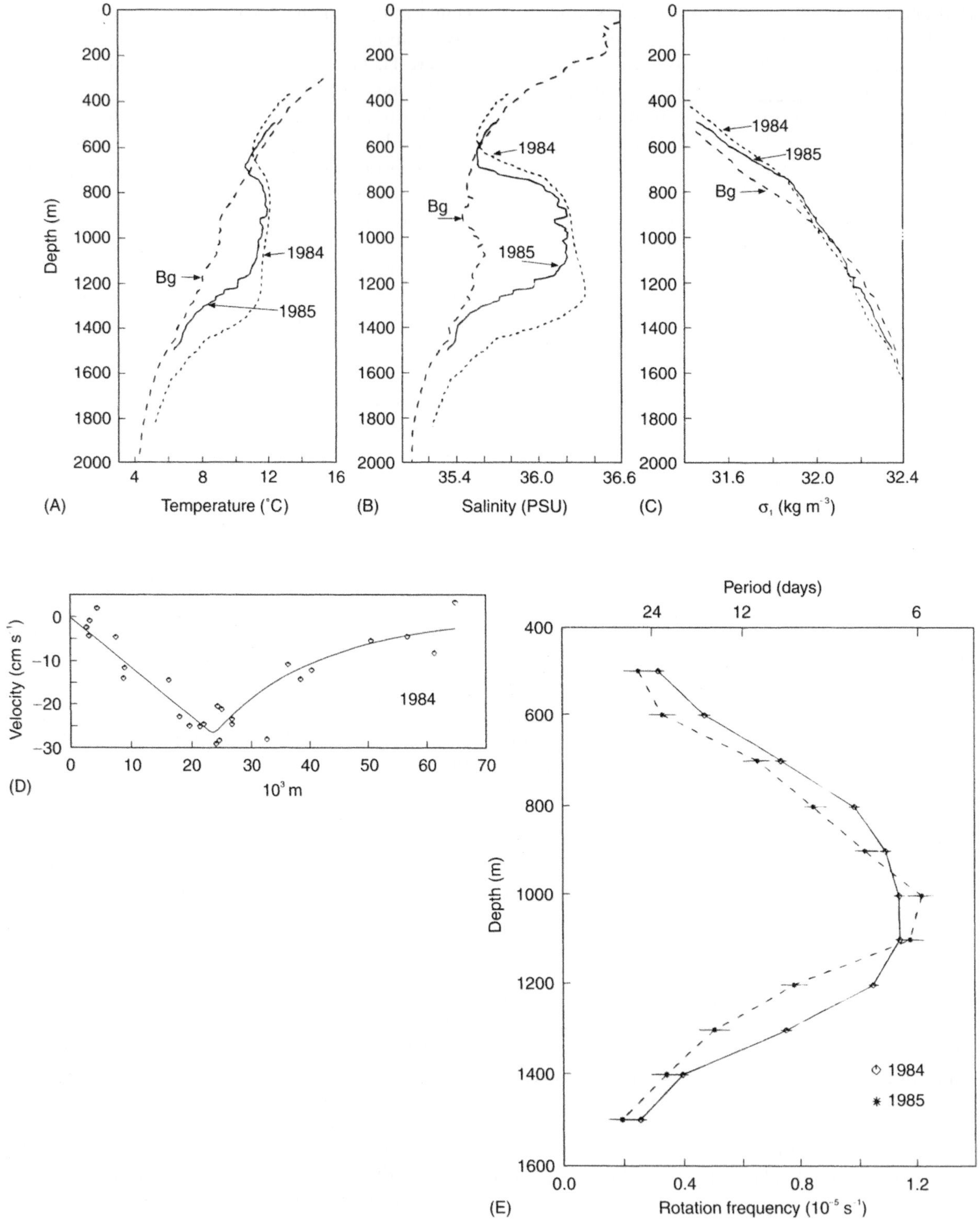

Figure 1 Temperature (A), salinity (B) and density (C) in the core of meddy 'Sharon' in October 1984 (solid), October 1985 (dotted) and background (dashed). Panel (D) shows azimuthal velocity in cm s^{-1} as a function of radius in 1984 and panel (E) shows angular velocity (bottom) and period of rotation (top) as a function of depth.

Between the visits to the meddy in 1984 and 1985, the strong core of undiluted salt had shrunk substantially as intrusions of fresher waters from the outside reached toward the center. This was not the case for the velocity field. Although it had shrunk in diameter, the core still evinced solid body rotation with a sharp transition to a region of radial decay beyond. Thus the dynamical structure is preserved even as the meddy loses its water mass anomaly. **Figure 2** shows cross-sections of the salinity for the two surveys. Note the substantial loss of core waters but retention of a well-defined velocity structure. Evidently small-scale leakage and diffusion along isopycnals can 'tunnel' through the larger-scale dynamics. But what maintains the sharp velocity transition at the velocity maximum (panel (D) in **Figure 1**)?

Fluid parcels do not remain at exactly the same radius as the meddy spins, but slowly wander in toward the center and out in an apparently random fashion. As SOFAR floats orbit around the meddy, they gradually drift in and out relative to the center, indicating a radial exchange of fluid within the core as it rotates. The very weak departures from solid body rotation within the core can be viewed alternatively as facilitating radial exchange and mixing or as the result thereof. Either way, this indicates a continuing process of homogenization of the core. This homogenization applies only to regions where the potential vorticity itself remains sufficiently uniform so as to allow the process to continue, i.e., within the core out to the radius of maximum velocity. The sharpness of the potential vorticity front for both years can be seen in **Figure 3**. Note how the radial boundary remains sharp despite the reduction in size. This stands in contrast to the continuing loss of salt inside the core of the meddy (**Figure 2**).

Meddy Formation

At the south-west corner of Portugal, Cape St. Vincent, the continental slope makes a sharp turn to the north. The Mediterranean outflow, a warm saline flow in geostrophic balance along the slope at about 1000 m depth, tends to follow the slope north. But sometimes, especially when the flow is strong, the current appears to overshoot and continues as an unbalanced flow to the west and curving to the north owing to the Coriolis force. If the curved motion is strong enough, it can fold back on itself, forming a closed loop and resulting in the genesis of a meddy. The negative relative vorticity of the meddy comes from the curvature of the flow and the negative lateral shear between the undercurrent and the bottom. The formation process was demonstrated by a series of RAFOS floats released into the Mediterranean Outflow over the period of a year. Some of these turned north along the bathymetry, but others exhibited the orbital motion we associate with meddies. These immediately broke away from the continent and started their drift to the south west. Other meddies were spawned farther north near the Estremadura Promontory, where the bathymetry also makes a sharp turn to the right. **Figure 4** shows an example of such a trajectory. The timescale for eddy formation is estimated at 3–7 days with, in all, about 15–20 meddies spawned per year.

Decay of the Meddy

It seems rather remarkable that these slender lenses with a core rotation period measured in days can last for hundreds of revolutions. Those that drift west to south-westward from Portugal toward the mid-Atlantic Ridge may reach 4–5 years of age. Meddy 'Sharon,' which was visited four times over a 2-year

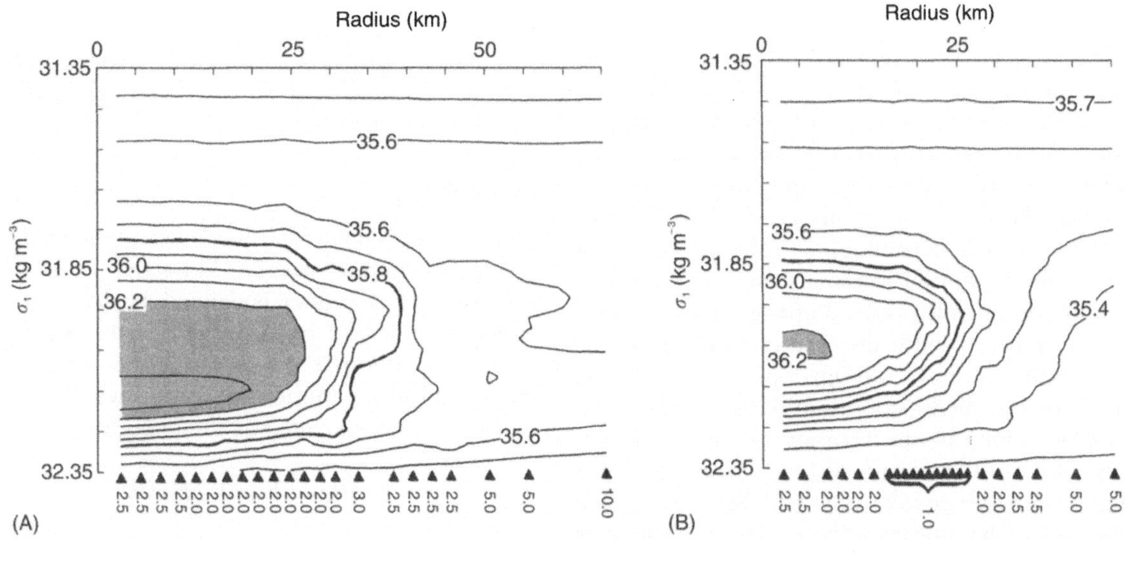

Figure 2 Radial σ_1 distribution of salinity (PSU) in meddy 'Sharon' for 1984 (A) and 1985 (B). σ_1 is density relative to a pressure of 1000 dbar.

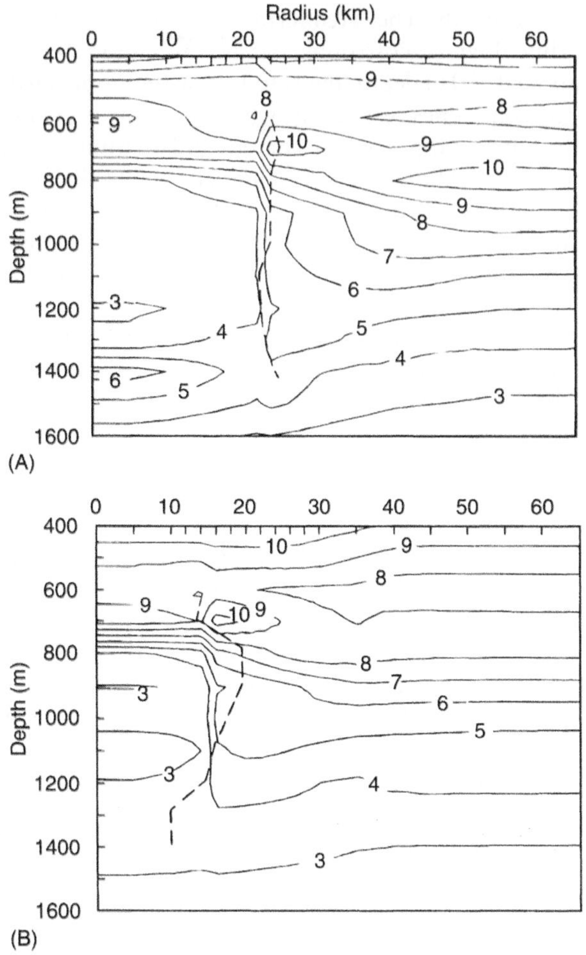

Figure 3 Potential vorticity distribution in meddy 'Sharon' for 1984 (A) and 1985 (B).

period during its 1100 km drift south, was probably at least a year old at the time of discovery. These repeat visits documented in detail both a radial and a vertical erosion. The organized orbital motion leads to substantial and sustained vertical shear across the large surfaces at the top and bottom of the meddy. However, the increased stability associated with the crowding of the isopycnals more than suffices to suppress shear flow instabilities. Double-diffusive processes, which depend upon the fact that heat diffuses more rapidly than salt, appear to play a more important role. Vertical erosion occurred both above and below, with the greater losses along the lower perimeter due to salt-fingering. Radial erosion, as indicated by the loss of salt in the core, appears to take place by means of intrusions along isopycnals. Between the first and last visit two years apart, the vertical and horizontal scales had been more than halved such that the identifiable volume had been reduced by a factor $(8\,\text{km}/30\,\text{km})^2 \times (300/700\,\text{m})$ to about 3% of its original size. The greater radial

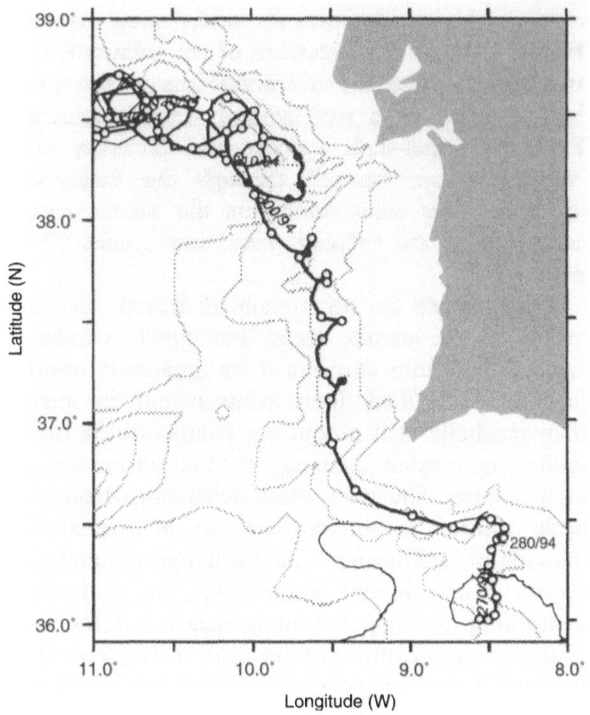

Figure 4 Formation of a meddy off Cape St. Vincent as indicated by the trajectory of a RAFOS float. (Reproduced from Bower *et al.* (1997) with permission of the American Meteorological Society.)

than vertical reduction points to some form of ablation at the perimeter rather than erosion at the top and bottom surfaces. The decrease in radial scale relative to the vertical might indicate an increase in Burger number (i.e., kinetic to potential energy ratio), but the uncertainties associated with this estimation process have left the matter unresolved. In any event, the faster radial than vertical erosion is testimony to the efficacy with which vertical stratification suppresses diapycnal exchange processes, even in the presence of enhanced vertical shear due to the rotation of the lens.

Energetics

Two measures define the energetics of the meddy, the available potential energy (APE) and its kinetic energy (KE). The former is defined as the energy that would be released by restoring all the density surfaces to a reference state at which all pressure gradients and hence motion associated with the meddy will vanish. It is defined as the integral

$$\text{APE} = \int_v 1/2\rho N^2 \pi^2 \mathrm{d}V \qquad [2]$$

While easy to state, the accuracy of the integration depends upon a determination of the background rest state, i.e., the accuracy with which the vertical displacement in the integral can be estimated. The N^2 term represents the vertical stratification. The corresponding KE integral

$$\mathrm{KE} = \int_v 1/2\rho v^2 \mathrm{d}V \qquad [3]$$

can be estimated fairly accurately. For most subsurface eddy studies the ratio of KE/APE, the energy Burger number, tends to be somewhat greater than unity, particularly as they age. This reflects the tendency for the aspect ratio of the lens to increase with age owing to the greater erosion around the perimeter than from above and below. For young meddies, APE and KE are of order 10^{14} J, which equals the output of a 1 GW electric utility plant for one day.

Sudden Death

Most meddies do not have the privilege of reaching a great age. Instead, there is a high probability of collision with one of the large number of seamounts in the eastern North Atlantic west of the Iberian Peninsula. Sometimes they fragment into smaller lenses that can continue for months to years. It has been estimated that perhaps 90% of all meddies eventually collide with a seamount, with an estimated average age at collision of 1.7 years. Those that do survive might live up to 5 years.

Apparently spontaneous breakup of meddies into smaller units has been observed, but the extent to which these occur owing to internal instabilities or result from interactions with other currents or eddies that shear them apart needs further study. Given the great age that meddies can reach, it would appear that the probability of spontaneous fission is quite small. Sharp lateral shear in the ambient flow could also wear at the meddy, but the large and organized relative vorticity of the meddy, typically 0.2–0.6 times the Coriolis parameter, renders it immune to the surrounding eddy field. Examples also exist of coalescence of smaller eddies into larger ones.

Significantly, almost all information about eddy interactions comes from the trajectories of SOFAR and RAFOS floats, which give us considerable spatial information as they drift about. Given the tight structure of the meddy, a single float will suffice to tell us the trajectory of the meddy, its collision with seamounts, and its possible demise. On the other hand, almost all our information on the mechanisms of aging comes from 'Sharon,' which, as noted above, shrank in volume by two orders of magnitude during the 2-year study. Curiously, meddies farther to the west appear to be much larger and vigorous at a comparable or greater age. Thus, it remains unclear how well 'Sharon' represents the meddy population as a whole. This is of more than passing interest because meddies have been suggested as a mechanism for maintaining the Mediterranean Salt Tongue (MST) that extends across the ocean near 30°N. If meddies are common in the eastern Atlantic, and it has been estimated that there might be of the order of 30 meddies at any given time, it seems remarkable that not a single meddy has been found west of the mid-Atlantic Ridge since the original meddy observation in the fall of 1976. **Figure 5** shows a summary of all meddy sightings in relation to the salinity anomaly of the Mediterranean Salt Tongue. Whereas many meddies drift south, note the conspicuous absence of meddies west of ~30°W along the axis of the salinity anomaly, the maintenance of which also remains an enigma.

Indeed, there is strong circumstantial evidence that the original meddy did not have a Mediterranean origin but came from quite far north in the northwest Atlantic. Near 50°N where the North Atlantic Current abruptly turns east as the Subpolar Front, a strong anticyclonic rotation is maintained by the current. It has been observed that this semidetached circulation can subduct and move south across the Newfoundland Basin as a subsurface eddy and continue west and south across the Sargasso Sea. With the help of the Gulf Stream recirculation system, the transit time may only be 2–3 years instead of 5 years, despite the large distance involved. Other subsurface warm-core lens sightings in the western Atlantic lend further support to this alternative origin.

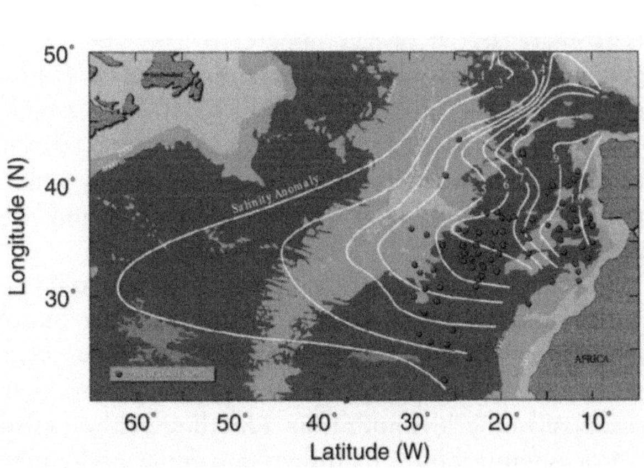

Figure 5 Summary of historical meddy observations. The diameter of the dots in the figure is about 50 km, somewhat smaller than the typical 100 km diameter of meddies. The contours show salinity anomaly relative to 35.01 PSU near a depth of 1100 m. (Reproduced from Richardson *et al.* (2000) with permission of Pergamon Press.)

Other Subsurface Lenses

Numerous other lenses have been observed and described. The structure of 'Sharon' seems to apply to others after appropriate scaling for size. Thus, an eddy about 50 km in diameter between 1000 and 2000 m depth located about 400 km west of Bermuda had a very similar structure. The isopycnals bend up and down forming a core region with weak stratification. The cold fresh waters in the core clearly point to a Labrador Sea origin, but where the lens itself was formed remains uncertain (analogous to the formation of meddies off Portugal containing waters originating in the Mediterranean Sea). One of the smallest yet very energetic subsurface eddies ever observed was tracked for 2 months in real time with a SOFAR float at 700 m depth west of Bermuda. A detailed hydrographic survey when the float was picked up indicated a diameter of $\sim$20 km and 300 m thickness. The core exhibited a distinct water mass anomaly with temperature, salinity, oxygen, and nitrate characteristics suggesting that the waters came from a low latitude, but where the eddy itself originated remains unclear; perhaps it was advected by the Gulf Stream north, perhaps it was formed by the meandering of the current. In any event, this lens, with a 1.5 : 100 aspect ratio, had a very fast rotation rate, about 3.5 days. An even faster rotation rate, 2 days, was observed for a small lens in the Gulf of Cadiz. This is very close to the theoretical limit where the relative vorticity of the lens exactly cancels the planetary vorticity. While the number of detailed lens studies remains limited, smaller lenses seem to have a higher rotation rate than larger ones. This suggests that, as the lenses age, they do so by decreasing their radius more rapidly than their height, so that their aspect ratio increases. The effect of this is to increase the Burger number of the lens. For a given pressure anomaly in the center, a smaller radius means a higher azimuthal velocity. The high angular velocity of these two small eddies compared to that of larger ones suggests a possible end fate in which the core remains intact as the ablation around the perimeter proceeds.

Curiously, almost all reports have focused on anticyclonic lenses despite the fact that we know from float observations that cyclonic lenses occur with near equal probability. Cyclonic lenses have received much less attention. For these, the density surfaces must bow in rather than out, inviting the description 'concave lenses.' The best-documented examples of these have been observed in the West European Basin. Interestingly, these also carry a positive salt anomaly, apparently to the north west toward the mid-Atlantic Ridge. Indeed, there is growing evidence that cyclonic eddies tend to drift poleward, whereas anticyclonic eddies drift equatorward. The classical argument for this is that as they age and lose their relative vorticity, they compensate for this by changing their latitude. In the case of meddy Sharon, the peak angular velocity actually increased, but the vertically averaged rotation rate clearly decreased (**Figure 1E**).

Discussion and Summary

The discovery of the meddy has a curious history. One of the first lenslike subsurface eddies to be identified as such was found just north-east of the Dominican Republic in the fall of 1976. It was nearly 150 km in diameter and 500 m thick, and the temperature–salinity characteristics of the core of the eddy and its proximity to the axis of the Mediterranean Salt Tongue suggested a Mediterranean origin. The report of this finding stimulated the search for similar lenses in the eastern Atlantic, and soon enough many others had been found. But what makes the original discovery all the more remarkable is that no other meddy has since been sighted west of the mid-Atlantic ridge. In addition, the probability of meddies getting that far decreases rapidly owing to the high risk of collision with seamounts. Even if a meddy did cross the mid-Atlantic ridge, the additional 2000 km distance to the original sighting makes that observation seem all the more extraordinary if not implausible. Instead, it now appears that the original 'meddy' actually originated in the north-west Atlantic where the North Atlantic Current turns east at 50°N. While the distance from that location almost matches that from Cape St. Vincent, Portugal, a lens originating in the North Atlantic Current (NAC) can be carried or advected rapidly to the south and west by the recirculating waters east of the NAC and south of the Gulf Stream, reducing the transit time to 2–3 years instead of 5 years. The decrease in latitude favors the anticyclonic eddy, but the nearly 50% reduction in Coriolis parameter suggests that the lens must undergo considerable adjustment. Perhaps the extraordinary width and flatness of the original meddy has its explanation here: As the Coriolis parameter f decreases, a decrease in thickness h would indicate a tendency to conserve its potential vorticity $(f + \zeta)/h$. On a more speculative note, if the lens did indeed flatten and widen, this could help explain the extraordinary diameter of the original 'meddy' and simultaneously give it an additional lease of life against radial erosion.

The fact that subsurface eddies larger than $\sim$100 km in diameter have not been observed suggests an upper limit at formation time set by inertia.

Meddies form at Cape St. Vincent at the south-west corner of Portugal where the Mediterranean Undercurrent must make a sharp turn to the north along the continental slope. Owing to inertia, the current may overshoot, becoming geostropically unbalanced where the bottom turns north. This causes the current to curve to the right owing to the Coriolis force. For faster than normal flow, this curving flow can almost fold back on itself, resulting in a closed loop leading to the genesis of a meddy. Given the frequent rate of formation of meddies at Cape St. Vincent site, this would be an excellent place to study the formation process in greater detail. Other sharp topographic features have been identified as sites for the formation of anticyclonic lenses. In contrast, remarkably little is known about how cyclonic lenses get spun up. Perhaps they result from instabilities of fronts and/or fission from larger eddies. No lower limit to the size of subsurface lenses has been established, but, at some limit, viscosity and double-diffusive processes will dissipate what is left. Before that limit is reached, however, the lenses can still remain remarkably energetic. But the very small pressure gradients needed to balance the cyclogeostrophic motion all but guarantees that they can only be detected and identified as such by Lagrangian means.

See also

Rossby Waves.

Further Reading

Bower AS, Armi L, and Ambar I (1997) Lagrangian observations of Meddy formation during a Mediterranean Undercurrent seeding experiment. *Journal of Physical Oceanography* 27: 2545–2575.

Hebert D, Oakey N, and Ruddick B (1990) Evolution of a Mediterranean salt lenss: calar properties. *Journal of Physical Oceanography* 20: 1468–1483.

Journal of Physical Oceanography March 1985. Special issue with numerous articles devoted to studies and observations of subsurface eddies.

Prater MD and Rossby T (1999) An alternative hypothesis for the origin of the 'Mediterranean' salt lens observed off the Bahamas in the fall of 1976. *Journal of Physical Oceanography* 29: 2103–2109.

Richardson PL, Bower AS, and Zenk W (2000) A census of Meddies tracked by floats. *Progress in Oceanography* 45: 209–250.

Robinson AR (ed.) (1983) *Eddies in Marine Science*. New York: Springer Verlag.

Schauer U (1989) A deep saline cyclonic eddy in the West European Basin. *Deep-Sea Research* 36: 1549–1565.

Schultz Tokos K and Rossby T (1991) Kinematics and dynamics of a Mediterranean salt lens. *Journal of Physical Oceanography* 21: 879–892.

TOPOGRAPHIC EDDIES

J. H. Middleton, The University of New South Wales, Sydney, New South Wales, Australia

Introduction

Topographic eddies in the ocean may have a range of scales, and arise from flow separation caused by an abrupt change in topography. This abrupt change may be of large scale, such as a major headland, in which case the topographic eddy is essentially a horizontal eddy of scale many tens of kilometers in a shallow coastal ocean. Eddies also occur at much smaller scales when ocean currents flow around small reefs, or over a rocky seabed. In this case the topographic eddies are perhaps only meters or centimeters in scale. A rule of thumb is that topographic eddies are generated at the same length scales as the generating topography.

Perhaps the earliest recorded evidence of a topographic eddy in the ocean comes from Greek mythology, where there is mention of a whirlpool occurring beyond the straits of Messina, between Sicily and Italy. Jason and the Argonauts in their vessel the Argo had to find the path between the Cliff known as Scylla, and the whirlpool having the monster Charybdis. The whirlpool still exists and occurs as the tides flood and ebb through the narrow straits. Another well-known tidal whirlpool, intensified at times by contrary winds and often responsible for the destruction of small craft, occurs off the Lofoten Islands of Norway. The Norwegian word maelstrom is associated with the whirlpool.

More recently, the fishing and marine lore of the Palau District of Micronesia, and the knowledge of ocean currents held apparently for hundreds, if not thousands, of years has been investigated. In fact, it was found 'The islanders had discovered stable vortex pairs and used them in their fishing and navigation long before they were known to science.' A sketch, drawn from the fishermen's description, is shown in **Figure 1**. The flow appears to comprise a stable vortex pair in the lee of the island, with identifiable zones of rougher water which would result from the conflicting directions of currents and waves, and calmer waters directly upstream from the island. A most interesting feature is the description of concentrations of tuna and flying fish, which clearly have a preference for congregating in certain zones, perhaps because there they find food more prevalent. This diagram underscores a fundamental importance of topographic eddies; they serve not only as a feature of the circulation, but also to provide preferential environments for the marine biota.

Although such whirlpools, eddies, and wakes had been well known for centuries by seafarers, in the early 1900s pilots and aerodynamicists 'rediscovered' eddies. The additional feature that they discovered was that eddies and wakes draw their energy from the mean flow, and provide a 'form drag' on the incident flow, tending to slow it down.

In the ocean, topographic eddies (or recirculating flows) comprise horizontal eddies generated by coastal currents flowing past coastal headlands, coral reefs, islands, or over undersea hills or ridges. They are important as they profoundly affect not only the

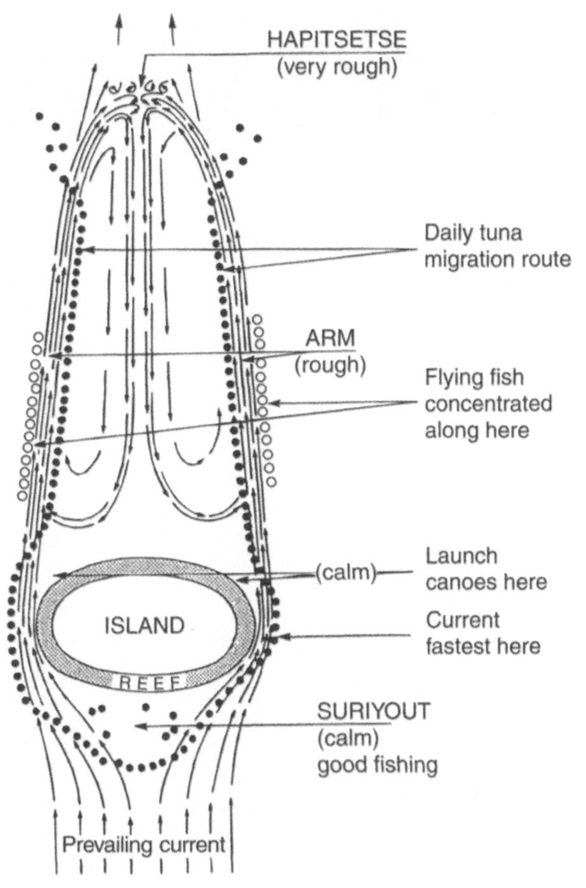

Figure 1 Island Wake of Tobi in Micronesia, showing flow patterns and concentrations of tuna and flying fish. (Reproduced from Johannes, 1981.) ●, Tuna concentrations; ○, Flying fish concentrations.

horizontal distribution of nutrients, pollutants, bottom sediments and biota through direct horizontal transport, but also the vertical distribution through the associated three-dimensional flow field. In addition, a cascade of turbulent energy to smaller scales provides a continued source of smaller-scale turbulence, which itself acts to further diffuse and transport such matters.

Topographic eddies may also be produced by smaller-scale reefs (submerged wholly or partly) with scales typical of the width and height of the reef, in which case the turbulent eddies are fully three-dimensional in nature. These eddies are also unstable and break down into turbulence of progressively smaller scale, stirring the ocean, and creating strong spatial gradients, which enhance mixing and diffusion of passive materials.

Perhaps it is appropriate now to discuss the terms turbulence and diffusion. Turbulence refers to a state of flow which is chaotic and random in its detail, such that any instantaneous state of flow will not ever be reproduced at any later time. However, there may be underlying physics which imply that measurements of properties made at any point will, after much averaging over time, produce an average which is predictable and reproducible. Diffusion refers to the stirring and mixing of waters as they flow in a turbulent manner. Diffusion tends to smear out or dilute unusually large concentrations of some property, such as a pollutant. The stronger the turbulence, the more rapid the diffusion.

Larger Scale Topographic Eddies

Many coastal headlands protrude several kilometers into coastal currents, where the ocean depth is often less than 100 m or so. The coastal currents may be tidally induced, changing over a period of 12 h or so, or may be relatively steady, changing perhaps only once every 7–10 days as a result of local synoptic scale atmospheric systems (*see* Wind Driven Circulation), or as a result of coastal trapped waves. Any resulting eddies are somewhat two-dimensional, with horizontal size many times that of vertical size, and occur downstream of the headland.

The generation of such recirculating flows or eddies has traditionally been considered to occur as follows. Flow separation occurring as a coastal current passes a headland is a result of the inability of the pressure field to allow the flow to follow the coastal contours, resulting in an adverse pressure gradient at the boundary. The separated flow has a very strong shear layer (with high vorticity), and a large-scale eddy may form, and either remain attached to the headland, or be carried downstream. In some cases, a string of eddies (known as a vortex street) may be generated. **Figure 2** shows a characteristic flow pattern behind Bass Point (near Sydney), with an overall larger-scale wake pattern, superimposed on which there are a number of smaller eddies.

There are several dimensionless numbers which represent various balances between physical processes, and hence terms in the equations of motion, which have been proposed in an attempt to simplify the physical balances which exist. For example, classical laboratory studies of the breakdown to turbulence have utilized the Reynolds number

$$R_e = UL/\nu$$

where U is the scale for the incident flow, L is the horizontal scale of the obstacle (reef or headland), and ν is the kinematic viscosity of the fluid. For example, Reynolds numbers between 4 and 40 for two-dimensional flows around a circular cylinder indicate a trapped and steady recirculating eddy-pair. For $R_e > 40$ the trapped eddy-pair maintains its presence, but the downstream wake begins to become unstable. At Reynolds numbers larger than $R_e = 80$, the eddy-pairs are swept downstream as a von Karman vortex street. Reports of such studies invariably cite the need to have no 'environmental noise' in the system to ensure reproducibility of the wake at low R_e, that is, a perfectly smooth incident flow upstream.

Field observations show some features which are at first sight similar to the laboratory observations, however, characterization of wakes and eddies based on the Reynolds numbers (and/or other simple dimensionless parameters) have often produced conflicting results. For field observations, relevant dimensional quantities include the incident current speed U, the distance the headland protrudes into the free stream L, the Coriolis parameter f and the water depth D. Vertical density stratification plays a role in deeper water, and the effects of the wind-driven surface layer and the frictional bottom boundary layer play a role in shallower water. Quantities arising from the flow itself are the horizontal eddy diffusivity ν_x and vertical eddy diffusivity ν_z associated with horizontal and vertical turbulent diffusion, respectively. Values of ν_x are usually several orders of magnitude greater than values of ν_z for most oceanic flows, indicating that horizontal diffusion dominates over vertical.

The assumption that turbulent Reynolds stresses are proportional to the mean velocity gradient allows an eddy-diffusivity approximation for the mean components of a turbulent flow. Reynolds numbers

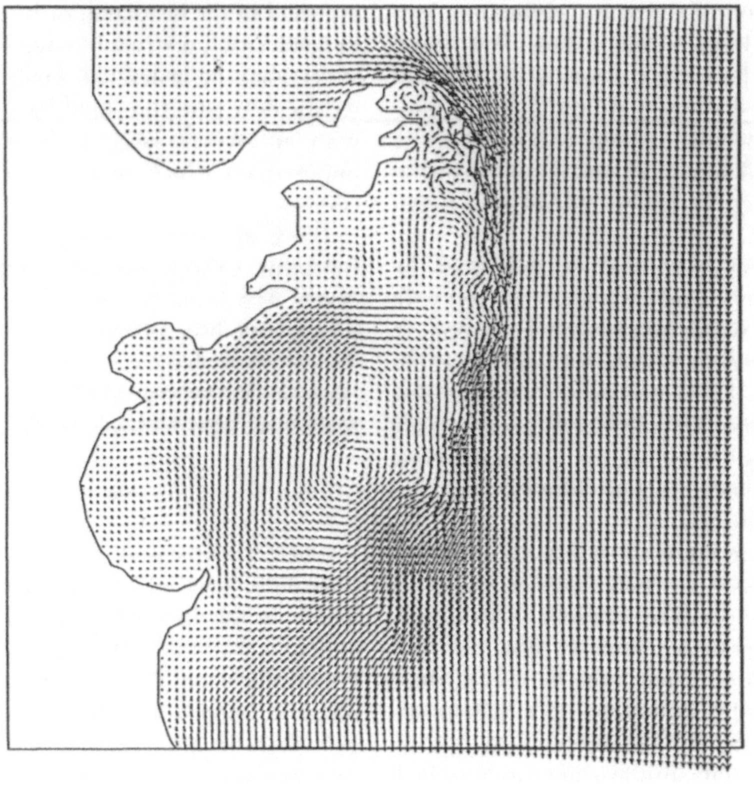

Figure 2 Attached eddies and vortex street in the wake of Bass Point as simulated by a computer model. (Reproduced from Denniss *et al.*, 1995.) The domain width is 6.67 km, and the maximum current vector, as indicated by the longest arrow, is 0.42 m s^{-1}.

for oceanic flows are then evaluated using the horizontal eddy diffusivity ν_x rather than the molecular viscosity. As an example, for flows around Bass Point Sydney, $R_e = UL/\nu_x \sim 1000$ using the overall headland width, and $R_e = 5$–10 for smaller-scale eddies produced by reefs at the tip of the headland, where $L \sim 100$ m and $\nu_x \sim 15\,\mathrm{m^2\,s^{-1}}$. Thus there are at least two different scales of topographic eddies in the recirculation processes depicted in **Figure 2**.

Other relevant dimensionless parameters include the Rossby number

$$R_o = U/fL$$

which gives a ratio of advective acceleration (non-linear) terms to Coriolis terms in the momentum equations. The Coriolis parameter f denotes the local rate of the earth's rotation about the vertical axis. Low R_o flows ($R_o \ll 1$) have the background rotation of the earth controlling the dynamics, with relatively slow flows, and a tendency to stable flow patterns. High R_o flows ($R_o \sim 1$) have a stronger tendency to produce eddies, as the non-linear terms which characterize energetic flows tend to dominate. Most larger-scale flows in the ocean have $R_o \ll 1$, indicating the rotation of the earth is a dominating effect, whereas for the Bass Point example described above, $R_o \sim 1$, indicating that the advective acceleration terms may be strong enough to produce eddies.

Derived parameters include the bottom boundary layer or Ekman layer thickness δ, which scales as

$$\delta \sim (\nu_z/f)^{1/2}$$

This height is a measure of the vertical extent above the bottom where the flow is affected by transfer of vertical stresses. This results in a deceleration of current from the free stream value U in the flow above to zero at the sea bed. In this bottom boundary layer, currents will change in direction, turning to the left (right) in the Northern (Southern) Hemisphere as the seabed is approached from above. If the bottom depth $D \gg \delta$, then the boundary layer provides a frictional decay on the overall flow. If, however, $D \sim \delta$, then the bottom turbulent layer dominates the entire water column, and somewhat different dynamics follow.

The vertical Ekman number giving the ratio of vertical momentum diffusion terms to the Coriolis term is

$$E = \nu_z/fH\delta$$

Thus E may be interpreted as a ratio indicating relative importance of bottom frictional effects and those due to the Earth's rotation. High E values are indicative of flows in which bottom friction dominates the flow (very shallow flows or flows with high vertical eddy viscosity), whereas $E \ll 1$ is indicative of deeper flows, or flows where bottom friction is less effective.

The importance of an island wake parameter P (or its square root) defined by

$$P = UD^2/\nu_z L$$

is discussed by several authors as being the relevant parameter (*see* Island Wakes) to describe a wake some distance downstream; it is essentially a Reynolds number based on vertical eddy diffusivity rather than horizontal. A survey of data from a range of island wakes indicates that for $P \ll 1$ the current simply flows around the headland with no recirculating eddies, for $P \sim 1$ the wake is steady and stable, and for $P \gg 1$ eddy shedding is observed.

For very shallow water flows where bottom stress is dominant, a summary of data show that a ratio of Rossby to Ekman numbers defined by

$$R_o/E_k = UD\delta/\nu_z L$$

is perhaps a better parameter than P with eddy shedding for large numbers ($R_o/E_k > 500$), steady eddy formation for $R_o/E_k \sim 100$ and fully attached eddies for $R_o/E_k < 10$. In shallow waters where $D \sim \delta$, these parameters (P and R_o/E_k) are essentially the same.

The use of dimensionless numbers to characterize flows is based on the assumptions that the essential processes are characterized by simple dynamical balances which will hold essentially throughout the domain of interest. However, for unsteady, nonlinear flows the balances are dependent on both location and time. Thus the Reynolds and Rossby numbers above are a measure of the flow balance in the upstream region. Reynolds numbers higher than some critical value imply that the resultant downstream flow is unsteady and chaotic, and is fundamentally different from the steady flow which occurs at lower Reynolds numbers. By contrast, the island wake parameters P and the R_o/E_k represent physical balances of the wake downstream of the headland. The bottom boundary layer thickness and Ekman numbers are properties of the vertical profile of the flow at any location.

A major feature, recognized in the early laboratory experiments was that flow stability at low Reynolds numbers was dependent on the absence of small-scale, rapidly changing background variability (referred to here as stochastic noise) occurring in the incident or upstream flow. However, it has been demonstrated theoretically that transition of the larger-scale flow to an unsteady chaotic system can be linked to the system's amplification of background stochastic noise. In the case of recirculating headland eddies, such stochastic noise might be due to variations in wind stress, wave activity (internal and surface), and nonlinear small-scale high-frequency turbulence caused by flow over or around bottom topography such as submerged or semisubmerged reefs.

Support for these ideas is provided by analyses of data from Bass Point. It is hypothesized that the turbulence generated by the smaller-scale reefs at the tip of Bass Point creates a turbulent horizontal shear layer. This pushes the flow separation point downstream, inhibiting the formation of a larger-scale attached eddy except under strong incident current conditions. Thus the smaller-scale turbulence at the tip of the Point has a substantial effect on the larger-scale wake flow. The small-scale turbulence is characterized in strength by a horizontal eddy diffusivity ν_x, and since ν_x is absent from the above wake parameters, the wake parameters cannot be definitive in terms of flow description. Thus it can be concluded that simulations or predictions of flow behavior based on dimensionless numbers alone, traditional stability analyses, or direct (nonlinear) numerical simulation may fail without exact knowledge of the background environmental stochastic noise.

The above description of flows is essentially applicable to cases of steady upstream inflows, however, in many coastal regions the tidal flood and ebb creates an alongshore current which floods and ebbs in opposite directions alongshore. In this case, there may be transient eddies generated at each half tidal cycle at every headland. The schematic diagram given here in **Figure 3** illustrates the point.

Eddies are generated on each cycle, and sit either side of the headland, with a residual flow always directed offshore at the tip of the point (**Figure 3A**); depth variability offshore ensures that rotational motions are generated by differential bottom friction (**Figure 3B**). A simple headland eddy is shown in **Figure 3C**, and a vortex street in **Figure 3D**.

The dynamics that allow generation of topographic eddies do not necessarily control their subsequent motion, the turbulent energy cascades or their ultimate dissipation. For topographic eddies in coastal waters, the presence of bottom boundary layers renders the flow partly three-dimensional, and so energy cascades to smaller and smaller scales as eddies break up. Vertical eddy viscosity in the bottom boundary layer, caused by friction at the seabed, also extracts energy from the system, creating a form drag on coastal currents.

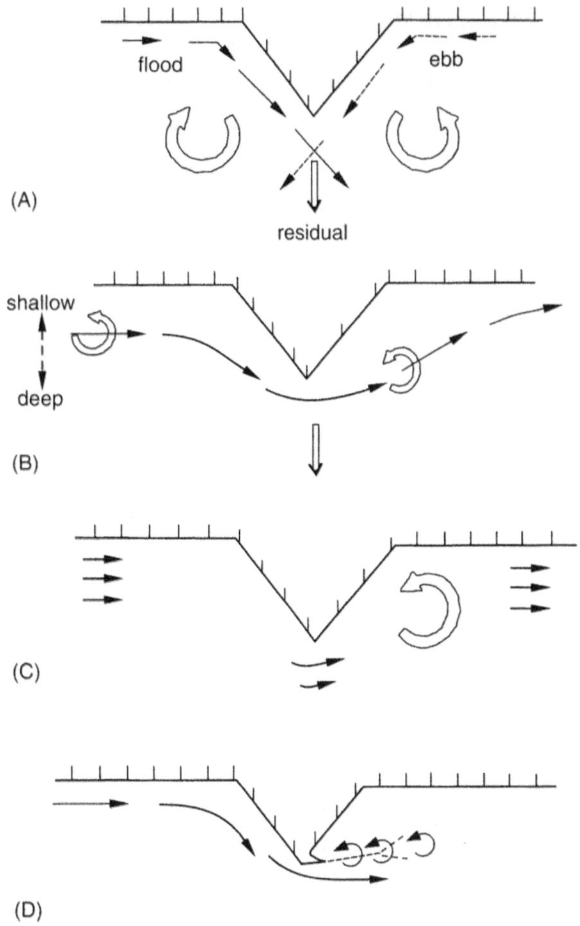

Figure 3 The possible mechanisms for the generation of headland eddies in a tidally cyclic flow (A), a steady flow where the depth increases offshore (B), and cases where a steady flow induces a simple attached eddy (C) or a train of separated eddies (D). (Reproduced from (Robinson 1975.)

Topographic Eddies Due To Bottom Topography in Stratified Flows

A dimensionless number which directly gives the ratio of current velocity U to the velocity of gravity waves in a current of depth D is known as the Froude number and is defined by

$$F_r = U/ND$$

The Froude number is the definitive number which divides a physical process where a disturbance may propagate upstream (called subcritical and denoted by $F_r < 1$), or a disturbance is swept downstream (called supercritical and denoted by $F_r > 1$). Flows over shallow sills, or coral reefs, in areas of strong tidal flow (e.g., north-western Australia) may sometimes be so rapid as to be supercritical. Hydraulic jumps, caused by a flow transition from rapid smooth flow upstream with $F_r > 1$ to slow turbulent flow downstream with $F_r < 1$, are also known to occur in strongly flowing rivers.

Flows which are density stratified are characterized by the flow speed U and the buoyancy frequency N, where N is defined by

$$N^2 = \frac{g d\rho}{\rho dz}$$

In flows of depth D, the dimensionless number which reflects the ratio of current speed to wave speed is

$$F_r = U/ND$$

Flows over and around obstacles depend not only on this internal Froude number, but also on the height H of the obstacle, its horizontal size and the steepness of the topography. Internal hydraulic jumps are also known to exist in the ocean where very strong tidal currents flow over steep topography, such as the Mediterranean outflow, or off the British Columbia coast (**Figure 4**). The subsequent turbulence is confined downstream causing a high level of mixing and turbidity, while upstream waters remain relatively placid and clear. Since such hydraulic jumps usually occur in irregularly shaped channels, they are often also the source of small topographic eddies.

Consideration of stratified flow around obstacles having an infinite value R_o (zero Coriolis parameter) provides many examples of the generation of topographic edies. These include hydraulic jumps, exchange flows, waves and recirculating flows over two and three-dimensional obstacles in finite depth and infinitely deep stratified flows. In water much deeper than the obstacle height H, the relevant height scale is H, and the relevant dimensionless parameter is NH/U, an inverse Froude number based on the obstacle height. An interesting case study is depicted in **Figure 5**, for stratified flow with parameter $NH/U = 5$. In this case the stratification is sufficiently strong to confine recirculation patterns to within about two obstacle heights of the seabed, with flow going both over and around the obstacle, and generating a pair of steady attached eddies. The dynamics are extremely complex, and consist of nodes where the flow separates, and stagnation points where the flow has zero current.

The flows described have zero background rotation (i.e., $f = 0$), and thus cannot describe a range of eddy-like flows, trapped above an obstacle in a current flow in a rotating reference frame and known as Taylor columns. For stratified flows over typical ocean seamounts in which the earth's rotation plays a

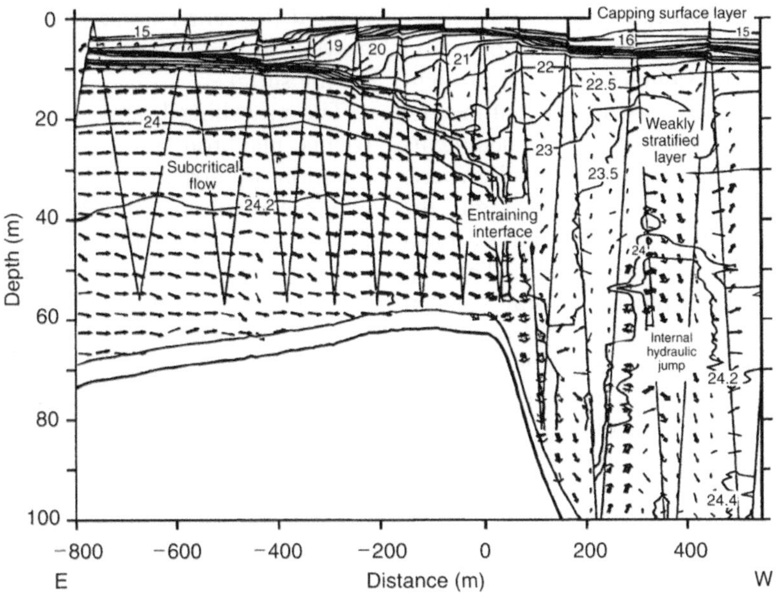

Figure 4 Schematic diagram of an internal hydraulic jump, in which the upstream flow (at left) is slow (with $F_r < 1/mn$) accelerates rapidly down a steep slope (with $F_r > 1/mn$), and flows into a turbulent hydraulic jump (with $F_r < 1/mn$) (Reproduced from Farmer and Armi, 1999.) The downstream turbulence cannot propagate up the steep slope as the stratification N is not sufficiently large for the internal wave speed ND to exceed U, and so the turbulent flow is confined downstream of the topographic slope.

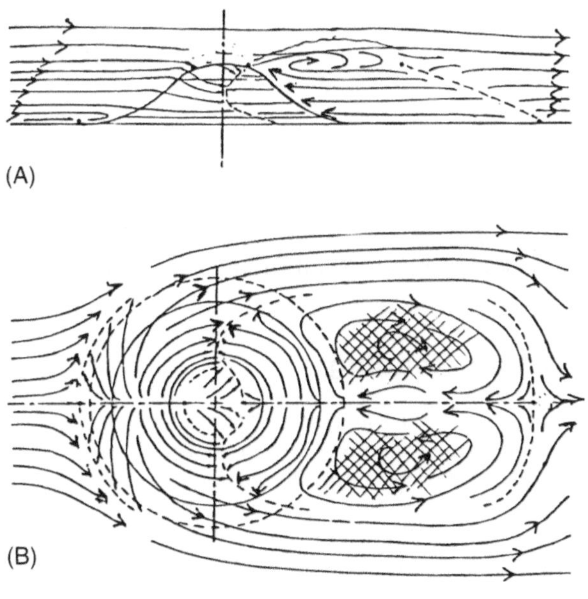

Figure 5 Topographic eddy in stratified flow over a three-dimensional obstacle with $NH/U = 5$ (reproduced from Baines, 1995), showing the side view through a cross-section along the line of symmetry (A), and the plan view showing the horizontal current components at a depth below the top of the obstacle (B). Also shown in (B) is the zone in which upwelling occurs at that same level (hatched).

role, the relevant scale of vertical disturbance above the seamount is fL/N, where L is the horizontal scale of the seamount. The ratio of the vertical scale of the obstacle H to this scale height is thus NH/fL, which in the form

$$B_u = (NH/fL)^2 = (R_o/F_r)^2$$

is known as the Burger number. B_u is also the ratio of Rossby number to Froude number squared. The Burger number is thus an indication of the balance of effects of stratification and earth's rotation, adjusted for the vertical aspect ratio of the seamount. High values of B_u tend to keep topographic disturbances more confined vertically, whereas low values permit taller Taylor columns, in which the effects of the obstacle extend higher in the water column. A full description of Taylor columns above seamounts in stratified flows is beyond the scope of this article.

Turbulence Due to Small-scale Bottom Topography and Reefs

For topography whose roughness scales are much smaller than the depth, and timescales are short, then the R_o number is high and Coriolis effects are negligible. Flow around such topography then has properties typical of those found in laboratory experiments, allowing for even smaller-scale turbulence to act as a stochastic noise at the inflow region. A number of topographic eddies may then be formed at different times and/or places, sometimes creating a wholly turbulent flow field over a limited region of the coastal ocean.

Such turbulence may be caused by strongly flowing tidal currents over a rough seabed or over and through coral reefs, for example, where the roughness scales may be as small as a few centimeters. These flows may scour the seabed, raising sediments off the seafloor and transporting them elsewhere. Such turbulence can also act to thoroughly mix the water column in areas where strong tidal currents flow over significant bottom topography. This occurs, for example, over submerged coral reefs at 80 m of water depth offshore from Hydrographers Passage in the Great Barrier Reef where phytoplankton multiply rapidly as a consequence of the combination of nutrient supply and light. Turbulence on these scales is still subject to the energy cascade phenomenon, whereby eddies continually break down to smaller and smaller eddies until fluid viscosity damps out the motions.

Biological Implications of Topographic Eddies

As depicted in **Figure 1**, there is clearly a relationship between the wake of Tobi Island and the fish concentrations, as described by the local fishermen. However, there are also some much more subtle responses. These are related to the vertical circulation which is necessarily part of a horizontally circulating eddy.

The physics is relatively straightforward. In a horizontal eddy, the eddy can only maintain its structure if the eddy center has low pressure, which exists by virtue of a reduced sea level in the eddy center. Throughout the main part of the water column (away from the seabed) the horizontal pressure gradient balances the centripetal force. However, in coastal waters the pressure gradient has an effect right down through the bottom boundary layer to the seabed. In this bottom boundary layer, the centripetal force is reduced because the velocity is reduced, and so the pressure gradient drives a flow toward the center along the seabed. This flow then upwells in the eddy center (**Figure 6**), and outflows on the surface.

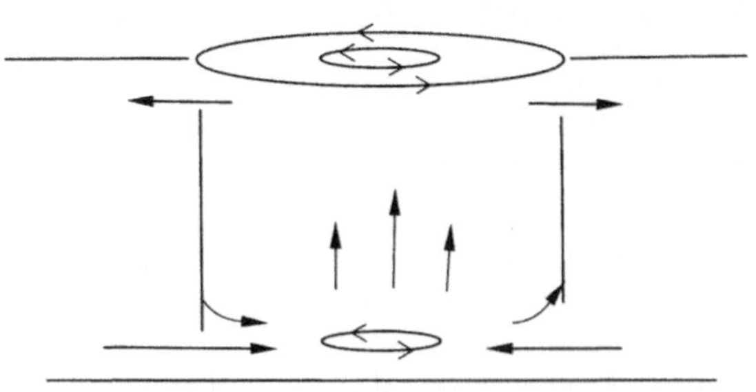

Figure 6 Topographic eddy in the coastal ocean showing inflow in the bottom boundary layer, upwelling in the eddy center, and outflow at the surface.

The upwelling brings with it fine sediments, nutrients, plankton, and perhaps larval fish. The combination of nutrients, plankton and greater light can enhance plankton growth. Thus eddies in shallow waters are intrinsically places of enhanced plankton growth, and perhaps enhanced concentrations of other elements of the marine food chain.

In deeper waters, where the ocean is stratified, the lower pressure at the eddy center also results in a general uplift of deeper nutrient-rich waters, creating the same effect as in shallower waters. Observations in the wake of Cato Reef off eastern Australia showed higher concentrations of nutrients, phytoplankton, and larval fish of better condition, and the principal effects of this increased productivity were attributed to the uplift in the wake.

Referring back finally to **Figure 1**, the diagram can now be interpreted. The arms along which the flying fish and tuna concentrate are zones in which upwelled nutrient-rich waters, now outflowing from the eddy centers, meet the nutrient poor free-stream currents. This is likely to be a zone where plankton in the free stream now have an opportunity to grow, and the larger fish are perhaps benefiting from the enhanced primary productivity.

See also

Island Wakes. Mediterranean Sea Circulation. Mesoscale Eddies. Wind Driven Circulation.

Further Reading

Baines PG (1995) *Topographic Effects in Stratified Flows*. Cambridge: Cambridge University Press.

Batchelor GK (1967) *An Introduction to Fluid Dynamics*. Cambridge: Cambridge University Press.

Boyer DL and Davies PA (2000) Laboratory studies in rotating and stratified flows. *Annual Reviews of Fluid Mechanics* 32: 165–202.

Coutis PF and Middleton JH (1999) Flow topography interaction in the vicinity of an isolated deep ocean island. *Deep-Sea Research* 46: 1633–1652.

Denniss T and Middleton JH (1994) Effects of viscosity and bottom friction on recirculating flows. *Journal of Geophysical Research* 99: 10183–10192.

Denniss T, Middleton JH, and Manasseh R (1995) Recirculation in the lee of complicated headlands; a case study of Bass Point. *Journal of Geophysical Research* 100: 16087–16101.

Farmer D and Armi L (1999) Stratified flow over topography; the role of small-scale entrainment and

mixing in flow establishment. *Proceedings of the Royal Society of London* A455: 3221–3258.
Farrell BF and Ioannou PJ (1996) Generalized stability theory. Part I: Autonomous operators. *Journal of the Atmospheric Sciences* 53: 2025–2040.
Huppert HE (1975) Some remarks on the initiation of internal Taylor columns. *Journal of Fluid Mechanics* 67: 397–412.
Johannes RE (1981) *Words of the Lagoon: Fishing and Marine Lore of the Palau District of Micronesia*. San Diego: University of California Press.
Kundu PK (1990) *Fluid Mechanics*. London: Academic Press.
Middleton JH, Griffin DA, and Moore AM (1993) Ocean circulation and turbulence in the coastal zone. *Continental Shelf Research*13143–13168.
Pattiaratchi C, James A, and Collins M (1986) Island wakes; a comparison between remotely sensed data and laboratory experiments. *Journal of Geophysical Research* 92: 783–794.
Rissik D, Taggart C, and Suthers IM (1997) Enhanced particle abundance in the lee of an isolated reef in the south Coral Sea; the role of flow disturbance. *Journal of Plankton Research* 19: 1347–1368.
Robinson IS (1975) Tidally induced residual flows. In: Johns B (ed.) *Physical Oceanography of Coastal and Shelf Seas*. Amsterdam: Elsevier.
Tennekes H and Lumley JL (1972) *A First Course in Turbulence*. Cambridge, MA: MIT Press.
Tomczak M (1988) Island wakes in deep and shallow water. *Journal of Geophysical Research* 93: 5153–5154.

ISLAND WAKES

E. D. Barton, University of Wales, Bangor, Menai Bridge, Anglesey, UK

Introduction

The 'island mass effect' has been documented for about half a century. This refers to a biological enrichment around oceanic islands in comparison to surrounding waters. Despite a relatively large body of evidence to support the existence of such an effect in the vicinity of islands, there have been few studies to investigate the underlying physical causes of this phenomenon. From a physical point of view the presence of an island in the background flow will disturb the flow regime to produce perturbations that ultimately must have biological consequences.

Two types of island disturbance have been investigated. The first takes place in shallow, stratified shelf seas with significant tidal regimes but no appreciable mean flow, where as the tide moves water back and forth the island acts as a stirring rod to enhance vertical mixing locally and break down the pycnocline. The second occurs in both shallow and deep water, where flow past an island generates eddies downstream and a wake of disturbed flow extends several island diameters away. This arises in the case of larger islands when a clear ambient flow dominates over tidal variability and also around islands small enough that the tidal stream itself can generate a similar effect. The scale of the eddies is typically close to the island diameter and their time scale will be several days for larger islands but only hours in the case of tidal flows. The nature of the wake and eddies may differ between the oceanic case where conditions can be considered quasigeostrophic and the shallow case where they are frictionally dominated by bottom stress.

Theory

Nonrotating Case

The simplest case is where flow past isolated oceanic islands is considered to be analogous to that of channel flow past a circular cylinder. The work of Batchelor presented the case of nonrotating flow in a homogeneous fluid. The form of the downstream disturbance or wake is related to the value of the Reynolds number

$$R_e = \frac{Ud}{\nu} \qquad [1]$$

where U is the free velocity upstream, d is the diameter of the cylinder and ν is the molecular viscosity. Although this formula appears simple, there are certain practical difficulties in applying it to even a nonrotating ocean of homogeneous character. Few islands are isolated or cylindrical, the upstream velocity is generally not well known and finally the molecular viscosity must be replaced by the horizontal eddy viscosity in the ocean, which is in general poorly known. Studies of Aldabra, an Indian Ocean atoll, indicate that the same current speed impinging on the island from different directions produces a different wake because of the asymmetrical form of the island. It is frequently the case that the upstream current in the ocean varies on a range of time scales, or may be subject to horizontal shear, both of which complicate the choice of a suitable value for the free stream velocity. Values of horizontal eddy viscosity coefficients in the ocean based on many experimental determinations increase with the length scale of interest, l, in a nonlinear fashion $K_h(m^2s^{-1}) = 2.2\times 10^{-4}l^{1.13}$. Typical values appropriate to oceanic islands vary between 10^2 and 10^5 m^2 s^{-1}.

The nature of the wake downstream of the obstacle changes as the Reynolds number increases (**Figure 1**). For low Reynolds numbers ($R_e < 1$) the flow pattern downstream is the same as that upstream and there is no perceptible wake. The flow remains attached to the sides of the cylinder and is laminar throughout the flow field. At Reynolds numbers between 1 and 40, the wake remains basically laminar away from the cylinder and two eddies are formed immediately behind the obstacle where they remain attached. At higher Reynolds number the wake becomes increasingly unstable and counter-rotating eddies form a vortex street. The eddies expand as they move away from the obstacle and gradually decay. For Reynolds numbers $R_e > 80$ eddies formed behind the island no longer remain trapped but are shed alternately into the vortex street. The frequency of eddy shedding n is related to another nondimensional number, the Strouhal number

$$S_t = \frac{nd}{U} \qquad [2]$$

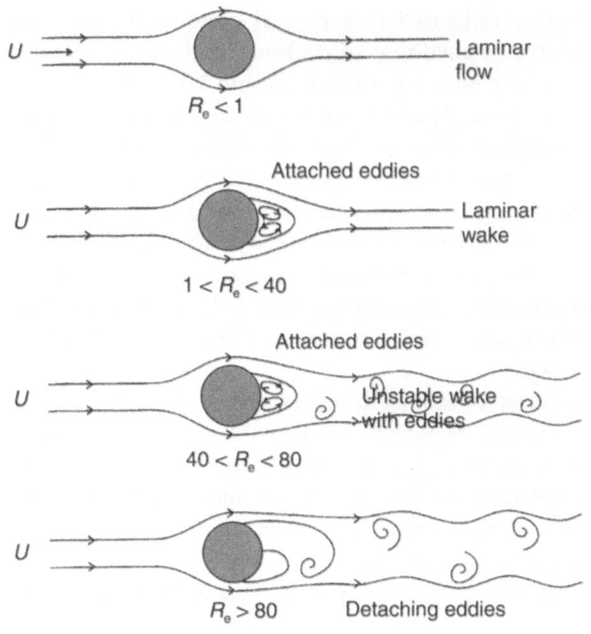

Figure 1 Reynold number regimes for flow past a cylinder.

This number approaches an asymptotic value of 0.21 at higher Reynolds numbers.

Effect of Rotation

The ocean of course is on the rotating Earth and so laboratory experiments have been carried out in rotating tanks to investigate the effect of this rotation. Two dimensionless quantities of importance here are the Rossby number

$$R_0 = \frac{U}{\Omega d} \qquad [3]$$

which represents the importance of rotation at rate Ω and the Ekman number

$$E_k = \frac{2\nu}{\Omega d^2} \qquad [4]$$

which determines the width of the wake. The ratio of Rossby number to Ekman number is proportional to the Reynolds number, generalizing this concept to the case of rotating flow. The Earth's rotation enhances the shedding of eddies in the same sense (cyclonic) so that in the Northern Hemisphere predominantly anticlockwise eddies are to be expected whereas in the Southern Hemisphere clockwise should be more common.

Because of the spherical shape of the Earth's surface the rate of rotation about the local vertical varies with latitude $f = 2\omega \sin\phi$, where ω is the rotation rate of the earth and ϕ the latitude. This variation of the Coriolis parameter f can be represented in terms of the 'beta plane' where $\beta = df/dy$ and y is distance north of the equator. If the dimensionless parameter

$$\beta' = \frac{\beta d^2}{4U} \qquad [5]$$

is small, the beta effect may be ignored and the rotation rate taken to be constant on the scale of the island in question. If on the other hand it is of order unity, differences occur from the case of uniform rotation. In particular flow separation is enhanced for flow towards the west and inhibited for flow towards the east.

Effects of Bottom Friction

Studies of flow patterns around small islands only a few kilometers in diameter in shallow shelf seas indicate that the Reynolds number as defined in eqn[1] overestimates the value at transition between the different cases of wake formation. It has been found that the island wake parameter

$$P = \frac{Uh^2}{K_z d} \qquad [6]$$

where U is the stream velocity, h is the water depth, K_z is the vertical eddy diffusivity and d is the dimension of the island, provided better agreement than the Reynolds number between observed and predicted wake parameters. However, P is actually a correct formulation for the Reynolds number when the effect of lateral and bottom frictional boundary layers is taken into account.

Observations

Until recently observations of island flow effects had been limited mainly to remote sensing reports of eddy production. Attempts to observe vortex production and development *in situ* behind oceanic islands have been largely unsuccessful because the background flow has been too weak or variable or the methods of observation insufficient to determine the flow regime adequately. However, a classic early report in 1972, based upon sparse observations of the drift of fishing gear and rudimentary surface current measurement, showed drift patterns downstream of Johnston Atoll in the Pacific Ocean in good agreement with a vortex street situation.

In the case of Aldabra Atoll, situated in the South Equatorial Current of the Indian Ocean, surveys made with acoustic Doppler current profiler indicated a single cyclonic (anticlockwise) eddy trapped behind the island in a low Reynolds number regime on two occasions. There was no evidence of continuous eddy or wake production downstream. The flow was variable during the experiment. The first trapped eddy was observed during westward background flow around 20 cm s^{-1}. Two subsequent rapid surveys, about ten days apart, of the current field showed predominantly northward and westward flows, respectively. In the first case the free stream flow was about 20 cm s^{-1} and impinged on the island's widest cross-section. No eddy was observed, only an asymmetrical deviation of currents behind the island. The later case found slightly stronger (30 cm s^{-1}) free stream velocity from the east impinging the narrow aspect of the island. This time a second weak eddy of diameter similar to the island width was indicated.

Another island showing highly variably flow regimes is Barbados. In Spring 1991, the flow seemed topographically steered around the Barbados ridge and there was no clear evidence of eddy production. The following spring, anticyclonic and cyclonic eddies of similar size to the island were found on either flank (**Figure 2**). Computer simulations of the flow regime indicated that typical conditions were conducive to shedding of cyclones and anticyclones alternately with a period of around 10 days. It was not possible to demonstrate the degree to which the observations were attributable to this type of vortex generation, however, and it was concluded that much more extensive observations would be needed to do so. One interesting indication of the simulation was that there was a continuous region of downstream reverse flow towards the island that could provide a return path for fish larvae swept away from the island.

Considerable evidence of recurrent eddy shedding has been reported recently in the Canary Island archipelago. There, the Canary Current flows southwestward at an average speed of 5 cm s^{-1}. Eddies of both signs have been reported (**Figure 3**) as being frequently spun off from the island of Gran Canaria. Cyclonic eddies of the same diameter as the island (50 km) and rotation period around 3 days have been observed to develop on the southwestern flank of the island and to move southwest at speeds of 5–15 cm s^{-1}. Almost as frequently, anticyclonic eddies have been observed to develop on the south east of the island. They have diameters up to twice

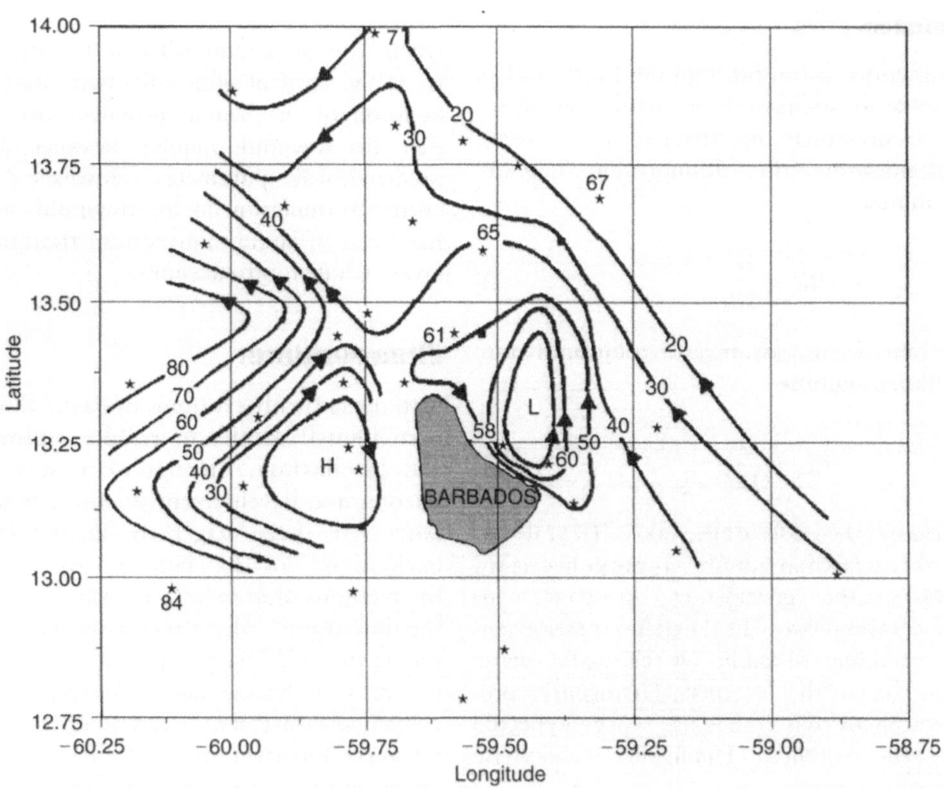

Figure 2 Sea surface topography in centimeters relative to 250 dbar 25 April–2 May 1991. Arrows denote the direction of surface flow. Note the overall northwestward flow and anticyclonic (H) and cyclonic (L) eddies either side of the island. (Adapted from Bowman *et al*., 1996.)

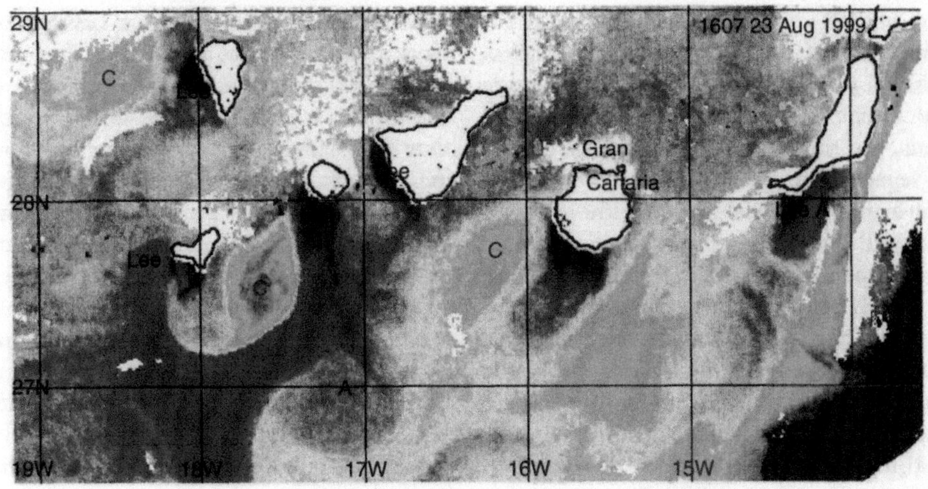

Figure 3 Sea surface temperature image showing multiple eddies shed from the Canary Islands during August 1999. Cyclones and anticyclones are labeled C and A, respectively.

the size of the cyclonic eddies, similar rotation rates and appear to persist for many months in the region.

One such eddy, seeded with drifters in June 1998, was observed to persist for at least seven months (**Figure 4**). It initially drifted slowly southwestward but later returned northwestward towards the outer islands of the archipelago. Other large anticyclones have been observed downstream of the island pair, Tenerife and La Gomera, trapped close to the islands for at least several weeks before moving away from the island. Smaller cyclones and anticyclones are frequently seen in sea surface temperature images being spun off from the flanks of the other smaller islands.

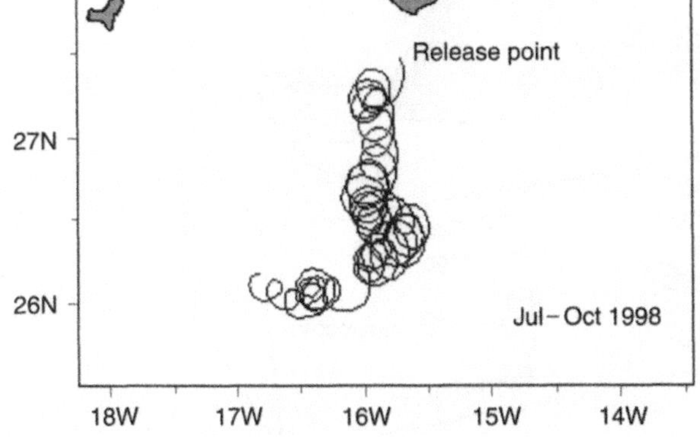

Figure 4 Path of a drifter released into an anticyclonic eddy shed from Gran Canaria in June 1998. The eddy persisted for 7 months though only the first three months are shown here (Courtesy of Dr Pablo Sangra, University of Las Palmas de Gran Canaria.)

The generation of these eddies may not be entirely a result of the oceanic flow past the islands. The Canaries are high volcanic islands situated in a regime of strong southwestward trade winds. The high island peaks block the flow of the trade winds to form extended lee regions downwind. These are bounded by localized horizontal shear in the wind field and so are locations of strong Ekman divergence and convergence. Ekman transport takes place in the near surface layer and is to the right of the wind in the Northern Hemisphere. At the western boundary of the lee, upwelling of deeper waters must compensate the divergence, while at the eastern boundary, sinking must occur. The upwelling and downwelling elevates or depresses, respectively, the pycnocline from its unperturbed depth. Because the downwind scale of the lee is limited, the elevation or depression of the pycnocline tends to form an eddy of the same sign as expected from the current past the island. The vertical motions expected from the horizontal wind shear on the lee boundaries are of the order of tens of meters per day, and so potentially could contribute significantly to the observed eddy production.

Despite the frequent reports of eddies spun off from the islands, it has proven difficult to obtain time series observations of eddy generation which would allow determination of basic eddy properties such as average shedding frequency, size, propagation speed or whether eddies are shed alternately from opposite island flanks. Remote sensing even in the subtropics is often blocked by cloud cover and time series *in situ* sampling is difficult to maintain.

A situation similar to the Canaries has been observed in the case of the Hawaiian archipelago, situated in the North Equatorial Current and trade winds of the Pacific. Downstream of these mountainous islands, the trade winds with speeds of 10–20 m s^{-1} are separated from the calmer lee by strong boundaries of high wind shear. Locally, the depth of the surface mixed layer depends on wind speed: in the channels between islands, deep mixed layers are observed; in the lee, stirring by the wind is too weak to distribute solar heating down below the surface layer and intense surface warming results during the day. Sharp surface temperature fronts up to 4°C, are often associated with these wind shear lines, as is also observed in the Canaries. The lee of islands in both archipelagos is often visible in sea surface temperature images as a significantly warmer triangular area extending downwind.

Ekman transports associated with the wind pattern produce pycnocline perturbations as in the Canaries, resulting in intense anticlockwise eddies under the northern shear lines, and less intense clockwise eddies under southern shear lines. The depth of the mixed layer in the lee of Hawaii can vary from less than 20 m in the counterclockwise eddy to more than 120 m in the clockwise eddy. **Figure** 5 shows a diagram of the Hawaiian island situation which applies equally well to the Canaries. Though the wind has long been viewed as an important generating mechanism for the Hawaiian eddies, it is still unclear how the variability of the wind field affects oceanic eddy generation. The wind itself has often been observed in satellite images of low level clouds to form wakes of counter-rotating atmospheric eddies behind Hawaii and the Canaries. The eddy-shedding period in this case is much shorter, about 10 h, than for oceanic eddies, which have a periodicity of many days. Presumably it is the wind field averaged over the timescale of the ocean eddies that is important. However, it is quite possible that atmospheric eddy shedding is intermittent; during periods when a trapped eddy regime dominates the wind shear lines are relatively stationary and able to feed energy into oceanic eddy production. Further observations are required to unravel the mechanisms at work in these island situations.

The North Equatorial Current impinging on the Hawaiian islands, of course, will also tend to produce eddies. The cumulative effect of the many eddies that are spun off is the formation of a mean large scale re-circulation behind the Hawaiian (cf. Barbados) chain that has been named the Hawaii Lee Counter Current. The longevity of individual eddies is further illustrated by one of their surface layer drifters which remained trapped in an anticyclonic vortex formed near the southern flank of the main island, drifting westward over 2000 km at 11 cm s^{-1}. This and other drifter tracks showed the remarkable phenomenon of vortex doubling, the process of merging of two vortices of the same sign. When two identical vortices merge, the radius of the merged eddy is $\sqrt{2}$ of the original and its period of rotation is doubled. The drifter appeared to show three such vortex-doubling episodes during its trajectory.

In the case of larger shallow sea islands in a tidal regime, the flow reverses before a wake can be properly set up. However, the relative motion between the island and surrounding body of water allows the island to act as a 'stirring rod' because of the increased flow speeds on the island flanks which produce vertical mixing within a tidal mixing front some distance off the shore. In a stratified region this

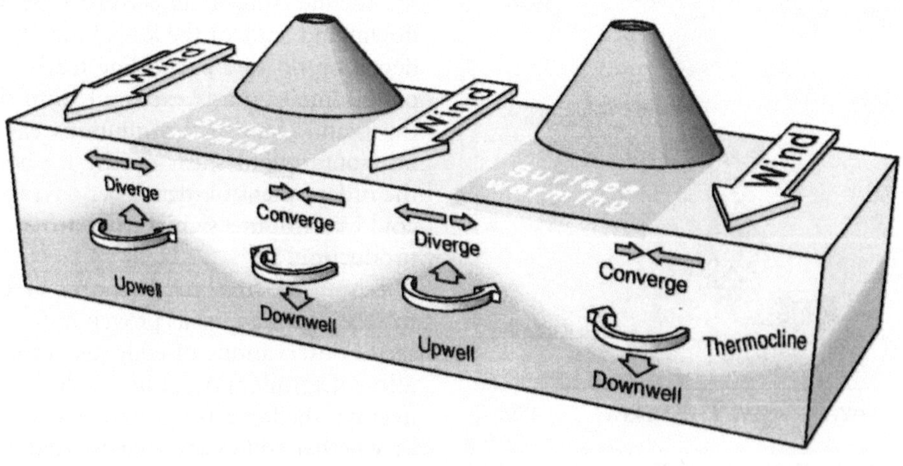

Figure 5 Hawaii schematic showing the mechanism of eddy generation by wind shear. Ekman surface layer transports lead to depression and elevation of the pycnocline in regions of convergence and divergence, respectively. This in turn generates oceanic anticyclones and cyclones. (Courtesy of Professor P. Flament, University of Hawaii.)

introduces nutrients from the lower layers into the surface layers so enhancing productivity around the island. In the case of the Scilly Islands mixing between low density surface layer and high density bottom layer waters produced intermediate density enriched water spreading out between the layers causing enhancement of the mid-depth chlorophyll maximum in the area around. Similar results were found around St Kilda off western Scotland.

Conclusions

Island wakes produced by eddy shedding have been observed in both deep ocean and shallow shelf sea situations. In some cases of shelf sea islands there is no wake as such, because of weak mean flows, but vigorous tidal currents can produce regions of well-mixed water in the surrounding area. In the oceanic case, though there are many examples of eddy and wake observations, there has yet to be a definitive study demonstrating the phenomenon over a range of flow conditions.

The effect of islands on the flow regime does have biological consequences, in that enhanced mixing related to shallow sea islands can increase primary production. It has often been argued that island shed eddies may provide a mechanism for retaining fish larvae in the vicinity of their spawning areas. Although the evidence for this is not yet convincing the existence of mean return circulation has been demonstrated in both model and drifter studies of oceanic islands.

It was suggested that energy dissipation caused by island flow disturbance could account for 10% of the wind kinetic energy input to the Pacific. However, it is also considered that general turbulent processes known within the deep ocean may be too weak by more than an order of magnitude to explain global redistribution of energy input. In this case vertical mixing must be greater at the ocean boundaries with land than previously considered and the role of islands and island chains may be greater than is presently perceived. Simulations of the Barbados wake indicated that the flow disturbance was extensive, reaching at least eight island diameters downstream. Both the Canaries and Hawaii archipelagoes clearly control physical conditions for large distances downstream, modifying water masses through mixing, enhancing productivity and shedding long-lived eddies.

See also

Canary and Portugal Currents. Ekman Transport and Pumping. Mesoscale Eddies. Pacific Ocean Equatorial Currents. Wind Driven Circulation.

Further Reading

Arístegui J, Tett P, Hernández-Guerra A, *et al.* (1997) The influence of island-generated eddies on chlorophyll distribution: a study of mesoscale variation around Gran Canaria. *Deep-Sea Research* 44(1): 71–96

Barkley RA (1972) Johnston Atoll's wake. *Journal of Marine Research* 30: 201–216.

Batchelor GK (1967) *An Introduction to Fluid Dynamics.* Cambridge: Cambridge University Press.

Bowman MJ, Dietrich DE, and Lin CA (1996) Observations and modelling of mesoscale ocean circulation near a small island. In: Maul G (ed.) *Small IslandsM: arine Science and Sustainable Development. Coastal and Estuarine Studies*, vol. 51, pp. 18–35. Washington: American Geophysical Union.

Chopra KP (1973) Atmospheric and oceanic flow problems introduced by islands. *Advances in Geophysics* 16: 297–421.

Flament P, Kennan SC, Lumpkin C, and Stroup ED (1998) The Atlas of Hawai'i. In: Juvik S and Juvik J (eds.) *The Ocean*, pp. 82–86. Honolulu: University of Hawai'i Press.

Simpson JH and Tett P (1986) Island stirring effects on phytoplankton growth. In: Bowman MJ, Yentsch M, and Peterson WT (eds.) *Lecture Notes on Coastal and Estuarine Studies*, vol. 17, *Tidal Mixing and Plankton Dynamics*, pp. 41–76. Berlin, Heidelberg: Springer-Verlag.

Wolanski E, Imberger J, and Heron ML (1984) Island wakes in shallow coastal waters. *Journal of Geophysical Research* 89C: 10533–10569.

ROSSBY WAVES

P. D. Killworth, Southampton Oceanography Centre, Southampton, UK

Introduction: What Are Rossby Waves?

Among the many wave motions that occur in the ocean, Rossby (or planetary) waves play one of the most important roles. They are largely responsible for determining the ocean's response to atmospheric and other climate changes; their energy dominates the ocean's energy spectrum at long timescales; they are responsible for setting up and maintaining the intense oceanic western boundary currents, and can be generated by those currents; they affect ocean color and biological interactions near the surface; and they moderate the ocean's behavior to decadal features such as El Niño and the North Atlantic Oscillation. The waves have a strong westward component in their phase speed (though short waves can propagate their energy eastward). There are two types of Rossby wave, as with many oceanic waves. The first is barotropic (independent of depth), which propagates rapidly (typically at 50 m s^{-1}), has variable space and time scales, and can cross an ocean basin in a few days. The second type of Rossby wave is baroclinic (varying with depth) and propagates fairly slowly (a few centimeters per second), has a long space scale (hundreds of kilometers), has a long period (of order a year), and takes a decade or more to cross an ocean basin westward. There is a countable infinity of baroclinic modes, but just one barotropic mode. Similar waves exist in the atmosphere, and are known to be responsible for most of the low-frequency variability observed there.

Figure 1 shows contours of the speed of the fastest propagating depth-varying Rossby wave within the ocean. This speed, as we shall see, increases dramatically near the equator, and decreases to become very slow at high latitudes.

All waves exist and propagate because of waveguides. Rossby waves owe their existence to an east–

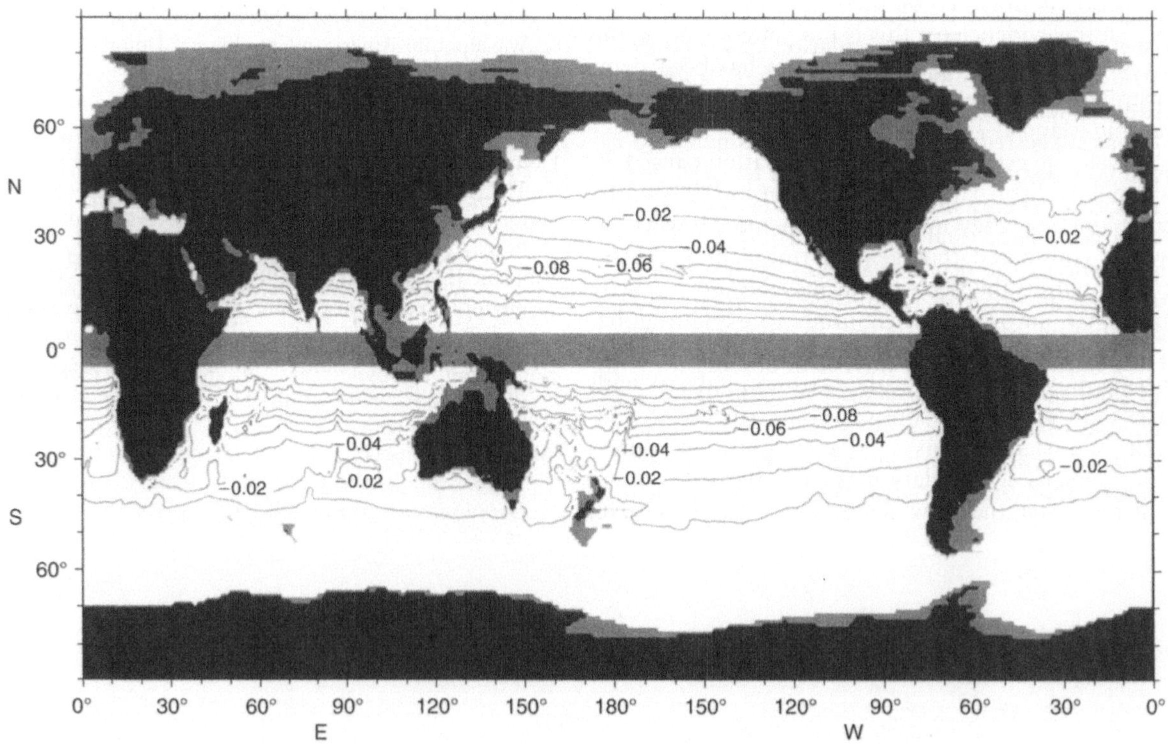

Figure 1 Contours of the speed of the fastest long baroclinic Rossby wave (m s^{-1}). Contour intervals are nonuniform: 0.3, 0.2, 0.15, 0.1, 0.08, 0.06, 0.04, 0.02, 0.01 m s^{-1}. Negative signs mean speed is westward. Values are masked within 5° of the equator where other theory holds, and for depths of less than 1000 m.

west waveguide that is present because the Coriolis parameter increases northward. In the simplest case of depth-independent two-dimensional flow (**Figure 2**), wave propagation occurs because of conservation of absolute vorticity of a fluid particle. (Absolute vorticity is the sum of the rotational, or relative, vorticity due to water motions plus the intrinsic, or planetary, vorticity due to the background earth rotation.) Imagine a line of particles oriented east–west (**Figure 2A**). These particles have no relative vorticity and no current shear. Now suppose the line is perturbed north–south in some manner (**Figure 2B**). The particles moved northward increase their intrinsic vorticity. To conserve their absolute vorticity, they must acquire negative rotational vorticity. The particles moved southward similarly acquire positive rotational vorticity. The rotational motions induced by these vorticity changes are shown in **Figure 2C**, and their effect on the particles in **Figure 2D**. The net effect is to move the original disturbed pattern of particle positions (B) westward.

In the more usual case of depth-varying flow, consider the ocean stratification as comprising a single layer of uniform depth H. When particles are displaced, they conserve their (potential) vorticity, which is now given for long length scales by f/H, where f is the Coriolis vector (the vorticity argument above continues to hold for short length scales). It is the variation of this quantity north–south that again induces a westward wave propagation. Particles displaced northward (**Figure 3A**) increase f and so increase H to compensate; particles moving southward decrease H. Thus the particle motions in **Figure 3C** give depth – and hence pressure – changes that lead to geostrophic flows (**Figure 3D**) that have similar effects on the displacements, moving the pattern westward. These schematic arguments also hold, with some modifications, near the equator.

For most practical purposes of measurement, Rossby waves propagate as a series of low and high pressures, which are constant normal to the direction of wave phase propagation. The resulting geostrophic flow is alternately in one direction and then in the opposite, perpendicular to the propagation direction. **Figure 4** shows an example of this.

The generation mechanisms for Rossby waves are unclear, though some form of surface forcing must be involved to induce changes in the upper ocean structure which can then propagate as waves. Thus, direct forcing by wind stress and to a lesser extent, by buoyancy forcing will both generate Rossby waves. Free waves must still be forced somewhere, and

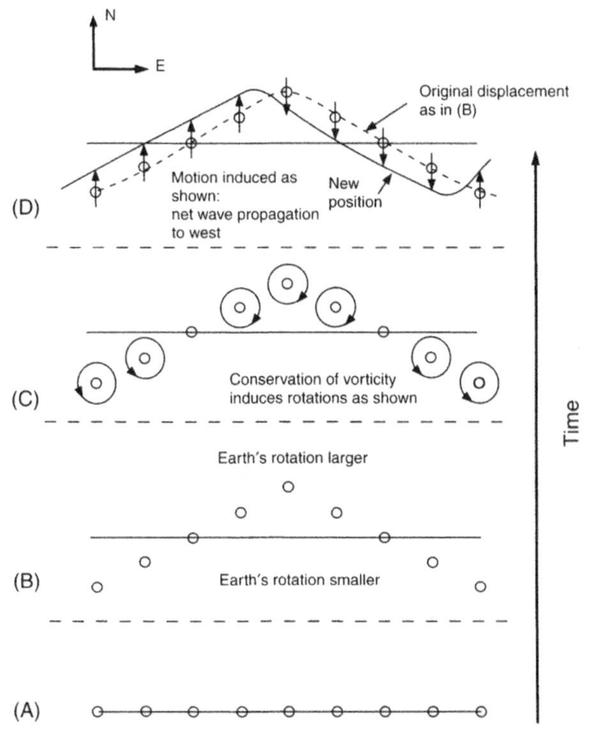

Figure 2 Schematic of depth-independent Rossby wave transmission; time reads up the diagram. An east–west line of particles (A) is displaced (B) and gathers vorticity owing to changes in the background Earth's rotation (C). This vorticity induces flow changes (D) that act to move the displacement pattern westward.

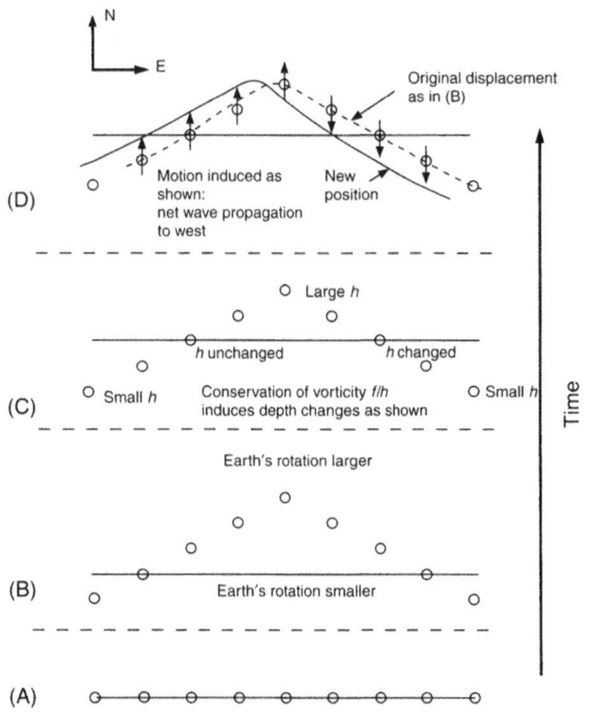

Figure 3 As **Figure 2**, but for a fluid layer. As the layer is displaced (B), it changes its depth (C) to conserve potential vorticity. These depth changes induce geostrophic flows (D) that act to move the displacement pattern westward.

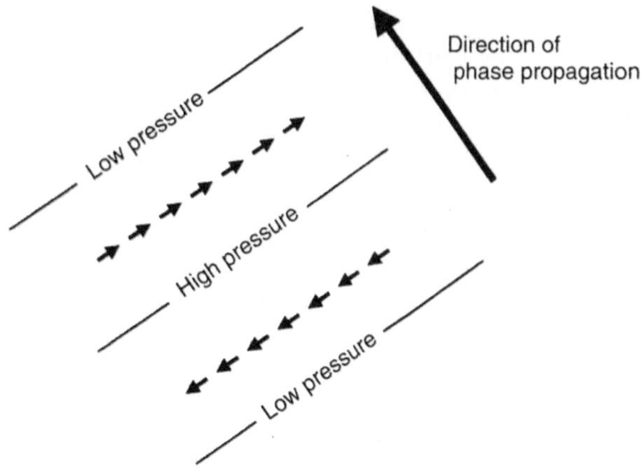

Figure 4 Schematic of long Rossby wave motions, which lie predominantly normal to the direction of phase propagation.

possible candidates include upwelling and downwelling on the eastern boundary induced by alongshore wind stress and topographic wave shedding over ocean ridges.

Observations of Rossby Waves

Rossby waves have been well observed in the atmosphere for decades. The large scale of oceanic Rossby waves necessitates an array of observations spanning a noticeable fraction of an ocean basin to distinguish phase variations, and the data obtained from cruises and ships of opportunity have been inadequate for the task, despite some valiant efforts. Rossby waves were, therefore, remarkably difficult to observe in the ocean until recently. (Observing barotropic Rossby waves is likely to continue to be difficult: their high speed makes the design of an observation network almost impossible.) Thus the theory of Rossby waves, discussed below, considerably predates their observation.

The launch of altimetric satellites changed all this dramatically. The barotropic Rossby waves remained essentially unobservable by satellites (the phases move so rapidly that they become indistinguishable between satellite passes). However, altimeter coverage proved ideal for the detection of surface signatures of baroclinic waves, which possessed surface height variations of a few centimeters and so were, at least for the TOPEX/Poseidon instrument, observable to the accuracy of the altimeter.

Satellite altimeters, to date, can only provide relative measures of sea surface height; in other words, they can report the variation of height accurately, but not the absolute value. For the purposes of observing wave propagation, this limitation presents no difficulties. Nonetheless, the ocean surface variation is made up of a superimposition of many different waves, direct responses to local forcing, and so on, so that the detection of Rossby waves still required massaging of the data.

The simplest approach used is to construct a time–space, or Hovmöller, diagram (**Figure 5**). A specific geographical line along which wave propagation is to be studied is chosen, typically a line of constant latitude. At successive times, determined by the repeat pattern of the satellite, the surface height – usually measured relative to some long-term average at each location so as to remove as much bias as possible – is plotted as a function of distance along the chosen line. These plots are then stacked perpendicular to each other, at separations proportional to the time interval. Wave propagation will then appear as contour lines across the diagram; the slope of these lines is a direct measure of the speed of the wave. Waves propagating at constant speed show up as lines of constant slope; those whose speed varies along the line will show a slope variation. In some cases more than one wave can be inferred, suggesting that various vertical modes are present simultaneously. There are hints that waves are sometimes generated near midocean ridges, though the reasons are as yet unclear. Certainly wavelike features can be identified in many parts of the ocean by this approach.

The Hovmöller diagram approach as in **Figure 5** is biased toward the direction of the chosen line (in this case toward east–west propagation). Here waves can be seen propagating 30° in longitude in about 60 cycles, with a cycle time of 10 days, this gives a speed of 0.06 m s^{-1}. Lines can be, and have been, oriented at other angles, but obviously a more systematic approach is necessary to locate the preferred orientation of the waves. Various methods have been used (e.g., Fourier transforms), with a popular approach being the Radon transform. This can be thought of as a simultaneous examination of Hovmöller diagrams at all possible angles, seeking for the orientation that gives the most energetic signal. So far, the preferred orientation for waves has been within a few degrees of westward, despite the fact that north–south propagation remains perfectly possible in theory. The reason remains unknown.

Detection is made more uncertain by the difficulty of disentangling other possible mechanisms that might have produced the given pattern. For example, a westward propagating eddy would appear as a wave on a Hovmöller diagram, while a propagating disturbance on a western boundary layer could appear as an eastward-oriented wave.

Recent increases in accuracy of other satellite instrumentation has meant that these too can be used

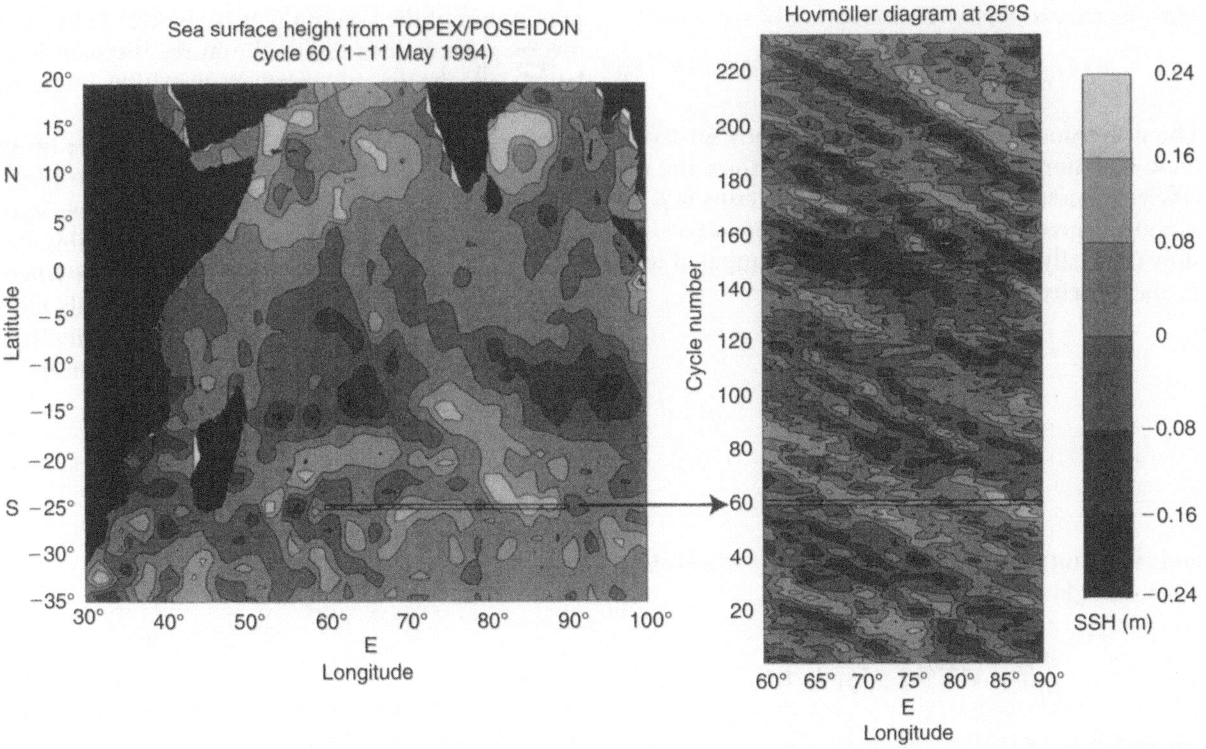

Figure 5 Hovmöller (or time–longitude) diagram, and its method of generation, for the South Indian Ocean. Snapshots of sea surface height (SSH) are taken along a given latitude at each satellite cycle (left-hand side of diagram) and assembled in sequence with time running upward (right-hand side of diagram). The signals of westward propagation then show clearly as bands tilted from upper left to lower right. One cycle is about 10 days. Note the amplitudes, typically about 10 cm and speed of waves, about 0.06 m s^{-1}.

to detect Rossby waves (and, more generally, large-scale anomaly propagation). Sea surface temperature measurements are a case in point. Their signal is dominated by the seasonal cycle. This can be removed in various ways (e.g., by removing the cycle explicitly at each location, or by taking the east–west derivative of the data at each point). The resulting signal also shows westward-propagating modes at similar, but not identical, speeds to the altimeter in most cases. This has been interpreted as the simultaneous presence of wave modes with different vertical structures (and so with different speeds). In the altimeter signal, one wave appears dominant; in the surface temperature signal, another mode is dominant. This is confirmed by theory, which shows that the relation between surface height and temperature variability depends strongly on the vertical structure of the wave, so that each wave could appear dominant when observed by one of the instruments.

Ocean color passive microwave instruments are now also used to elucidate large-scale wave propagation.

Theory of Rossby Waves

Theoretical predictions of Rossby waves have existed since the 1930s, and have long been known to give good predictions for atmospheric motions. The theory involves taking the three-dimensional problem and splitting it into two subproblems. The first involves the horizontal and time (as in the schematics above) and the second involves only the vertical structure. The latter is known as 'normal mode theory.' The idea is to find vertical structures that are maintained as the wave propagates, leaving the actual propagation behavior to be described by the pseudo-horizontal problem.

Horizontal Variation

It is logical to begin with the horizontal problem, for small motions in an ocean of uniform depth H, with a perturbation h. (An effective depth H will be determined later.) The momentum equations in the east (x) and north (y) directions for the velocity components (u, v, respectively) are

$$u_t - fv = -gh_x \quad [1]$$

$$v_t + fu = -gh_y \quad [2]$$

since the dynamic pressure (p/ρ_0, where p is pressure and ρ_0 the mean density of sea water) is given by gh.

Here h_x means $\partial h/\partial x$. Conservation of mass gives

$$h_t + H(u_x + v_y) = 0 \quad [3]$$

The dispersion relation for wave motions satisfying these equations involves a little work, since the coefficients possess y-variation due to the terms in f. To a good degree of approximation, if the waves are slow (formally, $\partial/\partial t$ is assumed small compared with f), the velocity components are

$$u \approx \frac{gh_y}{f} - \frac{gh_{xt}}{f^2} \quad [4]$$

$$v \approx \frac{gh_x}{f} - \frac{gh_{yt}}{f^2} \quad [5]$$

and substitution of these into mass conservation gives a single equation for h:

$$\frac{\partial}{\partial t}(h - a^2\nabla^2 h) - \frac{\beta C^2}{f^2}h_x \approx 0 \quad [6]$$

where $\nabla^2 = (\partial^2/\partial x^2) + (\partial^2/\partial y^2)$ and various small terms have been neglected, and

$$C^2 = gH \quad [7]$$

is the speed of long surface waves and

$$a = C/f \quad [8]$$

is the 'deformation radius,' or 'Rossby radius,' the natural length scale for the problem. Here β represents df/dy, the northward gradient of the Coriolis parameter. Equation[6] represents conservation of vorticity of the fluid column. Although its coefficients depend on latitude, because no more differentiation is required, it is traditional to 'freeze' the coefficients and treat f and β as if they are constant. We then pose

$$h \propto \exp i(kx + ly - \omega t) \quad [9]$$

which gives the dispersion relation connecting frequency ω with wavenumbers k, l as

$$\omega = -\beta a \frac{ak}{1 + (ak)^2} \quad [10]$$

where

$$K^2 = k^2 + l^2 \quad [11]$$

is the modulus of the wavenumber. Equation[10] has been written in a manner that emphasizes the importance of the size of the wavenumber in units of inverse deformation radius. From the dispersion relation all details of wave propagation may be determined.

To start with, [10] does not permit waves of all frequencies; there is a cutoff frequency of $|\beta a|^2$, above which Rossby waves may not propagate. Since a varies inversely with Coriolis parameter, this frequency becomes smaller as the poles are approached. (Baroclinic waves at the annual cycle can only exist for latitudes less than about 40–45°, for example.)

The phase velocity (c_x, c_y) is formally given by

$$c_x = \frac{\omega k}{K^2} = -\beta a^2 \frac{(ak)^2}{(ak)^2\left[1 + (ak)^2\right]} \quad [12]$$

$$c_y = \frac{\omega l}{K^2} = -\beta a^2 \frac{(ak)^2(al)}{(aK)^2\left[1 + (aK)^2\right]} \quad [13]$$

although these are not the propagation speeds of points of constant phase in the x and y directions, which are given by ω/k and ω/l, respectively. The first of these shows that waves propagate with crests moving westward (i.e., c_x is negative), with a maximum speed βa^2, when k and l are small (long waves). Since a varies inversely with Coriolis parameter, this speed will be a strong function of latitude, becoming infinite at the equator, where this midlatitude theory breaks down. The north–south phase velocity can take various values depending on the orientation of the wave crests.

The group velocity (c_{gx}, c_{gy}), i.e., the velocity at which the wave energy propagates, is given by

$$c_{gx} = \frac{\partial \omega}{\partial k} = -\beta a^2 \frac{1 + \left[(al)^2 - (ak)^2\right]}{\left[1 + (aK)^2\right]^2} \quad [14]$$

$$c_{gy} = \frac{\partial \omega}{\partial l} = 2\beta a^2 \frac{(ak)(al)}{\left[1 + (aK)^2\right]^2} \quad [15]$$

In general the group velocity is not the same as the phase velocity, so that Rossby waves are dispersive. If ak is sufficiently large (i.e., the waves are sufficiently short in the east–west direction), eqn[14] shows that c_{gx} can be positive: while crests move west, the wave energy moves east. The simplest case to discuss is when al is small, so that the waves are long in the north–south direction. Then the dispersion relation becomes

$$\omega = -\beta a \frac{ak}{1+(ak)^2};\ c_x = -\beta a^2 \frac{1}{1+(ak)^2}$$
$$c_{gx} = -\beta a^2 \frac{1-(ak)^2}{\left[1+(ak)^2\right]^2} \quad [16]$$

which is shown in **Figure 6**. This shows the following.

- Long waves ($ak \ll 1$) have the same phase and group velocity, $-\beta a^2 = -\beta C^2/f^2$, and so are nondispersive.
- Zero group velocity occurs when $ka = -1$.
- There is a maximum eastward group velocity of one-eighth the fastest westward group velocity.

Vertical Variation

We now return to the vertical structure of the waves. The idea is to seek a situation in which the waves may propagate without changing that structure. Write

$$\left(u, v, \frac{p}{\rho_0}\right) = \left(\tilde{u}, \tilde{v}, g\tilde{h}\right)(x, y, t)\cdot\hat{u}(z) \quad [17]$$

$$w = \tilde{w}(x, y, t)\cdot\tilde{w}(z). \quad [18]$$

Substitution into the two horizontal momentum and the divergence, buoyancy, and hydrostatic equations, linearized about a basic state of stratification $\bar{\rho}(z)$, where $N^2(z) = -g\bar{\rho}_z/\rho_0$ is the buoyancy frequency, gives respectively

$$(\tilde{u}_t - f\tilde{v})\hat{u} = -g\tilde{h}_x\hat{u} \quad [19]$$

$$(\tilde{v}_t + f\tilde{u})\hat{u} = -g\tilde{h}_y\hat{u} \quad [20]$$

$$(\tilde{u}_x + \tilde{v}_y)\hat{u} + \tilde{w}\hat{w}_z = 0 \quad [21]$$

$$-\frac{g\rho_t}{\rho_0} + \tilde{w}N^2\hat{w} = 0 \quad [22]$$

$$-\frac{g\rho}{\rho_0} = g\tilde{h}\hat{u}_z \quad [23]$$

(Note that the density ρ has not had a vertical structure defined but that its vertical structure is like $N^2\hat{w}(z)$). Eqs [19] and [20] already permit the cancellation of the common factor $\hat{u}(z)$ as required. Equation [21] will also permit cancellation of $\hat{u}(z)$ provided that we choose

$$\hat{w}(z) = \hat{u}_z(z) \quad [24]$$

Finally, combination of eqns[22] and [23] can only permit cancellation of the vertical structure if

$$\hat{u}_z(z) \propto N^2\hat{w} \quad \text{or} \quad \hat{w}_{zz} + \frac{N^2}{C^2}\hat{w} = 0 \quad [25]$$

Here C, with dimension of velocity, is an unknown constant of proportionality. It is given as an eigenvalue by solving eqn [25] with suitable boundary conditions at surface and floor.

There are a countable infinity of solutions, or normal modes. The zeroth (using a traditional numbering) is the barotropic mode, in which $\hat{u}$ is approximately independent of depth and $\hat{w}$ is linear with depth; C is given to an excellent degree of

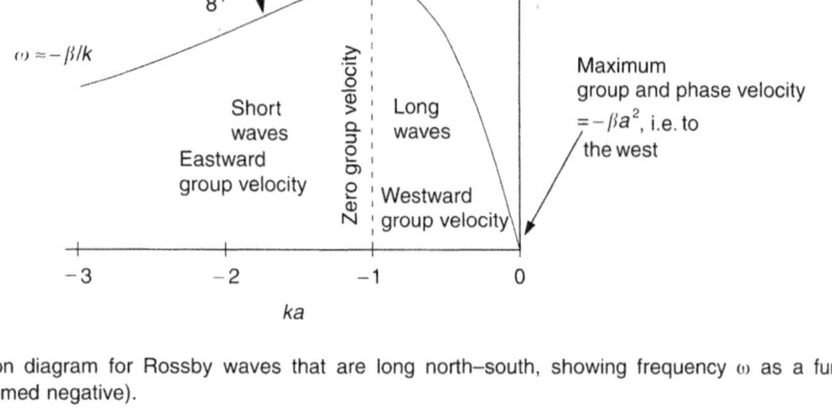

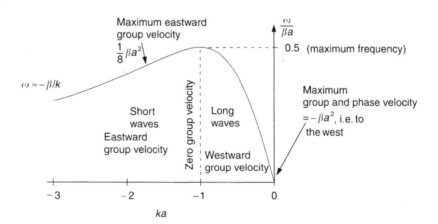

Figure 6 Dispersion diagram for Rossby waves that are long north–south, showing frequency ω as a function of east–west wavenumber k (assumed negative).

approximation by $C^2 = gH$, where H is the fluid depth. This gives values for C of around $200\,\mathrm{m\,s^{-1}}$ in ocean depths of 4000 m. This is just the long-wave speed of an unstratified fluid. The remaining solutions, called baroclinic, have much smaller eigenvalues C, and for these modes, to a good approximation, $\hat{w}$ vanishes at surface and floor. The nth vertical mode has $n-1$ zeros of $\hat{w}$ and n zeros of $\hat{u}$ in the fluid column, so that high modes oscillate strongly in the vertical. For various reasons, many not well understood, high modes are rarely found in observations. Typical values of C are 2–$3\,\mathrm{m\,s^{-1}}$ for the first mode, and steadily slower for high modes. **Figure 7** shows the vertical structure of $\hat{w}$ and $\hat{u}$ for the first two baroclinic modes, for the stratification (N) shown, which is an average over the major ocean basins.

Substitution back into the remaining (horizontal) part of the system gives for each normal mode in turn,

$$\tilde{u}_t - f\tilde{v} = -g\tilde{h}_x \qquad [26]$$

$$\tilde{v}_t + f\tilde{u} = -g\tilde{h}_y \qquad [27]$$

$$g\tilde{h}_t + C^2(\tilde{u}_x + \tilde{v}_y) = 0 \qquad [28]$$

If we define an effective depth H by

$$H = C^2/g \qquad [29]$$

then the system of eqns [26] and [28] reduces to [1] to [3], as required, though with H taking a different meaning. Thus each vertical mode has a separate horizontal behavior, characteristic speeds, and so on.

The fastest westward phase speed (i.e., for long waves), $-\beta a^2 = -\beta C^2/f^2$, for mode 1, the fastest baroclinic mode, is shown in **Figure 1**.

Time Variation: Ocean Spinup

The theory above is derived for free waves. When the waves are forced, the approach is to express the forcing as a sum of vertical normal modes in the same way, with coefficients varying with (x, y and t), and add the forcing on the right-hand-sides of eqns [26] and [28]. The simplest such problem is the response of an ocean to a wind that is suddenly turned on; for simplicity the wind does not vary east–west.

Several things happen immediately, indicated in the Hovmöller diagram in **Figure 8**. In the ocean interior, where no waves have yet reached, the ocean responds linearly in time to the local forcing. Near the eastern boundary, a Rossby wave is initiated,

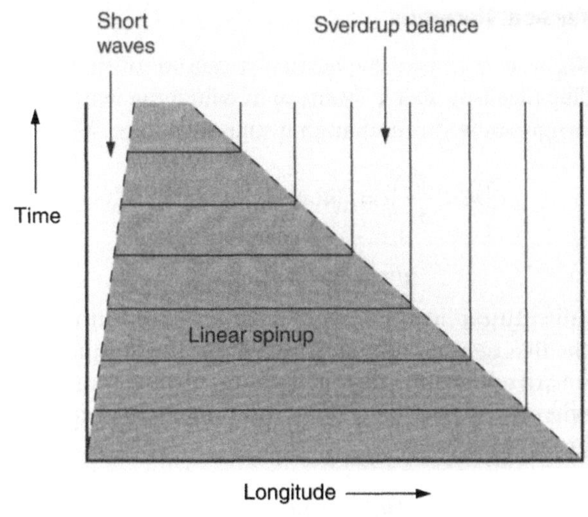

Figure 8 Schematic Hovmöller diagram showing the ocean's response to an applied steady wind stress. Long Rossby waves from the east and short, slower Rossby waves from the west move from their respective boundaries. Where they have not yet reached there is a linear spinup. After the wavefront has passed, a Sverdrup interior and a time-varying western boundary current remain.

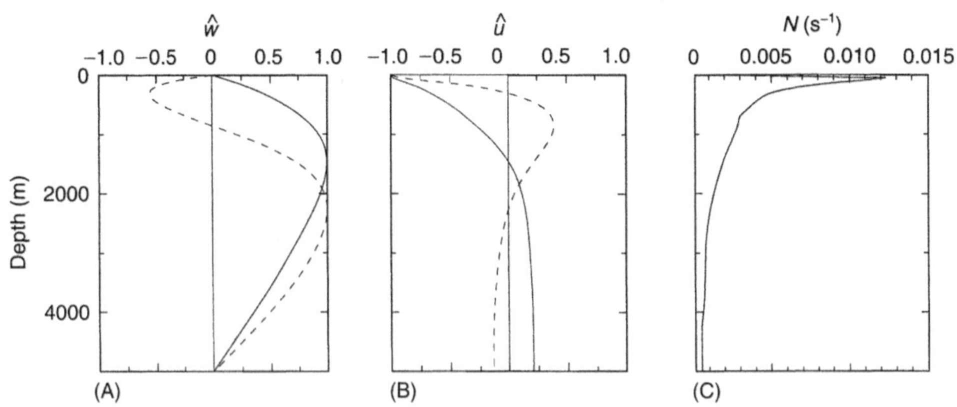

Figure 7 Vertical structure of the first two baroclinic modes, for the mean stratification of the major ocean basins: (A) $\hat{w}$; (B) $\hat{u}$; (C) N. Each mode is scaled to a maximum of unity.

carrying with it information that the ocean has an eastern boundary. This moves westward at the long-wave speed $-\beta a^2$. Behind it, the ocean becomes steady, in Sverdrup balance with the forcing. Near the western boundary, short waves, moving at speed $\beta a^2/8$, move eastward, conveying information about the western boundary to the fluid interior. (The solution left behind by these waves is complicated, and is heavily affected by any dissipative terms present.) Eventually the two wavefronts meet, giving a long-term solution with Sverdrup balance over most of the ocean, and a time-varying area near the western boundary that takes the form of an effective western boundary current.

Comparison with Observations

The satellite observations of Rossby waves discussed above showed that Rossby waves were found in many areas of the subequatorial and subpolar oceans. Their westward speeds were estimated using the Hovmöller diagram, or other approaches, and compared with the theory above (as indicated in **Figure 1**). The findings fall into three latitudinal bands (**Figure 9**). First, near the equator, observed wave speeds were usually somewhat less than predicted by theory. However, some of this area would be affected by equatorial wave theory, so making comparison difficult. Second, around latitudes of 10° to 20°, linear theory has succeeded admirably in estimating wave speeds. Third, for latitudes poleward of 20°, there appears to be a steady increase in the discrepancy between theory and observations, with theoretical speeds becoming as much as 2–3 times slower than observed speeds at high latitudes. (Some debate remains as to the magnitude of this shortfall, which depends to some extent on the longitude bands chosen for comparison. But the shortfall at high latitudes does seem to be unequivocal.) In areas such as the Antarctic Circumpolar Current (ACC), there is evidence of eastward propagation of Rossby waves. This is almost certainly caused by a combination of two factors: the natural speed of Rossby waves is very small at high latitudes, and the ACC is one of the few currents where the barotropic mean flow is large and eastward. Thus, the waves are simply swept eastward by the mean flow in a Doppler shift process.

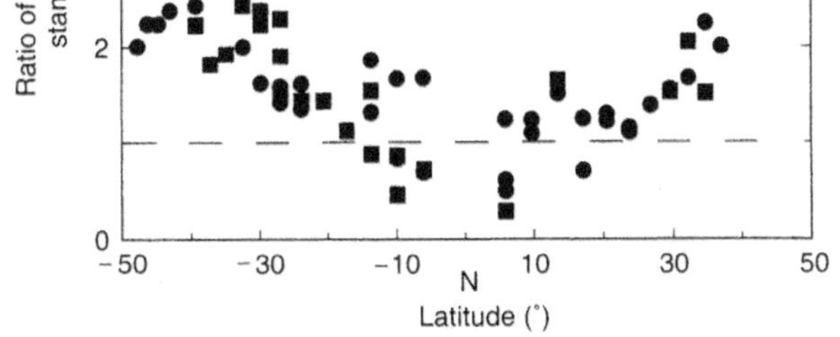

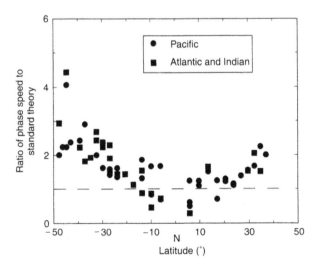

Figure 9 Ratio of observed Rossby wave speeds to linear theoretical predictions, as a function of latitude. Data from different ocean basins are indicated. The uncertainty in the ratio values is for less than the variation with latitude.

Improvements to Theory

Several suggestions have been made to explain the discrepancies, which assume either that the interpretation of the data is incorrect (i.e., that linear theory is in fact correct), or that some aspect of linear theory must be modified. The answer probably lies in a combination of these.

If the waves are forced by a wind which has (say) an annual cycle, then the ocean responds with a combination of free propagating waves (e.g., as sin $(kx - \omega t)$) and a forced response (e.g., as sin t). The sum of these two possesses a term in sin $(kx - 2\omega t)$, and it could be this term that apparently yields wave propagation at twice the predicted value. However, this term occurs multiplied by sin kx, and so the waves disappear at regular intervals east–west, which is not generally observed; in addition, Fourier and other methods of signal processing would show the two linear responses and not generate an erroneous doubling of the speed.

It is thus probable that some aspect of flat-bottom, dissipation-free linear theory has broken down. A prime candidate during the last 20 years has been that of varying ocean topography, which can generate a waveguide and permit the propagation of 'topographic Rossby waves' in which the bottom slope plays a role similar to that of the variations of Coriolis parameter. The resulting waves are frequently, but not always, trapped near the bottom; these waves can be faster or slower than their flat bottom relatives. Whether a topography that both rises and falls (e.g., over a midocean ridge) would generate any net speed increase remains unclear.

Active research is examining various options.

- If the waves are dissipative (either directly, or indirectly by heat losses at the surface), then the generation of a decay scale can induce an apparently different phase speed. This mechanism is particularly effective near an eastern boundary; if it is too successful, the dissipation stops the wave propagation completely.

- If the waves are of large enough amplitude, several things can happen. Pairs of waves can interact. Single large waves (sometimes known as 'solitons') can self-advect, at speeds that may differ from linear wave theory. Both these effects are beyond the scope of this article. Large waves can modify the ocean background stratification and increase their speed. However, the ubiquity of faster propagation at mid to high latitude would argue for almost permanent changes to the background stratification, which are not observed.
- The effects of ocean topography are being reexamined, with emphasis on the propagation of waves over a slowly varying floor (and concomitant normal mode change during the propagation). The results are as yet incomplete, but it looks as though the suggestion above that topography oriented in a variety of directions cannot yield a net speed increase continues to hold.
- The mid-latitude ocean is known to possess teleconnections with the equatorial ocean, so that there may be anomalous responses at mid latitude related to the faster-travelling equatorial systems. It is hard to see how this effect could propagate to higher latitudes, however, where there would be a severe mismatch in speeds.
- Finally, the background ocean is not at rest. This has three possible effects. First, a strong enough barotropic flow could simply sweep the waves westward with the speed of the flow. However, we do not believe that depth-averaged midocean flows are even one-twentieth of the speed necessary to achieve this; so the barotropic flow can be discounted. Second, mean flow could change the normal mode calculation in the vertical. However, oceanic motions are seldom as fast as the 2–3 m s^{-1} speed of the first vertical mode, so this can be discounted. Third, and more seriously, depth-varying ocean motions are as fast as the few centimeters per second speed of the fastest baroclinic Rossby wave. A background flow – produced geostrophically by density variations across the ocean basin – can strongly modify the northward potential vorticity gradient of the system. Direct calculations of the changes this produces in phase speed suggest that much, but not all, of the discrepancy between observations and linear theory is explained by the presence of such background flows.

Conclusions

The ocean appears to possess Rossby waves in most of its basins. Theory for such waves has existed for many years, but they have only recently been observed by satellite altimeters and other approaches. The theory gives predictions of the right order of magnitude for Rossby wave speeds, but at mid and high latitudes appears to underestimate the speeds by a factor of 2. Various theories have been put forward to explain this discrepancy, of which the most promising is the inclusion of background mean flow, not as a simple depth-independent advection but as a genuine interaction with the wave. None of these theories forms a complete explanation. It will probably be necessary to combine the processes (e.g., to include both topographic variation and a background mean flow) before the theory can be regarded as satisfactory.

Nomenclature

a	Rossby, or deformation, radius
c_x, c_y	Phase velocity
c_{gx}, c_{gy}	Group velocity
f	Coriolis parameter
g	Acceleration due to gravity
h	Perturbation to depth of a fluid layer
k	Eastward wavenumber
l	Northward wavenumber
p	Pressure
t	Time
u	Eastward velocity
v	Northward velocity
w	Vertical velocity
x	Eastward coordinate
y	Northward coordinate
z	Vertical coordinate
C	Modal wave speed
H	Ocean depth, or equivalent depth
K	Modulus of wavenumber
N	Buoyancy frequency of water
β	Rate of change of f with distance north
ρ_o	Mean density of sea water
ω	Frequency

See also

Ekman Transport and Pumping. Mesoscale Eddies. Wind Driven Circulation.

Further Reading

Chelton DB and Schlax MG (1996) Global observations of oceanic Rossby waves. *Science* 272: 234–238.

Dickinson RE (1978) Rossby waves — long-period oscillations of oceans and atmospheres. *Annual Reviews of Fluid Mechanics* 10: 159–195.
Gill AE (1982) *Atmosphere–Ocean Dynamics*. New York: Academic Press.
LeBlond PH and Mysak LA (1978) *Waves in the Ocean*. Amsterdam: Elsevier.
Rhines PB (1977) The dynamics of unsteady currents. In: Goldberg EG, McCave IN, O'Brien JJ, and Steele JH (eds.) *The Sea*, pp. 189–378. New York: Wiley.

EQUATORIAL WAVES

A. V. Fedorov and J. N. Brown, Yale University, New Haven, CT, USA

Introduction

It has been long recognized that the tropical thermocline (the sharp boundary between warm and deeper cold waters) provides a wave guide for several types of large-scale ocean waves. The existence of this wave guide is due to two key factors. First, the mean ocean vertical stratification in the tropics is perhaps greater than anywhere else in the ocean (**Figures 1** and **2**), which facilitates wave propagation. Second, since the Coriolis parameter vanishes exactly at 0° of latitude, the equator works as a natural boundary, suggesting an analogy between coastally trapped and equatorial waves.

The most well-known examples of equatorial waves are eastward propagating Kelvin waves and westward propagating Rossby waves. These waves are usually observed as disturbances that either raise or lower the equatorial thermocline. These thermocline disturbances are mirrored by small anomalies in sea-level elevation, which offer a practical method for tracking these waves from space.

For some time the theory of equatorial waves, based on the shallow-water equations, remained a theoretical curiosity and an interesting application for Hermite functions. The first direct measurements of equatorial Kelvin waves in the 1960s and 1970s served as a rough confirmation of the theory. By the 1980s, scientists came to realize that the equatorial waves, crucial in the response of the tropical ocean

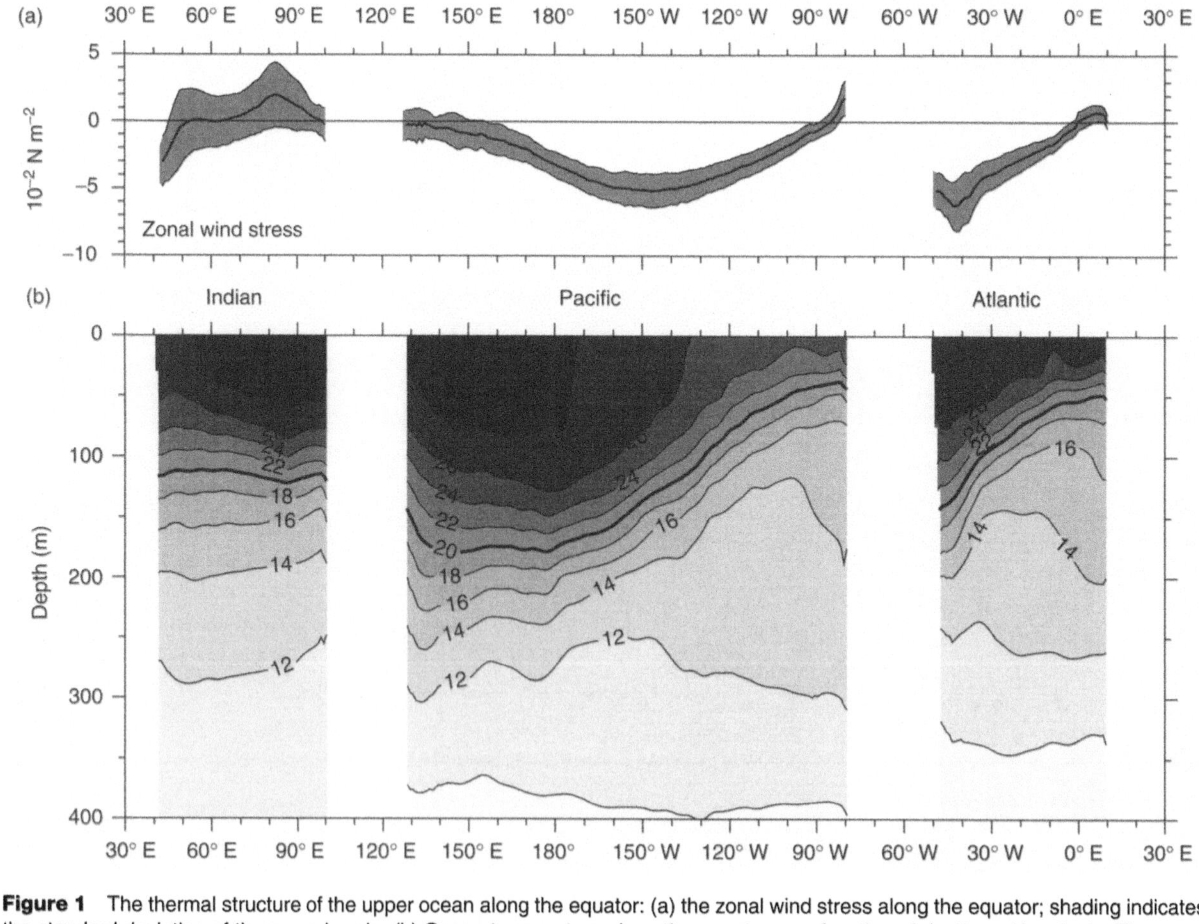

Figure 1 The thermal structure of the upper ocean along the equator: (a) the zonal wind stress along the equator; shading indicates the standard deviation of the annual cycle. (b) Ocean temperature along the equator as a function of depth and longitude. The east–west slope of the thermocline in the Pacific and the Atlantic is maintained by the easterly winds. (b) From Kessler (2005).

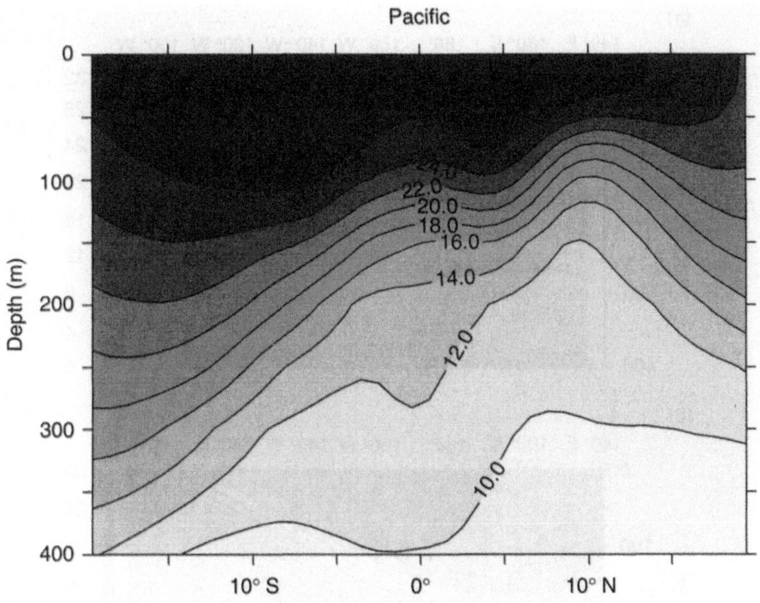

Figure 2 Ocean temperature as a function of depth and latitude in the middle of the Pacific basin (at 140° W). The thermocline is particularly sharp in the vicinity of the equator. Note that the scaling of the horizontal axis is different from that in **Figure 1**. Temperature data are from Levitus S and Boyer T (1994) *World Ocean Atlas 1994, Vol. 4: Temperature NOAA Atlas NESDIS4.* Washington, DC: US Government Printing Office.

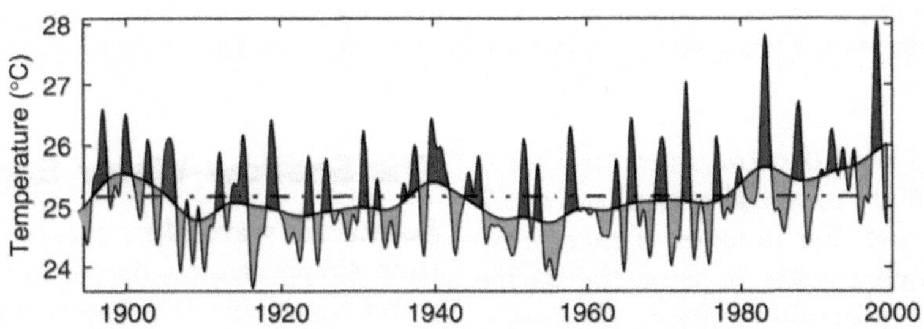

Figure 3 Interannual variations in sea surface temperatures (SSTs) in the eastern equatorial Pacific shown on the background of decadal changes (in °C). The annual cycle and high-frequency variations are removed from the data. El Niño conditions correspond to warmer temperatures. Note El Niño events of 1982 and 1997, the strongest in the instrumental record. From Fedorov AV and Philander SG (2000) Is El Niño changing? *Science* 288: 1997–2002.

to varying wind forcing, are one of the key factors in explaining ENSO – the El Niño-Southern Oscillation phenomenon.

El Niño, and its complement La Niña, have physical manifestations in the sea surface temperature (SST) of the eastern equatorial Pacific (**Figure 3**). These climate phenomena cause a gradual horizontal redistribution of warm surface water along the equator: strong zonal winds during La Niña years pile up the warm water in the west, causing the thermocline to slope downward to the west and exposing cold water to the surface in the east (**Figure 4(b)**). During an El Niño, weakened zonal winds permit the warm water to flow back eastward so that the thermocline becomes more horizontal, inducing strong warm anomalies in the SST (**Figure 4(a)**).

The ocean adjustment associated with these changes crucially depends on the existence of equatorial waves, especially Kelvin and Rossby waves, as they can alter the depth of the tropical thermocline. This article gives a brief summary of the theory behind equatorial waves, the available observations of those waves, and their role in ENSO dynamics. For a detailed description of El Niño phenomenology the reader is referred to other relevant papers in this encyclopedia and Pacific Ocean Equatorial Currents.

It is significant that ENSO is characterized by a spectral peak at the period of 3–5 years. The

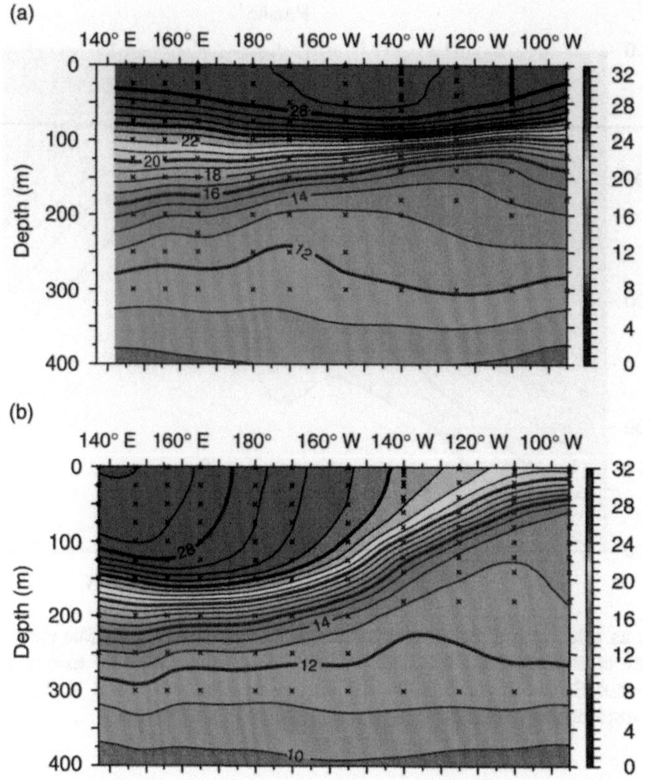

Figure 4 Temperatures (°C) as a function of depth along the equator at the peaks of (a) El Niño (Jan. 1998) and (b) La Niña (Dec. 1999). From the TAO data; see McPhaden MJ (1999) Genesis and evolution of the 1997–98 El Niño. *Science* 283: 950–954.

timescales associated with the low-order (and most important dynamically) equatorial waves are much shorter than this period. For instance, it takes only 2–3 months for a Kelvin wave to cross the Pacific basin, and less than 8 months for a first-mode Rossby wave. Because of such scale separation, the properties of the ocean response to wind perturbations strongly depend on the character of the imposed forcing. It is, therefore, necessary to distinguish the following.

1. Free equatorial waves which arise as solutions of unforced equations of motion (e.g., free Kelvin and Rossby waves),
2. Equatorial waves forced by brief wind perturbations (of the order of a few weeks). In effect, these waves become free waves as soon as the wind perturbation has ended,
3. Equatorial wave-like anomalies forced by slowly varying periodic or quasi-periodic winds reflecting ocean adjustment on interannual timescales. Even though these anomalies can be represented mathematically as a superposition of continuously forced Kelvin and Rossby waves of different modes, the properties of the superposition (such as the propagation speed) can be rather different from the properties of free waves.

The Shallow-Water Equations

Equatorial wave dynamics are easily understood from simple models based on the $1\frac{1}{2}$-layer shallow-water equations. This approximation assumes that a shallow layer of relatively warm (and less dense) water overlies a much deeper layer of cold water. The two layers are separated by a sharp thermocline (**Figure 5**), and it is assumed that there is no motion in the deep layer. The idea is to approximate the thermal (and density) structure of the ocean displayed in **Figures 1** and **2** in the simplest form possible.

The momentum and continuity equations, usually referred to as the reduced-gravity shallow-water equations on the β-plane, are

$$u_t + g'h_x - \beta y v = \tau^x/\rho D - \alpha_s u \qquad [1]$$

$$v_t + g'h_y + \beta y u = \tau^y/\rho D - \alpha_s v \qquad [2]$$

$$h_t + H(u_x + v_y) = -\alpha_s h \qquad [3]$$

These equations have been linearized, and perturbations with respect to the mean state are considered. Variations in the mean east–west slope of the

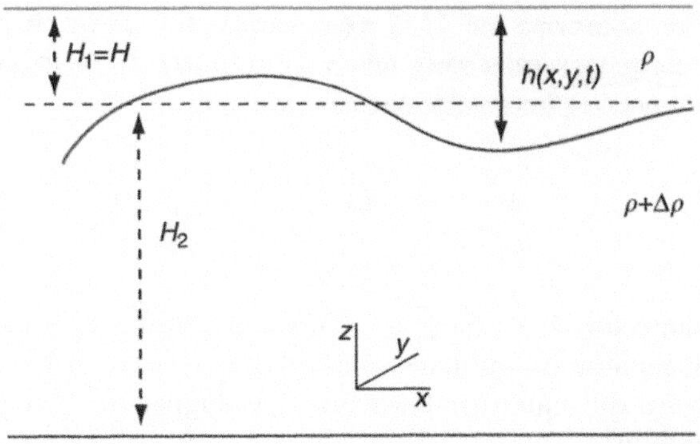

Figure 5 A sketch of the $1\frac{1}{2}$-layer shallow-water system with the rigid-lid approximation. $H_1/H_2 \ll 1$. The mean depth of the thermocline is H. The x-axis is directed to the east along the equator. The y-axis is directed toward the North Pole. The mean east–west thermocline slope along the equator is neglected.

thermocline and mean zonal currents are neglected. The notations are conventional (some are shown in **Figure** 5), with u, v denoting the ocean zonal and meridional currents, H the mean depth of the thermocline, h thermocline depth anomalies, τ^x and τ^y the zonal and meridional components of the wind stress, ρ mean water density, $\Delta\rho$ the difference between the density of the upper (warm) layer and the deep lower layer, $g' = g\Delta\rho/\rho$ the reduced gravity. D is the nominal depth characterizing the effect of wind on the thermocline (frequently it is assumed that $D = H$). The subscripts t, x, and y indicate the respective derivatives.

The system includes simple Rayleigh friction in the momentum equations and a simple linear parametrization of water entrainment at the base of the mixed layer in the continuity equation (terms proportional to α_s). Some typical values for the equatorial Pacific are $\Delta\rho/\rho = 0.006$; $H = 120$ m (see **Figure 1**); $D = 80$ m. The rigid-lid approximation is assumed (i.e., to a first approximation the ocean surface is flat). However, after computing h, one can calculate small changes in the implied elevation of the free surface as $\eta = -h\Delta\rho/\rho$. It is this connection that allows us to estimate changes in the thermocline depth from satellite measurements by measuring sea-level height.

The boundary conditions for the equations are no zonal flow ($u = 0$) at the eastern and western boundaries and vanishing meridional flow far away from the equator. The former boundary conditions are sometimes modified by decomposing u into different wave components and introducing reflection coefficients (smaller than unity) to account for a partial reflection of waves from the boundaries.

It is apparent that the $1\frac{1}{2}$-layer approach leaves in the system only the first baroclinic mode and eliminates barotropic motion (the first baroclinic mode describes a flow that with different velocities in two layers, barotropic flow does not depend on the vertical coordinate). This approximation filters out higher-order baroclinic modes with more elaborate vertical structure. For instance, the equatorial undercurrent (EUC) is absent in this model. Observations and numerical calculations show that to represent the full vertical structure of the currents and ocean response to winds correctly, both the first- and second-baroclinic modes are necessary. Nevertheless, the shallow-water equations within the $1\frac{1}{2}$-layer approximation remain very successful and are used broadly in the famous Cane-Zebiak model of ENSO and its numerous modifications.

For many applications, the shallow-water equations are further simplified to filter out short waves: in the long-wave approximation the second momentum equation (eqn [2]) is replaced with a simple geostrophic balance:

$$g'h_y + \beta y u = 0 \qquad [4]$$

The boundary condition at the western boundary is then replaced with the no-net flow requirement ($\int u dy = 0$).

It is noteworthy that the shallow-water equations can also approximate the mean state of the tropical ocean (if used as the full equations for mean variables, rather than perturbations from the mean state). In that case the main dynamic balance along the equator is that between the mean trade winds and the mean (climatological) slope of the thermocline (damping neglected)

$$g'h_x \sim \tau^x/\rho D \qquad [5]$$

This balance implies the east–west difference in the thermocline depth along the equator of about 130 m in the Pacific consistent with **Figure 1**.

Free-Wave Solutions of the Shallow-Water Equations

First, we consider the shallow-water eqns [1]–[3] with no forcing and no dissipation. The equations have an infinite set of equatorially trapped solutions (with $v \to 0$ for $y \to \pm\infty$). These are free equatorial waves that propagate back and forth along the equator.

Kelvin Waves

Kelvin waves are a special case when the meridional velocity vanishes everywhere identically ($v = 0$) and eqns [1]–[3] reduce to

$$u_t + g'h_x = 0 \quad [6]$$

$$g'h_y + \beta yu = 0 \quad [7]$$

$$h_t + Hu_x = 0 \quad [8]$$

Looking for wave solutions of [6]–[8] in the form

$$[u, v, h] = [\tilde{u}(y), \tilde{v}(y), \tilde{h}(y)]e^{i(kx-\omega t)} \quad [9]$$

$$\tilde{v}(y) = 0; \quad \omega > 0 \quad [10]$$

we obtain the dispersion relation for frequency ω and wave number k

$$\omega^2 = g'Hk^2 \quad [11]$$

and a first-order ordinary differential equation for the meridional structure of h

$$\frac{d\tilde{h}}{dy} = -\frac{\beta k}{\omega} y\tilde{h} \quad [12]$$

The only solution of [11] and [12] decaying for large y, called the Kelvin wave solution, is

$$h = h_0 e^{-(\beta/2c)y^2} e^{i(kx-\omega t)} \quad [13]$$

where the phase speed $c = (g'H)^{1/2}$ and $\omega = ck$ (h_0 is an arbitrary amplitude). Thus, Kelvin waves are eastward propagating ($\omega/k > 0$) and nondispersive. The second solution of [11] and [12], the one that propagates westward, would grow exponentially for large y and as such is disregarded.

Calculating the Kelvin wave phase speed from typical parameters used in the shallow-water model gives $c = 2.7\,\mathrm{m\,s^{-1}}$ which agrees well with the measurements. The meridional scale with which these solutions decay away from the equator is the equatorial Rossby radius of deformation defined as

$$L_R = (c/\beta)^{1/2} \quad [14]$$

which is approximately 350 km in the Pacific Ocean, so that at 5° N or 5° S the wave amplitude reduces to 30% of that at the equator.

Rossby, Poincare, and Yanai Waves

Now let us look for the solutions that have nonzero meridional velocity v. Using the same representation as in [9] we obtain a single equation for $\tilde{v}(y)$:

$$\frac{d^2\tilde{v}}{dy^2} + \left(\frac{\omega^2}{c^2} - k^2 - \frac{\beta k}{\omega} - \frac{\beta^2}{c^2}y^2\right)\tilde{v} = 0 \quad [15]$$

The solutions of [15] that decay far away from the equator exist only when an important constraint connecting its coefficients is satisfied:

$$\frac{\omega^2}{c^2} - k^2 - \frac{\beta k}{\omega} = (2n+1)\frac{\beta}{c} \quad [16]$$

where $n = 0, 1, 2, 3, \ldots$. This constraint serves as a dispersion relation $\omega = \omega(k,n)$ for several different types of equatorial waves (see **Figure 6**), which include

1. Gravity-inertial or Poincare waves $n = 1, 2, 3, \ldots$
2. Rossby waves $n = 1, 2, 3, \ldots$
3. Rossby-gravity or Yanai wave $n = 0$
4. Kelvin wave $n = -1$.

These waves constitute a complete set and any solution of the unforced problem can be represented as a sum of those waves (note that the Kelvin wave is formally a solution of [15] and [16] with $v = 0$, $n = -1$).

Let us consider several important limits that will elucidate some properties of these waves. For high frequencies we can neglect $\beta k/\omega$ in [16] to obtain

$$\omega^2 = c^2k^2 + (2n+1)\beta c \quad [17]$$

where $n = 1, 2, 3, \ldots$.

This is a dispersion relation for gravity-inertial waves, also called equatorially trapped Poincare waves. They propagate in either direction and are similar to gravity-inertial waves in mid-latitudes.

For low frequencies we can neglect ω^2/c^2 in [16] to obtain

$$\omega = -\frac{\beta k}{k^2 + (2n+1)\beta/c} \quad [18]$$

with $n = 1, 2, 3, \ldots$. These are Rossby waves similar to their counterparts in mid-latitudes that critically depend on the β-effect. Their phase velocity (ω/k) is always westward ($\omega/k < 0$), but their group velocity $\partial\omega/\partial k$ can become eastward for high wave numbers.

The case $n = 0$ is a special case corresponding to the so-called mixed Rossby-gravity or Yanai wave. Careful consideration shows that when the phase velocity of those waves is eastward ($\omega/k > 0$), they behave like gravity-inertial waves and [17] is satisfied, but when the phase velocity is westward ($\omega/k < 0$), they behave like Rossby waves and expression [18] becomes more appropriate.

The meridional structure of the solutions of eqn [15] corresponding to the dispersion relation in [16]

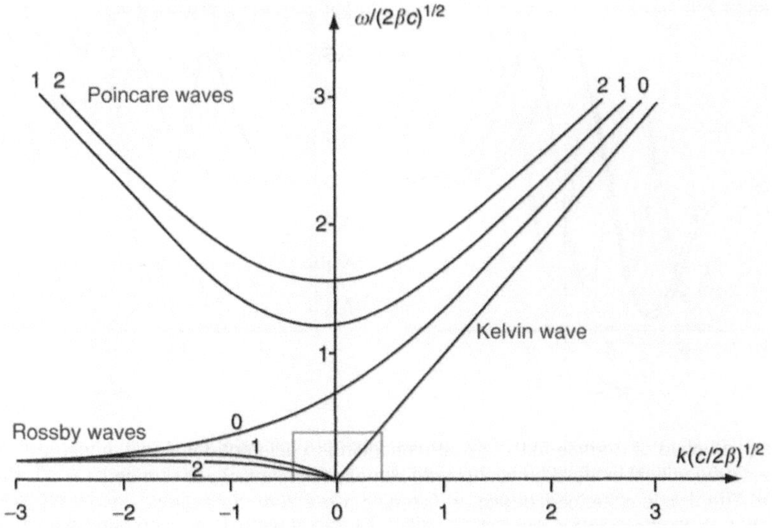

Figure 6 The dispersion relation for free equatorial waves. The axes show nondimensionalized wave number and frequency. The blue box indicates the long-wave regime. $n=0$ indicates the Rossby-gravity (Yanai) wave. Kelvin wave formally corresponds to $n=-1$. From Gill AE (1982) *Atmosphere-Ocean Dynamics*, 664pp. New York: Academic Press.

is described by Hermite functions:

$$\tilde{v} = \left(\sqrt{\pi}2^n n!\right)^{-1/2} H_n\left((\beta/c)^{1/2} y\right) e^{-(\beta/2c)y^2} \quad [19]$$

where $H_n(Y)$ are Hermite polynomials ($n = 0, 1, 2, 3, \ldots$),

$$H_0 = 1;\ H_1 = 2Y;\ H_2 = 4Y^2 - 2;$$
$$H_3 = 8Y^3 - 12Y; H_4 = 16Y^4 - 48Y^2 + 12; \ldots, \quad [20]$$

and

$$Y = (\beta/c)^{1/2} y \quad [21]$$

These functions as defined in [19]–[21] are orthonormal.

The structure of Hermite functions, and hence of the meridional flow corresponding to different types of waves, is plotted in **Figure 7**. Hermite functions of odd numbers ($n = 1, 3, 5, \ldots$) are characterized by zero meridional flow at the equator. It can be shown that they create symmetric thermocline depth anomalies with respect to the equator (e.g., a first-order Rossby wave with $n = 1$ has two equal maxima in the thermocline displacement on each side of the equator). Hermite functions of even numbers generate cross-equatorial flow and create thermocline displacement asymmetric with respect to the equator (i.e., with a maximum in the thermocline displacement on one side of the equator, and a minimum on the other).

It is Rossby waves of low odd numbers and Kelvin waves that usually dominate the solutions for large-scale tropical problems. This suggests that solving the equations can be greatly simplified by filtering out Poincare and short Rossby waves. Indeed, the long-wave approximation described earlier does exactly that. Such an approximation is equivalent to keeping only the waves that fall into the small box in **Figure 6**, as well as a remnant of the Yanai wave, and then linearizing the dispersion relations for small k. This makes long Rossby wave non-dispersive, each mode satisfying a simple dispersion relation with a fixed n:

$$\omega = -\frac{c}{2n+1} k, \quad n = 1, 2, 3, \ldots \quad [22]$$

Consequently, the phase speed of Rossby waves of different modes is $c/3$, $c/5$, $c/7$, etc. The phase speed of the first Rossby mode with $n = 1$ is $c/3$, that is, one-third of the Kelvin wave speed. It takes a Kelvin wave approximately 2.5 months and a Rossby wave 7.5 months to cross the Pacific. The higher-order Rossby modes are much slower. The role of Kelvin and Rossby waves in ocean adjustment is described in the following sections.

Ocean Response to Brief Wind Perturbations

First, we will discuss the classical problem of ocean response to a brief relaxation of the easterly trade winds. These winds normally maintain a strong east–west thermocline slope along the equator and their changes therefore affect the ocean state. Westerly wind bursts (WWBs) that occur over

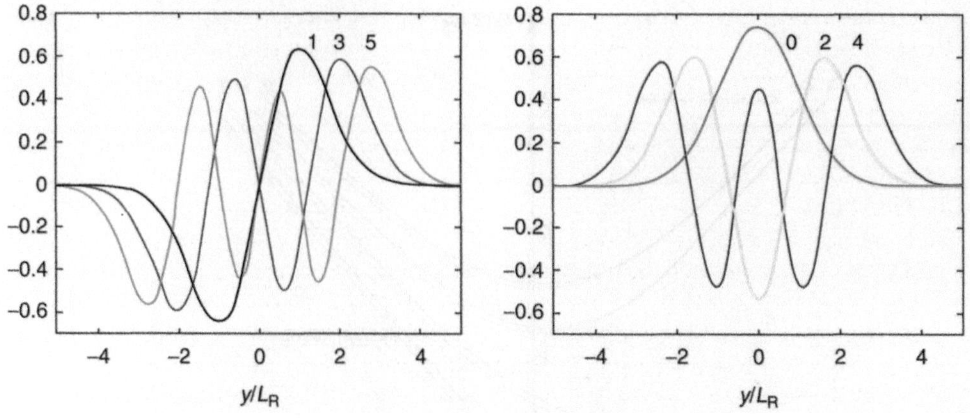

Figure 7 The meridional structure of Hermite functions corresponding to different equatorial modes (except for the Kelvin mode). The meridional velocity v is proportional to these functions. Left: Hermite functions of odd numbers ($n=1, 3, 5, \ldots$) with no meridional flow crossing the equator. The flow is either converging or diverging away from the equator, which produces a symmetric structure (with respect to the equator) of the thermocline anomalies. Right: Hermite functions of even numbers ($n=0, 2, 4, \ldots$) with nonzero cross-equatorial flow producing asymmetric thermocline anomalies.

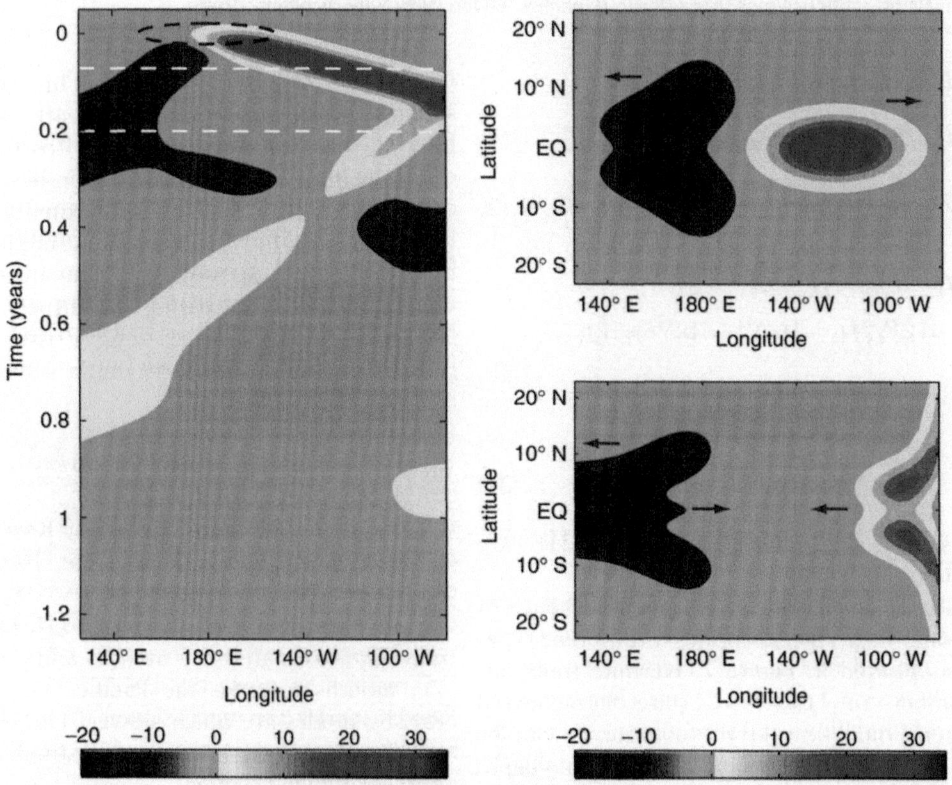

Figure 8 Ocean response to a brief westerly wind burst (WWB) occurring around time $t=0$ in a shallow-water model of the Pacific. Left: a Hovmoller diagram of the thermocline depth anomalies along the equator (in meters). Note the propagation and reflection of Kelvin and Rossby waves (the signature of Rossby waves on the equator is usually weak and rarely seen in the observations). Right: the spatial structure of the anomalies at times indicated by the white dashed lines on the left-side panel. The arrows indicate the direction of wave propagation. The wind-stress perturbation is given by $\tau = \tau_{\mathrm{wwb}} \exp[-(t/t_0)^2 - (x/L_x)^2 - (y/L_y)^2]$. Red corresponds to a deeper thermocline, blue to a shallower thermocline. The black dashed ellipse indicates the timing and longitudinal extent of the WWB.

the western tropical Pacific in the neighborhood of the dateline, lasting for a few weeks to a month, are examples of such occurrences. (Early theories treated El Niño as a simple response to a wind relaxation caused by a WWB. Arguably, WWBs may have contributed to the development of El Niño in 1997, but similar wind events on other occasions failed to have such an effect.)

As compared to the timescales of ocean dynamics, these wind bursts are relatively short, so that ocean adjustment occurs largely when the burst has already ended. In general, the wind bursts have several effects on the ocean, including thermodynamic effects modifying heat and evaporation fluxes in the western tropical Pacific. The focus of this article is on the dynamic effects of the generation, propagation, and then boundary reflection of Kelvin and Rossby waves.

The results of calculations with a shallow-water model in the long-wave approximation are presented next, in which a WWB lasting ~3 weeks is applied at time $t=0$ in the Pacific. The temporal and spatial structure of the burst is given by

$$\tau = \tau_{\mathrm{wwb}} e^{-(t/t_0)^2-(y/L_y)^2-(x-x_0)^2/L_x^2} \qquad [23]$$

which is roughly consistent with the observations. The burst is centered at $x_0 = 180°$ W; and $L_x = 10°$; $L_y = 10°$; $t_0 = 7$ days; $\tau_{\mathrm{wwb}} = 0.02\,\mathrm{N\,m^{-2}}$.

The WWB excites a downwelling Kelvin wave and an upwelling Rossby wave seen in the anomalies of the thermocline depth. **Figure 8** shows a Hovmoller diagram and the spatial structure of these anomalies at two particular instances. The waves propagate with constant speeds, although in reality Kelvin waves should slow down in the eastern part of the basin where the thermocline shoals (since $c = (g'H)^{1/2}$). The smaller slope of the Kelvin wave path on the Hovmoller diagram corresponds to its higher phase speed, as compared to Rossby waves.

The spatial structure of the thermocline anomalies at two instances is shown on the right panel of **Figure 8**. The butterfly shape of the Rossby wave (meridionally symmetric, but not zonally) is due to the generation of slower, high-order Rossby waves that trail behind (higher-order Hermite functions extend farther away from the equator, **Figure 7**).

The waves reflect from the western and eastern boundaries (in the model the reflection coefficients were set at 0.9). When the initial upwelling Rossby wave reaches the western boundary, it reflects as an equatorial upwelling Kelvin wave. When the downwelling Kelvin wave reaches the eastern boundary, a number of things occur. Part of the wave is reflected back along the equator as an equatorial downwelling Rossby wave. The remaining part travels north and

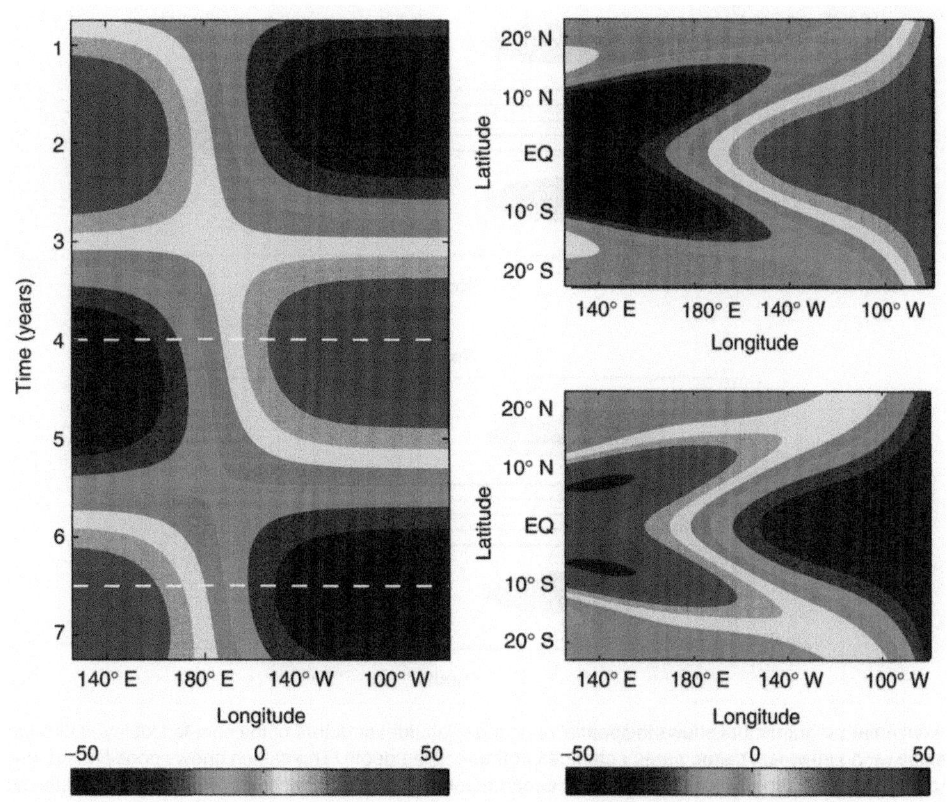

Figure 9 Ocean response to oscillatory winds in a shallow-water model. Left: a Hovmoller diagram of the thermocline depth anomalies along the equator. Note the different temporal scale as compared to **Figure 8**. Right: the spatial structure of the anomalies at times indicated by the dashed lines on the left-side panel. Red corresponds to a deeper thermocline, blue to a shallower thermocline. The wind-stress anomaly is calculated as $\tau = \tau_0 \sin(2\pi t/P)^* \exp[-(x/L_x)^2 - (y/L_y)^2]$; $P = 5$ years.

south as coastal Kelvin waves, apparent in the lower right panel of **Figure 8**, which propagate along the west coast of the Americas away from the Tropics.

Ocean Response to Slowly Varying Winds

Ocean response to slowly varying periodic or quasi-periodic winds is quite different. The relevant zonal dynamical balance (with damping neglected) is

$$(u_t - \beta y v) + g' h_x = \tau^x / \rho D \qquad [24]$$

It is the balance between the east–west thermocline slope and the wind stress that dominates the equatorial strip. Off the equator, however, the local wind stress is not in balance with the thermocline slope as the Coriolis acceleration also becomes important.

The results of calculations with a shallow-water model in which a periodic sinusoidal forcing with the period P is imposed over the ocean are presented in **Figure 9**. The spatial and temporal structure of the forcing is given by

$$\tau = \tau_0 \sin(2\pi t/P) e^{-(y/L_y)^2 - (x - x_0)^2 / L_x^2} \qquad [25]$$

where we choose $x_0 = 180°$ W; $L_x = 40°$; $L_y = 10°$; and $\tau_0 = 0.02\,\mathrm{N\,m^{-2}}$; and $P = 5$ years. This roughly approximates to interannual wind stress anomalies associated with ENSO.

Figure 9 shows a Hovmoller diagram and the spatial structure of the ocean response at two particular instances. The thermocline response reveals slow forced anomalies propagating eastward along the equator. As discussed before, mathematically they can be obtained from a supposition of Kelvin and Rossby modes; however, the individual free waves are implicit and cannot be identified in the response. At the peaks of the anomalies, the spatial structure of the ocean response is characterized by thermocline depression or elevation in the eastern equatorial Pacific (in a direct response to the winds) and off-equatorial anomalies of the opposite sign in the western equatorial Pacific.

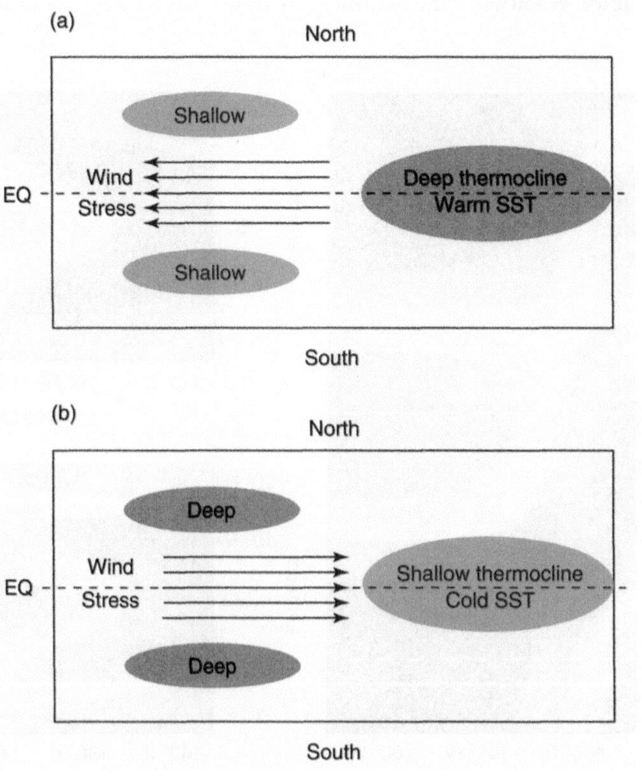

Figure 10 A schematic diagram that shows the spatial (longitude–latitude) structure of the coupled 'delayed oscillator' mode. Arrows indicate anomalous wind stresses, colored areas changes in thermocline depth. The sketch shows conditions during (a) El Niño and (b) La Niña. The off-equatorial anomalies are part of the ocean response to varying winds (cf. **Figure 9**). While the spatial structure of the mode resembles a pair of free Kelvin and Rossby waves, it is not so. The transition from (a) to (b) includes the shallow off-equatorial anomalies in thermocline depth slowly feeding back to the equator along the western boundary and then traveling eastward to reemerge in the eastern equatorial Pacific and to push the thermocline back to the surface. It may take, however, up to several years, instead of a few months, to move from (a) to (b). From Fedorov AV and and Philander SG (2001) A stability analysis of tropical ocean–atmosphere interactions: Bridging measurements and theory for El Niño. *Journal of Climate* 14(14): 3086–3101.

Conceptual Models of ENSO Based on Ocean Dynamics

So far the equatorial processes have been considered strictly from the point of view of the ocean. In particular, we have shown that wind variations are able to excite different types of anomalies propagating on the thermocline – from free Kelvin and Rossby waves generated by episodic wind bursts to gradual changes induced by slowly varying winds. However, in the Tropics variations in the thermocline depth can affect SSTs and hence the winds, which gives rise to tropical ocean–atmosphere interactions.

The strength of the easterly trade winds (that maintain the thermocline slope in **Figure 1**) is roughly proportional to the east–west temperature gradient along the equator. This implies a circular dependence: for instance, weaker easterly winds, during El Niño, result in a deeper thermocline in the eastern equatorial Pacific, weaker zonal SST gradient, and weaker winds. This is a strong positive feedback usually referred to as the Bjerknes

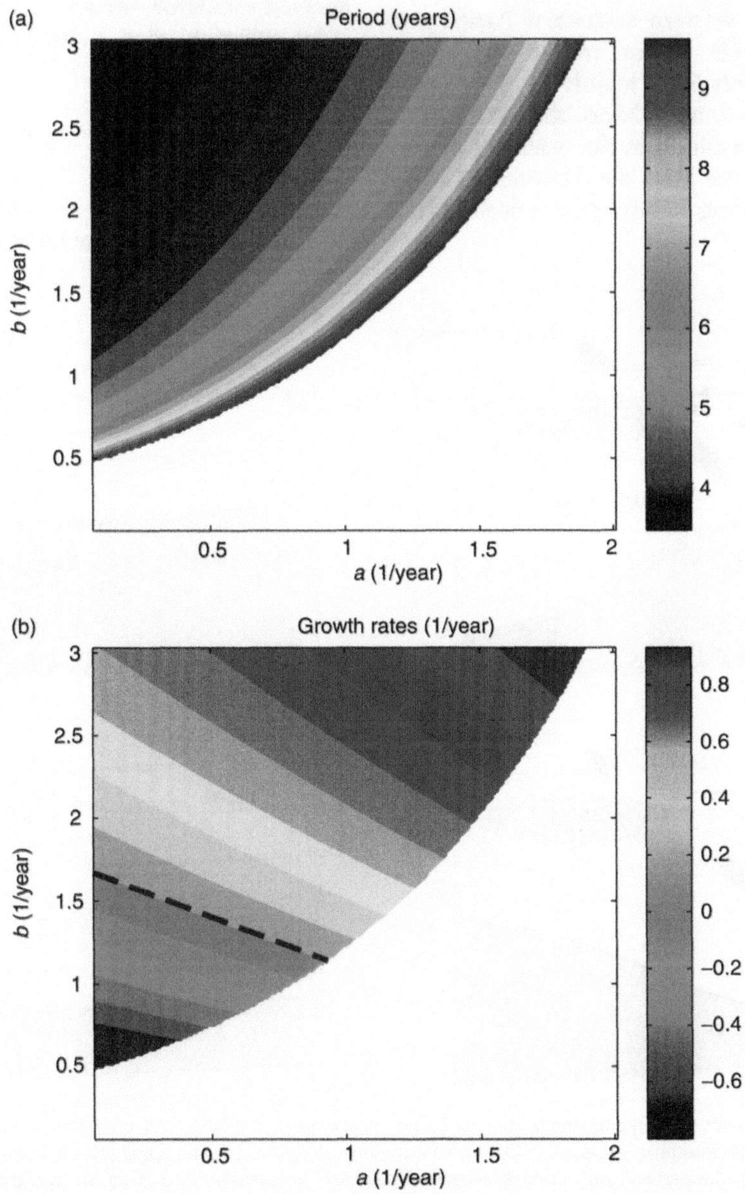

Figure 11 The period and the e-folding growth (decay) rates of the ENSO-like oscillation given by the delayed oscillator model $dT/dt = aT - bT(t - \Delta)$ as a function of a and b; for the delay $\Delta = 12$ months. The solutions of the model are searched for as $e^{\sigma t}$, where σ is a complex frequency. In the white area of the plot there are no oscillatory, but only exponentially growing or decaying solutions. At the border between the white and color areas the oscillation period goes to infinity, that is, imag$(\sigma) = 0$. The dashed line in (b) indicates neutral stability. Note that the period of the oscillation can be much longer than the delay Δ used in the equation.

feedback. On the other hand, the gradual oceanic response to changes in the winds (often referred to as 'ocean memory') provides a negative feedback and a potential mechanism for oscillatory behavior in the system. In fact, the ability of the ocean to undergo slow adjustment delayed with respect to wind variations and the Bjerknes feedback serve as a basis for one of the first conceptual models of ENSO – the delayed oscillator model.

Delayed Oscillator

Zonal wind fluctuations associated with ENSO occur mainly in the western equatorial Pacific and give rise to basin-wide vertical movements of the thermocline that affect SSTs mainly in the eastern equatorial Pacific. During El Niño, the thermocline in the east deepens resulting in the warming of surface waters. At the same time, the thermocline in the west shoals; the shoaling is most pronounced off the equator.

In this coupled mode, shown schematically in **Figure 10**, the response of the zonal winds to changes in SST is, for practical purposes, instantaneous, and this gives us the positive Bjerknes feedback described above. Ocean adjustment to changes in the winds, on the other hand, is delayed. The thermocline anomalies off the equator slowly feed back to the equator along the western boundary and then travel eastward, reemerging in the eastern equatorial Pacific, pushing the thermocline back to the surface, and cooling the SST. It may take up to a year or two for this to occur. This mode can therefore be called as a 'delayed oscillator' mode.

An equation that captures the essence of this mode is

$$T_t = aT + bT(t - \Delta) \qquad [26]$$

where T is temperature, a and b are constants, t is time, and Δ is a constant time lag. The first term on

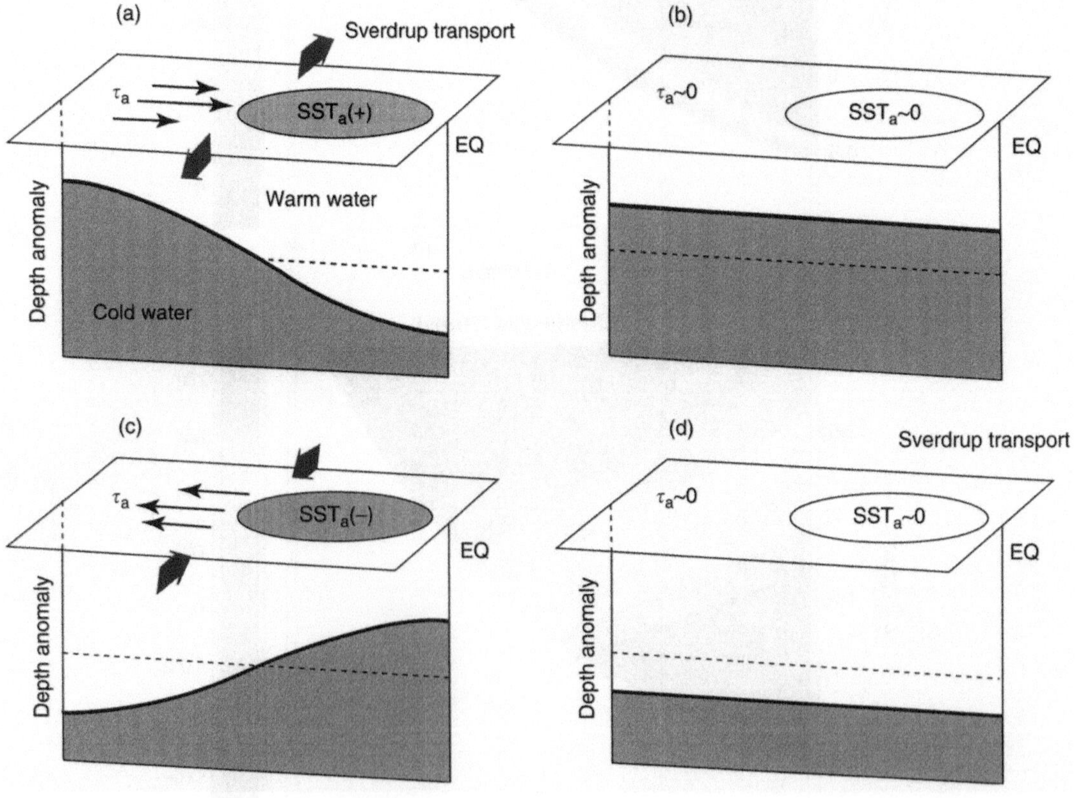

Figure 12 (a–d) A sketch showing the recharge-discharge mechanism of ENSO. All quantities are anomalies relative to the climatological mean. Depth anomaly is relative to the time mean state along the equator. Dashed line indicates zero anomaly; shallow anomalies are above the dashed line and deep anomalies are below the dashed line. Thin arrows and symbol τ_a represent the anomalous zonal wind stress; bold thick arrows represent the corresponding anomalous Sverdrup transports. SST$_a$ is the sea surface temperature anomaly. Oscillation progresses from (a) to (b), (c), and (d) clockwise around the panels following the roman numerals; panel (a) represents El Niño conditions, panel (c) indicates La Niña conditions. Note similarities with **Figure 10**. Modified from Jin FF (1997) An equatorial ocean recharge paradigm for ENSO. 1. Conceptual model. *Journal of the Atmospheric Sciences* 54: 811–829 and Meinen CS and McPhaden MJ (2000) Observations of warm water volume changes in the equatorial Pacific and their relationship to El Niño and La Niña. *Journal of Climate* 13: 3551–3559.

the right-hand side of the equation represents the positive feedbacks between the ocean and atmosphere (including the Bjerknes feedback). It is the presence of the second term that describes the delayed response of the ocean that permits oscillations (the physical meaning of the delay Δ is the time needed for an off-equatorial anomaly in the western Pacific to converge to the equator and then travel to the eastern Pacific).

The period of the simulated oscillation depends on the values of a, b, and Δ. Solutions of eqn [26] proportional to $e^{\sigma t}$ give a transcendental algebraic equation for the complex frequency σ

$$\sigma = a - be^{-\sigma t} \quad [27]$$

where

$$\sigma = \sigma_r + i\sigma_i \quad [28]$$

The solutions of eqn [27] are shown in **Figure 11** for $\Delta = 1$ year and different combinations of a and b.

Even though the term 'delayed oscillator' appears frequently in the literature, there is some confusion concerning the roles of Kelvin and Rossby waves, which some people seem to regard as the salient features of the delayed oscillator. The individual waves are explicitly evident when the winds change abruptly (**Figure** 6), but those waves are implicit when gradually varying winds excite a host of waves, all superimposed (**Figure** 7). For the purpose of deriving the delayed-oscillator equation, for instance, observations of explicit Kelvin (and for that matter individual Rossby) waves are irrelevant. The gradual eastward movement of warm water in **Figure** 7 (left panel) is the forced response of the ocean and cannot be a wave that satisfies the unforced equations of motion.

Recharge Oscillator

The delayed oscillator gave rise to many other conceptual models based on one or another type of the delayed-action equation (the Western Pacific, Advection, Unified oscillators, just to name a few, each emphasizing particular mechanisms involved in ENSO). A somewhat different approach was used by Jin in 1997 who took advantage of the fact that free Kelvin waves cross the Pacific very quickly, which allowed him to eliminate Kelvin waves from consideration and derive the recharge oscillator model.

The recharge oscillator theory is now one of the commonly used paradigms for ENSO. It relies on a phase lag between the zonally averaged thermocline depth anomaly and changes in the eastern Pacific SST.

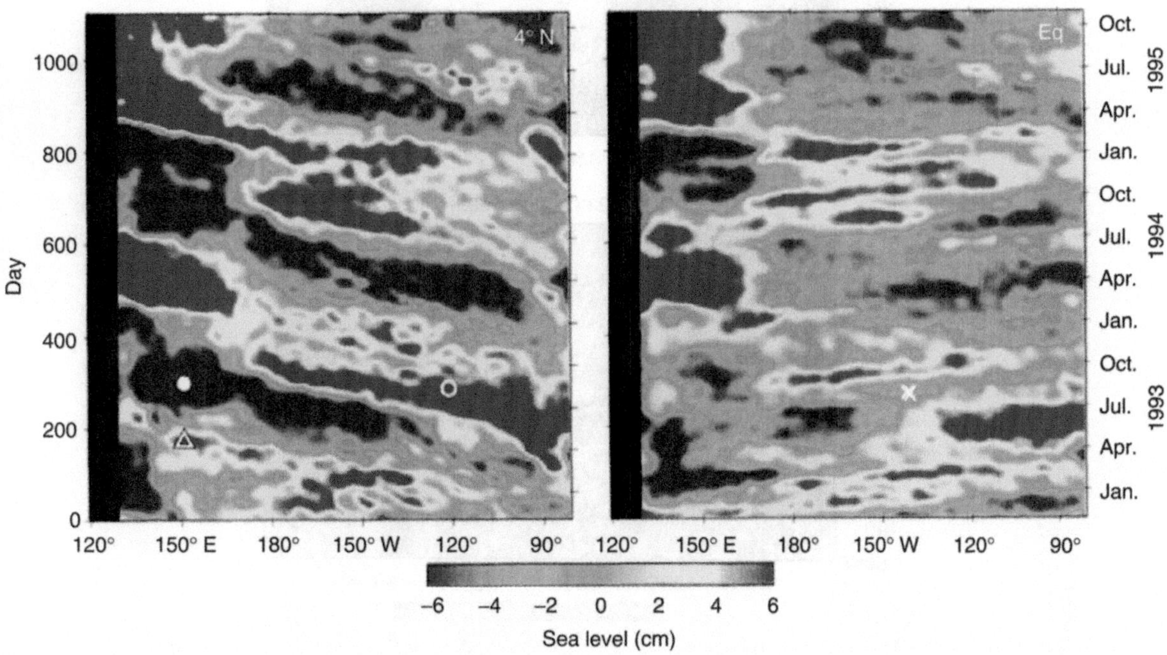

Figure 13 Observations of Rossby (left) and Kelvin (right) waves. Time-longitude sections of filtered sea level variations in the Pacific Ocean along 4° N and along the equator are shown. A section along 4° S would be similar to the 4° N section. The symbols (x, triangle, and circles) correspond to the times and locations of the matching symbols in **Figure 14**. Note that time runs from bottom to top. Obtained from TOPEX/POSEIDON satellite data; from Chelton DB and Schlax MG (1996) Global observations of oceanic Rossby Waves. *Science* 272(5259): 234–238.

Consider first a 'recharged' ocean state (**Figure 12(d)**) with a deeper than normal thermocline across the tropical Pacific. Such a state is conducive to the development of El Niño as the deep thermocline inhibits the upwelling of cold water in the east. As El Niño develops (**Figure 12(a)**), the reduced zonal trade winds lead to an anomalous Sverdrup transport out of the equatorial region. The ocean response involves a superposition of many equatorial waves resulting in a shallower than normal equatorial thermocline and the termination of El Niño (**Figure 12(b)**).

The state with a shallower mean thermocline (the 'discharged' state) is usually followed by a La Niña event (**Figure 12(c)**). During and after La Niña the enhanced trade winds generate an equatorward flow, deepening the equatorial thermocline and eventually 'recharging' the ocean (**Figure 12(d)**). This completes the cycle and makes the ocean ready for the next El Niño event.

Observations of Kelvin and Rossby Waves, and El Niño

Thirty years ago very little was known about tropical processes, but today an impressive array of instruments, the TAO array, now monitors the equatorial Pacific continuously. It is now possible to follow, as they happen, the major changes in the circulation of the tropical Pacific Ocean that accompany the alternate warming and cooling of the surface waters of the eastern equatorial Pacific associated with El Niño and La Niña. Satellite-borne radiometers and altimeters measure ocean temperature and sea level height almost in real time, providing information on slow (interannual) changes in the ocean thermal structure as well as frequent glimpses of swift wave propagation.

Figures 13 and **14** show the propagation of fast Kelvin and Rossby waves in the Pacific as seen in the satellite altimeter measurements of the sea level

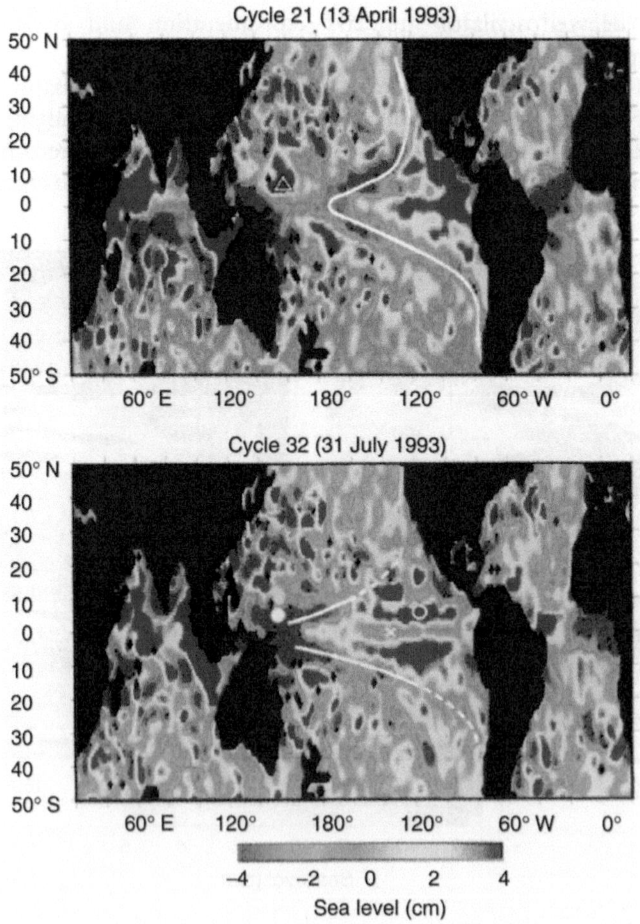

Figure 14 Observations of Rossby waves: global maps of filtered sea level variations on 13 April 1993 and $3\frac{1}{2}$ months later on 31 July. White lines indicate the wave trough. The time evolution of the equatorial Kelvin wave trough (x), the Rossby wave crest (open triangle and open circle), and the Rossby wave trough (solid circle) can be traced from the matching symbols in **Figure 13**. Obtained from TOPEX/POSEIDON satellite data, from Chelton and Schlax (1996).

height. The speed of propagation of Kelvin waves agrees relatively well with the predictions from the theory ($\sim 2.7\,\mathrm{m\,s^{-1}}$). The speed of the first-mode Rossby waves, however, is estimated from the observation to be $0.5\text{–}0.6\,\mathrm{m\,s^{-1}}$, which is somewhat lower than expected, that is, $0.9\,\mathrm{m\,s^{-1}}$. This appears to be in part due to the influence of the mean zonal currents.

Estimated variations of the thermocline depth associated with the small changes in the sea level in **Figure 13** are in the range of ± 5 m (stronger Kelvin waves may lead to variation up to ± 20 m). Note that the observed 'Rossby wave' in **Figure 14** is actually a composition of Rossby waves of different orders; higher-order waves travel much slower. The waves are forced by high-frequency wind perturbations, even though it seems likely that annual changes in the zonal winds may have also contributed to the forcing of Rossby waves.

As mentioned above, there is a clear distinction between free Kelvin waves and slow wave-like anomalies associated with ENSO. This is further emphasized by the measurements in **Figure 15** that contain evidence of freely propagating Kelvin waves (dashed lines in the left panel) but clearly show them to be separate from the far more gradual eastward movement of warm water associated with the onset of El Niño of 1997 (a dashed line in the right panel). This slow movement of warm water is the forced response of the ocean and clearly not a wave that could satisfy the unforced equations of motion. The characteristic timescale of ENSO cycle, several years, is so long that low-pass filtering is required to isolate its structure. That filtering eliminates individual Kelvin waves in the right-side panel of the figure. Whether the high- and low-frequency components of the signal can interact remains to be seen, even though it has been argued that the Kelvin waves in **Figure 15** may have contributed to the exceptional strength of El Niño of 1997–98.

The observations also provide confirmation of the recharge oscillator mechanism, which can be demonstrated, for example, by calculating the empirical orthogonal functions (EOFs) of the thermocline

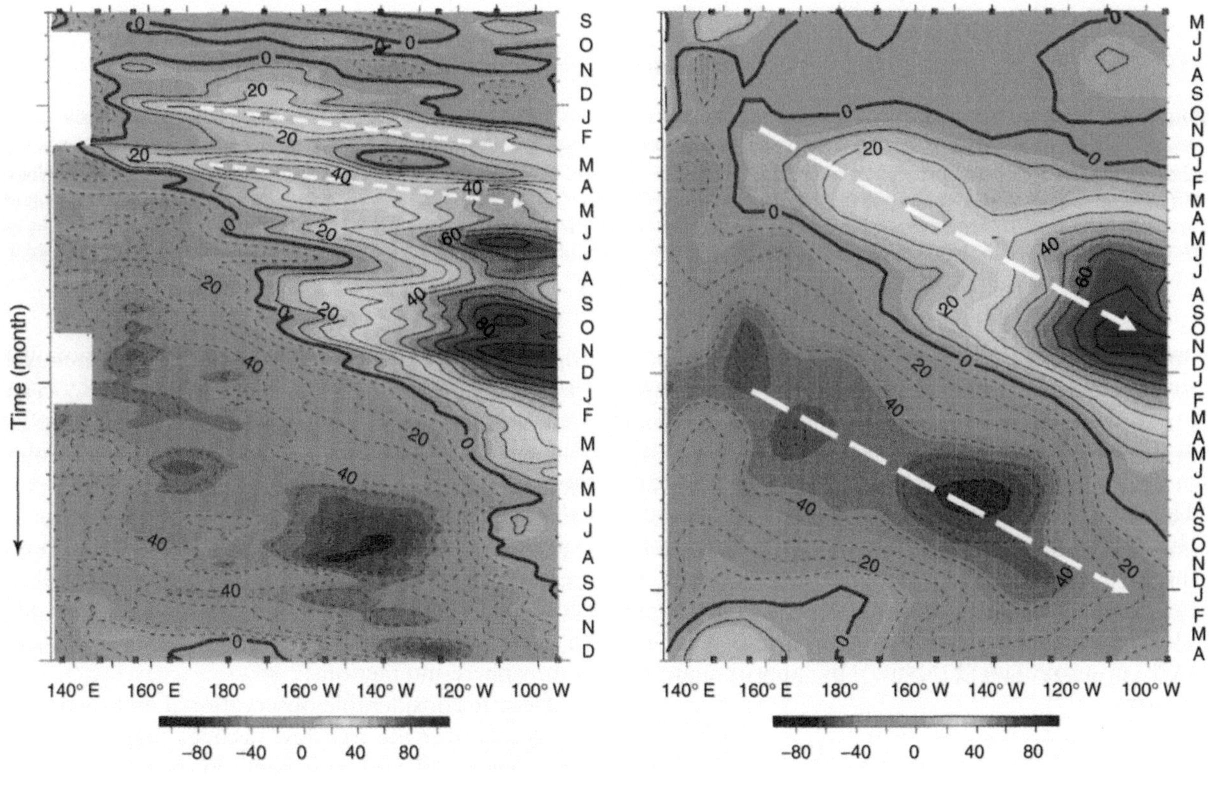

Figure 15 Observations of Kelvin waves and El Niño from the TAO array. Anomalies with respect to the long-term average of the depth of the 20 °C-degree isotherm are shown before and after the development of El Niño of 1997. Left: 5-day averages; the time axis starts in September 1996. Right: monthly averages; the time axis starts in May 1996 (cf. **Figures 8** and **9**). The dashed lines in the left-side panel correspond to Kelvin waves excited by brief WWBs and rapidly traveling across the Pacific. The dashed lines in the right-side panel show the slow eastward progression of warm and cold temperature anomalies associated with El Niño followed by a La Niña. The monthly averaging effectively filters out fast Kelvin waves from the picture leaving only gradual interannual changes. It has been argued that the Kelvin waves may have contributed to the exceptional strength of El Niño in 1997–98.

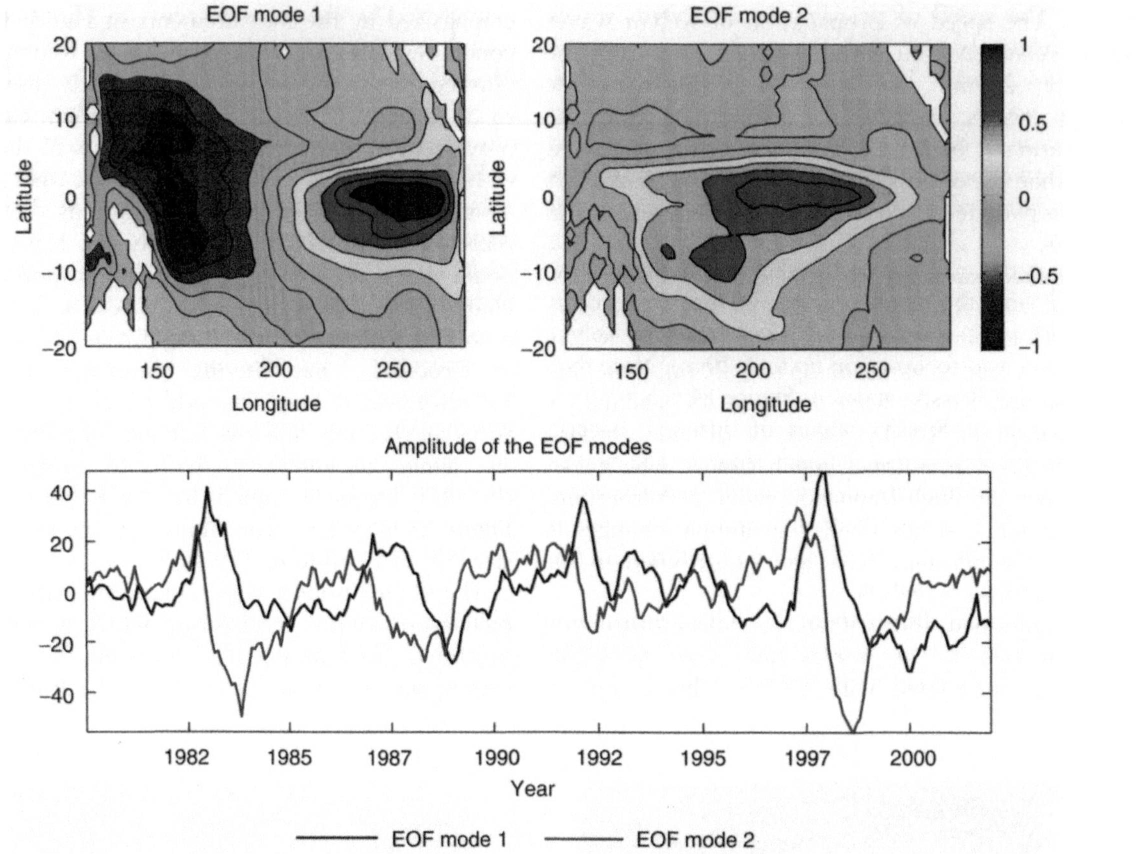

Figure 16 The first two empirical orthogonal functions (EOFs) of the thermocline depth variations (approximated as the 20 °C-degree isotherm depth) in the tropical Pacific. The upper panels denote spatial structure of the modes (nondimensionalized), while the lower panel shows mode amplitudes as a function of time (cf. **Figure 12**). The data are from hydrographic measurements combined with moored temperature measurements from the tropical atmosphere and ocean (TAO) array, prepared by Neville Smith's group at the Australian Bureau of Meteorology Research Centre (BMRC). Adapted from Meinen CS and McPhaden MJ (2000) Observations of warm water volume changes in the equatorial Pacific and their relationship to El Niño and La Niña. *Journal of Climate* 13: 3551–3559.

depth (**Figure 16**). The first EOF (the left top panel) shows the spatial structure associated with changes in the slope of the thermocline, and its temporal variations are well correlated with SST fluctuations in the eastern equatorial Pacific, or the ENSO signal. The second EOF shows changes in the mean thermocline depth, that is, the 'recharge' of the equatorial thermocline. The time series for each EOF in the bottom panel of **Figure 16** indicate that the second EOF (the thermocline recharge) leads the first EOF (a proxy for El Niño) by approximately 7 months.

Summary

Early explanations of El Niño that relied on the straightforward generation of free Kelvin and Rossby waves by a WWB have been superseded. Modern theories consider ENSO in terms of a slow oceanic adjustment which occurs as a sum of continuously forced equatorial waves. The concept of 'ocean memory' based on the delayed ocean response to varying winds has become one of the cornerstones for explaining ENSO cyclicity. It is significant that although from the point of view of the ocean a superposition of forced equatorial waves is a direct response to the winds, from the point of view of the coupled ocean–atmosphere system it is a part of a natural mode of oscillation made possible by ocean–atmosphere interactions.

Despite considerable observational and theoretical advances over the past few decades many issues are still being debated and each El Niño still brings surprises. The prolonged persistence of warm conditions in the early 1990s was as unexpected as the exceptional intensity of El Niño in 1982 and again in 1997. Prediction of El Niño also remains problematic. Not uncommonly, when a strong Kelvin wave crosses the Pacific and leaves a transient warming of

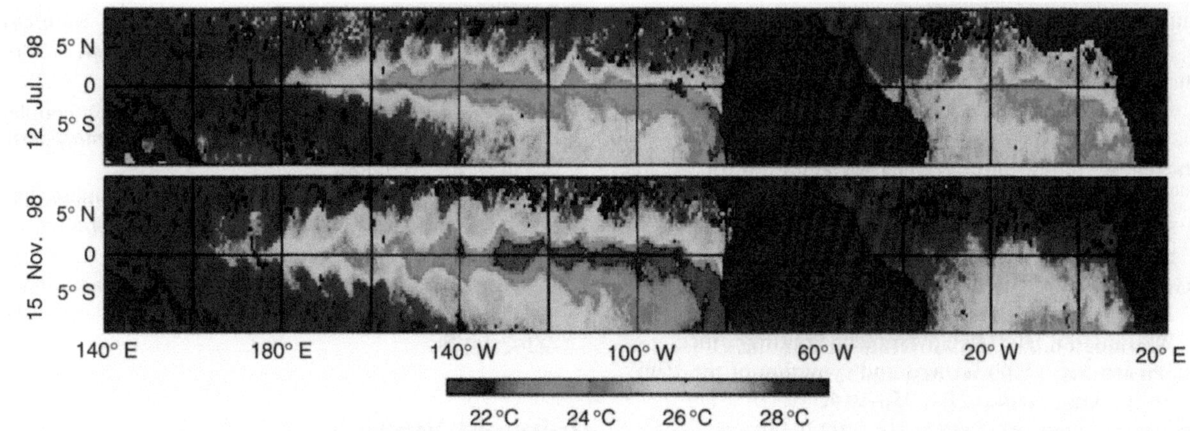

Figure 17 Tropical instability waves (TIWs): 3-day composite-average maps from satellite microwave SST observations for the periods 11–13 July 1998 (upper) and 14–16 November 1998 (lower). Black areas represent land or rain contamination. The waves propagate westward at approximately 0.5 m s^{-1}. From Chelton DB, Wentz J, Gentemann CL, de Szoeka RA, and Schlax MG (2000) Satellite microwave SST observations of trans-equatorial tropical instability waves. *Geophysical Research Letters* 27(9): 1239–1242.

1–2 °C in the eastern part of the basin, a question arises whether this might be a beginning of the next El Niño. To what degree, random transient disturbances influence ENSO dynamics remains unclear.

Many theoretical and numerical studies argue that high-frequency atmospheric disturbances, such as WWBs that excite Kelvin waves, can potentially interfere with ENSO and can cause significant fluctuations in its period, amplitude, and phase. Other studies, however, insist that external to the system atmospheric 'noise' has only a marginal impact on ENSO. To resolve this issue we need to know how unstable the coupled system is. If it is strongly damped, there is no connection between separate warm events, and strong wind bursts are needed to start El Niño. If the system is sufficiently unstable then a self-sustained oscillation is possible. The truth is probably somewhere in between – the coupled system may be close to neutral stability, perhaps weakly damped. Random atmospheric disturbances are necessary to sustain a quasi-periodic, albeit irregular oscillation.

Another source of random perturbations that affects both the mean state and interannual climate variations is the tropical instability waves (TIWs) typically observed in the high-resolution snapshots of tropical SSTs (**Figure 17**). These waves, propagating westward with typical phase speed of roughly 0.5 m s^{-1}, are excited by instabilities of the zonal equatorial currents with strong vertical and horizontal shear. The wave dynamical structure corresponds to that of cyclonic and anticyclonic eddies having maximum velocities near the ocean surface and penetrating into the ocean by a few hundred meters. The waves can affect the temperature of the equatorial cold tongue, and the properties of ENSO, by modulating meridional heat transport from the equatorial Pacific. Overall, the role of the TIWs remains a subject of intensive research which includes the effect of these waves on the coupling between the wind stress and SSTs and the interaction between the TIWs and Rossby waves.

See also

Pacific Ocean Equatorial Currents.

Further Reading

Battisti DS (1988) The dynamics and thermodynamics of a warming event in a coupled tropical atmosphere/ ocean model. *Journal of Atmospheric Sciences* 45: 2889–2919.

Chang PT, Yamagata P, Schopf SK, *et al.* (2006) Climate fluctuations of tropical coupled system – the role of ocean dynamics. *Journal of Climate* 19(20): 5122–5174.

Chelton DB and Schlax MG (1996) Global observations of oceanic Rossby waves. *Science* 272(5259): 234–238.

Chelton DB, Schlax MG, Lyman JM, and Johnson GC (2003) Equatorially trapped Rossby waves in the presence of meridionally sheared baroclinic flow in the Pacific Ocean. *Progress in Oceanography* 56: 323–380.

Chelton DB, Wentz J, Gentemann CL, de Szoeka RA, and Schlax MG (2000) Satellite microwave SST observations of trans-equatorial tropical instability waves. *Geophysical Research Letters* 27(9): 1239–1242.

Fedorov AV and Philander SG (2000) Is El Niño changing? *Science* 288: 1997–2002.

Fedorov AV and Philander SG (2001) A stability analysis of tropical ocean–atmosphere interactions: Bridging measurements and theory for El Niño. *Journal of Climate* 14(14): 3086–3101.

Gill AE (1982) *Atmosphere-Ocean Dynamics*, 664p. New York: Academic Press.

Jin FF (1997) An equatorial ocean recharge paradigm for ENSO. 1. Conceptual model. *Journal of the Atmospheric Sciences* 54: 811–829.

Kessler WS (2005) Intraseasonal variability in the oceans. In: Lau WKM and Waliser DE (eds.) *Intraseasonal variability of the Atmosphere-Ocean System*, pp. 175–222. Chichester: Praxis Publishing.

Levitus S and Boyer T (1994) *World Ocean Atlas 1994, Vol. 4: Temperature NOAA Atlas NESDIS4*. Washington, DC: US Government Printing Office.

McPhaden MJ (1999) Genesis and evolution of the 1997–98 El Niño. *Science* 283: 950–954.

Meinen CS and McPhaden MJ (2000) Observations of warm water volume changes in the equatorial Pacific and their relationship to El Niño and La Niña. *Journal of Climate* 13: 3551–3559.

Philander G (1990) *El Niño, La Niña, and the Southern Oscillation. International Geophysics Series*, 293p. New York: Academic Press.

Schopf PS and Suarez MJ (1988) Vacillations in a coupled ocean atmosphere model. *Journal of the Atmospheric Sciences* 45: 549–566.

Wang C, Xie SP, and Carton JA (2004) Earth's climate: The ocean–atmosphere interaction. *Geophysical Monograph 147, American Geophysical Union*, 405p.

Zebiak SE and Cane MA (1987) A model El Niño-southern oscillation. *Monthly Weather Review* 115(10): 2262–2278.

Relevant Website

http://www.pmel.noaa.gov
– Tropical Atmosphere Ocean Project, NOAA.

CIRCULATION AND RELATED MODELS

OCEAN CIRCULATION

N. C. Wells, Southampton Oceanography Centre, Southampton, UK

Introduction

This article discusses the following aspects of ocean circulation: what is meant by the term ocean circulation; how the ocean circulation is determined by measurements and dynamical processes; the consequences of this circulation on the Earth's climate.

What is Ocean Circulation?

The ocean circulation in its simplest form is the movement of sea water through the ocean, which principally transfers temperature and salinity, from one region to another. Temperature differences between regions cause heat transfers. Similarly, differences in salinity produce transfers of salt. On the time scale of the ocean circulation the inputs and exports of salt into and out of the ocean make a negligible contribution to overall salinity and so variations in salinity occur by the addition and removal of fresh water into and out of the ocean.

Two major processes control the ocean circulation: the action of the wind and the action of small density differences, produced by differences in temperature and salinity, within the ocean. The former process is the wind driven circulation (*see* Wind Driven Circulation) whereas the latter is the thermohaline circulation. Although it is useful to separate these two processes to better understand the ocean circulation, they are not independent from each other.

Ocean circulation is in reality a very complex system, as the flows are not steady in time or space. They are turbulent flows that show variability on scales from the largest scale of the ocean basins to the smallest scales where the energy is finally dissipated as heat. This turbulent structure of the ocean means there are fundamental limitations on the predictability of its behavior.

Because of this inherent complexity oceanographers have approached ocean circulation by using a combination of observational methods including ships, buoys Moorings and satellites, combined with the mathematical methods of dynamical oceanography (*see* Inverse Models). This integrated approach allows hypotheses to be made that can be tested by comparison with observations. Furthermore, mathematical models of the ocean circulation, based on the dynamical principles, can be constructed and tested against observations (*see* Regional and Shelf Models).

This article considers how ocean circulation is measured, how the major processes at work are determined and the consequences of the ocean circulation on the climate system.

How is the Ocean Circulation Determined?

The determination of ocean currents involves measurement of the displacement of an element of fluid over a measured time interval. The position of the measurement is defined mathematically in a Cartesian coordinate system (**Figure 1**) where x is positive eastward direction (lines of constant latitude), y is positive northward direction (geographic North Pole), and z is positive upwards; $z = 0$ corresponds to mean sea level. Without ocean currents and tides the sea level would be an equipotential surface, i.e., one of constant potential energy. The z coordinate is perpendicular to the equipotential surface. The origin is the intersection of the Greenwich meridian (Universal meridian) and the equator with mean sea level.

The coordinates of a parcel of water can be determined by the Global Positioning System (GPS). This satellite-based system provides a horizontal position with an accuracy of better than 100 m, which is sufficient for large-scale flows in the ocean. Large-scale flows are at least 10 km in spatial scale and have timescales of at least a day. A pressure device attached to the current meter normally determines the vertical position.

There are two mathematical methods for defining the displacement of the fluid. One is to measure the velocity of the fluid at a fixed point in the ocean, and the other is to follow the element of the fluid and to measure its velocity as it moves through the ocean. The first method is known as a Eulerian description and the second is a Lagrangian description of flow. In principle, the two methods are independent of each other. This means that a Eulerian measurement can not provide Lagrangian currents, and Lagrangian measurements can not provide Eulerian currents.

Having defined the two mathematical methods how the currents are measured in practice is now

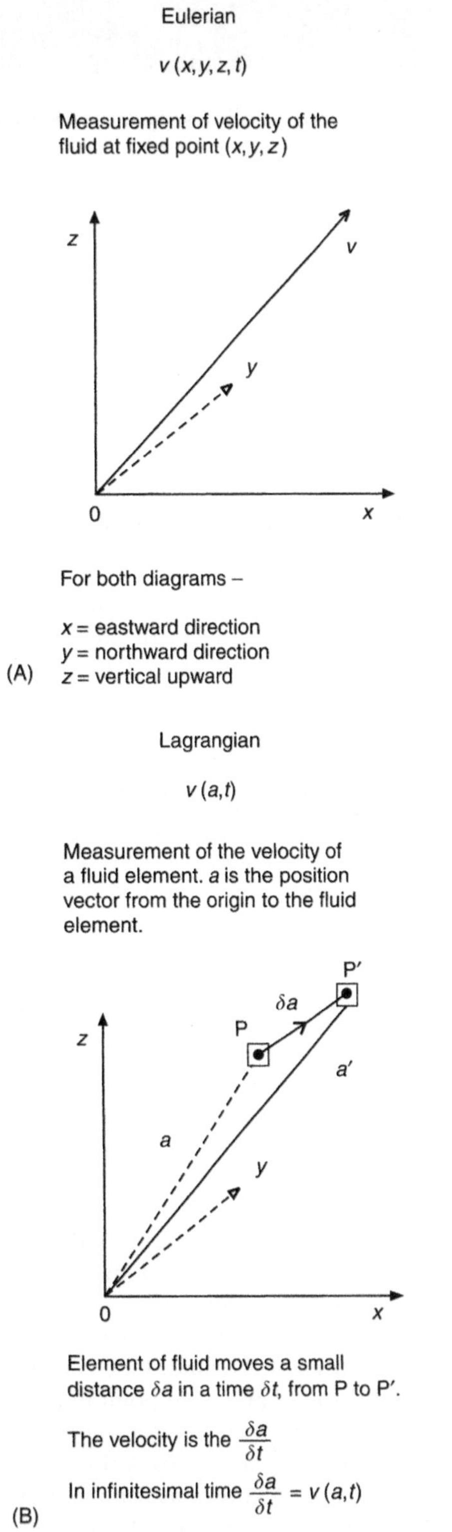

Eulerian
v (x, y, z, t)
Measurement of velocity of the fluid at fixed point (x, y, z)
z
v
y
0
x
For both diagrams –
x = eastward direction
y = northward direction
(A)
z = vertical upward
Lagrangian
v (a, t)
Measurement of the velocity of a fluid element. a is the position vector from the origin to the fluid element.
P′
δa
P
z
a′
a
y
0
x
Element of fluid moves a small distance δa in a time δt, from P to P′.
The velocity is the δa/δt
In infinitesimal time δa/δt = v (a, t)
(B)

Figure 1 Eulerian and Lagrangian specifications of flow.

considered. Initially, these methods will address only the measurement of the horizontal flow. The vertical flow is difficult to measure directly, and will be discussed later in this article.

First, the Eulerian method is considered. The measurement of the flow at a fixed point in the ocean is only straightforward when a fixed position can be maintained, for instance with a current meter attached to the bottom of the ocean or to a pier on the coast. Most measurements have to be made well away from land. This is achieved by attaching the current meter to a mooring which is attached to weights and then deployed (**Figure 2**). The position of a mooring can be determined by GPS. The current meter may be a rotary device or an acoustic device. The rotary current meter measures the number of revolutions over a fixed period, whereas the acoustic one measures the change in frequency of an emitted sound pulse caused by the ocean current (i.e., it uses the Doppler effect). Moorings may be deployed for periods of up to 2 years. In the analysis of the record it is normal to remove the high frequency variability of less than 1 day caused by tides by filtering the data.

A Lagrangian measurement of current can be determined by following an element of water with a float. The horizontal displacement of the water over a small interval of time defines the Lagrangian current. **Figure 3** shows typical float designs that are

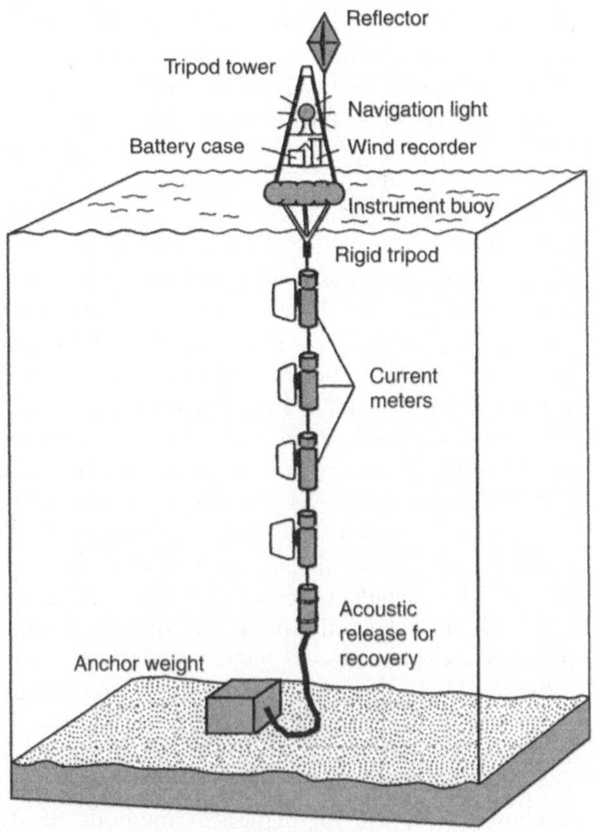

Figure 2 A typical current meter mooring.

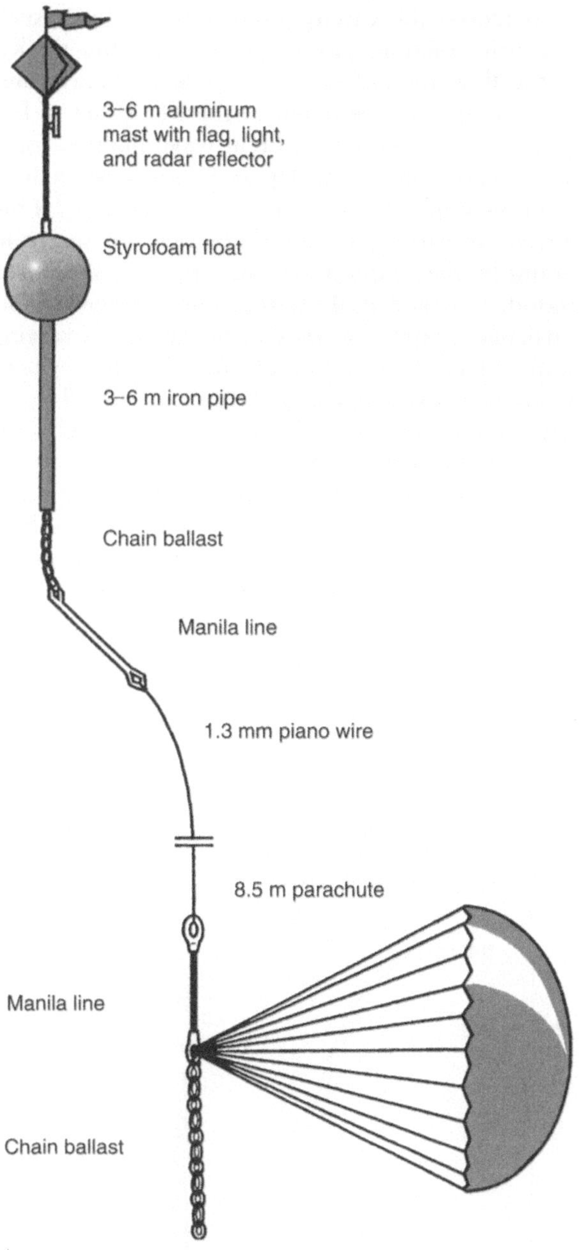

Figure 3 A typical drifter with a parachute drogue of a few meters below the surface. It will follow the current at the depth of the parachute.

used. The position of the float can be determined by two methodsA float that has a surface satellite transmitter/receiver can have its position determined by GPS, whereas a subsurface float would use an acoustic navigation system. Some floats can descend to a predetermined depth, maintain that depth for a few weeks and then return to the surface for a position fix. This technique allows the current to be measured down to depths of 1 km below the surface.

Each method gives different information on the flow field. A mooring will give a time series of the horizontal current, whereas a float will give the trajectory of the horizontal displacement of the parcel. It is worth remarking that most of the information on the surface ocean circulation has come from mariners' observations of the ships set, a method which has been used since the nineteenth century. However, these measurements have their limitations since they are neither eulerian nor lagrangian measurements and additional influences (e.g., wind effects on the ship) may cause errors.

This information can be analyzed in many different ways to discern the major current systems. From a set of moorings deployed for a few years across, say, the Gulf Stream, the mean flow (i.e., the average of all the current measurements) and its variability can be determined. The mean flow could be calculated over a particular time period. This time period is limited by the period of deployment, which is of the order of 2 years. This is rather short for a climatological mean, and a much longer period of 10 years is desirable. A few longer time series of currents have been determined for the Gulf Stream in the Florida Straits and for the Antarctic Circumpolar Current in the Drake Passage (*see* Antarctic Circumpolar Current).

Recall ocean currents are turbulent and therefore have variability on a whole range of timescales. Hence the mean flow gives no information on the variability of the flow. However, the statistics of the flow can be calculated, based on the kinetic energy (KE). The kinetic energy/unit volume is defined as:

$$\mathrm{KE} = \frac{1}{2}\rho\left[u^2 + v^2\right]$$

where ρ is the density of the sea water and u and v are the eastward and northward components of the horizontal flow, respectively.

If the time mean current is defined as $\bar{u}$ and u' as the deviation from $\bar{u}$ at any time, the mean kinetic energy (KEM) and eddy kinetic energy (EKE) can be defined by:

$$\mathrm{KEM} = \frac{1}{2}\rho\left[\bar{u}^2 + \bar{v}^2\right]$$

$$\mathrm{EKE} = \frac{1}{2}\rho\left[u'^2 + v'^2\right]$$

These two numbers give quantitative measures on the mean and variability of the flow respectively. The ratio EKE/KEM gives a measure of the relative variability of the flow. If the ratio is very much less than 1 then the flow is steady, whereas if the ratio is approximately equal to 1 then the flow is very variable.

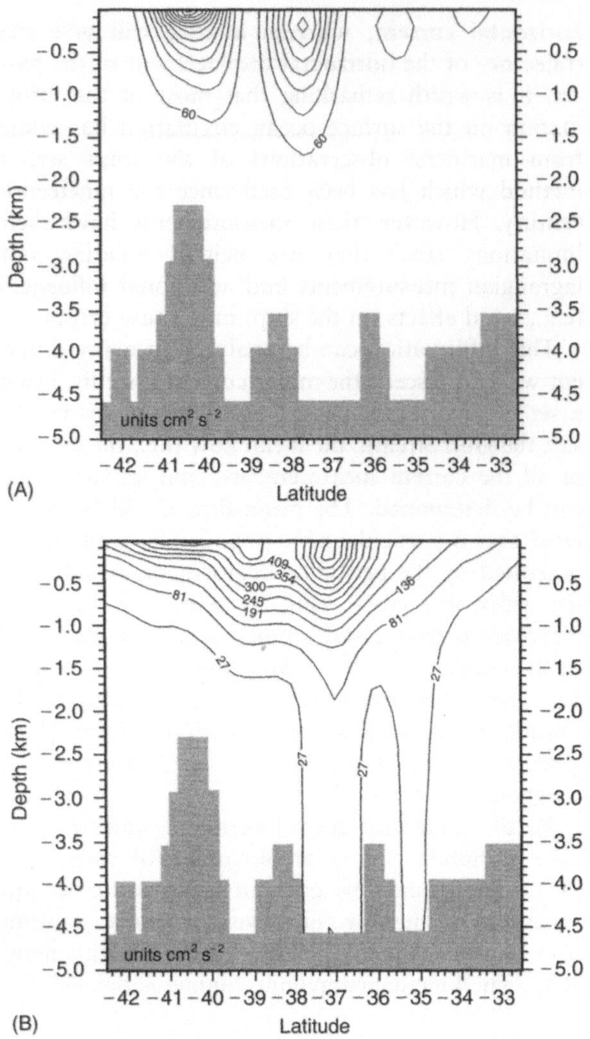

Figure 4 The mean kinetic energy (KEM) (A) and the eddy kinetic energy (EKE) (B) in a north–south slice through the Agulhas Current system at 14.4°E. The KEM maximum corresponds to the mean position of the Agulhas Return Current (Eastward flow) between 40° and 41°S, and the Agulhas Current (Westward flow) between 37°S and 38°S. The EKE distribution is much broader than KEM, which shows the large horizontal extent of the flow variability. The ratio of EKE/KEM is typically about a third, which indicates a very variable current system. (Reproduced from Wells NC, Ivchenko V and Best SE (2000) Instabilities in the Agulhas Retroflection Current system: A comparative model study. *Journal of Geophysical Research* 105: 3233–3246.)

Figure 4 shows the variability of the flow in the Agulhas Current (*see* Agulhas Current), which is an intense and highly variable current off the coast of South Africa.

Although this ratio gives a measure of the variability of the current, it does not give any idea of the exact time or space scales over which the current is varying. For example, the current may show a slow change from one season to another or it may show faster variation due to eddies.

To address this variation time series analysis, such as Fourier analysis, can be used to determine the KE of the flow for different time periods. Fourier analysis produces a spectrum of the KE, either in frequency for a time series, or in wave number for a spatial variation in flow. **Figure** 5 shows the analysis of a time series into its component frequencies. If the current is varying on all timescales the spectrum would be flat, but if there was only one dominant period, it would peak at that one frequency. This particular analysis shows that the current is varying at the tidal frequency and the inertial frequency both at the high frequency end of the spectrum. The inertial frequency is given by $2\Omega \sin \phi$ where Ω is the rotation rate of the earth and ϕ is the latitude. At the lower frequency end, which corresponds to the ocean circulation frequency, there is a broad band of high kinetic energy. This band is due to eddies (*see* Mesoscale Eddies) which cause fluctuations of currents on timescales of weeks to months.

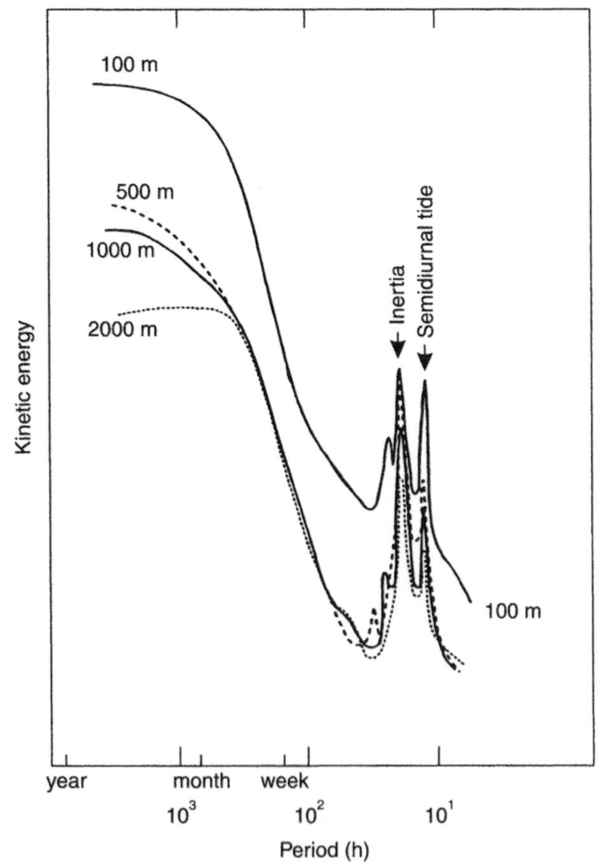

Figure 5 Frequency spectrum of kinetic energy from four depths at site D (39°N, 70°W), north of the Gulf Stream. Note the two high-frequency peaks, coinciding with the inertial period (19 h) and the semidiurnal tide (12.4 h). (Reproduced from Rhines PB (1971) A note on long-period motions at site D. *Deep Sea Research* 18: 21–26.)

For these mean climatological currents, our knowledge has been augmented by the application of the dynamic method. This method is based on the observation that large-scale ocean currents are in geostrophic balance, over large areas of the ocean. Geostrophic balance means that the Coriolis force balances the horizontal pressure gradient force. The geostrophic flow is a good approximation to the flow in the interior of an ocean outside the equatorial region. The horizontal pressure gradient is dependent on the slope of the ocean surface and the horizontal variation of the density distribution within the ocean. In the future, the former may be determined by satellite measurements of the sea surface height and the geoid[1] but at present we do not have an accurate geoid at sufficiently high resolution to measure the sea surface slope. The latter can be determined from temperature, salinity and pressure measurements that have been made over large ocean areas during the last century. The dynamic method allows the determination of the vertical shear of the horizontal geostrophic current, and therefore to determine the absolute geostrophic current, additional measurements are required. For example if the current has been measured at a particular depth then the dynamic method can be referenced to that depth and the vertical profile of current can be obtained.

The recent World Ocean Circulation Experiment (WOCE) hydrographic program has provided more measurements of the ocean than all previous hydrographic programs and will give the most comprehensive assessment of climatological horizontal ocean flow to date.

Recall that the vertical circulation of the ocean cannot be measured directly because it is technically too difficult. Current indirect methods used to determine the vertical circulation rely on the use of mathematical approaches, such as dynamical models, or the use of chemical tracers.

Observations of temperature and salinity can be inserted into a mathematical ocean general circulation model which allows the three-dimensional, circulation to be determined, subject to limitations in the accuracy of the model.

Chemical tracers have been inadvertently injected into the ocean from nuclear tests in the 1960s and from industrial processes (e.g., chlorofluorocarbons). Naturally occurring tracers such as ^{14}C also existThese tracers can be measured with high accuracy in a few laboratories around the world and from their distributions at different times, the three-dimensional circulation can be estimated. This method reveals the time history of the ocean circulation wherever the tracer is measured. This is very different information from that provided by the methods previously discussed, but nonetheless it can reveal unique aspects of the flow. For example, nuclear fallout deposited in the surface layer of the Nordic seas in the 1960s was located in the deep western boundary current 10 years later.

[1] The geoid is an equipotential surface, which would be represented by the sea level of a stationary ocean. Ocean currents cause deviations in sea level from the geoid.

An Ocean General Circulation Model

An ocean general circulation model is composed of a set of mathematical equations which describe the time-dependent dynamical flows in an ocean basin. The basin is discretized into a set of boxes of regular horizontal dimensions but variable thickness in the vertical dimension. The horizontal flow (northward and eastward components) is predicted by the momentum equation (**Figure 6A**) at the corners of each box (**Figure** 7).

The forcing for the flow may come from the wind stress (the frictional term in the momentum equation) and from the surface buoyancy fluxes, arising from heat and freshwater (precipitation + runoff–evaporation) exchange with the atmosphere and adjacent landmasses. These buoyancy fluxes change the temperature and salinity in the surface layer of the ocean. The surface water masses are then subducted into the interior of the ocean by the vertical and horizontal components of the flow, where they are mixed with other water masses.

The processes of transport and mixing are described by the temperature and salinity equations (**Figure 6B** and **C**), at the center of each ocean box (**Figure** 7). From these two equations the seawater density, and thence the pressure can be obtained for each box. The horizontal pressure gradient is then determined for the momentum equation, and the vertical velocity is calculated from the horizontal divergence of the flow. This set of time-dependent equations can then be used to describe all the dynamical components of the flow field, provided that suitable initial and boundary conditions are specified.

Wind-driven and Thermohaline Circulation

The wind-driven circulation (*see* Wind Driven Circulation) is considered first. The surface layer of the ocean is directly driven by the surface wind stress and is also subject to the exchange of heat and fresh water between ocean and atmosphere. This layer,

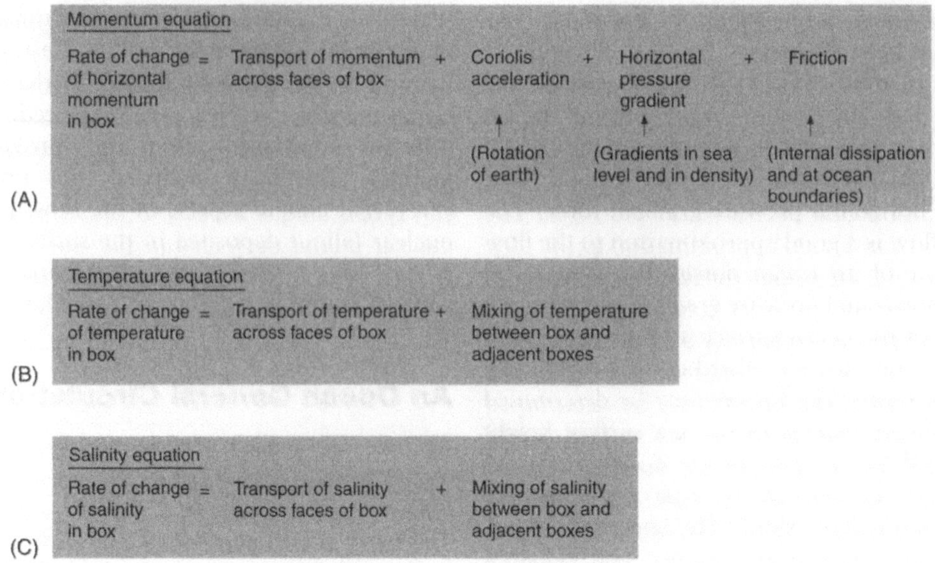

Figure 6 The basic equations for an ocean general circulation model. (A) Momentum equation; (B) temperature equation; (C) salinity equation. (Reproduced from Summerhayes and Thorpe, 1996.)

which is typically less than 100 m in depth, is referred to as the Ekman layer. That is a steady wind stress causes a transport of the surface water 90° to the right of the wind direction in the Northern Hemisphere and 90° to the left in the Southern Hemisphere. This is due to the combined action of the wind stress on the ocean surface and the Coriolis force. These Ekman flows can converge and produce a downwelling flow into the interior of the ocean. Conversely a divergent Ekman transport will produce an upwelling flow from the interior into the surface layer.

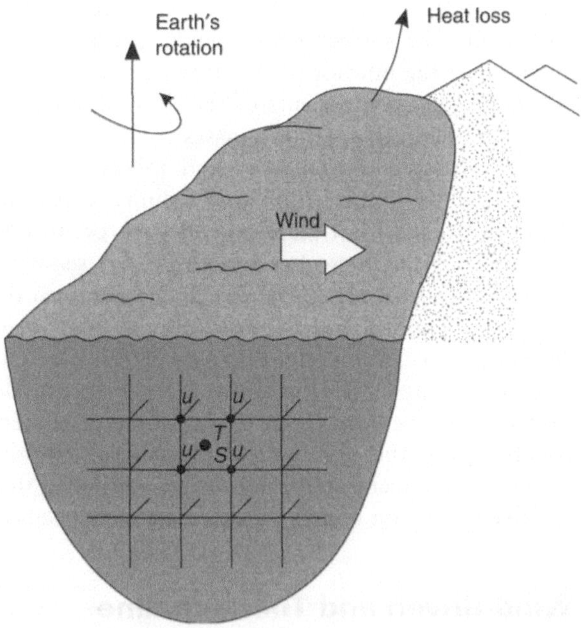

Figure 7 A schematic of the model boxes in an ocean general circulation model. The equations (**Figure 6**) for momentum are solved at the corners of the boxes (*u*), while the temperature (*T*), and salinity (*S*) equations are solved at the centers of boxes. The model is forced by climatological wind stress, surface heat fluxes, and freshwater fluxes. (Reproduced from Summerhayes and Thorpe, 1996.)

This type of flow is known as Ekman pumping, and is directly related to the Curl of the Wind Stress (*see* Wind Driven Circulation). It is of fundamental importance for the driving of the large-scale horizontal circulation, in the upper layer of the ocean. For example, between 30° and 50° latitude the climatological westerly wind, drives an Ekman flow equatorward, whereas between 15° and 30° latitude the trade winds drive an Ekman flow polewards. At about 30° latitude the flows converge and sink into the deeper ocean. Before discussing the influence of Ekman pumping on the interior ocean circulation the role of density is considered.

The density of sea water increases with depth. From hydrographic measurements of density, the horizontal variation of the depth of a chosen density surface can be mapped. These constant density surfaces are known as isopycnals. The flow tends to move along these surfaces and therefore the variations in the depth of these surfaces gives a picture of the horizontal flow in the deep ocean, away from the surface layer and benthic layer. The isopycnal surfaces dip down in the center of the subtropical gyre at about 30°. The formation of this lens of light warm water is related to the climatological distribution of surface winds, which produce a convergence of Ekman transport towards the center of the gyre, and a downwelling of surface waters into the interior of the ocean. At the center of the lens, the sea surface domes upwards reaching a height of 1 m above the sea surface at the rim. Due to hydrostatic forces the main thermocline is depressed downwards to depths of the order of 500–1000 m (**Figure 8**).

The surface horizontal circulation flows anticyclonically around the lens with the strongest currents on the western edge, where the slope of the density surface reaches a maximum. These are geostrophic currents, where there is a balance between the Coriolis force and the horizontal pressure gradient force. Generally, the circulation in the subtropical gyres is clockwise in the Northern Hemisphere and anticlockwise in the Southern Hemisphere. These large-scale horizontal gyres are ultimately caused by the climatological surface wind circulation and are found in all the ocean basins.

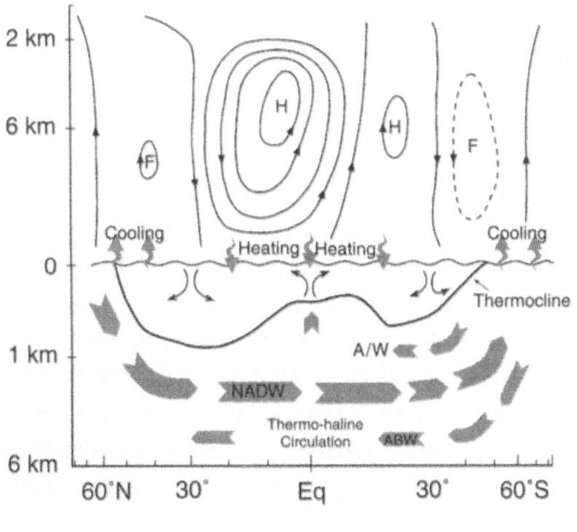

Figure 8 A representation of the meridional average section through the atmosphere for December–February in the Northern Hemisphere. The cells are the Hadley Cell (H) and Ferrel Cell (F). The strength of the cells is represented by the solid contours which are in units of 40 megatonnes/second, whilst the dashed contours are in units of 20 megatonnes/second. Note the predominantly downward motion at ~30 degree latitude, associated with the subtropical anticyclones, and the strong upward motion at equatorial latitudes which is associated with the Inter-Tropical Convergence Zone. A meridional transect through the Atlantic Ocean, showing the position of the main thermocline. The small arrows represent the wind driven downwelling (Ekman pumping) at ~30 degree latitude, and the equatorial upwelling, which occurs within and above the main thermocline. The North Atlantic Deep Water (NADW) is produced in the Labrador and Nordic Seas and is the predominant deep water mass by volume. The Antarctic Intermediate Water (AIW) is produced at ~50°S, and by virtue of its salinity is lighter than the NADW. In contrast Antarctic Bottom Water is the most dense water mass in the worlds ocean and is formed in the Weddell and Ross Seas. These deep flows form part of the thermo-haline circulation. The vertical scales are exaggerated in the lower troposphere and in the upper ocean. The horizontal scale is proportional to the area of the Earth's surface between latitude circles.

The surface layer is also subject to heating and cooling, and the exchange of fresh water between ocean and atmosphere, both of which will change the density of the layer. For example, heat is lost over the Gulf Stream on the rim of the light water lens of the subtropical gyre. Recall that flow tends to follow isopycnal layers and these layers will slope downwards towards the center of the gyre. Cooling of the waters in the Gulf Stream leads to the sinking of surface waters to produce a water mass known as 18°C water. This water, which is removed from the surface layer, will slowly move along the isopycnal layers into the thermocline. As it moves clockwise around the gyre it will be subducted in to the deeper layers of the thermocline, in a spiral-like motion (**Figure 9**). The deepest extent of the main thermocline is located in the subtropical gyre to the west of Bermuda on the eastern edge of the Gulf Stream rather than in the center of the ocean basin.

This asymmetry of the gyre is related to the beta effect, i.e., the change of the Coriolis parameter with latitude (*see* Agulhas Current).

The subtropical gyres are one of the most well-studied regions of the ocean, and our understanding is therefore most developed in these regions. These

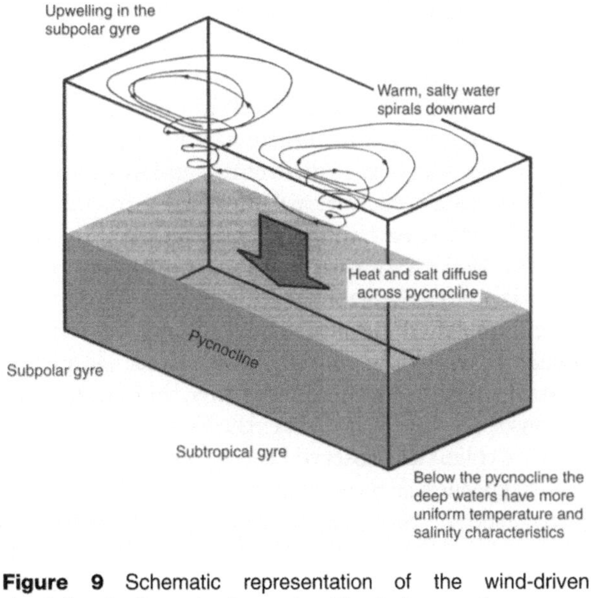

Figure 9 Schematic representation of the wind-driven circulation in the subpolar and subtropical gyre of an ocean basin. The wind circulation causes a convergence of Ekman transport to the center of the subtropical gyre and downwelling into the interior. Conversely in the subpolar gyre there is a divergence of the Ekman transport and upwelling from the interior into the surface layer. This Ekman pumping is responsible for the gyre circulations (see text for details). The western boundary currents are depicted by the closeness of the streamlines. They are caused by the poleward change in the Coriolis force known as the BETA effect. (Reproduced from Bean MS (1997) PhD thesis, University of Southampton.)

gyres occur in the surface and thermocline regions of the ocean and are primarily controlled by the wind circulation, with modifications due to heating and cooling of the surface. The question now arises of why thermoclines are seen in the ocean. For example, why is the warm water not mixed over the whole depth of the ocean and why is the average ocean temperature about 3°C.

To explain the observed behavior thermohaline circulation, which is generated by small horizontal differences in density, due to temperature and salinity, between low and high latitude is considered. How does it work? Consider an ocean of uniform depth and bounded at the equator and at a polar latitude. We will assume it has initially a uniform temperature and is motionless (for the moment the effect of salinity on density are ignored). This hypothetical ocean is then subject to surface heating at low latitudes and surface cooling at high latitudes. In the lower latitudes the warming will spread downwards by diffusion, whereas in high latitudes the cooling will spread downwards by convection which is a much faster process than diffusion. The heavier colder water will induce a higher hydrostatic pressure at the ocean bottom than will occur at low latitudes. The horizontal pressure gradient at the ocean bottom is directed from the high latitudes to the lower latitudes, and will induce an equatorward abyssal flow of polar water. The flow can not move through the equatorial boundary of our hypothetical ocean and therefore will upwell into the upper layer of the tropical ocean, where it will warm by diffusion. The flow will then return polewards to the high latitudes where it will downwell into deepest layers of the ocean to complete the circuit. It is found that the downwelling occurs in narrow regions of the high latitudes whereas upwelling occurs over a very large area of the tropical ocean. This hypothetical ocean demonstrates the key role of the deep horizontal pressure gradient, caused by horizontal variations in density, for driving the flow.

To explain the observed thermohaline circulation, this hypothetical ocean has to be modified to take into account the Coriolis force, which causes the deep abyssal currents to flow in narrow western boundary currents, the effect of salinity on the density (the haline component of the flow), asymmetries in the buoyancy fluxes between the Northern and Southern Hemispheres and the complex bathemetry of the ocean basins.

There follows a descriptive account of the thermohaline circulation. The deepest water masses in the ocean have their origin in the polar seasThese seas experience strong cooling of the surface, particularly in the winter seasons. In the North Atlantic, there are connections through the Nordic seas to the Arctic Ocean, from which sea ice flows. Heat energy melts the ice in the North Atlantic and the melt water gives rise to further cooling. There are two effects on the density of the water: cooling increases the density whereas surface freshening, due to ice melt, decreases the density of the water. The former process usually dominates and hence denser waters are produced. These dense cold waters flow into the Atlantic through the East Greenland and West Greenland Currents and then into the Labrador Current. These cold waters mix and sink beneath the warm North Atlantic Current.

In addition to surface polar currents there are also deep ocean currents (*see* Abyssal Currents). The cold saline water entering from the Nordic seas mixes as it sinks to the abyssal layers of the ocean and moves southward as a deep current along the western boundary of the Atlantic. This water mass is known as NADW (North Atlantic Deep Water) and it is the most prominent and voluminous of all the deep water masses in the global ocean. It flows into the Antarctic Circumpolar Current from where it flows into the Indian and Pacific Ocean. In addition to NADW, colder denser water, Antarctic Bottom Water (AABW) enters the Southern Ocean from the Antarctic shelf seas. It is not as voluminous as NADW but it flows northwards in the deepest layers into the Atlantic, where it can be distinguished as far north as 30°N. These deep flows upwell into the thermocline and surface waters where they return to the North Atlantic. This global thermohaline circulation has been termed the global conveyor circulation to signify its role in transporting heat and fresh water (**Figures 10** and **11**) around the planet.

How does this circulation explain the thermocline? The rate at which these cold deep abyssal waters are produced can be estimated and it is known for a steady state in the ocean that production has to be balanced by removal. A large-scale upwelling of the abyssal waters into the thermocline produces this removal. Our simple conceptual picture is of warm thermocline water mixing downwards, balanced by a steady upwelling of the cold abyssal layers. Without the upwelling, the warm waters would mix into the deepest layers of the ocean.

The Role of Fresh Water in Ocean Circulation

The present discussion has shown that the wind-driven circulation and the thermohaline circulation are major components of ocean circulation, which

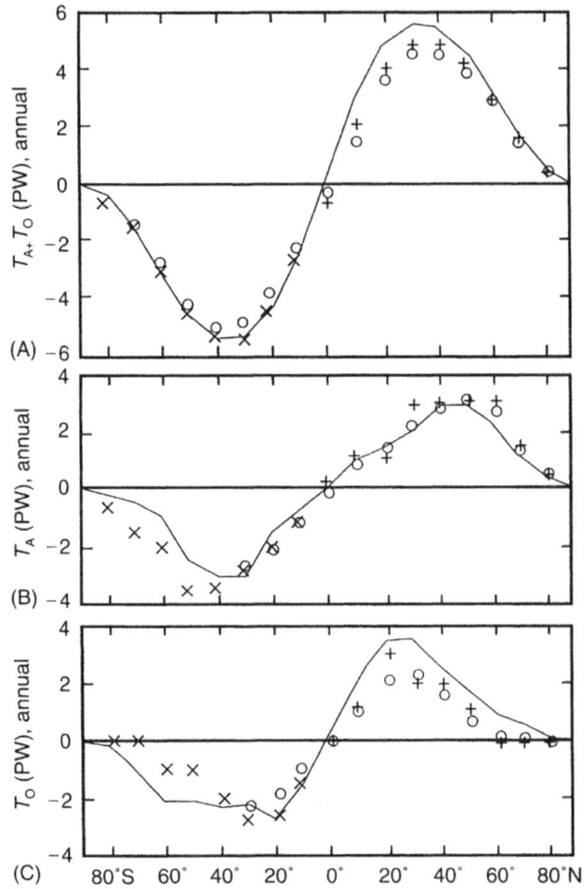

Figure 10 Poleward transfer of heat by (A) ocean and atmosphere together ($T_A + T_O$), (B) atmosphere alone (T_A), and (C) ocean alone (T_O). The total heat transfer (A) is derived from satellite measurements at the top of the atmosphere, that of the atmosphere alone (B) is obtained from measurements of the atmosphere, and (C) is calculated as the difference between (A) and (B). (1 Petawatt (PW) $= 10^{15}$ W.) Data compiled from three sources. (Reproduced from Carrissimo BC, Oort AH and Von der Harr TH (1985) Estimating the meridional energy transports in the atmosphere and ocean. *Journal of Physical Oceanography* 15: 52–91.)

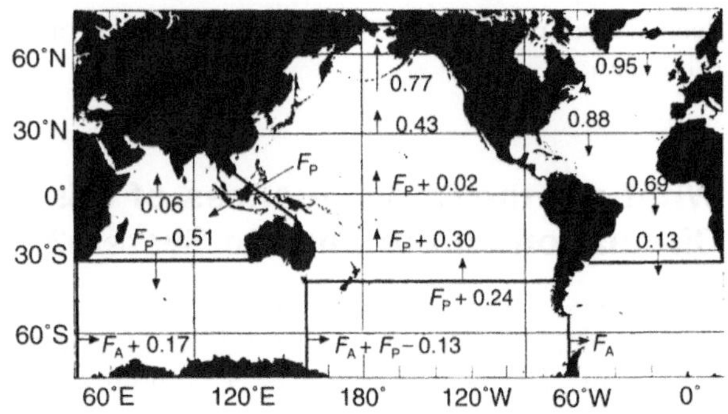

Figure 11 An estimate of the transfer of fresh water ($\times 10^9$ kg s^{-1}) in the world ocean. In polar and equatorial regions precipitation and river run-off exceed evaporation, and hence there is an excess of fresh water, whereas in subtropical regions there is a water deficit. A horizontal transfer of fresh water is therefore required between regions of surplus to regions of deficit. F_P and F_A refer to the freshwater fluxes of the Pacific–Indian throughflow and of the Antarctic Circumpolar Current in the Drake Passage, respectively. (Reproduced from Wijffels SE, Schmitt R, Bryden H and Stigebrandt A (1992) Transport of freshwater by the oceans. *Journal of Physical Oceanography* 22: 155–162.)

are ultimately driven by the surface wind stress and buoyancy fluxes. Buoyancy fluxes are the net effect of heat exchange and the freshwater exchange with the overlying atmosphere. It has been shown that heat exchange is a major process explaining existence of both the thermocline and the deep abyssal water but what is the role of the fresh water in ocean circulation?

In the subtropics there is net removal of fresh water by evaporation. This increases the salinity of the water which, in turn, increases the density of the thermocline waters. Normally this effect is opposed by heating, which lightens the water. However, in the Mediterranean and the Red Sea evaporation produces salient waters, which by virtue of their salinity and cooling in winter, sink to the deepest layers of the basins. At the Straits of Gibraltar (*see* Flows in Straits and Channels), this dense saline layer flows out beneath the incoming fresher and cooler Atlantic water. This Mediterranean water forms a distinct layer of high salinity water in the eastern Atlantic Ocean. Similar behavior occurs at Bab el Mandeb adjacent to the Gulf of Aden.

The influence of fresh water is more substantial in the polar oceans. A given amount of fresh water will have a greater effect on density at low temperatures than at high temperatures, because the thermal expansion of sea water decreases with decreasing temperature. At higher latitudes there is a net addition of fresh water into the oceans, which arises from the excess of precipitation over evaporation and the melting of sea ice moving towards the equator from the polar regions.

The addition of fresh water adds buoyancy to the surface layer while cooling removes buoyancy, therefore the fresh water will tend to reduce the effect of the cooling. In the Arctic Ocean the surface layer is colder but less dense than the warmer layer at ~100 m and therefore is in equilibrium. This stable halocline in the Arctic Ocean reduces the vertical heat flux in to the deep ocean.

In the subpolar oceans, the addition of fresh water reduces the density of the surface layer and can reduce the prevalence of deep convection. This happened in the late 1960s when fresh water, probably from excessive ice in the Arctic Basin, melted in the subpolar gyre. The effect on the thermohaline circulation is unknown, but it is believed from modeling studies that the decrease in the production of deep waters reduced the thermohaline circulation of the ocean.

What are the Consequences of this Circulation on the Climate System?

The effects of the ocean circulation on the climate can be understood in terms of the heat capacity of the ocean. The heat capacity of a column of sea water only 2.6 m deep is equivalent to that of a column of whole atmosphere and therefore the ocean heats and cools on a long timescale compared with the atmosphere.

It is known that there is a poleward gradient of temperature, which is driven by the thermal radiation imbalance between the low and high latitudes. In response to this temperature gradient there is a flow of heat from the warmer to cooler latitudes. Both the atmosphere and ocean circulations transfer this heat from low to high latitudes by a variety of mechanisms. In the low latitudes of the atmosphere there is the Hadley cell which transfers low temperature air in the lowest levels via the trade winds towards the equator and transfers warmer air poleward in the upper troposphere (**Figure 8**). At higher latitudes anticyclones and cyclones and their accompanying upper air jet streams transfer heat polewards. In the ocean, the wind-driven Ekman currents transfer heat as surface waters move across latitude circles. This water is returned deeper in the ocean at a different temperature from that of the surface water. The ocean gyres carry heat towards higher latitudes since the poleward flows of the western part of the gyres are warmer than the equatorward flows in the eastern parts of the gyre. Finally, and not least, is the contribution of the thermohaline circulation, which transports warm surface and thermocline waters to the highest latitudes and returns cold water to lower latitudes. **Figures 10** and **11** show the heat transport and fresh water transport in the ocean.

A major difference between the atmosphere and ocean is the relative speed of their circulation. The atmosphere circulation is a fast system, responding on timescales of days to weeks. For example, weather systems in temperate latitudes grow and decay on timescales of a few days. By contrast, the ocean tends to be slower in its response. The fastest part of the system are the surface Ekman layers which respond to changes in the surface wind circulation on a timescale of one or two days. Changes in wind circulation can cause planetary waves which will change sea level and surface temperature on monthly to seasonal timescales. In particular, the equatorial oceans respond to the surface wind stress on seasonal timescales, which allows a strong coupling between the ocean and atmosphere to take place. This gives rise to phenomena such as the El Niño Southern Oscillation. The subtropical gyres respond to changes in the wind circulation on decadal timescales, whereas the deep thermohaline circulation respond on millenial timescales. There is some evidence for rapid changes of local parts of the thermohaline circulation on timescales 50 years.

Observations of the deep ocean are far fewer in number than at the ocean and land surface. The longest continuous data set is a deep station at Bermuda that commenced operations in 1954. Observations from cruises in the earlier part of the century are of unknown quality and therefore it is difficult to know whether differences are due to the use of different instruments or to real changes in the ocean. It is only since the 1950s that such changes have been accurately measured. **Figure 12** shows changes in the temperature for that period of time across the Atlantic. These changes are of the order of a few tenths of a degree over periods of 15 years. As the heat

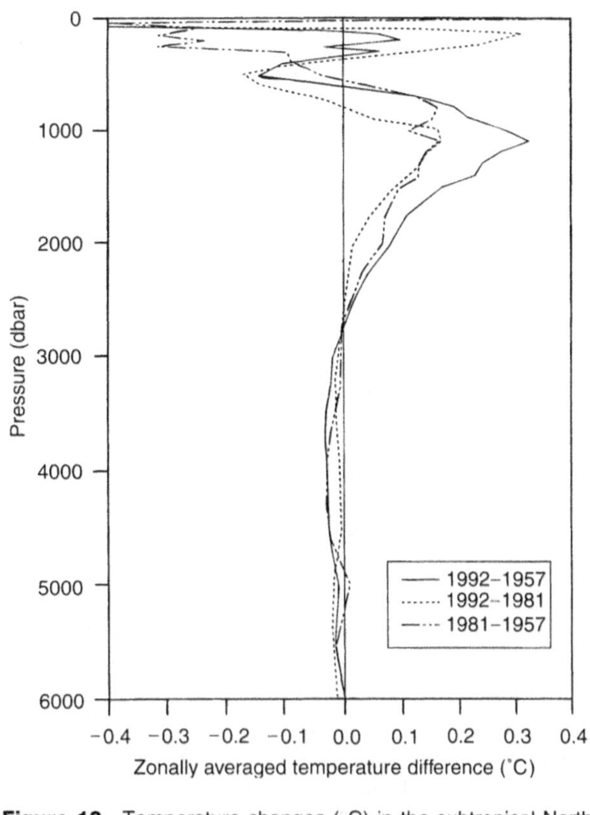

Figure 12 Temperature changes (C) in the subtropical North Atlantic (24 N), 1957–1992. The measurements have been averaged across 24 N between North Africa and Florida. (Reproduced from Parrilla G, Lavin A, Bryden H, Garcia M and Millard R (1994) Rising temperatures in the subtropical North Atlantic. *Nature* 369: 48–51.)

capacity of the oceans is very much larger than that of the atmosphere, these changes in temperature involve very significant changes in the heat content of the ocean. The World Ocean Circulation Experiment from 1990 to 1997 has provided measurements of ocean properties such as temperature, salinity, and chemical tracers as well as current measurements on a global scale. This set of high quality measurements will provide the baseline from which future changes in ocean circulation can be determined.

Despite the brief record of deep ocean observations, sea surface temperature measurements of reasonable quality go back to the late nineteenth century. These measurements can be used to assess changes in the surface layers. Salinity measurements are fewer and not as reliable but, nevertheless, changes can be still detected.

Salinity measurements in the Mediterranean over the last century have shown a warming of the Western Mediterranean Deep Water of 0.1°C and increase of 0.05 in salinity. The reasons for this change are not known, but it has been speculated that the change in salinity may be attributed to a reduction in the freshwater flow due to the damming of the Nile and of rivers flowing into the Black Sea.

An important recently identified question is the stability of the thermohaline circulation. The thermohaline circulation is driven by small density differences and therefore changes in the temperature and salinity arising from global warming may alter the thermohaline circulation. In particular, theoretical modeling of the ocean circulation has shown that the thermohaline circulation may be reduced or turned off completely when significant excess fresh water is added to the subpolar ocean. In the event of thermohaline circulation being significantly reduced or stopped, it may take many centuries before it returns to its present value.

In view of the current levels of uncertainty, it is necessary to continue to monitor the ocean circulation, as this will provide the key to the understanding of the present circulation and enhance our ability to predict future changes in circulation.

See also

Abyssal Currents. Agulhas Current. Antarctic Circumpolar Current. Current Systems in the Atlantic Ocean. Florida Current, Gulf Stream and Labrador Current. Flows in Straits and Channels. Flows in Straits and Channels. Inverse Models; Inverse Models. Regional and Shelf Models. Wind Driven Circulation.

Further Reading

Gill AE (1982) *Atmosphere–Ocean Dynamics*. London: Academic Press.

Siedler G, Church J, and Gould (2001) *Ocean Circulation and Climate*. London: Academic Press.

Summerhayes CP and Thorpe SA (1996) *Oceanography – An Illustrated Guide*. London: Manson Publishing.

Wells NC (1997) *The Atmosphere and Ocean: A Physical Introduction*, 2nd edn. Chichester: John Wiley.

WIND DRIVEN CIRCULATION

P. S. Bogden, Maine State Planning Office, Augusta, Maine
C. A. Edwards, University of Connecticut, Groton, CT

Introduction

Winds represent a dominant source of energy for driving oceanic motions. At the ocean surface, such motions include surface gravity waves, which are familiar as the waves that break on beaches. Winds are also responsible for small-scale turbulent fluctuations just beneath the ocean surface. Turbulent motions can be created by breaking waves or by the nonlinear evolution of currents near the air–sea interface. Subsurface processes such as these can lead to easily observed windrows or scum lines on the sea surface. Winds also generate other complex and varied small-scale motions in the top few tens of meters of the ocean. However, the surface/wind-driven circulation described here refers instead to considerably larger-scale motions that compare in size to the ocean basins and extend as much as a kilometer or more below the surface.

The textbook notion of the surface/wind-driven circulation includes most of the well-known surface currents, such as the intense poleward-flowing Gulf Stream in the western North Atlantic (**Figure 1**). Analogues of the Gulf Stream can be found in each of the major ocean basins, including the Kuroshio in the North Pacific, the Brazil Current in the South Atlantic, the East Australian Current in the South Pacific, and the Aghulas in the Indian. These 'western boundary currents' are not isolated structures. Rather, they represent the poleward return flow for basin-scale motions that occupy middle latitudes in all major oceans. Each of the major ocean basins has an analogous set of large-scale current systems. The western boundary currents are quite intense, reaching velocities in excess of 1 m s^{-1}, while the interior flow speeds are considerably smaller in magnitude.

The basin-scale patterns in the mid-latitude surface circulation are referred to as subtropical gyres. The gyres extend many hundreds of meters below the surface, reaching the bottom in some locations. Subtropical gyres rotate anticyclonically, that is, they rotate in a sense that is opposite to the sense of the earth's rotation (clockwise in the northern hemisphere and counterclockwise in the south). In the North Atlantic and North Pacific, subpolar gyres reside to the north of the subtropical gyres. They too include intense western boundary currents. However, the subpolar gyres rotate cyclonically, in the opposite sense of the subtropical gyres and in the same sense as the earth. Rotation of the wind-driven gyres is related to the rotation of the earth through a simple, though nonintuitive, physical mechanism. This mechanism is fundamental to understanding how the wind drives large-scale flows.

Our present understanding of the dynamics associated with the surface/wind-driven circulation developed largely during a 30-year period starting in the late 1940s. Before that time, oceanographers were aware of the gyre-scale features of the surface circulation. But it was not until the major theoretical advances in geophysical fluid dynamics beginning around 1947 that the surface circulation was conceptually linked to the winds.

Observations

Oceanic wind systems exhibit a large-scale pattern that is common to the major ocean basins (**Figure 2**). Near the equator, trade winds blow from east to west. Near the poles, westerly winds blow from west to east. The ocean gyres have similar distributions of east–west flow. But the reasons for this are quite subtle. Moreover, there are profound differences between oceanic and atmospheric motion. North–south flows in the ocean are much more strongly pronounced than they are in the atmosphere, and winds fail to exhibit analogues of the intense poleward western boundary currents found in the ocean.

Western boundary currents such as the Gulf Stream were evident in estimates of time-averaged surface circulation obtained over a century ago with 'ship-drift' data (**Figure 3**). Ship drift is the discrepancy between a ship's position as obtained with dead reckoning and that obtained by more accurate navigation. Thus, ship drift can be attributed at least in part to ocean currents. The patterns of the surface circulation that emerge after averaging large numbers of such measurements are qualitatively correct. In general, however, accurate measurements of currents are difficult to obtain, particularly below the surface.

Temperature and salinity measurements are relatively abundant, and they provide an alternative resource for estimating large-scale currents. Temperature and salinity determine seawater density. The density distribution provides information about the pressure field, which in turn can be used to diagnose currents.

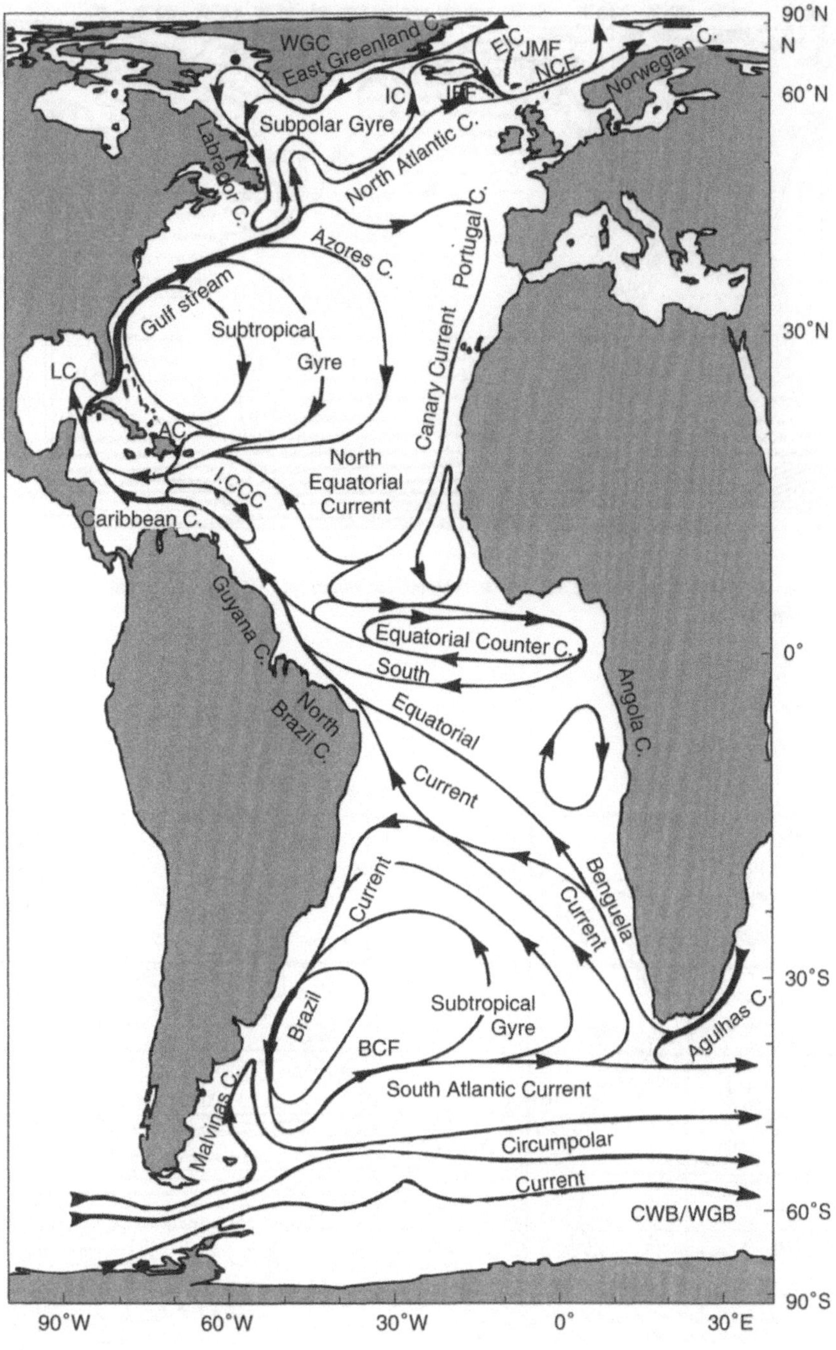

Figure 1 Schematic of large-scale surface currents in the Atlantic Ocean. (From Tomczak and Godfrey (1994) *Regional Oceanography: An Introduction.*)

For many decades, oceanographers have been routinely measuring vertical profiles of temperature and salinity. Consequently, detailed maps of the three-dimensional density structure exist for all the ocean basins.

Such maps clearly show a region of anomalously large vertical gradients in temperature, salinity, and density known as the main thermocline. The main thermocline divides two regions of relatively less stratified water near the surface and bottom. Thermocline depth varies substantially on the gyre scale, and can exceed 700 m depth in some regions. Lateral variations in the density field can be quite large as well, and these are associated with the currents that we refer to as the surface/wind-driven circulation.

The first step in estimating currents from density involves computation of dynamic height using the

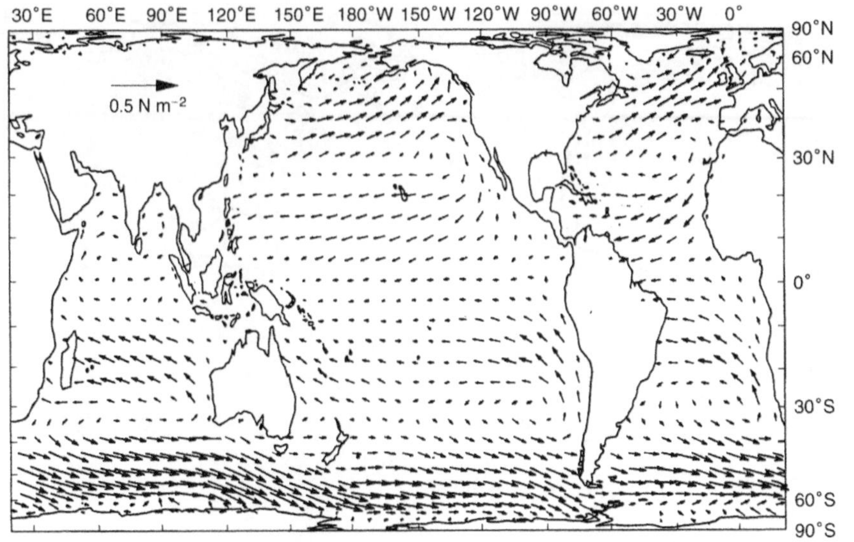

Figure 2 Global mean surface wind stress, which is related to wind [1]. (From Tomczak and Godfrey (1994) *Regional Oceanography: An Introduction.*)

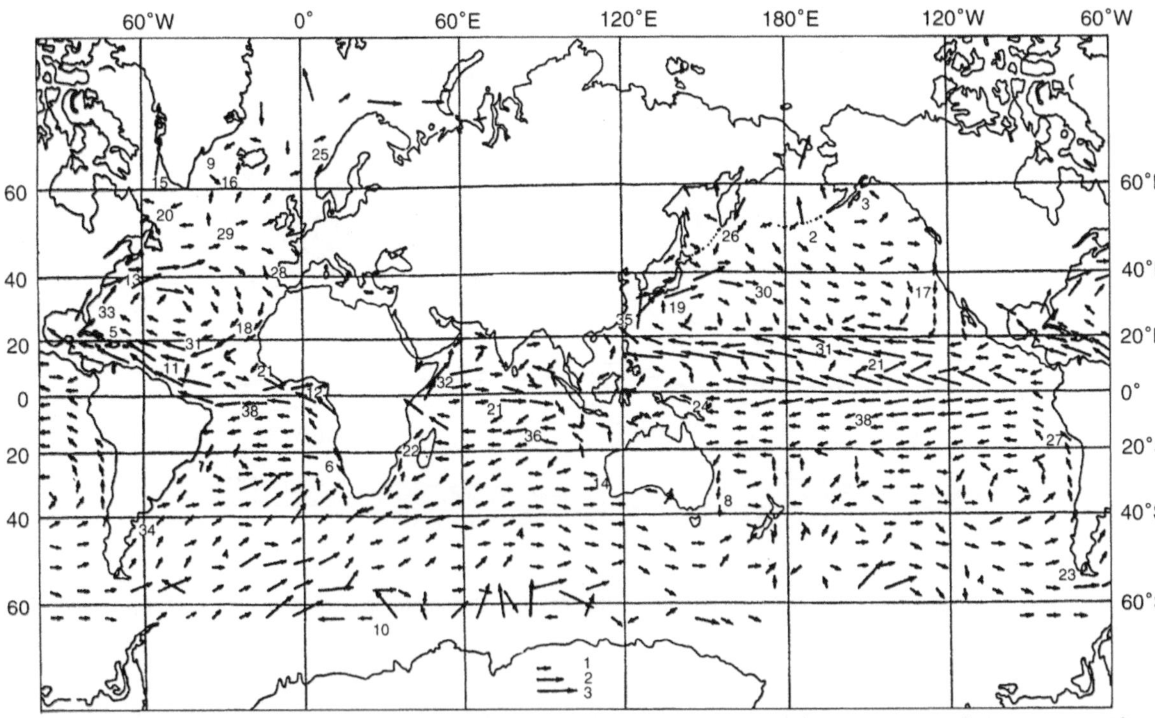

Figure 3 Surface currents inferred from ship-drift measurements. To simplify the presentation, there are three vector sizes in this figure indicated by the scale vectors at the bottom of the figure. A vector the size of vector 1 corresponds to flow speeds in the range 0–10 cm s^{-1}, vector 2 is 10–20 cm s^{-1}, vector 3 is more than 30 cm s^{-1}. While some of the values have questionable reliability, the vectors show the general patterns large-scale circulation at the ocean surface. From Stidd CK (1974) Ship Drift Components: Means and standard Deviations, SIO Reference Series 74-33 as appearing in Burkov VA 1980 *General Circulation of the World Ocean.* Gidrometeoizdat Publishers, Leningrad, published for the Division of Ocean Sciences, National Science Foundation, Washington, DC, by Amerind Publishing Co. Pvt. Ltd., New Delhi. 1993.

principle of isostasy. Isostasy describes, for example, the pressure field in a glass of ice water. An ice cube represents a region where water is slightly less dense than its surroundings. Buoyancy forces elevate the surface of the ice cube above the surface of the surrounding fluid. Similarly, a region in the ocean with less dense water than its surroundings will have a slightly elevated sea surface. In the ocean, this result involves the tacit assumption that currents in the abyssal ocean are weak relative to those nearer the surface, as is usually the case. The ocean's surface topography implied by the density field is referred to as dynamic height. **Figure 4** shows dynamic height computed from density between 2000 and 200 m depth. The variations of a meter or more are large enough to account for the pressure gradients that force the large-scale gyres.

The connection between pressure and large-scale currents involves the principle of 'geostrophy'. Geostrophic currents arise from a balance of the forces involving pressure gradient forces and Coriolis accelerations. This balance is a consequence of the large horizontal scales of the flow combined with the rotation of the earth. If the earth were not rotating, the sea surface elevations would accelerate horizontal flows down the pressure gradient, as occurs with smaller-scale motions such as surface gravity waves. With large-scale geostrophic flows, however, the Coriolis effect gives rise to currents that flow perpendicular to the pressure gradient, as indicated by the arrows in **Figure 4**.

Geostrophic currents such as those in **Figure 4** provide evidence of the surface/wind-driven circulation. This point requires some explanation, since the geostrophic flows are associated with large-scale pressure-gradient forces in the top kilometer of the ocean. As discussed below, winds directly drive motions in a relatively thin layer at the ocean surface known as the surface mixed layer. But these directly wind-driven flows give rise to other large-scale flows and, in turn, to the large-scale pressure gradients that can be estimated with dynamic height. Thus, it is accurate to refer to the large-scale geostrophic surface circulation as the wind-driven circulation because the pressure gradients would not exist without the wind.

Wind-Driven Surface Layer

Surface Mixed Layer

The surface mixed layer is loosely defined as a part of the water column near the surface where observed temperature and salinity fields are vertically uniform. In practice this layer extends from the ocean surface to a depth where stratification in temperature or density exceeds some threshold value. Typically, the underlying water is more strongly stratified. The mixed-layer depth often undergoes large diurnal and seasonal variations, varying between 0 and 100 m. However, the surface mixed layer rarely occupies more than 1% of the total water column.

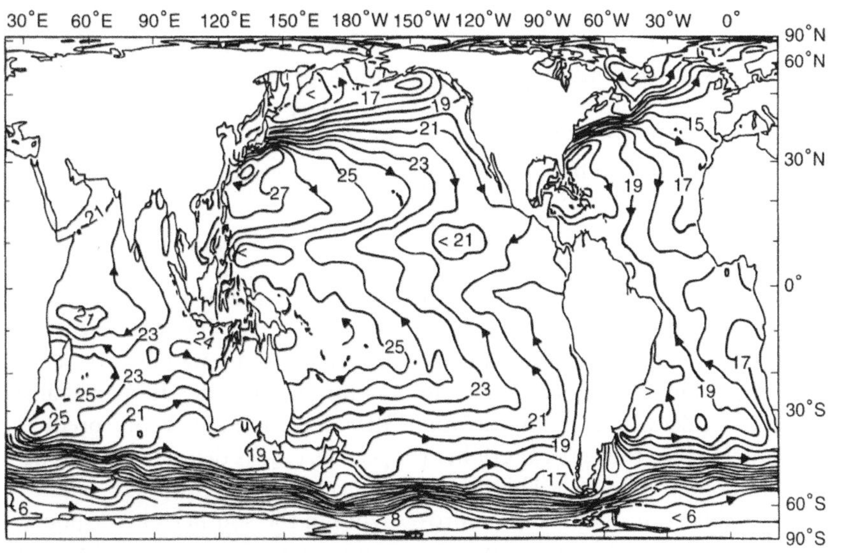

Figure 4 Dynamic height (m^2 s^{-2}) computed using the density field between 0 m and 2000 m depth, and assuming that the pressure field at 2000 m has no horizontal variation. Dynamic height is roughly proportional to the sea-surface height, in meters, multiplied by 10. (Based on Levitus (1982).)

Winds provide the primary source of mechanical forcing for the motions that homogenize water properties within the mixed layer. Mixing can also result from destabilizing effects of cooling and evaporation near the surface, as these can temporarily give rise to localized regions where the surface water overlies less dense water. Restratification of a stable water column involves solar heating from above, reduced salinity from precipitation, and other more subtle processes.

The detailed mechanisms by which wind generates small-scale motions (i.e., motions on scales smaller than the mixed-layer depth, such as breaking waves) are quite complex and incompletely understood. Nevertheless, the effect of wind on the surface mixed layer is commonly parameterized through a stress τ_w on the ocean surface. Wind stress has units of force per unit area. The standard empirical relation has the form of eqn [1].

$$\tau_w = \rho_{air} C_d u^2, \qquad [1]$$

where ρ_a is the density of air, $C_d \approx 10^{-3}$ is a drag coefficient that may depend on wind speed and atmospheric stability, and u is the wind speed 10 m above the sea surface. Ten meters is the standard height that commercial ships use to mount their anemometers, and ship reports still account for most of the direct measurements of wind over the ocean.

Ekman Dynamics

The small-scale motions that mix temperature and salinity also mix momentum. As a result, the momentum of the wind is efficiently transmitted throughout the mixed layer, thereby accelerating horizontal currents. The resulting motions have large horizontal length scales comparable to those of the wind systems that drive them. In 1905, V.W. Ekman developed a model revealing the influence of the earth's rotation on such large-scale flows. His dynamical model presumed that the force associated with a divergence in the vertical momentum flux is balanced by Coriolis accelerations associated with the horizontal flows. The vertical momentum flux in this wind-driven Ekman layer is the result of turbulent mixing processes that Ekman parametrized using Fick's law. Thus, the vertical turbulent flow of momentum is made proportional to the vertical gradient of the large-scale horizontal velocity. This parametrization involves an uncertain constant of proportionality called the vertical eddy viscosity A_v. Ekman's model predicts horizontal currents that simultaneously decrease and rotate with depth. Within this so-called Ekman spiral, currents decrease away from the surface with a vertical scale D known as the Ekman depth (eqn [2]).

$$D = (2A_v/f)^{1/2} \qquad [2]$$

This relation provides our first introduction to the Coriolis parameter $f = 2\Omega \sin\theta$, where θ is latitude, which appears here because of Coriolis accelerations in the Ekman dynamics. The angular velocity of the earth is a vector of magnitude $\Omega = 2\pi$ day that is parallel to the earth's axis of rotation. The Coriolis parameter equals twice the magnitude of the vector component that is parallel to the local vertical. The vertical component is the only component that creates horizontal Coriolis accelerations with horizontal flow. This dependence on the local vertical and the sphericity of the earth explain the $\sin\theta$ factor in the formula for f. Thus, for any given flow, Coriolis accelerations are strongest at the poles, negligible at the equator, and smoothly varying in between. As discussed below, this geometric detail has profound implications for the surface/wind-driven circulation.

Consider a typical mixed-layer depth of 30 m at mid-latitudes, where $f \approx 10^{-4}\,s^{-1}$. By relating these two quantities to an Ekman depth D, one deduces a vertical eddy viscosity $A_v \approx 0.05\ m^2\,s^{-1}$. This value is many orders of magnitude larger than the kinematic viscosity of water, $\nu \approx 10^{-6}\ m^2\,s^{-1}$. The large value of A_v indicates the efficiency of turbulent mixing compared with molecular diffusion. But A_v arises from the use of Fick's law to parametrize the turbulence, and Fick's law is an oversimplified model for turbulence. In fact, details of the wind-mixed layer that depend heavily on A_v, such as spiraling velocities, are rarely observed.

There is, nevertheless, one very important and robust conclusion from Ekman theory. The net mass transport (the Ekman transport) within the mixed layer, i.e., the vertical integral of the horizontal flow, has magnitude

$$U_{Ekman} = \tau/(\rho f) \qquad [3]$$

where ρ is the density of water. This result does not depend on A_v. Thus, while the details and vertical extent of the Ekman flow depend on the complexities of mixing, the net Ekman transport does not. Furthermore, Ekman theory predicts that the net transport U_{Ekman} is directed 90° to the right of the wind in the northern hemisphere and 90° to the left of the wind in the southern hemisphere. This result is quite contrary to what one would find if the earth were not rotating.

The Ekman transport describes the net horizontal motion in a thin surface mixed layer. Implications for flows in the interior of the ocean depend on the large-scale patterns in the wind stress, as shown in **Figure 2**. In particular, westward wind stress near the equator results in poleward Ekman mass transport and eastward wind stress at higher latitudes drives equatorward Ekman transport. This pattern results in a convergence of fluid that gives rise to an elevated sea surface at the center of the clockwise wind system. Ultimately, this horizontally convergent Ekman transport has only one direction in which to go – down. The resultant downward motion at the base of the mixed layer, called Ekman pumping, occurs in all mid-latitude ocean basins. Likewise, at higher latitudes, counterclockwise wind systems cause horizontally divergent Ekman transport, a depressed sea surface, and an upward motion known as Ekman suction.

In the classic wind-driven ocean circulation models discussed below, vertical Ekman flows drive the horizontal geostrophic flow. In fact, the net effect of all the complex motions in the mixed layer is often reduced to a simple prescription of the vertical Ekman-pumping velocity W_{Ekman} (eqn[4]).

$$W_{\mathrm{Ekman}} = \mathrm{curl}(\tau_w/\rho f) \qquad [4]$$

where $\mathrm{curl}(\tau_w/\rho f))$ represents the curl of the surface wind-stress vector divided by ρf. Thus, it is not simply the magnitude of the wind stress that determines the Ekman pumping velocities, but its spatial distribution. The Ekman-pumping velocity is often applied as a boundary condition at the sea surface associated with a negligibly thin mixed layer.

On average, Ekman-pumping speeds rarely exceed $1\ \mu\mathrm{m\ s}^{-1}$. Nevertheless, such minuscule vertical velocities give rise to the most massive current systems in the ocean. This remarkable fact reflects the enormous constraint that the earth's rotation plays in large-scale ocean dynamics.

Large-scale Dynamics

The directly wind-driven flows within the Ekman layer occupy only a small fraction of the total water column. In fact, the impact of the wind extends considerably deeper. The connection between the minute Ekman-pumping velocities and the tremendous horizontal flows associated with the surface/wind-driven circulation involves a balance of forces that is very different from that in the surface mixed layer.

Far from continental boundaries, and below the surface mixed layer, the basin-scale circulation varies on length scales measured in thousands of kilometers. The time-averaged horizontal velocities sometimes exceed $1\ \mathrm{m\ s}^{-1}$, but they more generally vary between 1 and $10\ \mathrm{cm\ s}^{-1}$. With these scales, flows are plausibly geostrophic. Furthermore, when the density is uniform, geostrophic flows exhibit no vertical variation. Rather, they behave like a horizontal continuum of vertical columns of fluid. It is reasonable to approximate the region between the mixed layer and the thermocline as a region of constant density. In this region the geostrophic columns of fluid span many hundreds of meters. The earliest models of the ocean circulation obtained remarkable predictive skills by assuming that columnar geostrophic flow extends from the top to the bottom of the ocean.

Downward Ekman-pumping velocities, as small as they may be, effectively compress the fluid columns. Under the influence of downward Ekman pumping, fluid columns below the mixed layer compress vertically and expand horizontally, as if they were conserving their total volume. Likewise, Ekman suction causes water columns to stretch vertically and contract horizontally. Because of the earth's rotation, the effect of Ekman pumping and suction on large-scale motions is related to the principle of angular momentum conservation in classical mechanics. For example, a water column that undergoes the stretching effect of Ekman suction is not unlike a rotating figure skater who draws in her arms, thereby decreasing her moment of inertia and rotating more rapidly. (Note: The analogy is incomplete because Ekman pumping is a consequence of external forcing, whereas the spinning skater is unforced. Nevertheless, the comparison is physically relevant.) Ekman pumping is then similar to a skater extending her arms, which causes a reduction of rate of spin.

The connection between water-column stretching and horizontal flow in the ocean involves one additional subtle point: Water columns on the earth rotate by virtue of their location on the earth's surface. Vertical fluid columns that appear stationary in the earth's frame of reference are actually rotating at a rate that is proportional to the vertical component of the earth's angular velocity. The absolute rotation rate is the sum of the earth's rotation plus the rotation relative to the earth. More precisely, the fluid's absolute vorticity includes a contribution from the planetary vorticity, which has magnitude equal to the Coriolis parameter f, plus a contribution from its relative vorticity.

Relative vorticity is measured from a frame of reference that is fixed on the earth's surface. The earth itself rotates at a rate of one revolution per day. In comparison, large-scale ocean currents progress around an ocean basin at average speeds much less than $1\ \mathrm{m\ s}^{-1}$, so the period of rotation associated with a complete circuit can be several years. Thus, the relative rotation rate of large-scale ocean currents is negligible compared with the rotation rate of the earth itself. For this reason, it is a very good approximation to neglect the relative vorticity and equate the vorticity of the large-scale circulation with f. The critically important result is that fluid columns change their vorticity by changing their latitude. Just as the skater who extends her arms starts to rotate more slowly, a water column undergoing the compression of Ekman pumping will travel toward the equator where the magnitude of f is smaller.

The physical mechanism that connects Ekman pumping and subsurface geostrophic flow was described mathematically by H.U. in 1947, and is summarized by the relation (eqn [5]).

$$\beta V = f(W_{\mathrm{Ekman}} - W_{\mathrm{deep}}) \qquad [5]$$

V represents the meridional (positive northward) velocity integrated over the depth H of the water column, $\beta = (1/R)\partial f/\partial \theta$ is the meridional variation in the Coriolis parameter, R is the radius of the earth, W_{w} is the Ekman pumping velocity, and W_{deep} is a vertical velocity at depth H. In this model, V is the depth-integrated geostrophic velocity. While prescription of W_{deep} is discussed further below, the early models of the wind-driven circulation assumed that $W_{\mathrm{deep}} = 0$ at some depth well below the main thermocline. Thus, V can be estimated using the Sverdrup relation with W_{Ekman} prescribed using eqn [1], eqn [4] and measurements of wind. V computed in this way agrees remarkably well with V computed from geostrophic flows estimated from the observed density field. The agreement is good everywhere except near the western boundaries of the ocean basins.

Westward Intensification

While the Sverdrup relation provides guidance for the ocean interior, it cannot describe the basin-wide circulation. From a mathematical viewpoint, the Sverdrup relation is a purely local relation between wind stress and meridional flow, so it does not determine east–west flow within the basin. Moreover, the Sverdrup relation predicts that V will be large only where W_{Ekman} is large. In fact, V is observed to be largest in the intense western boundary currents found in each ocean basin, such as the Gulf Stream. This is problematic because W_{Ekman} fails to exhibit the westward intensification, or a change in sign. This means that the Sverdrup relation predicts weak western boundary flow in the wrong direction! Thus, the Sverdrup balance must break down, at least in certain regions.

It is common to presume that Sverdrup theory holds everywhere in the ocean interior except near the western ocean boundary (and northern and southern boundaries if the wind-stress curl does not vanish there). Then, V can be integrated from east to west to determine the total transport required in a poleward western boundary current that returns the meridional Sverdrup transport back to its place of origin. This calculation comes close to predicting the observed transport in the poleward-traveling Gulf Stream at some locations. But models that predict the structure and location of the return flow require fundamentally different dynamics.

In 1948, H. Stommel developed a theory for the wind-driven circulation in which the ocean bottom exerts a frictional drag on the horizontal flow. In the ocean interior, the Stommel and Sverdrup dynamics are nearly indistinguishable, but bottom friction becomes important near the western boundary, allowing Stommel's model to predict a closed circulation for the entire ocean basin (**Figure 5**). The friction-dominated western boundary layer contains the intense poleward analogue of the Gulf Stream. Stommel showed the remarkable fact that westward intensification of the wind-driven gyres is fundamentally linked to the latitudinal variation of the Coriolis parameter. That is, the Gulf Stream and its western-boundary analogues in all the ocean basins exist because of the sphericity of the rotating earth (**Figure 6**).

Friction is the key to closing the circulation cell. In Stommel's model the friction parametrization was chosen for its simplicity, but it is ultimately unrealistic. In 1950, W. Munk developed a similar flat-bottom model with lateral viscosity, an entirely different form of dissipation. Nevertheless, Munk's model produces an intense western boundary current for the same reasons as does the Stommel model.

The primary results from these frictional models are robust. Both theories deduce the zonal flow within the basin, and share the central conclusion that the return flow for the interior Sverdrup transport occurs in a meridional current near the western edge of the basin. This current is an example of a boundary layer, a narrow region governed by different physical balances from those dominating the larger domain. In the western boundary layer, fluid columns can change latitude because dissipation changes their vorticity, thereby counteracting the effects introduced by the Ekman pumping or suction. Both models show that this dissipative mechanism can only occur in an intense western boundary layer.

Dissipation in both models actually parametrizes many interesting smaller-scale phenomena. This is evident in Munk's model. The horizontal viscosity needed to produce a realistic Gulf Stream is many orders of magnitude larger than molecular viscosity, larger even than Ekman's vertical eddy viscosity, A_{v}. Modern theories show that these viscous parametrizations for

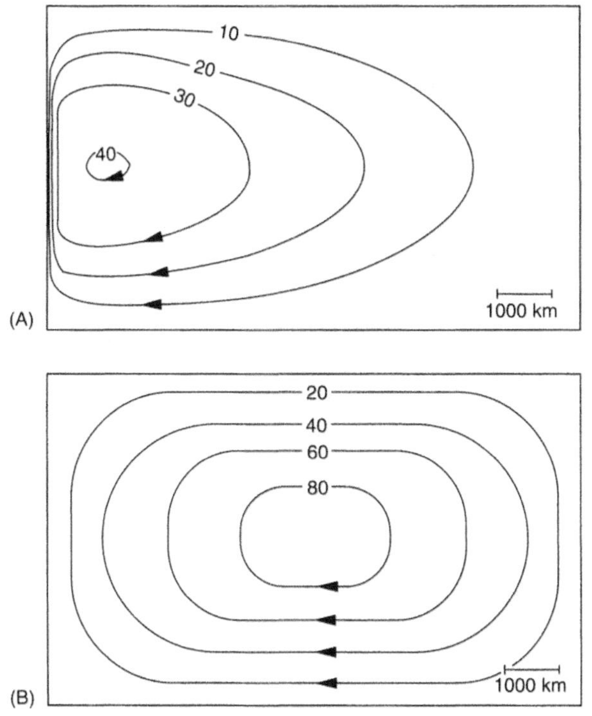

Figure 5 Streamlines from Stommel's model indicating the total flow in an idealized flat-bottom subtropical gyre. The flow is everywhere parallel to the streamlines in the direction indicated by the arrows. Flow intensity is greatest where the streamlines are closest together. (A) An idealized subtropical gyre for a rotating earth in which the Coriolis parameter varies with latitude. (B) Streamlines for a 'uniformly' rotating earth, that is, for a Coriolis parameter that does not vary with latitude. (From Stommel (1948).)

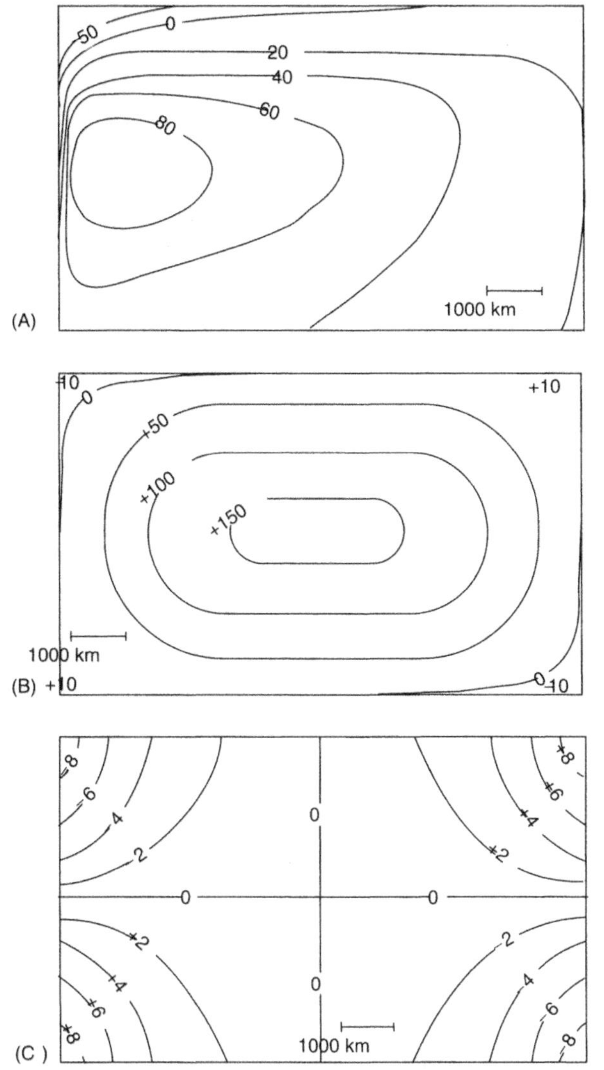

Figure 6 Contours of sea-surface height from Stommel's model. (A) Sea-surface height consistent with the stream function of **Figure 5**. Features in this figure can be directly compared with the dynamic height computed from data in **Figure 4**. (B) Sea-surface height for Stommel's model after setting the Coriolis parameter to a constant. This effectively removes the geometrical factor associated with sphericity. Stommel referred to this as the case of a 'uniformly rotation ocean'. (C) The sea-surface height for the same wind distribution as in (A) and (B), but for a nonrotating ocean. (From Stommel (1948).)

ocean turbulence greatly oversimplify the effect of small-scale motions on the large-scale circulation. More importantly, the Stommel and Munk models neglect the fact that the ocean has variable depth and density stratification.

Topography, Stratification, and Nonlinearity

The simplified Stommel and Munk models describe the wind-driven circulation for a rectangular ocean that has uniform density, a flat bottom, and vertical side walls. It remains to put these idealized models in context for an ocean that has density stratification, mid-ocean ridges, and continental slopes and shelves. The flat-bottom constant-density models clearly oversimplify the ocean geometry. Were the mid-ocean ridges placed on land, they would stand as tall as the Rockies and the Alps. The assumption of constant density turns out to be an oversimplification of comparable proportions.

In flat-bottom models, deep currents are unimpeded by topographic obstructions. With realistic bathymetry, however, flow into regions of varying depth can lead to large vertical velocities. For rotating fluid columns, these vertical velocities affect vorticity. Computer models that add realistic bathymetry and Ekman pumping to the Stommel or Munk models show that such vertical velocities can substantially alter the horizontal flow pattern, so much so that the flows in the center of the ocean no longer resemble the observed surface circulation. Thus, in idealized constant-density models, realistic bathymetry eliminates the most remarkable similarities between the models and the ocean observations.

This conundrum can be reconciled in a model that has variable density. In a constant-density ocean, geostrophic fluid columns extend all the way to the bottom. This allows bottom topography to have an unrealistically strong influence on the flow. Density stratification reduces the vertical extent of columnar motion. Conceptually, a stratified ocean behaves almost like a series of distinct layers, each with variable thickness and constant density. For example, the main thermocline may be considered the interface between one continuum of fluid columns in a surface layer and a second continuum of fluid columns in an abyssal layer. Generalizations of the Stommel and Munk models have often treated the ocean as two distinct layers of fluid.

The main thermocline varies smoothly compared with the ocean bottom. This means that there are fewer obstructions to the columnar flow above the thermocline than below. In this sense, the thermocline effectively isolates the ocean bathymetry from the surface circulation. In fact, observed currents above the main thermocline tend to be stronger. While the Sverdrup theory applies to the top-to-bottom transport, stratification allows the flow to be surface intensified. Smaller abyssal velocities reduce the influence of bottom topography. Flat-bottom models describe a limiting case where the topographic effects are identically zero.

Without question, the vertical extent of the large-scale wind-driven circulation is linked to density stratification. Realistic models of the large-scale circulation must include thermodynamic processes that affect temperature, salinity, and density structure. For example, atmospheric processes change the heat and fresh water content of the surface mixed layer. Large-scale motions can result when the water column becomes unstable, with more dense water overlying less dense water. The resulting motion is often referred to as the thermohaline circulation, as distinct from the wind-driven circulation, but the conclusion to be drawn from the more realistic ocean-circulation models is that the thermohaline circulation and the wind-driven circulation are inextricably linked.

Additional factors come into play in the more comprehensive ocean models. For example, the persistent temperature and salinity structure of the ocean indicates that many large-scale features in the ocean have remained qualitatively unchanged for decades, perhaps even centuries. But there are no simple (linear) theories that predict the existence of the thermocline. The transport and mixing of density by ocean currents are inherently nonlinear effects. Other classes of nonlinearities inherent to fluid flow add other types of complexity. Such nonlinear effects account for Gulf Stream rings, mid-ocean eddies, and much of the distinctly nonsteady character of the ocean circulation. Ocean currents are remarkably variable. Variability on much shorter timescales of weeks and months, and length scales of tens and hundreds of kilometers, often dominates the larger-scale flows discussed here. Thus, it is not appropriate to think of the ocean circulation as a sluggish, linear, and steady. Instead, it is more appropriate to think of the ocean as a complex turbulent environment with its own analogues of unpredictable atmospheric weather systems and climate variability. Nevertheless, the simplified theories of steady circulation illustrate important mechanisms that govern the time-averaged flows.

In closing, two ocean regions deserve special mention: the equatorial ocean and the extreme southern ocean. Equatorial regions have substantially different dynamics compared with models discussed above because Coriolis accelerations are negligible on the equator, where $f = 0$. The wind-related processes that govern El Niño and the Southern Oscillation, for example, depend critically on this fact. The southern ocean distinguishes itself as the only region without a western (or eastern) continental boundary. This absence of boundaries produces a circulation characteristic of the atmosphere, with intense zonal flows that extend around the globe. They represent some of the most intense large-scale currents in the world, and derive much of their energy from the wind. So they too represent an important part of the surface/wind-driven circulation.

See also

Atlantic Ocean Equatorial Currents. Benguela Current. Current Systems in the Atlantic Ocean. Florida Current, Gulf Stream and Labrador Current.

Further Reading

Henderschott MC (1987) Single layer models of the general circulation. In: Abarbanel HDI and Young WR (eds.) *General Circulation of the Ocean*. New York: Springer-Verlag.

Pedlosky J (1996) *Ocean Circulation Theory*. New York: Springer-Verlag.

Salmon R (1998) *Lectures on Geophysical Fluid Dynamics*. Oxford: Oxford University Press.

Stommel H (1976) *The Gulf Stream*. Berkeley: University of California Press.

Veronis G (1981) Dynamics of large-scale ocean circulation. In: Warren BA and Wunsch C (eds.) *Evolution of Physical Oceanography*. Cambridge. MA: MIT Press.

EKMAN TRANSPORT AND PUMPING

T. K. Chereskin, University of California, San Diego, La Jolla, CA, USA
J. F. Price, Woods Hole Oceanographic Institution, Woods Hole, Massachusetts, USA

Introduction

Winds blowing along the ocean's surface exert forces that set the oceans in motion, producing both currents and waves. Separating the wind force into the part that goes into making currents from that which goes into making waves is in fact very difficult. Conceptually, normal forces (i.e., think of the wind beating on the ocean surface like a drum) create waves, and tangential forces (i.e., frictional stresses exerted by the wind pulling on the sea surface) go into making currents. Although there are wind-generated currents that flow in a direction more or less downwind, the currents driven by the steady or slowly varying (compared to the period of the earth's rotation) wind stress flow in a direction that is quite different from the wind direction, sometimes by more than 90°, due to the combined effects of the wind force and the earth's rotation. These wind-driven currents, commonly called Ekman layer currents in recognition of the Swedish oceanographer W. Ekman who first described their dynamics, are the topic of this article, which has three themes: (1) the local dynamics of Ekman layer currents; (2) the spatial variation of wind stress and the resulting spatial variation of the Ekman layer currents; and (3) the effects of Ekman layer currents on the physical and biological environment of the oceans.

The effects of Ekman layer currents are quite far reaching despite the fact that the currents themselves are usually modest, typically no more than 0.05–0.1 m s^{-1} and smaller than many other currents. The importance of Ekman layer currents arises more from their horizontal spatial extent and variation than from their magnitude alone. Although small in magnitude and in vertical extent (typically the layer extends from the surface to about 100 m depth), the mass transport in the Ekman layer (integrated across the width of the ocean) can be substantial, comparable to the transport of major ocean currents such as the Gulf Stream or the Kuroshio. Spatial variation in the Ekman transport is caused by spatial variation in the wind (the wind stress curl) and results in an exchange of fluid with the ocean interior (Ekman pumping); this exchange of mass induces motion in the ocean interior in order to conserve angular momentum. In regions where the Ekman transport converges, conservation of mass requires that fluid be pumped from the surface Ekman layer into the ocean interior; in regions where the Ekman transport diverges, fluid is pumped into the Ekman layer from below. It is through the vertical velocity or pumping thus generated at the base of the Ekman layer that the wind ultimately forces the ocean interior circulation. Ekman pumping also transports material (nutrients, heat, etc.) from the upper thermocline toward the photic zone and sea surface. The pattern of converging and diverging Ekman layer currents is thus imprinted very strongly upon the patterns of biological productivity as well as upon the strength and shape of the major oceanic current gyres.

Ekman's Theory of the Wind-driven Currents

In 1905 Ekman published a simple but elegant theory to explain F. Nansen's observations of ice movements. Nansen was an oceanographer and Arctic explorer; he observed that wind blowing over ice floes caused them to drift at an angle of 20–40° to the right of the wind rather than downwind, and he correctly surmised that the earth's rotation was causing the ice to move at an angle to the wind. Ekman was the first to derive the equations that describe these surface-layer currents, and he used the solution to explain Nansen's observation.

Ekman assumed that the direct influence of the wind was confined to a surface layer, a frictional boundary layer approximately 10–100 m deep, where a steady wind stress was balanced by the Coriolis force. The Coriolis force is an apparent force due to the earth's rotation; it cannot set the fluid in motion, but it can act to change the motion over timescales of days or longer (i.e., timescales on the order of the earth's rotation period). Ekman's key assumptions were: (1) a steady wind blowing over the ocean surface, far from any coast, (2) a flat ocean surface, (3) a constant water density, and (4) a frictional force acting in the surface boundary layer that matched the wind stress at the sea surface and decayed to zero at the bottom of the surface layer. (An inviscid assumption, i.e., neglect of frictional forces, is valid for most of the ocean except near boundaries.) By making these assumptions Ekman arrived at a greatly simplified theoretical ocean, yet he retained the physics required to explain the wind-driven surface currents. For example, if the sea surface is horizontal and the density is uniform, the pressure at any depth is constant, and there will be no pressure-driven flows. In the real ocean, the tilt of the sea surface and the internal horizontal density gradients result in flows due to the pressure-gradient force, and separating these pressure-driven flows from wind-driven currents is the main challenge in direct testing of Ekman's theory from observations.

The Coriolis force is given by the vector cross-product

$$\text{Coriolis force} = \rho f \hat{\boldsymbol{k}} \times \vec{\boldsymbol{u}}; \quad f = 2\Omega \sin\phi, \qquad [1]$$

where ρ is the density of seawater, $\hat{\boldsymbol{k}}$ is the unit vector in the direction of the local vertical, and $\vec{\boldsymbol{u}} = (u, v)$ is the vector of east (u) and north (v) horizontal currents. The Coriolis parameter f is twice the magnitude of the vertical component of the Earth's rotation vector Ω (2π radians per sidereal day) and ϕ is the latitude. Ekman's steady momentum balance is between the Coriolis force and the vertical gradient of the frictional stress $\vec{\tau} = (\tau^x, \tau^y)$:

$$\rho f \hat{\boldsymbol{k}} \times \vec{\boldsymbol{u}} = \frac{\partial \vec{\tau}}{\partial z} \qquad [2]$$

At the sea surface the frictional stress equals the wind stress $\vec{\tau}_0$. Bulk formulae parameterize the wind stress in terms of a wind velocity at 10 m above the sea surface $\vec{U}_{10}$, the air density ρ_{air}, and a drag coefficient C_D:

$$\vec{\tau}_0 = \rho_{\text{air}} C_D \vec{U}_{10} \left| \vec{U}_{10} \right| \qquad [3]$$

A wind speed of 10 m s^{-1} corresponds to a stress of about 0.1 N m^{-2}.

Ekman's assumption that the frictional force acted only in the surface layer allowed him to estimate the volume transport within the layer by integrating from the sea surface, where the wind stress was known, to the unknown depth $-H$ where the frictional stress vanished by assumption. The Ekman transport relation is given by

$$M_x = \frac{\tau_0^y}{\rho f}; \quad M_y = -\frac{\tau_0^x}{\rho f} \qquad [4]$$

where M_x is the eastward component of Ekman transport and M_y is the northward component. One of the surprising results of the theory is that the net transport is at right angles to the wind direction and depends only on the wind stress at the sea surface. Most notably, the transport does not depend on the details of how the stress gets transferred through the Ekman layer. The northward wind component τ_0^y forces the eastward Ekman transport M_x, and the eastward wind component τ_0^x forces the northward Ekman transport M_y. The sign (to the right/left) of the wind depends on the sign of the Coriolis parameter f, which is positive/negative in the Northern/Southern Hemisphere, respectively. The transport is in units of m^2 s^{-1}, and it can be interpreted as the rate at which the volume (per unit width) of water in the Ekman layer is moving. The qualifier 'per unit width' emphasizes that this is the transport at a single point location; in practice, one is usually interested in integrating the Ekman transport over some distance, such as a latitude or longitude band, to see the total volume of water in the Ekman layer that is transported across that latitude or longitude.

Spatial variation in the wind and therefore in the Ekman transport results in local convergences and divergences within the surface layer. The vertical velocity (Ekman pumping) w_H that results from convergence or divergence in the Ekman layer is derived by integrating the mass conservation equation over the Ekman layer:

$$w_H = \frac{\partial(\tau_0^y/\rho f)}{\partial x} - \frac{\partial(\tau_0^x/\rho f)}{\partial y} \qquad [5]$$

Thus the Ekman pumping is given by the spatial derivative or curl of the wind stress (divided by ρf); it depends on the spatial variation of the wind rather than its magnitude.

Ekman's theory also predicted the currents within the surface layer. His solution for the velocity structure, the Ekman spiral, is the least robust of his results since it depends critically on how one assumes the stress that the wind exerts on the surface is transferred downward by friction and mixing. Ekman was the first to acknowledge that the wind momentum is transferred to ocean currents through turbulent mixing. He modeled it as a diffusion process, exactly analogous to molecular diffusion, but with an effective kinematic viscosity (eddy viscosity) many orders of magnitude larger than molecular viscosity. Turbulent eddies are much more efficient than molecular diffusion at stirring the fluid and hence mixing the wind momentum. For example, a wind of 10 m s^{-1} would require more than a day to penetrate the top meter of the surface layer if molecular diffusion were the sole process acting to mix the wind momentum. In fact, the wind momentum is observed to mix down by tens of meters within hours. Although turbulence is clearly the dominant process in creating the wind-mixed layer, the detailed structure of the turbulent boundary layer remains an active research question today.

To solve for the currents, Ekman assumed a constant eddy viscosity A_ν in place of the molecular viscosity ν. The value of ν for sea water is approximately 10^{-6} m^2 s^{-1} and applies for smooth laminar flow. The magnitude and structure of a turbulent eddy viscosity A_ν are not well known, but scaling arguments suggest that its magnitude may be as large as 10^{-1} m^2 s^{-1}. The current structure is a spiral (**Figure 1**) that decays in amplitude and rotates clockwise with increasing depth (z negative):

$$u = \exp(z/D_E)[V_+\cos(z/D_E) + V - \sin(z/D_E)]$$

$$v = \exp(z/D_E)[V_+\sin(z/D_E) - V - \cos(z/D_E)] \quad [6]$$

$$V_+ = \frac{\tau_0^x + \tau_0^y}{\rho\sqrt{(2A_\nu|f|)}}; \quad V- = \frac{\tau_0^x - \tau_0^y}{\rho\sqrt{(2A_\nu|f|)}}$$

The Ekman depth, $D_E = \sqrt{(2A_\nu|f|)}$, is the depth over which the amplitude decays by 1/e and over which the velocity vector rotates by one radian. Note that the predicted surface current lies at an angle of 45° to the wind direction.

Ekman spirals have been observed in the laboratory, where the appropriate viscosity is the molecular value ν. More limited observations are available in the open ocean, because of the difficulty in acquiring observations with the requisite vertical resolution and because of the difficulty in separating the wind-driven flow from pressure-driven flow. Ocean spirals have been observed (**Figure 1**); they tend to be flatter than Ekman's theory predicts, due to the effect of other processes such as stratification and the diurnal cycling of the mixed layer depth. Regardless of these details, integrating the Ekman spiral (eqn [6]) over the surface layer yields the more general Ekman transport result (eqn [4]).

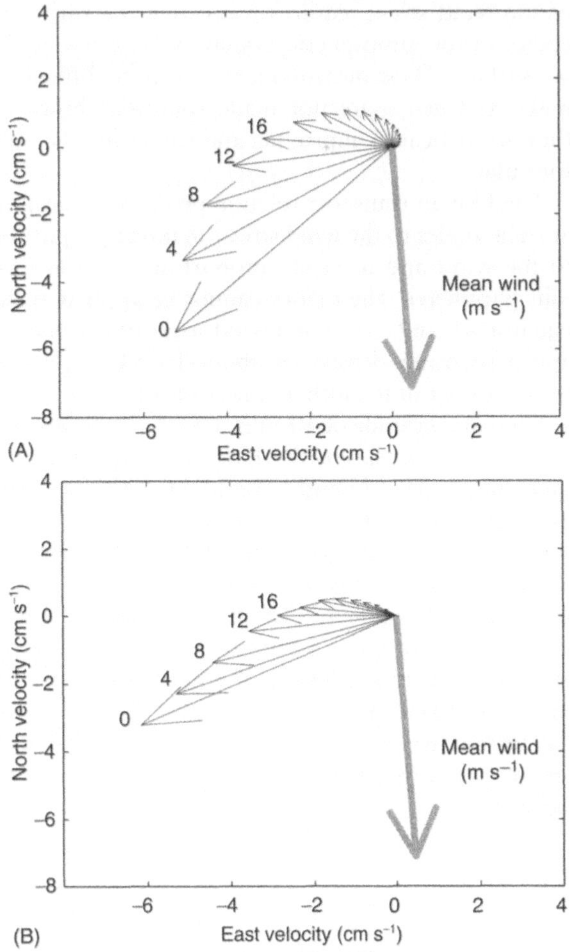

Figure 1 Comparison of an observed wind-forced velocity spiral and the theoretical Ekman spiral calculated for the same wind in the Northern Hemisphere. Each current vector is the time-averaged velocity at a particular depth; the first five depths are labelled. Depth units are m, current velocity units are cm s^{-1}, and wind velocity units are m s^{-1}. (A) The theoretical Ekman spiral: the surface current lies 45° to the right of the wind, and the rate of amplitude decay and the rate of rotation with depth are both set by the constant eddy viscosity (0.014 m^2 s^{-1}), chosen to match the rate of amplitude decay of the observed spiral. (B) Observed velocity spiral from averaged current observations from a location about 400 km offshore of the US west coast from the Eastern Boundary Currents Experiment. The currents spiral to the right of the wind direction and decay with depth, as predicted by the theory. However, the surface current lies more than the predicted 45° to the right of the wind, and the current amplitude decreases at a faster rate than it turns to the right, resulting in a flatter spiral than the theory predicts. These differences are due to unmodeled processes.

Ekman Transport and Pumping

One of the remarkable results of Ekman's theory is the Ekman transport relation (eqn [4]), which allows the surface layer transports to be predicted from the wind field. Ship observations of wind have been made over a much longer period of time and over a much greater area of the ocean than have direct measurements of ocean currents, and satellites presently provide global coverage of the wind field. Wind velocity is the quantity that is typically measured, at heights from 10 to 30 m above the sea surface. Wind stress is proportional to the wind speed squared and in the same direction as the wind velocity.

A first step in calculating wind stress from velocity measurements is to determine the wind speed at 10 m height above the sea surface, using a model of the wind profile versus height. The stress at the sea surface can then be calculated using eqn [3] that parameterizes the stress using the 10-m wind speed and a drag coefficient that depends on the speed and other properties of the air–sea interface such as air and sea temperature and relative humidity. Direct estimation of the wind stress requires measuring the turbulent eddies in the atmospheric boundary layer above the sea surface. These measurements are more difficult to make and hence are not made routinely; however, they are critical to improving and validating the bulk formulae.

The Ekman transport relation predicts a transport at right angles to the wind stress, in direct proportion to the wind and inversely proportional to the Coriolis parameter. The theory cannot be applied at the equator where $f = 0$. For a wind stress of 0.1 N m^{-2}, and a seawater density of about 1025 kg m^{-3}, the transport per unit width at a latitude of 30° is 1.3 m^2 s^{-1} and at a latitude of 10° it is 3.4 m^2 s^{-1}. Integrated across an ocean basin, the Ekman transport can be quite large. The Atlantic Ocean at 10°N is about 4000 km across. The mean wind stress is of order 0.1 N m^{-2}, and the predicted Ekman transport is 15×10^6 m^3 s^{-1}. This transport is about one-half the transport measured for the Gulf Stream where it passes through the Straits of Florida. Across the same latitude in the Pacific, the Ekman transport is estimated to be about 50×10^6 m^3 s^{-1}, not because the winds are stronger but because the Pacific is about three times the width of the Atlantic at that latitude.

The wind stress is communicated to the ocean interior through Ekman pumping, and therefore the spatial variation of the Ekman transport is just as important as its magnitude. The large-scale pattern of the wind is one of alternating bands of easterlies and westerlies (**Figure 2A**). The equator is spanned by a broad band of easterly winds known as the trade winds. The magnitude of these winds decreases with latitude, reaching a minimum about 30° or so away from the equator. Still further north/south lies the band known as the prevailing westerlies, with a maximum at about 50° latitude. The polar easterlies are the most poleward band of winds. This overall pattern dominates in both hemispheres of the Pacific and Atlantic Oceans. The Ekman transport relation implies a corresponding pattern of convergences and divergences in the Ekman transport that should be detectable through the Ekman pumping of ocean current gyres (**Figure 2C**). For example, the theory suggests that the persistent easterly trade winds in the tropics will force a poleward Ekman transport; the midlatitude westerlies will force an equatorward Ekman transport. The resulting Ekman transport convergence at subtropical latitudes should result in a thick warm layer of water above the main thermocline, an Ekman pumping that supplies fluid to

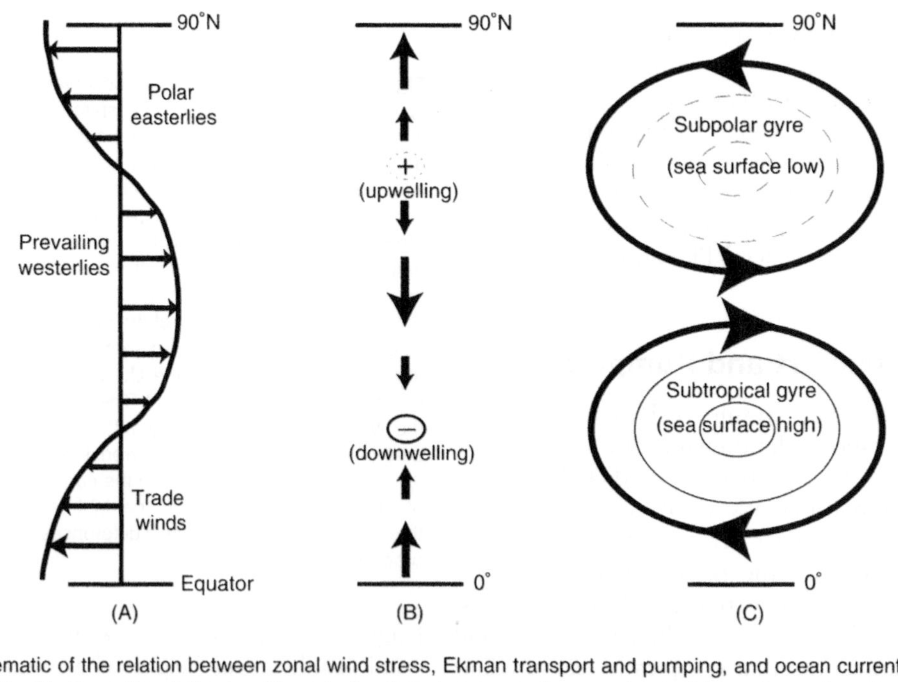

Figure 2 Schematic of the relation between zonal wind stress, Ekman transport and pumping, and ocean current gyres. (A) Zonal wind stress for the Northern Hemisphere, with easterly wind stress near the equator and the pole, and westerly wind stress at midlatitudes. (B) The meridional Ekman transport that results from the zonal wind stress. The magnitude is proportional to the wind stress, and the direction is orthogonal to the wind direction. The spatial variation in the wind stress results in Ekman convergences and divergences and Ekman pumping. (C) The wind-driven oceanic current gyres. Ekman convergence in the subtropics results in a sea-surface high and a clockwise-current gyre with downwelling at the gyre center. The Ekman divergence at subpolar latitudes results in a sea-surface low and an anticlockwise-current gyre with upwelling at the gyre center.

the ocean interior, and, through conservation of angular momentum, a clockwise/anticlockwise general circulation in the northern/southern hemisphere. Such gyres are major observed features of the subtropical oceans. Their existence is an indirect confirmation of Ekman's theory. It is through the pumping thus generated at the base of the Ekman layer that the wind ultimately forces the ocean interior. If the wind were spatially uniform, the wind-driven currents and mass transport would remain largely confined within the shallow surface Ekman layer and would have a negligible role in driving ocean circulation.

Note that using the wind to estimate the Ekman transport via eqn [4] gives no indication of how deep it extends. Direct verification, from measurements of both wind and currents, is required in order to determine the details of the turbulent mixing of the wind momentum, such as the maximum depth of frictional influence of the wind, and the importance of other processes such as time dependence and stratification. This direct confirmation eluded oceanographers until quite recently, because of the difficulty in making accurate measurements in the harsh surface zone and in separating the wind-driven flow from other pressure-driven flows. Also, since the wind is not steady, appropriate timescales for averaging need to be determined. Direct verification of the integrated Ekman transport has been made from measurements at tropical latitudes 8–11°N in the Atlantic, Pacific, and Indian Oceans, where the Ekman transport is large. High vertical resolution moored measurements at midlatitudes in the Atlantic and the Pacific have verified the Ekman transport relation for moderate winds, with transport per unit width of about $1\,m^2\,s^{-1}$.

Ekman transport and pumping have far-reaching implications, from the mechanism of momentum transfer from wind to water to the maintenance of the large-scale wind-driven oceanic current gyres. Although the original theory was developed for a steady wind, the phenomena of Ekman transport and pumping occur on much shorter timescales, e.g., the scale of synoptic storms. However, Ekman transport and pumping are most effective in driving large-scale ocean circulation when the wind is steady over long periods of time, because then the effects can accumulate.

Coastal Upwelling

Although spatial variation in the wind field is the cause of Ekman pumping in the open ocean, a spatially uniform wind blowing parallel to a coast can also cause Ekman pumping. In the Northern Hemisphere,

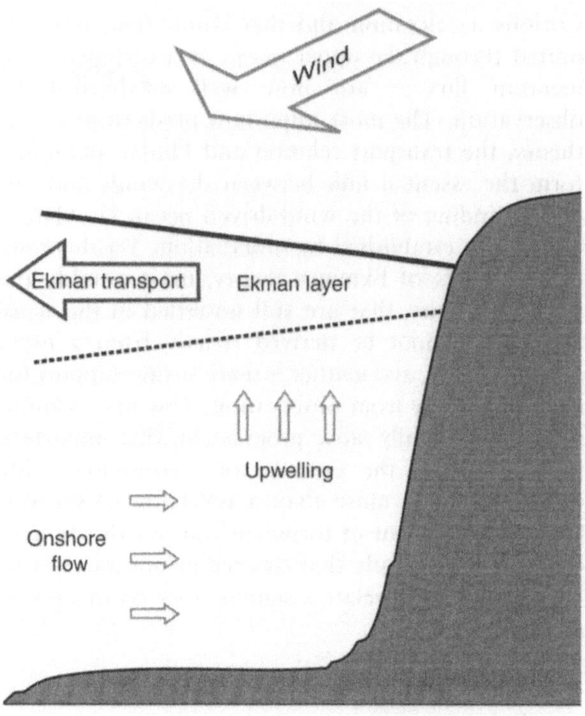

Figure 3 Diagram of upwelling (Ekman divergence) at a coast. The wind blows parallel to the coast (in a direction *out* of the diagram in the Northern Hemisphere), forcing a net Ekman transport offshore. This offshore transport is compensated by onshore flow at depth and upwelling.

Ekman divergence occurs when the wind blows parallel to a coastline on its left. For example, during spring and summer the mean winds along the west coast of North America are southward. Associated with these winds is a net westward Ekman transport. This offshore transport of mass in the surface layer is balanced by an onshore flow at depth and an Ekman pumping of water from depth into the surface layer (**Figure 3**). This phenomenon is known as coastal upwelling, and it occurs seasonally along the west coasts of continents in both hemispheres. Prolonged upwelling can provide a continuous source of cold, nutrient-rich water to the surface layer euphotic zone, replacing the nutrient-depleted water that is transported offshore. Many of the ocean's important fisheries are concentrated in upwelling zones. The reverse phenomenon, convergence from onshore Ekman transport and downward pumping at the coast also occurs and is called downwelling.

Summary

Ekman's theory of wind-driven currents, now almost 100 years old, is a cornerstone of physical oceanography. The two essential features of the theory – that the dominant momentum balance for steady wind-driven currents is between the wind stress and

Coriolis acceleration and that wind stress is transmitted through the upper ocean as a turbulent momentum flux – are now well established by observation. The most important predictions of the theory, the transport relation and Ekman pumping, form the essential link between the winds and our understanding of the wind-driven ocean circulation and is also established by observation. Yet there are other aspects of Ekman's theory, most notably the eddy diffusivity, that are still unsettled in the sense that they cannot be derived strictly from a more general theory, and neither is there strong support for eddy diffusivity from observation. This may seem to indicate unusually slow progress on this important point, and yet the uncertainty surrounding eddy diffusivity is no more than a reflection of the tortuous development of turbulent transfer theory generally. It seems likely that succeeding editions of this encyclopedia will relate a similar story on this point, at least.

See also

Meddies and Subsurface Eddies. Mesoscale Eddies. Wind Driven Circulation.

Further Reading

Chereskin TK (1995) Direct evidence for an Ekman balance in the California Current. *Journal of Geophysical Research* 100: 18261–18269.

Ekman VW (1905) On the influence of the earth's rotation on ocean-currents. *Arkiv för Matematik, Astronomi och Fysik* 2: 1–52.

Gill AE (1982) *Atmosphere–Ocean Dynamics*. New York: Academic Press.

Pond S and Pickard GL (1983) *Introductory Dynamical Oceanography*, 2nd edn. Oxford: Pergamon.

Price JF, Weller RA, and Schudlich RR (1987) Wind-driven currents and Ekman transport. *Science* 238: 1534–1538.

OCEAN CIRCULATION: MERIDIONAL OVERTURNING CIRCULATION

J. R. Toggweiler, NOAA, Princeton, NJ, USA

Introduction

The circulation of the ocean has been traditionally divided into two parts, a wind-driven circulation that dominates in the upper few hundred meters, and a density-driven circulation that dominates below. The latter was once called the 'thermohaline circulation', a designation that emphasized the roles of heating, cooling, freshening, and salinification in its production. Use of 'thermohaline circulation' has all but disappeared among professional oceanographers.

The preferred designation is 'meriodional overturning circulation', hereafter MOC. This usage reflects the sense of the time-averaged flow, which generally consists of a poleward flow of relatively warm water that overlies an equatorward flow of colder water at depth. It also reflects a recognition by oceanographers that most of the work done to drive the circulation, whether near the surface or at depth, comes directly or indirectly from the wind.

Like the thermohaline circulation of old, the MOC is an important factor in the Earth's climate because it transports roughly 10^{15} W of heat poleward into high latitudes, about one-fourth of the total heat transport of the ocean/atmosphere circulation system. Radiocarbon measurements show that it turns over all the deep water in the ocean every 600 years or so. Its upwelling branch is important for the ocean's biota as it brings nutrient-rich deep water up to the surface. It may or may not be vulnerable to the warming and freshening of the Earth's polar regions associated with global warming.

The most distinguishing features of the MOC are observed in the sinking phase, when new deep-water masses are formed and sink into the interior or to the bottom. Large volumes of cold polar water can be observed spilling over sills, mixing violently with warmer ambient water, and otherwise descending to abyssal depths. The main features of the upwelling phase are less obvious. The biggest uncertainty is about where the upwelling occurs and how the upwelled deep water returns to the areas of deep-water formation.

The Cooling Phase – Deep-Water Formation

The most vigorous overturning circulation in the ocean today is in the Atlantic Ocean where the upper part of the Atlantic's MOC carries warm, upper ocean water through the Tropics and subtropics toward the north while the deep part carries cold dense polar water southward around the tip of Africa and into the Southern Ocean beyond. The Atlantic's MOC converts roughly $15 \times 106\,\mathrm{m}^3\,\mathrm{s}^{-1}$ of upper ocean water into deep water. (Oceanographers designate a flow rate of $1 \times 106\,\mathrm{m}^3\,\mathrm{s}^{-1}$ to be 1 Sv. All the world's rivers combined deliver *c.* 1 Sv of fresh water to the ocean.)

The MOC in the Atlantic is often characterized as a 'conveyor belt' or more generally as a continuous current or ribbon of flow. It is shown following a path that extends through the Florida Straits and up the east coast of North America as part of the Gulf Stream. The ribbon then cuts to the east across the Atlantic and then northward closer to the coast of Europe. This is somewhat misleading because individual water parcels do not follow this kind of continuous path. The MOC is composed instead of multiple currents that transfer water and water properties along segments of the path. Individual water parcels loop around multiple times and may recirculate all the way around the wind-driven gyres while moving northward in stages.

As the MOC moves northward through the tropical and subtropical North Atlantic, it spans a depth range from the surface down to ~800 m and has a mean temperature of some 15–20 °C. During its transit through the Tropics and subtropics, the MOC becomes saltier due to the excess of evaporation over precipitation in this region. It also becomes warmer and saltier by mixing with the salty outflow from the Mediterranean Sea. By the time the MOC has crossed the 50° N parallel into the subpolar North Atlantic the water has cooled to an average temperature of 11–12 °C. Roughly half of the water carried northward by the MOC at this juncture moves into the Norwegian Sea between Iceland and Norway. Part of this flow extends into the Arctic Ocean.

The final stages of this process make the salty North Atlantic water cold and dense enough to sink. Sinking is known to occur in three main places. The densest sinking water in the North Atlantic is formed in the Barents Sea north of Norway where salty water from the Norwegian Current is exposed to the atmosphere on the shallow ice-free Barents shelf. Roughly 2 Sv of water from the Norwegian Current crosses the shelf and sinks into the Arctic basin after being cooled down to 0 °C. This water eventually flows out of the Arctic along the coast of Greenland at a depth of 600–1000 m. The volume of dense water is increased by additional sinking and open ocean convection in the Greenland Sea north of Iceland. The dense water then spills into the North Atlantic over the sill between Greenland and Iceland in Denmark Strait (at about 600-m depth) and through the Faroe Bank Channel between the Faroe Islands and Scotland (at about 800-m depth).

As 0 °C water from the Arctic and Greenland Seas passes over these sills, it mixes with 6 °C Atlantic water beyond the sills and descends the continental slopes down to a depth of 3000 m. The overflows merge and flow southward around the tip of Greenland into the Labrador Sea as part of a deep boundary current that follows the perimeter of the subpolar North Atlantic. A slightly warmer water mass is formed by open ocean convection within the Labrador Sea. The deep water formed in the Labrador Sea increases the volume flow of the boundary current that exits the subpolar North Atlantic beyond the eastern tip of Newfoundland.

Newly formed water masses are easy to track by their distinct temperature and salinity signatures and high concentrations of oxygen. The southward flow of North Atlantic Deep Water (NADW) is a prime example. NADW is identified as a water mass with a narrow spread of temperatures and salinities between 2.0 and 3.5 °C and 34.9 and 35.0 psu, respectively.

Figure 1 shows the distribution of salinity in the western Atlantic which tracks the southward movement of NADW. Newly formed NADW ($S > 34.9$) is easily distinguished from the relatively fresh intermediate water above ($S < 34.6$) and the Antarctic water below ($S < 34.7$). NADW exits the Atlantic south of Africa between 35° and 45° S and joins the eastward flow of the Antarctic Circumpolar Current (ACC). Traces of NADW reenter the Atlantic Ocean through Drake Passage (55–65° S) after flowing all the way around the globe.

The other major site of deep-water formation is the coast of Antarctica. The surface waters around Antarctica, like those over most of the Arctic Ocean, are ice covered during winter and are too fresh to sink. Deep water below 500 m around Antarctica, on the other hand, is relatively warm (1.5 °C) and fairly salty (34.70–34.75 psu). This water mass, known as Circumpolar Deep Water (CDW), penetrates onto the relatively deep continental shelves around

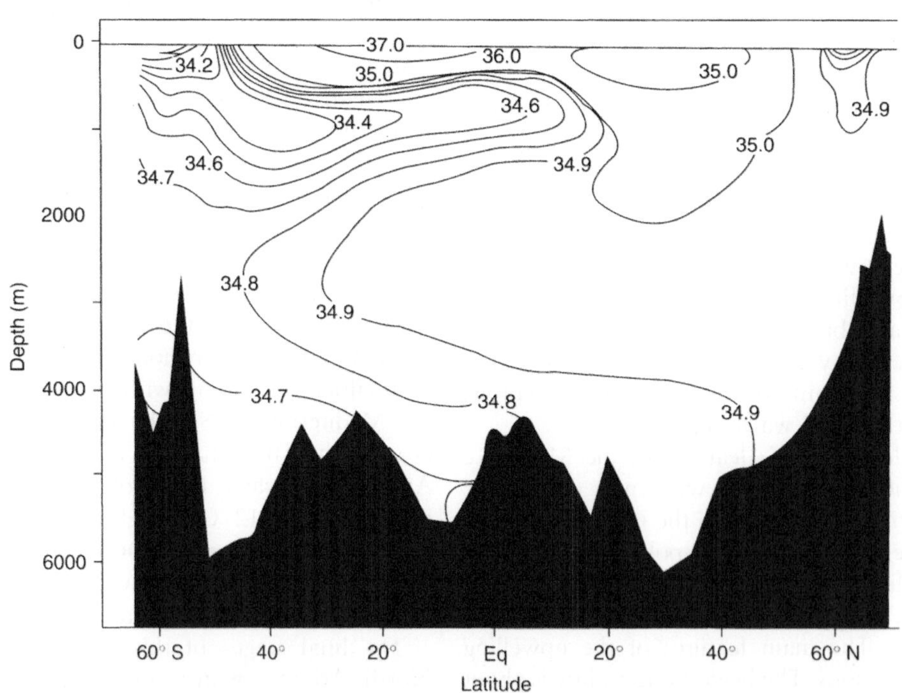

Figure 1 North–south section of salinity down the western Atlantic from Iceland to Drake Passage. The salinity distribution has been contoured every 0.1 psu between 34.0 and 35.0 psu to highlight NADW (34.9–35.0 psu).

Antarctica where it is cooled to the freezing point. Brine rejection from the formation of new sea ice maintains fairly high salinities on the shelf despite the freshening effects of precipitation and the input of glacial meltwater. Very cold shelf water (−1 °C, 34.6–34.7 psu) is then observed descending the continental slope to the bottom in the Weddell Sea. Bottom water is also observed to form off the Adelie coast south of Australia (~ 140° E).

The volume of new deep water formed on the Antarctic shelves is not very large in relation to the volume of deep water formed in the North Atlantic, perhaps 3–4 Sv in total. It however entrains a large volume of old Circumpolar Deep Water, as it sinks to the abyss. Even with the entrainment, the deep water formed around Antarctica is denser than the NADW formed in the north. Antarctic Bottom Water (AABW) occupies the deepest parts of the ocean and is observed penetrating northward into the Atlantic, Indian, and Pacific Oceans through deep passages in the mid-ocean ridge system.

Small quantities of deep water are also observed to form in semi-enclosed evaporative seas like the Mediterranean and Red Seas. These water masses are dense owing to their high salinities. They tend to form intermediate- depth water masses after exiting their formation areas in relation to the deep-water and bottom-water masses formed in the North Atlantic and around Antarctica.

The Warming Phase – Upwelling and the Return Flow

The upwelling of deep water back to the ocean's surface was thought at one time to be widely distributed over the ocean. This variety of upwelling was attributed to mixing processes that were hypothesized to be active throughout the interior. The mixing was thought to be slowly transfering heat downward, making the old deep water in the interior progressively less dense so that it could be displaced upward by the colder and saltier deep waters forming near the poles. Since the main areas of deep water formation are located at either end of the Atlantic, and since most of the ocean's area is found in the Indian and Pacific Oceans, it stood to reason that the warming of the return flow should be widely distributed across the Indian and Pacific. Schematic diagrams often depict the closure of the old thermohaline circulation as a flow of warm upper ocean water that passes from the North Pacific through Indonesia, across the Indian Ocean, and then around the tip of Africa into the Atlantic.

Observations made over the last 30 years point instead to turbulent mixing that is intense in some places but is not widespread. Attempts to directly measure the turbulent mixing in the interior have shown that there may only be enough mixing to modify perhaps 10% or 20% of the deep water formed near the poles. There is no indication that any deep water is actually upwelling to the surface in the warm parts of the Indian and Pacific Oceans. Vigorous mixing is found near the bottom, where it is generated by tidal motions interacting with the bottom topography. It is also found in the vicinity of strong wind-driven currents. Thus, the energy that is available to warm the deep waters of the abyss comes mainly from the winds and tides.

It now appears that most of the deep water sinking in the North Atlantic upwells back to the surface in the Southern Ocean. The upwelling occurs within the channel of open water that circles the globe around Antarctica. **Figure 2** shows schematically how this is thought to work. The curved lines in the background are isolines of constant density (also known as isopycnals). Their downward plunge to the north away from Antarctica reflects the flow of the ACC out of the page in the center of the figure. Salty dense water from the North Atlantic is found at the base of the plunging isopycnals. Westerly winds above the ACC (also blowing out of the page) drive the ACC forward and also push the cold fresh water at the surface away from Antarctica to the north. Dense salty water from the North Atlantic is drawn upward in its place. A mixture of the salty dense deep water and the cold fresh surface water is then driven northward out of the channel by the westerly winds and is forced down into the thermocline on the north side of the ACC.

In this way, the winds driving the ACC continually remove water with North Atlantic properties from the interior. Oceanographers are fairly certain that something like this is happening because the winds driving the ACC have become stronger over the last 40 years in response to global warming and the depletion of ozone over Antarctica. The subsurface water around Antarctica has become warmer, saltier, and lower in oxygen as upwelled water from the interior has displaced more of the cold fresh surface water, as shown in **Figure 2**. The thermocline water north of the ACC has also become cooler and fresher.

Numerical experiments with ocean general circulation models show that much of the water forced down into the thermocline around the open channel eventually makes its way into the Atlantic Ocean where it is converted again into deep water in the northern North Atlantic. Thus, the winds over the ACC, in drawing up deep water from the ocean's interior, have been shown to actively enhance the formation of deep water in the North Atlantic.

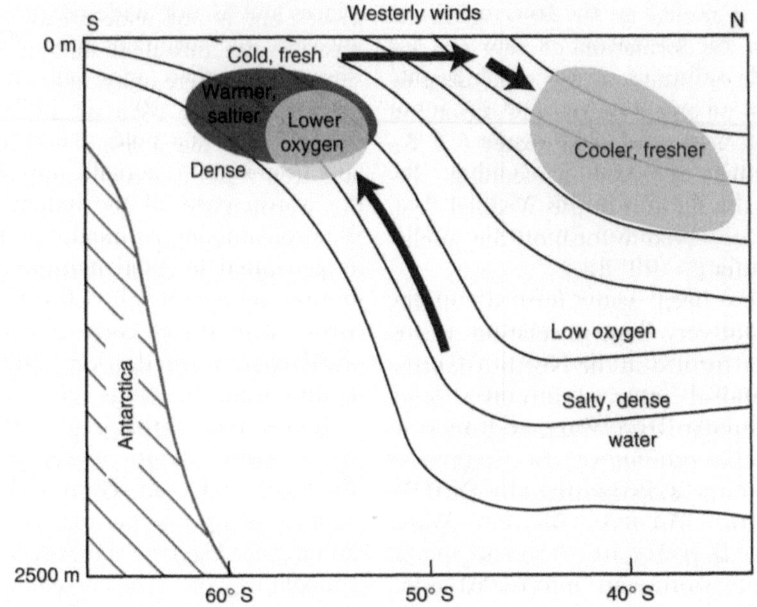

Figure 2 Schematic cross section of the ACC, showing how the winds driving the current in an eastward direction out of the page also drive an overturning circulation. Stronger winds over the last 40 years have drawn more deep water to the surface south of the ACC, which has caused the subsurface waters around Antarctica to become warmer, saltier, and lower in oxygen, and have produced more downwelling to the north, which has made the thermocline waters north of the ACC cooler and fresher. Adapted from Aoki S, Bindoff NL, and Church JA (2005) Interdecadal water mass changes in the Southern Ocean between 30 ° E and 160 ° E. *Geophysical Research Letters* 32 (doi:10.1029/2005GL022220) and reproduced from Toggweiler RR and Russell J (2008) Ocean Circulation in a warming climate. *Nature* 451 (doi:10.1038/nature06590).

The deep water drawn up to the surface around Antarctica is colder than the surface water that is forced down into the thermocline north of the ACC. As this cold water comes into contact with the atmosphere and is carried northward, it takes up solar heat that otherwise would be available to warm the Southern Ocean and Antarctica. The MOC then carries this southern heat across the equator into the high latitudes of the North Atlantic where it is released to the atmosphere when new deep water is formed. The MOC thereby warms the North Atlantic at the expense of a colder Southern Ocean and colder Antarctica. Indeed, sea surface temperatures at 60° N in the North Atlantic are on average *c.* 6 °C warmer than sea surface temperatures at 60° S.

If the warming phase of the MOC involved a downward mixing of heat in low and middle latitudes, as once thought, the MOC would carry tropical heat poleward into high latitudes. The warming phase of the Atlantic's MOC now seems to take place in the south instead. This means that the heat transport by the MOC through the South Atlantic is directed equatorward and is opposed to the heat transport in the atmosphere. The net effect is a weakening of the global heat transport in the Southern Hemisphere and a strengthening of the heat transport in the Northern Hemisphere.

General Theory for the MOC

Figure 3 is a north–south section showing the distribution of potential density through the Atlantic Ocean. Most of the northward flow of the Atlantic's MOC takes place between the 34.0 ($\sim$20 °C) and 36.0 ($\sim$8 °C) isopycnals, that is, within the main part of the thermocline, but some of the northward flow takes place among the denser isopycnals of the lower thermocline down to 36.6 $g\,kg^{-1}$ ($\sim$1000 m). The southward flow of NADW is located, for the most part, below the 37.0 isopycnal.

The isopycnals in the lower thermocline in **Figure 3** are basically flat north of 40° S but rise up to the surface between 40° and 60° S. The rise of the isopycnals marks the eastward flow of the ACC, as shown previously in **Figure 2**. The relatively deep position of these isopycnals north of 40° S reflects the mechanical work done by the winds that drive the ACC. The convergent surface flow forced by the winds north of the ACC pushes the relatively light lower thermocline water down in relation to the cold, dense water that is drawn up by the winds south of the ACC. Numerical experiments carried out in ocean global climate models (GCMs) suggest that all the isopycnals of the lower thermocline would be squeezed up into the main thermocline if the circumpolar channel were closed and the ACC eliminated.

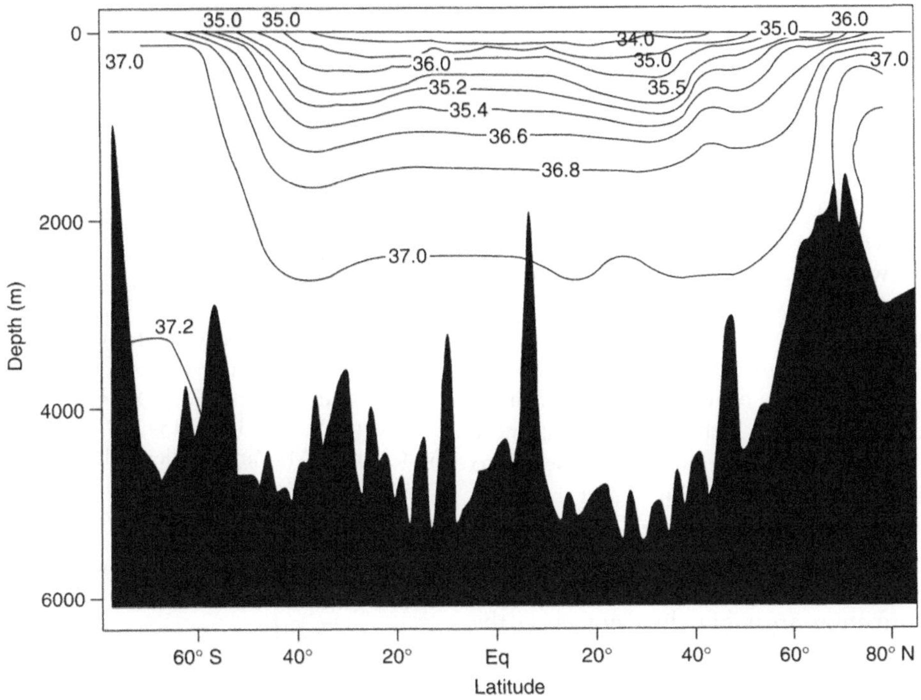

Figure 3 Density structure of the Atlantic thermocline. Seawater density is referenced to a depth of 2000 m and has been zonally averaged across the Atlantic basin (units $g\,kg^{-1}$). Contours have been chosen to highlight the lower part of the thermocline. The zonally averaged topography fails to capture the depths of the sills through the Greenland–Iceland–Scotland Ridge (65° N), which are at 600 and 800 m.

The relatively deep lower thermocline puts relatively warm water just to the south of the sills between Greenland, Iceland, and the Faroe Islands in the northern North Atlantic. This warm water is important because it sets up a sharp contrast with the cold water behind the sills, which ultimately drives the flow of dense water over the sills and into the deep Atlantic. In this way, the winds and the ACC in the south contribute as much, if not more, to the overflows in the north as the cooling that takes place at the surface.

The depths of the isopycnals in the lower thermocline lead one to a general theory for the MOC in the Atlantic. Deep-water formation in the North Atlantic converts relatively low-density thermocline water into new deep water. It thereby removes mass from the thermocline and allows the isopycnals of the thermocline to be squeezed upward. The winds in the south have the opposite effect as they draw deep water up to the surface and pump it northward into the thermocline. They convert dense water from the deep ocean into low-density thermocline water and cause the thermocline to thicken downward. In this way, the thermocline thickness reflects a balance between the addition of mass via winds in the south and the loss of mass by deep-water formation in the north. The strength of the MOC should in this sense be proportional to the thermocline thickness and the transformations in the north and south that convert light water to dense water and vice versa.

A different kind of theory is needed for the overturning of the bottom water formed around Antarctica because the winds cannot play the same role. The upwelling branch of this circulation is also associated with the ACC as in **Figure 2** but the upwelled water in this case flows poleward onto the adjacent Antarctic shelves where it is cooled and sinks back into the interior. The winds in this case help cool the upwelled water by exposing it to the atmosphere and by driving the fresh Antarctic surface waters away but the strength of the bottom water circulation would appear to be limited by the rate at which old bottom waters can be warmed by mixing with the overlying deep water.

Instability of the MOC

Cooling of the ocean in high latitudes makes polar surface waters denser in relation to warmer waters at lower latitudes. Thus cooling contributes to a stronger MOC. The salinity section through the Atlantic Ocean in **Figure 1** gives one a superficial impression that salinification also makes a positive contribution to the MOC. This is actually not true.

The cycling of fresh water between the ocean and atmosphere (the hydrological cycle) results in a net addition of fresh water to the polar oceans which reduces the density of polar surface waters. Thus, the haline part of the old thermohaline circulation is nearly always in opposition to the thermal forcing. The Earth's hydrological cycle is expected to become more vigorous in the future with global warming. This may weaken the MOC in a way which could be fairly abrupt and unpredictable.

NADW is salty because the upper part of the MOC flows through zones of intense evaporation in the Tropics and subtropics. If the rate at which new deep water is forming is relatively high, as it seems to be at the present time, the sinking water removes much of the fresh water added in high latitudes and carries it into the interior. The added fresh water in this case dilutes the salty water being carried into the deep-water formation areas but does not erase the effect of evaporation in low latitudes. Thus, NADW remains fairly salty and is able to export fresh water from the Atlantic basin.

If the rate of deep-water formation is relatively low or the hydrological cycle is fairly strong, the fresh water added in high latitudes can create a low-salinity lid that can reduce or annihilate the MOC. There seems to be a critical input of freshwater input for maintaining the MOC. If the freshwater input is close to this threshold, the overturning becomes unstable and may become prone to wild swings over time.

Coupled (ocean + atmosphere) climate models from the 1990s projected that a fourfold increase in atmospheric CO_2 would increase the hydrological cycle sufficiently that the MOC might collapse. More recent coupled models are predicting that the warming will, in addition, lead to stronger mid-latitude westerly winds that are shifted poleward with respect to their preindustrial position. This change in the westerlies puts stronger Southern Hemisphere westerlies directly over the ACC, which should make the MOC stronger. Thus, the MOC could become stronger or weaker depending on whether the winds or the hydrological cycle dominate in the future.

See also

Antarctic Circumpolar Current. Florida Current, Gulf Stream and Labrador Current. Ocean Circulation. Wind Driven Circulation.

Further Reading

Aoki S, Bindoff NL, and Church JA (2005) Interdecadal water mass changes in the Southern Ocean between 30° E and 160° E. *Geophysical Research Letters* 32: (doi:10.1029/2005GL022220).

Broecker WS (1991) The great ocean conveyor. *Oceanography* 4: 79–89.

Gnanadesikan A (1999) A simple predictive model for the structure of the oceanic pycnocline. *Science* 283: 2077–2079.

Kuhlbrodt T, Griesel A, Montoya M, Levermann A, Hofmann M, and Rahmstorf S (2007) On the driving processes of the Atlantic meridional overturning circulation. *Reviews of Geophysics* 45: RG2001 (doi:10.1029/2004RG000166).

Manabe S and Stouffer R (1993) Century-scale effects of increased atmospheric CO_2 on the ocean–atmosphere system. *Nature* 364: 215–218.

McCartney MS and Talley LD (1984) Warm-to-cold water conversion in the northern North Atlantic Ocean. *Journal of Physical Oceanography* 14: 922–935.

Munk W and Wunsch C (1998) Abyssal recipes II: Energetics of tidal and wind mixing. *Deep-Sea Research I* 45: 1977–2010.

Orsi A, Johnson G, and Bullister J (1999) Circulation, mixing, and production of Antarctic bottom water. *Progress in Oceanography* 43: 55–109.

Rahmstorf S (1995) Bifurcations of the Atlantic thermohaline circulation in response to changes in the hydrological cycle. *Nature* 378: 145–149.

Rudels B, Jones EP, Anderson LG, and Kattner G (1994) On the intermediate depth waters of the Arctic Ocean. In: Johannessen OM, Muench RD, and Overland JE (eds.) *Geophysical Monograph 85: The Polar Oceans and Their Role in Shaping the Global Environment*, pp. 33–46. Washington, DC: American Geophysical Union.

Russell JL, Dixon KW, Gnanadesikan A, Stouffer RJ, and Toggweiler JR (2006) The Southern Hemisphere westerlies in a warming world: Propping open the door to the deep ocean. *Journal of Climate* 19(24): 6382–6390.

Schmitz WJ and McCartney MS (1993) On the North Atlantic Circulation. *Reviews of Geophysics* 31: 29–49.

Toggweiler JR and Samuels B (1995) Effect of Drake Passage on the global thermohaline circulation. *Deep-Sea Research I* 42: 477–500.

Toggweiler JR and Samuels B (1998) On the ocean's large-scale circulation near the limit of no vertical mixing. *Journal of Physical Oceanography* 28: 1832–1852.

Toggweiler RR and Russell J (2008) Ocean Circulation in a warming climate. *Nature* 451: (doi:10.1038/nature 06590).

Tziperman E (2000) Proximity of the present-day thermohaline circulation to an instability threshold. *Journal of Physical Oceanography* 30: 90–104.

Wunsch C and Ferrari R (2004) Vertical mixing, energy, and the general circulation of the oceans. *Annual Reviews of Fluid Mechanics* 36: 281–314.

REGIONAL AND SHELF MODELS

J. J. Walsh, University of South Florida,
St. Petersburg, Florida, USA

Introduction

Simple biological models of nutrient (N), phytoplankton (P), and zooplankton (Z) state variables were formulated more than 50 years ago (Riley *et al.*, 1949), with minimal physics, to explore different facets of marine plankton dynamics. These N–P–Z models, usually in units of nitrogen or carbon are still used in coupled biophysical models of ocean basins, where computer constraints preclude the use of more complex ecological formulations of global biogeochemical budgets. At the regional scale (**Table 1**) of individual continental shelves (Steele, 1974; Walsh, 1988), however, pressing questions no longer focus on the amount of biomass of the total phytoplankton community, but are instead concerned with the functional types of algal species that may continue to support fisheries, generate anoxia, fix nitrogen, or form toxic red tides.

Regional models that presently address the outcome of such plankton competition must of course specify the rules of engagement among distinct functional groups of P, Z, and larval fish (F) living and dying within time-dependent physical (e.g., light, temperature, and current) and chemical (e.g., N) habitats. Successful ecological models are data-driven, distilling qualitative hypotheses and aliased field observations into simple analogs of the real world in a continuing cycle of model testing and revision. They are usually formulated as part of multidisciplinary field studies, in which the temporal and spatial distributions of the model's state variables, from water motion to plankton abundances, are measured to provide validation data. Here, examples of some state-of-the-art, complex regional biophysical models are drawn from specific field programs designed to validate them. Depending upon the questions asked of regional models, processes at smaller and larger time/space scales (**Table 1**) are variously ignored, parametrized, or specified as boundary conditions.

Statistical models elicit noncausal relationships among presumed independent variables with poorly known probability density functions. Deterministic simulation models in contrast assign cause and effect among the state variables and forcing functions described by a set of usually nonlinear ordinary or partial differential equations, whose solutions are obtained numerically. Within a fluid subjected to such external forcing, the vagaries of population changes of the embedded plankton, along measured spatial gradients, must be described in relation to both water motion and biotic processes.

For example, in a Lagrangian sense – following the motion of a parcel of fluid – the time dependence of larval fish (F) can be written simply as an ordinary differential equation (eqn [1]), where the total derivative is given by eqn [2].

$$\frac{\mathrm{d}F}{\mathrm{d}t} = (b - d)F \qquad [1]$$

$$\frac{\mathrm{d}}{\mathrm{d}t} = \frac{\partial}{\partial t} + u\frac{\partial}{\partial x} + v\frac{\partial}{\partial y} + w\frac{\partial}{\partial z} \qquad [2]$$

Mixing is ignored and biotic factors are just the linear birth, b, and death, d, rates of the fish. Because

Table 1 Domains of ecological and physical processes within a hierarchy of coupled models

Model	*Time domain*	*Space domain*	*Ecological focus*	*Physical focus*
Small scale	Seconds	mm–cm	Cellular metabolism	Dissipation, nutrient supply
Turbulent	Minutes	dm–m	Photoadaptation, cell quotas	Waves, Langmuir cells
Local	Hours	m	Plankton migration, cell division	Tides, inertial oscillations
Regional	Days–weeks	km	Food web interactions, populations, resuspension events, nutrient depletion, grazing controls	Trapped waves, storms, upwelling, frontal currents
Basin	Months	degree (latitude)	Seasonal succession, mammal migration, carbon sequestration	Gyre flows, changes of mixed layer depths
Global	Years	Planetary	Stock fluctuations	Thermohaline circulation
Geological	Centuries	–	Evolution, sedimentation	Climate change

drogues have difficulty following plankton patches, and current meters measure flows at a few fixed locations, it is usually more convenient to model circulation and the coupled biological fields in an Eulerian sense – over a grid mesh of spatial cells.

The underlying physical models (Csanady, 1982; Heaps, 1987; Mooers, 1998) have a rich history, since the development of calculus by Isaac Newton and Gottfried Leibnitz *c.* 1675, allowing expression of the rates of change, or derivatives, of properties over both time and space. After 1800, for example, Newton's second law of motion became transformed into the time (t)-dependent Navier–Stokes equations for the horizontal (u, v) and vertical (w) motions of a fluid, as a function of density (ρ), pressure (p), gravitational ($g\rho/\rho_0$), where ρ_0 is a reference density and Coriolis (fv, $-fu$, 0) forces, in a Cartesian coordinate system (x, y, z) as eqns [3]–[5].

$$\frac{\partial u}{\partial t} + u\frac{\partial u}{\partial x} + v\frac{\partial u}{\partial y} + w\frac{\partial u}{\partial z} = \frac{\partial}{\partial x}\left(K_x\frac{\partial u}{\partial x}\right) + \frac{\partial}{\partial y}\left(K_y\frac{\partial u}{\partial y}\right) + \frac{\partial}{\partial z}\left(K_z\frac{\partial u}{\partial z}\right) - \frac{1}{\rho}\frac{\partial \varphi_e}{\partial x} - \frac{1}{\rho}\frac{\partial \varphi_\rho}{\partial x} + fv \qquad [3]$$

$$\frac{\partial v}{\partial t} + u\frac{\partial v}{\partial x} + v\frac{\partial v}{\partial y} + w\frac{\partial v}{\partial z} = \frac{\partial}{\partial x}\left(K_x\frac{\partial v}{\partial x}\right) + \frac{\partial}{\partial y}\left(K_y\frac{\partial v}{\partial y}\right) + \frac{\partial}{\partial z}\left(K_z\frac{\partial v}{\partial z}\right) - \frac{1}{\rho}\frac{\partial \varphi_e}{\partial y} - \frac{1}{\rho}\frac{\partial \varphi_\rho}{\partial y} - fu \qquad [4]$$

$$\frac{\partial w}{\partial t} + u\frac{\partial w}{\partial x} + v\frac{\partial w}{\partial y} + w\frac{\partial w}{\partial z} = \frac{\partial}{\partial x}\left(K_x\frac{\partial w}{\partial x}\right) + \frac{\partial}{\partial y}\left(K_y\frac{\partial w}{\partial y}\right) + \frac{\partial}{\partial z}\left(K_z\frac{\partial w}{\partial z}\right) - \frac{1}{\rho}\frac{\partial \varphi_\rho}{\partial z} - \frac{g\rho}{\rho_0} \qquad [5]$$

subject to appropriate boundary conditions at the surface (s), bottom (b), and sides of the ocean.

For example, the surface wind and bottom friction stresses, $\tau_{s,b}$, are part of the boundary conditions for the third term of internal vertical Reynolds stresses, involving K_z on the right side of [3]–[5]. No water moves across the land interfaces of the first two terms, representing horizontal turbulent mixing at smaller length scales than those of the flow components (u, v, w) over longer timescales. The last three terms on the left side of eqns [3]–[5] are thus the nonlinear advection of momentum in the horizontal and vertical directions, while the first one is the local time change.

In terms of other forces exerting acceleration of the fluid on the right side of eqns [3]–[5], their last terms represent the impact of Coriolis (f) and gravitational (g) effects. The fourth and fifth terms of eqns [3] and [4] are the respective pressure forces exerted by the piled-up mass of sea water (ϕ_e) (i.e., the barotropic force from slope of the sea surface, e, and the baroclinic one (ϕ_ρ) from the density field, ρ, which in turn is a function of temperature, (T), salinity (S), and pressure. At the surface of the sea, it is given simply by eqn [6], where ρ_0 is a reference density of pure water and γ is a polynomial function of S and T.

$$\rho = \rho_0 + \gamma(S, T) \qquad [6]$$

As in the case of momentum, the time-dependent spatial distributions of these conservative parameters of temperature and salinity are described by equations [7] and [8].

$$\frac{\partial T}{\partial t} + u\frac{\partial T}{\partial x} + v\frac{\partial T}{\partial y} + w\frac{\partial T}{\partial z} = \frac{\partial}{\partial x}\left(K_x\frac{\partial T}{\partial x}\right) + \frac{\partial}{\partial y}\left(K_y\frac{\partial T}{\partial y}\right) + \frac{\partial}{\partial z}\left(K_z\frac{\partial T}{\partial z}\right) \qquad [7]$$

$$\frac{\partial S}{\partial t} + u\frac{\partial S}{\partial x} + v\frac{\partial S}{\partial y} + w\frac{\partial S}{\partial z} = \frac{\partial}{\partial x}\left(K_x\frac{\partial S}{\partial x}\right) + \frac{\partial}{\partial y}\left(K_y\frac{\partial S}{\partial y}\right) + \frac{\partial}{\partial z}\left(K_z\frac{\partial S}{\partial z}\right) \qquad [8]$$

The different forcings of net incident radiation, latent/sensible heat losses, precipitation, and evaporation at the sea surface are boundary conditions for the last terms on the right side of eqns [7] and [8]. Depending upon the time and space scales of interest (**Table 1**), the coefficients K_x, K_y, and K_z may represent either molecular diffusion, turbulence, viscosity, tidal stirring, or eddy mixing processes, while u, v, and w are obtained from solutions of eqns [3]–[5] and [9] (below).

Finally, to complete description of the physical habitat, conservation of momentum is invoked as the continuity equation [9] by assuming that sea water is an incompressible fluid.

$$\frac{\partial u}{\partial x} + \frac{\partial v}{\partial y} + \frac{\partial w}{\partial z} = 0 \qquad [9]$$

Thomas Malthus's consideration of exponential population growth in *c.* 1800 was placed in a marine context of eqn [1] during the 1920s, after recognition of the sources of variation of stock recruitment. Since then, quantitative description of

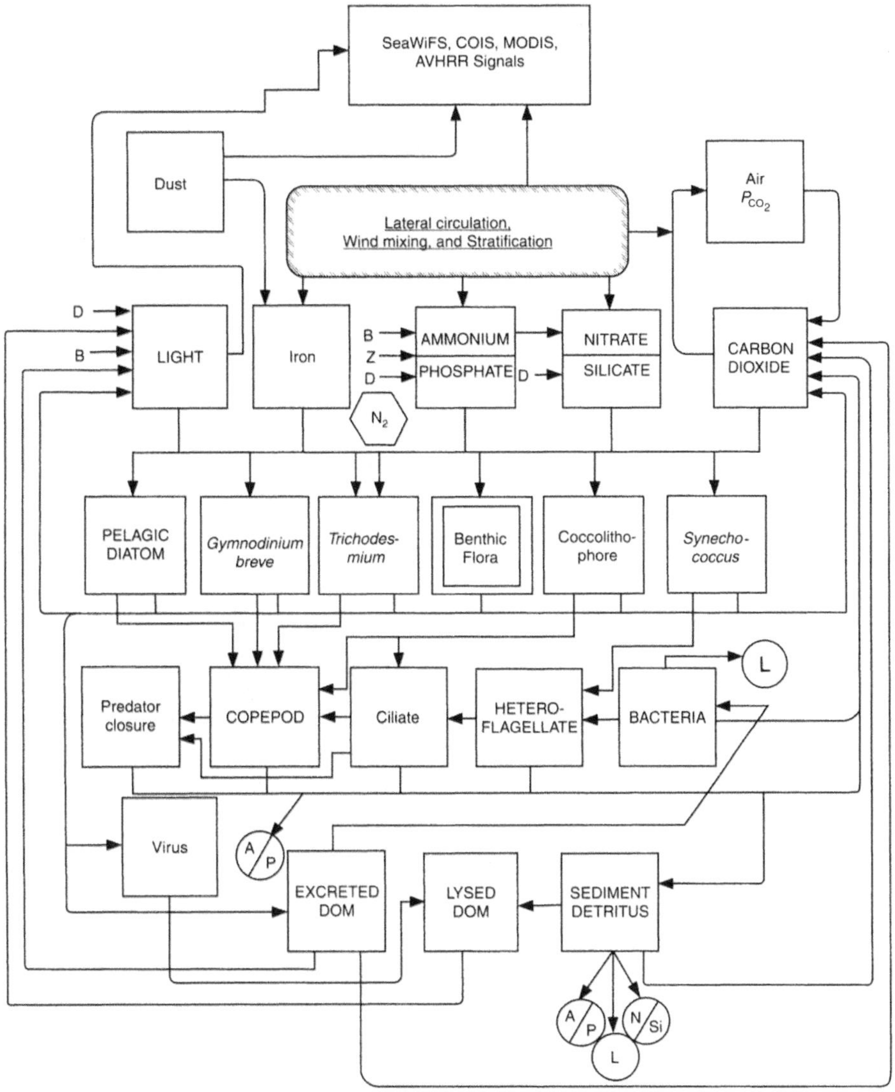

Figure 1 State variables of an ecologically complex numerical food web, coupled to a physical model of water circulation, for analysis of phytoplankton competition leading both to harmful algal blooms of *Gymnodinium breve* and to larval fish predators on the West Florida shelf.

the fluctuations of larval fish abundance have expanded to simple marine food webs (**Figure 1**) that support these vertebrates. State equation [10] for fish now has mixing and a nonlinear term for zooplankton prey, which are linked to N, P, bacteria (B), dissolved organic matter (D), and benthic microbiota (M) in eqns [11]–[16], where the advective and diffusive terms of eqns [10]–[16] are analogous to those of eqns [3]–[5].

$$\frac{\partial F}{\partial t} + u\frac{\partial F}{\partial x} + v\frac{\partial F}{\partial y} + w\frac{\partial F}{\partial z} = \frac{\partial}{\partial x}\left(K_x\frac{\partial F}{\partial x}\right) + \frac{\partial}{\partial y}\left(K_y\frac{\partial F}{\partial y}\right) + \frac{\partial}{\partial z}\left(K_z\frac{\partial F}{\partial z}\right) + (bZ - d)F \quad [10]$$

$$\frac{\partial Z}{\partial t} + u\frac{\partial Z}{\partial x} + v\frac{\partial Z}{\partial y} + w\frac{\partial Z}{\partial z} = \frac{\partial}{\partial x}\left(K_x\frac{\partial Z}{\partial x}\right) + \frac{\partial}{\partial y}\left(K_y\frac{\partial Z}{\partial x}\right) + \frac{\partial}{\partial z}\left(K_z\frac{\partial Z}{\partial z}\right) + (\alpha a\mathrm{P} + cB - e - \varepsilon_1 - \delta_1 - bF)Z \quad [11]$$

$$\frac{\partial P}{\partial t} + u\frac{\partial P}{\partial x} + v\frac{\partial P}{\partial y} + w\frac{\partial P}{\partial z} = \frac{\partial}{\partial x}\left(K_x\frac{\partial P}{\partial x}\right) + \frac{\partial}{\partial y}\left(K_y\frac{\partial P}{\partial x}\right) + \frac{\partial}{\partial z}\left(K_z\frac{\partial P}{\partial z}\right) + (gN - \varepsilon_2 - \delta_2 - aZ)\mathrm{P} + w_p\frac{\partial P}{\partial z} \quad [12]$$

$$\frac{\partial B}{\partial t}+u\frac{\partial B}{\partial x}+v\frac{\partial B}{\partial y}+w\frac{\partial B}{\partial z}=\frac{\partial}{\partial x}\left(K_x\frac{\partial B}{\partial x}\right)+\frac{\partial}{\partial y}\left(K_y\frac{\partial B}{\partial y}\right)+\frac{\partial}{\partial z}\left(K_Z\frac{\partial B}{\partial Z}\right)+(hD-\delta_3-cZ)B \quad [13]$$

$$\frac{\partial N}{\partial t}+u\frac{\partial N}{\partial x}+v\frac{\partial N}{\partial y}+w\frac{\partial N}{\partial z}=\frac{\partial}{\partial x}\left(K_x\frac{\partial N}{\partial x}\right)+\frac{\partial}{\partial y}\left(K_y\frac{\partial N}{\partial y}\right)+\frac{\partial}{\partial z}\left(K_Z\frac{\partial N}{\partial Z}\right)+\frac{\varepsilon_1}{\delta_1}Z+\delta_2P+\delta_3B+\frac{\partial}{\partial x}\left(K_x\frac{\partial M}{\partial z}\right)-gNP \quad [14]$$

$$\frac{\partial D}{\partial t}+u\frac{\partial D}{\partial x}+v\frac{\partial D}{\partial y}+w\frac{\partial D}{\partial z}=\frac{\partial}{\partial x}\left(K_x\frac{\partial D}{\partial x}\right)+\frac{\partial}{\partial y}\left(K_y\frac{\partial D}{\partial y}\right)+\frac{\partial}{\partial z}\left(K_z\frac{\partial D}{\partial z}\right)+(1-\alpha aP)+\varepsilon_2P+\frac{\partial}{\partial z}\left(K_z\frac{\partial M}{\partial z}\right)-hDB \quad [15]$$

$$\frac{\partial M}{\partial t}+w_z\frac{\partial eZ}{\partial z}+w_P\frac{\partial P}{\partial z}-\frac{\partial}{\partial z}\left(K_z\frac{\partial M}{\partial z}\right) \quad [16]$$

Thus far, F has represented various larval fish species subjected to the vagaries of the currents, of their supply of zooplankton prey (bZF), and of losses (dF) to predators. The simulated Z are a more diverse zooplankton group of protozoans, salps, copepods, and euphausiids, however, while P are numerous functional groups of light-regulated phytoplankton, e.g., the silicate-requiring diatoms, nitrogen-fixing diazotrophs, calcium-using coccolithophores, sulfur-releasing prymnesiophytes, and heavily-grazed chlorophytes, cyanophytes, and cryptophytes of the microbial food web.

Accordingly, the types of inorganic nutrients that have been modeled with eqn [14] are total carbon dioxide, nitrate, ammonium, dinitrogen gas, iron, silicate, calcium, and phosphate, while the dissolved organic matter (DOM) pools have been those of carbon, nitrogen, phosphorus, and sulfur, to provide sufficient algal niches. The particulate pools of the phytoplankton are usually chlorophyll (chl) biomass, or their elemental equivalents. Finally, B are ammonifying and nitrifying bacteria, and M represents both microflora and microfauna, as well as particulate and dissolved debris of the sediments.

The other biological terms of eqns [11]–[16] involve non-linear rate processes, in units of reciprocal time and associated with the various coefficients, such as in the temperature-dependent grazing of phytoplankton by zooplankton (aPZ). Their sloppy feeding is represented by α such that both the herbivores and the phytoplankton excretion (ε_2P) are sources of the DOM consumed by the bacteria (hDB), who in turn are eaten by other smaller zooplankton (cBZ). Some of the ingested food is released as egested fecal pellets (eZ) to sink (w_z), together with settling phytodetritus ($w_p\partial P/\partial z$) to the seafloor. Ammonium and phosphorus are excreted ($\varepsilon_1 Z$) by the respiring ($\delta_1 Z$) grazers as one source of recycled nutrients (including carbon dioxide), as well as those derived from respiring phytoplankton ($\delta_2 P$) and bacteria ($\delta_3 B$) and sediment releases $\partial/\partial z(K_z\,\partial M/\partial z)$. Finally, the temperature-dependent, light-regulated uptake of nutrients and subsequent growth (gNP) of the phytoplankton is usually a Liebig's law of the minimum formulation, involving saturation kinetics of the available resources, as function of cell quota and a ratio of the required elements.

The respective exchanges of biogenic gases, dissolved solutes, and bioturbated particulate matter across either the air–sea or water–sediment interfaces have been treated as part of the boundary conditions for the vertical diffusive term of these ecological state variables, similarly to the physical formulations of wind stress, bottom friction, and buoyancy fluxes. Further, since few elements escape the planet, a conservation of mass within marine food webs is also invoked as eqn [17], like that of eqn [9] for momentum.

$$\frac{\mathrm{d}F}{\mathrm{d}t}+\frac{\mathrm{d}Z}{\mathrm{d}t}+\frac{\mathrm{d}P}{\mathrm{d}t}+\frac{\mathrm{d}B}{\mathrm{d}t}+\frac{\mathrm{d}N}{\mathrm{d}t}+\frac{\mathrm{d}D}{\mathrm{d}t}+\frac{\mathrm{d}M}{\mathrm{d}t}=0 \quad [17]$$

Model Hierarchies

Despite extensive *in situ* and satellite validation data for today's nonlinear biophysical models, based on some or all of state equations [3]–[17], the models still have numerical problems, since any set of coupled partial differential equations is solved iteratively at each individual grid point of a spatial three-dimensional mesh (**Figure 2**), using time-dependent forcing functions. The smaller the spatial grid mesh and the larger the number of state variables (**Figure 1**), the more computer memory is required, such that any one model must be formulated to provide only a few answers to a hierarchy of possible questions (**Table 1**) – *ergo*, the reason for construction of regional models.

For example, 50 years after the pioneer studies of Riley and co-workers, the present general circulation models (GCMs) still have minimal ecological realism with just nitrate and the total phytoplankton community to allow spatial meshes of ~1/3° resolution of important physical processes of eddy mixing within basin simulations over ~37 depth levels.

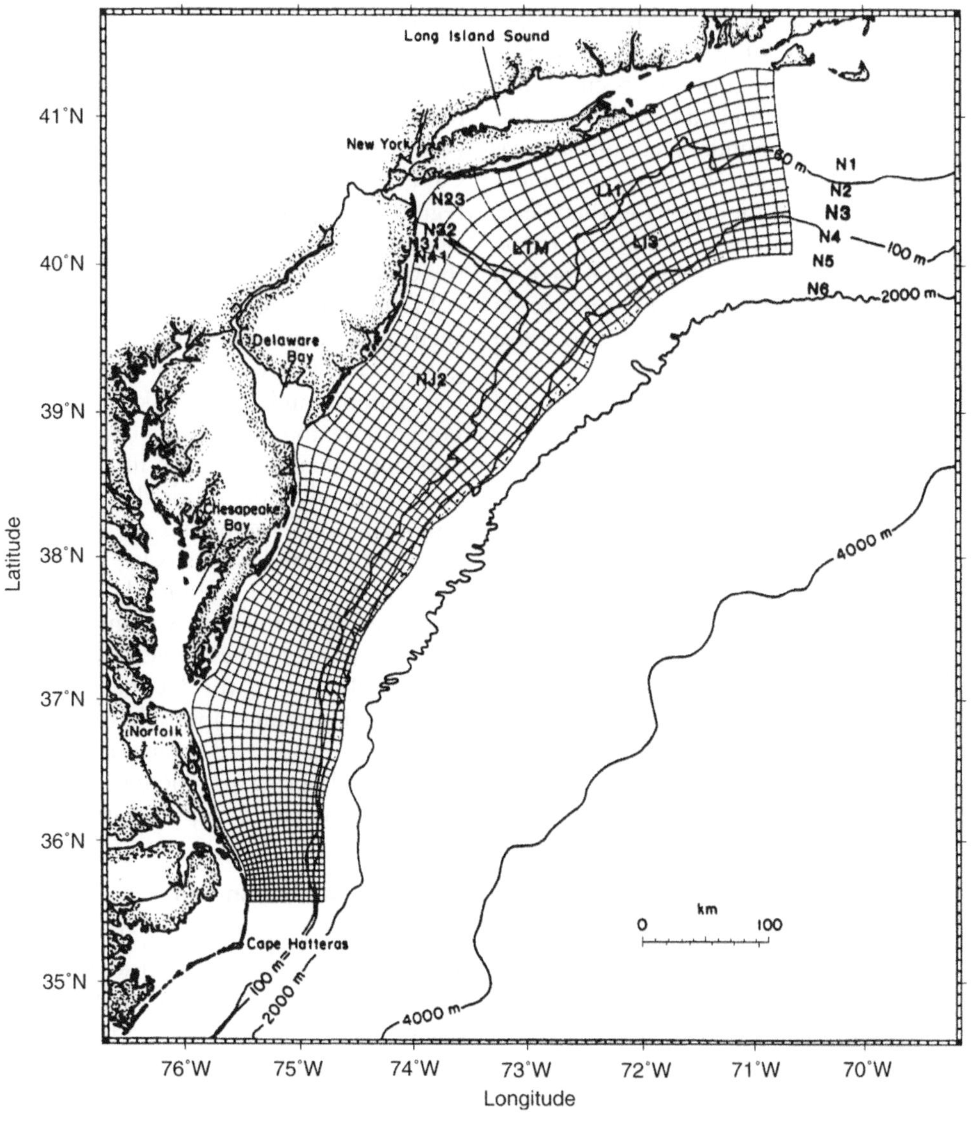

Figure 2 A curvilinear grid, showing every other grid line, of the first coupled physical-biochemical model of the Mid-Atlantic Bight, in relation to upstream (N1–N6) and interior (LI 1, LI 3, LTM, N23, N31–32. N41, and NJ2) current meter moorings.

Ideally, future GCMs may include more biogeochemical state variables to explicitly link carbon uptake during net photosynthesis with associated element cycles, e.g., nitrogen fixation, dimethyl sulfide evasion, organic phosphorus lability, calcification, and silica depletion in global budgets of climate change. But present questions of predicting the onset of harmful algal blooms, of the consequences of overfishing regional stocks, and of the fate of eutrophied coastal ecosystems require circulation models at smaller spatial resolution and biological models of greater ecological realism than those of GCMs (**Table 1** and **Figure 1**).

Regional models both incorporate local processes at the high end of their time/space scales of hours and decimeters and extend into the low end of basin models at resolution of months and degrees (**Table 1**). At a residual flow of $5\,\mathrm{cm\,s^{-1}}$, or $\sim 5\,\mathrm{km\,d^{-1}}$, it would take a plankton population about 8 weeks to drift $\sim 300\,\mathrm{km}$ across a region of $\sim 1 \times 10^5\,\mathrm{km}^2$ area, about the size of the Mid-Atlantic Bight, the southern Caribbean Sea, or the West Florida shelf. With a large set of ecological variables (**Figure 1**), however, present computer resources barely allow computation of the terms of [3]–[17] over the $\sim 4\,\mathrm{km}$ resolution of a regional circulation model (**Figure 2**), with ~ 10 depth layers. Thus the complexities of turbulent flows and cellular metabolism must be glossed over, while the results of the regional model cannot be extrapolated over $\sim 3 \times 10^7\,\mathrm{km}^2$ of an adjacent basin without the loss of additional realism.

At the global ocean level of $\sim 3.7 \times 10^8$ km^2, lateral boundary conditions of a GCM are not a problem, because land boundaries are specified and flows are computed throughout all sectors of the simulated ocean; surface boundary conditions are another story, requiring input from some atmospheric model! In a regional model of the shelf seas, however, three open boundary conditions must be specified, which determine the solution over the grid mesh of the region of interest. Various options are (1) to use data at some of the boundaries, (2) to move them far way with a telescoping grid mesh, or (3) to make some assumptions about the water flows and fluxes of elements across the boundaries from other models. Examples of results from coupled biophysical models of three shelf regions are shown, using each of these options.

Regional Case Studies

Mid-Atlantic Bight

The first set of coupled biophysical models of the continental shelf and slope of the Mid-Atlantic Bight (MAB) between Martha's Vineyard and Cape Hatteras used the same circulation model, which was wind-forced, with current meter data specifying flows at the upstream boundary (**Figure 2**). The horizontal grid mesh was an average of ~4 km, with analytical solutions yielding continuous vertical profiles of the barotropic flow fields, from the shallow water formulation of the depth-averaged Navier–Stokes equations. During spring conditions of minimal density structure, the barotropic flow is ~90% of the total transport, and the simulated flows of the model exhibited a vector error of 1–2 cm s^{-1} speed and 10–30° direction, compared to the observed currents at interior moorings (**Figure 2**). During a strong upwelling event (**Table 1**), when nutrient-rich subsurface water moves up into the sunlit euphotic zone, the model's vertical velocity, w from eqn [9], was 9–14 m d^{-1} for effecting both supply of unutilized nutrients in eqn [14] and resuspension of settled-out phytoplankton (**Figure 3A**) in eqn [12].

In the first version of a coupled biological model of the MAB, only three vertical levels of ecological interactions were computed within the spatial mesh of the circulation model, with a minimal realism of just nitrate and the whole phytoplankton community as state variables, forced in turn by incident light and time-dependent grazing stress. Within parts of the MAB, this study of 'new' nitrate-based production yielded similar chlorophyll stocks of phytoplankton on 10 April 1979 (**Figure 3B**), after such an upwelling event, to those seen concurrently by a color-sensing satellite (**Figure 3A**) (the Coastal Zone Color Scanner (CZCS)) and an *in situ* moored fluorometer, moored south of Long Island, i.e., between New Jersey and Martha's Vineyard, as well as off Cape Hatteras. At mid-shelf in the wider, central region of the MAB, however, the simulated surface stocks of the first model overestimated those seen by the CZCS.

In another version of the coupled model of both shelf and slope domains, temperature-dependent growth of two groups of phytoplankton (net plankton diatoms and nanoplankton flagellates) utilized both nitrate and ammonium over 10 depth intervals, allowing calculation of the total ('new' + 'recycled') primary production. With the more heavily grazed nanoplankton dominating on the outer shelf and slope, the simulated stocks on 10 April 1979 (**Figure 3C**) now more closely approximated those estimated from the CZCS (**Figure 3A**). The model's total net photosynthesis over March–April was then within 28–97% of the ^{14}C measurements of productivity, depending upon the area and month. A mean of 0.5 g C m^{-2} d^{-1} suggested that only ~13% of the total spring production was exported to the deep sea, compared to prior estimates of ~50%.

Southern Caribbean Sea

The second biophysical model of the Venezuelan shelf was again wind-forced with a mesh of ~4 km within 200 km of the land boundary, but the seaward boundary now extended ~1400 km out from the coast, intersecting Puerto Rico and Cuba, with a telescoping mesh of maximal size of ~110 km. A western boundary current was assumed to flow east–west along the model's shelf-break at a depth of ~100 m (**Figure 4**). Both barotropic and baroclinic flow fields were computed from a two-layered, linearized circulation model – i.e., again without advection of momentum and ignoring tidal forces. The ecological formulation over 30 depth intervals is also more complex, with the additional state variables of carbon dioxide, labile and refractory dissolved organic matter, bacteria, and zooplankton fecal pellets, as well as nitrate, ammonium, and light-regulated diatoms.

During spring upwelling of 8 m d^{-1} near the coast and 3 m d^{-1} offshore at a time-series site ('+' on **Figure 4**), the thermal fields of the physical model match satellite and ship estimates of surface temperature in February–April 1996–1997. Within the three-dimensional flow field, the steady solutions of detrital effluxes from a simple food web of diatoms, adult calanoid copepods, and the ammonifying/

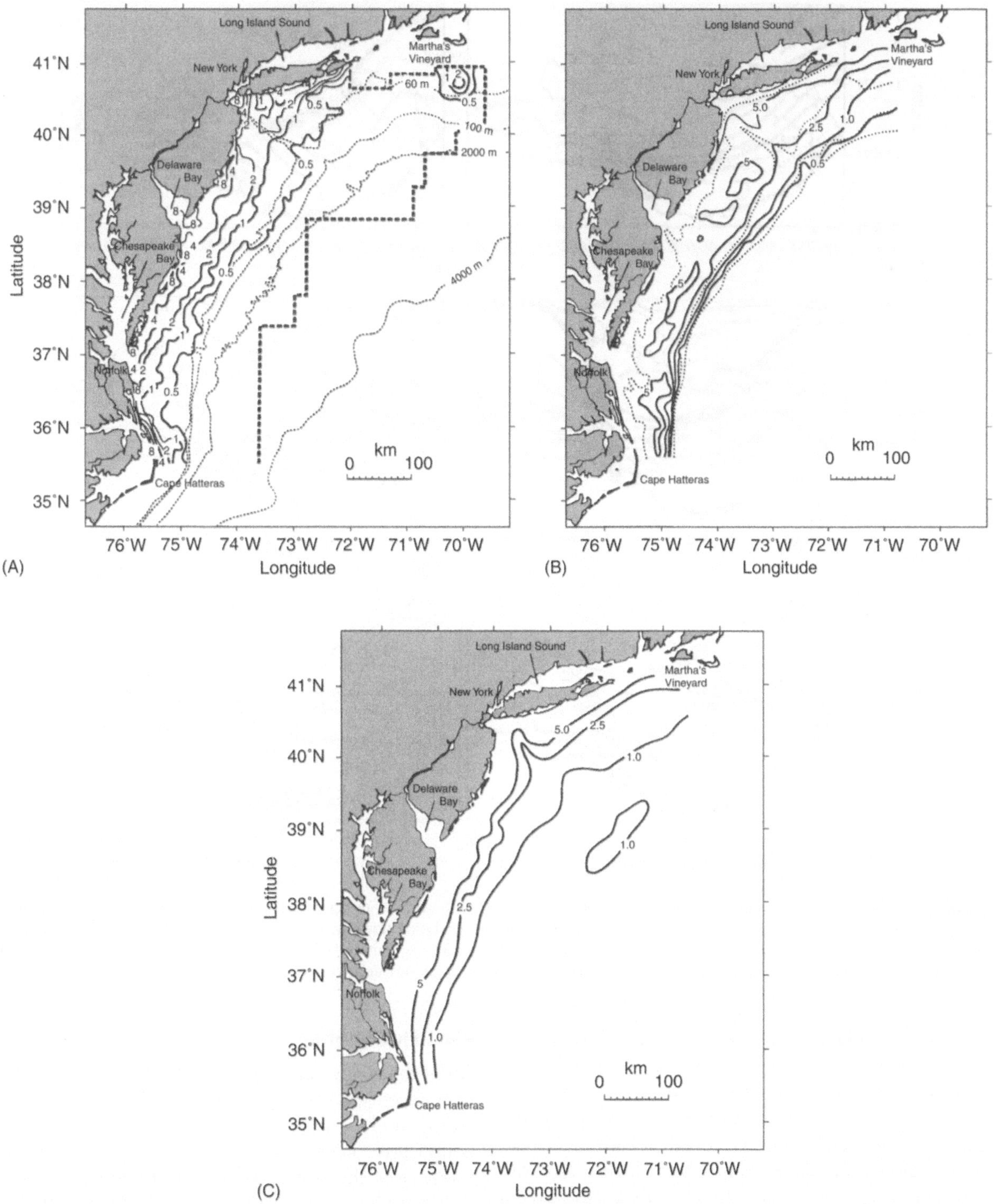

Figure 3 The observed (A) and simulated biomass (μg chl l^{-1}) at the surface of the Mid-Atlantic Bight during 10 April 1979, as seen by the Coastal Zone Color Scanner (CZCS) and modeled by simple food webs of (B) one and (C) two functional groups of phytoplankton within the same barotropic circulation model, under wind forcing measured at John F. Kennedy Airport.

nitrifying bacteria are also ~91% of the mean spring observations of settling fluxes of particulate carbon caught by a sediment trap at a depth of ~240 m, moored in the Cariaco Basin (+). At this station, the other state variables of the coupled models are within ~71% of the chlorophyll biomass (**Figure 4A**), ~89% of the average ^{14}C net primary production of $2.0\,g\,C\,m^{-2}\,d^{-1}$ (**Figure 4B**), ~80% of the light

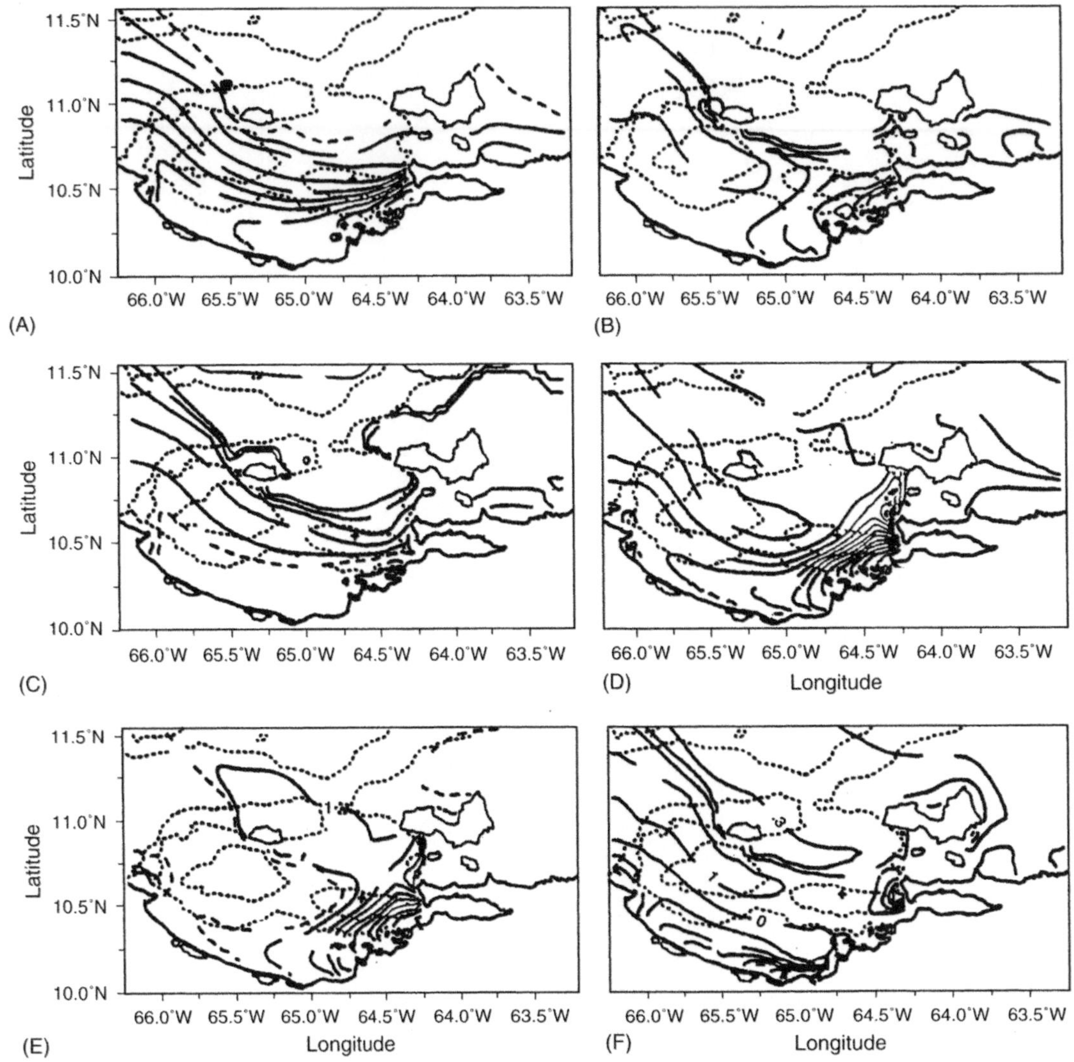

Figure 4 The simulated spring distributions within a two-layered baroclinic circulation model, under wind forcing measured during February–April 1996–1997 at Margarita Island, of (A) near-surface chlorophyll ($\mu g\,l^{-1}$), (B) water column net primary production ($mg\,C\,m^{-2}\,d^{-1}$), (C) 1% light depth (m), and, at ~25 m depth of (D) total CO_2 (μ mol DIC kg^{-1}), (E) nitrate (μ mol NO_3 kg^{-1}), and (F) ammonium (μ mol NH_4 kg^{-1}) around the CARIACO time-series site (+) on the Venezuelan shelf.

penetration (**Figure 4C**), and 97% of the total CO_2 stocks (**Figure 4D**). No data were then available for nitrate, ammonium, or satellite color.

When the Intertropical Convergence Zone of the North Atlantic winds moves north in summer, local winds along the Venezuelan coast slacken. Within one summer case of the model with weaker wind forcing, the simulated net primary production is only 14% of that measured in August–September, while the predicted detrital flux is then 30% of the observed. Addition of a diazotroph state variable (**Figure 1**), with another source of 'new' nitrogen via nitrogen fixation rather than upwelled nitrate, would remedy the seasonal deficiencies of the biological model, attributed to use of a single phytoplankton groups. We explore such a scenario in the last case.

West Florida Shelf

All of the terms eqns [3]–[5] and [9] were maintained in a recent series of wind-forced barotropic circulation models of the West Florida shelf, with horizontal grid meshes of ~4–9 km and 16–21 vertical levels over the water column. Open boundary conditions of flow at a depth, H, of ~1500 m in the Gulf of Mexico basin, as well as on the shelf, were of the radiation form [18], where C is the speed of gravity waves (i.e., $(gH)^{1/2}$) and u is determined from the adjacent interior grid points of the model.

$$\frac{\partial u}{\partial t} = C\frac{\partial u}{\partial x} \qquad [18]$$

When the additional constraint is imposed that the depth-integrated transport across these boundaries is zero, as in the shelf-break condition of the Mid-Atlantic Bight (**Figure 2**), fluxes of water within the surface and bottom Ekman layers cancel each other. Since biochemical variables exhibit nonuniform vertical profiles, however, the net influx of nutrient within the bottom layer of the embedded ecological model may instead be balanced by a net efflux of plankton within the surface layer. Prior one-dimensional models of element cycling by eight functional groups of phytoplankton on the West Florida shelf concerned most of the state variables of **Figure 1**, while the three-dimensional results, shown here, consider just the transport of a harmful algal bloom (**Figure 5**), with diel migration of the dinoflagellate *Gymnodinium breve*.

Once a red tide of *G. breve* is formed, after DON supplied from nitrogen-fixers of our one-dimensional model, its simulated trajectory over 16 vertical levels during December 1979 (**Figure 5A**) matches repeated shipboard and helicopter observations (**Figure 5B**) of this dinoflagellate bloom at the surface of the West Florida shelf, if one samples the model at sunrise after nocturnal convective mixing: at noon, the simulated red tide instead aggregates in a subsurface maximum, as observed during additional time-series studies.

Under the predominantly upwelling-favorable winds of fall/winter, the circulation model yields a positive w of ~ 0.5–$1.0\,\mathrm{m\,d^{-1}}$ within the red tide patch and of 1.0–$2.0\,\mathrm{m\,d^{-1}}$ at the coast. In another model case, without the vertical downward migration of *G. breve* at a speed of $\sim 1\,\mathrm{m\,h^{-1}}$ to avoid bright light, the model's surface populations then did not replicate the data; they were instead advected farther offshore than the *in situ* populations. It appears that in the 'real world' *G. breve* spent most of their time in the lower layers of the water column, before ascending to be sampled by ship and helicopters during daylight at the sea surface.

Furthermore, within the bottom Ekman layer, the simulated red tide is advected onshore, mimicking observations of shellfish bed closures on the barrier islands. Thus, the coupled models suggest that, upon maturation of a red tide from successful competition among functional groups of the phytoplankton community (**Figure 1**), vertical migration of *G. breve* in relation to seasonal changes of summer downwelling and fall/winter upwelling flow fields then determines the duration and intensity of red tide landfalls along the beaches of the west coast of Florida.

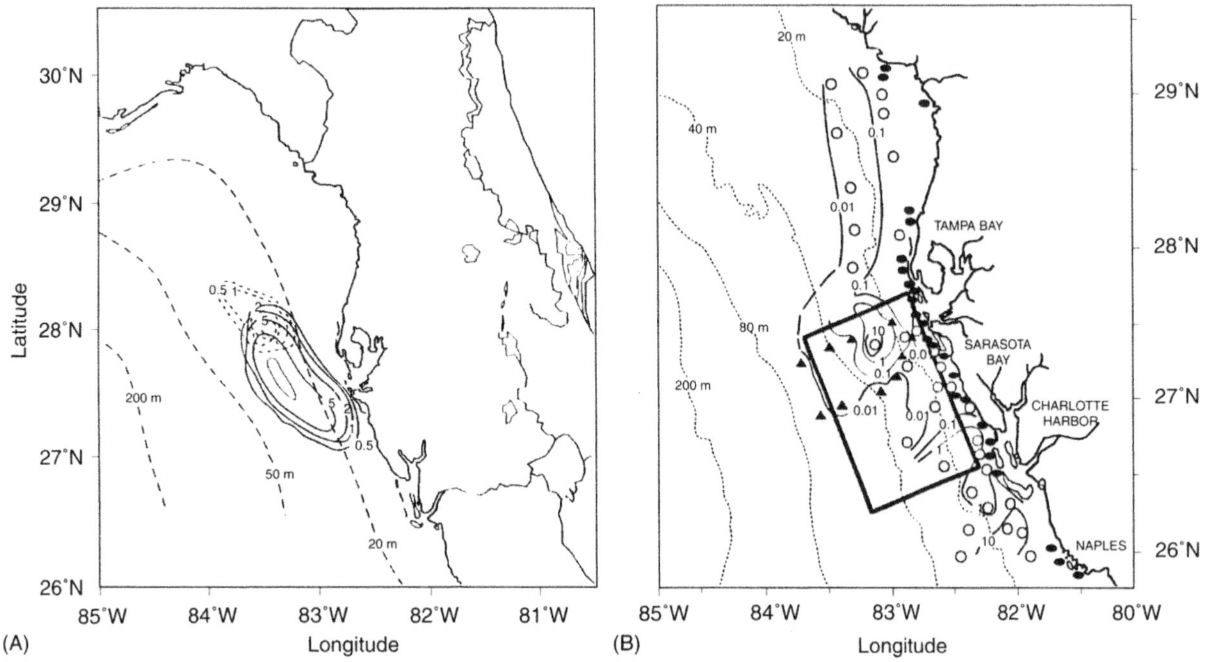

Figure 5 The (A) simulated and (B) observed daily trajectories on the West Florida shelf of a near-surface red tide (10^5 cells $\mathrm{l^{-1}} = 1\,\mu\mathrm{g\,chl\,l^{-1}}$) of *Gymnodinium breve*. The model's population is growing at a maximal rate of $\sim 0.15\,\mathrm{d^{-1}}$ with vertical migration and nocturnal convective mixing, during sunrise (06:00) on 20 December 1979, after initiation on 25 November (dashed contours), and under forcing of a barotropic circulation model by winds measured at the Tampa International Airport. The offshore observations were made during 10–11 December 1979 (solid triangles) and 19–21 December 1979 (open circles) in relation to landfalls of red tides sampled (solid circles) on piers and bridges during 11/25/79–2/8/80.

Prospectus

Other regional models of varying ecological and physical realism have been constructed for numerous shelf regions. They are mainly classic *N–P–Z* formulations, however, such that they may be improved with inclusion of a larger number of ecological state variables. Simply adding biochemical and physical variables for the next generation of coupled regional models is not sufficient, because the initial and boundary conditions will always be poorly known. Like models of the weather on land, such predictive models must be continually validated with data to correct for the poor knowledge of these conditions.

Given the expense of shipboard monitoring programs, a few bio-optical moorings (e.g., fluorometers or remote sensors (**Figure 1**)), are the most likely sources of such updates for the ecological models. Furthermore, the veracity of the underlying circulation models must be maintained with a complete suite of buoyancy flux measurements at the same moorings, to derive the baroclinic contributions important to the regional flow fields. For example, the barotropic calculations of the West Florida shelf case did not match current meter observations during summer on the outer shelf. The bio-optical implications of regional physical/ecological models driven by time-dependent density fields must be included in future simulation analyses.

See also

Inverse Models.

Further Reading

Csanady GT (1982) *Circulation in the Coastal Ocean.* Dordrecht: Riedel.

Heaps NS (1987) *Three-Dimensional Coastal Ocean Models.* Washington, DC: American Geophysical Union Coastal and Estuarine Series 4.

Mooers CN (1998) *Coastal Ocean Prediction.* Washington, DC: American Geophysical Union Coastal and Estuarine Series 56.

Riley GA, Stommel H, and Bumpus DF (1949) Quantitative ecology of the plankton of the western North Atlantic. *Bulletin of Bingham Oceanographic College* 12: 1–169.

Steele JH (1974) *The Structure of Marine Ecosystems.* Cambridge, MA: Harvard University Press.

Walsh JJ (1988) *On the Nature of Continental Shelves.* San Diego, CA: Academic Press.

COASTAL CIRCULATION MODELS

F. E. Werner and B. O. Blanton, The University of North Carolina at Chapel Hill, Chapel Hill, NC, USA

Introduction

Coastal environments are among the most complex regions of the world's oceans. They are the transition zone between the open ocean and terrestrial watersheds with important and disparate spatial and temporal scales occurring in the physical as well as biogeochemical processes. Coastal oceans have three major components, the estuarine and nearshore areas, the continental shelf, and the continental slope. The water column depth ranges from areas where flooding and drying of topography occurs over a tidal cycle at the landward boundary, to depths of thousands of meters seaward of the shelf break. The offshore extent of coastal environments can range from a few kilometers (off the Peru/Chile coast), to hundreds of kilometers (over the European or Patagonian shelf). The topography in coastal oceans can be relatively featureless, or it can be complex and include river deltas, canyons, submerged banks, and sand ridges.

Coastlines and coastal oceans span the globe, from near the North Pole to the Antarctic, and thus are subject to a full range of climatic conditions. Circulation in coastal regions is forced locally (for example by winds, freshwater discharges, formation of ice) or remotely (through interactions with the neighboring deep ocean, terrestrial watersheds, or large-scale atmospheric disturbances). Resulting motions include tides, waves, mean currents, jets, plumes, eddies, fronts, instabilities, and mixing events. Vertical and horizontal spatial scales of motion range from centimeters to hundreds of kilometers, and timescales range from seconds to interannual and longer.

Coastal regions can be very productive biologically and they support the world's largest fisheries. These regions are also preferred as recreational and dwelling sites for our increasing human population. There is evidence suggesting that changes in the coastal environment, such as degradation in habitat, water quality, as well as changes in the structure and abundance of fisheries have resulted from increases in commercial and residential development, agriculture, livestock, soil, and sediment loss. Therefore, although natural phenomena shaped the coastal environment in the past, in the future it will be defined jointly by natural and anthropogenic processes. To understand the coastal oceans, predict their future states, and reduce the human impact on the region through management strategies, it is necessary to develop a quantitative understanding of the processes that define the state of the coastal ocean. Coastal circulation models are tools rooted in mathematical and computational science formalism that allow the integration of measurements, theory and computational capability in our attempt to quantify the above processes.

Governing Equations and State Variables

The starting point for coastal circulation models is modified versions of the Navier–Stokes equations derived for the study of classical fluid mechanics. The fundamental differences are the inclusion of the Coriolis force associated with the Earth's rotation, and the inclusion of hydrostatic and Boussinesq approximations appropriate for a thin layer of stratified fluid on a sphere for circulation features of hundreds of meters and larger. Certain smaller-scale motions, such as convection and mixing, are not admitted by these approximations as they may be possibly nonhydrostatic. Additional departures from descriptions of other fluid motions are the consideration of temperature and salinity as thermodynamic variables, and a nonlinear equation of state.

Coastal ocean domains are subsets of the global ocean basins and are typically defined by a solid wall (landward) boundary and open (wet) boundaries which connect the region of interest to neighboring bodies of water. Islands inside the model domains are also considered as solid wall boundaries. The open boundaries are generally of two types: offshore boundaries along which the coastal domain exchanges information with the neighboring deep ocean, and cross-shelf boundaries along which the coastal domain receives and/or radiates information to regions up- or downstream of the study site. The sea surface and the bottom complete the definition of the model domain. With the model domain defined, solution of the governing equations is sought subject to specified initial and boundary conditions.

The simplest initial conditions specify the fluid to be at rest (no motion), and the sea level, temperature and salinity fields to be flat (no horizontal gradients). Boundary conditions are more problematic and must be specified on all model boundaries. They include stress conditions at the free surface where atmospheric winds are imposed and input energy into the coastal ocean, and at the bottom where frictional forces extract energy from the overlying motions. Heating and cooling are imposed through prescribed flux conditions at the surface (where the coastal ocean is in contact with the atmosphere), or along the model's lateral boundaries (where it is in contact with offshore regions). Additional buoyancy fluxes due to variations in salinity are imposed through prescribed evaporation or precipitation fluxes at the surface, along the open boundaries as a result of exchanges with the offshore and up- and downstream regions, and from either point- or line-sources representing riverine or larger watershed (terrestrial) inputs.

The proper specification of boundary conditions is one of the more difficult aspects of modeling coastal circulation. Although for most cases the mathematical approaches are well established, the data required to quantitatively specify the mass and momentum fluxes across the model boundaries are lacking. As a result, boundary conditions are generally idealized. In practice boundary conditions are a mixture of imposed observed quantities, derived values from larger-domain models, and conditions that minimize the uncertainties associated with the artificial nature of open (wet) boundaries.

The solution of the governing equations consists of the time-history in three-dimensions of the velocity field, the temperature, salinity, and density, and additional derived quantities describing mixing rates of mass, momentum, and other tracers. Analytic solutions, also known as closed form solutions, are not possible except for highly idealized cases. For example, topography and forcing must be simplified and certain nonlinear processes must be ignored. Nevertheless, even in these limiting cases, analytic approaches are desirable as they include in a single statement the solutions' dependence on a wide range of parameters over the entire model domain, allowing for a comprehensive understanding of the interaction between the fundamental processes. In the early 1980s, analytic solutions developed for coastal-trapped waves presented a breakthrough in the study of remotely forced currents in coastal regions.

Numerical approaches offer the possibility of retaining full dynamic and topographic complexity in the study of coastal circulation. In these approaches, the governing equations are discretized in space and time and the resulting algebraic discrete equations are solved using methods of numerical analysis. Spatial discretization in the horizontal is accomplished using the finite difference method with either structured regular grids, or, in cases where the shape of the coastline is highly irregular, curvilinear grids (**Figure 1**). The latter allow some degree of resolution and geometric flexibility. The finite element method uses unstructured grids and allows for greatest flexibility in capturing spatial heterogeneity and geometric complexity (**Figure 2**). Horizontal spatial discretization is important as the convergence of the models' solution is dependent on proper refinement of topography and flow structures associated with the presence of stratification, among others. The relative merits of structured versus unstructured meshes has not been fully addressed by the research community.

In the vertical, three approaches are commonly used in computing the depth-dependent structure of the circulation. The 'z'-coordinate computes the vertical structure along constant geopotential levels, the 'sigma' or σ-coordinate is bottom- or terrain-following and the solution is computed at the same number of points in the vertical regardless of the water column depth, and the isopycnal or density-coordinates in which the vertical structure is computed along the time-dependent location of the density surfaces. As in the case of horizontal discretization approaches, there is no optimum choice of vertical coordinate systems.

There are many algorithmic questions and mathematical formulations that are still not fully answered. Assuming smoothness in forcing functions, initial data, topography, etc., the choice of discretization method to the solution should not matter in the continuum limit of the equations. However, errors arising from solving the approximate forms of the governing equations display different behaviors due to discretization methods, and in some cases these solutions are spurious. Higher-order discretization schemes that reduce the truncation error although not significantly increasing the computational effort continue to be investigated. Similarly, many physical processes are not well understood, such as vertical mixing near the free surface, flow instabilities and horizontal mixing rates. The scales of these processes are frequently too small to be resolved in models and it is necessary to represent them by what is often 'ad hoc' parametrization. Thus, the development of coastal circulation models continues to be a specialized undertaking, with several approaches being developed by teams of investigators worldwide.

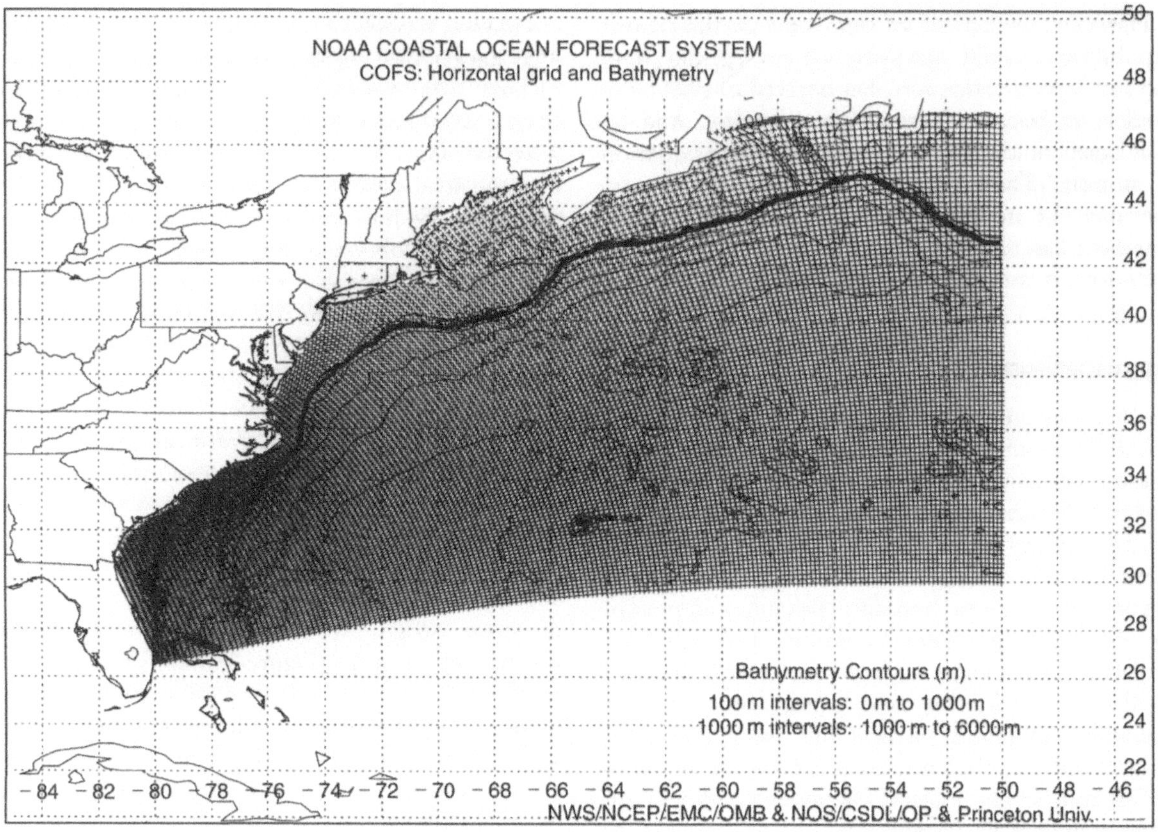

Figure 1 National Oceanic and Atmospheric Administration's Coastal Ocean Forecast System (COFS) curvilinear finite difference grid. Provided courtesy of National Weather Service's Environmental Modeling Center.

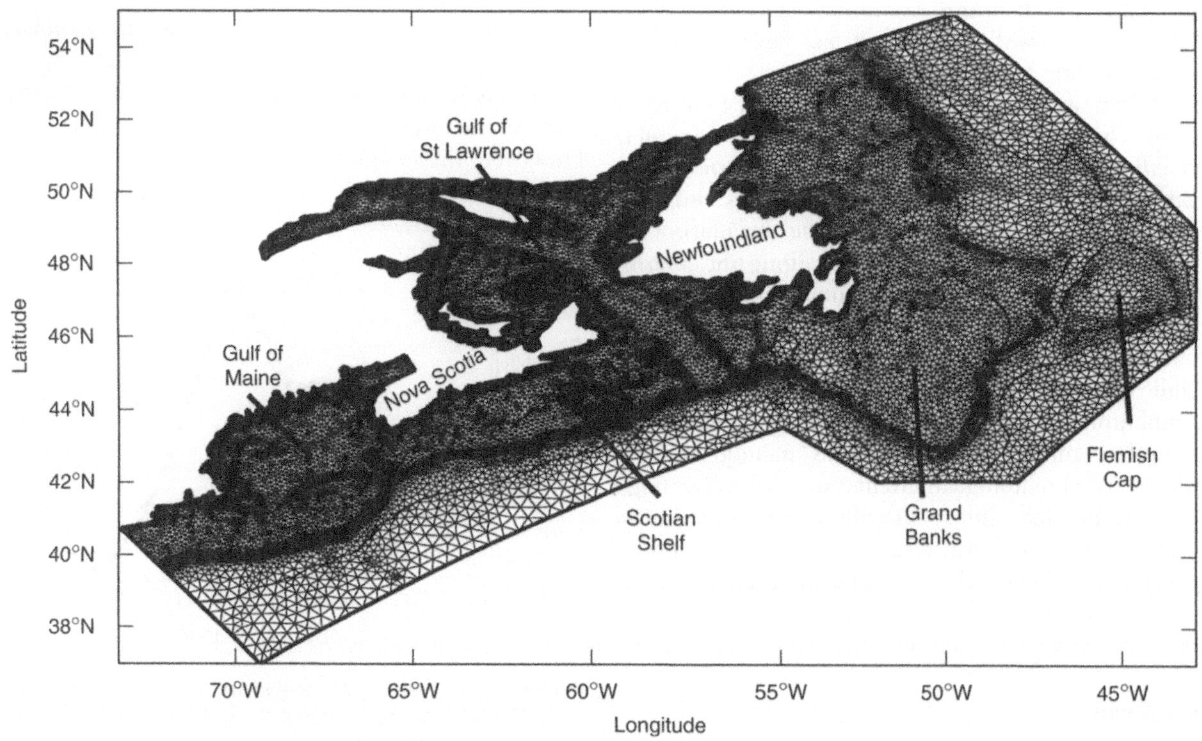

Figure 2 Northwest Atlantic shelf finite element domain. Flexibility of method increases resolution near sharp bathymetric gradients at the shelf break, deep channels, submerged banks, and along the coast.

However, the advent of significant computational capabilities (readily accessible on present-day desktop and laptop computers) has enabled coastal ocean models to become increasingly complex, and are now based on the fully stratified, nonlinear equations of motion. These advances coupled with the importance and interest in understanding coastal ocean processes has resulted in expansive growth and applications in some of the areas discussed next.

Applications

Applications of coastal circulation models can be broadly classified as process studies or regional studies. Process studies seek to identify the fundamental physical mechanisms responsible for observed features of the coastal ocean by idealizing complications of irregular shoreline geometry, time-dependent stochastic boundary forcing, and possibly simplifying the governing equations. Typically, these models retain effects such as the earth's rotation, idealized stratification and topography, idealized boundary conditions of heat flux and wind stress, and simplified turbulence closure. Early studies in the 1970s and 1980s focused on understanding large-scale wind-forced response of coastal regions including upwelling, and the nonlinear propagation of tides. Recently, with the increase in computing capabilities and improved mathematical formulations, the spatial and temporal resolution of process studies has also increased. The result has been a greater understanding of the detailed structure of phenomena such as the interaction of coastally trapped waves with bathymetric features and irregular coastlines (e.g., canyons, ridges, and capes), the generation of instabilities in the currents and formation of upwelling filaments, the formation of temperature and/or salinity fronts along the continental shelf break, and of river plume dynamics in the near-shore coastal and estuarine regions.

Regional coastal circulation studies attempt to include as much realism as possible into the numerical simulation of a specific region, including geometry and boundary conditions. These studies include the estimation of climatological circulation and tracer (e.g., temperature and salinity) distribution, fine resolution tidal simulations, storm surge analysis and prediction, transport of dissolved and particulate matter, coastal ocean prediction and forecasting, and coupled effects between estuaries, tidal inlets and the coastal ocean.

Sea Level

Many of the world's coastal regions are affected by large variations in sea level. The ability of coastal circulation models to accurately simulate coastal sea level has enabled the quantitative study of the impact of large tidal amplitudes and storm surges on low-lying coastal areas. Regional coastal sea-level models have become *de facto* components of emergency management systems in areas sensitive to sea-level variations. Robust and very good predictions of sea level can be obtained with horizontal two-dimensional models provided that accurate predictions of the surface wind field are available. The simplification from fully three-dimensional approaches is accomplished by averaging the governing equations along the vertical coordinate. Usually these models will also ignore effects of stratification, but include very high-resolution bottom topography and coastline features. The simplifications allow for significant speed-up of computations, which is necessary when issuing real-time forecasts. The rise of sea level associated with the passage of storm systems is known as storm surge. Accurate prediction of sea level during a storm surge (**Figure 3**) and its timing relative to the time of high tide are essential for the protection of property and life in low-lying coastal areas, such as the Dutch coast, the Gulf of Mexico, the east coast of the United States, and the southern Asian continent.

Engineering

Coastal circulation models are also used in engineering applications, such as in the design of ports, offshore platforms, and in the dredging of shipping lanes. These applications usually deal with the impact of the circulation on the structure being built. However, there are instances where the structure

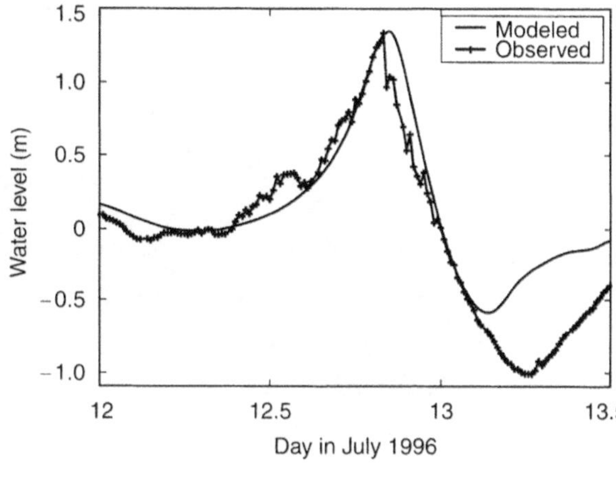

Figure 3 Observed and modeled water levels of New Bern, North Carolina, during the passage of Hurricane Bertha in July 1996. The modeled water level was computed using a two-dimensional circulation model, and accurately captures the magnitude and timing of the sea level response to the storm.

itself can have a significant effect on the coastal ocean, and circulation models are used in the quantitative assessment of its impact. The Bay of Fundy in the Gulf of Maine is known for its extreme (over 6 meter) tidal amplitudes, a result of the region's near-resonance with the principal lunar M_2 tide with period 12.4 h. The natural resonant period of the gulf-bay system is about 13.3 h. The large amplitude tide offers the opportunity to harness the tidal elevation and resulting potential energy to generate hydroelectric power by constructing dams. In 1987, a two-dimensional circulation model of the Gulf of Maine was used to investigate the tides, the effects of building tidal power plants, and their potential impact on the natural resonant period of the gulf-bay system. These studies showed that the tidal amplitude near the proposed barriers would decrease by about 25 cm due to the shortening of the bay's natural length and consequent decrease in the bay's resonant period. Furthermore, and perhaps somewhat unexpectedly the results also indicated that increases in tidal amplitude of 15–20 cm would occur in remote coastal areas, some of which are potentially sensitive to sea-level fluctuations and flooding. Predicted changes in circulation also suggested changes in sedimentation rates.

Coastal Ecosystems

The study of marine ecosystems requires that models of different systems be coupled to properly capture biological, geochemical, and hydrodynamic interactions across a wide range of temporal and spatial scales. Important biological processes are affected by transport mechanisms that can occur over hundreds of kilometers as well as turbulent mixing events that can occur on scales of several meters or less. Coastal circulation models have now achieved a level of sophistication and realism where new and significant opportunities for scientific progress in studying coupled physical–biological simulations are within reach. We are close to being able to construct spatially and temporally explicit models of the coastal physical environment, including the specification of velocity, hydrography, and turbulent fields, on scales relevant to biological processes. The investigation of ecosystem-level questions involving the role of hydrodynamics in determining the variability and regulation of planktonic and fish populations (**Figures 4** and 5) are now being attempted. There are now many case studies that have coupled the growth and feeding environment of planktonic and larval fish species with coastal circulation models. Examples include: the study of retention, survival, and

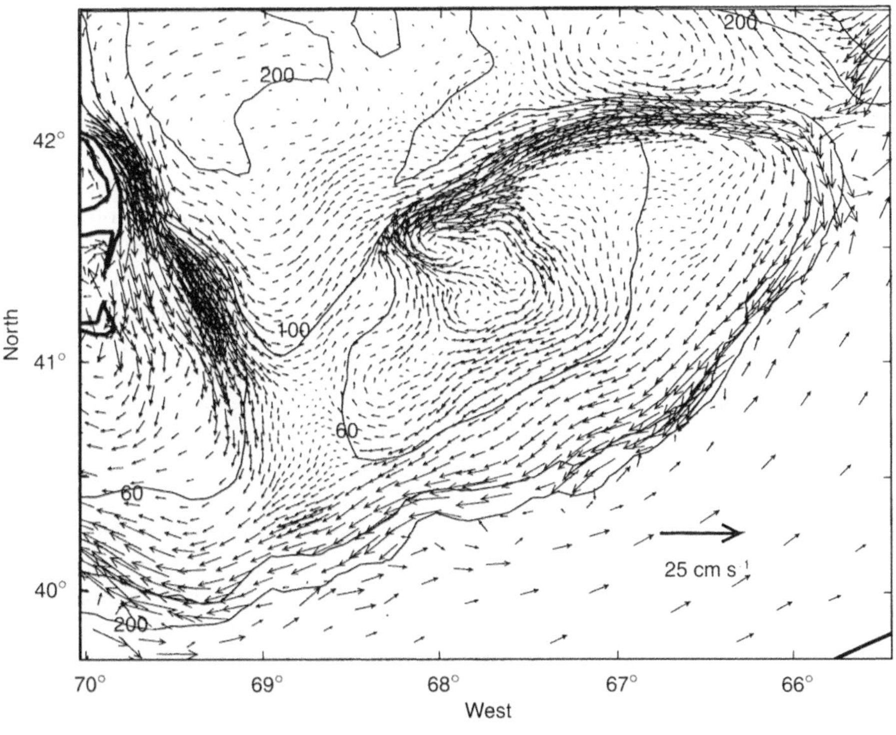

Figure 4 Modeled depth and time-averaged circulation on Georges Bank forced by tides, winds and upstream inflow. Adapted from Werner *et al* (1996).

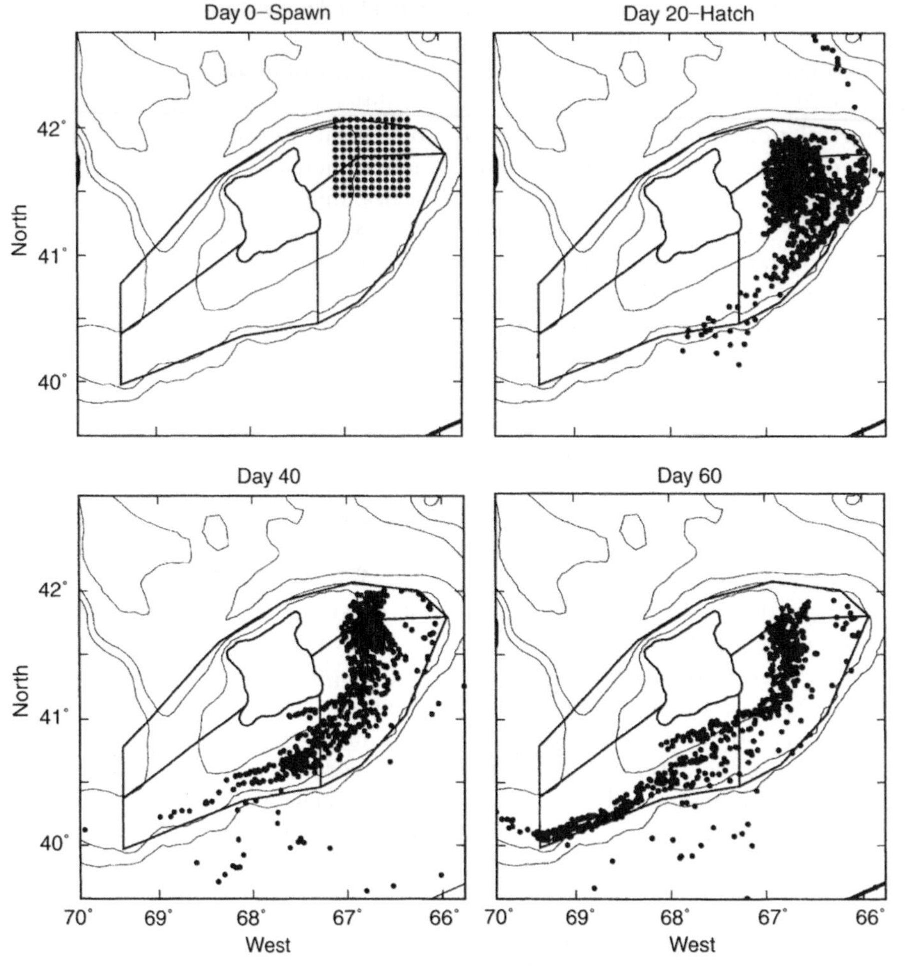

Figure 5 Simulated larval fish trajectories over Georges Bank at 20, 40, and 60 days post-spawn using flow fields from **Figure 4**. These trajectories are used to evaluate the on-bank retention versus off-bank loss. Adapted from Werner *et al.* (1996).

dispersal of larval cod, haddock and their prey in the Northwest Atlantic and North Sea; the transport of estuarine dependent fish from offshore coastal spawning regions to estuarine nursery habitats on the eastern US coast; the interannual recruitment variability of pollock in the Gulf of Alaska; and the dispersal of coral reef species. The development of management strategies used in the definition of marine sanctuaries or marine reserves can now look to circulation models for guidance in estimating population exchanges within and among neighboring coastal regions.

Operational Forecast Systems

A recent and evolving application of coastal circulation models is in operational coastal ocean prediction and forecasting systems. These systems are used to estimate in real-time the state of a particular region of the coastal ocean for the purposes of navigation, naval operations, search-and-rescue, oil spill impact assessment, or commercial and recreational fishing. The coastal circulation model is driven in part by forecasts of heat,moisture, and momentum from weather, tidal or large-scale ocean circulation models. However, to partially correct for erroneous (or imperfectly known) boundary and initial conditions and the resulting continuous accumulations of these errors, algorithms have been developed that allow assimilation of observed currents, water level, and/or hydrography within the domain and thereby improve model forecasts in real-time. The US National Oceanic and Atmospheric Administration's Coastal Ocean Forecast System (COFS) provides real-time forecasts of the coastal and open ocean state for the eastern US coast (**Figure 6**) by taking advantage of recent advances in coastal circulation models and observational systems. The coastal circulation model is forced at the surface by forecast surface flux fields of momentum, heat, and

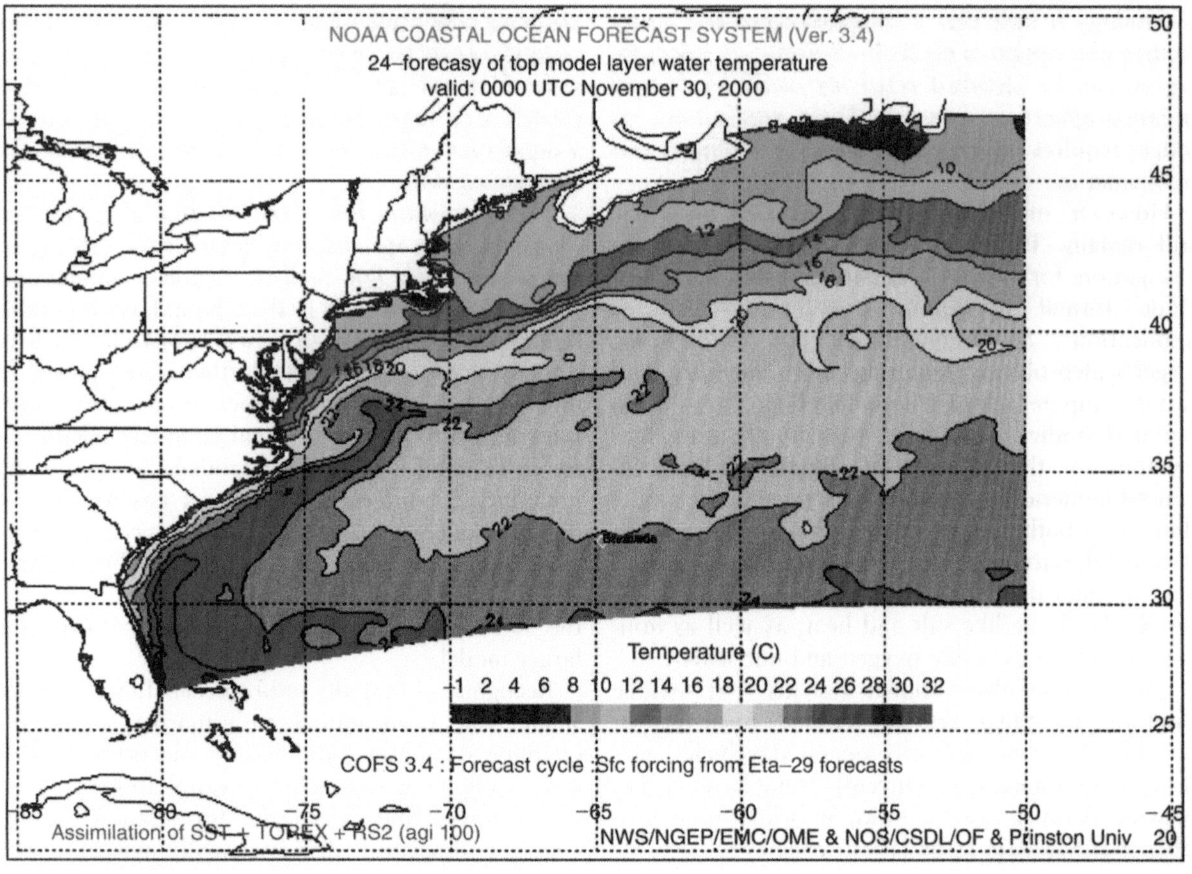

Figure 6 ECOFS 24-hour sea surface temperature forecast for 30 November 2000, computed on the grid in **Figure 1**. Note that a large portion of the deep water adjacent to the east coast shelf needs to be included in such a forecasting system. Provided courtesy of National Weather Service's Environmental Modeling Center.

moisture from a high-resolution weather forecast model. An assimilation system that incorporates both *in situ* and remotely sensed observations of surface and sysubsurface temperatures and sea-surface heights enables ECOFS to make relatively accurate 24-hour forecasts of Gulf Stream frontal position, water levels, three-dimensional currents, temperature, and salinity on a daily basis.

Sampling Design

Intense observational efforts focus on sampling physical and biogeochemical fields in the coastal ocean. The design of field sampling programs through Observational System Simulation Experiments (OSSEs) is a challenging modeling opportunity with extremely valuable results. Sampling strategies at sea can be difficult due to the evolving nature of the circulation and OSSEs provide a realistic site-specific simulation that can affect field protocols. Additionally, real-time limited-area forecast systems have been implemented on board research vessels to predict the transport of physical or biological tracers at sea. Thus, using a coastal circulation forecasting system the likely path of the tracer of interest can be predicted and help researchers in the field develop appropriate sampling schemes.

Conclusions and Future Directions

The ocean science community is currently presented with unprecedented opportunities and advanced technologies for understanding and managing coastal ecosystems. Rapid advancement of computer resources, observational systems and instruments, and numerical techniques are converging to enable real-time coastal observation systems for coastal monitoring and marine forecasting. In the past two decades, a variety of models for simulating the coastal ocean have emerged as significant tools for investigating processes and mechanisms as well as regional coastal ecosystems and environmental questions and issues. The application of these models to almost any regionin the world represents a remarkable scientific achievement. The state of the art of coastal oceancirculation models and computer technology is such that a comprehensive and quantitative description of the hydrodynamics in a specific region can be obtained relatively easily by coastal oceanographers in general. Their application no longer requires expertise in numerical techniques and mathematics.

However, many fundamental research questions still remain. There are several areas of active investigation for coastal circulation models that include formal development issues as well as applications. As computational power increases, larger-scaleproblems requiring more memory and faster computer speed will enable higher resolution regional studies as well as faster longer-term integrations for the purposes of climate studies. Advanced numerical methods for discretizing the model domain in both thehorizontal and vertical are being developed, particularly regarding mass conservation and the algorithms that transport scalar properties of the fluid volume like salt and heat, as well as nonconservative tracers like oxygen and nutrients.

Advances in observational systems that include satellite and radar remote sensing, fixed instrumented platforms, remotely operated vehicles, and moored instruments, are currently being harnessed to provide as much near real-time information as possible to use in data assimilation schemes for oceanic numerical models. The modeling community in general is striving to provide forecasted global ocean circulation fields in nereal-time that resolve basin-scale to coastal-scale features. Coastal circulation models will play an important role in: (1) communicating the open-ocean information to coastal and near-shore regions; (2) providing extensions to the basin-scale models to regions that are typically underresolved by the larger-scale models; and (3) providing realistic cross-shelf fluxes of mass and momentum to the bain-scale models.

Operational forecasting systems are being developed for site-specific, limited-area predictions of the coastal ocean. *In situ* and remotesensed data are being assimilated by these systems, driven by forecasting results frommeteorological and basin-scale ocean models. Mesoscale weather models are being used to provide spatially dense estimates of surface flux parameters to the coastal circulation models but this coupling is largely one-way; the forecasted weather parameters affect the coastal hydrodynamic evolution. However, it is well known that the ocean affects the atmosphere. Observations have shown, for example, that the surface heat flux between the ocean and atmosphere over Gulf Stream waters significantly affects the development and evolution of extratropical cyclones that routinely pass along the eastern United States seaboard. Effective two-way coupling that communicates surface fluxes from the coastal ocean to overlying atmosphere in coupled coastal ocean and regional weather forecasting models is currently being developed and will provide a significant enhancement to regional meteorological forecasting skill.

The open-water boundaries of coastal circulation models require specification of either the velocity or the water level. For realistic regional simulations, there exists uncertainty in these boundary conditions related to the sparsity of observations on which to directly deduce them. The further development of schemes for assimilating observations into model integrations to provide optimal boundary conditions for forecasting is critical. Obtaining the open water boundary conditions from a larger basin-scale prediction model is another method for specifying open boundary conditions in operational limited-area coastal prediction models. This, in effect, generates the smaller domain boundary conditions from the larger model.

Recognizing that the entire ocean functions as a single unit from global to estuarine scales, the coupling of coastal- and basin-scale ocean models will represent a significant advance toward global ocean forecasting. Since the formulations of the coastal and basin models are usually quite different, this poses an unsolved question of the communication between the two models.

The ability of coastal circulation models to integrate the governing equations and boundary conditions for the coastal environment is a powerful tool for exploring both questions of process and mechanism and for addressing realistic regional problems that include forecasting of the coastal ocean analogous to atmospheric weather prediction. As the need to understand and address the growing list of environmental concerns accelerates, broad interdisciplinary efforts that couple models of different physical, biological, chemical, and geological systems will be critical in addressing these issues.

See also

Data Assimilation in Models. Wind Driven Circulation.

Further Reading

Brink KH and Robinson AR (eds.) (1999) *The Sea*, vols 10 and 11. New York: John Wiley.

Crowder LB and Werner FE (eds) (1999) Fisheries oceanography of the estuarine-dependent fishes in the South Atlantic Bight. *Fisheries Oceanography* 8: 242pp.

Gill AE (1982) *Atmosphere–Ocean Dynamics*. New York: Academic Press.

Greenberg DA (1987) Modeling Tidal Power. *Scientific American* November: 128–131.

Haidvogel DB and Beckmann A (1998) Numerical models of the coastal ocean. In: Brink KH and Robinson AR (eds.) *The Sea*, vol 10, pp. 457–482. New York: John Wiley.

Heaps NS (ed.) (1987) *Three-dimensional Coastal Ocean Models*. Washington, DC: American Geophysical Union.

Lynch DR and Davies AM (eds.) (1995) *Quantitative Skill Assessment for Coastal Ocean Models*. Washington, DC: American Geophysical Union.

Malanotte-Rizzoli P (ed.) (1996) *Modern Approaches to Data Assimilation in Ocean Modeling*. New York: Elsevier.

Mooers CNK (ed.) (1998) *Coastal Ocean Prediction*. Washington, DC: Amercian Geophysical Union.

Werner FE, Perry RI, Lough G, and Naimie CE (1996) Trophodynamic amd advective influences on Georges Bank larval cod and haddock. *Deep-Sea Research II* 43: 1793–1822.

Werner FE, Quinlan JA, Blanton BO, and Luettich RA Jr (1997) The role of hydrodynamics in explaining variability in fish populations. *Journal of Sea Research* 37: 195–212.

Westerink JJ, Luettich RA Jr, Baptista AM, Scheffner NW, and Farrar P (1992) Tide and storm surge predictions using a finite element model. *ASCE Journal of Hydraulic Engineering* 118: 1373–1390.

INVERSE MODELS

C. Wunsch, Massachusetts Institute of Technology, Cambridge, MA, USA

Introduction

Inverse methods are formal procedures for making inferences about the ocean (or any other physical system) by using observations in combination with dynamical and kinematic models. As such, they are a part of the general methods of statistical inference done in the presence of known or assumed kinematic and dynamical constraints. Their first use in oceanography occurred in 1977 as a method for addressing the famous so-called level-of-no-motion problem in the oceanic general circulation, and the determination of the general circulation remains a major area of application. Subsequently, inverse methods became a central element of ocean acoustic tomography. More recently, they have begun to be applied widely in all areas of oceanography including biogeochemical problems.

In mathematical usage, 'inverse methods' often describe procedures directed at solving a variety of ill-posed problems, in the absence of observational noise. Although the terminology is much the same, noise-free observations exist only in textbooks, and this literature is useful, but tangential, to the oceanographic problem.

'Inverse methods' are often used in conjunction with, and thereby confused with, 'inverse models'. For historical reasons, and mathematical convention, one denotes many systems as being 'forward' or 'direct' problems or models. A simple example derives from a supposed theory that produces a rule for a variable, perhaps oceanic temperature, θ, as a function of z, in the form

$$\theta(z) = a_0 + a_1 z + a_2 z^2 + a_3 z^3 = \sum_{i=0}^{3} a_i z^3 \qquad [1]$$

If the theory tells us the a_i, we can calculate θ for any value of z. The theory is often labeled a 'forward' or 'direct model' and, more generally, may be either an analytical or a very complex numerical one. Equation [1] is called a 'forward solution'. If, on the other hand, $\theta(z)$ were known, but one or more of the a_i were not, one would have an inverse model, also given by equation [1], the label 'inverse' being employed only as a matter of convention – because there is a previously studied 'forward' version.

Another example comes from the classical advection/diffusion/decay equation for a concentration tracer, C,

$$w\frac{\partial C}{\partial z} - \kappa\frac{\partial^2 C}{\partial z^2} + \lambda C = q(z) \qquad [2]$$

where q is a source, w is the vertical advective velocity, κ is a diffusion coefficient and λ is a decay constant. All are known constants or functions of z. Given suitable boundary conditions, e.g., $C(z=0)C_0$, $C(z \to \infty) = 0$, one has a well-understood, well-posed problem, in which eqn [2] and its boundary conditions are labeled as 'forward.' But an equally compelling, and commonplace problem is the following: Given $C(z)$ and λ, what are w and κ? This type of problem occupies much of the mathematical literature on inverse problems. The oceanographers' version is, however, more likely to be: Given observations,

$$y_i = C(z_i) + n_i \qquad [3]$$

where n_i are noise, at a finite number of discrete positions z_i, what are w, κ? Many variations of this problem are possible; for example, given noisy or uncertain observations or estimates of C, w, κ, λ what was the boundary condition C_0? In this context, equation [2], along with any other available information, would now be described as the 'inverse model,' with the formal knowns and unknowns of the forward problem having in part, been interchanged.

Parameter determination from noisy observations goes back at least to Legendre and Gauss. The modern generalization, inverse theory, is often traced to the pioneering work of G. Backus and F. Gilbert in the solid-earth geophysics context. They initiated the study of model systems in which the number of formal unknowns greatly exceeds (perhaps infinitely so), the number of available data by exploiting the existence of an underlying differential or partial differential system. This form of inverse theory is rooted in functional analysis and is highly developed; see Parker (1994), or for oceanographic applications, Bennett (1992), in the Further Reading.

In oceanographic practice, even the simplest models are almost always reduced to some numerical, discrete form, either by expansion into a finite set of modes or by finite differences or related methods. The polynomial form eqn [1] is already discrete, with four parameters. Many other forms are possible. For example eqn [2], when written in finite differences, becomes

$$w_i \frac{C_{i+1} - C_i}{z_{i+1} - z_i} - \kappa_i \frac{C_{i+1} - 2C_i + C_{i-1}}{z_{i+1} - 2z_i + z_{i-1}} + \lambda C_i = q_i,$$
$$1 \leq i \leq M \quad [4]$$

which is a set of M simultaneous linear equations in the finite discrete set w_i, κ_i, λ. Alternatively, if w, κ are constant, an analytic solution, subject to a fixed surface concentration $C(z=0) = C_0$ is readily seen to be

$$C(z) = C_0 \exp\left\{ z \frac{w}{2\kappa}\left[1 + \left(1 + \frac{4\lambda\kappa}{w^2}\right)^{1/2}\right]\right\}, \quad z \leq 0 \quad [5]$$

which can be evaluated at $z = z_i$ to produce

$$C(z_i) = C_0 \exp\left\{ z_i \frac{w}{2\kappa}\left[1 + \left(1 + \frac{4\lambda\kappa}{w^2}\right)^{1/2}\right]\right\} \quad [6]$$

Because eqns [4] and [6] are algebraic equations in w, κ, λ the problem of determining them has been reduced from that of an infinite-dimensional Hilbert or Banach space to that of an ordinary finite-dimensional vector space. No matter how great the value of M, the corresponding mathematical simplifications render the methods of inverse theory much more transparent in this case. Reduction of the functional analysis methods of Backus and Gilbert to that of finite-dimensional spaces appears to begin with Wiggins, who used the singular value decomposition of Eckart and Young. In practice, almost all real inverse problems are solved on computers; they are thus automatically discrete and of finite dimension and this mathematical representation is the most useful one. Note that eqn [6] is nonlinear in the parameters, showing that the same problem can be rendered linear depending upon exactly how it is formulated.

Example

Assuming then, that observations always have a noise component, we can proceed to estimate whatever is known. For the simple power law of eqn [1], let us suppose that the 'truth' is

$$\theta(z) = 30 - 0.0005z^2, \quad -100 \leq z \leq 0 \quad [7]$$

but the theory says that it could actually be of the general form (1) so that the correct answer would be $a_0 = 30$, $a_2 = 5 \times 10^{-4}$, $a_i = 0$, otherwise. Defining $y_i = C_i + n_i$, the problem has become one of finding an estimate, $\tilde{\mathbf{x}}$, satisfying,

$$\mathbf{Ex} + \mathbf{n} = \mathbf{y} \quad [8]$$

Here the matrix vector notation $\mathbf{x} = [a_i]$ (called the 'state vector'), $\mathbf{n} = [n_i]$, $\mathbf{y} = [y_i]$ is being used, and $\mathbf{E}$ is the coefficient matrix. As stated, this is now a problem in polynomial regression theory, and much of inverse theory overlaps that branch of statistics.

The tracer problem (eqn [2]) can be reduced to this same form. Suppose that w_i, κ_i, are believed constant, independent of i, and that we have noisy measurements ξ_i [3] of C_i. Then one can readily make the substitution $C_i \rightarrow \xi_i$ in eqn [4], producing a set of simultaneous equations for w, κ_i, where the coefficient matrix E has elements depending upon

$$\frac{C_{i+1} - C_i}{z_{i+1} - z_i} \sim \frac{\xi_{i+1} - \xi_i}{z_{i+1} - z_i} \quad [9]$$

etc. and $\mathbf{y}$ now involves $\lambda\xi_i$, q_i, etc. $\mathbf{n}$ is a noise-vector representing the errors introduced by the observational noise, and any misrepresentation of the true full physics by eqn [2]. In practice, the structure of $\mathbf{n}$ can be a complicated function of the structure of $\mathbf{E}$, because equations such as (9) render the problem, in a rigorous sense, nonlinear, too. This nonlinearity is often ignored and there are many inverse problems where $\mathbf{E}$ is known exactly. Equation [8] can now be regarded as the 'model' instead of, or in addition to, eqn [2]. The state vector consists of the two unknowns $\mathbf{x}^{\mathrm{T}} = [w, \kappa_i]^{\mathrm{T}}$ (superscript T denotes a transpose; all vectors here are column vectors). The form of [8] obviously generalizes to any number of unknown elements, N, in $\mathbf{x}$.

A common method for dealing with equations of this type is to seek a least-squares solution that renders the noise vector n as small as possible:

$$J = \mathbf{n}^{\mathrm{T}}\mathbf{n} = (\mathbf{y} - \mathbf{Ex})^{\mathrm{T}}(\mathbf{y} - \mathbf{Ex}) \quad [10]$$

By taking the derivatives of the 'objective' or 'cost' function J with respect to the elements of $\mathbf{x}$ and setting them to zero, one obtains the 'normal equations.' Here, one finds, using this matrix notation, that the so-called normal equations are

$$\mathbf{E}^{\mathrm{T}}\mathbf{Ex} = \mathbf{E}^{\mathrm{T}}\mathbf{y} \quad [11]$$

whose solution is

$$\tilde{x} = \left(\mathbf{E}^{\mathrm{T}}\mathbf{E}\right)^{-1}\mathbf{y} \quad [12]$$

and which permits an estimate of the noise unknowns,

$$\tilde{\mathbf{n}} = \mathbf{y} - \mathbf{E}\tilde{x} = \mathbf{y} - \mathbf{E}\left(\mathbf{E}^{\mathrm{T}}\mathbf{E}\right)^{-1}\mathbf{y} \quad [13]$$

The forward model [1] was used to generate 'data' at seven observation points, and these values were then corrupted with noise having a standard deviation of 1. The least-squares solution produces a state vector $\tilde{\mathbf{x}}^T = [31, 0.02, 0.002, 6\times10^{-5}, 4\times10^{-7}]$, and the corresponding estimated is shown in **Figure 1**. It passes near the data points but is clearly not 'correct' in the sense of reproducing the known true values.

Although a very easy-to-use and common parameter-determining procedure, least-squares in this form is not an inverse method (statisticians call it 'curve-fitting'). The reasons for seeking a more powerful method are easy to see. At least as written, one can raise a number of questions about the solution:

1. Why should the smallest mean-square noise, $\mathbf{n}^T\mathbf{n}$, be regarded as the correct solution to choose?
2. What would happen if the inverse of $\mathbf{E}^T\mathbf{E}$ failed to exist? ($\mathbf{E}$ corresponding to [7] is a Vandermonde matrix, and known to be very badly conditioned.)
3. Suppose one knew that some of the noise values were likely to be much greater than others: Could that information be used?
4. Suppose one had a reasonable idea of the magnitude of the a_i, i.e., of $\mathbf{x}$: Cannot that information be used?
5. Just how reliable is the solution given the noisiness of the data, and the particular structure of the observations?
6. Are some observations more important than others in determining the solution?
7. Could one require $\kappa>0$ in the advection–diffusion problem?
8. In general, there are M eqns in (8) and N elements of $\mathbf{x}$. When $M>N$, the system appears to be comfortably overdetermined. But that comfort disappears when one recognizes that N noise unknowns have also been determined in eqn [13], producing a total of $M+N$ previously unknown values. Is it still sensible to term the problem 'overdetermined.'
9. Problems such as the advection-diffusion one involve a quantity C that is necessarily positive and that often renders the noise statistics highly non-Gaussian. How can one understand how that affects the best solution?

These and other issues lead one to find other means to make an estimate of $\mathbf{x}$. It is possible to modify the least-squares procedure so that it will produce solutions obtained by other methods; this correspondence has led, in the literature, to serious confusion about what is going on.

Two More Solution Methods

There are a number of methods available for estimating the elements of $\mathbf{x}$, $\mathbf{n}$ that produce useful answers to some or all of the questions listed above. We will briefly describe two such methods, leaving to the references listed in Further Reading (see Menke, 1989; Tarantola, 1987; Munk *et al.*, 1995; Wunsch, 1996) both the details and discussion of other approaches.

Gauss–Markov Method

We postulate some *a priori* knowledge of the elements of $\mathbf{x}$, $\mathbf{n}$ in a common, statistical, form. In particular, we assume that $\langle\mathbf{x}\rangle = \langle\mathbf{n}\rangle = 0$, where the brackets denote the expected value, and that there is some knowledge of the second moments, $\mathbf{S} = \langle\mathbf{x}\mathbf{x}^T\rangle, \mathbf{R} = \langle\mathbf{n}\mathbf{n}^T\rangle$.

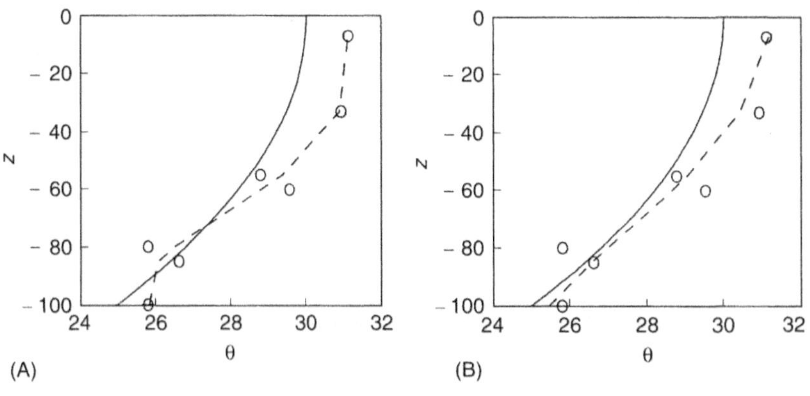

Figure 1 (A) A hypothetical temperature, θ, profile with depth, $-z$, drawn as a solid line. 'Observations' of that profile at discrete depths contaminated with unit noise variance errors are shown as open circles. Dashed line is the ordinary least-squares fit. (B) Same as in (A), except that the dashed line is the Gauss–Markov solution computed using *a priori* knowledge of the statistics of the noise and the solution itself. The fit is less structured, and a specific error estimate is available at every point, showing the true line to lie within the estimated uncertainty.

The so-called Gauss–Markov method (sometimes known as the 'stochastic inverse') produces a solution that minimizes the variance about the true value, i.e., it is a statistical method that minimizes not the sum of squares but, individually, all terms of the form $\langle (\tilde{x}_i - x_i)^2 \rangle$ where x_i is the true value, and $\tilde{x}_i$ is the estimate made. The solution is

$$\tilde{\mathbf{x}} = \mathbf{S}\mathbf{E}^{\mathrm{T}}\left(\mathbf{E}\mathbf{S}\mathbf{E}^{\mathrm{T}} + \mathbf{R}\right)^{-1}\mathbf{y} \qquad [14]$$

with solution uncertainty (error estimate)

$$\mathbf{P} = \langle (\tilde{\mathbf{x}} - \mathbf{x})(\tilde{\mathbf{x}} - \mathbf{x})^{\mathrm{T}} \rangle = \mathbf{S} - \mathbf{S}\mathbf{E}^{\mathrm{T}}\left(\mathbf{E}\mathbf{S}\mathbf{E}^{\mathrm{T}} + \mathbf{R}\right)^{-1}\mathbf{E}\mathbf{S} \qquad [15]$$

All of the available information has been used. A solution for the simple example is shown in **Figure 1**, where $\mathbf{S} = \mathrm{diag}\,([100,\ 1\times10^{-10},\ 1\times10^{-6},\ 1\times10^{-10},\ 1\times10^{-10}])$, $\mathbf{R} = \{\delta_{ij}\}$. One obtains,

$$\tilde{\mathbf{x}}^{\mathrm{T}} = [31 \pm 0.8, -1.4\times10^{-10} \pm 1\times10^{-5}, -6.1\times10^{-4} \pm 5.1\times10^{-4}, 4.5\times10^{-6} \pm 9.1\times10^{6}, 4.9\times10^{-8} \pm 6.9\times10^{-8}] \qquad [16]$$

that is, the correct answer now lies within two standard errors and, although it is not displayed here, the $\mathbf{P}$ matrix provides a full statement of the extent to which the noise elements in $\tilde{\mathbf{x}}$ are correlated with each other (which can be of the utmost importance in many problems).

Least-Squares by Singular Value Decomposition

As noted, least-squares can be modified so that it is more fully capable of producing answers to the questions put above. There are at least two ways to do this. The more interesting one is that based upon the singular value decomposition (SVD) and the so-called Cholesky decomposition of the covariance matrices, $\mathbf{S} = \mathbf{S}^{\mathrm{T}/2}\mathbf{S}^{1/2}$, $\mathbf{R} = \mathbf{R}^{\mathrm{T}/2}\mathbf{R}^{1/2}$.

One takes the original eqn [18] and employs the matrices $\mathbf{S}$, $\mathbf{R}$ to rotate and stretch $\mathbf{E}$, $\mathbf{n}$, $\mathbf{y}$, $\mathbf{x}$ into new vector spaces in which both observations and solution have uncorrelated structures:

$$\mathbf{R}^{-\mathrm{T}/2}\mathbf{E}\mathbf{S}^{\mathrm{T}/2}\mathbf{S}^{-\mathrm{T}/2}\mathbf{x} + \mathbf{R}^{-\mathrm{T}/2}\mathbf{n} = \mathbf{R}^{-\mathrm{T}/2}\mathbf{y} \qquad [17]$$

or

$$\mathbf{E}'\mathbf{x}' + \mathbf{n}' = \mathbf{y}' \qquad [18]$$

where

$$\mathbf{E}' = \mathbf{R}^{-\mathrm{T}/2}\mathbf{E}\mathbf{S}^{\mathrm{T}/2},\quad \mathbf{x}' = \mathbf{S}^{-\mathrm{T}/2}\mathbf{x},\quad \mathbf{n}' = \mathbf{R}^{-\mathrm{T}/2}\mathbf{n},\quad \mathbf{y}' = \mathbf{R}^{-\mathrm{T}/2}\mathbf{y} \qquad [19]$$

One then computes the singular value decomposition

$$\mathbf{E}' = \mathbf{U}\boldsymbol{\Lambda}\mathbf{V}^{\mathrm{T}} \qquad [20]$$

where $\boldsymbol{\Lambda}$ is a diagonal matrix, and $\mathbf{U}$, $\mathbf{V}$ are orthogonal matrices. Let there be K nonzero values on the diagonal of $\boldsymbol{\Lambda}$, and let the columns of $\mathbf{U}$, $\mathbf{V}$ be denoted $\mathbf{u}_i$, $\mathbf{v}_i$ respectively. Then it can be shown in straightforward fashion that the $\mathbf{u}_i^{\mathrm{T}}\mathbf{u}_j = \delta_{ij}, \mathbf{v}_i^{\mathrm{T}}\mathbf{v}_j = \delta_{ij}$ and that the solution to [18] is

$$\mathbf{x}' = \sum_{k=1}^{K} \mathbf{v}_k \frac{\mathbf{u}_k^{\mathrm{T}}\mathbf{y}'}{\lambda_k} + \sum_{k=K+1}^{N} \alpha_k \mathbf{v}_k \qquad [21]$$

where the α_k are completely arbitrary. The physical solution is obtained from $\tilde{\mathbf{x}} = \mathbf{S}^{\mathrm{T}/2}\tilde{\mathbf{x}}'$. A major advantage of this solution (and the uncertainly matrix, $\mathbf{P}$, can also be computed for it), is that it explicitly produces the solution in orthonormal structures, $\mathbf{v}_i$, in terms of orthonormal structures of the observations $\mathbf{u}_i$, and separates these elements from the second sum on the right of equations [21], which defines the so-called null space of the problem. The null space represents the structures possibly present in the true solution, about which the equations [18] carry no information. A null space is present too, in the Gauss–Markov solution, but the corresponding null space vectors are given coefficients α_k so as best to reproduce the *a priori* values of S. The SVD leads naturally to a discussion of what is called 'resolution' both of data and of the solution, and provides complete information about the solution. These issues, with examples, are discussed in Wunsch (1996) (see Further Reading).

A Hydrographic Example

The first use of inverse methods in oceanography was the hydrographic problem. In this classical problem, oceanographers are able to determine the density field, $\rho(z)$ at two nearby 'hydrographic stations' where temperature and salinity were measured by a ship as functions of depth. By computing the horizontal differences in density as a function of depth, and invoking geostrophic balance, an estimate could be made of the velocity field, up to an unknown constant (with depth) of integration. Although mathematically trivial, the inability to determine the integration constant between pairs of stations plagued oceanography for 100 years, and led to the employment of the *ad hoc* assumption that the flow field vanished at some specific depth z_0. This depth became known as the 'level-of-no-motion.' But by employing physical requirements such as mass and salt conservation (or any other constraint involving the absolute velocity), it is straightforward to reduce the problem to one of deducing the actual flow field at the level-of-no-motion (and which is thus better called the 'reference-level'). These constraints are readily written in the form of eqn [8], and solved with error estimates, etc.

An example of the practical application of this method is shown in **Figure 2**. In this calculation, constraints were written for mass and salt conservation in a triangular region bounded by the US coastline, and a pair of hydrographic sections in which the Gulf Stream flowed into the region across one section and out again in the other. Direct velocity measurements obtained from the ship permitted additional equations to be written for the unknown flow velocity at the reference level. The resulting estimated absolute flow at the reference level is shown, with error estimates, in the top panels. The lower panels show the total estimated absolute velocity. In general, any available information can be used to estimate the unknowns of the system as long as one has a plausible estimate of the error contained in the resulting constraint. For the hydrographic problem in particular, a number of variations on the constraints have been proposed, under the labels of 'beta-spiral,' 'Bernoulli-method,' etc., which we must leave to the references listed in Further Reading.

Extensions

Like eqn [6], many inverse problems are nonlinear. Tarantola (1987) provides some general background and specific oceanographic applications may be seen in Mercier (1989), Mercier *et al.* (1991), and Wunsch (1994). The use of inequality constraints leads to the general subject of mathematical programming, a part of the wide subject of optimization theory (see Arthnari and Dodge, 1981).

The Gauss–Markov solution method and the SVD version of least-squares have a ready interpretation, as minimum variance estimates of the true field. If the fields are all normally distributed, the solutions are also maximum-likelihood estimates, a methodology that is readily extended to non-Gaussian fields.

Time-dependent Problems

As originally formulated by Backus and Gilbert, and as exploited in most of the oceanographic literature

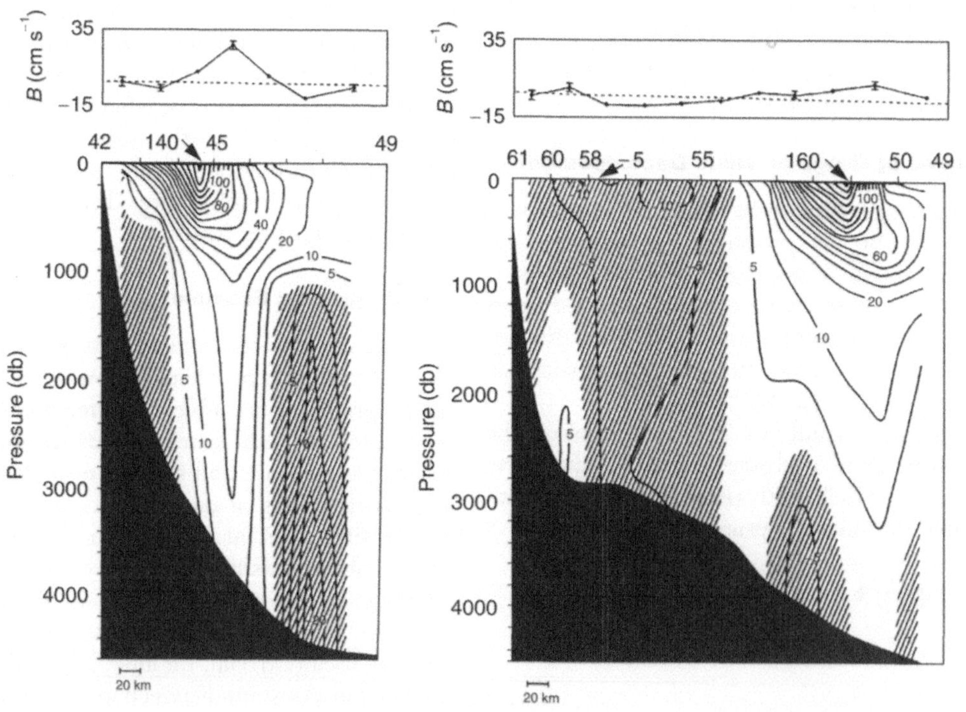

Figure 2 Example of the inversion for the reference level velocity in a triangular region bounded on two sides by hydrographic sections crossing the Gulf Stream, and on the third side by the US coastline. Velocity contours are in centimeters. They are the sum of the so-called thermal wind, which involves setting the velocity to zero at a reference depth. The actual flow at that depth (the 'reference level velocity') is shown in the top panels as estimated, with uncertainties, from the singular value decomposition solution, with $K=30$. (From Joyce *et al.*, 1986.) The 'columnar' structure, which is so apparent here, first appeared in inverse solutions, and was greeted with disbelief by those who 'knew' that oceanic flows were 'layered' in form. Acceptance of this type of structure is now commonplace.

to date, the problems have been essentially static, with time evolution not accounted for. If one has a system that evolves in time, the observations and state vector, $\mathbf{x}$, will also be time evolving. If one simply writes down the relationship between data and a model in which the two are connected by a set of equations, linear or nonlinear, one sees immediately that, mathematically, the problem is identical to that posed by eqn [8], in which the time of the observation or of the model calculation is just a bookkeeping index. The difficulty that arises is purely a practical one: the potentially extremely large growth in the number of equations that must be dealt with over long time spans. With a sufficiently large and fast computer, the most efficient way to solve such inverse problems would be the straightforward application of the same methods developed for static problems: Write out all of the equations explicitly and solve them all at once. In oceanographic practice however, one rapidly outstrips the largest available computers and the static methods become impractical if used naively.

Fortunately, time-evolving oceanographic models have a very special algebraic structure, which permits one to solve the corresponding normal equations [11] by methods not requiring storage of everything in the computer at one time. Numerous such special methods exist, and go by names such as sequential estimators (Kalman filter and various smoothers), adjoint equations (Pontryagin principle or method of Lagrange multipliers), Monte Carlo methods, etc. The generic terminology that has come into use is 'data assimilation' borrowed from meteorological forecasting terminology. Such methods are highly developed, if often abused or misunderstood, and require a separate discussion. What is important to know, however, is that they are simply algorithmically efficient solutions to the inverse problem as described here.

Common Misconceptions and Difficulties

Some chronic misunderstandings and difficulties arise. The most pernicious of these is the attempt to use inverse models, which are physically inconsistent with the known forward model or physics. This blunder corresponds, for example, in the simplest model above, to imposing a linear relationship (model) between θ and z, when it is clear or suspected that higher powers of z are likely to be present. Some writers have gone so far as to show that such an inversion does not reproduce a known forward solution, and then declared inverse methods to be failures. Another blunder is to confuse the inability to resolve or determine a parameter of interest, when the data are inadequate for the purpose, with a methodological or model failure. Inverse methods are very powerful tools. Like any powerful tool (a chain saw, for example), when properly used they are useful and even essential; when improperly used they are a grave danger to the user.

See also

Data Assimilation in Models.

Further Reading

Arthnari TS and Dodge Y (1981) *Mathematical Programming in Statistics*. New York: Wiley.

Bennett AF (1992) *Inverse Methods in Physical Oceanography*. Cambridge: Cambridge University Press.

Eckart C and Young G (1939) A principal axis transformation for non-Hermitian matrices. *Bulletin of the American Mathematical Society* 45: 118–121.

Joyce TM, Wunsch C, and Pierce SD (1986) Synoptic Gulf Stream velocity profiles through simultaneous inversion of hydrographic and acoustic doppler data. *Journal of Geophysical Research* 91: 7573–7585.

Lanczos C (1961) *Linear Differential Operators*. Princeton, NJ: Van Nostrand.

Liebelt PB (1967) *An Introduction to Optimal Estimation*. Reading, MA: Addison Wesley.

Menke W (1989) *Geophysical Data Analysis: Discrete Inverse Theory* 2nd edn. New York: Academic Press.

Mercier H (1989) A study of the time-averaged circulation in the western North Atlantic by simultaneous nonlinear inversion of hydrographic and current-meter data. *Deep-Sea Research* 36: 297–313.

Mercier H, Ollitrault M, and Le Traon PY (1991) An inverse model of the North Atlantic general circulation using Lagrangian float data. *Journal of Physical Oceanography* 23: 689–715.

Munk W, Worcester P, and Wunsch C (1995) *Ocean Acoustic Tomography*. Cambridge: Cambridge University Press.

Parker RL (1994) *Geophysical Inverse Theory*. Princeton, NJ: Princeton University Press.

Seber GAF (1977) *Linear Regression Analysis*. New York: Wiley.

Tarantola A (1987) *Inverse Problem Theory. Methods for Data Fitting and Model Parameter Estimation*. Amsterdam: Elsevier.

Van Trees HL (1968) *Detection, Estimation and Modulation Theory. Part I*. New York: Wiley.

Wiggins RA (1972) The general linear inverse problem: implication of surface waves and free oscillations for earth structure. *Reviews in Geophysics and Space Physics* 10: 251–285.

Wunsch C (1977) Determining the general circulation of the oceans: a preliminary discussion. *Science* 196: 871–875.

Wunsch C (1994) Dynamically consistent hydrography and absolute velocity in the eastern North Atlantic Ocean. *Journal of Geophysical Research* 99: 14071–14090.

Wunsch C (1996) *The Ocean Circulation Inverse Problem*. Cambridge: Cambridge University Press.

FORWARD PROBLEM IN NUMERICAL MODELS

M. A. Spall, Woods Hole Oceanographic Institution, Woods Hole, MA, USA

Introduction

The Forward Problem in numerical modeling describes a class of problems for which computers are used to advance in time discretized forms of a continuous set of equations of motion subject to initial and boundary conditions. Numerical models have emerged as powerful and versatile tools for addressing both fundamental and applied problems in the ocean sciences. This advance has been partly driven by increases in computing speed and memory over the past 30 years. However, advances in fundamental theories of ocean dynamics and a better description of the ocean currents as a result of a diverse and growing observational database have also helped to make numerical models both more realistic and more useful aids for understanding the basic physics of the ocean. Although many of the issues discussed in this article are relevant to interdisciplinary applications of forward models, the focus here is on using ocean models to understand the physics of the large-scale oceanic circulation.

One of the advantages of forward numerical models is that they provide dynamically consistent solutions over a wide range of parameter space and thus offer great flexibility for a variety of oceanographic problems. Forward models can be configured in quite realistic domains and subjected to complex, realistic forcing fields to produce simulations of the present-day ocean, past climates, or specific time periods or regions. However, the model physics and domain configuration can also be greatly simplified in order to address fundamental issues via process-oriented studies. The choice of model physics, numerical discretization, initial and boundary conditions, and analysis approach is strongly dependent on the application in mind.

More complete descriptions of the mean and time-dependent characteristics of the ocean circulation can provide valuable reference points for ocean models, targets towards which the scientist can aim. Discrepancies between models and data serve to identify deficiencies in the model solutions, which in turn lead to improved physics or numerics and, hopefully, a more faithful representation of the observations. The scientist can use the dynamical model to study the sensitivity of the ocean circulation to variations in model configuration, e.g., surface forcing, dissipation, or topography or to variations in model physics. Understanding how the solution depends on the fundamental parameters of the system leads to an increased understanding of the dynamics of the ocean circulation in general.

Analytic solutions to the equations of motion also provide valuable reference points for numerical models. Forward models can generally be applied to a wider range of problems than are typically accessible by purely analytic methods. However, the numerical models do not solve exactly the continuous equations of motion. They provide approximate solutions on a finite numerical grid and are generally subject to some form of truncation error, dissipation, and smoothing (discussed further below). It is often useful to configure the model such that direct comparisons with analytic solutions are possible in order to quantify the influences of the numerical method and to verify that, at least in the parameter space for which the analytic solution is valid, the model produces the correct solution. This starting point provides a useful reference for extending the model calculations into parameter space for which analytic solutions are not available. This typically involves increasing the nonlinearity of the system, introducing time dependence, and/or complexity of the domain configuration (topography, coastlines, stratification) and forcing.

This article provides an overview of the general issues relating to the use of forward numerical models for the study of the meso- to basin-scale general oceanic circulation. Although space does not permit a detailed discussion of each of the subject areas discussed below, it is intended that this introduction identify the major issues and concerns that need to be considered when using a numerical model to address a problem of interest. More detailed treatments of each of these topics can be found elsewhere in this text and in the Further reading list.

Equations of Motion

Forward models integrate discrete forms of a dynamically consistent set of equations of motion. The fundamental equations of motion for the oceanic circulation are the Navier–Stokes equations in a rotating coordinate system together with an equation of state that relates the density of the seawater to temperature, salinity, and pressure. However, it is not necessary to solve the full Navier–Stokes equations to study the large-scale, low frequency aspects of the oceanic circulation. Many general circulation models are based on the primitive equations, which form a subset of the Navier–Stokes equations by making the Boussinesq and hydrostatic approximations. The Boussinesq approximation neglects variations in the density everywhere in the momentum equations except where it is multiplied by the acceleration of gravity. This assumption is generally well satisfied in the ocean because changes in the density of sea water are much less than the density of sea water itself. The hydrostatic approximation neglects all vertical accelerations except that due to gravity. This assumption is valid as long as the horizontal scales of motion are much larger than the vertical scales of motion, and is well satisfied by the large-scale general circulation and mesoscale variability, but is violated in regions of active convection.

The horizontal momentum balance for the primitive equations is written in vector form as

$$\frac{\mathrm{d}\vec{u}}{\mathrm{d}t} + f\hat{k} \times \vec{u} = -\nabla P/\rho_0 + F_V \qquad [1]$$

where $\vec{u}$ is the horizontal velocity vector, $f = 2\Omega \sin \phi$ is the Coriolis term, ϕ is the latitude, P is pressure, ρ_0 is the mean density of sea water, and F_V represents horizontal and vertical subgridscale viscosity. The advection operator is defined as

$$\frac{\mathrm{d}}{\mathrm{d}t} = \frac{\partial}{\partial t} + \frac{u}{a \cos \phi} \frac{\partial}{\partial \lambda} + \frac{v}{a} \frac{\partial}{\partial \phi} + w \frac{\partial}{\partial z} \qquad [2]$$

where w is the vertical velocity, λ is longitude and a is the radius of the earth.

The vertical momentum equation is replaced by the hydrostatic approximation

$$\frac{\partial P}{\partial z} = -g\rho \qquad [3]$$

The fluid is assumed to be incompressible, so that the continuity equation reduces to

$$\nabla \cdot \vec{u} + \frac{\partial w}{\partial z} = 0 \qquad [4]$$

The density of sea water is a nonlinear function of temperature, salinity, and pressure. In most general form, the primitive equations integrate conservation equations for both temperature T and salinity S.

$$\frac{\mathrm{d}T}{\mathrm{d}t} = F_T \qquad [5]$$

$$\frac{\mathrm{d}S}{\mathrm{d}t} = F_S \qquad [6]$$

Subgridscale horizontal and vertical dissipative processes are represented as F_T and F_S. An equation of state is used to calculate density from T, S, and P. For process-oriented studies, it is often sufficient to solve for only one active tracer or, equivalently, the density,

$$\rho = \rho(T, S, P) \qquad [7]$$

The primary advantage of the primitive equations is that they are valid over essentially the entire range of scales appropriate for the large-scale, low-frequency oceanic motions. They are a good choice for climate models because they allow for spatially variable stratification, independent temperature and salinity influences on density and air–sea exchange, water mass transformations, large vertical velocities, and steep and tall topography. High resolution studies also benefit from a complete treatment of advection in regions of large Rossby number and large vertical velocities.

The main disadvantages of these equations are computational in nature. The equations admit high-frequency gravity waves, which are generally not of direct consequence for the large-scale physics of the ocean, yet can place computational constraints on the time step allowed to integrate the equations. Models based on the primitive equations must integrate four prognostic equations and one diagnostic equation, making them computationally more expensive than some lower order equations. Finally, for regional applications, the implementation of open boundary conditions is not well posed mathematically and, in practice, models are often found to be very sensitive to the details of how the boundary conditions on open boundaries (mainly on outflow and transitional points) are specified.

There are several additional subsets of dynamically consistent equations that may be derived from the primitive equations. The most common form used in forward numerical models are the quasi-geostrophic equations, which are a leading order asymptotic approximation to the primitive equations for small Rossby number, $R = V/f_0L$, and small aspect ratio $\delta = D/L$, where V is a characteristic velocity scale, D and L are characteristic vertical and horizontal length scales, and f_0 is the Coriolis parameter at the central latitude. The Coriolis parameter is assumed to vary linearly with latitude y, $f = f_0 + \beta y$. The mean stratification is specified and uniform over the entire model domain. Interface displacements due to the fluid motion are assumed to be small compared to the mean layer thicknesses. The quasigeostrophic limit allows for the length scale of motion to be the same order as the oceanic mesoscale, typically 10–100 km.

Quasigeostrophic models have been most often used for process studies of the wind-driven general circulation and its low-frequency variability. Their formulation in terms of a single prognostic variable, the quasigeostrophic potential vorticity, is often an advantage from a conceptual point of view. The equations are adiabatic by design, so that spurious diapycnal mixing is not a problem. The quasigeostrophic equations are generally more efficient to integrate numerically because there is only one prognostic equation and the time step can generally be larger than for comparable resolution primitive equation models because high-frequency Kelvin and gravity waves are not supported.

There are numerous drawbacks to the quasigeostrophic equations that make them less practical for large-scale realistic modeling studies. Several of these drawbacks stem from the assumption that the mean stratification is uniform throughout the model domain. This prohibits isopycnal surfaces from outcropping or intersecting the bottom. The influences of bottom topography are represented by a vertical velocity consistent with no normal flow through the sloping bottom, however it is imposed at the mean bottom depth. It is also not possible to represent the subduction of water masses, the process by which water is advected from the near surface, where it is in turbulent contact with the atmosphere, to the stratified interior, where it is shielded from direct influence from the atmosphere. Various higher order terms, such as the advection of relative vorticity and density anomalies by the nongeostrophic velocity field, are not represented. In addition, the equations do not consider temperature and salinity independently, and the geostrophic approximation breaks down near the equator.

Discretization Issues

Forward models solve for the equations of motion on a discrete grid in space and time. There are numerous considerations that need to be taken into account in determining what form of discretization is most appropriate for the problem of interest. Most numerical methods used in ocean general circulation models are relatively simple, often relying on finite difference or spectral discretization schemes, although some models have employed finite element methods. The convergence properties of discretization techniques are well documented in numerical methods literature, and the reader is referred there for a detailed discussion. Perhaps of most critical importance to the ocean modeler is the choice of vertical discretization. As will be clear, the following issues are most relevant to the primitive equation models because they arise as a consequence of spatially variable stratification or bottom topography. There are three commonly used vertical coordinates: z-coordinate or level models; sigma-coordinate or generalized stretched coordinates; and isopycnal coordinates. The ramifications of the truncation errors associated with each of these approaches differ widely depending on the problem of interest, so it is worth some discussion on the relative advantages and disadvantages of each method.

Level models solve for the dynamic and thermodynamic variables at specified depths which are fixed in time and uniform throughout the model domain. A schematic diagram of a surface intensified density gradient over a sloping bottom in level coordinates is shown in **Figure 1**. The advantages of this approach include: its relative simplicity, an accurate treatment of the horizontal pressure gradient terms, high vertical resolution in regions of weak stratification, and well-resolved surface boundary layers. The major disadvantage of the level models is in their treatment of bottom topography. In the simplest, and most commonly used, form the bottom depth must reside at one of the level interfaces. Thus, in order to resolve small variations in bottom depth one must devote large amounts of vertical resolution throughout the model domain (even on land points). As is evident in **Figure 1(A)**, the bottom slopes on horizontal scales of several model grid points will be only coarsely represented. There are two main errors associated with an inaccurate treatment of the bottom slope. The first is a poor representation of the large-scale stretching and/or compression associated with flow across a sloping bottom, which can affect the propagation characteristics of large-scale planetary waves. The second problem arises when dense water flows down steep topography, such as is found near sills connecting marginal seas to the open ocean. Standard level models excessively mix the dense water with the ambient water as it is advected downslope, severely compromising the properties of the dense water. In the ocean interior, care must be taken to ensure that most of the mixing takes place along isopycnal surfaces.

Terrain-following or, in more general terms, stretched coordinate models solve the equations of motion on a vertical grid that is fixed in time but varies in space. A schematic diagram of a surface intensified density gradient over a sloping bottom in terrain-following coordinates is shown in **Figure 1(B)**.

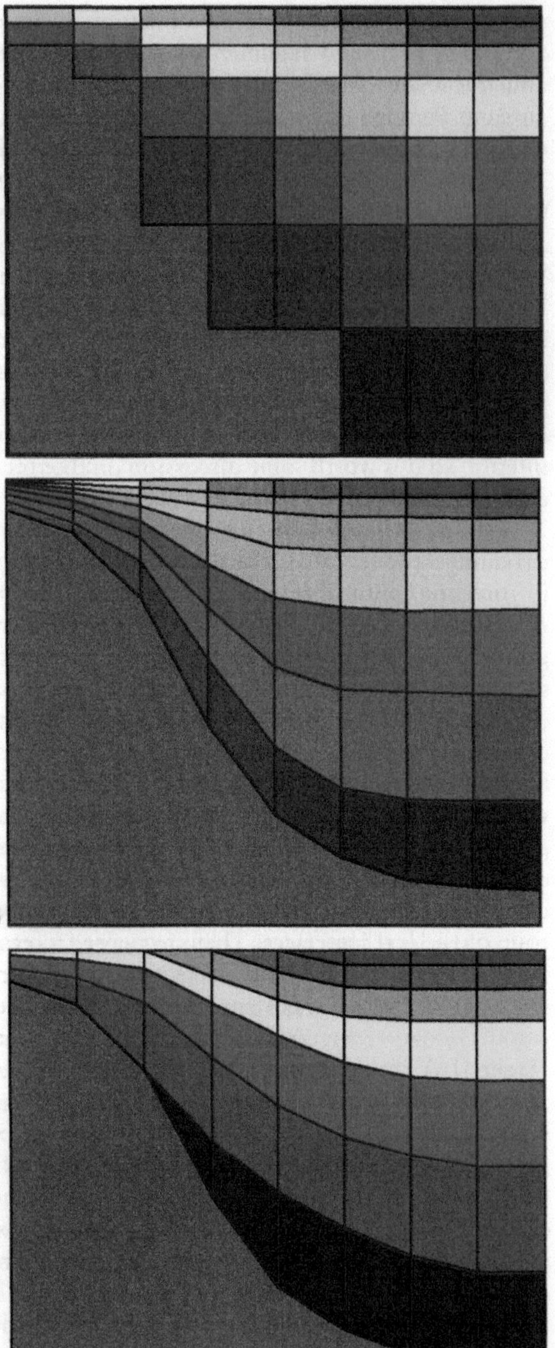

Figure 1 Schematic diagrams of a density front over a sloping bottom as represented in different model coordinate systems. Black lines denote boundaries of model grid cells, color indicates density (red light, blue dense). (A) Level coordinates; (B) terrain-following coordinates; (C) isopycnal coordinates.

The advantages of this approach are: accurate representation of topographic slopes, all model grid points reside in the ocean, and high resolution of surface and bottom boundary layers. Notice the smooth representation of the bottom slope and increase in vertical resolution in shallow water. The main disadvantage of the stretched coordinate approach is in calculating the horizontal pressure gradient in regions of steep topography. The calculation of the pressure gradient is prone to errors because the hydrostatic component of the pressure between adjacent grid points that are at different depths must be subtracted before calculating the (typically much smaller) dynamically significant lateral variation in pressure. As with level models, the parameterization of lateral subgridscale mixing processes must be carefully formulated to avoid spurious mixing across isopycnal surfaces. There can also be some computational constraints that arise when stretched coordinate models are extended into very shallow water as the vertical grid spacing becomes very small.

Isopycnal coordinate models solve the equations of motion within specified density layers, which are allowed to move vertically during the course of integration. A schematic diagram of a surface intensified density gradient over a sloping bottom in isopycnal coordinates is shown in **Figure 1**(C). Notice that the density is uniform along the model coordinate surfaces, and that these surfaces may intersect the surface and/or the bottom. This gives rise to a more discrete representation of the stratification than either level or stretched coordinate models. The advantages of this approach include: high vertical resolution in regions of large vertical density gradients; straightforward mixing along isopycnal surfaces; explicit control of diapycnal mixing; accurate treatment of lateral pressure gradients; smooth topography, and a natural framework for analysis. The primary disadvantages of the isopycnal discretization are: low resolution in weakly stratified regions (such as the mixed layer and high latitude seas); difficulty in the implementation of the nonlinear equation of state over regions of widely varying stratification such that static stability is uniformly assured; and the handling of the exchange of water of continuously varying density in the surface mixed layer with the discrete density layers in the stratified interior. Care must be taken so that the layer thicknesses can vanish in regions of isopycnal outcropping or intersection with topography. Although readily handled with higher order discretization methods, treatment of vanishing layers makes the models more computationally intensive and numerically complex.

Initial and Boundary Conditions

Forward numerical models integrate the equations of motion subject to initial and boundary conditions. Initial conditions for process-oriented studies are typically very idealized, such as a motionless, homogenous, stratified basin or statistically uniform, geostrophically balanced turbulence in a periodic basin. The initial conditions used for simulations in realistic basins can be more complex, depending on the application in mind. Long-term climate studies are often initialized with a motionless ocean of uniform temperature and salinity. Short-term basin-scale integrations are often initialized with a more realistic density field derived from climatological hydrographic databases. Although subject to much smoothing (particularly near narrow boundary currents and overflows), such renditions contain many realistic features of the large-scale general circulation. Some care should be taken so as not to introduce spurious water masses when averaging hydrographic data onto the model grid.

Forward models can also be used to produce short-term simulations or predictions of the oceanic state of a specific region and time. Although such calculations are technically feasible, they require large amounts of data to initialize the model state variables and force the models at the boundaries of the regional domain. The initialization of such models often makes use of direct *in situ* or remotely sensed observations together with statistical or dynamical extrapolation techniques to fill the regions void of data. The initial dynamic adjustment is greatly reduced if the velocity field is initialized to be in geostrophic balance with the density field. Higher order initializations can further reduce the high-frequency transients that are generated upon integration, although the effects of these transients are generally confined to the early integration period and are of most importance to short-term predictions.

Lateral boundary conditions are most straightforward at solid boundaries. Boundary conditions for the momentum equations are typically either no-slip (tangential velocity equals zero) or free-slip (no stress) and no normal flow. Tracer fluxes are generally no flux through the solid wall. Boundary conditions at the bottom are generally no normal flow for velocity and no flux for tracers. A stress proportional to $\vec{v}$ or $\vec{v}^2$ is often imposed as a vertical momentum flux at the bottom to parameterize the influences of unresolved bottom boundary layers.

Lateral boundary conditions are more problematic when the boundary of the model domain is part of the open ocean. The model prognostic variables need to be specified here but, in general, some of the information that determines the model variables on the open boundary will be controlled by information propagating from within the model domain and some will be controlled by information propagating from outside the domain. There is no general solution to this problem. Most regional models adopt a practical, rather than rigorous, approach through a combination of specifying the flow variables from 'observations' on inflow points and propagating information using simple advective equations on outflow points. Increasing the model dissipation near the open boundaries is sometimes required. The 'observations' in this case may be based on actual oceanic observations (in the case of regional simulations), climatology (basin-scale long-term integrations), or some idealized flow state (process studies). Basin-scale general circulation models often represent water mass exchanges that take place outside the model domain through a restoring term in the tracer equations that forces the model tracers towards the climatological tracer values near the boundaries. In their most simple application, however, these boundary conditions do not permit an advective tracer flux through the boundary ($\vec{v} = 0$) so that even though the tracer field may replicate the observed values, the tracer flux may be in error.

Primitive equation models explicitly integrate conservation equations for heat, salt, and momentum and thus require surface fluxes for these variables. It is commonly assumed that the flux of a property through the surface of the ocean is transported vertically away from the very thin air–sea interface through small-scale turbulent motions. Primitive equation models do not generally resolve such small-scale motions and rely on subgridscale parameterizations (see next section) to represent their effort on the large-scale flow. The vertical turbulent momentum and tracer fluxes are generally assumed to be downgradient and represented as a vertical diffusion coefficient times the mean vertical gradient of the quantity of interest. Thus, the surface flux of magnitude F for a model variable T is incorporated into the vertical diffusion term as

$$K_T \frac{\partial T}{\partial z} = F \qquad [8]$$

where K_T is a vertical diffusion coefficient.

Historically, climate models have represented the complex suite of heat flux components by simply restoring the model sea surface temperature towards a specified, spatially variable 'atmospheric' temperature with a given time scale (which itself may be a function of space and/or time). In its simplest form, the atmospheric temperature is taken to be the observed climatological sea surface temperature. This approach is simple and has the advantage that the ocean temperature will never stray too far away from the range specified by the boundary conditions. In the context of simulating the real ocean, however, the obvious drawback is that if the model ocean has the correct sea surface temperature, it also has zero heat flux, a condition that holds only over very limited regions of the ocean. Conversely, if the model ocean is being forced with the correct net surface heat flux it must have the wrong sea surface temperature. This approach also assumes that the atmosphere has an infinite heat capacity and thus can not respond thermodynamically to changes in the sea surface temperature.

A more realistic approach is to specify the surface heat flux to be the sum of the best estimate of the real net surface heat flux plus a restoring term that is proportional to the difference between the model sea surface temperature and the observed climatological sea surface temperature. An advantage of this approach is that the model is forced with surface heat fluxes consistent with the best observational estimates when the model SST agrees with climatology. This approach also does not allow the model SST to stray too far away from the specified climatology. The drawbacks include having to specify the surface heat flux (which is not well known) and enforcing at the surface the spatial scales inherent in the smoothed hydrographic climatology.

A dynamically more complex approach for climate is to include a simple active planetary boundary layer model to represent the atmosphere. In this case, the only specifications that are required are the incoming short-wave solar flux, which is reasonably well known, the temperature of the land surrounding the ocean basin, and the wind stress. The planetary boundary layer model calculates the atmospheric temperature and humidity and, with the use of bulk formulae, the net sensible, latent, and long-wave radiative heat fluxes. Although this approach is more complex than the simple restoring conditions, it has the advantages of thermodynamic consistency between the model physics and the surface heat fluxes, the spatial scales of the surface heat flux are determined by the model dynamics, and it allows the atmosphere to respond to changes in sea surface temperature.

The net fresh water flux at the surface is fundamentally different from and technically more problematic than the net heat flux. Primitive equation models integrate conservation equations for salinity (the number of grams of salt contained in 1 kg of sea water) yet the net salt flux through the sea surface is zero. The salinity is changed by exchanges of fresh water between the ocean and atmosphere. A major difference between the air–sea exchange of freshwater and the air–sea heat flux is that the freshwater flux is largely independent of salinity, the dynamically active tracer which it strongly influences. The change in salinity in the ocean that results from a net freshwater flux at the surface is typically represented as a virtual salt flux by changing the salinity of the ocean without a corresponding change in the volume of water in the ocean. Differences between the exact freshwater flux boundary condition and the virtual salt flux boundary condition are generally small. However, a serious challenge facing long-term climate integrations is to represent faithfully the surface boundary condition for salinity without allowing the model salinity to drift too far from the observed state. Most climate models parameterize the influence of the net freshwater flux at the surface by including at least a weak restoring of the sea surface salinity towards a specified spatially variable value.

Subgridscale Parameterizations

The large-scale general circulation in the ocean is influenced by processes that occur on spatial scales from less than a centimeter to thousands of kilometers. Temporal variations occur on time scales from minutes for turbulent mixing to millennia for climate variability. It is not possible now, nor in the forseeable future, to represent explicitly all of these scales in ocean models. It is also attractive, from a conceptual point of view, to represent the interplay between scales of motion through clear dynamically based theories. The important and formidable task facing ocean modelers is to parameterize effectively those processes that are not resolved by the space/time grid used in the model. The two main classes of motion that may be important to the large-scale circulation, and are often not explicitly represented in forward models, are small-scale turbulence and mesoscale eddy variability. Although a comprehensive review of the physics and parameterizations of these phenomena are beyond the scope of this article, a summary of the physical processes that need to be considered is useful.

Turbulent mixing of properties (temperature, salinity, momentum) across density surfaces takes place on spatial scales of centimeters and is driven by small-scale turbulence, shear and convective instabilities, and breaking internal waves. Although these processes take place on very small scales, they play a fundamental role in the global energy budget and are essential components of the basin- to global-scale thermohaline general circulation. Mixing across density surfaces is thought to be small over most of the ocean, however it can be intense in regions of strong boundary currents and dense water overflows, rough bottom topography, and near the ocean surface where direct atmospheric forcing is important. It will likely be a long time before these processes will be able to be resolved in large-scale general circulation models, so parameterizations of the turbulent mixing are necessary.

Intense diapycnal turbulent mixing is found in the boundary layers near the ocean surface and bottom. The planetary boundary layer near the ocean surface has received much attention from physical oceanographers because the ocean circulation is forced through this interface and because of its fundamental importance to air–sea exchange of heat, fresh water, and biogeochemically important tracers. Strong turbulent mixing is forced in the surface mixed layer as a result of shear and convective instabilities resulting from these surface fluxes. The bottom boundary layer is similar in some regards, but is primarily driven by the bottom stress with buoyancy fluxes generally negligible.

Parameterization approaches to the turbulent boundary layers generally fall into three categories. Bulk models assume that all properties are homogenized within the planetary boundary layer over a depth called the mixed layer depth. This depth is determined by a budget between energy input through the air–sea interface, turbulent dissipation, and entrainment of stratified fluid from below. These models are relatively simple and inexpensive to use, yet fail to resolve any vertical structure in the planetary boundary layer and assume that all properties mix in the same way to the same depth. Local closure models allow for vertical structure in the planetary boundary layer, yet assume that the strength of turbulent mixing is dependent only on the local properties of the fluid. A third class of planetary boundary layer models does allow for nonlocal influences on turbulent mixing.

Turbulent mixing in the ocean interior is generally much smaller than it is within the surface and bottom boundary layers. Nonetheless, water mass budgets in semi-enclosed abyssal basins indicate that substantial and important mixing must take place in the deep ocean interiors. Recent tracer release experiments and microstructure measurements suggest that elevated mixing may be found near and above regions of rough bottom topography, perhaps as a result of internal wave generation and subsequent breaking. Early subgridscale parameterizations of diapycnal mixing were very crude, and often dictated more by numerical stability constraints than by ocean physics. Downgradient diffusion is generally assumed, often with spatially uniform mixing coefficients for temperature and salinity. A few studies using a mixing coefficient that increases with decreasing stratification have been carried out, but there is still much more work to be done to understand fully the importance of diapycnal mixing distributions on the general oceanic circulation.

Moving to larger scales, the next category of motions that is likely to be important to the general circulation is the oceanic mesoscale. Mesoscale eddies are found throughout the world's oceans and are characterized by spatial scales of tens to hundreds of kilometers and time scales of tens to hundreds of days. This is much larger than the turbulent mixing scale yet still considerably smaller than the basin-scale. Variability on these space and time scales can result from many processes, such as wave radiation from distance sources, local instability of mean currents, vortex propagation from remote sources, and local atmospheric forcing. Although the oceanic mesoscale has been known to exist since the early 1960s, oceanographers still do not know in what proportion each of these generation mechanisms are important in determining the local mesoscale variability and, perhaps ever more daunting, their role in the general circulation. However, eddy-resolving modeling studies have shown that significant, large-scale mean circulations can be driven by mesoscale eddy motions so, in at least some regards, they are clearly important for the general circulation.

The diversity of generation mechanisms and complex turbulent dynamics of mesoscale eddies makes it very difficult to parameterize their influences on the large-scale circulation. Historical approaches have relied on downgradient diffusion of tracers with uniform mixing coefficients, often along the model coordinate surfaces. This approach is simple and produces smooth and stable solutions. Yet, in the case of coordinate surfaces that do not coincide with isopycnal surfaces, this approach can introduce excessive spurious diapycnal mixing, particularly in regions of strongly sloping isopycnal surfaces, such as near the western boundary current. More physically based subgridscale parameterizations of mesoscale eddies project the eddy-induced tracer fluxes along isopycnal surfaces. Note, however, that in the case where density is determined by only one tracer, there will be no effective tracer flux by the mesoscale eddies and some additional form of subgridscale mixing may be required for numerical stability.

Recent advances in the parameterization of mesoscale tracer transports have been based on the assumptions that (1) eddy tracer fluxes are primarily along isopycnal surfaces and (2) the strength of the eddy-induced tracer flux is proportional to the local gradient in isopycnal slope. This approach has several nice characteristics. First, it allows one to define a stream function for the eddy-induced velocity, thus ensuring adiabatic tracer advection and eliminating spurious diapycnal mass fluxes due to the parameterization. Second, the eddy fluxes work to relax sloping isopycnal surfaces and extract potential energy from the mean flow, crudely representing the effects of local baroclinic instability. Implementation of such parameterizations in global and basin-scale general circulation models has allowed for removal of horizontal diffusion and has resulted in much improved model simulations. Additional theories have been developed that relate the strength of the eddy-induced transport velocities to the local properties of the mean flow by making use of baroclinic instability theory. It should be pointed out that this approach makes the implicit assumption that the eddy flux divergence is proportional to the local mean flow so that it is not intended to parameterize the tracer transport by eddies that travel far from their source, either by self-propagation or advection by the mean flow.

Summary

Forward numerical models have emerged as a powerful tool in large-scale ocean circulation studies. They provide dynamically consistent solutions while allowing for much flexibility in the choice of model physics, domain configuration, and external forcing. When used together with observations, laboratory experiments, and/or theories of the ocean circulation, forward models can help extend our understanding of ocean physics into dynamically rich and complex regimes. However, the scientist must always be aware that forward models provide only approximate solutions to the equations of motion and are subject to various forms of smoothing and dissipation. Perhaps the most critical area for future development in forward models is in improving our ability to parameterize unresolved turbulent processes on scales from centimeters to the oceanic mesoscale.

See also

Mesoscale Eddies. Rossby Waves.

Further Reading

Chassignet EP and Verron J (eds.) (1998) *Ocean Modeling and Parameterization*. Dordrecht: Kluwer Academic Publishers.

Haidvogel DB and Beckmann A (1999) *Numerical Ocean Circulation Modeling*. London: Imperial College Press.

Iserles A (1996) *A First Course in the Numerical Analysis of Differential Equations*. Cambridge: Cambridge University Press.

Kantha LH and Clayson CA (2000) *Numerical Models of Oceans and Oceanic Processes*. San Diego: Academic Presss.

Roache PJ (1972) *Computational Fluid Dynamics*. Albuquerque: Hermosa Publishers.

DATA ASSIMILATION IN MODELS

A. R. Robinson and P. F. J. Lermusiaux,
Harvard University, Cambridge, MA, USA

Introduction

Data assimilation is a novel, versatile methodology for estimating oceanic variables. The estimation of a quantity of interest via data assimilation involves the combination of observational data with the underlying dynamical principles governing the system under observation. The melding of data and dynamics is a powerful methodology which makes possible efficient, accurate, and realistic estimations otherwise not feasible. It is providing rapid advances in important aspects of both basic ocean science and applied marine technology and operations.

The following sections introduce concepts, describe purposes, present applications to regional dynamics and forecasting, overview formalism and methods, and provide a selected range of examples.

Field and Parameter Estimation

Ocean science, and marine technology and operations, require a knowledge of the distribution and evolution in space and time of the properties of the sea. The functions of space and time, or state variables, which characterize the state of the sea under observation are classically designated as fields. The determination of state variables poses problems of state estimation or field estimation in three or four dimensions. The fundamental problem of ocean science may be simply stated as follows: given the state of the ocean at one time, what is the state of the ocean at a later time? It is the dynamics, i.e., the basic laws and principles of oceanic physics, biology, and chemistry, that evolve the state variables forward in time. Thus, predicting the present and future state of oceanic variables for practical applications is intimately linked to fundamental ocean science.

A dynamical model to approximate nature consists of a set of coupled nonlinear prognostic field equations for each state variable of interest. The fundamental properties of the system appear in the field equations as parameters (e.g., viscosities, diffusivities, representations of body forces, rates of earth rotation, grazing, mortality, etc.). The initial and boundary conditions necessary for integration of the equations may also be regarded as parameters by data assimilation methods. In principle the parameters of the system can be estimated directly from measurements. In practice, directly measuring the parameters of the interdisciplinary (physical-acoustical-optical-biological-chemical-sedimentological) ocean system is difficult because of sampling, technical, and resource requirements. However, data assimilation provides a powerful methodology for parameter estimation via the melding of data and dynamics.

The physical state variables are usually the velocity components, pressure, density, temperature, and salinity. Examples of biological and chemical state variables are concentration fields of nutrients, plankton, dissolved and particulate matter, etc. Important complexities are associated with the vast range of phenomena, the multitude of concurrent and interactive scales in space and time, and the very large number of possible biological state variables. This complexity has two essential consequences. First, state variable definitions relevant to phenomena and scales of interest need to be developed from the basic definitions. Second, approximate dynamics which govern the evolution of the scale-restricted state variables, and their interaction with other scales, must be developed from the basic dynamical model equations. A familiar example consists of decomposing the basic ocean fields into slower and faster time scales, and shorter and longer space scales, and averaging over the faster and shorter scales. The resulting equations can be adapted to govern synoptic/mesoscale resolution state variables over a large-scale oceanic domain, with faster and smaller scale phenomena represented as parameterized fluctuation correlations (Reynolds stresses). There is, of course, a great variety of other scale-restricted state variables and approximate dynamics of vital interest in ocean science. We refer to scale-restricted state variables and approximate dynamics simply as 'state variables' and 'dynamics'.

The use of dynamics is of fundamental importance for efficient and accurate field and parameter estimation. Today and in the foreseeable future, data acquisition in the ocean is sufficiently difficult and costly so as to make field and parameter estimates by direct measurements, on a substantial and sustained basis, essentially prohibitive. However, data acquisition for field and parameter estimates via data assimilation is feasible, but substantial resources must be applied to obtain adequate observations.

The general process of state and parameter estimation is schematized in **Figure 1**. Measurement models link the state variables of the dynamical model to the sensor data. Dynamics interpolates and extrapolates the data. Dynamical linkages among state variables and parameters allow all of them to be estimated from measurements of some of them, i.e., those more accessible to existing techniques and prevailing conditions. Error estimation and error models play a crucial role. The data and dynamics are melded with weights inversely related to their relative errors. The final estimates should agree with the observations and measurements within data error bounds and should satisfy the dynamical model within model error bounds. Thus the melded estimate does not degrade the reliable information of the observational data, but rather enhances that information content. There are many important feedbacks in the generally nonlinear data assimilation system or ocean observing and prediction system (OOPS) schematized in **Figure 1**, which illustrates the system concept and two feedbacks. Prediction provides the opportunity for efficient sampling adapted to real time structures, events, and errors. Data collected for assimilation also used for ongoing verification can identify model deficiencies and lead to model improvements.

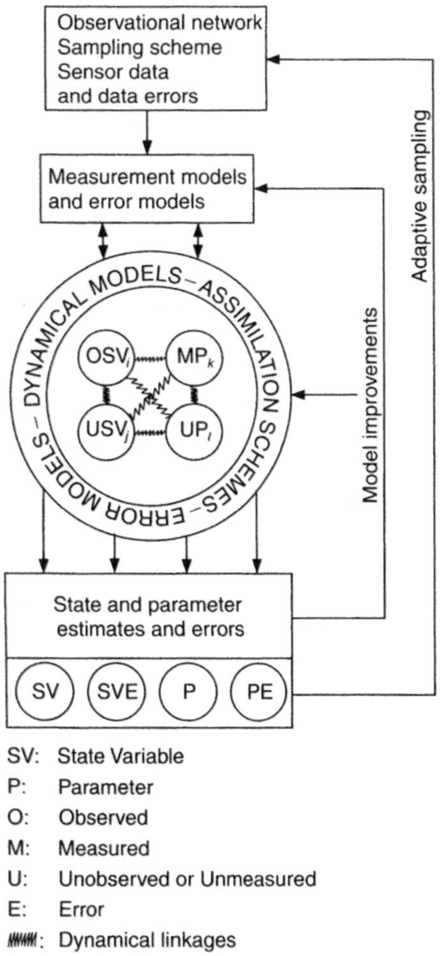

Figure 1 Data assimilation system schematic. Arrows represent the most common direction for the flows of information. The arrows between the measurement models and dynamical models double because measurement models can include operators that map state variables and parameters to the sensor data (e.g. interpolations, derivatives or integrals of state variables/parameters) and operators that transform sensor data into data appropriate for the model scales and processes (e.g. filtering, extrapolations or integrals of sensor data). The legend at the bottom explains abbreviations.

A data assimilation system consists of three components: a set of observations, a dynamical model, and a data assimilation scheme or melding scheme. Modern interdisciplinary OOPS generally have compatible nested grids for both models and sampling. An efficient mix of platforms and sensors is selected for specific purposes.

Central to the concept of data assimilation is the concept of errors, error estimation, and error modeling. The observations have errors arising from various sources: e.g., instrumental noise, environmental noise, sampling, and the interpretation of sensor measurements. All oceanic dynamical models are imperfect, with errors arising from the approximate explicit and parameterized dynamics and the discretization of continuum dynamics into a computational model.

A rigorous quantitative establishment of the accuracy of the melded field and parameter estimates, or verification, is highly desirable but may be difficult to achieve because of the quantity and quality of the data required. Such verification involves all subcomponents: the dynamical model, the observational network, the associated error models, and the melding scheme. The concept of validation is the establishment of the general adequacy of the system and its components to deal with the phenomena of interest. As simple examples, a barotropic model should not be used to describe baroclinic phenomena, and data from an instrument whose threshold is higher than the accuracy of the required measurement are not suitable. In reality, validation issues can be much more subtle. Calibration involves the tuning of system parameters to the phenomena and regional characteristics of interest. Final verification requires dedicated experiments with oversampling.

At this point it is useful to classify types of estimates with respect to the time interval of the data input to the estimate for time t. If only past and present data are utilized, the estimation is a filtering process. After the entire time series of data is available for $(0,T)$, the estimate for any time $0 \leq t \leq T$ is best based on the whole data set and the estimation is a smoothing process.

Goals and Purposes

The specific uses of data assimilation depend upon the relative quality of data sets and models, and the desired purposes of the field and parameter estimates. These uses include the control of errors for state estimates, the estimation of parameters, the elucidation of real ocean dynamical processes, the design of experimental networks, and ocean monitoring and prediction.

First consider ocean prediction for scientific and practical purposes, which is the analog of numerical weather prediction. In the best case scenario, the dynamical model correctly represents both the internal dynamical processes and the responses to external forcings. Also, the observational network provides initialization data of desired accuracy. The phenomenon of loss of predictability nonetheless inhibits accurate forecasts beyond the predictability limit for the region and system. This limit for the global atmosphere is 1–2 weeks and for the mid-ocean eddy field of the north-west Atlantic on the order of weeks to months. The phenomenon is associated with the nonlinear scale transfer and growth of initial errors. The early forecasts will accurately track the state of the real ocean, but longer forecasts, although representing plausible and realistic synoptical dynamical events, will not agree with contemporary nature. However, this predictability error can be controlled by the continual assimilation of data, and this is a major use of data assimilation today.

Next, consider the case of a field estimate with adequate data but a somewhat deficient dynamical model. Assimilated data can compensate for the imperfect physics so as to provide estimates in agreement with nature. This is possible if dynamical model errors are treated adequately. For instance, if a barotropic model is considered perfect, and baroclinic real ocean data are assimilated, the field estimate will remain barotropic. Even though melded estimates with deficient models can be useful, it is of course important to attempt to correct the model dynamics.

Parameter estimation via data assimilation is making an increasingly significant impact on ocean science via the determination of both internal and external parameter values. Regional field estimates can be substantially improved by boundary condition estimation. Biological modelers have been hampered by the inability to directly measure *in situ* rates, e.g., grazing and mortality. Thus, for interdisciplinary studies, internal parameter estimation is particularly promising. For example, measurements of concentration fields of plankton together with a realistic interdisciplinary model can be used for *in situ* rate estimation.

Data-driven simulations can provide four-dimensional time series of dynamically adjusted fields which are realistic. These fields, regarded as (numerical) experimental data, can thus serve as high resolution and complete data sets for dynamical studies. Balance of terms in dynamical equations and overall budgets can be carried out to determine fluxes and rates for energy, vorticity, productivity, grazing, carbon flux, etc. Case studies can be carried out, and statistics and general processes can be inferred for simulations of sufficient duration. Of particular importance are observation system simulation experiments (OSSEs), which first entered meteorology almost 30 years ago. By subsampling the simulated 'true' ocean, future experimental networks and monitoring arrays can be designed to provide efficient field estimates of requisite accuracies. Data assimilation and OSSEs develop the concepts of data, theory, and their relationship beyond those of the classical scientific methodology. For a period of almost 300 years, scientific methodology was powerfully established on the basis of two essential elements: experiments/observations and theory/models. Today, due to powerful computers, science is based on three fundamental concepts: experiment, theory, and simulation. Since our best field and parameter estimates today are based on data assimilation, our very perception and conceptions of nature and reality require philosophical development.

It is apparent from the above discussion that marine operations and ocean management must depend on data assimilation methods. Data-driven simulations should be coupled to multipurpose management models for risk assessments and for the design of operational procedures. Regional multiscale ocean prediction and monitoring systems, designed by OSSEs, are being established to provide ongoing nowcasts and forecasts with predictability error controlled by updating. Both simple and sophisticated versions of such systems are possible and relevant.

Regional Forecasting and Dynamics

In this section, the issues and concepts introduced in the preceding sections are illustrated in the context of real-time predictions carried out in 1996 for NATO naval operations in the Strait of Sicily and for interdisciplinary multiscale research in 1998 in Massachusetts Bay. The Harvard Ocean Prediction System (HOPS) with its primitive equation dynamical model was utilized in both cases. In the Strait of Sicily (**Figure 2**), the observational network with platforms

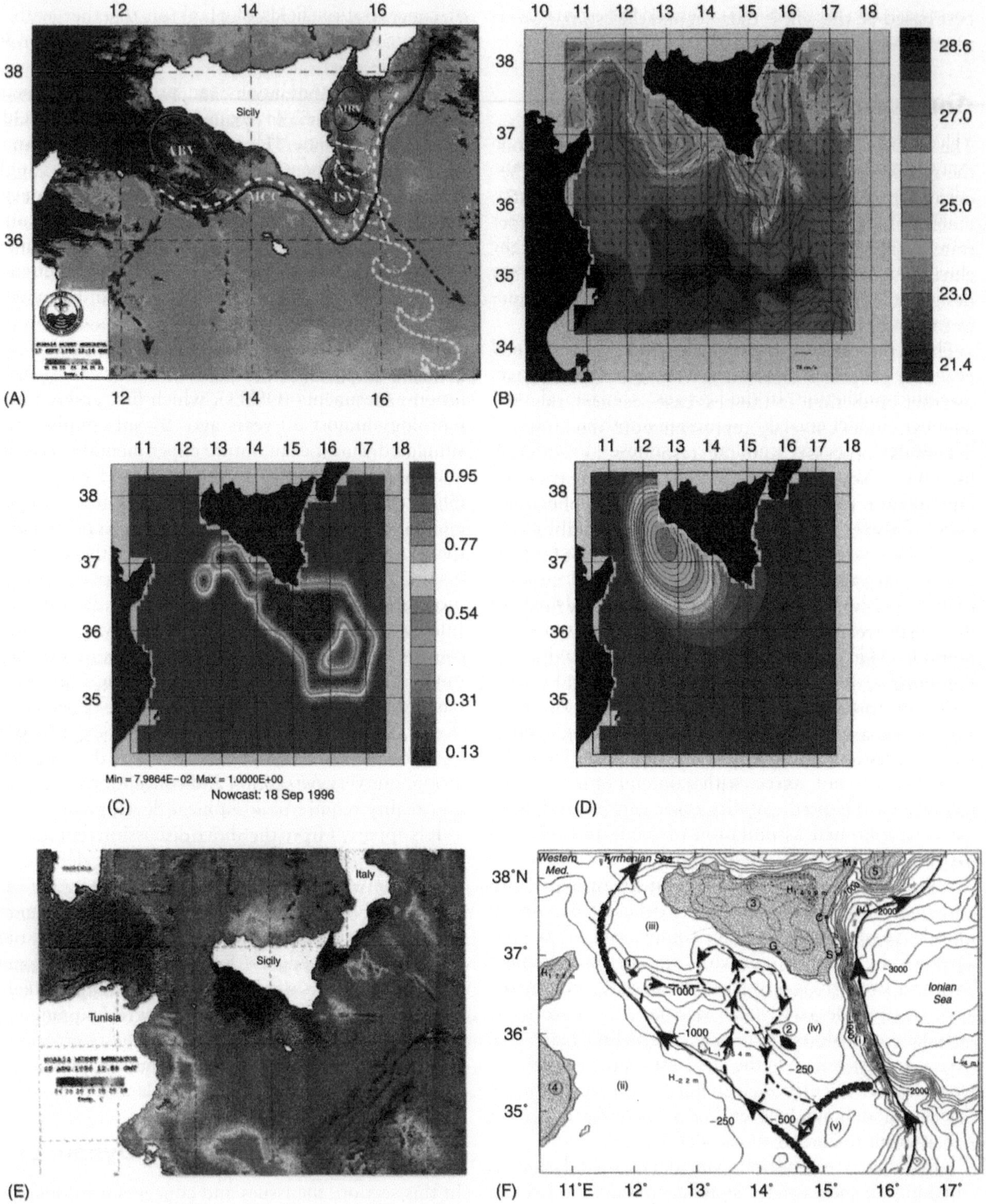

Figure 2 Strait of Sicily. (A) Schematic of circulation features and dominant variabilities. (B) Forecast of the surface temperature for 25 August 1996, overlaid with surface velocity vectors (scale arrow is 0.25 m s^{-1}). (C) Objectively analyzed surface standard error deviation associated with the aircraft sampling on 18 September 1996 (normalized from 0 to 1). (D) Surface values of the first nondimensional temperature variability mode. (E) Satellite SST distributions for 25 August 1996. (F) Main LIW pathways, features, and mixing on deep potential density anomaly iso-surface ($\sigma_\Theta = 29.05$), over bottom topography (for more details, see Lermusiaux, 1999, and Lermusiaux and Robinson, 2001).

consisting of satellites, ships, aircraft, and Lagrangian drifters, was managed by the NATO SACLANT Undersea Research Centre. In Massachusetts Bay (**Figure** 3), the observational network with platforms consisting of ships, satellites, and autonomous underwater vehicles, was provided by the Littoral Ocean Observing and Prediction System (LOOPS) project within the US National Ocean Partnership Program. The data assimilation methods used in both cases were the HOPS OI and ESSE schemes (see Estimation Theory below). In both cases the purposes of data assimilation were to provide a predictive capability, to control loss of predictability, and to infer basic underlying dynamical processes.

The dominant regional variabilities determined from these exercises and studies are schematized in **Figures 2A** and **3A**. The dominant near surface flow in the strait is the Atlantic Ionian Stream, AIS (black lines for the stream; smooth and meandering dashed lines for the common locations of fronts and wave patterns, respectively) and dominant variabilities include the location and shapes of the Adventure Bank Vortex (ABV), Maltese Channel Crest (MCC), Ionian Shelfbreak Vortex (ISV), and Messian Rise Vortex (MRV) with shifts and deformations 0(10–100 km) occurring in 0(3–5 days). The variability of the Massachusetts Bay circulation is more dramatic. The buoyancy flow-through current which enters the Bay in the north from the Gulf of Maine may have one, two or three branches, and together with associated vortices (which may or may not be present), can reverse directions within the bay. Storm events shift the pattern of the features which persist inertially between storms. Actual real-time forecast fields are depicted in **Figures 2B** and **3B**.

The existence of forecasts allows adaptive sampling, i.e., sampling efficiently related to existing structures and events. Adaptive sampling can be determined subjectively by experience or objectively by a quantitative metric. The sampling pattern associated with the temperature objective analysis error map (**Figure 2C**) reflects the flight pattern of an aircraft dropping subsurface temperature probes (AXBTs). The data were assimilated into a forecast in support of naval operations centered near the ISV (**Figure 2A**). The sampling extends to the surrounding meanders of the AIS, which will affect the current's thermal front in the operational region. The multiscale sampling of the Massachusetts Bay experiment is exemplified in **Figure 3C** by ship tracks adapted to the interactive submesoscales, mesoscales, bay-scales, and large-scales. Note that the tracks of **Figure 3D** are superimposed on a forecast of the total temperature forecast error standard deviation. The shorter track is objectively located around an error maximum. The longer track is for reduction of velocity error (not shown). Eigendecomposition of variability fields helps dynamical interpretations. This eigen-decomposition estimates and orders the directions of largest variability variance (eigenmodes) and the corresponding amplitudes (eigenvalues). The first temperature variability eigenmodes for the strait and the bay are depicted in **Figures 2D** and **3E** respectively. The former is associated with the dominant ABV variability and the latter with the location, direction, and strength of the inflow to the bay of the buoyancy current from the Gulf of Maine.

A qualitative skill score for the prediction of dominant variations of the ABV, MCC, and ISV indicated correct 2- to 3-day predictions of surface temperature 75% of the time. The scores were obtained by validation against new data for assimilation and independent satellite sea surface temperature data as shown in **Figure 2E** for the forecast of **Figure 2B**. An important kinematical and dynamical interconnection between the eastern and western Mediterranean is the deep flow of salty Levantine Intermediate Water (LIW), which was not directly measured but was inferred from data assimilative simulations (**Figure 2F**). The scientific focus of the Massachusetts Bay experiment was plankton patchiness, in particular the spatial variability of zooplankton and its relationship to physical and phytoplankton variabilities (**Figure 3B, G**). The smallest scale measurements in the bay were turbulence measurements from an AUV (**Figure 3F**), which were also used to research the assimilation in real time of subgridscale data in the primitive equation model.

Concepts and Methods

By definition (see Introduction), data assimilation in ocean sciences is an estimation problem for the ocean state, model parameters, or both. The schemes for solving this problem often relate to estimation or control theories (see below), but some approaches like direct minimization, stochastic, and hybrid methods (see below) can be used in both frameworks. Several schemes are theoretically optimal, while others are approximate or suboptimal. Although optimal schemes are preferred, suboptimal methods are generally the ones in operational use today. Most schemes are related in some fashion to least-squares criteria which have had great success. Other criteria, such as the maximum likelihood, minimax criterion or associated variations might be more appropriate when data are very noisy and sparse, and when probability density functions are multimodal (see Stochastic and hybrid models below). Parameters are assumed from here on to be

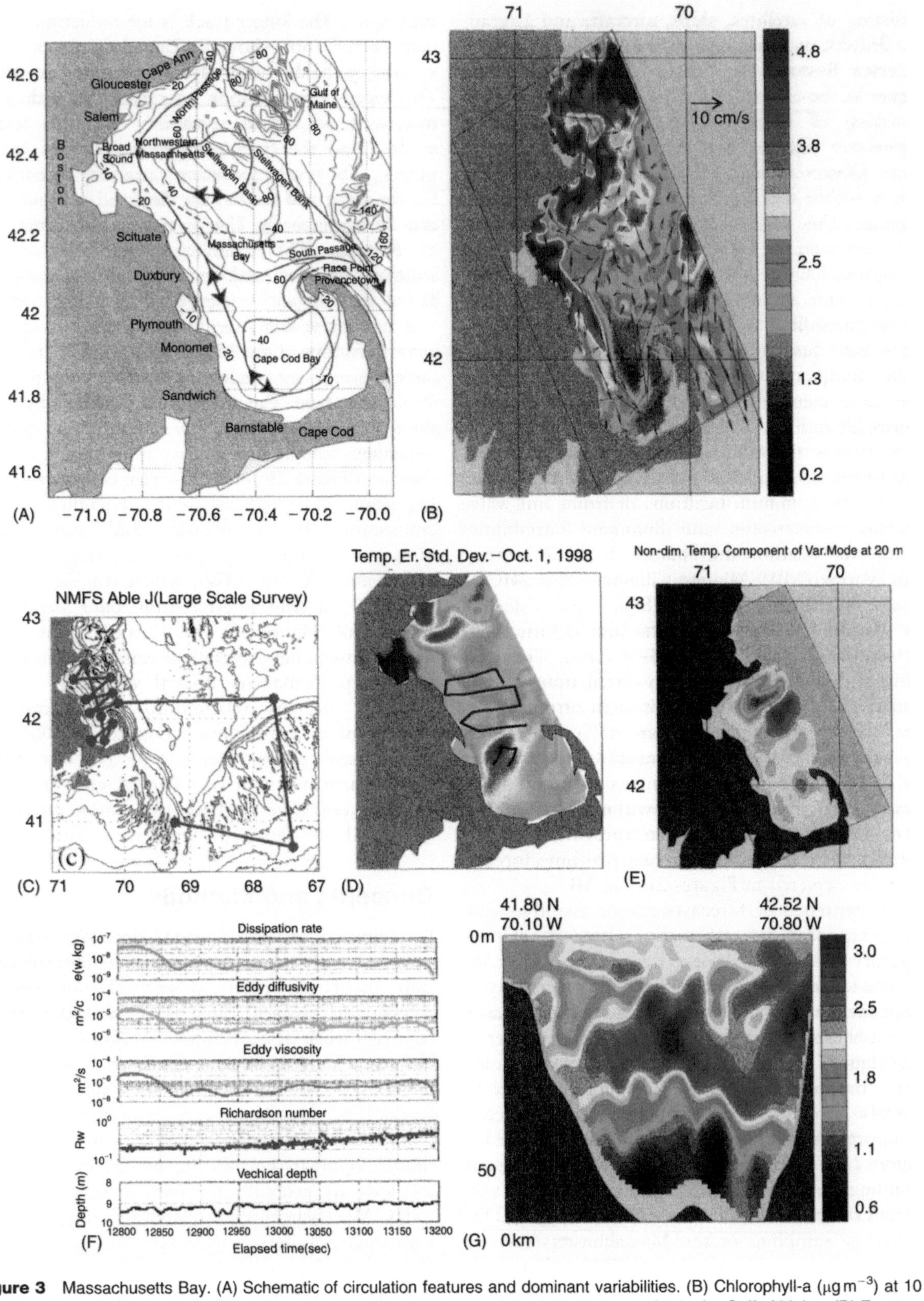

Figure 3 Massachusetts Bay. (A) Schematic of circulation features and dominant variabilities. (B) Chlorophyll-a ($\mu g\,m^{-3}$) at 10 m, with overlying velocity vectors. (C) Sampling pattern for the bay scales and external large-scales in the Gulf of Maine. (D) Forecast of the standard error deviation for the surface temperature (from 0°C in dark blue to a maximum of 0.7°C in red), with tracks for adaptive sampling. (E) 20 m values of the temperature component of the first nondimensional physical variability mode. (F) AUV turbulence data (Naval Underwater Warfare Center (NUWC)). (G) Vertical section of zooplankton ($\mu M\,m^{-3}$) along the entrance of Massachusetts Bay (for more details, see Robinson and the LOOPS group, 1999, and Lermusiaux, 2001).

included in the vector of state variables. For more detailed discussions, the reader is referred to the article published by Robinson *et al.* in 1998 (see Further Reading section).

Estimation Theory

Estimation theory computes the state of a system by combining all available reliable knowledge of the system including measurements and theoretical models. The *a priori* hypotheses and melding or estimation criterion are crucial since they determine the influence of dynamics and data onto the state estimate.

At the heart of estimation theory is the Kalman filter, derived in 1960. It is the sequential, unbiased, minimum error variance estimate based upon a linear combination of all past measurements and dynamics. Its two steps are: (1) the forecast of the state vector and of its error covariance, and (2) the data-forecast melding and error update, which include the linear combination of the dynamical forecast with the difference between the data and model predicted values for those data (i.e., data residuals).

The Kalman smoother uses the data available before and after the time of interest. The smoothing is often carried out by propagating the future data information backward in time, correcting an initial Kalman filter estimate using the error covariances and adjoint dynamical transition matrices, which is usually demanding on computational resources.

In a large part because of the linear hypothesis and costs of these two optimal approaches, a series of approximate or suboptimal schemes have been employed for ocean applications. They are now described, from simple to complex.

Direct insertion consists of replacing forecast values at all data points by the observed data which are assumed to be exact. The **blending** estimate is a scalar linear combination, with user-assigned weights, of the forecast and data values at all data points. The **nudging** or **Newtonian relaxation scheme** 'relaxes' the dynamical model towards the observations. The coefficients in the relaxation can vary in time but, to avoid disruptions, cannot be too large. They should be related to dynamical scales and *a priori* estimates of model and data errors.

In **optimal interpolation** (OI), the matrix weighting the data residuals, or gain matrix, is empirically assigned. If the assigned OI gain is diagonal, OI and nudging schemes can be equivalent. However, the OI gain is usually not diagonal, but a function of empirical correlation and error matrices.

The **method of successive corrections** performs multiple but simplified linear combination of the data and forecast. Conditions for convergence to the Kalman filter have been derived, but in practice only a few iterations are usually performed. Frequently, the scales or processes of interest are corrected one after the other, e.g., large-scale first, then mesoscale.

Control Theory

All control theory or variational approaches perform a global time-space adjustment of the model solution to all observations and thus solve a smoothing problem. The goal is to minimize a cost function penalizing misfits between the data and ocean fields, with the constraints of the model equations and their parameters. The misfits are interpreted as part of the unknown controls of the ocean system. Similar to estimation theory, control theory results depend on *a priori* assumptions for the control weights. The dynamical model can be either considered as a strong or weak constraint. Strong constraints correspond to the choice of infinite weights for the model equations; the only free variables are the initial conditions, boundary conditions and/or model parameters. A rational choice for the cost function is important. A logical selection corresponds to dynamical model (data) weights inversely proportional to *a priori* specified model (data) errors.

In an '**adjoint method**', the dynamical model is a strong constraint. One penalty in the cost function weights the uncertainties in the initial conditions, boundary conditions, and parameters with their respective *a priori* error covariances. The other is the sum over time of data-model misfits, weighted by measurement error covariances. A classical approach to solve this constrained optimization is to use Lagrange multipliers. This yields Euler-Lagrange equations, one of which is the so-called adjoint equation. An iterative algorithm for solving these equations has often been termed the adjoint method. It consists of integrating the forward and adjoint equations successively. Minimization of the gradient of the cost function at the end of each iteration leads to new initial, boundary, and parameter values. Another iteration can then be started, and so on, until the gradient is small enough.

Expanding classic inverse problems to the weak constraint fit of both data and dynamics leads to **generalized inverse problems**. The cost function is usually as in adjoint methods, except that a third term now consists of dynamical model uncertainties weighted by *a priori* model error covariances. In the Euler-Lagrange equations, the dynamical model uncertainties thus couple the state evolution with the adjoint evolution. The representer method is an algorithm for solving such problems.

Direct Minimization Methods

Such methods directly minimize cost functions similar to those of generalized inverse problems, but often without using the Euler-Lagrange equations. **Descent methods** iteratively determine directions locally 'descending' along the cost function surface. At each iteration, a minimization is performed along the current direction and a new direction is found. Classic methods to do so are the steepest descent, conjugate-gradient, Newton, and quasi-Newton methods. A drawback for descent methods is that they are initialization sensitive. For sufficiently nonlinear cost functions, they are restarted to avoid local minima.

Simulated annealing schemes are based on an analogy to the way slowly cooling solids arrange themselves into a perfect crystal, with a minimum global energy. To simulate this relatively random process, a sequence of states is generated such that new states with lower energy (lower cost) are always accepted, while new states with higher energy (higher cost) are accepted with a certain probability.

Genetic algorithms are based upon searches generated in analogy to the genetic evolution of natural organisms. They evolve a population of solutions mimicking genetic transformations such that the likelihood of producing better data-fitted generations increases. Genetic algorithms allow nonlocal searches, but convergence to the global minimum is not assured due to the lack of theoretical base.

Stochastic and Hybrid Methods

Stochastic methods are based on nonlinear stochastic dynamical models and stochastic optimal control. Instead of using brute force like descent algorithms, they try to solve the conditional probability density equation associated with ocean models. Minimum error variance, maximum likelihood or minimax estimates can then be determined from this probability density. No assumptions are required, but for large systems, parallel machines are usually employed to carry out Monte Carlo ensemble calculations.

Hybrid methods are combinations of previously discussed schemes, for both state and parameter estimation; for example, error subspace statistical estimation (ESSE) schemes. The main assumption of such schemes is that the error space in most ocean applications can be efficiently reduced to its essential components. Smoothing problems based on Kalman ideas, but with nonlinear stochastic models and using Monte Carlo calculations, can then be solved. Combinations of variational and direct minimization methods are other examples of hybrid schemes.

Examples

This section presents a series of recent results that serve as a small but representative sample of the wide range of research carried out as data assimilation was established in physical oceanography.

General Circulation from Inverse Methods

The central idea is to combine the equations governing the oceanic motion and relevant oceanic tracers with all available noisy observations, so as to estimate the large-scale steady-state total velocities and related internal properties and their respective errors. The work of Martel and Wunsch in 1993 exemplifies the problem. The three-dimensional circulation of the North Atlantic (**Figure 4A**) was studied for the period 1980–85. The observations available consisted of objective analyses of temperature, salinity, oxygen, and nutrients data; climatological ocean–atmosphere fluxes of heat, water vapor, and momentum; climatological river runoffs; and current meter and float records. These data were obtained with various sensors and platforms, on various resolutions, as illustrated by **Figure 4B**. A set of steady-state equations were assumed to hold *a priori*, up to small unknown noise terms. The tracers were advected and diffused. The advection velocities were assumed in geostrophic thermal–wind balance, except in the top layer where Ekman transport was added. Hydrostatic balance and mass continuity were assumed. The problem is inverse because the tracers and thermal wind velocities are known; the unknowns are the fields of reference level velocities, vertical velocity, and tracer mixing coefficients.

Discrete finite-difference equations were integrated over a set of nested grids of increasing resolutions (**Figure 4A**). The flows and fluxes at the boundaries of these ocean subdivisions were computed from the data (at 1° resolution). The resulting discrete system contained *c.* 29 000 unknowns and 9000 equations. It was solved using a tapered (normalized) least-squares method with a sparse conjugate-gradient algorithm. The estimates of the total flow field and of its standard error are plotted on **Figure 1C** and **D**. The Gulf Stream, several recirculation cells and the Labrador current are present. In 1993, such a rigorous large-scale, dense, and eclectic inversion was an important achievement.

Global versus Local Data Assimilation via Nudging

Malanotte-Rizzoli and Young in 1994 investigated the effectiveness of various data sets to correct and control errors. They used two data sets of different types and resolutions in time and space in the Gulf

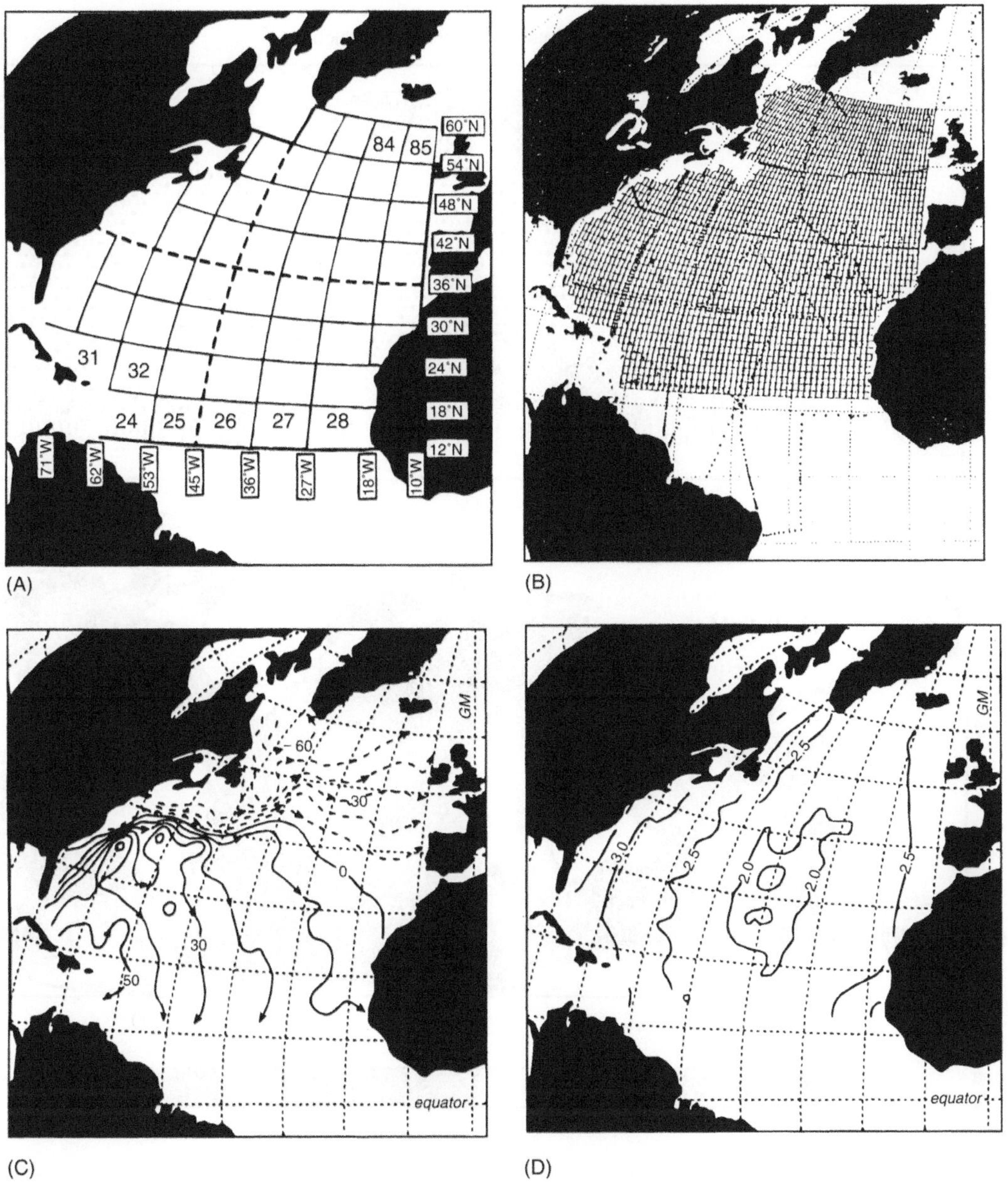

Figure 4 (A) Domain of the model used in the inverse computations. Weak dynamical constraints were imposed on the flow and tracers, and integrated over a set of nested grids, from the full domain (heavy solid lines) to an ensemble of successive divisions (e.g., dashed lines) reaching at the smallest scales the size of the boxes labeled by numbers. (B) Locations of the stations where the hydrographic and chemical component of the data set were collected (model grid superposed). (C) Inverse estimate of the absolute sea surface topography in centimeters (contour interval is 10 cm). (D) Inverse estimate of the standard error deviation (in cm) of the sea surface topography shown in (C). (Reproduced with permission from Martel and Wunsch, 1993.).

Stream region, at mesoscale resolution and for periods of the order of 3 months, over a large-scale domain referred to as global scale.

One objective was to assimilate data of high quality, but collected at localized mooring arrays, and to investigate the effectiveness of such data in improving the realistic attributes of the simulated ocean fields. If successful, such estimates allow for dynamical and process studies. The global data consisted of biweekly fields of sea surface dynamic height, and of temperature and salinity in three dimensions, over the entire region, as provided by the Optimal Thermal Interpolation Scheme (OTIS) of the US Navy Fleet Numerical Oceanography Center. The local data were daily current velocities from two mooring arrays. The dynamical model consisted of primitive equations (Rutgers), with a suboptimal nudging scheme for the assimilation.

The global and local data were first assimilated alone, and then together. The ‘gentle’ assimilation of

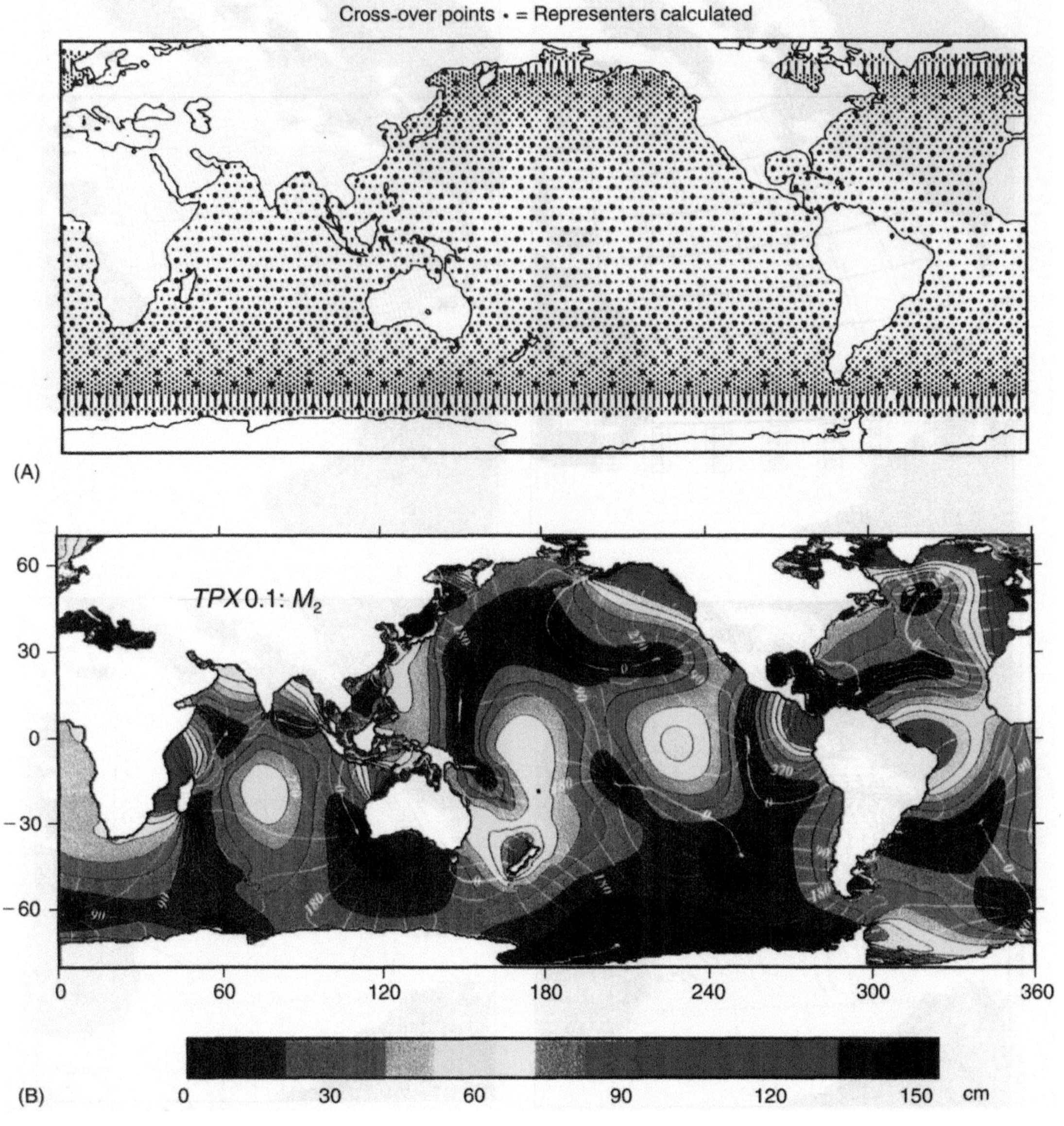

Figure 5 (A) TOPEX/POSEIDON crossover points and subsampling. Representers were calculated only for the windowed subset of 986 satellite crossover points (large filled dots), but differences from all 6355 crossover points (small dots and large filled dots) were included in the data-misfit penalty. (B) Generalized inverse estimate of the amplitude and phase of the M_2 tidal constituent. The phase isolines are plotted in white over color-filled contours of the amplitude. Contour interval is 10 cm for amplitude and 30° for phase. (Reproduced with permission from Egbert *et al.* 1994.).

the spatially dense global OTIS data was necessary for the model to remain on track during the 3-month period. The 'strong' assimilation of the daily but local data from the Synoptic Ocean Prediction SYNOP was required to achieve local dynamical accuracies, especially for the velocities.

Small-scale Convection and Data Assimilation

In 1994, Miller *et al.* addressed the use of variational or control theory approaches (see earlier) to assimilate data into dynamical models with highly nonlinear convection. Because of limited data and computer requirements, most practical ocean models cannot resolve motions that result from static instabilities of the water column; these motions and effects are therefore parameterized. A common parameterization is the so-called convective adjustment. This consists of assigning infinite mixing coefficients (e.g., heat and salt conductivities) to the water at a given level that has higher density than the water just below. This is carried out by setting the densities of the two parcels to a unique value in such a way that heat and mass are conserved. In a numerical model, at every time step, water points are checked and all statically unstable profiles replaced by stable ones.

The main issues of using such convective schemes with variational data assimilation are that: (1) the dynamics is no longer governed by smooth equations, which often prevents the simple definition of adjoint equations; (2) the optimal ocean fields may evolve through 'nonphysical' states of static instability; and (3), the optimization is nonlinear, even if the dynamics are linear. Ideally, the optimal fields should be statically stable. This introduces a set of inequality constraints to satisfy. An idealized problem was studied so as to provide guidance for realistic situations. A simple variational formulation had several minima and at times produced evolutions with unphysical behavior. Modifications that led to more meaningful solutions and suggestions for algorithms for realistic models were discussed. One option is the 'weak' static stability constraint: a penalty that ensures approximate static stability is added to the cost function with a very small error or large weight. In that case, static stability can be violated, but in a limited fashion. Another option is the 'strong constraint' form of static stability which can be enforced via Lagrange multipliers. Convex programming methods which explicitly account for inequality constraints could also be utilized.

Global Ocean Tides Estimated from Generalized Inverse Methods

In 1994, Egbert *et al.* estimated global ocean tides using a generalized inverse scheme with the intent of removing these tides from the data collected by the TOPEX/POSEIDON satellite and thus allowing the study of subtidal ocean dynamics.

A scheme for the inversion of the satellite crossover data for multiple tidal constituents was applied to 38 cycles of the data, leading to global estimates of the four principal tidal constituents (M_2, S_2, K_1 and O_1) at about 1° resolution. The dynamical model was the linearized, barotropic shallow water equations, corrected for the effects of ocean self-attraction and tidal loading, the state variables being the horizontal velocity and sea surface height fields. The data sets were linked to measurement models and comprehensive error models were derived.

The generalized inverse tidal problem was solved by the representer method. Representer functions are related to Green's functions: they link a given datum to all values of the state variables over the period considered. These representers were computed by solving the Euler-Lagrange equations in parallel. The size of the problem was reduced by winnowing out the full set of 6350 crossovers to an evenly spaced subset of 986 points (see **Figure 5A**). The resulting representer matrix was then reduced by singular value decomposition.

The amplitude and phase estimates for the M_2 constituent are shown in **Figure 5B**. The M_2 fields are qualitatively similar to previous results and amphidromes are consistent. However, when compared with previous tidal model estimates, the inversion result is noticeably smoother and in better agreement with altimetric and ground truth data.

Conclusions

The melding of data and dynamics is a powerful, novel, and versatile methodology for parameter and field estimation. Data assimilation must be anticipated both to accelerate research progress in complex, modern multiscale interdisciplinary ocean science, and to enable marine technology and maritime operations that would otherwise not be possible.

Acknowledgments

We thank Ms. G. Sweetland and Ms. M. Armstrong for help in preparing the manuscript, and the ONR for partial support.

See also

Coastal Circulation Models. Current Systems in the Mediterranean. Florida Current, Gulf Stream and Labrador Current. Forward Problem in Numerical Models. Inverse Models. Mesoscale Eddies. Arctic Ocean Circulation. Regional and Shelf Models.

Further Reading

Anderson D and Willebrand J (eds.) (1989) *Oceanic Circulation ModelsC: ombining Data and Dynamics*. Dordrecht: Kluwer Academic.

Bennett AF (1992) *Inverse Methods in Physical Oceanography. Cambridge Monographs on Mechanics and Applied Mathematics*. Cambridge: Cambridge University Press.

Brasseur P and Nihoul JCJ (eds.) (1994) *Data assimilationt: ools for modelling the ocean in a global change perspective. Series IG: lobal Environmental Change*, 19. Berlin: Springer-Verlag. NATO ASI Series.

Egbert GD, Bennett AF, and Foreman MGG (1994) TOPEX/POSEIDON tides estimated using a global inverse model. *Journal of Geophysical Research* 24: 821–824 852.

Haidvogel DB and Robinson AR (eds) (1989) Special issue on data assimilation. *Dynamics of Atmospheres and Oceans* 13: 171–517.

Lermusiaux PFJ (1999) Estimation and study of mesoscale variability in the Strait of Sicily. *Dynamics of Atmospheres and Oceans* 29: 255–303.

Lermusiaux PFJ (2001) Evolving the subspace of the three-dimensional multiscale ocean variability: Massachusetts Bay. *Journal of Marine Systems*. In press.

Lermusiaux PFJ and Robinson AR (1999) Data assimilation via error subspace statistical estimation, Part IT: heory and schemes. *Monthly Weather Review* 7: 1385–1407.

Lermusiaux PFJ and Robinson AR (2001) Features of dominant mesoscale variability, circulation patterns and dynamics in the Strait of Sicily. *Deep Sea Research*, Part I. In press.

Malanotte-Rizzoli P and Young RE (1995) Assimilation of global versus local data sets into a regional model of the Gulf Stream system: I. Data effectiveness. *Journal of Geophysical Research* 24: 773–796.

Malanotte-Rizzoli P (ed.) (1996) *Modern Approaches to Data Assimilation in Ocean Modeling, Elsevier Oceanography Series*. The Netherlands: Elsevier Science.

Martel F and Wunsch C (1993) The North Atlantic circulation in the early 1980s – an estimate from inversion of a finite-difference model. *Journal of Physical Oceanography* 23: 898–924.

Miller RN, Zaron EO, and Bennett AF (1994) Data assimilation in models with convective adjustment. *Monthly Weather Review* 122: 2607–2613.

Robinson AR, Lermusiaux PFJ, and Sloan NQ III (1998) Data assimilation. In: Brink KH and Robinson AR (eds.) *The SeaT: he Global Coastal Ocean I, Processes and Methods*, 10. New York: John Wiley and Sons.

Robinson AR (1999) Forecasting and simulating coastal ocean processes and variabilities with the Harvard Ocean Prediction System. In: Mooers CNK (ed.) *Coastal Ocean Prediction, AGU Coastal and Estuarine Study Series*, pp. 77–100. Washington: AGU Press.

Robinson AR and the LOOPS group (1999) *Real-time forecasting of the multidisciplinary coastal ocean with the Littoral Ocean Observing and Predicting System (LOOPS). Third Conference on Coastal Atomspheric and Oceanic Prediction and Processes*. New Orleans, LA: American Meterological Society, 30*35. (3-5 Nov 1999).

Robinson AR and Sellschopp J (2000) Rapid assessment of the coastal ocean environment. In: Pinardi N and Woods JD (eds.) *Ocean ForecastingC: onceptual Basis and Applications*. London: Springer-Verlag.

Wunsch C (1996) *The Ocean Circulation Inverse Problem*. Cambridge: Cambridge University Press.

APPENDICES

APPENDIX 1. SI UNITS AND SOME EQUIVALENCES

Wherever possible the units used are those of the International System of Units (SI). Other "conventional" units (such as the liter or calorie) are frequently used, especially in reporting data from earlier work. Recommendations on standardized scientific terminology and units are published periodically by international committees, but adherence to these remains poor in practice. Conversion between units often requires great care.

The base SI units

Quantity	*Unit*	*Symbol*
Length	meter	m
Mass	kilogram	kg
Time	second	s
Electric current	ampere	A
Thermodynamic temperature	kelvin	K
Amount of substance	mole	mol
Luminous intensity	candela	cd

Some SI derived and supplementary units

Quantity	*Unit*	*Symbol*	*Unit expressed in base or other derived units*
Frequency	hertz	Hz	s^{-1}
Force	newton	N	$kg\,m\,s^{-2}$
Pressure, stress	pascal	Pa	$N\,m^{-2}$
Energy, work, quantity of heat	joule	J	N m
Power	watt	W	$J\,s^{-1}$
Electric charge, quantity of electricity	coulomb	C	A s
Electric potential, potential difference, electromotive force	volt	V	$J\,C^{-1}$
Electric capacitance	farad	F	$C\,V^{-1}$
Electric resistance	ohm	ohm (Ω)	$V\,A^{-1}$
Electric conductance	Siemens	S	Ω^{-1}
Magnetic flux	weber	Wb	V s
Magnetic flux density	tesla	T	$Wb\,m^{-2}$
Inductance	henry	H	$Wb\,A^{-1}$
Luminous flux	lumen	lm	cd sr
Illuminance	lux	lx	$lm\,m^{-2}$
Activity (of a radionuclide)	becquerel	Bq	s^{-1}
Absorbed dose, specific energy	gray	Gy	$J\,kg^{-1}$
Dose equivalent	sievert	Sv*	$J\,kg^{-1}$
Plane angle	radian	rad	
Solid angle	steradian	sr	

*Not to be confused with Sverdrup conventionally used in oceanography: see SI Equivalences of Other Units.

SI base units and derived units may be used with multiplying prefixes (with the exception of kg, though prefixes may be applied to gram $= 10^{-3}$kg; for example, 1 Mg $= 10^6$ g $= 10^6$kg)

Prefixes used with SI units

Prefix	*Symbol*	*Factor*
yotta	Y	10^{24}
zetta	Z	10^{21}
exa	E	10^{18}
peta	P	10^{15}
tera	T	10^{12}
giga	G	10^{9}
mega	M	10^{6}
kilo	k	10^{3}
hecto	h	10^{2}
deca	da	10
deci	d	10^{-1}
centi	c	10^{-2}
milli	m	10^{-3}
micro	μ	10^{-6}
nano	n	10^{-9}
pico	p	10^{-12}
femto	f	10^{-15}
atto	a	10^{-18}
zepto	z	10^{-21}
yocto	y	10^{-24}

SI Equivalences of Other Units

Physical quantity	*Unit*	*Equivalent*	*Reciprocal*
Length	nautical mile (nm)	1.85318 km	km = 0.5396 nm
Mass	tonne (t)	10^3 kg = 1 Mg	
Time	min	60 s	
	h	3600 s	
	day or d	86 400 s	s $= 1.1574 \times 10^{-5}$ day
	y	3.1558×10^7 s	s $= 3.1688 \times 10^{-8}$ y
Temperature	°C	°C = K − 273.15	
Velocity	knot (1 nm h^{-1})	0.51477 m s^{-1}	m s^{-1} = 1.9426 knot
		44.5 km d^{-1}	
		16 234 km y^{-1}	
Density	gm cm^{-3}	tonne m^{-3} = 10^3 kg m^{-3}	
Force	dyn	10^{-5} N	
Pressure	dyn cm^{-2}	10^{-1} N m^{-2} = 10^{-1} Pa	
	bar	10^5 N m^{-2} = 10^5 Pa	
	atm (standard atmosphere)	101 325 N m^{-2} = 101.325 kPa	
Energy	erg	10^{-7} J	
	cal (I.T.)	4.1868 J	
	cal (15°C)	4.1855 J	
	cal (thermochemical)	4.184 J	J = 0.239 cal

(*Note*: The last value is the one used for subsequent conversions involving calories.)

Energy flux	langley (ly) min^{-1} = cal cm^{-2} min^{-1}	697 W m^{-2}	W m^{-2} = 1.434 × 10^{-3} ly min^{-1}
	ly h^{-1}	11.6 W m^{-2}	W m^{-2} = 0.0860 ly h^{-1}
	ly d^{-1}	0.484 W m^{-2}	W m^{-2} = 2.065 ly d^{-1}
	kcal cm^{-2} y^{-1}	1.326 W m^{-2}	W m^{-2} = 0.754 kly y^{-1}
Volume flux	Sverdrup	10^6 m^3 s^{-1} 3.6 km^3 h^{-1}	
Latent heat	cal g^{-1}	4184 J kg^{-1}	J kg^{-1} = 2.39 × 10^{-4} cal g^{-1}
Irradiance	Einstein m^{-2} s^{-1} (mol photons m^{-2} s^{-1})		

*Most values are taken from or derived from *The Royal Society Conference of Editors Metrication in Scientific Journals*, 1968, The Royal Society, London.

The SI units for pressure is the pascal (1 Pa = 1 N m^{-2}). Although the bar (1 bar = 10^5 Pa) is also retained for the time being, it does not belong to the SI system. Various texts and scientific papers still refer to gas pressure in units of the torr (symbol: Torr), the bar, the conventional millimetre of mercury (symbol: mmHg), atmospheres (symbol: atm), and pounds per square inch (symbol: psi) – although these units will gradually disappear (see Conversions between Pressure Units).

Irradiance is also measured in W m^{-2}. Note: 1 mol photons = 6.02 × 10^{23} photons.

The SI unit used for the amount of substance is the mole (symbol: mol), and for volume the SI unit is the cubic metre (symbol: m^3). It is technically correct, therefore, to refer to concentration in units of molm^3. However, because of the volumetric change that sea water experiences with depth, marine chemists prefer to express sea water concentrations in molal units, mol kg^{-1}.

Conversions between Pressure Units

	Pa	kPa	bar	atm	Torr	psi
1 Pa =	1	10^{-3}	10^{-5}	9.869 23 × 10^{-6}	7.500 62 × 10^{-3}	1.450 38 × 10^{-4}
1 kPa =	10^3	1	10^{-2}	9.869 23 × 10^{-3}	7.500 62	0.145 038
1 bar =	10^5	10^2	1	0.986 923	750.062	145.038
1 atm =	101 325	101.325	1.013 25	1	760	14.6959
1 Torr =	133.322	0.133 322	1.333 22 × 10^{-3}	1.315 79 × 10^{-3}	1	1.933 67 × 10^{-2}
1 psi	6894.76	6.894 76	6.894 76 × 10^{-2}	6.804 60 × 10^{-2}	51.715 07	1

psi = pounds force per square inch.
1 mmHg = 1 Torr to better than 2 × 10^{-7} Torr.

APPENDIX 2. USEFUL VALUES

Molecular mass of dry air, $m_a = 28.966$
Molecular mass of water, $m_w = 18.016$
Universal gas constant, $R = 8.31436\,\mathrm{J\,mol^{-1}\,K^{-1}}$
Gas constant for dry air, $R_a = R/m_a = 287.04\,\mathrm{J\,kg^{-1}\,K^{-1}}$
Gas constant for water vapor, $R_v = R/m_w = 461.50\,\mathrm{J\,kg^{-1}\,K^{-1}}$
Molecular weight ratio $\varepsilon \equiv m_w/m_a = R_a/R_v = 0.62197$
Stefan's constant $\sigma = 5.67 \times 10^{-8}\,\mathrm{W\,m^{-2}\,K^{-4}}$
Acceleration due to gravity, $g\,(\mathrm{m\,s^{-2}})$ as a function of latitude φ and height $z\,(\mathrm{m})$

$$g = (9.78032 + 0.005172\sin^2\varphi - 0.00006\sin^2 2\varphi)(1 + z/a)^{-2}$$

Mean surface value, $\bar{g} = \int_0^{\pi/2} g\cos\varphi\,d\varphi = 9.7976$
Radius of sphere having the same volume as the Earth, $a = 6371\,\mathrm{km}$ (equatorial radius = 6378 km, polar radius = 6357 km)
Rotation rate of earth, $\Omega = 7.292 \times 10^{-5}\,\mathrm{s^{-1}}$
Mass of earth = $5.977 \times 10^{24}\,\mathrm{kg}$
Mass of atmosphere = $5.3 \times 10^{18}\,\mathrm{kg}$
Mass of ocean = $1400 \times 10^{18}\,\mathrm{kg}$
Mass of ground water ≑ $15.3 \times 10^{18}\,\mathrm{kg}$
Mass of ice caps and glaciers = $43.4 \times 10^{18}\,\mathrm{kg}$
Mass of water in lakes and rivers = $0.1267 \times 10^{18}\,\mathrm{kg}$
Mass of water vapor in atmosphere = $0.0155 \times 10^{18}\,\mathrm{kg}$
Area of earth = $5.10 \times 10^{14}\,\mathrm{m^2}$
Area of ocean = $3.61 \times 10^{14}\,\mathrm{m^2}$
Area of land = $1.49 \times 10^{14}\,\mathrm{m^2}$
Area of ice sheets and glaciers = $1.62 \times 10^{13}\,\mathrm{m^2}$
Area of sea ice = $1.9 \times 10^{13}\,\mathrm{m^2}$ in March and $2.9 \times 10^{13}\,\mathrm{m^2}$ in September (averaged between 1979 and 1987)

APPENDIX 10. BATHYMETRIC CHARTS OF THE OCEANS

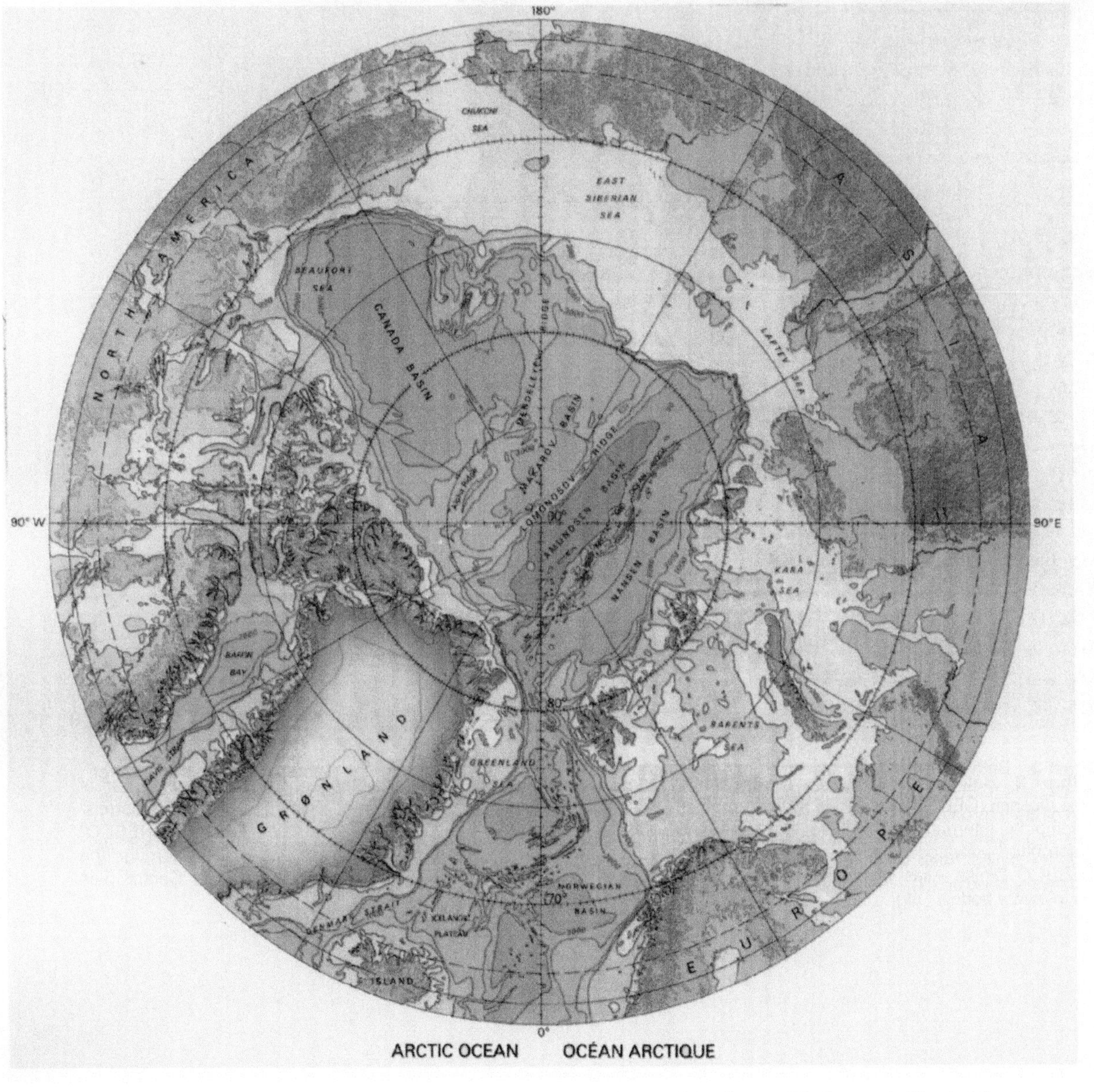

Chart 1 Bathymetric chart of the Arctic Ocean. Based on the World Sheet of the General Bathymetric Chart of the Oceans (GEBCO), published by the Canadian Hydrographic Service, Ottawa, Canada, 1984; reproduced with permission of the International Hydrographic Organization and the Intergovernmental Oceanographic Commission (of UNESCO) (see http://www.ngdc.noaa.gov/mgg/gebco). The GEBCO bathymetry is currently maintained and updated through the GEBCO Digital Atlas which is published periodically on CD-ROM by the British Oceanographic Data Centre (see http://www.bodc.ac.uk).

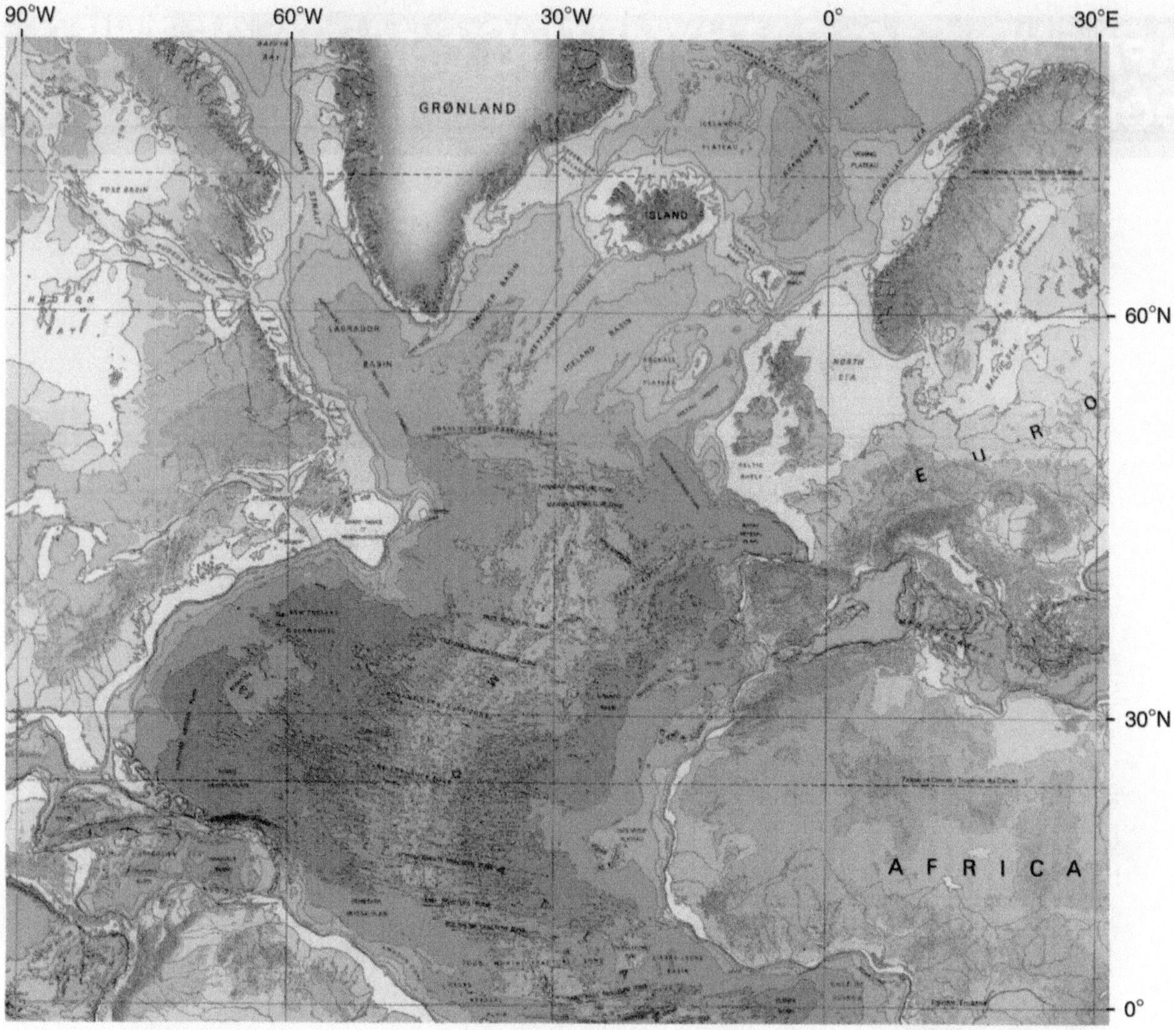

Chart 2 Bathymetric chart of the North Atlantic Ocean. Based on the World Sheet of the General Bathymetric Chart of the Oceans (GEBCO), published by the Canadian Hydrographic Service, Ottawa, Canada, 1984; reproduced with permission of the International Hydrographic Organization and the Intergovernmental Oceanographic Commission (of UNESCO) (see http://www.ngdc.noaa.gov/mgg/gebco). The GEBCO bathymetry is currently maintained and updated through the GEBCO Digital Atlas which is published periodically on CD-ROM by the British Oceanographic Data Centre (see http://www.bodc.ac.uk).

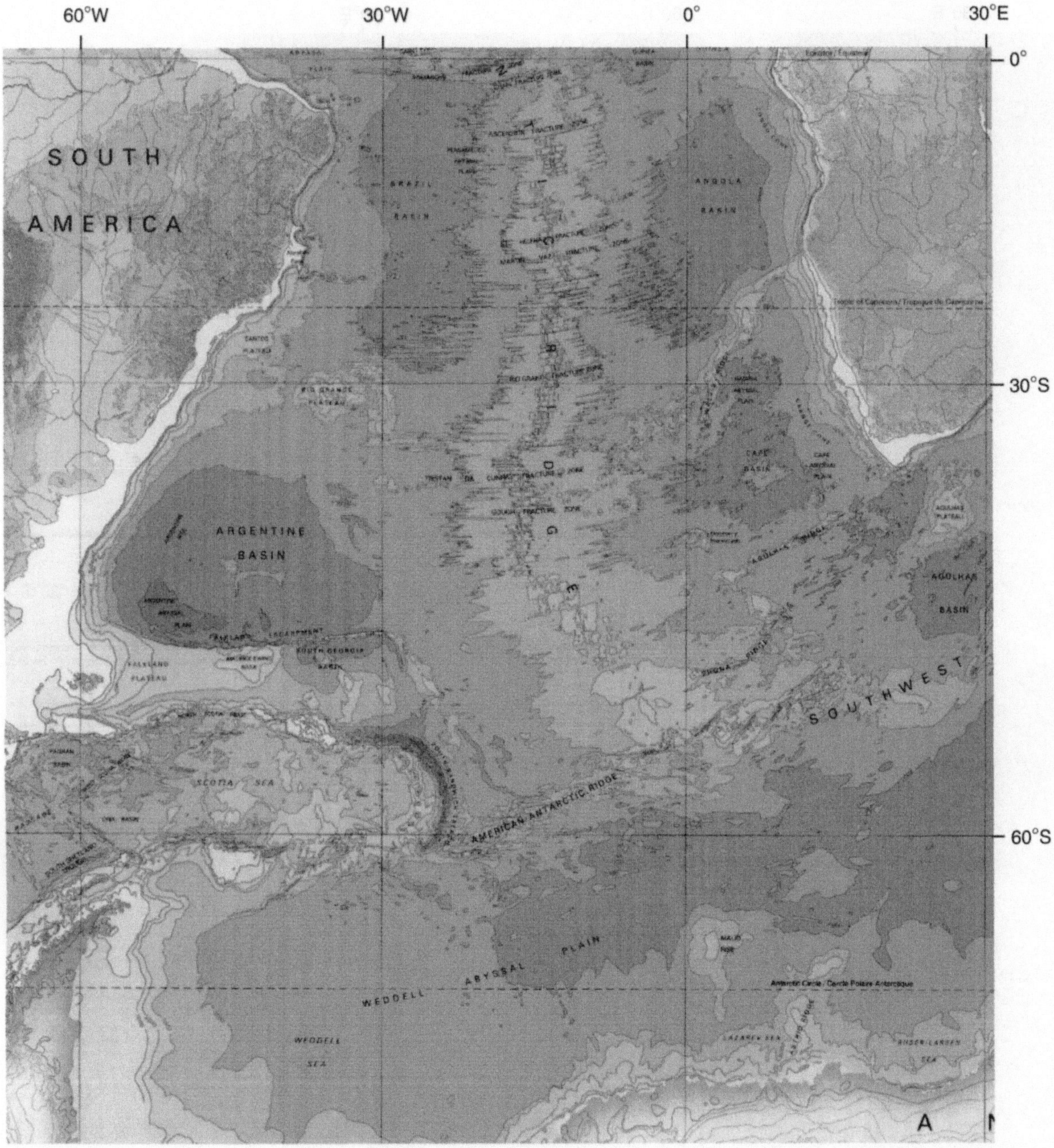

Chart 3 Bathymetric chart of the South Atlantic Ocean. Based on the World Sheet of the General Bathymetric Chart of the Oceans (GEBCO), published by the Canadian Hydrographic Service, Ottawa, Canada, 1984; reproduced with permission of the International Hydrographic Organization and the Intergovernmental Oceanographic Commission (of UNESCO) (see http://www.ngdc.noaa.gov/mgg/gebco). The GEBCO bathymetry is currently maintained and updated through the GEBCO Digital Atlas which is published periodically on CD-ROM by the British Oceanographic Data Centre (see http://www.bodc.ac.uk).

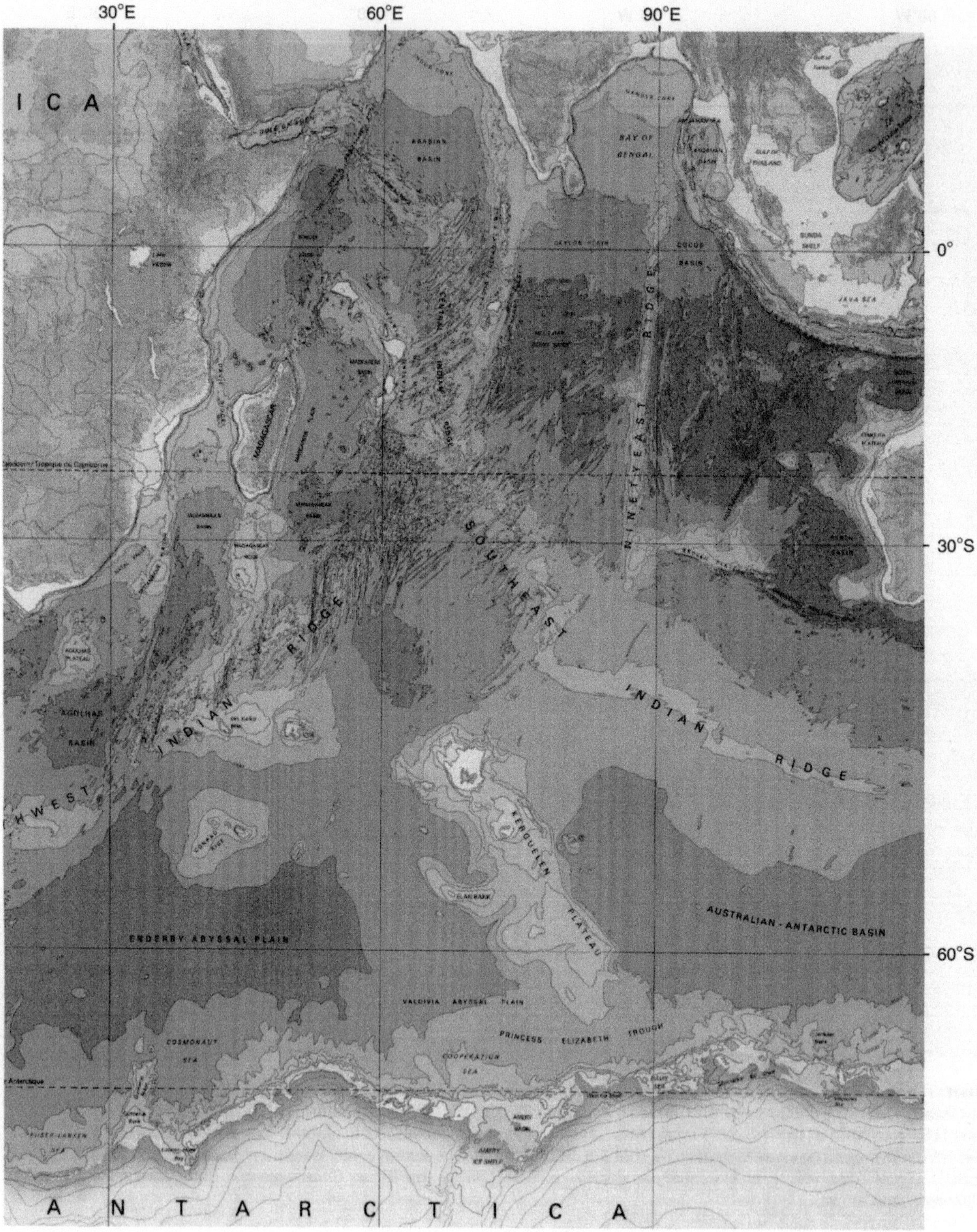

Chart 4 Bathymetric chart of the Indian Ocean. Based on the World Sheet of the General Bathymetric Chart of the Oceans (GEBCO), published by the Canadian Hydrographic Service, Ottawa, Canada, 1984; reproduced with permission of the International Hydrographic Organization and the Intergovernmental Oceanographic Commission (of UNESCO) (see http://www.ngdc.noaa.gov/mgg/gebco). The GEBCO bathymetry is currently maintained and updated through the GEBCO Digital Atlas which is published periodically on CD-ROM by the British Oceanographic Data Centre (see http://www.bodc.ac.uk).

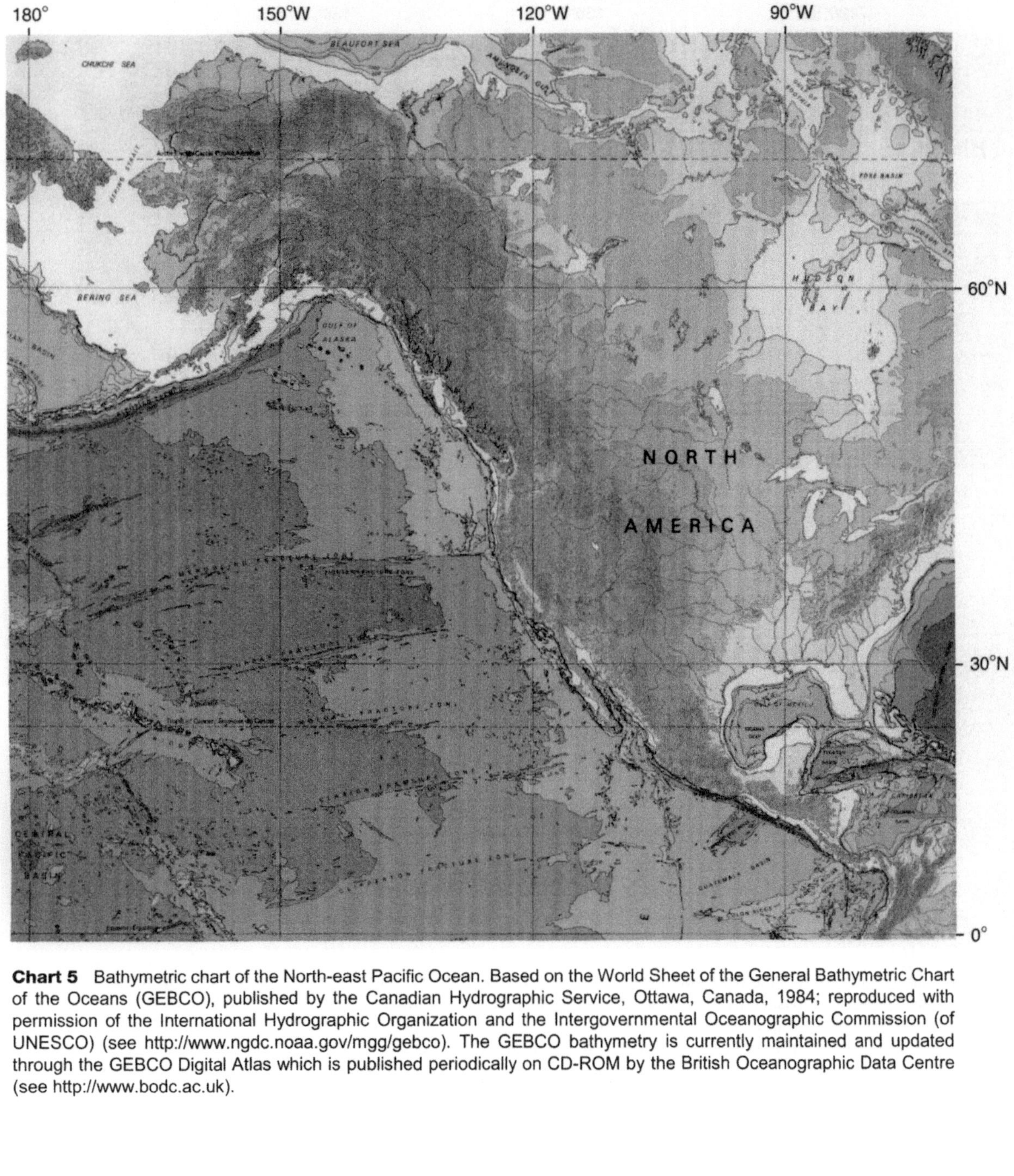

Chart 5 Bathymetric chart of the North-east Pacific Ocean. Based on the World Sheet of the General Bathymetric Chart of the Oceans (GEBCO), published by the Canadian Hydrographic Service, Ottawa, Canada, 1984; reproduced with permission of the International Hydrographic Organization and the Intergovernmental Oceanographic Commission (of UNESCO) (see http://www.ngdc.noaa.gov/mgg/gebco). The GEBCO bathymetry is currently maintained and updated through the GEBCO Digital Atlas which is published periodically on CD-ROM by the British Oceanographic Data Centre (see http://www.bodc.ac.uk).

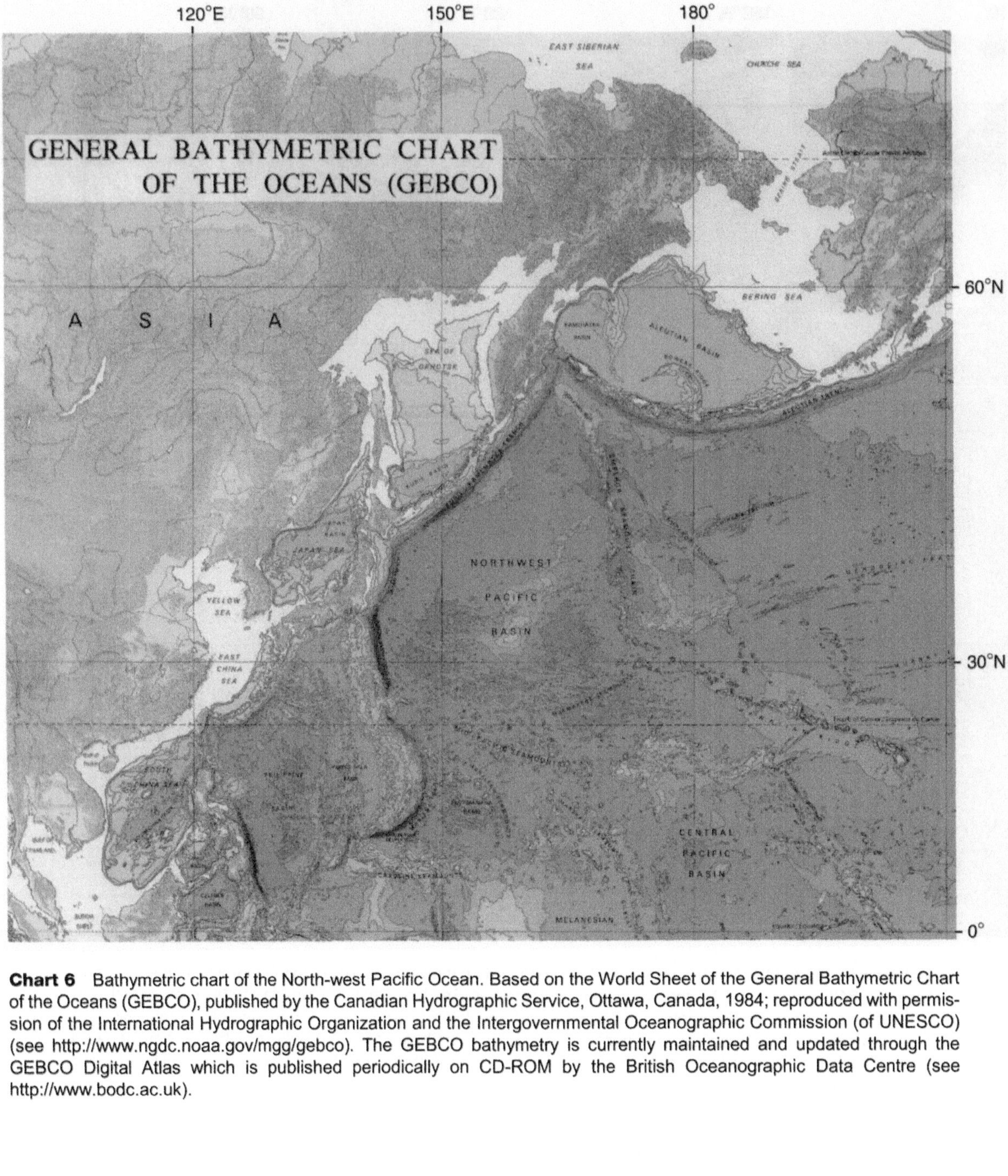

Chart 6 Bathymetric chart of the North-west Pacific Ocean. Based on the World Sheet of the General Bathymetric Chart of the Oceans (GEBCO), published by the Canadian Hydrographic Service, Ottawa, Canada, 1984; reproduced with permission of the International Hydrographic Organization and the Intergovernmental Oceanographic Commission (of UNESCO) (see http://www.ngdc.noaa.gov/mgg/gebco). The GEBCO bathymetry is currently maintained and updated through the GEBCO Digital Atlas which is published periodically on CD-ROM by the British Oceanographic Data Centre (see http://www.bodc.ac.uk).

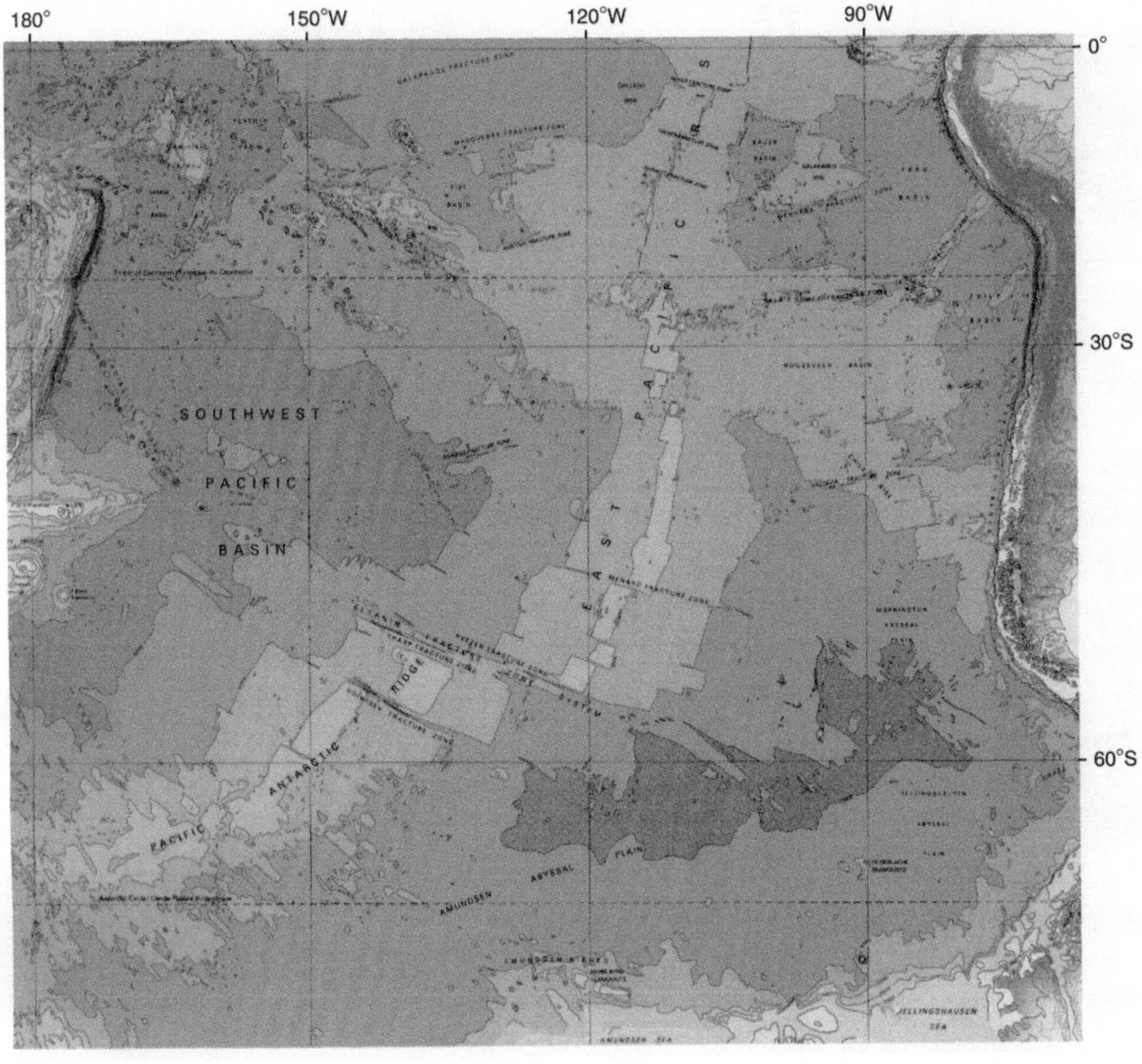

Chart 7 Bathymetric chart of the South-east Pacific Ocean. Based on the World Sheet of the General Bathymetric Chart of the Oceans (GEBCO), published by the Canadian Hydrographic Service, Ottawa, Canada, 1984; reproduced with permission of the International Hydrographic Organization and the Intergovernmental Oceanographic Commission (of UNESCO) (see http://www.ngdc.noaa.gov/mgg/gebco). The GEBCO bathymetry is currently maintained and updated through the GEBCO Digital Atlas which is published periodically on CD-ROM by the British Oceanographic Data Centre (see http://www.bodc.ac.uk).

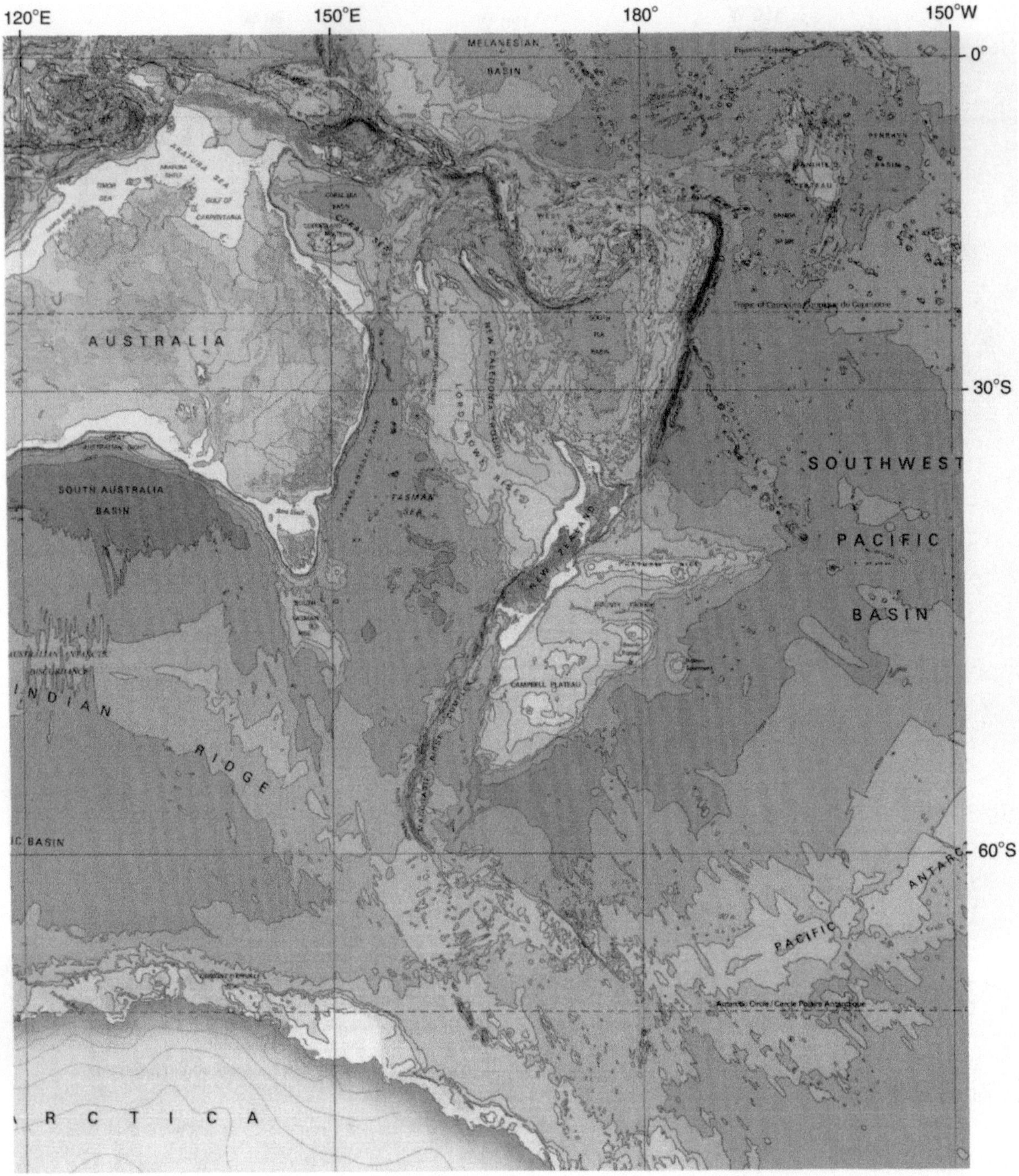

Chart 8 Bathymetric chart of the South-west Pacific Ocean. Based on the World Sheet of the General Bathymetric Chart of the Oceans (GEBCO), published by the Canadian Hydrographic Service, Ottawa, Canada, 1984; reproduced with permission of the International Hydrographic Organization and the Intergovernmental Oceanographic Commission (of UNESCO) (see http://www.ngdc.noaa.gov/mgg/gebco). The GEBCO bathymetry is currently maintained and updated through the GEBCO Digital Atlas which is published periodically on CD-ROM by the British Oceanographic Data Centre (see http://www.bodc.ac.uk).

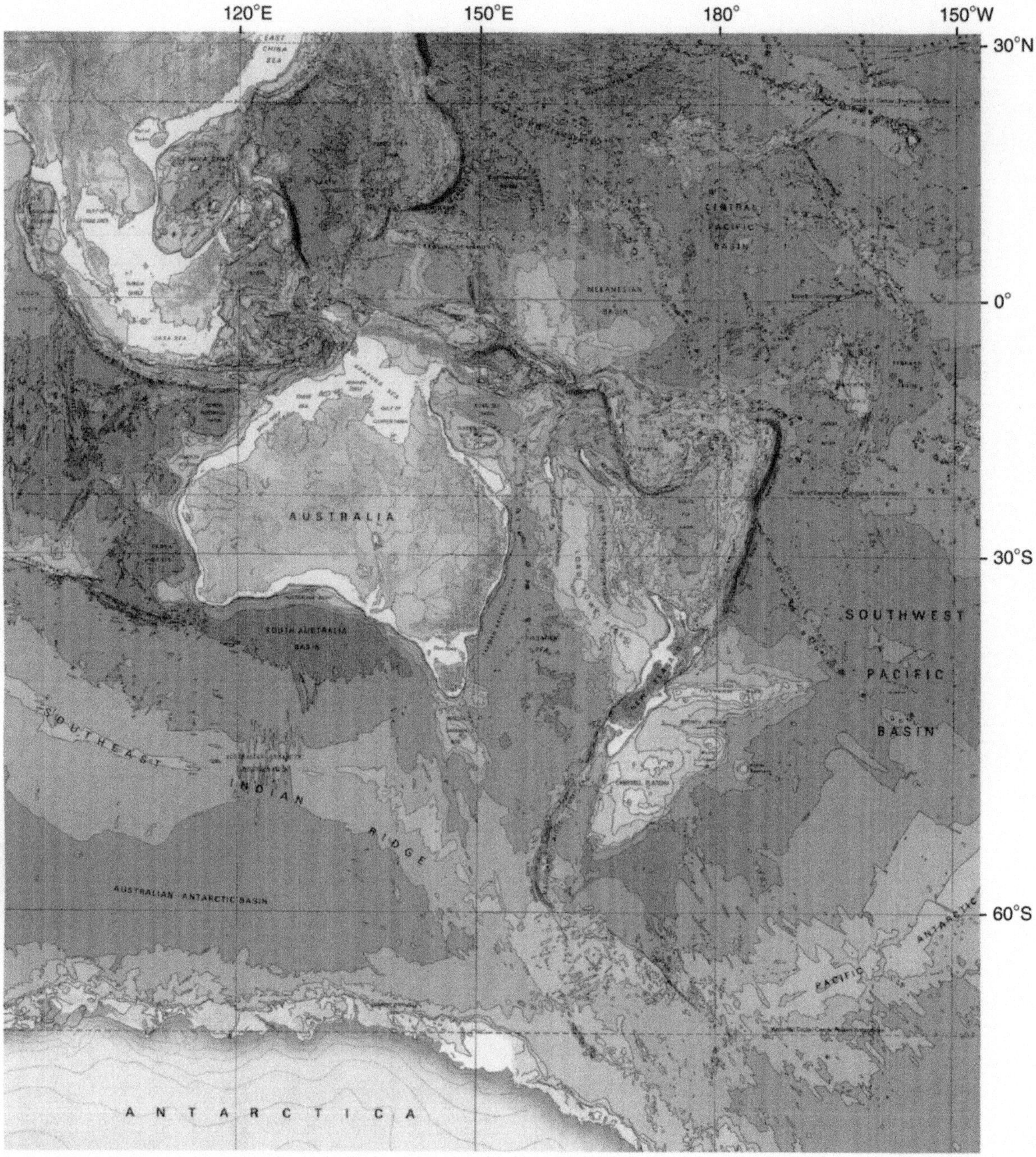

Chart 9 Bathymetric chart of Ocean surrounding Australia and New Zealand. Based on the World Sheet of the General Bathymetric Chart of the Oceans (GEBCO), published by the Canadian Hydrographic Service, Ottawa, Canada, 1984; reproduced with permission of the International Hydrographic Organization and the Intergovernmental Oceanographic Commission (of UNESCO) (see http://www.ngdc.noaa.gov/mgg/gebco). The GEBCO bathymetry is currently maintained and updated through the GEBCO Digital Atlas which is published periodically on CD-ROM by the British Oceanographic Data Centre (see http://www.bodc.ac.uk).

INDEX

Notes

Cross-reference terms in italics are general cross-references, or refer to subentry terms within the main entry (the main entry is not repeated to save space). Readers are also advised to refer to the end of each article for additional cross-references - not all of these cross-references have been included in the index cross-references.

The index is arranged in set-out style with a maximum of three levels of heading. Major discussion of a subject is indicated by bold page numbers. Page numbers suffixed by T and F refer to Tables and Figures respectively. vs. indicates a comparison.

This index is in letter-by-letter order, whereby hyphens and spaces within index headings are ignored in the alphabetization. For example, 'oceanography' is alphabetized before 'ocean optics.' Prefixes and terms in parentheses are excluded from the initial alphabetization.

Where index subentries and sub-subentries pertaining to a subject have the same page number, they have been listed to indicate the comprehensiveness of the text.

Abbreviations used in subentries

ENSO - El Niño Southern Oscillation
MOC - meridional overturning circulation
NADW - North Atlantic Deep Water
SST - sea surface temperature

Additional abbreviations are to be found within the index.

A

B

C

D

F

G

H

J

K

N

Q

R

S

T

X

Y

Z

Printed and bound by CPI Group (UK) Ltd, Croydon, CR0 4YY
03/10/2024
01040311-0007